Forensic Neuropathology

— A Practical Review of the Fundamentals —

Forensic Neuropathology

A Practical Review of the Fundamentals

Hideo H. Itabashi, MD

Former Head, Neuropathology, Harbor-UCLA Medical Center, Torrance, California
Professor Emeritus, Department of Pathology and Neurology, UCLA School of Medicine
Neuropathology Consultant, Department of Coroner, County of Los Angeles

John M. Andrews, MD

Deputy Medical Examiner and Neuropathology Consultant, Department of Coroner, County of Los Angeles

Uwamie Tomiyasu, MD[†]

Clinical Professor of Pathology, Department of Pathology, UCLA School of Medicine
Neuropathologist, Department of Pathology, Veterans Affairs Medical Center, West Los Angeles

Stephanie S. Erlich, MD

Clinical Assistant Professor of Pathology and Neurology, KECK School of Medicine, USC
Deputy Medical Examiner, Neuropathology Consultant, Department of Coroner, County of Los Angeles

Lakshmanan Sathyavagiswaran, MD, FRCP(C), FACP, FCAP

Clinical Professor, KECK School of Medicine, USC
Clinical Assistant Professor, UCLA School of Medicine
Chief Medical Examiner-Coroner, Department of Coroner, County of Los Angeles

[†] deceased

AMSTERDAM • BOSTON • HEIDELBERG • LONDON
NEW YORK • OXFORD • PARIS • SAN DIEGO
SAN FRANCISCO • SINGAPORE • SYDNEY • TOKYO

Academic Press is an imprint of Elsevier

Acquisitions Editor: Jennifer Soucy
Assistant Editor: Kelly Weaver
Marketing Manager: Diane Jones
Project Manager: Phil Bugeau
Cover Designer: Dennis Schaefer

Academic Press is an imprint of Elsevier
30 Corporate Drive, Suite 400, Burlington, MA 01803, USA
525 B Street, Suite 1900, San Diego, California 92101-4495, USA
84 Theobald's Road, London WC1X 8RR, UK

This book is printed on acid-free paper. ∞

Copyright © 2007, Elsevier Inc. All rights reserved.

No part of this publication may be reproduced or transmitted in any form or by any means, electronic or mechanical, including photocopy, recording, or any information storage and retrieval system, without permission in writing from the publisher.

Permissions may be sought directly from Elsevier's Science & Technology Rights Department in Oxford, UK: phone: (+44) 1865 843830, fax: (+44) 1865 853333, E-mail: permissions@elsevier.com. You may also complete your request on-line via the Elsevier homepage (http://elsevier.com), by selecting "Support & Contact" then "Copyright and Permission" and then "Obtaining Permissions."

Library of Congress Cataloging-in-Publication Data

Forensic neuropathology : a practical review of the fundamentals / Hideo H. Itabashi . . . [et al.].
    p. ; cm.
  Includes bibliographical references and index.
  ISBN-13: 978-0-12-058527-4 (hard cover : alk. paper)
  ISBN-10: 0-12-058527-8 (hard cover : alk. paper) 1. Forensic neurology. 2. Nervous system—Pathophysiology. I. Itabashi, Hideo H.
  [DNLM: 1. Forensic Pathology—methods. 2. Central Nervous System—pathology. 3. Trauma, Nervous System—pathology. W 750 F7127 2007]
  RA1147.F6773 2007
  614'.1—dc22

2007006301

British Library Cataloguing-in-Publication Data

A catalogue record for this book is available from the British Library.

ISBN: 978-0-12-058527-4

For information on all Academic Press publications
visit our Web site at www.books.elsevier.com

Printed in China
07 08 09 10   9 8 7 6 5 4 3 2 1

Working together to grow
libraries in developing countries

www.elsevier.com | www.bookaid.org | www.sabre.org

ELSEVIER   BOOK AID International   Sabre Foundation

Dedication

Uwamie Tomiyasu, M.D.

December 30, 1922–February 29, 2004

An admirable human being, esteemed colleague and teacher, loved by students, residents, fellow workers and all privileged to know her.

Contents

Preface ix
Foreword xi
Acknowledgments xiii

Chapter 1
THE FORENSIC NEUROPATHOLOGY AUTOPSY
I: SELECTED GROSS AND MICROSCOPIC EXAMINATION CONSIDERATIONS 1

Chapter 2
THE FORENSIC NEUROPATHOLOGY AUTOPSY
II: DEVELOPMENTAL CONSIDERATIONS 27

Chapter 3
DATING/AGING OF COMMON LESIONS IN NEUROPATHOLOGY 49

Chapter 4
NEUROPATHOLOGY OF PREGNANCY AND DELIVERY: MOTHER AND CHILD 123

Chapter 5
MALFORMATIONS AND OTHER CONGENITAL CENTRAL NERVOUS SYSTEM LESIONS 151

Chapter 6
BLUNT FORCE HEAD INJURY 167

Chapter 7
THE SUSPECTED CHILD ABUSE CASE 199

Chapter 8
INJURIES DUE TO FIREARMS AND OTHER MISSILE-LAUNCHING DEVICES 211

Chapter 9
SUDDEN UNEXPECTED DEATH 255

Chapter 10
RESPONSES OF THE CENTRAL NERVOUS SYSTEM TO ACUTE HYPOXIC/ISCHEMIC INJURY AND RELATED CONDITIONS 289

Chapter 11
VASCULAR DISEASES OF THE CENTRAL NERVOUS SYSTEM 307

Chapter 12
INFECTIONS OF THE CENTRAL NERVOUS SYSTEM 335

Chapter 13
BRAIN TUMORS 359

Chapter 14
NEURODEGENERATIVE DISORDERS 395

Chapter 15
DEMYELINATING DISORDERS 413

Chapter 16
PERIPROCEDURAL COMPLICATIONS 423

Chapter 17
MISCELLANEOUS TOPICS 443

Appendix 465
Index 471

Preface

The primary goal of this book is to provide visual examples from our experience of the more commonly encountered conditions in forensic neuropathology and, in the form of a manual, answer some of the most frequently asked questions that arise regarding neuropathologic findings in our consultations. Restating the obvious, all experienced pathologists know that diagnoses printed on documents are only convenient shorthands for communication purposes, medicolegal or otherwise, and they fail to convey the complexity and possible variations that might exist within the descriptors. One cannot escape the fact that every case is unique. And with this caveat, we present a practical and much simplified review of some of the more common conditions encountered in a forensic office. General pathology residents, forensic pathology and neuropathology fellows, general pathologists, clinicians involved in referred cases, and individuals in allied fields, such as law enforcement officers and attorneys, may find this book useful.

As Harrison notes, "It is difficult to think like someone other than yourself."[1] In attempting to do so, reliance is placed on the large volume and wide range of pathologic materials examined over the past 28 years at the Department of Coroner, County of Los Angeles, to guide this highly selective and undoubtedly biased choice of material. Clearly, no claim can be made to comprehensive coverage. As the book title indicates, the topics covered are highly selective, some subjects are necessarily omitted, and others, such as gunshot wounds, are covered in some detail. More time is spent on the most frequently encountered conditions, and case studies and clinical-pathologic correlations are used to illustrate key points. Were one to adhere to strict criteria presently proposed for "evidence-based medicine," such experience and correlations would be largely discounted. However, we believe they are useful so long as their limitations are understood.

Information has been included from some older published sources that we find helpful in present-day consultation work, since those sources have been largely displaced in many modern texts and journals in order that the latest imaging techniques, immunohistochemical methods, or insights derived from molecular biology methods be included. Results of a survey of readers of the *Journal of Neuropathology and Experimental Neurology* in 2002, for example, indicated that degenerative diseases ranked highest and trauma ranked lowest in the list of topics desired by readers in future issues of the journal. The forensic neuropathologist is likely to rearrange these priorities.

The importance of newer techniques is not in question, but most of them do not help us deal with issues we presently encounter, such as dating/aging of a variety of neuropathologic lesions, and whether dark neurons might be confused with hypoxic/ischemic neuronal injury. We found it useful to review, for example, existing guidelines to help us decide whether or not a particular gunshot wound to the head was immediately incapacitating. In the same vein, we want to know the errors to avoid when answering hypothetical questions about the body position of a decedent at the instant a particular gunshot wound was sustained. Many such issues are more likely to be encountered in forensic work than in an academic or community neuropathology consultation practice.

Our intent is to provide a practical summary of selected forensic neuropathology topics to supplement the excellent neuropathology and forensic pathology textbooks currently available. We do not cover, for example, most degenerative and toxic metabolic disorders rarely seen by medical examiners. Not included are most aspects of scene investigation, medicolegal aspects of composing death certificates, and other subjects readily found and thoroughly discussed in well-known textbooks or monographs. References cited reflect the practical impossibility of including every author who has written on every statement in this text. In some instances, an observation we thought was original was subsequently found in some earlier publication, and the latter has been consistently credited as the source for the statement.

Forensic neuropathology is a field of study with many challenging components. The need for careful and detailed documentation in words, diagrams, and photographs of gross and microscopic findings is more critical than in many other pathoanatomic disciplines, largely because of potential future allegations and as support in court testimony. Routine light microscopy with routine H&E stains are something more than a preliminary guide to a battery of expensive special stains. More attention is paid to correlation of information from a wider range of disciplines than virtually any other medical subspecialty. For example, scene investigators, witness interrogators, tool mark experts on patterned injuries, anthropologists, entomologists, and veterinary pathologists often play an important role in the final analysis.

We remain always mindful that the resultant work product may have an impact on the living equal to that in any other medical specialty with more direct patient involvement.

Reference

1. Harrison JH Jr: Pathology informatics questions and answers from the University of Pittsburgh pathology informatics rotation. Arch Pathol Lab Med 2004;128:71–83.

Foreword

It is my pleasure to write the foreword for this book based on the experiences of the Neuropathology Service of the Department of Los Angeles County Chief Medical Examiner-Coroner's Office over many decades.

The Department has been fortunate in having the services of generations of outstanding neuropathologists. First was Professor Cyril B. Courville, who was the Chief of Neuropathology at Los Angeles County General Hospital and Professor of Neurology and Pathology at the Loma Linda University School of Medicine at the same time. Of his numerous important publications, a major contribution by Professor Courville was his textbook on forensic neuropathology (*Forensic Neuropathology: Lesions of the Brain and Spinal Cord of Medico-legal Importance.* Illinois: Callaghan and Company, 1964), one of the first textbooks on forensic neuropathology that we used throughout our era. A number of residents also studied forensic neuropathology under Dr. Courville and, following his death, one of them, the late Dr. Abraham T. Lu, Professor of Pathology (Neuropathology), Loma Linda University School of Medicine, took over his service in forensic neuropathology. Dr. Lu's legacy remains in the Neuropathology Wet Tissue Medical Museum housed at the Loma Linda University School of Medicine.

In 1967, I became the Chief Medical Examiner-Coroner for the County of Los Angeles, and I felt that instead of sending the brain specimens to the Los Angeles County General Hospital for analyses, it would be best to incorporate the neuropathologic studies on forensic cases into the residency training program in forensic pathology at the Chief Medical Examiner's Office. In May 1972, the Los Angeles County Medical Examiner's Office moved to a new Forensic Science Center Building on the grounds of the Los Angeles County/USC Medical Center. In 1977, I asked Dr. Hideo H. Itabashi, Professor of Pathology and Neurology at UCLA and the Head of Neuropathology at Harbor-UCLA Medical Center in Torrance, to join us as the Chief Forensic Neuropathologist.

Dr. Itabashi's superb service led to considerable increase in cases referred for neuropathologic consultation by the deputy medical examiners. At about the same time, the late Dr. Uwamie Tomiyasu, Clinical Professor of Pathology at UCLA and Neuropathologist, Veterans Affairs Medical Center, West Los Angeles, came to the Office and provided many years of invaluable volunteer assistance to the Department as the service work expanded. My successor, Dr. Lakshmanan Sathyavagiswaran, co-author of this book, and current Chief Medical Examiner-Coroner, expanded the consultative services, especially in neuropathology, by adding full time staff neuropathologists Dr. John M. Andrews and Dr. Stephanie S. Erlich, also co-authors of this book.

I believe the authors have not only met but surpassed their goals stated in the Preface. Their analyses of a large collection of a wide range of carefully documented cases combined with extensive reference sources for forensic neuropathology will aid practicing medical examiners and forensic neuropathologists, and the book will also be a most useful resource for general pathologists, other physicians, attorneys, and investigator professionals.

It was a privilege to review this book and write this foreword.

Thomas T. Noguchi, M.D.
Professor Emeritus of Forensic Pathology
University of Southern California Keck School of Medicine and Chief Medical Examiner-Coroner (Ret)
County of Los Angeles, California

January 31, 2007

Acknowledgments

The authors would be greatly remiss not to mention and thank our colleagues in the Department of Coroner of the County of Los Angeles for their support and cooperation. We thank Christopher Rogers, M.D., Chief of the Forensic Medicine Division, and Senior Physicians Eugene Carpenter Jr., M.D., James K. Ribe, J.D., M.D., and William E. Sherry, M.D., for their willingness to share their extensive experience in forensic cases. Francis V. Hicks provided administrative assistance. Deputy Medical Examiners Juan J. Carrillo, M.D., Chanikaren Changsri, M.D, Ogbonna Chinwah, M.D., Raffi S. Djabourian, M.D., Paul V. Gliniecki, M.D., Irwin L. Golden, M.D., Jeffrey P. Gutstadt, M.D., Vladimir Levicky, M.D., Pedro M. Ortiz-Colom, M.D., Ajay J. Panchal, M.D., Louis A. Pena, M.D., Vadims Poukens, M.D., Solomon L. Riley, Jr., M.D., Lisa A. Scheinin, M.D., Stephen Scholtz, M.D., Susan F. Selser, M.D., Yulai Wang, M.D., and David B. Whiteman, M.D., provided us with many interesting and unusual cases, and submitted questions leading to many of the topics emphasized in this text. Cases in the pediatric age range were primarily referred through the courtesy of James K. Ribe, J.D., M.D., and David B. Whiteman, M.D. We thank Steven J. Dowell, B.S., for sharing his special expertise in patterned injuries, electron probe analysis, unusual case presentations, and other areas.

Special thanks also go to Carol L. Andrews, M.D., Clarissa De La Torre, Mauricio Molina, Roy Fernandez, Rudolfo Enriquez, George McDowell, and John Marsden for extra photographic assistance, and to Maria Diaz, M.D., for radiographic assistance.

Adriana Flores was invaluable for her tireless efforts in transcription of our notes and dictations, without whom this project would not have been possible.

The team from Elsevier/Academic Press have our gratitude for their patience and encouragement. Thanks go to Jennifer Soucy, Kelly Weaver, Diane Jones, and Phil Bugeau.

The Forensic Neuropathology Autopsy

I: Selected Gross and Microscopic Examination Considerations

INTRODUCTION 1
DECOMPOSED CASES 1
THE ROUTINE FORENSIC NEUROPATHOLOGY
 AUTOPSY 1
LEGAL ISSUES, INCLUDING POTENTIAL ERRORS IN
 THE PERFORMANCE OF THE FORENSIC
 NEUROPATHOLOGY CONSULTATION 7
TOPOGRAPHIC APPROACH TO
 NEUROPATHOLOGY 8
SUDDEN UNEXPECTED DEATH IN INFANTS AND
 CHILDREN DUE TO NATURAL CAUSES, AND
 THE SUSPECTED SUDDEN INFANT DEATH
 SYNDROME CASE 8

TIME SINCE DEATH 9
CRANIOCEREBRAL RELATIONSHIPS 10
MENINGES 10
BRAINSTEM 12
SPINAL CORD 13
NECK DISSECTION 15
DISCOLORATION OF SKULL, MENINGES, AND
 BRAIN 15
SELECTED ARTIFACTS 17
EFFECT OF FORMALIN FIXATION ON CNS
 TISSUE 22
PINEAL GLAND 22
REFERENCES 23

Introduction

The references cited at the end of this chapter include a review of forensic pathology issues, including autopsy procedures, designed primarily for attorneys[56]; forensic pathology textbooks and articles that include sections on differences between routine hospital and forensic autopsy procedures and/or techniques more commonly used in forensic autopsies[36,39,45,101,112,129,140,144,148]; and articles and monographs more specifically directed to neuropathology aspects of the autopsy.[38,123,166]

On occasion, however, it becomes necessary to explore a particular topic relevant to the forensic neuropathology consultation in greater detail. The goal of this chapter is to highlight selected sources and issues we found helpful in the categories that follow.

Decomposed Cases

There is an understandable tendency to dispense with examination of decomposed tissues as unnecessary or unproductive activity in any busy office. It has been our experience, however, that in cases with "Swiss-cheese" type change, with centrally located pink-colored softening due to poor penetration of formalin, or with decomposition that has progressed to the point that central nervous system (CNS) tissue consistency was just barely firm enough to be cut into a block for embedding, a surprising degree of histologic detail sometimes remained. Others have had a similar experience,[118] and making the attempt is worthwhile if an important neuropathologic question might be clarified. Even brain tissue decomposed to the consistency of soft paste can be submitted for toxicologic screening, and often yields useful qualitative, if not quantitative, results. Common artifacts in forensic material are seen in Figs. 1.1–1.5.

The Routine Forensic Neuropathology Autopsy

Our role as forensic neuropathology consultants places us in the position of having our portion of the case

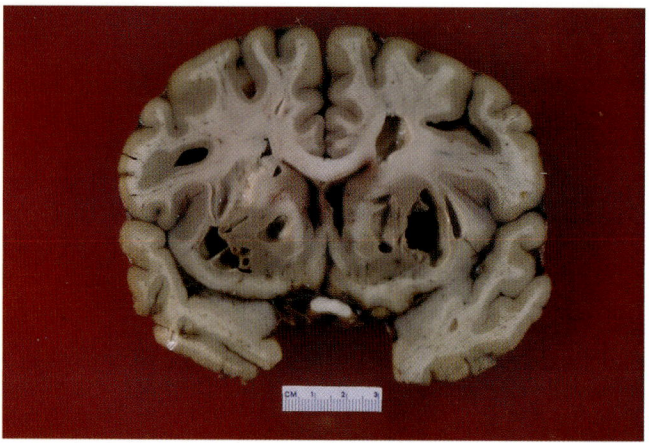

Fig. 1.1. Coronal section of brain with multiple cystic spaces due to postmortem overgrowth of gas forming bacteria; so-called "Swiss-cheese" artifact.

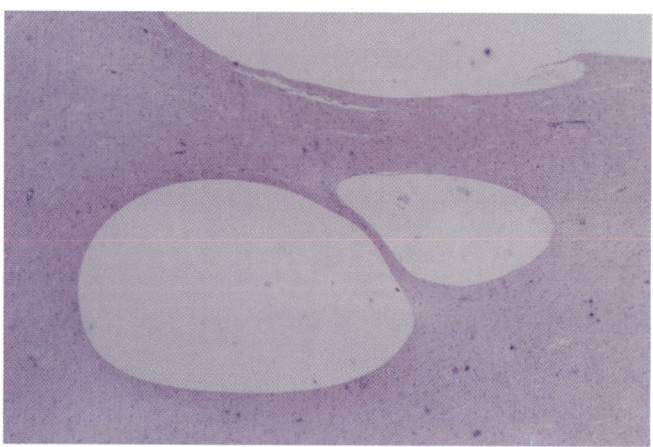

Fig. 1.3. Low-power micrograph of "Swiss-cheese" artifact. No evidence of tissue reaction. At higher power, bacteria will be evident (usually rod-shaped or mixed flora) in cyst wall and nearby parenchyma. (H&E.)

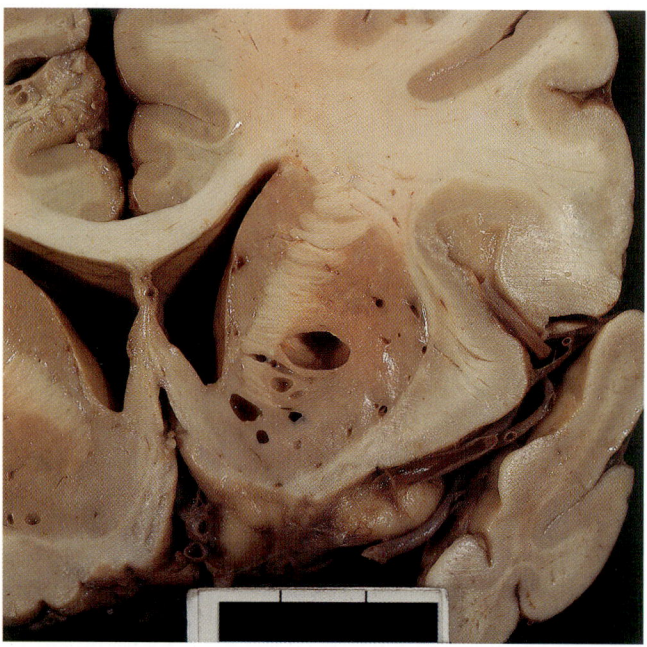

Fig. 1.2. Detail of cysts due to gas forming organisms. Smooth walls, no tissue reaction, punched-out appearance.

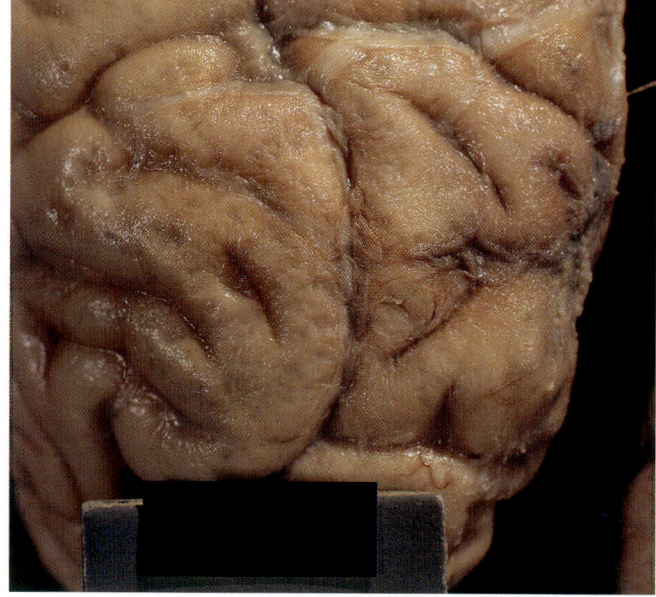

Fig. 1.4. Roughened cerebral cortical surface due to superficial cortical avulsion. This artifact occurs when fresh or fixed cortex moves against, or is excessively indented by, rough or textured surfaces such as surgical caps (when brain is suspended in fixative) or paper towels, and is to be distinguished from granular cortical atrophy.

evaluation compromised if specimens provided us contain significant preventable artifact, since we are not present at the time of autopsy in most cases. Direct communication with the medical examiner facilitates the extra care necessary for successful brain and/or spinal cord removal with the least amount of damage.

Most readers will already have arrived at a protocol for the CNS portion of the general autopsy with which they are comfortable and proficient, and descriptions of various techniques of brain and spinal cord removal in routine adult autopsies are easily found. Finkbeiner et al.[45] and Ludwig[101] provide suggested CNS techniques in greater detail than many other texts on autopsy techniques, and the College of American Pathologists publishes several relevant manuals dealing specifically with forensic autopsy issues.

Our department uses narrative gross examination and microscopic reports in most consultation cases. A short-

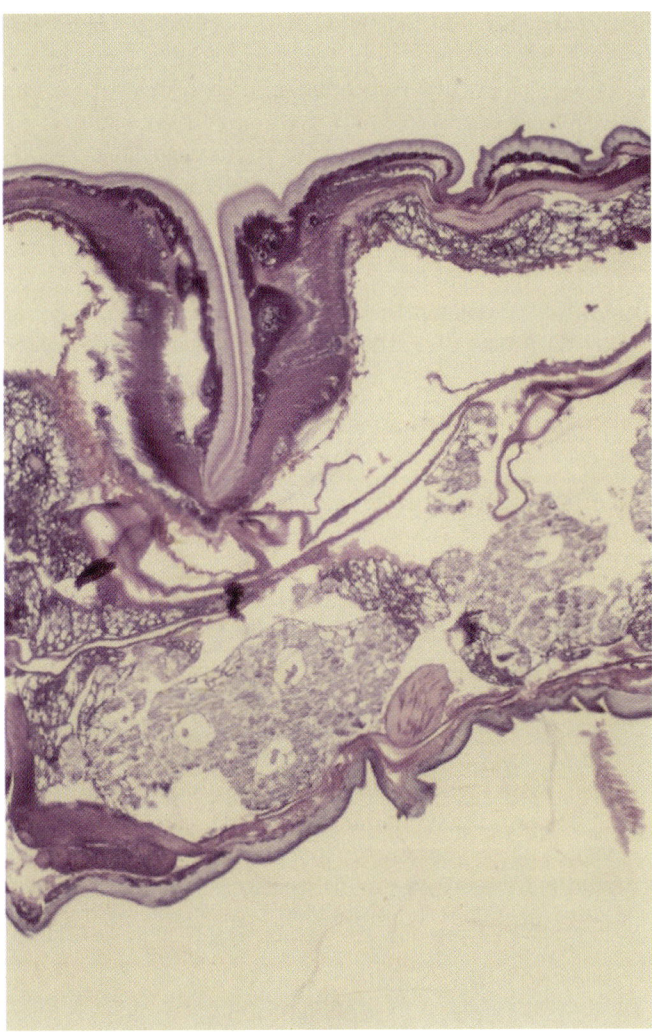

Fig. 1.5. Portions of longitudinal section of fly larva (maggot), encountered in cases with more advanced decomposition (and on some forensic board examinations). (H&E.)

form neuropathology protocol is also available for use by deputy medical examiners (DMEs) who have themselves performed the gross and microscopic examination of the CNS in their case, or which we use when they seek later consultation from us for some limited aspect of the CNS portion of their case.

We encourage suspension of the fresh brain in 10% formalin after weighing, either by a string under the basilar artery (not recommended in brains very soft for any reason) or within a porous (usually green) surgical cap suspended by string within the formalin-filled container. A complete change of formalin after 24 hours and toward the end of the first week greatly improves fixation. We do not employ perfusion fixation for CNS tissue in forensic cases as a means to shorten turnaround time. This is primarily because of the potential for dislodgement of intravascular pathology and for inconsistent perfusion (and thus fixation) with the death-to-autopsy delays often encountered in forensic cases.

With rare exception, cerebral hemispheres are separated from the brainstem by a transverse section of brainstem at the superior pontine sulcus, and hemispheres are sectioned in a coronal plane and the brainstem and cerebellum in a transverse plane.

Brain tissue may be immature and soft, or soft and friable in the fixed state for other reasons, such as in a respirator brain. Sections of routine thickness in such brains would undergo fragmentation if handled in the usual manner. Rectangles of paper towels, moistened with water and placed on the flat brain surface created by the initial cut, can form a support for each subsequent slice. This method is facilitated by an assistant who can provide mild pressure against the paper towel on the cut brain surface, and gently guide the brain slice to the cutting board as the cut is made. An assistant is required because it is not possible for one person to simultaneously support the brain at the angle necessary for a proper coronal section, cut the brain, and control the cut soft brain slice as it collapses and falls to the cutting board. In the absence of an assistant, after the initial cut the flat cut brain surface is placed against the moist paper towel on the cutting board. The brain is sectioned at the appropriate thickness parallel to the cutting board. The uncut portion of brain is carefully moved to the next prepositioned moist paper towel, sectioned, and the process repeated. This method allows transfer of the sections by lifting the paper towel that supports the section. Both sides of the slice can be easily examined by transferring it from one moist paper towel to another, and one can easily position the tissue slice for cutting blocks for histology by manipulating the paper towel rather than the slice per se. Selected slices, paper towel included, are subsequently placed in the save bottle at the conclusion of the gross examination, minimizing the chance of distortion and fragmentation of the saved slices.

For estimating degree of occlusion of vessels, Fig. 1.6 has been helpful to us.[24] In cases of spontaneous subarachnoid hemorrhage, search for possible berry (saccular) aneurysm is conducted in the fresh state after photography of the undissected specimen. Soft blood clot is removed from major basal vessels with the aid of a gentle stream of water. Extension of vessel examination for aneurysms in this manner much beyond 1.0 to 1.5 cm peripheral to the circle of Willis is a low-yield endeavor that becomes increasingly destructive to brain parenchyma, so further search for a bleeding source is deferred to the fixed specimen.

When identification of laterality of a tissue block submitted for histology is necessary, a simple routine is to "V-notch" a nonessential margin on blocks from the right side only. The histology laboratory should be instructed to not excessively "face" the submitted block (always using the "down" side of each block) unless otherwise

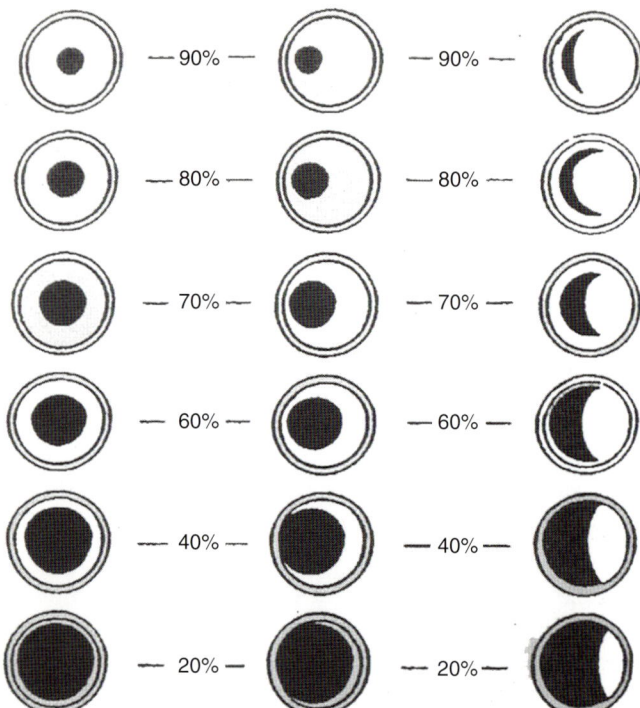

Fig. 1.6. Schematic diagram representing the percentage reduction in cross-sectional area of coronary (or cerebral) artery vessels. The outer circle represents vessel exterior, inner circle represents the elastic lamina, and black area represents the lumen (from Champ CS, Coghill SB. Visual aid for quick assessment of coronary artery stenosis at necropsy. J Clin Pathol 1989;42:887–888, with permission).

instructed, since an essential lesion may not extend throughout the depth of the block. One of the coronal cuts is made through the midmammillary bodies. At the conclusion of the gross examination, a narrative description of the "Gross Impressions" is listed.

As much tissue as will reasonably fit in the save bottle is retained; a full bottle uses up no more storage space than a partially empty bottle. Sections are saved from every lobe, visual and sensorimotor cortex, the frontal/temporal/occipital pole, at least three levels of deep gray matter (including mammillary bodies, amygdala, and adjacent structures), orbitofrontal cortex, both hippocampi, the entire brainstem not already submitted for histology, cerebellar vermis and lateral hemisphere (including dentate nucleus), several levels of cervical/thoracic/lumbosacral cord and attached roots, and samples of cranial and spinal dura. The pineal gland is often avulsed and lost at general autopsy, but, if received, it is also placed in the save bottle.

Sections submitted for histology include the usual representative samples for specific protocols (e.g., SIDS protocol), and samples of lesions and of selective grossly normal areas that are relevant to the case history or are useful for "mirror-image" opposite hemisphere comparison purposes. A block designation different from the routine general autopsy blocks (e.g., using a letter of the alphabet rather than numbers), and cassettes of a different color are useful for neuropathology tissue submitted to the histology laboratory. This minimizes the possibility that two different histologic sections from the same case (e.g., one of spleen and one of brain) will have identical cassette color and number designations.

In the microscopic report, each slide is listed with its anatomic source. At the conclusion of the microscopic description narrative, the "Final Neuropathologic Diagnosis" items are listed, with or without a comment.

Potential Problem Areas in the CNS Portion of the Autopsy

1. **Removal, Fixation, and Gross Examination of the Immature Brain.**
   a. *Removal.* In addition to recommendations in standard autopsy texts as noted earlier,[45,101] various brain removal techniques for perinatal cases are described by Valdés-Dapena and Huff[162] and others.[52,53,124] A simplified head immersion technique for brain removal in the perinatal period credited to Dimensten uses an approximately 7 × 5 × 4-inch plastic container from which a semicircular portion is cut on the rim aspect, and the container filled with water. The infant neck is placed in the concavity of the cut such that the head is in the water, and the water is used to support the brain as it is removed.[60] Leestma[94] and Gilles[53] suggest different approaches to facilitate removal of severely hydrocephalic infant brains, to which the reader is referred for details.
   b. *Fixation.* Various fixative combinations, such as formalin followed by alcohols[132,136] or use of 20% formalin, have been suggested as ways to harden immature brain tissue to facilitate brain cutting. Although it should be noted that the Occupational Safety and Health Administration and budgetary restrictions may limit its use, 20% formalin is preferred.
   c. *Gross examination of soft, immature brains* is aided by the moist paper towel method described earlier, which is less problematic for soft brain slices than use of a spatula.
   d. *Investigation of stillbirths (and deaths in the early postnatal period),* while aided by CNS examination, is incomplete unless it also includes examination of the placenta.[104]

2. **The High-Risk Autopsy.** This term is generally applied to autopsies on decedents with infectious diseases that may be transmitted to autopsy room personnel, including but not limited to viral hepatitis, tuberculosis, human immunodeficiency virus (HIV), and prion diseases. Recommendations to maximize the safety of individuals exposed to the autopsy or to tissues from such cases continue to evolve. Autopsies on cases of poisoning by a variety of toxins (e.g., cyanide, organo-

phosphates, metallic phosphides), formalin spills, high doses of radiation, and certain medical devices such as implantable cardioverter-defibrillators have the potential to pose risk to autopsy personnel, and detailed recommendations for precautions to take under these various circumstances are well described in the literature.[45,49,52,74,77,96,114] Neuropathology consultants are sometimes called to the autopsy table for an immediate opinion, and should be aware of possibilities of general autopsy risks. Usual scalpel cutting injury risks,[126] needle injuries including neck needle foreign bodies in intravenous drug addicts,[72] potential for sharp missile fragments in brain wounds (see Chap. 8), and potential aerosol exposure to infectious disease (such as when opening the skull with the autopsy saw) may also involve the neuropathologist assisting with brain, spinal cord, or *en bloc* pituitary fossa removal in certain cases.

In our department, despite dealing with a population in which the HIV incidence is in the range of 2.5% of cases, known disease transmission to staff has been largely limited to tuberculosis.[154] Much remains uncertain regarding the duration of survivability of various organisms, both in unembalmed bodies and in formalin-fixed tissue,[7,28,113,114] so universal precautions must be respected. DMEs are urged to alert consultants if known or suspected hazards exist in specimens referred for consultation.

3. Intraosseous Carotid Artery Dissection. A procedure for dissection of the intraosseous portion of the carotid artery has been described.[92] A slight modification of this procedure by narrowing the coronal diameter of the skull base quadrangle removed can be useful in cases in which the primary focus is the pituitary fossa and immediately adjacent structures (Case 1.1).

4. In-Custody Deaths. Protocols in this category of cases differ to some extent from the routine case, primarily by an increased emphasis on detailed diagrams and photographic documentation of the entire external body surface, including layer-by-layer dissection photographs, especially in the head and neck area, not dissimilar to the United Nations protocol for the investigation of deaths of questioned origin[160] (see later). As in all homicide, or suspected homicide, cases, clothing is carefully examined. Neuropathology consultation is obtained in all cases in this category, even in the absence of abnormalities seen at the time of the general autopsy. DiMaio[35] has recently reviewed the essentials of the medical examiner's role in in-custody deaths.

5. Investigation of Extralegal, Arbitrary, and Summary Executions. Forensic autopsies involving political assassinations, torture, executions without due process, acts of genocide, and so on understandably attract considerable public, political, and professional attention. Primary responsibility for such cases often falls under the jurisdiction of forensic pathologists in a government agency, but civilian forensic pathologists

Case 1.1. (Figs. 1.7–1.10)

An elderly man underwent transsphenoidal surgery for a clivus bone tumor situated posteroinferior and slightly to the left of the pituitary gland. Attempted biopsy resulted in massive hemorrhage that required packing for hemostasis, and surgical closure. Postoperative imaging studies revealed a left internal carotid/cavernous sinus fistula and pseudoaneurysm in the posterior left cavernous sinus region. Left internal carotid stent placement and coil embolization to the area of the fistula and pseudoaneurysm was performed on the fifth postoperative day. Four days later death occurred from a sudden, massive hemorrhage at the operative site.

Neuropathologic examination included an *en bloc* resection of the sellar and perisellar skull base. Specimen x-rays (Figs. 1.7–1.9) following specimen fixation and decalcification were used to guide a near-midline sagittal section (Fig. 1.10), avoiding encounter with embolized metallic coils and resultant mechanical artifact in the histologic areas of interest. Diagrams and photographs made at the time of autopsy dissection allowed unequivocal identification of anatomic features on the subsequent specimen (Fig. 1.10). These procedural precautions made possible the identification of the clival mass as a rare form of pituitary adenoma: an ectopic clival prolactinoma.

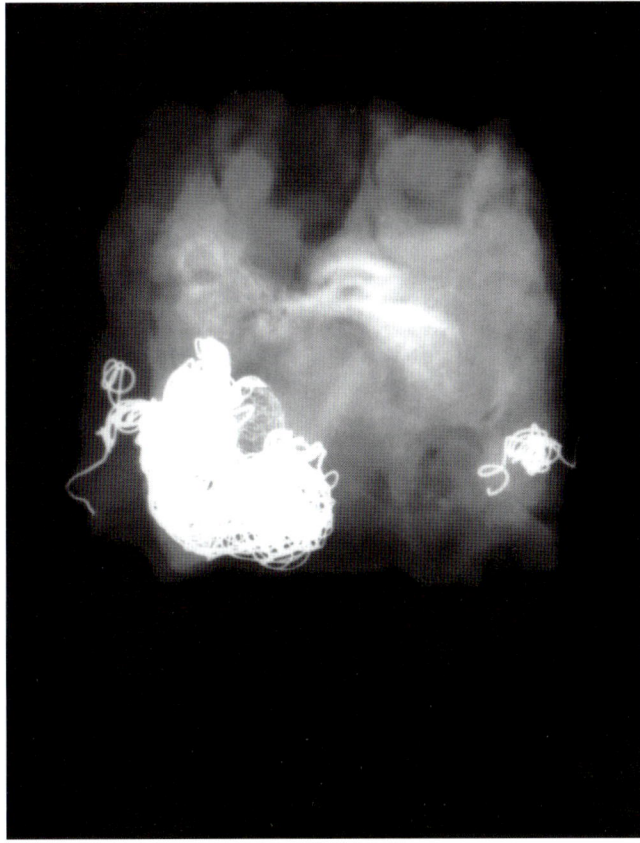

Fig. 1.7. Anteroposterior x-ray of sellar block. Dense wire coils are evident.

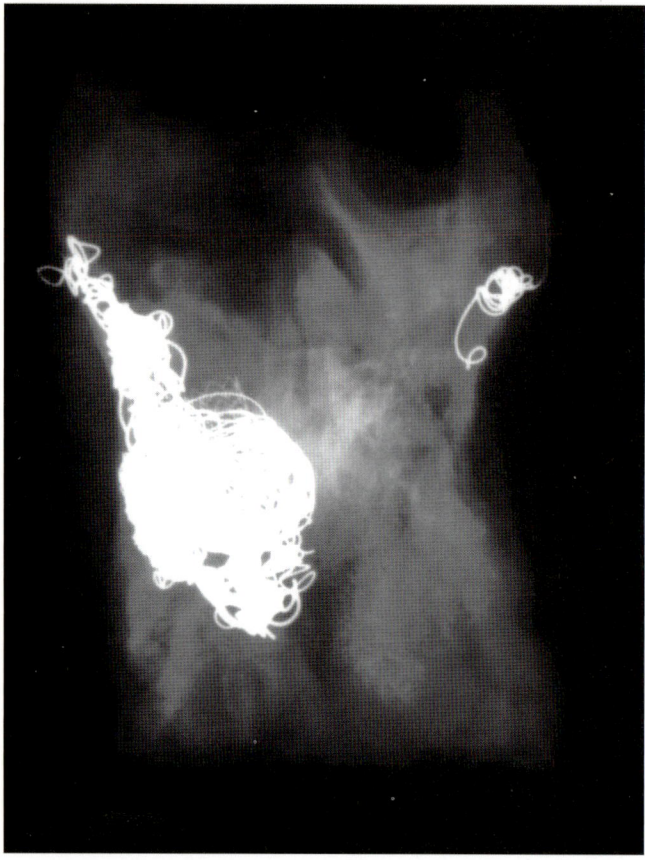

Fig. 1.8. Dorsal (horizontal plane) x-ray of sellar block.

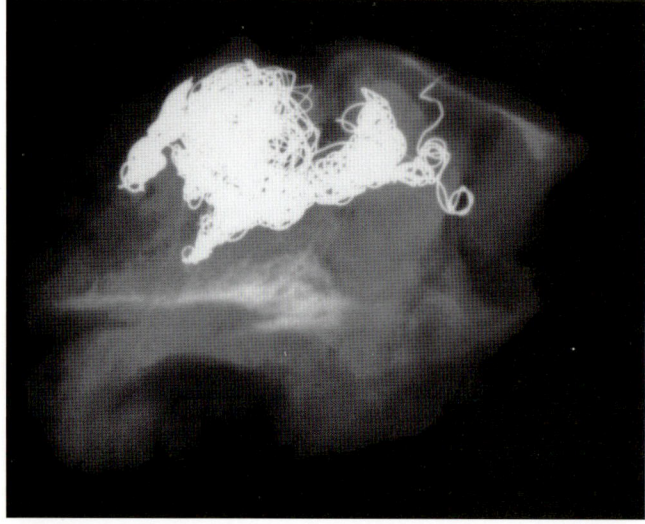

Fig. 1.9. Lateral x-ray view of sellar block.

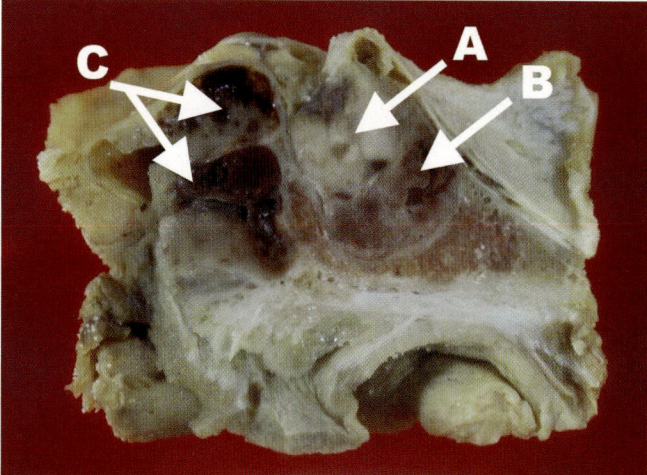

Fig. 1.10. Sellar block from Case 1.1 sectioned in mid-sagittal plane between wire coils. A, pituitary; B, tumor; C, blood in sphenoid sinus.

are sometimes recruited for such studies. Useful background information concerning autopsy protocols for such cases can be obtained from the United Nations.[160] The recommendations differ in some details from routine procedures used in many departments, and familiarity with torture techniques used in some parts of the world is also helpful as additional background.[10,18,95,110,122]

6. Report Length. Narrative reports can run several pages in complex cases, which may evoke sighs from the impatient. However, there should be no apology for a lengthy report full of relevant, irreproducible descriptions. No volunteers will come forth to offer to testify in defense of an abbreviated, incomplete neuropathology consultation report.

One will be very grateful for having a detailed documentation during testimony occurring years after the consultation was completed. Our view is that having more information that is later found to attract directed scrutiny is less of a problem than the alternative. One need not document all findings not present, but using common sense to list particularly relevant negative findings can be helpful. For example, one might indicate that no source of bleeding in a case of subarachnoid hemorrhage was found, either on external examination of major surface or basal vessels, or by careful dissection of vessels made accessible by subsequent coronal sections of the brain and transverse sections of the brainstem and cerebellum.

Other approaches include having some documentation of one's standard gross CNS examination procedure, which one uses in every case, available as a baseline to which one may refer. One may also use a template, checklist report form similar to those being proposed for certain standardized surgical pathology specimens.

Legal Issues, Including Potential Errors in the Performance of the Forensic Neuropathology Consultation

Most of the legal issues regarding performance of an autopsy, or any restrictions upon the autopsy that may apply, have already been resolved well before the consulting neuropathologist examines the fixed brain or other CNS tissue, a fact for which the neuropathologist can be thankful. The myriad social, religious, cultural, and ever-changing legal forces influencing autopsy performance, tissue and organ transplantation, and so on[13,88] typically require the ongoing attention of a group of attorneys in the office of the County Counsel in our jurisdiction, in order to advise our department on special autopsy procedures.

It follows that the neuropathologists should have a basic understanding of the laws regulating autopsy performance in the area in which they work, carefully review each file prior to performing the consultation for any documentation that indicates, for example, that all brain tissue not used for histologic sections must be returned to the mortuary for burial at the family's request, or that standard procedure should be otherwise altered. If any question arises based on the decedent's religion or other factors noted during the file review, the neuropathologist should contact the DME prior to the consultation to determine whether specific restrictions exist. The DME should notify all others involved in the case concerning restrictions that apply, but may not have done so.

It is recommended that consulting forensic neuropathologists read certain articles on the most common potential errors in the performance of forensic autopsies, which have been summarized by experienced forensic pathologists and briefly described later. These articles are emphasized in forensic pathology, but not generally in neuropathology, training programs.

Many of these points are directly applicable to the neuropathology examination and report. Examples by Moritz (Table 1.1)[111] and Sturner (Table 1.2)[151] demonstrate a degree of overlap, indicating that many of these errors persist over time. Petty (Table 1.3)[121] lists common medicolegal misconceptions that may be held by nonphysicians as well as physicians without forensic experience. Indeed, our experience is that Petty's misconception number 3 is alive and well, in that some colleagues continue to simply request a forensic neuropathology consultation, but provide no information and simply ask that "any abnormalities found" be reported.

Even cases with prior "biopsy proven" diagnoses may be found to have incorrect diagnoses, with many surgical pathology diagnoses now being rendered on extremely small needle biopsy samples that may or may not be representative of the underlying lesion, or which are difficult to interpret for a variety of other reasons.[19]

The challenges to the forensic pathologist and forensic consultant are ongoing and evolving,[31] and consulting neuropathologists will benefit from knowledge of current departmental legal policies relevant to their area of involvement.

Table 1.2. Common Errors in Forensic Pediatric Pathology

Incomplete or absent scene investigation.
Inadequate or insufficient photography.
Improper postmortem internal assessment.
Inadequate or incomplete medical records.
Incomplete preautopsy study.
Inadequate gross autopsy examination.
Incomplete microscopic examination.
Incomplete laboratory studies.
Failure to document and differentiate artifacts.
Failure to give appropriate importance and significance to findings.

Data from Sturner WQ. Common errors in forensic pediatric pathology. Am J Forensic Med Pathol 1998;19:317–320.

Table 1.1. Classical Mistakes in Forensic Pathology

Not being aware of the objective of the medicolegal autopsy.
Performing an incomplete autopsy.
Permitting the body to be embalmed before performing a medicolegal autopsy.
Mistakes resulting from nonrecognition or misinterpretation of postmortem changes.
Failure to make an adequate examination and description of external abnormalities.
Confusing the objective with the subjective sections of the protocol.
Not examining the body at the scene of the crime.
Not making adequate photographs of the evidence.
Not exercising good judgment in the taking or handling of specimens for toxicologic examination.
Permitting the value of the protocol to be jeopardized by minor errors.

Data from Moritz AR. Classical mistakes in forensic pathology. Am J Clin Pathol 1956;26:1383–1397.

Table 1.3. The Devil's Dozen: Popular Medicolegal Misconceptions

1. That the time of death can be precisely determined by the examination of the body.
2. That the autopsy always yields the cause of death.
3. That the autopsy can properly be carried out without a "history."
4. That the autopsy is over when the body leaves the autopsy room.
5. That embalming will not obscure the effects of disease and trauma.
6. That only true and suspected homicide victims need examination.
7. That the cause and manner of death are the only results of autopsy.
8. That any pathologist is qualified.
9. That the autopsy must be immediate.
10. That the poison is always detected by the toxicologists.
11. That all physicians are good death investigators.
12. That the medicolegal autopsy is criminally or prosecution oriented.

Data from Petty CS. The devil's dozen: Popular medicolegal misconceptions. South Med J 1971;64:919–923.

An additional comment made by experienced forensic pathologists, such as those listed earlier, is to avoid the temptation to speculate on conclusions prior to having all relevant data available for review. A wide-ranging discussion at the autopsy table of differential diagnostic possibilities, speculation on survival period post-injury, and so on, while perhaps useful in an academic setting when teaching, is counterproductive in a forensic setting. One should avoid any statements that, should you find them attributed to you on page one of tomorrow morning's local newspaper, would make you uncomfortable.

Topographic Approach to Neuropathology

A list of useful tissue blocks one might select for sampling is, in our opinion, an individual choice based on experience, the disorder(s) suspected, the technical facilities for histologic study available to the examiner, and budgetary limitations. Lists of tissue blocks recommended for one or another entity are widely available in the neuropathology literature, and frequently differ from one another. In some cases (e.g., SIDS), department-mandated sampling protocols are followed. In others, it is left to the discretion of the examiner. There is no need to add our personal preference list to the already numerous published lists available. It is recommended that whatever the extent of initial sampling performed, large sampling of the remainder of the CNS be placed in the save bottle for future use if needed (see earlier section, "The Routine Forensic Neuropathology Autopsy"). In some cases, more than one save bottle or even the entire CNS specimen not used for initial screening blocks should be retained in formalin until the slide review with routine stains is completed. At that point a decision is made as to whether additional histologic sampling, special stains, and/or reducing the gross specimen to conserve limited storage space is appropriate.

Sudden Unexpected Death in Infants and Children Due to Natural Causes, and the Suspected Sudden Infant Death Syndrome Case

Chapter 9, on Sudden Unexpected Death, includes neuropathologic conditions that may involve a wide range of age groups. Byard's[20] monograph on this topic also provides a wealth of useful information.

Forensic neuropathology consultation is performed in all sudden unexpected death (SUD), and all suspected sudden infant death syndrome (SIDS) cases, in our department. The suspected SIDS case differs from other referrals in that sections of pontomedullary junction, pons, midbrain, hippocampus, frontal lobe, cerebellum, and choroid plexus are obtained in every case, as recommended in the Standardized Autopsy Protocol for the CNS evaluation of SIDS cases approved by the California Department of Health Services.[23] The recommended discretionary microscopic sections of cervical spinal cord and basal ganglia are also taken routinely, and thalamus is generally included in the block for "basal ganglia" due to its vulnerability to hypoxic-ischemic injury in this age group. Initial screening is performed using only routine hematoxylin-eosin (H&E) stain.

A portion of the suggested SIDS protocol to which our department policy does not presently conform is to provide fresh and fixed tissue weights separately for supratentorial and infratentorial intracranial CNS structures. The information gained by additional manipulation and sectioning of the soft, immature brain does not outweigh the likelihood of mechanical artifact introduced (particularly to the midbrain area). We also note that, although figures for differential supra- and infratentorial weights are available for the 10- to 41-week gestational age period,[59] such figures are rare, difficult to locate, and virtually ignored in contemporary texts and articles on brain weights in the age groups most relevant to SIDS cases.[15] Such figures are also minimally available for adults,[15,97] and no definitive abnormality in SIDS brain weights has been found in studies to date.[147,157]

If large case series providing differential supra- and infratentorial brain weights exist for the SIDS age group, they have escaped our notice.

SIDS continues to generate considerable controversy due to its frequency, evolving concepts regarding its definition and diagnosis, and varied suggestions as to its underlying cause(s). In addition to the monographs[21,161] and chapters in books,[20,52,84] numerous journal articles are devoted to this topic. A brief overview for the consulting forensic neuropathologist with minimal prior experience in this category of cases, since most cases in this country fall under the jurisdiction of the medical examiner and are not usually autopsied at teaching institutions or private hospitals, might include not only references such as those noted previously, but also some representative samples of diagnostic and definitional issues.[11,46,54,71,89,145,153] A review might also include examples of the importance of examining not only expressed opinions[55,87] but also sources for critiques of such opinions[14,130] in an effort to assess the pros and cons of a given concept.

From the neuropathologist's standpoint, the California Department of Health Services SIDS protocol includes recommendations for the DME to document scalp abnormalities, head circumference, calvarial abnormalities (sutures, fontanels, head shape, etc.), foramen magnum appearance, meninges, fresh brain weight, and spinal cord appearance and abnormalities. Routine brain gross and microscopic examination is recommended, including the mandatory and discretionary microscopic sections noted previously. Blood is routinely screened for acylcarnitine profile and galactose (Gal) and galactose 1-phosphate (Gal-1-P), both of which are directly relevant to SUD. Blood is also routinely screened for 17-hydroxyprogesterone (for congenital adrenal hyperplasia) and thyroid-stimulating hormone (for congenital hypothyroidism), which are useful but less likely to be a cause of SUD.

By definition, neuropathology examination does not reveal a cause of death in SIDS cases, and it is the role of the consultant to exclude occult, potentially fatal CNS disorders as well as provide an opinion on the significance (or lack thereof) of subtle abnormalities that may be found.

A growing literature regarding neurophysiologic and neuropathologic findings in SIDS cases was reviewed by Kinney and Filiano in 2001,[82] Sparks and Hunsaker in 2002,[147] and Kinney and Paterson in 2004.[84] Neurophysiologic studies have largely centered on neurotransmitters, and anatomic studies have tended to focus on subtle changes in brain maturation and on brainstem autonomic nuclear group differences from controls as revealed by quantitative histologic and immunohistochemical studies. Additional neuropathologic studies on SIDS cases have been published since the review articles noted earlier.[55,75,81,83,102,107,155,158,159]

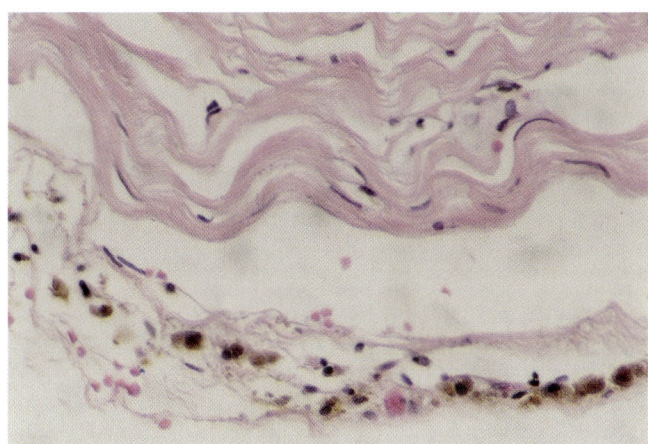

Fig. 1.11. Example of the very thin subdural neomembranes that are grossly represented by a faintly visible tan discoloration on the inner dura mater, and microscopically typically consist of loosely organized collagen fibers, a few capillary-sized blood vessels, and some scattered macrophages containing hemosiderin-like pigment.[135] Native dura, inner surface, is in upper one-half of photo. (H&E.)

At present there is no reliable anatomic marker for the diagnosis of SIDS discernable by commonly employed routine neuropathologic gross and microscopic study.

As noted elsewhere in this text, subtle subdural resolving neomembranes that could date to birth do not exclude a diagnosis of SIDS (Fig. 1.11).[135]

Increased evidence of skull asymmetry in children who die of various causes might be expected in the next few years as a consequence of the evident success of the American Academy of Pediatrics program encouraging the supine sleeping position as a means to reduce unexplained infant deaths.[70] This may not develop if the recommendations of Peitsch et al.[119] succeed in preventing this apparent cause-and-effect relationship between supine position sleeping and skull asymmetry. Clinical studies in several areas, such as the search for early chemical markers of infants at risk for SIDS[105,146] and clinical clues derived from cases of breath-holding spells in infants,[16,17,37,78,90,100] may gradually reduce not only the incidence of this category of death, but also yield insights regarding yet-undiscovered mechanisms of sudden unexpected death in this age group.

Time Since Death

Time since death based on gross and histologic examination in stillbirths is briefly discussed in Chapter 4. Otherwise, the forensic neuropathologist is unlikely to address such issues unless questions arise concerning persistence of certain spontaneous movements after brain death (see Chapter 3, section on brain death). The examination of the eye as an aid to time since death (see Chapter 7), and the testing of certain reflexes as an aid to determine the time of death have also been described in the literature. To our knowledge, the latter are confined to occasional research studies[103] or found in sources that are more of historical interest,[141] and find no widespread practical application currently in the U.S. forensic community.

Craniocerebral Relationships

During routine autopsies, brains in the fresh (unfixed) state have a consistency such that gravity-induced alterations in shape and craniocerebral relationships significantly differ from conditions while the person was alive. This soft parenchymal consistency also results in considerable distortion of the brain when sectioned in the fresh state at the autopsy table. The result is that, although most neuropathologic findings such as the path of projectiles or sharp force injury can be determined by major anatomic landmarks, subtle observations may be lost due to tissue fragmentation and stretching. There may be a need to make rather thick sections to allow handling of sections with less distortion or fragmentation. Many other examples of information lost by sectioning of unfixed brains at the time of autopsy could be given. The point is, with rare exceptions, if neuropathologic information is considered relevant to issues that may arise in a given case, fixation prior to sectioning is preferable. A portion of brain can be removed at the autopsy for viral or other special studies, for example, and the remainder of the unsectioned brain placed in fixative for later study.

Craniocerebral relationships are even more difficult to assess in decomposed or burned brains, owing to CNS liquefaction, dehydration, and other resultant anatomic distortions.

The question thus arises regarding the degree to which one can expect to accurately extrapolate the position of certain brain structures during life from external landmarks on the skull. Obviously it would be useful, for example, to be able to describe what specific internal brain structures were involved in a gunshot wound path if one knew only the exact location of the entrance and exit wounds in the skull. Craniocerebral relationships are typically commented upon rather briefly in older[57,69] and more recent textbooks of anatomy. The problem is minimal in more localized areas of the cranial vault such as the posterior fossa or pituitary fossa. More often, the issue arises in the cerebral hemispheres.

One of the first detailed efforts to determine craniocerebral relationships prior to the modern imaging era was that of Symington in 1903.[152] Others examining this topic found that "neither the position nor the inclination of the coronal suture is constant with respect to other skull landmarks" and that that the Freeman-Watts landmark relative to the longitudinal, central, and lateral fissures of the brain "showed wide variation" (e.g., up to several centimeters in some cases).[138] The position of the asterion relative to the underlying transverse sinus is also variable,[32] and the distance of the motor cortex from the coronal suture increased with patient age in a study of 17 cases ranging in age from 10 months to 14.6 years.[134] Many other examples could be cited. For the neurosurgeon, these issues have been largely resolved by modern imaging and stereotactic surgical techniques.

These are among considerations leading to DMEs' being instructed to fix brains prior to sectioning if there is any question of later neuropathology issues arising.

In cases where careful CNS anatomic study did not occur, premortem CNS imaging studies, if performed, can be obtained for review (a forensic radiology consultant is available to our department). Investigative reports and witness statements can be reviewed, and the data analyzed and compared with skull wound measurements and with cross-sectional radiology atlases such as that of Cahill et al.[22] This combined approach often, but not always, allows conclusions as to at least the major brain structures involved in a given wound path with reasonable medical certainty.

Meninges

Questions pertaining to meninges are relatively limited. A few points of interest are the following:

- Cranial dura mater is composed of two layers, periosteal and meningeal. Spinal dura mater has one layer, meningeal, surrounded by epidural fat, venous plexus, etc.
- Conspicuous meningeal arteries are a good marker for the external layer of the cranial dura mater (Figs. 1.12–1.13).
- Leptomeningeal vessels often appear congested at autopsy, even when thoracoabdominal organs are removed prior to intracranial examination (Fig. 1.14).
- Spinal arachnoid diverticulae, although uncommon, are considered a major diagnostic criterion for Marfan's syndrome. They may also be an isolated idiopathic finding, and have been described in association with congenital pigmented nevi, diastematomyelia, multiple sclerosis, and syringomyelia.[27,34] Other than Marfan's syndrome and diastematomyelia, it seems the other speculated associations have not held up over time.
- White leptomeningeal plaques occur that are composed of fibrous tissue, typically in a lamellar pattern, and at times with a pia-arachnoid cell surface layer (Figs. 1.15–1.17). They are more commonly encountered in older persons, and are occasionally more

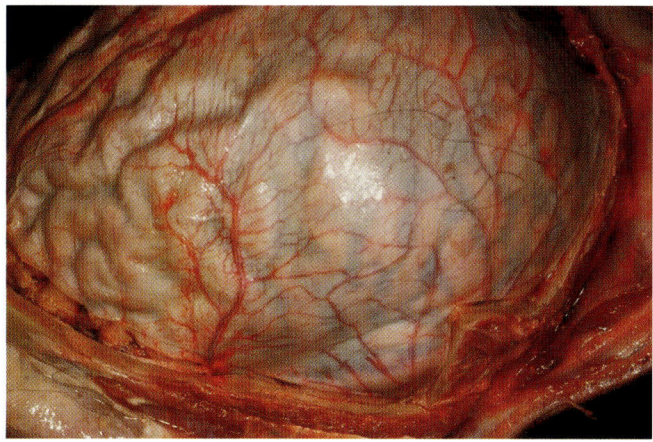

Fig. 1.12. Normal appearance of external aspect of cranial dura mater after calvarium removed. Middle meningeal artery branches are more visible on the external surface of the dura mater.

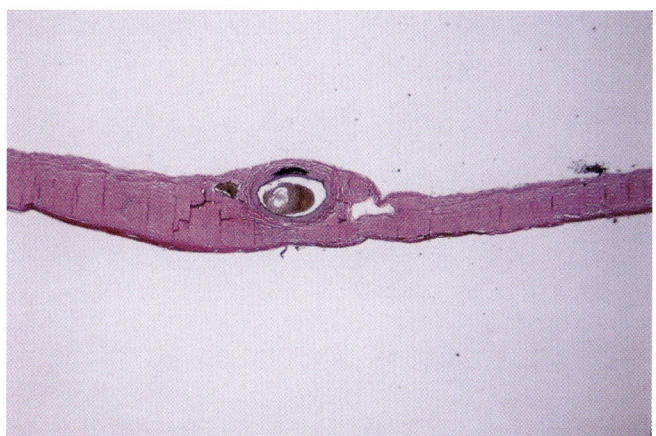

Fig. 1.13. Low-power micrograph of cranial dura, with middle meningeal artery branch in the external (periosteal) layer of dura; this is a useful landmark to distinguish the external from the internal layer of cranial dura mater. (H&E.)

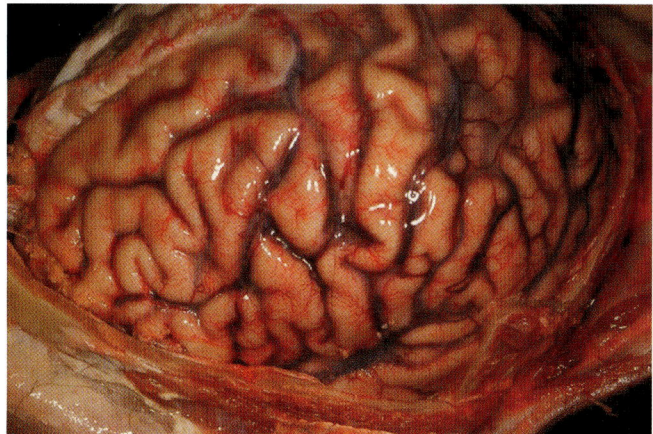

Fig. 1.14. Usual autopsy appearance of brain leptomeninges after removal of dura mater. Mild to moderate leptomeningeal vascular congestion, as seen here, is common.

Fig. 1.15. Gross appearance of leptomeningeal white plaques on dorsal aspect of lumbosacral spinal cord.

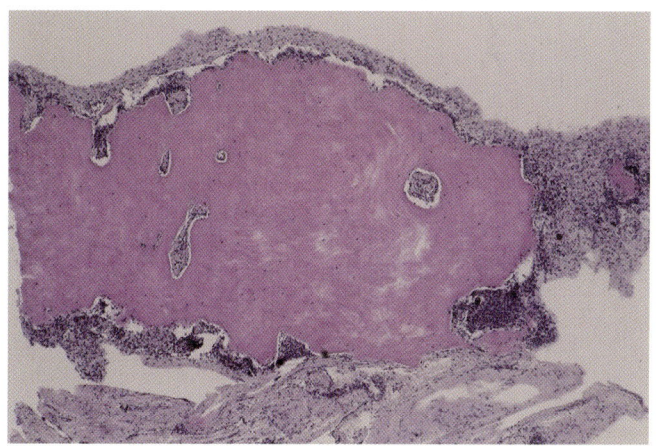

Fig. 1.16. Low-power micrograph of leptomeningeal white plaques, primarily composed of paucicellular dense fibrous tissue. Arachnoid cells are evident in and around the plaque. (H&E.)

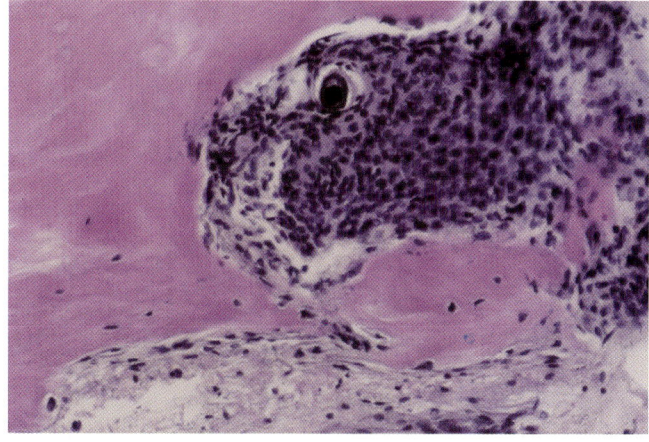

Fig. 1.17. Medium-power micrograph of leptomeningeal white plaque, with psammoma body within an arachnoid cell aggregate. Calcification in the fibrous tissue may occur in some cases. (H&E.)

focally prominent in the vicinity of other spinal pathology. They often have a hyalinized appearance, and may contain areas of calcification or ossification.[85] They are typically found on the dorsal leptomeninges of the lower thoracic and lumbar spinal cord. Fuller and Burger[48] also indicate that hyaline plaques may occur in cerebral leptomeninges. No importance is attached to these plaques except for clinicians who may encounter their resistance during a lumbar puncture procedure.

- Mild to moderate lymphocytic cuffing is sometimes seen around a rare blood vessel in leptomeninges, or in the Virchow-Robin space, particularly in and around the inferior olivary nucleus in cases devoid of any other abnormality. It is encountered commonly enough in isolation that it is probably not of significant diagnostic importance.
- Leptomeningeal melanocytes are common, particularly in dark-skinned individuals, and should not be mistaken for hemosiderin-containing macrophages or other abnormality (see later section, "Discoloration of Skull, Meninges, and Brain").

Most other common questions are well described in standard textbooks, such as by Fuller and Burger.[48]

Brainstem

Certain features of brainstem anatomy lead to occasional questions, answers to which, while in the literature, may be somewhat difficult to find.

Neuronal lipofuscin accumulation with increasing age is widespread, but occurs preferentially in certain brain locations. It is particularly prominent in inferior olivary nuclei, dentate nuclei, anterior horn cells, and dorsal root ganglion cells. Distinction from the pathologic accumulation of lipofuscin in disorders such as the neuronal lipofuscinoses should present no problem. Purkinje cells usually contain relatively mild lipofuscin accumulation.[66]

Neuromelanin-containing cells occur not only in the substantia nigra (Fig. 1.18) and locus ceruleus, where the neuromelanin is grossly visible in more mature brains, but also in the dorsal motor nucleus of the vagus, neurons of the red nucleus capsule, dorsal root ganglia, sympathetic ganglia, and scattered neurons in the roof of the fourth ventricle.[66] Other nuclei that may contain neuromelanin include the mesencephalic nucleus of cranial nerve V and the nucleus parabrachialis medialis.[117] In the substantia nigra, neuromelanin may be seen using routine light microscopy by 2 to 5 years of age, and is grossly visible by the early to mid second decade, gradually increasing with aging.

Brainstem nuclei that normally may demonstrate more conspicuous glial satellitosis around neurons, not to be mistaken for neuronophagia, include the nucleus of the superior and inferior colliculus, nucleus intercollicularis, red nucleus, nucleus parvocellularis, and nucleus cuneiformis.[117]

Somas of most CNS neurons contain Nissl substance distributed throughout the cytoplasm (Fig. 1.19). A number of brainstem nuclei, however, contain neurons that normally demonstrate more peripherally situated Nissl substance, with a relatively pale perinuclear zone (Fig. 1.20). Familiarity with this appearance allows one to avoid mistaking neurons with such nuclei for a neuron with true central chromatolysis. Such normal cells are seen in many nuclei, with a few of the more easily recognized examples being the dorsal motor nucleus of cranial nerve X, mesencephalic nucleus of cranial nerve V, Edinger-Westphal nucleus, nucleus gracilis and cuneatus, and Clark's column in the spinal cord.[17] In true chromatolysis, perikaryal pallor is typically more exaggerated (Fig. 1.21), accompanied by peripheral nuclear displacement and peripheral margination, with altera-

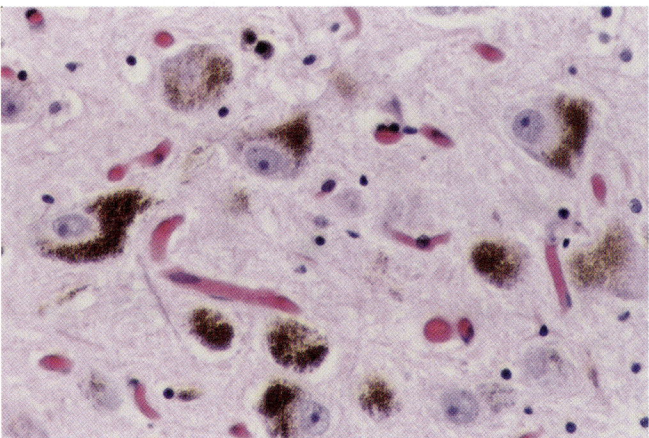

Fig. 1.18. Normal substantia nigra neuromelanin-containing neurons. (H&E.)

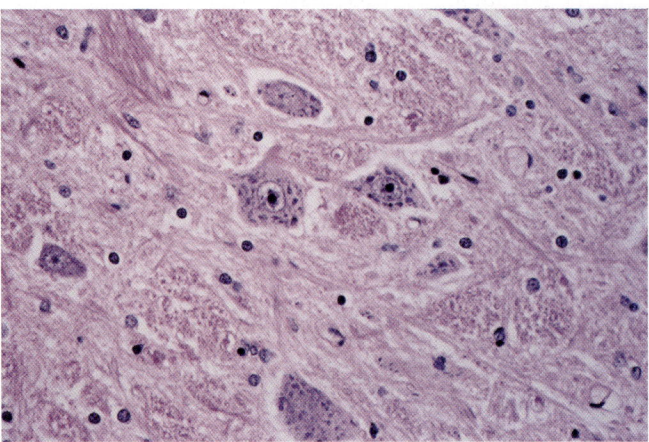

Fig. 1.19. Normal nucleus ambiguus neurons demonstrating diffusely distributed Nissl substance, as is commonly seen in most CNS neurons. (H&E.)

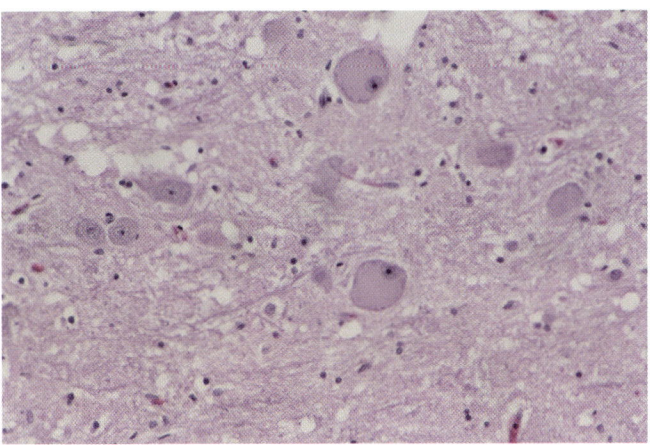

Fig. 1.20. Normal lateral cuneate nucleus neurons, demonstrating peripherally distributed chromatin producing a "pseudochromatolysis" effect (see text). (H&E.)

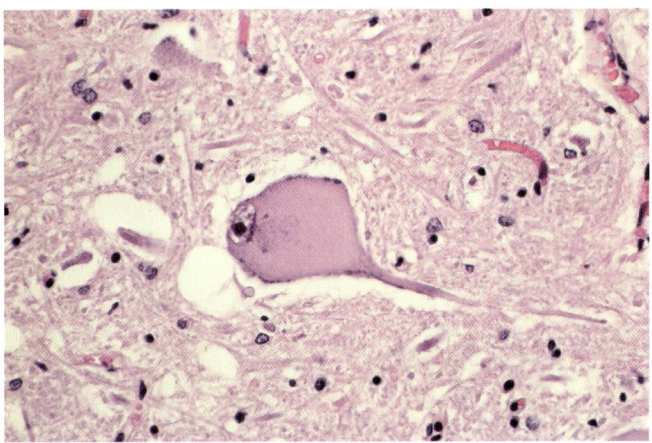

Fig. 1.21. Abnormal chromatolytic anterior horn cell from a case of Guillain-Barré syndrome. (H&E.)

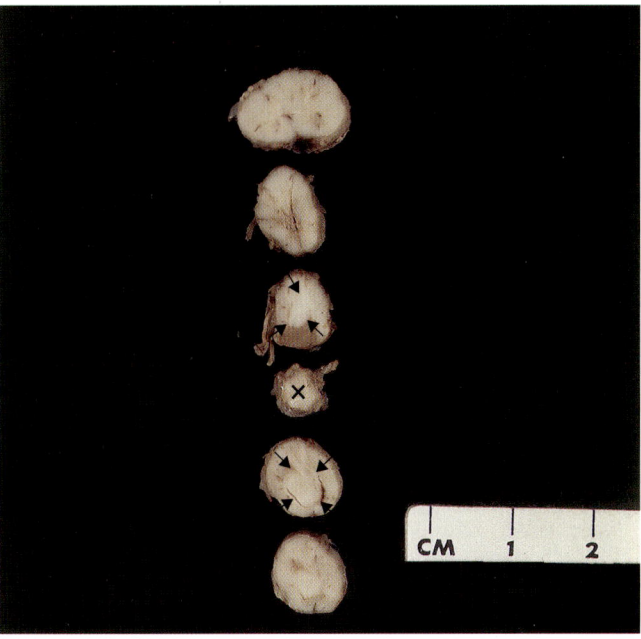

Fig. 1.22. Spinal cord "toothpaste" artifact. X marks the zone of maximum compression artifact, with cord parenchyma at this level forced upward and downward to create round masses (*arrows*) mimicking a tumor. A common artifact in spinal cord removal involving excessive compression or traction.

tion of Nissl substance progressing to an extent that Nissl substance becomes undetectable by routine light microscopy, and additional abnormalities indicating the cause of the chromatolytic reaction will be present.

Measurement of brainstem diameter attracted relatively little attention until neuroradiologists began efforts to add imaging criteria to the clinical diagnostic features in disorders such as parkinsonism, progressive supranuclear palsy, and olivopontocerebellar atrophy. The sparse data we include here is for unusual situations in which some quantitative data may be considered useful.

In adult males, the transverse diameter (width) of the medulla has been measured to be approximately 17.8 to 18.2 mm, and the width of the pons from 27 to 34 mm (mean, 30.6 mm).[15] Imaging study measurements of brainstem differ somewhat from autopsy specimen measurements, as also occurs in the case of spinal cord measurements (see later). Radiologically, the medulla width has been measured as approximately 14 to 18 mm, and the pons transverse diameter from 29 to 39 mm.[128]

The widest adult midbrain (at the level of the red nuclei) measurement range in one imaging study was 2.7 to 5.3 cm (mean 4.03 ± 0.6 cm).[40] Our informal measurements of midbrain width are nearly always in the range of 3.5 to 4.0 cm in the absence of general atrophy or midbrain compression secondary to brain swelling.

It should be noted that imaging studies tend to emphasize sagittal measurements, or measurements at angles other than strictly anterior-posterior or transverse, and attention to methodology is necessary when attempting to correlate imaging measurements with gross specimen measurements.[40,86,164] Information on brainstem measurements by imaging techniques in children is also included in the study by Raininko *et al.*,[128] including some data on coronal widths in the pons and medulla.

Spinal Cord

Preferably, whenever possible, the spinal cord should be removed by the posterior approach,[45,101] which is less likely to result in artifact when properly performed (Fig. 1.22). The spinal cord might not be removed in the absence of any suspicion that it will be other than normal. The rapid cord traction removal method described by

Spitz[149] and by Dolinak and Matshes[38] cannot be recommended due to inevitable artifact and incomplete specimens, in our experience.

Whether the spinal cord is too large or too small in diameter is often determined by visual approximations, which vary with the examiner's prior experience. Rather consistently, adult spinal cord diameters measure very close to 1.5 cm in lateral diameter and 1.0 cm in anterior-posterior (AP) diameter at the C6 spinal segment, 1.0 cm in lateral diameter and 0.8 cm in AP diameter at the high thoracic level, and 1.0 cm in lateral diameter and 1.0 cm in AP diameter at the lumbosacral enlargement. Focal swelling, atrophy, hyperemia, or discoloration should be self-evident. Data exist on adult spinal cord diameters, with some sources providing considerable detail. Formalin fixation was not found to alter spinal cord measurements more than approximately 0.5 mm according to Elliott.[42] Measurements made by radiologic imaging studies, including those in embalmed cadavers,[25,51,115,137,143,156] and measurements of histologic sections[76] tend to be smaller than those made in fixed gross specimens. This should be considered if one compares radiographic measurements of spinal cords in infants and children,[131] spinal cords examined at autopsy, or cord examination following formalin fixation. If a large case series of infant or child lateral and AP measurements performed on formalin-fixed spinal cords has been published, it has not yet come to our attention.

Cord length is variable, and not necessarily related to body height.[108] Average lengths are therefore of limited value (i.e., 44.79 cm in adult males, and 41.8 cm for adult females). Average changes in stature have also likely occurred in the population since that early study was performed. During development, the conus becomes progressively more rostral in relationship to the vertebral column, reaching the L1-L2 interspace between approximately 35 weeks' gestation to 2 months postnatal age.[64] A conus level below the L2-L3 interspace beyond 5 years of age is considered abnormal (sometimes referred to as a tethered cord).[64] In adults, McCotter (in a study of 234 cases)[108] found the maximum range of the lower level of the spinal cord to be from the middle of the body of the T12 vertebra to the inferior border of the L2 vertebra. The conus medullaris and filum terminale are infrequently examined in a forensic autopsy setting, and may contain astrocytic and ependymal cell clusters, degenerative neurologic elements, or other features that may suggest pathology if one is not familiar with the normal histologic features of this region.[26]

It is not unusual in a forensic setting to receive less than an entire spinal cord for neuropathology consultation. The general region (cervical, thoracic, or lumbosacral) is easily determined by cord size and/or cross-sectional gray and white matter contour, but accurate segmental level identification is not always possible. If the cervical-thoracic junction area is included with the specimen, however, one can be more definite as to segmental level (of a lesion not already determined at the time of autopsy) by a method familiar to neuropathologists.[29] The brachial plexus is typically formed primarily from contributions from spinal segment C5-T1, with minimal contributions from C4 and T2 (excluding the infrequent instances of prefixed or postfixed plexus). The result is a conspicuously larger T1 spinal root compared to the T2 spinal root, allowing one to use the discrepancy in size between T1 and T2 as an easily identifiable gross landmark (e.g., the lowest large root in the cervical enlargement–upper thoracic region) from which to count roots rostrally or caudally.

Histologically, an easy means to determine whether one is examining an anterior or posterior spinal rootlet is that anterior rootlets contain a rather uniform population of large nerve fibers (Figs. 1.23–1.24), whereas posterior rootlets contain a much broader spectrum of admixed large and small nerve fibers (Figs. 1.25–1.26).

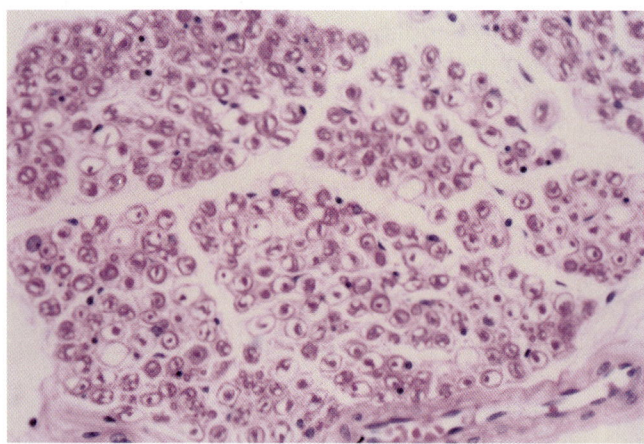
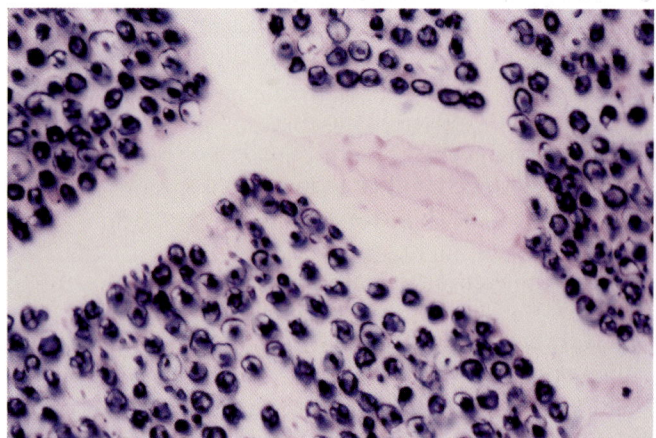

Figs. 1.23–1.24. *Left* (Fig. 1.23; H&E) and *right* (Fig. 1.24; Weil). Anterior spinal root cross-section. Relatively uniform axon diameters are present throughout the root.

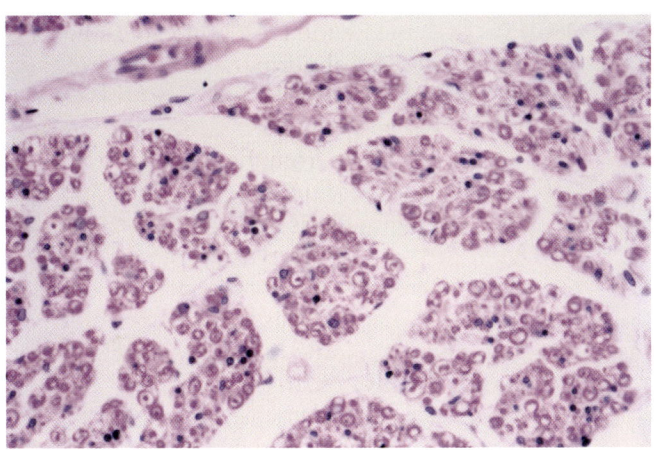 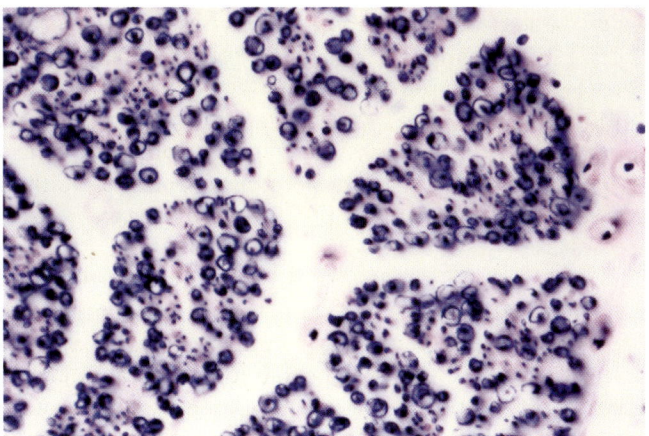

Figs. 1.25–1.26. Left (Fig. 1.25; H&E) and right (Fig. 1.26; Weil). Posterior spinal root cross-section, demonstrating marked variation in axon diameters.

Neck Dissection

Infrequent autopsy table consultations by the neuropathologist involve neck dissections, which are performed after thoracoabdominal viscera and intracranial contents have been removed. Experienced medical examiner colleagues routinely do careful layer-by-layer neck dissections in cases with known or suspected neck trauma (strangulations, etc.), and examination of the neck of special neuropathologic interest should be coordinated with their examination.

Sources of additional information helpful in circumstances where such guidance is less available include the following:

- Technique details for anterior and posterior neck dissection.[1,2]
- Special en bloc techniques for examination of the craniocervical junction and cervical spinal cord.[12,50]
- Technique for examining the intraosseous portion of the carotid artery.[92]
- Technique for vertebral artery dissection.[79]
- Technique for minimizing extravasated blood artifact in neck dissections by in situ rather than en bloc dissections.[125]
- Need for caution in interpreting hemorrhage in retroesophageal and cervical paravertebral areas as trauma-induced rather than as artifact secondary to congestion and rupture of pharyngolaryngeal plexus vessels.[36]
- Need for caution in interpreting the significance of cervical and upper thoracic epidural space hemorrhages in infants less than 1 year of age.[63]

This last article[63] described either focal or confluent epidural hemorrhage, usually in the dorsal or dorsolateral quadrant of the epidural space, in 15 of 19 infants without evidence of trauma. Most cases were also associated with pulmonary edema and/or thoracic petechiae in cases of sudden unexpected death, and evidence of respiratory tract infection was present in most cases. The methods section of that report indicated that the en bloc removal of the vertebral column with the spinal canal and its contents was performed after removal of neck organs, but does not specify whether intracranial and thoracoabdominal organs had previously been removed; the reputation of the investigators would lead us to believe that they were. Additional evidence of trauma would be advisable before concluding that trauma played a role in this specific finding.

Discoloration of Skull, Meninges, and Brain

Bodies undergoing decomposition/putrefaction, or in contact with soil, may develop a variety of soft tissue and bone color changes, primarily of interest to the forensic anthropologist and general forensic pathologist.[61,140]

Embalming procedures may cause brain pallor (in well-perfused brains), irregular decomposition (in inadequately perfused brains), or increased pink to reddish color (due to dyes used to simulate natural skin tones).[62] Perfusion during the embalming procedure will often cause some enlargement of the brain, with flattening of gyral crowns and grooving of uncus and cerebellar tonsilar/biventer lobule areas, superficially mimicking brain swelling with herniation syndrome. Formalin fixation may impart a grayish discoloration to tissues in some cases,[62] and decolorize the yellow color of the basal ganglia and other affected areas of the brain in kernicterus. However, formalin does not cause significant fading of the cherry red color caused by CO intoxication. Yellow jaundice converts to green jaundice by oxidation. Some color changes of interest to the forensic neuropathologist are presented in Table 1.4.

Table 1.4. Possible Sources of Skull, Meninges, or Brain Discolorations

Color	Possible Associations
Skull color changes	
Yellowish skull (Figs. 1.27–1.28)	Remote chronic tetracycline treatment,[165] diabetes[44]; minocycline treatment
Yellow/green/black skull	Minocycline treatment
Meninges color changes	
Yellow to green dura mater, leptomeninges (or brain, in areas of blood–brain barrier breakdown) (Fig. 1.29)	Jaundice
Black discoloration of dura	Ochronosis[98]
Widespread leptomeningeal rust-brown color	Superficial siderosis[91]
Focal leptomeningeal/cortical rust-brown color	Remote contusion
Foci of brown/gray/or black leptomeningeal discoloration (Figs. 1.30–1.33)	Leptomeningeal melanosis
Leptomeningeal congestion with white to yellow to greenish infiltrates	Meningitis (*note:* in cryptococcal meningitis, the brain surface feels mucoid or slimy, but is not discolored)
Brain color changes	
Dusky pink	Congestion
Cherry red (Fig. 1.34)	Carbon monoxide poisoning; cyanide poisoning
Dusky gray, congested and soft	Respirator brain
Slate gray (Fig. 1.35)	Cerebral malaria[43]; may also occur in patchy distribution with postmortem overgrowth of bacteria
Grayish to white to faintly yellowish, slightly translucent areas	Chronic gliosis
Greenish-gray to greenish-purple (fades with formalin fixation)	Hydrogen sulfide intoxication[3,6]
Bright yellow color of several nuclear groups, mainly globus pallidus, subthalamic nucleus, and hippocampus (fades with formalin fixation)	Kernicterus[47]
Globus pallidus and substantia nigra rust brown to yellow brown color	Hallervorden-Spatz disease[43,99]
Globus pallidus ± putamen atrophy and gray to brown discoloration	Chronic sequelae of any cause of severe hypoxic-ischemic injury; Marchiafava-Bignami disease; Leigh's disease; infantile bilateral striatal necrosis; and some mitrochondrial disorders[43]
Brownish to reddish-brown or brick-red atrophic globus pallidus and predominantly middle one-third of putamen (Fig. 1.36)	Wilson's disease[41,43]
Slate gray to gray-green putamen	Striatonigral degeneration[43,99]
Depigmentation of substantia nigra, usually also locus ceruleus	Parkinsonism (idopathic, postencephalitic, and secondary to MPTP toxicity); progressive supranuclear palsy; corticobasal degeneration[43]
Pink spots with white centers in deep hemispheric areas in formalin-fixed brains	Present in some cases with infectious bacterial disease[80]

MPTP, 1-methyl-4-phenyl-1,2,3,6-tetrahydropyridine.

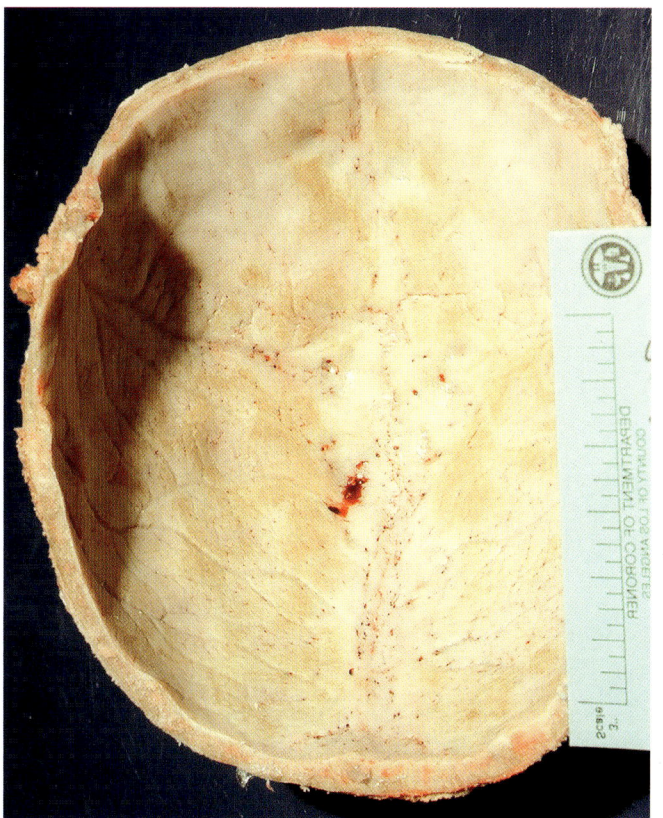
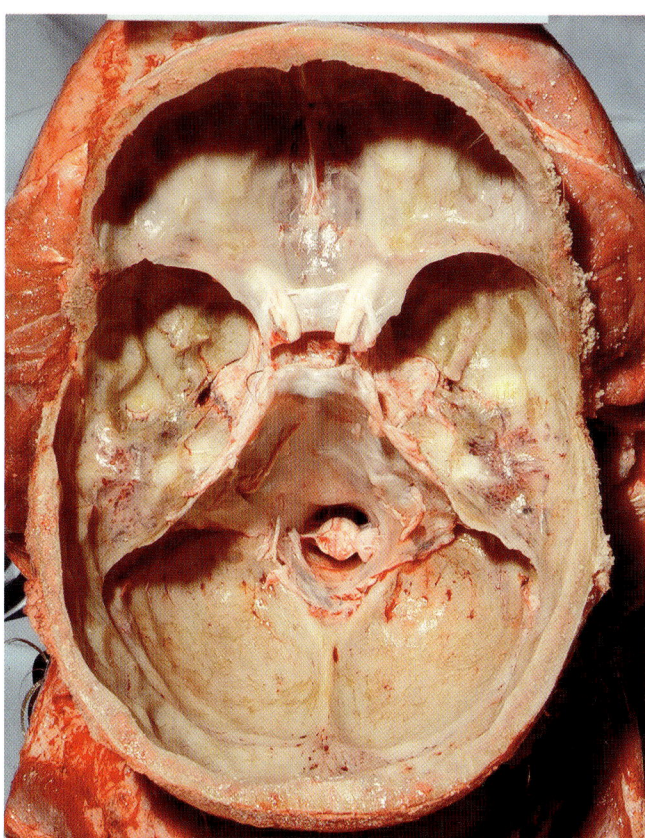

Figs. 1.27–1.28. *Left* (Fig. 1.27), inner calvarium, and *right* (Fig. 1.28), skull base. Yellow skull discoloration; probable remote tetracycline treatment.

Leptomeningeal melanosis is very commonly encountered in locations such as the ventral aspect of the medulla and cervical spinal cord, particularly in racial groups with dark skins. Less frequently it may occur in a patchy, at times symmetric distribution elsewhere, such as on the dorsal or ventral, mainly frontal, cerebral hemispheres (Figs. 1.30–1.33). A case in which an initial examiner misinterpreted such changes as foci of cortical bruising related to child abuse was later readily disproved by microscopic examination demonstrating that the discoloration was due to leptomeningeal melanosis without underlying cortical abnormality.

Selected Artifacts

A variety of selected artifacts are described in standard atlases and textbooks.[167] Normal CNS findings sometimes misinterpreted as pathology are also reviewed by Fuller and Burger.[48] A few additional comments follow.

1. *Gravity-Related Congestion.* Body position influences postmortem location of blood in livor mortis, and can cause not only prominent congestion in dependent areas but also petechial or larger blood leakages in scalp or intracranial areas. The latter may appear as

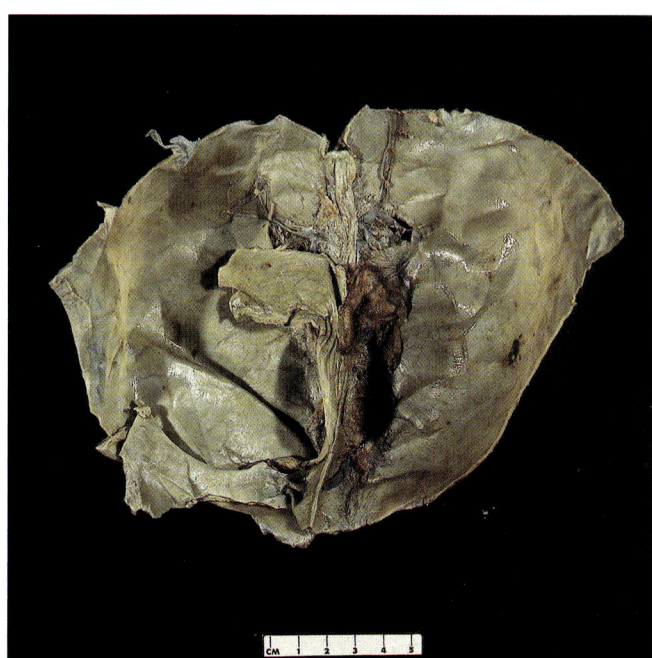

Fig. 1.29. Green dura mater in case of jaundice.

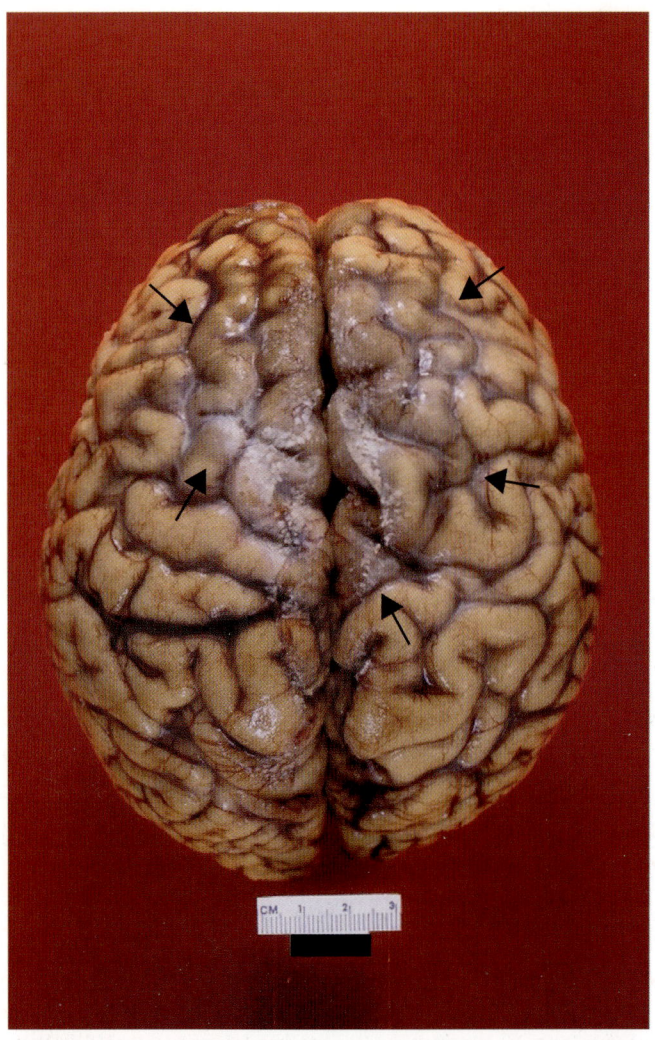

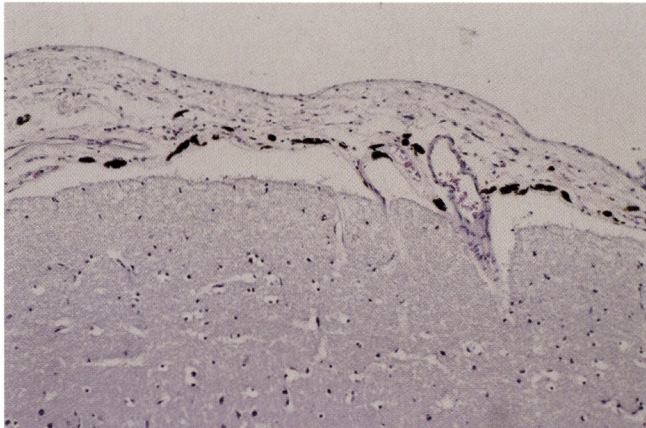

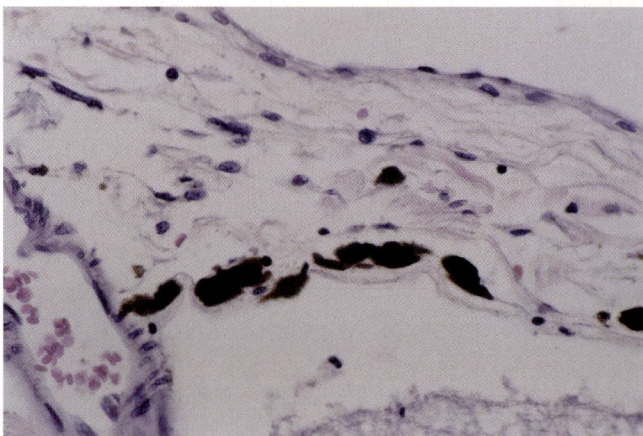

Figs. 1.32–1.33. *Top* (Fig. 1.32; low-power) and *bottom* (Fig. 1.33; medium-power) micrographs of leptomeningeal melanosis from case seen in Figs. 1.30 and 1.31. Slender to somewhat more plump spindle cells packed with brown melanin granules. (H&E.)

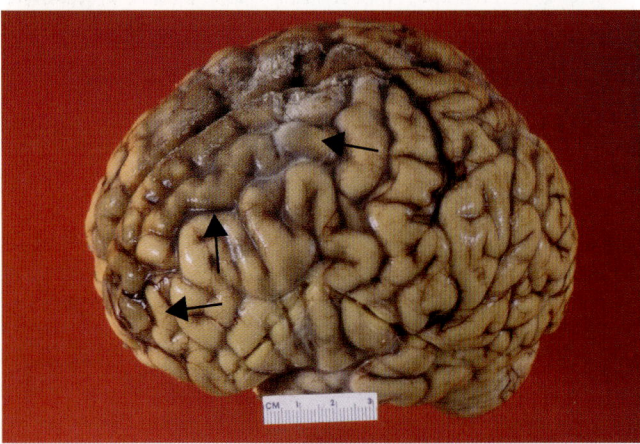

Figs. 1.30–1.31. *Top* (Fig. 1.30; dorsal view) and *bottom* (Fig. 1.31; lateral view). Cerebral leptomeningeal melanosis (*arrows*).

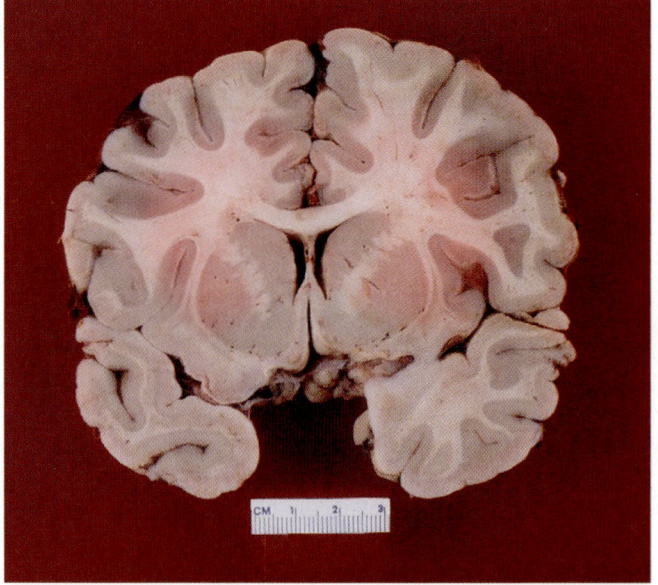

Fig. 1.34. Cherry red brain discoloration in case of carbon monoxide intoxication.

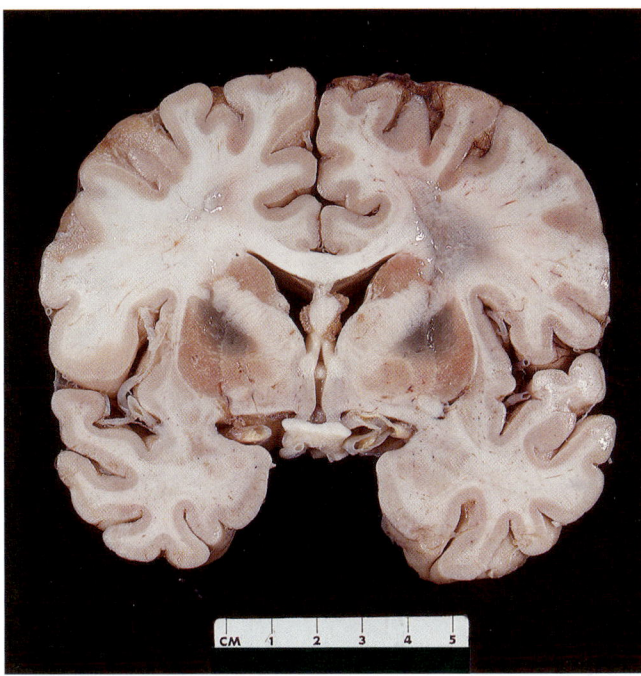

Fig. 1.35. Patchy gray discoloration in gray and white matter due to postmortem bacterial overgrowth without gas formation.

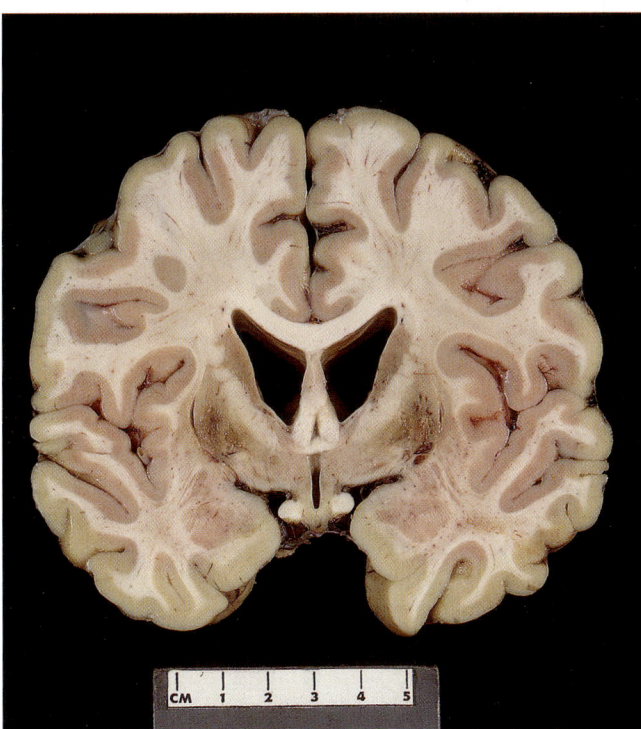

Fig. 1.36. Brownish discoloration of medial lenticular nuclei in Wilson's disease.

localized subdural or subarachnoid hemorrhages,[36] indicating the importance of scene examination and correlation of neuropathologic findings with descriptions of more superficial structures in the general autopsy report.

2. *Diagnostic Procedures.* In addition to the usual artifacts produced by antemortem therapeutic efforts, postmortem diagnostic procedures, such as cisternal puncture to obtain a cerebrospinal fluid (CSF) sample, can produce local vascular injury and hemorrhage.[65] When one person performs a cisternal puncture at autopsy, and another prepares the CNS examination weeks later, the potential for interpretive error increases.

3. *Selective Vulnerability to Autolysis.* Minute vacuoles in the CNS are so common in autopsy material because of death to autopsy time lag, that diagnosis of the earliest stages of hypoxic-ischemic injury or edema is compromised compared with perfusion-fixed animal experimental material. For example, to diagnose early edema at a lesion margin we require a distinct difference in frequency and size of clear vacuoles near the lesion, and a gradation to the lesser degree of vacuolization elsewhere in the neuropil as distance from the lesion increases.

The most common CNS site of selective postmortem autolysis is the cerebellar granular layer, an artifact that is important to distinguish from the several degenerative, toxic, and metabolic diseases that may preferentially involve this neuronal population[73] (Figs. 1.37–1.39). It is a common finding in respirator brain death cases.[116] Grossly, the presence of a light gray line in the deeper cerebellar cortex is seen in advanced cases (see Fig. 2.21). Studies indicate that this is a postmortem artifact,[5,73] which increases in frequency with increasing interval between death and autopsy. It may be more likely to occur in patients with terminal acidosis.[4]

4. *Pupillary Inequality.* Unequal pupils, a useful clinical finding in the neurologic examination of patients, may also occur as a postmortem artifact due to asymmetric rigor mortis of involuntary muscles in the iris.[120]

Brain Weight Measurements

Brain and other organ weights are routinely recorded in autopsy protocols. Many studies attempting to determine normal brain weight ranges in varying age groups exist, and current textbooks on autopsy methods or on neuropathology demonstrate considerable differences of opinion on which study or studies on this topic deserve to be cited. Certain studies that are rather widely quoted, and those on which we have tended to place more emphasis, are summarized in the following (in order of increasing age range). This section concludes with some comments by the authors.

1. Maroun and Graem[106] published a report in 2005 in which they studied 1800 cases ranging in gestational age (GA) from 12 to 43 weeks. Advantages of this study are its attention to degree of maceration of the fetus, specifying that the brain weights are based on fresh specimens (i.e., nonformalin-fixed), its use of 1-week intervals, and its more recent publication in comparison

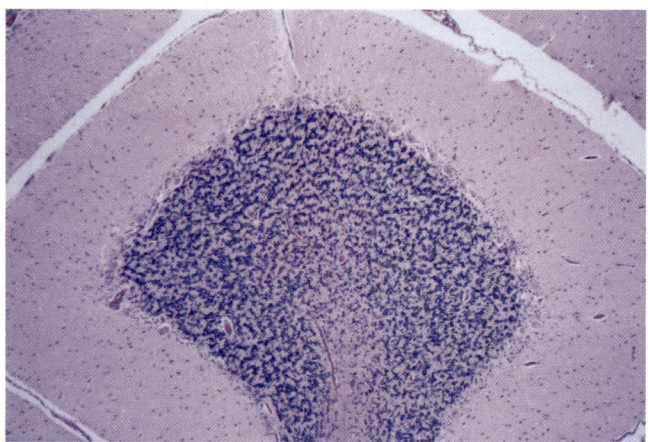

Fig. 1.37. Normal cerebellar cortex. Note granule cell packing density and degree of basophilia. (H&E.)

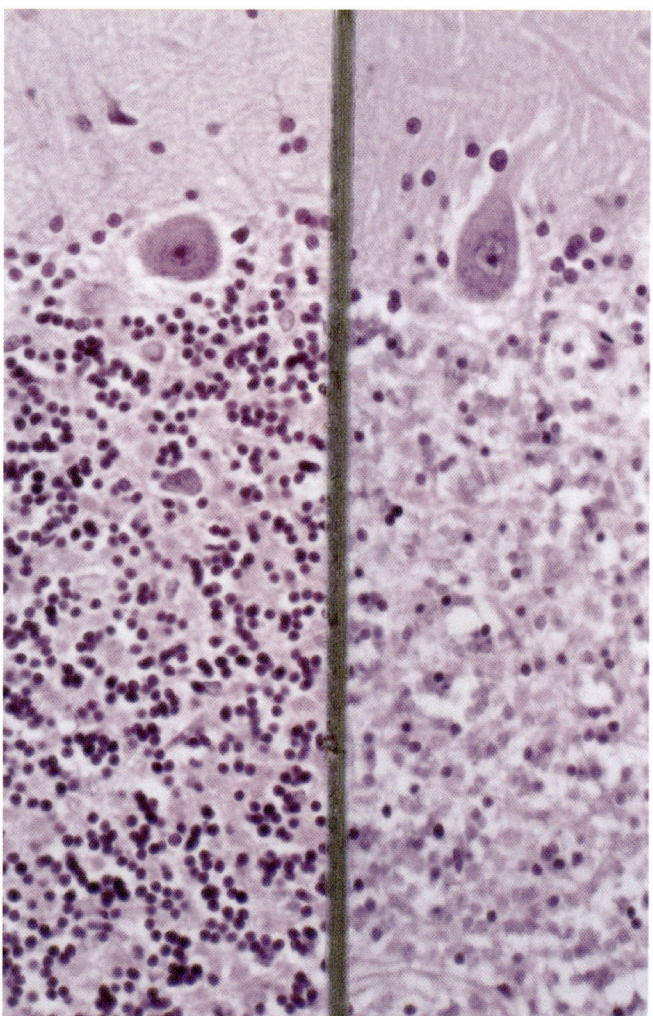

Fig. 1.39. Side-by-side comparison at medium power of normal (left half of photo) and autolyzed (right half of photo) granule cells. Intact Purkinje cells and lower margin of molecular layer are seen at top of photos. (H&E.)

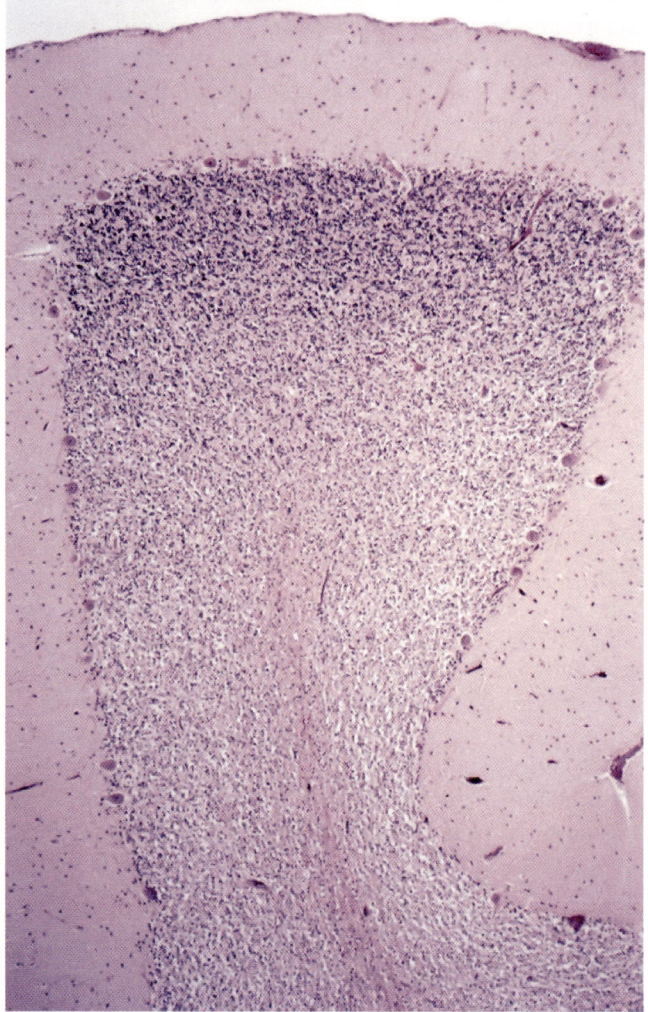

Fig. 1.38. Postmortem autolysis of granule cells, represented by pale granule cell zone, least evident at crown of folium. (H&E.)

to the more widely quoted 1960 report by Gruenwald and Minh.[58] The latter authors reported findings on 1355 newborn infants, including stillborns and those surviving for 3 days or less, from GA 24 to 44 weeks, at 2-week intervals. Their study demonstrated a trend toward lower brain weights (presumably fresh), most evident in the later stages of gestation, in comparison to the recent findings of Maroun and Graem.[106] The trend toward higher brain weights in normals in more recent studies is also demonstrated by Thompson and Cohle,[157] noted later.

2. Guihard-Costa and Larroche,[59] in 1990, reported brain weights in 298 fetuses from GA 8 to 41 weeks, at 2-week intervals. It is included here for two main reasons. First, it documents both fresh and fixed brain weights. Second, it provides the infrequently reported separate weights for the infratentorial components in 114 of

the 298 cases studied. They demonstrated acceleration of cerebellar growth relative to the brainstem after 20 weeks' GA. Parenthetically, it is not until approximately 2 years of age that cerebellar size relative to cerebrum achieves an essentially adult appearance.

3. One of the most quoted brain weight reports is that of Schulz et al.,[142] published in 1962 and based on 638 fetuses and newborn infants ranging from 5 months' GA to 1 week post-term. Measurements were at 1-month intervals up to 1 week post-term, and at 1-month intervals in an additional 701 infants from 1 month to 1 year of age. Their tables (presumably based on fresh brains) are used by the State of California for their SIDS protocol brain weights.

4. Thompson and Cohle,[157] in 2004, published a study of 453 cases from 30 weeks' GA to 12 months of age, including stillborns. It is included here because it is more recent, it has a reasonable number of cases, and it indicates that (presumably fresh) organ weights in this more recent population are increased in comparison to older studies such as that of Schulz et al.[142]

5. Coppoletta and Wolbach[30] published a 1933 study providing data derived from 1198 cases, in which "each selected autopsy contributed a normal weight (presumably fresh) of at least one vital organ." Although not helpful for newborns (organ weights were given without birth weights or GA specified), it extended previously available data by adding brain weights at from ½- to 2- or 3-week intervals up to 9 weeks of age, at 1-month intervals up to 1 year of age, at 2-month intervals from 1 to 2 years of age, and at 1-year intervals from 2 to 12 years of age. Thus, it also added more time intervals to the later study by Dekaban and Sadowsky[33] in the 12 years and under age group (see later). Although widely quoted, Ribe[133] has indicated reasons why this study is less applicable to current children under 2 years of age than more recent studies, despite its widespread use as a source for brain weights.

6. Dekaban and Sadowsky, in 1978, reported fresh brain weights in 2603 males and 1848 females ranging from 0 to 10 days of age to 86+ years of age at varying intervals, thus extending the age range of available data well beyond the limits provided by others.[33] The decreasing brain weight with more advanced age demonstrated by this study was further supported by the subsequent reports of Ho et al., who examined 1261 fresh brains between the ages of 25 and 80 years.[67,68]

Comment on Brain Weight Tables

The brain weights in the tables described earlier should be considered estimates. In most cases, use of grossly normal brains was specified, but in the GA range these studies are largely on premature stillbirths and, in contrast to the report by Maroun and Graem,[106] the degree of maceration present is not always specified. Cause of death, duration of terminal illness, and terminal state of hydration are not usually specified. Comparisons of these various studies in similar age groups reveals a tendency for brain weights to be somewhat greater in more recent studies compared with older studies in fetuses and children up to 12 months of age.[106,157] Also, a study in cases presenting to a forensic laboratory demonstrated somewhat higher brain weights (weight fresh, with leptomeninges and CSF not removed). In this latter study, however, the authors ascribed the increased weights to varying degrees of cerebral swelling, documented in their report and presumed related to the acute terminal illnesses present in the majority of the population studied.[163]

Another consideration is that most reports do not specify death to autopsy interval. Variation in the latter may account for at least some of the difference between studies, and between cases of similar age range in any single study, since Appel and Appel[8,9] noted that mean brain weights in 1544 white males, age range approximately 57 to 63 years, increased by approximately 9% over the course of measurements taken from the 0- to 11-hour to the 84- to 146-hour postmortem interval. The most rapid increase occurred in the first 12 hours. The observations by Appel and Appel[8,9] also provide an additional possible basis for the postmortem decrease in ventricular and sulcal size described subsequently in a study using serial computed tomography scans.[139]

Lindboe[97] examined relative contributions of cerebrum, cerebellum, brainstem, upper spinal cord, leptomeninges, and intraventricular CSF to total brain weight in eight adult cases. He advocated published brain weights being more specific as to exactly which anatomic structures are being weighed, as well as whether the weight is based on fresh or fixed organs.

In summary, based on the several considerations noted earlier, we presently favor the following brain weight sources as a general guide in the following age groups, noting that some age group overlaps occur.

1. Estimated gestational age (EGA) 12 to 43 weeks: Maroun and Graem.[106]
2. EGA 30 weeks to 1-year-old infant: Thompson and Cohle[157] would seem to be a logical choice to use in this age range, except that their figures for one standard deviation (SD) are so broad at several age intervals (e.g., approximately 40% of the mean at age 5 months) that a single SD could easily encompass weights of both significantly atrophic and significantly swollen brains, and possibly create an illusion of a discrepancy in the report. Possibly for this reason, the figures by Schulz et al.[142] are currently used as normal values for this age group by the State of California, as noted earlier.
3. Newborn to 86+ years of age: Dekaban and Sadowsky.[33]

4. When found useful to separately weigh supra- and infratentorial structures, or separately weigh other components of the intracranial contents, some guidance in the 8- to 41-week estimated gestational age fetus can be obtained from Guihard-Costa and Larroche,[59] in 9-day-old infants to 20-year-old individuals (consult Tables 119 and 120 in Blinkov et al.)[15] and in adults aged 48 to 82 years old in Lindboe.[97]

5. One should carefully describe any evidence of brain atrophy or swelling in a narrative report. Brain weight, after all, is only a crude indicator of these and other abnormalities except at the extremes of normal values. Discrepancies between brain weight recorded at autopsy and the brain received for neuropathology consultation could also alert one to the possibility of specimen mislabeling.

Effect of Formalin Fixation on CNS Tissue

Formalin fixation (typically in "10%" solutions, which are actually 3.7% formaldehyde) for times varying from several days to several weeks in different studies results in weight changes in intracranial tissues, compared with fresh weights, with estimates varying somewhat in different studies, as follows:

- An 8.8% gain in whole brain weight in adults[45,101]
- A 10% gain in whole brain weights in children[93]
- A 14.1% gain in whole brain weights in children[101]
- A 3.9% decrease in cranial dura mater weights[101]
- Up to 33% gain in whole brain weight in larger, late gestation fetuses[109]
- 10 to 15% gain in whole brain weight in fetuses after 28 weeks' gestation[59]
- A 4 to 5% gain in infratentorial CNS weight (cerebellum plus brainstem) in fetuses after 28 weeks' gestation[59]

Quester and Schröder[127] considered the effects not only of fixation (in 4% formalin) but also effects of embedding CNS tissues. They reported:

- A 7 to 13.4% gain in weight and volume of cerebrum, with the maximum weight gain between 1 and 5 days' fixation, a greater effect with lower formalin concentrations, followed by a gradual decrease in weight in subsequent weeks and months (although usually not back to original fresh weight, according to Stevenson).[150]
- A 12% gain in cerebellar weight was noted.
- Less distinct weight changes in brains with formalin fixation described in elderly brains in some, but not all, reports of elderly brains.

With combined formalin fixation and paraffin embedding, Quester and Schröder described a linear shrinkage of 12% in medullary olivary breadth and 11% in medullary pyramid breadth, compared with previously reported figures of 13.7 to 34.4% in cerebral tissue.[127]

These studies on combined effects of formalin fixation and paraffin embedding involved relatively small numbers of cases, and indications are that there is not only considerable variation in results between individuals but also between different structures within the same brain.

Pineal Gland

The insular architecture and tendency to develop cysts and concretions by the normal pineal gland (Figs. 1.40–1.41) periodically result in a slide that has inadvertently included this structure being brought to us to "rule out

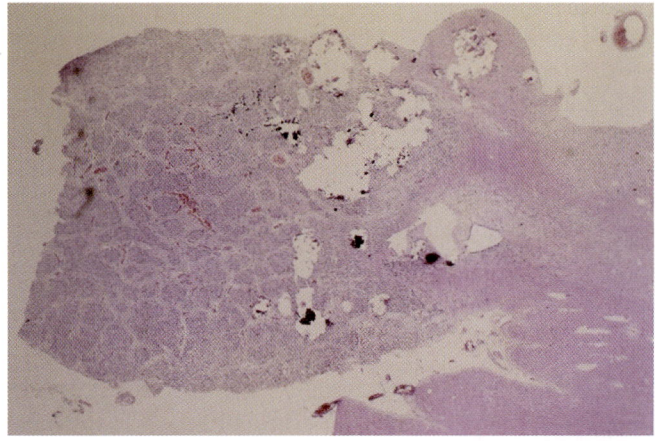

Fig. 1.40. Low-power micrograph of normal pineal gland. Tissue tears are due to knife artifact from intrapineal calcification. (H&E.)

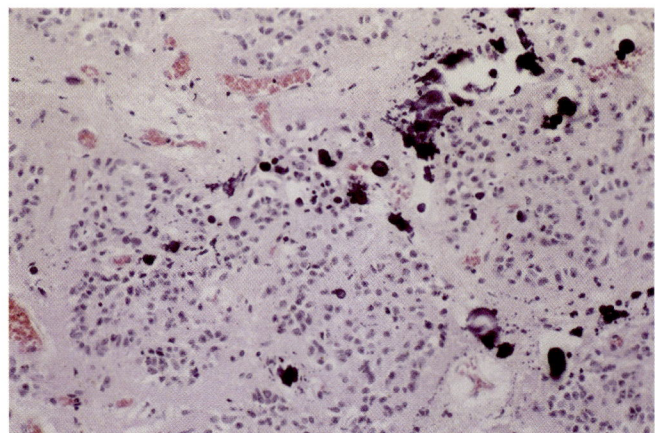

Fig. 1.41. Medium-power micrograph, normal pineal gland. Pinealocytes form nests and cords resulting in an insular architectural pattern. Calcium deposits are present, often with psammoma body features. (H&E.)

brain tumor." It also tends to appear with fair regularity as a board examination question. A simple solution is to have a slide of typical pineal gland in one's personal teaching set for comparison purposes.

References

1. Adams VI. Autopsy technique for neck examination. I. Anterior and lateral compartments and tongue. Pathol Annu 1990;25: 331–349.
2. Adams VI. Autopsy technique for neck examination. II. Vertebral column and posterior compartment. Pathol Annu 1991;26: 211–226.
3. Adelson L, Sunshine I. Fatal hydrogen sulfide intoxication: Report of three cases occurring in a sewer. Arch Pathol 1966;81:375–380.
4. Albrechtsen R. The pathogenesis of acute selective necrosis of the granular layer of the human cerebellar cortex. Acta Neuropathol (Berl) 1977;37:31–34.
5. Albrechtsen R. The incidence of the so-called acute selective necrosis of the granular layer of cerebellum in 1000 autopsied patients. Acta Pathol Microbiol Scand [A] 1977;85:193–202.
6. An TL. Hydrogen sulfide intoxication due to coal tar distillation column accident (abstract). Proc Am Acad Forensic Sci 1998;4: 155–156.
7. Anhalt JP, Witebsky FG. Is there evidence that *Mycobacterium tuberculosis* survives in routine formalin fixatives used in surgical pathology? CAP Today 1996;Mar:75–76.
8. Appel FW, Appel EM. Intracranial variation in the weight of the human brain. Hum Biol 1942;14:48–68.
9. Appel FW, Appel EM. Intracranial variation in the weight of the brain (concluded). Hum Biol 1942;14:235–250.
10. Asirdizer M, Yavuz S, Sari H, et al. Unusual torture methods and mass murders applied by a terror organization. Am J Forensic Med Pathol 2004;25:314–320.
11. Bajanowski T, Vennemann M, Bohnert M, et al., for the GeSID Group. Unnatural causes of sudden unexpected deaths initially thought to be sudden infant death syndrome. Int J Legal Med 2005;119:213–216.
12. Berzlanovich AM, Sim E, Muhm MA. Technique for dissecting the cervical vertebral column. J Forensic Sci 1998;43:190–193.
13. Bierig JR. A potpourri of legal issues relating to the autopsy. Arch Pathol Lab Med 1996;120:759–762.
14. Blair PS. Sudden infant death syndrome: A commentary on the review by Goldwater. Arch Dis Child 2003;88:1031.
15. Blinkov SM, Glezer II. The Human Brain in Figures and Tables: A Quantitative Handbook. New York: Plenum Press and Basic Books, 1968.
16. Breningstall GN. Breath-holding spells. Pediatr Neurol 1996;14: 91–97.
17. Breukels MA, Plötz FB, van Nieuwenhuizen O, van Diemen-Steenvoorde JAAM. Breath holding spells in a 3-day-old neonate: An unusual early presentation in a family with a history of breath holding spells. Neuropediatrics 2002;33:41–42.
18. Brogdon BG, Vogel H, McDowell JD. A Radiologic Atlas of Abuse, Torture, Terrorism, and Inflicted Trauma. Boca Raton, FL: CRC Press, 2003.
19. Bruner JM, Inouye L, Fuller GN, Langford LA. Diagnostic discrepancies and their clinical impact in a neuropathology referral practice. Cancer 1997;79:796–803.
20. Byard RW. Sudden Death in Infancy, Childhood and Adolescence, ed 2. Cambridge: Cambridge University Press, 2004.
21. Byard RW, Krous HF (eds). Sudden Infant Death Syndrome: Problems, Progress and Possibilities. London: Arnold, 2001.
22. Cahill DR, Orland MJ, Miller GM. Atlas of Human Cross-Sectional Anatomy: With CT and MRI Images, ed 3. New York: Wiley-Liss, 1995.
23. California Department of Health Services: Standardized Autopsy Protocol for the Evaluation of Sudden Unexpected Infant Death. CDHS Maternal, Child, and Adolescent Health/Office of Family Planning Branch, Epidemiology and Evaluation Section, P.O. Box 997420, MS 8304, Sacramento, CA 95899-7420. Also available at: http://www.mch.dhs.ca.gov/epidemiology, or at www.californiasids.com.
24. Champ CS, Coghill SB. Visual aid for quick assessment of coronary artery stenosis at necropsy. J Clin Pathol 1989;42:887–888.
25. Choi D, Carroll N, Abrahams P. Spinal cord diameters in cadaveric specimens and magnetic resonance scans, to assess embalming artifacts. Surg Radiol Anat 1996;18:133–135.
26. Choi BH, Kim RC, Suzuki M, Choe W. The ventriculus terminalis and filum terminale of the human spinal cord. Hum Pathol 1992; 23:916–920.
27. Cilluffo JM, Gomez MR, Reese DF, et al. Idiopathic ("congenital") spinal arachnoid diverticula: Clinical diagnosis and surgical results. Mayo Clin Proc 1981;56:93–101.
28. Claydon SM. The high-risk autopsy: Recognition and protection. Am J Forensic Med Pathol 1993;14:253–256.
29. Clemente CD (ed). Gray's Anatomy, 30th American edition. Philadelphia: Lea & Febiger, 1985, p 1192.
30. Coppoletta JM, Wolbach SB. Body length and organ weights of infants and children: A study of the body length and normal weights of the more important vital organs of the body between birth and twelve years of age. Am J Pathol 1933;9:55–70.
31. Davis JH. The future of the medical examiner system. Am J Forensic Med Pathol 1995;16:265–269.
32. Day JD, Tschabitscher M. Anatomic position of the asterion. Neurosurgery 1998;42:198–199.
33. Dekaban AS, Sadowsky D. Changes in brain weights during the span of human life: Relation of brain weights to body heights and body weights. Ann Neurol 1978;4:345–356.
34. De Paepe A, Devereux RB, Dietz HC, et al. Revised diagnostic criteria for the Marfan syndrome. Am J Med Genet 1996;62: 417–426.
35. DiMaio V. Deaths in custody investigations. In Ross DL, Chan TC (eds): Sudden Deaths in Custody. Totowa, NJ: Humana Press, 2006, pp 167–172.
36. DiMaio VJ, DiMaio D. Forensic Pathology, ed 2. Boca Raton, FL: CRC Press, 2001.
37. DiMario FJ Jr. Breath-holding spells and pacemaker implantation. Pediatrics 2001;108:765–766.
38. Dolinak D, Matshes E. Medicolegal Neuropathology: A Color Atlas. Boca Raton, FL: CRC Press, 2002.
39. Dolinak D, Matshes EW, Lew EO: Forensic Pathology: Principles and Practice. Amsterdam: Elsevier Academic Press, 2005.
40. Doraiswamy PM, Na C, Husain MM, et al. Morphometric changes of the human midbrain with normal aging: MR and stereologic findings. AJNR Am J Neuroradiol 1992;13:383–386.
41. Duchen LW, Jacobs JM. Nutritional deficiencies and metabolic disorders. In Adams JH, Corsellis JAN, Duchen LW (eds): Greenfield's Neuropathology, ed 4. New York: John Wiley & Sons, 1984, pp 573–626.
42. Elliott HC. Cross-sectional diameters and areas of the human spinal cord. Anat Rec 1945;93:287–293.
43. Ellison D, Love S, Chimelli L, et al. Neuropathology: A Reference Text of CNS Pathology. London: Mosby, 1998.
44. Fierro MF. Identification of human remains. In Spitz WU (ed): Spitz and Fisher's Medicolegal Investigation of Death: Guidelines for the Application of Pathology to Crime Investigation, ed 3. Springfield, IL: Charles C. Thomas, 1993, pp 71–117.
45. Finkbeiner WE, Ursell PC, Davis RL. Autopsy Pathology: A Manual and Atlas. Philadelphia: Churchill Livingston, 2004.
46. Fleming PJ, Blair PS. How reliable are SIDS rates? The importance of a standardized, multiprofessional approach to "diagnosis."

(Commentary on the paper by Sheehan et al: Arch Dis Child 2005;90:1082–1083.) Arch Dis Child 2005;90:993–994.
47. Friede RL. Developmental Neuropathology, ed 2. Berlin: Springer-Verlag, 1989.
48. Fuller GN, Burger PC. Central nervous system. In Sternberg SS (ed): Histology for Pathologists. New York: Raven Press, 1992, pp 145–167.
49. Galloway A, Snodgrass JJ. Biological and chemical hazards of forensic skeletal analysis. J Forensic Sci 1998;43:940–948.
50. Geddes JF, Gonzalez AG. Examination of spinal cord in diseases of the craniocervical junction and high cervical spine. J Clin Pathol 1991;44:170–172.
51. Gellad F, Rao KCVG, Joseph PM, Vigorito RD. Morphology and dimensions of the thoracic cord by computer-assisted metrizamide myelography. AJNR Am J Neuroradiol 1983;4:614–617.
52. Gilbert-Barness E, Debich-Spicer DE. Handbook of Pediatric Autopsy Pathology. Totowa, NJ: Humana Press, 2005.
53. Gilles FH. Perinatal neuropathology. In Davis RL, Robertson DM (eds): Textbook of Neuropathology, ed 3. Baltimore: Williams & Wilkins, 1997, pp 331–385.
54. Goldwater PN. Sudden infant death syndrome: A critical review of approaches to research. Arch Dis Child 2003;88:1095–1100.
55. Grafe MR, Kinney HC. Neuropathology associated with stillbirth. Semin Perinatol 2002;26:83–88.
56. Graham MA, Hanzlick R. Forensic Pathology in Criminal Cases. Carlsbad, CA: Lexis Law Publishing, 1997.
57. Gray H (Lewis WH, ed). Anatomy of the Human Body. Philadelphia: Lea & Febiger, 1942.
58. Gruenwald P, Minh HN. Evaluation of body and organ weights in perinatal pathology. I. Normal standards derived from autopsies. Am J Clin Pathol 1960;34:247–253.
59. Guihard-Costa A-M, Larroche J-C. Differential growth between the fetal brain and its infratentorial part. Early Hum Dev 1990;23:27–40.
60. Haber SL. Innovations in pathology. CAP Today 2004;Sept:114.
61. Haglund WD, Sorg MH (eds): Forensic Taphonomy: The Postmortem Fate of Human Remains. Boca Raton, FL: CRC Press, 1997.
62. Hanzlick R. Embalming, body preparation, burial, and disinterment: An overview for forensic pathologists. Am J Forensic Med Pathol 1994;15:122–131.
63. Harris LS, Adelson L. "Spinal injury" and sudden infant death: A second look. Am J Clin Pathol 1969;52:289–295.
64. Harwood-Nash D, Fitz CR: Neuroradiology in Infants and Children. St. Louis: CV Mosby, 1976, p 1138.
65. Henry TE. Case FR-01 (2001 FR-A). In American Society for Clinical Pathology: Forensic Pathology. Northfield, IL: College of American Pathologists, 2001; pp FR-A:1–2.
66. Hirano A. Neurons and astrocytes. In Davis RL, Robertson DM (eds): Textbook of Neuropathology, ed 3. Baltimore: Williams & Wilkins, 1997, pp 1–109.
67. Ho K, Roessmann U, Straumfjord JV, Monroe G. Analysis of brain weight. I. Adult brain weight in relation to sex, race and age. Arch Pathol Lab Med 1980;104:635–639.
68. Ho K, Roessmann U, Straumfjord JV, Monroe G. Analysis of brain weight. II. Adult brain weight in relation to body height, weight and surface area. Arch Pathol Lab Med 1980;104:640–645.
69. Hollinshead WH. Anatomy for Surgeons, Vol 1. The Head and Neck. New York: Hoeber-Harper Book, 1954.
70. Huang C-S, Cheng H-C, Lin W-Y, et al. Skull morphology affected by different sleep positions in infancy. Cleft Palate Craniofac J 1995;32:413–419.
71. Hunt CE, Hauck FR. Sudden infant death syndrome. Cam Med Assoc J 2006;174:1861–1869.
72. Hutchins KD, Williams AW, Natarajan GA. Neck needle foreign bodies: An added risk for autopsy pathologists. Arch Pathol Lab Med 2001;125:790–792.
73. Ikuta F, Hirano A, Zimmerman HM. An experimental study of post-mortem alterations in the granular layer of the cerebellar cortex. J Neuropathol Exp Neurol 1963;22:581–593.
74. Johnson MD, Schaffner W, Atkinson J, Pierce MA. Autopsy risk and acquisition of human immunodeficiency virus infection: A case report and reappraisal. Arch Pathol Lab Med 1997;121:64–66.
75. Kadhim H, Kahn A, Sébire G. Distinct cytokine profile in SIDS brain: A common denomination in a multifactorial syndrome? Neurology 2003;61:1256–1259.
76. Kameyama T, Hashizume Y, Sobue G. Morphologic features of the normal human cadaveric spinal cord. Spine 1996;21:1285–1290.
77. Kappel TJ, Reinartz JJ, Schmid JL, et al. The viability of *Mycobacterium tuberculosis* in formalin-fixed pulmonary autopsy tissue: Review of the literature and brief report. Hum Pathol 1996;27:1361–1364.
78. Kelly AM, Porter CJ, McGoon MD, et al. Breath-holding spells associated with significant bradycardia: Successful treatment with permanent pacemaker implantation. Pediatrics 2001;108:698–702.
79. Kessler SC, Evans RJ. Through the tight canal swiftly—A review of the technique for evaluating vertebral artery trauma at autopsy (abstract). Proc Am Acad Forensic Sci 2001;7:230–231.
80. Kibayashi K, Ng'walali PM, Honjyo K, et al. Pink spots of Hedley-White in the brain: Evaluation of the significance in the forensic autopsy. Legal Med 2000;2:88–92.
81. Kinney HC, Armstrong DL, Chadwick AE, et al. Seizures, cerebral edema and hippocampal anomalies in sudden unexplained death in children (SUDC): Report of a series (abstract). J Neuropathol Exp Neurol 2004;63:556.
82. Kinney HC, Filiano JJ. Brain research in sudden infant death syndrome. In Byard RW, Krous HF (eds): Sudden Infant Death Syndrome: Problems, Progress and Possibilities. London: Arnold, 2001, pp 118–137.
83. Kinney HC, Myers MM, Belliveau RA, et al. Subtle autonomic and respiratory dysfunction in sudden infant death syndrome associated with serotonergic brainstem abnormalities: A case report. J Neuropathol Exp Neurol 2005;64:689–694.
84. Kinney HC, Paterson DS. Sudden infant death syndrome. In Golden JA, Harding BN (eds): Pathology and Genetics: Developmental Neuropathology. Basel: ISN Neuropath Press, 2004, pp 194–203.
85. Knoblich R, Olsen BS. Calcified and ossified plaques of the spinal arachnoid membranes. J Neurosurg 1966;25:275–279.
86. Koehler PR, Haughton VM, Daniels DL, et al. MR measurement of normal and pathologic brainstem diameters. AJNR Am J Neuroradiol 1985;6:425–427.
87. Koehler SA, Ladham S, Shakir A, Wecht CH. Simultaneous sudden infant death syndrome: A proposed definition and worldwide review of cases. Am J Forensic Med Pathol 2001;22:23–32.
88. Kohr RM. Autopsies on executed federal prisoners (letter to editor). J Forensic Sci 2003;48:1203.
89. Krous HF, Beckwith JB, Byard RW, et al. Sudden infant death syndrome and unclassified sudden infant deaths: A definitional and diagnostic approach. Pediatrics 2004;114:234–238.
90. Kuhle S, Tiefenthaler M, Seidl R, Hauser E. Prolonged generalized epileptic seizures triggered by breath-holding spells. Pediatr Neurol 2000;23:271–273.
91. Kumar N, Cohen-Gadol AA, Wright RA, et al. Superficial siderosis. Neurology 2006;66:1144–1152.
92. Langlois NEI, Little D. A method for exposing the intraosseous portion of the carotid arteries and its application to forensic case work. Am J Forensic Med Pathol 2003;24:35–40.
93. Larroche J-C. Developmental Pathology of the Neonate. Amsterdam: Excerpta Medica, 1977.
94. Leestma JE. Forensic neuropathology. In Duckett S (ed): Pediatric Neuropathology. Baltimore: Williams & Wilkins, 1995, pp 243–283.

95. Leth PM, Banner J. Forensic medical examination of refugees who claim to have been tortured. Am J Forensic Med Pathol 2005;26:125–130.
96. Li L, Zhang X, Constantine NT, Smialek JE. Seroprevalence of parenterally transmitted viruses (HIV-1, HBV, HCV, and HTLV-I/II) in forensic autopsy cases. J Forensic Sci 1993;38:1075–1083.
97. Lindboe CF. Brain weight: What does it mean? Clin Neuropathol 2003;22:263–265.
98. Liu W, Prayson RA. Dura mater involvement in ochronosis (alkaptonuria). Arch Pathol Lab Med 2001;125:961–963.
99. Lowe JS, Leigh N. Disorders of movement and system degenerations. In Graham DI, Lantos PL (eds): Greenfield's Neuropathology, ed 7. London: Arnold, 2002;2:325–430.
100. Lucet V, deBethmann O, Denjoy I. Paroxysmal vagal overactivity, apparent life-threatening event and sudden infant death. Biol Neonate 2000;78:1–7.
101. Ludwig J. Handbook of Autopsy Practice. Totowa, NJ: Humana Press, 2002.
102. Machaalani R, Waters KA. NMDA receptor 1 expression in the brainstem of human infants and its relevance to the sudden infant death syndrome (SIDS). J Neuropathol Exp Neurol 2003;62:1076–1085.
103. Madea B, Krompecher T, Knight B. Muscle and tissue changes after death. In Henssge C, Knight B, Krompecher T, et al. (eds): The Estimation of the Time Since Death in the Early Postmortem Period. London: Edward Arnold, 1995, pp 138–220.
104. Magee JF. Investigation of stillbirth. Pediatr Develop Pathol 2001;4:1–22.
105. Malloy M. SIDS—A syndrome in search of a cause. N Engl J Med 2004;351:957–959.
106. Maroun LL, Graem N. Autopsy standards of body parameters and fresh organ weights in nonmacerated and macerated human fetuses. Pediatr Develop Pathol 2005;8:204–217.
107. Matturri L, Ottaviani G, Alfonsi G, et al. Study of the brainstem, particularly the arcuate nucleus, in sudden infant death syndrome (SIDS) and sudden intrauterine unexplained death (SIUD). Am J Forensic Med Pathol 2004;25:44–48.
108. McCotter RE. Regarding the length and extent of the human medulla spinalis. Anat Rec 1916;10:559–564.
109. McLennan JE, Gilles FH, Neff RK. A model of growth of the human fetal brain. In Gilles FH, Leviton A, Dooling EC (eds): The Developing Human Brain: Growth and Epidemiologic Neuropathology. Boston: John Wright, PSG, 1983, pp 43–58.
110. Moreno A, Grodin MA. Torture and its neurological sequelae. Spinal Cord 2002;40:213–223.
111. Moritz AR. Classical mistakes in forensic pathology. Am J Clin Pathol 1956;26:1383–1397.
112. National Association of Medical Examiners. Forensic autopsy performance standards. Am J Forensic Med Pathol 2006;27:200–225.
113. Nolte KB. Survival of *Mycobacterium tuberculosis* organisms for 8 days in fresh lung tissue from an exhumed body. Hum Pathol 2005;36:915–916.
114. Nolte KB, Taylor DG, Richmond JY. Biosafety considerations for autopsy. Am J Forensic Med Pathol 2002;23:107–122.
115. Nordqvist L. The sagittal diameter of the spinal cord and subarachnoid space in different age groups: A roentgenographic post-mortem study. Acta Radiol [Diagn] 1964;(Suppl)227:1–96.
116. Ogata J, Yutani C, Imakita M, et al. Autolysis of the granular layer of the cerebellar cortex in brain death. Acta Neuropathol (Berl) 1986;70:75–78.
117. Olszewski J, Baxter D. Cytoarchitecture of the Human Brain Stem, ed 2. Basel: S. Karger, 1982.
118. Omalu BI, Mancuso JA, Cho P, Wecht CH. Diagnosis of Alzheimer's disease in an exhumed decomposed brain after twenty months of burial in a deep grave. J Forensic Sci 2005;50:1453–1458.
119. Peitsch WK, Keefer CH, LaBrie RA, Mulliken JB. Incidence of cranial asymmetry in healthy newborns. Pediatrics 2002;110(6). Available at: http://www.pediatrics.org/cgi/content/full/110/6/e72. Accessed 5/9/06.
120. Perper JA. Time of death and changes after death, Part 1: Anatomical considerations. In Spitz WU (ed): Spitz and Fisher's Medicolegal Investigation of Death: Guidelines for the Application of Pathology to Crime Investigation, ed 3. Springfield, IL: Charles C. Thomas, 1993, pp 14–49.
121. Petty CS. The devil's dozen: Popular medicolegal misconceptions. South Med J 1971;64:919–923.
122. Piwowarczyk L, Moreno A, Grodin M. Health care of torture survivors. JAMA 2000;284:539–541.
123. Powers JM, Autopsy Committee of College of American Pathologists. Practice guidelines for autopsy pathology: Autopsy procedures for brain, spinal cord, and neuromuscular system. Arch Pathol Lab Med 1995;119:777–783.
124. Prahlow JA, Ross KF, Salzberger L, et al. Immersion technique for brain removal in perinatal autopsies. J Forensic Sci 1998;43:1056–1060.
125. Prinsloo I, Gordon I. Post-mortem dissection artifacts of the neck: Their differentiation from ante-mortem bruises. S Afr Med J 1951;25:358–361.
126. Pritt BS, Waters BL. Cutting injuries in an academic pathology department. Arch Pathol Lab Med 2005;129:1022–1026.
127. Quester R, Schröder R. The shrinkage of the human brain stem during formalin fixation and embedding in paraffin. J Neurosci Methods 1997;75:81–89.
128. Raininko R, Autti T, Vanhanen SL, et al. The normal brain stem from infancy to old age: A morphometric MRI study. Neuroradiology 1994;36:364–368.
129. Randall BB, Fierro MF, Froede RC, for the Members of the Forensic Pathology Committee, College of American Pathologists. Practice guideline for forensic pathology. Arch Pathol Lab Med 1998;122:1056–1064.
130. Reece RM. Editor's note Re: Koehler SA, et al. Simultaneous sudden infant death syndrome. Am J Forensic Med Pathol 2001;22:23–32. Child Abuse Q Med Update 2001;VIII(3):13–14.
131. Resjö IM, Harwood-Nash DC, Fitz CR, Chuang S. Normal cord in infants and children examined with computed tomographic metrizamide myelography. Radiology 1979;130:691–696.
132. Reske-Nielsen E, Oster S, Reintoft I. Astrocytes in the prenatal nervous system. Acta Pathol Microbiol Immunol Scand [A] 1987;95:339–346.
133. Ribe JK. Reviewer's note. Quarterly Update 2005;XII(4):33–34.
134. Rivet DJ, O'Brien DF, Park TS, Ojemann JG. Distance of the motor cortex from the coronal suture as a function of age. Pediatr Neurosurg 2004;40:215–219.
135. Rogers CB, Itabashi HH, Tomiyasu U, Heuser ET. Subdural neomembranes and sudden infant death syndrome. J Forensic Sci 1998;43:375–376.
136. Rorke LB, Riggs HE. Myelination of the Brain in the Newborn. Philadelphia: JB Lippincott, 1969.
137. Rosenbloom S, Cohen WA, Marshall C, Kricheff II. Imaging factors influencing spine and cord measurements by CT: A phantom study. AJNR Am J Neuroradiol 1983;4:646–649.
138. Rowland LP, Mettler FA. Relation between the coronal suture and cerebrum. J Comp Neurol 1948;89:21–40.
139. Sarwar M, McCormick WF. Decrease in ventricular and sulcal size after death. Radiology 1978;127:409–411.
140. Saukko P, Knight B. Knight's Forensic Pathology, ed 3. London: Arnold, 2004.
141. Schleyer F. Determination of the time of death in the early postmortem interval. In Lundquist F (ed): Methods of Forensic Science. London: Interscience Publishers, 1963, II, pp 253–293.
142. Schulz DM, Giordano DA, Schulz DH. Weights of organs of fetuses and infants. Arch Pathol 1962;74:244–250.

143. Seibert CE, Barnes JE, Dreisbach JN, et al. Accurate CT measurement of the spinal cord using metrizamide: Physical factors. AJNR Am J Neuroradiol 1981;2:75–78.
144. Sheaff MT, Hopster DJ. Post Mortem Technique Handbook. London: Springer, 2001.
145. Sheehan KM, McGarvey C, Devaney DM, Matthews T. How reliable are SIDS rates? Arch Dis Child 2005;90:1082–1083.
146. Smith GCS, Wood AM, Pell JP, et al. Second-trimester maternal serum levels of alpha-fetoprotein and the subsequent risk of sudden infant death syndrome. N Engl J Med 2004;351:978–986.
147. Sparks DL, Hunsaker JC III. Neuropathology of sudden infant death (syndrome): Literature review and evidence of a probable apoptotic degenerative cause. Childs Nerv Syst 2002;18:568–592.
148. Spitz WU (ed). Spitz and Fisher's Medicolegal Investigation of Death: Guidelines for the Application of Pathology to Crime Investigation, ed 3. Springfield, IL: Charles C. Thomas, 1993.
149. Spitz WU. Selected procedures at autopsy. In Spitz WU (ed): Spitz and Fisher's Medicolegal Investigation of Death: Guidelines for the Application of Pathology to Crime Investigation, ed 3. Springfield, IL: Charles C. Thomas, 1993, pp 776–797.
150. Stevenson GS. Stabilizing brain tissue during fixation. Arch Neurol Psychiatry 1923;9:763–768.
151. Sturner WQ. Common errors in forensic pediatric pathology. Am J Forensic Med Pathol 1998;19:317–320.
152. Symington J. Observations on the relationships of the deeper parts of the brain to the surface. J Anat Physiol 1903;37:241–250 (and Plates XXIV–XXIX).
153. Task Force on Sudden Infant Death Syndrome. The changing concept of sudden infant death syndrome: Diagnostic coding shifts, controversies regarding the sleeping environment, and new variables to consider in reducing risk. Pediatrics 2005;116:1245–1255.
154. Templeton GL, Illing LA, Young L, et al. The risk of transmission of *Mycobacterium tuberculosis* at the bedside and during autopsy. Ann Intern Med 1995;122:922–925.
155. Thach BT. The brainstem and vulnerability to sudden infant death syndrome. Neurology 2003;61:1170–1171.
156. Thijssen HOM, Keyser A, Horstink MWM, Meijer E. Morphology of the cervical spinal cord on computed myelography. Neuroradiology 1979;18:57–62.
157. Thompson WS, Cohle SD. Fifteen-year retrospective study of infant organ weights and revision of standard weight tables. J Forensic Sci 2004;49:575–585.
158. Tolcos M, McGregor H, Walker D, Rees S. Chronic prenatal exposure to carbon monoxide results in a reduction in tyrosine hydroxylase-immunoreactivity and an increase in choline acetyltransferase-immunoreactivity in the fetal medulla: Implications for sudden infant death syndrome. J Neuropathol Exp Neurol 2000;59:218–228.
159. Tonkin SL, Gunn TR, Bennet L, et al. A review of the anatomy of the upper airway in early infancy and its possible relevance to SIDS. Early Hum Dev 2002;66:107–121.
160. United Nations Office at Vienna, Centre for Social Development and Humanitarian Affairs. Manual on the Effective Prevention and Investigation of Extra-Legal, Arbitrary and Summary Executions. Publication Sales No. E.91.IV.1. New York: United Nations, 1991.
161. Valdéz-Dapena M, McFeeley PA, Hoffman HJ, et al. Histopathology Atlas for the Sudden Infant Death Syndrome. Washington, DC: Armed Forces Institute of Pathology, 1993.
162. Valdés-Dapena M, Huff D. Perinatal Autopsy Manual. Washington, DC: Armed Forces Institute of Pathology, 1983.
163. Voigt J, Pakkenberg H. Brain weight of Danish children: A forensic material. Acta Anat 1983;116:290–301.
164. Warmuth-Metz M, Naumann M, Csoti I, Solymosi L. Measurement of the midbrain diameter on routine magnetic resonance imaging: A simple and accurate method of differentiating between Parkinson disease and progressive supranuclear palsy. Arch Neurol 2001;58:1076–1079.
165. Wetli CV, Mittleman RE, Rao VJ. An Atlas of Forensic Pathology. Chicago: ASCP Press, 1999, pp 242–243.
166. Whitwell HL (ed). Forensic Neuropathology. London: Hodder Arnold, 2005.
167. Yeh I-T, Brooks JSJ, Pietra GG. Atlas of Microscopic Artifacts and Foreign Materials. Baltimore: Williams & Wilkins, 1997.

The Forensic Neuropathology Autopsy

II: Developmental Considerations

INTRODUCTION 27
LARGE AND SMALL HEADS; LARGE AND SMALL BRAINS 27
HYDROCEPHALUS 28
FONTANEL SIZE 29
CENTRAL NERVOUS SYSTEM DEVELOPMENT: SELECTED ASPECTS 29
REFERENCES 45

Introduction

In Chapter 1, many of the topics discussed were equally applicable to mature as well as immature central nervous system (CNS) tissue. This chapter focuses on a few selected aspects of CNS development that may be useful in the examination of immature brain and spinal cord tissue.

Forensic cases involving CNS malformations frequently include questions of whether some incident during pregnancy or delivery was responsible for the abnormality present. For this reason, neuropathology problems encountered in pregnant women, the fetus, and in early childhood, including malformations, are further discussed in Chapters 4 and 5.

Large and Small Heads; Large and Small Brains

Definitions of large or small heads or brains vary in different textbooks, and the definitions used are briefly summarized here. All brains questioned to be unusually large or small should be referred for neuropathologic evaluation to resolve any future allegations.

Megalocephaly, or macrocephaly, is a term used for a large head/cranial vault of any cause. A variety of cranial measurements have been described in the literature, but head circumference is the most commonly used. A value exceeding two standard deviations (SD) over the mean for age group and sex is a basis for further evaluation, but serial measurements over time are more useful than a single measurement, since normal variations in growth rate and other factors can result in transient size variations in a given age range.[66] The same principle, of course, applies to heads that are too small. Skull size is rather incidental to the neuropathologist, who has the brain itself to examine. References to tables for head circumference are included, since these data are occasionally requested by medical examiners. Department protocols include such measurements in the general autopsy data required in children. The following selected references include head circumference tables in age ranges from fetuses to adulthood,[45,66] and additional tables are available in several general pediatric textbooks and from the Centers for Disease Control and Prevention. Pediatricians favor tables relating head circumference to chronologic age rather than to height and weight, even in dwarfs, since head and brain size are more closely related than head and body size.[8,69]

Megalencephaly, or macrencephaly, refers to an enlarged, abnormally heavy brain. In any age group, a brain exceeding 1600[38] to 1800 g[68] is considered abnormal. In younger age groups, it is variably defined as a brain weight greater than 2 SD (or above the 90th to 98th percentile) above the mean for age group and sex,[30,68] or a brain weight greater than 2.5 SD above the mean for age group and sex.[48,100] Megalencephaly can be unilateral (hemimegalencephaly), so anatomic features in addition to weight must be considered. The differential diagnostic list for megalencephaly is extensive.[30,101] A case of unilateral megalencephaly with high output cardiac failure has also been reported.[104] Left–right asymmetry of several structures is recognized as a normal feature in the human brain,[18] and some evidence suggests that the molecular events underlying such asymmetries occur sometime between 6 and 20 weeks' gestational age.[39] Whether insults in this same time frame can result in hemimegalencephaly is not, to our knowledge, established, although some have indicated a time frame of the third to fourth month of gestation for events producing both this malformation and megalencephaly.[40]

Microcephaly refers to a smaller than normal head/cranial vault of any cause, and has been defined as a head circumference "2 or 3" SD below normal for age group.[68] This term is often (mis)used to denote a small brain.[36,48]

Microencephaly, or micrencephaly, refers to a small brain, variously considered in children to be > 2 SD[48] or > 3 SD[102] below the mean for age group and sex, or a brain weight of less than 900 g in an adult, regardless of cause.

In head circumference measurements as well as in brain weights (see Chap. 1), varied and evolving ranges of "normal" values would support a policy of citing one's source for such figures, if used in formal reports.

Hydrocephalus

Implicit in the term *hydrocephalus* are enlarged lateral ventricles. It may be noncommunicating, communicating, "normal pressure," or "*ex vacuo*" in type. It is so regularly discussed in depth in standard neuropathology textbooks, including its occasional relationship to sudden unexpected death in children[92] and adults (see Chap. 9), that it needs only brief mention here. One aspect of consequence in hydrocephalus not frequently emphasized in the literature, but mentioned by Friede,[38] is what has sometimes been termed redundant gyration, or stenogyria, a common feature of Arnold-Chiari (Chiari II) malformation. This refers to a more complex pattern of gyri on the surface of the hydrocephalic brain, superficially resembling polymicrogyria at times (Fig. 2.1). Histologically, the cortical architecture is unremarkable except for changes related to mechanical stretching or atrophy in severe cases. The primary usefulness of this observation from a forensic standpoint is that its presence strongly favors the etiology of the hydrocephalus having occurred early in life, certainly before the age of 5 years. Hydrocephalus developing after 5 years of age does not show this type of complex gyral pattern. The term *external hydrocephalus* is applied to excessive cerebrospinal fluid (CSF) accumulation in the subarachnoid

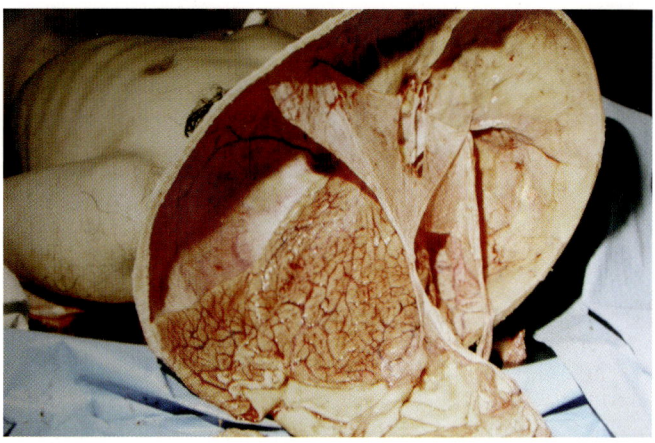

Fig. 2.1. This 35-year-old man had developed hydrocephalus shortly after birth. Parents refused medical intervention. The collapsed thin rim of cerebral cortex demonstrates an excessively complex gyral pattern (redundant gyration, or stenogyria).

space, with the so-called benign form demonstrating normal or only slightly enlarged ventricular size and prominent basilar cisterns.[63] Most descriptions of external hydrocephalus have been in children less than 2 years of age.

In the normal fetus the extracerebral space in early gestation tends to be widest over the posterior parietal area, but by term the widest space is frontal in location, with or without associated enlargement of frontal horns of the lateral ventricle.[88]

In infants, an enlarged pericerebral subarachnoid space (SAS) is usually encountered during the course of imaging studies prompted by serial head circumference measurements suggesting megalocephaly. If there is associated megalencephaly (see earlier), the evaluation proceeds toward the latter differential diagnostic considerations. If additional studies establish that the enlarged space between brain and calvarial dura is due to subdural fluid collections, one is led to a different set of possible causes such as subdural hematoma or hygroma. If the SAS is enlarged exclusively, or disproportionately compared with lateral ventricles, but the enlargement follows hypoxic-ischemic injury, subarachnoid hemorrhage, meningitis, or venous hypertension,[15] the term *external hydrocephalus* has sometimes been applied. However, the use of this term in situations where there is some obvious predisposing factor such as those mentioned earlier is considered an inappropriate use of the term by some.[70] If there is no apparent etiology for the SAS enlargement, it has been referred to as idiopathic benign external hydrocephalus.[3,72] In some cases there appears to be a genetic predisposition, with relatives of the proband sometimes having either macrocephaly[3,58] or microcephaly.[2] A case of external hydrocephalus was reported in only one of a set of monochorionic twins presumed to be monozygotic, but more recent evidence that dizygous monochorionic twins can occur might lend more substance to the nongenetic causes discussed by the author.[74,94]

Most cases of external hydrocephalus appear to be sporadic, and to follow a benign, self-limiting course, resolving by the second year of life.[4,96] Persistence beyond age 2 is considered a basis for further evaluation,[96] and a few cases evolve into symptomatic progressive internal communicating hydrocephalus requiring CSF shunting.[15]

The primary forensic interest in this condition, probably best viewed as a syndrome with varied etiologies, is some preliminary evidence raising the question of whether children with external hydrocephalus may have a greater than average tendency to develop subdural hematomas.[72,73,75] Undoubtedly this issue will be further explored by those involved in the differential diagnosis of child abuse, in an effort to establish the presence or absence of such an association.

Cardoso and Schubert[15] described "external hydrocephalus" as a transient phase preceding internal hydrocephalus following spontaneous or traumatic subarachnoid hemorrhage, and resolving following CSF shunting in adults, but these authors seem to use the term to include extracerebral fluid (including blood) collections that include subarachnoid, subdural, and possibly extradural/subcutaneous operative site fluid collections. The typical benign form of idiopathic external hydrocephalus seen in infants has not, to our knowledge, been described in adults.

Fontanel Size

Fontanel measurements are often included in pediatric autopsy protocols. Once taken, however, what do they tell us? Most of the time, they do not aid the determination of cause and manner of death beyond what the rest of the autopsy findings reveal. In some cases, however, they may suggest a direction of investigation that is helpful in arriving at the correct diagnosis.

Menkes and Sarnat[62] indicate that the posterior fontanel closes by 3 months of age, and the anterior fontanel by 20 months of age. This is likely based in part on an earlier study of 530 apparently normal children in whom the mean age of anterior fontanel closure, determined radiologically, occurred at 17.9 months in boys and 19.7 months in girls, and, by clinical examination, occurred at 16.3 months in boys and 18.8 months in girls.[1] However, the extremes of ages at which the anterior fontanel closed by clinical criteria were from 6 months to 2½ years in boys, and from 1 to 4½ years in girls in that study.[1] No correlation was found between fontanel closure and hand/wrist skeletal maturity, number of deciduous teeth erupted, or skull circumference in these normal children.[1]

The range of normal size of anterior and posterior fontanels has been described during the first year of life in 201 subjects (figures given are based on the average of length plus width in millimeters) by Popich and Smith.[76] Their original article with its useful tables should be reviewed for details; the most useful table is also reproduced in Jones.[50] Briefly, there is a tendency for the anterior fontanel size to enlarge up to about 2½ months to 3 months of age, followed by a gradual decrease in size and eventual closure. The posterior fontanel was smaller, with only 3% of full-term newborns demonstrating a mean size exceeding 0.5 cm (the largest being 2.0×2.1 cm).[76]

Fontanel size may provide clues to the presence of a variety of disorders. Unusually small anterior fontanel size may occur in microcephaly, craniosynostosis involving sagittal or coronal sutures, or accelerated osseous maturation such as in hyperthyroidism.[76] Given the broad age range of normal closure, however, a small anterior fontanel as an isolated finding (i.e., normal head size and shape, absence of cranial suture ridging, normal brain weight) is not significant.[76]

Delayed closure of fontanels may result from a broader spectrum of disorders. In addition to its occurrence in cases of increased intracranial pressure from various causes, Jones[50] lists 43 malformation syndromes in which delayed fontanel closure is either frequent or occasional. Additional conditions in which delayed fontanel closure can be seen include achondroplasia, congenital hypothyroidism,[93] aminopterin-induced syndrome, Kenny's syndrome, vitamin D-deficiency rickets, malnutrition, and rubella syndrome.[76] In achondroplasia, the delay in fontanel closure may reflect hydrocephalus, seen in some of these cases as a result of a narrow foramen magnum and partial blockage of CSF effluence from the fourth ventricle, rather than as a delay in osseous maturation.[50]

Unusually large anterior and posterior fontanels for age group may, by indicating a prenatal delay in osseous maturation, suggest the possibility of congenital hypothyroidism before myxedema and growth retardation become clinically evident.[93]

Central Nervous System Development: Selected Aspects

Numerous sources for information on this complex subject exist, and many readers will undoubtedly have already settled on those they consider most useful to them. For others, we can suggest a few selected sources, several of which also contain more extensive citations to original studies on specific topics. A brief introduction to developmental topics is included in several general neuropathology and pediatric autopsy texts,[38,40,51,52,89] although it should be noted that some sources clearly specify that they extrapolate to humans to some extent from animal experimental material.[10]

In the following sections, a few sources representative of the more extensive literature in each category are discussed.

Cerebral Cortex Development

Determining whether or not cerebral cortical development is appropriate to gestational age can be helpful in the decision of whether to submit additional histologic sections to confirm the presence of congenital malformations, including the more subtle forms of cortical dysplasia. References helpful in this regard include those with either sequential photographs or diagrams of cortical gyral development[16,29,32,42,55,56] and more recent books on brain development, such as the series represented by the text of Bayer and Altman[10] of the external brain appearance at various developmental stages.

There is an extensive literature on cytoarchitectonic development of the cerebral and cerebellar cortex, with sources particularly useful on prenatal development being that of Larrouche.[55,56,57] For postnatal development of the cerebral cortex, useful are the series of eight volumes by Conel[20–27] or the more recent series of Bayer and Altman,[10] and the study of the birth to 1-year age group by Dekaban.[29] These sources bracket the age ranges of cortical development most often needed by the pathologist.

Most neurons are generated in the embryo between day 40 and 125,[69] and the thickness of the subependymal germinal matrix is maximum at approximately 26 to 30 weeks' gestation,[42] then rapidly decreases. Neuronal migration assumes various forms, including columns of cells oriented either perpendicular (sunburst or radiating pattern) or parallel (onionskin pattern) to the ependymal surface. Migrating cells also form perivascular aggregates or nestlike aggregates that are not vessel centered. One helpful distinction between migrating neuroblasts and/or spongioblast cells versus chronic inflammatory cells is the presence in the former group of rod-shaped cells. Another useful guide is to compare such cell nests with the "internal control" of residual, clearly germinative cells in subependymal areas in the same brain. Glial precursors begin to outnumber neuronal precursors in the third trimester.[51,52] These neuronal and glial precursors have, for the most part, disappeared by about 1 year postpartum, although rare nests of cells can be seen in some cases (usually in perivascular or periventricular white matter sites). The earlier migrating neurons form deeper cerebral cortical layers, with subsequent cells migrating to more superficial cortical layers ("inside out" developmental process). Distinction between the various cortical layers gradually becomes more evident with time, with different cytoarchitectonic areas maturing at different rates. The six-layered motor cortex is evident well before birth, for example.[56]

From a practical standpoint, sufficient familiarity with normal development to appreciate significant deviation from normal is necessary (Figs. 2.2–2.3), with attention given particularly to abnormalities such as heterotopic neuronal and/or glial aggregates; a "windblown" or "helter-skelter" disarray of cortical neurons (Figs. 2.4–2.5); presence of radial cell columns rather than a laminar pattern in mature frontal lobes; foci of relative cell depopulation within a given layer or layers; clusters of neurons in cortex which are ovoid aggregates, separated from one another by cortical areas less well populated with neurons (normal in certain areas of the medial temporal lobe but not elsewhere); and unusually large, bizarre, multinucleated or swollen-appearing neurons and/or glia (Fig. 2.5). In the hippocampus, cell gaps or dispersion of the dentate fascia layer is also a helpful clue to give careful attention to the presence of possible cytoarchitectural and other anomalies, assuming that the section is at the level of the lateral geniculate body. Sections of hippocampus in more anterior or posterior portions do not normally show the same classical orientation and cell layer regularity. Such abnormalities on routine sections suggest the need for carefully oriented blocks of cortex cut perpendicular to the pial surface at 10- to 12-μm thickness and stained with the Nissl method for confirmation of suspected subtle cortical dysplasia.

Additional comments on some of the aforementioned clues are as follows. A radial orientation of cortical neurons in regions where they would not be expected beyond 1 year of age, such as in the frontal lobe, warrants further study. It is normal to see radial linear neuronal orientation in other areas, such as the paracalcarine occipital cortex, para-acoustic temporal lobe, and medial temporal isocortex. Also, in the frontal lobe there should not be obvious gaps between cortical neuronal groups, but rather a continuous laminar pattern is the norm.

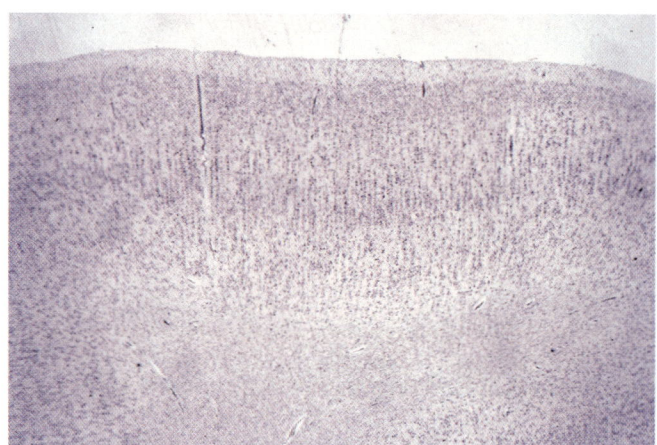

Fig. 2.2. Low-power micrograph of angular gyrus, with normal cortical cytoarchitecture. Orderly laminar pattern is seen. (Nissl.)

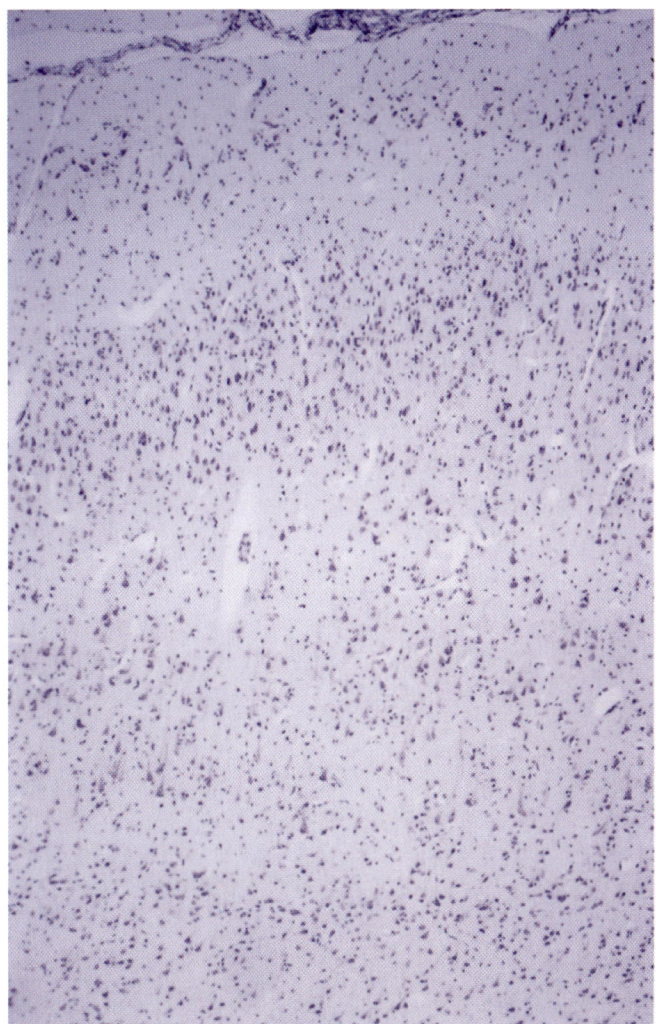

Fig. 2.3. Slightly higher-power micrograph than Fig. 2.2 of frontal lobe dysplastic cortex, with increased neuronal packing density and thickness most apparent in layer 2. Layers 3 and 4 show sparse, irregular neuronal distribution. Layer 5 is somewhat more organized. (Nissl.)

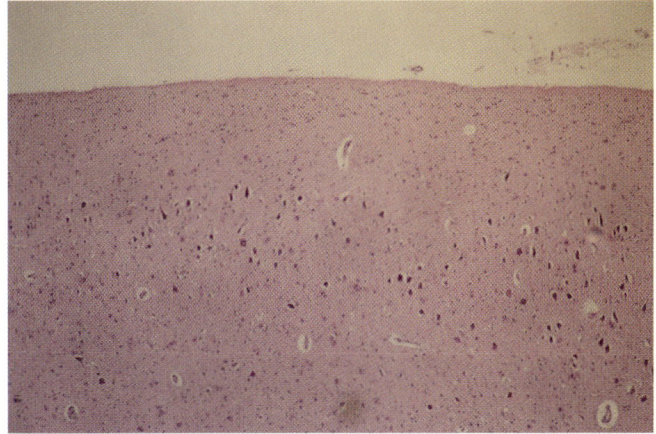

Fig. 2.4. Low-power micrograph of dysplastic cortex, with "windblown" neuronal disarray and severely disturbed laminar pattern. This section is from the thickened cortex in Fig. 2.7. (H&E.)

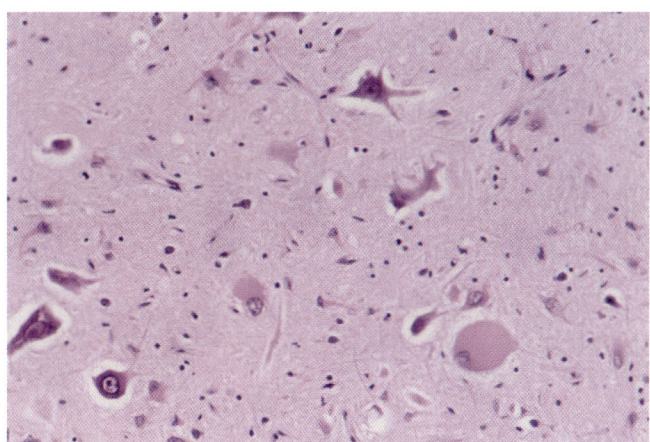

Fig. 2.5. High-power micrograph of atypical, dysplastic neurons and astrocytes from cortex seen in Fig. 2.4. (H&E.)

The cortical thickness may provide a clue to cortical cytoarchitectonic abnormalities (Figs. 2.6–2.7). In the mature brain, normal measurements for cortical width of the frontal lobe and most areas of isocortex average about 2.5 mm. The precentral motor cortex may measure up to about 4.5 mm, but primary visual cortex is thin, measuring only about 1.5 mm. Another clue to possible dysplasias is the depth and complexity of gyri and sulci as viewed on coronal sections of the cerebral hemispheres. In the posterior parietal area it is not unusual to see several isolated tongues of cortex in the depths of sulci, surrounded by white matter, sometimes referred to as cortical undercutting. If seen in coronal sections in other areas of the brain to this same degree, careful examination is indicated to rule out true heterotopic islands of cortex or irregularities of cortical thickness suspicious for dysplasia.

One should be cautious in diagnosing cerebral cortical dysplasia if cortex is not sampled perfectly perpendicular to the surface. Assessment of cytoarchitecture is best made along banks of sulci and not at the crown or base of the sulcus. Normal areas of cortex can also be a clue to the diagnosis between areas suspicious for dysplasia. Usually, dysplastic cortex is present over a broad zone rather than minute isolated zones. Recent reviews and suggested classifications of cortical dysplasias are available.[65,71,77,99]

Noncortical Regions

Other potentially problematic areas can be encountered in infant brains. Infants may normally demonstrate larger numbers of leptomeningeal macrophages[42] or lymphocytes than would be considered within the normal range for adults (analogous to the age differences in normal CSF cell counts in infants versus adults).[5,54] Gilles et al.[42] describe large numbers of microglia-like cells in telencephalic structures as a normal finding in infant brains.

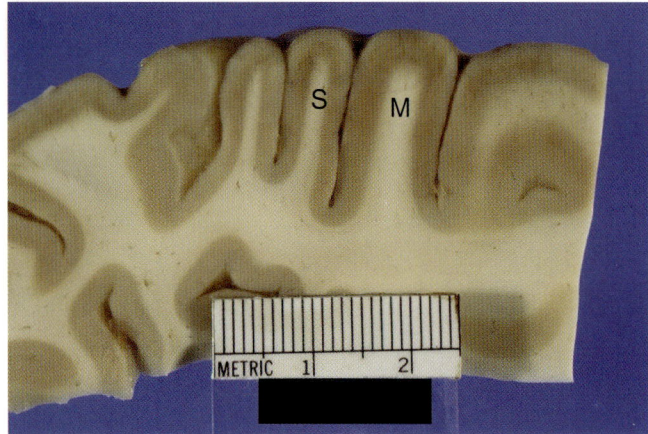

Fig. 2.6. Normal brain section perpendicular to central sulcus, with upper portion of photo showing sensory cortex (S) (left side) merging with thicker motor cortex (M) (right side).

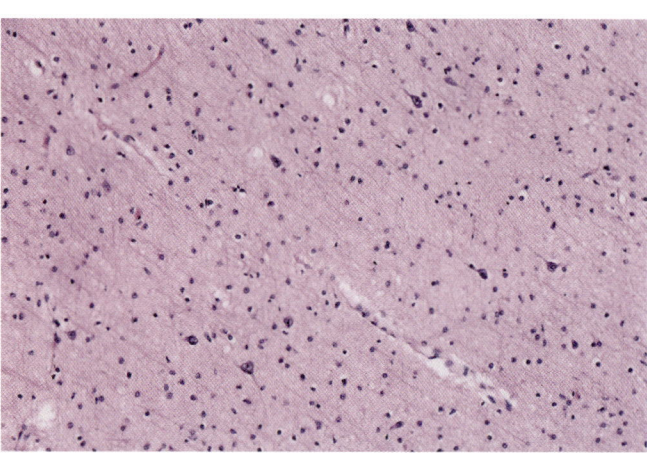

Fig. 2.8. Low-power micrograph. Pyramidal-shaped larger cells are neurons present in deep cerebral subcortical white matter. (H&E.)

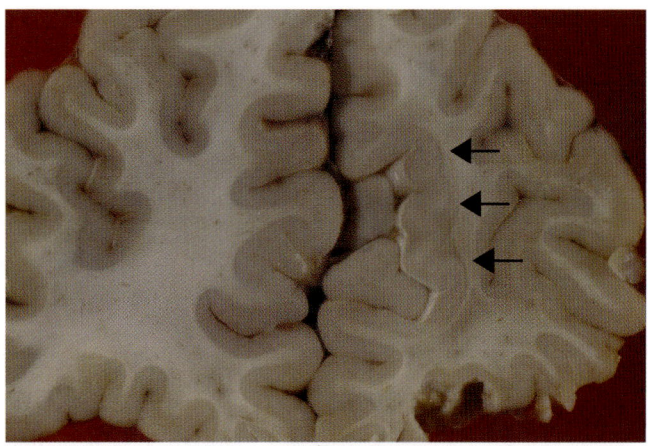

Fig. 2.7. The right medial frontal lobe in this brain contains a zone of abnormally thick cortex and slightly discolored subcortical white matter (*arrows*), suspicious for cortical dysplasia (also see Figs. 2.4 and 2.5).

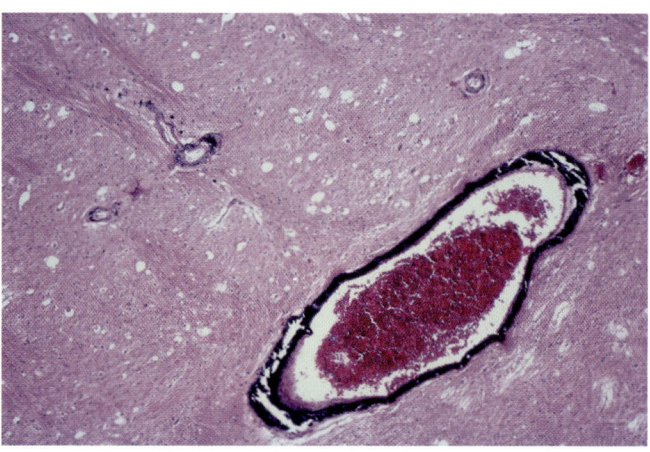

Fig. 2.9. Low-power micrograph of calcification in globus pallidus vessel wall and, to a less extent, in parenchyma in middle-aged adult. (H&E.)

Possibly some of these are residual neuroblasts or spongioblastic elements. Scattered single neurons in white matter (Fig. 2.8) as an isolated finding, especially in the immediately subcortical zone, are not an unusual finding in infant brains, and they should be interpreted as pathologic only with caution.[38,69] Nonetheless, brains with this finding should be carefully examined for subtle abnormalities, since it seems to us to be more frequent in the presence of other CNS congenital abnormalities. These individual heterotopic neurons have also been described as increased in number in certain forms of epilepsy.[60,61,98] Rojiani *et al*.[83] have demonstrated that heterotopic white matter neurons are more common in temporal lobes than in occipital or frontal lobes in normal brains.

Micronodular mineralization in the developing corpus striatum is seen in normal brains, and is also of uncertain significance.[38] As with the amphophilic globules described by Gilles,[41] they merit mention in the microscopic report and should alert one to search for associated abnormalities. No further comment on them is made in the final diagnoses if they are unassociated with other significant findings. More extensive basal ganglia mineralization in vessel walls and parenchyma, most common in the globus pallidus, is often present with aging, usually as an incidental finding but occasionally associated with calcium–phosphorus metabolic disorders (Fig. 2.9).

Subependymal gliosis, accompanied by ependymal rosettes and prominence of small blood vessels, is a normal finding at the tip of the occipital horn of the lateral ventricle.

Cerebellum Development

The cerebellar cortex appearance during the postnatal period is a source of some variation of opinion in the literature. In part, this undoubtedly reflects individual differences in the sample population, and the difficulty of distinguishing outliers from the majority of the population. The practical importance of this issue is that, occasionally, cases are referred of the cerebellar external granular layer (Fig. 2.10) misdiagnosed as, for example, chronic meningitis, lymphoma, leukemia, or medulloblastoma. Estimates in the literature vary as to the thickness of the external granular layer at different ages, and the age at which it finally disappears. There is less disagreement on the appearance of recognizable Purkinje cells (i.e., at about 25 weeks' gestation, initially in the vermis), and when a recognizable four-layered cerebellar cortex appears (i.e., external granular layer, molecular layer, Purkinje cell layer, and internal granular layer, at approximately 26 weeks' gestation). Only during the 21st to 32nd week of gestation is the cerebellar cortex five-layered, during the life span of the lamina desiccans.[89] A detailed review of the histologic features of the cerebellar cortex during brain development is beyond the scope of this text, but is available in the references cited.

In our experience, the cerebellar external granular layer, which first appears in the 9th week of gestation and reaches maximum thickness at approximately 24 weeks' gestation, does not exceed 6 to 7 cell layers in thickness in an otherwise normal full-term infant. It decreases in thickness over the next several months, becoming a discontinuous single layer by approximately 9 to 10 months postnatally. Various authors cite the age of complete disappearance of the external granular layer as varying from 8 to 10 months postnatally[56] to up to 18 months postnatally.[102] In our experience, the external granular layer essentially disappears by 14 months of age, aside from rare isolated cells that may persist up to 16 months of age. Typically, the lateral cerebellar hemisphere is sampled for this assessment. Lateral folia mature later than the vermis, which loses its external granular layer as early as 4 to 12 months' postpartum.[89]

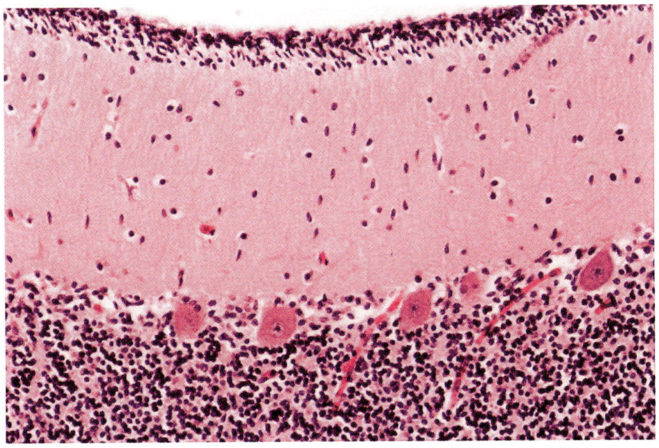

Small cerebellar neuronal heterotopias, particularly large cell type (Fig. 2.11), are frequent in apparently normal infants.[69,86] They are normally encountered much less frequently in later childhood and in mature brains, but larger collections are present in association with certain trisomies (trisomy 13, 15, 18, and 21).[28,97] Cerebellar neuronal heterotopias characterized by large nodules of mixed cells (i.e., mature neurons of Purkinje cell type, small granule cells, and spindle cells) are the most likely to be associated with other brain or visceral malformations[57] (Figs. 2.12–2.14). Less frequently, other cellular components such as ependymal cells and increased vascular elements may be present in such heterotopias,[12,13] and cerebellar cortical heterotopias and dysplasias in various patterns occur (Figs. 2.15–2.16). Neuronal heterotopias of small cell type are not limited to the cerebellum, but may occur in other sites such as the dorsal cochlear nucleus (Fig. 2.17).

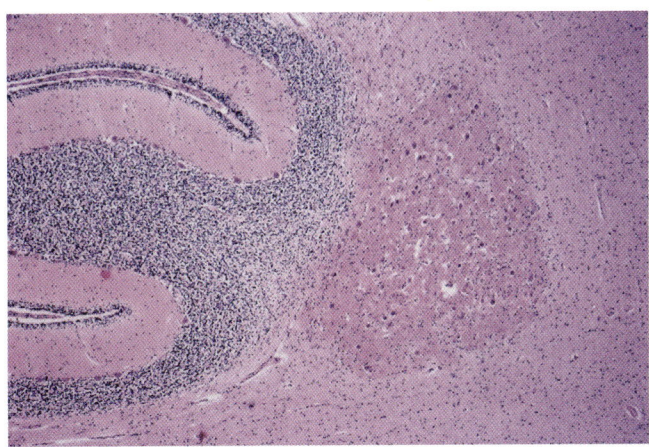

Fig. 2.11. Low-power micrograph of cerebellar subcortical large cell neuronal heterotopia. (H&E.)

Fig. 2.10. Medium-power micrograph. Normal subpial cerebellar external granular layer (see text). (H&E.)

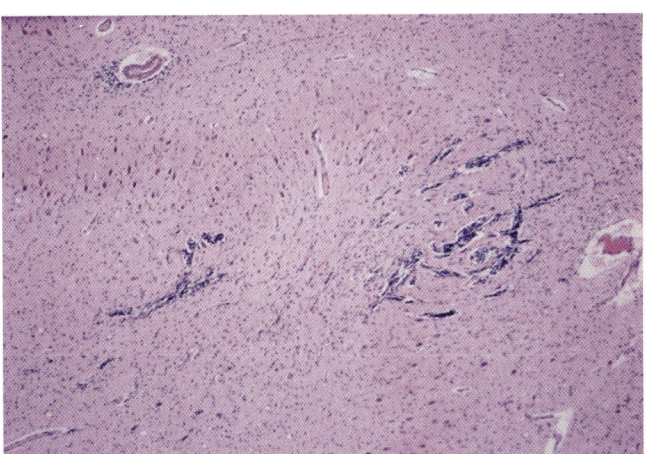

Fig. 2.12. Low-power micrograph of mixed large and small cell cerebellar white matter neuronal heterotopia in region of dentate nucleus. (H&E.)

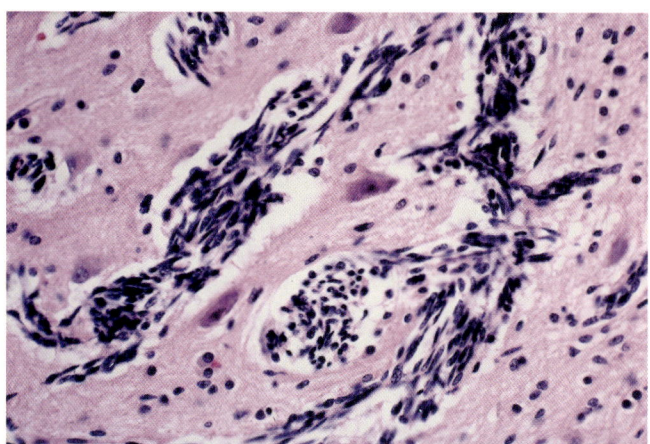

Fig. 2.13. High-power micrograph of case seen in Fig. 2.12. Streams of elongate and round small cells resembling neuroblasts with hyperchromatic nuclei and sparse cytoplasm, interspersed with larger heterotopic neurons. (H&E.)

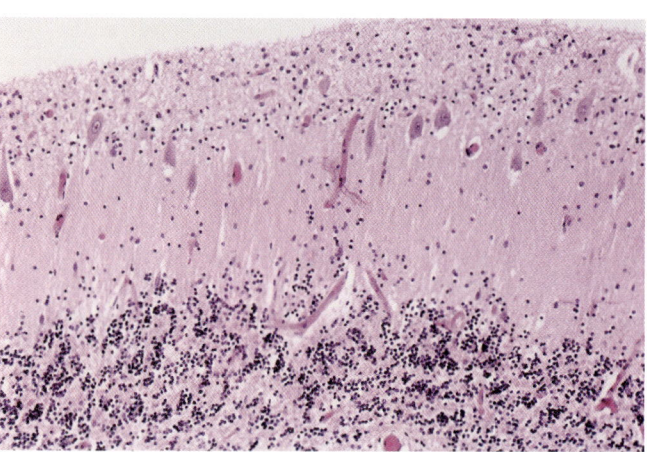

Fig. 2.15. Medium-power micrograph demonstrates unusual form of cerebellar cortical dysplasia, with granule cells and Purkinje cells in upper molecular layer. (H&E.)

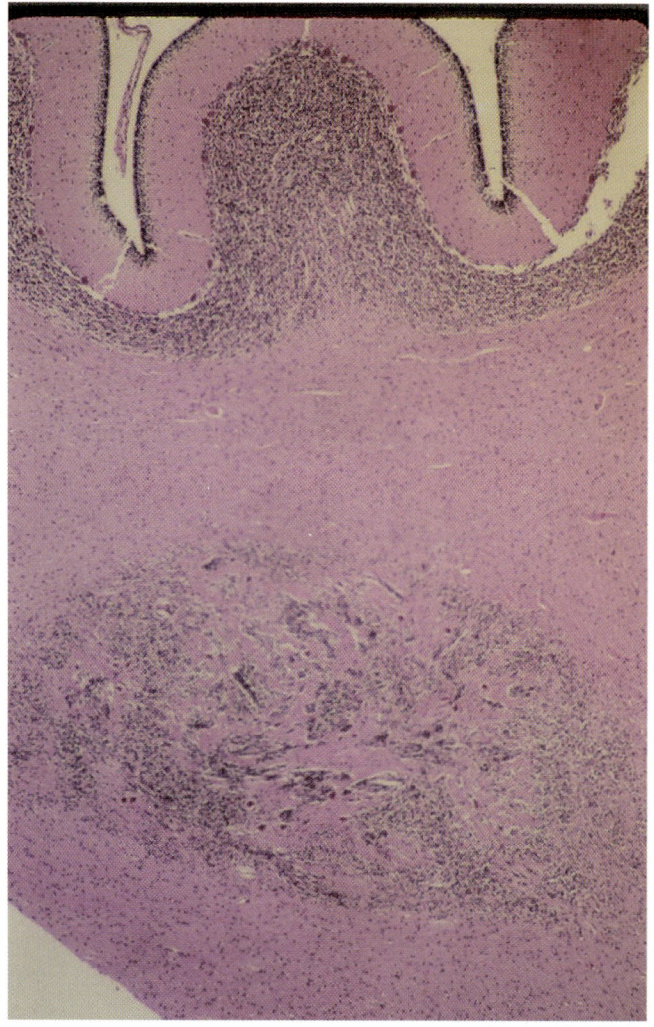

Fig. 2.14. Low-power micrograph. Mixed large and small cell neuronal heterotopia in cerebellar white matter. This infant had gastroschisis and died shortly after birth. Normal cortical external granular layer is present. (H&E.)

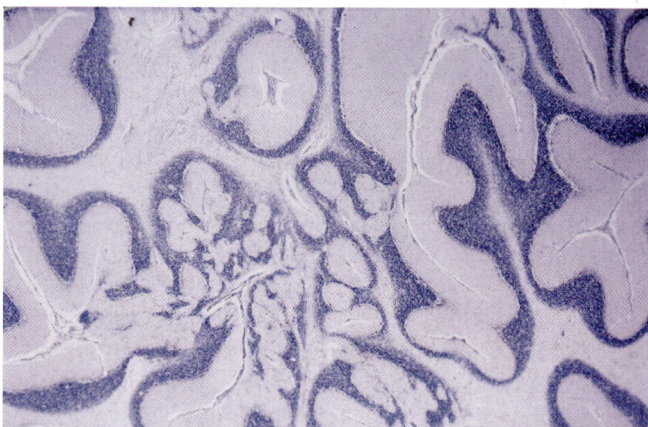

Fig. 2.16. Low-power micrograph. Dysplastic, irregularly layered cerebellar cortex within deeper subcortical white matter. (H&E.)

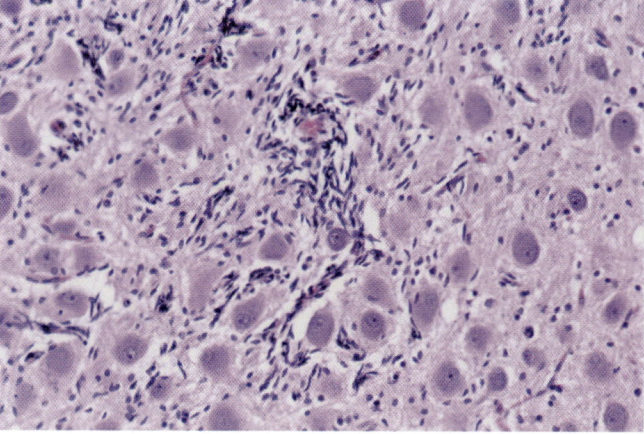

Fig. 2.17. Medium-power micrograph of small cell neuronal heterotopia in dorsal cochlear nucleus. (H&E.)

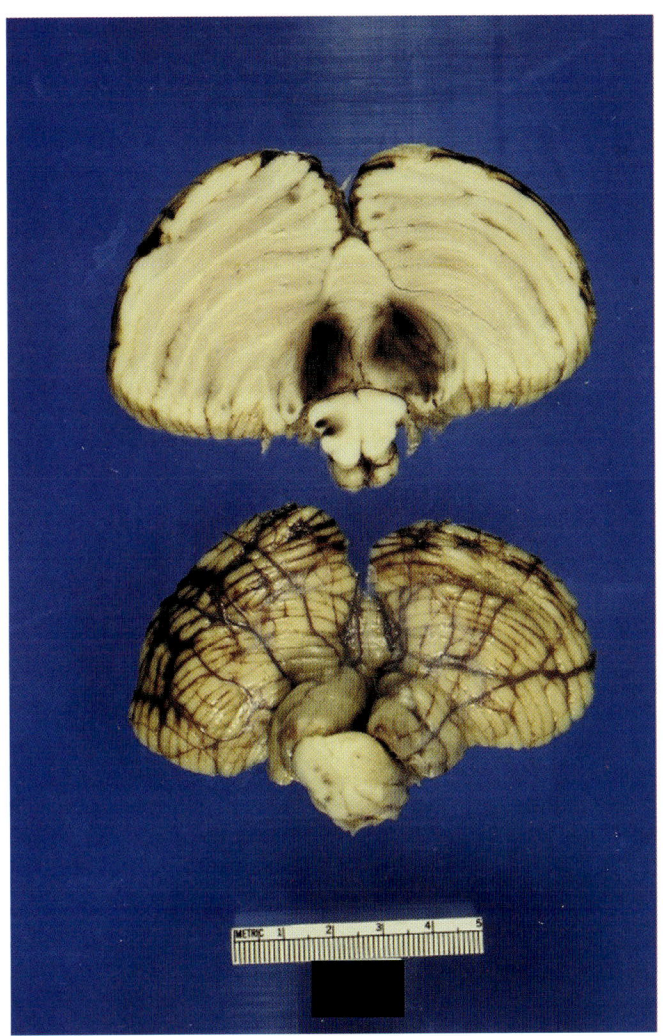

Fig. 2.18. Cerebellar "tonsillar" herniation. Biventer lobule is deeply grooved bilaterally (lower image). Tissue medial to the groove is discolored and soft, and a section (upper image) reveals parenchymal hemorrhage (herniation contusion) in biventer–tonsillar area and adjacent left medulla. Tonsils are not seen on the external examination at the stage of herniation present in this case.

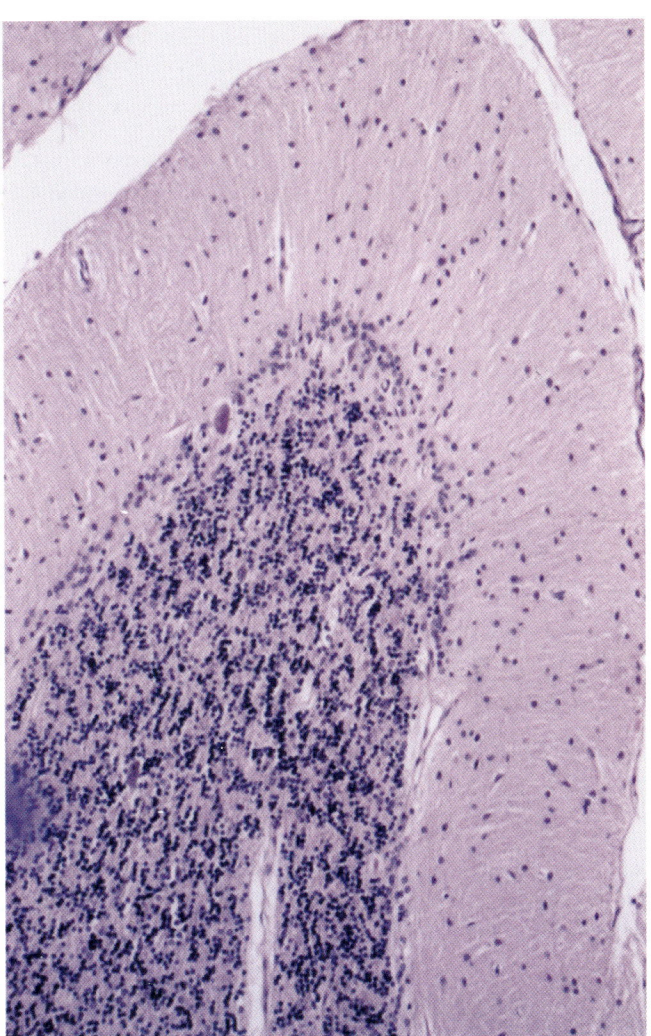

Fig. 2.19. Medium-power micrograph of atrophic cerebellar cortex. Purkinje cell dropout, with only a single Purkinje cell present in what is otherwise a layer of conspicuous Bergmann gliosis. (H&E.)

A few further comments on the cerebellum are in response to recurring questions raised. Cerebellar tonsillar herniation actually includes the medial portion of the biventer lobule of the inferior cerebellum, lateral to the tonsil. In the normal state, the tonsil faces medially and is not regularly exposed on the inferior aspect of the cerebellum unless sufficient herniation effect occurs (Fig. 2.18). Grooving of the biventer lobule is a common finding of postmortem settling of the brain or with the brain enlargement seen as a result of perfusion pressure in embalmed bodies. It does not imply herniation unless there is focally increased vascular congestion and discoloration of the biventers, especially at the impressions created at the margins of the foramen magnum, and/or softening in comparison to adjacent brain tissue. The term "tonsillar herniation" should perhaps be reserved for true caudal displacement of the tonsils with its invariable compression of the medulla. Some make no distinction between herniation and biventer molding in the expression "cerebellar pressure cone." The important point is that molding of the biventer alone devoid of the changes mentioned here is not properly called tonsillar herniation. The latter is of utmost pathologic importance.

Astrocyte nuclei in the region of the dentate nucleus not infrequently are slightly larger, and more often display irregularities in nuclear outline (e.g., clefts, focal protuberances) than other areas (aside from the globus pallidus) in normal brains.

One should be wary of diagnosing Bergmann's gliosis in the cerebellar cortex in the absence of a definite increased number and prominence of Bergmann glia nuclei (Fig. 2.19) and associated dropout of Purkinje cells. In later stages of atrophy, eosinophilic glial processes radially extend straight up from the Bergmann glia layer to the pia mater (Fig. 2.20). Lastly, on gross examination of sections of the cerebellum, cortical surfaces of adjacent folia should be closely approximated. If space exists between folia (enlarged sulci), sample the area for atrophy (Fig. 2.21).

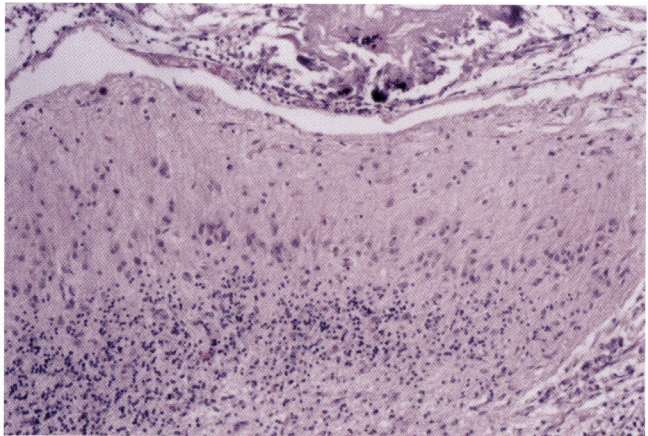

Fig. 2.20. Medium-power micrograph of atrophic cerebellar cortex. Granule cell depopulation, complete loss of Purkinje cells, and hypertrophy of Bergmann glia with radiating glial processes extending toward pia mater. Area of pressure atrophy from dermoid cyst of fourth ventricle (note squamous debris and foreign body giant cells in overlying leptomeninges). (H&E.)

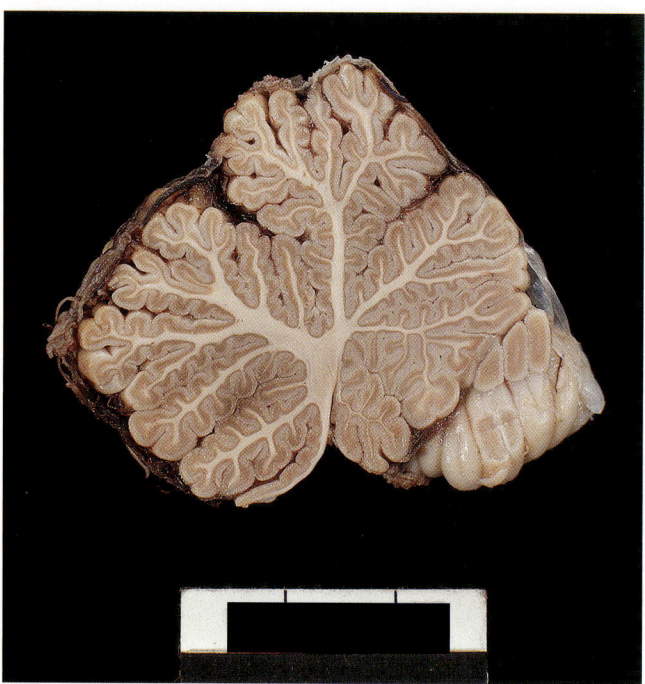

Fig. 2.21. Cerebellar cortex on section is atrophic (wide sulci) at top and at upper left of photo. Other regions are within normal limits on gross inspection, aside from gray discoloration typical of granule cell autolysis (see Chap. 1, Figs. 1.37–1.39).

Appearance of Reactive Cells in the Developing Brain

The appearance of macrophages, astrocytes, and the like in the developing brain transforms the nature of the CNS response to insults to those that are so familiar in more mature brains. Prior to their appearance, sequelae of a brain insult may simply be represented by absence of a normal structure, architectural abnormality, neuronal depopulation, cystic change, and/or mineralization without presence of macrophages, gliosis, or other reactive stigmata. Estimates as to the timing of these reactive capabilities vary in the literature, and results from several sources are summarized in Table 2.1. These should be regarded as estimates only, as the variation between sources suggests. Also, the first appearance of astrocytes varies in different regions of the developing brain, which may account for some of the estimated variations seen,[79] and continued morphologic (and presumably functional) changes in astrocytes continue well into postnatal life.[78]

In addition to the timetable for brain parenchyma summarized in Table 2.1, some data also exist for other specific areas. Gilles[41] summarized reports indicating that a histocytic response may be seen in the developing cavum septi pellucidi at "about 12 weeks'" gestation, and in the leptomeninges at "about 20 weeks'" gestation.

Cavum Septi Pellucidi and Cavum Vergae

The cavum septi pellucidi, a space between the two layers of the septum pellucidum at the level of the descending columns of the fornix, represents a normal stage in brain development. It is invariably seen in the premature brain. The cavum Vergae is a cavity posterior to the cavum septi pellucidi between the psalterium (consisting of hippocampal commissure and pillars of the fornices) and the splenium. It mostly communicates with the cavum septi pellucidi, forming anterior and posterior sectors of the cavity. Both (especially the cavum Vergae) become increasingly obliterated late in the third trimester, but residual cavum septi pellucidi can be seen in up to 97% of term infants, becoming obliterated at about 2 months of age[38,91] in the majority of individuals. It persists to varying degree in up to 20% of adults (Case 2.1, Fig. 2.22).[48] The nature of the inner surface layer of the cavum septi pellucidi varies with age, but in adults is most commonly a single layer of flattened or cuboidal cells, at times including some ciliated cells.[59] Only in rare instances do cavi present a clinical problem.

Hippocampus

The hippocampus warrants specific mention because of the importance that it be sampled and examined in such a wide spectrum of age groups and disorders, including but not limited to hippocampal sclerosis (e.g., including types related to hypoxic-ischemic injury, trauma,

Table 2.1. Selective Reactive Cellular Landmarks in the Developing Central Nervous System*

Gestational Period	Event
8 weeks' gestation[79]	First GFAP-positive astrocytes appear in midthoracic spinal cord, and in subventricular zone of mesial cerebrum.
12 weeks' gestation[79]	First GFAP-positive astrocytes appear in brainstem.
Early in 2nd trimester[34]	Macrophages appear in CNS.
15–16 weeks' gestation[68]	Macrophages appear in CNS.
15 weeks' gestation[82]	Mature astrocytes appear in CNS.
17 weeks' gestation[79]	First GFAP-positive astrocytes appear in cerebellum (vermis).
17–20 weeks' gestation[80]	Astrocytes become capable of forming glial scar.
After 20 weeks' gestation[34]	Fiber-forming astrocytes appear.
20–27 weeks' gestation[52]	First appearance of reactive astrocytes (variable, depending on site).
20 weeks' gestation[38,81]	GFAP-positive reactive astrocytes first detected in CNS.
After 20 weeks' gestation[34]	GFAP-positive fiber-forming astrocytes appear.
After 20 weeks' gestation[68]	Gliosis can be seen.
23 weeks' gestation[38,81]	GFAP-positive gemistocytic astrocytes first seen.
Slightly before 24 weeks' gestation[102]	Macrophages appear in CNS.
24 weeks' gestation[102]	GFAP-immunoreactive gliosis seen (rare prior to this period).
"Midgestation" plus[42]	Probable earliest appearance of plasma cells in CNS.
After 35 weeks' gestation[52]	Well-differentiated (i.e., well-ramified) microglia present.
Perinatal injury[84]	Alzheimer type II glia may be seen in striatum.
Beyond 6 months' postnatal, and readily seen after 1–2 years of age[38]	Alzheimer type II glia first seen.

*These generalizations apply primarily to human telancephalic structures and also reflect the lack of uniformity currently in the literature. Areas other than forebrain are specified.
GFAP, glial fibrillary acidic protein.

excitotoxicity, and epilepsy); changes of aging; degenerative diseases of a variety of types (e.g., Alzheimer's disease, Pick's disease, intranuclear hyaline inclusion disease, various system degenerations such as multiple system atrophy and progressive supranuclear palsy, diffuse Lewy body disease, and argyrophilic grain disease); other disorders of memory and cognition; infectious diseases (e.g., Creutzfeldt-Jakob disease, herpes simplex encephalitis); toxins (e.g., domoic acid); and congenital malformations (e.g., malrotation, cortical dysplasia). It is also a location that tends to develop certain tumors (e.g., hamartomas, gangliogliomas).

Screening for these various disorders, especially the more esoteric degenerative disorders, is beyond the spectrum of activity in many forensic pathology departments, primarily due to time, and technical and budgetary restraints. The need for resident training in autopsy techniques and forensic pathology on the part of academic institutions, and the benefit of increased access to clinical and pathology subspecialty consultants and more extensive laboratory facilities by forensic pathology departments, can lead to formal or informal arrangements beneficial to both parties.[31] Selected forensic cases may, for example, be referred for more extensive neuropathologic study to individuals in academic institutions with a special interest in the disease category in question.

Evaluation of the hippocampus is facilitated by sampling it in a coronal section taken at the level of the lateral geniculate body. The anterior one-half of the hippocampus does not display the classic appearance of the various pyramidal cell layer sectors (e.g., CA1-4) due to the greater irregularity of the cortical band and dentate fascia anteriorly, and the more posterior hippocampus also becomes comparatively less well organized.

The hippocampus has a vertical orientation at 15 weeks' gestation, and gradually rotates to become a horizontally oriented structure with formation of the hippocampal fissure by 15 to 19 weeks' gestation.[6] It clearly has its mature brain overall profile by 23 weeks' gestation, suggesting that abnormalities leading to hippocampal malrotations occur at some time prior to approximately 20 to 23 weeks' gestation (Case 2.2, Figs. 2.23–2.26).

There is nonsynchronous maturation of pyramidal layer neurons in the hippocampus, which may lead to misinterpretations in the immature brain. Early in gestation, CA1 neurons take the lead in developmental matu-

Case 2.1. (Fig. 2.22)

A 23-year-old man was playfully boxing on a public sidewalk when he sustained several blows and collapsed, becoming comatose. Transported to a hospital, he arrived in full arrest and died within 45 minutes of his collapse. Brain showed moderate swelling. Clinical diagnosis of intracranial hemorrhage was not confirmed.

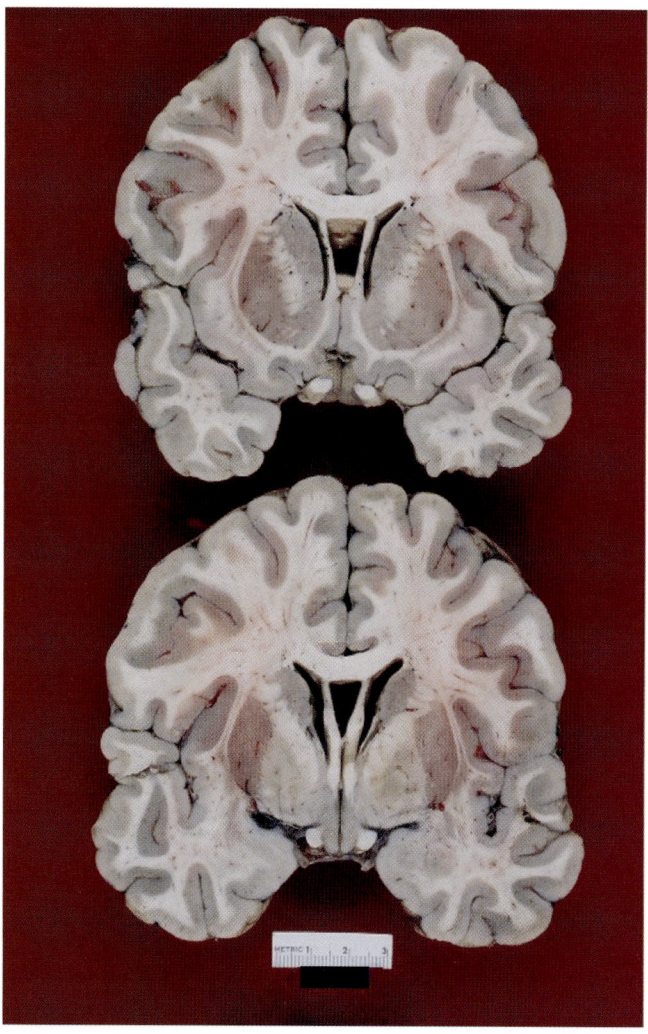

Fig. 2.22. Coronal sections of frontal lobes reveal a broad cavum septi pellucidi, an incidental finding, with symmetrically narrow but open frontal horns of lateral ventricles. Anterior striatum appears somewhat larger than average, but was without diagnostic abnormality.

Case 2.2. (Figs. 2.23–2.26)

A 61-year-old woman was battered in domestic violence by her husband and sent to a hospital. Formal evaluation, if any, of her mental capacity was not known. She was said to be prone to bouts of yelling and screaming. For "convulsions," Dilantin had been prescribed. She was unexpectedly found dead in a convalescent home where she was sent following discharge from the hospital.

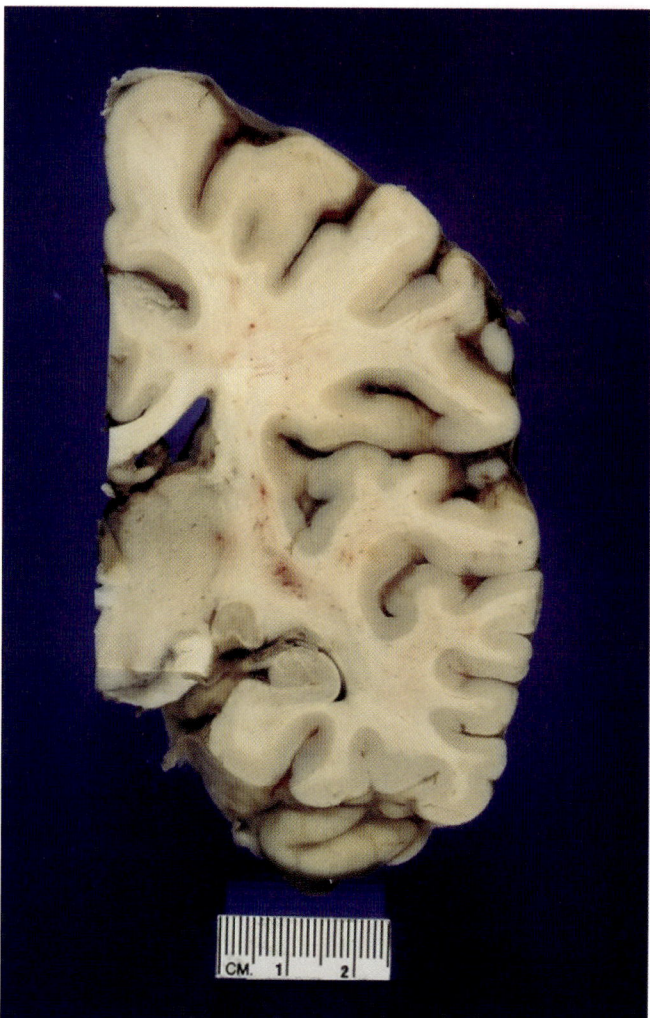

Fig. 2.23. Right hemisphere, normal brain, with normal rotation of hippocampus at level of lateral geniculate body, for comparison with Fig. 2.24.

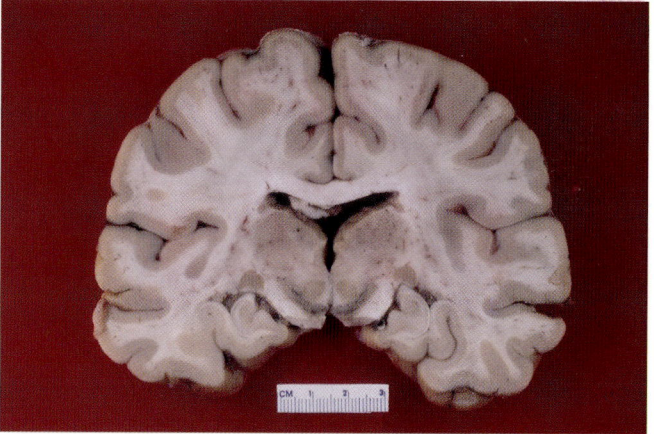

Fig. 2.24. Coronal section of cerebrum of Case 2.2 at level of lateral geniculate bodies shows bilateral "external rotation" of hippocampi. Microscopically, hippocampi were intact, including Sommer's sector. Cerebellum showed severe diffuse folial atrophy with retrograde inferior olivary atrophy. Postfixation brain weight was 1139 g. Frontal and parietal convolutions were simplified, with paucity of tertiary sulci.

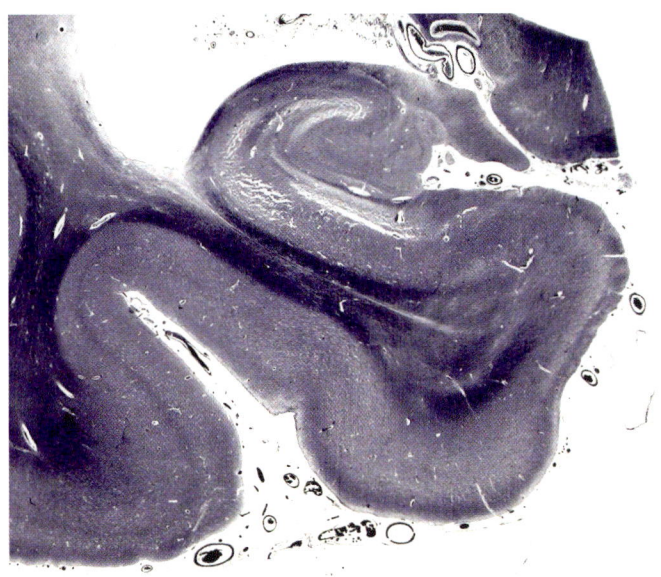

Fig. 2.25. Normal hippocampal position. Alveus is superior to ventricular surface of left hippocampus, leading to fimbria and fornix medially (to right of photograph). (Weil.)

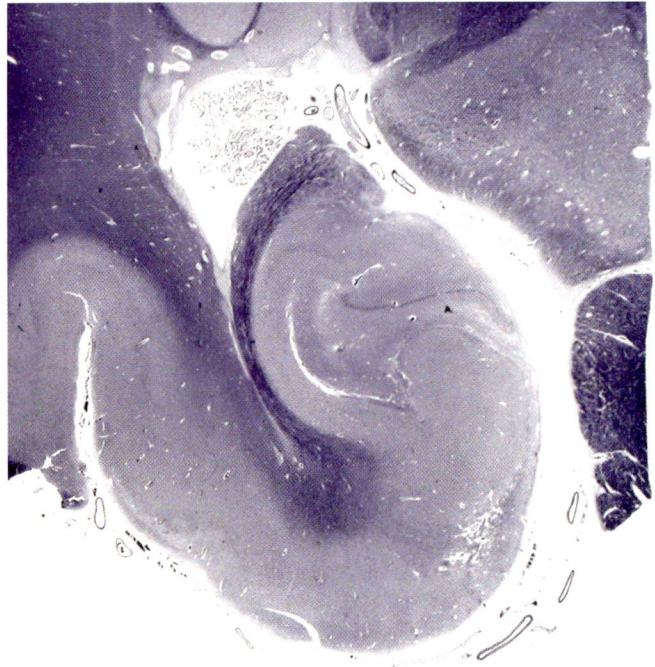

Fig. 2.26. Hippocampus from Case 2.2. Incomplete rotation of left hippocampus. Alveus is lateral (to left on photograph) to hippocampus, with fimbria superiorly situated. (Weil.)

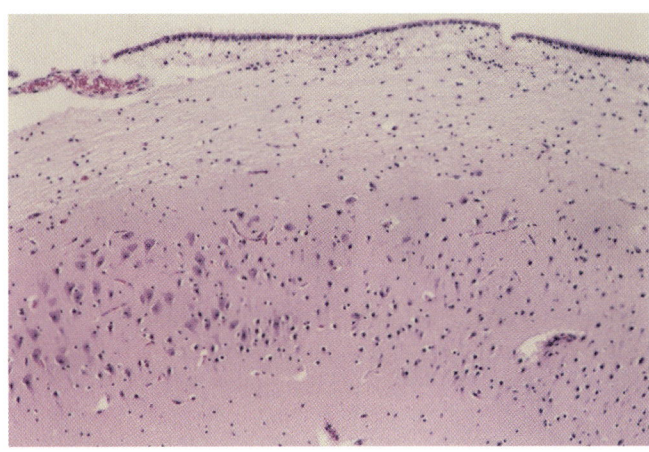

Fig. 2.27. Low-power micrograph of hippocampus in 3-month-old boy. Larger pyramidal cells of CA2 (left side) merge with smaller pyramidal cells of CA1 (right side) near center of photograph. (H&E.)

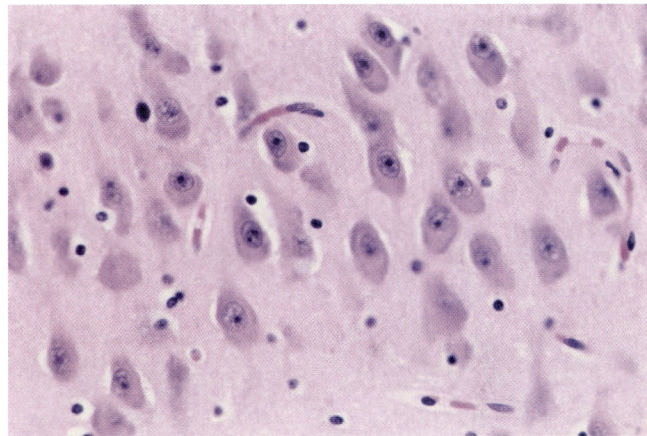

Fig. 2.28. High-power micrograph. CA2 neurons from case seen in Fig. 2.27. (H&E.)

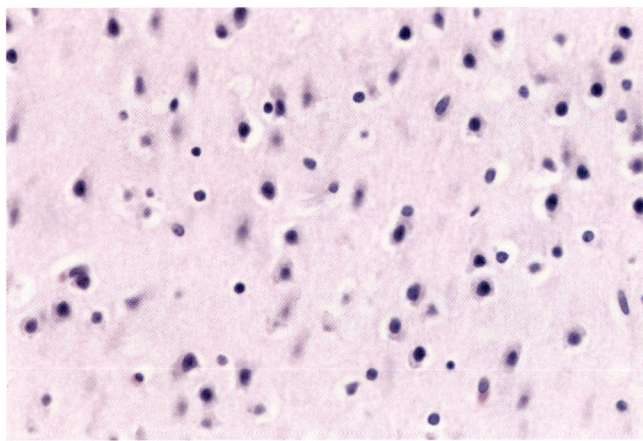

Fig. 2.29. High-power micrograph of CA1 neurons from case seen in Figs. 2.27–2.28. Taken at same magnification as Fig. 2.28. Autolysis is somewhat more evident in these CA1 neurons compared to CA2 neurons. (H&E.)

rity appearance, but between 25 and 32 weeks' gestation CA1 neuronal maturity is surpassed by a more rapid development of pyramidal neurons in CA2-4.[6] The result is, at term, the cells of CA1 are normally smaller and may appear more dense in routine stains, giving a false impression of pathologic pyknosis, apoptosis, or necrosis (Figs. 2.27–2.29). Because pyknosis is the more characteristic type of neuronal necrosis in neonates, rather than the classic red neuron change seen in larger mature neurons, it is important to compare CA1 cells with similar-sized neurons in clearly unaffected areas of the brain as an aid in distinguishing immature neurons from those undergoing necrosis.[51,85] CA1 neurons achieve an essentially mature appearance by approximately 2 years of age,[51] although hippocampal neurons continue to increase in size slightly up to 16 years of age.[6]

An incidental benefit of the CA1 neuronal immaturity in infants is that one can easily delineate the CA1 field from adjacent CA2 and subiculum, respectively, and thus develop greater accuracy in judging its location and extent in the mature brain. In maturity, CA1 neurons still remain perceptibly smaller than those of CA2.

Central Nervous System Myelination

Most textbooks and articles[11,53] on pediatric neuropathology and CNS developmental landmarks contain some reference to timetables for myelin development in the normal human brain and spinal cord. (Also see references in the previous section on general information sources on CNS developmental landmarks.) Most of these tables are based on microscopic analysis of myelin-stained sections and are very useful if one routinely does myelin stains in a number of CNS regions in all cases. An admittedly more crude approach is more practical in our setting, and it consists of gross examination of coronal sections of the fixed neonatal and infant brain. Some guidance is available in the literature as to which brain areas should contain myelin that is readily discernible to the unaided eye at various ages, thus allowing one some discretion as to whether special myelin stains may be indicated.

Gilles et al.[43] and Dekaban[29] provide information which, combined with our experience and comparison with illustrations of brains at various stages of development in other textbooks cited in this chapter's references, has resulted in the following general rules concerning areas in which evidence of myelination is grossly apparent in nearly all normal brains.

1. Term infant
   - Cerebrum (Figs. 2.30–2.31): Posterior limb of internal capsule, ansa lenticularis, and optic chiasm and tracts.
   - Midbrain and pons (Figs. 2.32–2.33): Medial longitudinal fasciculus, medial and lateral lemniscus, and superior cerebellar peduncle.

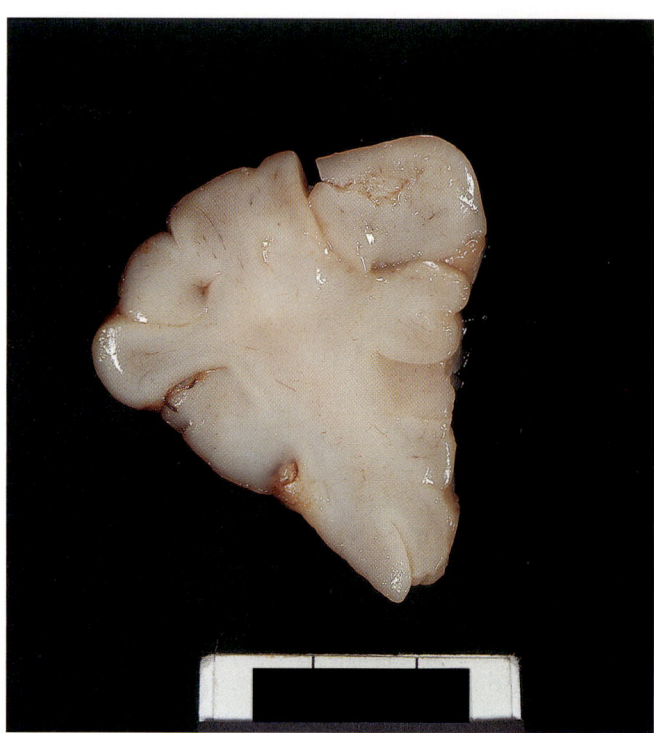

Fig. 2.30. Term newborn, anterior frontal lobe without grossly evident myelin.

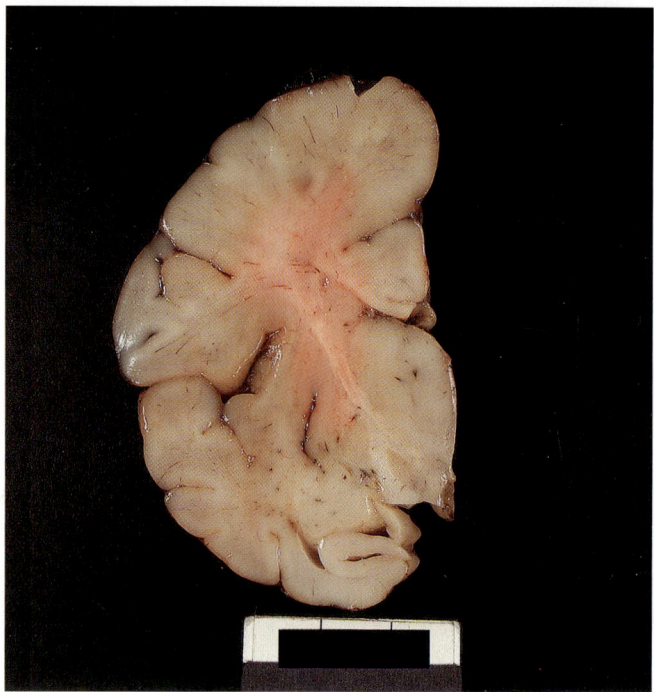

Fig. 2.31. Term newborn, posterior frontal lobe with internal capsule myelin visible.

   - Medulla: Medial longitudinal fasciculus, medial lemniscus, inferior cerebellar peduncle, usually parasagittal and peridentate nucleus cerebellar areas, and at least some peri-inferior olivary nucleus fibers.

- Cerebellum (Fig. 2.33): White matter surrounding dentate nucleus, and in portions of the vermis.
- Spinal cord: Posterior columns.

2. Three-month-old infant: Midportion of the corpus callosum (at the level of the precentral gyrus); frontal lobe centrum semiovale more than other portions of the centrum semiovale; and fibers surrounding the inferior olivary nucleus in the medulla.
3. Six- to 12-month-old infant: Progressively more widespread myelin is visible, approaching the appearance and consistency of the mature brain at 12 months of age.

Should more detailed microscopic analysis of possible retarded myelination be indicated, additional data for comparison are available in several references.[29,42,52,55,56,87,89,103] The reference by Rorke and Riggs[87] includes a large number of illustrations.

During active myelination stages, still relatively sparse numbers of oligodendrocytes typically display somewhat more open-faced nuclei and more conspicuous, lightly eosinophilic cytoplasm. At the same time, exposed astrocytes appear prominent. Their similar appearance is rather eye-catching in the otherwise relatively paucicellular and sparsely myelinated white matter, leading to the occasional use of the misnomer "myelination gliosis" (Figs. 2.34–2.35). Mistaking this for reactive astrocytic gliosis is a potential pitfall, as is well illustrated and described in the references cited.

Cerebral Ventricular Size, Configuration, and Cerebrospinal Fluid Production: Developmental Aspects

In premature brains the superior angle of the lateral ventricle is rounded rather than sharp, and the degree of roundness decreases toward term, as any review of coronal sections in immature brains demonstrates.

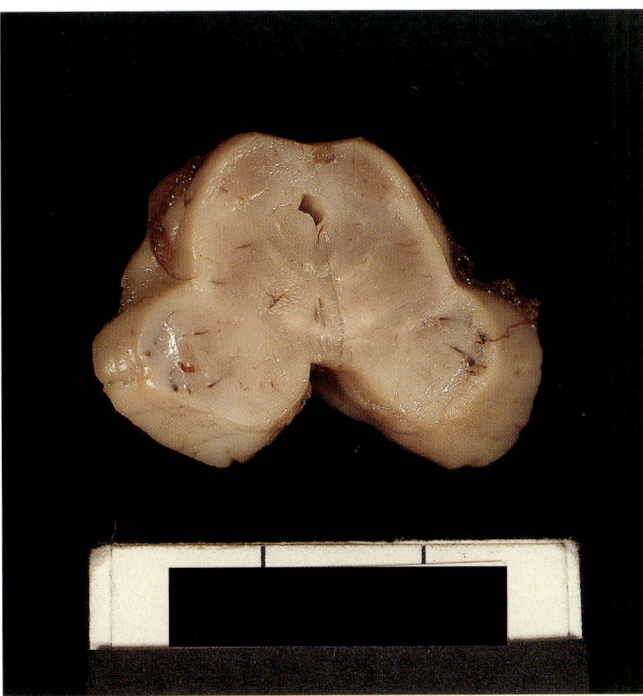

Fig. 2.32. Term newborn, midbrain. Myelin is visible in medial longitudinal fasciculus, medial and lateral lemniscus, and decussation of superior cerebellar peduncle.

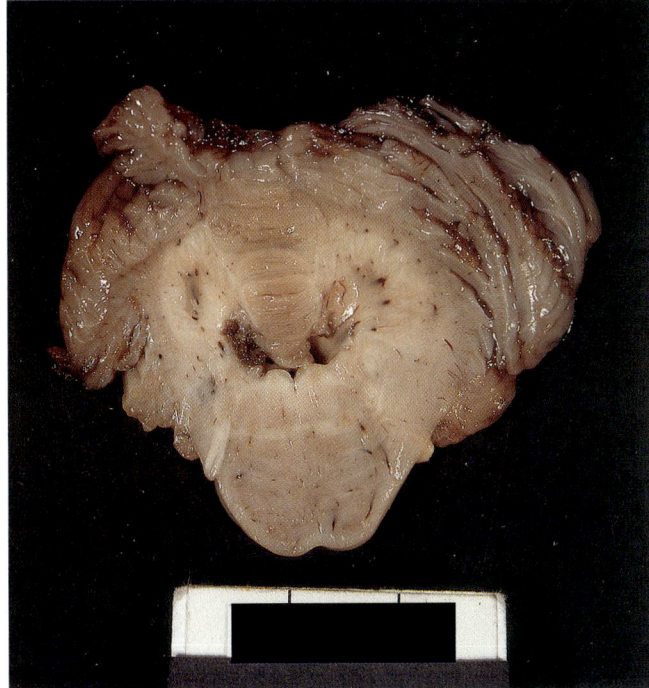

Fig. 2.33. Term newborn, pons and cerebellum. Myelin is visible in cerebellar vermis white matter and white matter surrounding dentate nucleus of cerebellum, medial longitudinal fasciculus, medial lemniscus, and trigeminal nerve trunk.

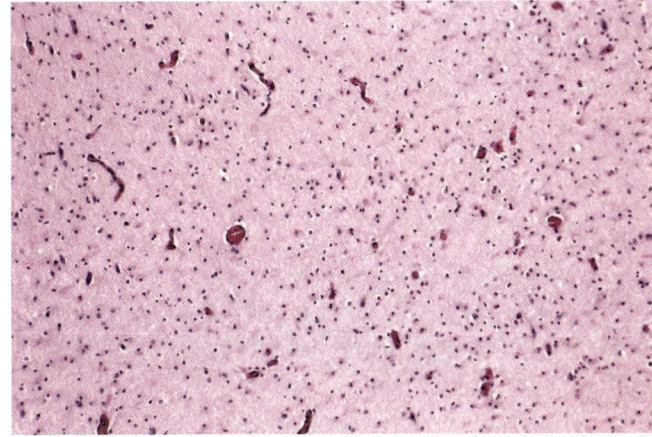

Fig. 2.34. Low-power micrograph. Oligodendrocytes and astrocytes in sparsely myelinated neonatal white matter (see text). (H&E.)

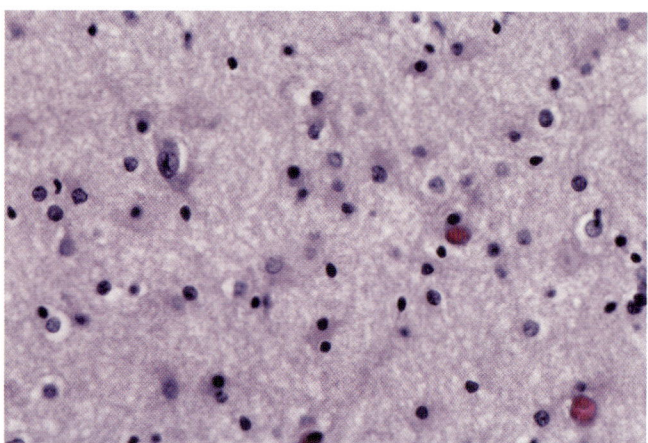

Fig. 2.35. High-power micrograph. Detail of oligodendrocytes and astrocytes in white matter undergoing active myelination. The larger cell in upper left quadrant may be a neuron, not an unusual finding in immature brain subcortical white matter. (H&E.)

Sonography during prenatal care includes assessment of ventricular size due to the relationship of ventriculomegaly to malformations, developmental delay, and increased mortality rates.[14,67,95] If clearly rounded superior angles of lateral ventricles persist into the latter part of the first year of life, it should prompt a careful search for other evidence of abnormality, such as hydrocephalus and periventricular leukomalacia, for example.

It has been demonstrated sonographically that the width of the lateral ventricular atrium does not change significantly during the second and third trimesters,[14] but, following vaginal birth, the ventricles are collapsed in 80% of newborns.[67] Thereafter, the ventricles open gradually, with the median time from birth to partial opening approximating 1.5 to 2.5 days. Only 1% of the ventricles in this study were completely open within 12 hours of birth. The authors concluded that, in the absence of other imaging abnormalities, sonographic evidence of collapsed ventricles in the immediate postnatal period is not to be equated with cerebral edema, as is the usual interpretation of such a finding in most age groups.[67]

These results are difficult to reconcile with published rates of CSF production. For example, in adults the average volume of the brain ventricular system is reportedly 22.4 ml (range, 7.4–56.6 ml),[19,37] and it has been estimated that the average CSF production rate (approximately 0.5 ml per minute) approaches the ventricular volume (at least in most young adult brains) in 20 to 30 minutes.[64] Studies in adult communicating hydrocephalus have demonstrated CSF flow rates from 0.05 to 0.78 ml per minute (3–43.8 ml per hour).

Data for infants and children are rather sparse, but a few studies are available.[33,105,106] These have indicated that the mean ventricular volume at 12 months of age was 17 cm³ (ml) (20 cm³ for boys, and 15 cm³ for girls). The age differential disappeared after 6 years of age. CSF production in hydrocephalic infants from 1 month to 8 years of age has been estimated as 0.25 ml per minute (15 ml per hour), and studies indicate that CSF production remains normal in hydrocephalus.[106] This would result in an estimated time of 1.13 hours to replace the approximately 17-ml intraventricular volume of a 12-month-old child.

In an even younger age group, a cerebral ventricular volume study using imaging methods in live neonates[49] revealed right ventricular (RV) volume (in ml) of 1.17 ± 0.39 and left ventricle (LV) volume of 1.00 ± 0.32 in the first week after birth, increasing in the second week to LV, 1.36 ± 0.46 ml and RV, 1.12 ± 0.38 ml. The third and fourth ventricular volumes were not measured in this study but were included in the total ventricular volume of 6 to 7 cm³ determined in two children 1 month of age by Xenos et al.[105] Application of the CSF production rate in hydrocephalic children noted earlier (i.e., 15 ml per hour) to a 1-week-old neonate ventricular volume indicates the ventricles should be completely refilled by CSF after removal of head compression in the vaginal canal well within 1 hour, rather than the "partial opening" over 1.5 to 2.5 days observed by Cardoza et al.[14] unless (1) CSF production is substantially less than 15 ml per hour; (2) CSF absorption is more rapid in this age group than in more mature infants; or (3) some degree of cerebral swelling unaccompanied by imaging study or clinical abnormalities exists in the first few days after vaginal birth. The evidence at this time does not, therefore, exclude the possibility of some postdelivery transient brain swelling.

Choroid Plexus

During choroid plexus development, serial ultrasound techniques have described the choroid plexus as containing a loose parenchymal stroma with cysts devoid of epithelial lining when examined pathologically between 17 and 28 weeks' gestation, with spontaneous regression of this feature occurring by approximately 24 weeks' gestation in approximately 80% of cases.[17,35] Chitkara et al.[17] classified the choroid plexus cysts as small (5 mm or less), medium (6–10 mm), or large (11 + mm). About 90% of cysts resolved by 28 weeks' gestation, and the maximum size range was 3 to 20 mm in a series of 41 cases, with only four cases exceeding 10 mm. These authors[17] recommended chromosomal studies in cases in which cystic change exceeded 11 mm and persisted beyond 20 to 22 weeks' gestation.

It is our practice in infant brains to routinely examine choroid plexus in all sudden infant death syndrome cases (choroid plexus being one of the recommended sampling sites), and it is frequently sampled in other categories of cases. The choroid plexus of the temporal horn usually appears in our hippocampal sections, but additional samples of the glomus might be obtained. In the glomus of the choroid plexus, most neonates demonstrate abundant, loosely organized choroid plexus stroma, the edematous-appearing mesenchymal stroma containing smooth-walled cysts of various size, and variable areas of meningothelial elements with or without sparse ependymal cells and a few lymphocyte-like cells (Figs. 2.36–2.40). The maximum dimension of such stromal enlargements is typically in the range of approximately 1.5 to 2 cm in length by 0.5 to 1.0 cm in diameter, but it is unlikely they would appear cystic on sonography in most infants due to the small size of the contained cysts (i.e., usually no more than 0.2–0.3 mm in maximum diameter). The cystic change may, in occasional cases, be fairly conspicuous, and an aggregate of small cysts could possibly be interpreted as a single large cyst by sonography. Giving them the descriptive term of "hydropic meningothelial islands," they gradually "involute" to the typical appearance of the choroid plexus in other portions of the brain, and, in more mature brains, to the usual sparse fibrovascular stromal appearance in nearly every case by 1 to 2 years of age. Only subtle reminders of their presence can be seen in some mature brains, such as small cysts, or single or small groups of arachnoid cells in the choroid plexus stroma along with psammoma bodies. This is such a consistent finding in the glomus of the choroid plexus that it is considered a normal developmental stage without pathologic significance.

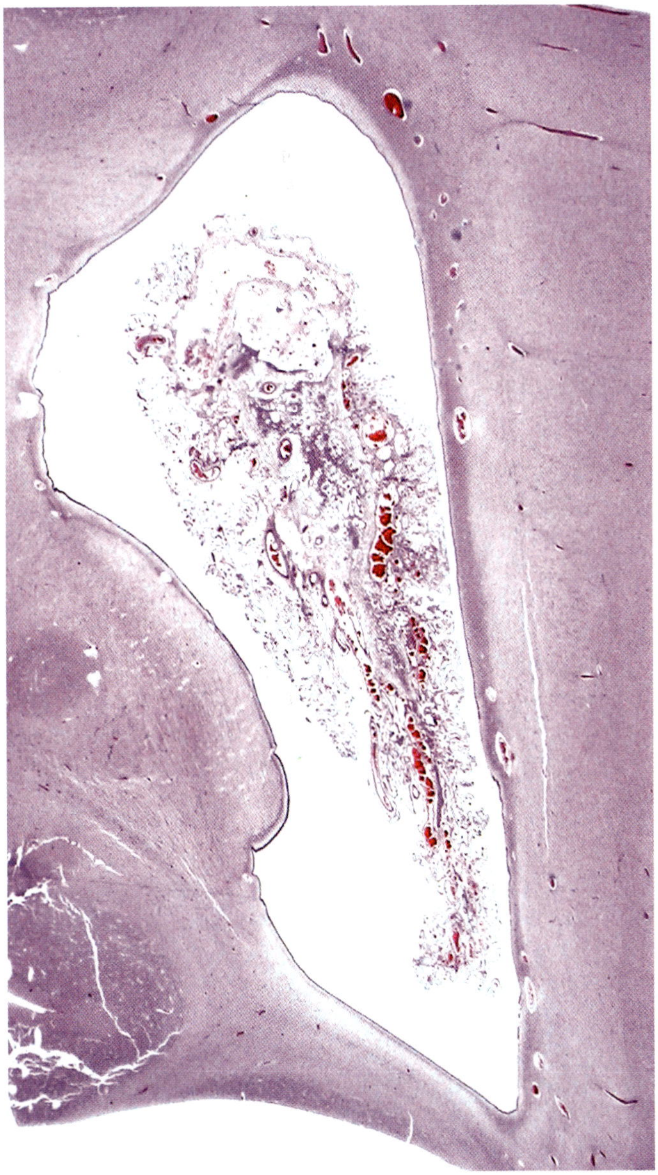

Pituitary Gland

A normal developmental stage in the anterior pituitary gland includes parenchymal microcalcifications with psammoma body-like laminated concretions. These increase in the first 2 weeks of life, then decrease in number and disappear by approximately 6 months of age.[46,47] This age-related site-specific pattern, the exact opposite of that seen in concretions that increase with advancing age in other intracranial sites (meninges, pineal gland, choroid plexus), is a potential source of error in interpretation if not recognized as a normal but transient phenomenon. Other common occurrences in pituitary glands that can result in requests for review of slides include questions concerning normal trophic cell aggregates versus microadenomas, salivary gland rests, squamous cell nests, and foci of "basophil invasion" of

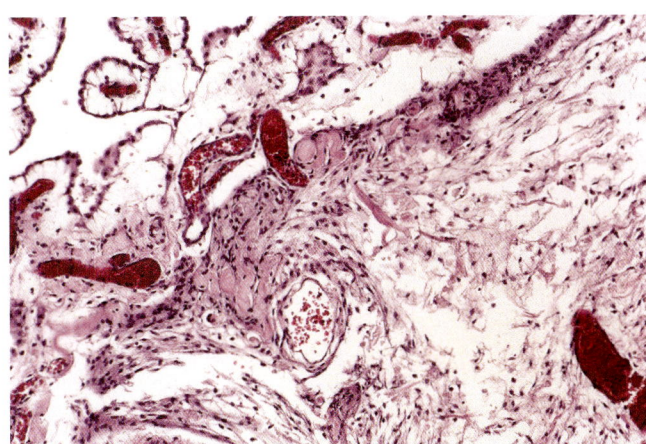

Fig. 2.37. Medium-power micrograph with microcysts in hydropic meningothelial island (*bottom*), merging with more typical choroid plexus with sparse and swollen stroma (*top*) in neonate. At least some of the swelling in the more typical choroid plexus zone here probably represents autolysis. The cystic change in the hydropic meningothelial island portion is present in many cases devoid of significant autolysis. (H&E.)

Fig. 2.36. Very low power micrograph. Glomus of choroid plexus with hydropic meningothelial island in 4-month-old child. (H&E.)

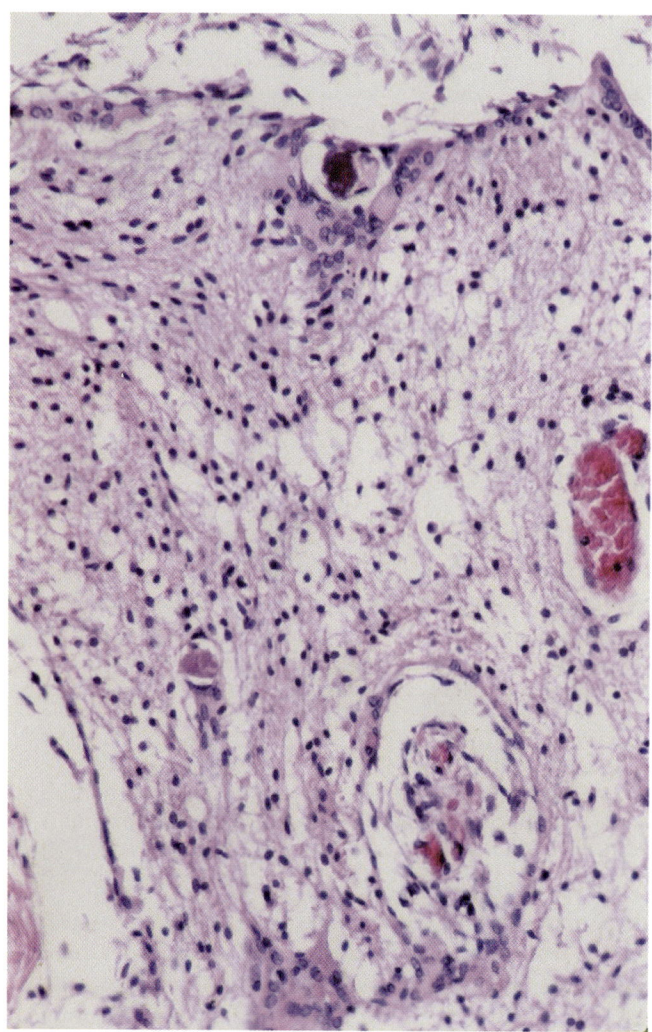

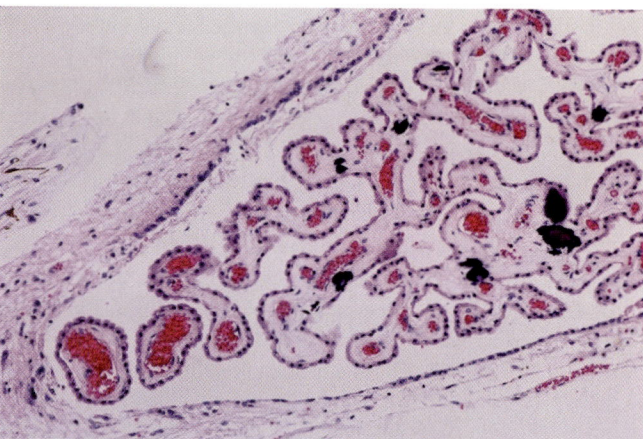

Fig. 2.40. Normal middle-aged adult choroid plexus with calcification, for comparison with Figs. 2.36–2.39. (H&E.)

the pars nervosa versus neoplasia. Such findings, while more often described in adults, can sometimes be found in infants and are well described in the surgical pathology literature.[7,44,90]

Final Comments

A brief summary of some of the more common questions encountered in the evaluation of immature brains, discussed in the previous sections, is as follows:

1. Normal migrating neuroblast/spongioblast cells mistaken for inflammatory cells.
2. Cerebellar external granular layer mistaken for meningitis, lymphoma, or medulloblastoma.
3. Sommer's sector neuronal immaturity mistaken for necrosis or ischemic change.
4. Mildly increased subarachnoid lymphocytes and macrophages normal in first few months of life, mistaken for meningitis.
5. Oligodendrocytes engaged in active myelination during normal development mistaken for pathologic gliosis.
6. Overinterpretation of significance of scattered single neurons in cerebral white matter in infants and children, in absence of associated abnormalities.
7. Overinterpretation of significance of cerebellar neuronal heterotopias, especially small collections of large cell type, in immature brains.
8. Overinterpretation of sparse microcalcification in developing corpus striatum.
9. Normal presence of hydropic meningothelial islands in the glomus of the choroid plexus, mistaken for abnormal cysts or neoplasms.
10. Lack of taking into consideration the fact that an inflammatory response in the CNS is not seen before the necessary cellular components are present and are sufficiently mature to respond.

Fig. 2.38. Medium-power micrograph of hydropic meningothelial island of 5-month-old. Vacuoles are seen in stroma, which is also rich in blood vessels. A calcified meningothelial whorl (psammoma body) is present in mid upper margin. (H&E.)

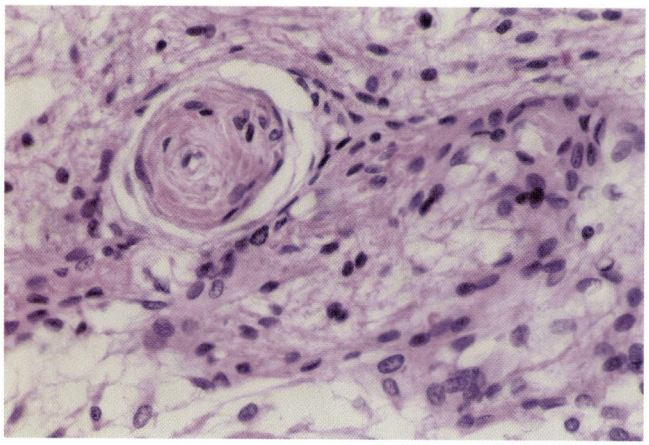

Fig. 2.39. High-power micrograph of meningothelial whorl and cells in hydropic meningothelial island of 1-month-old child. (H&E.)

References

1. Acheson RM, Jefferson E. Some observations on the closure of the anterior fontanelle. Arch Dis Child 1954;29:196–198.
2. Akabushi I, Ikeda T, Yoshioka S. Benign external hydrocephalus in a boy with autosomal dominant microcephaly. Clin Genet 1996;9:160–162.
3. Alvarez LA, Maytal J, Shinnar S. Idiopathic external hydrocephalus: Natural history and relationship to benign familial macrocephaly. Pediatrics 1986;77:901–907.
4. Al-Saedi SA, Lemke RP, Debooy VD, Casiro O. Subarachnoid fluid collections: A cause of macrocrania in preterm infants. J Pediatr 1996;128:234–236.
5. Andrews JM, Schumann GB: Neurocytopathology. Baltimore: Williams & Wilkins, 1992.
6. Arnold SE, Trojanowski JQ. Human fetal hippocampal development: Cytoarchitecture, myeloarchitecture, and neuronal morphologic features. J Comp Neurol 1996;367:274–292.
7. Asa SL. Tumors of the Pituitary Gland. Atlas of Tumor Pathology. Third Series, Fascicle 22. Washington, DC: Armed Forces Institute of Pathology, 1998.
8. Bartholomeusz HH, Courchesne E, Karns CM. Relationship between head circumference and brain volume in healthy normal toddlers, children, and adults. Neuropediatrics 2002;33:239–241.
9. Bayer SA, Altman J. The Human Brain During the Third Trimester, Vol 2. Boca Raton, FL: CRC Press, 2003.
10. Bayer SA, Altman J, Russo RJ, Zhang X. Embryology. In Duckett S (ed): Pediatric Neuropathology. Baltimore: Williams & Wilkins, 1995:54–107.
11. Brody BA, Kinney HC, Kloman AS, Gilles FH. Sequence of central nervous system myelination in human infancy. I. An autopsy study of myelination. J Neuropathol Exp Neurol 1987;46:283–301.
12. Brzustowicz RJ, Kernohan JW. Cell rests in the region of the fourth ventricle. I. Their site and incidence according to age and sex. AMA Arch Neurol Psychiatry 1952;67:585–591.
13. Brzustowicz RJ, Kernohan JW. Cell rests in the region of the fourth ventricle. II. Histologic and embryologic consideration. AMA Arch Neurol Psychiatry 1952;67:592–601.
14. Cardoza JD, Goldstein RB, Filly RA. Exclusion of fetal ventriculomegaly with a single measurement: The width of the lateral ventricular atrium. Radiology 1988;169:711–714.
15. Cardoso ER, Schubert R. External hydrocephalus in adults: Report of three cases. J Neurosurg 1996;85:1143–1147.
16. Chi JG, Dooling EC, Gilles FH. Gyral development of the human brain. Ann Neurol 1977;1:86–93.
17. Chitkara U, Cogswell C, Norton K, et al. Choroid plexus cysts in the fetus: A benign anatomic variant or pathologic entity? Report of 41 cases and review of the literature. Obstet Gynecol 1988;72:185–189.
18. Cohen MM Jr. Asymmetry: Molecular, biologic, embryopathic, and clinical perspectives. Am J Med Genet 2001;101:292–314.
19. Condon BR, Patterson J, Wyper D, et al. A quantitative index of ventricular and extraventricular intracranial CSF volumes using MR imaging. J Comput Assist Tomogr 1986;10:784–792.
20. Conel JL. The Postnatal Development of the Human Cerebral Cortex, Vol 1: The Cortex of the Newborn. Cambridge, MA: Harvard University Press, 1939.
21. Conel JL. The Postnatal Development of the Human Cerebral Cortex, Vol II: The Cortex of the One-Month Infant. Cambridge, MA: Harvard University Press, 1941.
22. Conel JL. The Postnatal Development of the Human Cerebral Cortex, Vol III: The Cortex of the Three-Month Infant. Cambridge, MA: Harvard University Press, 1947.
23. Conel JL. The Postnatal Development of the Human Cerebral Cortex, Vol IV: The Cortex of the Six-Month Infant. Cambridge, MA: Harvard University Press, 1951.
24. Conel JL. The Postnatal Development of the Human Cerebral Cortex, Vol V: The Cortex of the Fifteen-Month Infant. Cambridge, MA: Harvard University Press, 1955.
25. Conel JL. The Postnatal Development of the Human Cerebral Cortex, Vol VI: The Cortex of the Twenty-Four-Month Infant. Cambridge, MA: Harvard University Press, 1959.
26. Conel JL. The Postnatal Development of the Human Cerebral Cortex, Vol VII: The Cortex of the Four-Year-Old Child. Cambridge, MA: Harvard University Press, 1963.
27. Conel JL. The Postnatal Development of the Human Cerebral Cortex, Vol VIII: The Cortex of the Six-Year-Old Child. Cambridge, MA: Harvard University Press, 1967.
28. Costa C, Hauw J-J. Pathology of the cerebellum, brain stem and spinal cord. In Duckett S (ed): Pediatric Neuropathology. Baltimore: Williams & Wilkins, 1995:217–238.
29. Dekaban A. Neurology of Early Childhood. Baltimore: Williams & Wilkins, 1970.
30. DeMyer W. Megalencephaly: Types, clinical syndromes, and management. Pediatr Neurol 1986;2:321–328.
31. Djabourian R, Sathyavagiswaran L, Fishbein MC. Forensic autopsy in a pathology training program. Arch Pathol Lab Med 1998;122:750–751.
32. Dorovini-Zis K, Dolman CL. Gestational development of brain. Arch Pathol Lab Med 1977;101:192–195.
33. Drake JM, Sainte-Rose C, DaSilvia M, Hirsch J-F. Cerebrospinal fluid flow dynamics in children with external ventricular drains. Neurosurgery 1991;28:242–250.
34. Ellison D, Love S, Chimelli L, et al. Neuropathology: A Reference Text of CNS Pathology. London: Mosby, 1998.
35. Encha-Razavi F. Fetal neuropathology. In Duckett S (ed): Pediatric Neuropathology. Baltimore: Williams & Wilkins, 1995:108–122.
36. Ferrer I, Armstrong J. Microcephaly. In Golden JA, Harding BN (eds): Pathology and Genetics: Developmental Neuropathology. Basel: ISN Neuropath Press, 2004:26–31.
37. Fishman RA. Cerebrospinal Fluid in Diseases of the Nervous System, ed 2. Philadelphia: WB Saunders, 1992.
38. Friede RL. Developmental Neuropathology, ed 2. Berlin: Springer-Verlag, 1989.
39. Geschwind DH, Miller BL. Molecular approach to cerebral laterality: Development and neurodegeneration. Am J Med Genet 2001;101:370–381.
40. Gilbert-Barness E, Debich-Spicer DE. Handbook of Pediatric Autopsy Pathology. Totowa, NJ: Humana Press, 2005.
41. Gilles FH. Perinatal neuropathology. In Davis RL, Robertson DM (eds): Textbook of Neuropathology, ed 3. Baltimore: Williams & Wilkins, 1997:331–385.
42. Gilles FH, Leviton A, Dooling EC. The Developing Brain: Growth and Epidemiologic Neuropathology. Boston: John Wright, PSG Inc, 1983.
43. Gilles FH, Shankle W, Dooling EC. Myelinated tracts: Growth patterns. In Gilles FH, Leviton A, Dooling EC (eds): The Developing Brain: Growth and Epidemiologic Neuropathology. Boston: John Wright, PSG Inc, 1983.
44. Goldberg GM, Eshbaugh DE. Squamous cell nests of the pituitary gland as related to the origin of craniopharyngiomas. Arch Pathol 1960;70:293–299.
45. Greenwood Genetic Center (compiled by Saul RA, Geer JS, Seaver LH, Phelan MC, Sweet KM, Mills CM). Growth References: Third Trimester to Adulthood. Greenville, SC: Keys Printing, 1998.
46. Groisman GM, Amar M, Polak-Charcon S. Microcalcifications in the anterior pituitary gland of the fetus and newborn: A histochemical and immunohistochemical study. Hum Pathol 1999;30:199–202.
47. Groisman GM, Kerner H, Polak-Charcon S. Calcified concretions in the anterior pituitary gland of the fetus and newborn: A light and electron microscopic study. Hum Pathol 1996;27:1139–1143.

48. Harding BN, Copp AJ. Malformations. In Graham DI, Lantos PL (eds): Greenfield's Neuropathology, ed 7. London: Arnold, 2002;1:357–483.
49. Ichihashi K, Takahashi N, Honma Y, Momoi M. Cerebral ventricular volume assessment by three-dimensional ultrasonography. J Perinatal Med 2005;33:332–335.
50. Jones KL. Smith's Recognizable Patterns of Human Malformation, ed 5. Philadelphia: WB Saunders, 1997.
51. Kinney HC, Armstrong DD. Perinatal neuropathology. In Graham DI, Lantos PL (eds): Greenfield's Neuropathology, ed 6. London: Arnold, 1997;1:535–599.
52. Kinney HC, Armstrong DD. Perinatal neuropathology. In Graham DI, Lantos PL (eds): Greenfield's Neuropathology, ed 7. London: Arnold, 2002;1:519–606.
53. Kinney HC, Brody BA, Kloman AS, Gilles FH. Sequence of central nervous system myelination in human infancy. II. Patterns of myelination in autopsied infants. J Neuropathol Exp Neurol 1988;47:217–234.
54. Krieg AF, Kjeldsberg CR. Cerebrospinal fluid and other body fluids. In Henry JB (ed): Clinical Diagnosis and Management by Laboratory Methods. Philadelphia: WB Saunders, 1991:445–473.
55. Larroche J-C. Part II: The development of the central nervous system during intrauterine life. In Falkner F (ed): Human Development. Philadelphia: WB Saunders, 1966:257–276.
56. Larroche J-C. Developmental Pathology of the Neonate. Amsterdam: Excerpta Medica, 1977.
57. Larroche J-C. Malformations of the nervous system. In Adams JH, Corsellis JAN, Duchen LW (eds): Greenfield's Neuropathology, ed 4. New York: John Wiley, 1984:385–450.
58. Laubscher B, Deonna T, Uske A, vanMelle G. Primitive megalencephaly in children: Natural history, medium-term prognosis with special reference to external hydrocephalus. Eur J Pediatr 1990;149:502–507.
59. Liss L, Mervis L. The ependymal lining of the cavum septi pellucidi: A histological and histochemical study. J Neuropathol Exp Neurol 1965;23:355–367.
60. Meencke HJ. The density of dystrophic neurons in the white matter of the gyrus frontalis inferior in epilepsies. J Neurol 1983;230:171–181.
61. Meencke HJ, Janz D. The significance of microdysgenesis in primary generalized epilepsy: An answer to the considerations of Lyon and Gastaut. Epilepsia 1985;26:368–371.
62. Menkes JH, Sarnat HB (ed). Child Neurology, ed 6. Philadelphia: Lippincott Williams & Wilkins, 2000:352.
63. Ment LR, Duncan CC, Geehr R. Benign enlargement of the subarachnoid spaces in the infant. J Neurosurg 1981;54:504–508.
64. Milhorat TH: Clark RG, Hammock MK. Experimental hydrocephalus. Part 2: Gross pathological findings in acute and subacute obstructive hydrocephalus in the dog and monkey. J Neurosurg 1970;32:390–399.
65. Morris EB III, Parisi JE, Buchhalter JR. Histopathologic findings of malformations of cortical development in an epilepsy surgery cohort. Arch Pathol Lab Med 2006;130:1163–1168.
66. Nellhaus G. Head circumference from birth to eighteen years: Practical composite international and interracial graphs. Pediatrics 1968;41:106–114.
67. Nelson MD Jr, Tavaré CJ, Petrus L, et al. Changes in the size of the lateral ventricles in the normal-term newborn following vaginal delivery. Pediatr Radiol 2003;33:831–835.
68. Norman MG, Ludwin SK. Congenital malformations of the nervous system. In Davis RL, Robertson DM (eds): Textbook of Neuropathology, ed 3. Baltimore: Williams & Wilkins, 1997:265–329.
69. Norman MG, McGillivray BC, Kalousek DK, et al. Congenital Malformations of the Brain: Pathologic, Embryologic, Clinical, Radiologic and Genetic Aspects. New York: Oxford University Press, 1995.
70. Odita JC. The widened frontal subarachnoid space: A CT comparative study between macrocephalic, microcephalic, and normocephalic infants and children. Childs Nerv Syst 1992;8:36–39.
71. Palmini A, Najm I, Avanzini G, et al. Terminology and classification of the cortical dysplasias. Neurology 2004;62(Suppl 3): S2-S8.
72. Papasian NC, Frim DM. A theoretical model of benign external hydrocephalus that predicts a predisposition towards extra-axial hemorrhage after minor head trauma. Pediatr Neurosurg 2000;33:188–193.
73. Piatt J Jr. A pitfall in the diagnosis of child abuse: External hydrocephalus, subdural hematoma, and retinal hemorrhages. Neurosurg Focus 1999;7(4):1–6. Available at: http://www.medscape.can/viewarticle/413309. Accessed 11/12/03.
74. Piatt JH Jr. Monozygotic twins discordant for external hydrocephalus. Pediatr Neurosurg 2001;35:211–215.
75. Pittman T. Significance of a subdural hematoma in a child with external hydrocephalus. Pediatr Neurosurg 2003;39:57–59.
76. Popich GA, Smith DW. Fontanels: Range of normal size. J Pediatr 1972;80:749–752.
77. Prayson RA. Some thoughts on the classification of malformations of cortical development. Arch Pathol Lab Med 2006;130:1101–1102.
78. Reske-Nielsen E, Gregersen M, Lund E. Astrocytes in the postnatal central nervous system: From birth to 14 years of age. An immunohistochemical study on paraffin-embedded material. Acta Pathol Microbiol Immunol Scand [A] 1987;95:347–356.
79. Reske-Nielsen E, Oster S, Reintoft I. Astrocytes in the prenatal central nervous system: From 5th to 28th week of gestation. An immunohistochemical study on paraffin-embedded material. Acta Pathol Microbiol Immunol Scand [A] 1987;95: 339–346.
80. Roessmann U. Congenital malformations. In Duckett S (ed): Pediatric Neuropathology. Baltimore: Williams & Wilkins, 1995: 123–148.
81. Roessmann U, Gambetti P. Pathological reaction of astrocytes in perinatal brain injury: Immunohistochemical study. Acta Neuropathol (Berl) 1986;70:302–307.
82. Roessmann U, Gambetti P. Astrocytes in the developing human brain: An immunohistochemical study. Acta Neuropathol (Berl) 1986;70:308–313.
83. Rojiani AM, Emery JA, Anderson KJ, Massey JK. Distribution of heterotopic neurons in normal hemispheric white matter: A morphometric analysis. J Neuropathol Exp Neurol 1996;55:178–183.
84. Rorke LB. Pathology of Perinatal Brain Injury. New York: Raven Press, 1982.
85. Rorke LB. Perinatal brain injury. In Adams JH, Duchen LW (eds): Greenfield's Neuropathology, ed 5. New York: Oxford University Press, 1992:639–709.
86. Rorke LB, Fogelson MH, Riggs HE. Cerebellar heterotopia in infancy. Dev Med Child Neurol 1968:10:644–650.
87. Rorke LB, Riggs HE. Myelination of the Brain in the Newborn. Philadelphia: Lippincott, 1969.
88. Rutherford MA. Magnetic resonance imaging of injury to the immature brain. In Squire W (ed): Acquired Damage to the Developing Brain: Timing and Causation. London: Arnold, 2002: 166–192.
89. Sarnat HB. Microscopic criteria to determine gestational age of the fetal and neonatal brain. In Garcia JH (ed): Neuropathology: The Diagnostic Approach. St. Louis: Mosby, 1997:529–540.
90. Schochet SS Jr, McCormick WF, Halmi NS. Salivary gland rests in the human pituitary. Light and electron microscopical study. Arch Pathol 1974;98:193–200.
91. Shaw CM, Alvord EC Jr. Cava septi pellucidi et vergae: Their normal and pathological states. Brain 1969;92:213–224.
92. Shemie S, Jay V, Rutka J, Armstrong D. Acute obstructive hydrocephalus and sudden death in children. Ann Emerg Med 1997;29:524–528.

93. Smith DW, Popich G. Large fontanels in congenital hypothyroidism: A potential clue toward earlier recognition. J Pediatrics 1972;80:753–756.
94. Souter VL, Kapur RP, Nyholt DR, et al. A report of dizygous monochorionic twins. N Engl J Med 2003;349:154–158.
95. Squier W. Pathology of fetal and neonatal brain damage: Identifying the timing. In Squier W (ed): Acquired Damage to the Developing Brain: Timing and Causation. London: Arnold, 2002:110–127.
96. Suara RO, Trouth A-J, Collins M. Benign subarachnoid space enlargement of infancy. J Natl Med Assoc 2000;93:70–73.
97. Sumi SM. Brain malformations in the trisomy 18 syndrome. Brain 1970;93:821–830.
98. Takao M, Ghetti B, Murrell JR, et al. Ectopic white matter neurons, a developmental abnormality that may be caused by the PSEN1 S169 L mutation in a case of familial AD with myoclonus and seizures. J Neuropathol Exp Neurol 2001;60:1137–1152.
99. Thom M. Epilepsy. Part 1: Cortical dysplasia. In Golden JA, Harding BN (eds): Pathology and Genetics: Developmental Neuropathology. Basel: ISN Neuropath Press, 2004:61–66.
100. Thom M. Epilepsy. Part 1: Hemimegalencephaly. In Golden JA, Harding BN (eds): Pathology and Genetics: Developmental Neuropathology. Basel: ISN Neuropath Press, 2004:67–72.
101. Townsend JJ, Nielsen SL, Malamud N. Unilateral megalencephaly: Hamartoma or neoplasm? Neurology 1975;25:448–453.
102. Vinters HV, Farrell MA, Mischel PS, Anders KH. Diagnostic Neuropathology. New York: Marcel Dekker, 1998.
103. Volpe JJ. Neurology of the Newborn, ed 3. Philadelphia: WB Saunders 1995.
104. Walters BC, Burrows PE, Musewe N, et al. Unilateral megalencephaly associated with neonatal high output cardiac failure. Childs Nerv Syst 1990;6:123–125.
105. Xenos C, Sgouros S, Natarajan K. Ventricular volume change in childhood. J Neurosurg 2002;97:584–590.
106. Yasuda T, Tomita T, McLone DG, Donovan M. Measurement of cerebrospinal fluid output through external ventricular drainage in one hundred infants and children: Correlation with cerebrospinal fluid production. Pediatr Neurosurg 2002;36:22–28.

Dating/Aging of Common Lesions in Neuropathology

INTRODUCTION 49
SCALP INJURIES 52
SKULL FRACTURE 54
MINERALIZATION IN CENTRAL NERVOUS SYSTEM LESIONS 58
HETEROTOPIC OSSIFICATION FOLLOWING CENTRAL NERVOUS SYSTEM INJURY 60
EPIDURAL HEMATOMA 62
SUBDURAL HEMATOMA 63
SUBARACHNOID HEMORRHAGE 67
OBSERVATIONS ON POSTMORTEM CEREBROSPINAL FLUID EXAMINATION 70
BRAIN CONTUSIONS 72
AXONAL SPHEROIDS, DIFFUSE AXONAL INJURY, AND β-APP 74
TRAUMATIC AXONAL INJURY 77
PUNCTURE WOUNDS IN THE BRAIN 79
INTRACEREBRAL HEMORRHAGE 81
SPINAL CORD TRAUMATIC LESIONS 82
SCHWANNOSIS 84

SPINAL SHOCK 84
SPINAL CONCUSSION 87
ADULT CEREBRAL INFARCTION 88
GENERALIZED BRAIN HYPOXIC-ISCHEMIC INJURY 91
SPINAL CORD INFARCTION 92
BRAIN DEATH AND RESPIRATOR BRAIN 92
CEREBROSPINAL FLUID CYTOLOGY IN STROKE 98
MALIGNANT CEREBRAL EDEMA 99
WALLERIAN DEGENERATION 100
PERIPHERAL NERVE REGENERATION 103
TRANSNEURONAL DEGENERATION 104
ASTROCYTE TERMINOLOGY 107
BLOOD CELLS IN CENTRAL NERVOUS SYSTEM LESIONS 108
DO REACTIVE CELL CHANGES OF IMMATURE VERSUS MATURE BRAIN DIFFER? 109
DENERVATION OF MUSCLE 112
MUSCLE REINNERVATION 114
REFERENCES 114

Introduction

Questions concerning the timing of lesions that arise in the courtroom are not usually emphasized in current textbooks on neuropathology. The purpose of this chapter is to summarize our present understanding of dating/aging issues (some of which are discussed in other chapters in this text) that may be encountered by the forensic neuropathology consultant, and to do this in a format that also highlights areas in which data are not sufficient to answer some frequently asked questions.

Issues of timing of tissue responses receive a great deal more emphasis in forensic pathology in comparison to general pathology because timing conclusions based on histologic studies may help to confirm or refute certain witness and advocate statements. In general pathology, etiologic and mechanistic issues understandably will often take precedence over detailed timing of the earliest appearance, maximum expression, or longest duration of a given histologic finding.

For the consulting neuropathologist involved in forensic work the question is usually not about "time of death" so often as it is about "time interval between central nervous system (CNS) insult and death." This question is perhaps most often asked in the context of child abuse cases in an attempt by counsel to include or exclude involvement by a given caretaker. It can also occur in a variety of other scenarios.

This subject is a controversial one in some areas of the forensic literature. There is much to be gained by careful attention to the pace and sequence of cellular events following various types of nervous system insults, but large gaps in our knowledge of many aspects of such events require a cautious and objective approach when providing opinions in a forensic setting.

The dating/aging data in this text, in this and other chapters, have been selected with criteria that reflect our conceptual filters, or gradually evolving working hypotheses, at this point in time. These "timetables" continue to be revised as indicated as additional data are collected and examined and by reviewing pertinent publications that become available. More data must be gathered before a general consensus is likely to be reached in many of these parameters.

Several factors make evaluation of much of the published literature on dating/aging of various lesions potentially confusing or problematic, given current hypotheses. If such issues are not addressed, however, there are concerns that they could adversely impact some forensically critical conclusions. These concerns have influenced the format chosen to present the bulk of the dating/aging data in this and other chapters. *By citing a variety of sources without the exclusion of outliers, the reader can decide whether or not a given observation meets their own criteria for usefulness.*

It should be emphasized that the following discussion is not intended as a critique of the authors cited. Their work has collectively provided the foundation for this discussion by publishing research findings yielding useful data. The question is whether conclusions derived for the most part from studies designed for nonforensic purposes might also be found applicable to human forensic cases.

1. *Species studied.* Some neuropathology textbooks in common use provide rather detailed descriptions of dating/aging of lesions such as, for example, cerebral infarction. However, the data presented may be derived from combined studies of experimental animals and humans,[115] or are provided without any references as to the information sources used.[112] Given the well-established variations in numerous pathologic processes between different species, lesion timetable opinions may have an adverse influence on court decisions unless opinions are confined to data derived from studies on human tissue. If a nonhuman data source is used, indicating the species studied may be relevant.

For example, secondary axotomy following traumatic axonal injury requires a survival period of 12 hours or more in humans, but only 2 to 4 hours in rat and cat.[60] The same point applies to other lesions, such as cerebral contusions, in which cited references sometimes provide timetables that combine human and experimental animal data.[150] Hardmann,[146] in his discussion of subarachnoid hemorrhage cellular responses, references data derived from experimental studies in dogs.[168] Another example is the detailed timetable of histologic events in cerebral infarction in a 1988 publication[392] citing the source as a 1971 article that discusses cerebral arteriosclerosis in humans, but is without histologic data.[391] Further investigation reveals that a nearly identical timetable of histologic events following cerebral infarction by the same author published in 1971 was based on studies in cats, not humans, and was very clearly stated as such in the text of this publication.[390] The implication that the 1988 publication was based on human data was obviously due to a typographical error (citing the source as "1971b," rather than the correct reference, which was "1971a") in the 1988 article by Zülch and Hossmann.[392]

We do not imply that such experimental animal references are not useful for examining mechanisms of disease, guiding experiments to determine whether similar mechanisms occur in humans, and the like, but especially careful examination of the original source for lesion dating/aging data that seamlessly combines human and animal data may be necessary in order to allow them to be separated into their respective categories for human forensic work.

2. *Age.* Many studies of a given lesion type are based on populations combining a wide age range, such as from infancy to old age. Given the well-known differences in reaction to pathologic processes in different age groups, an attempt is made in this chapter to separate data that are entirely or predominantly derived from the mature nervous system from those derived from the immature nervous system. Developmental age-confined lesions such as status marmoratus provide justification for attention to this variable (see Chaps. 4 and 5).

3. *Focal versus diffuse lesions.* The spectrum of cellular reactions and their pace may differ, depending on the volume of nervous tissue affected (e.g., middle cerebral artery territory infarction versus global hypoxic-ischemic encephalopathy due to cardiac arrest.[99]

4. *Preexisting conditions.* Preexisting health status of the individual, and/or the duration of the agonal period,[203,215] may affect the nature and pace of reactive changes, and they are often not specifically addressed in articles on dating/aging of lesions.

5. *Single versus sequential insults in the same individual.* In forensic cases, where the timeline of events in the clinical history may be inaccurate, incomplete, or purposely misleading, this possibly should be routinely considered. An inaccurate clinical history also can introduce apparent inconsistency in dating/aging of histologic events following single insults.

6. *Histologic staining method variations.* Staining method influences timetable data. The most commonly employed stain in routine forensic work is the hematoxylin-eosin (H&E) stain, and the majority of the dating/aging data in this text is based on that method. A widely recognized difference is the initial recognition of axonal spheroids in diffuse axonal injury (DAI) being earlier with β-amylase precursor protein (β-APP) immunostaining than with routine H&E stains (see discussion of axonal injury, later).

7. *Combined forms of injury, primary versus secondary effects of injury, and artifact.* It is uncommon for studies on human material to be "pure" lesions unadulterated by additional factors. As one example, are the pathologic changes seen in a case of blunt force cerebral trauma the direct and immediate result of the trauma itself, or rather do they reflect superimposed secondary events, such as complications of post-traumatic apnea, cerebral swelling and herniation syndromes, consequences of neurogenic pulmonary edema, post-traumatic catecholamine surge, post-traumatic coagulopathy, post-traumatic excitotoxicity effects, hematoma mass effects, or infection (discussed in greater detail in Chap. 6)? Further, one needs to factor into the interpretation of findings whether or not medical intervention, such as artificial respiration, has allowed the development of so-called respirator (or "nonperfused") brain changes that can slow or abort the usual progress of cellular reactions that might otherwise be expected for the same insult to death interval if unaccompanied by an extended period of nonperfused brain effects. Finally, the death to autopsy interval, by allowing the development of autolysis and possibly putrefactive change, can further compromise interpretation of histologic findings.

8. *Heterogeneity of population studied.* In some reports, insults of various types (e.g., infarction and blunt force trauma) may be combined in the patient population studied, potentially obscuring differences in cellular reaction to different forms of brain injury.

9. *Design of the study.* In a forensic pathology setting, it is rarely possible to control variables to the degree achievable in bench research. Service requirements, budgetary limitations, legal restrictions, religious customs, and a host of other factors act in concert to limit one's approach to solving certain questions.

Data that can be helpful in elucidating questions concerning the aging/dating of lesions might include the following: What is the earliest survival time studied, and the latest? How many case examples were studied at each time interval? Did such factors influence the data as to when a particular cellular change was first seen? Is any quantitative estimate as to the degree of cellular reaction given, and at what intervals? What area of the CNS was examined? How specific is the terminology employed? For example, does the author's use of the term "reactive astrocyte" mean that the astrocyte is simply swollen, or a gemistocytic astrocyte, or an astrocyte with fibrillar processes, or is the reader left to speculate as to the nature of the astrocyte to which the author refers? In our summary of dating/aging parameters, the authors' own terminology is used, or occasionally a revision is given based on the article's illustrations.

10. *Individual variability in response due to other undetermined factor or factors.* While attempting to apply considerations as noted previously to screen the cited reference material for inclusion in the dating/aging data in this text, some useful human data may have been inadvertently excluded. Readers are encouraged to bring such omissions to our attention, and, hopefully, others will be prompted to publish their own material, which can fill gaps that are obvious in this subject.

This general approach has been helpful to begin to appreciate how, for example, careful workers could have such apparently discrepant opinions regarding the earliest appearance of "red neurons" following cerebral insults. Red neurons (often, but not accurately, assumed to be synonymous with neuronal hypoxic-ischemic injury) have been observed to initially appear as early as less than 1 hour,[16] to 1 to 2 hours (but generally 6–12 hours),[198] to 24 to 48 hours after brain insult.[181,182] Such variation could lead one to employ a very wide time bracket in attempting to date/age the onset of this cellular change, but it should be noted that one of the earlier mentioned reports was based on trauma cases only,[16] one on hypoxic-ischemic injury in the perinatal age group only,[181,182] and one on hypoxic-ischemic injury in (predominantly or exclusively mature) brains in a variety of situations.[198] In our experience, red neurons are present within a very few millimeter zone at the margin of a traumatic lesion (such as cerebral contusion), and not elsewhere in the section, in adults within a 1- to 2-hour survival period. Moreover, we have not seen red neurons in global hypoxic-ischemic brain insults (such as those due to cardiac arrest) prior to about 6 hours' survival. It has become clear that the chemical differences accompanying neuronal necrosis, as reflected by the "red neuron" appearance, may occur by more than one mechanism[19] (also see Chap. 10). These considerations have influenced our decision to deal with most of the dating/aging data in this text according to lesion etiology and duration, rather than by an individual cellular reaction.

The most accurate answer to some questions regarding the dating/aging of lesions may be that there is insufficient data to give an opinion on some specific point, or that a given finding is either consistent with, or not consistent with, a given scenario or hypothetical question, with reasonable medical certainty. Reasonable medical certainty is considered here to be analogous to having no good reason to seriously consider another possibility; or sufficient certainty to act on definitively for diagnosis and/or treatment, if one were in a clinical setting. We have also repeatedly convinced ourselves of the wisdom expressed by Davis[74] concerning the usefulness of reviewing all available information from reports generated by first responders, police, medical examiner department investigator, medical records, and general autopsy findings when constructing event timelines and other factors pertinent to the forensic neuropathology conclusions. All available information relevant to lesion age (radiologic, biochemical, etc.) should also be compared to gross and histologic findings, and reasons for any apparent discrepancy explored.

For a general overview of the various classification systems for types of hypoxia, hypoxic versus ischemic events, and other general background information, the reader is referred to any of several excellent general textbooks on neuropathology.[75,88,100,114,174,175,319] Once references are eliminated that combine animal experimental data with human data, or those in which it is difficult for us to determine whether or not this has been done, the sources cited for the dating/aging sections become relatively limited. If nonhuman sources have been used to illustrate a specific point, they are so indicated.

Scalp Injuries

Skin differs considerably in thickness of its various layers, component adnexae, vascularity, and degree of pigmentation in various parts of the body and between different individuals. It also changes its appearance as a consequence of aging, prior injury or disease, and a variety of other factors.[68] We have not encountered forensic studies specific to scalp skin aging/dating of injuries. Articles and illustrations concerning dating/aging of skin wounds tend to focus on presumably aseptic injuries in nonscalp areas, with the implication that changes with time are not site-specific. Thus, the following discussion is based on an assumption, rather than data, that scalp wounds (despite the increased vascularity in scalp compared with most other skin sites) behave in a fashion approximately similar to that of nonscalp skin wounds.

Based on currently available literature, certain general conclusions appear reasonable. It is hoped that future, well-designed studies in humans will allow more definitive statements to be made.

Color Changes in Skin Bruises

Saukko and Knight[319] offer a general, rather cautious, timetable of estimated age of a bruise based on visible color changes, but indicate that bruises may appear fresh (i.e., less than 24 hours old) for periods well beyond 2 days in aged persons, even for "the remainder of their lives" (p. 147). Subungual hemorrhages may persist for "several months, without changing greatly."[369]

A recent literature review concerning the potential value of dating/aging of skin bruises in individuals less than 18 years of age by assessment of color changes (in vivo and from photographs) concluded that accurate aging cannot be done by these means.[216] The data on which prior suggested timetables for aging of skin wounds by color changes have been based were relatively restricted, and the degree of variation between these few studies showed that any color (i.e., red, blue, purple, yellow, brown, green) could be seen in fresh (less than 48 hours), intermediate (48 hours–7 days), and old (greater than 7 days) bruises. Also, not all colors appear during the evolution of a single bruise over time. Yellow, when it did occur in a bruise, was not seen in the cases studied before about 24 to 48 hours, but the sparse data on which this conclusion was based were considered insufficient for an acceptable level of accuracy.[216]

When tempted to offer an opinion that bruises were sustained at different times based on color differences, one should note that Maguire et al.[216] described one child with a blue bruise on the arm and a green-yellow bruise on the leg that were sustained at the same time. Depth of hemorrhage can influence perceived color.[34]

Histologic Dating/Aging of Scalp Wounds

As noted earlier, studies on which histologic aging of skin wounds are based are not specific to the scalp, but rather emphasize nonscalp areas. The general sequence and timing of histologic events following skin wounds compiled by Saukko and Knight[319] emphasize that individual variation occurs based on age, type of wound (bruise, abrasion, laceration), depth of hemorrhage, wound size, degree of pigmentation of the individual, and other factors. The following is a brief summary of their conclusions, and the reader is referred to the original text for a more thorough description of these changes. Sources for some differing estimates of event timing are also noted.

Histologic Dating/Aging of Scalp Wounds
Survival Period
- 30 minutes to 4 hours
  — Fibrin deposition, mast cell degranulation, and margination of neutrophils and small vessels. Some extravascular migration of neutrophils may occur near the end of this time frame.[319]
  — Others have described a more rapid cellular response. In surgical (aseptic) incisions in human abdominal skin, Ojala et al.[267] found intravascular neutrophil clumping within 15 to 30 minutes exceptionally, and intra- or extravascular neutrophils within 60 minutes in the majority, in 30 cases studied. Within 2 hours, granulocytes formed a zone at the wound skin margin and were "often flattened between the collagen fibers." Macrophages were found at the skin wound edge after 3 hours. The response was more rapid in subcutaneous fat. Granulocytes were seen in the fat in two-thirds of cases within 30 minutes and in all cases within 60 minutes, marginating on the endothelium or extravascularly between adipocytes. Macrophages containing lipid droplets were present within 3 hours in subcutaneous fat.[267]

- Yet others describe no histologic response in less than 4 hours.[369]
- 4 to 12 hours
  - Definite neutrophil and possible mononucleated inflammatory cell extravascular migration more likely. Possible edema and endothelial swelling. Small skin wounds may show basal epidermal layer regeneration.[319]
  - Perivascular neutrophils at 4 hours; neutrophils > macrophages (5:1) and activated fibroblasts present at 8 to 12 hours; imminent necrosis in central zone.[369]
- 12 to 24 hours
  - Leukocytes (neutrophils and a greater proportion of mononucleated cells including macrophages) form marginal layer around wound. Fibroblast mitoses seen from about 15 hours plus. Epidermis begins to spread into wound.[319]
- 16 to 24 hours
  - Relative increase in macrophages (neutrophil to macrophage ratio now 0.4:1). Prior to 16 hours fibrin stains yellow, but after 16 hours fibrin stains bright red (with Martius scarlet blue stain).[369]
- 24 hours
  - Maximum neutrophils and fibrin presence; remains at this level for 2 to 3 days; cytoplasmic processes at cut edge of epidermis.[369]
- 24 to 48 hours
  - Migration of epidermis toward wound center; necrosis present at 32 hours and after; earliest hemosiderin seen at 24 to 48 hours.[369]
- 15 to 30 hours
  - Erythrophagocytosis seen at 15 to 17 hours.[369]
  - Some leukocytes infiltrating the wound are becoming necrotic.[169,369]
- 48 hours
  - Leukocyte infiltration peaks.[319]
  - Macrophage infiltration peaks.[369]
- 2 to 4 days
  - Fibroblasts migrate into wound periphery.[369]
- 72 hours
  - Fibroblasts more conspicuous, and capillaries have begun to bud from nearby vessels into wound.[319]
- 3 to 4 days
  - Capillary buds appear.[369]
- 3 to 6 days
  - Hemosiderin stains positive with Perl's method; collagen appears; foreign body giant cells may appear. Note that the earliest appearance of hemosiderin staining depends on the method employed.[169]
- 4 to 8 days
  - First new collagen fibers seen; profuse new capillaries up to 8 days; maximum lymphocytes at wound periphery at 6 days.[369]
- 8 to 10 days
  - Decrease in inflammatory cells, fibroblasts, and capillaries; increased number and size of collagen fibers; earliest hematoidin seen at 9 days.[369]
- > 12 days
  - Decreasing cellularity, vascularity; decreasing fibroplasia after 14 days; maturation of collagen; basement membrane restored in epithelium.[369]
- 10 to 15 days plus
  - Cellularity and vascularity gradually decreases. Fibroblast activity continues.[319]
- Weeks to months
  - Eventual mature scar formation.[319]

Antemortem Versus Postmortem Scalp Wounds

The amount of blood extravasation into adjacent tissue is not necessarily a reliable indicator of antemortem versus postmortem bruising, especially in dependent areas, and in areas of particular interest to the neuropathologist. The latter include basilar skull fractures leading to periorbital hemorrhages, scalp injuries, and postmortem corneal harvesting for transplantation (especially if the face has been in a dependent position after death).[47]

Histologic changes which might lead one to conclude that an injury was antemortem have been described as postmortem events in rare instances. For example, Saukko and Knight[319] indicate that margination of neutrophils within vessel walls and fibrin deposition within a wound can occur within a few minutes in antemortem wounds, or in wounds sustained during the early postmortem period.

Can extravascular migration of neutrophils alone in the margins of a wound be considered definitive evidence of an antemortem wound? Gardner[117] studied gastric content entry into the lungs by aspiration during life and by gravity flow after death, and briefly commented that cellular reactive changes may continue after death. The described changes included intra-alveolar erythrocytes, macrophages, and a few neutrophils. These histologic findings were neither illustrated nor described in more detail in his article, however.

It has been stated that the earliest appearance of neutrophils in an antemortem wound (whether by intravascular margination or extravascular infiltration was unspecified) is 20 to 30 minutes, but that the nature of the agonal state of the individual can significantly delay this estimate.[215] The authors described a chronically ill 82-year-old man who died after a 2½- to 3-hour agonal period following a self-inflicted left wrist incision involving the radial artery and overlying tissue. No local inflammatory reaction was present in the wrist wound.[215] Robertson and Mansfield made similar observations decades earlier.[304] On the other hand, experiments in rats in which postmortem abdominal wall and skin injections of chemotactic materials were performed demonstrated that neutrophils retain motility for about 6 to 8 hours after death.[8,9]

The question is whether is such observations in human lung, and in experimental animals, can be applied to human skin wounds. Saukko and Knight[319] seem to imply that such data are directly applicable to humans, and that extravascular neutrophilic infiltrates in wound margins are not a reliable guide to antemortem versus postmortem wounding. Madea and Grellner,[215] however, attempted to reproduce Ali's[8,9] experiment in mice and in pigs, but with much less dramatic results. They found relatively sparse (judging by their Figures 16 and 17) perivascular accumulation of mononucleated cells and a few neutrophils in only 3 of 30 specimens, mainly younger animals.

These observations, together with the well-known risks of direct application of animal experimental results to humans without further confirmation, suggest that it would be premature to conclude that mild to moderate extravascular neutrophilic infiltrates in tissues in and around wounds have no value in determining antemortem versus postmortem timing of the wound, or that the absence of such infiltrates rules out antemortem wounding up to 2½ to 3 hours prior to death in certain types of prolonged agonal state situations. Reactive changes that go beyond mild to moderate extravascular neutrophil infiltrates (e.g., presence of mononucleated cells, macrophages, edema, reactive fibroblasts, endothelial changes), if present, simplify the interpretation assuming that proper criteria for more subtle changes such as edema, endothelial reactive changes, and so on are employed in order to exclude autolysis as their basis.

Histochemical studies may someday contribute to accurate wound dating/aging, but are still in an experimental stage and have not yet reached widespread acceptance as a practical tool in the forensic community.[9,68,153,264,284,319] Determination of the frequency of apoptosis has been suggested as a means to establish early vital changes following skin injury.[345] An alternative light source (such as Wood's lamp) combined with digital photography has been found useful in detecting the presence of skin and subcutaneous hemorrhage that is undetectable or indistinct to visual inspection, but has not yet aided dating/aging of bruises to our knowledge.[372]

Skull Fracture

Many orthopedic, radiologic, and forensic textbooks discuss healing of fractures. However, there is an appropriate reluctance by the majority of authors to be very specific about providing timetables for dating/aging fractures because of the well-recognized variables that affect the pace of bone healing. These variables include the specific bone and site within the bone involved, patient age, degree of local trauma, whether the fracture is open or closed, vascularity of the fracture fragment(s), method of treatment, degree of bone loss or displacement of fractured fragments, degree of immobilization, presence of infection, associated bone pathology at the fracture site (e.g., neoplasia, radiation injury), gender, hormonal factors, presence or absence of contiguous joint involvement, and presence of repeated local injury or repeated fracture.[245,283,298,307,314]

We offer a brief summary, being well aware of the considerable complexity of the subject of fractures. Current texts even differ in how the stages of fracture healing are categorized. For example, one major multi-volume text on bone and joint disorders describes fracture healing as occurring in three indistinctly separated phases. These are the inflammatory, reparative, and remodeling phases, representing approximately 10%, 40%, and 70%, respectively, of the total healing time.[298] Another current major text describes fracture healing in four stages: induction, soft callus, hard callus, and remodeling.[263] Both of these sources describe an essentially similar sequence of histologic events, as they occur typically in uncomplicated long bone shaft fractures. The initial hematoma at the fracture site forms a clot, and necrosis occurs at the fracture site and traumatized adjacent soft tissues. An inflammatory response develops with vasodilation, exudate, and cellular infiltration (including neutrophils, histiocytes, and mast cells). Granulation tissue formation follows, with organization of the clot. Osteoclasts and, later, osteoblasts occur; and cartilage, osteoid, woven bone and, later, lamellar bone develop as the inflammatory phase recedes. Sufficient callus of new bone formation stabilizes the fracture, and during subsequent remodeling the medullary canal reforms, the cortex returns to its original diameter, and bony deformities gradually become corrected. Klotzbach et al.[186] mention a few histologic findings at the early stages of fracture healing in long bones. The pace of healing is faster, and the completeness of the remodeling process is generally greater, in children compared with adults. For example, one report describes "ossification" within 15 days in the newly formed granulation tissue on the dura mater bordering an epidural hematoma in an 8½-month-old boy,[167] and comments that such findings tended to be more conspicuous in younger patients in that series of 21 cases. Physeal fractures tend to heal faster than shaft fractures. For a more detailed account of the aging/dating of fracture healing in long bones, the cited references should be consulted.[263,298] A recent review of radiologic dating of fractures in children emphasizes the need for further studies on this complex subject.[283]

As noted earlier, when estimates of aging/dating of fractures based on radiologic or histologic data are given, it usually is in the context of implied or specified long bone (especially rib)[245] fractures or in a particular age group, so that extrapolating such estimates to other age groups or bones is not necessarily appropriate.[263,288,292] This caution certainly applies to skull fractures, which is the emphasis in this section. Some authors do provide

Table 3.1. Appropriate Dating Aging of Skull Fractures Based on Skull Radiographs

Age at Time of Linear Skull Fracture	Fracture Line Indistinct	Fracture Line No Longer Visible
Birth[185]	2 months	6 months
Infancy and early childhood[354]		3–6 months
Children 5–12 years old[354]		Usually heal within a year
Children (not otherwise specified as to age)[28]		A few months, possibly a year average time
Adolescents[28]		More than a year average
Adults[354]		Average time 2–3 years, but fracture line may even persist throughout life

general radiologic guidelines for the rate of skull fracture healing in various age groups, primarily as a guide to clinicians.[28,185,354] Some generalizations that have been suggested for healing rates of linear skull fractures in the usual, uncomplicated case, based on radiologic criteria, are provided in Table 3.1.

Awareness of the individual variations that may alter the general rules can, in our opinion, become more important in the rather unforgiving environment of court testimony in forensic cases. Normal and premature closure of cranial sutures has been studied in humans and in animals for many years, yet its mechanisms are not clearly understood, and a study that compares and contrasts the detailed morphologic features of normal cranial suture closure and skull fracture healing has not come to our attention.[48,201,221,389] In newborn skulls there exist normal variants that may be misinterpreted as fractures or other abnormalities. Examples include parietal thinning, parietal fissures, parietal foramina, the interparietal fontanel, persistent mendosal sutures and other developmental delays of suture closure, unusually prominent venous channels, and prominent intrafontanel and intrasutural (wormian) bones.[58,104,149,185,332] Figure 1.58 in Hardwood-Nash and Fitz[149] is a diagram of infant sutural and synchondrosis anatomy that can be a useful guide at the autopsy table.

Some of these variants can persist into adulthood, and (typically asymptomatic) conditions such as hyperostosis frontalis interna[80] and sinus pericranii; dural, pineal, habenular, and choroid plexus calcifications may also sometimes be confusing to nonradiologists.[159,353]

Skull fractures may heal by bony union or fibrous union, with examples of both mechanisms sometimes seen within different portions of the same fracture line.[330] Cartilaginous callous may be absent, whether the portion of skull fractured was originally of membranous or of cartilaginous origin, according to Sevitt.[330] Fractures may widen over time, particularly when occurring in patients under 3 years of age.[108] Experimental studies suggest that this is most likely to occur when a dural tear accompanies the fracture, allowing dura and possibly arachnoid tissue to be interposed between the fracture edges, presumably by thus exposing the fracture margins to cerebrospinal fluid (CSF) pulsations.[108,354]

Neurosurgical burr holes may heal by fibrous tissue or by bone filling the defect. In a case reported by Sauer and Dunlap[318] a 16-year-old female disappeared who, 6 years previously, had a neurosurgical procedure in which bilateral burr holes were placed near the skull vertex, each 7 to 8 mm in diameter and approximately 3 cm lateral to the midline. The left burr hole encircled a portion of the coronal suture; the right burr hole did not. Three years later her skull was found in a wooded area by a hiker. It demonstrated a unilateral, left-sided burr hole 7 mm in maximum diameter, with minimal evidence of healing at its margin. On the right side, the prior burr hole was externally represented only by a shallow depression and interruption of the coronal suture line, and on the inner skull suture by an abrupt depression or pit. The asymmetry was so pronounced that the left-sided defect was suspected to possibly be a bullet hole by some initial observers.

The reason for this asymmetric healing is unclear. Burr holes often heal with new bone in children, but are much less likely to do so in adults. There is no evidence that burr holes overlying cranial sutures heal differently than those not involving sutures.[300] The historical avoidance of cranial sutures when opening the skull was based primarily on fear of causing uncontrollable hemorrhage from a major dural venous sinus, not concern that the burr hole would heal less completely, and has been rendered invalid by modern neurosurgical techniques.[21] The asymmetric burr hole healing is probably best explained by a theory based on integrity of the dura suggested by one of our neurosurgical colleagues.[23] The likelihood exists that, during surgery, hemorrhage from the incised dura occurring in the left-sided burr hole required extensive cautery of that dural margin. Because the dura is also the periosteum of the inner skull surface, its destruction on the left (but not the right, in this hypothesis) burr hole site would impair subsequent bone regeneration ipsilaterally. This hypothesis is supported by some clinical observations as well as experimental studies in animals.[201]

Skull and dura alterations may also result from neurosurgical procedures unrelated to defects from burr holes or skull flaps. Marked skull thickening, premature suture closure, and thickening of the dura with associated meningeal vascular proliferation have been described in patients with chronic ventricular shunts placed in early childhood for the treatment of hydrocephalus.[15,212] We have also seen the development of this sequence of events, but differ in our interpretation of the dural findings in that we regard both the dural changes in Figure 3B of Lucey et al.[212] and in our case as the presence of a chronic subdural neomembrane resulting from prior subdural hemorrhage (see Case 3.1, Figs. 3.1–3.6). The exact mechanism for the postshunt skull thickening is unclear. The differential diagnosis for skull thickening in this age group not only includes the possible effect of low postshunt intracranial pressure, but also cerebral atrophic processes early in life (including unilateral thickening of the skull as seen in cerebral hemiatrophy), diphenylhydantoin therapy,[176] rickets, severe chronic anemias, craniometaphyseal dysplasia, osteogenesis imperfecta, and microcephaly. In somewhat older age groups (usually early adolescence or later), skull thickening may also be associated with fibrous dysplasia, renal osteodystrophy, hyperostosis frontalis interna, or Paget's disease.[15,139,176,212,354,382] A more complete review of various syndromes, metabolic disorders, and skeletal dysplasias that may result in skull thickening (for which 67 conditions are listed), as well as conditions resulting in a variety of other skull abnormalities (e.g., defective closure, excessive wormian bones), can be found in Taybi and Lachman.[355]

Intradiploic meningoencephalocele may occur as an infrequent post-traumatic lesion.[272] In the absence of

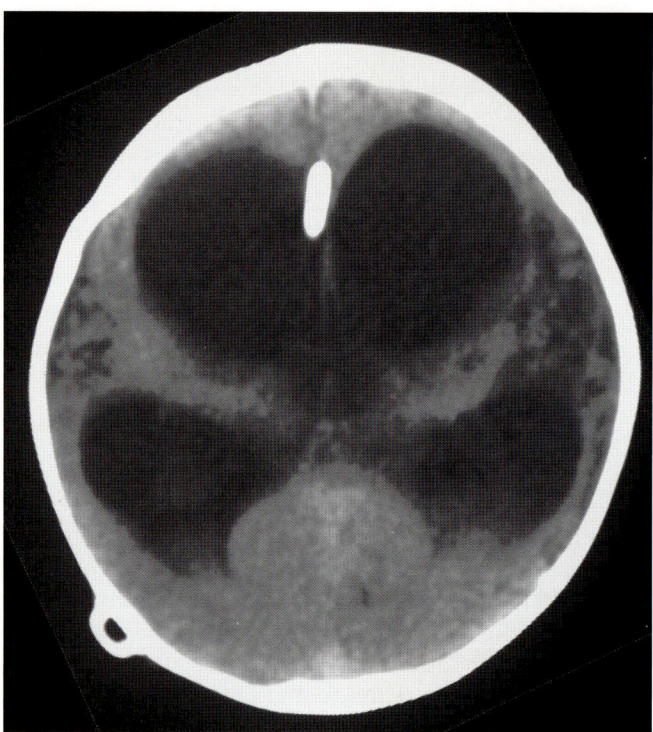

Fig. 3.1. Head CT of Case 3.1 at age 8 months. Axial image at the level of lateral ventricles with frontal ventricular shunt crossing anterior falx from left to right (as viewed from below). A second ventricular shunt valve is noted in right parietal calvarium. Marked hydrocephalus is present.

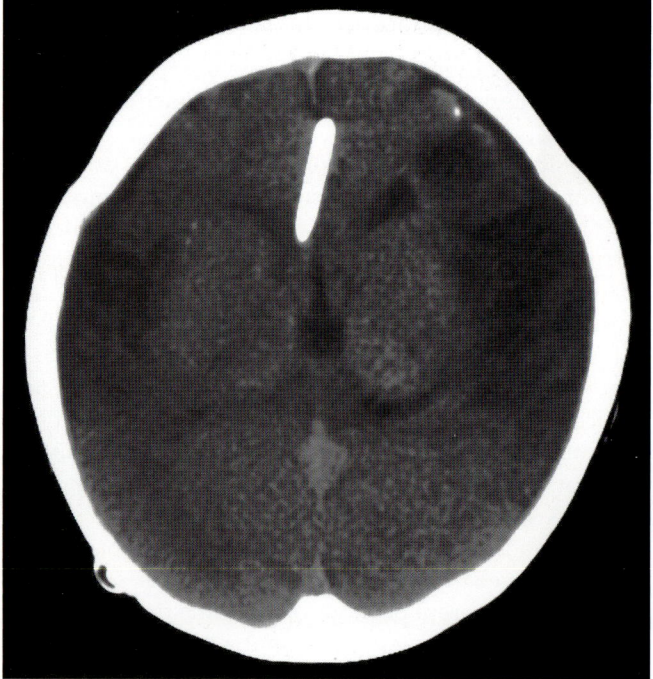

Fig. 3.2. Head CT of Case 3.1, 4 years later, at a level similar to that seen in Fig. 3.1. The same frontal ventricular shunt and right parietal shunt valve are seen. Interval resolution of hydrocephalus but irregularly distributed cerebral atrophy and prominent cranial vault thickening are now evident.

> **Case 3.1.** (Figs. 3.1–3.6)
>
> A 5-year-old girl developed pneumococcal meningitis at age 3 months with severe neurologic sequelae that included hydrocephalus, respiratory difficulties, convulsive disorder, and psychomotor retardation requiring daily nursing assistance. A ventriculoperitoneal shunt was placed for the treatment of severe hydrocephalus at approximately age 8 months (Fig. 3.1). A repeat computed tomography (CT) head scan at a similar plane of section 4 years later demonstrated interval improvement in hydrocephalus, bihemispheric irregularly distributed cerebral atrophy, and skull thickening (Fig. 3.2). She was found unresponsive without preceding recent change in clinical status at age 5 years, and pronounced dead at the scene. Autopsy revealed acute pneumonia, skull thickening of up to 1.0 cm in areas, bihemispheric thin chronic subdural neomembranes (Figs. 3.3–3.4), severe cerebral atrophy (particularly in the middle cerebral artery territory) (Figs. 3.5–3.6), and hippocampal sclerosis. The skull was not sampled for histology at autopsy.

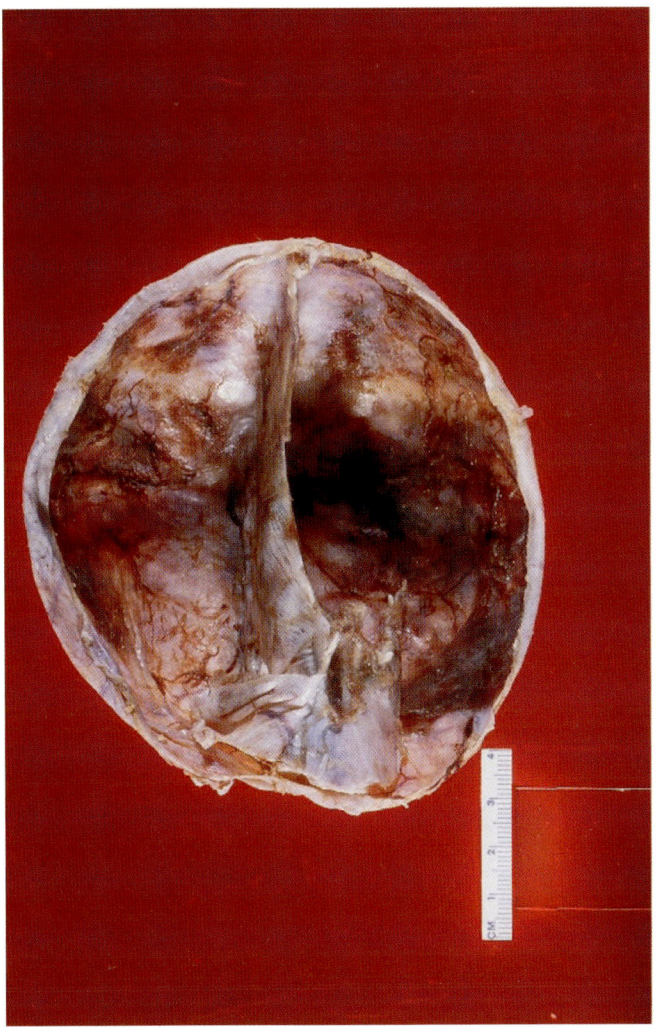

Fig. 3.3. Inner surface of cranial dura mater, frontal aspect at top of photo. Bihemispheric chronic subdural neomembranes represented by tan to brown areas of discoloration.

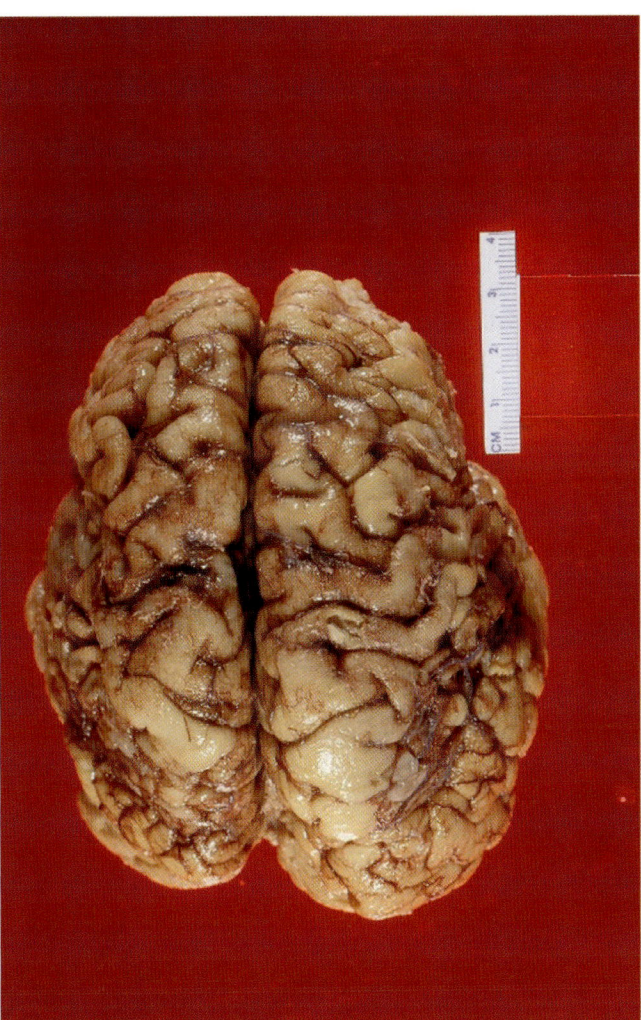

Fig. 3.5. Dorsal view of brain from Case 3.1, with atrophy primarily in distribution of middle cerebral artery.

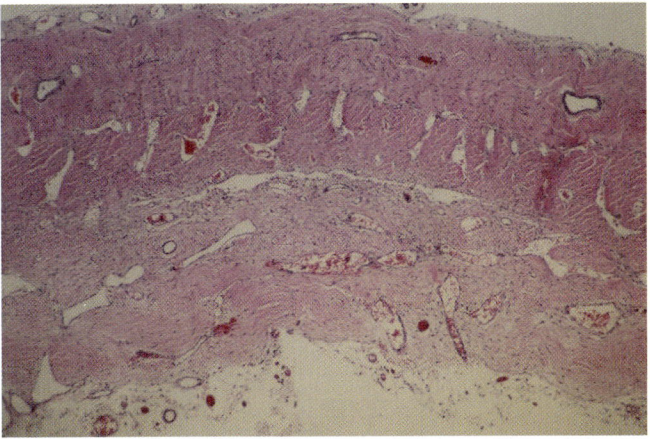

Fig. 3.4. Low-power micrograph of representative portion of chronic subdural neomembrane. Native dura in upper one-half of photo is more eosinophilic, and both its outer and inner layers are easily distinguished and are not thickened. The chronic subdural neomembrane is in the lower one-half of the photo, slightly less eosinophilic, well collagenized, with areas of hyalinization and with moderate vascularity. Scattered macrophages with hemosiderin-like pigment were present (not visible at this magnification). The neomembrane is approximately equal in thickness to the native dura. (H&E.)

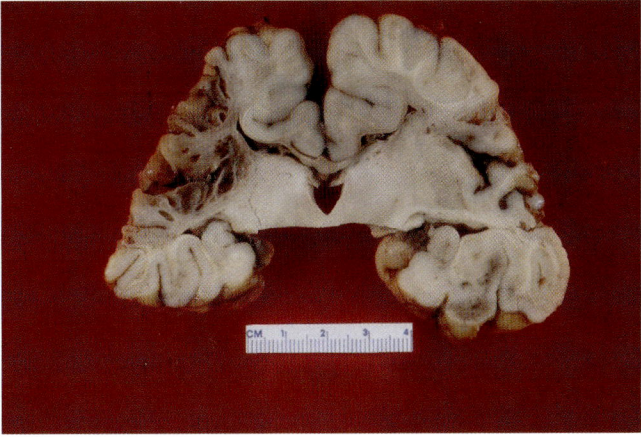

Fig. 3.6. Coronal section of brain from Case 3.1. Atrophy primarily, but not exclusively, in middle cerebral artery territory.

trauma, this appearance can be mimicked rather well by small occipital meningoencephaloceles, the preoperative diagnosis in one of our cases being "bone cyst."

In summary, while the degree of skepticism expressed by Saukko and Knight[319] regarding the unreliability of presently available criteria for dating/aging of fractures is perhaps not shared by other workers cited herein, it should be kept in mind so that unrealistic precision in dating/aging estimates for skull (and other site) fractures is avoided when expressing opinions on this subject. Clinical, radiologic, and histologic data combined are more likely to yield useful information than reliance on any single parameter, especially when questions arise concerning whether multiple fractures occurred simultaneously. Milgram[245] provides a more extensive set of H&E-stained photographs of healing fractures than other single texts we have reviewed, which can be helpful for comparison in certain cases.

Separation of nonfused cranial sutures as a result of increased intracranial pressure (sutural diastasis) is not uncommon in children, usually below the age of 12 years and rarely seen after 16 years of age. Radiologically, a separation of greater than 2.0 mm is excessive, and after the age of 16 years is more likely to be traumatic diastasis (diastatic feature).[354] Measurement of suture width at autopsy should be prior to removal of the skull cap to avoid artifact.

Calvarial bone is initially formed as a single plate, with a double plate arrangement separated by diploë generally occurring at an age range variously estimated as about 4 years[54] to (in the case of the parietal bone) 8 years.[63]

Mineralization in Central Nervous System Lesions

Mineralization of CNS tissues occurs in a wide variety of circumstances in both mature and immature brains, as evidenced by even a cursory review of current neuropathology textbooks and periodical literature.[108,137,200,302,355] Less frequently mentioned, although potentially useful as an additional parameter for estimating the age of a particular lesion, is the time required for light-microscopically visible mineralization to develop.

Ferrugination of neurons, sometimes referred to by other terms such as mummified, mineralized or fossilized neurons, is a condition in which deeply basophilic to black neuronal profiles are present in routine H&E stains (Figs. 3.7–3.8), and stains for iron (e.g., Prussian blue) and/or calcium (e.g., von Kossa) are positive. Other elements (i.e., phosphorous, sulfur) may also be present.[95]

Ferruginated neurons are seen in a wide variety of (generally chronic) lesions in both immature and mature brains, perhaps most commonly in the thalamus of infants

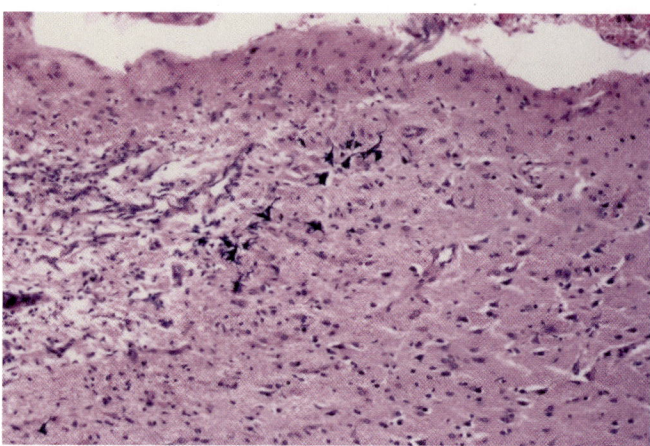

Fig. 3.7. Low-power micrograph. Margin of chronic cystic/malacic zone in midcerebral cortex (chronic pseudolaminar necrosis lesion). Dark cells at the border with more intact cortex are ferruginated neurons. (H&E.)

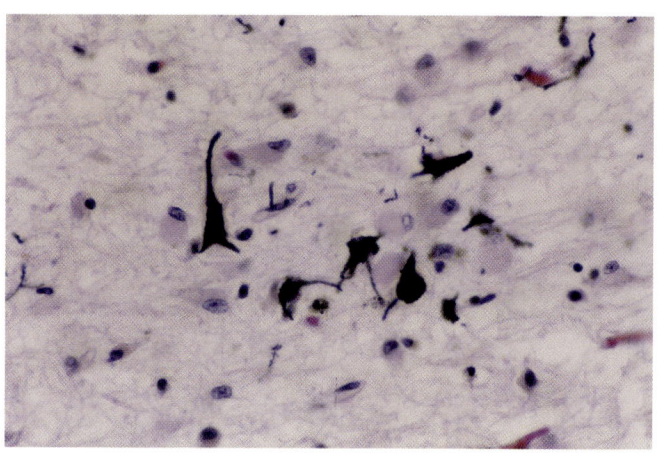

Fig. 3.8. High-power micrograph of ferruginated neurons in gliotic cortex. Cells with pink cytoplasm are reactive astrocytes. (H&E.)

who have experienced hypoxic-ischemic injury in utero.[310] This would infer that the responsible injury to thalamus occurs at some time after the peak period of thalamic development, which is estimated as occurring at approximately the second and third months of gestation.[24,126,373]

The association of ferruginated neurons and/or mineralization of other CNS components (e.g., neuropil, capillaries), in many cases with lesions which are chronic (i.e., do not come to autopsy for at least several months to several years after the insult to the CNS), makes it difficult to determine how quickly such changes may develop. In cerebral infarcts, ferruginated neurons have been described as soon as 10 days' post-insult.[61] Once present, there seems to be general agreement that such changes may persist for many years.

Granular encrustations of neurons and capillaries (which are PAS positive and are occasionally also positive for iron or calcium salts) have been described at the margins of necrotic areas due to cerebral infarction of "several weeks" duration in neonatal brains,[309] and are inferred to develop in infarcts older than several months in adults.[100]

Banker and Larroche[22] described the appearance of basophilic fibers on H&E-stained sections at the periphery of periventricular leukomalacia lesions believed to be secondary to intrauterine hypoxic/ischemic injury. These fibers were interpreted as encrusted axons. They occasionally, but not invariably, also were positive for calcium (von Kossa stain) and, more frequently, for iron (Prussian blue stain). These basophilic fibers were not seen in the first two weeks post-insult, but exactly how quickly they made their appearance in cases with survival periods beyond 2 weeks is not specified in this report. Sarnat[117] describes mineralization in cyst walls which develop from 2 to 24 weeks following white matter infarction in preterm and term neonates, but details of the shortest time period in which such mineralization occurs are not specified.

Rather striking neuronal ferrugination in the margin of a neonatal necrotic brain lesion was illustrated within 12 days following injury.[339] The stain illustrated (plate 16) is luxol fast blue/cresyl violet. The author comments also that minerals (particularly calcium) are "readily laid down in areas of damaged tissue in the fetal brain,"[339] and comments elsewhere[340] that mineralization may be seen within 10 to 12 days in macrophages, on axons, and within neuronal cell bodies in hypoxic-ischemic lesions in immature brains. A recent case at this Office supports the rapid development of mineralization in an immature brain (Case 3.2, Fig. 3.9).

Development of tissue mineralization in the CNS occurring within days is perhaps not too surprising when one recalls that (1) marked myometrial calcification in an adult sufficient to be evident in a CT scan and on gross examination may occur in less than 4 weeks following a postpartum hypotensive episode[232]; (2) myocardial fibers in transplanted hearts may calcify within 9 days post transplantation[271]; and (3) calcium deposits may develop in subcutaneous tissues surrounding polyHEMA sponge material within 14 to 21 days, and around suture material in the urinary tract within 3 days or less.[383]

The fact that strongly basophilic granular material is seen in H&E stains does not necessarily mean that calcium or iron is present.[22,129,309] If this distinction is considered important, special stains or other methods (e.g., electron probe analysis) can be employed for verification of the presence and type of mineralization. For example, Perl's prussian blue reaction reportedly stains ferric iron,[214] the Turnbull blue reaction stains ferrous iron,[219] and Lillie's method stains both ferric and ferrous iron.[214] Exceptionally, intraparenchymal CNS calcification can be an artifact produced by rinsing formalin-fixed brains in calcium-rich (i.e., 78–125 mg Ca/liter) tap water.[296]

> **Case 3.2.** (Fig. 3.9)
>
> This 11-month-old child was reportedly accidentally dropped onto a concrete pavement by her grandfather, rapidly became unconscious and developed seizures. Hospital evaluation revealed sagittal suture diastasis, subarachnoid hemorrhage, right hip fracture, and groin pseudoaneurysm, and a left abdominal wall hematoma. Clinical criteria for brain death were met, and organ donation followed on the sixth postinjury day. Neuropathologic examination revealed acute epidural and intradural hemorrhage adjacent to the sagittal suture diastasis, acute hypoxic-ischemic encephalopathy, acute subarachnoid hemorrhage, and respirator brain changes. Fine powdery subpial calcification was present in the basis pontis, but was particularly evident in localized areas of the dense dendritic arborization of Purkinje cells, where the beaded mineralization was positive with Prussian blue and von Kossa stains (Fig. 3.9).

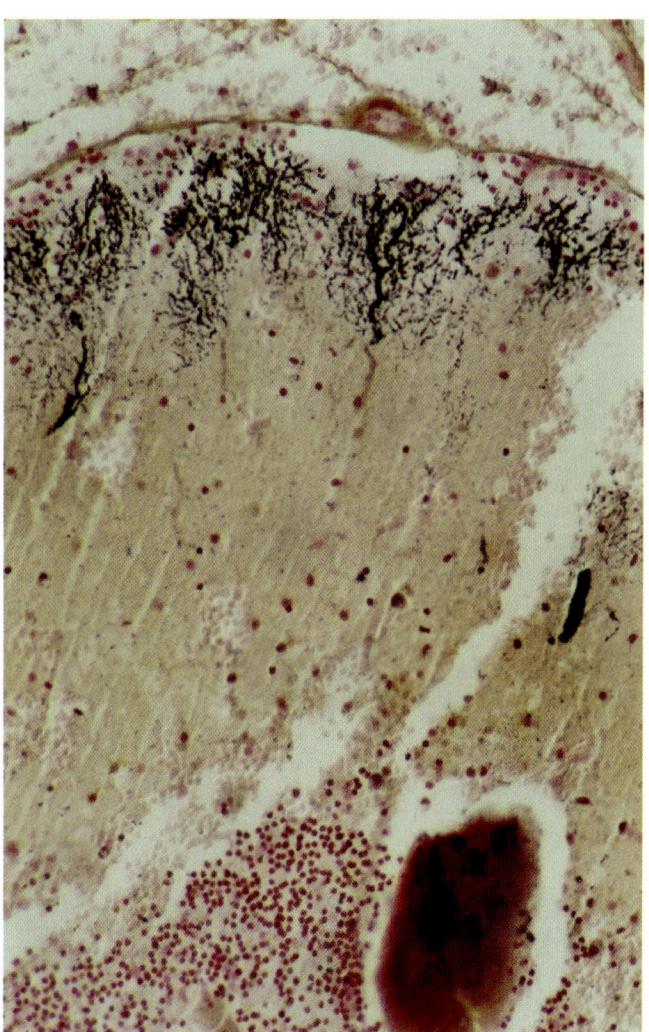

Fig. 3.9. Medium-power micrograph. Heavily mineralized superficial portion of Purkinje cell dendritic arborization is seen as black granules, 6 days after a severe acute hypoxic-ischemic insult in this 11-month-old child. (Prussian blue.)

Heterotopic Ossification Following Central Nervous System Injury

A propensity for development of heterotopic bone formation in patients with CNS injury, while well-described in neurologic, orthopedic, radiologic, and rehabilitation medicine literature, appears to be less appreciated in the forensic literature. Since it may be seen in patients more prone to develop decubiti, or in circumstances which come to the attention of the medical examiner as a consequence of referrals for suspected or alleged elder abuse or caretaker neglect, it is important for medical examiners to be aware of this condition so that its presence is correctly interpreted.

Following CNS lesions, ossification (true bone formation) may occur in soft tissues of weakened or paralyzed limbs, often in a para-articular location with or without a bony connection to the adjacent, typically otherwise normal, joint or long bone. In some cases, it may progress to form significant mass lesions, restricted range of motion of the joint, or adjacent joint ankylosis.*

This neurogenic form of heterotopic calcification is to be distinguished from that formed secondary to neoplasms, from local trauma, from other rare idiopathic forms of soft tissue ossification (such as myositis ossificans progressiva), and from ectopic calcification (in which mineralization, but no true trabecular bone formation, occurs in soft tissue).[286,297,376] The development of heterotopic bone formation in survivors of tetanus[140,213] is also likely based on local soft tissue injury rather than the CNS involvement per se. A case of periarticular ossification following pharmacologically induced paralysis,[3] and a case in which spinal meningeal (dura and arachnoid) ossification followed spinal cord trauma,[26] may represent unusual variants of heterotopic ossification.

The question as to whether preexisting CNS lesions influence healing of subsequent fractures in paralyzed limbs, either by retarding or accelerating the pace or extent of the healing process, remains unclear.[118,202,270] Similarly, a review of reported cases tends to provide some evidence favoring[178,228,312] and other evidence discounting[148,303] associated soft tissue injury as an etiologic factor, over and above CNS influences, in the process of heterotopic calcification.

Certain generalities relevant to heterotopic calcification following CNS injury in potential forensic cases may include the following. Most cases occur in adults, and it is somewhat more common in males. The sites most commonly affected are para-articular soft tissues of the hip, elbow, shoulder, and knee.[270] Hand or spine involvement is rare.[107,228,377] Although the location of the heterotopic bone may vary, certain sites and patterns predominate, and they differ somewhat depending on whether brain or spinal cord is the site of the CNS insult.[270,297]

The neurologic disorders cited as predisposing to heterotopic ossification include a variety of brain (encephalitis, trauma, cerebrovascular lesions causing hemiplegia, hypoxic-ischemic encephalopathy, neoplasia), spinal cord (multiple sclerosis, poliomyelitis, neoplasia, epidural abscess, trauma, tabes dorsalis, syringomyelocele, arachnoiditis) (Case 3.3, Figs. 3.10–3.13), and combined brain and spinal cord (encephalomyelitis, multiple sclerosis) conditions.[14,113,128,286,297,311,342,364] The development of heterotopic ossification may be seen as early as 19–21 days or delayed as long as 19 years following CNS lesions, although in most cases becomes apparent within 2 to 6 months.[119,148,228,257,297,377] It occurs in up to 22% of brain-injured patients and up to 53% of spinal cord-injured patients in various series,[270,297] in both flaccid and spastic limbs. Clinical symptoms may be essentially absent, or early stages may resemble acute arthritis, synovitis, tumor, infection, or thrombophlebitis.[268,297] Heterotopic ossification occurs less commonly (i.e., 3.3% of 152 patients devoid of other complicating factors) in children with brain or spinal cord lesions, and tends to be delayed in onset in the pediatric population (average delay of 14 months after CNS injury, range 3 to 36 months). It also tends to be less symptomatic in children

---

Case 3.3. (Figs. 3.10–3.13)

A 36-year-old man sustained a gunshot wound to the thoracic spine, resulting in paraplegia. He had no direct trauma to the pelvis. He moved to California 18 months later, where medical evaluation revealed limited lower extremity joint passive range of motion and no prior physical therapy. CT imaging, obtained on arrival in California, revealed heterotopic ossification in the paralyzed musculature and fascial planes about the pelvic girdle. Initial scout localizer film (Fig. 3.10) shows bullet fragments in the lower thoracic spine as well as increased bone density around the sacroiliac joints and bilateral hips. Lines are drawn at levels corresponding to axial images A (Fig. 3.11), B (Fig. 3.12), and C (Fig. 3.13). Axial image A (Fig. 3.11) at the lower thoracic spine reveals bullet fragments (*arrow*) traversing the right pedicle and extending into the T11 vertebral body. Axial image B (Fig. 3.12) obtained at the lower sacroiliac joint demonstrates mature heterotopic ossification (*white arrow*) arising from the ventral surface of the right iliac bone and less mature, evolving ossification along the left anterior inferior iliac spine (*arrowheads*). Axial image C (Fig. 3.13) obtained at the proximal femurs demonstrates heterotopic ossification at various stages of maturation. A mass containing developing immature ossification in the margins (*arrowheads*) is located anterior to the right femoral neck, and more mature ossification (*arrows*) is attached to and surrounding the ventral left femoral neck.

(Case courtesy of Jennifer A. Hill, M.D.)

---

*References 14, 71, 107, 113, 118, 119, 120, 128, 160, 161, 241, 246, 247, 268, 311, 312, 342.

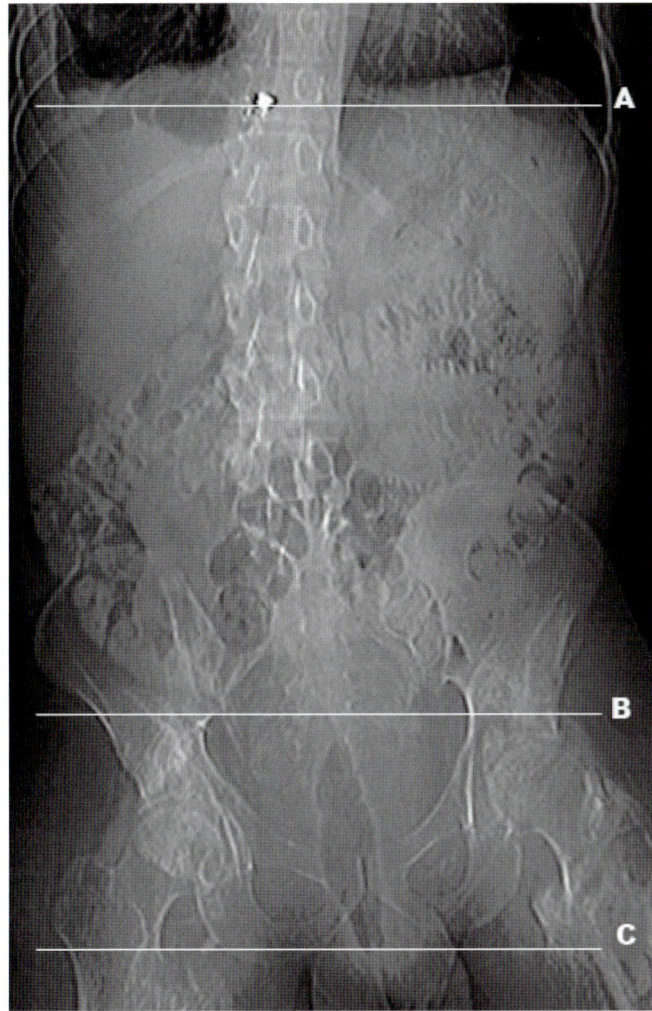

Fig. 3.10. Scout localizer CT film from Case 3.3.

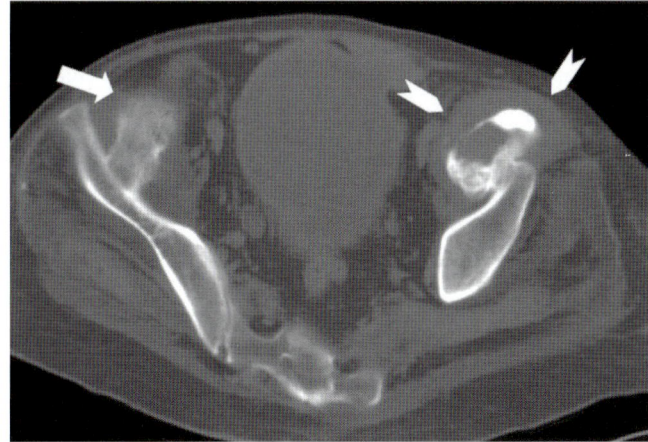

Fig. 3.12. Horizontal level B in Fig. 3.10.

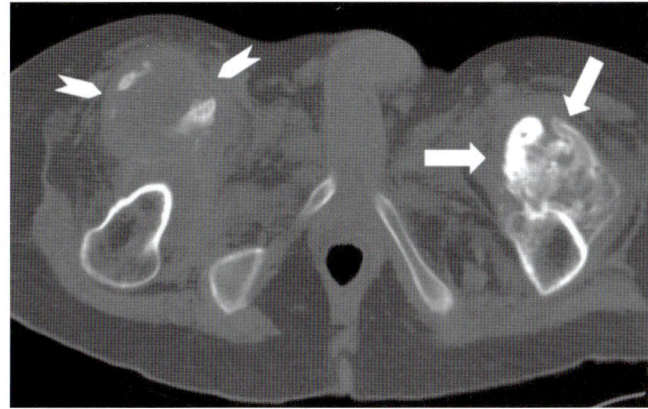

Fig. 3.13. Horizontal level C in Fig. 3.10.

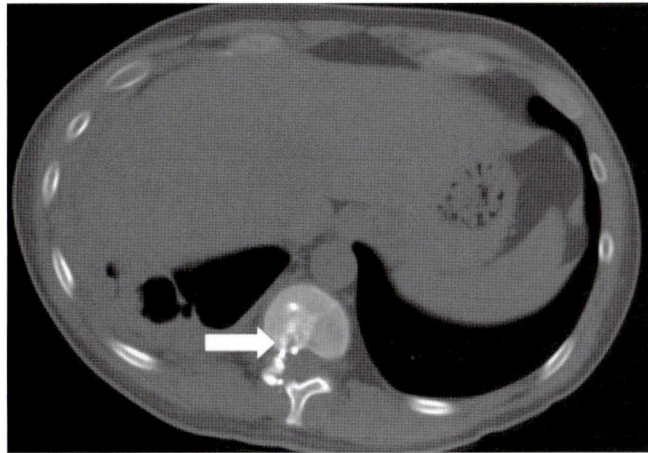

Fig. 3.11. Horizontal level A in Fig. 3.10. (See Case 3.3 text for explanation.)

(most consistent sign was decreased range of motion) and has more tendency to spontaneously resorb than in adults.[120] The location of the pediatric heterotopic ossification in this series was hip, and much less commonly peridiaphysial femur or shoulder.[120] In 10 additional cases in this same patient population, modifying local factors such as decubiti, hip dislocation, surgery, and acute local trauma were felt to be precipitating events leading to heterotopic ossification; these cases were not included in the group considered to be solely the consequence of the CNS lesions.[118]

Paralyzed patients tend to develop varying degrees of bone demineralization considered secondary to disuse. This in turn predisposes them to fractures that may even occur during the course of, for example, physical therapy to prevent contractures and other complications.[1] There is some evidence to suggest that heterotopic ossification is more likely to occur following such fractures in paralyzed limbs as opposed to nonparalyzed limbs, particularly when the paralysis is due to brain injury rather than spinal cord injury.[118] A high incidence of decubiti occurs in paralyzed limbs with heterotopic calcification,[359] but it is not clear from published studies whether the decubiti set into motion a cascade of events leading to the underlying heterotopic ossification, or whether the soft tissue ossification antedates and predisposes to overlying skin breakdown. It seems reasonable to conclude that either or both mechanisms may occur in a given patient. Knowledge of the rather characteristic patterns of bone deposition seen in many of these CNS-injured patients, as noted earlier, may be helpful in making such distinctions in individual cases.

Epidural Hematoma

The etiology of the great majority of epidural hematomas (EDH) is trauma, with or without associated skull fracture (including diastatic fractures). In cases of hypoxic-ischemic encephalopathy that occur prior to fusion of cranial sutures, resultant brain swelling may produce sutural diastasis sufficient to cause at least mild parasutural epidural hemorrhage (Case 3.4, Fig. 3.14). Other causes, such as blood dyscrasias, vascular malformations, neoplasia, or complications of neurosurgical procedures[144] are infrequent. Bleeding sources include meningeal arteries and veins, diploic veins, emissary veins, and dural venous sinuses.[205] In nontraumatic hematomas, histologic sampling to exclude occult neoplasia or other underlying cause is recommended.

Size of Epidural Hematoma in Relation to Clinical Symptoms

Clinical symptoms are not rigidly predictable from hematoma size alone. Variables include head size; brain size relative to skull size (e.g., cerebral atrophy); how rapidly the hematoma enlarges; whether cranial sutures are fused; functional margin of safety in individual cases (e.g., borderline dementia cases may develop increased mental symptoms with a smaller clot size than a person with normal mentation, other factors being equal); and presence or absence of other simultaneous intracranial mass lesions (e.g., brain swelling, tumor). Nevertheless, in forensic practice it can be useful to document the volume of intracranial hematomas in any location.

Hardman[146] considered 25 to 50 ml of EDH "significant," 75 to 100 ml "lethal," and indicated the "maximum" size to be approximately 300 ml of blood. Lindenberg[205] indicated 75 cm^3 or more was generally required to produce clinical symptoms in the absence of other space occupying lesions, and 75 cm^3 "may already endanger vital functions" in the presence of simultaneous brain swelling. Although Lindenberg[205] cited Vance[367] as the source of a maximum EDH size of 450 g, review of the original article revealed that the maximum EDH size in Vance's series was 250 g. (Given the near inevitability of similar typographical errors in large manuscripts, it is time well spent for the reader to consult original references if the data are to be used in testimony, and we

Case 3.4. (Fig. 3.14)

A 7-month-old boy was found unresponsive in bed with "blankets wrapped around his head." Resuscitative efforts restored a cardiac rhythm, but coma and the need for respiratory support persisted until brain death was pronounced 5 days later. Investigation revealed no evidence of trauma, and no suspicion of foul play. Neuropathologic examination revealed severe sutural diastasis that was accompanied by a midline sagittal external skull surface periosteal laceration and immediately adjacent patchy thin periosteal hemorrhage. The external surface of the underlying dura mater demonstrated a thin layer of epidural blood corresponding in distribution to the region of the severely diastatic sagittal suture line. Subarachnoid hemorrhage, a very swollen soft brain, and a brain weight just in excess of 2 standard deviations above the mean for age group and sex were present.

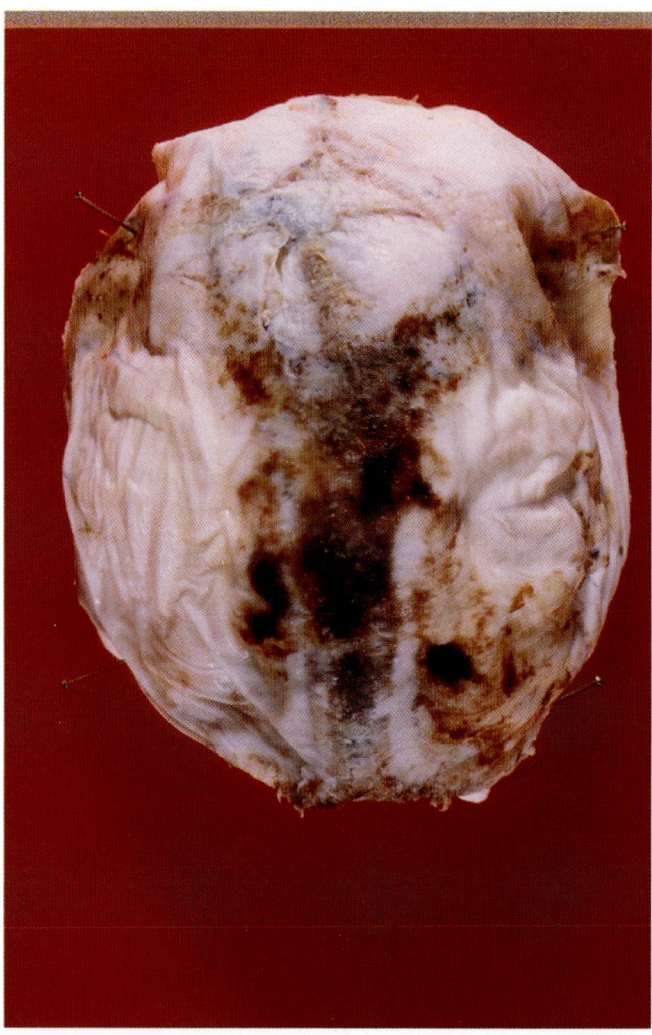

Fig. 3.14. Dorsal view of external aspect of cranial dura mater. Thin epidural hemorrhage largely parallels the region of the diastatic sagittal suture.

recommend this rule also apply to the present text.) The figures given refer to adults (i.e., neonates and children are not specified in the discussions), and it follows that smaller quantities may be significant in infancy and early childhood. It is also apparent that all associated findings such as hematomas in other head compartments (subdural, etc.), brain contusions, lacerations, swellings, diffuse axonal injury, and so on must be considered in the overall assessment of clinical-pathologic correlation and mechanism of death in a given case.

Dating/Aging of Epidural Hematoma

In comparison to the gross or microscopic dating/aging of subdural hematoma (SDH) (see later), the literature regarding EDH is relatively sparse. The general impression is that the sequence of cellular reactions is slightly slower in EDH compared with SDH, but detailed studies are not available to verify this impression. One source cites the first sprouting of fibroblasts from the outer layer of the dura into the epidural blood clot as occurring after 2 to 3 days, and indicates that it can be pronounced after 6 days.[205] In SDH (see later), by comparison, by day 4 there may be a fibroblast cell layer 2 to 4 cells in thickness, with fibroblasts entering the clot.

Until shown otherwise, the general consensus seems to be that no differences in cellular response pace exist between EDH and SDH that exceeds that seen in SDH between different individuals. Figures given for SDH (and implied for EDH), as in other timetables for cellular responses in this book, are to be regarded as general guidelines and not rigid rules.

Chronic Epidural Hematoma

Most EDHs are of arterial origin and present clinically within hours or a few days of their initiation. Exceptionally, EDHs may, as occurs in many SDHs, become more chronic. Chronicity is variously defined, but most cases described under this term had been several days to several weeks (rarely several years) in duration. These chronic EDHs are more likely to be extratemporal in location, more likely to contain clotted than liquid blood (or both, rather than liquid blood only), and more likely to be nonarterial (e.g., calvarial diploë, dural sinus) in origin, or due to delayed rupture of a middle meningeal arterial pseudoaneurysm.[158,167,227] The newly formed membrane bordering the dural aspect of these chronic EDHs is also more likely (compared with SDH neomembranes) to show areas of ossification within survival periods as brief as 15 days.[167] This is more likely to occur in children, and is not unexpected since the outer dura also serves as the inner skull periosteum.

Spontaneous Epidural Hematoma Resolution

From a clinical standpoint, most traumatically induced EDHs that produce sufficient symptomatology to result in imaging studies revealing their presence will require operative treatment. Much less frequently, EDHs may spontaneously gradually, or even rapidly, resolve.

Spontaneous resolution of EDH without surgical intervention following accidental cranial trauma with skull fracture and hemorrhage has been documented in a few case reports.[18,190] The hematomas measured up to 1.5 cm in thickness. Evidence of resolution in the volume of hematoma was recorded as early as 2 hours' postinjury in one case,[190] and within 3 and 10 hours in two others.[18] Curiously, associated epicranial hematomas enlarged as the EDH was resolving.[18]

Such cases may provide an explanation for apparent discrepancies between antemortem hospital imaging study diagnoses and autopsy findings. Other potential sources of such discrepancies are discussed in Chapter 8 on gunshot wounds, and in the neuroradiology section of Chapter 17.

Subdural Hematoma

As with EDH, most SDHs are secondary to trauma, with other causes (blood dyscrasias, severe dehydration, neoplasia, aneurysms, vascular malformations, etc.) being much less common.[57] Subdural fluid collections largely devoid of formed blood elements (such as secondary to liquefaction of an SDH or due to escape of cerebrospinal fluid (CSF) into the subdural space) are often referred to as subdural hygromas (although some workers reserve that term specifically for fluid collections composed of spinal fluid, and not those previously containing blood). Bleeding sources include bridging cortical veins entering major dural sinuses (most frequently the superior sagittal sinus in hematomas over the dorsal hemispheric convexity; less frequently the lateral sinus in subtemporal hematomas)[229,385]; bridging veins plus or minus arteries entering the dura remote from a major sinus and reaching the sinus through an intradural route directly or via an ectopic pacchionian granulation; small cerebral arteries and/or veins ruptured with an arachnoid tear, with or without cortical contusions; and ruptures of carotid or vertebral arteries or the great vein of Galen as they pass through the arachnoid layer lining the dura.[205,367] Predisposing factors may also include arachnoid cysts,[125,282] ventriculoperitoneal or other shunts,[111,253] and anticoagulant therapy.

Size of Subdural Hematoma in Relation to Clinical Symptoms

As with EDH, several variables determine the size of an SDH that will produce clinical symptomatology (see EDH, earlier).

One author considered greater than 50 ml of subdural blood to be "significant," greater than 100 ml to be "lethal," and the "maximum size" to be approximately 300 ml (again, presumably in adults).[146] Another indicated that an acute SDH in an adult becomes life-threatening when it reaches approximately 50 ml in size, and in infants a smaller amount may be life-threatening.[82] A third regarded 30 to 50 ml as "significant" in children 1 to 3 years of age.[273]

The importance of measuring, or at least estimating the weight/volume of the subdural hemorrhage, cannot be overemphasized. It is also useful to save the clot for chemical analysis, at least until the need for such testing is found to be unnecessary.[222]

Anatomy of Subdural Hematoma

Although the terms "subdural hematoma" and "subdural neomembranes" are both misnomers, they are so entrenched in the current pathology literature that they continue to be used in their commonly understood meanings in this text. For the sake of clarification, however, ultrastructural studies have revealed that SDHs actually form within the inner layer of the dura mater formed by "dural border cells" and "arachnoid barrier cells," in a potential space created by opening relatively weak intercellular attachments of the two layers. The "neomembrane" is the result of proliferation and thickening of the already present normal layer of dural border cells still adherent to the dura mater proper, accompanied by vascular sprouting into the layer of proliferative fibroblasts and SDH proper from the inner dura vasculature.[109,143,321] The innermost "neomembrane," closest to the brain, consists of a very thin layer of dural border cells and arachnoid. Compared with the outer neomembrane, a longer period is necessary before sufficient cellular proliferation forming the inner neomembrane has progressed to the stage when it is readily identifiable at the light microscope level of resolution. If the SDH resolves within a few days, an inner neomembrane is not seen.

In submitting sections of dura mater with SDHs for histologic examination, strips of dura and attached hematoma no longer than between 1.5 and 2.0 cm, and approximately 0.3 cm in width, allow histotechnologists to embed on edge without twisting or curling. The dura should be cut perpendicular to the major meningeal arteries in the immediate area of interest, and a grossly evident meningeal artery should be included in this section wherever possible. In adults, the middle meningeal artery branches are in the outer one-half of the dura (i.e., in the outer of the two major layers of collagen fibers comprising the dura mater), a useful guideline in distinguishing inner and outer dural surfaces in sections (see Chap. 1, Figs. 1.12–1.13). In infants the position of at least some obvious arteries in the dura can be more variable. Usually they are limited to the outer one-half, but occasionally are found in mid-dura or even slightly within the inner one-half of the dura. With increasing age, however, the rule that most or all of the prominent arteries are in the outer dural layer is reliable. Another useful guide to inner versus outer dura is the periosteal layer on the outer dural surface, which typically has a more loosely organized stroma, is more amphophilic (compared with the eosinophilic dense collagen in the dura proper), and which may contain a few osteoblasts, osteoclasts, and sparse mononucleated chronic inflammatory cells, particularly in infancy and early childhood. A few macrophages with hemosiderin (positive with Perl's stain for iron) or "pigmented osteoblasts" mimicking hemosiderin are also commonly seen in early infancy,[54,55,209] and we occasionally see small islands of extramedullary hematopoiesis even in infants without similar evidence in other organs.

Aging/Dating of Subdural Hematoma

Some consider any attempt to age/date subdural hemorrhages to be misleading.[108] Certain generalizations can be reasonably made, keeping in mind the need to avoid dogmatism and use special care when there is evidence of repeated traumatic events or other factors, such as questionable reliability of the history provided in forensic cases by caretakers, and so on. It can be useful, for example, to approximate the age of the most chronic portions of the neomembrane in a case that also has multiple other acute injuries and a large fresh subdural hematoma, since this may support evidence for repeated episodes of head trauma (see discussion on child abuse in Chap. 7).

Gross Appearance of Subdural Hematoma (from Lindenberg[205] and other sources)

Survival Period
- Day 1 to 3
  — Clot appears dark purplish and is easily washed from the dura by a gentle stream of water.
- Day 4 plus
  — Some blood remains adherent to the dura when washed by a gentle stream of water.
- Day 6 to 10
  — Blood often (but not invariably; see Munro and Merritt[250]) becomes partially clotted, crumbly, and develops a dark brown color.
  — An outer neomembrane is visible and easily stripped from the dura.
- After 2 to 3 weeks
  — Thinner blood collections may already be enclosed by an inner neomembrane.
  — Blood develops a milk chocolate to dirty motor oil color.
  — Outer neomembrane is thicker and more resistant to tearing.
- 4 weeks[233–235]
  — Clot usually liquefied.
- 8 weeks
  — Larger hematomas are generally entirely enclosed by neomembranes by this time, and tend to be gray-brown to rust brown in color.
- 2 to 3 years

- Neomembrane may be represented only by a faint yellow-brown mottled discoloration on the inner dura.
- Small foci of rebleeding in the interval may result in various shades of purple, black, brown, rust, and/or yellow colors.

Microscopic Appearance of Subdural Hematoma (From Various Sources[13,146,147,205,233–235,250] and Personal Observations)

Survival Period
- First 24 hours
  - Intact red blood cells (RBCs), fibrin at dura-clot interface and on arachnoid side, and at times as layers within the clot.
  - ± Few neutrophils in native dura inner layer, usually in perivascular location.
  - ± Dilated dural capillaries and/or slightly enlarged capillary endothelium adjacent to hematoma.[205]
- 24 to 48 hours
  - Neutrophils often, but not invariably, more conspicuous in inner dura.
  - ± Activated fibroblasts at dura-clot interface.
  - ± Rare small macrophages at dura-clot interface.
  - Intact RBCs.
- 2 to 3 days
  - ± Early evidence of endothelial hypertrophy, ± endothelial proliferation.[205]
- 3 to 4 days
  - Macrophages with and without hemosiderin at dura-clot interface (i.e., first appearance of hemosiderin positive with Perl's stain for iron).[13]
  - During this period, fibroblast layer of neomembrane is typically 3 to 4 cell layers in thickness and beginning to enter the clot; endothelial hyperplasia and hypertrophy first seen (according to most authors other than Lindenberg).[205]
  - A few erythrocytes may already demonstrate degenerative changes, such as pallor and loss of normal contour.
- 4 to 5 days
  - Early lysis of some RBCs. Fibrin on arachnoid side.[233-235]
- 5 days
  - Fibroblast layer typically about 3 to 7 cell layers in thickness, and more are entering clot.
  - More conspicuous hemosiderin-containing and foamy macrophages.
- 5 to 10 days
  - First capillaries enter clot; more degenerating RBCs and laked RBCs.
  - More macrophages with and without pigment.
  - Fibroblast layer up to approximately 12 to 15 cell layers in thickness by day 8; on the arachnoid side, still only fibrin and sparse fibroblasts ± hemosiderin-laden macrophages.
- 7 days
  - Laking of RBCs.
  - Angiofibroblastic invasion of clot.[233-235]
- 11 days plus
  - Clot starting to be broken into islands of degenerating RBCs by fibroblasts.
- 13 to 17 days
  - Neomembrane thickness variable; often from ⅓ to 1½ times the native dural thickness.
  - Many pigment-laden macrophages.
  - Brain side (i.e., inner) neomembrane evident in peripheral portions of the clot (in thin SDH, it may consist of only a few fibroblasts in thickness, ± a few capillaries).[233-235]
  - Blood in thin SDH nearly completely absorbed, with only a few RBCs and variable numbers of pigment-containing macrophages remaining.
  - Giant capillaries (vascular sinusoids) apparent at 7 days.
- 13 to 20 days
  - Capillary formation in the clot is more conspicuous, ± giant "sinusoidal" capillary profiles.
  - Up to this stage, fibroblast layer is rather loosely organized, and amphophilic.
- 20 to 25 days
  - Clot often liquefied; few intact RBCs remain from the original clot; ± foci of fresh rebleeding may be present.
  - Increased numbers of typical capillaries as well as giant capillaries.
  - Early collagenation is first seen in many cases during this time frame (collagenation refers to a definite pink, not amphophilic, color to the connective tissue fibers in the neomembrane with H&E stain).
  - Numerous pigment-laden macrophages present; neomembrane "about" the thickness of native dura (but quite variable, sometimes exceeding twice the native dural thickness).
  - Inner (brain side) neomembrane may be completed at this stage; it is essentially avascular in our experience, and thinner than the neomembrane at the native dura interface of the clot.
- 25 to 30 days
  - Neomembrane is well formed, the stroma more compact, and becoming well collagenized and less cellular.
  - Clot usually liquefied.
  - Brain-side neomembrane is well formed but relatively avascular.[233-235]
- 30 days plus
  - Arteries may sometimes occur in the neomembrane by this stage.[205]
- 1 to 3 months
  - Hyalinization of both dura and brain side neomembranes; often secondary hemorrhages.[233-235]
- 40 days plus

- Capillaries may be prominent.
- Neomembrane is beginning to resemble native dura to a greater degree (i.e., early hyalinization; hyalinization refers to paucicellular, usually closely packed pink collagen fibers resembling native dura in degree of cellularity and overall appearance).
- 60 days plus
  - Neomembrane thickness continues to be variable (i.e., less than 1 time up to about 2 times native dural thickness).
  - Progressive hyalinization of neomembrane.
  - Giant capillaries can persist up to 90 days plus.
  - Thinner SDHs may be largely or completely absorbed by this time.
- 3 to 12 months
  - Neomembranes usually fuse; consist of more mature fibrous tissue; contain scattered pigment-laden macrophages.[233–235]
- 6 months plus
  - A few hemosiderin-laden macrophages, and clefts with blood debris or cholesterol clefts may be seen.
  - Blood vessels typically quite sparse, most often consisting of small capillaries, but may persist in some neomembranes even for years.[205]
- 1 to 2 years
  - Thin neomembrane almost indistinguishable from native dura may be seen, recognized because the orientation of collagen fibers (compared with native dura) is altered.
  - A few pigment-laden macrophages may persist.

Additional Comments on Subdural Hematoma

If the neomembrane is well hyalinized (i.e., very similar in density, cellularity, and tinctorial characteristics to native dura in H&E-stained material) and also demonstrates calcification or ossification, our cases with this appearance have all been in excess of 1-year duration. Hardman[146] and McCormick[233-235] describe calcification or ossification as occurring after approximately 3 years. This is probably not an invariable aging/dating observation, however, for reasons indicated elsewhere in this text (see various factors influencing the appearance of mineralization in tissues elsewhere in this chapter). If the SDH is small, it may become organized and absorbed from the dural side without the formation of an inner neomembrane. If present, the arachnoid-side neomembrane is first apparent at the light microscopic level of resolution at the clot peripheral margin, and typically remains thicker at the margin even when fully formed. A number of factors determine the eventual fate of a given SDH (e.g., shrinkage and resolution, progressive enlargement, degree of admixture with cerebrospinal fluid), and are as yet incompletely understood.[54,163,223,385–387]

Very thin, chronic (i.e., several months plus) subdural neomembranes in early infancy, especially when located in the dorsal (vertex) posterior frontal/anterior parietal areas, seem not to have the same degree of forensic implications for suspicious nonaccidental injury as do more laterally placed subdural neomembranes or acute SDH in an infant several months of age. They correlate poorly with other stigmata of child abuse.[306] Our policy is to no longer routinely take histologic sections of these particular minor stains on the inner dura in this age group, but rather provide a gross description and place representative tissue in the "save" bottle in the event that sections might subsequently be required.

Eosinophils and mast cells are infrequently found in subdural neomembranes, usually as sparse isolated cells. On very rare occasions, moderate numbers of eosinophils occur in neomembranes that do not correlate with any specific systemic process. A remarkable frequency of eosinophilia in SDH of 60% was reported in an Indian population series, perhaps unique to that population.[316] Although not emphasized in the literature to the same extent as other cellular reactions described earlier, small numbers of mononucleated chronic inflammatory cells in subdural neomembranes, usually as single cells or small groups of small lymphocytes, are occasionally seen. Less often there are moderate numbers of lymphocytes with or without sparse plasma cells in cases with no evidence of superimposed infection, and in which acid-fast, Gomori methenamine silver (GMS), Gram, and Giemsa stains are negative. In our material, these chronic inflammatory cells began to appear at about the same time frame as macrophages, and can persist for months.

Spontaneous Subdural Hematoma Resolution

There are occasional opinions expressed in the neuropathology literature indicating that SDH in children can resolve more rapidly than in adults.[56,197] No large case series documenting this impression exists, to our knowledge. Duhaime et al.[98] described four such cases (ages 10 months, 9 months, 4 and 10 years), but indicated that this phenomenon has more often been reported in adults.

Polman et al.[280] described a 26-year-old woman with head trauma whose initial CT head scan revealed an acute SDH 1.0 cm in thickness that had resolved (at the resolution level of a repeat CT scan) 6 hours later. A magnetic resonance image (MRI) 3 days later revealed only a very thin layer of more widely distributed blood, and clinical recovery progressed without surgery. Nagao et al.[252] reported the case of a 3-year, 10-month girl with two separate episodes of head trauma resulting in an acute SDH. The initial episode demonstrated that the SDH had completely disappeared on a repeat CT scan 2 days after the initial CT scan. In a second episode, she developed an SDH after a fall, which substantially decreased in volume on a repeat CT scan 17 hours later. Kuroiwa et al.[190] described a 17-year-old man who sustained a linear fracture of the right occipital bone, an epidural hematoma in the right posterior fossa, and a subdural hematoma in the left frontotemporal parietal region after a fall, demonstrated in a CT scan 1 hour postinjury. A repeat CT scan 2 hours postinjury revealed a clear reduction in volume of both hematomas, and a third CT scan 12 hours postinjury revealed near-total disappearance of both the epidural and the subdural hematoma. Surgery was deferred, and the clinical recovery was excellent. Erol et al.[101] reported the case of a 13-year-old boy involved in an auto accident who sustained a scalp laceration, linear left frontotemporal skull fracture, right frontotemporal SDH, and small left EDH on initial CT scan. A repeat CT scan 24 hours later revealed a larger but less dense SDH with some compression of adjacent brain, and brain edema. Thirty-six hours after admission, a repeat CT scan showed disappearance of the SDH, but a left occipital lobe intracerebral hematoma had developed. The latter had resolved on a CT scan performed 1 month later.

The aforementioned cases make the point that clinical information such as CT and MRI reports may not always coincide with the autopsy report. The finding of less SDH volume at autopsy, for example, compared with that estimated by the radiologist in an antemortem imaging study, does not necessarily mean that the scan report was an error, or that the pathologist made a gross underestimation of the volume of SDH.

Subarachnoid Hemorrhage

The most common cause of subarachnoid hemorrhage (SAH) is trauma to any vessel traversing the subarachnoid space, including many of those listed as subdural hemorrhage sources (see earlier), and those upon or within simultaneously injured brain tissue. Intracerebral hemorrhages may also gain access to the subarachnoid space by broaching the pia mater, or the ependymal lining of the ventricular system. Potential predisposing factors for subarachnoid hemorrhage from relatively mild trauma, not ordinarily expected to cause hemorrhage, might include cortical/dural adhesions related to prior traumatic or inflammatory disorders, anticoagulant therapy, and preexisting bleeding diatheses. Nontraumatic causes include intracranial aneurysms and vascular malformations, primary vascular diseases such as arteriosclerosis, infectious vasculitis, noninfectious angiopathies, blood dyscrasia, vein and dural venous sinus thrombosis, neoplasia, and a variety of less frequent conditions.[360] The rate and amount of hemorrhage will determine whether the clinical result is catastrophic and leads rapidly to death, or results in less dramatic or insignificant symptoms.

A useful clue as to whether subarachnoid hemorrhage is from an antemortem event or from a brain removal artifact is the appearance of dural arachnoid granulations alongside major dural sinuses such as the superior sagittal sinus. Their normal color is similar to that of the dura (Fig. 3.15). In antemortem diffuse subarachnoid hemorrhage, they become dark red-brown to black bilaterally (Fig. 3.16).[235] Unilateral hemispheric SAH will result in blood-discolored arachnoid granulations along the ipsilateral superior sagittal sinus. Similar asymmetry of arachnoid granulation discoloration can occur ipsilateral to a unilateral subdural hematoma.

In reviewing the dating/aging timetables for *gross* appearance of SAH that follow, some caveats should be considered. First, findings described are, unless otherwise stated, derived from studies obtained by lumbar puncture, and do not apply to CSF obtained by ventricular puncture.[226] Second, Tourtellotte et al.[363] studied clinical factors that might correlate with speed of clearance of erythrocytes (RBCs) from CSF (with clearance defined as CSF containing less than 100 RBCs/mm^3) in patients with spontaneous SAH. The time from onset of SAH to clear CSF varied from 6 days to between 20 and 30 days in their series of 62 patients. There tended to be a slower clearing of RBCs from CSF in association with older age, past or family history of diabetes mellitus or vascular disease, the presence of permanent neurologic deficits, and possibly with the size of the spontaneous hemorrhage. However, exceptions occur, emphasizing the difficulty of making predictions in individual cases. While it is not possible with present techniques to determine with precision the age of the SAH in a particular case with no known acute symptomatic onset, most individuals studied exhibit findings that lead to the following generalizations.

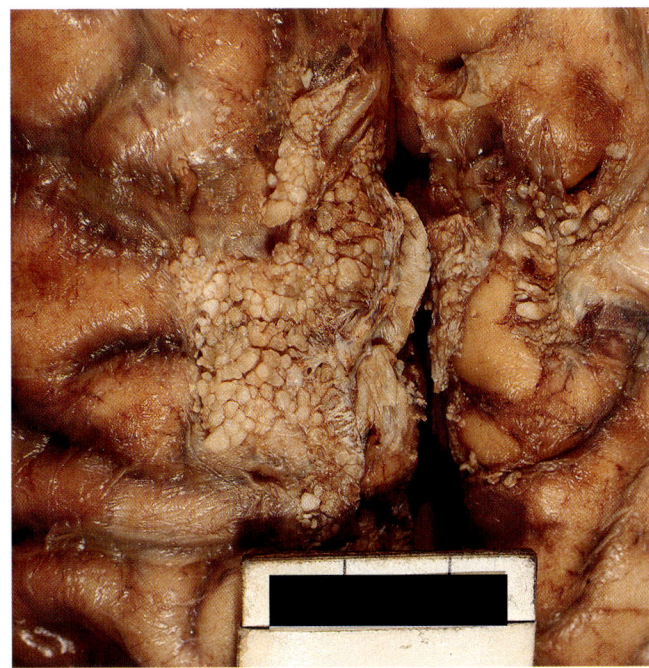

Fig. 3.15. Detail of arachnoid granulations in parasagittal vertex area of hemisphere in an adult. Normal light tan color is apparent.

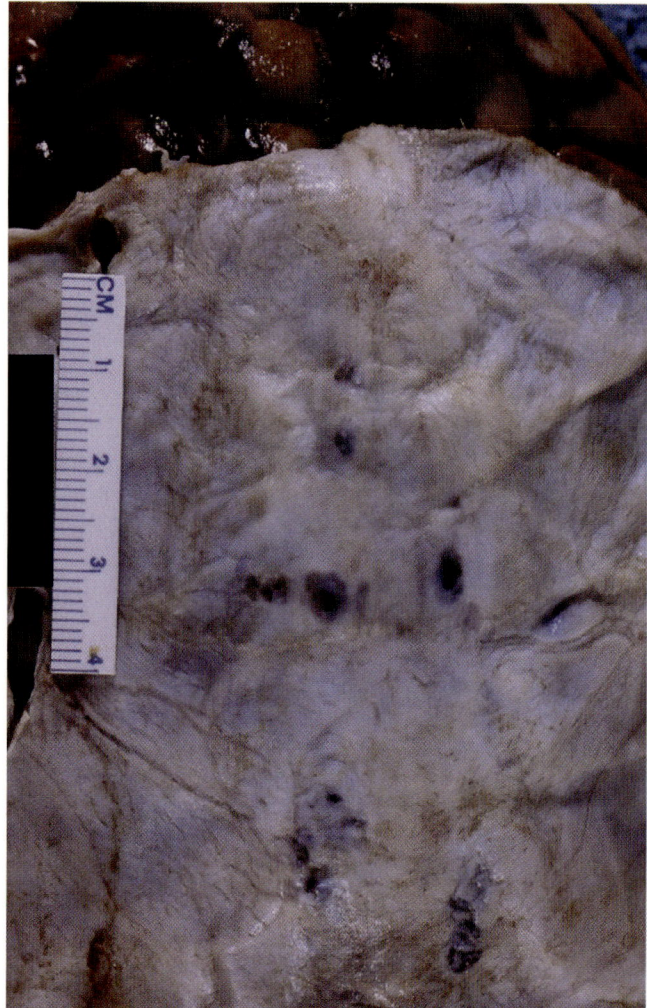

Fig. 3.16. External surface of dura mater in case of subarachnoid hemorrhage. Note dark brown to black discoloration of arachnoid villi which are adjacent to the superior sagittal sinus at sites of thinner dura. The discoloration is due to subarachnoid blood being filtered by the arachnoid villi. Subarachnoid hemorrhage is seen overlying the frontal lobe at top of photograph.

Cerebrospinal Fluid Gross Appearance in Subarachnoid Hemorrhage

Postmortem CSF can be readily obtained in adults via a cisternal (magna) puncture using an 18-gauge needle slid along the midline base of the occipital bone to a depth of about 6 cm from the skin.[106] Aseptic technique is obviously mandatory in a case examined for infection. Tonsillar herniation, including that with respirator brain, precludes the cisternal route, and one needs to revert to the lumbar route.

Initiation of lysis of RBCs in CSFs following SAH requires at least 2 to 4 hours, at which point xanthochromia of supernatant in centrifuged CSF develops.[106] Persistence of xanthochromia varies with the amount of initial hemorrhage and other undetermined factors, including the possibility of recurrent blood leakage from the original source of hemorrhage.[106,363] Xanthochromia has been observed in 70% of patients 3 weeks following SAH.[106] Adams and Sidman [7] indicate that the CSF does not become colorless for approximately 15 to 30 days following SAH. In contrast, it typically clears within approximately 2 to 5 days following a traumatic lumbar puncture.

The differential diagnosis of xanthochromia includes the turbidity associated with any cause of elevated CSF protein (generally requiring greater than 150 mg/dl), CSF leukocytosis; aspiration of epidural fat during the lumbar puncture (LP); use of radiographic contrast media; jaundice (yellowish, but rarely with a faintly greenish tint); food faddists with dietary hypercarotenemia (orange tint); malignant leptomeningeal melanomatosis (shades of brown tint); and rifampin treatment (orange-red tint).[106,151]

Gross Appearance of Brain in Subarachnoid Hemorrhage

Survival Period
- "Acute"[106,360]
  — Clear CSF (if < 360 RBC/mm^3 present).
  — Pink to mildly xanthochromic CSF (if > 1000 RBC/mm^3, after ~ 4 hours).
  — Red blood-stained CSF (at ~ 6000 RBC/mm^3).
- First few days
  — Gradual change to purplish-black blood (methemoglobin).
- 1 week
  — Early appearance of brownish color (purplish-black color persists longer in larger SAH).
  — Maximum xanthochromia after ~ 1 week.[360]
- Many weeks to months
  — Gradual change to rust and eventually yellowish color.
- 6 to 12 months
  — Usually no gross discoloration related to a single prior SAH event, unless associated with cortical contusion.

Microscopy of Subarachnoid Hemorrhage

The average aneurysmal SAH is said to release 7 to 10 ml of blood into the CSF.[360] SAH, a foreign material in the subarachnoid space, elicits reactive changes over time that may result in CSF leukocyte counts of up to several hundred cells/mm^3; elevated white blood cell (WBC) counts may persist on average up to 4 weeks.[360] The initial predominance of neutrophilic reaction is replaced by mononucleated cells (lymphocytes, monocytes, and macrophages) over several days. Experience is usually that duration of the abnormal erythrocyte to leukocyte ratio in CSF, or the impression gained from microscopy of the subarachnoid space, is not sufficient to raise the question of associated purulent meningitis in the great majority of cases. In fatal cases, postmortem cultures and/or stains for infectious agents in tissue sections can be employed if an unusually vigorous leukocytosis has occurred in response to SAH.

In the past, air introduced into the SAS for pneumoencephalography was well known to cause an increase in CSF cell count, and traumatic pneumocephalus can be expected to produce a similar response. Anesthetic agents, chemotherapeutic drugs, antibiotics, radiographic contrast media, and a host of other substances intentionally introduced into the SAS can and have produced cellular meningitis reactions,[106] in some cases leading to more serious arachnoiditis. A good history will resolve any mystery of increased inflammatory cells in the SAS at autopsy in those circumstances. Modern methods of cell identification can also be used to determine the components of the reactive cell population with greater precision.

Microscopy of Brain and Subarachnoid Space Tissue Sections in Subarachnoid Hemorrhage

The summary that follows is based on varying and overlapping patient survival periods following SAH, as studied by several different workers who bracketed their results based on available patient survival times and other undetermined factors. This point is emphasized, since a similar format of overlapping survival periods is used in some other timetables of histologic findings in this text. It more accurately reflects the data from the source cited, avoids oversimplification of complex events, and allows one to appreciate that the reaction patterns to SAH are more consistent in some parameters than in others. Each of the time frames indicated subsequently signifies the survival period of the patient from the onset of subarachnoid hemorrhage, whether the SAH was traumatic or nontraumatic in origin.

Microscopic Examination of Subarachnoid Hemorrhage

Survival Period
- < 10 minutes
  — Fresh blood in the SAS. No reactive changes.[145]
  — Presence of crenation does not distinguish SAH from a traumatic LP in CSF samples from surviving patients.[363]
- 1 to 4 hours
  — Fresh blood in the SAS.
  — A few neutrophils in perivascular location in pia mater.[145]
  — Erythrocytes begin to lyse in 2 to 4 hours.[251]
- 1 hour to 8 days
  — RBCs within the SAS, and do not pass into the cortical perivascular spaces of Virchow-Robin.[162] (Note that the anatomy of human spinal meninges is similar to that over the human cerebral cortex, so that similar results can be expected in that location.[256])
- "Few hours"
  — Intact RBCs plus an increase in leukocytes (type of leukocytes not otherwise specified).[225]
  — Increased neutrophils first seen.[147]
- 4 to 16 hours
  — More intense neutrophil reaction that is more diffuse in the SAS. The vigor of response is not clearly related to the amount of blood in the SAS.
  — Lymphocytes are starting to increase in number as early as 4 hours' survival, seen first in perivascular spaces.[145]
- < 12 hours
  — Fresh and crenated RBCs present; slight fibrin; slight hemosiderin crystals; very occasional macrophages with evidence of phagocytosis; and an increase in leukocytes.[11] (Leukocyte reaction varied throughout this series of cases, in appearance and number, and the response did not correlate with the size of the hemorrhage. *Note*: The presence of "hemosiderin" was an opinion based on routine stains such as H&E and connective tissue stains. No iron stain was employed in this study.)
- 12 hours
  — Early arachnoid cell nuclear swelling, becoming more obvious at 24 to 48 hours.[205]
  — Occasional macrophages appear.[7]
- 12 to 24 hours
  — Increased numbers of leukocytes, and fresh and crenated RBCs present.
  — Hemosiderin, fibrin strands, and phagocytosis (more prominent than in any of the cases seen at less than 12 hours' survival).[11] (See comment regarding hemosiderin in < 12 hour-survival period, earlier.)
- 16 to 32 hours
  — Increased numbers of neutrophils and lymphocytes, present more diffusely in the SAS than at the 4- to 16-hour stage, but the numbers of neutrophils are never comparable to that seen in purulent meningitis.
  — From approximately 24 hours on, at least some of the macrophages appear to be derived from pia-subarachnoid cells lining the SAS.
  — First appearance of brown pigment and iron positivity in this case group.[145]
- < 1 day
  — These authors imply that neutrophils were more prominent than lymphocytes in the SAS portion of the arachnoid villi, and that erythrocytes were present in arachnoid villi.[225]
- 1 to 2 days
  — Erythrophagocytosis first seen in CSF specimens from living patients.[151]
- 1 to 3 days
  — Neutrophil reaction peaks at 3 days.[360]
  — Neutrophils begin to be replaced by lymphocytes, and macrophages (including lipid-filled macrophages) are prominent and remain so for another few days.[147]
- Within 2 days

- Early lysis of erythrocytes, first macrophages appear, and the latter may also contain iron-positive pigment.[205]
- 2 to 4 days
  - Macrophages with hemosiderin first seen in CSF in living patients.[151]
- 2 to 5 days
  - Erythrocytes mainly in clumps.
  - Fibrin and hemosiderin more marked than in previous survival periods.
  - Phagocytosis well developed.
  - Early but definite fine reticulin fiber organization seen with Laidlaw's stain, arising from pia-arachnoid and from adventitia of meningeal blood vessels.
  - Early separation of RBCs into islands by reticulin fibers; slight collagen fiber formation.
  - Increased numbers of leukocytes.[11]
- >2 to 3 days
  - Degenerating erythrocytes first noted in arachnoid villi; increased mitotic activity in arachnoid villi in arachnoid cap cells as early as 24 to 48 hours' survival.[225]
- 3 days
  - Siderophages first seen in CSF samples from living patients.[106]
  - First fibroblast proliferation in meninges seen.[147]
  - All reactive cell forms increased in number; neutrophil reaction at its height but constitutes less than 50% of the total reactive cells in the SAS (outnumbered by phagocytes and possibly lymphocytes).
  - Many phagocytes, which contain RBCs, pigment, and iron; and degenerating leukocytes also present.[145]
  - Quite variable cellular component preponderance from area to area even within the same case.[145]
  - Earliest appearance of hemosiderin within arachnoid villi.[225]
- 6 to 7 days
  - First appearance of hemosiderin in macrophages.[7]
- 7 days
  - Neutrophilic response subsided (unless rebleeding occurred); approximately equal numbers of lymphocytes and macrophages; increased amounts of pigment and iron compared with prior time frames.
  - Some intact RBCs still present.[145] Some intact RBCs persist even in the absence of evidence of rebleeding.[205]
  - Prominent hemosiderin in arachnoid villi.[225]
- 9 to 20 days
  - Erythrocytes are completely lysed within this period.[251]
- 10 days
  - First definitive fibrosis in leptomeninges is patchy, and focally begins to obliterate the SAS; fibrosis varies considerably in amount from case to case; present in only about 50% of 53 cases at 10+ days' survival, but present in all cases that survived 7 weeks or longer.[145] No relation of amount of fibrosis to age of patient.[145]
  - Neutrophils decreasing in arachnoid villi; focal thickening of arachnoid cap cells of villi.[225]
- 10 to 14 days
  - Macrophages with hematoidin pigment first seen in CSF samples from living patients.[106]
- 12 to 35 days
  - RBCs mainly in clumps with few fresh or crenated cells; secondary hemorrhage occurred in 3 of 7 cases in this time frame; hemosiderin present.
  - Fibrin strands present.
  - Phagocytosis well developed.
  - Progressive connective tissue organization.
  - Collagen prominent at 35 days' survival.[11]
- 2 weeks to 6 months
  - Evidence of rebleeding (in all cases in this time frame).
  - Fibrosis from 7 weeks on.
  - Variable numbers of macrophages, leukocytes, pigment, and iron.[145]
  - Mesothelial cell reactive changes, free phagocytes and lymphocytes present as long as any blood or blood breakdown products persisted in the SAS (neutrophils were present after 1 week only if recurrent hemorrhage occurred).[145]
- 2 to 4 weeks
  - Maximum cellular reactive changes in the SAS.[205]
- 6 to 8 weeks
  - Subarachnoid space reactive changes slowly clear.[205]
- 2 months
  - More pronounced thickening (10–15 layers, compared with normal range of 3–5 layers) of arachnoid cap cells in arachnoid villi, with some neovascularization.
  - Hemosiderin in arachnoid cap cells.[180,225]
- Several months
  - Macrophages with hemosiderin may persist in CSF samples from living patients.[151,360]
- Years
  - Increased numbers of macrophages and iron-positive pigment may persist for years in the SAS;[205] for months to years in the SAS and Virchow-Robin spaces (but possibly disappear more rapidly in children).[147]

Observations on Postmortem Cerebrospinal Fluid Examination

Interpretation of postmortem CSF results, even if obtained rapidly after death, requires the usual precautions employed in living patients as well as consideration of certain additional factors. Chemical changes occur rapidly in postmortem CSF, particularly elevations of potassium and magnesium.[106]

Studies of leukocyte survival in CSF samples obtained from living patients and placed in sterile plastic tubes kept at room temperature may also be relevant to these considerations, such as the studies performed by Steele et al.[341] They found that neutrophils rapidly lysed, being reduced to 91 ± 2% (of the original neutrophil count) at 30 minutes, 68 ± 10% at 1 hour, and 50 ± 12% at 2 hours post-lumbar puncture. Both lymphocytes and monocytes were less rapidly lysed, with 88 ± 10% of lymphocytes and 80 ± 8% of monocytes surviving at 2 hours, and 69 ± 7% lymphocytes and 66 ± 7% monocytes surviving at 3 hours.[341] Leukocyte degenerative changes occurred much more rapidly in CSF than those described in blood samples taken from nonrefrigerated cadavers, or samples taken from living patients and placed in ethylenediaminetetraacetic acid (EDTA)-treated glass containers, in which no morphologic changes occur until after 6 hours in vivo or in vitro.[85]

In contrast, Platt et al.[278] studied postmortem CSF samples from pediatric cases obtained by lumbar puncture, and postmortem CSF samples from adult cases obtained by lumbar, cisternal, or ventricular aspiration from either hospital cases or medical examiner cases. They demonstrated a postmortem pleocytosis in all pediatric cases and in all but one adult case. CSF samples were analyzed in the fresh state (all pediatric cases and some adult cases) or after RBC lysis with glacial acetic acid followed by 1:1 dilution with formalin. In 26 sudden infant death syndrome (SIDS) cases, CSF examined from 2 to 28 hours postmortem demonstrated cell counts of from 37 to 3250 cells/mm^3. In 24 non-SIDS pediatric cases, CSF studies performed from 1½ to 22 hours postmortem revealed cell counts of from 0 to 593/mm^3. In 14 adult cases studied 5 to 48 hours postmortem, CSF cell counts were from 1 to 108 cells/mm^3. CSF cells were characterized as greater than 90% mononuclear in the SIDS group, and "most were mononuclear" in the non-SIDS pediatric and the adult groups.[278] CSF cultures were negative in all cases, and histology of brain sections showed no inflammation. Only two cases, both in the SIDS category, were examined at two separate postmortem intervals. In one case, CSF sampling revealed a cell count of 175 cells/mm^3 at 9 hours, and 1400 cells/mm^3 at 27 hours postmortem. In the second case, the cell count was 118 cells/mm^3 at 2 hours and 180 cells/mm^3 at 25 hours postmortem. The authors speculated that the findings might be secondary to postmortem diapedesis of lymphocytes and monocytes across the choroid plexus and arachnoid blood vessels.

It seems reasonable to speculate that leukocytes might rapidly lyse in untreated postmortem CSF in situ (even in bodies refrigerated soon after death) as they do in test tubes kept at room temperature for even brief periods, as studied by Steele et al.[341] If so, how does one reconcile the rapid lysis of leukocytes in CSF in vitro[341] and the postmortem (mainly mononuclear leukocytic) pleocytosis and presumed persistent functional mobility in vivo?[278] Using fresh peripheral blood leukocytes as controls for the various leukocyte immunohistochemical stains employed by Platt et al.[278] might not exclude the possibility of misidentification of postmortem cells in CSF for the same reason that similar stains in tissue undergoing autolysis or necrosis are often difficult to interpret (e.g., either due to nonstaining, or due to nonspecific staining).

Another possible explanation for the results of Platt et al.[278] is suggested by the studies of McGarrey et al.,[236] who performed a postmortem cytologic study of CSF obtained by cisternal puncture. A portion of the CSF was examined without further manipulation, and to another portion of the CSF obtained by the same cisternal puncture was added cells gently swabbed with a cotton-tipped applicator from the pia-arachnoid membrane, ependyma, or choroid plexus from subsequently exposed brain in the same case (in an effort to have a population of cells of known source as an aid to cell identification). Stains employed were H&E, and a modified Papanicolaou method. The study was primarily qualitative, with only subjective quantitative assessments. Postmortem intervals were not specified, age range of cases was 2 weeks to 89 years, and cases with- and other cases without-CNS pathology were examined. The only CSF cells encountered in the postmortem cases without CNS pathology were lymphocytes, pia-arachnoid cells, ependymal cells, choroid plexus epithelium, and accidentally introduced contaminants from overlying tissues (such as normal epidermal squamous cells, capillaries, erythrocytes, adipose tissue, fibrous tissue, and skeletal muscle cells) or underlying structures (nerve cells and glia). They commented that ependymal cells could not be recognized in most samples unseeded by cotton swabs "probably because of rapid disintegration of cytoplasm and pyknosis of nuclei after exfoliation," and that "in this degenerated state these bare nuclei were indistinguishable from those of lymphocytes." They also noted that degenerating choroid plexus cells "bore a striking resemblance to macrophages." Increased numbers of CSF cells were only described in cases with CNS pathology in their study.[236]

Desquamation of ependymal cells, choroid plexus cells, and granule cells and Purkinje cells from disintegrating cerebellar cortex all increase in frequency in cerebral ventricles or in the SAS as autolysis progresses with increasing postmortem intervals. Perhaps these and other degenerating intracranial cell populations contribute to what could appear as an increasing postmortem CSF (predominantly "mononuclear") pleocytosis, but not a true leukocytosis. Future studies with more modern techniques should clarify this question.

Brain Contusions

Brain contusions are surface bruises in which the leptomeningeal layer on the surface overlying the brain remains intact. If brain substance is actually disrupted, the lesion is termed a laceration.[234] Lacerations, it follows, are essentially always accompanied by contusions. Most frequently, contusions occur on the surface of the cerebral hemispheres and, less frequently, on the cerebellum. The various types of contusion (e.g., coup, contrecoup, fracture, herniation, gliding contusional tear) are discussed in most general pathology and neuropathology textbooks, as well as in the older literature often cited.[76,205,206,234]

One should resist the temptation to be too dogmatic regarding the direction, amount, and type of force that produced a given brain contusion based on its location(s). Any such opinions, if offered, should be based on not only the brain lesions but also the general autopsy findings in the scalp and skull, evidence of any underlying coagulopathy, and all other available information concerning the case. It is increasingly evident that the location of a contrecoup contusion, for example, is not necessarily a good indicator as to the direction or mechanism of force by which it was produced.[135,136,248] Contusions related to gunshot wounds are discussed separately (see Chap. 8).

Fresh contusions may appear only as closely aggregated petechiae in apices of the superficial cortex. Criteria useful for differentiating somewhat more superficial contusions from circulatory infarction, as summarized by Lindenberg,[205] are that contusions

1. Are located at the crests of gyri.
2. Have a wedge shape on cross-section, with a broad base at the surface.
3. Show evidence of associated SAH and participation of leptomeningeal elements in the healing process.
4. Are devoid of a residual cortex layer 1 "glial membrane."
5. Demonstrate persistence of intact cortex in the depths of sulci when two or more adjacent gyral contusions are present.

It will become quickly evident that CNS contusions have attracted more dating/aging attention from investigators than most other CNS lesions, particularly during the early stages of reactive changes. This allows inclusion of the first appearance of certain cellular reactions (e.g., neuronophagia, neuronal incrustation, and ferrugination), which are data difficult to find in other sources despite the common presence of these changes in many other traumatic and nontraumatic lesions. Whether their initial appearance and life span in other types of lesions are essentially the same as in contusions remains to be determined.

It will also be evident that the pace of certain cellular reactions is more rapid in trauma than in ischemic circulatory lesions, and that a more vigorous cellular response accompanies larger and more hemorrhagic lesions.

Gross Appearance of Brain Contusions

Survival Period

- Seconds to minutes[205]
  — Focal subpial red hemorrhage at contusion site ± SAH (dark brown to black post-formalin fixation).
- Several days[205]
  — Color more purplish-black.
- 3 weeks[205]
  — Color beginning to change to dark brown.
- 3 to 4 weeks[205]
  — Color lighter brown.
- 6 weeks plus
  — Color usually golden orange-brown (i.e., plaque jaune)[7] to rust brown,[205] although some will show persistence of dark black/brown color for extended periods, possibly due to rebleeding.[205]
- "Chronic" stage[7]
  — Underlying white matter may be grayish due to myelin loss (associated with wallerian degeneration) and gliosis.
  — Dura-pia/arachnoid adhesions may form (especially with associated laceration/contusions).

Microscopic Appearance of Brain Contusions

For this dating/aging section, references on the studies of brain contusions are limited by intention. An exception is one study[16] that was based on cases of traumatic brain lesions (age range of patients 2–88 years) resulting from motor vehicle accidents, blunt force homicides, gunshot wounds, and falls. Focus was on the histologic changes in the first 48 hours of survival. The authors' implication was that no significant differences in reaction pace or patterns occurred in this group of lesions (with the exception of their motor vehicle accident group having 1 of 28 cases demonstrating neuronal incrustation in the first 6 hours of survival, not seen in any of the 12 gunshot wound cases examined during the same time period). We are including their data under this heading with the understanding that their timetable represents an aggregate of histologic reactions seen in contusions, lacerations, intraparenchymal hemorrhages, and infarctions. Our review of dating/aging timetables for more "pure" case series of cerebral contusion comprises the bulk of the following information, however. We deal separately with infarction, lacerations, intracerebral hemorrhage, and gunshot wounds elsewhere in this text. Since observation sources are specified in the following information, readers may select those observations most suitable to their own cases.

Microscopic Examination of Brain Contusions

Survival Period
- Within 1 hour
  — Minimal edema in some cases,[146,266] lasting up to 6[146] to 9 days.[264] Others do not include early edema in evaluation parameters due to the difficulty of determining its presence in postmortem material.[16]
  — Hemorrhage (initially mainly SAH and parenchymal perivascular hemorrhage); RBCs seen in the lesion up to 5 months' survival (presumed rebleeding).[264]
  — "Degenerating" neurons (eosinophilic cytoplasm not specified), persisting up to 6 months' survival.[264]
  — Red neurons present in lesion area.[16,210] Red neurons may persist for greater than 15 days.[210]
  — Swollen eosinophilic glia present (mainly subpial, subependymal, and bordering hemorrhages);[16] persist for somewhat greater than 4 to 5 days' survival.[210]
  — "Dark" neurons present, and may persist for greater than 48 hours.[16] These may represent reactive changes in dying neurons rather than artifact, according to the studies cited[16,210] (also see Chap. 10).
  — Neutrophils in lesion area: marginating along vascular endothelium; in perivascular neuropil (rare); and around corpora amylacea (mainly subpial).[16,210] Neutrophils not seen around corpora amylacea beyond 2 days' survival.[16]
  — Swollen axons/spheroids present; persisting greater than 48 hours (seen with H&E and with Garvey's silver axon stain earlier than with β-APP immunostaining).[16]
- 1 to 3 hours
  — First appearance of neutrophils at 2 hours, 10 minutes; present up to survival periods of 28 days.[264]
  — Astrocyte swelling, especially perivascular podocyte processes, decreasing at 5 days' survival, but with increasing glycogen content; electron microscope study.[44]
  — Nerve fiber changes, as demonstrated by silver (but not H&E) stains in this study, were present to a mild degree as early as 1¼ hours' survival, but rapidly became more conspicuous, with typical spheroids seen at 26 hours and beyond (seen up to 7 months' survival in a case of traumatic intracerebral hemorrhage).[290]
- 3 to 6 hours
  — Neuronal encrustation first seen, and persisted greater than 48 hours.[16]
  — Neuronal encrustation first seen at greater than 5 hours' survival, and is prominent at 36 to 48 hours (using Nissl stain).[206]
- 6 to 12 hours
  — Maximal edema and vascular congestion; first appearance of neutrophils and macrophages in lesion.[146]
  — Glial cell necrosis (with eosinophilic nuclei) seen in ⅓ of 65 cases).[210]
  — Erythrophagocytosis first seen at 8 hours' survival.[264]
  — Minimal axonal swelling appears; progresses in time to typical axonal spheroids, which can persist greater than 12 months.[146]
  — Minimal necrosis of neurons and glia appears; necrosis increases and persists greater than 1 month, but less than 2 months.[146]
- Within 24 hours
  — Hypoxic-ischemic neuronal injury present; electron microscope study.[44]
- 12 to 24 hours
  — First appearance of red neurons and neuronal incrustation.[205]
  — Early astrocyte "reactive" changes; some with pyknotic nuclei.[205]
  — Macrophages first seen at 14 hours, and persisted up to 58 years' survival.[264]
  — Macrophages first seen at 19 hours.[16]
  — RBCs seen moving through fenestrations in endothelium to leave vessel lumen by diapedesis; electron microscope study.[44]
  — Neuronophagia first seen at 14 hours, and persisted up to 5 days' survival.[264]
  — Endothelial swelling first seen.[210]
  — Mild neutrophil infiltrates, moderate axonal swellings, prominent necrosis.[146]
- 24 to 48 hours
  — Persistence of changes as previously described.
  — Thrombosis of small intracortical vessels first seen, subsequently increasing in frequency.[210]
  — Granulocyte karyorrhexis first seen from the end of day 2, and present in all 65 cases studied thereafter up to and including their 10- to 15-day survival group.[210]
  — Axonal swelling first seen at 31 hours, and persisted up to 28 years' survival.[264]
  — Macrophages first seen at 36 to 48 hours, and rare karyorrhexis.[205]
- 2 to 3 days
  — Early RBC degeneration.[205]
  — Axonal spheroids first seen at lesion margin at 2 to 5 days' survival, and in most cases thereafter.[210]
  — Macrophages containing hemosiderin and infiltration of lymphocytes first seen at 71 hours; both persisted in cases up to 44 years' survival.[264]
  — First appearance of macrophages containing hemosiderin.[197]
- 2 to 6 days

- First appearance of endothelial proliferation and reactive astrocytes.[146]
- 3 to 4 days
  - True sprouting of capillaries first seen.[210]
  - First appearance of foamy macrophages on day 3, and of erythrophagocytosis on day 4.[343]
  - Foamy macrophages seen at lesion margins in most cases after 3 to 4 days, and in all 63 cases after 4 days' survival.[210]
  - Vascular endothelial swelling and proliferation first seen.[165]
  - Pyknosis and shrinkage of both neurons and glia, and lysis of neuronal nuclei first seen; neuronophagia first seen (electron microscope study).[44]
  - "Increase in vessels" first seen at 94 hours, and still present in a case surviving 31 years.[264]
- 4 to 6 days
  - Reactive, swollen astrocytes with more prominent pink cytoplasm present.[197]
  - Hemosiderin-containing macrophages first seen at 4 to 5 days.[210]
  - Astrocytes with swollen nuclei appear from day 5 onward.[210]
  - "Protoplasmic" astrocytes first seen at 101 hours, persisting in a case surviving 26 years.[264]
  - First collagen fiber deposition seen at 5 to 11 days; electron microscope study.[44]
- 6 days
  - "Piloid" astrocytes first seen at 6 days, persisting in cases surviving up to 58 years.[264]
  - Fibroblast/fibrocyte reaction first seen at 6 days; persisted in cases surviving up to 8 months.[264]
- 7 to 10 days
  - Probable increase in numbers of astrocytes, as well as reactive astrocyte changes.[197]
  - Siderin-containing astrocytes first seen at day 8; may persist up to 44 years' survival.[264]
- 7 to 14 days
  - Progressive reduction of edema, hemorrhage, vascular congestion, neutrophil infiltrate, and degenerating neurons and glia.[146]
- 10 to 15 days
  - Macrophages containing hematoidin first seen at 10 days[344] to 14 days.[343]
  - Macrophages containing hematoidin first seen at 12 days, and present up to 12 months' survival.[264]
  - Gemistocytic astrocytes first seen at 10 to 15 days' survival.[210]
  - Ferrugination of neurons first seen at 9 to 10 days, and persisted in survival periods up to 28 years.[264]
- 21 to 23 days
  - Ferrugination of neurons first seen at greater than 21 days; can persist for "years."[291]
  - Coagulation necrosis first seen at 23 days, and persisted up to 53 days' survival.[264]
- 6 to 12 months
  - Moderate numbers of macrophages and reactive astrocytes may persist. Maximum glial scar, which persists greater than 12 months' survival.[146]
  - Gemistocytic astrocytes have largely evolved to become fibrillary astrocytes by approximately 1-year survival.[210]
- > 12 months
  - Persistent findings for varying periods, as noted above.

Axonal Spheroids, Diffuse Axonal Injury, and β-APP

Axonal Spheroids

Various terms have been applied to describe the ovoid enlargements seen in injured or regenerating axons, the ultrastructure of which was well described in the experimental study by Lampert.[192] The authors prefer the term "axonal spheroids." The spheroids have been referred to in the English language literature as axonal retraction balls, retraction bulbs, dystrophic axons, axonal torpedos (if located in Purkinje cells), axonal balloons, axonal swellings, and axonal varicosities. Some writers use the terms "axonal swelling" and "axonal varicosities" to refer to more subtle changes, such as variations in axon caliber that remain in continuity throughout the injured area, and that fall short of well-developed, focal axonal spheroids associated with secondary axotomy.[230]

Diffuse Axonal Injury

Diffuse axonal injury (DAI) was the term used for several years for widespread axonal injury resulting from trauma to the CNS, and represented in severe acute cases by a characteristic distribution of petechial hemorrhages (the latter, when more widespread, more recently referred to as diffuse vascular injury).[87,277] In cases surviving beyond approximately 18 to 24 hours,[82] one finds a characteristic distribution of axonal spheroids when examined by H&E stain or by certain silver stains. Milder injuries might still result in widespread axonal spheroids, but with few or no petechiae, since axons are more vulnerable to acceleration-deceleration injury than blood vessels.[5]

These earlier concepts have continued to evolve, with unqualified use of the DAI designation more recently falling out of favor. This is primarily because of increasing recognition that widespread—and considered by some authors to be essentially indistinguishable patterns—of axonal injury may be seen in nontraumatic conditions. It has thus been suggested that trauma-induced widespread brain injury be termed "diffuse traumatic brain injury," or "diffuse traumatic axonal injury," based on whether the preponderant pathology is limited to axons or also includes other brain lesions (e.g., hemorrhages, lacerations).[87] As newer technologies (e.g., β-APP immunostaining; see later) have evolved, the frequency and extent of axonal spheroids not only in traumatic brain injury, but also in a variety of nontraumatic CNS insults, have become more widely appreciated. Contemporary neuropathology textbooks and periodicals continue to expand the list of conditions in which axonal spheroids are found.[86,90,96,100,124,136,137,238] Table 3.2 is a brief summary of several conditions in which axonal spheroids may be found. The common denominator in this rather nonspecific histologic finding is considered to be a disturbance of rapid axonal transport mechanisms. Of these various conditions, the most likely to cause interpretation difficulty in forensic cases with regard to determining the relative contributions of each are traumatic diffuse axonal injury, hypoxic-ischemic axonal injury, and brain swelling with associated increased intracranial pressure and herniation syndromes. Two or more of these three conditions may be present simultaneously in the same forensic case, and all three are known to independently produce axonal injury (AI).[91]

Careful attention to the clinical history, associated CNS pathology, and especially the distribution of axonal spheroids in the CNS, will usually allow the other conditions listed in Table 3.2 to be readily distinguished from traumatic diffuse axonal injury[89,239] (also see the discussion of this topic in Chap. 6). Another approach which has been suggested is to separate axonal injury into focal AI (generally related to local mechanisms, such as hemorrhage, hypoxic-ischemic focal injury, and focal trauma) and nonfocal AI (such as seen with diffuse traumatic axonal injury, diffuse hypoxic-ischemic encephalopathy).[2] It is also recommended that traumatic axonal injury (diffuse or focal) be so designated, and that the basis for any form of nontraumatic AI be specifically indicated in the report (e.g., hypoxic-ischemic DAI), if identifiable.[293]

β-Amyloid Precursor Protein Immunostains

The last issue to consider in this section is the use of β-APP immunostains in the diagnosis of AI.

For many years, axonal spheroids were diagnosed primarily with H&E-stained sections or silver stains for axons. By H&E methods, axonal spheroids due to trauma could usually be recognized at anywhere from 12 to 24 hours' post-trauma interval,[82] to 18 to 24 hours' post-trauma interval.[87] With silver stains, the earliest axonal spheroid recognition has been estimated as from 8 hours[31] to 15 to 18 hours' post-trauma,[82] although axonal varicosities may be recognized earlier (12–24 hours).[100] Sparse references, such as Anderson and Opeskin,[16] expressed the opinion that H&E stains, together with silver axon stains (such as Garvey's), are more useful than β-APP in the diagnosis of early traumatic AI, indicating the former two methods demonstrated axonal swelling and spheroids in post-trauma survival periods of less than 1 hour. The results of most investigators, however, are consistent with the previously cited general time frames for the earliest recognition of axonal spheroids by H&E or silver stains.

Attempts to diagnose AI with greater accuracy or at earlier stages of evolution have included several approaches over the years, such as using antibodies against neurofilament protein,[31,127,132,331] α-synuclein,[255] and ubiquitin.[127,265] Efforts to use chromogranin A, synaptophysin, anti-cathepsin D, SNAPP-25, GAP-43, and tau either did not stain injured axons, or proved to have no advantage over the previously listed stains.[334]

The stain that has received the most attention throughout these various efforts is β-APP. Proponents of using this method to diagnose the earliest stages of AI tend to emphasize the following points:

Table 3.2. Conditions in Which Axonal Spheroids May Be Found

1. Reactive and degenerative axons (e.g., post-traumatic).
2. Regenerative axons (e.g., growth cones of proximal axon sprouts during peripheral nerve regeneration).
3. Dystrophic axons.
    a. Physiologic, seen increasingly as the brain ages, most commonly in the gracile and cuneate nuclei, zona reticulata of the substantia nigra, inner segment of globus pallidus, spinal cord anterior horns, sympathetic ganglia, and dorsal root ganglia.
    b. Degenerative diseases primarily represented by axonal dystrophy. This group includes conditions such as Hallervorden-Spatz disease; Nasu-Hakola disease; giant axonal neuropathy; and the various infantile, juvenile and adult forms of neuroaxonal dystrophy.
4. Miscellaneous disorders that include axons in one or another of the above, more distinct categories include a variety of systemic degenerations (e.g., amyotrophic lateral sclerosis, inherited and acquired cerebellar degenerations, Huntington's disease), numerous disorders producing necrosis (e.g., infectious [including malaria], multifocal necrotizing leukoencephalopathy), demyelinating diseases (e.g., multiple sclerosis), inherited and acquired metabolic disorders (e.g., hypoglycemia, vitamin E deficiency, Niemann-Pick disease type C), hypoxic-ischemic injury (including focal disorders such as various types of emboli; global hypoxic-ischemic encephalopathy), a wide variety of neurotoxins, other conditions (e.g., Menke's disease), and radiation injury.

1. This method, or some variation of this method, can identify AI earlier than other methods, particularly when AI is defined simply as positivity with β-APP in axons without other evident pathology by light microscopy. Methodologic differences, together with defining AI as axonal β-APP positivity only, may account for AI being diagnosed as early as 30 minutes' post-trauma in spinal cord injuries,[64] and within 35 to 60 minutes' postinjury using an antigen retrieval method.[12] AI has been detected within 1.75 to 2 hours' postinjury when defined as β-APP positivity as well as presence of axonal swelling in continuity, and axonal spheroids detected as early as 3 to 6 hours' postinjury with β-APP.[135,136,230,237]
2. β-APP is claimed to be more sensitive than other methods in detecting AI, in that more axons demonstrate positivity with this method than other methods applied to a given case.[124]
3. It can be useful to confirm post-trauma survival for "at least 2 to 3 hours' postinjury."[124]
4. Absence of β-APP staining in cases with greater than 3 hours' survival "suggests there has been no significant axonal damage."[124]
5. β-APP stains only injured axons, and can be seen in injured axons that are normal in size, before axonal swellings and axonal spheroids develop.[2]

The β-APP method is an interesting and useful approach in experimental neuropathology. Our personal experience with β-APP immunostaining is limited. If conclusions in testimony in forensic cases are to be based on the results of this method, however, certain issues need further clarification:

1. Some proponents of this method diagnose AI based solely on β-APP immunostain positivity of axons normal in every other way, or with only subtle enlargements in diameter. Axons may demonstrate progressive swelling as postmortem interval increases, when studied by silver stains.[66] Since β-APP accumulation within axons sufficient to result in positivity with β-APP immunostains is considered to be an energy-requiring process, and thus an indicator of antemortem AI, postmortem increase in axonal diameter is thus not considered by some investigators to be a confounding factor.[234]

Perhaps approaches using β-APP positivity as the sole criterion for AI would inspire greater confidence if studies attempting to produce β-APP axonal positivity during the early postmortem period analogous to, for example, the extensive experiments of Cammermeyer in attempting to produce dark neurons, were available (see discussion of dark neurons in Chap. 10). For example, studies using silver stains have demonstrated that enlarged, sinusoidal axons resembling those seen following closed head injury can be readily produced by postmortem stretching of brain tissue;[370] could such altered axons also demonstrate increased affinity for β-APP stains? A second concern is that it is not known whether β-APP positivity in otherwise normal-appearing axons, or even in axons with mild swelling or varicosities, inevitably progresses to secondary axotomy. Could such early indicators of AI be reversible and not contribute significantly to clinical symptomatology or fatal outcome?

2. If β-APP staining of isolated, scattered, or small groups of axons in the absence of axonal spheroids, and only very limited (i.e., two slides) sampling of the CNS, is used as an indicator of AI as occurs in some publications, it allows AI to be diagnosed in an individual who died "instantly" after a gunshot wound to the temple.[258] Most investigators recommend a wider spectrum of areas sampled in the brain, with a requirement for at least some axonal abnormality present in addition to positive β-APP staining, in order to make a diagnosis of DAI.

3. Given the various time frames and criteria used for AI, it is clear that using β-APP staining as an indicator of survival period should be done cautiously, with very precisely defined criteria with regard to staining method used, presence or absence of additional abnormalities besides β-APP positivity, etc.[124] Careful attention to the possibility of prior injury (i.e., more than one injury, of varying duration) includes review of witness statements as well as histologic data.

4. If the only reliable method to determine significant axonal injury leading to secondary axotomy by β-APP also requires identification of definite axonal spheroids by this method,[5,122,123] the relative benefit of β-APP compared with H&E stains is somewhat diminished, since the earliest development of definite axonal spheroids in the view of most investigators is substantially longer than the earliest evidence of β-APP positivity in axons.

5. Axonal injury following a single traumatic episode may not be a simultaneous, all-or-none phenomenon. There is evidence indicating continued recruitment of damaged axons for at least 99 days' postinjury.[230] Limited staining at an early stage after injury could, in this instance, underestimate the degree of AI.

6. As with other staining methods used to detect axonal injury, β-APP does not distinguish between axonal injury of various causes (e.g., trauma, hypoxic-ischemic injury, hypoglycemia).[124] Differences in location and pattern of β-APP positivity are helpful in distinguishing, for example, traumatic causes from hypoxic-ischemic or hypoglycemic injury, but the distinction is not always possible when survival periods allow the effects of secondary complications to combine with the effects of the primary brain insult.[91,123,124,191,265,293,295]

7. As noted in the section on primary and secondary consequences of head injury (see Chap. 6), medical examiners are unlikely to see "pure" AI in isolation from the numerous other sequelae that may occur following head trauma. Whether or not AI is demonstrated in short survival cases, there is typically brain swelling, herniation syndromes, hemorrhage, prominent pulmonary edema, myocardial contraction band necrosis, or other findings

sufficient to determine cause or manner of death. Trauma sufficient to cause death so rapid that there is not an opportunity for axonal injury to be demonstrated will usually provide other anatomic markers of injury. If diffuse traumatic brain injury is sufficient to cause rapid death, the majority of these cases will also demonstrate diffuse traumatic vascular injury. In those cases where post-traumatic apnea or catecholamine surge leading to a fatal cardiac arrhythmia was the responsible mechanism for rapid death, does the ability to demonstrate some β-APP positivity in otherwise normal axons in the brain aid the determination of cause or manner of death? β-APP studies specifically addressing these latter conditions are not yet sufficiently persuasive for or against its routine use under such circumstances.

8. β-APP is not quite as specific as implied in many publications. Unpredictable and variable degrees of β-APP positivity have been described in pediatric cases (where questions of AI in nonaccidental trauma frequently arise). In such instances, β-APP positivity has been described in the neuron soma, in circulating blood cells within vessels and cells comprising blood vessel walls, in meningothelial cells, choroid plexus cells, ependymal cells, tanycytes, astrocytes, oligodendroglia, dorsal root ganglia neurons and satellite cells, peripheral myelin, and cells within the cerebellar internal granular layer.[294] These investigators also demonstrated a perivascular axonal reactivity pattern for β-APP, but did not consider it to be evidence of trauma. Others have described β-APP positivity in neuronal perikarya and in senile plaques.[123]

If continuing studies verify that only significantly injured axons decorate with β-APP, and not axons with only minor and reversible injury unaccompanied by physiologic conduction abnormality, the technique may yet be generally accepted as a routine part of the evaluation of certain forensic cases. Some believe that it has already achieved such status.[124] Others,[78] including ourselves, have not quite reached that conclusion. From a practical standpoint, factors other than our interest in this subject and desire to employ this and other interesting techniques on a wide variety of our forensic cases must be balanced against the reality that any use of expensive immunostains must compete with other demands on limited funds, such as toxicology studies, DNA studies for identification of remains, and so on. Over the years we have witnessed the introduction of many special neuropathology stains originally thought to be reliable and specific in some fashion, only to have further experience prove disappointing (e.g., neuron-specific enolase). Continuing studies on β-APP and other new developments are anticipated.

Traumatic Axonal Injury

Estimates of the timeline of the histologic sequence of events in AI vary, no doubt in part due to variables as outlined in the introduction to this chapter. In human studies, a major variable is the type of staining method used (see preceding section on β-APP). In the following outline for dating/aging of traumatic AI, the stain and/or method employed for a given survival period is specified. The data in this outline are preliminary, since it is based on relatively few cases. However, it is representative of the extremes of time frames published.

None of the following gross or microscopic findings are pathognomic of traumatic DAI. It is the history, general autopsy findings, and totality (and especially the topography) of neuropathologic findings that allow one to diagnose traumatic AI or DAI.

Gross Appearance of Diffuse Axonal Injury

Macroscopic abnormalities are seen only in more severe grades of traumatic DAI.

Gross Appearance of Traumatic Axonal Injury

Survival Period

- Acute (minutes to a few days)[87,135]
  — Hemorrhages in more severe forms of traumatic DAI are most typically seen in anterodorsal frontal and anterior temporal white matter, corpus callosum, posterior limb of internal capsule, superior cerebellar peduncle, and dorsolateral quadrants of rostral brainstem.
  — Gliding contusions may also occur.
- "Several days"[135]
  — Lesions previously hemorrhagic become granular and less obvious grossly.
- 2 to 3 months
  — May develop a decrease in white matter bulk in anterior cerebral hemispheres, thin corpus callosum, and compensatory ventricular enlargement.[136]
- Chronic
  — Lesions may become shrunken and sometimes cystic (illustration used in cited reference is of a case at 21 months' survival)[135]; faint hemosiderin staining may occur.

Microscopic Findings in Axonal Injury, Including Observations in Diffuse Axonal Injury

Survival Period

- 30 minutes[64]
  — Earliest positive staining of otherwise normal but presumably injured axons following spinal cord trauma (β-APP-positive material seen "up to 50 years'" post-trauma in this series) (β-APP stain).
- 35 to 60 minutes[12]
  — Earliest positive staining of injured axons in brain (whether or not other axonal morphologic abnormality was required for the diagnosis is not described) (β-APP antigen retrieval method).

- < 1 hour[16]
  — Axonal injury changes positive with H&E, silver stain, and β-APP stain.
- 1 hour[381]
  — Positive axons present without other axon morphologic abnormality (β-APP antigen retrieval method).
- 1.75 to 2 hours[230]
  — Earliest staining of injured axons when axonal swelling in continuity is also required for the definition of axonal injury (β-APP stain).
- ~ 2 hours[86]
  — Axonal injury positive (β-APP stain).
- 2.5 hours[381]
  — β-APP positivity of axons with additional finding of axonal swellings first seen (no distinction was made between axonal swellings and bulbs).
- 2 to 3 hours[124,136]
  — Axonal injury first positive with β-APP in all cases with 2 to 3 hours' survival. β-APP remained positive in axonal spheroids for about a month, although granular β-APP deposits (not definitely in axons) could be seen with "longer survival."
- 3 to 5 hours[237]
  — First evidence of β-APP positivity in axonal bulbs (spheroids) in traumatic axonal injury.
- 4 to 5 hours[100]
  — First evidence of β-APP positivity (and positivity with other distally transported proteins) in axonal injury.
- 5 hours[123]
  — Focal linear positivity with small axonal bulbs (β-APP stain).
- ~ 5 hours[122]
  — Axonal varicosities and bulbs (spheroids) first seen with β-APP stain.
- 6 hours
  — Axonal injury first seen with ubiquitin stain.[82]
  — Focally enlarged immunoreactive axons with axonal infolding or disordered neurofilaments (using immunostain for neurofilament subunits, light microscopy of plastic embedded sections, and electron microscopy).[60]
- 12 hours[60]
  — Axonal disconnections seen in some axons, with focally distended and swollen axonal segments lacking apparent continuity with their distal process (neurofilament subunit stain, light microscopy of plastic-embedded sections, and electron microscopy).
- 12 to 18 hours[124]
  — Axonal spheroids first positive with silver stains.
- 12 to 24 hours
  — Axonal spheroids first seen by H&E stain.[82] Axonal varicosities first seen with silver stains.[100]
- 18 to 24 hours
  — Axonal spheroids first seen by H&E.[86]
  — Axonal spheroids first seen by H&E and silver stains.[122]
- 15 hours
  — First detection of axonal injury by silver stains.[237]
  — First detection of axonal injury by H&E and silver stains.[135]
- 15 to 18 hours
  — Numerous axonal swellings adjacent to focal traumatic lesions by silver stain.[136]
- 24 hours
  — Axonal bulbs large, and strongly positive with β-APP.[123]
  — Axonal spheroids first seen by H&E and silver stains, and may be seen up to 2 months' postinjury.[100]
  — Axonal spheroids identified by β-APP positivity tend to increase in number up to about this time, and decrease in number thereafter, being identifiable up to about 3 months' postinjury.[136]
- 30 hours (~ 1¼ days) and 60 hours (~ 2.5 days)[60]
  — Axonal disconnections frequent. Further enlargement of reactive axon tip, surrounded by myelin sheath in some profiles (neurofilament subunit stain, light microscopy of plastic-embedded sections, and electron microscopy).
- ~ 48 hours[124]
  — Microglial hyperplasia first seen (CD68 stain for microglia).
- 3 days[123]
  — Axonal bulbs remain as large as when seen at 24 hours (β-APP).
  — Subjective increase in size and number of microglia first seen (CD68).
- 85 hours (~ 3½ days)[381]
  — Survival period at which mean size of axonal swellings reached maximum diameter, and plateaued thereafter. The size variations suggested to these authors possible sequential axonal injury from secondary complications following the original injury, or differences due to original size of injured axons, or variations in development phase of secondary axotomy in injured axons. They note that this observation should always be "assessed with other evidence" in the forensic setting, with regard to dating/aging issues.
- 88 hours (~ 3.7 days)[60]
  — Changes similar to those described at 30 and 60 hours, but with more heterogeneity of severity of changes in various axons (mild to severe) (neurofilament subunit stain, light microscopy of plastic-embedded sections, and electron microscopy).
- "Days"[4,5]
  — Large numbers of eosinophilic and azurophilic swellings on nerve fibers (referring to axonal

spheroids), and coarse varicosities in axons (H&E, silver stain, cresyl fast violet, and Marchi stain).
- 5 days[123]
  — Aggregation of microglia around axonal bulbs first seen (CD 68).
  — No astrocyte response seen up to 5 days' survival (glial fibrillary acidic protein [GFAP]).
- "After ~ 1 week"[124]
  — Axonal spheroids consistently show reduced β-APP positivity, but adjacent varicose axons continue to be strongly positive.
- 8 days[123]
  — Some axonal bulbs are pale or unstained (β-APP).
- "Beyond 8-day survival"[123]
  — Marked reactive astrocytosis, but no aggregation of astrocytes around individual axonal bulbs or areas of damage (GFAP).
- "~ 10 to 11 days"[123,124]
  — Microglial clusters first seen. These were only occasionally present, but by 14 days' survival were present in considerable numbers (also referred to as microglial stars) (CD68).
- 10 to 14 days[122]
  — Marked variation in intensity of β-APP staining in different parts of the same brain. One case at 14 days had β-APP-negative axonal bulbs in the pons, but strongly positive axonal bulbs in the substantia nigra.
- ~ 30 days[124]
  — All β-APP staining in axons disappears, with only occasional small granules of positive β-APP material seen in white matter.
- "Few weeks"[4,5]
  — Many small clusters of microglia present throughout involved white matter; damaged axons and myelin sheaths are fragmented, and axonal spheroids are rarely identifiable (H&E, silver stain, cresyl fast violet, Marchi stain).
- "After a few weeks"
  — Large numbers of small clusters of microglia present throughout injured white matter.[136]
  — Lipid-containing macrophages first seen in long tracts undergoing secondary degeneration (stain unspecified).[124]
- 7 weeks[123]
  — Occasional axonal bulbs still detectable with β-APP, and occasional small granular clumps of β-APP-positive material, some in macrophages, present between nonstaining axons.
  — Microglial clusters (stars) seen in all cases of survival between 7 weeks and 5 months, but not seen in a case surviving 18 months (CD68).
- 7 weeks to 18 months[123]
  — Abundant foamy macrophages (stain unspecified) and diffuse microglial proliferation in corpus callosum and long tracts (CD68).

- > 2 to 3 months[136]
  — Wallerian degeneration in long tracts present (Marchi method).
- Beyond 3 months[123]
  — Marked diffuse astrocytic gliosis in corpus callosum, hemispheric white matter, and long tracts (GFAP).
- 99 days (~ 3½ months)[32,33]
  — β-APP positivity persisted (implied to be in axonal "bulbs").
- >6 months[123]
  — No β-APP positivity remained except in one case with 3-year survival, demonstrating scattered granules of β-APP positivity in severely atrophic corpus callosum.
- Beyond 1 year[122]
  — Only occasional granular extracellular deposits of β-APP remain.

Puncture Wounds in the Brain

To our knowledge, only one article specifically deals with dating/aging of human brain puncture wounds. The report by Baggenstoss et al.[20] was based on the study of surgical, sterile ventricular needle puncture wounds performed for diagnosis and localization of brain tumors. The authors noted that cerebral swelling, and in some cases herniation syndrome, was present. They examined a total of 70 cases, with wound tracts evaluated in both longitudinal and cross-sectional views in both gray and white matter. Patients ranged in age from the first through the sixth decades. Stains used included H&E and a wide variety of special stains for individual CNS cell types, including Cajal's gold sublimate method for astrocytes, modifications of Hortega's silver impregnation method for oligodendroglia and microglia, and several other special methods. The reader is directed to their report for details. Compensating to some extent for our inability to locate other studies specifically attending to the dating/aging reactive changes following sharp force injury in the human brain is the attention of these authors to detail.

Microscopic Examination of Puncture Wounds in the Brain

Survival Period
- 1 day or less (12 cases)
  — Neurons became acidophilic, with karyolysis or pyknosis; swelling and fragmentation of myelin sheaths and axons.
  — Central zone of hemorrhage and necrosis, with swelling and disintegration of glial cells.
  — Surrounding 1.5- to 2.0-mm zone of edema, dilated capillaries, and many focal hemorrhages, often perivascular.

- Occasional necrotic vessel walls with perivascular fibrin.
- Neutrophils vary in number and appear to be involved in phagocytosis of debris.
- Swollen oligodendroglia.
- A few large mononuclear cells present but no compound granular corpuscles (i.e., no definite macrophages).

• 2 days (5 cases)
- Marginal zone demonstrated more prominent edema, vascular congestion, and perivascular hemorrhages.
- Swollen endothelial cells frequently present.
- Border zone with normal brain often abrupt.
- Increased numbers of neutrophils; few mononuclear cells with lipid droplets contained therein.
- Spherical and irregular acidophilic masses interpreted as probable remnants of nerve cells and myelin sheaths.

• 3 days (5 cases)
- First definite appearance of compound granular corpuscles (macrophages), few in number, and generally in perivascular areas.
- Neutrophils still prominent.
- Microglia not obviously reactive at this point.
- Some mononuclear cells present.
- Occasional mitotic figures; cell type uncertain.

• 4 days (6 cases)
- Definite capillary endothelial proliferation, with mitotic figures.
- First appearance of gemistocytic astrocytes.
- Increased compound granular corpuscles, often closely related to blood vessels.
- Some hypertrophy of microglia.
- Increased numbers of neutrophils.

• 5 to 7 days (12 cases)
- Findings as above; more prominent capillary endothelial proliferation; numerous macrophages and neutrophils.
- Occasional lymphocytes and rare fibroblast.
- More prominent gemistocytic astrocytes.
- Many nerve fibers swollen and fragmented; some axons showing bulbous ends, corkscrew deformities, and other irregularities.
- Up to this point, the wounds demonstrate two main zones: central hemorrhage and necrotic zone; peripheral edema with perivascular hemorrhages.

• 8 to 10 days (8 cases)
- More prominent endothelial cell and capillary proliferation.
- First appearance of significant numbers of fibroblasts and collagen fibers.
- Many macrophages.
- Decreased edema in marginal zone.
- Prominent neutrophils and macrophages persist.
- Definite increased numbers of lymphocytes, usually perivascular.
- More prominent astrocytic hypertrophy, including multinucleated forms.
- Oligodendroglial swelling persists.
- Neuronal degenerative changes persist, including satellitosis and neuronophagia.

• 12 to 14 days (8 cases)
- Zone of edema and perivascular hemorrhages now largely replaced by a zone of capillaries and increased numbers of proliferating fibroblasts together with macrophages, more prominent in gray matter than in white matter.
- Maximum number of macrophages observed at approximately 10- to 14-day survival.
- Neutrophils decreased in number; lymphocytes increased in number.
- Plasma cells first seen, occasionally binucleate.
- Prominent astrocytic hypertrophy.
- Nerve cell degenerative changes now less conspicuous.

• 19 to 25 days (5 cases)
- Three wound zones now more obvious: central, partly cystic zone with necrotic debris being removed by macrophages; middle zone of capillaries with the overall appearance of this zone resembling granulation tissue, with many fibroblasts and reticular and collagen fibrils, macrophages, lymphocytes, some plasma cells, and sparse neutrophils; and peripheral zone of reactive astrocytes.

• 30 to 58 days (4 cases)
- Less endothelial cell and fibroblastic proliferation but more reticular and collagen fibrils.
- As in shorter survival times, reactive changes more prominent in gray than in white matter.
- Near-complete removal of necrotic debris in most cases.
- Persistent macrophages.
- One wound (58 days) showed persistent islands of necrotic tissue and neutrophils.
- Macrophages and lymphocytes the predominant cell types in the capillary zone; few plasma cells persist.
- Hypertrophied astrocytes (including gemistocytes) persist in outer zone.
- Degenerative changes (unspecified) in neurons at margin of wound still rarely seen.
- Ependymal cells at wound margin flattened or absent (also true in wounds of shorter survival periods).

• 75 to 80 days (2 cases)
- Somewhat narrower capillary zone.
- White matter portion of wounds show almost no connective tissue response, in contrast to gray matter.

- 196 to 206 days (2 cases)
  — Increased numbers of hypertrophic astrocytes in peripheral zone, often commingled with collagen fibers; some astrocytes multinucleated; others piloid in appearance.
  — White matter demonstrates cystic spaces surrounded by hypertrophied astrocytes, with "hardly any evidence" of mesodermal reaction.
  — Some macrophages still present.
- 7 years (1 case)
  — Central portion of wound loosely filled with collagen fibers.
  — No macrophages, lymphocytes, or plasma cells.
  — No capillary zone apparent; no obvious increase in astrocytes.
  — Wound margin hypocellular and atrophic.

The authors anticipated the presence of, and specifically searched for, differences in the reactive changes in children compared with adults. No significant differences were found based on age range, however.[20]

Review of dating/aging studies, as outlined in various sections of this text, reveals minor variations for many of the cellular components in one type of lesion compared with another. Some of these discrepancies, however, cannot be easily reconciled. For example, "piloid" astrocytes are described as initially occurring at 6-day survival in contusions,[264] but not until 196 days or more in ventricular needle wounds[20] (see later section, "Astrocyte Terminology" for further discussion of this point).

Intracerebral Hemorrhage

For a discussion of the etiology of intracerebral hemorrhages, the reader is referred to Chapter 6 on blunt force head injury, and to Chapter 11 on vascular diseases of CNS.

Volume of Intracerebral Hemorrhage

Adams and Sidman[7] indicate that a putamen hemorrhage may reach a volume of 100 ml, but in the brainstem a 5- to 10-ml hemorrhage can be fatal. Of course, the degree and pace of the secondary events will also influence whether the patient survives the primary hemorrhage. A very small hemorrhage (or infarction) involving a cardiac or respiratory center in the brainstem may be fatal with less than 5 ml in volume. A putaminal hemorrhage that ruptures into the ventricle may be rapidly fatal.

Histologic Examination of Intracerebral Hemorrhage

Studies specifically commenting on the sequence and timing of cellular reactions to human intracerebral hemorrhage are relatively sparse in comparison to those available for subarachnoid hemorrhage, cerebral infarction, and cerebral contusion. It should also be noted that many infarcts contain some degree of hemorrhagic component, and the brain at the margin of a cerebral hemorrhage will demonstrate varying degrees of secondary effects, such as infarction and edema due to vascular compromise from the mass effect of the hemorrhage. Thus, one is not necessarily dealing with the response to a pure ischemic or pure hemorrhagic lesion in such cases. Review of the small amount of data available in the literature addressing this specific circumstance does suggest, however, some minor differences in the cellular reactions to lesions that are primarily characterized by a sudden extravasation of blood into otherwise intact or only mildly bruised cerebral parenchyma, and those in which contused or ischemic brain tissue comprises the primary component of the lesion (also see discussion later in the section on the dating/aging of adult cerebral infarction).

Microscopic Examination of Intracerebral Hemorrhage

Survival Period
- Hours to a few days
  — Edema present in brain parenchyma surrounding hemorrhage.[116]
  — Swollen oligodendrocytes in immediate vicinity of hemorrhage (using Penfield's method for oligodendroglia and microglia).[289]
  — The pace of development of clotting in the initial liquid hemorrhage, or pace and completeness of subsequent liquefaction of clot, is quite variable from person to person.[116]
- 1 day
  — Necrosis of neurons and glia present at hemorrhage margin.[174,175]
- 2 days
  — First appearance of neutrophils (said to be slower in intracerebral hemorrhage than in infarction).[174,175]
- 3 to 4 days plus
  — Definite macrophage response is not seen until after 2 to 3 days, and seems to be delayed until some lysis of RBCs begins. Both RBC fragments and intact RBCs are phagocytosed. Once present, macrophages may persist for greater than 7 months.[289]
  — Inflammatory cell infiltrates appear in response to RBC lysis at approximately 4 to 10 days (mainly neutrophils and "activated microglia").[116]
  — Hemosiderin in macrophages (positive Prussian blue stain) seen about 1 day after initiation of phagocytosis, and may persist in both macrophages and astrocytes in chronic lesions.[174,175]
- 1 week
  — Astrocyte proliferation seen.[174,175]

Other general observations include the clot being resorbed at an estimated pace of "0.7 ml/day."[174,175] Ferrugination is seen in neurons in the margins of some chronic brain wounds, but its earliest appearance in hemorrhagic lesions is not specified other than the general statement that it is not seen until after 3 weeks' survival[291] (see earlier section on dating/aging of brain contusions for additional data that may apply to primarily hemorrhagic lesions).

In summary, the sparse data available on primarily hemorrhagic human intracerebral lesions, traumatic or spontaneous, suggest that the first appearance of both neutrophils and macrophages may be slower by anywhere from 1 to 3 days in comparison to infarcts and contusions. Studies addressing this specific issue in more detail are needed before such conclusions can be used with confidence in a forensic setting.

Spinal Cord Traumatic Lesions

The database for dating of traumatic lesions of the human spinal cord is much more limited compared with cerebral contusion. No single report cited provides a comprehensive overview of this topic but, collectively, they do allow for some generalizations.

The 65 cases studied by Bruce et al.[41] varied in age from 3 months to 86 years, varied in survival time from 3 hours to 24 years, and varied in severity from grossly intact cords to anatomically complete cord lesions. Croul and Flanders[67] examined eight cases, all adults, with cord injury resulting from vertebral fractures or fracture-subluxation. Their patient population varied in survival time from 5 days to 5 months, and varied in severity from unilateral dorsal horn involvement to complete anatomic transection. Kakulas[173] examined 564 cases, with lesion severity varying from grossly intact to completely transected cords, and survival times (when specified) varying from dead on arrival at the hospital to 38 years' survival. Quencer et al.[287] studied three cases, ages 65, 67, and unspecified, with survival times of 3 days, 6 weeks, and 7 months. Their cases were selected on the basis of clinically meeting criteria for a "central cord syndrome," and no gray matter lesions or hemorrhage were found in any of their cases. The injury mechanism was believed to be traumatic hyperextension in patients with preexisting spinal stenosis. Finally, Norenberg et al.[261] examined 180 cases of spinal cord blunt force trauma, age range 8 months to 92 years, and with survival periods ranging from "almost instantaneous" death to 51 years' survival. Cord lesion severity varied from grossly normal cords to near-complete cord transection.

Microscopic Examination of Spinal Cord Trauma

Survival Period
- 1 to 2 hours
  — Hemorrhage, vasodilation, and congestion[261] (Case 3.5, Figs. 3.17–3.22).
- > 2 hours to 3 days[261]
  — Hemorrhage, edema (gray greater than white matter).
  — Inflammation (mild influx of neutrophils within 1 day, reaching peak numbers at 2 days, and most gone at 3 days).
  — Acidophilic neurons present.
  — Axonal spheroids present as early as 1-day survival.
  — Myelin breakdown, oligodendrocyte apoptosis, and necrosis.
- < 24 hours
  — Axonal damage detected by β-APP (irregular varicose axis cylinder enlargement).[173]
- 48 to 72 hours
  — Macrophages first seen.[173]
- 3 days
  — Edema in white matter, with early myelin and axon destruction, centered in the lateral columns (especially corticospinal tracts).[287]
- 4 to 5 days
  — Astrocytes appear, and "multiply" in 5 to 7 days.[173]
- 5 days[67]
  — Hemorrhage (either gray matter, white matter, or both); focal petechiae to widespread cord involvement of hemorrhage.
  — Necrosis within, and peripheral to, hemorrhage.
  — Inflammatory cell infiltrates with lymphocytes greater than neutrophils in two cases, and mononuclear cells noted only in gray matter in one case.
  — Edema present.
- Days to weeks[261]
  — Microglial activation occurs as early as 1-day survival (enlarged, pale, with pleomorphic nuclei on H&E), and gradually transform to macrophages over the "next several days." Macrophages, once present, can persist from weeks to months.
  — Hypertrophy of astrocytes begins in "several days," peaks at 2 to 3 weeks, and gradually thereafter astrocytes develop more extended, fibrillar processes leading to a glial scar.
  — Increased numbers of blood vessels in lesion by 7 to 10 days, with hypertrophic vessel walls.
- 13 days (1 case)[67]
  — Hemorrhage, with necrosis in margin.
  — Edema.
  — Lymphocytes present.

Case 3.5. (Figs. 3.17–3.22)

A 77-year-old man slipped and fell while taking a shower. His wife thought he fell on his back or buttocks. He lodged no complaint and was able to take his dog out for a walk. Returning home and upstairs, he was heard to collapse. He was found flat on his back and was pronounced dead at the scene.

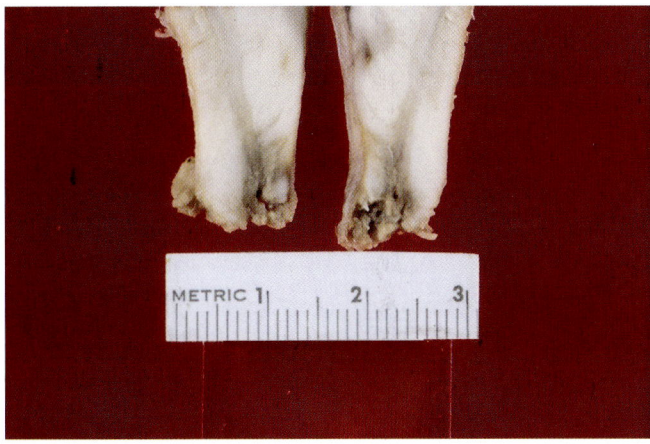

Fig. 3.17. Closeup view of midsagittal section at cervicomedullary junction laid apart shows faintly hemorrhagic necrosis at medial (anterior) halves of cord.

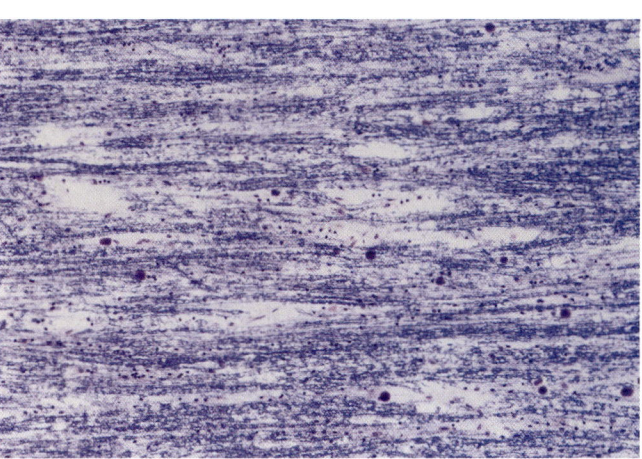

Fig. 3.19. Medium-power micrograph of internal control of myelinated tract rostral to level marked by arrows in Fig. 3.18. Note uniformly straight myelinated fibers. (Luxol fast blue.)

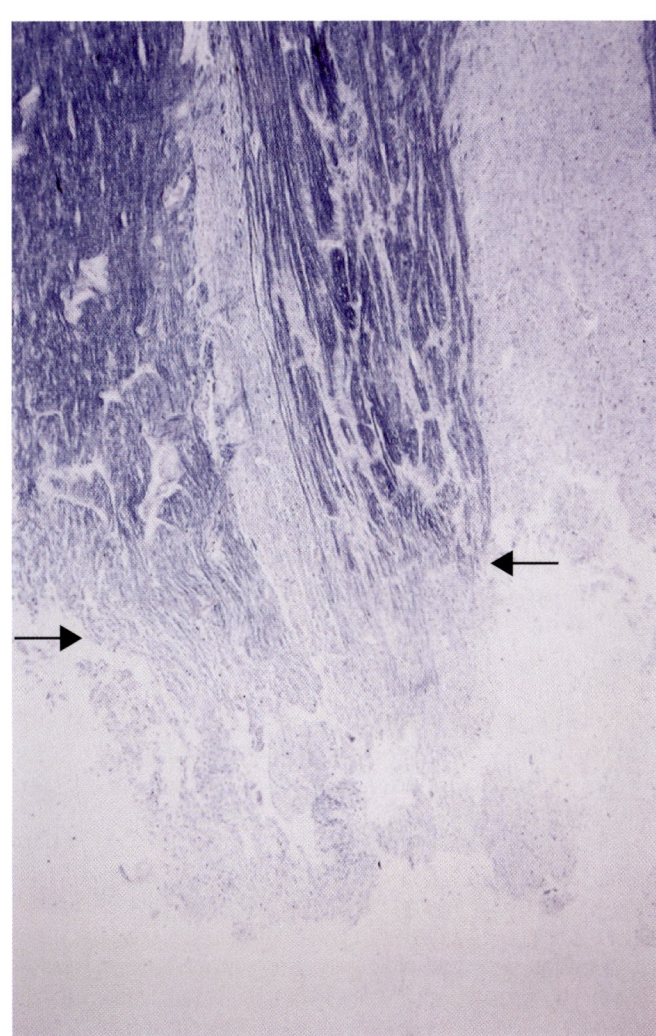

Fig. 3.18. Low-power micrograph shows loss of stainability of myelin caudal to arrows near the point of transection. (Luxol fast blue.)

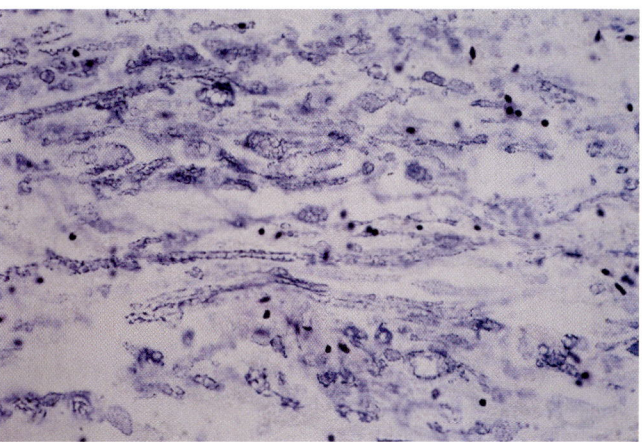

Fig. 3.20. Lower half of Fig. 3.18 (below arrows) taken at high power shows myelinated fibers in disarray at the level of transection, but free of inflammation. (Luxol fast blue.)

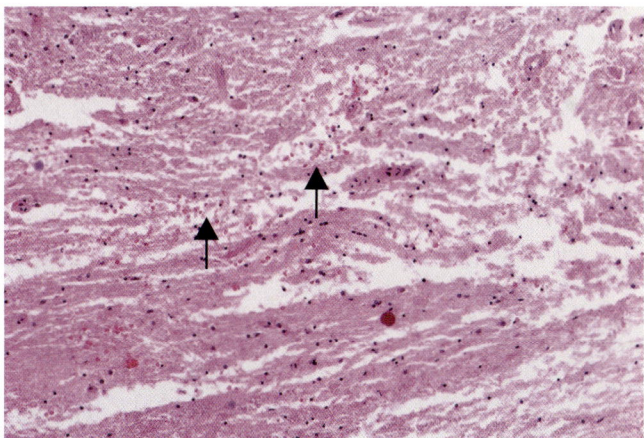

Fig. 3.21. Medium-power micrograph of area of transection shows a few scattered RBCs (*arrows*). (H&E.)

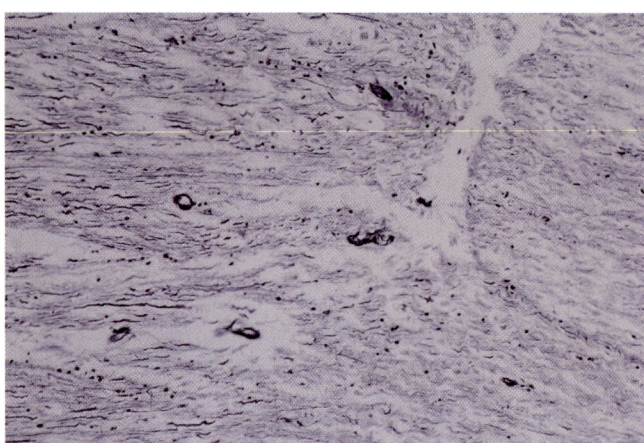

Fig. 3.22. Silver stain of the area shown in Fig. 3.20 shows absence of axonal swelling. (Bodian.)

- 24 days (1 case)[67]
  — No edema.
  — Residual hemosiderin present.
  — Necrosis persists in gray matter and white matter.
  — Macrophages present.
- 1 month (1 case)[67]
  — No hemorrhage, hemosiderin, or edema described.
  — Gray matter necrosis persists.
  — Macrophages persist.
- 6 weeks[287]
  — Disruption of posterior, anterior, and lateral columns, with myelin breakdown and marked axon breakdown most evident in lateral columns.
- 2 months[67]
  — Hemosiderin, macrophages, and cavitation present.
- 5 months[67]
  — Cavitation present, with persistent macrophages.
- 6 to 12 months[173]
  — Wallerian degeneration becomes apparent within 6 to 12 months (see later discussion of CNS wallerian degeneration).
- Weeks to months/years[261]
  — Wallerian degeneration develops.
  — Development of mesenchymal scar, with minor contribution of arachnoid cells.
  — Cystic change may occur, with cyst contents including fluid, residual macrophages, small bands of connective tissue, and blood vessels.
  — Syrinx formation may occur.
- 7 months[287]
  — Loss of myelin and axons in lateral columns (larger axons more affected than smaller diameter axons).

Schwab and Bartholdi[326] suggest that there is a slower evolution of necrotic changes in white matter compared with gray matter, but no further details are specified in their discussion of human material.

Schwannosis

Schwannosis is a general term applied not only to focal intraparenchymal proliferation of Schwann cells and peripheral nerves in patients with von Recklinghausen's disease, but also more frequently applied to proliferation of peripheral nerves and Schwann cells, associated with fibrous connective tissue, entering spinal cord traumatic lesions. Kakulas,[173] although examining a total of 564 cases of spinal cord injury, searched for schwannosis in only 27 of these cases, and found it to be present in 16 of the 27 cases. In this subgroup, the age range was 12 to 52 years and the survival time was from 1 to 52 years.

Bruce et al.[41] examined 65 cases of spinal cord injury varying in age from 3 months to 86 years, with survival times varying from 3 hours to 24 years. Schwannosis was not seen in any case with survival periods of 4 months or less, but did occur in 82% of the cases that had survival times from 6 months' to 24 years' postinjury. They were not able to demonstrate any consistent correlation between the degree of schwannosis and site, severity, or type of cord injury; extent of meningeal or parenchymal fibrosis; or patient age. All of their cases with greater than 4 months' survival were adults.

In a series of 180 spinal cord injury cases, schwannosis was demonstrated in more than 90% of cases with survival periods exceeding 4 months.[261] A trichrome stain can be quite useful in demonstrating this phenomenon (Case 3.6, Figs. 3.23–3.27).

Spinal Shock

Spinal shock is defined in a medical dictionary as a "transient depression or abolition of reflex activity below the level of an acute spinal cord injury or transection."[152] Tator[352] indicates that "spinal shock consists of the loss of somatic motor, sensory, and sympathetic autonomic function due to spinal cord injury."

In order for spinal shock to occur, it is implied if not specifically stated that the responsible spinal cord lesion is either a complete or incomplete, but definite and permanent, anatomic disruption of spinal cord descending pathways that leads to isolation from suprasegmental

Case 3.6. (Figs. 3.23–3.27)

This 45-year-old man died from complications (renal insufficiency and sepsis) following a gunshot wound to the L2 spinal level that rendered him paraplegic 5 years previously.

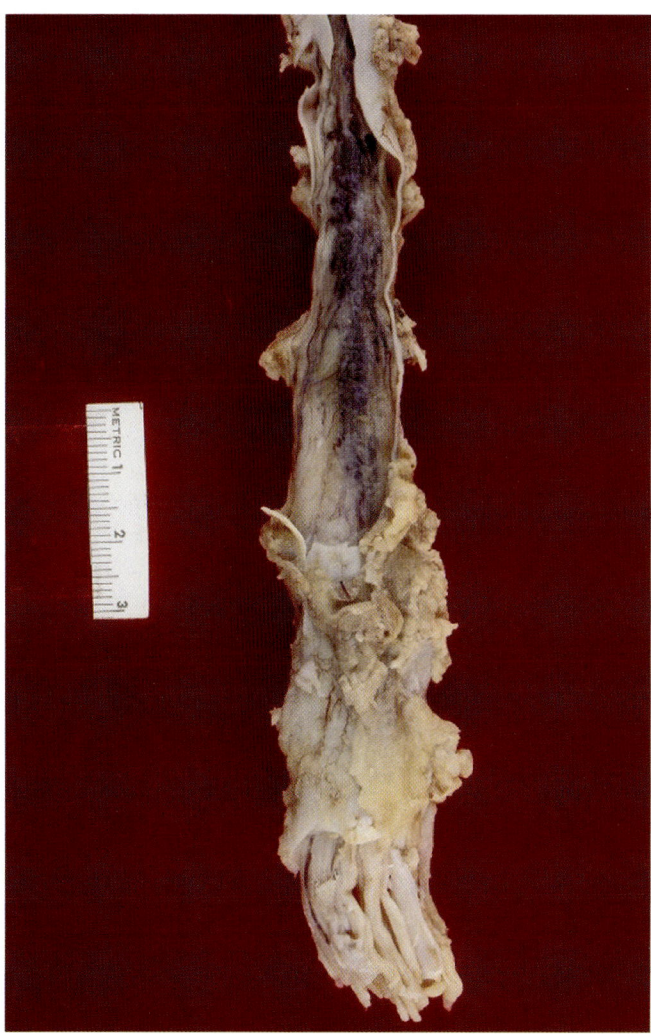

Fig. 3.23. Leptomeningeal hypervascularity and spinal cord atrophy at L1–2 level.

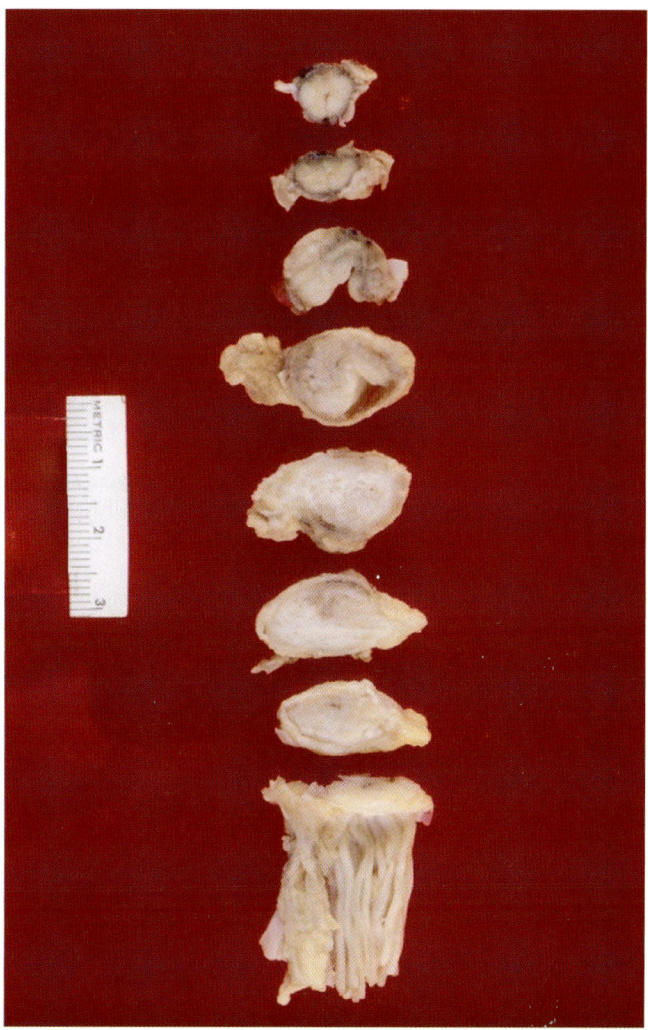

Fig. 3.24. Transverse sections of spinal cord in atrophic region demonstrate meningeal fibrosis and spinal cord atrophy and distortion.

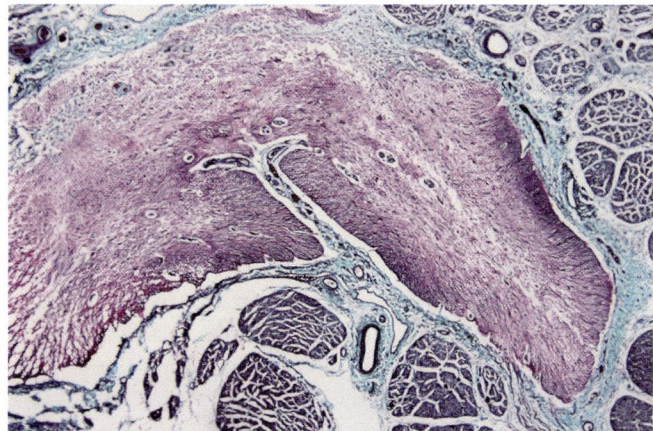

Fig. 3.25. Very low power micrograph of spinal cord, roots, and leptomeninges. Leptomeningeal fibrosis, severe spinal cord atrophy and gliosis, and patchy fibrosis and schwannosis (most readily visible as green tissue invading posterior spinal cord region at upper portion of photo). (Trichrome.)

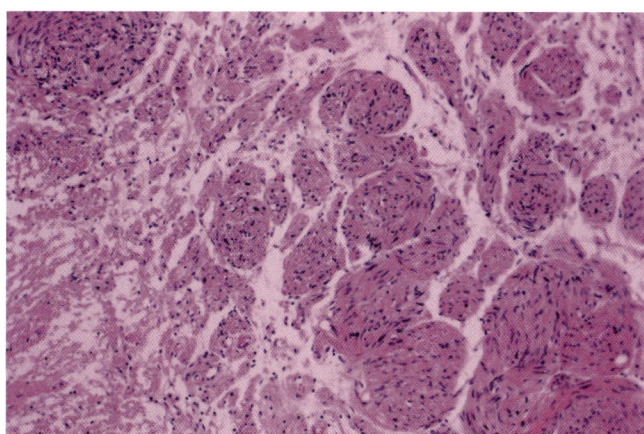

Fig. 3.26. Low-power micrograph. Intramedullary spinal schwannosis in another case of chronic spinal cord trauma. Schwann cell whorls with associated peripheral nerve and fibrous tissue (right and center portions of photo) invade degenerated neural tissue (left). (H&E.)

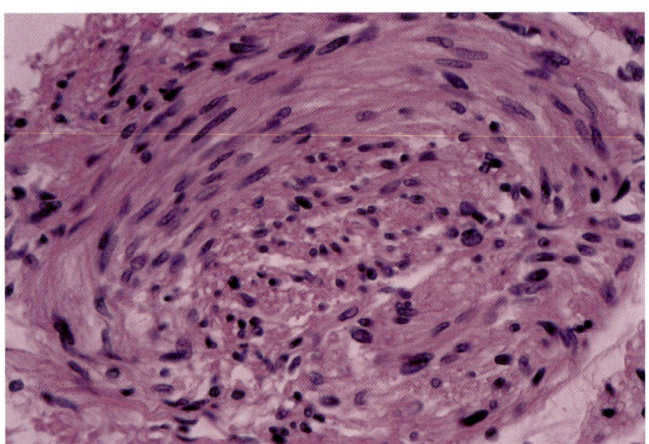

Fig. 3.27. High-power micrograph of case seen in Fig. 3.26. Abundant Schwann cells with more centrally located peripheral neuronal elements. (H&E.)

functional input to spinal cord segments caudal to the lesion. If the lesion is a complete anatomic spinal cord transection, its level determines whether the consequence is permanent paraplegia or quadriplegia. Indeed, most of the earlier studies on spinal shock dealt only with cases exhibiting persistent paraplegia or quadriplegia. More recent studies also include patients with much less severe neurologic deficits. For example, the series of 70 cases of spinal shock reported by Ko et al.[187] included 12 cases with an initial emergency room neurologic examination demonstrating American Spinal Injury Association (ASIA) Impairment Scale grade of D (i.e., preserved motor function caudal to the lesion, with a majority of key muscles below the lesion exhibiting active movements against gravity but not against any resistance).[352] The percentage, if any, of patients with eventual full clinical recovery of motor and sensory function and return to normal reflex status after an initial period of "spinal shock" is unstated in our literature review. This "gray area" of uncertainty of where to place timetable brackets around the category of spinal shock seems to be filled in the acute stage with the term "spinal concussion" (see later discussion).

Even a brief review of the relevant literature reveals marked differences of opinion concerning both intensity and duration of spinal shock. Many of these apparent discrepancies may be explained as follows:

1. Species variation is considerable, with spinal shock tending to increase in prominence as cerebral development increases.[141]
2. Recovery of reflex function after spinal injury may be delayed by other medical problems in the patient, such as urinary tract infections, effects of nonneural associated injuries, persistent shock, or the presence of infected decubiti.[25]
3. There is evidence indicating that age may play a role in the intensity and duration of spinal shock. Allen[10] indicates that spinal shock in the newborn "may last for several months." But, in children (defined in this reference as under the age of 13 years), Burke[46] indicated that the period of spinal shock is much shorter compared with "the usual 3–4 week period seen in adults." He indicated that in many cases (presumably with complete paraplegia) these children have normotonic legs, active reflexes, and even hyperactive withdrawal reflexes with plantar stimulation, with or without abnormal plantar reflexes, within a few hours' postinjury.[46] It should also be kept in mind, when reviewing medical records in childhood cases, that extensor plantar responses are normally present in infants up to 4 to 6 months of age, and, according to some reports, up to 2½ years of age.[243]
4. Spinal shock patterns may vary with the level of the lesion, the completeness of the lesion (partial versus complete transection at a given level), and the fact that even lesions causing a complete transection may vary considerably in their degree of extent of rostral/caudal destructiveness (e.g., gunshot wound versus knife wound, lesion extension due to edema, hemorrhage, vascular compromise with infarction, syrinx formation, infection).[141,171,208,352]
5. It is not sufficient to know simply that recovery of a reflex occurred, perhaps interpreted by a given author as the beginning of a recovery phase of spinal shock. Rather, for forensic questions, we believe that it is helpful to know exactly which reflex or cluster of reflexes were tested, and at what intervals, postinjury. This may allow a more informed opinion as to whether or not it is possible to suggest the timing of an injury in a given case, particularly when correlated with autopsy findings. At a minimum, consultants who seek to provide such opinions may find it helpful to be particularly familiar with the references by Gutmann,[141] Ko et al.,[187] and Ditunno et al.,[83] and also search for subsequent related studies in the literature. Consultation with clinical colleagues in neurology, neurosurgery, or rehabilitation medicine who have experience with spinal shock can also be very helpful in selected cases.

As a generalization, if initially absent in the early stage of spinal shock, recovery of the cremasteric reflex often precedes the return of ankle jerk, the deep plantar response often precedes the bulbocavernosus reflex and Babinski sign, and the Babinski sign generally precedes the recovery of ankle and knee jerks.[187] The evolution of recovery of these and other reflexes after spinal cord injury may vary, however, and the patterns may have some prognostic significance for ambulation potential.[187]

Occasionally, the medical examiner who performs the autopsy will request consultation from the forensic neuropathologist regarding spinal shock. Some questions relate to the likelihood of spinal shock developing after certain types of spinal cord injury, and other questions relate to spinal shock duration (Case 3.7).

Case 3.7.

A 34-year-old man, upon hearing a car backfire, stated that he was "shot" and suddenly began to run around a supermarket parking lot trying to open doors of parked cars, jumped into and then fell out of the bed of a passing pickup truck, then entered a supermarket. He jumped over the market counter and attacked several employees. He was eventually subdued by supermarket employees, and dragged from the building through the front entrance. One employee sat on the buttock area of the prone attacker as he assisted in restraint efforts until police arrived and handcuffed the assailant. Paramedics transported him to an emergency room (ER) nearby. Police and paramedics noted no evidence of paralysis in the field. ER notes described the patient as able to move all extremities, thrashing and kicking, and mumbling incomprehensible words. In the midst of this activity in the ER, the patient suddenly became apneic and pulseless, and resuscitation efforts resulted in recovery of a pulse and blood pressure, but the patient remained comatose. He did not respond to treatment measures and eventually met clinical criteria for brain death. He was pronounced after a 9-day survival period following the cardiopulmonary arrest.

Initial ER blood tests were positive for cocaine and cannabinoid. A neurologic consultation on the day of admission, post-cardiac arrest and resuscitation, listed findings that included coma, withdrawal to painful stimuli in all four extremities, "flaccid upper extremities," and increased tone in both lower extremities with bilateral ankle and patellar clonus. Response to plantar stimulation (e.g., toe flexion versus Babinksi sign) was not stated. The neurologic consultant indicated "rule out spinal cord trauma" among several other initial clinical impressions. Same-day CT scans of the brain and the cervical, thoracic, and lumbar spine were performed and were negative. Follow-up CT head scans revealed the development of severe cerebral edema.

General autopsy findings included bronchopneumonia, hepatomegaly (2550 g) with portal triaditis, minor healing abrasions of the extremities, and no evidence of internal injuries (including no evidence of spinal fracture or cervical soft tissue injury).

Neuropathologic examination of specimens submitted (cranial dura mater, brain, portion of cervical spinal cord, and a portion of rostral spinal column including the odontoid process and vertebral bodies, but devoid of lateral processes or posterior elements) revealed no abnormality in the submitted spinal column elements. Brain and spinal cord findings were consistent with spinal cord infarction with superimposed respirator brain changes. The findings included cervical spinal cord infarction with marginal zone cellular reactive changes including vascular endothelial hyperplasia, macrophages, a few necrotic neutrophils, a small central hematomyelia, and a few scattered neurons with acute neuronal hypoxic-ischemic injury.

Comment. One might wonder why a question of "duration of spinal shock" even arose in this case. Months after the autopsy, the medical examiner and neuropathology consultant were asked to review a surveillance videotape that showed the decedent being pulled out of the supermarket by two employees, one holding each arm, and the decedent's legs dragging behind him. The surveillance tape alone provides no information as to whether this leg appearance is a consequence of voluntary passive resistance or due to another cause. However, interested parties had been reviewing medical records and the videotape, had apparently focused on the neurologist's initial impression that spinal cord injury should be excluded, and concluded that police brutality causing spinal injury might potentially be claimed. Clearly, it is appropriate for clinicians to exclude a possible spinal cord injury with even the most meager of suspicion, considering the consequences of missing such findings. The leap in logic to the assumption of police brutality, however, requires selective amnesia for evidence in the medical records that the decedent exhibited vigorous muscular strength in all four extremities immediately preceding the cardiopulmonary arrest, the fact that there was no subsequent radiologic or autopsy evidence of spinal column injury or cervical soft tissue injury, and that the cervical spinal cord findings were perfectly consistent with those that may occur following cardiac arrest. It was the opinion of the neuropathology consultant that questions regarding duration of spinal shock, absent any evidence of precardiopulmonary arrest spinal injury, were thus irrelevant in this case. Had it been a relevant question, the references cited earlier would be useful in forming an opinion or, at the very least, in aiding the consultant in determining whether the information needed to form an opinion was available (e.g., detailed timeline of events, exactly which reflexes and functions were tested, results of serial examinations, and so on).

Spinal Concussion

Spinal concussion, in contrast to spinal shock, is a clinical term referring to a partial or complete impairment of spinal cord function following trauma to the spine that is followed by rapid and complete recovery of function. This term is also sometimes applied to the recovery of function at levels rostral to a known cord destructive lesion that is more rapid (e.g., within hours or days) than the recovery of spinal shock signs caudal to the same lesion. Thus, this more dramatic early improvement is assumed to relate to a transient physiologic block of function rather than an anatomic destructive lesion.[141]

As with many terms used in medicine, "rapid" has been variously defined in the past as anywhere from 24 hours to 20 days.[171] Recent reports tend to restrict the time frame for complete recovery to no more than 48 to 72 hours' postinjury in order to qualify the case as one of spinal cord concussion.[77,393] Zwimpfer and Bernstein[393] further defined their cases of spinal cord concussion by including only those in which trauma immediately preceded the onset of neurologic deficits, and the neurologic deficits were consistent with spinal cord involvement at the level of injury. Although spinal stenosis was not a predisposing factor in their 19 reported cases, all occurred at either the cervical or thoracolumbar area, the two most mobile areas of the vertebral column.

The neuropathologic substrate for spinal concussion is not clear due to the low likelihood of these cases having a fatal outcome. It is likely that this current absence of data concerning the pathologic substrate of spinal concussion will, as cases continue to be examined by newer techniques, yield findings similar to those seen in studies in human cerebral concussion. Blumbergs and colleagues[32,33] have described evidence of axonal injury using immunostaining for amyloid precursor protein in six patients with what they refer to as "concussive syndromes" or mild head injury. In their most recent study,[33] postinjury unconsciousness duration ranged from 1 minute to less than 10 minutes in five of these cases, and was less than 40 minutes in the sixth case. Postinjury survival periods in cases varied from 3.75 hours to 28 days, with death due to systemic problems unrelated to the head injury. Four of the six cases were devoid of any associated neuropathologic abnormalities, one case demonstrated some hypoxic-ischemic damage, and one case had a small gliding contusion. Axonal injury was present in the corpus callosum and fornices in all six cases, and in cerebral hemispheric white matter in five of the six cases. A similar study on a series of human spinal cord concussion cases has not yet been published, to our knowledge.

Adult Cerebral Infarction

A detailed discussion of the numerous causes of cerebral infarction is beyond the scope of this chapter's focus (see also Chap. 11). A more thorough review can be obtained from general neuropathology texts and from references such as Meyer et al.[244] The latter authors group cerebral infarction causes under the following general categories: cardiac causes (e.g., embolism, cardiac arrhythmias), noncardiogenic disturbances of perfusion pressure (e.g., hypotension, hypertension, steal syndromes), arterial obstructions (e.g., arteriosclerosis, emboli, vasculitis), changes in blood constituents (e.g., polycythemia, sickle cell disease, disseminated intravascular coagulation, paraproteinemias), venous obstructions (including dural venous sinuses), and hypoxic and metabolic causes (e.g., hypoxia, hypoglycemia).

From the pathology viewpoint, cerebral infarction may be ischemic, partially ischemic and partially hemorrhagic (i.e., hemorrhagic infarction), a global reduction in blood flow, or a focal reduction in blood flow.[72] The term "infarction" is usually used to described brain injury limited to the territory of a single major brain artery, with brain tissue bordering that region maintaining its circulation. With focal reductions in blood flow, that portion of the brain surrounding the lesion area may have normal or abnormal perfusion. Alternatively, the brain tissue surrounding the original infarction may initially be well perfused, but with time (and lesion area swelling, increasing inflammation, secondary hemorrhage, etc.) become itself hypoperfused and infarcted. The presence of hemorrhage, according to some observers, also tends to stimulate a more vigorous inflammatory response than a pure ischemic lesion.[72] For this reason, where the information has been provided by the source cited, the additional presence of hemorrhage in a predominantly ischemic infarct has been noted in our timetables.

Gross Examination of Central Nervous System Ischemic Infarction

Survival Period
- 8 to 12 hours[197]
  — ± Slightly more congested in infarct zone.
  — ± Slightly darker color than normal brain (gray matter more so than white matter).[233]
  — ± Slightly softer than surrounding normal brain tissue (especially postfixation). May not detect infarct by gross examination for up to 24 hours in unfixed brain.[174,175]
- 12 hours plus[197]
  — ± Early swelling.
- 12 to 24 hours
  — Soft, pale.[105]
- 18 to 24 hours[233]
  — Typical ischemic infarction becomes clearly demarcated from normal brain.
  — A few petechiae are often present, especially in gray matter and at lesion margin.
  — (*Comment*: Determination of when extravasated blood is sufficient to be designated a hemorrhagic infarction is a subjective one. Adams and Sidman[7] consider it "an abundance of petechial hemorrhages up to 1 to 2 mm in size").
- 24 to 48 hours[105]
  — Swollen, indistinct gray/white matter junction, darker.
- 1½ to 3 days[233]
  — Separation of zone of infarct from more intact adjacent tissue begins at about 36 to 48 hours, but is usually not marked in less than 72 hours.
- 2 to 10 days[105]
  — Yellowish color, sharper margins, usually triangular or wedge-shaped, progressive tissue breakdown.
- 3 to 4 days[197]
  — Infarct zone easily distinguished by dusky gray-brown color. Injection of small pial capillaries. Postinfarction swelling tends to peak at between 3 and 5 days, then gradually decreases.[233]
- 4 to 5 days[233]
  — First liquefaction visible. Earliest evidence of cavitation may be seen, depending on infarct size.

- 7 days[197]
  — Infarct clearly soft and easily fragmented (before and after fixation).
- 1 to 2 weeks[197]
  — Infarct area easily breaks away from unaffected tissue.
- 10 days to 3 weeks[105]
  — Tissue liquefaction progressing to cavity lined by dark gray tissue.
- 2 to 3 weeks[197]
  — Infarct area more spongelike, and trend toward decreased ease of fragmentation begins.
  — Necrotic debris removed at about the rate of 1.0 ml every 2 months.
- 3 weeks to months[105]
  — Irregular cavity forms, containing clear, pale yellow fluid.
- ~ 2 months plus[197]
  — Gradual evolution to more spongy, ± sunken, ± cystic zone of infarction.
  — ± Some tan to brown pigment (depending on the amount of blood present in infarct area).
  — A cavity of 2 to 3 cm in diameter may take 6 to 8 months to form.[233]

Microscopic Examination of Central Nervous System Ischemic Infarction

Survival Period
- \> 1 hour[174,175]
  — Neuronal microvacuoles (swollen mitochondria) followed by perineuronal vacuolation (swollen astrocyte process).
  — *Comment*: We cannot reliably distinguish this early change from the artifact of postmortem autolysis in human forensic cases, and do not base a diagnosis of acute neuronal hypoxic-ischemic injury on these criteria alone.
- ~ 2 hours[197]
  — First appearance of red neurons (under certain special circumstances, such as prior good health and prompt rescue but failed resuscitation efforts in hanging or cardiac arrest cases).
- 2 to 5 hours[184]
  — First appearance of red neurons.
  — Astrocyte swelling present.
- 4 to 12 hours[36]
  — First appearance of red neurons in neocortex.
- 4 to 12 hours[174,175]
  — First appearance of red neurons.
  — Disappearance of Nissl bodies.
  — Neuronal nuclei pyknotic and nucleoli no longer visible.
- 6 to 8 hours[233]
  — "May" see edema, capillary endothelial swelling, red neurons, swollen oligodendroglia (Fig. 3.28), RBC diapedesis, neutrophil margination ± focal extravascular migration, and paleness of neuropil.
- Within 6½ hours[37]
  — First appearance of red neurons.
- Within 6 to 8 hours[110,371]
  — First appearance of red neurons.
- After 6 hours[133]
  — First appearance of red neurons.
- 6 to 8 hours[260]
  — First appearance of red neurons.
- 6 to 8 hours[233]
  — First appearance of axonal swellings, which become conspicuous at 24 hours.
- 8 to 12 hours[197]
  — First appearance of red neurons (under most circumstances).
- ~ 12 to 20 hours[233]
  — First appearance of macrophages, and there may be well developed macrophage response by 24 hours.
- 12 to 24 hours[197]
  — Neuropil vacuoles present, most conspicuous at marginal zone of infarct.
  — Infarct area ± slightly more pale than surrounding brain on H&E stain.
- 15 to 24 hours[174,175]
  — Neutrophil infiltration first seen.
- 15 to 28 hours[207]
  — \> 10-fold increase in neutrophils in infarct zone, mostly marginating along inner vessel walls in the 15-hour survival period (*note*: using CD15 immunostaining to identify neutrophils).
- "During first 24 hours"[211]
  — Moderate neuronal and cytoplasmic shrinkage or indentation.

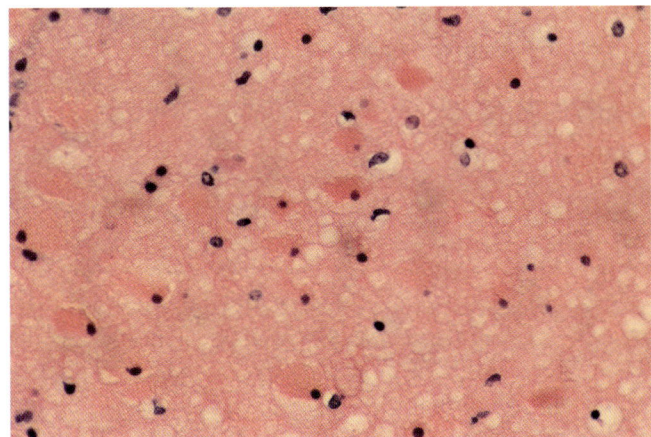

Fig. 3.28. Medium-power micrograph. Cells with small round dense nuclei and abundant pink cytoplasm are swollen oligodendroglia, often seen within a few hours of insult in infarctions or at the margins of hemorrhagic or traumatic lesions. (H&E.)

- — Perineuronal, followed by more generalized, neuropil vacuolization.
- — More triangular nuclear profile in some neurons.
- "By 24 hours"[211]
  - — Neuronal nuclei oval or triangular.
  - — Some neurons with chromatin starting to disperse.
  - — Many neuronal nuclei amphophilic rather than simply pyknotic.
  - — Neuron nucleolus usually just discernable.
  - — Increased neuronal cytoplasmic eosinophilia.
  - — Neutrophils usually present (margination, and in perivascular neuropil).
- 1 day[327]
  - — Hypoxic neuronal damage.
  - — Spongiosis.
  - — Discrete, rare granulocytes present.
  - — Endothelial swelling.
- 1 day[281]
  - — Infarct zone demarcated by neutrophil infiltration (in a case with hemorrhage also present).
  - — Infarct zone with scattered neutrophils (in a case with slight hemorrhage present).
- 1 day plus[165]
  - — Axonal spheroids may be found in and around infarct zone.
- 1 to 2 days[7]
  - — Typical red neurons seen.
- 1 day, 14 hours to 2 days, 5 hours[207]
  - — Maximal neutrophil response in ischemic infarct (both marginating within vessels and present in neuropil) (based on CD15 immunostaining).
- 24 to 36 hours[7]
  - — Neutrophils in perivascular areas and scattered in infarct long before first macrophage seen. Neutrophils may be extensive, mimicking infection; they disintegrate within a few days.
- 1 to 2 days[327]
  - — Neutrophils marginating on inner vessel walls and in neuropil (hemorrhage present in this case).
  - — Spongiosis.
  - — Endothelial swelling.
  - — First appearance of macrophages and capillary proliferation toward end of this period.
- ~ 1 to 2 days[7]
  - — Macrophages appear (after neutrophils present), and become lipid-laden (Gitter cells) in 48 hours, thereafter increasing in number in 5 to 30 days.
- 2 days[211]
  - — Many neurons with glassy, hyaline, brightly eosinophilic cytoplasm.
  - — Some neurons with less eosinophilia and granular cytoplasm.
- 2½ days[327]
  - — First appearance of macrophages.
  - — First appearance of capillary proliferation.
- 3 days[7]
  - — Earliest astrocyte proliferation.
- 1 to 4 days[61]
  - — Red neurons (present in < 24 hours); minimal poorly staining ("pale") neurons beginning at day 2 to 3.
  - — Mild decrease in myelin staining.
  - — Neutrophils present by 1-day survival, maximum at 2 to 3 days' survival.
- 5 days[174,175]
  - — Neutrophil infiltration ceases.
- 5 to 7 days[165]
  - — Necrotic Purkinje cells have largely disappeared (if cerebellum involved), but many red neurons persist in other areas, such as Ammon's horn, neocortex, brainstem, and/or spinal cord.
- 5 to 7 days[61]
  - — Red and pale neurons in equal numbers; necrotic and "ghost" oligodendrocytes present.
  - — Myelin markedly decreased.
  - — "Slight" axonal changes.
  - — Decreasing neutrophils.
  - — Scant macrophages appear.
  - — Hemosiderin first seen at day 7.
  - — Neovascularization first appears at day 5.
- 1 week[233]
  - — Heavy infiltrate of macrophages.
  - — Astrocytic proliferation and hypertrophy at infarct border.
- ~ 1 week plus[174,175]
  - — Reactive astrocytes appear at infarct margin.
  - — Capillary density increased.
- ~ 1 week plus[211]
  - — Some dying neurons still recognizable.
- 7 to 10 days[7]
  - — Faint outlines of necrotic neurons may persist.
  - — Capillary wall thickening first seen at 7 days.
- 8 to 14 days[61]
  - — Fewer red neurons; pale neurons more frequent, peaking in week 2; scant myelin.
  - — Some axonal fragmentation.
  - — Neutrophils absent.
  - — Prominent vessels, increased in number.
  - — Scant reactive astrocytes with a few gemistocytes (first seen at day 10).
  - — Numerous macrophages.
  - — Ferruginated neurons first seen at day 10.
  - — First cavitation seen at day 13.
- Up to 15 days[110]
  - — Residual red neurons may still be found in some brain regions (i.e., hippocampal pyramidal cell layer in sectors CA1 and CA2, and periamygdaloid [pyriform] cortex). Not seen in these regions or elsewhere at 27 days' survival in this study.

- 15 to 27 days[61]
  — No red neurons remain; ghost neurons present.
  — No myelin seen.
  — Scant axons remaining.
  — Prominent macrophages, vessels, and proliferating astrocytes (with an increase in gemistocytes).
- 17 to 18 days[207]
  — Degree of macrophage influx and phagocytosis of neutrophils prevented reliable counts of neutrophils (based on CD15 and HAM56 immunostaining).
- 26 days
  — Ferrugination of neurons first seen (after "asphyxial" episode).[291]
- "Weeks"[327]
  — Moderate to dense macrophage presence.
  — Rests of necrotic tissue present.
  — Moderate to prominent astrocytosis.
  — Fibrotic vessels.
  — Capillary proliferation persists.
- "Several weeks"[281]
  — Macrophages persist.
  — Capillary sprouting persists.
  — Reactive gliosis.
  — Cystic reorganization.
- 1 month[233]
  — Prominent margin-zone gliosis.
  — Inflammatory response has subsided.
  — Macrophages may persist.
  — Cavitation may be obvious.
- "Months"[326]
  — Macrophages present.
  — Glial scar.
  — Capillary proliferation.
- Months to years[174,175]
  — Foamy macrophages may persist in infarct zone.

We also note variable findings in chronic infarcts, to some degree depending on the infarct size, which may include micro- or macrocystic change, neuronal ferrugination, Rosenthal fibers, neuroglial mineralization and/or phagocytosis of hemosiderin, and vascular wall sclerosis.

One of the aforementioned studies[61] noted some differences in reactive changes when they compared hemorrhagic and ischemic cerebral infarctions. They found, in the regions of hemorrhagic infarction:

- Neutrophils first appeared in the 1- to 4-day period, as in ischemic infarction, but the response was more robust and more persistent (up to week 4) compared with ischemic infarction, in which neutrophils disappeared by day 8.
- A slower, less vigorous macrophage response (not seen until the 8- to 14-day period) in hemorrhagic infarction, compared with onset in the 5- to 7-day interval in ischemic infarction.
- Less conspicuous breakdown of neuropil in hemorrhagic infarction areas compared with anemic infarction areas.
- Later appearance of neovascularization (first seen in the 8- to 14-day period in hemorrhagic infarctions, but seen "at the end of the first week" in anemic infarctions).
- Hemosiderin was not seen until the second week in hemorrhagic infarctions, but was seen at 7 days in anemic (ischemic) infarction.
- Their study was based on 15 cases of hemorrhagic infarction and 16 cases of ischemic infarction,[61] so that the rather subtle changes in neovascularization and earliest hemosiderin formation, in particular, would seem to warrant further studies for confirmation.

Generalized Brain Hypoxic-Ischemic Injury

There are occasional suggestions in the literature that the cellular reaction to diffuse cerebral hypoxic-ischemic insult may be somewhat less vigorous than to localized infarction with well-perfused surrounding brain. For example, Graham,[134] within a single chapter, indicated that typical neuronal ischemic cell change occurs within 4 to 12 hours in generalized hypoxic-ischemic insults, and within 4 to 6 hours in focal infarction. On the other hand, Brierley et al.[37] described red neurons developing within 6½ hours' survival subsequent to an open heart surgery procedure, and Norenberg and Bruce-Gregorios[260] described the first appearance of red neurons in anoxic-ischemic encephalopathy occurring within 6 to 8 hours. These estimates are well within the most frequent time frame in which other authors describe red neurons first appearing in infarction. It is our impression that, assuming a respirator brain syndrome does not become inserted into the clinical situation while criteria for brain death are being established, the margin between widely infarcted brain and well-perfused brain demonstrates a sequence of histologic changes very similar to that which occurs in more localized infarctions, wherever the margin between well-perfused and nonperfused brain CNS tissue exists (e.g., even if that margin is at a brainstem level).

Additional Comments on Infarction

Neuronal incrustations (dark specks on neuronal perikarya and dendrites by light microscopy; seen on electron microscopy as electron-dense surface projections of neuronal cytoplasm bordered by swollen pale astrocyte processes indenting the neuronal surface[94,95]) are commonly seen in acute CNS infarction. They may be seen in experimental models as early as 30 minutes (rat) to 90 minutes (monkey) and persist for 48 hours,[94,95] but the earliest occurrence in human infarcts related to circulatory disturbance alone is less well described. In human traumatic brain injury (including post-traumatic infarction), however, neuronal incrustation was first seen at 3 hours' survival, and persisted up to 48 hours' survival.[16] Other investigators have indicated the first appearance of neuronal incrustation to be greater than 5 hours, and becoming prominent at 36 to 48 hours (with Nissl stain),[206] or being first seen in the time frame of 12 to 24 hours' survival.[205]

Spinal Cord Infarction

Only one article was located that specifically dealt with cellular events following an essentially pure spinal cord infarction. Most reports on spinal cord infarction either refer to the sequence of reactive changes in very general terms or describe infarction related to trauma. Toro *et al.*[362] reported a single human case of spinal cord infarction due to nucleus pulposus embolism, and their report included a review of the literature of 31 previously reported cases of similar etiology. Their brief summary of the microscopic neuropathologic sequence of events follows.

Microscopic Examination of Spinal Cord Infarction

Survival Period
- "Few hours"
  — Multiple microscopic hemorrhages, predominantly in gray matter.
- > 48 hours
  — Acute ischemic necrosis of Spielmeyer (i.e., red neurons) seen primarily in anterior horn cells and ventral one-third of spinal cord, in sulcal-commissural region.
  — No microglial or astrocyte reaction seen.
  — Early spongiosis in white matter.
- > 1 week (Case 3.8, Figs. 3.29–3.32)
  — Frank necrosis, neuronal loss.
  — Macrophages, capillary proliferation, and astroglial reaction present.
- "Advanced" cases
  — Cavitation, glial scar, and long tract degeneration (see later section on CNS wallerian degeneration).

Examination of a larger series of patients with specific attention to cellular reaction timetables will be necessary in order to determine whether or not infarction in the spinal cord differs significantly in cellular reaction dating/aging in comparison to cerebral infarction. The time brackets noted earlier are too general to allow firm conclusions in this regard.

Brain Death and Respirator Brain

Plum and Posner[279] defined brain death as a state in which "irreversible brain damage is so extensive that the organ enjoys no potential for recovery and can no longer maintain the body's internal homeostasis, that is, normal respiratory and cardiac function, normal temperature control, normal gastrointestinal function, and so on." Current medical practice is designed to provide the patient with all reasonable supportive measures until such time that the clinical criteria establishing irreversibility of functional loss and a hopeless prognosis for brain recovery are confirmed. Only then is a diagnosis of brain death made. Criteria for determination of brain death vary in different jurisdictions,[240] and will continue to evolve as experience and technologic developments accumulate. It follows that brain death is a clinical, not a pathologic, diagnosis.

The history of brain death concepts is summarized in monographs on this topic,[375,378,379] where the reader can also find a more detailed review of gross and microscopic findings.

This brief review of the topic will focus on the most common neuropathologic questions asked, and a few findings less commonly described in such cases. The neuropathology of brain death, generally termed "respirator brain," varies in detail from case to case but has certain shared features. Factors such as underlying cardiovascular status; duration and degree of hypoxic-ischemic insult; duration of respirator support; extent, if any, of local reperfusion phenomena; and extent of secondary phenomena, such as brain swelling and herniation effects, coagulopathies, hyperthermia, and so on may influence the ultimate findings.

Gross Examination of Brain Death

Leestma[197,199] has outlined the evolution of gross examination findings in typical respirator brain cases, which may be briefly summarized as follows. The brain is swollen, typically in sufficient degree to show secondary herniation effects. It is congested, unusually soft despite adequate fixation time, and has a grayish to gray-brown tint referred to as a "dusky" color on the surface and also present to varying degree on cut sections (Figs. 3.33–3.36). These changes are visible on gross inspection in about 12 to 16 hours in most cases, but may first appear as early as 6 hours' or as late as 24 hours' post-circulatory arrest.[197]

Additional findings may include subarachnoid hemorrhage and/or intraparenchymal areas of softening and/or small hemorrhages. Intraparenchymal hemorrhages, often petechial in size, tend to favor gray matter, and are influenced by brain necrosis-associated coagulopathy[380] and by the presence and extent of reperfusion phenomena at the border zones of nonperfused and perfused tissue. Thrombosis may be found in cortical veins and/or in dural venous sinuses (the latter in approximately 10–15% of cases).[375] The caudal brainstem and spinal cord (rostral greater than caudal) may have surface structures obscured by tan to gray, soft adherent

Case 3.8. (Figs. 3.29–3.32)

A 76-year-old man was found down at home incontinent of urine and feces, and he was transported to the hospital where he was alert and oriented to name and place only. He was unable to stand. While in the emergency room, he fell from bed and was noted to be quadriplegic. X-rays revealed fracture of the C4 vertebra. He remained quadriplegic and died 18 days after being found.

Fig. 3.29. Anterior aspect of cervical spinal cord with dura reflected shows dusky discoloration of cord heightened at C7 and C8 levels by what appears to be transverse compressions at those points, but are postmortem bending artifacts.

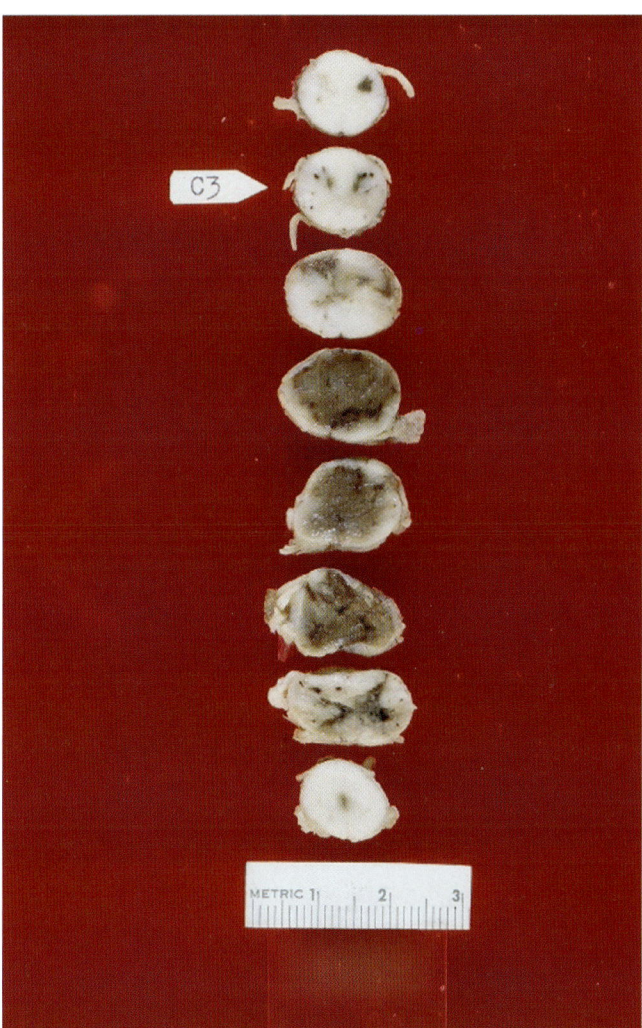

Fig. 3.30. Transverse sections of spinal cord at segmental levels show brown necrosis greatest at C5 but involving virtually the entire length of cervical cord.

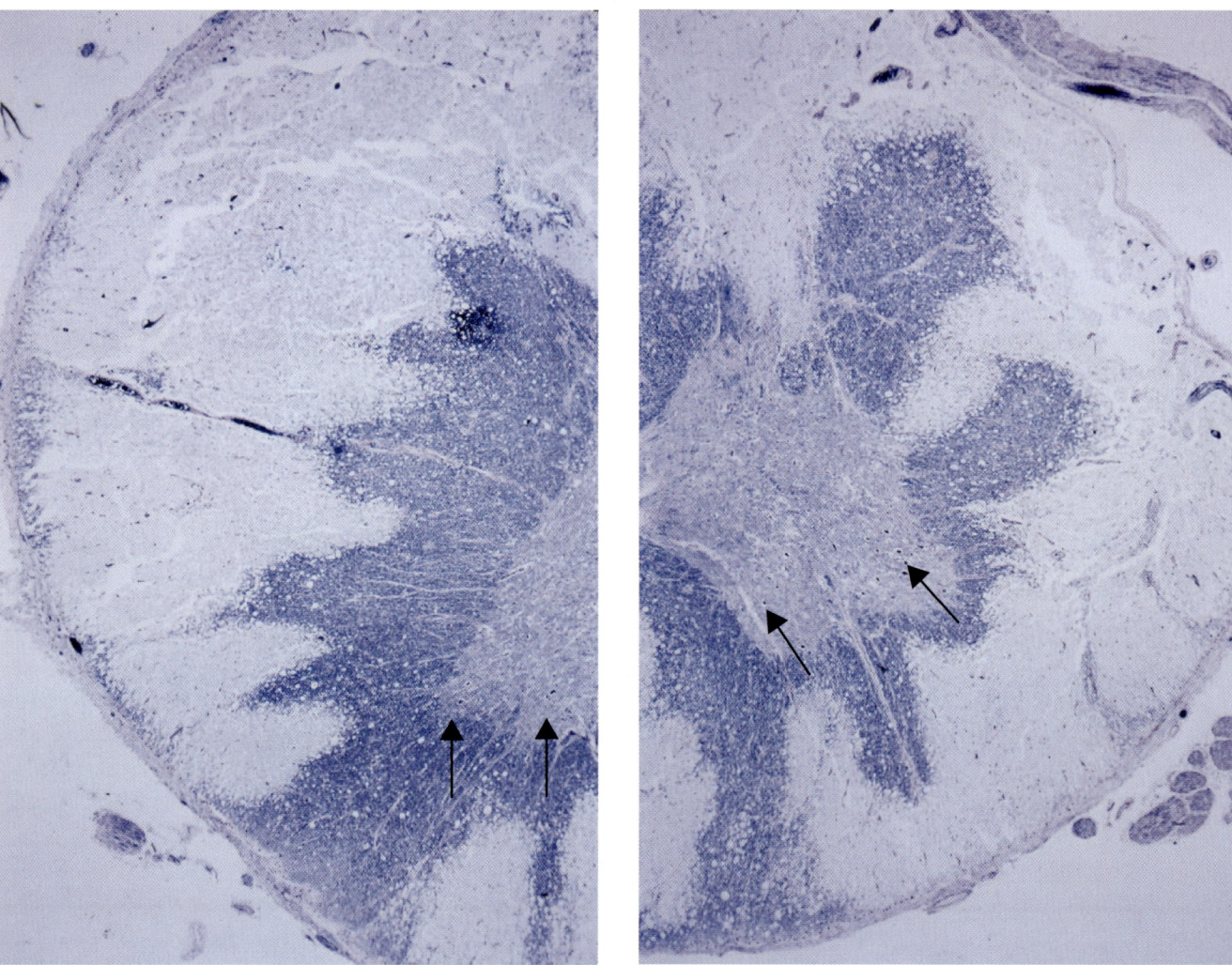

Figs. 3.31–3.32. Low-power micrographs of the two halves of cervical spinal cord show geographic loss of myelin staining representing areas of acute infarction, at peripheral cord on both sides in territories of penetrating arteries of pial plexus arising from medullary ("radicular") arteries. Central white matter around central gray matter, including anterior horns, is spared. Intact motor neurons were present, but are barely visible at this magnification (*arrows*). (Luxol fast blue.)

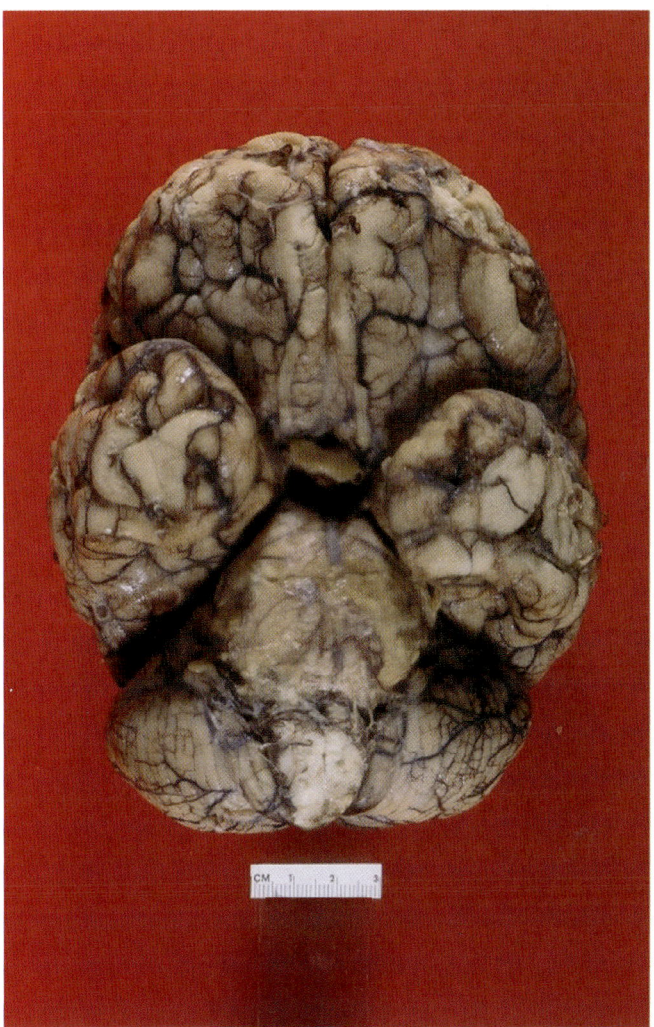

Fig. 3.33. Basal view of respirator brain.

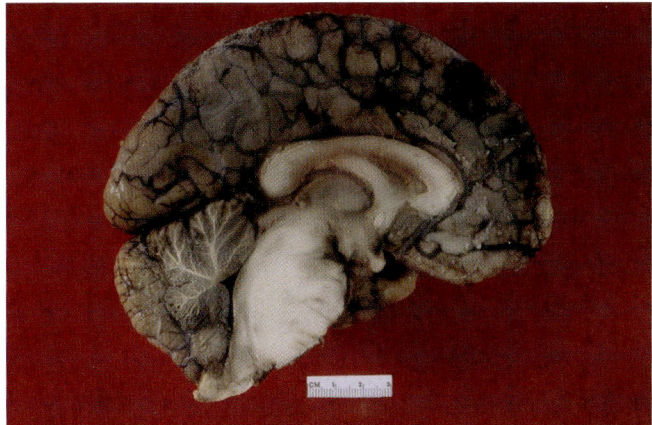

Fig. 3.34. Midsagittal view of respirator brain. Pons is edematous; no neoplasm or inflammation was present.

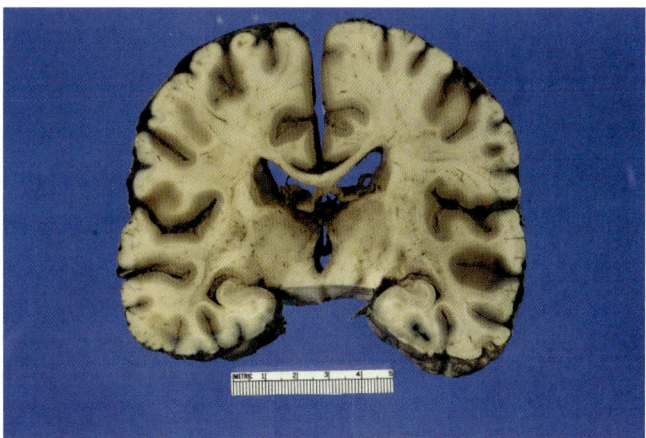

Fig. 3.35. Respirator brain, coronal section, demonstrating swelling, congestion, and tan-brown to grayish, "dusky" discoloration.

Fig. 3.36. Respirator brain. Transverse section of cerebellum through rostral pons. Dusky gray discoloration is seen to better advantage on the more superficial areas of cerebellar cortex. A light-gray line is present in deep cortex, representing classic gross appearance of granular layer autolysis (see Chap. 1).

material derived primarily from fragments of necrotic cerebellar cortex (Figs. 3.37–3.38). The degree of diencephalic, brainstem, and spinal cord involvement is also variable.

In occasional cases, the brunt of the pathology involves the caudal brain, brainstem, and cerebellum more than forebrain structures.[199]

Microscopic Examination of Brain Death

In the majority of brain death cases, microscopy demonstrates an important discrepancy compared with perfused brains. That is, the nature and extent of certain reactive changes expected in response to a severe hypoxic-ischemic event in the CNS, based on the interval between the insult and criteria for brain death having

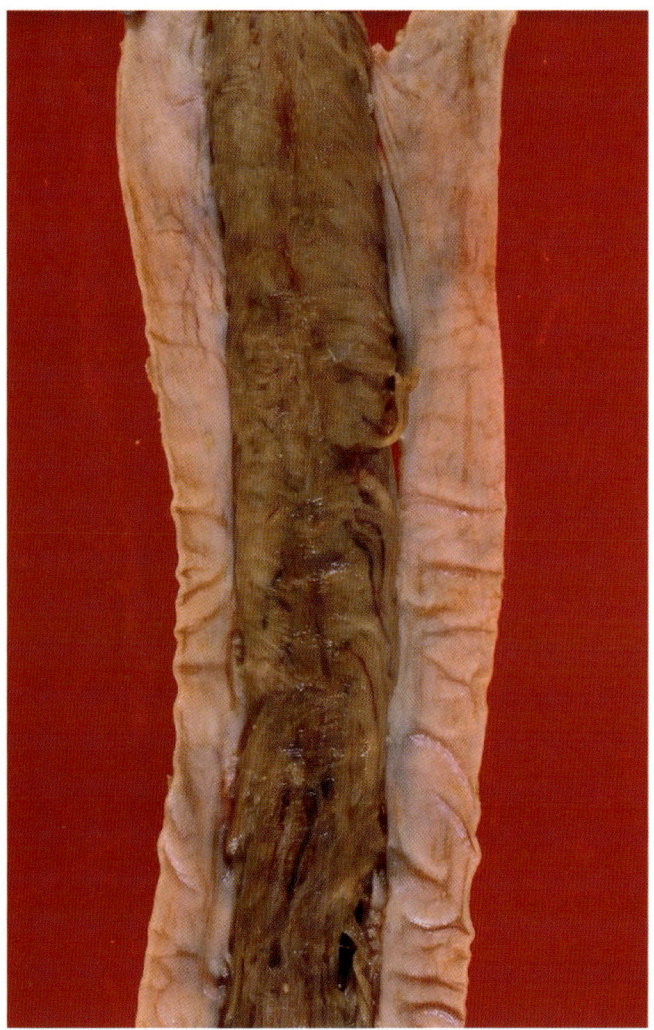

Fig. 3.37. Respirator brain. External spinal cord landmarks and rootlets are obscured by accumulated necrotic cerebellar tissue, which was most pronounced in upper one-third of spinal cord.

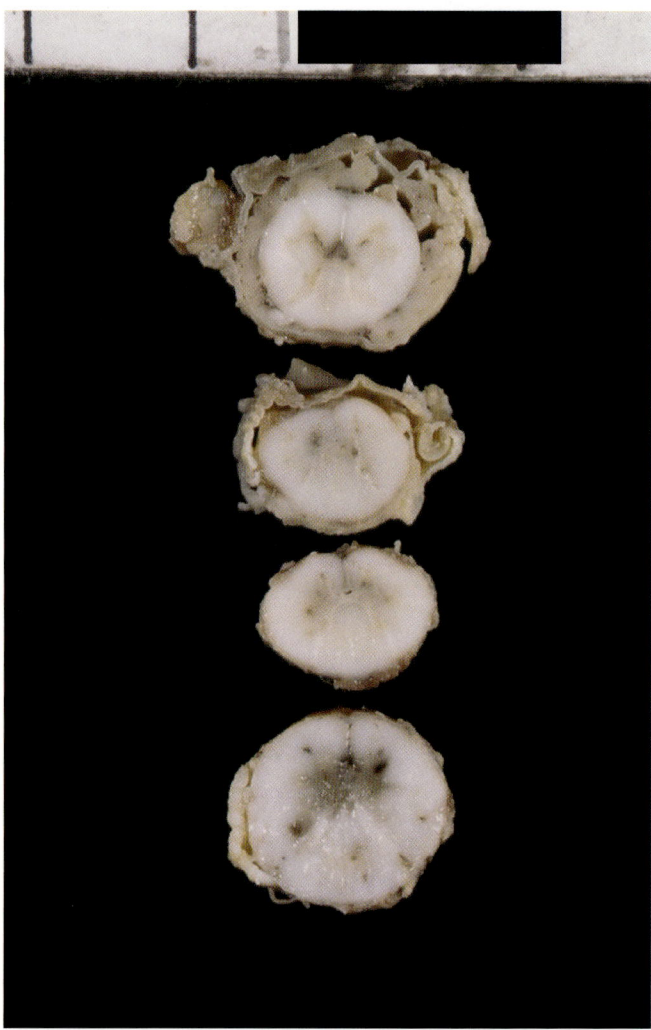

Fig. 3.38. Respirator brain. Transverse sections of spinal cord with necrotic cerebellar tissue in subarachnoid space, most easily seen in upper two profiles.

been documented, is much less than would be expected in the margin of, for example, a localized infarction or hemorrhage of similar duration in an otherwise intact patient (see previous section on cerebral infarction regarding timetables of cellular responses). More specifically, degenerative changes in already-present nerve cells (e.g., red neurons) proceed as expected, but the reactive cellular changes such as an increase in rod cells, infiltrates of neutrophils and mononucleated chronic inflammatory cells, development of foamy and/or hemosiderin-laden macrophages, glial response, endothelial hypertrophy and proliferation, and so on, are not seen in the absence of ongoing local reperfusion following the initial insult.

Probably at least in part related to the realization that time of cessation of brain perfusion need not necessarily coincide exactly with time on respirator,[375] it can be appreciated that any suggested timetable for histologic events developing in brain death are general estimates and not to be interpreted dogmatically.

It follows that one is not comfortable making a diagnosis of findings consistent with respirator brain in cases with less than 24 hours' survival after pronouncement of brain death. This is the outside limit, in our experience, when one might reasonably expect neutrophils to appear in the periphery of infarcted parenchyma in cases of hypoxic-ischemic encephalopathy in which perfusion is restored by the initial resuscitative efforts. If neutrophils are absent beyond 24 hours' survival, respirator brain becomes the more appropriate diagnosis. One should be reluctant to pathologically diagnose nonperfused brain in the absence of at least patchy acute neuronal hypoxic-ischemic injury in the brainstem. Routine sampling in such cases should include the neocortex, dorsal thalamus, hippocampus, cerebellum, midbrain, pons, and medulla. The diagnosis of brain death is a clinical determination, with the gross and microscopic neuropathologic examination being simply supportive. Thus, reports will commonly indicate "findings consistent with respirator (i.e., nonperfused) brain," rather than "brain death." There usually is not equivocation in the diagnosis of respirator brain in cases maintained on life support for greater than 24 hours after diagnosis of brain death.

Infarction and/or hemorrhage of the pituitary gland may or may not be present, since it derives some of its blood supply from extracranial sources, and its circulation may also be compromised by the degree of brain swelling and herniation.

Spinal cord involvement is also variable in severity and extent, with most reference sources indicating it is limited, when present, to rostral cervical levels.[375,378] Rare reports[325] and a very few of the authors' cases have provided exceptions to this rule of thumb, with caudal cervical cord involvement, and, in rare cases, more widespread spinal cord involvement seen.

With its extracranial blood supply source, cranial dura mater reactive changes may occur in the form of inflammatory cells infiltrating the outer greater than the inner layers of the dura mater. This is likely secondary to a relative hypoperfusion of the inner dura related to the increased intracranial pressure present in brain death cases.

Necrotic cerebellar fragments displaced into the spinal canal are seen in brain death cases maintained for longer periods (Figs. 3.37–3.38). Such displaced tissue has been seen as early as 9 hours' postinsult (Schröder, quoted by Sayer et al.[320]), but usually takes longer to appear. In autopsy cases, necrotic cerebellar and other CNS tissue may be seen in the subdural space (especially dorsally), or in the subarachnoid space.[154,375] Such deposits of necrotic tissue may stimulate a local meningeal inflammatory response by granulocytes and lymphocytes, and some indicate that meningeal blood vessels may also become involved in this inflammatory process, leading to thrombosis.[325] These authors[325] hypothesize that subpial spinal cord edema and radial perivenous hemorrhages seen in some cases with longer survival periods are secondary to such leptomeningeal vessel involvement, which itself is secondary to a foreign body-type reaction to the displaced necrotic cerebellar tissue. They suggest that mechanical compromise of the spinal venous drainage by accumulated cerebellar tissue may be a contributory factor to such subpial spinal lesions. Others[154] indicate that similar cord lesions occur in some of their cases despite an absence of meningeal inflammation; they consider mechanical compromise of cord circulation by accumulated displaced necrotic CNS tissue to be the most likely basis for such lesions. Although we do not discount this latter possibility, we note that the myriad of secondary effects that occur following brain death (e.g., a vicious circle of brain and cord swelling with marginal-zone progressive vascular compromise, coagulopathies, excitotoxic mechanisms, progressive hypoxic and other chemical abnormalities that accompany terminal multisystem failure) may play an as yet poorly understood role in the eventual neuropathologic histologic picture.

Cerebrospinal fluid findings are variable in brain death cases, as expected from the differences seen in the degree of meningeal inflammatory reaction from case to case. In several reported cases with varying survival periods, the major findings are a significant pleocytosis that may include neutrophilia; mononucleated cells (usually interpreted as leukocytes but possibly including desquamated cerebellar cortex granule cells); eosinophils; large neurons (usually resembling Purkinje cells); macrophages that may demonstrate erythrophagocytosis or leukophagocytosis or that may contain hemosiderin granules; rare nuclei of degenerate cells; and corpora amylacea[320] (also see earlier section discussing observations on postmortem CSF examination).

Spontaneous and Reflex Movements in Brain Death Patients

Questions occasionally arise in forensic cases concerning extremity and other body movements that occur during the period when clinical criteria for brain death are being documented. Such questions may result from, for example, witness statements that investigators consider possibly inconsistent with pathologic findings, or by family members of the decedent who observe movements that cause them to question a clinical diagnosis of brain death. Such movements are well described in the literature and are considered to be of spinal origin.[156,166,172,279,308,315,379] Movements described in these references include spontaneous finger jerks and facial myokymia, and reflex movements in response to stimuli. The latter include various tonic neck reflexes, withdrawal responses of an extremity, persistence of cutaneous or deep tendon reflexes, and more complex combinations of extremity and truncal movement, such as symmetric or asymmetric elevation of the arms with touching together of hands, or touching of the chin, followed by arm extension and replacement of arms alongside the body; arching of the back on one or the other side; and trunk flexion causing a partial sitting posture. Consultation with neurology or neurosurgery colleagues can be very useful, particularly when the forensic neuropathologist has not had clinical experience with such cases.

Case 3.9 illustrates how an initial question submitted to the neuropathology consultant can lead to what is more appropriately a multidisciplinary approach in a given case.

Dating/Aging of Common Lesions in Neuropathology

> **Case 3.9.**
>
> A 25-year-old woman died from a single gunshot wound to the head interpreted as a close range, but not contact, wound. The bullet passed through both temporal lobes and the rostral pons. The bullet path within the pons involved the ventral tegmentum and dorsal basis pontis bilaterally. Relevant additional findings included remote contusions of frontal, temporal, and occipital lobes; and remote fractures of the crista galli, the cribriform plate of the ethmoid bone, and the orbital roof plate of the frontal bone bilaterally.
>
> The weapon was a Walther PPK .380 semiautomatic pistol, with a single ejected cartridge case. The witness claimed that the decedent had committed suicide with the weapon, with the witness present. The confounding factor was that the weapon found on the floor near the decedent demonstrated the trigger in a forward position, and the hammer in a forward (decocked) position. Had a single shot been fired from this type of weapon, the trigger should normally have been in its rear position, and the hammer cocked in its rear position. Even with the magazine removed, when a round in the chamber is fired the hammer should remain cocked in its rear position. Firearms analysts had tested the weapon and found that dropping it from various heights up to 5 feet onto a carpeted surface (as was the case at the scene) numerous times with the safety off and the hammer cocked did not cause the hammer to fall. Suspicion then fell on the witness for having possibly staged a suicide scene in order to conceal a homicide. It was concluded by the firearms analysis that the way in which the weapon was found could only have happened if: (1) the weapon had been fired without a magazine in place, and after firing had been manually decocked or the trigger pulled again; or (2) the weapon had been fired, and after firing the safety lever was used to decock the weapon, and then this safety lever was returned to the "fire" position (thus returning the trigger to the forward position).
>
> The issues involved were analyzed from the neuropathologist's perspective as follows. The preponderance of data concerning the gunshot wound favored immediate incapacitation, based on the gunshot wound path as well as the evidence of secondary peak overpressure effects (see Chap. 8). Any subsequent acts by the decedent would necessarily, therefore, be of an involuntary nature. In the latter category would be spontaneous or reflex movements, including convulsive activity.
>
> The neuropathologist consulted in this case requested additional testing on the weapon involved. It was found that 17½ to 18 lb effort was required to pull the slide back on the weapon and 4½ to 4¾ lb effort was necessary to manually cock the hammer. The trigger pull required to discharge the weapon was 5½ lb single action and 17 lb double action. Based on this information, it was considered at least theoretically possible that an involuntary convulsive muscular spasm could generate sufficient power to accomplish decocking the firearm. For example: (1) the authors recently reviewed a surveillance camera videotape in which a homicidal gunshot wound to the upper cervical spinal cord of the victim caused a fleeting startle-type motion immediately followed by collapse without further movement; and (2) exceptionally, seizures can cause fractures, and electroconvulsive therapy has used muscle relaxants to minimize this complication of uninhibited muscle spasm. However, the eyewitness in the case in question did not describe any convulsive-type movements by the decedent.
>
> It was considered much less likely that movements of the type described in brain death cases would have occurred in the moments immediately following this gunshot wound. Had they occurred, informal tests revealed that approximately 5 lb of resistance was required to bring a hand from alongside the body up to the face, with the body in the spine position. Through courtesy of Dr. J. A. Bueri[43] it was learned that the more commonly encountered finger jerks seen in brain death cases, while not measured as to strength with any device during his published studies,[315] are typically isolated, tiny, low-amplitude flexion jerks of one or two fingers.
>
> Speculation as to whether such spontaneous or reflex movements combined with the fall to the floor by the decedent, leading to various "cannot exclude" hypotheses, became unnecessary when the last request by the neuropathologist was carried out. Familiarity with a case in another jurisdiction in which repeated testing of a firearm prior to trial revealed no malfunction, only to have the malfunction claimed by the defendant occur in front of the jury when the identical procedure was performed during subsequent testimony by the firearms expert, led to a request that the firearm "drop on the carpet" test be repeated with the neuropathologist present. Shortly after the repeat test began, the weapon malfunctioned on more than one trial and the hammer dropped to the forward (decocked) position at impact despite numerous previous negative similar tests. Subsequent disassembly of the weapon revealed a worn sear. These results supported the account of events given by the witness.
>
> This case serves as a reminder that, although questions raised in some cases can lead to interesting conjecture about complex neural mechanisms for which few quantitative data are available, revisiting the possibility of a simpler explanation can be helpful.

Cerebrospinal Fluid Cytology in Stroke

CSF examination is of very limited value in the diagnosis of stroke type (e.g., ischemic versus hemorrhagic infarction, lobar intracerebral hemorrhage), size, or location (superficial versus deep). Fortunately, neuroradiologic imaging studies have largely solved such questions.

Nevertheless, familiarity with spinal fluid cytologic studies in stroke can be helpful when one examines stroke histopathology in tissue sections. The forensic neuropathologist, unlike an academic researcher who is often performing a detailed examination of a very restricted portion of the spectrum of changes in a given lesion, must examine the totality of abnormalities in a given case. It is necessary, for example, to determine whether the stroke patient has subarachnoid space cellular changes consistent with, or atypical for, the brain parenchymal lesion present. Inconsistencies must necessarily lead to additional studies. Published descriptions of the pathology of brain infarction rarely, if ever, comment in any significant detail on associated cellular changes in the subarachnoid space. CSF studies in stroke patients are an indirect source of data that can reasonably be extrapolated to determine what one may encounter in stroke case histologic sections. With such data as an initial guide, and increasing experience, one gains judgment on what is, and what is not, to be expected in a given stroke setting.

Cerebrospinal fluid studies in stroke patients suggest the following generalizations:

1. Most patients with ischemic infarction have normal spinal fluid cytology and total WBC counts with standard counting and differential counting techniques.[106]
2. If special techniques are used (e.g., sedimentation of CSF in order to study a larger number of cells, even in cases with normal total cell counts), up to 60% of stroke patients have greater than 5% neutrophils in the differential count.[338]
3. When patients with ischemic infarction have elevated CSF leukocyte counts, they are generally less than 100 WBCs/mm^3.
4. In stroke patients with elevated CSF cell counts, the initial prominent cell is the neutrophil, with peak counts at about day 3 to 4 following the stroke, and a gradual return to normal counts by about 2 weeks.[338]
5. In stroke patients with elevated CSF cell counts, monocyte/macrophage elements become the predominant cell population between the fourth and fifth day post stroke, as neutrophils began to diminish in number.[338]
6. Exceptionally, some patients with strokes develop CSF leukocyte counts suggestive of infectious meningitis. CSF WBC counts of from several hundred up to 1000 neutrophils/mm^3 have been reported in infarction,[106,338] and up to 21,000 mm^3 in intracerebral hematomas.[338] Evidence of subarachnoid blood was present in 38% of the intracerebral hematoma group in the latter study.[338] Such cases necessitate further studies to exclude infection (e.g., sepsis, endocarditis).[251]
7. Elevated CSF WBC counts, in general, are much more likely to occur in patients with hemorrhagic infarction and intracerebral hematoma than in patients with ischemic infarction.[251,338] One report suggested that ischemic strokes related to infectious or autoimmune vasculitis are more likely to be associated with elevated CSF leukocyte counts than are those related to arteriosclerosis.[79]
8. A more recent study did not find a correlation between ischemic infarct size and CSF WBC count, but did find a positive correlation between increased infarct size and an elevated peripheral blood neutrophil count.[351] The patients studied were all considered to have latent syphilis, and CSF samples were taken at different intervals in different patients, so that comparison to previous cited studies is difficult.

Malignant Cerebral Edema

Malignant cerebral edema has been described under various terms, including malignant cerebral swelling and hyperacute cerebral edema. The term *edema*, in this context, is generally used to indicate an increase in brain volume due to an increase in tissue water, whereas brain swelling implies an increase in blood volume as the main contributing factor to the brain volume increase. In our discussion, we use the term *malignant cerebral edema*, without implying that it is restricted to tissue water content (see later).

Malignant cerebral edema is a clinical term that is most often used to describe brain volume increase which is more rapid than, for example, that occurring in a typical case of regional cerebral infarction. Although the term *malignant middle cerebral artery edema syndrome* has been suggested for massive infarction related to internal carotid artery or middle cerebral artery occlusion that can cause a fatal herniation syndrome,[51] the bulk of the literature does not so restrict the implications of the term, and it is more often applied to post-traumatic events.

Following regional cerebral infarction, brain swelling is generally maximal in various studies in the time range of 2 to 3 days, 3 to 5 days, or up to 7 to 10 days following infarction, during which time herniation syndromes are most likely to develop.[84,254,333] The discrepancy between the duration of the initial cause of the swelling (e.g., infarction, blunt force trauma) and the development of secondary herniation effects may sometimes be apparent in neuropathologic examination by the relative prominence of marginal reactive changes associated with the original insult compared with the paucity of reactive changes at the sites of herniation, contusions, and Duret's hemorrhages, if the interval between initial insult and death is several days.

It follows that malignant cerebral edema develops "faster" than ordinarily expected, using the majority of postinfarction, postcerebral hemorrhage, or post-traumatic insults as a general guideline. We are occasionally asked, therefore, how fast can it develop?

Some authors apply the term to indicate cases in which symptoms and/or loss of consciousness develop "almost immediately or very quickly" following trauma, although the case cited in this particular report was also complicated by recent alcohol ingestion.[81] Lindenberg[204] described a 4-year-old boy who became "pale, then confused and soon unconscious," within a few minutes of sustaining a minor injury to his forehead, with death following 20 minutes later from respiratory failure. Autopsy revealed a minor bruise of the forehead and severe brain swelling, the bulk of which appeared to be white matter enlargement.

While generally considered to occur most commonly in infants and children,[40] malignant cerebral edema also occurs in adults.[193] Two cases known to us with fortuitous rapid documentation by imaging studies serve to illustrate the speed with which malignant cerebral edema may occur.

In the first case, a 23-year-old woman was struck by an automobile while riding her bicycle in front of a hospital. Serial clinical examinations initiated approximately 3 to 4 minutes' postinjury demonstrated unconsciousness and a progressive deterioration of neurologic status. Initial CT head scan, performed within 20 minutes of injury, revealed massive swelling of the right cerebral hemisphere with a right to left midline shift, with no evidence of intracranial hematoma. Treatment resulted in clinical and CT image study improvement by the second day postinjury, and she was discharged approximately 3 weeks' postinjury with cognitive functions unchanged from preaccident status and minimal motor findings on neurologic examination.[188]

The second case was a 20-year-old male baseball player struck by a baseball in the left retroauricular area. There was no loss of consciousness, and he was immediately placed in a chair. A few minutes later, headache and vomiting developed, and he was taken to the emergency room of a nearby hospital. CT head scan 30 minutes following the injury revealed massive swelling of the left cerebral hemisphere with a left to right midline shift and collapse of the left lateral ventricle. No intracranial hematoma or cerebral contusion was evident. Skull x-rays revealed no fracture. Treatment produced clinical improvement in the brain swelling by CT scan repeated 6 hours' postinjury. He had no sequelae at discharge 1 week after injury.[374]

These cases provide examples useful in forensic evaluations, indicating the rapidity with which life-threatening brain volume increases may sometimes unpredictably occur by still incompletely understood mechanisms. We do not see a reason to restrict the term to cases of diffuse, as apposed to unilateral, brain volume increase. Either event can be rapidly fatal within minutes to a few hours untreated, and may not always respond favorably to rapid, currently accepted treatment. The literature cited suggests that the brain volume increase in malignant cerebral edema is initially predominantly the result of blood volume increase (i.e., swelling), with an increase in brain water content (i.e., edema) becoming a significant component at a somewhat slower pace.

Wallerian Degeneration

The considerable variables involved in determining the time course of wallerian degeneration are almost entirely derived from experimental work in animals, and include the conditions of the experiment, species, individual variation within a species, age, axon diameter, temperature, severity of the injury (i.e., neurotmesis, axonotmesis, or neurapraxia[35]), proximity of the injury to the cell body, whether the nerve fiber is myelinated or unmyelinated, functional characteristics (e.g., motor versus sensory),[59] and whether the lesion is in the CNS or peripheral nervous system (PNS). It is generally accepted that wallerian degeneration occurs relatively more rapidly in the PNS than in the CNS; more rapidly in larger axons compared with smaller axons; in unmyelinated axons compared with myelinated axons; in younger animals compared with adults; and in mammals compared with cold-blooded animals.[67,177,347] Interspecies variations in these studies indicate that dating/data aging from one species may not be applicable to another species.

It follows that estimates in dating/aging of human cases, while not necessarily ignoring the abundant experimental studies, should primarily rely on the relatively few human studies available and not be extrapolated to include all age groups, should not assume that data from one specific nerve applies to nerves in all body sites, and similar concerns.

The general sequence, if not the pace, of morphologic events in wallerian degeneration is generally agreed upon, consisting first of axon and myelin sheath degeneration, clearing of cellular debris by macrophages, proliferation of Schwann cells in the PNS, and gliosis in the CNS.

Wallerian Degeneration in the Peripheral Nervous System

Axonal beading, seen early in the course of wallerian degeneration in the PNS, is regarded as a valid indicator of pathology in this setting irrespective of the demonstration that a similar change can be produced by nerve stretch and other experimental manipulations in animal studies.[262] The internal control used would be whether or not such changes occurred in anatomic sites beyond those under the influence of the lesion site.

In the PNS, motor nerve terminals cease functioning approximately 2 to 3 days faster than sensory nerve terminals, based on electrophysiologic studies in humans. Thus, axoplasm remains at least functionally intact for several days after axonotmesis. In the facial nerve,[27] evoked potentials cease within 5 to 7 days following axonotmesis, and in extremity nerves within approximately 9 days for motor fibers and 11 days for sensory fibers.[51] The axon is said to appear morphologically intact for approximately 3 to 5 days,[301] then begins to disintegrate, accompanied by myelin sheath degeneration and phagocytosis of debris both by Schwann cell and macrophages. The axonal breakdown, once begun, is so rapid that a spatial sequence (that is, proximal versus distal) is not clear in humans.[59] The morphologic sequence of events subsequently appears somewhat slower in humans than in several of the animal species studied experimentally. For example, in animal experiments the axon and myelin debris is generally removed from the endoneurial tube by the end of the fifth to eighth week, although a few macrophages may be seen up to 3 months after nerve injury.[347] Foamy macrophages containing myelin debris can be found in humans as long as 7 months after nerve transection.[59]

Wallerian Degeneration in the Central Nervous System

Wallerian degeneration in the CNS in humans has been studied primarily in long tracts distant from a given primary lesion, such as trauma, hemorrhage, or infarction. This avoids confounding variables such as more rapid access of blood-borne macrophages into the area as a consequence of vascular damage in the vicinity of such primary lesions. The following sequence of events, therefore, refers to events in long tracts beyond both the point of injury and the immediate margins of the primary lesion.

Somewhat more detailed information is available on wallerian degeneration in the human CNS than in the PNS. The following description is subdivided under the headings of axon, myelin sheath, microglia, and astrocytes, rather than by survival time groups used elsewhere in this text (Figs. 3.39–3.42).

Axon

No evidence of CNS wallerian degeneration of axons is described prior to 12 days' survival following a spinal cord lesion, or middle cerebral artery territory vascular lesions (in most cases described, the latter consists of middle cerebral artery territory infarction rather than hemorrhage). This summary of axon injury in the following references[49,285] combines data derived from neurofilament immunohistochemical staining, routine H&E stains, and silver stains for axons. Axonal degeneration, as determined by axonal swelling and/or fragmentation as well as by increased spacing between axons, is the earliest noted change. This has been noted first as early as 8 days' postinjury survival in dorsal columns (six segments removed from an injury site in the spinal cord), although the lateral columns remained normal in appearance until 4 weeks' postinjury.[285] Buss et al.,[49] however, first noted axonal degeneration in the corticospinal tract in the lumbar cord 12 days following a T6-level injury; no corticospinal tract axonal degeneration was seen by these same workers in another similar case at 11 days' survival.

By 14 days' survival after cerebral infarction, axonal injury was seen in the corticospinal tract in cervical, but not thoracic or lumbar, segments.

Twenty-four days following cord lesions, ascending pathways related to the cord lesion site were nearly

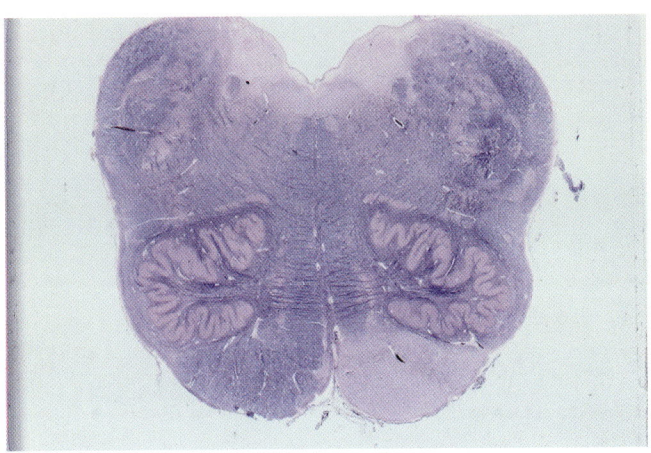

Fig. 3.39. Chronic CNS wallerian degeneration. Transverse section of medulla with myelin pallor of right pyramidal tract. (Luxol fast blue/H&E.)

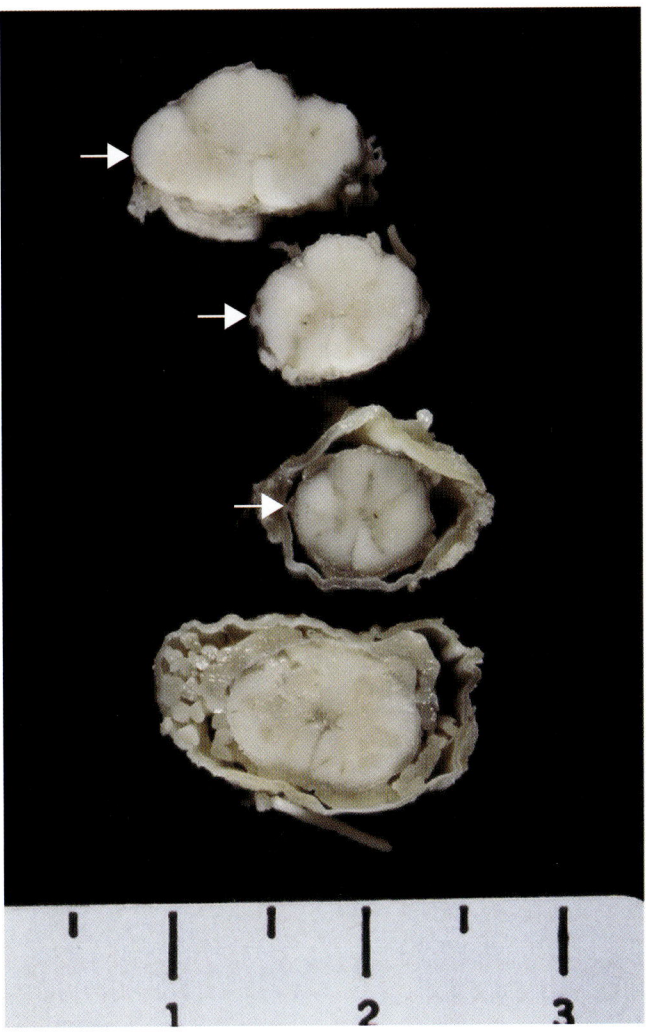

Fig. 3.40. Chronic CNS wallerian degeneration. Lateral (and ventral) corticospinal tracts are pale (and slightly more firm) bilaterally, but more obviously on left side of sections (*arrows*).

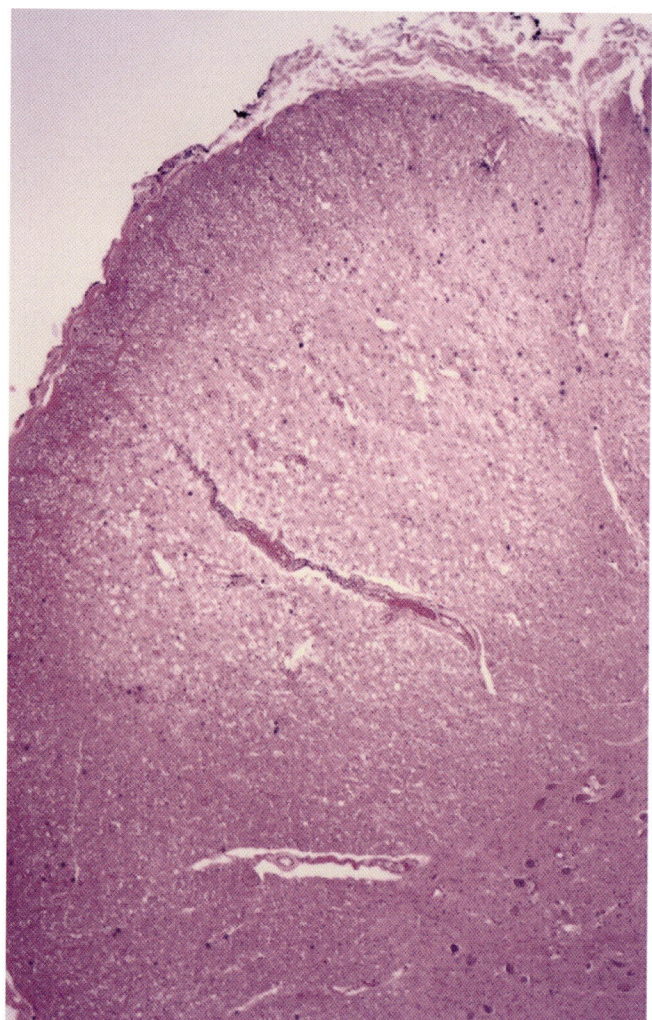 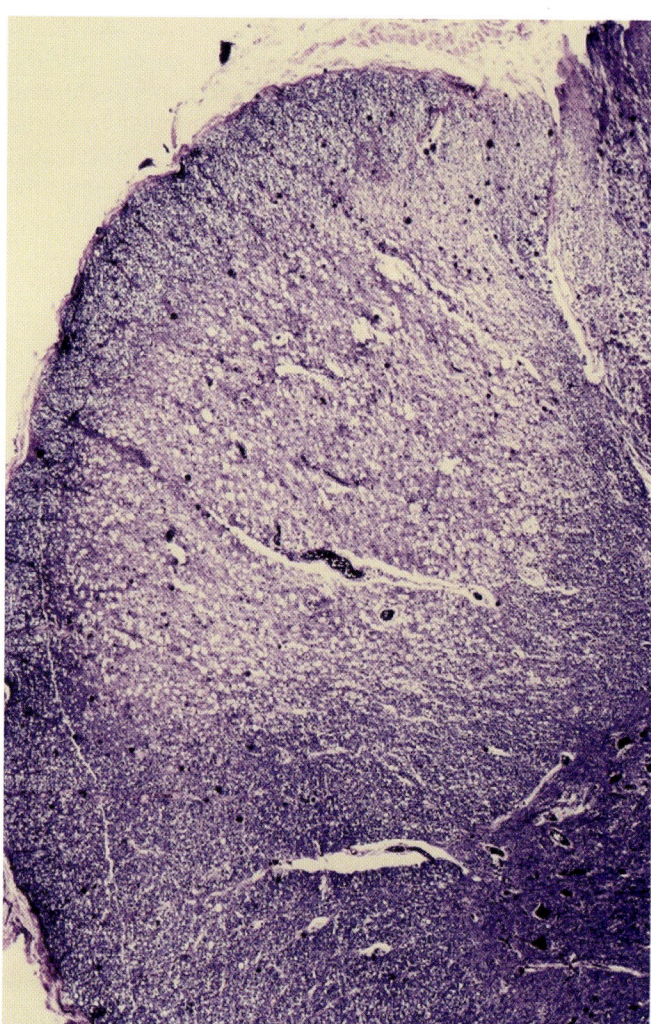

Figs. 3.41–3.42. Fig. 3.41 *(left)* (H&E) and Fig. 3.42 *(right)* (Luxol fast blue/H&E). Chronic CNS wallerian degeneration. Low-power micrographs of dorsolateral spinal cord with pallor due to axon and myelin loss with gliosis in lateral corticospinal tract.

devoid of neurofilament staining, with remaining axons showing an irregular and/or swollen appearance.[49]

Five weeks after a cerebral infarction, degenerating axons in the lateral corticospinal tract were seen as far distally as the lumbar spinal cord.

Four months' postinjury, axonal degeneration was sufficient to result in fluid and macrophage-filled spaces in involved tracts, surrounded by GFAP-positive processes.[49] One year or more following the initial lesion, only rare neurofilament-positive elements were found in the related ascending and/or descending pathways.

Myelin Sheath

Evidence of myelin degeneration was seen in the dorsal columns as early as 8 days' survival, six segments from the injury site in the spinal cord, although the lateral columns were unremarkable with regard to myelin sheaths until 4 weeks' postinjury.[285] Using myelin basic protein immunohistochemistry as an indicator of myelin breakdown, Buss *et al.*[49] found no evidence of myelin sheath breakdown until 14 days after spinal cord lesions.

Four months after spinal injury or cerebral infarction, corticospinal tracts contained swollen and irregularly shaped rings and amorphous structures, and reduced (but not absent) normal-appearing myelin rings based on myelin basic protein staining.[49]

Up to 3 years' postinjury, a few myelin basic protein weakly positive ringlike structures were present in the involved tracts.

Eight years after the injury, no myelin basic protein-positive structures were present in the involved tracts.

Microglia

The following observations were based on examination of microglia using leukocyte common antigen (LCA), major, histocompatibility class II (MHC II) (HLA-DR in humans), and CD68 immunohistochemistry staining.

There was no evidence of microglial activation in one case of spinal cord trauma 6 hours' postinjury.[323,324]

Within 3 days' postinjury, LCA and MHC II intensely positive cells were present in perineural intermediate gray matter and ventral horn gray matter. This finding was present bilaterally in spinal cord injury cases, and unilaterally in the middle cerebral artery territory infarction cases. In the cord lesion, this gray matter positivity was present as far caudal as the lumbar spinal cord. In the stroke cases, the gray matter positivity was more pronounced at cervical and thoracic areas than in lumbar areas. The morphology of the microglia was that of ramified cells. A few of the microglia were also positive for CD68 at 4 to 5 days, and more so at 10 to 14 days, but the CD68-positive cells were always fewer than the MHC II-positive cells. In one case at 5 weeks' survival, microglial nodules were also seen.

Between 5 weeks' and 1-year survival, LCA- and MHC II-positive cells decreased in gray matter, and showed a strong increase in the corticospinal tracts. The enhanced corticospinal tract staining intensity decreased to normal levels after 2 years. Most of the microglia resembled foamy macrophages. As with the shorter survival times, the corticospinal tract abnormalities were bilateral in the cord injury cases and unilateral in the stroke cases, and were also more prominent in cervical and thoracic areas than in lumbar areas in the stroke cases. Most of the microglia expressed both MHC II as well as CD68 positivity.

Two years after the injury, microglia were seen only in perivascular areas.[285] Slightly thickened blood vessel walls were also noted in the degenerating areas.

Astrocytes

No evidence of astrocytic response was found in CNS wallerian degeneration prior to survival periods of 4 months.[49,285] This included studies of survival starting at 3 days, and included cases with survival periods of 1, 4, 5, and 7 weeks. The first evidence of increased GFAP reactivity at 4 months' survival was in the tracts involved in wallerian degeneration in both spinal cord lesions and stroke cases, and included increased staining of both soma and processes of astrocytes.

At 1 year or more survival periods, dense GFAP-positive processes were present in the involved tracts,[49] and 3 years' postinjury, there was less intense staining of astrocyte processes. It was considered difficult to see the astrocyte soma based on GFAP positivity.[285] By 2 years' postinjury, GFAP staining in degenerating tracts was actually less than that of surrounding neuropil, and remained so in cases examined up to 23 years following injury.[285]

These findings suggest that, in CNS wallerian degeneration, the earliest degeneration occurs in presynaptic terminals and terminal axons, and subsequently proceeds centrifugally.[49]

Peripheral Nerve Regeneration

Following peripheral nerve wallerian *degeneration*, the pace and degree of nerve regeneration depends on a number of variables, several of which are noted earlier at the beginning of the section on wallerian degeneration. Additional variables specific to PNS *regeneration* include presence and nature of surgical intervention (for example, removal of scar and reanastomosis versus insertion of nerve graft; the state of the nerve stumps at the suture line; time interval between injury and surgery; degree of nerve stretch required for reanastomosis); the relationship between the site of injury and major branching points of the nerve; presence or absence of postinjury infection; and level of lesion in relation to the neuron soma and the muscle or skin termination point.[35,328,329]

The most consistent human data obtained, to our knowledge, relate to studies in cases of axonotmesis, despite the variation and severity of nerve injury encompassed by this term.[329] With regard to recovery of motor function, Seddon et al.[329] gave an estimate of recovery rate (roughly correlated with axon growth rate) of the radial nerve of 1.6 ± 0.2 mm per day after nerve suture, and 1.5 ± 0.1 mm per day after axonotmesis not treated by surgery. All motor nerves studied (including radial, ulnar, median, and perineal nerves) gave an average rate of 1.5 ± 0.2 mm per day after suture and 1.4 ± 0.1 mm per day after unoperated axonotmesis. The rate of recovery of sensory nerves, based on eliciting Tinel's sign after suture, was found to be 1.7 mm per day.[329]

These averaged nerve recovery rate estimates should be considered together with the additional knowledge that the pace of recovery slows as the recovery period is lengthened. An initial recovery rate as high as 3.0 mm per day fell to less than 1.0 mm per day by approximately 100 days after recovery began.[329] There is also evidence to suggest that the recovery rate is faster in children than in adults.[329] Both of these conclusions are supported by subsequent electrophysiologic studies.[42]

A final consideration is the extent of latent period to add to these recovery dates in order to allow for an initial delay in axon growth tips bridging the lesion area and commencing their more unimpeded period of growth down the still-intact endoneurial tubes, and the delay required to establish functional contact with muscle or skin receptors at their termination. Seddon[328] points out the lack of data governing such estimates, but based on clinical experience suggested an estimate of an additional 40 or 50 days as a general guideline. Bowden and Goodman[35] suggested adding an additional 20 days to the estimate for a mild intraneural scar in cases of axonotmesis, and an additional 50 days after operative reanastomosis. This latter figure should probably be doubled in a case of a nerve graft placement, since two suture lines must be traversed by the growing axons.[42]

Transneuronal Degeneration

In a number of unusual, and in some locations common, situations, it becomes important to consider the possibility of transneuronal degeneration. This phenomenon may provide a basis for proper interpretation of otherwise confounding anatomic findings that are sometimes quite remote from the primary pathology. Issue may also be raised as to the time required for these changes to develop.

Some groups of neurons undergo degenerative changes discernable by routine histologic methods when deprived of afferent nerves, or of the neurons to which they project. Animal studies support certain general conclusions regarding this form of degenerative change, which crosses synaptic boundaries.[65] The change tends to be more prominent in immature animals, and it varies from species to species in the same neuronal (nuclear) group. In most areas studied, the likelihood of observing transneuronal degenerative changes is proportional to the degree of deafferentation produced by a lesion. The degeneration in at least some neuronal systems appears to be biphasic, with an earlier stage of more rapid change followed by a chronic phase of slower change.

In most areas, the nature of the neuronal change is one of neuronal atrophy and, usually to a less extent, cell loss. Examples in humans include the visual, cochlear, trigeminal, olfactory, cerebrocortical, and limbic systems.[65,361] An exception is the dentatoolivary system reaction to deafferentation, in which inferior olivary nucleus neurons undergo hypertrophy for extended periods and gliosis is conspicuous. There is some evidence to suggest that transneuronal degeneration may occur in the spinal cord in humans from limb amputations, but not from corticospinal tract lesions.[350,356]

Dating/aging of transneuronal degeneration is largely unexplored in humans, and the following discussion is limited to those reports on human neuronal systems in which at least preliminary data concerning the pace of degenerative change is included. That is, case series with varying survival periods are included in which some quantitative data were presented, or single case reports with relatively brief survival periods that demonstrate some abnormality. The need is apparent for more data of this type in order to reach firm conclusions.

Visual System

Following unilateral eye enucleation or severe macular lesions, primary anterograde transneuronal degeneration occurs in the appropriate laminae of the lateral geniculate body (LGB) (i.e., laminae 2, 3, and 5 ipsilaterally and laminae 1, 4, and 6 contralaterally). Atrophy of LGB neurons, as measured by shrinkage of mean cell body area (in μm^2), was observed to be from 15.2 to 26% at 1-month survival, 23.3 to 45% at 2 years' survival, and did not exceed 51% in any LGB laminae in cases with survival times of 9, 56, 62, or 73 years following eye lesions.[189] No conspicuous LGB neuronal loss was appreciated within the first 2 years following enucleation (i.e., no more than 10% cell loss estimated), and a cell loss of only 20 to 50% could be determined in survival periods of 9 to 73 years. Significant individual differences in degree of transneuronal degeneration was apparent in this series of only seven cases. No obvious gliosis was seen, in contrast to that which accompanies retrograde degeneration in the LGB following lesions of the optic radiations.[189]

A serially examined single case suggested that some individuals may exhibit secondary retrograde transneuronal degeneration.[142] A soldier sustained a gunshot wound involving both occipital lobes at approximately 21 years of age. At surgery, near-complete loss of the left occipital lobe and severe, but less extensive, damage to the right occipital lobe were observed. Initially he was completely blind, but vision gradually recovered to a correctible 5/200 in each eye, with a right homonymous hemianopsia. On serial ophthalmic examinations, no optic atrophy was noted up to 11¼ months' postinjury. At the next eye examination, 3½ years' postinjury, temporal pallor of both optic discs was described and a diagnosis of early bilateral nerve atrophy was made. Reexamination at 5½ years' postinjury revealed severe optic atrophy, with chalky pallor of both optic discs and a visible lamina cribrosa.

Cases of unilateral or bilateral anopthalmia describing the LGB appearance have been reported, but the degree to which the abnormalities relate only to transneuronal effects is unclear.[97,249]

Trigeminal System

Penman and Smith[276] described alterations consistent with both primary and secondary anterograde transneuronal degeneration in a 73-year-old man who survived approximately 3¾ months following alcohol injection of the right gasserian ganglion for tic douloureux. Loss of pain and deep pressure sense in the right forehead and cheek was accomplished, and persisted until death from metastatic neoplasia. Histologic examination of the peripheral trigeminal (cranial nerve V) pathway revealed swelling, segmentation, and decreased number and disorganization in nerve fibers, together with fibrous tissue scarring in the sensory root of cranial nerve V at the injection site. Decreased myelin, increased numbers of Schwann cells, granular debris, an increase in leukocytes, and a relatively mild loss of ganglion cells were also noted. Silver stains demonstrated thickened pericellular networks around remaining ganglion cells, with some nerve endings being bulbous. Varicosities were present in some nerve fibers. In the central trigeminal pathway, the cranial nerve V root entering the pons was abnormal, with marked reduction in axis cylinders and myelin sheaths. Clumped and granular myelin debris and a definite reduction in neurons were seen in the chief sensory nucleus with many of the remaining neurons being dark and shrunken. A microglial reaction was present in this nucleus. The spinal tract of cranial nerve V was degenerating, especially in more distal portions in the medulla. Many of the medium-sized neurons of the spinal nucleus of cranial nerve V demonstrated swelling, peripheral displacement of the nucleus, and chromatolysis. A few of the smaller neurons in this nucleus were shrunken, with dark nuclei. Microglial groups were seen, suggesting possible neuronophagia. Myelin staining revealed pallor of the secondary spinal fibers in the hilum of the spinal nucleus of cranial nerve V. No abnormality was seen in the mesencephalic or motor nuclei of cranial nerve V. These findings were interpreted by the authors as transneuronal degeneration in the chief sensory nucleus of cranial nerve V and the nucleus of the spinal tract of cranial nerve V.

Dentatoolivary System

A very unusual form of degenerative change, often referred to as inferior olivary hypertrophy, may occur following lesions in the dentatoolivary pathway, typically supraolivary lesions.[194] The responsible lesion, often either a vascular occlusion with infarction or trauma, is usually found in the ipsilateral central tegmental tract, the contralateral dentate nucleus, or at the junction of the dentate's projection with the central tegmental tract in the region of the internal and dorsal surfaces of the red nucleus.[170,194] This has led to general agreement that these changes in the inferior olivary nucleus are due to anterograde transneuronal degeneration. Additionally, a topographic relationship between the primary lesion location and the location of the transneuronal degeneration within the inferior olive nuclear complex in humans has been confirmed by the examination of such cases.[170]

Often, though not invariably, such lesions are clinically associated with palatal myoclonus.[155,170] Likewise, not all cases with palatal myoclonus demonstrate olivary hypertrophy.[388] The onset of palatal myoclonus in relation to the causal lesion is usually a matter of weeks, but has ranged from 1 day to 30 months.[194] When unilateral, the palatal myoclonus is ipsilateral to the cerebellar lesion, but contralateral to the central tegmental tract lesion.[121] Palatal myoclonus is not seen with inferior cerebellar peduncle lesions of olivodentate fibers,[194] nor with focal lesions within the inferior olivary nucleus.[337]

To our knowledge, olivary hypertrophy is currently the only form of transneuronal degeneration in which long-standing enlargement of the targeted nuclear group occurs. Whether the swollen neurons described in the spinal nucleus of the trigeminal nerve 3¾ months after alcohol injection of the gasserian ganglion[276] would persist, enlarge, become vacuolated, or become associated with sufficient gliosis to cause visible enlargement on gross examination with longer survival periods has

never, to our knowledge, been described. Peña[274] speculated on several explanations for the conspicuous neuronal hypertrophy resembling central chromatolysis (but without neuronal loss or mammillary body hypertrophy) in rare cases of Wernicke's encephalopathy, including a possible transneuronal effect. We are not aware of any supporting evidence for this possibility.

Histologic changes resulting in inferior olivary nucleus hypertrophy as anterograde transneuronal degeneration include[121,337] enlargement of some neurons that may have a bizarre shape and cytoplasmic vacuoles (Fig. 3.43); some admixed neurons that are atrophic; reduced numbers of neurons; increased astrocyte number and size, some with thick processes and/or binucleation (Fig. 3.44); fine fibrillary gliosis in later stages and myelin pallor in the panniculus and intraolivary fibers. Empty spaces and argentophilic tangles, termed "residual bodies," are interpreted as markers of

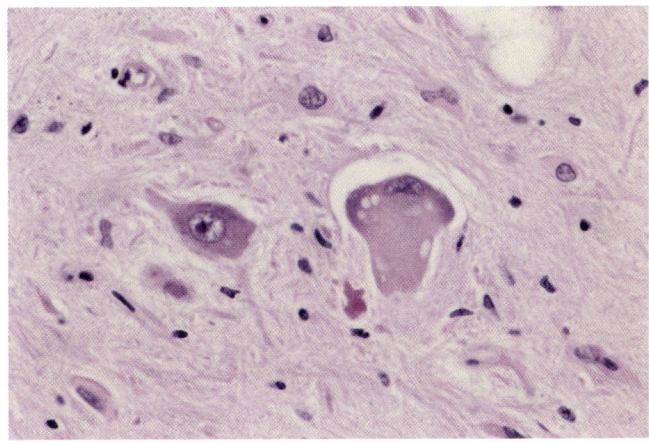

Fig. 3.43. High-power micrograph. Inferior olivary hypertrophy. Neuronal hypertrophy with cytoplasmic vacuoles. (H&E.)

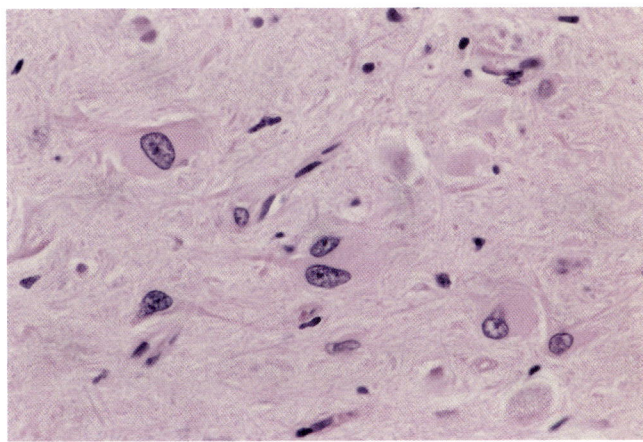

Fig. 3.44. High-power micrograph. Inferior olivary hypertrophy. Florid astrocytosis in inferior olivary nucleus. A binucleate astrocyte is present. (H&E.)

dead neurons. In at least some cases surviving beyond 8 to 9 months, there is a clear loss of neurons and large, bizarre neurons are less frequently found. However, collapse of the dense glial network or shrinkage of a previously enlarged olivary nucleus has not, to our knowledge, been documented.[121,130]

These histologic changes may be focal, multifocal, diffuse, unilateral, or bilateral; symmetric or asymmetric. Medial or dorsal accessory olivary nuclei are less frequently involved, and not necessarily ipsilateral to the main inferior olivary nucleus involvement. Gautier and Blackwood[121] ascribed the inferior olivary nucleus enlargement as most likely resulting from a combination of enlarged neurons and their processes, and the increase in size and number of astrocytes. The latter is the dominant component in long-term survivors, since neurons are often substantially reduced in number in the latter, and giant neurons much less frequent.

The pace of development of grossly apparent olivary hypertrophy has been difficult to determine, and nearly all described cases have been in adults. Rorke[309] indicated that she had not seen classic olivary hypertrophy in infants. Based on a review of the personal cases described in her book, it does not appear that lesion location in some instances, or survival periods in other instances, would have been sufficient to allow this lesion to develop even if that possibility exists in immature brains (see later). One of Jellinger's reported cases of olivary hypertrophy was a 12-year-old boy.[170]

Gautier and Blackwood[121] described classic gross and histologic changes of inferior olivary hypertrophy in cases with postlesion survival periods of 4 months (one case) and approximately 4 to 5 months (one case). The earliest histologic evidence of developing olivary hypertrophy reported is that of Jellinger,[170] who described "enlargement of olivary neurons with vacuolar degeneration of the cytoplasm and astroglial hyperplasia" in cases surviving 12 to 20 days' postlesion. Goto and Kaneko[130] studied eight cases of olivary hypertrophy with varying survival periods following pontine hemorrhage involving one or both central tegmental tracts. They found no evidence of olivary hypertrophy with survival periods of 24 hours, or of 2, 3, 5, or 7 days (decedent age range, 46–65 years). At 21 days' survival, a 51-year-old woman demonstrated microscopic evidence of inferior olive neuronal hypertrophy. Some neurons showed central chromatolysis, and others had minute invaginations containing an "eosinophilic corpuscle" that they referred to as an "insect bite appearance." They did not describe neuronal vacuoles or bizarre neurons in this case, and no glial reaction was seen. They also described similar small eosinophilic bodies ("corpuscle") within vacuoles in some of the vacuolated neurons encountered in the only two cases with longer survival periods in their study (i.e., 8½ months and 9½ months, respectively). The latter two cases demonstrated the classic findings described by Gautier and Blackwood[121] in their cases with 4-month, and 4- to 5-month survival periods, respectively. Goto and Kaneko's[130] morphometric analysis of the inferior olive nuclear area demonstrated some subtle increase in size at 21 days' survival, probably insufficient to appreciate by visual inspection of the brainstem, and more obvious enlargement at 8½ and 9½ months' survival. One case of ours with a unilateral ischemic central tegmental tract lesion surviving 21 days had, on H&E staining, only a suspicion of minimal neuronal enlargement of a few neurons in the ipsilateral inferior olivary nucleus with no neuronal vacuoles or chromatolysis. We acknowledge that, had we not been aware of the central tegmental tract lesion from a separate section of pons, the changes in the inferior olivary nucleus by routine methods in this case were probably too subtle to have alerted us to the presence of a central tegmental tract lesion.

MRI studies of cases with lesions likely to produce olivary hypertrophy tend to support the timetable suggested by the limited histologic studies noted earlier. Birbamer et al.[30] described two cases with increased signal intensity without hypertrophy in the inferior olivary nucleus in T2 and proton density-weighted images at 4 weeks' and 7 weeks' survival, respectively, following dentatoolivary pathway lesions. In another small case series, measurable enlargement of the inferior olive by MRI was not seen at 2 months' survival (three cases), but it was seen in cases surviving 4 months to 4 years.[365] One case examined by serial MRI showed no inferior olivary abnormality at 10 days' postlesion, but a gradual development of olivary hypertrophy was seen in this same case in MRIs performed at 6 months' and 10 months' postlesion.[357] No further increase in size of the inferior olives in this case was noted on repeat MRIs at 20 months' and 24 months' survival.

To summarize, the cases described earlier support preliminary conclusions that MRI imaging may, at least in some cases, demonstrate a hyperintense signal in an inferior olivary nucleus destined to develop olivary hypertrophy as early as 4 weeks postlesion. This hyperintense signal can persist at least 4 years postlesion,[365] but may become less intense and more irregular in serially examined cases by 20 months postlesion.[357] Measurable enlargement of the inferior olivary nucleus can be detected by MRI as early as 4 months postlesion.[365] The earliest histologic changes of developing olivary hypertrophy have been noted at 12 to 20 days postlesion.[170] The diagnosis of olivary hypertrophy is based on histologic criteria. Gautier and Blackwood[121] make the point that the "actual increase in size of the olivary nucleus is often of moderate degree," and in many cases "does not take place." It follows that when lesions of the dentatoolivary pathway are present, one should routinely submit at least one section of medulla that includes the inferior olivary nuclei.

Large lesions involving a cerebellar hemisphere, which not only include the dentate nucleus but also the remainder of a hemisphere, result in retrograde degeneration of the contralateral inferior olive due to the extensive olivocerebellar pathway injury. This will mask any potential for olivary hypertrophy. Rather, there is a contralateral atrophic, gliotic inferior olivary nucleus with extensive neuronal loss.[336]

In our experience, chronic lesions were discovered in a number of cases in the dentatoolivary pathway that were too small to be appreciated grossly, when lesions characteristic of inferior olivary hypertrophy were observed in an initial section of medulla. This finding should prompt submission of additional slides from the reserve material in the "save" bottle to search for lesions in the appropriate, more rostral portions of the dentatoolivary pathway.

Astrocyte Terminology

A number of terms have been used over the years to describe normal astrocytes, such as protoplasmic,[218] fibrous,[29,218] type 1,[138,259] type 2,[138,259] Bergmann glia in the cerebellum,[218] and Müller cells in the retina.[218] Other terms are used to describe reactive astrocytes, namely, gemistocytic,[133] reactive fibrous,[259] piloid,[275] Nissl-plump,[259] multinucleated, swollen, hypertrophied, bizarre, Alzheimer type I cell,[138] Alzheimer type II cell (Fig. 3.45),[138,259] and Creutzfeldt cell (Fig. 3.46),[45,62] Yet other terms refer to changes in processes of astrocytes such as Rosenthal fibers,[133] and corpora amylacea.[133] Increasingly appearing in the current literature are descriptive terms for astrocyte inclusions in various diseases, demonstrated by specific immunostains, such as for tau protein,[133] which are beyond the scope of this discussion.

Most of the aforementioned terms are well described in standard textbooks and review articles, such as those cited earlier. Two of these terms, one old (piloid astrocytes) and one newer (Creutzfeldt cell), however, provoke more questions and are briefly reviewed.

"Piloid" astrocyte is a term that, because it is the term used by the authors cited, is in some of the dating/aging sections in this text (e.g., cerebral contusions, needle puncture wounds). Yet, it is apparent that all authors using this term are not describing the same type of astrocyte. This becomes relevant in discussions of dating/aging of CNS lesions. For example, Oehmichen and Raff[266] described the first appearance of piloid astrocytes (as a reactive change) in human brain contusions at 6 days' postinjury. Baggenstoss et al.,[20] examining ventricular needle tracks in human brains, described a variety of astrocyte reactive changes over time. They mentioned hypertrophied astrocytes with eccentric nuclei, binucleate forms, and gemistocytes, but they did not report the appearance of the "piloid astrocytes of Penfield" until

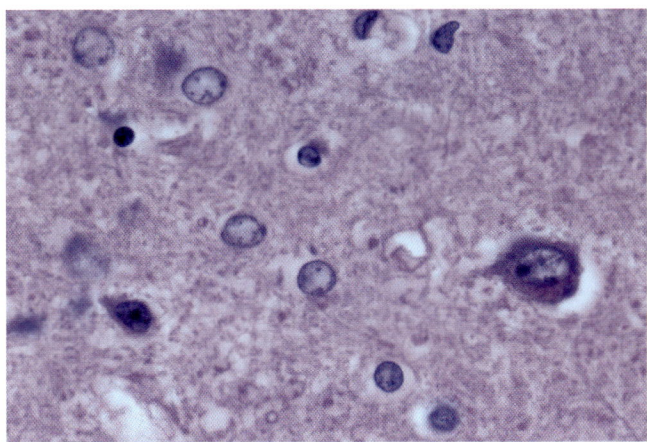

Fig. 3.45. High-power micrograph. Alzheimer type II cells are represented by the enlarged astrocyte nuclei with pale centers (nuclear chromatin is peripherally situated). Nuclear shape may be variable (round, elongate, kidney-shaped, indented). In forensic material they are most often encountered in cases accompanied by liver failure. (H&E.)

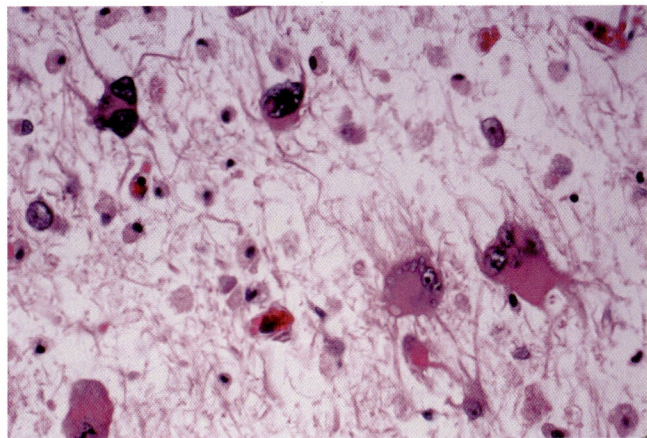

Fig. 3.46. Creutzfeldt cells in a case of progressive multifocal leukoencephalopathy. The two large cells in the right lower quadrant of the photo are the most typical examples (see text for details). (H&E.)

196 to 206 days' survival. Such astrocytes were not described in their 75- to 80-day survival group.[20]

Penfield's description of the piloid astrocyte refers to the origins of the word (*pilus*, hair, and *oides*, like) and indicates its "outstanding characteristic" to be the "length and slenderness of the fibers." He further described the cell bodies as "elongated," and describes this "hair-fibered gliosis" as occurring in (among other lesions) "brain wounds of long standing and in cases of cerebral destruction from birth injury."[275]

It seems evident that Penfield's piloid astrocyte is what has been termed by more recent authors as the fine fibrillar astrocyte of chronic, not acute to subacute, lesions.[157] If seen in acute to early subacute lesions, it is either an acute lesion superimposed on a more chronic lesion that produced fine fibrillar gliosis, or an application of the term in some fashion other than that of Penfield. Fortunately, investigators such as Oehmichen and Raff[266] assist us by defining their terminology. They apply the term "piloid" to any astrocyte with Holzer-positive fibrils. Thus defined, their piloid astrocyte would include much earlier reactive astrocytic changes such as hypertrophied astrocytes and gemistocytes, which may show Holzer- and phosphotuagstic acid hematoxylin (PTAH)-positive fibrils, but do not resemble the piloid astrocyte as Penfield[275] and others (including Baggenstoss et al.,[20] and ourselves) apply the term.

Parenthetically, if stains for astrocytes other than GFAP are used, it has been suggested that a modified form of Mallory's PTAH method is more glial fiber-specific and less toxic than the Holzer stain.[220]

The Creutzfeldt cell (i.e., granular mitosis) (Fig. 3.46) is an astrocyte described as a "large, glassy, eosinophilic cell with nuclei fragmented into multiple, small, and seemingly viable micronuclei of varying size."[45] It is also described as an enlarged astrocyte with bizarre morphology and nuclear fragmentation.[62] It has thus far been found in cases of multiple sclerosis, glial neoplasms, and "reactive lesions" not otherwise specified.[45,62] It is distinct from karyorrhexis and mitosis.[45] Such cells have also been observed at the margin of subacute hemorrhages, and at the margin of necrotic lesions from various infectious agents. In an experimental study of stab wounds in rat brains, Creutzfeldt cells were observed in a subacute stage of the lesions.[165]

Blood Cells in Central Nervous System Lesions

Among the questions we are occasionally asked is "How long do blood cells that leave the intravascular compartment survive in CNS tissues?" Or, as another approach, "Are disintegrating neutrophils useful in dating/aging the lesion in which they are found?" To some extent we have addressed the relative increase and decrease in blood elements in the discussion of reactive changes in several of the dating/aging sections in this text. But a definitive answer to this question is more elusive, as the following examples demonstrate.

Erythrocytes (Red Blood Cells)

Mature RBCs in the circulating blood have a life span of approximately 120 ± 20 days.[139] In extravascular sites, such as CSF or subdural hematomas, their survival may be considerably less. As just one example, some RBCs entering CSF may demonstrate evidence of hemolysis in less than 12 hours.[360] Erythrophagocytosis may be seen as early as 1 to 2 days after SAH,[151] and RBCs are cleared from CSF within 20 to 30 days following SAH.[363]

An incidental observation which may be useful in occasional cases is that RBCs that have entered subcutaneous tissue in the perimortem period undergo degeneration less rapidly than RBCs that remain in intravascular spaces, in cases with more advanced postmortem autolysis/putrefaction change.[299] A trichrome stain highlights the more intact extravascular RBCs in contrast to the autolyzed intravascular RBCs more distinctly than H&E staining. Whether this observation applies to other tissues, including brain, is unknown.

Neutrophils (Polymorphonuclear Leukocytes)

Mature PMNs reportedly have the shortest life span of all leukocytes, variously estimated as from a few hours up to about 5 days.[39,269,322] In the normal situation, they may leave the circulation to enter tissues and body cavities to perform various functions or become excreted (e.g., via intestinal tract, salivary glands).[39] The period of time spent by the PMN in peripheral blood does not reflect its total survival time or life span potential in various tissues, either in a normal state or in lesions.[53] Recent studies, primarily examining human neutrophils in vitro and neutrophils in experimental animals, have identified several factors (including reactive oxygen species, and a variety of inflammatory mediators) that can either shorten or prolong neutrophil life span.[195,217,322] Further research will be needed in order to clarify the determinants of PMN survival in various types of CNS lesions in which they participate (e.g., whether persistence in lesions beyond about 5 days is due to prolongation of life span, or continued recruitment from the circulation, and so on).

Monocyte/Macrophages

The mature monocyte, after approximately 12 hours in the peripheral blood circulation, enters tissues and differentiates into the macrophage type dictated by the tissue it inhabits.[39] Its fate and life span in tissues are quite variable, depending on a number of poorly understood factors. It continues to be capable of cell division. A number of animal experimental studies support the current consensus that, particularly in traumatic CNS lesions, the bulk of the cerebral macrophages are derived from circulating monocytes rather than from intrinsic resting microglia.[92] Data from human bone marrow transplantation studies also tend to support this concept, concluding that at least small numbers of donor marrow-derived monocytes can be found in a perivascular location within the CNS in cases with post-transplantation survival periods from 5 to 28 months.[366] One possible case of CNS graft-versus-host disease in a bone marrow transplant recipient has been reported, also suggesting that the lymphoid and monocytoid cells found in the CNS were donor marrow-derived.[313]

How long individual macrophages survive in cerebral lesions is less clear. In human cerebral contusions, "hemosiderophages" were noted in lesions up to 44 years after the injury, and "scavenger cells" were found in lesions 58 years after the injury.[264] Whether these cells included a portion of the population that entered the acute lesion, or rather reflected cell turnover of successive generations of replacement macrophages (either by local macrophage replication or by bone marrow-derived monocytes) is unclear.

This question is perhaps less illogical than it might initially appear. The monocyte life span is, as in the case of PMNs, altered in experimental studies by various tissue factors, including some inflammatory mediators.[349] Experimental studies in mice have indicated no significant replacement of peritoneal macrophages in a steady state (i.e., after injection of PKH-1, a dye that selectively labels resident peritoneal macrophages, but without further treatment) over a 49-day period (the duration of the study).[242] However, mouse peritoneal macrophages involved in an acute inflammatory response were largely (an estimated 73% plus) derived from circulating monocytes rather than by local proliferation of resident peritoneal macrophages.[368] Human bone marrow transplantation studies have indicated that pulmonary alveolar macrophages under these conditions have a life span of approximately 81 days, being replaced by donor marrow-derived macrophages.[358] Persistence of smoker's inclusions in pulmonary alveolar macrophages were present for at least 2 years following transplantation of a smoker's lung to a nonsmoker's lung.[224] These latter authors noted, however, that their data could be explained by several factors, including persistence of inclusion-bearing macrophages initially present in the transplanted lung; by alveolar donor lung macrophages that divided in the lung and kept their inclusions; or by particles released from old alveolar macrophages and rephagocytosed by young macrophages derived from the recipient's marrow.

We conclude that macrophage life span can vary in different tissues and under different pathologic conditions, that their appearance in very chronic lesions can be theoretically accounted for without necessarily requiring that the macrophage life span is identical to the age of the lesion, and that we are largely ignorant as to the life span of individual macrophages in most CNS lesions. Given these variables, the first appearance of macrophages is currently more useful for dating/aging of a lesion than their total population or persistence.

Lymphocytes

There is evidence, both in experimental animals and in humans, of two major populations of lymphocytes with respect to life span. One population, estimated to be at about 15% of the lymphocyte population, has a short life span of approximately 3 to 4 days; the majority of lymphocytes are able to survive 4 years or more, with some estimated to be capable of surviving up to at least 10 years.[39,131] The influence of various immune and inflammatory factors on lymphocyte survival in various tissues, including CNS, is not yet well understood, and likely varies in different types of lesions.

Other Leukocytes

Studies on the life span of human eosinophils are sparse. One study suggests a mean blood half-life of 12.4 ± 2 hours in normal subjects, but this does not necessarily reflect survival time in tissues in normal or pathologic states.[70] An *in vitro* study of human basophils suggested a survival period of no more than approximately 2 weeks, although some life span modifications could be induced by certain hemopoietic-growth factors.[384] Their life span in various CNS lesions is unknown to us.

Do Reactive Cell Changes of Immature Versus Mature Brain Differ?

The answer to this question varies with the developmental age chosen for comparison with mature brains. In Chapter 2, reactive cellular developmental landmarks were reviewed in Table 2.1, which emphasized the point that the response to a given CNS insult cannot include a given cell type until that particular cell appears in the CNS and has matured enough to respond to an insult. On this point there is no particular disagreement in the literature, although estimates as to the exact gestational age at which a given cell type becomes capable of reacting varies in different studies, as discussed in Chapter 2.

There is less information and, therefore, more controversy on whether there are any significant differences in the nature or pace of CNS reactive cell changes in the brains of the late gestational age fetus, neonate, or young child compared with mature brains.

Somewhat more data concerning this question exist regarding hypoxic-ischemic injury in the immature brain compared with that available for hemorrhagic lesions. This review of selected sources is in a format similar to that which has been used elsewhere in this text for lesions of similar etiology in mature brains, to facilitate comparison.

Since dating/aging of certain cell reactions in traumatic and hemorrhagic lesions differs from that seen in hypoxic-ischemic lesions in adults, it was arbitrarily chosen to separately list sources in which results for ischemic and hemorrhagic lesions are seamlessly combined in the summarized results. For example, a table (Table 16.6) in one text provides a timetable of cellular and tissue reactions for hypoxic-ischemic lesions in neonates,[17] but its source is indicated as modified from Ellis et al.[99] The latter study was based on seamlessly combined hypoxic-ischemic and hemorrhagic lesions in neonates, and differences, if any, in reactions to these two types of lesions could thus be obscured.

Squier[339] indicates that data provided are from the fetus and neonate, although the precise fetal gestational ages are not given. The same text elsewhere defines the neonatal period as less than 28 weeks from birth.[231] Further, the cellular reactions described are not specified as to lesion etiology, but rather appear to be generalizations based on data derived from hypoxic-ischemic as well as hemorrhagic lesions, and possibly hypoglycemic and other etiologies.

Such combined lesion results are cited only if a particular reaction type is not mentioned in studies limited to nonhemorrhagic or hemorrhagic lesions, respectively. This section concludes with a brief summary on whether there are any convincing differences in the nature and/or timing of reactive changes of forensic significance in the data reviewed, when comparing immature and mature brains.

Hypoxic-Ischemic Lesions (Predominantly)

Sarnat's[317] data refer to reactions in the periventricular and subcortical white matter in preterm and term neonates. The data of Kinney and Armstrong[181,182] refer to the terms "neonate" and "infant," not otherwise specified (NOS). Volpe's[373] data are for the "perinatal period." One of Fawer et al.'s cases (Case 15) of periventricular leukomalacia (PVL) was 37 weeks' estimated gestational age (EGA).[103] The PVL case cited by Rodriguez et al.[305] was 27 weeks' postmenstrual age (PMA). The study of cases of PVL by Banker and Larroche[22] was based on 51 infants aged 6 hours to 13 months, of which 40.9% were premature (defined as birth weight of < 5½ lb/2.5 kg). All of their cases were believed to have suffered severe "anoxia," and the earliest EGA studied was 23 weeks.[22] Five of their cases had subependymal hemorrhage with intraventricular rupture, and they were not separately described in their results. The extent to which this could have influenced the earliest appearance of a given cell reaction is thus unclear. The specific case origins for the timetable data provided by Kinney, Haynes, and Folkerth[183] are not specified. These considerations should be kept in mind when reviewing the results discussed.

Microscopic Examination of Hypoxic-Ischemic Lesions in Immature Brains

Survival Period
- 3 hours
  — Microglia accumulate at periphery of infarct.[317]
  — Coagulation necrosis present, with axonal "balls" at lesion margin.[22]
- 6 hours
  — Coagulation necrosis; axonal retraction balls at lesion margin; pallor on H&E stain.[317]
- 8 hours
  — Astrocyte degeneration within infarct zone.[22,317]
  — Activated microglia at margin of infarct.[22]
- 12 hours plus
  — Reactive astrocytes and microglia at infarct margin.[22,317]
  — Capillary endothelial hyperplasia.[22]
- 24 hours
  — Endothelial cell hyperplasia at infarct margin.[317]
  — *After* 24 hours, endothelial cell hyperplasia.[22]
  — Coagulative necrosis (e.g., hypereosinophilia of all cell types, nuclear pyknosis, and axonal spheroids present.)[183]
- 24 to 36 hours
  — Red neurons appear.[373]
- 24 to 48 hours
  — Initial evidence of ischemic injury: red neurons, loss of Nissl substance, and pyknotic or karyorrhectic neuronal nuclei (presumably in term infants).[181,182]
- 1 to 3 days
  — Prominent endothelial swelling.[317]
- 3 to 5 days
  — Initial appearance of macrophages.[181,182]
  — Hypertrophic astrocytes appear.[373]
- 5 days plus
  — Macrophages appear in the infarct zone.[317]
  — Macrophages appear in clusters.[22]
- 5 to 7 days
  — Neovascular proliferation.[317]
- 1 week
  — Organized necrotic foci with infiltrating macrophages and reactive astrocytes at margin.[183]
- 7 to 14 days
  — "Porencephalic cyst" can develop in area of ischemic infarct in neonates within this period.[102]
- 8 to 12 days
  — Subependymal and deep white matter microcysts and small cavities; ependymal cell loss (although the latter may be minimal).[317]
- 10 days
  — Cystic change (in PVL) first seen by ultrasound.[103]
- 14 days
  — Infarct zone and cavities filled with macrophages.
  — Glial proliferation at margin.[317]
  — Neovascularization appears.[22]
- "Within a few weeks"
  — Cavitation and periventricular cysts with, in some cases, subsequent cyst collapse and glial scar, typically with lipid-laden macrophages and mineralized axons.[183]
- 2- to 24-week period
  — White matter atrophy, cavitation, and microscopic pseudocyst formation.
  — Subependymal glial nodules.

- Mineralization of cyst walls.
- Multicystic encephalomalacia.[317]
- Cavitation increases.
- Axonal basophilia (occasionally positive for iron and/or calcium).[22]
* 15 weeks
  - Periventricular cysts (in PVL) up to 0.9 cm in diameter (per ultrasound at 7 weeks of age) are no longer present at autopsy at 15 weeks of age.
  - Glial scar present at the site.[305]

Comment. Much of the aforementioned information is based on studies in PVL lesions, and a curious feature is the apparent lack of participation of neutrophils in the reactive changes.[182]

Hemorrhagic Lesions

Cases of spontaneous subependymal/intraventricular hemorrhage form the basis for the following cellular reaction timetable in neonates. Sherwood et al.[335] studied 27 cases, EGA from 23 to 36 weeks, and survival periods from 4 to 59 days.

Gross Appearance of Hemorrhagic Lesions in Immature Brains

Survival Period
* < 14 days
  - Hemorrhage purplish-red.[335]
* > 14 days
  - Tendency for hemorrhage to develop a brownish tint "to greater or lesser extent."[335]

Microscopic Changes in Hemorrhagic Lesions in Immature Brains

Survival Period
* 5 days
  - First macrophage response (in a 36-week EGA infant).[335]
* 8 days
  - First capillary proliferation seen (in 28-week EGA infant).[335]
* 10 days
  - First astrocyte response seen in subependymal zone near hemorrhage, but not around hemorrhage itself (in 28-week EGA infant).[335]
* 13 days
  - First macrophage response (in a 28-week EGA infant).
  - First evidence of extramedullary hematopoiesis in the subependymal area (seen in only 4 of 27 cases in this series).
  - First capillary proliferation seen at margin of hematoma (in 28-week EGA infant).[335]
* 14 days
  - First astrocyte reactivity that includes slight fiber production at margin of hemorrhage (in 29-week EGA infant).[335]
* 59 days
  - No hematoidin was encountered at this longest survival period, or in any of the shorter survival periods studied.[335]
* Chronic
  - Hemosiderin is often eventually cleared completely following milder grades of periventricular hemorrhage in premature infants in contrast to adult brain hemorrhagic lesions, where it may persist for years.[317]

Comment: Sherwood et al.[335] noted the difficulty in identifying astrocytes with the stains employed (prior to GFAP availability). Furthermore, they described the astrocytic response to hemorrhage in the subependymal plate region to be slower, with less fibrillar astrocyte component, than in adults or in other areas of neonatal brains of the same age group. In the one case in their series in which they described significant simultaneous hypoxic injury, the astrocyte response at 59 days' survival was "more mature" and fibrillar than in the subependymal areas, suggesting regional differences in astrocyte response to insults.

The timetable format for premature brains used by Darrow et al.[73] was based on a division into seven stages determined by sonographic appearance. Each of these stages encompassed a rather broad survival time frame from a minimum of 20 days to a maximum of 119 days for stages 1 through 6. A single case was at stage 7, which had a survival time of 290 days. There is difficulty in converting their gross and microscopic data to our preferred format. For example, it is indicated that grossly the hemorrhagic clot is solid in stages 1 through 4, encompassing survival periods varying from 1 through 61 days, and becomes partially liquefied in stage 5, which includes survival periods from 1 to 120 days. One thus cannot determine whether the earliest survival period in which the clot becomes partially liquefied is at 1 day's or 120 days' survival, or somewhere between these extremes. Similarly, Darrow et al.[73] indicated that stages 4 and 5 demonstrated prominent macrophages with hemosiderin and hematoidin pigment, with the main difference between the two stages being that hemosiderin predominated in stage 4 and hematoidin predominated in stage 5. The survival periods listed for stage 4 are 41 to 61 days, and for stage 5 are 1 to 120 days. Clearly, the data were intended to relate histologic findings to sonographically determined stages, and not to clarify earliest survival period in which a given histologic change occurred.

Other data, however, can be derived from the observations of Darrow et al.[73] that are more relevant to our area of emphasis. First, no mention is made of a polymorphonuclear response in their cases. Second, they found no significant difference in macrophage response to cerebral hemorrhage in premature infants compared with adult cases of intracerebral hemorrhage that they studied, at least up to the stage (stage 3) in which they described hemosiderin-laden macrophages surrounding the hemorrhage. They described more hemosiderin in adult macrophages, possibly related to larger hemorrhages occurring in adult brains, and more conspicuous acquisition of iron pigment in astrocytes in adults compared with the premature brains. In later stages, the main difference they described in mature versus immature brains appears to be the greater prominence of the hematoidin pigment in the adult brain hemorrhagic lesions, even though its earliest appearance was in sonographic stage 4 (41- to 61-day survival period) in both the prematures and the adults. Their original article should be reviewed for further details.[73]

Miscellaneous Immature Brain Reactions, Unspecified As to Lesion Etiology

Mineralization

Ferrugination of the neuronal soma is said to occur within 10 to 12 days'[339,340] to 21 days'[99] postinsult in immature brains. Ferrugination of axons is said to occur in the 8- to 14-day range in the literature review of Ellis et al.,[99] but their material revealed this change in only three of their cases of "greater than 14 day" survival. Calcification, not otherwise specified, was noted in one case in the 8- to 10-day period.[99] (See also Case 3.2, Fig. 3.9, earlier in this chapter.)

Use of iron or calcium stains to confirm the presence of either of these minerals is not always specified,[339] and dark basophilic neurons or axons are not always positive with these stains.[22,309]

Other Reactions

Red neurons are said to occur about "5 to 6 hours'" postinjury, microglial response "within 2 to 3 days," and neonatal brain swelling is said to develop "within 1 hour" and "usually subsides by 7 days."[339] Central wallerian degeneration in the corticospinal tract is described as much more rapid in immature CNS than in the adult, "approaching the rates observed in peripheral nerves."[200] This latter study was based on sections stained by oil-red-O and by the Marchi method. The only reference cited in support of this observation is an abstract.[196] A detailed report of this apparently dramatic difference in immature versus mature brain reaction to a specific stimulus, accompanied by photographs, is not known to us.

Conclusion

Too few immature brains have been studied to reach firm conclusions regarding differences in the nature and pace of most cellular reactions in premature, term, and neonatal brains compared with mature brains. For example, the study by Banker and Larroche[22] in PVL, by our review, seems to be the source of most later statements concerning some of the cellular responses in immature brains being faster than responses in mature brains. As noted earlier, 5 of their 51 cases included hemorrhagic lesions, and we cannot exclude the possibility that this may have biased their results toward a more rapid response to some extent, similar to that seen in our experience of adult hemorrhagic lesions secondary to trauma. If their results are confirmed in subsequent case studies in which both "pure" ischemic versus hemorrhagic lesions are separately examined, data more persuasive in a forensic setting would then be available.

The presently available data do suggest a few directions for further study:

- Prior to 28 weeks' gestation, and for an ill-defined period after 28 weeks' gestational age, astrocytes present in the subependymal zone specifically may be less capable of developing a conspicuous fibrillar gliosis response to insults compared with astrocytes in other brain areas.
- Macrophages in premature brains seem more able to convert hemoglobin to hemosiderin than to hematoidin.
- Neutrophils do not appear to play a significant role in the reactive changes in evolving lesions of PVL in the studies to date, and they are not described in the study of periventricular/intraventricular hemorrhagic lesions by Sherwood et al.[335] in the 23 to 36 weeks' gestational age group. The emphasis of the latter authors was on macrophages and astrocytes, however.
- The statements in the pediatric neuropathology literature regarding the increased speed of development of microglial responses, axonal retraction bulbs, reactive astrocytes, endothelial hyperplasia and hypertrophy, mineralization, and possibly cystic change are intriguing, but further studies are indicated for confirmation.

Denervation of Muscle

Routine histologic changes in human muscle following denervation have received relatively little attention since the era of human biopsy and necropsy studies on poliomyelitis and peripheral nerve injuries many decades ago. Since direct effects of poliovirus infection on muscle could conceivably alter some of the histologic changes described,[164] more reliance is placed on changes described in muscle following peripheral nerve trauma as a guide to dating/timing of denervation issues. For forensic cases, we further confine our information sources to human material, since there is considerable variation in the nature and pace of the histologic changes in denervated muscle in animals compared with humans, in different individuals within a species, in different muscles and muscle fiber types in the same animal, and depending on mechanical, thermal, and electrical factors to which the subject is exposed.[35,347,348]

Following muscle denervation in humans, the major changes in muscle evident by routine H&E staining in one series of 140 biopsies from 86 cases[35] are emphasized in the following paragraphs. Supplemental information from other selected sources is included.

First 3 Months Following Denervation

Bowden and Gutmann[35] described no major change in appearance of muscle fibers during this time interval. Minor changes included some loosening of fibrillar structure within myofibers, increased waviness of the myofiber as seen in longitudinal sections, more conspicuous spaces between myofibers, and slight increase in fine dark granules at the edges of "Q stripes" (Q stripes are an older terminology for what corresponds to the combined A band and H zone in more recent studies[61]). Potential sources for this granular appearance are several, based on electron microscopic studies in denervated muscle.[50] Subsarcolemmal nuclei appeared more frequently in groups of two, three, or more rather than singly, and central nuclei became evident. Nucleoli of subsarcolemmal nuclei were somewhat more numerous (from one to four rather than the usual one to two), and end-plate nuclei were slightly more closely packed. No hyaline, granular, or fatty degeneration was seen, and cross-striations were maintained.

Intramuscular arteries appeared thickened, with hypertrophic media and some lumen narrowing. Veins were frequently dilated and appeared congested. Capillaries appeared more intensely stained, possibly due to thickened endothelium. There was little increase in connective tissue or fat, and that which did occur appeared more prominent around larger blood vessels more obviously than between myofibers. Mononucleated cells, mainly lymphocytes or possibly including some fibroblasts, were increased around smaller vessels.

From 4 Months to 1 Year Following Denervation

Development of marked fiber atrophy occurred by the end of this period, with the mean diameter of 100 myofibers in the flexor carpi ulnaris muscle being 20 μm, and many no more than 5 to 10 μm in diameter. For comparison, the mean muscle fiber diameters from normal adult human limb muscles varies from 61 to 69 μm in males and 42 to 53 μm in females.[93] In children, mean muscle fiber diameter is 15 μm at birth, gradually increasing to adult size by approximately 13 years of age.[38]

The degree of muscle atrophy was quite variable in various fibers within the muscle, and hypertrophic fibers appeared in some cases (the latter possibly representing, at least in part, some reinnervation effects; see later). Cross-striations were still usually present, but were sometimes more difficult to see in routine stains. There was a more striking arrangement of subsarcolemmal nuclei in rows, often with increased size and number of nucleoli. On cross-section, atrophic fibers were often angulated in outline.

Larger intramuscular nerve trunks were recognizable, and some contained macrophages. Smaller nerves were not well seen, probably obscured by associated fibrosis. A more conspicuous increase in fibrous tissue and fat between myofibers might be present, but the degree of increase in these elements showed considerable individual variation even at similar periods of denervation. Artery media thickening with lumen narrowing became more obvious, and capillaries were more irregular in appearance.

From 1 to 3 Years Following Denervation

More conspicuous atrophy was seen, although still quite variable in amount from myofiber to myofiber within a muscle (many myofibers were in the range of 3–5 μm in diameter). Striations were still intact, but special methods for visualization (e.g., polarization) were required on some cases. Granularity was present within some shrunken sarcolemmal fibers. Nuclear chains of subsarcolemmal nuclei were present, with fewer nucleoli. It was difficult to identify end-plate nuclei, and there was a further increase in connective tissue and/or fat in some cases. There was a further tendency for arterial thickening.

Beyond 3 Years Following Denervation

There was continued, unequal myofiber atrophy together with apparent splitting or fragmentation of myofibers. Occasional elongate sarcolemmal nuclei resembling nuclei of fibroblasts were present, and other sarcolemmal nuclei were pyknotic and in chains. Only an occasional large nucleus was seen. Beaded myofibers as well as some round or oval myofiber fragments ringed by nuclei could be seen. The latter often contained granular material, vacuoles, and droplets of eosinophilic material referred to as hyaline degeneration. The latter changes were more likely in cases of many years' duration. Cross-striations were still present in some myofibers 24 years after denervation. Only the largest nerve trunks were identifiable. Arterial walls were thickened with narrowed, or even completely occluded, lumens.

Prominent increase in connective tissue was present in some cases. In other cases, the muscle tissue was largely replaced by fat. After 30 years of denervation, no muscle fibers were seen by the methods used. Only connective tissue, fat, blood vessels, and some large empty nerve trunks were identified.

The aforementioned pattern differs in some specifics with that encountered in some more recent texts, but at least some of the latter sources either combine human and experimental animal data (judging by the cited references), or do not specify whether the data are derived from human or animal material.

It is recommended that when any of the aforementioned variations in muscle histology are anticipated in cases in which timing and/or nature of the muscle atrophy is in question, appropriate tissue be saved frozen and some also saved in glutaraldehyde for possible special study. This is due to the lack of specificity of many of the changes described in routine stains. Small angular fibers, fiber type grouping, and many other abnormalities will be more easily diagnosed with ancillary methods. If treatment measures have included functional electrical stimulation of the denervated muscle, maintenance of more normal myofiber diameter or even enlargement of previously atrophic myofibers may invalidate even these crude guidelines for dating/aging of muscle denervation described earlier.[179]

Muscle Reinnervation

Muscle reinnervation (which may be admixed with denervation changes, especially in more chronic conditions) may include such changes as grouped atrophy, fiber type grouping (requiring special stains), target fibers, and fiber hypertrophy. As noted earlier, distinction of these alterations from those produced by some primary myopathies, particularly in end-stage disease, is often not possible without special studies.

Evaluation of unusual cases of muscle weakness and atrophy in modern medical practice generally requires a team approach using methods such as serial clinical examination, electrodiagnostic tests, histochemistry and immunohistochemistry, biochemical studies, electron microscopy, genetic study, and the like, which are beyond the scope of this brief review.

Recovery of Muscle Function with Reinnervation

This question is occasionally raised in a forensic setting. Sunderland[346] reviewed 10 human cases of peripheral nerve surgical reanastomosis in which: injury was confined to one nerve; there was no direct injury to relevant muscles or periarticular structures; the period of denervation exceeded 6 months; there was sufficient recovery of post-treatment (e.g., surgical anastomosis of severed nerves and/or removal of scar tissue with reanastomosis) muscle function to warrant inclusion in the report; and patients were observed sufficiently long to confirm a final end result. He found that "complete or very good" restoration of function could result after up to 12 months' duration of denervation if sufficient nerve fibers reached the muscle, and if appropriate physical therapy had been used in that interval. Variables considered that might alter end results included degree of retrograde degeneration, fiber misdirection at the suture line, and degree of muscular atrophy present by the time the growing nerve reestablished end-organ contact.

Carter et al.[52] indicate that muscles may recover clinically significant strength if reinnervation occurs within up to 18 to 24 months after injury, possibly reflecting the interval progress in surgical technique.

We have not found specific information regarding the percentage of the original axon population needed to reinnervate a denervated muscle in humans that is sufficient to result in a clinically significant recovery of strength. A possible clue is that, in postpolio patients, clinically normal muscle strength recovery can occur in the presence of up to 50% loss of lower motor neurons.[69] These postpolio cases demonstrated extensive axonal branching from still-viable intramuscular axons.[69]

References

1. Abramson AS. Bone disturbances in injuries to the spinal cord and cauda equina (paraplegia). J Bone Joint Surg (Am) 1948;30:982–987.
2. Abou-Hamden A, Blumbergs PC, Scott G, et al. Axonal injury in falls. J Neurotrauma 1997;14:699–713.
3. Ackman JB, Rosenthal DI. Generalized periarticular myositis ossificans as a complication of pharmacologically induced paralysis. Skeletal Radiol 1995;24:395–397.
4. Adams JH. Head injury. In Adams JH, Duchen LW (eds): Greenfield's Neuropathology; ed 5. New York: Oxford University Press, 1992, pp 106–152.
5. Adams JH, Doyle D, Ford I, et al. Diffuse axonal injury in head injury: Definition, diagnosis and grading. Histopathology 1989;15:49–59.
6. Adams RD, Denny-Brown D, Pearson CM. Diseases of Muscle: A Study in Pathology, ed 2. New York: Harper and Row, 1962.
7. Adams RD, Sidman RL. Introduction to Neuropathology. New York: McGraw-Hill, 1968.
8. Ali TT. The role of white blood cells in post-mortem wounds. Med Sci Law 1988;28:100–106.
9. Ali TT. Post-mortem autolytic changes and the role of white blood cells in wounds post-mortem. PhD Thesis, Leeds: The University of Leeds, Department of Forensic Medicine, 1989.
10. Allen JP. Spinal cord injury at birth. In Vinken PJ, Bruyn GW, Braakman R (eds): Injuries of the Spine and Spinal Cord. Part I. Amsterdam: North-Holland, 1976;25:155–173.
11. Alpers BJ, Forster FM. The reparative processes in subarachnoid hemorrhage. J Neuropathol Exp Neurol 1945;4:262–268.
12. Al-Sarraj ST, Hortobagyi T, Wise S. Beta amyloid precursor protein (Beta APP) immunohistochemistry detects axonal injury in less than 60 minutes after human brain trauma (abstract). J Neuropathol Exp Neurol 2004;63:534.
13. Al-Sarraj S, Mohamed S, Kibble M, Rezaie P. Subdural hematoma (SDH): Assessment of macrophage reactivity within the dura mater and underlying hematoma. Clin Neuropathol 2004;23:62–75.
14. An HS, Ebraheim N, Kim K, Jackson WT, Jane JT. Heterotopic ossification and pseudoarthrosis in the shoulder following encephalitis: A case report and review of the literature. Clin Orthop 1987;219:291–298.
15. Anderson R, Kieffer SA, Wolfson JJ, et al. Thickening of the skull in surgically treated hydrocephalus. AJR Am J Roentgenol 1970;110:96–101.
16. Anderson R McD, Opeskin K. Timing of early changes in brain trauma. Am J Forensic Med Pathol 1998;19:1–9.
17. Armstrong DD. Neonatal encephalopathies. In Duckett S (ed): Pediatric Neuropathology. Baltimore: Williams & Wilkins, 1995, pp 334–351.
18. Aoki N. Rapid resolution of acute epidural hematoma. J Neurosurg 1988;68:149–151.

19. Auer RN, Benveniste H. Hypoxia and related conditions. In Graham DI, Lantos PL (eds): Greenfield's Neuropathology, ed 6. London: Arnold, 1997;1:263–314.
20. Baggenstoss AH, Kernohan JW, Drapiewski JF. The healing process in wounds of the brain. Am J Clin Pathol 1943;13:333–348.
21. Bakay L. The danger of trephining through the cranial sutures. Surg Neurol 1982;18:284–285.
22. Banker BQ, Larroche J-C. Periventricular leukomalacia of infancy: A form of neonatal anoxic encephalopathy. Arch Neurol 1962;7:386–410.
23. Batzdorf U. Personal communication, 2006.
24. Bayer SA, Altman J, Russo RJ, Zhang X. Embryology. In Duckett S (ed): Pediatric Neuropathology. Baltimore: Williams & Wilkins, 1995, pp 54–107.
25. Bedbrook GM. Injuries of the thoracolumbar spine with neurological symptoms. In Vinken PJ, Bruyn GW, Braakman R (eds): Injuries of the Spine and Spinal Cord. Part I. Amsterdam: North-Holland, 1976;25:437–466.
26. Bell RB, Wallace CJ, Swanson HA, Brownell AKW. Ossification of the lumbosacral dura and arachnoid following spinal cord trauma: Case report. Paraplegia 1995;33:543–546.
27. Bendet E, Maranta C, Vajtai I, Fisch U. Rate and extent of early axonal degeneration of the human facial nerve. Ann Otol Rhinol Laryngol 1998;107:1–5.
28. Bergeron RT, Rumbaugh. Skull trauma. In Newton TH, Potts DH (eds): Radiology of the Skull and Brain. The Skull, Vol 1, Book 2. St Louis: CV Mosby, 1971, pp 763–818.
29. Berry M, Butt AM, Wilkin G, Perry VH. Structure and function of glia in the central nervous system. In Graham DI, Lantos PL (eds): Greenfield's Neuropathology, ed 7. London: Arnold, 2002;1:75–121.
30. Birbamer G, Buchberger W, Kampfl A, Aichner F. Early detection of post-traumatic olivary hypertrophy by MRI. J Neurol 1993;240:407–409.
31. Blumbergs PC, Jones NR, North JB. Diffuse axonal injury in head trauma. J Neurol Neurosurg Psychiatry 1989;52:838–841.
32. Blumbergs PC, Scott G, Manavis J, et al. Staining of amyloid precursor protein to study axonal damage in mild head injury. Lancet 1994;344:1055–1056.
33. Blumbergs PC, Scott G, Manavis J, et al. Topography of axonal injury as defined by amyloid precursor protein and the sector scoring method in mild and severe closed head injury. J Neurotrauma 1995;12:565–572.
34. Bohnert M, Baumgartner R, Pollak S. Spectrophotometric evaluation of the colour of intra- and subcutaneous bruises. Int J Legal Med 2000;113:343–348.
35. Bowden REM, Gutmann E. Denervation and re-innervation of human voluntary muscle. Brain 1944;67:273–313.
36. Brierley JB, Graham DI. Hypoxia and vascular disorders of the central nervous system. In Adams JH, Corsellis JAN, Duchen LW (eds): Greenfield's Neuropathology, ed 4. New York: John Wiley & Sons, 1984, pp 125–207.
37. Brierley JB, Meldrum BS, Brown AW. The threshold and neuropathology of cerebral "anoxic-ischemic" cell change. Arch Neurol 1973;29:367–373.
38. Brooke MH, Engel WK. The histographic analysis of human muscle biopsies with regard to fiber types: Four children's biopsies. Neurology 1969;19:591–605.
39. Brown BA. Hematology: Principles and Procedures, ed 6. Philadelphia: Lea & Febiger, 1993.
40. Bruce DA, Alavi A, Bilaniuk L, et al. Diffuse cerebral swelling following head injuries in children: The syndrome of "malignant cerebral edema." J Neurosurg 1981;54:170–178.
41. Bruce JH, Norenberg MD, Kraydieh S, et al. Schwannosis: Role of gliosis and proteoglycan in human spinal cord injury. J Neurotrauma 2000;17:781–788.
42. Buchthal F, Kühl V. Nerve conduction, tactile sensibility, and the electromyogram after suture or compression of peripheral nerve: A longitudinal study in man. J Neurol Neurosurg Psychiatry 1979;42:436–451.
43. Bueri JA. Personal communication, April 2000.
44. Bullock R, Maxwell WL, Graham DI, et al. Glial swelling following human cerebral contusion: An ultrastructural study. J Neurol Neurosurg Psychiatry 1991;54:427–434.
45. Burger PC, Scheithauer BW, Vogel FS. Surgical Pathology of the Nervous System and Its Coverings, ed 4. New York: Churchill Livingston, 2002.
46. Burke DC. Injuries of the spinal cord in children. In Vinken PJ, Bruyn GW, Braakman R (eds): Injuries of the Spine and Spinal Cord. Part I. Amsterdam: North-Holland, 1976;25:175–195.
47. Burke MP, Olumbe AK, Opeskin K. Postmortem extravasation of blood potentially simulating antemortem bruising. Am J Forensic Med Pathol 1998;19:46–49.
48. Burke MJ, Winston KR, Williams S. Normal sutural fusion and the etiology of single sutural craniosynostosis: The microspicule hypothesis. Pediatr Neurosurg 1995;22:241–247.
49. Buss A, Brook GA, Kakulas B, et al. Gradual loss of myelin and formation of an astrocytic scar during Wallerian degeneration in the human spinal cord. Brain 2003;127:34–44.
50. Carpenter S, Karpati G. Pathology of Skeletal Muscle. New York: Churchill Livingstone, 1984.
51. Carter BS, Rabinov JD, Pfannl R, Schwamm LH. Case 5-2004: A 57-year-old man with slurred speech and left hemiparesis. N Engl J Med 2004;350:707–716.
52. Carter GT, Robinson RL, Chang VH, Kraft GH. Electrodiagnostic evaluation of traumatic nerve injuries. Hand Clin 2000;16:1–12.
53. Cartwright GE, Athens JW, Wintrobe MM. The kinetics of granulopoiesis in normal man. Blood 1964;24:780–803.
54. Case MES. Head injury in a child. Check Sample, Forensic Pathology No. FP97-6 (FP-227). New York: ASCP Press, 1997;39:79–95.
55. Case MES. Irresponsible medical testimony in cases of abusive head injury in children (abstract). In Annual Meeting of the National Association of Medical Examiners, Richmond, VA, 2001, p 24.
56. Case ME, Graham MA, Handy TC, et al. N.A.M.E. Ad Hoc Committee on shaken baby syndrome: Position paper on fatal abusive head injuries in infants and young children. Am J Forensic Med Pathol 2001;22:112–122.
57. Cave WS. Acute, nontraumatic subdural hematoma of arterial origin. J Forensic Sci 1983;28:786–789.
58. Chasler CN. The newborn skull: The diagnosis of fracture. Am J Roentgenol Radium Ther Nucl Med 1967;100:92–99.
59. Chaudhry V, Glass D, Griffin JW. Wallerian degeneration in peripheral nerve disease. Neurol Clin 1992;10:613–627.
60. Christman CW, Grady MS, Walker SA, et al. Ultrastructural studies of diffuse axonal injury in humans. J Neurotrauma 1994;11:173–186.
61. Chuaqui R, Tapia J. Histologic assessment of the age of recent brain infarcts in man. J Neuropathol Exp Neurol 1993;52:481–489.
62. Colodner KJ, Montana RA, Anthony DC, et al. Proliferative potential of human astrocytes. J Neuropathol Exp Neurol 2005;64:163–169.
63. Cormack DH. Ham's Histology, ed 9. Philadelphia: JB Lippincott, 1987, p 279.
64. Cornish R, Blumbergs PC, Manavis J, et al. Topography and severity of axonal injury in human spinal cord trauma using amyloid precursor protein as a marker of axonal injury. Spine 2000;25:1227–1233.
65. Cowan WM. Anterograde and retrograde transneuronal degeneration in the central and peripheral nervous system. In Nauta WJH, Ebbeson SDE (eds): Contemporary Research Methods in Neuroanatomy. New York: Springer, 1970, pp 217–251.

66. Crooks DA, Scholtz CL, Vowles G, Greenwald S. Axonal injury in closed head injury by assault: A quantitative study. Med Sci Law 1992;32:109–117.
67. Croul SE, Flanders AE. Neuropathology of human spinal cord injury. Adv Neurol 1997;72:317–323.
68. Daily JC, Bowers CM. Aging of bitemarks: A literature review. J Forensic Sci 1997;42:792–795.
69. Dalakas MC. Pathogenetic mechanisms of post-polio syndrome: Morphological, virological, and immunological correlations. Ann N Y Acad Sci 1995;753:167–185.
70. Dale HC, Hubert RT, Fauci A. Eosinophile kinetics in the hypereosinophilic syndrome. J Lab Clin Med 1976;87:487–495.
71. Damanski M. Heterotopic ossification in paraplegia: A clinical study. J Bone Joint Surg (Br) 1961;43:286–299.
72. Danton GH, Dietrich WD. Inflammatory mechanisms after ischemia and stroke. J Neuropathol Exp Neurol 2003;62:127–136.
73. Darrow VC, Alvord EC Jr, Mack LA, Hodson WA. Histologic evolution of the reactions to hemorrhage in the premature human infant's brain: A combined ultrasound and autopsy study and a comparison with the reaction in adults. Am J Pathol 1988;130:44–58.
74. Davis JH. The future of the medical examiner system. Am J Forensic Med Pathol 1995;16:265–269.
75. Davis RL, Robertson DM (eds). Textbook of Neuropathology, ed 2. Baltimore: Williams & Wilkins, 1991.
76. Dawson SL, Hirsch CS, Lucas FV, Sebek BA. The contrecoup phenomenon: Reappraisal of a classic problem. Hum Pathol 1980;11:155–166.
77. Del Bigio MR, Johnson GE. Clinical presentation of spinal cord concussion. Spine 1989;14:37–40.
78. Denton S, Miluesnic D. Response to letter from Dr. Omalu (Omalu BI: Diagnosis of traumatic diffuse axonal injury. Am J Forensic Med Pathol 2004;25:270). Am J Forensic Med Pathol 2004;25:270–271.
79. DeReuck J, DeCoster W, Vander Eecken H. Cerebrospinal fluid cytology in acute ischemic stroke. Acta Neurol Belg 1985;85:133–136.
80. Devriendt W, Piercecchi-Marti M-D, Adalian P, et al. Hyperostosis frontalis interna: Forensic issues. J Forensic Sci 2005;50:143–146.
81. DiCarlo FJ, Ross DJ. Forensic Pathology, Check Sample No. FP04-8 (FP-299), Chicago: Am Soc Clin Pathol, 2004;46:91–106.
82. DiMaio VJ, DiMaio D. Forensic Pathology, ed 2. Boca Raton, FL: CRC Press, 2001.
83. Ditunno JF, Little JW, Tessler A, Burns AS. Spinal shock revisited: A four-phase model. Spinal Cord 2004;42:383–395.
84. Djang WT, Gray L, Drayer BP. Intracranial occlusive vascular disease. In Taveras JM, Ferrucci JT (eds): Radiology: Diagnosis-Imaging-Intervention. Neuroradiology. Philadelphia: Lippincott-Raven, 1997;3:1–23.
85. Dokgöz H, Arican N, Elmas I, Fincanci SK. Comparison of morphological changes in white blood cells after death and in vitro storage of blood for the estimation of postmortem interval. Forensic Sci Int 2001;124:25–31.
86. Dolinak D. Case 2005 NPB-05. Surveys 2005, NP-B Neuropathology Program Participant Summary. Northfield, Ill.: College of American Pathologists, 2005, pp 2–4.
87. Dolinak D, Matshes EW. Forensic Neuropathology. In Dolinak D, Matshes EW, Lew EO (eds): Forensic Pathology: Principles and Practice. Amsterdam: Elsevier, Academic Press, 2005, pp 423–465.
88. Dolinak D, Matshes EW, Lew EO. Forensic Pathology: Principles and Practice. Amsterdam: Elsevier, Academic Press, 2005.
89. Dolinak D, Reichard R. An overview of inflicted head injury in infants and young children, with a review of β-amyloid precursor protein immunochemistry. Arch Pathol Lab Med 2006;130:712–717.
90. Dolinak D, Smith C, Graham DI. Hypoglycemia is a cause of axonal injury. Neuropathol Appl Neurobiol 2000;26:448–453.
91. Dolinak D, Smith C, Graham DI. Global hypoxia per se is an unusual cause of axonal injury. Acta Neuropathol 2000;100:553–560.
92. Dolman CL. Microglia. In Davis RL, Robertson DM (eds): Textbook of Neuropathology, ed 2. Baltimore: Williams & Wilkins, 1991, pp 141–163.
93. Dubowitz V, Brooke MH. Muscle Biopsy: A Modern Approach. London: WB Saunders, 1973.
94. Duchen LW. General pathology of neurons and neuroglia. In Adams JH, Corsellis JAN, Duchen LW (eds): Greenfield's Neuropathology, ed 4. New York: John Wiley & Sons, 1984, pp 1–52.
95. Duchen LW. General pathology of neurons and glia. In Adams JH, Duchen LW (eds): Greenfield's Neuropathology, ed 5. New York: Oxford University Press, 1992, pp 1–68.
96. Duchen LW, Jacobs JM. Nutritional deficiencies and metabolic disorders. In Adams JH, Duchen LW (eds): Greenfield's Neuropathology, ed 5. New York: Oxford University Press, 1992, pp 811–880.
97. Duckworth T, Cooper ERA. A study of anophthalmia in an adult. Acta Anat 1966;63:509–522.
98. Duhaime A-C, Christian C, Armonda R, et al. Disappearing subdural hematomas in children. Pediatr Neurosurg 1996;25:116–122.
99. Ellis WG, Goetzman BW, Lindenberg JA. Neuropathologic documentation of prenatal brain damage. Am J Dis Child 1988;142:858–866.
100. Ellison D, Love S, Chimelli L, et al. Neuropathology: A Reference Text of CNS Pathology. London: Mosby, 1998.
101. Erol FS, Kaplan M, Topsakal C, et al. Coexistence of rapidly resolving acute subdural hematoma and delayed traumatic intracerebral hemorrhage. Pediatr Neurosurg 2004;40:238–240.
102. Evans D, Levene M. Clinical assessment of the neonate. In Squire W (ed): Acquired Damage to the Developing Brain: Timing and Causation. London: Arnold, 2002, pp 139–165.
103. Fawer C-L, Calame A, Perentes E, Anderegg A. Periventricular leukomalacia: A correlation study between real-time ultrasound and autopsy findings. Periventricular leukomalacia in the neonate. Neuroradiology 1985;27:292–300.
104. Fenton LZ, Sirotnak AP, Handler MH. Parietal pseudofracture and spontaneous intracranial hemorrhage suggesting nonaccidental trauma: Report of 2 cases. Pediatr Neurosurg 2000;33:318–322.
105. Finkbiner WE, Ursell PC, Davis RL. Autopsy Pathology: A Manual and Atlas. Philadelphia: Churchill Livingstone, 2004.
106. Fishman RA. Cerebrospinal Fluid in Diseases of the Nervous System, ed 2. Philadelphia: WB Saunders, 1992.
107. Freiberg JA. Para-articular calcification and ossification following acute anterior poliomyelitis in an adult. J Bone Joint Surg (Am) 1952;34:339–348.
108. Friede RL. Developmental Neuropathology, ed 2. Berlin: Springer-Verlag, 1989.
109. Friede RL, Schachenmayr W. The origin of subdural neomembranes. II. Fine structure of neomembranes. Am J Pathol 1978;92:69–84.
110. Fujikawa DG, Itabashi HH, Wu A, Shinmei SS. Status epilepticus-induced neuronal loss in humans without systemic complications or epilepsy. Epilepsia 2000;41:981–991.
111. Fukuhara T, Vorster SJ, Luciano MG. Critical shunt-induced subdural hematoma treated with combined pressure-programmable valve implantation and endoscopic third ventriculostomy. Pediatr Neurosurg 2000;33:37–42.
112. Fuller GN, Goodman JC. Practical Review of Neuropathology. Philadelphia: Lippincott Williams & Wilkins, 2001, pp 226–244.
113. Furman R, Nicholas JJ, Jivoff L. Elevation of serum alkaline phosphatase coincident with ectopic-bone formation in paraplegic patients. J Bone Joint Surg (Am) 1970:52:1131–1137.

114. Garcia JH (ed). Neuropathology: The Diagnostic Approach. St Louis: Mosby, 1997.
115. Garcia JH, Anderson ML. Circulatory disorders and their effects on the brain. In Davis RL, Robertson DM (eds): Textbook of Neuropathology, ed 2. Baltimore: Williams & Wilkins, 1991, pp 621–718.
116. Garcia JH, Mena H. Vascular diseases. In Garcia JH (ed): Neuropathology: The Diagnostic Approach. St Louis: Mosby, 1997, pp 263–320.
117. Gardner AMN. Aspiration of food and vomit. Q J Med 1958;27:227–242.
118. Garland DE. Clinical observations on fractures and heterotopic ossification in the spinal cord and traumatic brain injured populations. Clin Orthop 1988;233:86–101.
119. Garland DE, Blum CE, Waters RL. Periarticular heterotopic ossification in head-injured adults: Incidence and location. J Bone Joint Surg (Am) 1980;62:1143–1146.
120. Garland DE, Shimoyama ST, Lugo C, et al. Spinal cord insults and heterotopic ossification in the pediatric population. Clin Orthop 1989;235:303–310.
121. Gautier JC, Blackwood W. Enlargement of the inferior olivary nucleus in association with lesions of the central tegmental tract or dentate nucleus. Brain 1961;84:341–361.
122. Geddes JF. What's new in the diagnosis of head injury? J Clin Pathol 1997;50:271–274.
123. Geddes JF, Vowles GH, Beer TW, Ellison DW. The diagnosis of diffuse axonal injury: Implications for forensic practice. Neuropathol Appl Neurobiol 1997;23:339–347.
124. Geddes JF, Whitwell HL, Graham DI. Traumatic axonal injury: Practical issues for diagnosis in medicolegal cases. Neuropathol Appl Neurobiol 2000;26:105–116.
125. Gelabert-Gonzáles M, Fernández-Villa J, Cutrin-Prieto J, et al. Arachnoid cyst rupture with subdural hygroma: Report of three cases and literature review. Childs Nerv Syst 2002;18:609–613.
126. Gilles FH. Telencephalon medium and the olfacto-cerebral outpouching. In Gilles FH, Leviton A, Dooling EC (eds): The Developing Human Brain: Growth and Epidemiologic Neuropathology. Boston: John Wright PSG, 1983, pp 59–86.
127. Gleckman AM, Bell MD, Evans RJ, Smith TW. Diffuse axonal injury in infants with nonaccidental craniocerebral trauma. Enhanced detection by β-amyloid precursor protein immunohistochemical staining. Arch Pathol Lab Med 1999;123:146–151.
128. Goldberg MA, Schumacher HR. Heterotopic ossification mimicking acute arthritis after neurologic catastrophes. Arch Intern Med 1977;137:619–621.
129. Gore I, Arons W. Calfication of the myocardium. Arch Pathol 1949;48:1–12.
130. Goto N, Kaneko M. Olivary enlargement: Chronologic and morphometric analyses. Acta Neuropathol 1981;54:275–282.
131. Gowans JL. Life span, recirculation, and transformation of lymphocytes. Int Rev Exp Pathol 1966;5:1–24.
132. Grady MS, McLaughlin MR, Christman CW, et al. The use of antibodies targeted against the neurofilament subunits for the detection of diffuse axonal injury in humans. J Neuropathol Exp Neurol 1993;52:143–152.
133. Graeber MB, Blakemore WF, Kreutzberg GW. Cellular pathology of the central nervous system. In Graham DI, Lantos PL (eds): Greenfield's Neuropathology, ed 7. London: Arnold, 2002;I:123–191.
134. Graham DI. Hypoxia and vascular disorders. In Adams JH, Duchen LW (eds): Greenfield's Neuropathology, ed 5. New York: Oxford University Press, 1992, pp 153–268.
135. Graham DI, Gennarelli TA. Trauma. In Graham DI, Lantos PL (eds): Greenfield's Neuropathology, ed 6. London: Arnold, 1997;I:197–262.
136. Graham DI, Gennarelli TA, McIntosh TK. Trauma. In Graham DI, Lantos PL (eds): Greenfield's Neuropathology, ed 7. London: Arnold, 2002;1:823–898.
137. Graham DI, Lantos PL (eds). Greenfield's Neuropathology, ed 7. London: Arnold, 2002.
138. Greenfield JG, Meyer A. General pathology of the nerve cell and neuroglia. In Blackwood W, McMenemey WH, Meyer A, et al. (eds): Greenfield's Neuropathology, ed 2. London: Arnold, 1963, pp 1–70.
139. Griscom NT, Oh KS. The contracting skull: Inward growth of the inner table as a physiologic response to diminution of intracranial content in children. AJR Am J Roentgenol 1970;110:106–110.
140. Gunn DR, Young WB. Myositis ossificans as a complication of tetatnus. J Bone Joint Surg (Br) 1959;41:535–540.
141. Gutmann L. Spinal shock. In Vinken PJ, Bruyn GW, Braakman E (eds): Injuries of the Spine and Spinal Cord. Part II. Amsterdam, North-Holland. 1976;26:243–262.
142. Haddock JN, Berlin L. Transsynaptic degeneration in the visual system: Report of a case. Arch Neurol Psychiatry (Chicago) 1950;64:66–73.
143. Haines DE. On the question of a subdural space. Anat Rec 1991;230:3–21.
144. Hamlat A, Heckly A, Doumbouya N, et al. Epidural hematoma as a complication of endoscopic biopsy and shunt placement in a patient harboring a third ventricle tumor. Pediatr Neurosurg 2004;40:245–248.
145. Hammes EM Jr. Reaction of the meninges to blood. Arch Neurol Psychiatry (Chic) 1944;52:505–514.
146. Hardman JM. Microscopy of traumatic central nervous system injuries. In Perper JA, Wecht CH (eds): Microscopic Diagnosis in Forensic Pathology. Springfield, IL: Charles C Thomas, 1980, pp 268–326.
147. Hardman JM. Cerebrospinal trauma. In Davis RL, Robertson DM (eds): Textbook of Neuropathology, ed 3. Baltimore: Williams & Williams, 1997, pp 1179–1232.
148. Hardy AG, Dickson JW. Pathological ossification in traumatic paraplegia. J Bone Joint Surg (Br) 1963;45:76–87.
149. Harwood-Nash DC, Fitz RC. Neuroradiology in Infants and Children. St Louis: CV Mosby, 1976;1:1–70.
150. Hausmann R. Timing of cortical contusions in human brain injury: Morphological parameters for a forensic wound-age estimation. In Tsokos M (ed): Forensic Pathology Reviews. Totowa, NJ: Humana Press, 2004;1:53–75.
151. Henry JB. Clinical Diagnosis and Management by Laboratory Methods. Philadelphia: WB Saunders, 1991.
152. Hensyl WR (ed). Stedman's Medical Dictionary, ed 25. Baltimore: Williams & Wilkins, 1990.
153. Hernandez-Cueto C, Girela E, Sweet DJ. Advances in the diagnosis of wound vitality: A review. Am J Forensic Med Pathol 2000;21:21–31.
154. Herrick MK, Agamanolis DP. Displacement of cerebellar tissue into spinal canal. Arch Pathol 1975;99:565–571.
155. Hermann C Jr, Brown JW. Palatal myoclonus: A reappraisal. J Neurol Sci 1967;5:473–492.
156. Heytens L, Verlooy J, Gheuens J, Bossaert L. Lazarus sign and extensor posturing in a brain-dead patient: Case report. J Neurosurg 1989;71:449–451.
157. Hirano A. Neurons and astrocytes. In Davis RL, Robertson DM (eds): Textbook of Neuropathology, ed 3. Baltimore: Williams & Wilkins, 1997, pp 1–109.
158. Hirsh LF. Chronic epidural hematomas. Neurosurgery 1980;6:508–512.
159. Hodges FJ III. Pathology of the skull. In Taveras JM (ed): Radiology: Diagnosis-Imaging-Intervention, Vol 3. Philadelphia: Lippincott-Raven, 1997, pp 3.1–3.21.
160. Hossack DW, King A. Neurogenic heterotopic ossification. Med J Aust 1967;1:326–328.

161. Hsu JD, Sakimura I, Stauffer ES. Heterotopic ossification around the hip joint in spinal cord injured patients. Clin Orthop 1975;112:165–169.
162. Hutchings M, Weller RO. Anatomical relationships of the pia mater to cerebral blood vessels in man. J Neurosurg 1986;65:316–325.
163. Hymel KP, Jenny C, Block RW. Intracranial hemorrhage and rebleeding in suspected victims of abusive head trauma: Addressing the forensic controversies. Child Maltreatment 2002;7:329–348.
164. Illa I, Leon-Monzon M, Agboatwalla M, et al. Role of muscle in acute poliomyelitis infection. Ann N Y Acad Sci 1995;753:58–67.
165. Itabashi HH. Unpublished observations.
166. Ivan LP. Spinal reflexes in cerebral death. Neurology 1973;23:650–652.
167. Iwakuma T, Brunngraber CV. Chronic extradural hematomas: A study of 21 cases. J Neurosurg 1973;38:488–493.
168. Iwanowski L, Olszewski J. The effects of subarachnoid injections of iron-containing substances on the central nervous system. J Neuropathol Exp Neurol 1960;19:433–448.
169. Janssen W. Forensic Histopathology. Berlin: Springer-Verlag, 1984.
170. Jellinger K. Hypertrophy of the inferior olives: Report on 29 cases. Z Neurol 1973;205:153–174.
171. Jellinger K. Neuropathology of cord injuries. In Vinken PJ, Bruyn GW, Braakman R (eds): Injuries of the Spinal Cord. Part I. Amsterdam: North-Holland, 1976;25:43–121.
172. Jorgensen EO. Spinal man after brain death: The unilateral extension-pronation reflex of the upper limb as an indication of brain death. Acta Neurochirurg 1973;28:259–273.
173. Kakulas BA. A review of the neuropathology of human spinal cord injury with emphasis on special features. J Spinal Cord Med 1999;22:119–124.
174. Kalimo H, Kaste M, Haltia M. Vascular diseases. In Graham DI, Lantos PL (eds): Greenfield's Neuropathology, ed 6. London: Arnold, 1997;1:315–396.
175. Kalimo H, Kaste M, Haltia M. Vascular diseases. In Graham DI, Lantos PL (eds): Greenfield's Neuropathology, ed 7. London: Arnold, 2002;1:281–355.
176. Kattan KR. Calvarial thickening after dilantin medication. AJR Am J Roentgenol 1970;110:102–105.
177. Kazui S, Kuriyama Y, Sawada T, Imakita S. Very early demonstration of secondary pyramidal tract degeneration by computed tomography. Stroke 1994;25:2287–2289.
178. Keret D, Harcke HT, Mendez AA, Bowen R. Heterotopic ossification in central nervous system-injured patients following closed nailing of femoral fractures. Clin Orthop 1990;256:254–259.
179. Kern H, Boncampagni S, Rossini K, et al. Long-term denervation in humans causes degeneration of both contractile and excitation-contraction coupling apparatus, which is reversible by functional electrical stimulation (FES): A role for myofiber regeneration? J Neuropathol Exp Neurol 2004;63:919–931.
180. Kida S, Yamashima T, Kubota T, et al. A light and electron microscopic and immunohistochemical study of human arachnoid villi. J Neurosurg 1988;69:429–435.
181. Kinney HC, Armstrong DD. Perinatal neuropathology. In Graham DI, Lantos PL (eds): Greenfield's Neuropathology, ed 6. London: Arnold, 1997, pp 535–599.
182. Kinney HC, Armstrong DD. Perinatal neuropathology. In Graham DI, Lantos PL (eds): Greenfield's Neuropathology, ed 7. London: Arnold, 2002;1:519–606.
183. Kinney HC, Haynes RL, Folkerth RD. White matter lesions in the perinatal period. In Golden JA, Harding BN (eds): Pathology and Genetics: Developmental Neuropathology. Basel: ISN Neuropath Press, 2004, pp 156–170.
184. Kitamura O. Immunohistochemical investigation of hypoxic/ischemic brain damage in forensic autopsy cases. Int J Legal Med 1994;107:69–76.
185. Kleinman PK, Barnes PD. Head trauma. In Kleinman PK: Diagnostic Imaging of Child Abuse, ed 2. St Louis: Mosby, 1998, pp 285–342.
186. Klotzbach H, Delling G, Richter E, Sperhake JP. Post-mortem diagnosis and age estimation of infants' fractures. Int J Legal Med 2003;117:82–89.
187. Ko H-Y, Ditunno JF Jr, Graziani V, Little JW. The pattern of reflex recovery during spinal shock. Spinal Cord 1999;37:402–409.
188. Kobrine AI, Timmins E, Rajjoub RK, et al. Demonstration of massive traumatic brain swelling within 20 minutes after injury: Case report. J Neurosurg 1977;46:256–258.
189. Kupfer C. The distribution of cell size in the lateral geniculate nucleus of man following transneuronal cell atrophy. J Neuropathol Exp Neurol 1965;24:653–661.
190. Kuroiwa T, Tanabe H, et al. Rapid spontaneous resolution of acute extradural and subdural hematomas: Case report. J Neurosurg 1993;78:126–128.
191. Lambri M, Djurovic V, Kibble M, et al. Specificity and sensitivity of β-APP in head injury. Clin Neuropathol 2001;20:263–271.
192. Lampert PW. A comparative electron microscopic study of reactive, degenerating, regenerating, and dystrophic axons. J Neuropathol Exp Neurol 1967;26:345–368.
193. Lang DA, Teasdale GM, Macpherson P, Lawrence A. Diffuse brain swelling after head injury: More often malignant in adults than children? J Neurosurg 1994;80:675–680.
194. Lapresle J. Palatal myoclonus. Adv Neurol 1986;43:265–273.
195. Lee A, Whyte MKB, Haslett C. Inhibition of apoptosis and prolongation of neutrophil functional longevity by inflammatory mediators. J Leukoc Biol 1993;54:283–288.
196. Leech RW, Alvord EC Jr: Wallerian degeneration in the premature human (abstract). J Neuropathol Exp Neurol 1975;34:92.
197. Leestma JE. Forensic Neuropathology. New York: Raven Press, 1988.
198. Leestma JE. Forensic neuropathology. In Garcia JH (ed): Neuropathology: The Diagnostic Approach. St Louis: Mosby, 1997, pp 475–527.
199. Leestma JE. Neuropathology of brain death. In Wijdicks EFM (ed): Brain Death. Philadelphia: Lippincott Williams & Wilkins, 2001, pp 45–60.
200. Lemire RJ, Loeser JD, Leech RW, Alvord EC Jr. Normal and abnormal development of the human nervous system. Hagerstown, MD: Harper and Row, 1975.
201. Lenton KA, Nacamuli RP, Wan DC, et al. Cranial suture biology. Curr Top Dev Biol 2005;66:287–328.
202. Liberson M. Soft tissue calcifications in cord lesions. JAMA 1953;152:1010–1013.
203. Lindenberg R. Morphometric and morphostatic necrobiosis: Investigations on nerve cells in the brain. Am J Pathol 1956;32:1147–1177.
204. Lindenberg R. Patterns of CNS vulnerability in acute hypoxaemia, including anesthetic accidents. In Schadé JP, McMenemey WH (eds): Selective Vulnerability of the Brain in Hypoxaemia. Philadelphia: FA Davis, 1963, pp 189–209.
205. Lindenberg R. Trauma of the meninges and brain. In Minckler J (ed): Pathology of the Nervous System. New York: McGraw-Hill, 1971;2:1705–1765.
206. Lindenberg R, Freytag E. Morphology of cortical contusions. Arch Pathol 1957;63:24–42.
207. Lindsberg PJ, Carpén O, Paetau A, et al. Endothelial ICAM-1 expression associated with inflammatory cell response in human ischemic stroke. Circulation 1996;94:939–945.
208. Lipschitz R. Stab wounds of the spinal cord. In Vinken PJ, Bruyn GW, Braakman R (eds): Injuries of the Spine and Spinal Cord. Part I. Amsterdam: North-Holland, 1976;25:197–207.
209. Little D, Stevens MP. The significance of hemosiderin in the dura of infants (abstract). Proceedings of the Annual Meeting of the National Association of Medical Examiners, Richmond, VA, 2001, pp 12–13.

210. Løberg EM, Torvik A. Brain contusions: The time sequence of the histological changes. Med Sci Law 1989;29:109–115.
211. Love S, Barber R, Wilcock GK. Neuronal death in brain infarcts in man. Neuropathol Appl Neurobiol 2000;26:55–66.
212. Lucey BP, March GP, Hutchins GM. Marked calvarial thickening and dural changes following chronic ventricular shunting for shaken baby syndrome. Arch Pathol Lab Med 2003;127:94–97.
213. Luisto M, Zitting A, Tallroth K. Hyperostosis and osteoarthritis in patients surviving after tetanus. Skeletal Radiol 1994;23:31–35.
214. Luna LG (ed). Manual of Histologic Staining Methods of the Armed Forces Institute of Pathology, ed 3. New York: McGraw-Hill, 1968.
215. Madea B, Grellner W. Vitality and supravitality in forensic medicine. In Oehmichen M, Kirchner H (eds): The Wound Healing Process: Forensic Pathological Aspects. Lübeck: Schmidt-Römhild, 1996, pp 259–282.
216. Maguire S, Mann MK, Sibert J, Kemp A. Can you age bruises accurately in children? A systematic review. Arch Dis Child 2005;90:187–189.
217. Maianski NA, Maianski AN, Kuijpers TW, Roos D. Apoptosis of neutrophils. Acta Haematol 2004;111:56–66.
218. Malhotra SK, Shnitka TK, Elbrink J. Reactive astrocytes—A review. Cytobios 1990;61:133–160.
219. Mancuso M, Davidzon G, Kurlan RM, et al. Hereditary ferritinopathy: A novel mutation, its cellular pathology, and pathogenetic insights. J Neuropathol Exp Neurol 2005;64:280–294.
220. Manlow A, Munoz DG. A non-toxic method for the demonstration of gliosis. J Neuropathol Exp Neurol 1992;51:298–302.
221. Manzanares MC, Goret-Nicaise M, Dhem A. Metopic sutural closure in the human skull. J Anat 1988;161:203–215.
222. Mariya F, Hashimoto Y. Medicolegal implications of drugs and chemicals detected in intracranial hematomas. J Forensic Sci 1998;43:980–984.
223. Markwalder T-M. Chronic subdural hematomas: A review. J Neurosurg 1981;54:637–645.
224. Marques LJ, Teschler H, Guzman J, Costabel U. Smoker's lung transplanted to a nonsmoker: Long-term detection of smoker's macrophages. Am J Respir Crit Care Med 1997;156:1700–1702.
225. Massicotte EM, Del Bigio MR. Human arachnoid villi response to subarachnoid hemorrhage: Possible relationship to chronic hydrocephalus. J Neurosurg 1999;91:80–84.
226. Mathiesen T, Lefvert AK. Cerebrospinal fluid and blood lymphocyte subpopulations following subarachnoid haemorrhage. Br J Neurosurg 1996;10:89–92.
227. Mathur PPS, Dharker SR, Agarwal SK, Sharma M. Fluid chronic extradural haematoma. Surg Neurol 1980;14:81–82.
228. Maury M, Bidart Y. The para-osteo-arthropathies. In Vinken PJ, Bruyn GW, Braakman R (eds): Handbook of Clinical Neurology. Injuries of the Spine and Spinal Cord. Part II. Amsterdam: North-Holland, 1976;26:501–519.
229. Maxeiner H. Demonstration and interpretation of bridging vein ruptures in cases of infantile subdural bleedings. J Forensic Sci 2001;46:85–93.
230. Maxwell WL, Povlishock JT, Graham DI. A mechanistic analysis of nondisruptive axonal injury: A review. J Neurotrauma 1997;14:419–440.
231. McCarthy GT. Cerebral palsy: The clinical problem. In Squier W (ed): Acquired Damage to the Developing Brain Timing and Causation. London: Arnold, 2002, pp 3–25.
232. McGluggage WG, Sloan JM, Traub AI. Postpartum diffuse calcification of the myocardium. Int J Gynecol Pathol 1996;15:82–84.
233. McCormick WF. Vascular diseases. In Rosenberg RN (ed): Clinical Neurosciences: Neuropathology. New York: Churchill Livingstone, 1983;III:35–83.
234. McCormick WF. Trauma. In Rosenberg RN (ed): Clinical Neurosciences: Neuropathology. New York: Churchill Livingstone, 1983;III:241–283.
235. McCormick WF. Pathology of closed head injury. In Wilkins RH, Rengachary SS (eds): Neurosurgery. New York: McGraw-Hill, 1985, pp 1544–1570.
236. McGarry P, Holmquist ND, Carmel A. A postmortem study of cerebrospinal fluid with histologic correlation. Acta Cytol 1969;13:48–52.
237. McKenzie KJ, McLellan DR, Gentleman SM, et al. Is β-APP a marker of axonal damage in short-surviving head injury? Acta Neuropathol 1996;92:608–613.
238. Medana IM, Day NP, Hien TT, et al. Axonal injury in cerebral malaria. Am J Pathol 2002;160:655–666.
239. Medana IM, Esiri MM. Axonal damage: A key predictor of outcome in human CNS diseases. Brain 2003;126:515–530.
240. Mejia RE, Pollack MM. Variability in brain death determination practices in children. JAMA 1995;274:550–553.
241. Melamed E, Robinson D, Halperin N, et al. Brain-injury related heterotopic bone formation: Treatment strategy and results. Am J Phys Med Rehabil 2002;81:670–674.
242. Melnicoff MJ, Horan PK, Breslin EW, Morahan PS. Maintenance of peritoneal macrophages in the steady state. J Leukoc Biol 1988;44:367–375.
243. Menkes JH, Sarnat HB, Moser FG. Introduction: Neurologic examination of the child and infant. In Menkes JH, Sarnat HB (eds): Child Neurology, ed 6. Philadelphia: Lippincott Williams & Wilkins, 2000, pp 1–32.
244. Meyer JS, Imai A, Shinohara T. Causes of cerebral ischemia and infarction. In Toole JF (ed): Handbook of Clinical Neurology. Vascular Diseases, Part I. Elsevier, 1988;53(9):155–173.
245. Milgram JW. Radiologic and Histologic Pathology of Nontumorous Diseases of Bones and Joints. Northbrook: Northbrook Publishing, 1990;1:215–254.
246. Miller LF, O'Neill CJ. Myositis ossificans in paraplegics. J Bone Joint Surg (Am) 1949;31:283–294.
247. Money RA. Ectopic paraarticular ossification after head injury. Med J Aust 1972;1:125–127.
248. Morrison AL, King TM, Korell MA, et al. Acceleration-deceleration injuries to the brain in blunt force trauma. Am J Forensic Med Pathol 1998;19:109–112.
249. Moskowitz N, Noback CR. The human lateral geniculate body in normal development and congenital unilateral anophthalmia. J Neuropathol Exp Neurol 1962;21:377–382.
250. Munro D, Merritt HH. Surgical pathology of subdural hematoma: Based on a study of one hundred and five cases. Arch Neurol Psychiatry 1936;35:64–75.
251. Nadis SM, Klawans HL. Cerebrospinal fluid in stroke. In Toole JF (ed): Handbook of Clinical Neurology. Vascular Diseases, Part II. New York: Elsevier Science, 1989;54(10):195–202.
252. Nagao T, Aoki N, Mizutani H, Kitamura K. Acute subdural hematoma with rapid resolution in infancy: Case report. Neurosurgery 1986;19:465–467.
253. Nakamizo A, Inamura T, Inoha S, et al. Occurrence of subdural hematoma and resolution of gait disturbance in a patient treated with shunting for normal pressure hydrocephalus. Clin Neurol Neurosurg 2002;104:315–317.
254. Nedergaard M, Klinken L, Paulson OB. Secondary brain stem hemorrhage in stroke. Stroke 1983;14:501–505.
255. Newell KL, Boyer P, Gomez-Tortosa E, et al. α-Synuclein immunoreactivity is present in axonal swellings in neuroaxonal dystrophy and acute traumatic brain injury. J Neuropathol Exp Neurol 1999;58:1263–1268.
256. Nicholas DS, Weller RO. The fine anatomy of the human spinal meninges: A light and scanning electron microscopy study. J Neurosurg 1988;69:276–282.
257. Nicholas JJ. Ectopic bone formation in patients with spinal cord injury. Arch Phys Med Rehabil 1973;54:354–359.
258. Niess C, Grauel U, Toennes SW, Bratzke H. Incidence of axonal injury in human brain tissue. Acta Neuropathol 2002;104:79–84.

259. Norenberg MD. Astrocyte responses to CNS injury. J Neuropathol Exp Neurol 1994;53:213–220.
260. Norenberg MD, Bruce-Gregorios J. Nervous system manifestations of systemic disease. In Davis RL, Robertson DM (eds): Textbook of Neuropathology, ed 3. Baltimore: Williams & Wilkins, 1997, pp 547–626.
261. Norenberg MD, Smith J, Marcillo A. The pathology of human spinal cord injury: Defining the problems. J Neurotrauma 2004;21:429–440.
262. Ochs S, Pourmand R, Jersild RA Jr, Friedman RN. The origin and nature of beading: A reversible transformation of the shape of nerve fibers. Prog Neurobiol 1997;52:391–426.
263. O'Connor JF, Cohen J. Dating fractures. In Kleinman PK: Diagnostic Imaging of Child Abuse, ed 2. St Louis: Mosby, 1998, pp 168–177.
264. Oehmichen M, Kirchner H (eds). The Wound Healing Process: Forensic Pathological Aspects. Lübeck: Schmidt-Römhild, 1996.
265. Oehmichen M, Meissner C, Schmidt V, et al. Axonal injury—A diagnostic tool in forensic neuropathology? A review. Forensic Sci Int 1998;95:67–83.
266. Oehmichen M, Raff G. Timing of cortical contusion: Correlation between histomorphologic alterations and post-traumatic interval. Z Rechtsmed 1980;84:79–94.
267. Ojala K, Lempinen M, Hirvonen J. A comparative study of the character and rapidity of the vital reaction in the incised wounds of human skin and subcutaneous adipose tissue. J Forensic Med 1969;16:29–34.
268. Orzel JA, Rudd TG. Heterotopic bone formation: Clinical, laboratory, and imaging correlation. J Nucl Med 1985;26:125–132.
269. Osgood EE, Krippaehne ML. Comparison of the life span of leukemic and non-leukemic neutrophils. Acta Haematal 1955;13:153–160.
270. Pape HC, Marsh S, Morley JR, et al. Current concepts in the development of heterotopic ossification. J Bone Joint Surg (Br) 2004;86:783–787.
271. Pardo-Mindán FJ, Herreros J, Marigil MA, et al. Myocardial calcification following heart transplantation. J Heart Transplant 1986;5:332–335.
272. Patil AA, Etemadrezaie H. Post-traumatic intradiploic meningo-encephalocele: Case report. J Neurosurg 1996;84:284–287.
273. Pearl GS. Traumatic neuropathology. Clin Lab Med 1998;18:39–64.
274. Peña CE. Wernicke's encephalopathy: Report of seven cases with severe nerve cell changes in the mammillary bodies. Am J Clin Pathol 1969;51:603–609.
275. Penfield W. Neuroglia: Normal and pathological. In Penfield W (ed): Cytology and Cellular Pathology of the Nervous System. New York: Hafner, 1965;II:423–479.
276. Penman J, Smith MC. Degeneration of the primary and secondary sensory neurons after trigeminal injection. J Neurol Neurosurg Psychiatry 1950;13:36–46.
277. Pittella JEH, Gusmão SNS. Diffuse vascular injury in fatal road traffic accident victims: Its relationship to diffuse axonal injury. J Forensic Sci 2003;48:626–630.
278. Platt MS, McClure S, Clarke R, et al. Postmortem cerebrospinal fluid pleocytosis. Am J Forensic Med Pathol 1989;10:209–212.
279. Plum F, Posner JB. The Diagnosis of Stupor and Coma, ed 3. New York: Oxford University Press, 2000.
280. Polman CH, Gijsbers CJ, Heimans JJ, et al. Rapid spontaneous resolution of an acute subdural hematoma. Neurosurgery 1986;19:446–448.
281. Postler E, Lehr A, Schluesener H, Meyermann R. Expression of the S-100 proteins MRP-8 and -14 in ischemic brain lesions. Glia 1997;19:27–34.
282. Prabhu VC, Bailes JE. Chronic subdural hematoma complicating arachnoid cyst secondary to soccer-related head injury: Case report. Neurosurgery 2002;50:195–197 discussion 197–198.
283. Prosser I, Maguire S, Harrison SK, et al. for the Welsh Child Protection Systematic Review Group. How old is this fracture? Radiologic dating of fractures in children: A systemic review. AJR Am J Roentgenol 2005;184:1282–1286.
284. Psaroudakis K, Tzatzarakis MN, Tsatsakis AM, Michalodimitrakis MN. The application of histochemical methods to the age evaluation of skin wounds: Experimental study in rabbits. Am J Forensic Med Pathol 2001;22:341–345.
285. Puckett WR, Hiester ED, Norenberg MD, et al. The astroglial response to Wallerian degeneration after spinal cord injury in humans. Exp Neurol 1997;148:424–432.
286. Puzas JE, Miller MD, Rosier RN. Pathologic bone formation. Clin Orthop 1989;245:269–281.
287. Quencer RM, Bunge RP, Egnor M, et al. Acute traumatic central cord syndrome: MRI-pathological correlations. Neuroradiology 1992;34:85–94.
288. Raekallio J. Histological estimation of the age of injuries. In Perper J, Wecht CH (eds): Microscopic Diagnosis in Forensic Pathology. Springfield, IL: Charles C Thomas, 1980, pp 3–16.
289. Rand CW, Courville CB. Histologic changes in the brain in cases of fatal injury to the head. III. Reactions of microglia and oligodendroglia. Arch Neurol Psychiatry 1932;27:605–644.
290. Rand CW, Courville CB. Histologic changes in the brain in cases of fatal injury to the head. V. Changes in the nerve fibers. Arch Neurol Psychiatry 1934;31:527–555.
291. Rand CW, Courville CB. Iron encrustation of nerve cells in the vicinity of old traumatic lesions of the cerebral cortex. Bull Los Angeles Neurol Sci 1945;10:95–106.
292. Reece RM. Child Abuse: Medical Diagnosis and Management, ed 2. Philadelphia: Lippincott Williams & Willkins, 2001.
293. Reichard RR, Smith C, Graham DI. The significance of β-APP immunoreactivity in forensic practice. Neuropathol Appl Neurobiol 2005;31:304–313.
294. Reichard RR, White CL III, Hladik CL, Dolinak D. Beta-amyloid precursor protein staining in nonhomicidal pediatric medicolegal autopsies. J Neuropathol Exp Neurol 2003;62:237–247.
295. Reichard RR, White CL III, Hladik CL, Dolinak D. Beta-amyloid precursor protein staining of nonaccidental central nervous system injury in pediatric autopsies. J Neurotrauma 2003;20:347–355.
296. Reske-Nielsen E, Lund E, Gregersen M. Central nervous system calcifications following tap water rinsing in autopsy material from children: A pitfall. Acta Pathologica Microbiologica et Immunologica Scandinavia 1991;99:765–768.
297. Resnick D. Neuromuscular disorders. In Resnick D (ed): Diagnosis of Bone and Joint Disorders, ed 4. Philadelphia: WB Saunders, 2002;4:3490–3493.
298. Resnick D, Goergen TG. Physical injury: Concepts and terminology. In Resnick D (ed): Diagnosis of Bone and Joint Disorders, ed 4. Philadelphia: WB Saunders, 2002;3:2627–2782.
299. Ribe JK. Personnal communcation, 2006.
300. Rich JR. Personal communication, 2006.
301. Richardson EP Jr, DeGirolami U. Pathology of the peripheral nerve. In Livolsi V (ed): Major Problems in Pathology. Philadelphia: WB Saunders, 1995;35:9.
302. Rickert CH, Rieder H, et al. Neuropathology of Raine syndrome. Acta Neuropathol 2002;103;281–287.
303. Roberts PH. Heterotopic ossification complicating paralysis of intracranial origin. J Bone Joint Surg (Br) 1968;50:70–77.
304. Robertson I, Mansfield RA. Ante-mortem and post-mortem bruises of the skin: Their differentiation. J Forensic Med 1957;4:2–10.
305. Rodriguez J, Claus D, Verellen G, Lyon G. Periventricular leukomalacia: Ultrasonic and neuropathological correlations. Dev Med Child Neurol 1990;32:347–355.
306. Rogers CB, Itabashi HH, Tomiyasu U, Heuser ET. Subdural neomembranes and sudden infant death syndrome. J Forensic Sci 1998;43:375–376.

307. Rogers LF, Hendrix RW. Fracture healing. In Rogers LF (ed): Radiology of Skeletal Trauma, ed 2. New York: Churchill Livingstone, 1992;1:197–221.
308. Ropper AH. Unusual spontaneous movements in brain-dead patients. Neurology 1984;34:1089–1092.
309. Rorke LB. Pathology of Perinatal Brain Injury. New York: Raven Press, 1982.
310. Rorke LB. Perinatal brain damage. In Adams JH, Duchen LW (eds): Greenfield's Neuropathology, ed 5. New York: Oxford University Press, 1992, pp 639–708.
311. Rosin AJ. Periarticular calcification in a hemiplegic limb—A rare complication of a stroke. J Am Geriatr Soc 1970;18:916–921.
312. Rosin AJ. Ectopic calcification around joints of paralyzed limbs in hemiplegia, diffuse brain damage, and other neurological diseases. Ann Rheum Dis 1975;34:499–505.
313. Rouah E, Gruber R, Shearer W, et al. Graft-versus-host disease in the central nervous system. A real entity? Am J Clin Pathol 1988;89:543–546.
314. Salter RB. Birth and pediatric fractures. In Heppenstall RB (ed): Fracture Treatment and Healing. Philadelphia: WB Saunders, 1980, pp 189–234.
315. Saposnik G, Bueri JA, Mauriño J, et al. Spontaneous and reflex movements in brain death. Neurology 2000;54:221–223.
316. Sarkar C, Lakhtakia R, Gill SS, et al. Chronic subdural haematoma and the enigmatic eosinophile. Acta Neurochir 2002;144:983–988.
317. Sarnat HB. Perinatal hypoxic/ischemic encephalopathy: Neuropathological features. In Garcia JH (ed): Neuropathology: The Diagnostic Approach. St Louis: Mosby, 1997, pp 541–580 (Table 13.1).
318. Sauer NJ, Dunlap SS. The asymmetrical remodeling of two neurosurgical burr holes: A case study. J Forensic Sci 1985;30:953–957.
319. Saukko P, Knight B. Knight's Forensic Pathology, ed 3. London: Arnold, 2004.
320. Sayer H, Wiethölter H, Oehmichen M, Zentner J. Diagnostic significance of nerve cells in human CSF with particular reference to CSF cytology in the brain death syndrome. J Neurol 1981;225:109–117.
321. Schachenmayr W, Friede RL. The origin of subdural neomembranes. I. Fine structure of the dura-arachnoid interface in man. Am J Pathol 1978;92:53–68.
322. Scheel-Toellner D, Wang K, Craddock R, et al. Reactive oxygen species limit neutrophil life span by activating death receptor signaling. Blood 2004;104:2557–2564.
323. Schmitt AB, Brook GA, Buss A, et al. Dynamics of microglial activation in spinal cord after cerebral infarction are revealed by expression of MHC class II antigen. Neuropathol Appl Neurobiol 1998;24:167–176.
324. Schmitt AB, Buss A, Breuer S, et al. Major histocompatibility class III expression by activated microglia caudal to lesions of descending tracts in the human spinal cord is not associated with a T cell response. Acta Neuropathol 2000;100:528–536.
325. Schneider H, Matakas F. Pathological changes of the spinal cord after brain death. Acta Neuropathol 1971;18:234–247.
326. Schwab ME, Bartholdi D. Degeneration and regeneration of axons in the lesioned spinal cord. Physiol Rev 1996;76:319–370.
327. Schwab JM, Nguyen TD, Meyermann R, Schluesener HJ. Human focal cerebral infarctions induce differential lesional interleukin-16 (IL-16) expression confined to infiltrating granulocytes, CD8+ T lymphocytes and activated microglia/macrophages. J Neuroimmunol 2001;114:232–241.
328. Seddon HJ. Three types of nerve injury. Brain 1943;66:237–288.
329. Seddon HJ, Medawar PB, Smith H. Rate of regeneration of peripheral nerves in man. J Physiol 1943;102:191–215.
330. Sevitt S. Bone Repair and Fracture Healing in Man. New York: Churchill Livingstone, 1981, pp 231–243.
331. Shannon P, Smith CR, Deck J, et al. Axonal injury and the neuropathology of shaken baby syndrome. Acta Neuropathol 1998;95:625–631.
332. Shapiro R. Anomalous parietal sutures and the bipartite parietal bone. AJR Am J Roentgenol 1972;115:569–577.
333. Shaw C-M, Alvord EC Jr, Berry RG. Swelling of the brain following ischemic infarction with arterial occlusion. Arch Neurol 1959;1:161–177.
334. Sherriff FE, Bridges LR, Gentleman SM, et al. Markers of axonal injury in post mortem human brain. Acta Neuropathol 1994;88:433–439.
335. Sherwood A, Hopp A, Smith JF. Cellular reactions to subependymal plate haemorrhage in the human neonate. Neuropathol Appl Neurobiol 1978;4:245–261.
336. Smith MC. Histologic findings after hemicerebellectomy in man: Anterograde, retrograde and transneuronal degeneration. Brain Res 1975;95:423–442.
337. Sohn D, Levine S. Hypertrophy of the olives: A report on 43 cases. In Zimmerman HM (ed): Progress in Neuropathology. New York: Grune and Stratton, 1971;1:202–217.
338. Sörnäs R, Östlund H, Müller R. Cerebrospinal fluid cytology after stroke. Arch Neurol 1972;26:489–501.
339. Squier W. Pathology of fetal and neonatal brain damage: Identifying the timing. In Squier W (ed): Acquired Damage to the Developing Brain: Timing and Causation. London: Arnold, 2002, pp 110–127.
340. Squier W. Gray matter lesions. In Golden JA, Harding BN (eds): Pathology and Genetics: Developmental Neuropathology. Basel: ISN Neuropath Press, 2004, pp 171–175.
341. Steele RW, Marmer DJ, O'Brien MD, et al. Leukocyte survival in cerebrospinal fluid. J Clin Microbiol 1986;23:965–966.
342. Storey G, Tegner WS. Paraplegic para-articular calcification. Ann Rheum Dis 1955;14:176–182.
343. Strassmann G. Hemosiderin and tissue iron in the brain, its relationship, occurrence and importance: A study of ninety-three human brains. J Neuropathol Exp Neurol 1945;4:393–401.
344. Strassmann G. Formation of hemosiderin and hematoidin after traumatic and spontaneous cerebral hemorrhages. Arch Pathol 1949;47:205–210.
345. Suárez-Peñaranda JM, Rodríguez-Calvo MS, Ortiz-Rey JA, et al. Demonstration of apoptosis in human skin injuries as an indicator of vital reaction. Int J Legal Med 2002;116:109–112.
346. Sunderland S. Capacity of reinnervated muscles to function efficiently after prolonged denervation. Arch Neurol Psychiatry 1950;64:755–771.
347. Sunderland S. Nerves and Nerve Injuries, ed 2. Edinburgh: Churchill Livingstone, 1978.
348. Sunderland S, Ray LJ. Denervation changes in mammalian striated muscle. J Neurol Neurosurg Psychiatry 1950;13:159–177.
349. Suttles J, Evans M, Miller RW, et al. T cell rescue of monocytes from apoptosis: Role of the CD40-CD40L interaction and requirement for CD40-mediated apoptosis—Role of the induction of protein tyrosine kinase activity. J Leukoc Biol 1996;60:651–657.
350. Suzuki H, Oyanagi K, Takahashi H, Ikuta F. Evidence for transneuronal degeneration in the spinal cord in man: A quantitative investigation in the intermediate zone after long-term amputation of the unilateral upper arm. Acta Neuropathol 1995;89:464–470.
351. Suzuki S, Kelley RE, Reyes-Iglesias Y, et al. Cerebrospinal fluid and peripheral white blood cell response to acute cerebral ischemia. South Med J 1995;88:819–824.
352. Tator CH. Classification of spinal cord injury based on neurological presentation. In Narayan RK, Wilberger JE, Povlishock JT (eds): Neurotrauma. New York: McGraw-Hill, 1996, pp 1059–1073.
353. Taveras JM. Anatomy and examination of the skull. In Taveras JM (ed): Radiology: Diagnosis-Imaging-Intervention. Philadelphia: Lippincott-Raven, 1997;3:2.1–2.9.

354. Taveras JM, Wood EH. Diagnostic Neuroradiology. Baltimore: Williams & Wilkins, 1964;1:764.
355. Taybi H, Lachman RS. Radiology of Syndromes, Metabolic Disorders, and Skeletal Dysplasias, ed 4. St Louis: Mosby, 1996.
356. Terao S, Li M, Hashizume Y, et al. Upper motor neuron lesions in stroke patients do not induce anterograde transneuronal degeneration in spinal anterior horn cells. Stroke 1997;28:2553–2556.
357. Terao S, Sobue G, Shimada N, et al. Serial MRI of olivary hypertrophy: Long-term follow-up of a patient with the "top of the basilar" syndrome. Neuroradiology 1995;37:427–428.
358. Thomas ED, Ramberg RE, Sale GE. Direct evidence for a bone marrow origin of the alveolar macrophage in man. Science 1976;192:1016–1018.
359. Tibone J, Sakimura I, Nickel VL, Hsu JD. Heterotopic ossification around the hip in spinal cord-injured patients. J Bone Joint Surg (Am) 1978;60:769–775.
360. Toole JF, Robinson MK, Mercuri M. Primary subarachnoid hemorrhage. In Toole JF (ed): Handbook of Clinical Neurology. Vascular Diseases, Part III. New York: Elsevier Science, 1989;11(55): 1–39.
361. Torch WC, Hirano A, Solomon S. Anterograde transneuronal degeneration in the limbic system: Clinical-anatomic correlation. Neurology 1977;27:1157–1163.
362. Toro G, Roman GC, Navarro-Roman L, et al. Natural history of spinal cord infarction caused by nucleus pulposus embolism. Spine 1994;19:360–366.
363. Tourtellotte WW, Metz LN, Bryan ER, DeJong RN. Spontaneous subarachnoid hemorrhage: Factors affecting the rate of clearing of the cerebrospinal fluid. Neurology 1964;14:301–306.
364. Trentz OA, Handschin AE, Bestmann L, et al. Influence of brain injury on early posttraumatic bone metabolism. Crit Care Med 2005;33:399–406.
365. Uchino A, Hasuo K, Uchida K, et al. Olivary degeneration after cerebellar or brainstem hemorrhage: MRI. Neuroradiology 1993;35:335–338.
366. Unger ER, Sung JH, Manivel JC, et al. Male donor-derived cells in the brains of female sex-mismatched bone marrow transplant recipients: A Y-chromosome specific in situ hybridization study. J Neuropathol Exp Neurol 1993;52:460–470.
367. Vance BM. Ruptures of surface vessels on cerebral hemispheres as a cause of subdural hemorrhage. Arch Surg 1950;61:992–1006.
368. Van Furth R, Diesselhoff-Den Dulk MMC, Mattie H. Quantitative study on the production and kinetics of mononuclear phagocytes during an acute inflammatory reaction. J Exp Med 1973;138: 1314–1330.
369. Vanezis P. Interpreting bruises at necropsy. J Clin Pathol 2001;54:348–355.
370. Vanezis P, Chan KK, Scholtz CL. White matter damage following acute head injury. Forensic Sci Int 1987;35:1–10.
371. Vinters HV, Farrell MA, Mischel PS, Anders KH. Diagnostic Neuropathology. New York: Marcel Dekker, 1998.
372. Vogeley E, Pierce MC, Bertocci G. Experience with Wood lamp illumination and digital photography in the documentation of bruises on human skin. Arch Pediatr Adolesc Med 2002;156:265–268.
373. Volpe JJ. Neurology of the Newborn, ed 3. Philadelphia: WB Saunders, 1995.
374. Waga S, Tochio H, Sakakura M. Traumatic cerebral swelling developing within 30 minutes after injury. Surg Neurol 1979; 11:191–193.
375. Walker AE. Cerebral Death, ed 3. Baltimore-Munich: Urban and Schwarzenberg, 1985.
376. Wang D, Shurafa MS, Acharya R, et al. Chronic abdominal pain caused by heterotopic ossification with functioning bone marrow: A case report and review of the literature. Arch Pathol Lab Med 2004;128:321–323.
377. Wharton GW, Morgan TH. Ankylosis in the paralyzed patient. J Bone Joint Surg (Am) 1970:52:105–112.
378. Wijdicks EFM (ed). Brain Death. Philadelphia: Lippincott Williams & Wilkins, 2001.
379. Wijdicks EFM. Clinical diagnosis and confirmatory testing of brain death in adults. In Wijdicks EFM (ed): Brain Death. Philadelphia: Lippincott Williams & Wilkins, 2001, pp 61–90.
380. Wijdicks EFM, Atkinson JLD. Pathophysiologic responses to brain death. In Wijdicks EFM (ed): Brain Death. Philadelphia: Lippincott Williams & Wilkins, 2001, pp 29–43.
381. Wilkinson AE, Bridges LR, Sivaloganathan S. Correlation of survival time with size of axonal swellings in diffuse axonal injury. Acta Neuropathol 1999;98:197–202.
382. Wolf DA, Falsetti AB. Hyperostosis cranii ex vacuo in adults: A consequence of brain atrophy from diverse causes. J Forensic Sci 2001;46:370–373.
383. Woodward SC. Mineralization of connective tissue surrounding implanted devices. Trans Am Soc Artif Intern Organs 1981;27: 697–702.
384. Yamaguchi M, Hirai K, Morita Y, et al. Hemopoietic growth factors regulate the survival of human basophils in vitro. Int Arch Allergy Immunol 1992;97:322–329.
385. Yamashima T, Friede RL. Why do bridging veins rupture into the virtual subdural space? J Neurol Neurosurg Psychiatry 1984;47: 121–127.
386. Yamashima T, Yamamoto S. How do vessels proliferate in the capsule of a chronic subdural hematoma? Neurosurgery 1984;15:672–678.
387. Yamashima T, Yamamoto S, Friede RL. The role of endothelial gap junctions in the enlargement of chronic subdural hematomas. J Neurosurg 1983;59:298–303.
388. Yokota T, Hirashima F, Furukawa T, et al. MRI findings of inferior olives in palatal myoclonus. J Neurol 1989;236:115–116.
389. Zimmerman B, Moegelin A, de Souza P, Bier J. Morphology of the development of the sagittal suture of mice. Anat Embryol 1998;197:155–165.
390. Zülch KJ. Hemorrhage, thrombosis, embolism. In Minckler J (ed): Pathology of the Nervous System. New York: McGraw-Hill, 1971;2:1499–1536.
391. Zülch KJ. Quelques observations sur l'artériosclérose intracranienne en allemagne de l'ouest. Afr J Med Sci 1971;2:301–318.
392. Zülch KJ, Hossmann V. Patterns of cerebral infarctions. In Toole JF (ed): Handbook of Clinical Neurology. Vascular Diseases. Part I. New York: Elsevier Science Publishing Co., 1988;53(9): 175–198.
393. Zwimpfer TJ, Bernstein M. Spinal cord concussion. J Neurosurg 1990;72:894–900.

Neuropathology of Pregnancy and Delivery: Mother and Child

4

INTRODUCTION 123
THE PREGNANT WOMAN 124
STILLBORN INFANTS 129

LABOR AND DELIVERY 130
THE NEWBORN INFANT 135
REFERENCES 146

Introduction

Cerebral palsy can be used to illustrate a general approach to cases submitted for neuropathologic consultation that involve the pregnant woman, fetus, or young child. Cerebral palsy, a clinical term, has been defined as "a group of conditions that are characterized by chronic disorders of movement or posture; it is cerebral in origin, arises early in life, and is not the result of progressive disease."[83] The neuropathologic substrate of cerebral palsy is quite variable from case to case, because a number of lesions may result in this spectrum of clinical features.[98] Furthermore, recent experimental and clinical studies also suggest that brain insults sustained early in development may result in secondary, progressive structural (e.g., cerebral cortical dysplasia) and presumably functional (e.g., seizure propensity) changes that continue to evolve in surviving brain tissue for extended periods beyond the healing phase of the original tissue destruction produced by the primary insult.[24,75–78] It follows that, from a neuropathologic standpoint, cerebral palsy is a syndrome rather than a disease entity and may well have a progressive underlying structural component when the brain is studied by special techniques, albeit not necessarily obvious clinically.

Only a few years ago, consensus held that most cases of cerebral palsy were the result of hypoxic-ischemic injury during labor and delivery.[104] More recent studies suggest that intrapartum hypoxia cannot be implicated as a cause in an estimated 90% of cases of cerebral palsy, and that "in the remaining 10% the intrapartum signs compatible with damaging hypoxia may have had antenatal or intrapartum origins."[72] Developmental abnormalities of various etiologies (genetic or other), prematurity, effect of multiple pregnancies, recognized associated cardiac and brain malformative disorders as well as central nervous system (CNS) lesions secondary to isolated cardiac malformations and/or therapy thereof,[61] metabolic abnormalities, autoimmune and alloimmunization disorders, coagulation disorders, infections, trauma, toxins,[12] low birth weight, twin-twin transfusion syndrome all must be included within the etiologic considerations and known associations in a given case, in addition to the possibility of perinatal hypoxic-ischemic insults related in time to labor and delivery. Furthermore, the possible relationship of umbilical cord (e.g., nuchal cord) and placental abnormalities to hypoxic-ischemic insults requires examination.[57,63,92,105]

Such considerations apply to various other hemorrhagic and nonhemorrhagic neuropathologic disorders that are either typically seen in, or unique to, the premature infant, the term infant, and the first few months of life, as has been stressed by Gilles.[42]

These considerations should be kept in mind, despite the fact that most current textbooks of neuropathology group many of these conditions (e.g., periventricular leukomalacia, multicystic encephalopathy, status marmoratus) under the category of hypoxic-ischemic lesions. That is not to say that a hypoxic-ischemic factor may not play an important role in many of these conditions, but rather it may simply be a final common pathway mechanism that has been instigated by an antecedent chain of events leading to fetal distress (or "nonreassuring fetal status")[72] during labor and delivery, which in turn provides indications for obstetric interventional procedures. Rarely, sudden onset of abnormal fetal movements variously characterized as epileptiform, or as abnormal decerebrate movements, have called attention to fetal distress.[54] The brain in the latter case was reported as normal,[54] but the extent of neuropathologic investigation is unclear.

A reasonable approach under such circumstances is to evaluate each case with the aid of guidelines such as those recommended in the consensus statement of the International Cerebral Palsy Task Force,[72] keeping in mind that the legal profession's approach to such issues may not be identical.[93]

In this chapter, conditions in mother and child that are more likely to be encountered by the forensic neuropathologist are emphasized. Thus, fatal conditions such as cerebrovascular catastrophes are considered, but other neurologic conditions simply potentially aggravated by pregnancy that are less likely to present to a medical examiner than to a clinician (e.g., multiple sclerosis) are generally omitted. Reviews of pregnancy-related deaths that include non-CNS causes have recently appeared in the forensic literature, and provide a useful perspective for the medical examiner.[19,94]

The term "prematurity," as used throughout this text, is defined as an infant born after less than 37 weeks' gestation, irrespective of birth weight. In this chapter, the perinatal conditions emphasized are related to events likely to precipitate medical examiner referral, and include some neuropathologic lesions that may be useful in dating an insult to a particular developmental stage (also see Chap. 5).

The Pregnant Woman

Stroke

Pregnancy confers no immunity to any of the usual differential diagnostic considerations in the stroke patient. The most common type of cerebrovascular event in pregnancy is ischemic infarction due to arterial occlusion, the incidence of which is nearly evenly divided among each trimester. Intracranial infarction or hemorrhage may also potentially occur in the pregnant or puerperant woman as a result of varied conditions, such as aortic or vertebral artery dissection[87,116] or peripartum cardiomyopathy (emboli from mural thrombi with subsequent hemorrhage), cerebral phlebothrombosis, sickle cell crisis, or pregnancy-induced states such as eclampsia (see later).[11,16]

Cerebral venous thrombosis is a potential complication of pregnancy, usually presenting in the postpartum period. The usual differential diagnosis for hypercoagulable states, such as dehydration, eclampsia, protein S deficiency, and factor V Leiden deficiency, should be explored.

Subarachnoid hemorrhage in pregnant women under 25 years of age is most commonly due to antecedent arteriovenous malformations of the CNS, usually in primiparas, and most commonly occurs during the second trimester or during labor.[28] Subarachnoid hemorrhages in pregnant women aged 25 or older are most often due to antecedent saccular aneurysms, which tend to rupture with progressively increasing incidence as the third trimester progresses.[28] Remaining cases include a long list of less frequent underlying causes of subarachnoid hemorrhage that also are encountered in nonpregnant patients.

Blood loss during delivery, if not controlled or replaced in a timely fashion, can lead to maternal hypoxic-ischemic encephalopathy and its complications (Case 4.1, Figs. 4.1–4.3).

Epilepsy

Seizure frequency increases in some epileptic patients during pregnancy. Factors implicated include hormonal effects, sleep deprivation arising from a variety of pregnancy-associated causes, noncompliance with anticonvulsant therapy, salt and water retention, mild compensated respiratory alkalosis, and increased plasma clearance of anticonvulsants during pregnancy.[28] The last requires clinical monitoring of anticonvulsant levels and dosage during pregnancy and after delivery in order to prevent overdose toxicity as drug clearance returns to nongravid levels. Theoretically, increased seizure frequency or severity could also predispose to an increased rate of maternal trauma and sudden unexpected death (see Chap. 9). Vitamin K supplements in epileptic mothers receiving certain anticonvulsants that depress vitamin K levels may be useful in reducing the potential for newborn intracerebral hemorrhages,[60,61] although recent data suggest factors other than maternal hepatic enzyme-inducing anticonvulsants (e.g., prematurity, alcohol use during pregnancy) may be more likely causes of such coagulopathies.[55]

Case 4.1. (Figs. 4.1–4.3)

This 37-year-old woman's pregnancy was complicated by development of diabetes mellitus and hypertension. Her vaginal delivery was further complicated by a vaginal laceration and hemorrhage from the vagina into the retroperitoneal space. As she was being prepared for emergency exploratory laparotomy, she had a cardiorespiratory arrest. She was resuscitated but remained comatose until death 7 days later. Neuropathologic examination revealed severe hypoxic-ischemic encephalopathy with herniation syndrome, greater on the left than the right side.

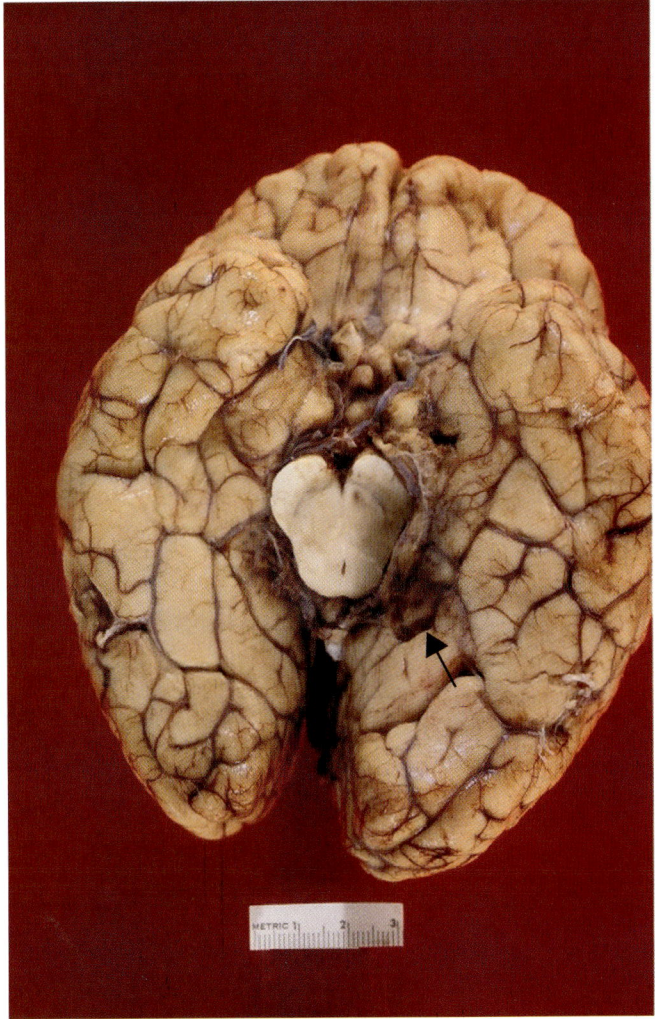

Fig. 4.1. Basal view of brain shows left greater than right side grooving from tentorial herniation. Some softening and hemorrhage are present medial to the herniation groove bilaterally, more pronounced posteriorly (*arrow*).

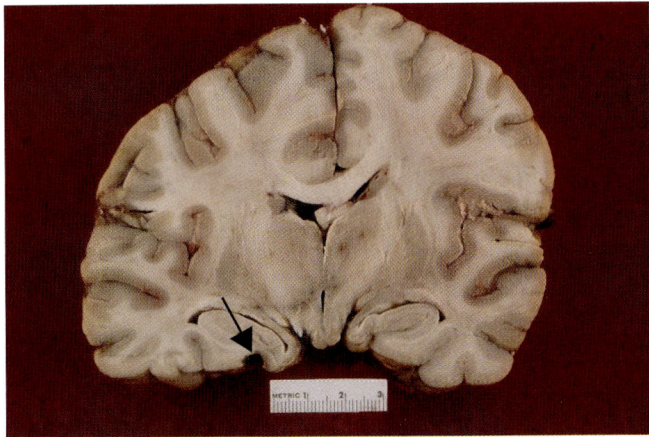

Fig. 4.2. Coronal section of brain at level of anterior thalamus. Left greater than right tentorial herniation with "herniation contusion" seen on left (*arrow*).

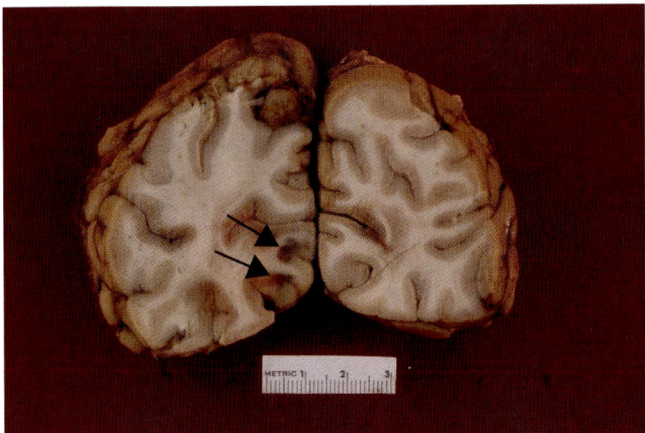

Fig. 4.3. Coronal section of brain at level of occipital lobes shows congestion and gray matter hemorrhage in distribution of left posterior cerebral artery due to its compromise at left tentorial herniation site (*arrows*). A smaller, similar lesion is seen in the depths of a sulcus in the right medial hemisphere.

Case 4.2. (Fig. 4.4)

This 21 week estimated gestational age male stillborn fetus was delivered spontaneously 2 days after the mother was assaulted by her common-law husband, who dragged her by the hair and repeatedly kicked her in the abdomen with a shod foot. Due to persistent abdominal pain, the mother was evaluated by ultrasound, which revealed no fetal heart motion. The external fetal examination (i.e., extent of skin desquamation and discoloration) was consistent with the time of death of the fetus being 24 or more hours prior to delivery.

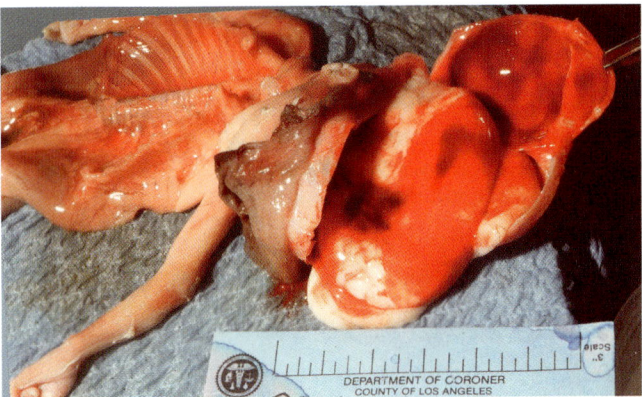

Fig. 4.4. Calvarium of fetus opened, demonstrating subdural and subarachnoid hemorrhage.

Trauma

Blunt force trauma, whether by falls, assaults (Case 4.2, Fig. 4.4), motor vehicle accidents, or other mechanisms, poses some risks unique to the pregnant woman. The most common cause of fetal death in maternal trauma is maternal death, sustained hypotension, or placental abruption.[5,28,69,95] Traumatic splenic rupture is more likely in gravid females, presumably due to hypervolemia of pregnancy.

Uterine rupture as a direct result of blunt force trauma is rare (Case 4.3, Fig. 4.5). Direct trauma to the unborn fetus may result in skull fracture and/or intracranial hemorrhage.[31] Fetal intrauterine skull fractures are often parietal, and probably result from pressure of the fetal head against the maternal sacral promontory or pubic ramus. Thus, they are more likely to occur late in gestation when the fetal head is within the pelvis.[4]

Penetrating wounds of the uterus are infrequently reported, and include stab wounds and gunshot wounds (Case 4.4, Figs. 4.6–4.10). Fetal gunshot wounds may have atypical features related to, for example, effects of intermediate targets, shoring of wounds, and skin immaturity.[17]

Amniocentesis may also result in fetal head injury.[26,107] Folk medicine practices, such as traditional uterine massage techniques, have also been suggested as contributors to fetal injury.[3,47]

Electric Shock

The fetus is more susceptible to death from electric shock than the mother, with a fetal mortality of 73% in a series of 15 cases reviewed by Fatovich.[35] Most reported cases have involved household direct current, and typical current paths were from hand or upper body to foot. Somogyi and Tedeschi[106] have suggested criteria by which to establish a positive correlation between the electric shock and subsequent fetal (and, in some instances, maternal) death, as follows:

Case 4.3. (Fig. 4.5)

A 32-year-old woman, recently released from a psychiatric facility, committed suicide by jumping from the roof of a 16-story building. She died of multiple traumatic injuries that included pelvic fracture, and lacerated uterus and placenta. A dead fetus, estimated gestational age 23 weeks, was found free in the peritoneal cavity.

Case 4.4. (Figs. 4.6–4.10)

A 37 week estimated gestational age fetus was the victim of an intrauterine 20-gauge shotgun wound sustained by the mother while at a party. Eight pellet wounds were present in the mother's left buttock, and abdominal and kidney areas. No fetal activity was found on ultrasound examination in the emergency room, and an emergency C-section delivered a stillborn fetus. Examination revealed shotgun pellet wounds of the head and left hand.

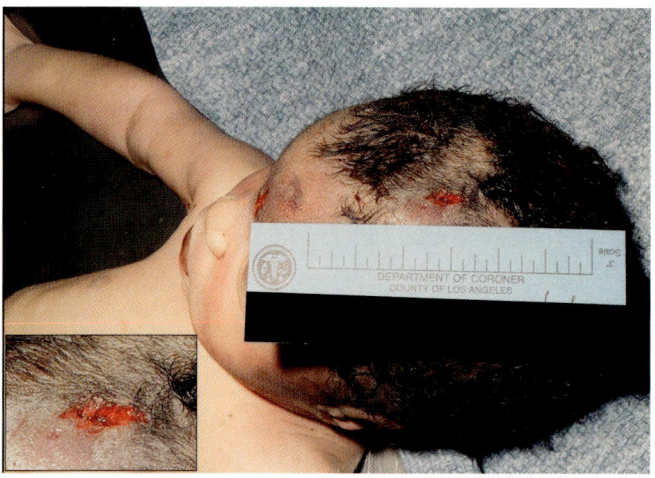

Fig. 4.6. Shotgun pellet entrance wound seen in upper left frontal scalp. *Inset.* Detail of angled entrance wound.

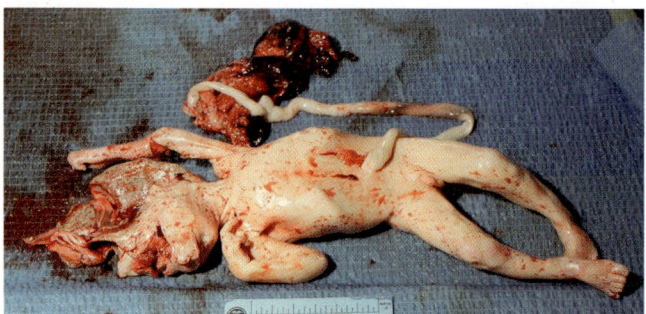

Fig. 4.5. Fetus with portion of attached placenta found lying free in the peritoneal cavity, demonstrating massive craniofacial trauma (and upper extremity deformity due to fractures).

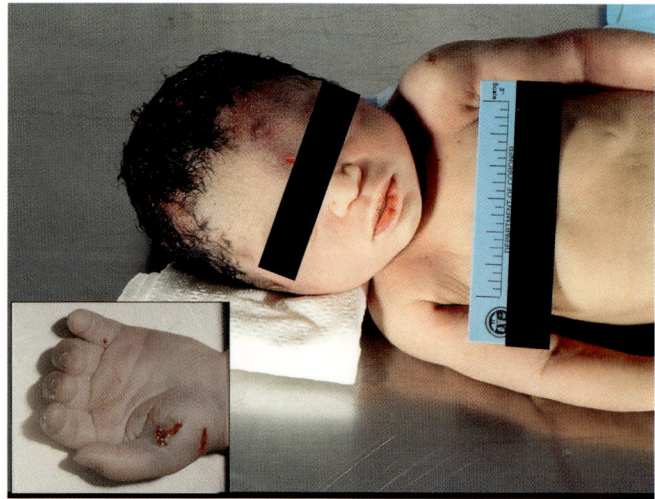

Fig. 4.7. Shotgun pellet exit wound in left glabella area. *Inset.* Shotgun pellet wound in left thenar eminence region.

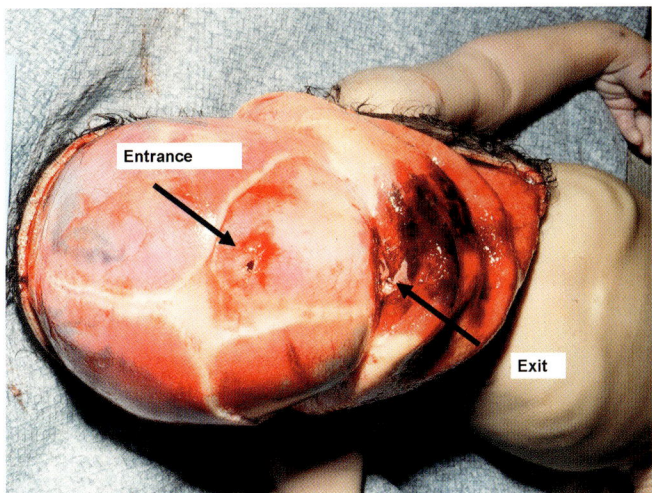

Fig. 4.8. Skull entrance and exit wounds revealed after reflection of hemorrhagic frontal scalp.

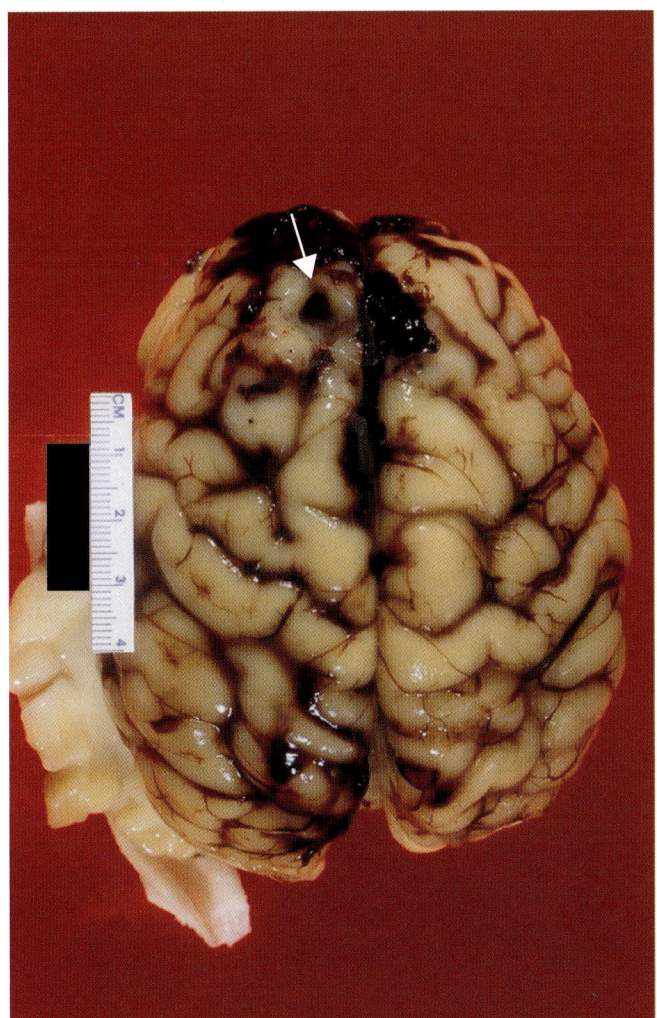

Fig. 4.9. Dorsal view of cerebral hemispheres, with subarachnoid hemorrhage and left medial frontal lobe entrance wound (*arrow*).

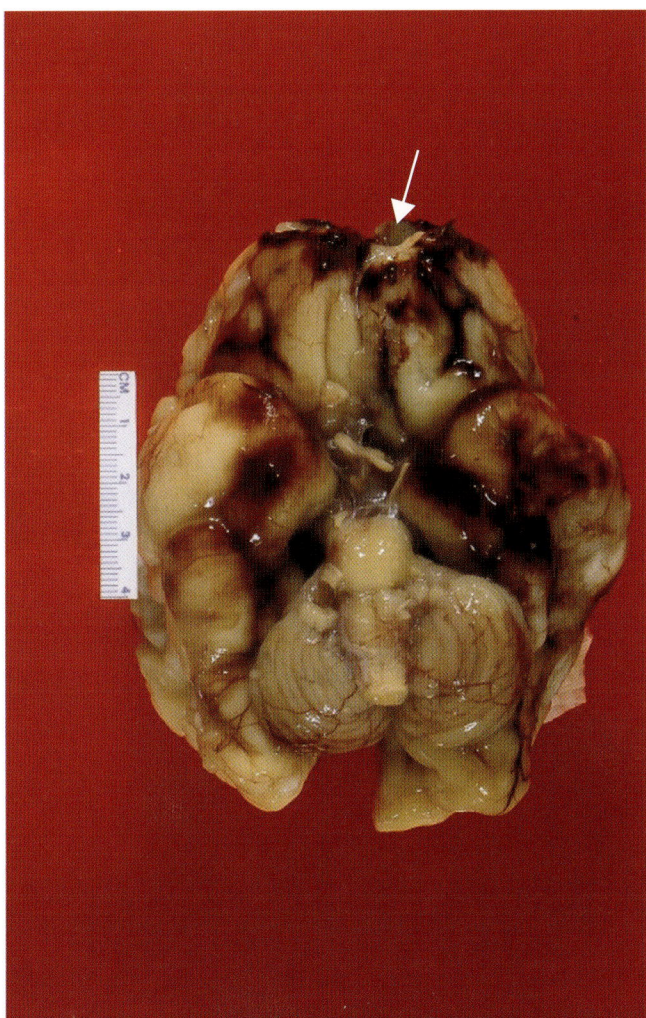

Fig. 4.10. Basal view of brain with exit wound at inferior left frontal pole (*arrow*).

1. There has been a normal pregnancy (pregnancies) and delivery (deliveries) in the past gestational history of the mother.
2. The course of the pregnancy was uneventful until the occurrence of the accident.
3. The accident has been proved.
4. No other factor or factors can be advocated to explain the loss of the fetus or the death of the mother.
5. The stage of development of the dead fetus corresponds to the stage of pregnancy, and the degree and extent of its maceration are proportional to the time that has elapsed between the accident and the death of the fetus.

Sublethal consequences to the fetus following electric shock to the mother may include intrauterine growth retardation, decreased movement, or oligohydramnios.[1]

In a medical examiner setting, issues arising during the course of arrest and/or in-custody use of force may arise, such as whether use of electrical devices (e.g., the taser) may cause miscarriage. An initial report suggesting the taser as a cause of miscarriage appeared in 1992.[80] This question is also discussed briefly in Chapter 8.

Burns
The fetal prognosis in burns tends to parallel maternal prognosis, without evidence of significant differential susceptibility.[1] Thermal injury to the nervous system is discussed in Chapter 17.

Asphyxia
The pregnant woman is reportedly more vulnerable to CNS hypoxic-ischemic injury compared to her unborn child and the nonpregnant woman (see Chaps. 10 and 11). In contrast, in cellular hypoxia induced by carbon monoxide (CO) poisoning the fetus is more vulnerable than the pregnant woman, due to altered kinetics in fetal versus maternal hemoglobin.[34]

Poisonings
The fetus tends to be more susceptible than the mother to toxic effects of acetaminophen, salicylates, and CO.[1] Neurotoxicology is also briefly discussed in Chapter 17.

Intracranial Tumors
Many types of brain tumors undergo accelerated enlargement during pregnancy, ascribed to various factors such as high tumor vascularity associated with increased cardiac output during pregnancy, or to hormonal effects. Meningiomas, neurofibromas, and capillary hemangioblastomas are among the frequently described tumors showing this propensity. Pituitary adenomas may also show significant increase in size during pregnancy. Mass effects of intracranial tumors combined with labor may precipitate a variety of neurologic signs and symptoms, including a herniation syndrome.

Of the pregnancy-associated systemic tumors, choriocarcinoma is the most likely to metastasize to the brain, and it may present as an acute intracerebral hemorrhage. This tumor usually presents months after a pregnancy that was associated with a hydatidiform mole or miscarriage, but it may also follow a normal pregnancy.

Eclampsia
Eclampsia may present during gestation, during delivery, or up to 7 days' postpartum. Together with HELLP syndrome, amniotic fluid embolism, and neurologic manifestations of tocolytic agents, eclampsia and these conditions are the most common causes of admission of the pregnant patient to the intensive care unit.[122]

Neuropathologic manifestations of eclampsia are primarily maternal cerebral hemorrhage (Case 4.5, Fig. 4.11) and ischemic stroke, with hemorrhages occurring in the CNS locations commonly encountered in hypertension from other causes. It has been postulated that eclampsia

Case 4.5. (Fig. 4.11)
This 25-year-old woman with a 28-week pregnancy presented to the emergency room with hypertension, and right hemiparesis of acute onset. An emergency cesarean section was performed, with delivery of a viable 1225-g male infant. The mother developed cardiorespiratory arrest during surgical closure, and could not be resuscitated.

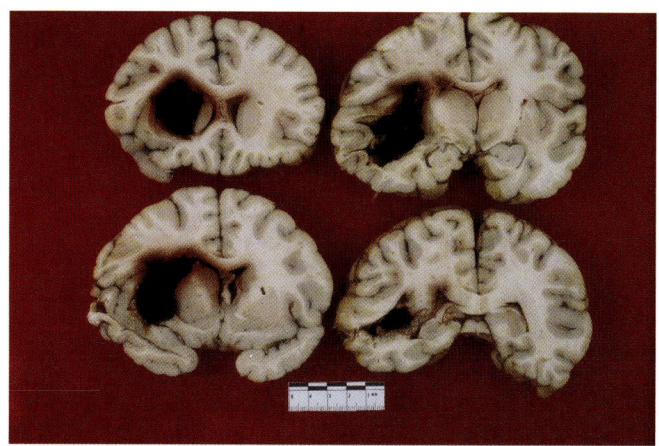

Fig. 4.11. Autopsy revealed a massive, acute left hemisphere intracerebral hemorrhage centered in putamen with hemorrhagic rupture into adjacent lateral ventricle and diffuse cerebral swelling with herniation syndrome, evident on coronal sections.

may be a risk factor for the development of hippocampal sclerosis and temporal lobe epilepsy.[65]

HELLP Syndrome
This acronym refers to the essential manifestations of this complication of eclampsia, namely, *H*emolysis, *E*levated *L*iver enzymes, and *L*ow *P*latelets.[115,122] Neuropathologic manifestations, when present, primarily consist of subarachnoid or intracerebral hemorrhage, or hypoxic-ischemic effects such as cerebral edema secondary to associated multisystem failure, disseminated intravascular coagulation, or postpartum uterine hemorrhage.[114]

Amniotic Fluid Embolism
The neuropathologic manifestations of this disorder, in which maternal mortality approaches 80% and fetal mortality is approximately 70%, are mainly the result of hypoxic-ischemic events caused by maternal hypotension, cardiac arrest, acute respiratory failure, postpartum hemorrhage, or disseminated intravascular coagulation.[20,122]

Spontaneous amniotic fluid embolism may develop anytime from the first trimester to 48 hours postpartum, and onset has also been described in association with delivery (including illegal abortion; see later), abdominal trauma, and amnioinfusion.[74] The latter procedure is sometimes advocated as an aid in the management of conditions such as meconium-stained amniotic fluid, oligohydramnios (with or without gastroschisis), preterm rupture of membranes, variable fetal heart rate decelerations suggesting umbilical cord compression, or chorioamnionitis. One of us (H.H.I.) has seen fetal squames in brain blood vessels in cases of amniotic fluid embolism, although we have not yet seen trauma-precipitated chorionic villi embolization in brain tissue.[59]

Tocolytic Agents

These agents, used to inhibit premature labor, are unlikely to produce a fatal outcome but may cause alarming symptoms such as ptosis, diploplia, photophobia, precipitation of migraine headaches, and pulmonary edema. Wijdicks[122] provides a brief review of commonly used agents, and Graber[46] has also discussed potential risks of this class of agents.

Air Embolism

This condition, occasionally pregnancy-related, has been reported in association with orogenital sex (cunnilingus) in pregnant (and rarely in nonpregnant) women,[22] and in the context of cesarean section, placenta previa, and illegal abortion.[27] Neurologic symptoms may be transient in milder cases, or can result in permanent neurologic defects or death. The neuropathologic lesions may relate to diffuse hypoxic-ischemic effects due to the cardiopulmonary consequences of air embolism, or may be more localized due to air emboli in brain or spinal cord arteries leading to microinfarcts.

Stillborn Infants

Risk factors for increased fetal loss include such diverse conditions as maternal diabetes mellitus, Cushing's syndrome, immune disorders (e.g., systemic lupus erythematosus, antiphospholipid syndrome), infections (e.g., listeriosis), and a variety of other disorders.[16] Screening for these risk factors is typically performed by the clinicians involved in the case and/or the medical examiner performing the general autopsy.

A stillbirth is defined as delivery of an infant with no sign of life between 20 weeks' gestation and term. Previable fetal death is delivery of a dead fetus between 11 and 19 weeks' gestation.[73]

Forensic neuropathology issues in these two categories most often arise in the stillborn infant, and most often deal with questions related to the timing of intrauterine death, and/or whether abnormalities present in the CNS can be theoretically linked causally to a particular event (e.g., trauma, exposure to drugs) that occurred during gestation. For example, it has been recommended that an abnormal gyral pattern on prenatal ultrasound in stillborn cases should prompt rapid postdelivery collection of the types of specimens needed for evaluation of perioxisomal and other metabolic disorders.[73] In the absence of congenital anomalies of the CNS (discussed in Chapter 5), the question most often asked relates to possible effects of prepartum or intrapartum trauma as a cause of stillbirth. Timing and dating of CNS lesions encountered thus becomes an important aspect of the neuropathology evaluation (see Chapters 3 and 5).

Fetal Artifacts

An artifact, which reportedly is largely confined to macerated, aborted fetuses at developmental gestational age of 9 to 16 weeks, is the presence of autolyzed cerebral tissue variously found in the retroperitoneal space, under the pleura, in the neck, in paraspinal areas, in systemic and pulmonary vessels, in cardiac chambers, placenta, adrenal medulla, and inguinal areas.[56,64,85,100] This may be confused with a primitive neuroectodermal tumor in macerated fetuses. Such autolyzed CNS material is believed to gain access to these various sites via squeezing of the autolyzed brain tissue *in utero* or at delivery into, for example, the spinal canal and along spinal nerves entering these various locations.

This artifact in macerated, aborted fetuses is to be distinguished from the embolism of brain tissue (usually cerebellar cortex) to lungs, placenta, coronary arteries, pial veins, choroid plexus, and meningeal, cerebral, and/or renal arteries that may occur in term or preterm neonates sustaining fatal injury during complicated deliveries, and in children following head trauma.[10,30]

Death to Delivery Interval in Stillborn Infants

This question is occasionally posed to the neuropathology consultant. As Case 4.6 illustrates, the CNS examination may yield useful information, although not necessarily related to the question asked.

Comparison of Stillborn and Liveborn Infant Neuropathologic Lesions

In considering possible cause-and-effect relationships between particular traumatic events and presence of intracranial hemorrhage in the neonate, consideration must also be given to studies that have demonstrated the relative frequency and location of intracranial hemorrhage in stillborn infants. Considering large, macroscopic hemorrhages only, for example, Dooling and Gilles[29] examined stillborns of greater than 20 weeks' gestation compared to liveborn infants surviving less than 7 days. They found subarachnoid (leptomeningeal) hemorrhages in 43% of stillborns and 56% of liveborn infants, intraventricular hemorrhage in 9% of stillborns and 20% of liveborn infants, intracerebral white matter hemorrhage in 3% of stillborns and 5% of liveborn infants, and germinal matrix hemorrhage in 5% of stillborns and 19% of liveborn infants. The incidence of smaller, microscopic

Case 4.6. (Figs. 4.12–4.13)

A 28-year-old woman was involved in a physical altercation with her sister approximately 2 hours prior to delivery of a stillborn fetus (estimated gestational age 30 to 31 weeks). Skin slippage was noted on the fetus. The brain had a semiliquid consistency, and only scattered foci of retained recognizable cerebral gyri. Within the pasty material was an ovoid blood clot. Histologic sections of more intact regions revealed cerebral cortical cytoarchitectural abnormalities, subarachnoid blood clot, leptomeningeal vascular congestion, and multiple cerebellar neuronal heterotopias, mixed cell type (Figs. 4.12–4.13). The degree of autolysis microscopically was much less impressive than the gross appearance of the brain suggested. This variety of cerebellar neuronal heterotopia is more likely to be associated with nonneural organ anomalies (see Chap. 2).

Comment. Histologic appearance of brain tissue is a poor indicator of time of fetal death to time of delivery interval in stillborn infants. Nuclear basophilia in cerebral cortical neurons may persist at least focally in fetuses with death to delivery intervals of 8 weeks or more.[39] Cerebral cortical nuclear basophilia even persisted in a fetus papyraceous whose death was documented 25 weeks before delivery.[39] Histologic examination of other (non-CNS) organs (e.g., kidney, liver, myocardium, lung, gastric intestinal tract, adrenal),[39] of the placenta and umbilical cord,[40,110] and even external examination of the fetus for degree of skin slippage or desquamation[41] provides more accurate estimates of death to delivery interval than does examination of CNS histology or examination for the presence or absence of calvarial collapse. In our case, the degree of skin maceration (desquamation) clearly indicated that intrauterine death preceded any trauma inflicted during the fight between sisters, and the main value of the neuropathologic study was to provide at least one potential etiology for the intrauterine demise (i.e., that of CNS developmental anomalies). This case also illustrates the point that the degree of cellular and architectural detail seen in histologic sections is sometimes considerably better than one might suspect based on gross appearances, especially in immature brain tissue.

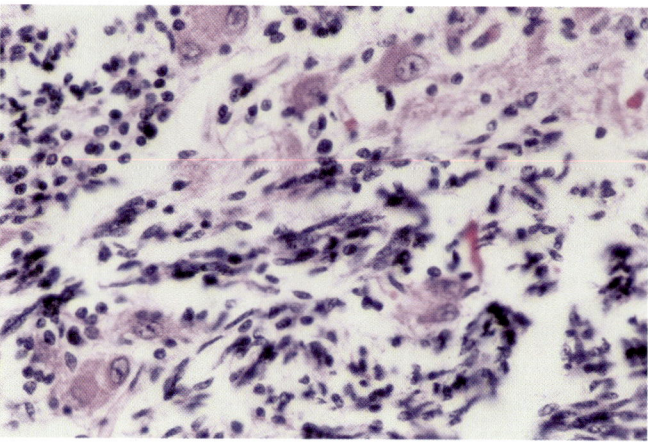

Fig. 4.13. Medium-power micrograph of cerebellar neuronal heterotopia seen in Fig. 4.12, demonstrating mixed population of larger neurons with pink cytoplasm and smaller, immature forms with sparse cytoplasm and dense round to elongate basophilic nuclei (H&E.)

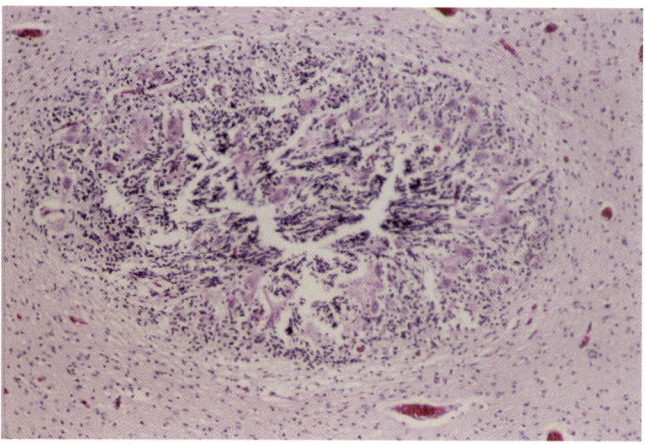

Fig. 4.12. Low-power micrograph of neuronal heterotopia in cerebellar subcortical white matter, mixed large and small cell type. (H&E.)

hemorrhages in both groups was even greater. Singer[104] has described germinal matrix hemorrhage in macerated stillborn fetuses as early as the 13th week of gestation. Such observations, as Gilles[42] succinctly states, "lend support to the suspicion that some of the significant antecedents of neonatal intracranial hemorrhage may antedate birth."

In later stages of gestation, intrauterine subdural hemorrhages may also occur in the absence of known trauma. Unlike posterior fossa subdural hemorrhages which are most often ascribed to traumatic delivery, they are usually supratentorial in location.[3,47] Spontaneous resolution of fetal bilateral subdural hematoma prior to delivery has also been reported, the hematoma in this instance possibly related to minor blunt force maternal abdominal trauma.[13]

In cases that come to autopsy, one looks for clinical record as well as autopsy evidence of coagulopathies, toxins, infection, metabolic disorders, hypoxic-ischemic episodes, and so on. The focus of the neuropathologist is primarily to exclude identifiable anatomic lesions predisposing to hemorrhage such as vascular anomalies, vasculitis, phlebothrombosis, embolism, neoplasia, and infection, and to determine the age of the lesions present based on accompanying cellular responses (see Chaps. 2, 3, and 11). Correlation of all available historical and anatomic findings with the timing and nature of proposed etiologic events can then be attempted.

Labor and Delivery

Spinal or epidural anesthetic procedures employed during labor and delivery may infrequently be accompanied by untoward effects, a topic discussed in Chapter

16. It is worth noting again, however, that association in time with delivery does not prove a cause-and-effect relationship, and all available data should be fully evaluated before reaching conclusions concerning such relationships. For example, several of the groups of neuropathologic findings reported in the newborn in association with a particular obstetric procedure, such as forceps or vacuum-assisted delivery, should be interpreted with the knowledge that such interventions are almost exclusively employed for deliveries in which some problem is already manifest or in the process of developing. To minimize errors in interpretation, it is prudent to search for any possible preexisting problem leading to the intervention, its possible causes, and its possible relationship to any neuropathologic lesions subsequently found.

Vaginal Delivery

A variety of neuropathologic findings occur in infants born by vaginal delivery. Their exact incidence is difficult to assess, but it is evident that some complications are more likely to occur when delivery is complicated by either a prolonged or an unusually short labor, suboptimal presentations, varying degrees of cephalopelvic disproportion, need for more than usual traction on presenting parts, or other untoward events short of clear indications for intervention in the form of cesarean section or other types of instrument-assisted delivery.

Caput succedaneum, a result of edema with or without associated hemorrhage in the subcutaneous tissue of the scalp, provides—by its location—an indication of mode of presentation of the head during delivery.

A subaponeurotic (subgaleal) hemorrhage (Case 4.7, Fig. 4.14) involves a rather large potential space from orbital ridge to nape of neck. Although most are not large enough to require transfusion, this potential space is sufficient to allow a fatal degree of blood volume loss in severe cases,[124] or to produce mass effects resulting in symptoms related to increased intracranial pressure.[6] An estimated 28% of newborn subgaleal hematomas occur after spontaneous vaginal delivery. Vessel injury may result from suture diastasis and/or overriding of skull bones, skull fracture, or emissary vein injury. Potential risk factors include coagulopathies, prematurity, rapid delivery, macrosomia, and male gender.[123,124]

Cephalhematomas are hemorrhages in the subperiosteal space (Fig. 4.15). Limited as they typically are by attachment of the periosteum to the unfused suture lines at the skull bone margin (most often overlying the parietal bone), they do not pose a risk for life-threatening blood loss. They are associated with linear or depressed skull fractures in 10 to 25% of cases (Hovind, quoted by Leestma[66]). Rarely, a linear skull fracture may occur in spontaneous vaginal delivery uncomplicated by cephalopelvic disproportion or extrinsic trauma.[51] In most instances, however, skull fractures occur in deliveries

Case 4.7. (Fig. 4.14)

Induced labor in this term female infant delivery became complicated by shoulder dystocia, leading to a vacuum extraction procedure. The neonate was in cardiorespiratory arrest upon arrival in the pediatric intensive care unit and did not respond to resuscitation efforts. Autopsy findings included extensive subgaleal hemorrhage, overriding sutures, cranial subperiosteal hemorrhage, subdural and subarachnoid hemorrhage, and cervico-occipital distraction with posterior epidural hemorrhage at the C2 vertebra.

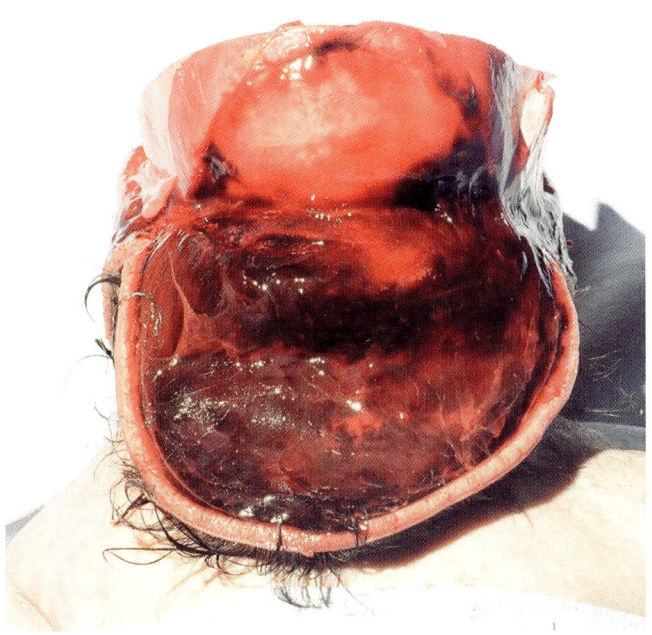

Fig. 4.14. Reflected scalp with extensive subgaleal hemorrhage.

requiring medical intervention.[82] Petechiae within the falx cerebri or tentorium cerebelli are commonly found in both premature and term infants, and are considered by Friede[37] to be more likely of asphyxial than of traumatic origin.

Epidural, subdural, and subarachnoid hemorrhage also may occur, with or without associated skull fracture. Other conditions mimicking such lesions must also be excluded. While infrequent, subdural hemorrhage is seen more frequently in term than in premature infants.[18,118] Primary subarachnoid hemorrhage is seen more frequently in prematures.[118] Symptoms associated with subdural hemorrhage may be subtle[18] and it is likely that minor, asymptomatic subdural hemorrhages occur more frequently following uneventful vaginal delivery than is generally appreciated, based on a pathologic study in sudden infant death syndrome infants.[96] A recent clinical study using magnetic resonance imaging (MRI) has documented unilateral and bilateral subdural hemorrhages

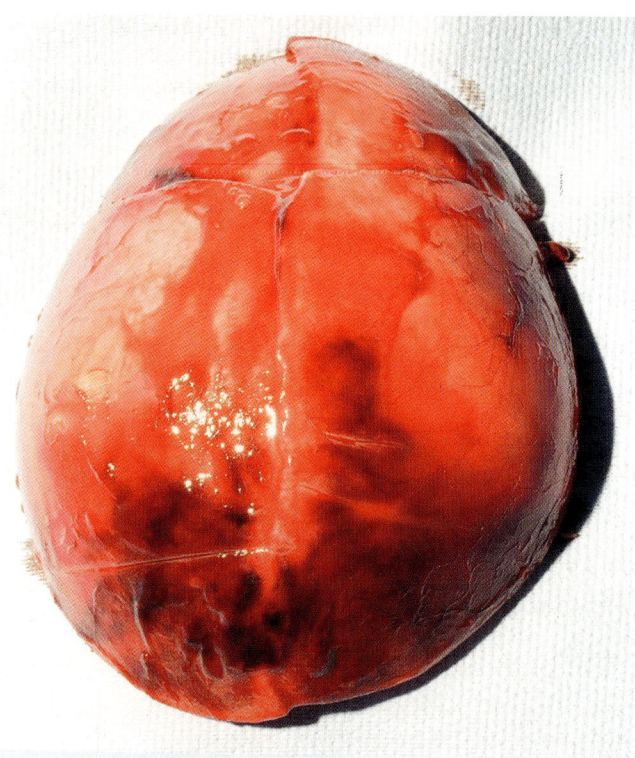

Fig. 4.15. Widespread subperiosteal hemorrhage on external surface of calvarium, more pronounced posteriorly (lower portion of photo). With severe cranial trauma, periosteal hemorrhage may involve more than one of the cranial bones, as seen here.

Direct (i.e., not secondary to brainstem vascular or traumatic lesions) cranial nerve injury during vaginal delivery is usually limited to the facial nerve, and is more likely to occur with forceps-assisted delivery. Infrequently, the abducens and/or oculomotor nerves may be compressed during delivery, probably in the region of the tip of the petrous portion of the temporal bone, and most often associated with forceps delivery.[81,112] These cranial nerve injuries may accompany other lethal injuries, and hemorrhagic foci within cranial nerve trunks are relevant nonlethal observations to include in the consultation report.

Brain tissue emboli may occur in both systemic and intracranial vessels in infants that have sustained intracranial injury in spontaneous vaginal delivery or in forceps-assisted vaginal deliveries.[14]

Brainstem and spinal cord injuries in newborns infrequently occur in spontaneous vaginal delivery, and if present may be accompanied by cervical fractures. Approximately 75% occur in the context of breech presentations. Head extension, leg traction while the head is held by an intrauterine contraction, or longitudinal stretching of the vertebral canal are thought to be the most likely mechanisms by which cord traction, laceration, or vascular compromise occurs. Cohen[21] states that the vertebral column of the newborn can be stretched up to 2 inches, but the spinal cord can only be stretched up to approximately ¼ inch without injury; the evidence supporting these figures is not detailed. Injuries by such mechanisms are most often in the lower cervical and upper thoracic regions.[118]

With cephalic presentations, torsional forces occurring during head rotation or shoulder dystocia resulting in cord traction via stretching of the brachial plexus are suggested mechanisms of cord injury; the resultant lesions tend to localize in mid to upper cervical and/or caudal medulla regions.[66,118]

The brachial plexus, originating from spinal roots C5 through T1 in most individuals, is especially subject to birth injury when exposed to extreme lateral traction. The resultant forces usually involve the rostral roots, and less commonly the plexus as a whole. As with many of the other injuries discussed, this complication is more likely in the presence of large fetal size, shoulder dystocia, breech or abnormal cephalic presentations, or in fetal distress situations necessitating prompt assisted completion of delivery. The traction on spinal roots may cause associated spinal cord injury. It may also be accompanied by sympathetic nervous system injury (T1, Horner's syndrome) or fracture of the clavicle or humerus.[118] A prior Horner's syndrome may be suspected at autopsy in the presence of heterochromia iridis, characterized by persistence of a blue color (as a result of sympathetic denervation) in the affected iris for months or years, while the contralateral iris becomes progressively pigmented.[118]

in asymptomatic babies following normal vaginal delivery or instrument-assisted delivery; all hematomas had resolved by 4 weeks' postpartum.[120] The gross appearance of epidural hemorrhage may also be simulated by epidural extramedullary hematopoiesis (see Chap. 17).

Cerebral contusion of the type seen in mature brains is uncommon in the perinatal period up to approximately 6 months of age. Contusions during that developmental period more frequently appear as slitlike defects in the subcortical white matter, particularly in the orbital frontal, temporal, and superior frontal gyrus regions.[70] Similar lesions (i.e., "gliding" contusions) may, of course, be seen in adults, generally in the context of diffuse axonal injury (see Chap. 6).

The incidence of hemorrhage within the neonate brain parenchyma in vaginal deliveries also varies as to site. Intraventricular and intracerebellar hemorrhages are seen more frequently in premature than in term infants.[118] Hemorrhages linked to traumatic delivery are briefly discussed later in this chapter in the section "The Newborn Infant." As in other age groups, histologic sections should be taken to exclude local causes such as neoplasm and vascular malformations or inflammatory causes of hemorrhage, and to screen for (maternal and infant) coagulopathies.

Phrenic nerve injury, with ipsilateral diaphragmatic paralysis, may be seen as an isolated peripheral neuropathy or in association with brachial plexus injuries during delivery.[118] The mechanism of injury is most likely extreme lateral traction involving cervical roots 3, 4, and/or 5, with most of the fibers of the phrenic nerve usually contributed by root C4.

Branches of the laryngeal nerve may be injured during birth, probably as a result of head rotation accompanied by lateral flexion. Compression of the superior branch of the laryngeal nerve between the thyroid cartilage and hyoid bone can lead to swallowing difficulty, and compression of its recurrent branch between the thyroid and cricoid cartilages can result in vocal cord palsy and dyspnea.[118]

More distal upper extremity peripheral neuropathies in isolation are infrequently seen as a result of birth injury, but when present they may be associated with fractures (e.g., humerus, clavicle). Lumbosacral plexus injury even during complicated deliveries is rare, and preexisting problems such as occult lumbosacral midline closure defects should be excluded.[118]

Both the neuropathologist and the forensic pathologist performing the general autopsy should have a general understanding of the spectrum of injuries that may follow complicated deliveries, including extremity fractures similar to those seen in nonaccidental infant trauma, in order to minimize interpretive error.[71]

Unwitnessed births in an outside hospital, with mother and/or neonate subsequently evaluated at another hospital for evolving difficulties, can be diagnostically challenging. A recent report describes respiratory distress and hypoxic-ischemic encephalopathy in neonates delivered by water-birth, a history of which was not initially disclosed by the mother in two of the four cases reviewed.[84]

With regard to unwitnessed births outside the hospital, DiMaio and DiMaio[27] cite evidence suggesting that infant skull fractures from precipitous delivery with the mother in the standing position are rare, even with the infant falling to the ground.

Assisted Delivery

Complications during usual vaginal delivery may require obstetric intervention in an effort to expedite completion of the birth process and improve the likelihood of survival with minimum untoward consequences for mother and infant. Thus, the setting in which such procedures are often employed makes it inevitable that a higher percentage of certain birth-related abnormalities will be encountered in this population. For example, in some published series the rate of intracranial hemorrhage is higher in infants delivered by forceps, vacuum extraction, or cesarean section performed *during* labor than in infants born by spontaneous vaginal delivery.[113] In other reports, subdural hemorrhage was equal in frequency in spontaneous delivery as compared with forceps-assisted delivery.[18] However, the rate is not higher in cesarean section performed *before* labor, suggesting that the common risk factor is abnormal labor rather than the method of delivery.[113] A complication associated with a given intervention may be concrete and visible, in contrast to the far more serious (and possibly fatal) theoretical and invisible outcome prevented by such intervention. This, combined with a common parental attitude in today's society (i.e., every child with a defect must be someone else's fault until proven otherwise), results in many such cases being referred to the medical examiner or other consultants in the course of litigation. This having been said, however, does not imply that there are not certain consequences that, on occasion, can be related to the procedure itself, whether properly or improperly applied.

1. **Forceps-Assisted Delivery.** Injury to the mother with forceps are localized to cervix, vagina, and perineum, but may include lumbosacral plexus involvement due to the closely associated nerves.

    Consequences to the infant, including common and minor skin erythema or ecchymosis, may include facial or abducens nerve palsy, cephalhematoma, and/or subdural and intracerebral hemorrhage with or without skull fracture, especially if improper forceps application or excessive traction is used. The incidence of a transient facial palsy is greater with midforceps delivery than in spontaneous vaginal delivery or vacuum extraction.[118] The incidence of subgaleal hemorrhage is less in midforceps delivery (14%) than in either vacuum extraction (49%) or spontaneous vaginal delivery (28%), but exceeds that in cesarean section (9%).[123]

    In a comparison of forceps and polyethylene vacuum cup-assisted delivery, Williams *et al.*[124] found an increased rate of retinal hemorrhage in vacuum deliveries (38%, compared to 17% for forceps), and a higher rate of facial nerve injury in forceps delivery (18%, compared to 2% for vacuum delivery).

2. **Vacuum-Assisted Delivery.** Experience with vacuum-assisted delivery, augmented by data subsequent to the 1998 Food and Drug Administration health advisory encouraging voluntary reporting of adverse events,[98] is still under study. The following comments should be considered preliminary. Reported vacuum extraction-related injuries have included chignon (scalp edema beneath the vacuum cup), subgaleal hemorrhage, cephalohematoma, skull fracture (including growing skull fracture), growing fontanelle, epidural hemorrhage, subdural hemorrhage, subarachnoid hemorrhage, retinal hemorrhage, and diffuse hypoxic-ischemic encephalopathy.[53,88,89,99,123,124] In comparison to unassisted delivery, available data suggest that vacuum-assisted delivery may increase the inci-

dence of cephalhematoma, subgaleal hemorrhage, and retinal hemorrhage. Whether or not the rate of intracranial hemorrhage (e.g., posterior fossa subdural and subarachnoid hemorrhage, intraparenchymatous brain hemorrhage) is increased in vacuum delivery remains controversial.[50,86,123] Infrequently, more severe complications may occur (Case 4.8, Fig. 4.16).

3. **Cesarean Section.** Intracranial hemorrhage frequency is similar in infants born by spontaneous vaginal delivery and those born by cesarean section with no labor, suggesting that the increased incidence of intracranial hemorrhage with cesarean section performed under other circumstances (e.g., to terminate a long second stage of labor or arrest of labor) is due to the dysfunctional labor rather than the operative intervention.[113] The incidence of subgaleal hemorrhage (9%) and retinal hemorrhage (7%) (see later) is less in cesarean section than in other forms of assisted delivery.

4. **Perimortem Cesarean Section.** Emergency cesarean section in a pregnant woman who develops cardiopulmonary arrest, if performed promptly, may not only result in a normal infant, but also improve the likelihood of response by the mother to ongoing cardiopulmonary resuscitation efforts.[58] The relative susceptibility of mother and fetus to the associated hypoxic-ischemic insult is discussed in Chapter 10.

5. **Animal Horn Cesarean Section.** Savage[102] cites several reports of so-called animal horn cesarean sections, first reported in 1647, in which a pregnant woman was gorged by a bovine horn with subsequent expulsion of the fetus through the wound. Fetus expulsion occurred immediately or after a short interval. This form of "cesarean section" clearly should be categorized under uterine trauma rather than assisted delivery, but is included here due to the misnomer applied.

6. **Kidnapping by Cesarean Section.** A 2002 report reviewed cases in which pregnant mothers were caused to undergo cesarean section through deception or by forcible abduction and attack, in order for the perpetrator(s) to obtain the desired child.[15] Methods of extracting the fetus varied from cesarean section performed by a physician in one case, to another case where the pregnant woman was strangled to unconsciousness, tied to a tree, and a car key substituted for a scalpel. In most cases reported, the mother was murdered in the process and efforts were made to hide her body. Many of the babies survived.

Illegal Abortion

It is not unexpected that several complications of pregnancy and delivery would be more likely to occur under conditions that prevail in the setting of suboptimal medical care[52] or that of illegal abortion. Air embolus or cardiac arrhythmia may occur during dilatation of the cervical os, or as a result of douching or syringing techniques. Syringing may also lead to perforation of the uterus or vagina. Sepsis, hemorrhage, thrombotic emboli, and anesthetic complications may occur.[27] Unsuccessful attempts to abort a pregnancy illegally may result in fetal abnormalities discovered after birth. The pregnant mother involved in such activities, in our experience, may also have a history of illicit drug use or other unhealthy practices, making it difficult to impossible to assign relative importance to the several risk factors they exhibit for the subsequent fetal neuropathologic lesions encountered (Case 4.9, Figs. 4.17–4.18).

> **Case 4.8.** (Fig. 4.16)
>
> A 39½-week-old estimated gestational age male infant was delivered by vacuum extraction when the delivery was complicated by shoulder dystocia. Death occurred 4 days after delivery. Autopsy findings included cervical fracture-dislocation, C6 spinal cord transection, acute spinal epidural and subdural hemorrhage, and diffuse subarachnoid hemorrhage.

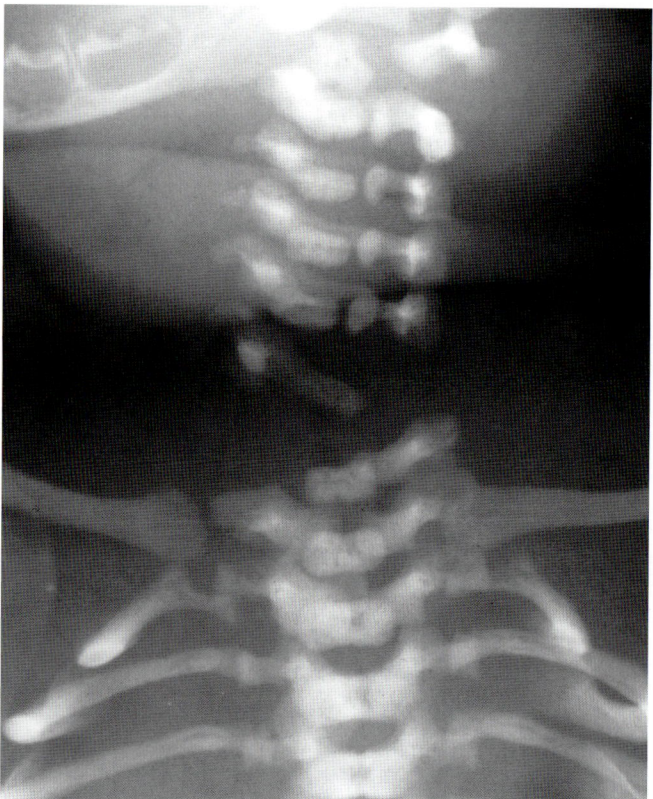

Fig. 4.16. Anteroposterior cervical radiograph demonstrates C6 fracture-dislocation with distraction across the C5-6 disk.

Case 4.9. (Figs. 4.17 and 4.18)

A 21-year-old G3, P2 mother who was positive for hepatitis C virus surface antigen, cytomegalovirus (CMV), herpes simplex type I and II antibodies, as well as cocaine, delivered a 32-week estimated gestational age (EGA) female fetus. The mother had made efforts to induce an abortion earlier during the pregnancy with unknown types of drugs. The maternal history also included heroin use. The infant remained institutionalized until its death at age 3 years, when it suddenly became apneic and could not be resuscitated. Clinical studies and therapy had revealed multiple cutaneous hemangiomas, mild pulmonary stenosis, gastroesophageal reflux, apneic episodes, bilateral congenital hip subluxation (surgically repaired), atrial and ventricular septal defects (surgically repaired), scoliosis, and severe mental retardation. Neuropathologic examination revealed a brain weight of 660 g (estimated normal brain weight for height, 1170 g; and for age, 1141 g), bilateral asymmetric porencephaly (left greater than right side) (Figs. 4.17–4.18), partial fusion of the frontal lobes obliterating the interhemispheric fissure focally, extensive convolutional pattern and cytoarchitectural abnormalities (including a radial orientation of gyri at the margin of the porencephalic cysts, micropolygyria, nodular neuronal heterotopias, and cortical dysplasia); corticospinal tract atrophy, hypoplastic optic nerves, and evidence of acute hypoxic-ischemic injury in the right Sommer's sector.

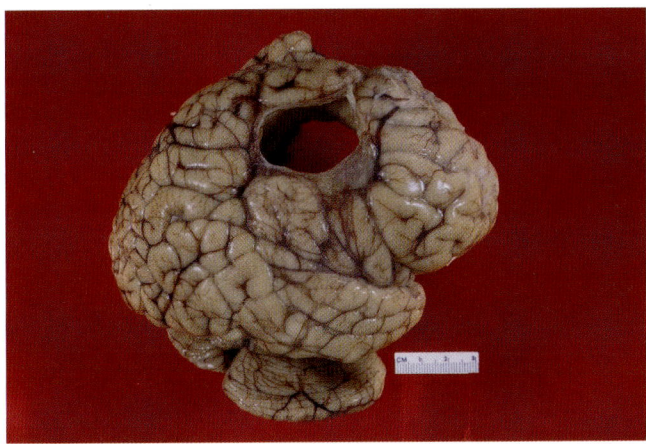

Fig. 4.18. Lateral view of right hemisphere with smaller porencephalic cyst, convolutional abnormalities, and widely exposed, abnormally formed insula.

The Newborn Infant

Several other neuropathologic conditions are associated in the literature with the "perinatal" period of development. The dictionary definition of "perinatal" refers to "before delivery from the 28th week of gestation to the first 7 days after delivery," and "neonatal" refers to "the period immediately succeeding birth and continuing through the first 28 days of life."[109] Such terms in the literature may be used interchangeably or may be applied to differing time frames. Of more importance to the forensic neuropathologist is evidence that conditions sometimes considered characteristic of the perinatal period may, in some instances, originate from insults extending well beyond the time frame just defined. This latter point is discussed in subsequent sections, and is also dealt with in another chapter (see Chap. 5). Although several of the possible sequelae of insults to the fetus *in utero* and during delivery have been discussed earlier in this chapter, the following section comments on selected lesions not conveniently grouped under prior settings. We note at the onset that separation of these lesions into such groups is primarily for descriptive convenience, because they may occur either in isolation or in various combinations in any given case. Most of the following lesions might best be considered under the category of "neonatal encephalopathy," since many of them are neither exclusive to the more narrow definitions of the perinatal period nor of proven or exclusive hypoxic-ischemic origin.

Even in cases with hypoxic-ischemic insults, our focus in this text on CNS sequelae should not distract one from the important multisystem involvement occurring in such cases, discussed elsewhere in this text and by others,[33] or the several mechanisms by which an insult may lead to neuronal injury.[36]

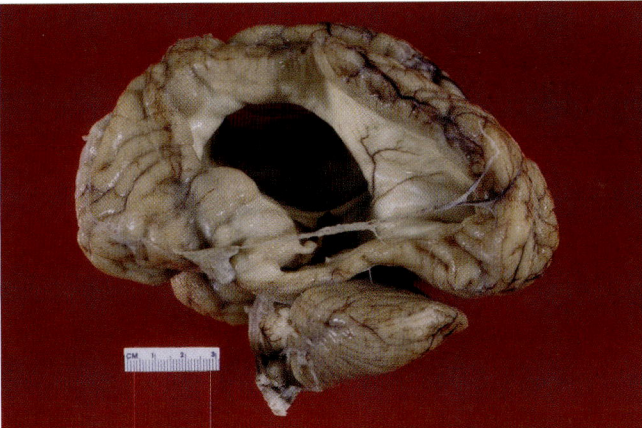

Fig. 4.17. Lateral view of left hemisphere, with large porencephalic cyst with gliotic margins and abnormal surrounding convolutional pattern.

This review of hemorrhagic and nonhemorrhagic lesions arising from insults during the late gestational and early postnatal period of life is brief, in keeping with our emphasis on the forensic aspects of these conditions. Recently published reviews of cerebrovascular disease in the perinatal and early childhood period are available.[25,62,108,117] These and other references cited provide the interested reader with a wealth of information regarding details of the clinical manifestations, prognosis, and pathologic findings. A recent series of articles also provides insight into the detailed cellular events that accompany many of these lesions, and that have evolving structural and functional repercussions extending well beyond the period of apparent completed repair as judged by routine histologic methods.[75–77]

Subependymal (Germinal Matrix) Hemorrhage

The subependymal zone of the lateral ventricle is the most common intracranial hemorrhagic site in prematures less than 32 to 34 weeks' gestation (most commonly in the 28- to 33-week gestational stage),[60,61] and tends to develop within the first 72 hours' postpartum. The source is generally conceded to be vessels such as the vena terminalis within the subependymal germinal matrix. The context in which germinal matrix hemorrhage occurs is considered by most authors to include a hypoxic-ischemic component.[60,61]

Intraventricular Hemorrhage

Intraventricular hemorrhage (IVH) occurs in highest incidence in the same developmental stage, under similar conditions as described earlier for germinal matrix hemorrhage, and is probably most often secondary to rupture of the latter through the ependymal lining of the ventricle (Figs. 4.19–4.20). Intracerebral (including intraventricular) hemorrhages in premature infants are graded 1 to 4 regarding severity (Figs. 4.19–4.20).[48]

Choroid plexus vessels (usually in the region of the glomus) may be the source of IVH in some cases, more often in term than in premature infants.[44,118] Exceptionally, IVH may occur in term newborns, and one report described an IVH in a term neonate that may have resulted from abdominal compression at 13 days' postpartum.[119] In stillbirths with liquefied brains, careful washing of the cranial contents may reveal a still-firm ventricular cast of clotted blood that provides clear evidence of a perimortem intraventricular hemorrhage (Fig. 4.21).

Cerebellar Hemorrhage

This occurs most often in approximately the same developmental stage as subependymal and intraventricular hemorrhage (i.e., 24 to 40 weeks' gestational age),[60,61] but much less frequently. It is more likely to be associated with difficult delivery, and findings may include occipital osteodiastasis (see later). It may also occur spontaneously, prior to labor, in apparently normal pregnancies.[49]

In the latter case, histologic examination may aid in determining lesion duration, particularly in cases with brief postpartum survival. It has also been noted to occur in higher incidence in premature infants in whom older methods of mask ventilation (resulting in cranial deformity) were employed.[90]

Delayed development of hydrocephalus is a recognized sequelae of intracranial hemorrhage, especially when significant amounts of free blood enter the ventricles and subarachnoid space. The likely mechanism relates to aqueductal blockage and a cerebrospinal fluid (CSF) production–absorption imbalance secondary to effects of erythrocyte trapping in CSF absorption sites.[79]

Neonatal Sinovenous Thrombosis

Thrombosis within the venous structures of the brain and its coverings in the neonatal period typically is seen in the context of one or more maternal or neonatal major

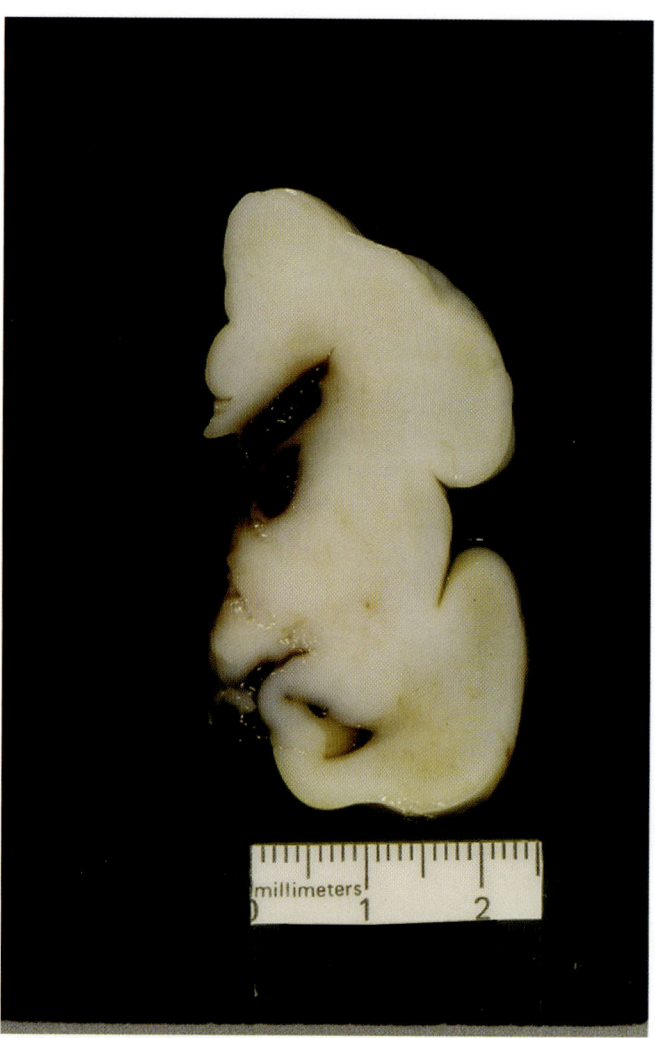

Fig. 4.19. In the grading of premature infant intracranial hemorrhage, hemorrhage confined to the subependymal germinal matrix is grade 1. If extension into the ventricle occurs without dilatation of the ventricle, as in this case, it is grade 2.

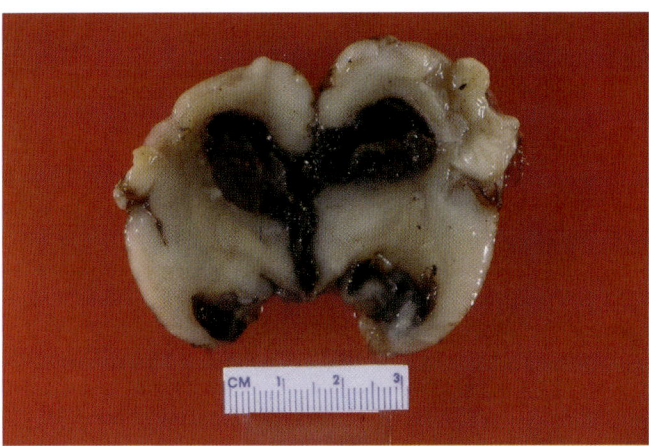

Fig. 4.20. This more extensive hemorrhage has dilated the cerebral ventricles (grade 3), but the extension into adjacent brain parenchyma also present (most evident at this coronal level in the region of the right temporal horn) upgrades this to a grade 4 hemorrhage.

Fig. 4.21. Blood clot retrieved from cranial cavity in a stillborn infant with liquefied brain parenchyma has a configuration typical of a ventricular mold from the trigone region, indicating a perimortem intraventricular hemorrhage.

medical problems. Examples of risk factors include maternal hypertension, diabetes mellitus, chorioamnionitis, and/or neonatal problems such as congenital heart disease, disseminated intravascular coagulation, congenital diaphragmatic hernia, meconium aspiration, sepsis, polycythemia, severe dehydration, a genetic thrombophilia, or ECMO therapy (see later).[125] Neuropathologic findings may include venous infarction, intraventricular hemorrhage, and deep parenchymal hemorrhages (e.g., in the caudate nucleus and thalamus).

Retinal Hemorrhage

Spontaneous vaginal delivery is the most common cause of neonatal retinal hemorrhage.[32,68] They are seen in 19 to 50% of deliveries in various series. Retinal hemorrhages occur in up to 75% of vacuum-assisted deliveries, 33% of spontaneous vaginal deliveries, less often in forceps delivery, and least often in cesarean section (0.8–7%).[67,123,124] There is no clear association between the presence of retinal hemorrhages and brain injury in this context.[67] Conditions associated with an increased incidence of neonatal retinal hemorrhage also include acidosis, intrauterine growth retardation, and short second stage of labor.[123,124]

In view of the emphasis placed on presence of retinal hemorrhages in suspected child abuse cases, it is noted that dot-blot and flame hemorrhages are the most common types seen in newborns after delivery. Subinternal limiting membrane hemorrhage in the macula, and subhyaloid, subretinal, or choroidal hemorrhages are rare. Following vaginal delivery, flame hemorrhages reportedly resolve within 1 week postpartum (usually by 3 days), and dot-blot hemorrhages resolve within 6 weeks (usually by 2–4 weeks).[68] Thus, retinal hemorrhage in infants over 1 month of age should prompt exploration for causes other then birth trauma.[32,45] See also the retinal hemorrhage section in Chapter 7.

Perinatal Telencephalic Leukoencephalopathy

Under this term is described white matter lesions consisting of varying patterns of hypertrophic astrocytes, amphophilic globules, acutely damaged glia (possibly oligodendrocyte precursors), and sometimes foci of necrosis.[43,44] Proposed etiologic factors have included hypoxic-ischemic insults, nutritional deficiencies, or endotoxemia.[43,118] The exact timing of the insult leading to such changes is unclear, although the timing of the insult in some cases with predominant white matter involvement does include the perinatal period. We have seen several cases of disproportionate white matter atrophy in long-term survivors with a history of "cerebral palsy." There are different degrees of severity. In milder cases, white matter bulk may be only moderately reduced, but it is firm and sclerotic (gliotic). In more severe cases, white matter bulk is severely reduced (Case 4.10, Figs. 4.22–4.23), or may show cystic changes.

Periventricular Leukomalacia

The period of vulnerability for this lesion to be caused by a CNS insult is the perinatal period up to and including term infants. Late gestation (from approximately 24 to about 32–35 weeks) has been suggested as the period of greatest risk.[60–62] The term is typically used to refer to necrotic lesions centered in white matter dorsal and lateral to the superior angles of the lateral ventricles, involving the centrum semiovale with or without involvement of optic and acoustic radiations.[118] It is more common in prematures. Gaffney[38] is of the opinion that injury at term does not produce lesions typical of periventricular leukomalacia. Many authors consider the commonest etiology to be hypoxic-ischemic injury possibly related to venous infarction.[111] In the latter instance, the periventricular lesion may be quite hemorrhagic and

> **Case 4.10.** (Figs. 4.22 and 4.23)
>
> A 15-year-old male choked on food while he was eating his lunch, became unresponsive, and did not respond to resuscitation. He had a past history of cerebral palsy, spina bifida, seizures, and mental retardation. Neuropathologic examination revealed predominantly central cerebral atrophy, with massive ventricular enlargement and descending long tract degeneration (including bilateral corticospinal tract degeneration). Thalamic sclerosis was interpreted as retrograde degeneration, and the inferior olivary nucleus atrophy as retrograde type due to the severe cerebellar atrophy present.

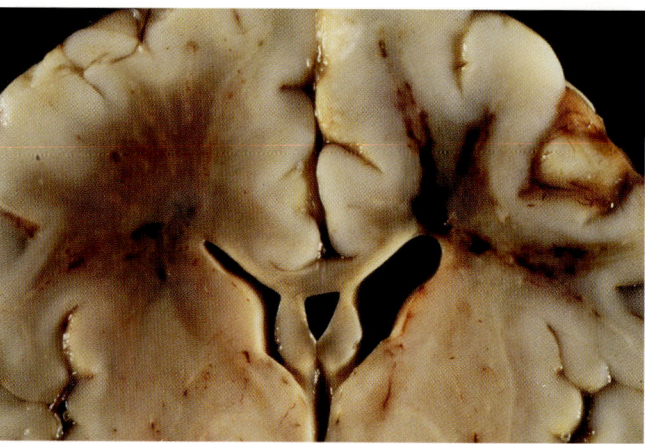

Fig. 4.24. Periventricular leukomalacia, acute stage with hemorrhagic component (see text).

> **Case 4.11.** (Fig. 4.25)
>
> This 2½-year-old boy's brain demonstrated multifocal cystic degeneration of cerebral white matter, and descending corticospinal and corticopontine tract degeneration. The appearance is consistent with an insult during the perinatal period.

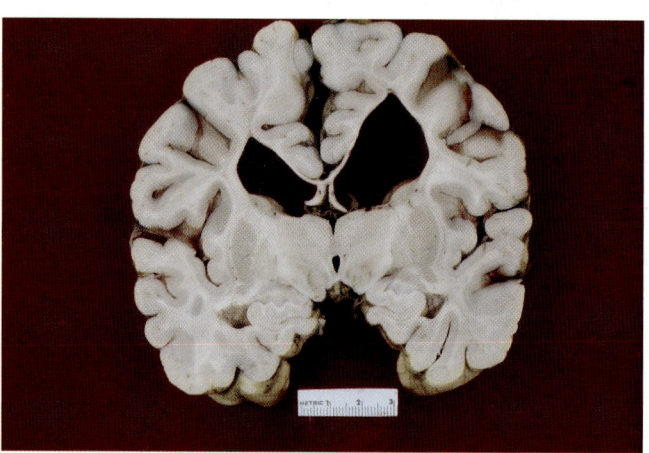

Fig. 4.22. Extensive periventricular atrophy is seen in this coronal section, primarily at the expense of white matter.

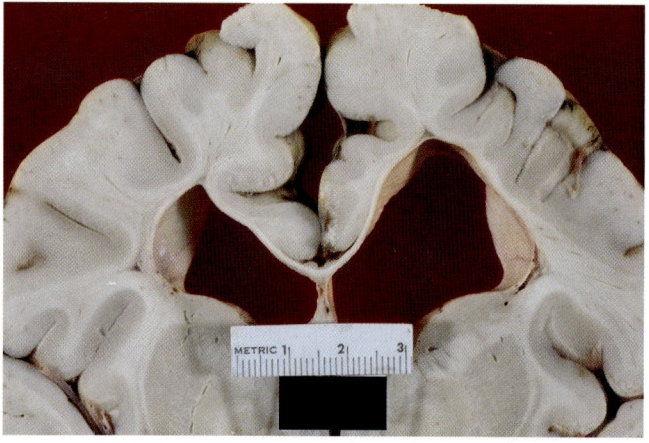

Fig. 4.23. Detail of a level just anterior to Fig. 4.22, coronal section. Note very thin corpus callosum and markedly reduced centrum semiovale white matter.

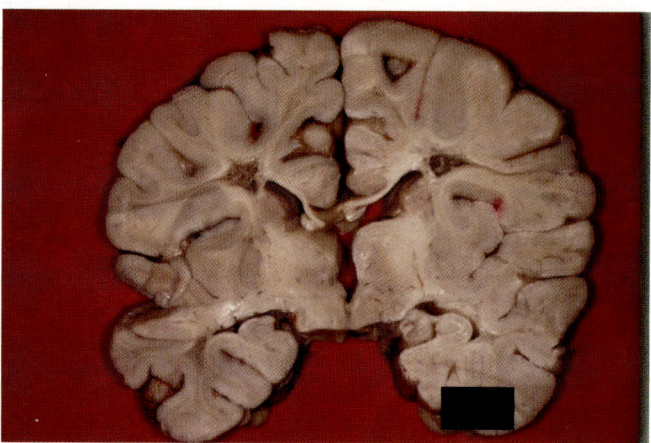

Fig. 4.25. Coronal section of the brain demonstrates the most prominent cystic change at this level is in the white matter adjacent to the superior angles of the lateral ventricles, the typical location of periventricular leukomalacia.

extend into the surrounding centrum semiovale and subcortical white matter (Fig. 4.24). More frequently, periventricular gliosis or cystic-malacic lesions are seen as chronic sequelae in long-term survivors (Case 4.11, Fig. 4.25).

In addition to infarct, however, Friede[37] notes that porencephaly, multicystic encephalopathy, postmeningitic encephalopathy, and residua of periventricular hemorrhage must be included in the differential diagnosis. We also distinguish the subependymal cystic lesions situated in the ventromedial body and tail of the caudate nucleus, which Shaw and Alvord[103] termed subependymal germinolysis, from periventricular leukomalacia (Case 4.12, Figs. 4.26–4.27).

Case 4.12. (Figs. 4.26 and 4.27)

A 5-week-old infant born at an estimated gestational age of 37 weeks was cosleeping with both parents. The infant was found unresponsive 1½ hours after her last feeding and did not respond to resuscitative efforts. The past history was remarkable for a nuchal cord (without apparent sequelae) and neonatal jaundice believed to be due to ABO-incompatibility. The latter was treated briefly with phototherapy, before the child was signed out of the hospital against medical advice. Neuropathologic examination revealed unilateral subependymal germinolysis, in the absence of perceptible atrophy or macrophages with hemosiderin pigment in the periventricular white matter.

Multicystic Encephalopathy

This pattern of injury is characterized by multiple foci of cavitary necrosis of varying size, typically bihemispheric, with involvement of both white and gray matter (Case 4.13, Figs. 4.28–4.38). The main period of vulnerability for this lesion is the occurrence of CNS insult from as early as 30 weeks' gestation up through the early postnatal period (i.e., approximately 1 month postpartum).[60,61,118] Associations, but not necessarily etiologies, include hypoxic-ischemic injury of various types (e.g., difficult labor, cyanosis at birth, cord prolapse), anaphylactic shock from a bee sting, abdominal trauma during gestation, and infections such as CMV, herpes simplex, neonatal encephalitis, and neonatal meningitis.[37]

Ulegyria

This lesion, believed to be associated with intracortical hypoxic-ischemic injury during a particular developmental time frame, consists of one or more mushroom-shaped gyri resulting from the bulk of the

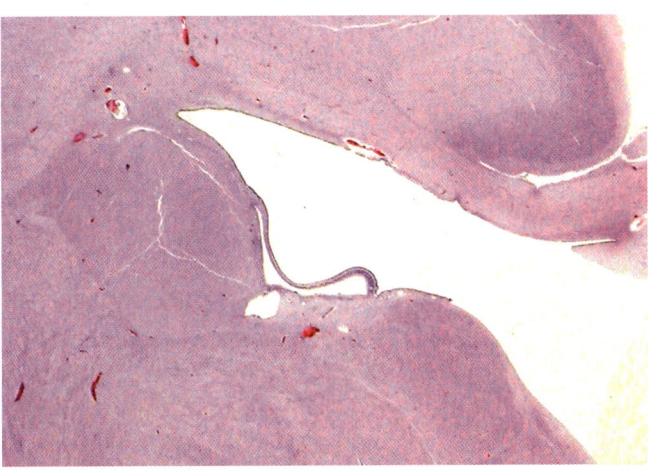

Fig. 4.26. Low-power micrograph of superior angle of lateral ventricle with slitlike cystic area beneath ependyma inferiorly. The adjacent rounded space is a vein. (H&E.)

Case 4.13. (Figs. 4.28–4.38)

A 16-year-old G4, P1, Ab 2 woman had routine prenatal care and an uneventful pregnancy until term, when her spontaneous-onset labor failed to progress. Vaginal delivery with forceps was attempted but was unsuccessful, and a crash cesarean section was performed. A male infant was born apneic, cyanotic, atonic, and with a heart rate of 100. Apgar scores were 2 and 2 at 1 and 5 minutes, respectively. Two days' postpartum a cranial ultrasound revealed small ventricles consistent with diffuse cerebral swelling, and a suspected small cystic area near one caudate nucleus. Three days' postpartum a noncontrast computed tomography (CT) brain scan revealed diffuse lucency of supratentorial structures, with posterior fossa contents being normal in appearance. No cystic change was described.

The infant's subsequent course was characterized by seizures, need for respiratory assistance, gradual increase in tone, and development of clonus. There was some subsequent improvement with supportive care including anticonvulsant therapy, to the point where the infant was discharged home. Five days after discharge he was found unresponsive in bed at home and pronounced dead at the scene, 5 weeks' postpartum. Neuropathologic examination revealed severe cystic encephalopathy, without evidence of mechanical trauma. Microscopically, cystic degeneration extended caudally to the level of the rostral pons. Neuronal loss and gliosis was also seen in the cerebellum, and long tract degeneration was present. It is noted that this extensive pathology, with presumed insult occurring 5 weeks previously, raises questions concerning the speed with which these various gross and microscopic features may develop (see Chap. 3 on aging and dating of neurological lesions).

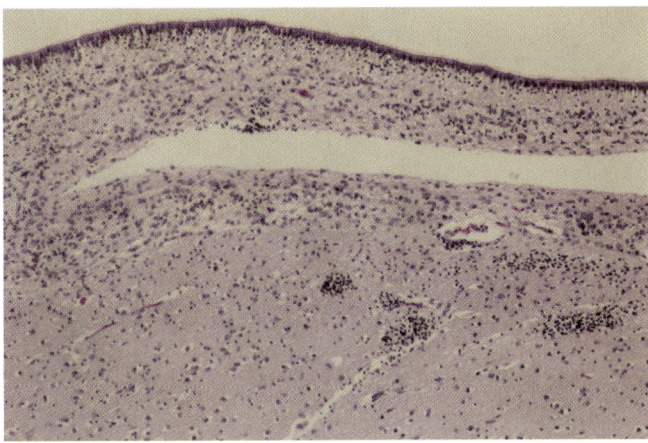

Fig. 4.27. Medium-power micrograph of supependymal germinolysis cavity (SG). Islands of small dark nuclei in lower right portion of photo are migrating neuroblast–spongioblast cellular elements. (H&E.)

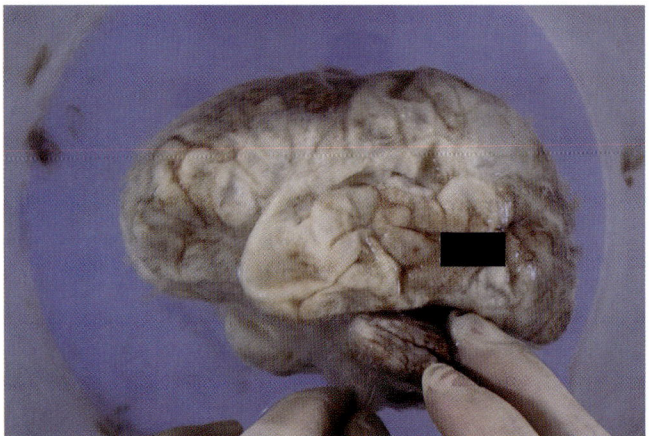

Fig. 4.28. Lateral view of brain immersed in water demonstrates maintained general configuration of hemispheres despite extensive cystic change.

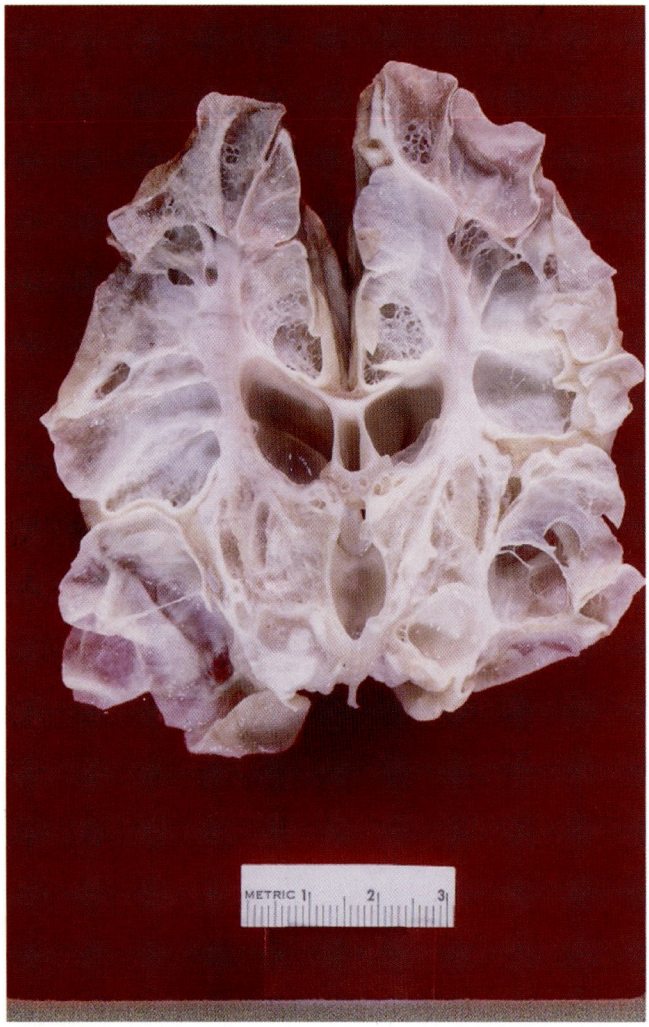

Fig. 4.29. Coronal section of brain at mid third ventricle demonstrates severe cystic change and parenchymal atrophy.

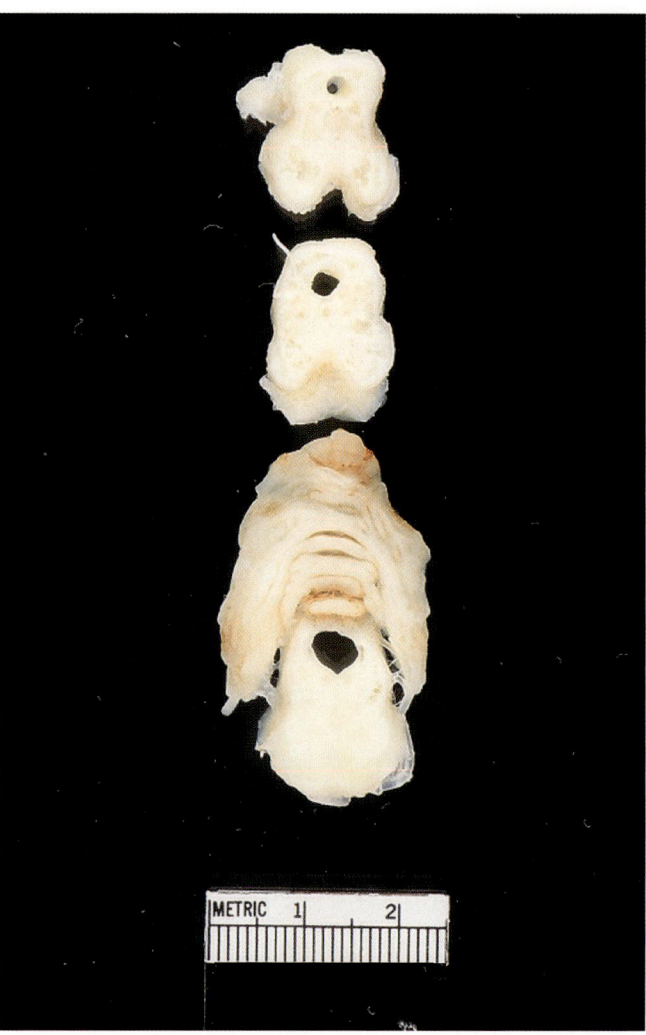

Fig. 4.30. Transverse sections of brainstem and cerebellum demonstrate severe, diffuse atrophy.

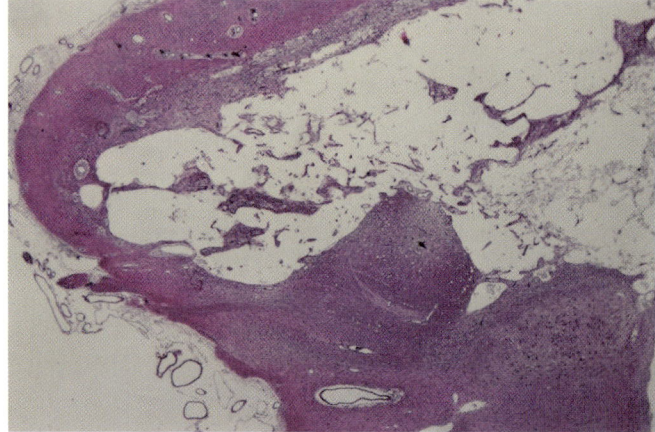

Fig. 4.31. Very low power micrograph of cerebral gyrus with cystic change. (H&E.)

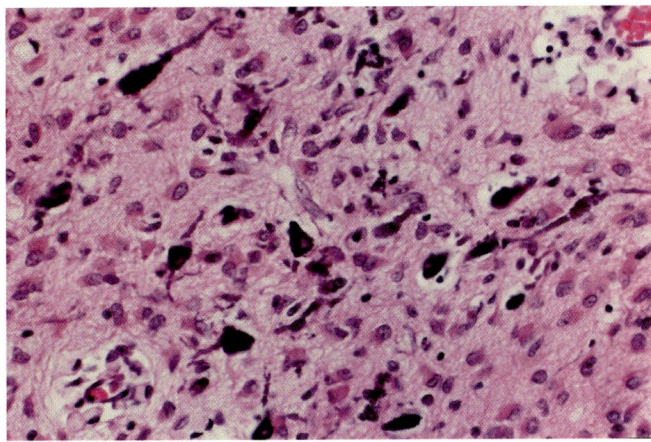

Fig. 4.32. Medium-power micrograph. Black ferruginated neurons in gliotic neuropil at marginal zone surrounding cystic areas in Fig. 4.31. (H&E.)

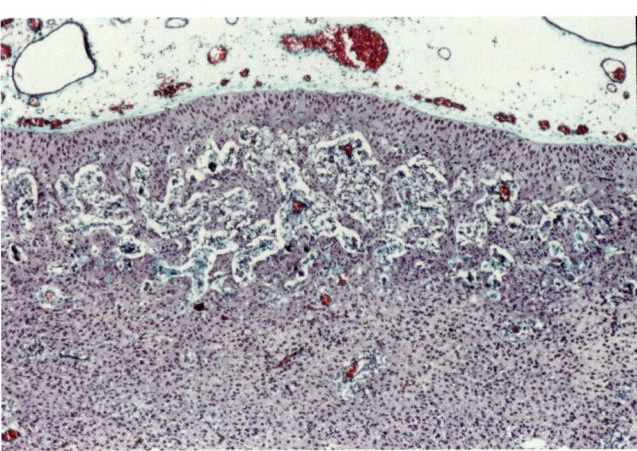

Fig. 4.33. Low-power micrograph of cerebral cortex with less extensive cystic change compared to Fig. 4.31. (Trichrome.)

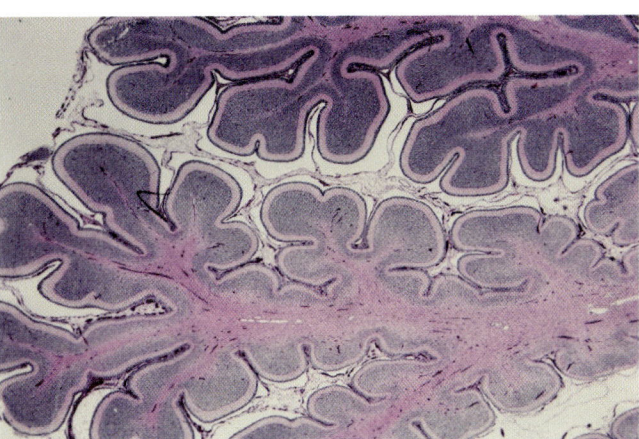

Fig. 4.35. Very low power micrograph of atrophic cerebellum, with narrow folia and wide sulci between folia. (H&E.)

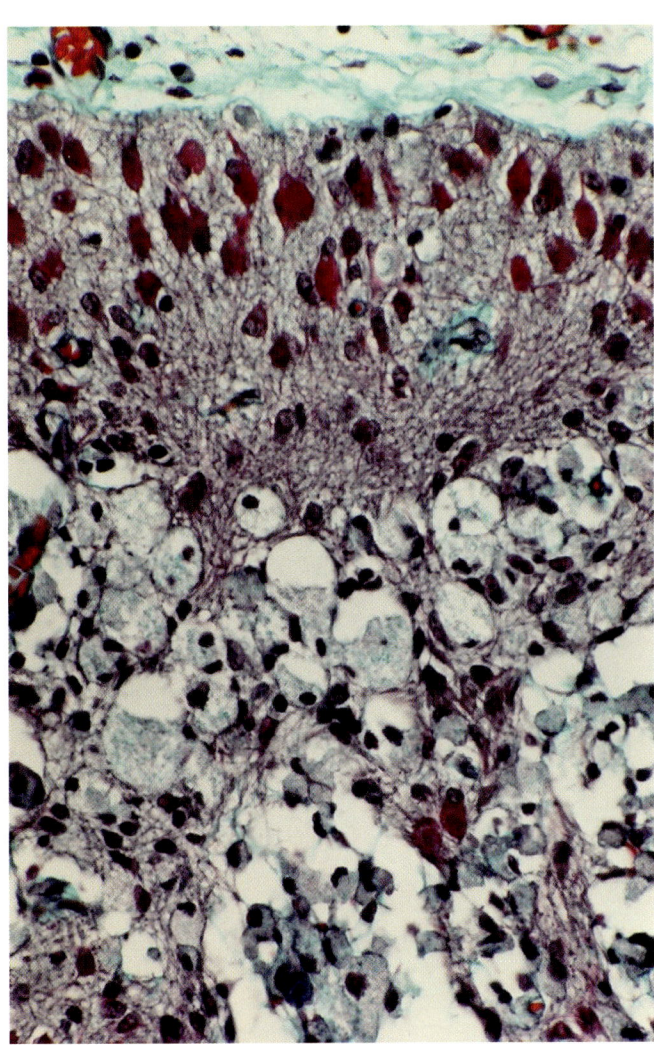

Fig. 4.34. Medium-power micrograph of superficial cortex with pia mater above, gliotic layer 1, and cystic layers with foamy macrophages below. (Trichrome.)

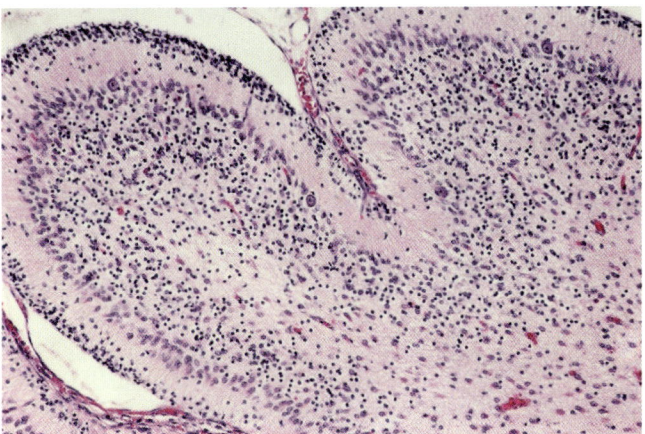

Fig. 4.36. Low-power micrograph of cerebellum. Near-complete absence of Purkinje cells, marked Bergmann gliosis, moderately severe depopulation of granule cells, and some depopulation of external granule cell layer. (H&E.)

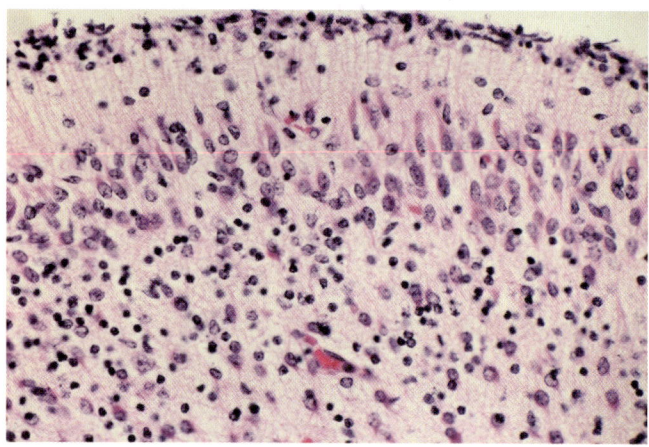

Fig. 4.37. Detail of Fig. 4.36. The external granule cell layer is depopulated, since it normally is six to seven cell layers in thickness in this age group. (H&E.)

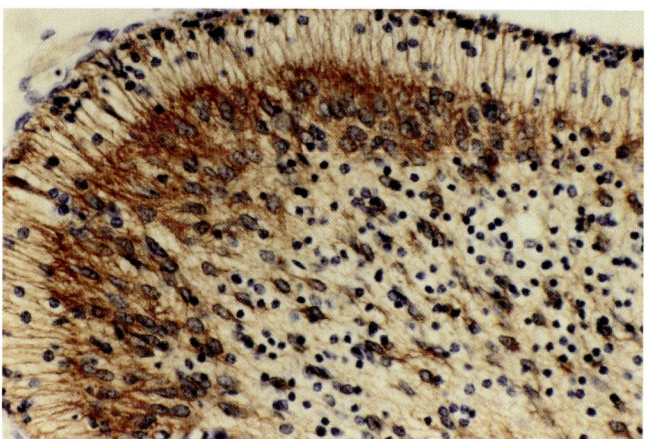

Fig. 4.38. Medium-power micrograph of cerebellar cortex with reddish-brown florid Bergmann gliosis emphasized. (Glial fibrillary acid protein immunostain.)

ischemic-induced neuronal loss and gliosis involving the cortex in the depths of the sulci, with relative sparing of the crowns of gyri (Case 4.14, Figs. 4.39–4.43). When present, it also serves as a marker for the insult having occurred during the perinatal period or sometime during approximately the first 6 to 9 months of postnatal life.[7,37,77]

Status Marmoratus

This lesion consists of a patterned neuronal loss and gliosis, at times with some ferrugination of neurons, accompanied by hypermyelination in the gliotic areas. It occurs in the corpus striatum (lateral greater than medial portion), and may also be found in the thalamus, head of caudate nucleus, and globus pallidus. The unique feature is the hypermyelination within the marbled zone of neuronal loss and gliosis (Figs. 4.44–4.47). The usual

insult is hypoxic-ischemic in nature, but cases have been described as a consequence of a postnatal acute febrile illness. According to most authors, the period of vulnerability is from the peripartum period up to 6 to 9 months' postpartum.[7,37,60,61,101] After this approximate age the hypermyelination component is absent, and lesions from similar insults are not always distinguishable from those seen in more mature brains.

Paracentral Cortical Atrophy

Attention was first directed to an unusual pattern of paracentral cortical atrophy in infants sustaining hypoxic-ischemic insults during the perinatal period by Azzarelli and colleagues in 1980 (Case 4.15, Figs. 4.48–4.50).[8] This appearance is sometimes referred to as a "rubber band" brain (e.g., as if mid dorsolateral hemispheres had been constricted by a rubber band). This finding is conspicuous on external examination of the brain by the focal atrophy of precentral and postcentral gyri, usually bilaterally. It is associated with ulegyria and evidence of more widespread neuronal loss and gliosis (e.g., status marmoratus of the basal ganglia and thalamus, cerebellar cortical atrophy, and infrequent atrophy of selected brainstem nuclei (V, VI, and VII) and selected spinal cord neurons). Azzarelli et al.[8] speculated that this pattern of cortical atrophy resulted from the vulnerability of areas of primary myelination most active at that stage of development having the greatest metabolic need. Areas that

> **Case 4.14.** (Figs. 4.39–4.43)
>
> This 42-year-old man from a chronic care institution had a history of mental retardation and seizure disorder, with episodic status epilepticus. He was found dead in his bed. Neuropathologic examination revealed focal cerebral cortical atrophy with features of ulegyria involving frontal, parietal, and occipital lobes.

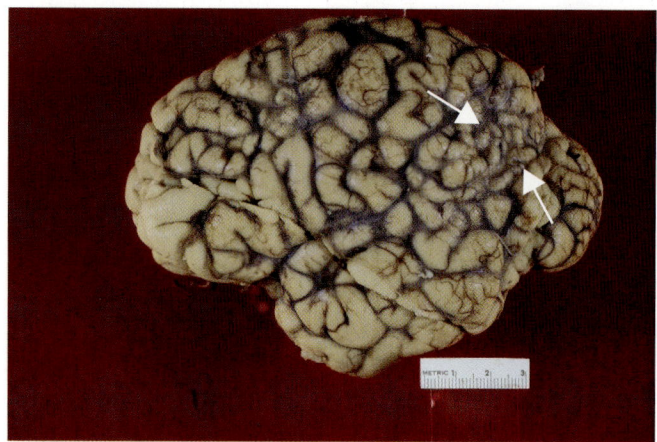

Fig. 4.39. Lateral view of left cerebral hemisphere with focal zone of ulegyria (*arrows*).

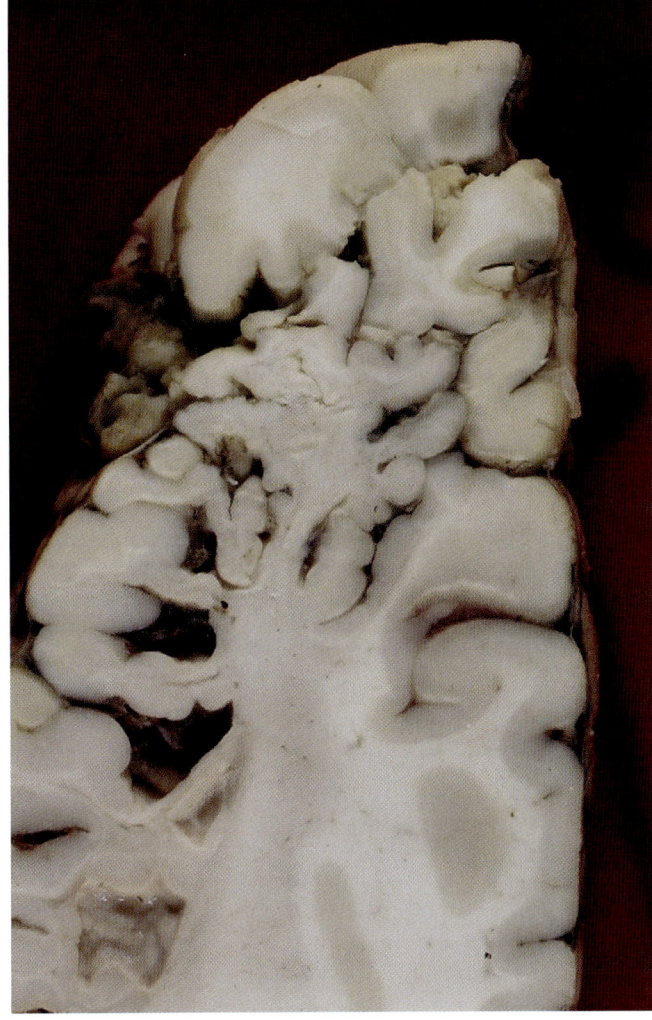

Fig. 4.40. Coronal section in zone of ulegyria, left parietal lobe.

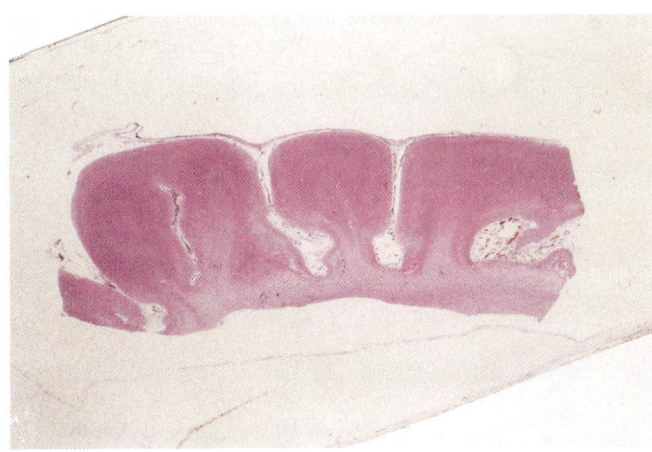

Fig. 4.41. Whole mount of ulegyria, with mushroom-shaped gyri. (H&E.)

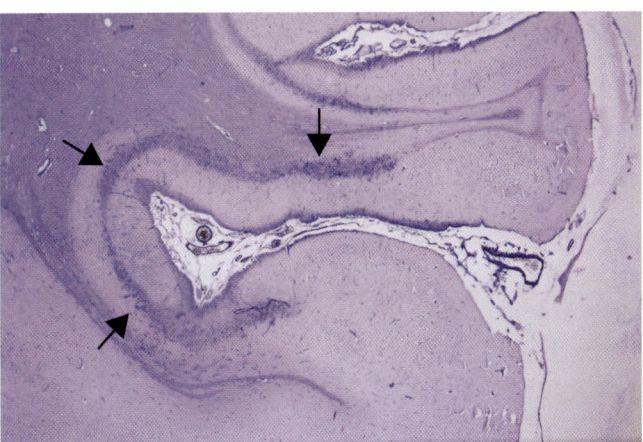

Fig. 4.42. Ulegyria. Gliosis (dark blue stain) involves cortex in deeper portions of sulcus (*arrows*) but essentially spares crowns of gyri. (Holzer.)

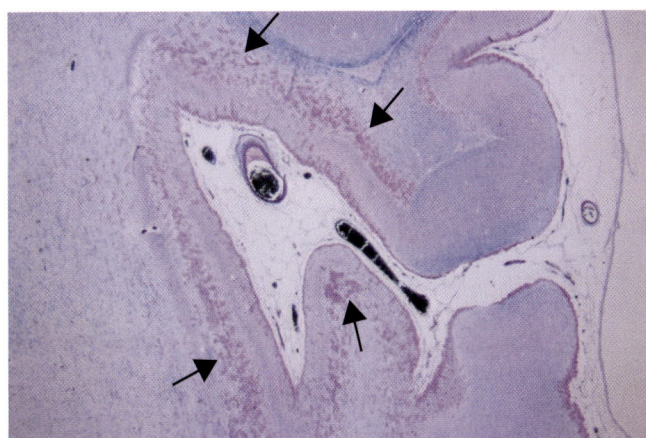

Fig. 4.43. Ulegyria with severe myelin loss. Red-stained corpora amylacea are in the expected subpial region, but are also markedly increased in the zones of cortical injury (*arrows*). Focal prominence of corpora amylacea in the vicinity of CNS parenchymal lesions is a nonspecific finding in several types of chronic lesions. (Luxol fast blue-periodic acid-Schiff.)

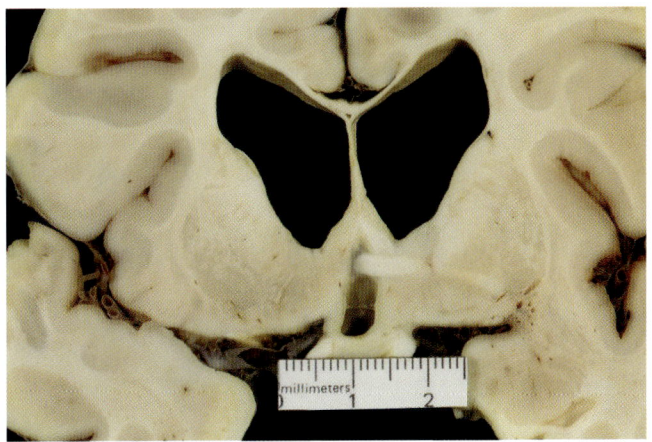

Fig. 4.44. Status marmoratus. Whitish blotchy pallor of basal ganglia in coronal section at level of anterior commissure. Additional findings included enlarged ventricles, very thin corpus callosum, and reduced bulk of centrum semiovale.

144 Neuropathology of Pregnancy and Delivery: Mother and Child

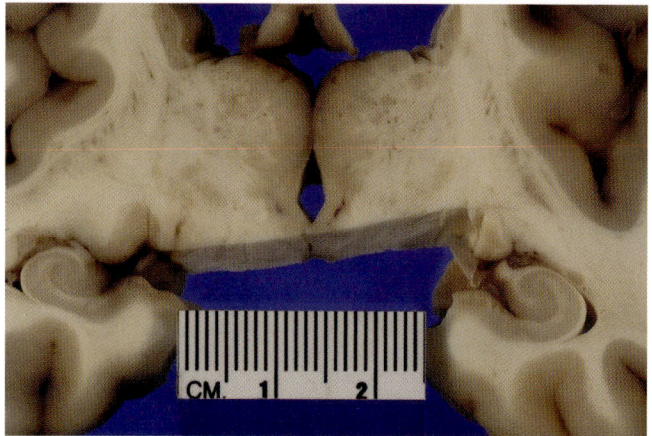

Fig. 4.45. Status marmoratus in another case, showing mottled pallor in thalamus in this coronal section at the level of the lateral geniculate body.

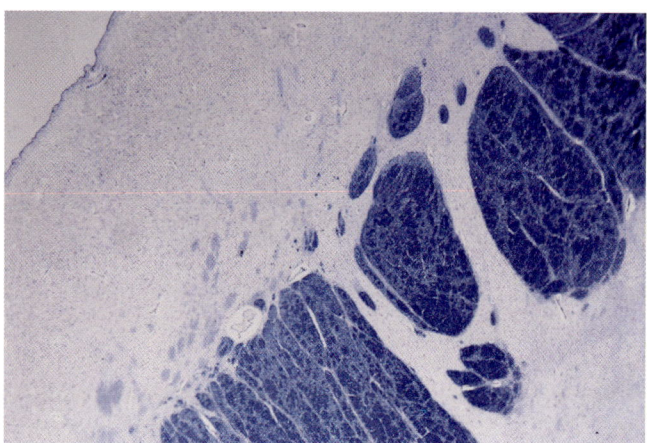

Fig. 4.46. Light blue-stained myelin in abnormal, patchy distribution in thalamus of case seen in Fig. 4.45. The large darker blue myelinated fiber bundles are internal capsule fibers. (Luxol fast blue/cresyl violet.)

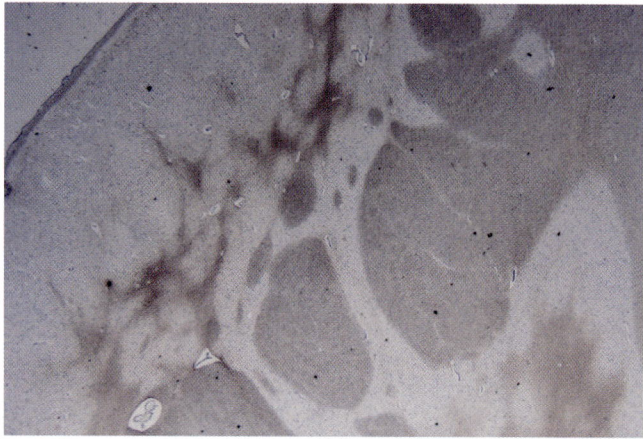

Fig. 4.47. Reddish-brown staining of gliosis in thalamus in section adjacent to that in Fig. 4.46, showing a distribution of gliosis similar to that of the abnormally located myelin in Fig. 4.46. (Holzer.)

Case 4.15. (Figs. 4.48–4.50)

This 16-month-old male became limp and unresponsive while being fed by his mother, and arriving paramedics found him in asystole. Past history was notable for cerebral palsy and seizure disorder since birth. The brain weight was 470 g (expected normal for age and body size, 944 g). Neuropathologic examination revealed chronic sequelae of perinatal injury, including microcephaly; ulegyria; atrophy of thalamus, basal ganglia, hippocampus, and cerebellum; retrograde degeneration of inferior olivary nuclei; and cortical atrophy that was particularly severe in the paracentral area.

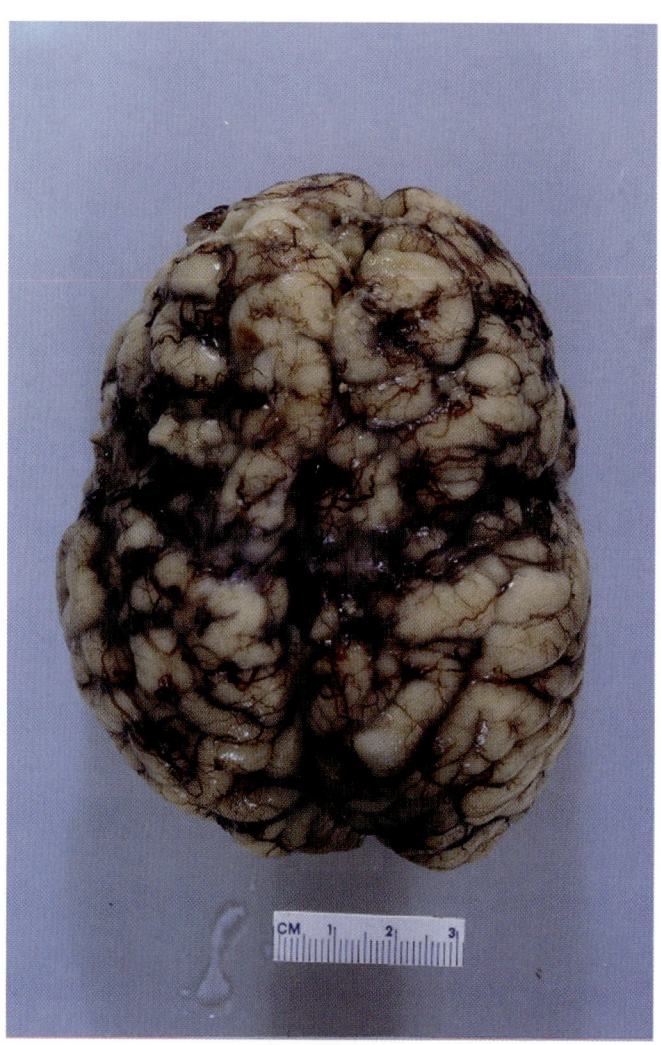

Fig. 4.48. Dorsal view of brain, frontal lobe above, shows a bandlike zone of atrophy marked by prominent leptomeningeal vessels in the vicinity of the central sulcus (fissure of Rolando).

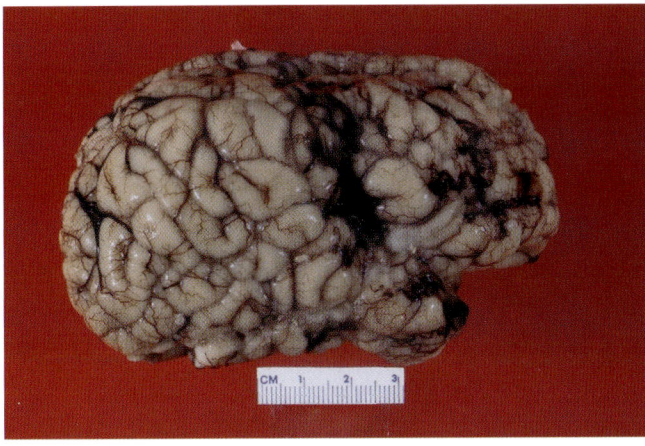

Fig. 4.49. Lateral view of right hemisphere. Prominence of leptomeningeal vessels in region of atrophic pre- and postcentral gyri.

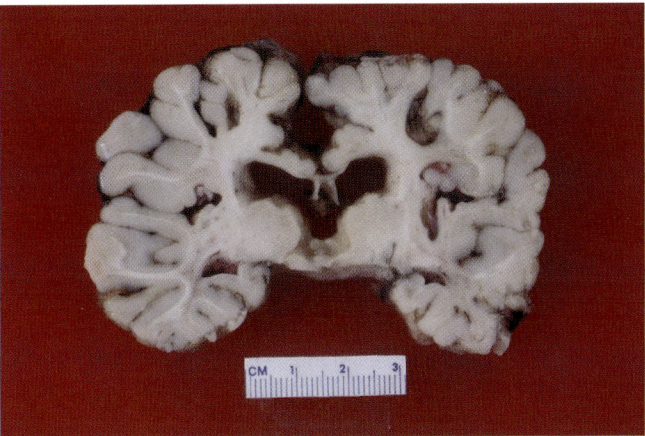

Fig. 4.50. Coronal section at level of atrophic cortex shows cortical and white matter atrophy, with ulegyria and ventricular enlargement.

myelinated later in the postnatal period are undamaged. A follow-up study by Azzarelli and colleagues published 16 years later, using newer techniques of neuroimaging and positron emission tomography, has provided further data supporting this hypothesis by indicating that the areas most affected are those with a higher rate of oxygen-glucose utilization[9] in this developmental period. The gestational age range of the reported cases at the time of the hypoxic-ischemic insult was from 35 to 44 weeks. The latter case was born at 36 weeks' estimated gestational age, sustained ventricular fibrillation for 7 minutes during abdominal surgery 2 months' postpartum, and remained comatose until death at age 4 months' postpartum.

Pontosubicular Necrosis

The main period of vulnerability for insults to produce this combination of lesions is from approximately 22 to 31 weeks' gestation up to 2 months' postpartum.[7,61] Hypoxic-ischemic insults as well as hyperoxygenation have been suggested as causal factors.

A few other patterns of selective neuronal involvement have been suggested as favoring the dating of the etiologic event to the late gestation to early childhood time frame. In our opinion, they are less reliable for purposes of court testimony on such matters. They are described further in Chapter 10.

Kernicterus

This condition, now infrequently seen, is due to hyperbilirubinemia of the nonconjugated form, arising from overload of enzymatic conjugation. It can occur, for example, in association with excessive hemolysis, inborn enzyme deficiencies, toxic effects of drugs, and vitamin K overload. The time frame of vulnerability extends from premature infants (primarily from approximately 20 weeks' gestation) to approximately 3 months' postpartum. The areas of CNS that may be stained by bilirubin, and involved by the neuronal necrosis that accompanies it, can include the globus pallidus, subthalamic nucleus, mammillary bodies, thalamus, hypothalamus, hippocampus, subiculum, induseum griseum, substantia nigra, cranial nerve nuclei, brainstem reticular formation, dentate nuclei, inferior olivary nuclei, uncus, flocculus, and spinal cord anterior horn cells.[37,97] Premature infants may vary somewhat from this pattern in that lateral thalamus billirubin staining may predominate, and locus ceruleus and Purkinje cells may also stain, but the hypothalamus in premature infants is unstained.[37] The icteric color is apparent in fresh tissue and frozen sections, but disappears in formalin-fixed tissue. Residual pigment or crystals may, but usually do not, survive in paraffin-embedded sections.[37,118]

Skull Congenital Anomalies

Some of the less common skull anomalies, including persistent sutures, may be misinterpreted as evidence of trauma. These are discussed further in Chapter 3.

Occipital Osteodiastasis

This condition consists of a traumatic separation of the cartilaginous joint between the squamous and lateral portions of the occipital bone, and may result from birth injury.[121] The vulnerability of this site exists because this synchondrosis does not ossify until approximately the fourth or fifth year of life.[23] In the presence of bilateral tears of the synchondrosis, the inferior margin of the squamous occipital bone is able to slip forward and above the lateral portions of the occipital bone. The result, depending on the degree of displacement, may include tears of the dura and dural venous sinuses, posterior fossa subdural and/or subarachnoid hemorrhage, cerebellar contusions or lacerations, and cerebellar or medullary compression due to resultant narrowing of the posterior fossa and foramen magnum (the posterior border of the foramen magnum is the squamous occipital bone). Wigglesworth[121] has observed this condition in gestations from 27 weeks up to term. Milder, nonfatal cases are also seen as a result of birth trauma, and if suspected clinically may be diagnosed by neuroimaging techniques demonstrating the development of callus formation bridging the innominate synchondrosis in subsequent weeks.[23] Perhaps most important from the standpoint of the forensic specialist, given the fact that severe forms of this lesion are now rare as a complication of birth trauma, is that the same lesion may be seen during postnatal life (prior to ossification of this synchondrosis) as a result of injuries due to, for example, child abuse or motor vehicle accidents.[23] Routine autopsy examination for excess motion between cranial bones will prevent this unusual injury from being overlooked.

Infantile Atlantooccipital Instability

Gilles and colleagues,[43] in 1979, suggested a possible mechanism for some unexplained neonatal and infant deaths, and Gilles has more recently again raised this question.[42] They described the relative immaturity of development of the lateral mass of the atlas, which is most pronounced during the first 6 months of postnatal life, and suggested that this may result in atlantooccipital joint motion sufficient in some infants to compromise vertebral artery circulation. Their hypothesis is based on a cadaver angiographic study and careful dissection. They also commented on the discrepancy between the size of the (larger) foramen magnum and the (smaller) atlantal arch openings in infants, which may predispose to rostral displacement (i.e., "atlantal arch inversion") of the atlas in the presence of sufficient ligament laxity (seen in 10 of 17 cadavers examined).[43] While we are not aware of subsequent studies providing proof of this mechanism as a cause of death, we believe the questions raised by these authors justify their recommendation that caretakers avoid extremes of head extension without immediate respiratory support available in neonates, and that further studies are needed to determine whether such a theoretical mechanism could underlie some unexplained infant deaths.

Extracorporeal Membrane Oxygenation (ECMO)

ECMO is a form of mechanical cardiopulmonary support that has been found useful in a variety of conditions in infants, children, and adults. In near-term and term infants (the focus of this chapter), it has been used in respiratory distress syndrome, meconium aspiration, persistent pulmonary hypertension, congenital diaphragmatic hernia, sepsis, cardiac dysfunction following cardiac surgery for congenital heart disease, and as a bridge to heart or lung transplantation.

The technique, as usually performed, requires ligation of the right common carotid artery and right jugular vein as well as systemic heparinization. In some cases, venovenous ECMO may be used.

We include this topic here in view of the age group involved, and because the reported neuropathologic complications of ECMO include several of the intracranial and intraocular lesions described earlier. It is not unexpected that stresses accompanying such intervention, while life-saving in many instances, may in some cases aggravate or be coincidentally associated with pathologic lesions that occur spontaneously in critically ill premature and term infants.

Reported neuropathologic findings in infants receiving ECMO have included sinovenous thrombosis, massive cerebral infarction (at times with subsequent development of cystic periventricular leukomalacia), smaller infarctions in a major cerebral artery distribution, thalamic infarctions, cerebral parenchymal hemorrhages, and pontine and olivary necrosis and/or gliosis.[60,61,101,118,125] The massive cerebral infarction may be ipsilateral or contralateral to the carotid and jugular ligation and cannulation. Intracranial hemorrhages in this setting have been described in the subarachnoid space, all cerebral lobes, basal ganglia, subependymal germinal matrix, intraventricular space, cerebellum, and brainstem.[118] Laryngeal nerve injury may occur during cannulation of neck vessels or due to injury accompanying intubation.[118] Retinal hemorrhages may also be seen following ECMO.[68,91] Varying degrees of neurodevelopmental impairment may occur in surviving infants.[2]

References

1. Abbott JT. Emergency management of the obstetric patient. In Burrow GN, Duffy TP (eds): Medical Complications During Pregnancy, ed 5. Philadelphia: WB Saunders, 1999, pp 225–236.
2. Ahmad A, Gangitano E, Odell RM, et al. Survival, intracranial lesions, and neurodevelopmental outcome in infants with congenital diaphragmatic hernia treated with extracorporeal membrane oxygenation. J Perinatol 1999;19:436–440.
3. Akman CI, Cracco J. Intrauterine subdural hemorrhage. Dev Med Child Neurol 2000;42:843–846.
4. Alexander E Jr, Kushner J. Intrauterine head injuries. In Vinken PJ, Bruyn GW, Braakman R (eds): Handbook of Clinical Neurology, Part I: Injuries of the Brain and Skull. Amsterdam: North-Holland, 1975, pp 471–476.
5. Alley JR Jr, Yahagi Y, Mencure MM, Strickler JC. A case of in utero fetal brain trauma after motor vehicle collision. J Trauma 2003; 55:782–785.
6. Amar AP, Aryan HE, Meltzer HS, Levy ML. Neonatal subgaleal hematoma causing compression: Report of two cases and review of the literature. Neurosurgery 2003;52:1470–1474.
7. Armstrong DD. Neonatal encephalopathies. In Duckett S (ed): Pediatric Neuropathology. Baltimore: Williams & Wilkins, 1995, pp 334–351.
8. Azzarelli B, Meade P, Muller J. Hypoxic lesions in areas of primary myelination: A distinct pattern in cerebral palsy. Childs Brain 1980;7:132–145.
9. Azzarelli B, Caldemeyer KS, Phillips JP, DeMyer WE. Hypoxic-ischemic encephalopathy in areas of primary myelination: A neuroimaging and PET study. Pediatr Neurol 1996;14:108–116.

10. Baergen RN, Castillo MM, Mario-Singh B, et al. Embolism of fetal brain tissue to the lungs and the placenta. Pediatr Pathol Lab Med 1997;17:159–167.
11. Bateman BT, Schumacher HC, Bushnell CD, et al. Intracerebral hemorrhage in pregnancy: Frequency, risk factors, and outcome. Neurology 2006;67:424–429.
12. Bates SR, Scalzo FM, Fowler KK. Drugs and neurotoxicants. In Duckett S (ed): Pediatric Neuropathology. Baltimore: Williams & Wilkins, 1995, pp 465–478.
13. Barozzino T, Sgro M, Toi A, et al. Fetal bilateral subdural hemorrhages: Prenatal diagnosis and spontaneous resolution by time of delivery. Prenat Diag 1998;18:496–503.
14. Böhm N, Keller KM, Kloke WD. Pulmonary and systemic cerebellar tissue embolism due to birth injury. Virchows Arch [Pathol Anat] 1982;398:229–235.
15. Burgess AW, Baker T, Nahirny C, Rabun JB Jr. Newborn kidnapping by cesarean section. J Forensic Sci 2002;47:827–830.
16. Burrow GN, Duffy TP (eds). Medical Complications During Pregnancy, ed. 5. Philadelphia: WB Saunders, 1999.
17. Catanese CA, Gilmore K. Fetal gunshot wound characteristics. J Forensic Sci 2002;47:1067–1069.
18. Chamnanvanakij S, Rollins N, Perlman JM. Subdural hematoma in term infants. Pediatr Neurol 2002;26:301–304.
19. Christiansen LR, Collins KA. Pregnancy-associated deaths: A 15-year retrospective study and overall review of maternal pathophysiology. Am J Forensic Med Pathol 2006;27:11–19.
20. Clark SL, Hankins GDV, Dudley DA, et al. Amniotic fluid embolism: Analysis of the national registry. Am J Obstet Gynecol 1995;172:1158–1169.
21. Cohen M. Case FR-11. Forensic Pathology Survey FR-B 2001. Northfield, Ill: College of American Pathologists, 2001, pp FR-B:13–14.
22. Conrad MR, Smith CH. Cerebral air embolus following oral-genital sex: First reported case in a non-pregnant patient. Female Patient 1988;13:19–20.
23. Currarino G. Occipital osteodiastasis: Presentation of four cases and review of the literature. Pediatr Radiol 2000;30:823–829.
24. Deguchi H, Antal FFY, Armstrong DD. Neural stem cell reduction in the ventricular zone of perinatal injured brains. Abstracts of the 80th Annual Meeting of the American Association of Neuropathologists, Cleveland, 2004. (Abstract 556.)
25. Del Bigio MR. Hemorrhagic lesions. In Golden JA, Harding BN (eds): Pathology and Genetics: Developmental Neuropathology. Basel: ISN Neuropath Press, 2004, pp 150–155.
26. DeLong GR. Mid-gestational right basal ganglia lesion: Clinical observations in two children. Neurology 2002;59:54–58.
27. DiMaio VJ, DiMaio D. Forensic Pathology, ed 2. Boca Raton, FL: CRC Press, 2001.
28. Donaldson JO. Neurologic complications. In Burrow GN, Duffy TP (eds): Medical Complications During Pregnancy, ed. 5. Philadelphia: WB Saunders, 1999, pp 401–414.
29. Dooling EC, Gilles FH. Intracranial hemorrhage: Topography. In Gilles FH, Leviton A, Dooling EC (eds): The Developing Human Brain: Growth and Epidemiologic Neuropathology. Boston: Wright PSG, 1983, pp 193–203.
30. Drut R. Embolism of fetal brain tissue to the lungs (letter to editor). Pediatr Pathol Lab Med 1997;17:987.
31. Dyer I, Barclay DL. Accidental trauma complicating pregnancy and delivery. Am J Obstet Gynecol 1962;83:907–926.
32. Emerson MV, Piermici DJ, Stoessel KM, et al. Incidence and rate of disappearance of retinal hemorrhage in newborns. Ophthalmology 2001;108:36–39.
33. Evans D, Levene M. Clinical assessment of the neonate. In Squier W (ed): Acquired Damage to the Developing Brain: Timing and Causation. London: Arnold, 2002, pp 139–165.
34. Farrow JR, Davis GJ, Roy TM, et al. Fetal death due to nonlethal maternal carbon monoxide poisoning. J Forensic Sci 1990;35:1448–1452.
35. Fatovich DM. Electric shock in pregnancy. J Emerg Med 1993;11:175–177.
36. Ferriero DM. Neonatal brain injury. N Engl J Med 2004;351:1985–1995.
37. Friede RL. Developmental Neuropathology, ed 2. Berlin: Springer-Verlag, 1989.
38. Gaffney G. Etiology of fetal and neonatal brain damage. In Squier W (ed): Acquired Damage to the Developing Brain: Timing and Causation. London: Arnold, 2002, pp 39–55.
39. Genest DR, Williams MA, Greene MF. Estimating the time of death in stillborn fetuses. I. Histologic evaluation of fetal organs: An autopsy study of 150 stillborns. Obstet Gynecol 1992;80:575–584.
40. Genest DR. Estimating the time of death in stillborn fetuses. II. Histologic evaluation of the placenta: A study of 71 stillborns. Obstet Gynecol 1992;80:585–592.
41. Genest DR, Singer DB. Estimating the time of death in stillborn fetuses. III. External fetal examination: A study of 86 stillborns. Obstet Gynecol 1992;80:593–600.
42. Gilles FH. Perinatal neuropathology. In Davis RL, Robertson DM (eds): Textbook of Neuropathology, ed 3. Baltimore: Williams & Wilkins, 1997, pp 331–385.
43. Gilles FH, Bina M, Sotrel A. Infantile atlantooccipital instability: The potential danger of extreme extension. Am J Dis Child 1979;133:30–37.
44. Gilles FH, Leviton A, Golden JA, et al. Groups of histopathologic abnormalities in brains of very low birth weight infants. J Neuropathol Exp Neurol 1998;57:1026–1034.
45. Goetting MG, Sowa B. Retinal hemorrhage after cardiopulmonary resuscitation in children: An etiologic reevaluation. Pediatrics 1990;85:585–588.
46. Graber EA. Dilemmas in the pharmacological management of preterm labor. Obstet Gynecol Surv 1989;44:512–517.
47. Greene PM, Wilson H, Ramaniuk C, et al. Idiopathic intracranial hemorrhage in the fetus. Fetal Diagn Ther 1999;14:275–278.
48. Gross GW, Goldberg BB. Neurosonography of the fetal, neonatal, and infant brain and spine. In Duckett S (ed): Pediatric Neuropathology. Baltimore: Williams & Wilkins, 1995, pp 830–881.
49. Hadi HA, Finley J, Mallett JQ, Strickland D. Prenatal diagnosis of cerebellar hemorrhage: Medicolegal implications. Am J Obstet Gynecol 1994;170:1392–1395.
50. Hanigan WC, Morgan AM, Stahlberg LK, Hiller JL. Tentorial hemorrhage associated with vacuum extraction. Pediatrics 1990;85:534–539.
51. Heise RH, Srivatsa PJ, Karsell PR. Spontaneous intrauterine linear skull fracture: A rare complication of spontaneous vaginal delivery. Obstet Gynecol 1996;87:851–854.
52. Horn KD, Nolte KB. An unusual maternal death. Forensic Pathology Check Sample. No. FP03-1 (FP-282). Chicago: American Society of Clinical Pathology, 2003;45:1–15.
53. Huisman TA, Fischer J, Willi UV, et al. "Growing fontanelle": A serious complication of difficult vacuum extraction. Neuroradiology 1999;41:381–383.
54. James SJ. Fetal brain death syndrome: A case report and literature review. Aust N Z J Obstet Gynaecol 1998;38:217–220.
55. Kaaja RJ, Kaaja EH, Hiilesmaa VH. Enzyme-inducing antiepileptic drugs in pregnancy and the risk of bleeding in the neonate (abstract). Neurology 2004;62(Suppl 5):A351.
56. Kalousek DK. Pathology of abortion: The embryo and the previable fetus. In Gilbert-Barness E (ed): Potter's Pathology of the Fetus and Infant. St Louis: Mosby, 1997;1:106–127.
57. Kaplan CG. Forensic aspects of the placenta. In Dimmick JE, Singer DB (eds): Forensic Aspects in Pediatric Pathology. Perspect Pediatr Pathol Basel: Karger, 1995;19:20–42.
58. Katz VL, Dotters DJ, Droegemueller W. Perimortem cesarean delivery. Obstet Gynecol 1986;68:571–576.

59. Kingston NJ, Baillie T, Chan YF, et al. Pulmonary embolism by chorionic villi causing maternal death after a car crash. Am J Forensic Med Pathol 2003;24:193–197.
60. Kinney HC, Armstrong DD. Perinatal neuropathology. In Graham DI, Lantos PL (eds): Greenfield's Neuropathology, ed 6. London: Arnold, 1997;I:535–599.
61. Kinney HC, Armstrong DD. Perinatal neuropathology. In Graham DI, Lantos PL (eds): Greenfield's Neuropathology, ed 7. London: Arnold, 2002;1:519–606.
62. Kinney HC, Haynes RL, Folkerth RD. White matter lesions in the perinatal period. In Golden JA, Harding BN (eds): Pathology and Genetics: Developmental Neuropathology. Basel: ISN Neuropath Press, 2004, pp 156–170.
63. Kraus FT, Acheen VI. Fetal thrombotic vasculopathy in the placenta: Cerebral thrombi and infarcts, coagulopathies, and cerebral palsy. Hum Pathol 1999;30:759–769.
64. Langlois NEI, Gray ES. Scatterbrain fetus. J Pathol 1992;168:347–348.
65. Lawn N, Laich E, Ho S, et al. Eclampsia, hippocampal sclerosis, and temporal lobe epilepsy: Accident or association? Neurology 2004;62:1352–1356.
66. Leestma JE: Forensic neuropathology. In Duckett S (ed): Pediatric Neuropathology. Baltimore: Williams & Wilkins, 1995, pp 243–283.
67. Levin AV: Retinal hemorrhages and child abuse. In David TJ (ed): Recent Advances in Paediatrics, Vol 18. New York: Churchill Livingston, 2000, pp 151–219.
68. Levin AV. Ophthalmic manifestations of inflicted childhood neurotrauma. In Reece RM, Nicholson CE (eds): Inflicted Childhood Neurotrauma. Elk Grove Village; IL: American Academy of Pediatrics, 2003, pp 127–159.
69. Lifschultz BD, Donoghue ER. Fetal death following maternal trauma: Two case reports and a survey of the literature. J Forensic Sci 1991;36:1740–1744.
70. Lindenberg R, Freytag E. Morphology of brain lesions from blunt trauma in early infancy. Arch Pathol 1969;87:298–305.
71. Lysack JT, Soboleski D. Classic metaphyseal lesion following cephalic version and cesarean section. Pediatr Radiol 2003;33:422–424.
72. MacLennan A (for the International Cerebral Palsy Task Force). A template for defining a causal relation between acute intrapartum events and cerebral palsy: International consensus statement. BMJ 1999;319:1054–1059.
73. Magee JF. Investigation of stillbirth. Pediatr Dev Pathol 2001;4:1–22.
74. Maher JE, Wenstrom KD, Hauth JC, Meis PJ. Amniotic fluid embolism after saline amnioinfusion: Two cases and review of the literature. Obstet Gynecol 1994;83:851–854.
75. Marín-Padilla M. Developmental neuropathology and impact of perinatal brain damage. I. Hemorrhagic lesions of the neocortex. J Neuropathol Exp Neurol 1996;55:758–773.
76. Marín-Padilla M. Developmental neuropathology and impact of perinatal brain damage. II. White matter lesions of the neocortex. J Neuropathol Exp Neurol 1996;56:219–235.
77. Marín-Padilla M. Developmental neuropathology and impact of perinatal brain damage. III. Gray matter lesions of the neocortex. J Neuropathol Exp Neurol 1999;58:407–429.
78. Marin-Padilla M, Parisi JE, Armstrong DL, et al. Shaken infant syndrome: Developmental neuropathology, progressive cortical dysplasia, and epilepsy. Acta Neuropathol 2002;103:321–332.
79. McComb JG: Cerebrospinal fluid, hydrocephalus, and cerebral edema. In Davis RL, Robertson DM (eds): Textbook of Neuropathology, ed 3. Baltimore: Williams & Wilkins, 1997, pp 225–251.
80. Mehl LE. Electrical injury from tasering and miscarriage. Acta Obstet Gynecol Scand 1992;71:118–123.
81. Menkes JH, Sarnat HB. Perinatal asphyxia and trauma. In Menkes JH, Sarnat HB (eds): Child Neurology, ed 6. Philadelphia: Lippincott Williams & Wilkins, 2000, pp 401–466.
82. Nadas S, Gudinchet F, Capasso P, Reinberg O. Predisposing factors in obstetrical fractures. Skel Radiol 1993;22:195–198.
83. Nelson KB. Can we prevent cerebral palsy? N Engl J Med 2003;349:1765–1769.
84. Nguyen S, Kuschel C, Teele R. Water birth—A near-drowning experience. Pediatrics 2002;110:411–413.
85. Nichols MM, Gelman BB, Gilbert-Barness E. Pathological case of the month. Autolyzed brain tissue beneath dermis of thigh and back. Arch Pediatr Adolesc Med 1996;150: 997–998.
86. Odita JC, Hebi S. CT and MRI characteristics of intracranial hemorrhage complicating breech and vacuum delivery. Pediatr Radiol 1996;26:782–785.
87. O'Gara PT, Greenfield AJ, Afridi NA, Houser SL. Case 12–2004: A 38-year-old woman with acute onset of pain in the chest. N Engl J Med 2004;350:1666–1674.
88. Okuno T, Miyamoto M, Itakura T, et al. [A case of epidural hematoma caused by a vacuum extraction without any skull fractures and accompanied by cephalohematoma.] No Shinkei Geka 1993;21:1137–1141. (Article in Japanese; abstract translated by Pub Med.)
89. Papaefthymiou G, Oberbauer R, Pendl G. Craniocerebral birth trauma caused by vacuum extraction: A case of growing skull fracture as a perinatal complication. Childs Nerv Syst 1996;12:117–120.
90. Pape KE, Armstrong DL, Fitzhardinge PM. Central nervous system pathology associated with mask ventilation in a very low birthweight infant: A new etiology for intracerebral hemorrhages. Pediatrics 1976;58:473–483.
91. Pollack J, Tychsen L. Prevalence of retinal hemorrhage in infants after extracorporeal membrane oxygenation. Am J Ophthalmol 1996;121:297–303.
92. Pourbabak S, Rund CR, Crookston KP. Three cases of massive fetomaternal hemorrhage presenting without clinical suspicion. Arch Pathol Lab Med 2004;128:463–465.
93. Powers MJ. Causation-legal proof. In Squier W (ed): Acquired Damage to the Developing Brain: Timing and Causation. London: Arnold, 2002, pp 209–217.
94. Prahlow JA, Barnard JJ. Pregnancy-related maternal deaths. Am J Forensic Med Pathol 2004;25:220–236.
95. Ribe JK, Teggatz JR, Harvey CM. Blows to the maternal abdomen causing fetal demise: Report of three cases and a review of the literature. J Forensic Sci 1993;38:1092–1096.
96. Rogers CB, Itabashi HH, Tomiyasu U, Heuser ET. Subdural neomembranes and sudden infant death syndrome. J Forensic Sci 1998;43:375–376.
97. Rorke LB. Kernicterus. In Golden JA, Harding BN (eds): Pathology and Genetics: Developmental Neuropathology. Basel: ISN Neuropath Press, 2004, pp 206–208.
98. Rosen RS, Armbrustmacher V, Sampson BA. Mortality in cerebral palsy (CP): The importance of the cause of CP on the manner of death. J Forensic Sci 2003;48:1144–1147.
99. Ross MG, Fresquez M, El-Haddad MA. Impact of FDA advisory on reported vacuum-assisted delivery and morbidity. J Matern Fetal Med 2000;9:321–326.
100. Rushton I. Letter to the editor (Re: Langlois NEI, Gray ES: Scatterbrain fetus. J Pathol 1992;168:347–348). J Pathol 1993;170:211.
101. Sarnat HB: Perinatal hypoxic/ischemic encephalopathy: Neuropathological features. In Garcia JH (ed): Neuropathology: The Diagnostic Approach. St. Louis: Mosby, 1997, pp 541–580.
102. Savage JE. Discussion. Am J Obstet Gynecol 1962;83:926.
103. Shaw C-M, Alvord EC Jr. Subependymal germinolysis. Arch Neurol 1974;31:374–381.
104. Singer DB. Expectations and reality: Untoward outcome of gestation. In Dimmick JE, Singer DB (eds): Forensic Aspects in Pediatric Pathology. Perspectives in Pediatric Pathology. Vol 19. Basel: Karger, 1995, pp 59–75.
105. Singh V, Khanum S, Singh M. Umbilical cord lesions in early intrauterine fetal demise. Arch Pathol Lab Med 2003;127:850–853.

106. Somogyi E, Tedeschi CG. Injury by electrical force. In Tedeschi CG, Eckert WG, Tedeschi LG (eds): Forensic Medicine: A Study in Trauma and Environmental Hazards. Philadelphia: WB Saunders, 1977;I:645–676.
107. Squier M, Chamberlain P, Zaiwalla Z, et al. Five cases of brain injury following amniocentesis in mid-term pregnancy. Dev Med Child Neurol 2000;42:554–560.
108. Squier W. Gray matter lesions. In Golden JA, Harding BN (eds): Pathology and Genetics: Developmental Neuropathology. Basel: ISN Neuropath Press 2004, pp 171–175.
109. Stedman's Medical Dictionary, ed 25. Baltimore: Williams & Wilkins, 1990.
110. Szulman AE. Embryonic death: Pathology and forensic implications. In Dimmick JE, Singer DB (eds): Forensic Aspects in Pediatric Pathology. Perspectives in Pediatric Pathology, Vol 19. Basel: Karger, 1995, pp 43–58.
111. Takanashi J, Barkovich AJ, Ferriero DM, et al. Widening spectrum of congenital hemiplegia: Periventricular venous infarction in term neonates. Neurology 2003;61:531–533.
112. Thompson JP. Forceps deliveries. Clin Perinatol 1995;22:953–972.
113. Towner D, Castro MA, Eby-Wilkins E, Gilbert WM. Effect of mode of delivery in nulliparous women on neonatal intracranial injury. N Engl J Med 1999;341:1709–1714.
114. Tsokos M. Pathological features of maternal death from HELLP syndrome. In Tsokos M (ed): Forensic Pathology Reviews. Totowa, NJ: Humana Press, 2004;1:275–290.
115. Tsokos M, Longauer F, Kardošová V, et al. Maternal death in pregnancy from HELLP syndrome: A report of three medico-legal autopsy cases with special reference to distinctive histopathological alterations. Int J Legal Med 2002;116:50–53.
116. Tuluc M, Brown D, Goldman B. Lethal vertebral artery dissection in pregnancy: A case report and review of the literature. Arch Pathol Lab Med 2006;130:533–535.
117. Vinters HV. Acquired vascular lesions in children. In Golden JA, Harding BN (eds): Pathology and Genetics: Developmental Neuropathology. Basel: ISN Neuropath Press, 2004, pp 176–183.
118. Volpe JJ. Neurology of the Newborn, ed 3. Philadelphia: WB Saunders, 1995.
119. Wehberg K, Vincent M, Garrison B, et al. Intraventricular hemorrhage in the full-term neonate associated with abdominal compression. Pediatrics 1992;89:327–329.
120. Whitby EH, Griffiths PD, Rutger S, et al. Frequency and natural history of subdural hemorrhages in babies and relation to obstetric factors. Lancet 2004;363:846–851.
121. Wigglesworth JS. Perinatal pathology. In Bennington JL (ed): Major Problems in Pathology, Vol 15. Philadelphia: WB Saunders, 1984.
122. Wijdicks EFM. Neurologic Complications of Critical Illness, ed 2. New York: Oxford University Press, 2002, pp 238–247.
123. Williams MC. Vacuum-assisted delivery. Clin Perinatol 1995;22:933–952.
124. Williams MC, Knujppel RA, O'Brien WF, et al. A randomized comparison of assisted vaginal delivery by obstetric forceps and polyethylene vacuum cup. Obstet Gynecol 1991;78:789–794.
125. Wu YW, Miller SP, Chin K, et al. Multiple risk factors in neonatal sinovenous thrombosis. Neurology 2002;59:438–440.

Malformations and Other Congenital Central Nervous System Lesions

5

TIMING OF CNS DEVELOPMENTAL ABNORMALITIES AND PERINATAL INJURIES 154

REFERENCES 165

The forensic neuropathology consultant may be called on to investigate central nervous system (CNS) developmental disorders from several sources. All patients who expire in county governmental institutions are under the jurisdiction of our department, and include individuals who have been classified as having mental illness, mental retardation, cerebral palsy, behavioral disorders, developmental delay, seizure disorders, or other relatively nonspecific diagnostic categories. Not only are various neurocutaneous syndromes and degenerative disorders encountered in this group, but many types of previously undiagnosed CNS abnormalities grouped under the general heading of malformations are seen. In this chapter we use the term "malformations" in a general sense, without restricting it to a specific etiology for reasons discussed further on.

Abnormally formed brains are also sometimes encountered in suspected natural deaths such as possible sudden infant death syndrome (SIDS) cases, stillbirths, neonatal deaths, and in infant cases referred to a medical examiner when the attending physician is unwilling to sign the death certificate. In some cases, there is no information available concerning gestational age or family history.

In several of these contexts, ability to determine whether a given gross or microscopic abnormality is or is not consistent with origin in a given time frame of alleged causative factor(s) is of considerable importance in the resolution of medicolegal questions. Considerations of this type have, therefore, influenced our choice of emphasis on topics that could otherwise be considered under this heading.

Sources of information helpful in the identification of the various types of CNS malformations can provide a starting point for further study. Many of these sources, including those on topics such as CNS development, brain weights, and so on have been reviewed in Chapters 1 and 2. Cited sources in those chapters include several texts devoted entirely to CNS malformations. A recent monograph on developmental neuropathology[15] includes data on genetic and molecular mechanisms associated with these disorders.

The dilemma presented to the forensic pathologist often centers on whether a particular event occurring during an individual's gestation, delivery, or early childhood was, or was not, related causally to a given CNS malformation or other lesion found in that individual. Several well-known observations can complicate the analysis of such issues. For example, abnormalities similar in appearance may result from diverse etiologic factors acting at the same developmental period. The same etiologic factor, acting at different stages of development, may produce different abnormalities. It can be difficult or impossible with available knowledge to distinguish primary from secondary events during the evolution of a developmental disorder. Although global features may be similar in different cases, each case has its unique features. Older nomenclature for a particular abnormality may infer causation not supported by more recent studies. More than one type of simultaneous insult may act in concert to produce a particular abnormality. Alternatively, one must consider the possibility of adverse events occurring on more than one occasion during gestation, the resulting abnormality representing the aggregate result of a sequence of similar or dissimilar etiologies.

There is a natural, though often inaccurate, tendency to link the etiology of an abnormality to a clinically dramatic event, such as illness or trauma during pregnancy, fetal distress precipitating emergency delivery, and so on. It is important to consider why the fetal distress or other emergency near parturition developed in the first place, such as whether it was secondary to a preexisting abnormality developing spontaneously or due to an earlier, clinically inapparent, adverse event. Finally, there are surely individual differences in susceptibility to similar insults, judging by experience in virtually all other aspects of pathology.

Situations occur in legal proceedings in which one is asked to disprove something being stated, but for which no proof has been put forth. Fackler[10] has quite lucidly stated the problem and its obvious solution: "You will save yourself a lot of time, and possibly nudge a few toward valid self-education, by demanding evidence for the things that are asserted, rather then responding to an irrational request-disproof of the unproven."

We have collected isolated data from a variety of dispersed sources that offer suggestions as to when, during the course of CNS development, a given insult of some type may produce a particular type of CNS abnormality. *Our attempt to place such information in an easily accessible format (see Table 5.2) has the potential for misinterpretation if it is taken too literally. We see this data as being useful if viewed as an index to the literature that can provide a starting point for the interested reader, to be supplemented by further study as needed.* The omission of specific etiologies in place of "period of insult" in the Table 5.2 is purposeful, for reasons outlined earlier. Consideration of potential contributing factors, such as cardiovascular disease, other causes of hypoxic-ischemic events, hypervitaminosis A, nutritional deficiency, chromosomal abnormalities, maternal metabolic disorders, infections, drugs, toxins, hormonal abnormalities, trauma, antigen-antibody interactions, irradiation, placental or umbilical cord abnormalities, and so on, may be necessary on a case-by-case basis.

The suggested timetable of events that may relate to a given type of abnormality should be viewed as approximations from the cited sources, with the degree of overlap in opinion from several sources hinting at varying degrees of consensus. The present authors have not personally done research that would either support or disprove these cited time estimates.

Although this summary is intended to be helpful as a brief overview, we would caution that the cited sources sometimes differ in their use of terms, or use terms without specifically defining them. For example, a dictionary[8] indicates the perinatal period to be variously defined as from 20 to 28 weeks' gestation to 7 to 28 days after birth. Gilles[13] indicates that this range can theoretically be extended up to 2 years of age. We have the impression that, defined or not, most authors use the term "perinatal period" to refer to the more restricted age range of from 20 weeks' gestational age to 28 days after birth, unless otherwise specified herein.

Several of the CNS deformities included in this chapter are obvious on cursory inspection of the brain. In Chapter 2, some clues on gross examination that suggest the need to search for more subtle developmental abnormalities are discussed. Other nonspecific clues include minor abnormalities in cerebral convolutional pattern, persistence of an exposed insula, a disproportionately small cerebellum for age group, an atypical contour of the cerebellum (e.g., tented cerebellum) or cerebrum (e.g., steep occipital slope), or presence of the infrequently mentioned rostrum orbitale (Fig. 5.1). A rostrum orbitale consists of a very conspicuous "beak" shape to the midline rostral orbitofrontal gyri. The basis for this malfor-

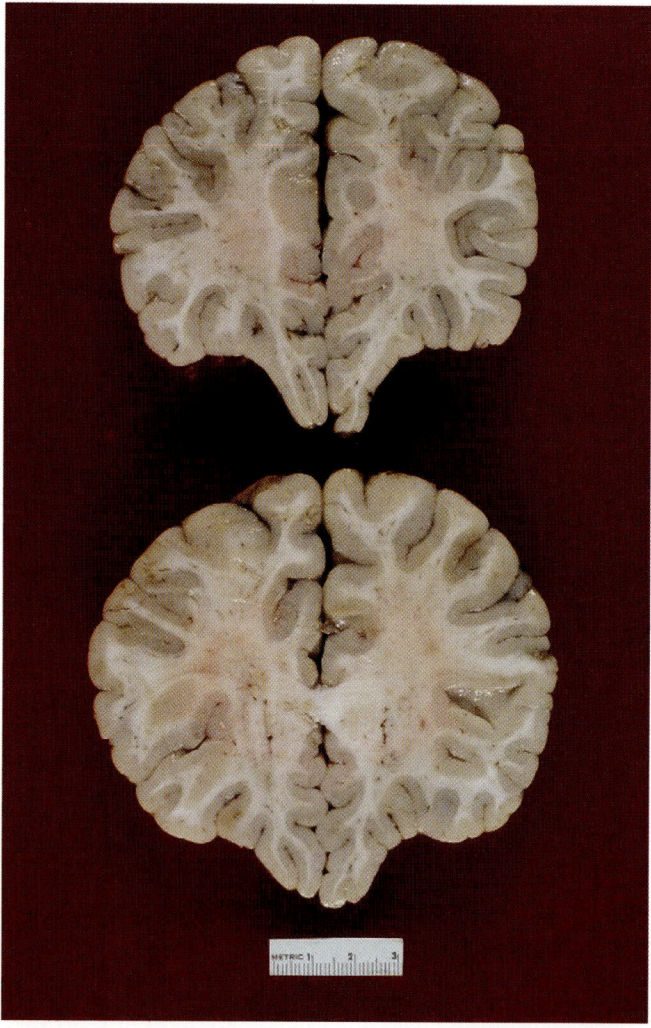

Fig. 5.1. Coronal sections at pregenu (*upper*) and anterior genu (*lower*) frontal lobe, with prominent inferior protrusion of anterior medial orbitofrontal gyri, termed "rostrum orbitale."

mation is unknown, but its presence is an indicator that one should search more carefully for other brain anomalies. It may be seen in Down syndrome and a variety of other brain malformations, often associated with clinical evidence of mental retardation.

A suggested manner of approaching this subject might be illustrated by briefly reviewing agenesis of the corpus callosum. Table 5.2 suggests that this condition, as an isolated abnormality, may develop prior to about 20 weeks' gestation. To determine with more confidence, however, that this is a reasonable assumption, one must also be aware that agenesis of the corpus callosum has been described in at least 70 different conditions, is associated with other CNS abnormalities in 80% of cases (some of which may alter one's estimated timing of a potential causative factor), and is associated with malformations in non-CNS organ systems in 62% of cases.[30,31] Furthermore, under this term is grouped both partial (hypoplastic) (Case 5.1, Figs. 5.2–5.3) and complete (Fig. 5.4) absence of the corpus callosum, and this term does not distinguish whether the corpus callosum is absent as a primary, isolated event or secondary to another cause. For example, holoprosencephaly "has inherently no corpus callosum,"[11] and its association in some cases with encephaloclastic lesions infers callosal destruction after prior formation.

An additional potential pitfall in this specific abnormality (agenesis of the corpus callosum) is the fragile

> **Case 5.1.** (Figs. 5.2 and 5.3)
>
> A 6½-month-old girl was found dead in her crib. A single seizure at age 4 months led to a magnetic resonance scan that reported partial agenesis of the corpus callosum with "hypogenesis" of the anterior corpus callosum.

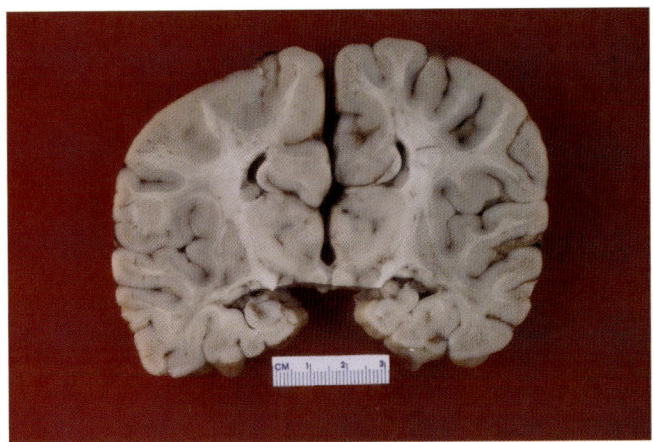

Fig. 5.3. Coronal section at mid thalamus in Case 5.1. Absent posterior one-half of corpus callosum.

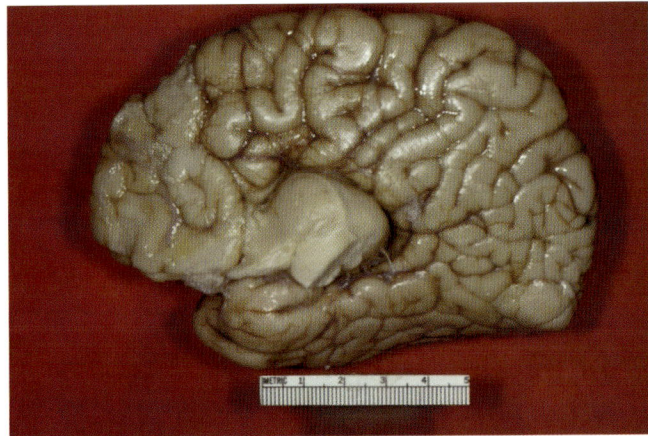

Fig. 5.4. Sagittal section of another brain, in this case with complete agenesis of corpus callosum. Note absence of identifiable cingulate sulcus and gyrus.

nature of the unmyelinated normal corpus callosum in immature brains, rendering it susceptible to avulsion during the course of brain removal. Norman and Ludwin[30] indicate that a helpful clue, particularly in the 12- to 24-week gestational age group in which the soft cerebral hemispheres can easily be separated by rupture of the corpus callosum during brain removal, is to look for a cingulate sulcus (Fig. 5.4). The presence of the latter indicates that a corpus callosum was formed, and this can later be confirmed in coronal sections of the fixed brain.

Many other examples exist to reinforce the point that finding a single malformation does not signal that the brain examination is complete (i.e., the "satisfaction of search" syndrome). Rather, it is an incentive to carefully search for additional abnormalities. For example, Jones[19] lists 2 frequent and 9 occasional syndromic associations with encephalocele, 2 frequent and 19 occasional syndromic associations with Dandy-Walker malformation, and 5 frequent and 47 occasional syndromic associations for hydrocephalus, and so on.

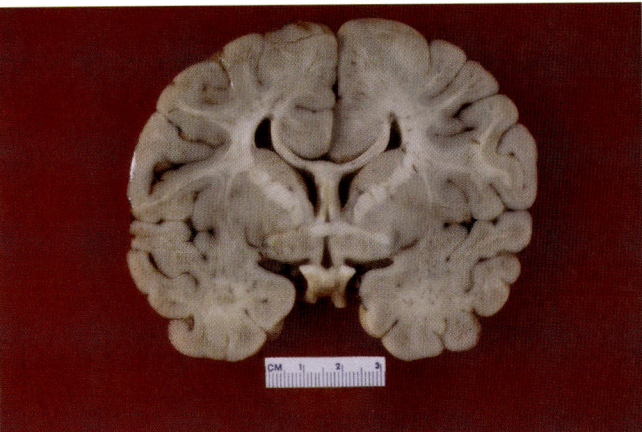

Fig. 5.2. Coronal section of brain in Case 5.1 at level of anterior commissure reveals a thin corpus callosum with upswept lateral extensions bordering a classic "butterfly wing" or "bat wing" profile of the lateral ventricles, characteristic of this anomaly.

It follows that careful correlation of a given finding with the individual family history, maternal history, autopsy findings in other organs including placenta, the relative severity of a given abnormality, the pathologic "company it keeps" in the CNS (e.g., appearance of adjacent structures, such as gyri and sulci in a defect involving the cortex, presence or absence of macrophage activity or glial scar, related long tract agenesis versus degeneration), and relevant experimental data are among the many factors to be evaluated in attempting to reach conclusions "to a reasonable degree of medical certainty" in these complex and challenging cases. A suggested sequence of evaluation of these cases is summarized in Table 5.1. We believe that it is especially important during step 7 in Table 5.1 to consider carefully the evidence put forth by an author to support a suggested cause-and-effect relationship between an event occurring during gestation and subsequent appearance of a CNS abnormality. Scientific investigators will offer hypotheses regarding such correlations that may be suggested—but not proved—by their findings in a sincere effort to advance knowledge by prompting other investigators to design studies that will confirm or refute their speculation. Such hypothetical correlation should not be taken out of context and used in court testimony, purposely or inadvertently, as if it were reliable and reproducible fact. Aspects of this problematic area are discussed by Bird.[5]

In Table 5.2, we have used the developmental stage terminology as it is indicated in the cited source, since our assumption of an estimated age may or may not correspond to the intent of the source. As pointed out by Norman and Ludwin,[30] there is more than one way to calculate gestational age, and the method used may not be specified in a given reference. For example, menstrual age, based on the mother's last menstrual period onset, yields a gestation period of 40 weeks, but the postconception gestational period is only 38 weeks.

Table 5.1. Suggested Steps in the Evaluation of CNS Malformations

1. Clinical history (including extended family, maternal, gestational history, peripartum and postpartum history).
2. Placenta pathology report.
3. Gross and microscopic CNS abnormalities tabulated, as well as abnormalities in other organ systems. Does the aggregate of CNS and non-CNS abnormalities conform to a previously described syndrome?
4. Sequential sorting of CNS abnormalities by approximate gestational age timing of origin of abnormality, including considerations of primary versus secondary abnormalities, presence or absence of evidence of reactive cellular changes, etc. (see Chap. 2).
5. Correlation of results of steps 4 and 5 with timing of suggested or claimed etiologic associations by participants in the case.
6. Literature review to determine whether there is precedent for any potential cause-and-effect relationship in situations where there is some coincidence in timing after steps 5 and 6 have been completed, based on information from human case reports as well as from animal experimentation (while carefully separating data derived from human versus nonhuman species).
7. Formulate final opinion based on objective data and reasonable degree of medical certainty.

Table 5.2. Timing of CNS Developmental Abnormalities and Perinatal Injuries[a]

Period of Insult	1. Holoprosencephaly Group — Abnormality Produced
Before or at 18 days' postovulation[30,31]	Holoprosencephaly with cyclops
20 days' postovulation[30]	Holoprosencephaly with fused double eye
At or before 4 weeks' postovulation[30]	Telencephalon fails to form hemispheres
Before 4th week of gestation[14,28,31]	Holoprosencephaly NOS* (Case 5.2, Figs. 5.5–5.6)
4th–6th week of gestation[11]	Holoprosencephaly NOS

Period of Insult	2. Sacral Agenesis — Abnormality Produced
May arise as early as 2½–3 weeks' gestational age[31]	Sacral agenesis (i.e., caudal regression syndrome)

Period of Insult	3. Neurenteric Cysts — Abnormality Produced
First 3 weeks' gestation[6]	Neurenteric cysts (enterogenic cysts)

Table 5.2. Continued

4. Midline Defect Group

Period of Insult	Abnormality Produced
18 days' postovulation[29]	Split spinal cord malformations (e.g., diastematomyelia, diplomyelia)
24 days' gestation[12]	Anencephaly; myeloschisis
No later than 24 days' gestation[41]	Anencephaly; myeloschisis
18–26 days' postovulation[19,29,30]	Anencephaly
Onset no later than 19–23 days' gestation[24]	Complete dysraphia (craniorachischisis)
After 23 days to?[27]	Anencephaly
Onset no later than 23–26 days' gestation[24]	Anencephaly
Probably no later than 26 days' gestation[41]	Myelomeningocele (Figs. 5.7–5.8)
Approximately 2½–3½ weeks' postovulation[29,30]	Lumbosacral myeloschisis portion of Chiari type II malformation (Fig. 5.9)
No later than 20–22 days' gestation[12,41]	Craniorachischisis totalis
26–28 days' gestation[12]	Meningomyelocele
After 26 days' gestation to?[27]	Cranium bifidum, spina bifida cystica, spina bifida occulta
Earliest onset could be 26–30 days' gestation[24]	Encephalocele; meningomyelocele
Prior to 28 days' gestation[6,19]	Meningomyelocele; meningocele
On or before 28 days' gestation[36]	All neural tube fusion defects
From 3–7 or 8 weeks' postovulation[31]	Occipital encephalocele (Fig. 5.10)
Approximately 26 days' gestation or shortly thereafter[41]	"Severe" encephaloceles
26 days' gestation[12]	Encephalocele, NOS
Onset after 26–30 days' gestation, and possibly during the 28- to 32-day gestational age period[24]	Lower spinal myelocystocele (localized cystic dilatation of central canal)
Probably between 50 and 70 gestational days[24]	Lower spine meningocele (dorsal prolapse of meninges without spinal cord involvement, with or without vertebral fusion abnormalities)

5. Forebrain Group

Period of Insult	Abnormality Produced
Probably no later than beginning of 2nd month of gestation[41]	Aprosencephaly; atelencephaly

6. Micrencephaly Group

Period of Insult	Abnormality Produced
No later than early in 2nd month of gestation[12,41]	Radial microbrain
From approximately 6–18 weeks' gestation[41]	Microcephaly vera
Approximately 18 weeks' gestation[12]	Microcephaly vera

7. Ventricular Adhesions

Period of Insult	Abnormality Produced
Approximately 9th gestational week to birth, plus postnatal period[17]	Adhesions of walls of lateral ventricles

8. Status Verrucosus Simplex

Period of Insult	Abnormality Produced
Probably prior to 10 weeks' gestation[17,22]	Status verrucosus simplex (Status pseudoverrucosus)

Malformations and Other Congenital CNS Lesions

Table 5.2. Continued

9. Polymicrogyria Group

Period of Insult	Abnormality Produced
8–24 weeks' gestation[29]	Polymicrogyria NOS (Figs. 5.11–5.14)
13–16 weeks' gestation[40]	Polymicrogyria with unlayered cortex
Approximately 3rd through 5th month gestation[18]	Polymicrogyria NOS
Prior to 16–22 weeks' gestation[9]	Polymicrogyria (with or without neuronal heterotopias)
No later than 4th–5th month gestation[41]	Polymicrogyria with unlayered cortex
3rd–4th month gestation[17]	Polymicrogyria NOS
20–24 weeks' gestation[39,41]	Polymicrogyria with 4-layered cortex
20–24 weeks' gestation "and beyond"[12]	Polymicrogyria (layered)
Between 2nd and 5th month of gestation[7]	Polymicrogyria NOS
Near 5th month or before 6th month of gestation[11]	Polymicrogyria NOS
Before, but not after, 28 weeks' gestation[37]	Polymicrogyria NOS
Perinatal (after 21 weeks' gestation)[26]	Polymicrogyria NOS

10. Porencephaly Group

Period of Insult	Abnormality Produced
Event prior to approximately 20–24 weeks' gestation[17]	Porencephaly, NOS (Figs. 5.15–5.18)
Approximately 22–24 weeks' gestation (peak incidence)[20]	Porencephaly, NOS
Approximately 5th fetal month[11]	Porencephaly[b] with polymicrogyria, with or without cortical lamination defects and heterotopias
As early as 20–27 weeks' gestation[41]	Porencephaly, NOS
Comment:	
Porencephalic-producing event prior to development of adjacent secondary and tertiary gyri/sulci pattern (i.e., before about 26–28 weeks' gestation)	Gyri at porencephalic margin have radial orientation to cyst opening (i.e., perpendicular to porencephalic cyst margin). (Rarely this may be mimicked by postnatal injuries, but will be devoid of polymicrogyria or heterotopia; see Case 5.3, Fig. 5.19).[11]
Porencephalic-producing event after development of adjacent secondary/tertiary gyri/sulci pattern (e.g., after about 26–28 weeks' gestation)	Gyri at porencephalic margin may have essentially normal orientation and convolutional pattern, aside from atrophy and gliosis.

11. Mobius Syndrome

Period of Insult	Abnormality Produced
Possibly 33–40 days' gestational age[17]	Möbius syndrome
Also, late fetal life by a destructive lesion[34,35]	Möbius syndrome

12. Dandy-Walker Malformation

Period of Insult	Abnormality Produced
By 6–7 weeks' gestation[30]	Dandy-Walker malformation (Fig. 5.20)
By 7–8 weeks' gestation[29]	Dandy-Walker malformation
Before 3rd month of gestation[11,17,41]	Dandy-Walker malformation

13. Fowler's Syndrome

Period of Insult	Abnormality Produced
Prior to week 7 of gestation[17]	Fowler's syndrome (proliferative vasculopathy and hydranencephaly-hydrocephaly)

Table 5.2. Continued

Period of Insult	14. Cerebellar/Brainstem Abnormality Group Abnormality Produced
1st trimester[11,16]	Olivary heterotopias/dysplasias (Figs. 5.21–5.27)
1st trimester[16]	Dentate nucleus dysplasias
Possibly at end of 3rd month of gestation[17]	Pontocerebellar hypoplasia
Before 20 to 30 weeks' gestation[37]	Brainstem neuronal heterotopias
Possibly 2nd trimester[17]	Olivary and dentate nucleus dysplasia
Possibly between 5th and 6th months of gestation[11]	Cerebellar granular layer aplasia

Period of Insult	15. Lissencephaly Abnormality Produced
Prior to 11–13 weeks' gestational age[17]	Lissencephaly (Figs. 5.28–5.29)
Prior to 3rd or 4th month of gestation[7,12,41]	Lissencephaly

Period of Insult	16. Agenesis of Corpus Callosum Abnormality Produced
Probably prior to 10–12 weeks' gestation[22]	Agenesis of corpus callosum (Figs. 5.2–5.4)
No later than 9–20 weeks' gestation[41]	Agenesis of corpus callosum

Period of Insult	17. Cerebral Neuronal Heterotopias and Dysplasia Group Abnormality Produced
Probably during 1st trimester[17]	Nodular neuronal heterotopias (Fig. 5.30)
Prior to 20 weeks' gestation[31,41]	Neuronal heterotopias, NOS
Before 21 weeks' gestation[37]	Cerebral hemisphere neuronal heterotopias
During 2nd trimester of gestation[11]	Periventricular heterotopic gray matter
Perinatal period[26]	Cerebral cortical dysplasias

Period of Insult	18. Neuronal and/or Glial Leptomeningeal Heterotopias Abnormality Produced
Any factor preventing normal pial-glial border development by approximately 7–8 weeks' gestation, or disrupting it after its formation (including the perinatal period)[25,31]	Leptomeningeal glial and glioneuronal heterotopias

Period of Insult	19. Hydranencephaly Group Abnormality Produced
15–16 weeks' gestation[9]	Hydranencephaly, NOS
Between 4th and 6th months of gestation[17]	Hydranencephaly, NOS
Before 5th fetal month[11]	Hydranencephaly with agenesis of lateral corticospinal tracts
5th–6th fetal month[11]	Hydranencephaly with relatively small lateral corticospinal tracts
Approximately gestational weeks 22–27[11]	Hydranencephaly with polymicrogyria
Can develop as early as 20–27 weeks' gestation[41]	Hydranencephaly, NOS
After 6th fetal month[11]	Hydranencephaly with normal cord configuration, but with corticospinal tract degeneration
1st week postnatal[42]	"Hydranencephaly"[c]

Period of Insult	20. Cerebro-Ocular Dysplasia Abnormality Produced
Prolonged disruptive process active during 2nd and 3rd trimesters[17]	Cerebro-ocular dysplasias

Table 5.2. Continued

21. Necrotic/Gliotic/Malacic Injury Pattern Group

Period of Insult	Abnormality Produced
After 16–22 weeks' gestation[9]	Laminar or total cortical plate necrosis
Perinatal[11]	Periventricular leukomalacia (Chap. 4, Fig. 4.24; Case 4.11, Fig. 4.25)
Approximately 24–40 weeks' gestation (mainly 24–32 weeks) but does occur in full-term infants[20]	Periventricular leukomalacia
Premature and term newborn[13]	Thalamic-brainstem injury
Perinatal period, including early infancy, up to 6–9 months of age[11,26,39]	Ulegyria (Chap. 4, Case 4.14, Figs. 4.39–4.43)
Between approximately 29 and 30 weeks' gestation to 2nd postnatal month (including some stillborns),[11,20,35] possibly as early as 22 weeks' gestation[11]	Pontosubicular necrosis (as the predominant lesion)
Approximately 32 weeks' gestational age[32]	Prenatal symmetric thalamic degeneration (symmetric thalamic degeneration in infancy)
"Perinatal," usually in term infants[20]	Parasagittal cerebral injury
Premature and term infants and first 4 postnatal months (major risk period is 28–35 weeks' gestation)[21]	Diffuse white matter gliosis (Chap. 4, Case 4.10, Figs. 4.22–4.23)
After 36 weeks' gestation[20]	Selective necrosis of Purkinje cells
Premature infants[41]	Selective necrosis of cerebellar granule cells
Postnatal period[34]	Selective necrosis of cerebellar granule cells
Early postpartum to 18 months of age[11]	Global hemispheric necrosis in infants
Perinatal[2,11,17,30,38]	Multicystic encephalopathy (i.e., multicystic encephalomalacia) (Chap. 4, Case 4.13, Figs. 4.28–4.38)
"Most cases" occur from insults between 30 and 44 weeks' gestation[37]	Multicystic encephalopathy
Approximately 30 weeks' gestation to first postnatal month[20,41]	Multicystic leukoencephalopathy

22. Status Marmaratus

Period of Insult	Abnormality Produced
Possibly as early as 5th month of gestation[34]	Status marmaratus (with other abnormalities) (Chap. 4, Figs. 4.44–4.47)
Approximately 35 weeks' gestation to 5th postnatal month[20]	Status marmaratus
Perinatal, and from insults up to 3 months postnatally[35]	Status marmaratus
Peripartum, and up to 9 months postpartum[2,11,20,35]	Status marmaratus
First apparent at about 6 months of age[37]	Status marmaratus

23. Fetal Ventriculomegaly

Period of Insult	Abnormality Produced
May be seen by 17–20 weeks' gestation[37]	Fetal ventriculomegaly (may be associated with other CNS or systemic malformations)

[a]Sequence by approximate gestational age of insult.
[b]Defined here as smooth-walled cystic communication between subarachnoid space and lateral ventricle.
[c]This case probably corresponds more appropriately to Rorke's[34] "encephaloclastic" lesions (rather than true hydranencephaly) occurring in the perinatal period, and to Friede's[11] "global hemispheric necrosis" in infants, characterized by some residual pattern of cortical convolutions, which may be seen due to insults occurring from the first few days postpartum up to 18 months of age.
*NOS—not otherwise specified.

> **Case 5.2.** (Figs. 5.5 and 5.6)
>
> A 12-year-old boy with a diagnosis of cerebral palsy and an estimated functional age of a 3-month-old was cared for at home. He was found unresponsive and was pronounced dead on arrival at the hospital.

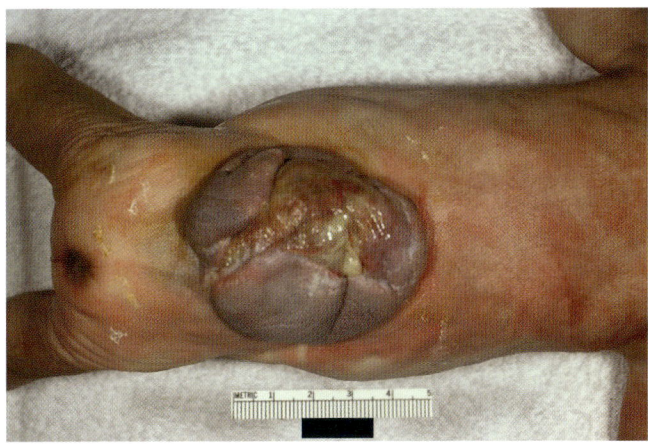

Fig. 5.7. Meningomyelocele, external view. Ulceration and infection, seen here, are potential complications.

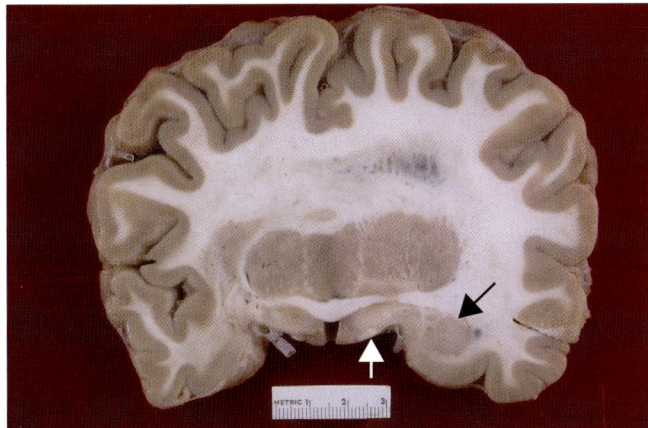

Fig. 5.5. Coronal section of frontal lobes shows midline fusion of basal ganglia overlying the anterior commissure. A shallow interhemispheric fissure is present, but cortex crosses the midline in place of corpus callosum. A sylvian fissure is present on the left, demarcating temporal lobe. A rudimentary amygdala (*black arrow*) and optic tract (*white arrow*) are present on the right.

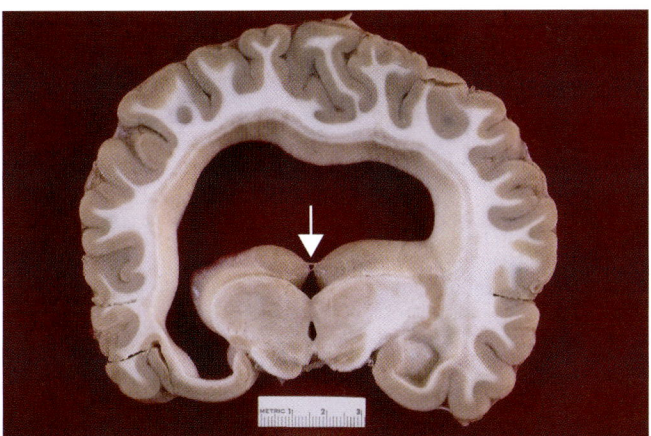

Fig. 5.6. Coronal section at post-thalamic level shows a bit of separated thalamus (*arrow*). Cerebral hemispheres are fused dorsally. A massive common ventricle is confluent with temporal horn. Hippocampi are rudimentary. Note the thin cerebral peduncles, especially medially (frontopontine tracts).

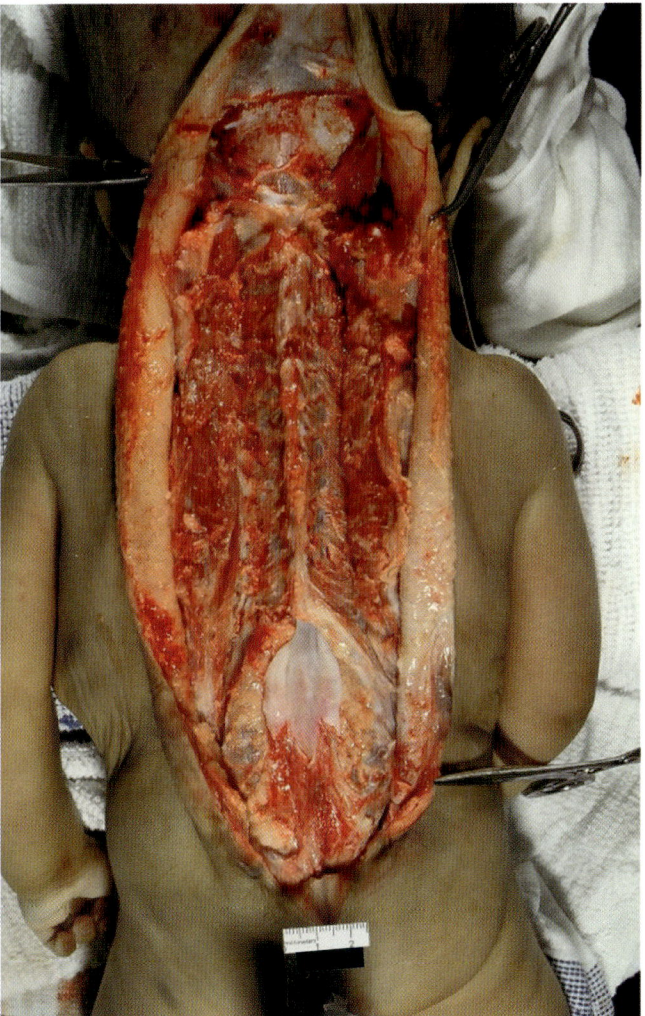

Fig. 5.8. Another case of meningomyelocele, with wide lumbosacral spinal defect exposed.

160 Malformations and Other Congenital CNS Lesions

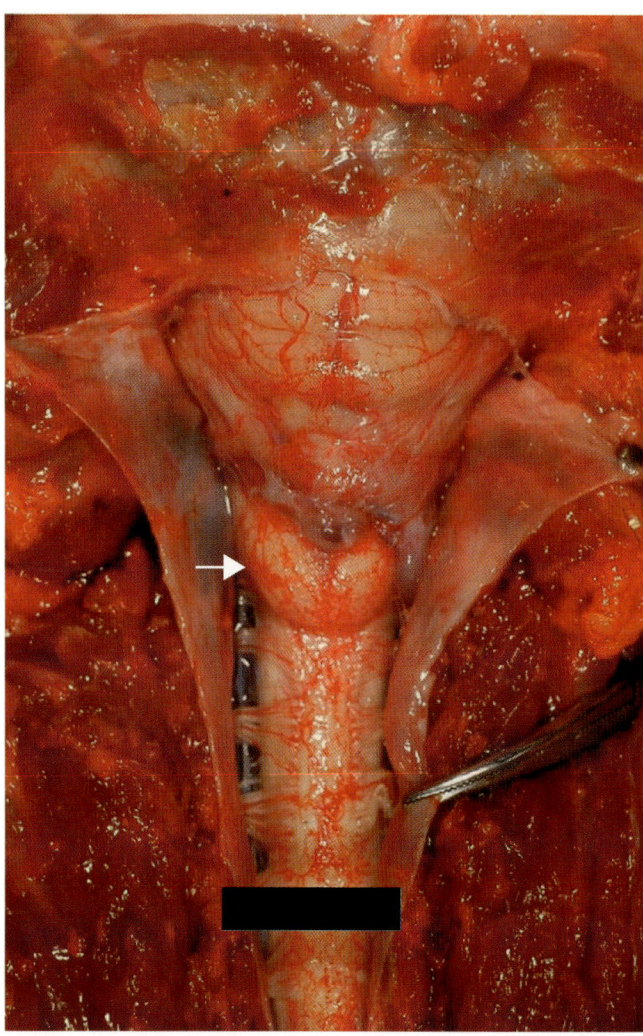

Fig. 5.9. Same case as Fig. 5.10, showing a common associated anomaly in meningomyelocele cases: Chiari type II malformation with prominent medullary knob (*arrow*).

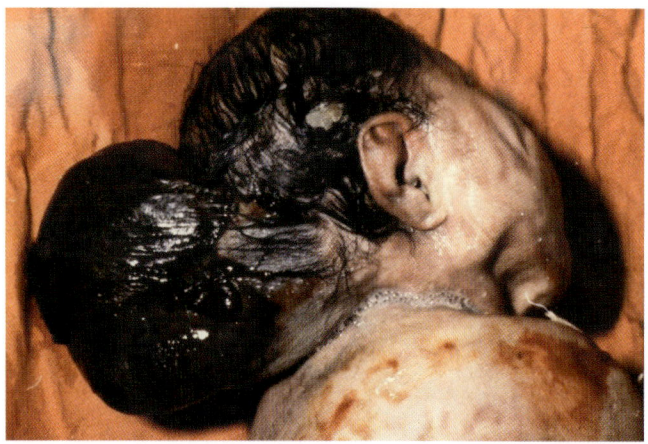

Fig. 5.10. Occipital encephalocele.

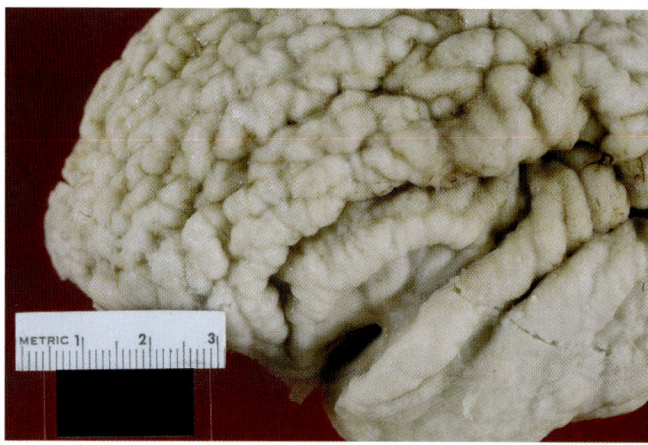

Fig. 5.11. Lateral view of frontotemporal area of brain with extensive polymicrogyria, sometimes referred to as a "corn cob" brain. Temporal area has a paucity of deeper sulci.

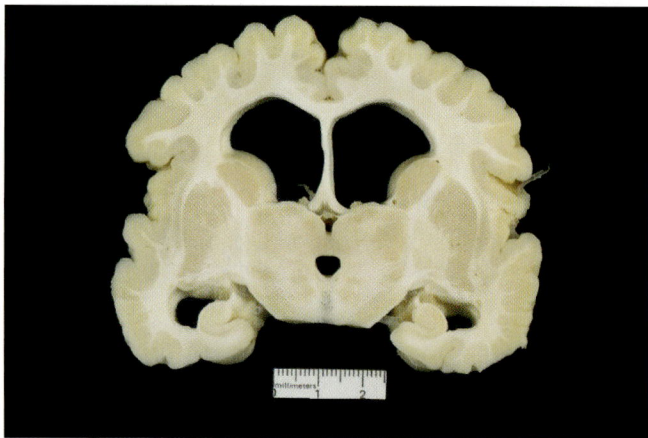

Fig. 5.12. Coronal section of brain seen in Fig. 5.11. Variable cortical thickness and surface irregularity and enlarged ventricles are the most conspicuous abnormalities.

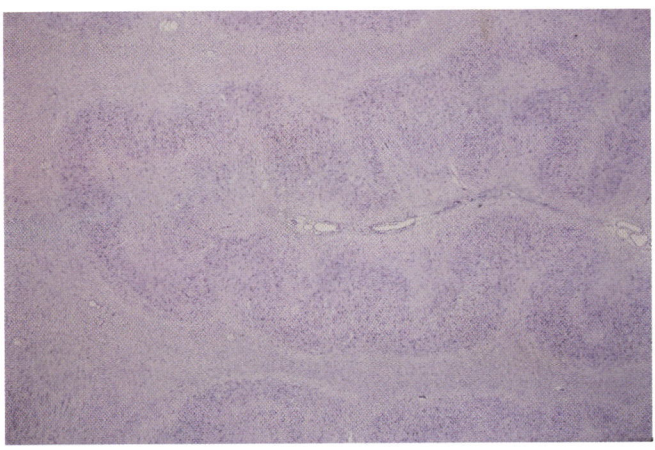

Fig. 5.13. Low-power micrograph of cortex of brain seen in Fig. 5.11. Plications of cortex are seen within what appears to be a single gyrus. (H&E.)

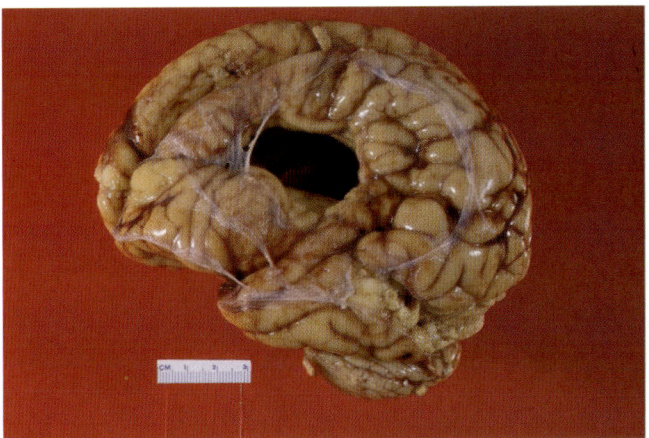

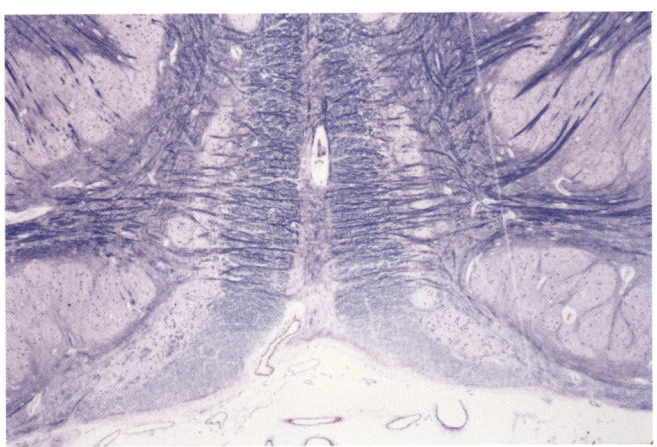

Fig. 5.14. Transverse section of ventral medulla at level of inferior olivary nucleus in case seen in Fig. 5.11 shows virtual absence of pyramids. (Luxol fast blue/H&E.)

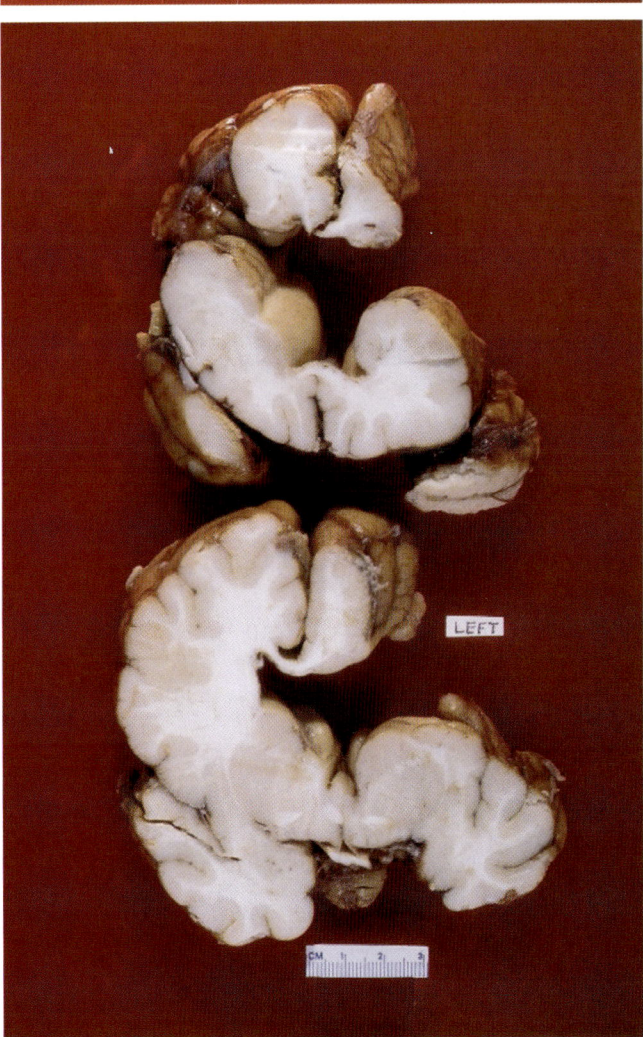

Figs. 5.15–5.16. *Top* (Fig. 5.15; lateral view) and *bottom* (Fig. 5.16; coronal view) show bilateral congenital porencephaly in a common location, the vicinity of the sylvian fissure. Dome of cyst was ruptured with brain removal.

162 Malformations and Other Congenital CNS Lesions

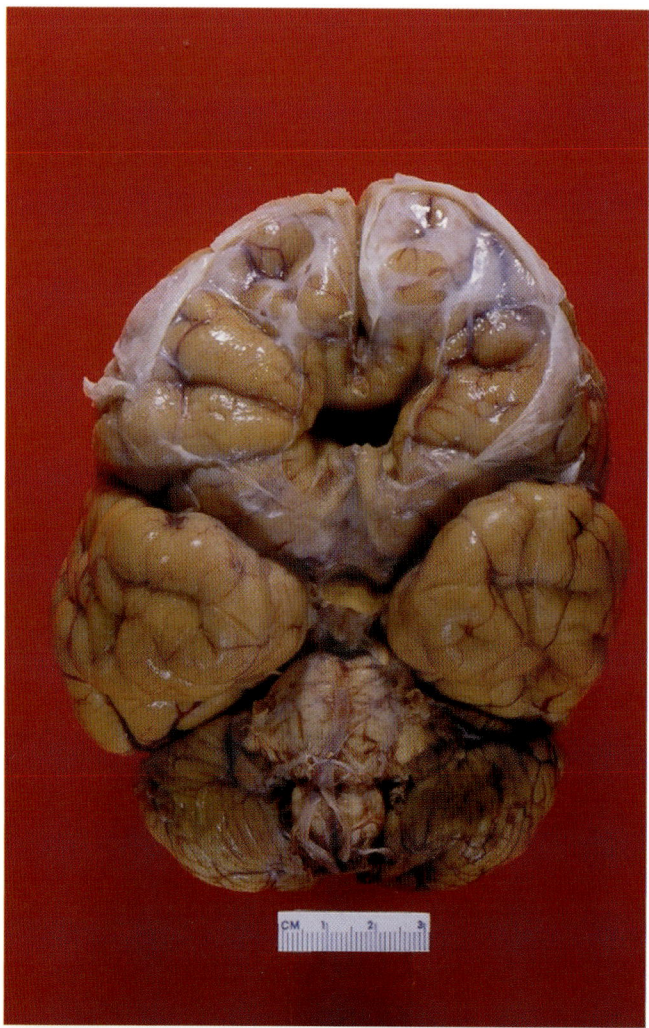

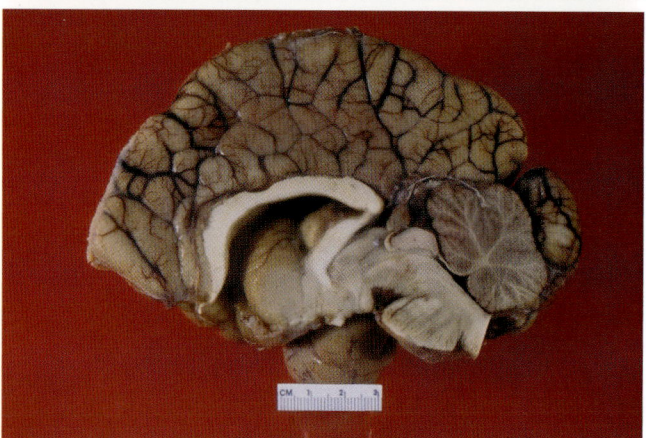

Figs. 5.17–5.18. *Top* (Fig. 5.17; basal view) and *bottom* (Fig. 5.18; sagittal view) show congenital porencephaly in an uncommon location, the orbitofrontal cortex. Dome of cyst ruptured at brain removal, but white margins of cyst are evident.

Case 5.3. (Fig. 5.19)

This 37-year-old man was 7 years old when struck by a car as he was walking across the street. Neurosurgery was performed to elevate a depressed fracture of the left frontal area, and neurologic sequelae included need for inpatient care for the next 1½ years, and subsequently had outpatient care for assistance with activities of daily living, severe memory impairment, mental capacity resembling that of a "10- to 11-year-old," mild right hemiparesis, and seizure disorder.

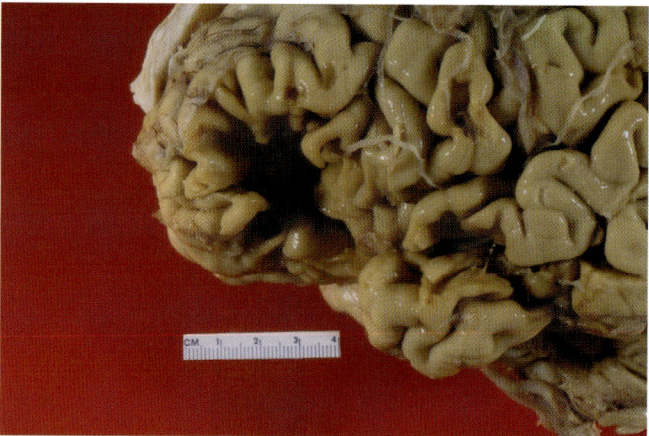

Fig. 5.19. Lateral view of left frontotemporal region demonstrates a large cystic/malacic post-traumatic defect with radial orientation of some bordering gyri. Cyst margin gliosis and chronic sequelae of diffuse axonal injury were also present. No cortical dysplasia or congenital anomalies were present.

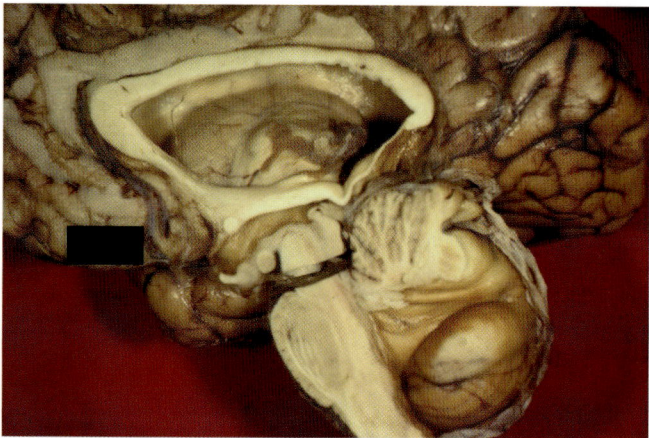

Fig. 5.20. Midsagittal section of Dandy-Walker malformation. Hydrocephalus, including enlarged fourth ventricle, and hypoplastic cerebellar vermis are visible in the photo. Findings also included posterior fossa enlargement with upward displacement of tentorium and transverse sinus.

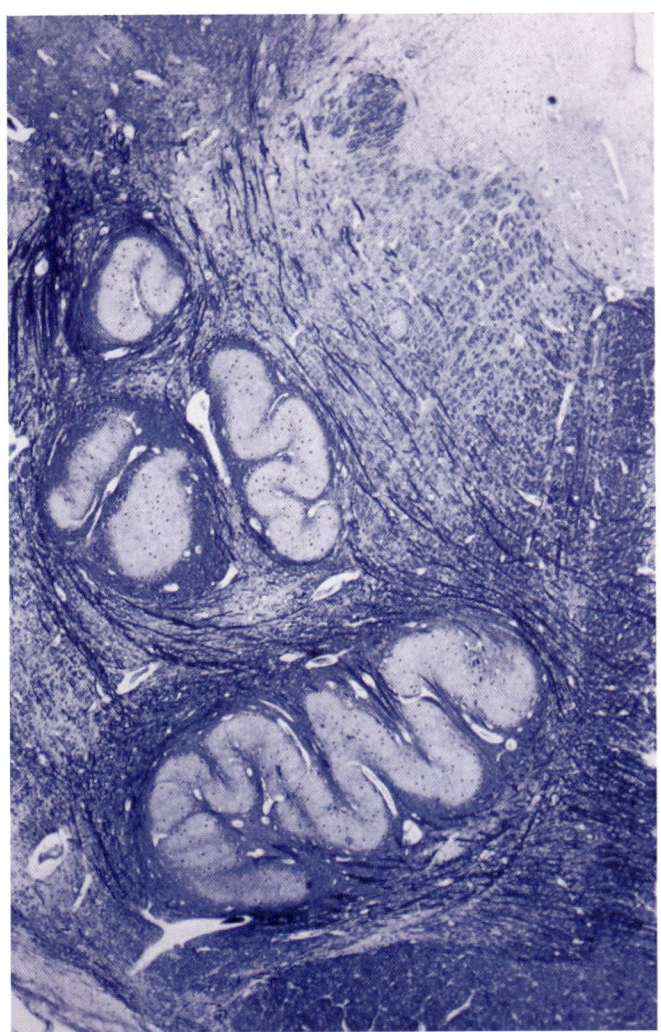

Fig. 5.21. Dysplastic inferior olivary nucleus, with abnormal heterotopic islands of olivary nucleus scattered along their developmental migration route from the rhombic lip to the eventual location of the normal inferior olivary nucleus. (Luxol fast blue/cresyl violet.)

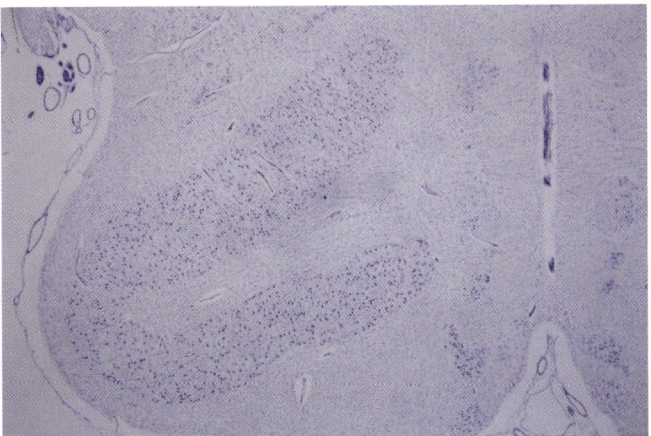

Fig. 5.22. Horseshoe-shaped inferior olivary nucleus with marked simplification of plicae. This degree of dysplasia is often associated with other significant CNS malformations. (Nissl.)

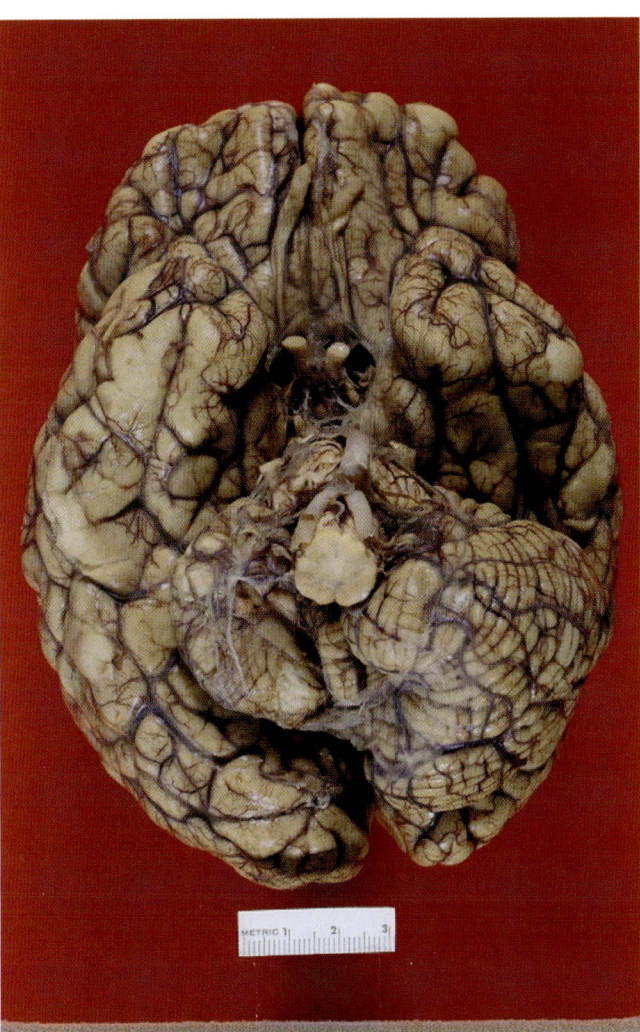

Fig. 5.23. Basal view of right cerebellar hemisphere hypoplasia with grossly normal left cerebellar hemisphere.[1,23]

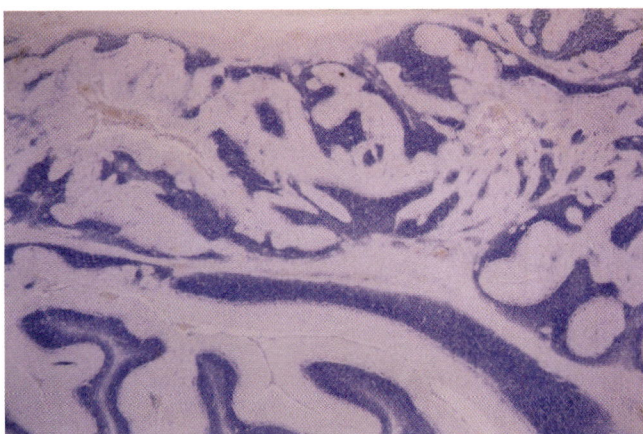

Fig. 5.24. Dysplastic cerebellar cortex in hypoplastic right cerebellar hemisphere from case seen in Fig. 5.23. (Luxol fast blue/cresyl violet.)

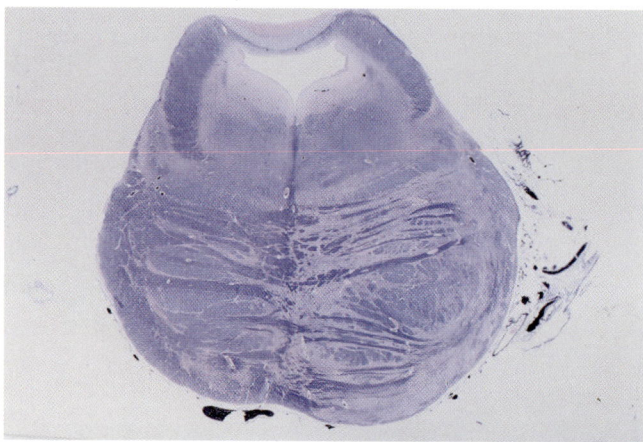

Fig. 5.25. Transverse section of pons from case seen in Fig. 5.23, demonstrating contralateral basis pontis hypoplasia due to reduced pontine nuclei and pontocerebellar fibers. (Luxol fast blue/cresyl violet.)

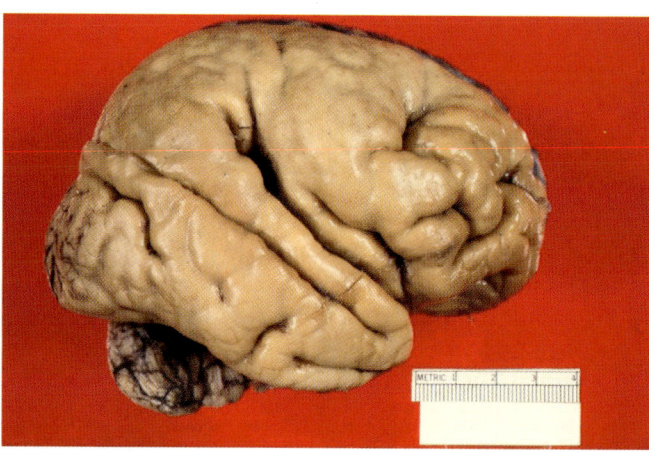

Fig. 5.28. Lissencephaly. Lateral view of right hemisphere, with marked simplification of convolutional pattern resulting in large zones of "smooth brain" appearance.

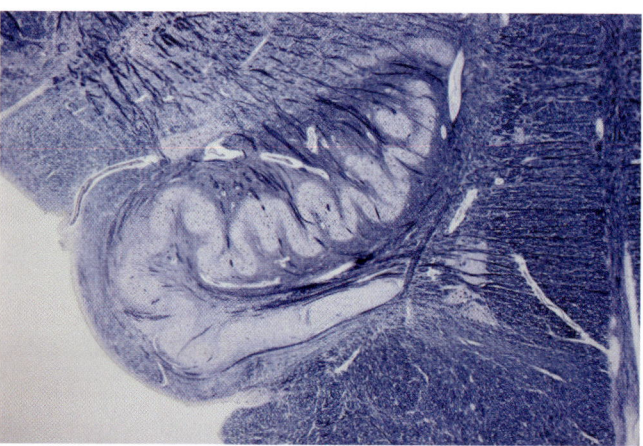

Fig. 5.26. Dysplastic left inferior olivary nucleus from case seen in Fig. 5.23, a common finding contralateral to the unilateral cerebellar hypoplasia anomaly.

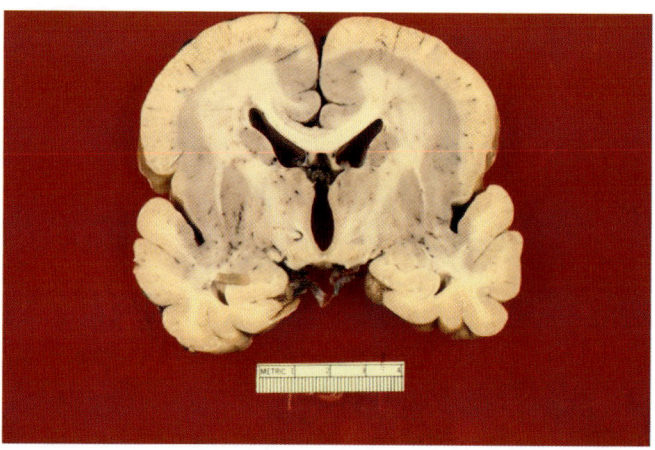

Fig. 5.29. Lissencephaly. Coronal section of brain seen in Fig. 5.28, at level of lenticular nucleus, shows nonconvoluted dorsolateral hemispheres.

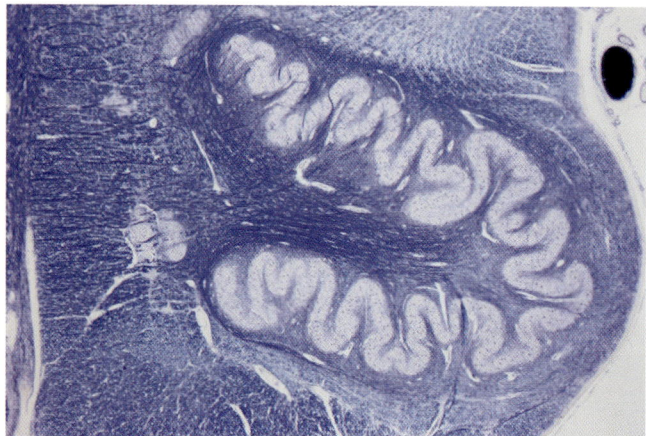

Fig. 5.27. Normal right inferior olivary nucleus contralateral to the normal left cerebellar hemisphere seen in Fig. 5.23.

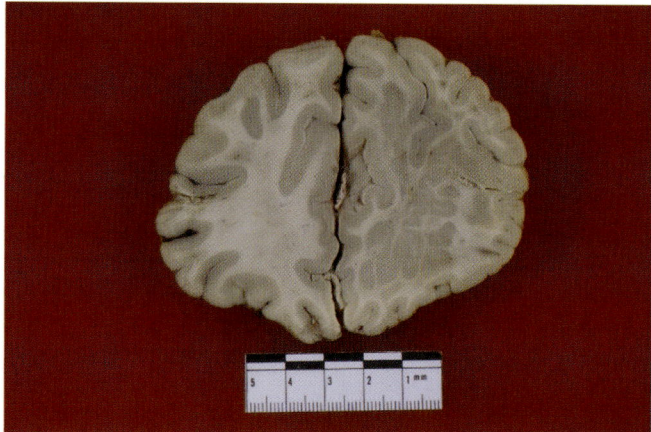

Fig. 5.30. Coronal section of pregenu cerebral hemispheres. White matter of right frontal lobe is largely occupied by gray matter from nodular neuronal heterotopias.

Table 5.2 sorts the heterogeneous conditions we have elected to include under the general heading of developmental abnormalities by advancing gestational age, starting with earlier stages of development and progressing toward later stages of development. It is immediately evident that many types of malformations and other abnormalities found in infant brains examined at this office are not listed in this table, due to the pertinent timetable relationships not having been suggested in the English-language literature, or due to our inability to locate the information. Further data from readers concerning such correlations would be welcomed by us. Doubtless our grouping of abnormalities will not be considered satisfactory to all readers, which in part reflects the lack of consensus regarding the best classification system for this heterogeneous and complex group of disorders.[3,4] Advances in molecular biology, in particular, are likely to clarify many current controversies on origin and classification of CNS malformations.[33]

References

1. Arcudi G, DiCorato A, D'Agostino G, Marella GL. Cerebellar agenesis in a suicide. Am J Forensic Med Pathol 2000;21:83–85.
2. Armstrong DD. Neonatal encephalopathies. In Duckett S (ed): Pediatric Neuropathology. Baltimore: Williams & Wilkins, 1995, pp 334–351.
3. Barkovich AJ, Kuzniecky RI, Jackson GD, et al. Classification system for malformations of cortical development: Update 2001. Neurology 2001;57:2168–2178.
4. Barkovich AJ, Kuzniecky RI, Jackson GD, et al. A developmental and genetic classification for malformations of cortical development. Neurology 2005;65:1873–1887.
5. Bird SJ. Scientific certainty: Research versus forensic perspectives. J Forensic Sci 2001;46:978–981.
6. Costa C, Hauw J-J. Pathology of the cerebellum, brain stem and spinal cord. In Duckett S (ed): Pediatric Neuropathology. Baltimore: Williams & Wilkins, 1995, pp 217–238.
7. Dekaban A. Neurology of Early Childhood. Baltimore: Williams & Wilkins, 1970.
8. Dorland's Illustrated Medical Dictionary, ed 27. Philadelphia: WB Saunders, 1988.
9. Encha-Razavi F. Antenatal disruptive lesions. In Golden JA, Harding BN (eds): Pathology and Genetics: Developmental Neuropathology. Basel: ISN Neuropath Press 2004, pp 144–147.
10. Fackler M. Editorial comment. Wound Ballistics Rev 2001;5(1):6–7.
11. Friede RL. Developmental Neuropathology, ed 2. Berlin: Springer-Verlag, 1989.
12. Gilbert-Barness E, Debich-Spicer DE. Handbook of Pediatric Autopsy Pathology. Totowa, NJ: Humana Press, 2005.
13. Gilles FH. Perinatal neuropathology. In Davis RL, Robertson DM (eds): Textbook of Neuropathology, ed 3. Baltimore: Williams & Wilkins, 1997, pp 331–406.
14. Golden JA. Holoprosencephaly: A defect in brain patterning. J Neuropathol Exp Neurol 1998;57:991–999.
15. Golden JA, Harding BN (eds) Pathology and Genetics. Developmental Neuropathology. Basel: ISN Neuropath Press, 2004.
16. Harding BN. Brainstem malformations. In Golden JA, Harding BN (eds): Pathology and Genetics: Developmental Neuropathology. Basel: ISN Neuropath Press, 2004, pp 105–108.
17. Harding B, Copp AJ. Malformations. In Graham DI, Lantos PL (eds): Greenfield's Neuropathology, ed 6. London: Arnold, 1997;1:397–533.
18. Harding BN, Pilz DT. Polymicrogyria. In Golden JA, Harding BN (eds): Pathology and Genetics: Developmental Neuropathology. Basel: ISN Neuropath Press, 2004, pp 49–51.
19. Jones KL. Smith's Recognizable Patterns of Human Malformation, ed 5. Philadelphia: WB Saunders, 1997.
20. Kinney HC, Armstrong DD. Perinatal neuropathology. In Graham DI, Lantos PL (eds): Greenfield's Neuropathology, ed 7. London: Arnold, 2002, pp 519–606.
21. Kinney HC, Haynes RL, Folkerth RD. White matter lesions in the perinatal period. In Golden JA, Harding BN (eds): Pathology and Genetics: Developmental Neuropathology. Basel: ISN Neuropath Press, 2004, pp 156–170.
22. Larroche J-C. Developmental Pathology of the Neonate. Amsterdam: Excerpta Medica, 1977.
23. Leestma JE, Torres JV. Unappreciated agenesis of cerebellum in an adult: Case report of a 38-year-old man. Am J Forensic Med Pathol 2000;21:155–161.
24. Lemire RJ, Loeser JD, Leech RW, Alvord EC Jr. Normal and Abnormal Development of the Human Nervous System. Hagerstown, MD: Harper and Row, 1975.
25. Marin-Padilla M. Developmental neuropathology and impact of perinatal brain damage. I. Hemorrhagic lesions of neocortex. J Neuropathol Exp Neurol 1996;55:758–773.
26. Marin-Padilla M. Developmental neuropathology and impact of perinatal brain damage. III. Gray matter lesions of neocortex. J Neuropathol Exp Neurol 1999;58:407–429.
27. Menkes JH, Sarnat HB. Malformations of the central nervous system. In Menkes JH, Sarnat HB (eds): Child Neurology, ed 6. Philadelphia: Lippincott Williams & Wilkins, 2000, pp 305–400.
28. Ming JE, Golden JA. Midline patterning defects. In Golden JA, Harding BN (eds): Pathology and Genetics: Developmental Neuropathology. Basel: ISN Neuropath Press, 2004, pp 14–25.
29. Norman MG. Malformations of the brain. J Neuropathol Exp Neurol 1996;55:133–143.
30. Norman MG, Ludwin SK. Congenital malformations of the nervous system. In Davis RL, Robertson DM (eds): Textbook of Neuropathology, ed 3. Baltimore: Williams & Wilkins, 1997, pp 265–329.
31. Norman MG, McGillivray BC, Kalousek DK, et al. Congenital Malformations of the Brain: Pathologic, Embryologic, Clinical, Radiologic and Genetic Aspects. New York: Oxford University Press, 1995.
32. Parisi JE, Collins GH, Kim RC, Crosley CJ. Prenatal symmetrical thalamic degeneration with flexion spasticity at birth. Ann Neurol 1983;13:94–97.
33. Pilz D, Stoodley N, Golden JA. Neuronal migration, cerebral cortical development, and cerebral cortical anomalies. J Neuropathol Exp Neurol 2002;61:1–11.
34. Rorke LB. Pathology of Perinatal Brain Injury. New York: Raven Press, 1982.
35. Sarnat HB. Perinatal hypoxic/ischemic encephalopathy: Neuropathological features. In Garcia JH (ed): Neuropathology: The Diagnostic Approach. St Louis: Mosby, 1997, pp 541–580.
36. Squier W. Brain development: Normal and abnormal. In Squier W (ed): Acquired Damage to the Developing Brain: Timing and Causation. London: Arnold, 2002, pp 101–109.
37. Squier W. Pathology of fetal and neonatal brain damage: Identifying the timing. In Squier W (ed): Acquired Damage to the Developing Brain: Timing and Causation. London: Arnold, 2002, pp 110–127.
38. Towbin A. Cerebral hypoxic damage in fetus and newborn: Basic patterns and their clinical significance. Arch Neurol 1969;20:35–43.
39. Vinters HV: Vascular diseases. In Duckett S (ed): Pediatric Neuropathology. Baltimore: Williams & Wilkins, 1995, pp 302–333.
40. Vinters HV, Farrell MA, Mischel PS, Anders KH. Diagnostic Neuropathology. New York: Marcel Dekker, 1998.
41. Volpe JJ. Neurology of the Newborn, ed 3. Philadelphia: WB Saunders, 1995.
42. Weiss MH, Young HF, McFarland DE. Hydranencephaly of postnatal origin: Case report. J Neurosurg 1970;32:715–720.

Blunt Force Head Injury

6

INTRODUCTION   167
SCALP ABRASION, CONTUSION, AND
   LACERATION   167
SUBCUTANEOUS SCALP AND SUBGALEAL
   HEMORRHAGES   168
SKULL FRACTURES   168
EPIDURAL HEMATOMA   169
SUBDURAL HEMATOMAS   170
SUBARACHNOID HEMORRHAGE   174
BRAIN CONTUSIONS, LACERATIONS, AND
   HEMATOMAS   177

DIFFUSE AXONAL INJURY   183
BRAIN STEM AVULSION   190
VENTRICULAR HEMORRHAGE   191
TRAUMATIC CRANIAL NEUROPATHIES   191
LOCKED-IN SYNDROME   192
BOXING AND OTHER SPORTS   192
CONSEQUENCES OF HEAD INJURY   195
REFERENCES   197

Introduction

Traumatic brain injury (TBI) has been called the "silent epidemic." Conservative estimates report 2 million injured annually in the United States, a half million with injury serious enough for hospitalization.[19] Mortality from TBI is said to comprise 1 to 2% of deaths from all causes. One-third to one-half of all traumatic deaths are due to head injury.[20] Of those who survive, the majority are left with important disabilities, including 3% in a vegetative state, with only about 30% making a good recovery.[20] The importance of TBI as a public health problem in this country cannot be overstated, and the medical examiner's office plays a major role in identifying the specific brain injuries resulting from each circumstance.

Head injuries can be broadly divided into blunt force and penetrating, or closed and open injuries, simple in concept but infinitely variable in individual cases. The types of injuries are limited in number, but extremely complex in each case. Types of head injuries are listed as follows and reviewed in turn with illustrated case examples.

Scalp abrasion and laceration
Subcutaneous hemorrhage
Subgaleal hemorrhage
Skull fracture
Epidural hematoma
Subdural hematoma
Diffuse axonal injury
Brain avulsion
Ventricular hemorrhage
Traumatic cranial
   neuropathies
Locked-in syndrome
Subarachnoid hemorrhage
Cerebral cortical contusions
   and lacerations
Boxing and other sports
Secondary consequences
   of head injury

Scalp Abrasion, Contusion, and Laceration

The scalp provides an ideal anatomic arrangement for development of abrasions, contusions, and lacerations due to scalp soft tissue overlying the hard, rounded bony prominences of the cranium. The hair, depending on its thickness, provides a certain amount of protection that may also mask detection of scalp abrasions and contusions. Abrasions resulting from occipital and parietal scalp impacts may be missed without close inspection and careful shaving of the areas when trauma to the head is suspected. Determination of impact point(s) may become important to bear out witness statements and/or to correlate with possible underlying brain injuries. Locations of abrasion sites should be documented by photographs, diagrams, and measurements to the center of the abrasion from landmarks such as the inion, ear, midline, vertex, and orbital ridge. Histologic evaluation of an abrasion is important for dating the injury and correlating it with brain injury dating. Abrasions from frontal impacts are usually in plain view and are not overlooked. Scalp lacerations, because of their usually profuse and sometimes fatal bleeding, are rarely missed. The form of the laceration, its size, and its special shape and wound deposit, if any, give clues to the inflicting instrument.

Subcutaneous Scalp and Subgaleal Hemorrhages

Traumatic subcutaneous hemorrhage of the scalp may be broadly localizing, but when it underlies an abrasion, contusion, or laceration, it would be an indicator of the impact point. The volume of hemorrhage may be considerable and impressive but not necessarily life-threatening. Estimated external blood loss at the scene of the incident is also a factor to consider. The galea aponeurotica, or epicranial aponeurosis, is a sheet of dense fibrous connective tissue covering the upper cranium which forms attachments to the frontalis and occipitalis muscles. Laterally, it is continuous with the temporal fascia. Subgaleal hemorrhage may be a marker of severe head trauma, or an imprecise locator of impact site. Beneath the galea is the periosteum, which is tightly adherent to the cranium. Periosteal and subperiosteal hemorrhages may accompany skull fractures. This combination is more often seen in skull fractures in infants and very young children.

Skull Fractures

The biomechanics of the mechanisms at play in skull fractures are relevant in assessing types and directions of forces occurring in any injury. However, outside of experimental conditions, rarely can the exact forces involved in the TBI be determined. The science is complicated, and the reader is referred to several references.[12,13,21–23] As might be intuitively evident, a loose correlation exists between the degree of a severity of a skull fracture and the underlying TBI. For a detailed discussion of fractures likely to occur with forces applied to different points of the skull, the reader is referred to the brief but clear exposition of the subject in DiMaio and DiMaio.[12] Some basic information should be emphasized. The frontal bone in the young adult male has the greatest thickness, about 7.70 ± 1.82 mm; the parietal bone is about 6.57 ± 1.31 mm in thickness,[67] and the temporal squama is about 4 mm thick.[42] However, as noted by Knight,[42] the thickness of the skull, unless very unusual, has little relevance in the court setting, because lethal injuries can occur without a skull fracture. It has been stated that a skull covered with soft tissue has a threshold to fracture of 1100 lb/in^2 in the frontal lobe, 550 lb/in^2 in the parietal lobe, and 225 lb/in^2 at the zygoma.[16]

Skull fractures can be broadly classified as linear or depressed. Their presence is made known by routine x-rays and other imaging studies if the victim survives long enough for hospitalization. Fractures can be confirmed by visual inspection at autopsy after stripping the periosteum and dura mater. It is not unusual to receive a clinical report of a skull fracture not confirmed at autopsy and, conversely, to discover a fracture not found by x-ray or computed tomography (CT) scan. Nondepressed fractures of the cranial vault are linear or slightly curving and vary in length, occurring within frontal, parietal, temporal, and occipital plates, but often crossing into adjacent bones (via fused cranial sutures) and extending into the floor of the temporal fossa, along or across the petrous ridge of the temporal bone, and toward or into the foramen magnum. Linear fractures occur with a head impact of sufficient velocity on a hard, unyielding flat surface that is large enough in area to avoid penetration of the cranium. By itself, no special importance can be ascribed to a single linear fracture other than the fact that the cranium had been subjected to trauma. The trauma may be as mild as a nonaccelerated ground-level fall, for example, without neurologic symptoms. On the other hand, it may be an indicator of serious underlying inflicted TBI. Bilateral linear skull fractures and fractures involving extracranial bones are highly suspicious for inflicted trauma. A linear fracture interrupted at another fracture line is second in the sequence of occurrence (Puppe's rule).[42] A linear fracture of the parietal bone and temporal squama is a potential cause of laceration of the middle meningeal arteries and a source for epidural hematoma (see later).

Skull fractures of the base of the skull are usually an indirect effect of impact on the cranial vault and are usually accompanied by a calvarial (vault) fracture. The skull base can be defined as those bony structures below a plane drawn from the glabella to the inion. Bones of the skull base are relatively fragile and more prone to fracture than those of the vault. Fractures of the frontal vault may extend toward the skull base into the orbital roof and the frontal sinus, the latter creating a potential pathway for intracranial infections. Periorbital ecchymosis or "raccoon eyes" should alert the autopsy physician to the possible presence of an orbital fracture. Fractures extending medially across the thin cribriform plate of the ethmoid may result in cerebrospinal fluid (CSF) rhinorrhea, creating a potential pathway for meningitis. Vault fractures commonly extend into the temporal fossa where fracture of the petrous ridge is said to occur in 75% of all basal fractures. Petrous ridge fracture running longitudinally along the external auditory meatus may result in CSF leakage mixed with blood and seen as blood coming from the ear canal or as postauricular ecchymosis (Battle's sign).

Hinge fracture of the base of the skull results from severe lateral and vertex blows of high energy, such as in traffic accidents and falls from great heights. The fracture crosses the temporal fossae from side to side along or crossing the petrous ridge and through the pituitary fossa, often causing transection of the pituitary stalk and diabetes insipidus. Immediately overlying a hinge fracture is the hypothalamus, which is commonly contused, potentially causing impairment of life-sustaining autonomic functions.

A ring fracture at the base of the skull is descriptive of a circular fracture around the foramen magnum. It is said to occur most often from a fall from a great height onto the feet or buttocks, resulting in the cranial base around the foramen magnum collapsing down upon the spine. The vertex impact of a heavy weight may have a similar result. Such fractures, with their associated cord–brainstem injury, are lethal.

Diastatic fracture is to be differentiated from separation of an unfused cranial suture, which occurs most commonly in infants and in the young up through the mid-teens due to severe increased intracranial pressure (for further discussion, see Chap. 3). In adults, diastatic fracture most often accompanies an adjacent vault fracture. Persistent metopic (frontal midline), occipitomastoid, and parietomastoid sutures and other skull anomalies are not to be confused with fractures.

A fall from a moderate height with head impact, such as from the second story of a building, or a fall upon other than a flat surface, for example, upon a small object such as a rock on a hard surface, can be expected to produce a stellate, depressed skull fracture. Random fracture lines radiate from the point of impact. The inbending fractured inner table may lacerate the underlying dura and penetrate the cortex to varying degree. High-velocity falls onto a flat surface can result in a circular fracture at the margins of the inbending cranium. Blows from small objects on the skull, such as with a hammer, will typically produce a punched-out small circular fracture accompanied by a wide range of underlying injuries depending on the severity of the blow. Depending on the instrument and intensity of the homicidal attack, a wide variety of scalp and/or skull tool marks may be produced that can provide instrument class or even specific weapon identification.

Epidural Hematoma

Epidural hematoma is a collection of blood expanding the space between the cranium and dura mater, usually as a result of laceration of a meningeal artery from an overlying cranial fracture (Fig. 6.1). The cranial dura is a tough, densely collagenous membrane of variable thickness, but typically of about 1 mm in thickness, consisting of a fused double layer, an outer periosteal and

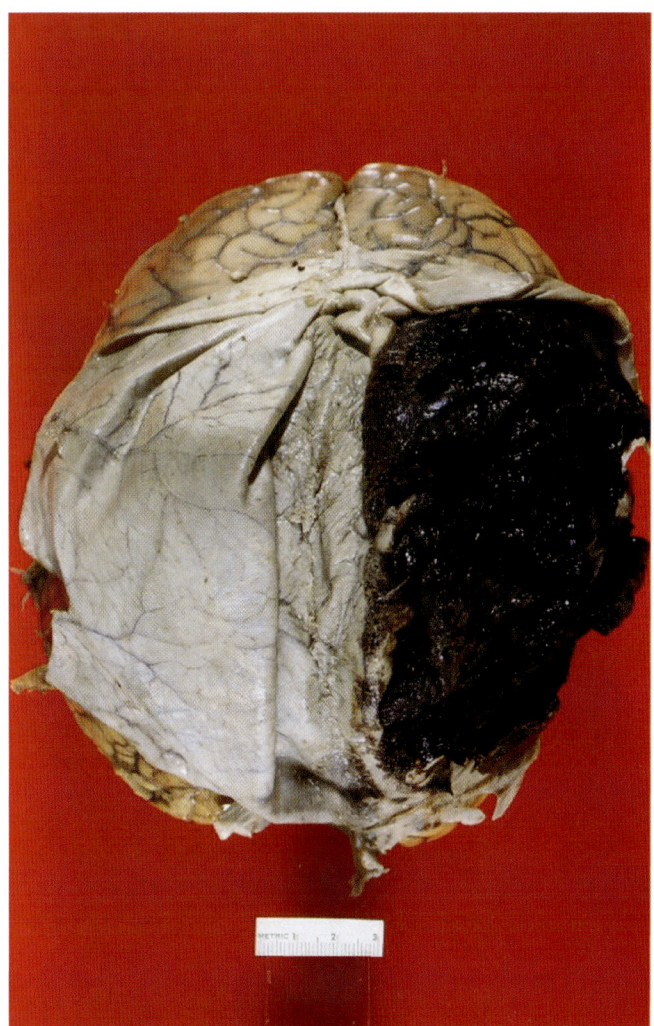

Fig. 6.1. An unusually large fresh epidural hematoma covers virtually the entire dorsal aspect of right hemisphere.

an inner meningeal layer. Meningeal arteries lie within the periosteal layer, and within a groove or even a tunnel of overlying skull bone, making them more prone to injury from cranial fracture. The outer dural layer forms the periosteum of the inner table of the cranium, and is tightly adherent to the cranium, especially in the developing cranium in infancy and very early childhood and the crania of many very elderly individuals. For this reason, epidural hematoma is less likely to form in either of these extremes of age group. Hendrick et al.[27] found epidural hematomas in only 40 of 4465 children hospitalized for head trauma. Another study found epidural hematomas of sufficient size to present a surgical problem in only 3% of head injuries. Figures range from 1 to 5%. Nevertheless, epidural hematomas are potentially lethal lesions in a high percentage of cases in which they do occur. They form focal lentiform collections of blood limited in spread by adherent dura (particularly at cranial suture boundaries) and, for that reason, behave as a rapidly developing focal hematoma under arterial pressure.

Although the source of hemorrhage in the majority of cases is a lacerated meningeal artery (Fig. 6.2), other causes include tears in a dural vein or dural sinus. Without immediate surgical intervention, the cascading secondary effects of cerebral edema secondary to mass effect lead to a shift of the cerebral hemisphere across midline, cingulate and uncal–parahippocampal herniations, midbrain compression, and finally Duret's hemorrhage in the midbrain and pons. Depending on the size of the hematoma, the underlying cortical surface will show variable smooth focal depression. In contrast to subdural hematoma (see later), which produces irregular focal cortical depressions and some maintenance of open sulci, cortical surfaces compressed by an epidural hematoma will show apical gyral flattening and effacement of sulci.

Because arterial laceration and bleeding are the usual source of epidural hemorrhage, the clinical course is typically short with a brief lucid interval between impact and onset of clinical symptoms. The interval varies due to several biologic variables, but it is said that clinically important hemorrhages may form in as short a time as one half-hour. Most become apparent in a few to several hours. This contrasts with a relatively longer lucid interval seen with subdural hematomas, which are usually of venous origin. Lucid intervals vary considerably in individual cases, and only a range is usually stated. "Coma within hours" is a common statement. An altered level of consciousness accompanies significant brain injury, and a true "lucid period" due to hematoma enlargement per se may thus be concealed.

At autopsy, on removal of the skull cap, photographic documentation of the usually clotted hematoma is essential, and volume should be measured. Saving the blood for toxicologic analysis may assist in determining if the person was under the influence of alcohol or drugs at the time of injury. Gross or histologic estimation of dating of epidural hematoma is not commonly an issue, and data on this topic are sparse (see Chap. 3).

Subdural Hematomas

Subdural hematomas (SDHs) are hemorrhages that, upon gross examination, appear to form between the dura and the arachnoid membrane (Fig. 6.3). Electron microscopic studies have shown a cell layer forming an

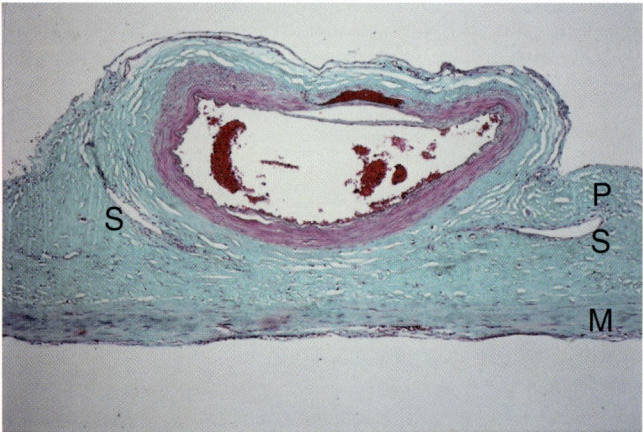

Fig. 6.2. Transverse section of meningeal artery in cranial dura mater. The vessel lies in (outer) periosteal layer (P), which is fused with inner meningeal layer (M). Paired small dural sinuses bracket the artery (S). Laceration of meningeal artery is usual source of epidural hematoma caused by overlying skull fracture. Such lacerations may also shunt high-pressure arterial blood into low-pressure venous sinuses to form "railroad tracking" beyond the point of injury, as demonstrated by contrast angiography (Gomori trichrome). Also see Fig. 1.13.

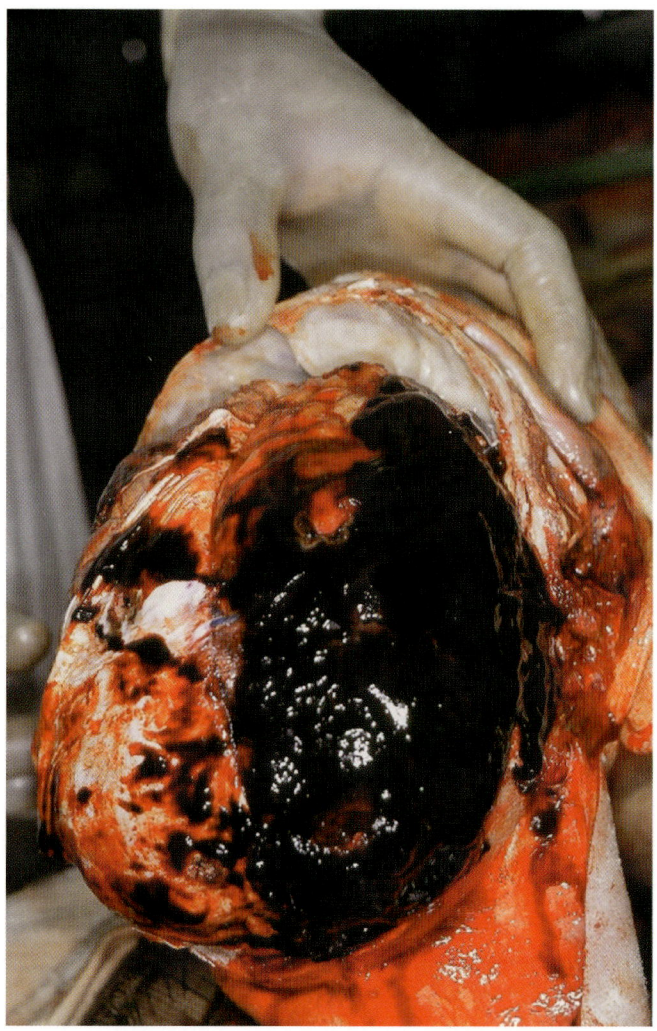

Fig. 6.3. Acute subdural hematoma over the right cerebral hemisphere. Dura mater over the hematoma has been reflected onto the left hemisphere to expose the hematoma.

"arachnoid barrier layer" loosely fused to a layer of "dural border cells" with no preexisting space at the interface. Therefore, a subdural hematoma is formed by insinuation of hemorrhage into the separated interface of these dural layers (i.e., anatomically the hematoma is actually intradural).[71] Support for this construct at a gross level can be found in those rare instances in which the brain had been carefully removed from the calvarium and the brain fixed with the dorsal hemispheric dura left in place undisturbed. In such cases, a minimal degree of adhesion can be experienced in pulling up the dura from the free edge away from the arachnoid surface. Unlike epidural hematomas, which are confined to the fractured skull bone by dural adhesion at cranial sutures, subdural hematomas spread widely over the cerebral hemisphere through the created space, filling and maintaining sulcal openings as well as forming shallow pools with depression of cortical surfaces. The CT scan image is that of layering of hemorrhage of variable thickness overlying the cerebral hemispheres. The typical appearance is one of relative maintenance of sulcal openings on the side of the hematoma, with the contralateral cortical surface ironed smooth by cerebral edema. The source of a subdural hematoma is most commonly a tear of so-called "bridging veins," which are superficial cortical veins draining toward the dural sinuses and perforating the arachnoid membrane before entering the sinuses (Fig. 6.4). For a short segment, draining veins lie free in the "virtual subdural space," leaving them vulnerable to shear forces as the brain shifts within the calvariun with impact. Electron microscopic study has shown the wall of this segment of vein to be very thin, as thin as $10\,\mu M$, whereas the subarachnoid portions of the vein measure 50 to $200\,\mu M$ in wall thickness.[78] Subdural hematomas are usually the result of acceleration–deceleration injuries with impact, although subdural hematoma in shaken baby syndrome is thought to occur without impact (see Chap. 7). Whiplash injury without impact is said to have caused subdural hematomas in adults.[59] Controversy continues to exist as to whether subdural hematoma can occur in infancy without impact. Clinically, onset of subdural hematoma symptoms may be rapid, with cerebral edema and brainstem compression and Duret's hemorrhage forming in as little as 30 minutes.[12] Typically, cited lucid interval from impact to onset of neurologic symptoms is given as several hours, but the interval range is so great that it is of little value and misleading to provide a specific interval. It may be days or longer, especially in the elderly, attributed in part to larger volumes of subarachnoid CSF in the elderly that could be displaced before brain compression ensues.

Accurate assessment of volume of acute subdural hematoma is very important information. Hematomas should be carefully collected in a catch basin at autopsy and measured in a volumetric flask. If firmly clotted, measurement can be made by displacement of a measured volume of water. A collection as small as 35 ml in adults may cause neurologic symptoms.[1] A hematoma measuring 50 ml is regarded as dangerous by some, while others tolerate as much as 100 ml before serious effects occur, probably depending on the volume of intracranial CSF of the individual (also see Chap. 3). Intracranial CSF volume varies considerably from individual to individual. Condon et al.[7] using magnetic resonance imaging (MRI) technique reported a wide range of total intracranial CSF volume of 57 to 286 ml in 64 volunteers between the ages 18 and 64 years. The findings included anticipated increasing volume with age. A larger volume of displaceable CSF in the elderly could accommodate a hematoma of large volume before symptoms.

Subdural hematomas are often classified as acute, subacute, and chronic but not always with agreement as to definition of time span of each type. Gross visual assessment of the appearance of a subdural hematoma at the time of surgery may be of value in evaluation of its age (see Chap. 3). Solid clotted blood would indicate acute SDH; mixed fluid and clotted blood, subacute SDH; and a completely fluid collection, chronic SDH. The color distinction between the dark red-black clots of acute SDH and the dark chocolate brown of subacute SDH is also useful. These generalized observations, however, do not always correlate well with those of neuroimaging (see the section on forensic neuroradiology in Chap. 17). Application of Gelfoam and similar hemostatic materials in the treatment of subdural hematoma may lead to the misleading impression of residual subdural hematoma (Case 6.1, Figs. 6.5–6.7). Microscopic examination of the collection will quickly resolve any question (Fig. 6.7). When reliable clinical information is available, the need for dating a subdural hematoma is obviated in the absence of prior trauma. The estimated volume of hema-

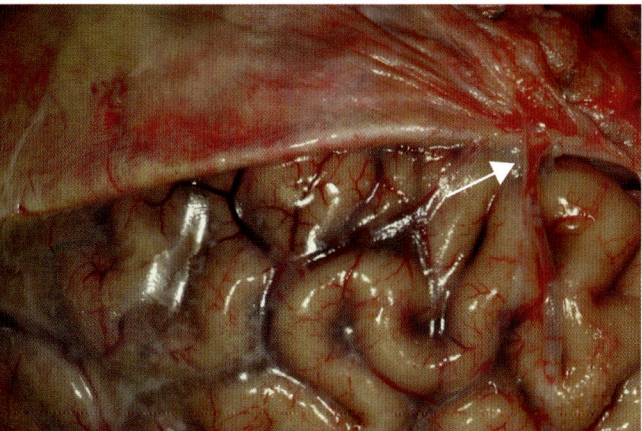

Fig. 6.4. Reflected dura mater exposes normal cortex with a bridging vein (*arrow*). Traction on such vessels is regarded as a common cause of rupture into the potential "subdural" space and formation of a subdural hematoma.

toma at surgery would be important for correlation with outcome, but it is rarely recorded. SDHs that become neurologically symptomatic after head impact within a few days are acute; those with neurologic symptoms occurring in 2 to 3 weeks are subacute; and those manifesting neurologic symptoms beyond 3 weeks are chronic, as these nonspecific terms are applied by the authors. It

Case 6.1. (Figs. 6.5–6.7)

A 79-year-old man fell and struck his head in his bathroom. After an unknown period, his daughter noted his unsteady gait, and he fell several more times before he saw a physician, who sent him home. He was admitted to the hospital 2 days later where CT scan revealed an occipital subdural hematoma that was evacuated on the same day. He died several weeks following surgery.

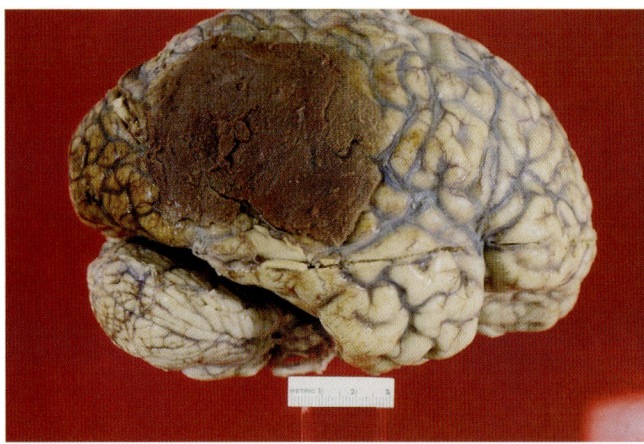

Fig. 6.5. What appears to be a large chocolate brown subacute subdural hematoma overlying right parieto-occipito-temporal lobe is a thin layer of Gelfoam suffused with blood.

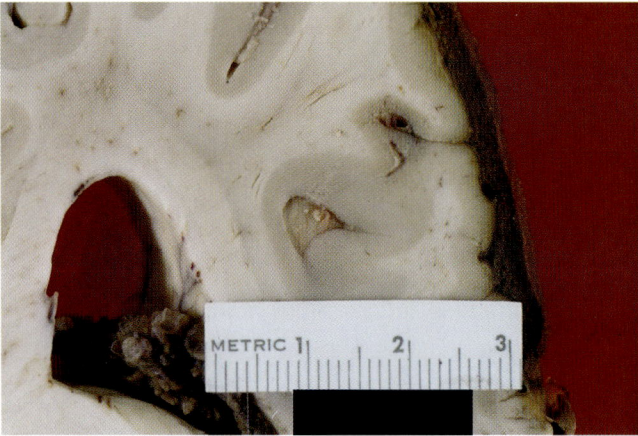

Fig. 6.6. Closeup view of coronal section at atrium of lateral ventricle demonstrates a layering of Gelfoam with a pale thinner outer layer and a thicker dark, blood-filled inner layer. Leptomeninges and underlying cortex are compressed.

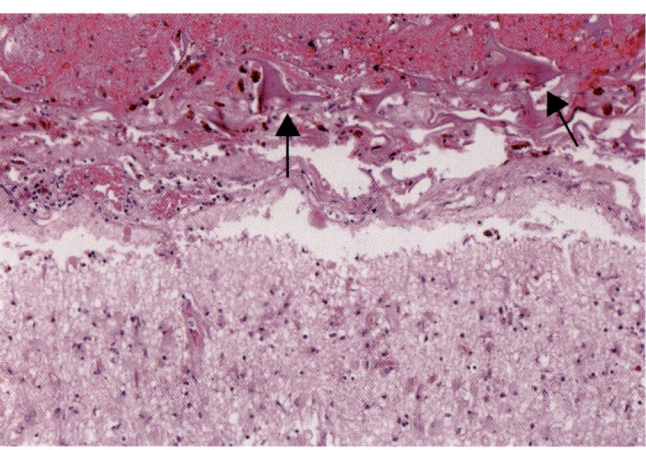

Fig. 6.7. Medium-power micrograph shows flattened leptomeninges between Gelfoam (*arrows*) packed with red blood cells in the interstices above and atrophic cortex below. Pia is artifactually separated away from cortex. (H&E.)

is then a matter of correlating history with autopsy findings for confirmation.

Histologic Dating of Subdural Hematoma

An early attempt at histologic dating of subdural hematoma by examination of the dura overlying the hematoma was that of Munro and Merritt,[56] and several subsequent publications have appeared (for a detailed overview of these sources, see Chap. 3).

Source of Neomembrane

Identifying the source of cells responsible for formation of a subdural neomembrane has been controversial, with the general assumption being activation of dural fibroblasts at the interface of hematoma with the dura as it appears on routine hematoxylin-eosin (H&E) preparations. However, as noted earlier, the electron microscopic study of Friede and Schachenmayr[14] demonstrated that neomembranes form by proliferation of dural border cells, creating a multilayered sheet with extracellular spaces containing collagen, elastic fibers, and amorphous material produced by the same border cells. Capillaries sprouting from the inner surface of the dura become integrated between proliferated border cells.

Chronic Subdural Hematoma

Chronic subdural hematomas are usually defined as those clinically manifested approximately 1 month after head trauma, and pathologically defined as liquefied hemorrhagic remnants within neomembranes. The liquid is often likened to "machinery oil." The fluid is straw-colored late, that is, months to years later. Discovery of hematoma may be by examination after persistent symptoms or sometimes serendipitously in a not-so-symptomatic patient. Not uncommonly, smaller chronic subdural hematomas are incidental findings at autopsy.

Usually, smaller subdural hemorrhages resolve as a single neomembrane (Case 6.2, Figs. 6.8–6.10). Hematomas are most often unilateral, but bilateral cases are not unusual (Case 6.3, Figs. 6.11–6.12). Autopsy discloses a thick, well-developed outer membrane sometimes equal to or exceeding the thickness of the overlying dura. The thinner inner membrane is more difficult to preserve in most cases but is easily demonstrated at the margins, and histologic sections taken at the border will clearly demonstrate inner and outer membranes (Case 6.4, Figs.

Case 6.2. (Figs. 6.8–6.10)

A 67-year-old man was found down in his residence after being drunk and falling. He had a laceration on his forehead and abrasions of the extremities. Examination revealed facial fractures and pneumocephalus. Initially lethargic and confused, his mental status further declined, and he died 5 weeks after his fall. No prior head trauma was known.

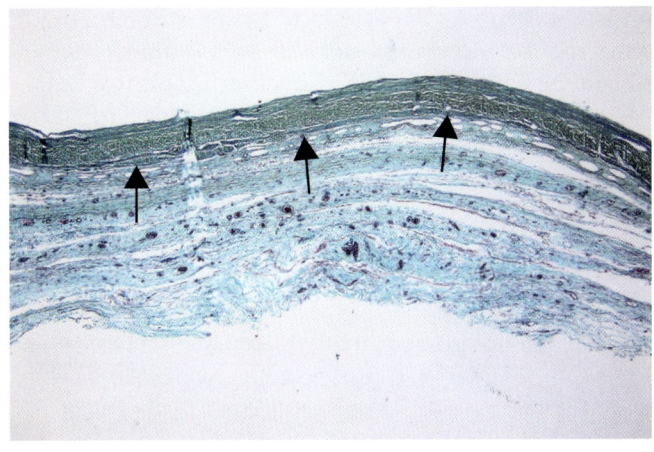

Fig. 6.9. Low-power micrograph demonstrates thick chronic subdural neomembrane several times the thickness of overlying dark green band (*arrows*), which is native dura. Neomembrane vascularized to its free edge makes its fusion with an inner membrane unlikely. (Gomori trichrome.)

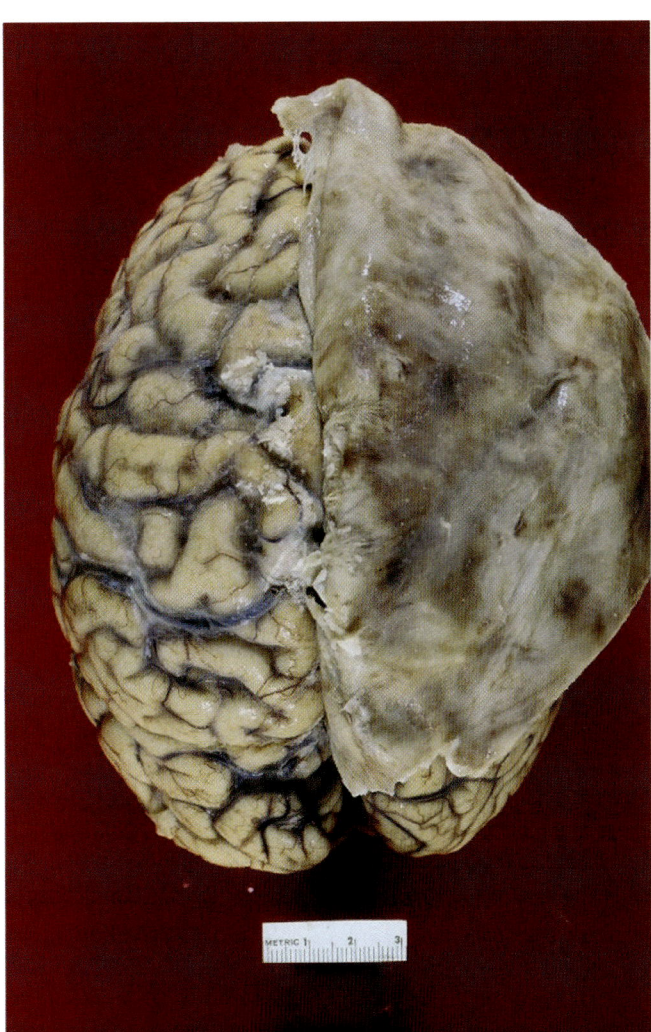

Fig. 6.8. Vertex view of brain with dura mater of left convexity turned over to the right exposing a gray-tan, well-formed subdural neomembrane. The membrane was bilateral.

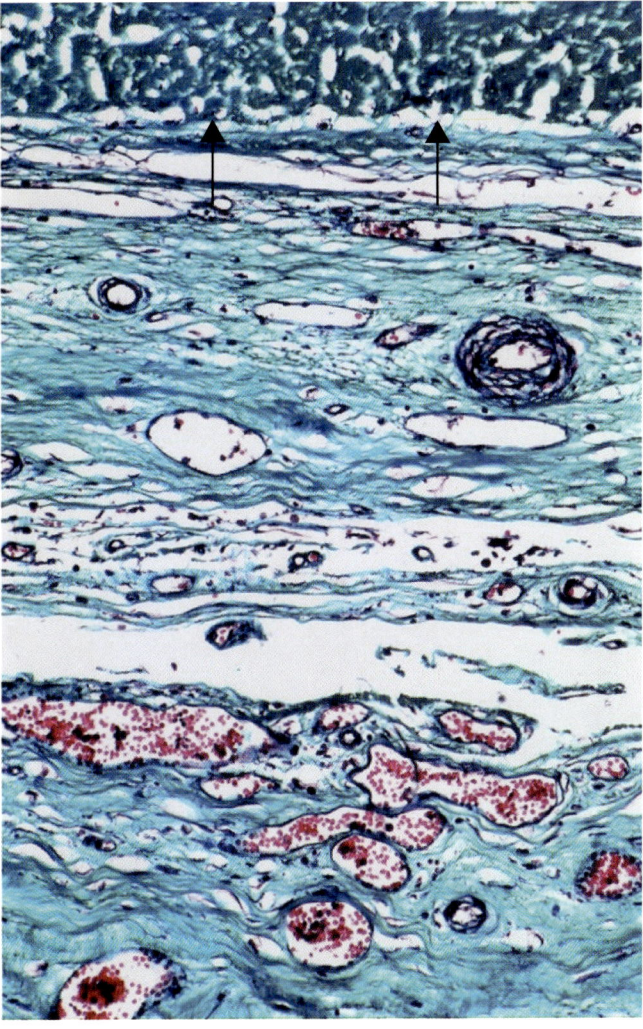

Fig. 6.10. High-power micrograph shows collagenized neomembrane of low cellularity and persistent capillaries. Arrows indicate interface of neomembrane with dura. (Gomori trichrome.)

6.13–6.14; Figs. 6.15–6.17). Early in the chronic period, scattered red blood cells and hemosiderin-laden macrophages remain along with giant capillaries, but only a collagenized hyaline membrane remains late, after several months or more. Grossly, depending on the size of the hematoma, the brain may show a surprising degree of compression distortion, usually over the dorsal hemisphere, with relatively minor or no neurologic deficit. Relatively silent chronic subdural hematomas may eventually become symptomatic with fresh bleeding into an old collection with or without a history of new trauma.

Subarachnoid Hemorrhage

Subarachnoid hemorrhage (SAH) is the most common indicator of TBI (see Fig. 6.18). A minor degree of SAH, especially located at the frontoparietal paramedian cortex, may be difficult to distinguish from artifact resulting from poor handling at autopsy with tearing of subarachnoid blood vessels. This potential problem may be avoided in cases with a history of trauma by inspection of the meningeal surfaces while the brain is still *in situ* by careful inspection while turning the dural flap. SAH found in this manner is likely not to be artifactual.

> ### Case 6.3. (Figs. 6.11 and 6.12)
> A 68-year-old man had a history of chronic alcoholism and multiple arrests over several years for public drunkenness. After being taken into custody for drunkenness, he was found down in his cell and was transported to a hospital. Examination revealed gastrointestinal bleeding from esophageal varices, anemia, and end-stage liver disease. Chronic bilateral subdural hematomas were discovered and surgically evacuated. Pneumonia complicated the postoperative period, and he died 4 weeks after surgery.

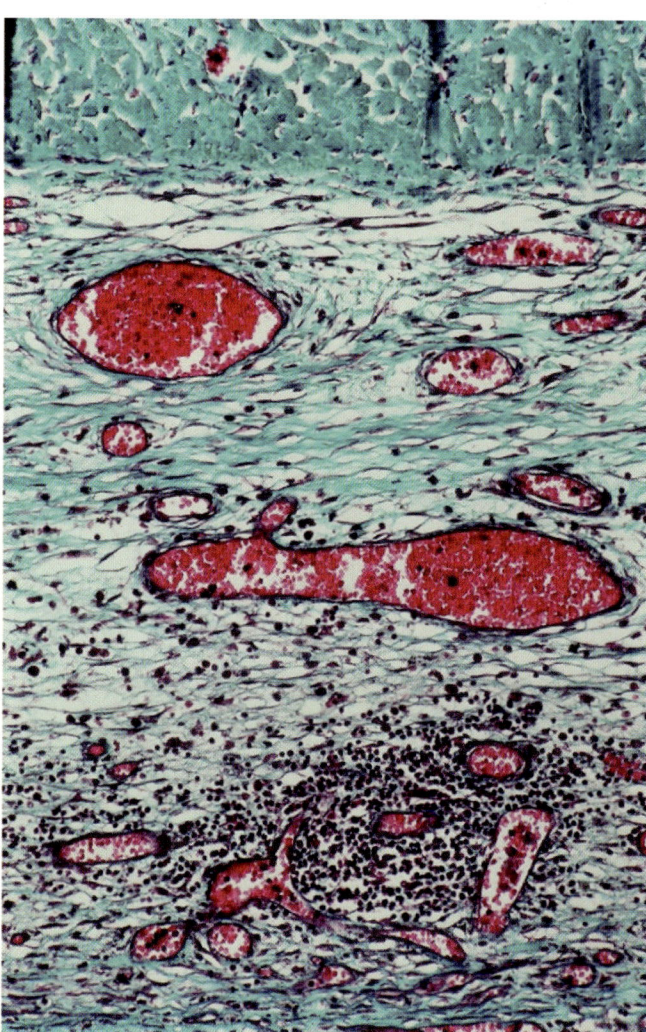

Fig. 6.12. Medium-power photomicrograph of dura mater shows portion of dark green-stained native dura (*top*) beneath which is the loosely organized fibrotic outer layer of chronic subdural neomembrane containing giant capillaries and focal chronic inflammation (*bottom*). (Trichrome.)

> ### Case 6.4. (Figs. 6.13 and 6.14)
> An 87-year-old man slipped and fell, striking his head, in a supermarket. He was treated at the scene by paramedics, and he drove home alone. One day, 3 weeks later, and again on the following day, he fell in his bedroom in his retirement home. After the second fall, he was admitted to the hospital and was found to have hypertension, episodic confusion, weakness, and difficulties with speech, coordination, and gait. CT scan revealed a left subdural hematoma that was evacuated on the same day. He died 15 days later.

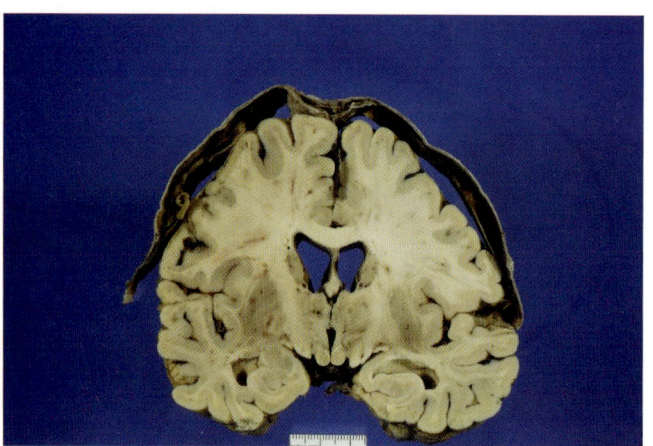

Fig. 6.11. Bilateral chronic subdural hematomas are seen as thick dark brown layering beneath the thin white line of dura mater.

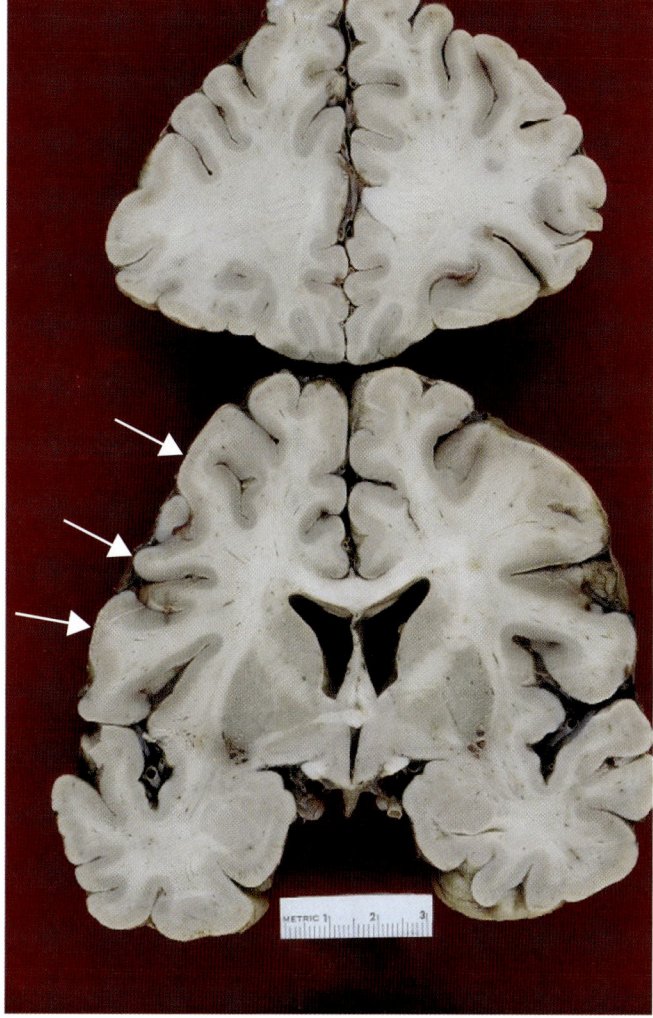

Fig. 6.13. Coronal sections of frontal lobes show slight flattening of left dorsal convexity (*arrows*) due to bilateral chronic subdural hematomas, greater on the left.

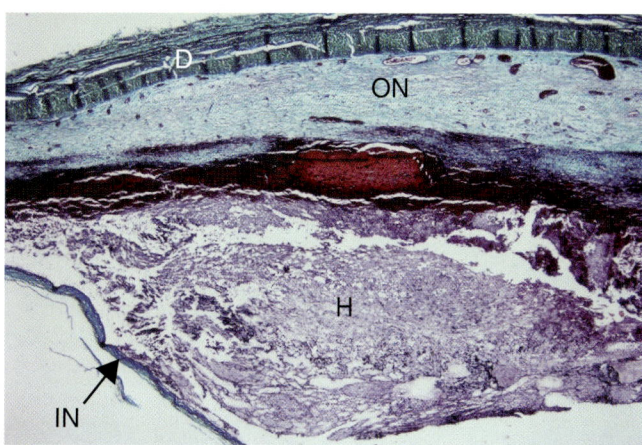

Fig. 6.15. Low-power micrograph from another case demonstrates thick outer and very thin inner neomembranes of a chronic subdural hematoma between which necrotic remnants of hematoma and small amount of recent hemorrhage remain. D, dura mater; ON, outer neomembrane; IN, inner neomembrane (*arrow*); H, hematoma. (Gomori trichrome.) Figs. 6.16 and 6.17 are from this same lesion.

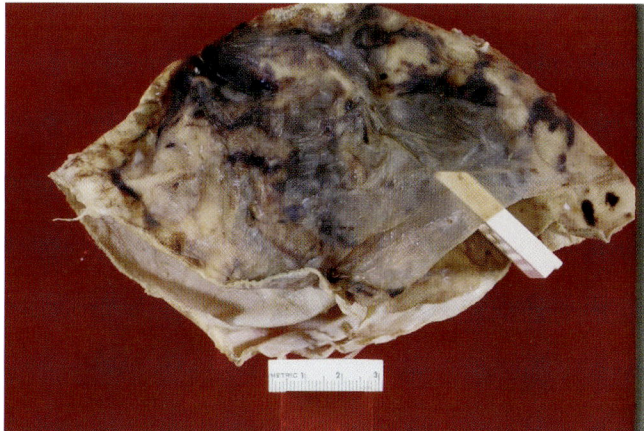

Fig. 6.14. View of subdural surfaces with chronic subdural hematoma and irregular collections of superimposed recent hemorrhage. Right side of photo is from left hemisphere, and a tissue cassette elevates a semitransparent inner membrane.

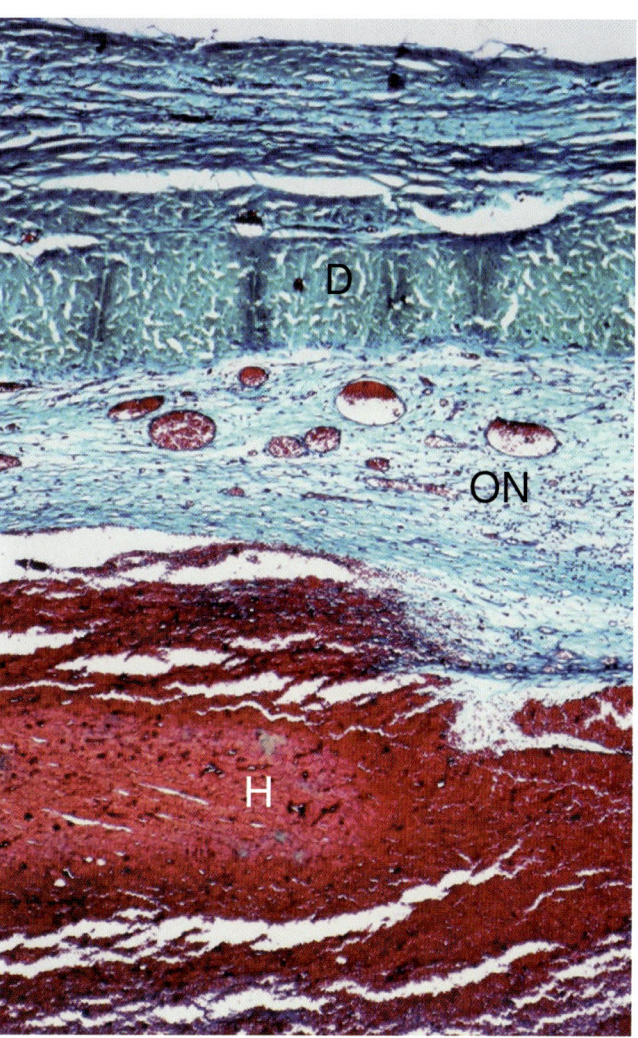

Fig. 6.16. Medium-power micrograph shows detail of outer neomembrane with loose fibrosis in which "giant" capillaries persist. (Gomori trichrome.)

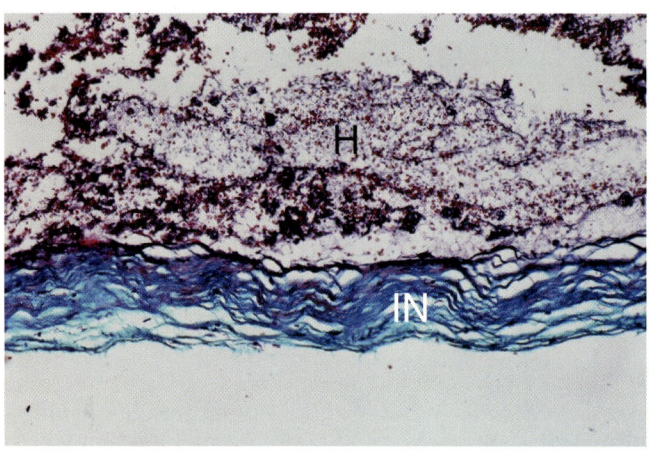

Fig. 6.17. Thin fibrotic inner membrane (IN) is devoid of capillaries. H = hematoma. (Gomori trichrome.)

Another indicator of true antemortem SAH is dark hemorrhagic congestion of epidurally exposed arachnoid granulations (see Fig. 3.16).

With TBI, it is most unlikely that a small amount of SAH alone can lead to death. However, it is indisputable that a large-volume SAH can lead to death, for example, with a ruptured berry aneurysm. Associated conditions such as diffuse axonal injury, hypoxic-ischemic injury, and cerebral edema are likely to be the true cause of death. Small or large amounts of localized SAH consistently accompany acute cerebral cortical contusions. The amount of SAH overlying minor contusions may be minimal and the contusions visible, but severe contusions may be obscured by thick layering of overlying SAH and exposed only on sectioning.

Significant SAH may underlie acute subdural hemorrhage, secondary to simultaneous leakage of ruptured subarachnoid cortical and bridging veins. Scattered small amounts of acute SAH in a widespread area is occasionally encountered at autopsy in cases of sudden death associated with head trauma. The amount is often barely sufficient to fill sulcal openings and stain gyral crowns. In the absence of subdural hemorrhage and evidence of diffuse axonal injury, the cause of death becomes problematic when one is left with only acute SAH. A small amount of SAH is usually not a factor in immediate cause of death. However, it is an indicator of TBI, with or without microscopic confirmation of underlying diffuse axonal injury and the presence of cerebral edema. Post-traumatic apnea or consequences of catecholamine surge are frequently hypothesized as the underlying mechanism of death in cases where the intracranial anatomic findings are limited to minor degrees of SAH (see later section on consequences of head injury).

Acute SAH around the lower brainstem with reflux into the fourth ventricle may be lethal. The mechanism is unknown, but speculation centers on the irritant effects of hemorrhage on the floor nuclei of the fourth ventricle, in particular, dorsal vagal nuclei and the nucleus of the tractus solitarius. Vasospasm of arteries bathed in acute subarachnoid hemorrhage secondary to ruptured aneurysm often causes infarction and other ischemic complications. However, the true role of SAH as the cause of death in traumatic head injury is not always apparent (see earlier). SAH in the posterior fossa with or without a traumatic history necessitates a thorough search for a ruptured aneurysm as a cause. The rupture of a preexisting aneurysm may be precipitated by an abrupt elevation of blood pressure during altercations and thus may be indirectly due to trauma. Acute SAH from traumatic avulsion of an otherwise normal intracranial vertebral artery may occur, or it may occur at the site of atherosclerosis. The precipitating traumatic event may be surprisingly minor.

Contostavlos,[9] Mant,[48] and others described a circumstance and a mechanism by which relatively minor trauma to the neck at the base of the skull may lead to massive SAH at the base of the brain. A minor blow to the high neck, such as with a fist, critically localized just below the mastoid process and behind the mandible, can fracture the transverse process of the atlas, causing damage to the wall of the vertebral artery within the foramen transversarium. This can lead to hemorrhage dissecting along the wall of the artery and eventually rupturing into the posterior fossa. Only a careful dissection of the high posterior neck and exposure of the vertebral artery in its extracranial course over the arch of the atlas will disclose the true cause of the hemorrhage. Sudden death of four ice hockey players with massive basilar SAH was attributed to presumed injury of the vertebral artery due to a blow by a puck driven at high velocity to the high neck,[49] a mechanism similar to that of Contostavlos and Mant, described earlier. In the same report, another player collapsed and died when struck with a fist in an altercation.

Survivors of traumatic SAH may later develop a triad of dementia, gait ataxia, and urinary incontinence, a syndrome sometimes called normal pressure hydrocephalus. It is a type of communicating hydrocephalus presumed to be caused by hemorrhagic blockage of resorption of CSF at arachnoid granulations and impaired CSF flow secondary to leptomeningeal and arachnoid granulation fibrosis. An occasional unexpected finding at autopsy, it is rarely the immediate cause of death.

Microscopically, the effect of SAH on the underlying cerebral cortex is minimal, resulting in pial fibrosis and scattered deposition of hemosiderin in the subarachnoid space. The brain parenchyma rarely shows any abnormality beyond mild subpial gliosis. Clearing of SAH is efficient and thorough, leaving behind only these mild brain changes, but gross staining of leptomeninges may sometimes persist.

Brain Contusions, Lacerations, and Hematomas

Cerebral cortical contusions are by far the most common markers of TBI. They are bruises found at the apices of gyri occurring at typically recurring locations. Contusions occur at the moment of impact against overlying bony prominences: anterior rectus gyri against crista galli (Case 6.5, Figs. 6.18–6.19; Figs. 6.20–6.21), frontal orbital cortex against the ridges of the roof of the orbit (Case 6.6, Fig. 6.22), posterior orbital cortex (Case 6.7, Figs. 6.23–6.24) and medial temporal pole against the lesser wing of the sphenoid bone (see Fig. 6.21), the midlevel of the inferior temporal gyrus against the petrous ridge of the temporal bone, and the anterior parasylvian cortices against the greater wing of the sphenoid (Case 6.8, Figs. 6.25–6.26; Case 6.9, Fig. 6.27). Isolated contusions in hidden areas such as the posterior

Case 6.5. (Figs. 6.18 and 6.19)

A 28-year-old man was one of three victims accosted by five gang suspects in a robbery attempt, and a fight ensued. The decedent was punched and kicked, and he fell to the ground unconscious. He was found by his cohorts who had initially fled, and later returned to the scene. They poured water on his face to revive him and walked him home. En route, a police car stopped to ask if the victim was OK. Fearing retaliation, they told the police that the victim was drunk. At home, he complained of neck pain but did not want medical attention. Before going to bed, he drank a quarter pint of vodka. He was found unresponsive in the morning and confirmed dead at the scene.

Autopsy revealed a nondepressed midline fracture of the frontal bone and a small acute epidural hematoma over the superior sagittal sinus. Small Duret's hemorrhages were found in the midbrain and rostral pons.

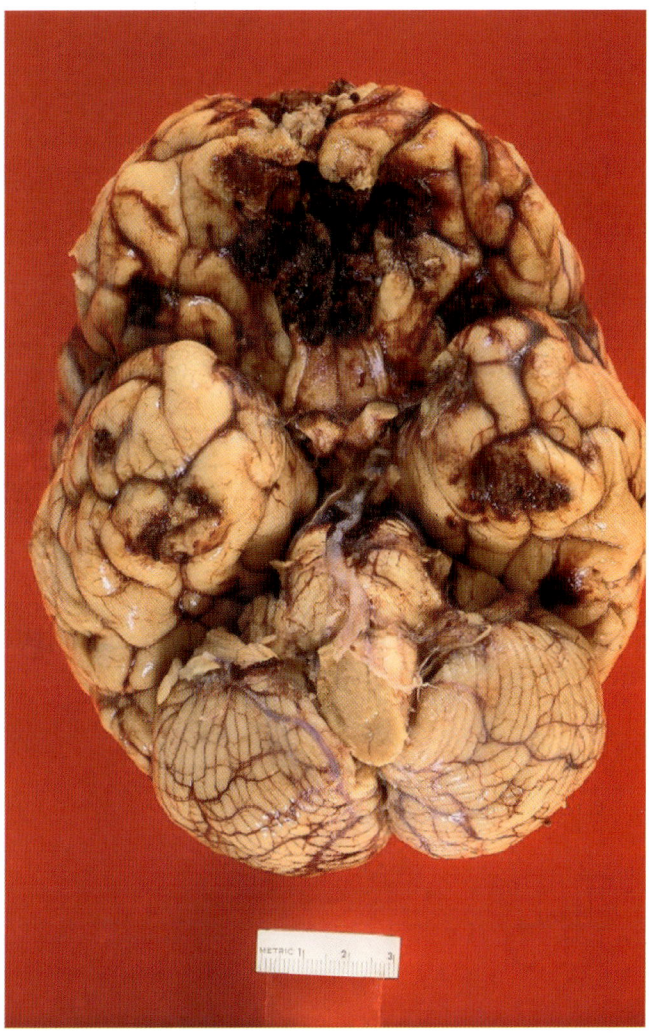

Fig. 6.19. Base of brain shows acute subarachnoid hemorrhage marking sites of acute contusions involving rectus gyri, posterior orbital cortex, and temporal lobes.

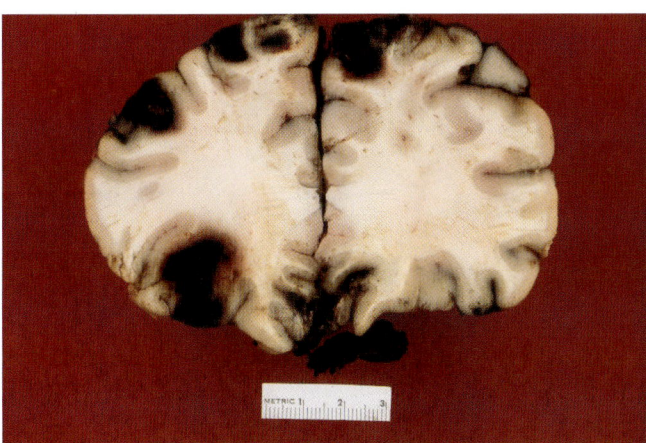

Fig. 6.18. Pregenu frontal lobes contain acute contusions with cortical/subcortical hematoma of rectus gyri, orbital cortex, parasagittal cortex, and an atypically located contusion of middle frontal gyrus.

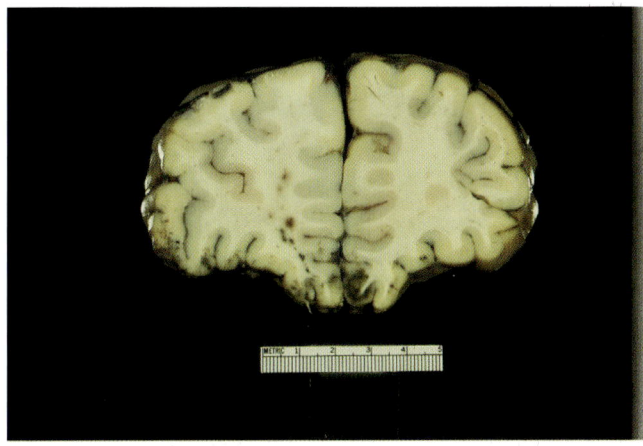

Fig. 6.20. Coronal section of frontal lobes demonstrate typical acute cortical contusions of rectus gyri.

178 Blunt Force Head Injury

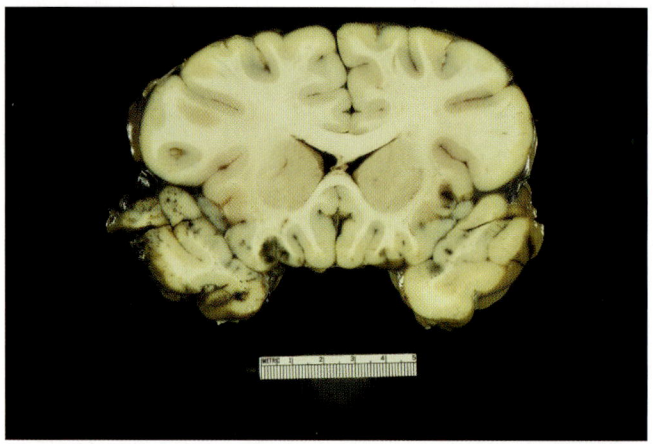

Fig. 6.21. Posterior orbital cortex, and medial temporal poles facing lesser wing of sphenoid bone are also common sites of contusions. Effaced sulci and ventricles give evidence of edema.

> **Case 6.6.** (Fig. 6.22)
>
> A 31-year-old male with evidence of head trauma of unknown circumstance was found down and taken by paramedics to the hospital where he arrived unresponsive with a "high" blood alcohol level. CT head scan revealed a right occipital skull fracture and a left frontal intracerebral hemorrhage. A decompressive craniotomy was performed on the second hospital day, but he remained comatose and died on the 18th hospital day.

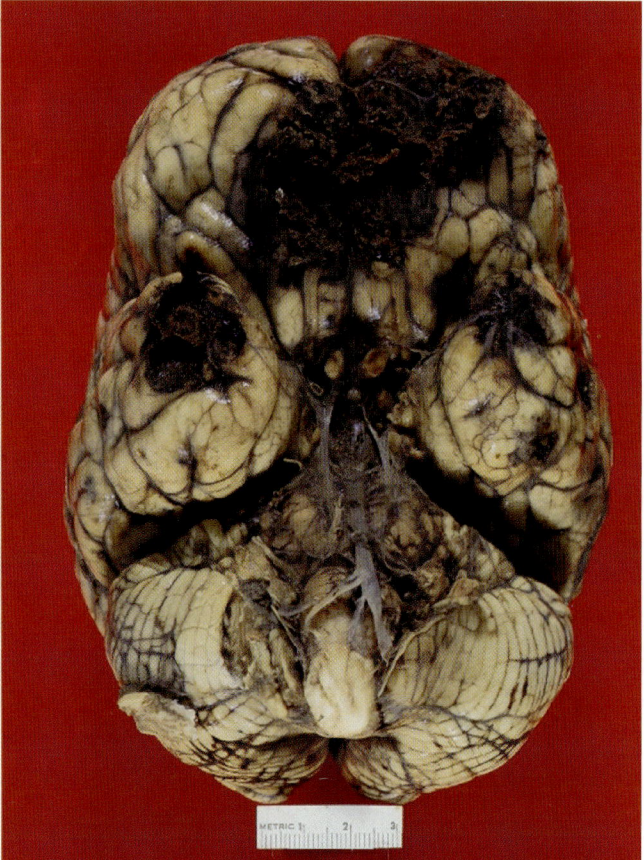

Fig. 6.22. Acute cortical contusions most pronounced in orbital frontal lobes but also involving temporal poles and inferior temporal gyri.

orbital cortex may be missed unless specifically sought for (see Fig. 6.23). Contusions of the tips of the frontal poles are less common, and those of the occipital poles are even more uncommon. Cortical contusions may be found over the hemispheric convexity beneath focused impacts (Case 6.10, Figs. 6.28–6.30) or depressed

> **Case 6.7.** (Figs. 6.23 and 6.24)
>
> A 34-year-old man was found in a field in early September in coma and in full arrest with a rectal temperature of 110°F. Resuscitative efforts failed, and he was pronounced dead soon after arrival in the ER. Cause of death was attributed to heat stroke subsequent to being left in the field after an inflicted head injury.

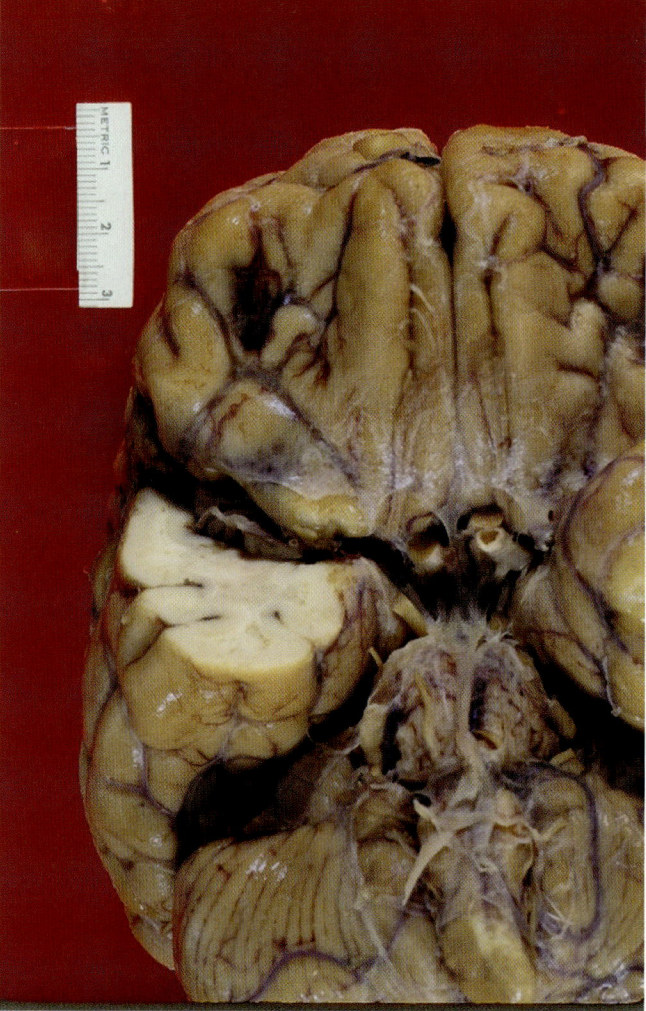

Fig. 6.23. Partial base view of brain shows two small dark discolorations of gyral apices of right orbital cortex (left in photo) (*arrows*), one of posterior orbital cortex hidden and only exposed by amputation of temporal lobe.

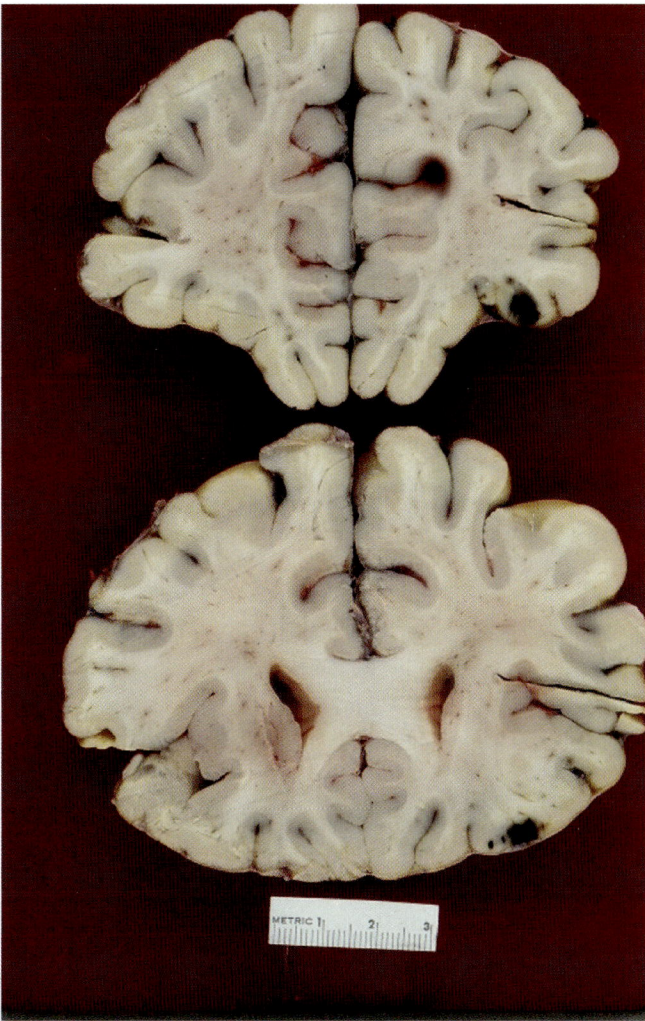

Fig. 6.24. Coronal sections display acute contusions noted externally, and focal hemorrhagic infarction in right medial frontal lobe.

> **Case 6.8.** (Figs. 6.25 and 6.26)
>
> A 59-year-old man was found dead at the scene. Cause of death was attributed to liver disease, complicated by leg phlebitis and multiple drug intoxication, including drugs for pain, anxiety, sleep, and anticonvulsants. He had a history of epilepsy that followed remote head injury.

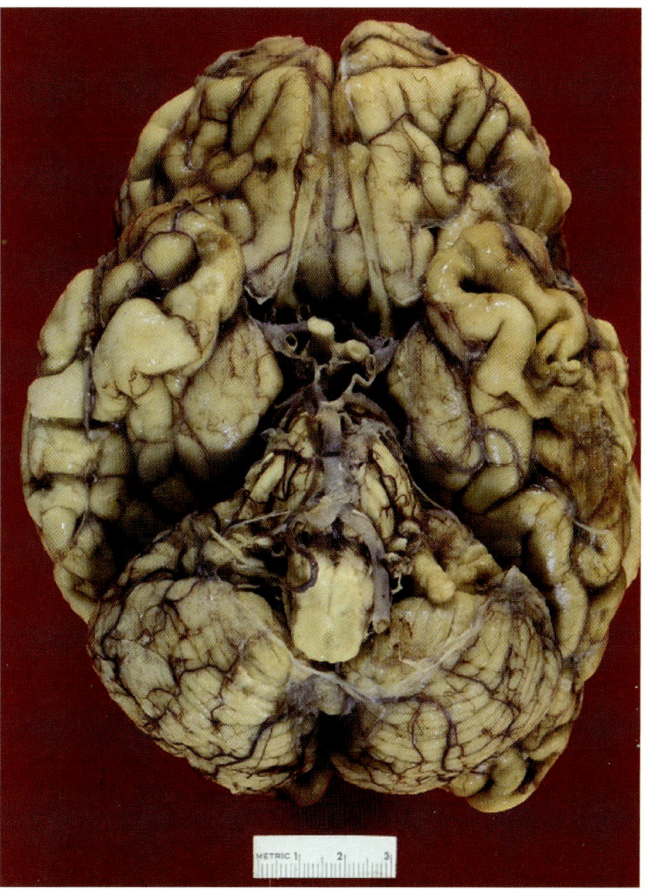

Fig. 6.25. External view of brain shows chronic cortical contusions over a wide extent of anterior half of middle and inferior temporal gyri. Reportedly, the decedent attempted to retrieve a dog from a moving vehicle, fell, and struck his head years before.

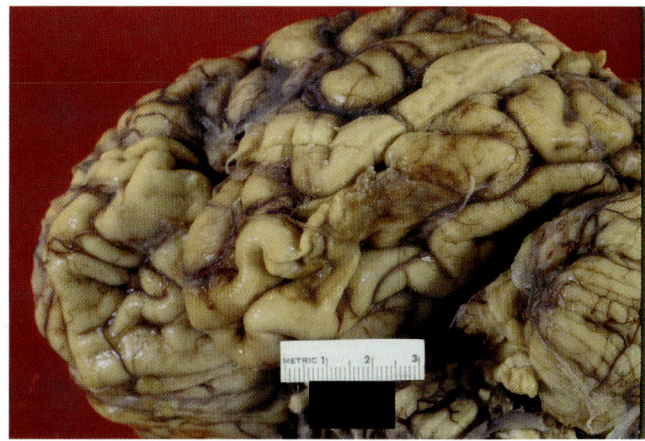

Fig. 6.26. Closeup view of left inferior temporal lobe shows a chronic contusion centered at inferior temporal gyrus.

Case 6.9. (Fig. 6.27)

A 67-year-old man fell and struck his head in a syncopal attack, and he was taken to a community hospital where he was lethargic with alcohol on his breath. Refusing an electrocardiogram (ECG) he was transferred to a tertiary hospital where hemotympanum was noted, and CT head scan revealed a left frontal subdural hematoma and ipsilateral temporal contusions. Burr holes, placed almost 4 weeks after the fall, exposed a chronic subdural hematoma that was drained. He died 2½ weeks later, 6 weeks after the fall.

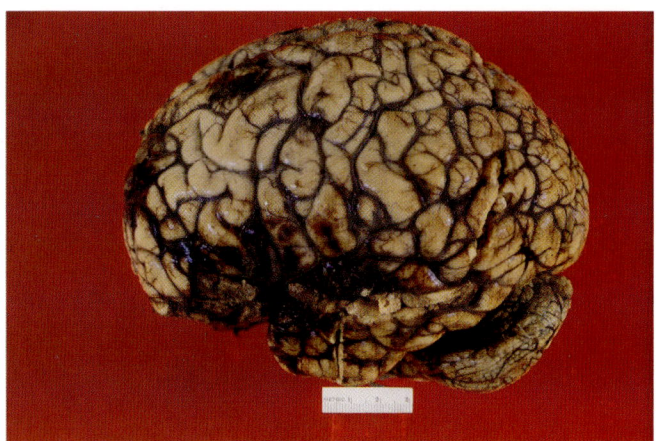

Fig. 6.28. Left lateral hemisphere shows a small amount of acute subarachnoid hemorrhage in anterior parasylvian region with contusions of inferior frontal gyrus and temporal pole.

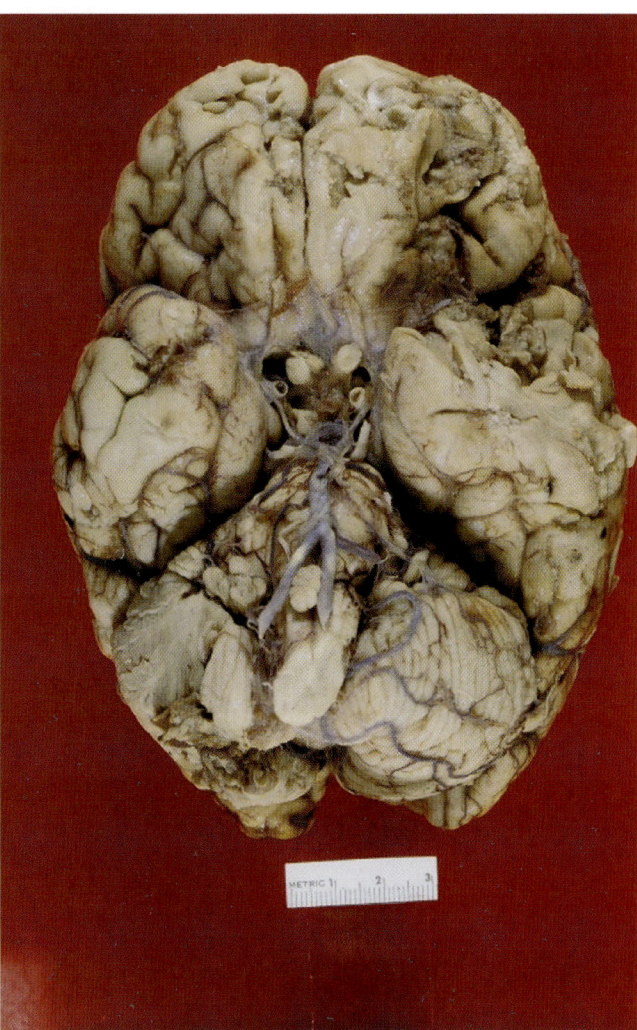

Fig. 6.27. Base view of brain shows chronic contusions of lateral orbital frontal lobe and anterior temporal lobe.

Case 6.10. (Figs. 6.28–6.30)

A 22-year-old man was assaulted during a robbery. Later that night, he was found unresponsive and taken to a hospital where subdural hematoma, fractured ribs, and pneumothorax were found. He remained in coma and died 2 days later. Autopsy revealed a right occipital nondepressed linear skull fracture.

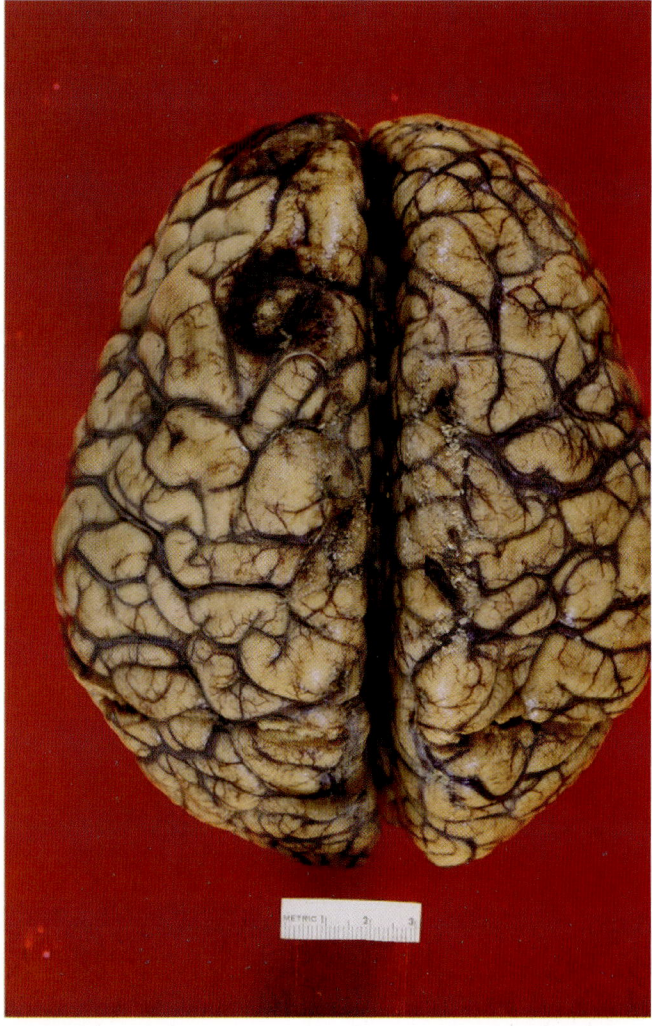

Fig. 6.29. Vertex view displays two foci of discrete round subarachnoid hemorrhage of superior frontal gyrus of possible impact contusion origin.

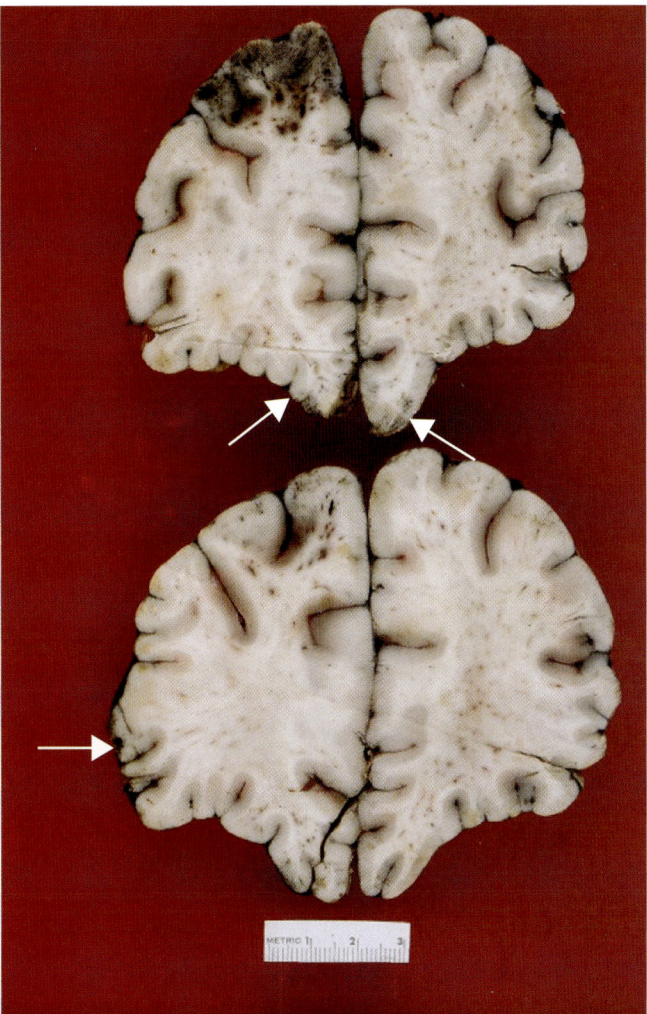

Fig. 6.30. Coronal sections of anterior frontal lobes show acute hemorrhagic necrosis of cortex into white matter localized to the left superior frontal gyrus. An impact contusion was a consideration, and hemorrhagic infarction secondary to some vascular event was considered unlikely. Minimal cortical contusions of rectus gyri and left inferior frontal gyrus are seen (*arrows*). Petechiae of frontal white matter suggested diffuse axonal injury.

Case 6.11. (Fig. 6.31)

A 36-year-old man, a known alcoholic with cirrhosis of the liver, had recent deterioration of motor function before a traffic accident. He drove his vehicle into a parked car and was arrested for DUI. Following release from jail, he was hospitalized for deteriorating medical condition. A flapping tremor and altered mental status were noted at admission, attributed to hepatic encephalopathy. CT head scan revealed a left temporoparietal hematoma with left lateral ventricular hemorrhage. He died 12 days after the accident.

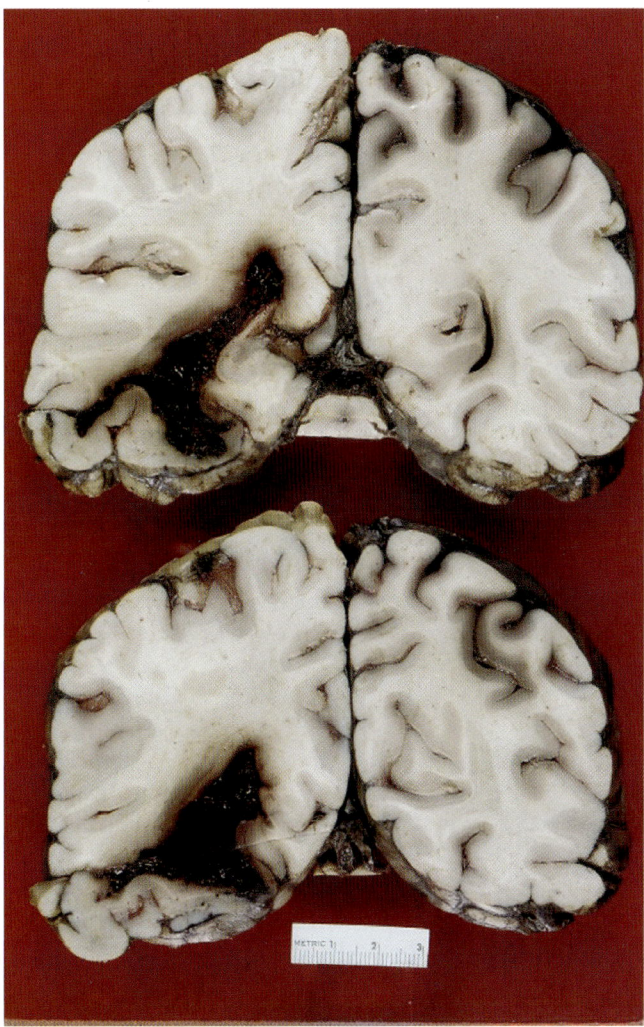

Fig. 6.31. Postsplenial coronal sections reveal left posterior temporal subcortical hematoma dissecting into adjacent atrium and posterior horn of lateral ventricle. Cortex overlying hematoma appears contused. Microscopic examination showed Alzheimer type II cells of hepatic encephalopathy in thalamus.

fractures, or under graze (tangential) gunshot wounds. Isolated fresh superficial cortical microinfarcts, from whatever cause, may be hemorrhagic and can mimic minimal contusion on gross inspection. However, contusions always occur at the crowns of gyri and at the characteristic anatomic locations mentioned earlier. Infarcts have more random distribution, are not limited to gyral apices, and demonstrate relative preservation of layer I of the cortex.

Cerebral cortical contusions may be accompanied by cortical/subcortical hematomas of variable volume. Small ones may remain intracortical. Large hematomas intrude deeper into an subcortical white matter, and may dissect into an adjacent ventricle (Case 6.11, Fig. 6.31). Follow-up CT scans occasionally discover an intracerebral hematoma not present on the admission scan. These delayed traumatic intracerebral hematomas often correlate with sudden worsening of the victim's clinical course,[79] but they also may remain clinically silent and are discovered only on imaging study or at autopsy.

Microscopically, very acute contusions in victims found dead at the scene are seen as focally clustered petechiae at the apex of the gyrus, more superficially located than deep, with no cellular reaction other than "dark" neurons amid petechiae. Later cellular reaction progresses in accordance with the length of survival (see Chap. 3). Chronically, cortical contusions are discrete wedge-shaped defects broad at the pial surface and most of the time opening into the subarachnoid space. On the deep side, the lesion is pointed toward and into subcortical white matter to variable depths. Pial margins of the defect often show hemosiderin granules, hemosiderin-decorated astrocytes, and occasionally ferruginated neurons.

Contusions are most often the result of acceleration–deceleration forces and, as expected, a loose correlation exists between the force of impact and extent of contusions. Common scenarios are unbroken ground-level falls on the back of the head, especially in the inebriated, on one end of the scale, and high-speed motor vehicle accidents on the other. When witnessed or analyzed by the scalp impact marks, and so on, a frequent mechanism is contre coup, that is, the contusions are located at a point opposite the point of impact. As a typical example, a fall on the right occiput results in contusions of the left frontal orbital cortex and the temporal pole. Coup or impact contusions, as might be expected, can occur subjacent to the impact site, occasionally in isolation at the frontal pole with frontal impact. Compared with the contre coup contusions, coup lesions are relatively small in closed head injury. Coup cortical contusion can be extensive under depressed fracture.

Unilateral hemorrhage in the basal ganglia is sometimes a head trauma autopsy finding seeking a mechanism. It has been hypothesized that an impact causing head rotation around a vertical axis of the body in a defined direction may cause such a hemorrhage by stretch and injury of the lateral lenticulostriate arteries arising from the horizontal middle cerebral artery.[47] These cases of Maki *et al.*[47] in children probably represented isolated examples of deep vascular injury of a type of deep traumatic hemorrhagic injury in brain termed "intermediary coup" by Lindenberg,[44] and it likely represents a larger hemorrhagic lesion of an injury now termed "diffuse axonal injury." The intermediary coup lesions described by Lindenberg included those in the corpus callosum that are now regularly sought for in the investigation of TBI.

"Gliding contusion" is a type of contusion occurring primarily in the superior frontal gyrus and superior parietal lobule, but may also occur in the anterior temporal lobes. It may be seen only as a faint bruise at the surface, but with a larger hemorrhagic component extending into subcortical white matter (Case 6.12, Figs. 6.32–6.33). This form of contusion is an important marker in accidental and inflicted head injury in infants and very young chil-

Case 6.12. (Figs. 6.32 and 6.33)

A 22-year-old workman apparently backed into an open elevator shaft and fell 2½ stories to the bottom. He was not found for some 2 hours, and he was taken to the hospital where multiple traumatic injuries included a fractured skull. He remained hospitalized in a coma until he died almost 20 months after the accident.

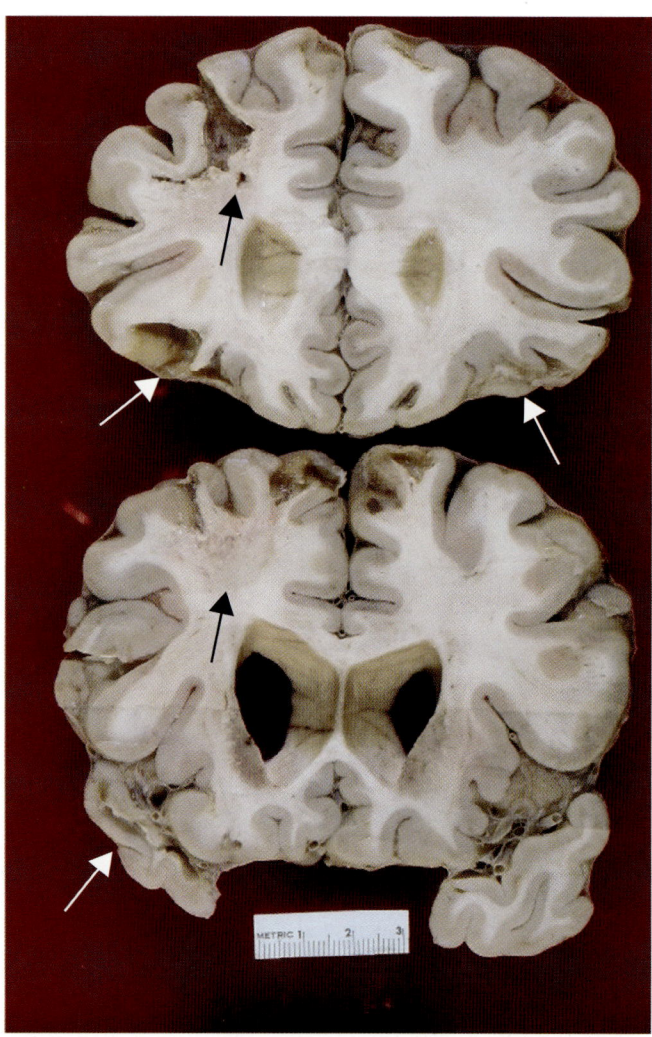

Fig. 6.32. Coronal sections of frontal lobes show chronic cystic lesions involving cortical apices of superior frontal gyri consistent with gliding contusions. Upper tier of middle frontal gyrus shows similar cystic cortical atrophy which, however, is accompanied by underlying necrosis of deep white matter of border zone infarction (*black arrows*). Chronic contusions are seen in orbital cortex and left temporal pole (*white arrows*). Corpus callosum is grossly intact, and lateral ventricles are enlarged.

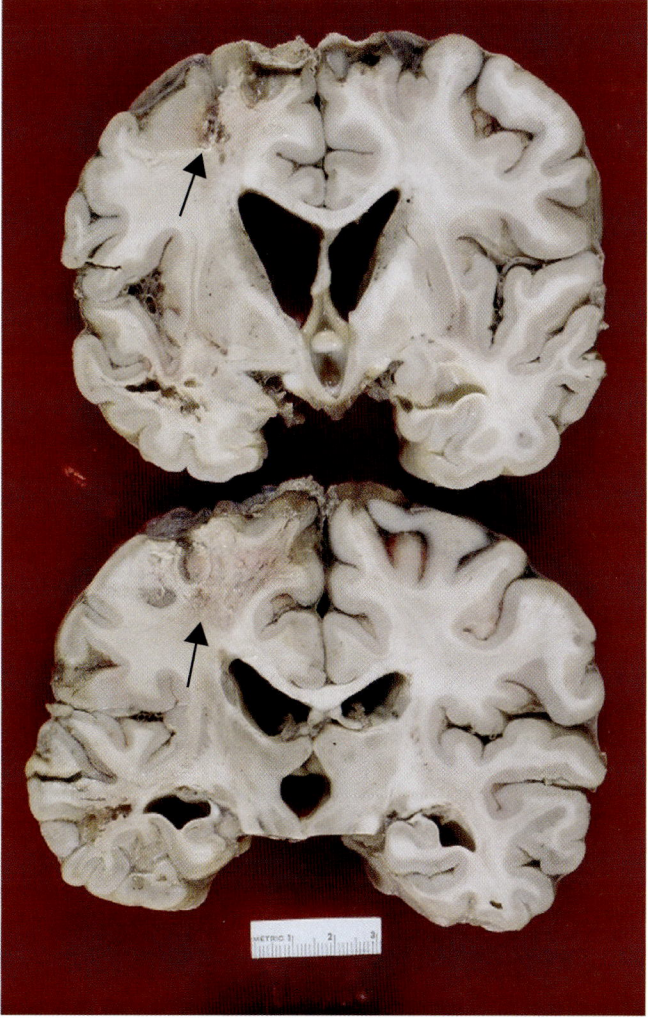

Fig. 6.33. Coronal sections of cerebral hemispheres at more posterior levels show extension of left subcortical white matter necrosis that merges with necrosis of parasagittal cortex. Hippocampi are atrophic. Left thalamus is shrunken, probably secondary to retrograde degeneration. No more than patchy bilateral semitransparent filmy subdural neomembranes were present.

Case 6.13. (Fig. 6.34)

A 20-year-old woman was broadsided by a speeding passenger car on the driver's side as she drove away from a gas station. Although she was seat-belted, her car was not equipped with a side-impact airbag, and her head went through the side window to impact the other vehicle's hood. A large left craniotomy was performed to evacuate a subdural hematoma. Other injuries included pelvic fracture, pulmonary contusions, and internal bleeding requiring multiple transfusions. She died 6 hours after the accident without gaining consciousness.

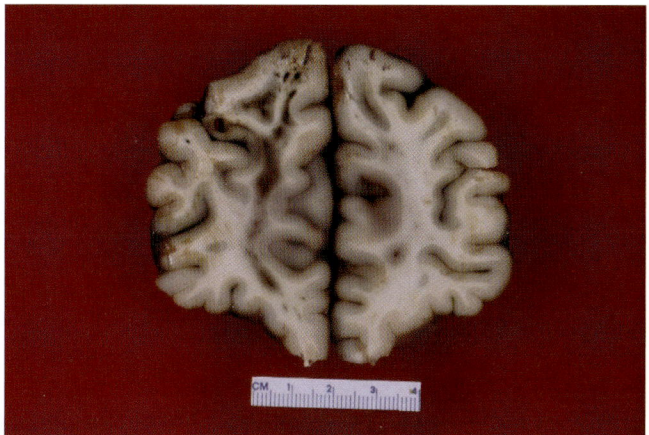

Fig. 6.34. Coronal section of pregenu frontal lobe shows characteristic vertical curvilinear array of hemorrhage through center of white matter on the left. Minor cortical and subcortical petechiae of superior frontal gyri are typical of acute gliding contusions.

dren. The same contusion pattern is also encountered in adults in traffic accidents and other major injuries, including inflicted trauma. "Herniation contusion" refers to the occurrence of superficially localized petechiae involving the pyriform cortex, including the uncus, adjacent hypothalamus, medial parahippocampal gyrus, and folia of biventer of cerebellum at the moment of impact compression of the brain by a force applied to a static head or as a result of shifting of the brain secondary to edema. On occasion, in the presence of TBI, deep-lying hemorrhages may occur in cerebral central white matter, basal ganglia, thalamus, and subcortical insula. Lindenberg suggested the term intermediary coup for such lesions (see earlier). These hemorrhages are often found in the midposition between the point of impact on the cranium and contre-coup cortical contusions. The mechanism for intermediary coup remains uncertain. Other causes of brain hemorrhage in trauma cases may mimic intermediary coup hemorrhage, including secondary coagulopathy.

Diffuse Axonal Injury

Diffuse axonal injury (DAI) is a term applied to pathologic findings following head trauma brought to attention by Sabrina Strich in 1956,[73] although, as noted by Adams et al.,[2] the concept of diffuse brain injury in humans and experimental models was known.[28] As the term implies, the injury to axons is in widespread areas, but it is not diffusely present throughout the brain. There are many spared areas. The injury tends to occur in certain selective sites such as the rostral levels of dorsal frontal white matter (Case 6.13, Fig. 6.34; Case 6.14, Fig. 6.35; Case 6.15, Figs. 6.36–6.38); white matter of temporal pole (Case 6.16, Figs. 6.39–9.40); corpus callosum at different anterior to posterior levels (Case 6.17, Figs. 6.41–

Case 6.14. (Fig. 6.35)

A 9-year-old girl was a seat-belted passenger in a car involved in a head-on accident, and she remained in the vehicle. She was taken to the hospital in a coma with minimal vital signs. A superficial abrasion of the right frontal area was noted at autopsy. Other injuries included ruptured spleen and fractured femur. She died 3 days after the accident.

Case 6.15. (Figs. 6.36–6.38)

A 49-year-old male motorcycle rider ran a red light and broadsided a car at an intersection. CT head scan revealed "a very small subdural hematoma, diffuse subarachnoid hemorrhage, and edema." He sustained multiple limb fractures and a closed head injury. Bilateral small craniotomies were performed for unstated indications. He was admitted in a coma and remained in a coma until he died 2½ months from the time of the accident.

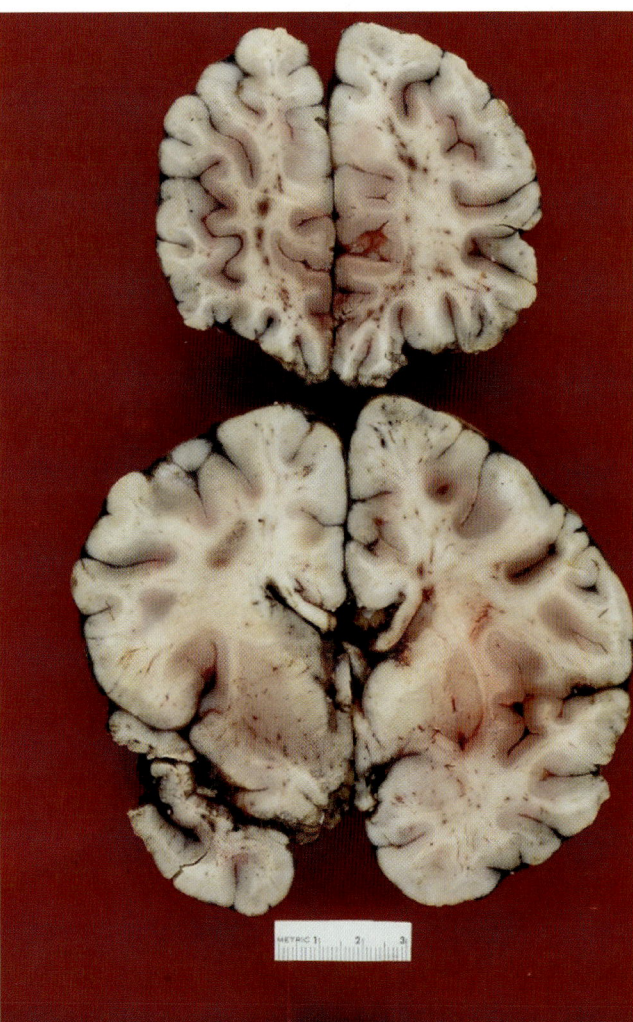

Fig. 6.35. Coronal sections of frontal lobes show bilateral, vertically oriented curvilinear arrays of petechiae in white matter in the anterior of two levels marking foci of diffuse axonal injury. A gliding contusion is seen in the right superficial frontal gyrus in the lower section. A few petechiae are also present in left temporal white matter. Separated corpus callosum is likely a postmortem artifact, but traumatic avulsion was not ruled out. Bilateral subdural hematomas were also present. The widespread red discolorations are due to incomplete fixation and not hemorrhage.

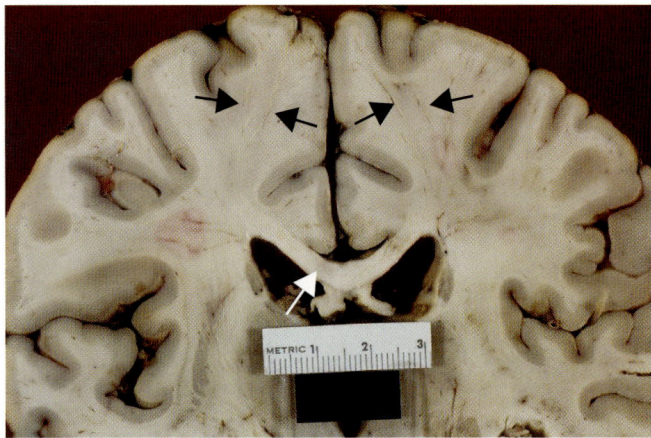

Fig. 6.36. Closeup view of coronal sections at the level of motor cortex reveals subtle, fine, linear streaks of necrosis in bilateral dorsal white matter typical of diffuse axonal injury (*black arrows*). Note a very subtle lesion in corpus callosum (*white arrow*).

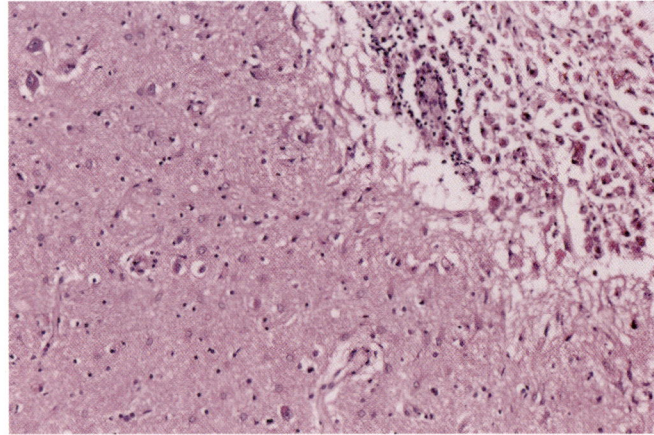

Fig. 6.37. Medium-power micrograph of one white matter lesion pointed out in Fig. 6.36 shows a uniform field of fibrillary astrogliosis. Focus of necrosis is shown in upper right corner. (H&E.)

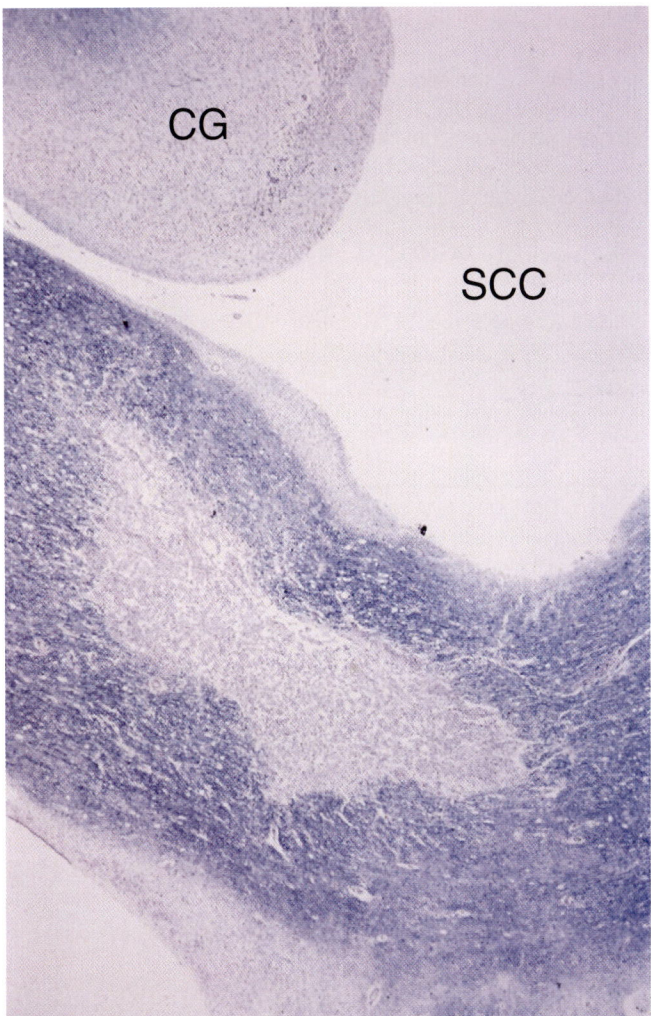

Fig. 6.38. Low-power micrograph of lesion pointed out in corpus callosum in Fig. 6.36 demonstrates focal necrosis typical, but not specific, for diffuse axonal injury. SCC, supracallosal cistern; CG, cingulate gyrus. (Luxol fast blue.)

> **Case 6.16.** (Figs. 6.39 and 6.40)
>
> A 15-year-old girl was a front seat passenger in an automobile parked in a gas station that was struck by an out-of-control vehicle. She was unconscious when found and remained in a coma for 11 weeks before she died. No abnormality was detected on the initial CT brain scan, and a repeat scan 8 weeks later showed only mild dilatation of the ventricles.

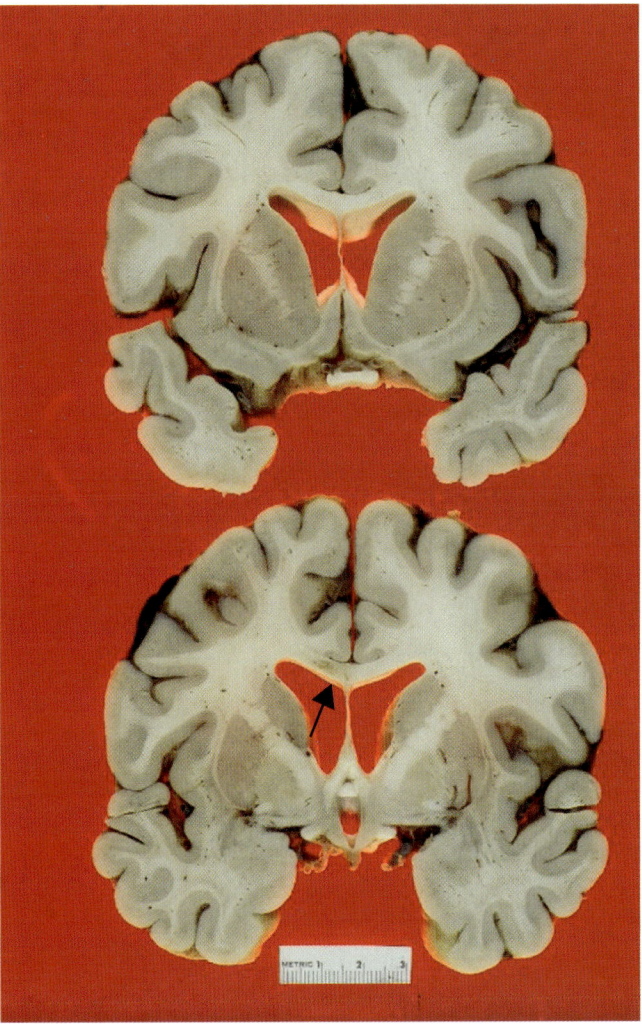

Fig. 6.39. Coronal sections of frontal lobes confirm mild but definite dilatation of lateral ventricles; otherwise of essentially normal appearance. A single important exception is a small faint discoloration in the corpus callosum just to the left of midline, typical of diffuse axonal injury (*arrow*). Although white matter otherwise appears normal grossly, it showed diffuse astrogliosis microscopically.

6.43; Fig. 6.44); posterior limb of internal capsule (Fig. 6.43) (especially at the mesencephalic transition); white matter of the basis pontis (greater in descending corticospinal tracts and much less so in pontocerebellar tracts); cerebral peduncles and medial lemnisci in the midbrain; tracts of the rostral pontine tegmentum (especially the superior cerebellar peduncle (Fig. 6.45), central tegmental bundle, and medial lemniscus) (Case 6.18, Figs. 6.46–6.47; Figs. 6.48–6.50).

Axonal spheroids can also be seen in cerebellar central white matter. Outside of cases of direct impact to the medulla, axonal retraction balls are rarely seen in the medulla in adults. Occipital white matter and basal ganglia usually escape, at least by H&E assessment. Strich's view was that injury was caused by shear forces acting at predictable anatomic loci quite independent of expectations in hypoxic-ischemic injury. The

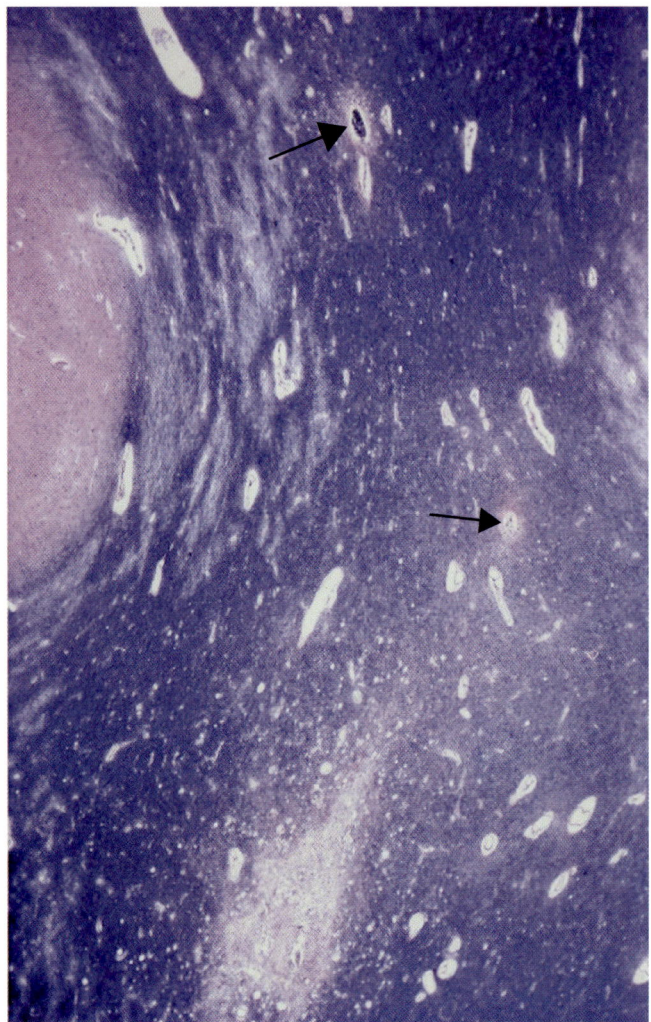

Fig. 6.40. Low-power micrograph of temporal white matter shows a stacked array of white lines of subcortical demyelination parallel to deep cortex. They represent lines of atrophy in this interesting pattern suggesting strain lines. A focus of necrosis is seen in deeper white matter along with minute perivascular atrophy (*arrows*). (Luxol fast blue.)

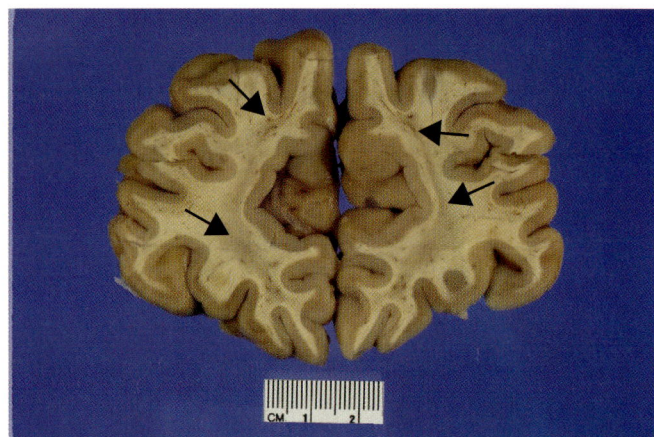

Fig. 6.41. Pregenu coronal section demonstrates vertical curvilinear gray and faintly stained white matter lesions characteristic but not necessarily diagnostic of old diffuse axonal injury (DAI). Vertically oriented arcs of myelin loss through frontal white matter are typical for DAI. Postplenial lesions are less common but do occur.

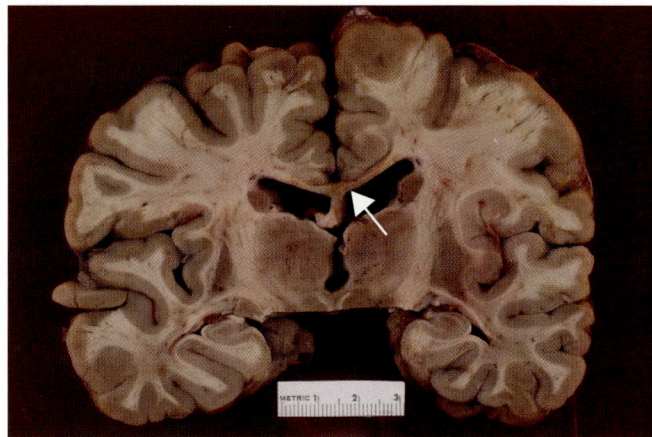

Fig. 6.42. Same case as Fig. 6.41 shows focal atrophy of corpus callosum to right of midline.

Case 6.17. (Figs. 6.41–6.43)

A 50-year-old man was found unresponsive in a convalescent home which was on a watch list. Solid food was obstructing his airway, and he was pronounced dead at the hospital. He had sustained head injury 4 years earlier in a motor vehicular accident, and he was in a coma for 14 days before regaining consciousness.

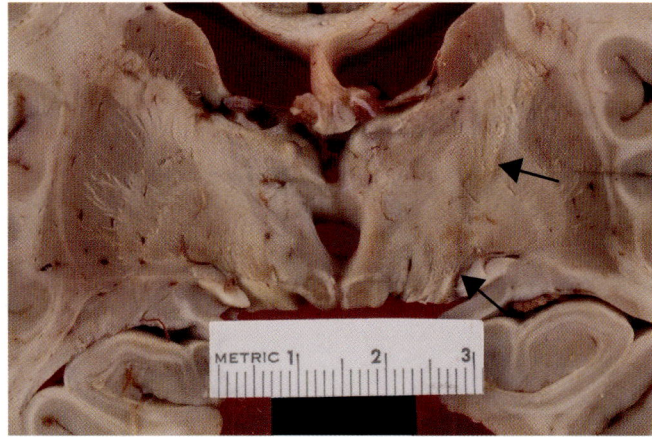

Fig. 6.43. Closeup view of posterior limbs of internal capsules shows characteristic diffuse axonal injury lesions (*arrows*).

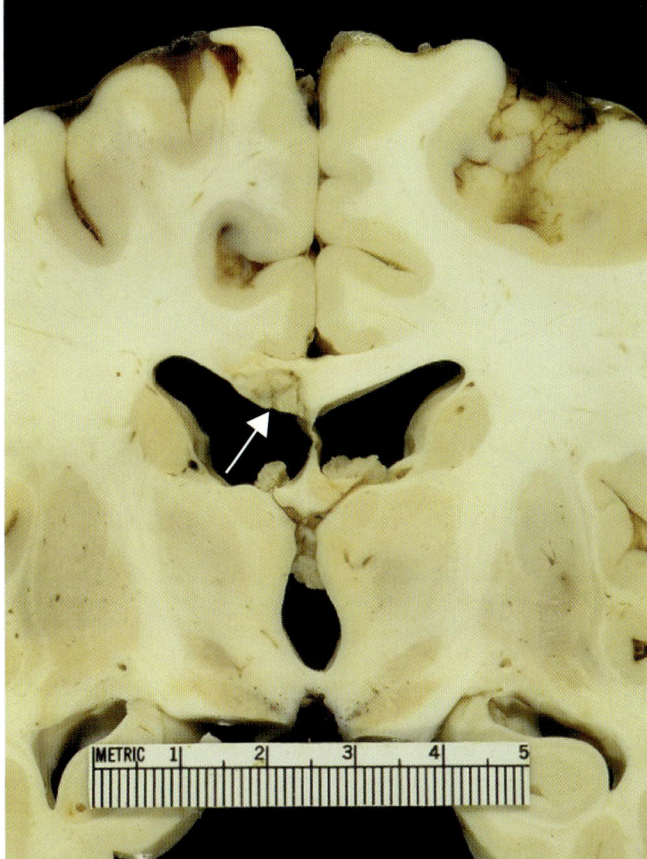

Fig. 6.44. Closeup view of coronal section of another case at a thalamic level shows focal chronic necrosis (*arrow*) on the ventricular surface of corpus callosum, characteristically just to one side of midline.

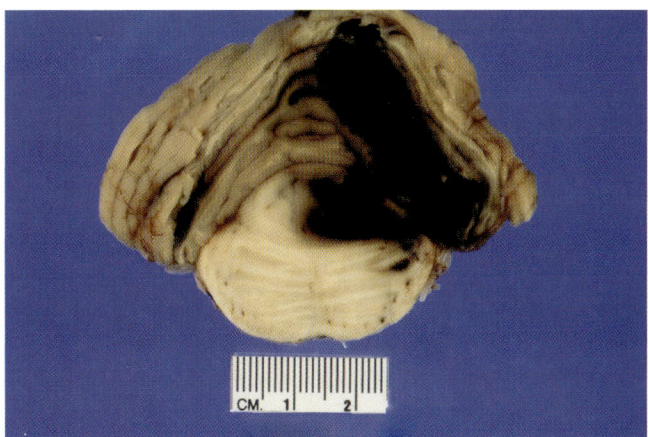

Fig. 6.46. Fresh hemorrhage of entire left pontine tegmentum (right side of photo) and overlying cerebellum is consistent with primary impact injury against the ipsilateral posterolateral margin of tentorium cerebelli. Other findings included a large left frontal cortical contusion/subcortical hematoma. Chronic liver disease suggested the presence of an underlying bleeding diathesis to account for the size of the hematomas.

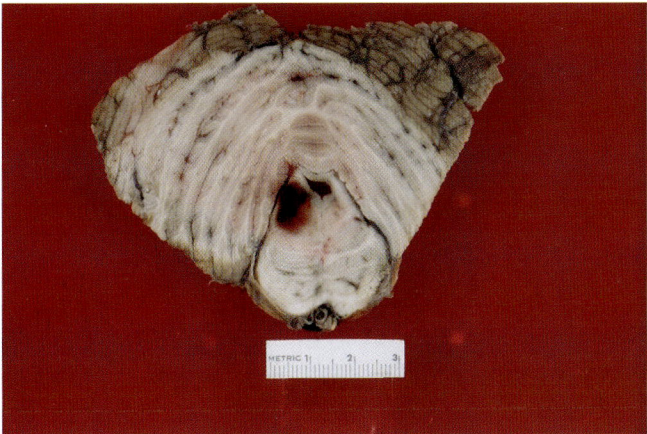

Fig. 6.45. Example of typical hemorrhage of diffuse axonal injury in dorsolateral pontine tegmentum involving superior cerebellar peduncle and adjacent tracts.

Case 6.18. (Figs. 6.46–6.47)

A 49-year-old man with a history of alcohol abuse was found at the bottom of a hotel stairwell with several abrasions. He had difficulty righting himself. After being assisted to bed, he was heard snoring loudly, but he was found expired in bed about 10 hours later.

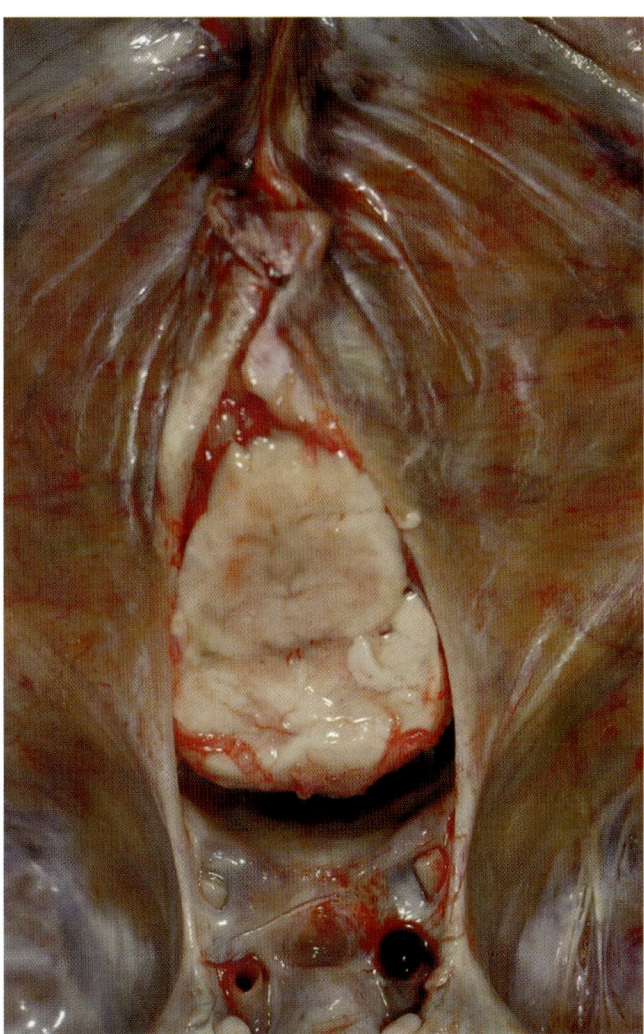

Fig. 6.47. Photo shows transverse section of brainstem at the pontomesencephalic transition in its relation to the margins of tentorium at the incisura. Note proximity of posterolateral incisurial margin to the brainstem tegmentum.

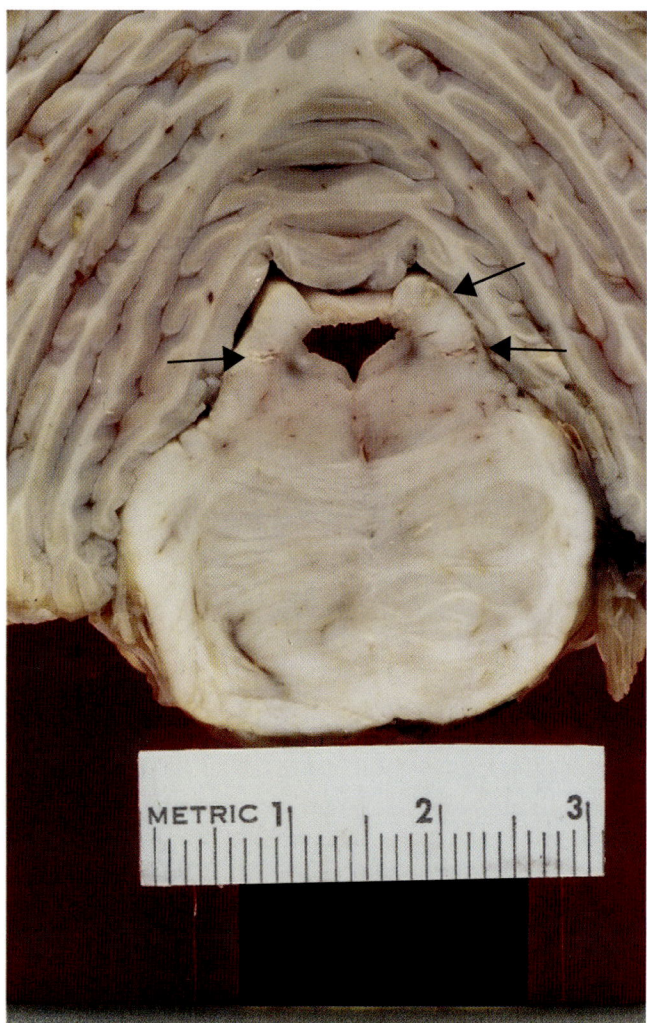

Fig. 6.48. Closeup view of rostral pons shows subtle but detectable necrosis in superior cerebellar peduncle on both sides (*black arrows*).

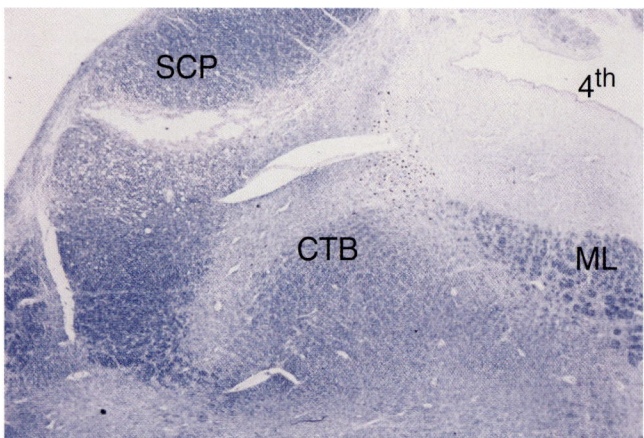

Fig. 6.49. Low-power micrograph of half of pontine tegmentum shows focal necrosis of superior cerebellar peduncle. SCP, superior cerebellar peduncle; CTB, central tegmental bundle; ML, medial longitudinal fasciculus; 4th, fourth ventricle. (Luxol fast blue.)

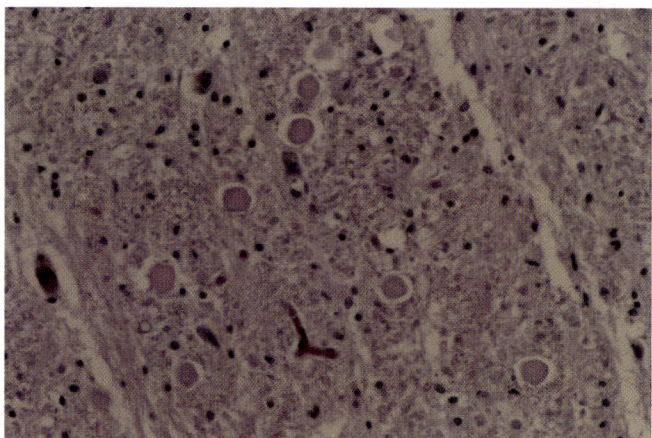

Fig. 6.50. Medium-power micrograph of central tegmental bundle in pons shows several axonal retraction balls, which are swollen axons seen as round eosinophilic bodies in transverse section. (H&E.)

critical histologic finding are axonal swellings, called axonal retraction balls, bulbs, or axonal spheroids (AS), readily detected by H&E stains in severe cases after a survival period of about 15 hours[20] (Fig. 6.50). Using a sensitive immunocytochemical method for β-amyloid precursor protein (β-APP), axonal spheroids may be detected in those surviving 3 hours or more[20] (see also Chap. 3).

Axonal spheroids by themselves are not specific for DAI, because similar bodies are found wherever axonal injury has occurred, for example, around cerebral hemorrhage and infarct examined after a certain period of survival. Axonal injury may follow hypoxic-ischemic injury and other unrelated conditions, such as multiple sclerosis and AIDS, among others, which may cloud the diagnosis of TBI. However, other than the presence of coincidental multiple diagnoses, critical differentiation remains only between trauma and hypoxic-ischemic injury in a trauma setting.[17] Changes of DAI in the basis pontis with selective involvement of axons with AS in descending corticospinal tracts, while sparing the transversely oriented pontocerebellar axons, as often illustrated in publications[20] (Figs. 6.51–6.52; Fig. 6.53), leaves little doubt of traumatic cause rather than ischemic injury. In less severe cases of TBI in which axonal damage is detected only by β-APP immunostain in only a few of many areas sampled, a definite etiologic diagnosis may not be possible. Nevertheless, hypoxic-ischemic encephalopathy has its own histopathologic features, and correlation of β-APP results with H&E and any other stains employed is indicated.

The more common sites prone to axonal spheroids have been noted earlier, along with the recommended areas of sampling for AS. Occasionally, within these areas, axonal injury can be focused to exceedingly limited sites, such as the midbrain ventral tegmentum where rostral-caudally oriented small nerve fibers adjacent to

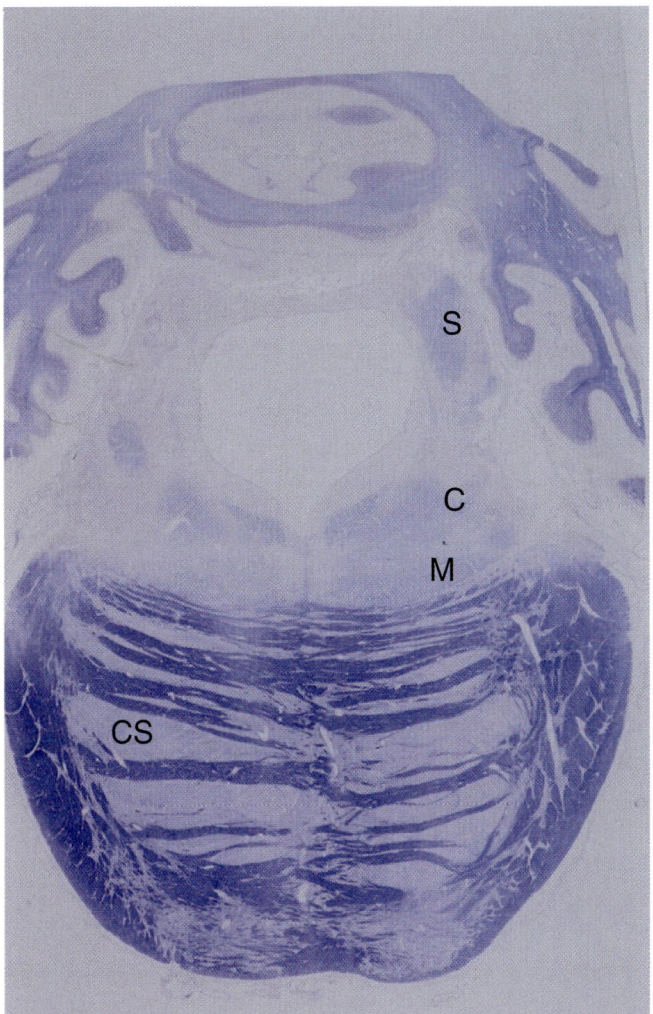

Fig. 6.51. Low-power micrograph of chronic sequelae of severe diffuse axonal injury shows transverse section of rostral pons with virtual absence of myelin staining of tracts of pontine tegmentum. Only faint islands of myelin remain in superior cerebellar peduncles, central tegmental bundles, and medial lemnisci with medial longitudinal fasciculi best preserved. Fourth ventricle is grossly enlarged. In basis pontis, corticospinal tracts are pale and totally devoid of myelin while crossing pontocerebellar tracts are normally stained S, superior cerebellar peduncle, C, central tegmental bundle; M, medial lemniscus; CS, corticospinal tracts. (Luxol fast blue/cresyl violet.)

transversely passing larger third cranial nerve fiber bundles, show axonal injury (Figs. 6.54–6.56). The arrangement is similar to juxtaposition of corticospinal tract versus pontocerebellar fibers.

Beyond the first 24 hours, AS remain for a time until they gradually diminish, and they may remain demonstrable for some 4 weeks to about 3 months.[20] During this interval, small aggregates of microglia called microglial stars, macrophagic infiltration, and astrogliosis will ensue (see Chap. 3 for further details).

The clinical correlate of severe DAI is usually one in which the victim is rendered unconscious at impact and

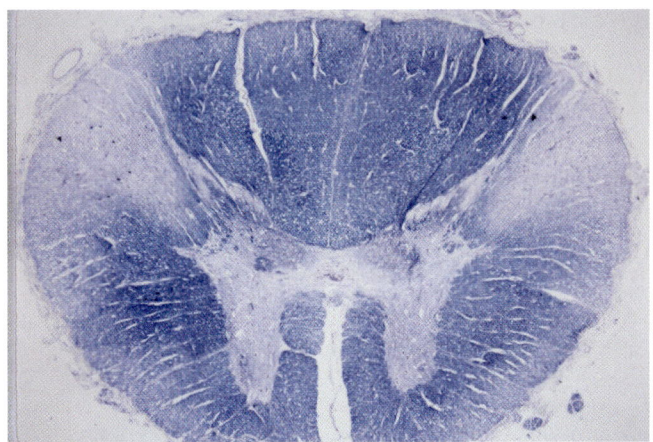

Fig. 6.52. Low-power micrograph of transverse section of thoracic spinal cord of the same case seen in Fig. 6.51 shows atrophy of descending degeneration of lateral corticospinal tracts. (Luxol fast blue/cresyl violet.)

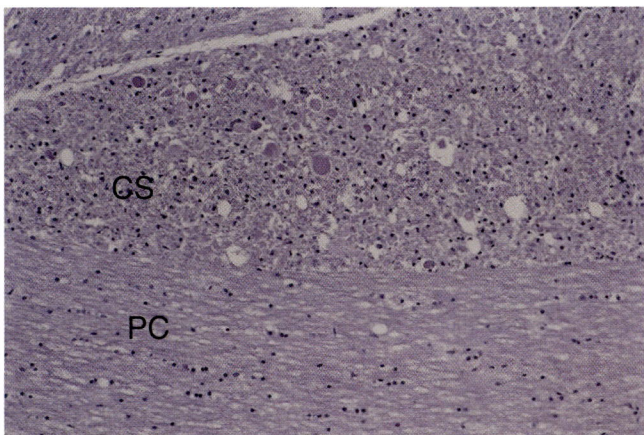

Fig. 6.53. Medium-power micrograph of basis pontis shows axonal spheroids in corticospinal tract (CS), while transverse pontocerebellar (PC) fibers are free of axonal spheroids. (H&E.)

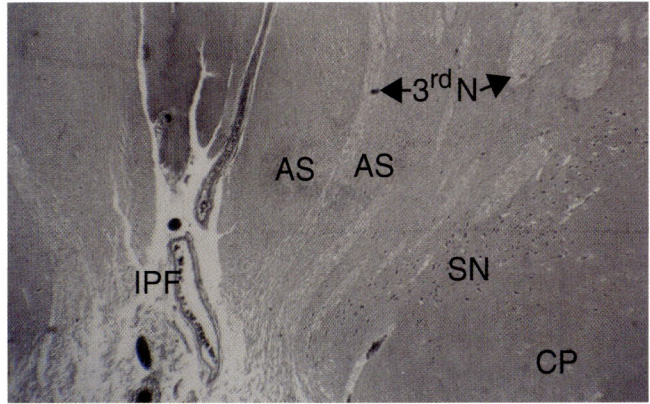

Fig. 6.54. Low-power micrograph of ventral midbrain adjacent to interpeduncular fossa (IPF) of another case shows curvilinear profiles of intramedullary course of third cranial nerve (3rd N) next to which very focal groupings of axonal spheroids (AS) were found. CP = cerebral peduncle, SN = substantia nigra. (Bodian.) Also see Figs. 6.55 and 6.56 from this same case.

remains in a coma to eventuate in severe neurologic disability or in a vegetative state. Having made this statement, as might be expected there are degrees of DAI, and the so-called grade 1 cases may have a lucid period after which a coma may develop. Other cases recover to varying degrees. No universally accepted grading system exists, but one suggested by Adams *et al.*[1] is encountered in several reports. They defined grades of DAI in a series of 122 cases in which 83 (68%) were considered grade 3 (most severe), and, of the 83, 59% were macroscopically apparent, presumably as focal hemorrhage or necrosis in the corpus callosum and midbrain or rostral pons. That is a relatively high percentage of cases in which a diagnosis of DAI can be suspected with some confidence by gross examination alone (see Fig. 6.45). The same mechanism responsible for DAI in the rostral pons may produce a large hemorrhage when accompanied by an underlying disorder predisposing to coagulopathy, such as chronic alcoholism (see Fig. 6.46).

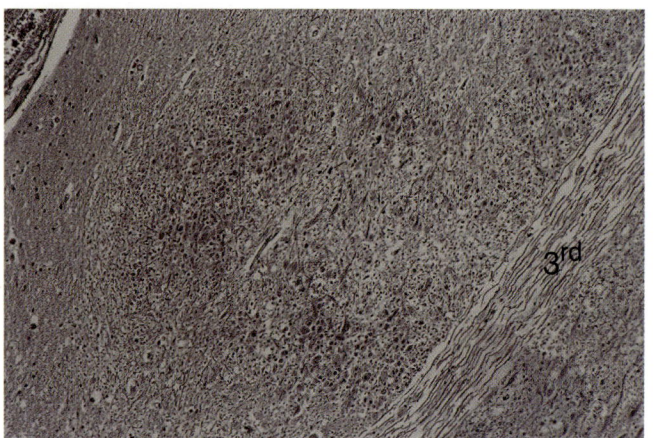

Fig. 6.55. Medium-power micrograph shows third cranial nerve (3rd) in the right corner and localized area of axonal spheroids in the midfield. (Bodian.)

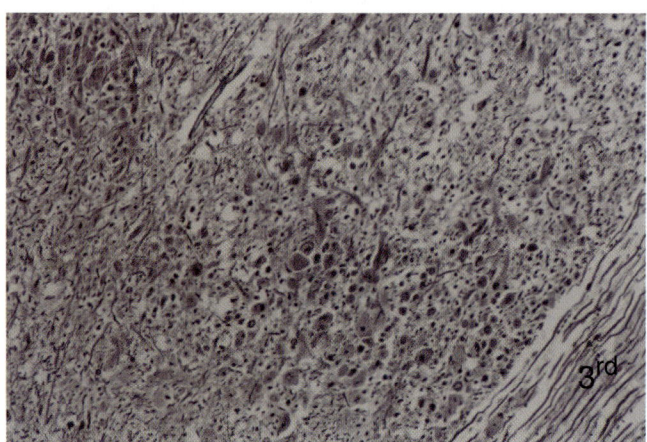

Fig. 6.56. High-power micrograph of the same field as in Fig. 6.55 shows myriads of small axonal spheroids in fine nerve fibers cut transversely. (Bodian.)

Brainstem Avulsion

Avulsion of the brainstem is a form of head injury occasionally encountered in motorcycle and automobile accidents in which the victim is found without signs of life by those immediately on the scene. That is, death is almost instantaneous. Autopsy may reveal an almost bloodless transverse separation of the brainstem by hyperextension through the inferior pontine sulcus beginning at the pyramids.[45] The separation is often complete with distraction of the avulsed brainstem (Fig. 6.57; Case 6.19, Figs. 6.58–6.59). The separated margins may show only minimal petechiae, more on the side proximal to the heart. A small amount of surrounding subarachnoid hemorrhage may be present. One mechanism of injury in motorcyclists is hyperextension of the lower brainstem and spinal cord. Scene investigation often determines that the victim was catapulted off the motorcycle upon collision at an intersection, and a part of the torso, such as the shoulder, was arrested by something unforgiving, like a telephone pole or windshield post of a car, leaving the head free to exert its inertial force. In that circumstance, the brainstem is consistently transversely avulsed at the pontomedullary transition through the inferior pontine sulcus. Retrohyperextension of the head on the neck with impact, such as against the wind-

Fig. 6.57. Midsagittal section of brainstem and cerebellum shows an example of traumatic hyperextension avulsion with total separation at the pontomedullary junction. Note the congestion in the caudal medulla but not in the pons rostral to the separation.

Case 6.19. (Figs. 6.58 and 6.59)

A 28-year-old female pedestrian was struck by a pickup truck and transported to the hospital comatose and apneic. Neurologic examination showed slight arm extension and leg flexion in response to deep pain stimuli. She survived 9 months in a vegetative state with quadriplegia.

shield in a frontal crash, and other forces can produce a similar lesion (Case 6.19). One needs to keep in mind that rough extraction of the brain at autopsy by forceful traction on the brainstem may produce an artifact mimicking the aforementioned lesion.

Ventricular Hemorrhage

Intraventricular hemorrhage as an isolated lesion in death due to head trauma is a very rare occurrence. Furthermore, identification of the source of such hemorrhage is nearly impossible. Potential sources include sheared blood vessels of the septum pellucidum, fornices, and choroid plexus (especially at the glomus), and a rent in the lateral ventricular wall through the vena terminalis, taenia choroidea,[5] and roof of the atrium of the ventricle. The amount is usually relatively small, forming only loose blood clots within the lateral ventricles. Reflux SAH through the fourth ventricle may also be a source of intraventricular hemorrhage with or without trauma. The mechanism of death in such cases is unknown, but a search for diffuse axonal injury is mandatory in such cases even though survival times may be too brief to allow typical histologic changes to develop.

Traumatic Cranial Neuropathies

Clinical investigation indicates that olfactory, facial, and audiovestibular nerve functions are most often affected in blunt force head trauma, whereas lower cranial nerves are rarely involved.[36,37] Severe hyperextension head injury with pontomedullary avulsion may be accompanied by avulsions of the abducens, seventh and eighth, and possibly ninth and tenth cranial nerves, but these nerve injuries would be only pathologic curiosities in lethal brainstem avulsions. Contusions commonly involving the rectus-orbital cortex account for contusion/maceration of the olfactory bulbs, the most commonly damaged cranial nerve in blunt force head trauma. Facial and audiovestibular nerve injuries are less readily confirmable due to the often forceful manner in which the brain is removed from the cranium at autopsy. Routine identification and section of individual cranial nerves, including the lower ones, before transecting the spinal cord and lifting the brain from the cranium is not the norm in most forensic autopsies. Uncal herniation with compression and contusion of the oculomotor nerve in its course over the margin of the tentorium at the incisura and its clinical manifestation of a dilated pupil, nonreactive to light, is familiar to most. However, oculomotor nerve damage and injuries about the orbit may present a variety of third cranial nerve dysfunctions far too

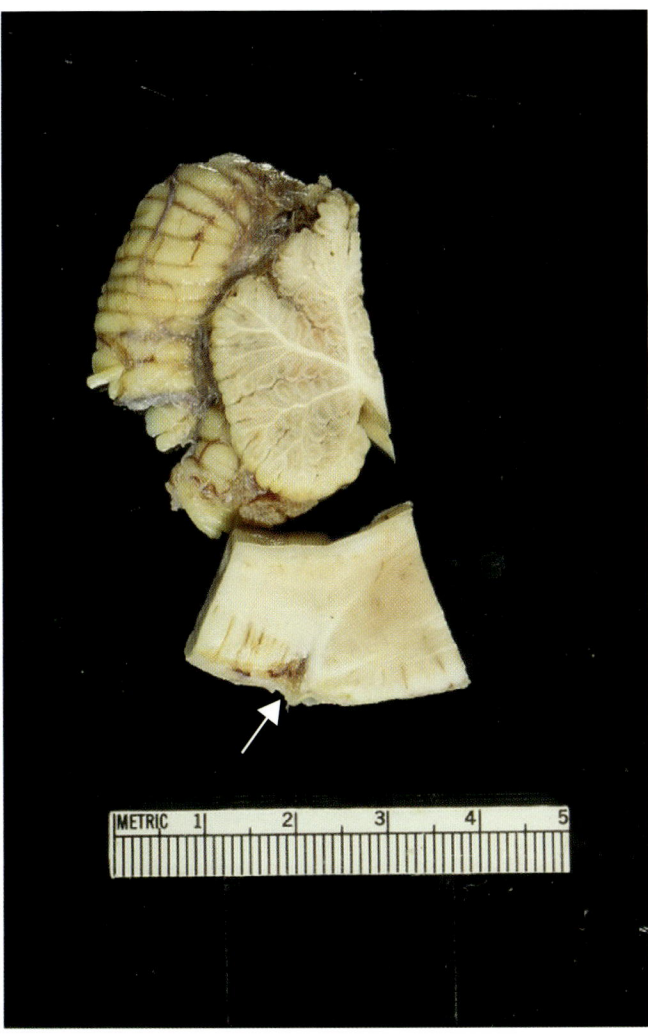

Fig. 6.58. Midsagittal section of a portion of brainstem and cerebellum shows a slight tear of the pontomedullary junction at the inferior pontine sulcus (*white arrow*).

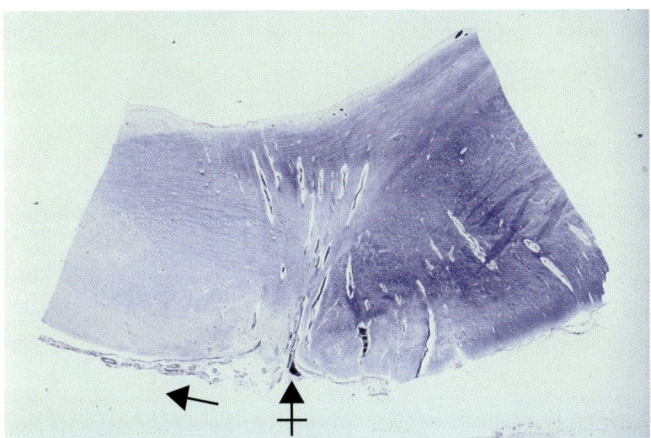

Fig. 6.59. Low-power micrograph of brainstem shows loss of myelin staining caudal to inferior pontine sulcus (*crossed arrow*; *arrow to left* points in caudal direction). Pontine tegmental atrophy of diffuse axonal injury was also present. (Luxol fast blue/cresyl violet.)

complicated to try to establish its pathologic basis in the usual autopsy. More readily found is infarction in the territory of the posterior cerebral artery, compressed with the third cranial nerve during herniation.

Locked-in Syndrome

The term "locked-in syndrome" was coined by Plum and Posner in 1966[62] to describe a condition of paralysis of all four extremities and loss of function of the lower cranial nerves, leaving the victim unable to communicate by gestures or words. Consciousness is retained, and communication is still possible for the victim by coded eye movements, such as a single blink for "yes" and many blinks for "no." The condition had been previously beautifully portrayed in the character of Monsieur Noirtier de Villeforte, in Alexandre Dumas' novel, *The Count of Monte Cristo* in 1844.[77] The pathoanatomic basis for the syndrome is found in lesions destructive of corticospinal and corticobulbar tracts in the midbrain or pons while sparing the midbrain and pontine tegmentum, leaving oculomotor nuclei intact (Case 6.20, Figs. 6.60–6.65). In the event of loss of abducens nerves, only vertical eye movements are possible. Cerebrovascular disease with pontine infarction,[25] pontine tumor, pontine hemorrhage, central pontine myelinolysis, and head injury with injury of the brainstem[36,38] have been reported as causes of the syndrome[62] (also see discussion in Chap. 11).

Boxing and Other Sports

Professional and amateur boxing, a sport with intent and license to do harm, has been long recognized as a cause of traumatic head injury, mainly in the form of acute subdural hematoma and the devastating long-term effect of chronic traumatic encephalopathy (CTE), also known as "punch drunk syndrome"[50] and dementia pugilistica.[54] The reports of acute subdural hematoma are small in number but with a high risk of lethal outcome (Case 6.21, Figs. 6.66–6.68). The number is said to be no greater than with other sports.[10] Although treatable, lethal outcome or disability (50% and 36%, respectively) from subdural hematoma, even in good hands, still leaves much to be desired.[33] Because these SDHs develop acutely in most fatal cases, the probability of concomitant brain injury is high and likely contributes to the high fatality rate. However, the chance of recovery exists. Such is not the case with CTE, a delayed-onset disorder of brain characterized by progressive personality and behavioral abnormalities, parkinsonism, tremor, ataxia, and cerebellar signs.[68] Onset of the disorder was delayed an average of 16 years (6–40 years) after starting boxing in one study.[11] In a clinical study of a selected group of professional and amateur, active and retired boxers,

> **Case 6.20.** (Figs. 6.60–6.65)
>
> A 28-year-old man was assaulted in a robbery attempt and was hit with a blunt instrument over the right forehead, losing consciousness temporarily. When his forehead was sutured, he was alert and oriented. There was no skull fracture, and he was released. He was returned to the hospital 3 hours later by family who witnessed a generalized seizure. He was comatose with a dilated right pupil. CT head scan revealed a marked midline shift due to a right epidural hematoma that was evacuated. Postoperatively, he was thought to be in a coma, but he blinked to visual threat in the right visual field but not the left, and he followed objects visually, indicating a locked-in-syndrome. He died 6½ months later.

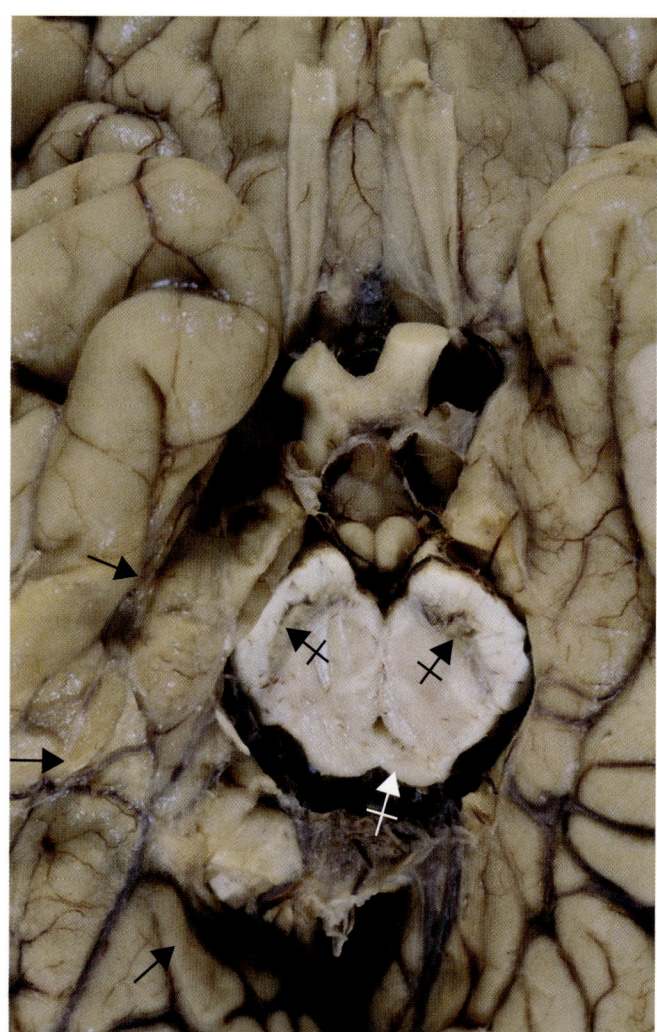

Fig. 6.60. Right (left in photo) uncal/parahippocampal infarction (*black arrows*) lies within territory of anterior temporal branch of posterior cerebral artery. Midbrain contains minute rusty brown necrosis of middle sector of cerebral peduncle (corticospinal tract) on both sides (*crossed black arrows*). Periaqueductal gray and midtegmentum appear intact with only a minute focus of necrosis in left tectum (*crossed white arrow*).

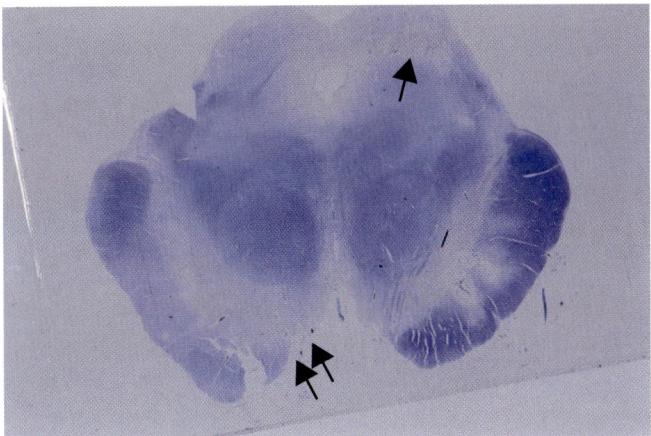

Fig. 6.61. Patchy infarctions of midportions of cerebral peduncles seen as loss of myelin staining, greater on the right (notched). Small infarct of tectum is on the left (*arrow*). Note loss of oculomotor nerve on the right (*double arrows*). Oculomotor nerve on the left is intact. (Luxol fast blue/cresyl violet.)

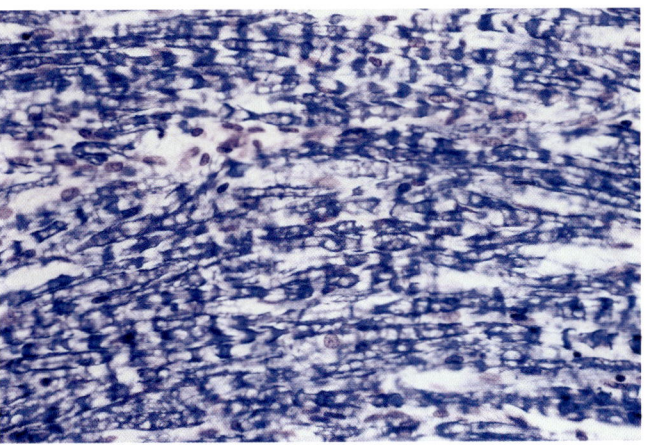

Fig. 6.64. Normal left oculomotor nerve. (Luxol fast blue/cresyl violet.)

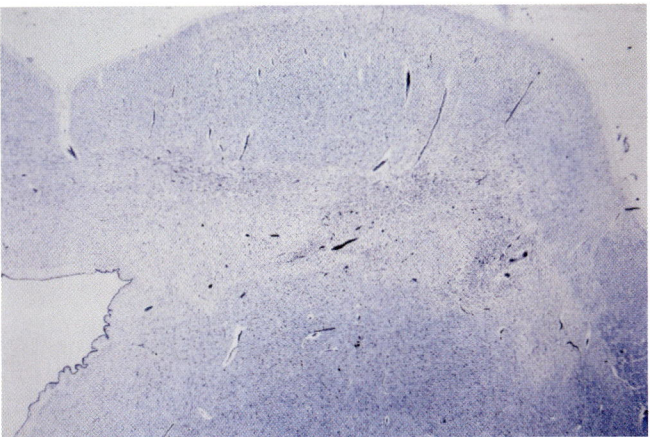

Fig. 6.62. Closeup of necrosis of left tectum. (Luxol fast blue/cresyl violet.)

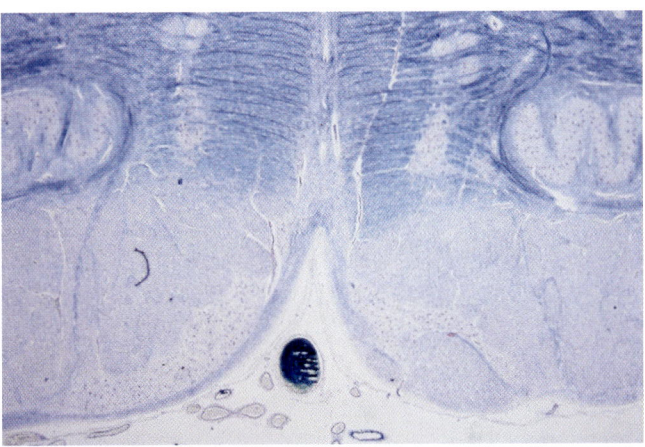

Fig. 6.65. Atrophic pyramids reflect severe descending tract degenerations seen as lost myelin staining. (Luxol fast blue/cresyl violet.)

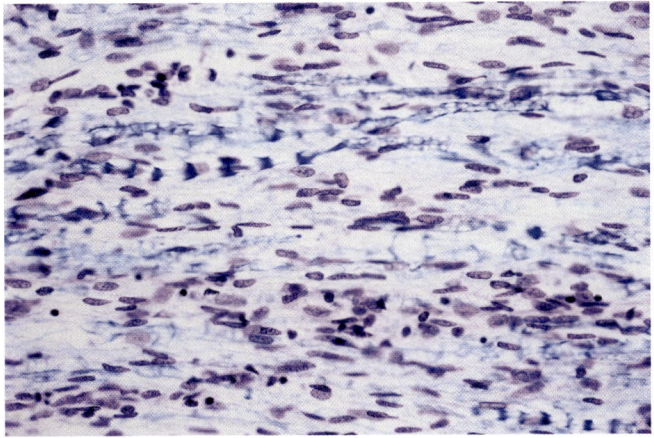

Fig. 6.63. Atrophic right oculomotor nerve.

studies have shown high percentages of abnormalities on neurologic examination, electroencephalogram, psychometric testing, and CT scan.[6,52] One study showed that 87% of the boxers had evidence of brain damage based upon these tests, including 45% with abnormal CT scans. Three of eight boxers with abnormal CT scans had cavum septi pellucidi.[6] Victims of this syndrome are not limited to boxers, as cases of CTE have been noted in participants in other sports such as soccer,[51] rugby, wrestling, parachuting, steeple chase,[11] NFL football,[58] and in traumatic brain injury unrelated to sports, such as in a female victim of spousal abuse.[65] The case of Omalu et al.[58] is of special interest, as it allegedly represents the first neuropathologically documented case of CTE in an NFL football player, a lineman. Studies of concussion in NFL players found wide receivers, kick return carriers, running backs, and quarterbacks more at risk.[61]

CTE is characterized pathologically by cerebral atrophy with neurofibrillary tangles (NFTs) in the absence of senile plaques,[8] although some plaques might be expected in older survivors as a change of aging. NFTs are found diffusely in the neocortex in conformity with the topographic distribution of Alzheimer's disease and in brainstem areas such as the substantia nigra[20] (Case 6.22, Figs. 6.69–6.70). Diffuse (non-neuritic) amyloid plaques may be demonstrated by the immunohistochemical method for β-APP, but not by Congo red stain.[20] Other findings included broadly torn and fenestrated cavum septi pellucidi, pale substantia nigra with loss of pigmented neurons and presence of neurofibrillary tangles, and cerebellar folial atrophy with loss of Purkinje cells. Omalu's case of a football player is of interest in that neurofibrillary tangles were sparse in the neocortex and absent in the hippocampus proper and entorhinal cortex, but diffuse amyloid plaques were frequent.[58] A recent case encountered by the authors of a 68-year-old professional football player, a running back, with personality change and memory impairment is apropos (Case 6.22, Figs. 6.69–6.70). Quite opposite to the Omalu case, neurofibrillary tangles were frequent in the hippocampus proper with considerable loss of neurons in CA2 (Fig. 6.69), and subtle in the neocortex. Diffuse plaques were present but rare in the cerebral cortex. Moderate dropout of neurons in the substantia nigra was accompanied by a few neurofibrillary tangles.

Case 6.21. (Figs. 6.66–6.68)

A 20-year-old male professional boxer collapsed in the ring during a match and was unresponsive. Studies a few hours after the fight revealed a small right-sided subdural hematoma and massive swelling of the right cerebral hemisphere. A partial frontal lobectomy was performed, but he remained in a coma postoperatively and died 2½ days after collapse.

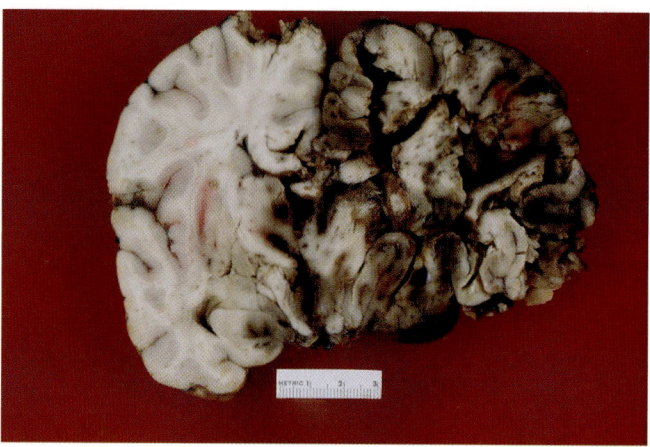

Fig. 6.67. Coronal section at anterior thalamic level shows massive hemorrhagic infarction of entire right hemisphere, also involving contralateral basal ganglia and thalamus.

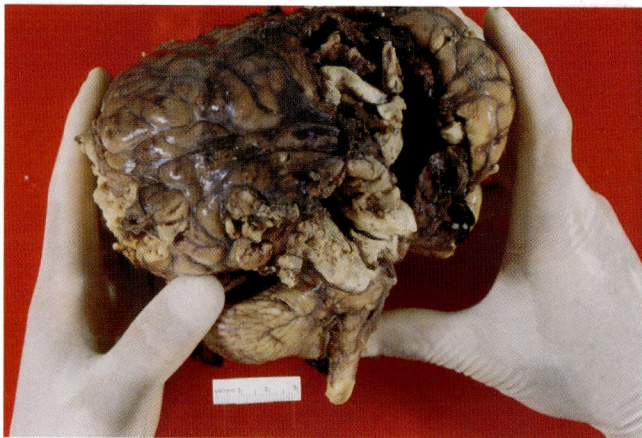

Fig. 6.66. Frontal lobe is to the right of photo, and it shows a large hemorrhagic cavitation at site of lobectomy.

Fig. 6.68. Tegmental hemorrhage of midbrain and pons has the appearance of Duret's hemorrhage more than primary impact injury, although underlying diffuse axonal injury cannot be ruled out on gross inspection.

> **Case 6.22.** (Figs. 6.69 and 6.70)
>
> A 68-year-old professional football running back, a legend in his time, had a 16-year career in the NFL. His friends noted a personality change and declining memory in his last two years. No formal psychological testing was available.

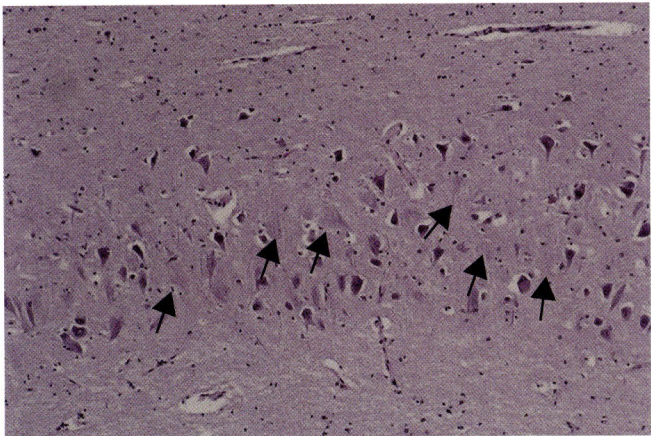

Fig. 6.69. Low-power photomicrograph of hippocampal formation shows loss of neurons in CA2 with a few neurofibrillary tangles and numerous ghost tangles (*arrows*) in the absence of senile plaques. (H&E.)

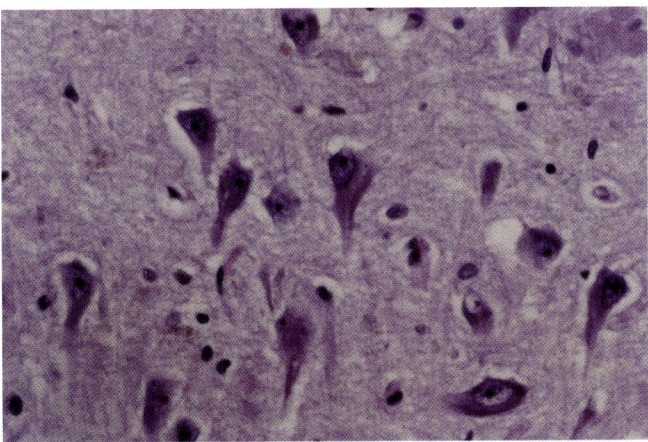

Fig. 6.70. High-power micrograph of hippocampus proper demonstrates neurofibrillary tangles in this case of chronic traumatic encephalopathy. (H&E.)

Consequences of Head Injury

Head injury potentially results in a wide variety of consequences that interact in complex ways. A common method of categorizing these consequences is to divide them into primary and secondary effects.

Primary Effects

Primary effects result from direct injury to central nervous system (CNS) parenchymal and supporting structures. The instigating injury may be traumatic (e.g., blunt force, sharp force, missile injury) or nontraumatic (infarction, hemorrhage, or insults such as hypoglycemia or infection). Some regard herniation syndromes as primary and others consider them as secondary. Direct traumatic injury to neurons, glia, vascular elements, and covering membranes will, depending on the volume of tissue involved and a variety of other factors (e.g., age, general health, involvement of other organ systems), determine the immediate clinical neurologic symptomatology depending on the site(s) and severity of insult. Review of serial clinical and imaging studies can be helpful in attempting to distinguish primary from secondary injury effects. If such studies are not available, and the survival period allows the development of secondary effects, anatomic markers of secondary effects may obscure the primary injury or injuries sufficiently to make distinction between the two categories more difficult. From a more practical standpoint, however, such distinctions are less important than being able, with reasonable medical certainty, to explain how the original injury initiated an unbroken chain of events that led to the individual's demise. If the primary injury or injuries do not cause immediate death, longer survival allows development of secondary effects.

Secondary Effects

The order in which these secondary effects are discussed is arbitrary, and it does not necessarily reflect their relative frequency or importance. It is also apparent that many of these potential complications may interact and induce or enhance one another in complex ways.

Secondary vascular effects may include events such as external compression due to brain parenchymal swelling or hematoma, hydrocephalus, an increase in intracranial pressure, impaired reactivity to physiologic stimuli such as O_2 and CO_2 levels, vascular dilatation; vascular spasm, thrombosis or embolism, infarction, abnormal vascular permeability with tendency for edema and/or hemorrhage, and traumatic aneurysms or arteriovenous fistulae.[18,63,76] Such events may occur rapidly after injury (e.g., minutes to hours), or in some cases may appear after a delay of days to weeks (e.g., rupture of traumatic aneurysm; delayed intracerebral hemorrhage). Intracranial hemorrhages may include epidural hematoma, subdural hematoma, subarachnoid hemorrhage, and/or brain parenchymal hemorrhage. These may occur singly or in combination, with acute or delayed onset. The reader is referred to the discussion of these separate lesions elsewhere in this text.

Brain swelling is a common consequence of a wide variety of CNS insults, including trauma, and is extensively discussed in most current neuropathology and forensic pathology textbooks, to which the reader is referred. Recent human and animal studies emphasize ionic and neurotransmitter disturbances as key mechanisms in its production.[30,64] Malignant cerebral swelling is discussed elsewhere in this text (see Chap. 3). Consequences of brain swelling, with or without other post-traumatic mass lesions such as hematomas, can lead to increased intracranial pressure, herniation syndrome (Case 6.23, Figs. 6.71–6.72), cranial nerve palsies, respirator (nonperfused) brain, and hypothalamic and/or pituitary compromise leading to the syndrome of inappropriate antidiuretic hormone secretion and other complications.

The role of immediate post–head injury apnea as a contributory factor in subsequent events, leading to hypoxic-ischemic CNS injury, has been increasingly appreciated. Levine et al.[43] briefly described two cases, at least one of which strongly suggested that traumatically induced apnea of significant degree may occur even when the CNS injury producing it was largely or completely recoverable. Subsequent studies have reinforced the importance of early treatment of post-traumatic apnea in determining prognosis.[4,34,40]

Catecholamine surge plays an important role in the outcome of head injury. Experimental studies in animals have confirmed clinical impressions in humans that sympathomimetic effects of head injury (e.g., elevated blood pressure and pulse rate) are a consequence of rapid and massive increases in factors such as plasma epinephrine and norepinephrine.[4] This catecholamine surge is also considered to be an important component in development of certain other secondary complications such as coagulopathy, hyperglycemia, gastric mucosal ulceration, myocardial injury, neurogenic pulmonary edema, and the potential for systemic inflammatory response syndrome (SIRS).[4,39,46]

Cardiac complications of acute CNS injury can be found in a wide variety of circumstances. Catecholamine surge is believed responsible for at least a majority of such consequences. Hypokinesis ("stunned" myocardium) with various degrees of cardiac output failure; cardiac wall motion abnormalities; myocardial injury with elevated cardiac enzymes; and a variety of cardiac rhythm disturbances (e.g., bradycardia, asystole, ventricular fibrillation), and conduction disturbances (e.g., QT interval prolongation; QRS, ST segment, and T-wave abnormalities; and virtually every other static ECG abnormality) have been reported.[46,57,70] Histologic evidence of myocardial contraction band necrosis is found in some cases.[57]

Neurogenic pulmonary edema is considered to be, at least in large part, a consequence of catecholamine surge

Case 6.23. (Figs. 6.71 and 6.72)

A 38-year-old man was witnessed in an argument with a suspect who struck the decedent and caused him to fall and strike his head on the pavement. CT head scan revealed a left frontotemporoparietal subdural hematoma with midline shift and obliteration of the third ventricle and basal cisterns. Ventriculostomy was performed on the day of admission, but there was no craniotomy. He remained comatose and died 6 days later.

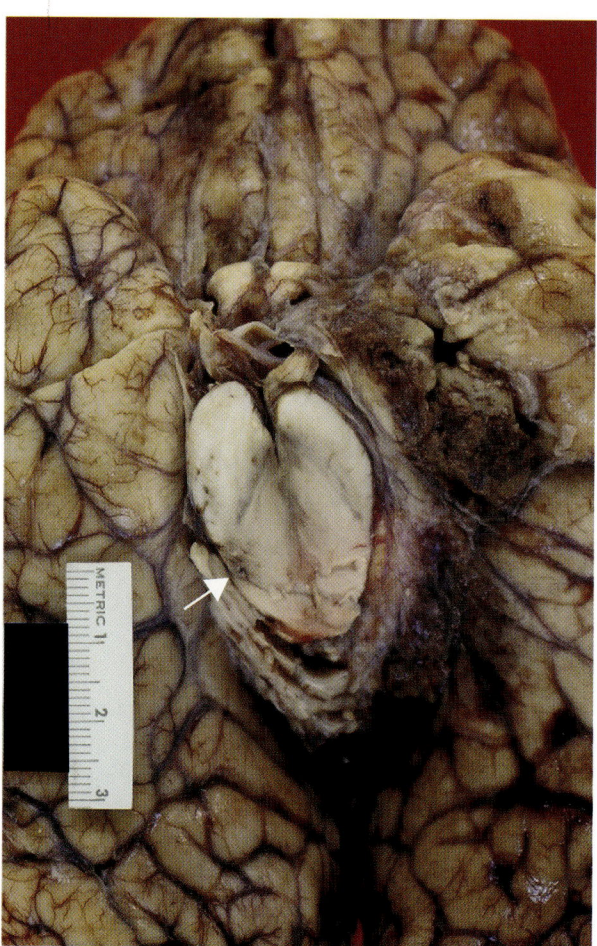

Fig. 6.71. Closeup view of midbrain shows deep notch of marked, slightly hemorrhagic, uncal-parahippocampal herniation across which blood vessels (including left posterior cerebral artery) pass. Note side-to-side compression of midbrain but no Duret's hemorrhage. A few minute petechiae along right lateral tegmentum suggest Kernohan's notch effect (*arrow*).

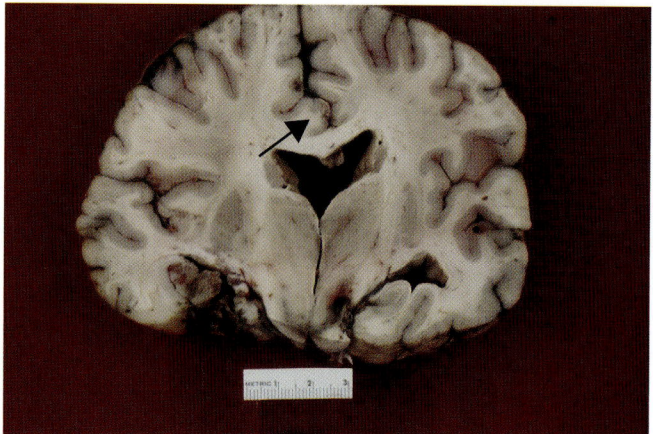

Fig. 6.72. Coronal section at midthalamic level reveals hemorrhagic infarction, including hippocampus, in territory of anterior temporal branch of left posterior cerebral artery. Note left cingulate gyrus herniation (*arrow*).

and its associated cardiac effects. It may be a major contributory factor to death following a variety of CNS insults such as head trauma, spontaneous subarachnoid hemorrhage, cerebral embolism, cerebral infarction secondary to vertebral artery dissection, inflicted suffocation, bacterial meningitis, epilepsy, brain tumors, medullary hemorrhage, increased intracranial pressure, acute hydrocephalus, and the use of some anesthetic agents.[15,32,41,46,53,60,69,72,75] Other pulmonary complications including impaired breathing, impaired cough and gag reflex, and need for assisted ventilation secondary to brain injury may lead to other, nonspecific pulmonary complications, such as aspiration pneumonia, adult respiratory distress syndrome, and pulmonary embolism.[15,24,66] Direct chest trauma accompanying the CNS trauma may add conditions such as pneumothorax, hemothorax, flail chest, and empyema to the patient's problem list.

Cerebral parenchymal injury may result in coagulopathy, including disseminated intravascular coagulation. This sequence of events is important to recognize, so that an intracerebral hemorrhage due to trauma is not incorrectly assumed to be necessarily a result, and not the cause, of the subsequently discovered coagulopathy.[29,31,39]

Systemic inflammatory response syndrome (SIRS) is still incompletely understood, but there is an overexuberant response to a variety of infectious and/or noninfectious insults, the latter category including trauma. This leads to release into the circulation of excessive amounts of the various molecules ordinarily involved in local inflammatory reactions. The end result in the more severe forms of SIRS may include septic encephalopathy (a potentially reversible condition)[26] or the more serious and potentially fatal multiple-organ dysfunction syndrome (MODS).[3,55]

The suggestion that excitotoxicity-induced acute or chronic neuronal injury follows head trauma or other CNS insults such as hypoglycemia is based primarily on animal experimental studies, but is persuasive.[74]

Convulsive disorders of a variety of types, and hydrocephalus in either noncommunicating or communicating forms, may result following head trauma. Either or both can occur during the acute stage or after a brief, moderate, or considerable delay.[18,20] Impaired immune responsiveness may occur following severe trauma or systemic illness, further reducing the victim's resistance to complications such as infection.[35] Impaired homeostatic mechanisms may lead to fluid and electrolyte imbalance (e.g., syndrome of inappropriate antidiuretic hormone secretion, hyperglycemia, and other metabolic disorders). Irreversible shock is a not uncommon component of terminal events preceding death in these cases.

A careful review of all available information, including first responder and hospital records, correlated with general autopsy findings, is essential in order to reconstruct the most logical sequence of primary versus secondary events.

References

1. Adams JH, Doyle D, Ford I, et al. Diffuse axonal injury in head injury: Definition, diagnosis and grading. Histopathology 1989; 15:49–59.
2. Adams JH, Mitchell DE, Graham DI, Doyle D. Diffuse brain damage of immediate impact type: Its relationship to primary brain-stem damage in head injury. Brain 1971;100:489–502.
3. Anderson WR. Forensic Sciences in Clinical Medicine: A Case Study Approach. Philadelphia: Lippincott-Raven, 1998.
4. Atkinson JLD. The neglected prehospital phase of head injury: Apnea and catecholamine surge. Mayo Clin Proc 2000;75:37–47.
5. Berry K, Rice J. Traumatic tear of tela choroidea resulting in fatal intraventricular hemorrhage. Am J Forensic Med Pathol 1994; 15:132–137.
6. Casson IR, Siegel O, Sham R, et al. Brain damage in modern boxers. JAMA 1984;251:2663–2667.
7. Condon BR, Patterson J, Wyper D, et al. A quantitiative index of ventricular and extraventricular intracranial CSF volumes using MR imaging. J Comput Assist Tomogr 1989;10:784–792.
8. Constantinidis J, Tissot R. Lesions neurofibrillaires d'Alzheimer generalisees sans plaque seniles. Arch Suisse Neurol Neurochirug Psychiatrie 1967;100:117–130.
9. Contostavlos DL. Massive subarachnoid hemorrhage due to laceration of the vertebral artery associated with fracture of the transverse process of the atlas. J Forensic Sci 1971;16:40–56.
10. Corsellis JAN. Boxing and the brain. BMJ 1989;298:105–109.
11. Critchley M. Medical aspects of boxing, particularly from a neurological standpoint. BMJ 1957;1:357–366.
12. DiMaio DJ, DiMaio VJM. Forensic Pathology. New York: Elsevier, 1989.
13. Evans FG, Lissner HR, Lebow M. The relation of energy, velocity, and acceleration to skull deformation and fracture. Surg Gynecol Obstet 1958;107:593–601.
14. Friede RL, Schachenmayr W. The origin of subdural neomembranes. II. The fine structure of neomembranes. Am J Pathol 1978;92:69–84.
15. Friedman JA, Pichelmann MA, Piepgras DG, et al. Pulmonary complications of aneurysmal subarachnoid hemorrhage. Neurosurgery 2003;52:1025–1032.
16. Gadd CW, Nahum AM. Head and facial bone impact tolerances. Proceedings Automobile Safety Seminar, Milford, 1968.
17. Geddes JF, Whitwell HL, Graham DI. Traumatic axonal injury: Practical issues for diagnosis in medicolegal cases. Neuropathol Appl Neurobiol 2000;26:105–116.
18. Gilles EE, Nelson MD Jr. Cerebral complications of nonaccidental head injury in childhood. Pediatr Neurol 1998;19:119–128.
19. Goldstein M. Traumatic brain injury: A silent epidemic. Ann Neurol 1990;27:327.
20. Graham DI, Lantos PL (eds). Greenfield's Neuropathology, ed 6. London: Arnold, 1997.
21. Gurdjian ES, Webster JE, Lissner HR. The mechanism of skull fracture. Radiology 1950;54:313–338.
22. Gurdjian ES, Webster JE, Lissner HR. The mechanism of skull fracture. J Neurosurg 1950;7:106–114.
23. Gurdjian ES, Webster JE, Lissner HR. Studies on skull fracture with particular reference to engineering factors. Am J Surg 1984;98: 736–742.
24. Hammond FM, Meighen MJ. Venous thromboembolism in the patient with acute traumatic brain injury: Screening, diagnosis, prophylaxis, and treatment issues. J Head Trauma Rehabil 1998;13: 36–50.
25. Hawkes CH. "Locked-in" syndrome: Report of seven cases. BMJ 1974;4:379–382.
26. Hemphill JC. Critical care neurology. In Kasper DL, Braunwald E, Fauci AS, et al (eds): Harrison's Principles of Internal Medicine, ed 16. New York: McGraw-Hill, 2005, pp 1631–1637.

27. Hendrick EB, Hardwood-Hash DCF, Hudson AR. Head injuries in children: A survey of 4465 consecutive cases at the Hospital for Sick Children, Toronto, Canada. Clin Neurosurg 1963;11:46–65.
28. Holbourn AH. Mechanics of head injury. Lancet 1943;245:438–441.
29. Huber A, Dorn A, Witzmann A, Cervós-Navarro J. Microthrombi formation after severe head trauma. Int J Leg Med 1993;106:152–155.
30. Hubschmann OR. Potassium and intracranial pressure (letter to editor). J Neurosurg 2001;94:1025–1027.
31. Hymel K, Abshire T, Luckey D, Jenny C. Coagulopathy in pediatric abusive head trauma. Pediatrics 1997;99:371–375.
32. Inobe J, Mori T, Ueyama H, Kumamoto T, Tsuda T. Neurogenic pulmonary edema induced by primary medullary hemorrhage: A case report. J Neurol Sci 2000;172:73–76.
33. Jennett B, Teasdale G. Management of head injuries. Philadelphia: FA Davis, 1981.
34. Johnson DL, Boal D, Baule R. Role of apnea in nonaccidental head injury. Pediatr Neurosurg 1995;23:305–310.
35. Kampalath B, Cleveland RP, Chang C-C, Kass L. Monocytes with altered phenotypes in post-trauma patients. Arch Pathol Lab Med 2003;127:1580–1585.
36. Keane JR. Neurologic signs following motorcycle accidents. Arch Neurol 1989;46:761–762.
37. Keane JR, Baloh RW. Post-traumatic cranial neuropathies. In Evans RW (ed): Neurology and Trauma. Philadelphia: WB Saunders, 1996.
38. Keane JR, Itabashi HH. Locked-in syndrome due to tentorial herniation. Neurology 1985;35:1647–1649.
39. Kearney TJ, Bentt L, Grode M, et al. Coagulopathy and catecholamines in severe head injury. J Trauma 1992;32:608–612.
40. Kemp AM, Stoodley N, Cobley C, et al. Apnea and brain swelling in nonaccidental head injury. Arch Dis Child 2003;88:472–476.
41. Kerr GW. Neurogenic pulmonary edema. J Accid Emerg Med 1998;15:275–276.
42. Knight B. Forensic Pathology, ed 2. London: Arnold, 1996.
43. Levine JE, Becker D, Chun T. Reversal of incipient brain death from head-injury apnea at the scene of accidents (letter to editor). N Engl J Med 1979;301:109.
44. Lindenberg R. Trauma of meninges and brain. In Minckler J (ed): Pathology of the Nervous System, Vol 2. New York: McGraw-Hill, 1971.
45. Lindenberg R, Freytag E. Brainstem lesions characteristic of traumatic hyperextension of the head. Arch Pathol 1970;90:509–515.
46. Macmillin CSA, Grant IS, Andrews PJD. Pulmonary and cardiac sequelae of subarachnoid haemorrhage: Time for active management? Intensive Care Med 2002;28:1012–1023.
47. Maki Y, Akimoto H, Enomoto T. Injuries of basal ganglia following head trauma in children. Child's Brain 1980;7:113–123.
48. Mant AK. Traumatic subarachnoid hemorrhage following blows to the neck. J Forensic Sci Soc 1972;12:567–572.
49. Maron BJ, Poliac LC, Ashare AB, Hall WA. Sudden death due to neck blows among amateur hockey players. JAMA 2003;290:599–601.
50. Martland HS. Punch drunk. JAMA 1928;91:1103–1107.
51. Matser JT, Kessels AGH, Jordan BD, et al. Chronic traumatic brain injury in professional soccer players. Neurology 1998;51:791–796.
52. McLatchie G, Brooks N, Galbraith S, et al. Clinical neurological examination, neuropsychology, electroencephalography and computed tomographic head scanning in active amateur boxers. J Neurol Neurosurg Psychiatry 1987;50:96–99.
53. McManis P, Lee C, Morgan M, Stewart D. Neurogenic pulmonary edema (letter to editor). Aust N Z J Med 2000;30:514.
54. Millspaugh JA. Dementia pugilistica. U S Naval Bull 1937;35:297–302.
55. Munford RS. Severe sepsis and septic shock. In Kasper DL, Braunwald E, Fauci AS, et al (eds): Harrison's Principles of Internal Medicine, ed 16. New York: McGraw-Hill, 2005, pp 1606–1612.
56. Munro D, Merritt HH. Surgical pathology of subdural hematoma: Based on a study of one hundred and five cases. Arch Neurol Psychiatry 1936;35:64–75.
57. Naidech AM, Kreiter KT, Janjua N, et al. Cardiac troponin elevation, cardiovascular morbidity, and outcome after subarachnoid hemorrhage. Circulation 2005;112:2851–2856.
58. Omalu BI, Dekosky ST, Minster RL, et al. Chronic traumatic encephalopathy in a National Football League player. Neurosurgery 2005;57:128–134.
59. Ommaya AK, Yarnell P. Subdural haematoma after whiplash injury. Lancet 1969;2:237–239.
60. Orme RML'E, McGrath NM, Rankin RJ, Frith RW. Extracranial vertebral artery dissection presenting as neurogenic pulmonary oedema (letter to editor). Aust N Z J Med 1999;29:824–825.
61. Pellman EJ, Viano DC, Casson IR, et al. Concussion in professional football: Repeat injuries—Part 4. Neurosurgery 2004;55:860–876.
62. Plum F, Posner JB. The Diagnosis of Stupor and Coma, ed 3. Philadelphia: FA Davis, 1980, p 9.
63. Ransom GH, Mann FA, Vavilala MS, et al. Cerebral infarct in head injury: Relationship to child abuse. Child Abuse Negl 2003;27:381–392.
64. Reinert M, Khaldi A, Zauner A, et al. High level of extracellular potassium and its correlates after severe head injury: Relationship to high intracranial pressure. J Neurosurg 2000;93:800–807.
65. Roberts GW, Whitwell HL, Acland PR, Bruton CJ. Dementia in a punch-drunk wife. Lancet 1990;335:918–919.
66. Rogers FB, Shackford SR, Trevisani GT, et al. Neurogenic pulmonary edema in fatal and nonfatal head injuries. J Trauma 1995;39:860–866; discussion 866–868.
67. Ross AH, Jantz RL, McCormick WF. Cranial thickness in American females and males. J Forensic Sci 1998;43:267–272.
68. Rowland LP, Sciarra D. Head injury. In Rowland LP (ed): Merritt's Textbook of Neurology, ed 8. Philadelphia: Lea & Febiger, 1989.
69. Rubin DM, McMillan CO, Helfaer MA, Christian CW. Pulmonary edema associated with child abuse: Case reports and review of the literature. Pediatrics 2001;108:769–775.
70. Sakr YL, Ghosn I, Vincent JL. Cardiac manifestations after subarachnoid hemorrhage: A systematic review of the literature. Prog Cardiovasc Dis 2002;45:67–80.
71. Schachenmayr W, Friede RL. The origin of subdural neomembranes. I. Fine structure of the dura-arachnoid interface in man. Am J Pathol 1978;92:53–68.
72. Smith WS, Matthay MA. Evidence for a hydrostatic mechanism in human neurogenic pulmonary edema. Chest 1997;111:1326–1333.
73. Strich SJ. Diffuse degeneration of cerebral white matter in severe dementia following head injury. J Neurol Neurosurg Psychiatry 1956;19:163–185.
74. Whetsell WO Jr. Current concepts of excitotoxicity. J Neuropathol Exp Neurol 1996;55:1–13.
75. Wiercisiewski DR, McDeavitt JT. Pulmonary complications in traumatic brain injury. J Head Trauma Rehabil 1998;13:28–35.
76. Wilkins RH. Intracranial vascular spasm in head injuries. In Vinken PJ, Bruyn GW, Braakman R (eds): Injuries of the Brain and Skull. Handbook of Clinical Neurology. Amsterdam: North-Holland, 1975;23(1):163–189.
77. Williams AN. Cerebrovascular disease in Dumas' The Count of Monte Cristo. J R Soc Med 2003;96:412–414.
78. Yamashima T, Friede RL. Why do bridging veins rupture into the virtual subdural space? J Neurol Neurosurg Psychiatry 1984;47:121–127.
79. Young HA, Gleave JRW, Schmidek HH, Gregory S. Delayed traumatic intracerebral hematoma: Report of 15 cases operatively treated. Neurosurgery 1984;14:422–425.

The Suspected Child Abuse Case

INTRODUCTION 199
PRELIMINARY COMMENTS 199
INFORMATION SOURCES 203
TADD SYNDROME 204
TIN EAR SYNDROME 204
HOW UNIQUE IS THE MECHANISM OF HEAD INJURY IN YOUR CASE? 204
TERSON'S SYNDROME 204
SHAKEN ADULT SYNDROME 205
CRUSH INJURIES TO THE HEAD 205
SUBDURAL HEMORRHAGE IN CHILDHOOD 205
FOLK REMEDIES AND CUSTOMS 205
SPINAL INJURY IN CHILD ABUSE 206
RETINAL HEMORRHAGE 206
REFERENCES 208

Introduction

Cases of blunt force trauma in infants and children lead to more questions directed to the forensic neuropathology consultant by medical examiners, attorneys, and other interested parties than any other single case category, in our experience. It is also our impression that this single topic generates more monthly publications, and more controversial opinions, than any other single topic in forensic pathology at the present time. Indeed, it seems at times that one can find some published article to "support" almost any viewpoint on this topic, no matter how contrary to common sense or to one's own experience. Opinions are far easier to find than facts. Under such circumstances, how might one approach cases in this category (or any other controversial area) with the goal in mind of generating a final report that is detailed, objective, and concludes with an opinion that will withstand careful scrutiny from a variety of interested parties?

In this section, some general principles, hypothetical case examples, and a few selected citations chosen solely for the purpose of illustrating a given point are used to suggest methods we have found helpful in the analysis of suspected child abuse cases. It is not intended to be a comprehensive literature review, and, by no means, a rigid template that will answer all pending issues or even be a guide to the only logical approach to such cases. We urge the forensic neuropathologist new to child abuse case work to develop a basic foundation of knowledge concerning the main areas of agreement and disagreement among workers in this field, keeping in mind, however, the reality that it takes years to develop a reputation as a trustworthy expert witness, but it can be destroyed in minutes.

Preliminary Comments

Several of the general points made elsewhere in this text are applicable to suspected child abuse cases. The general forensic autopsy and pediatric autopsy methods discussed in Chapters 1 and 2 are applicable to these cases. Protocols more specific to suspected child abuse cases are also available,[40,58] and they contain valuable suggestions. Documentation by photography and by diagrams becomes very important, not only to assist the autopsy surgeon writing the report, but also to aid other involved parties (criminal investigators, additional expert witnesses involved in the case, etc.) in their analysis.

How much emphasis to place on finding a few extravasated erythrocytes in cranial dura mater histologic sections, for example, continues to generate controversy. The firm adherence of the dura to the infant skull, in comparison to most young and middle-aged adults, often requires application of considerable traction by forceps when stripping the dura. Dura submitted in such cases is often received in multiple fragments. *In the absence of gross evidence of hemorrhage prior to stripping the dura*, it seems reasonable to discount microscopic evidence of a few extravasated epidural, intradural, or subdural erythrocytes in dura that has been subjected to traction sufficient to have produced postmortem transmural dural tears.

The immature brain, in comparison to mature brains, is also somewhat more likely to develop gliding contusions with blunt force trauma (also see Chap. 6). While more often seen in an acute stage, evidence of unexplained prior gliding contusions in children allegedly found "dead in the crib" can guide investigators to more thoroughly explore the possibility of child abuse (Case 7.1, Figs. 7.1–7.3).

Case 7.1. (Figs. 7.1–7.3)

This 3-month-old infant was allegedly found dead in bed by the caretaker 7 hours after the last feeding. There was no history of trauma.

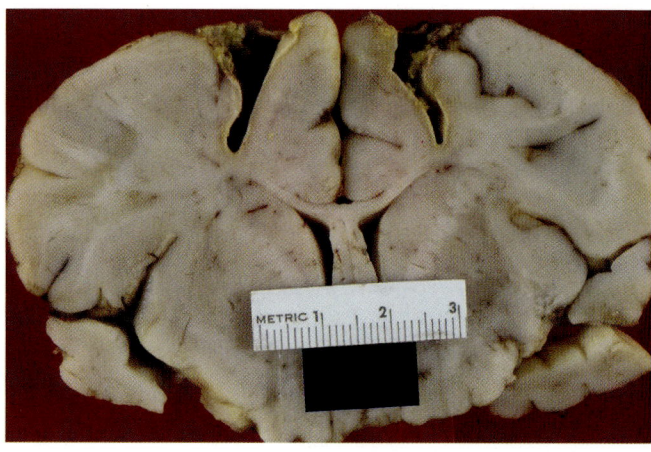

Fig. 7.2. Coronal section of posterior frontal lesions at level of corpus striatum. Chronic sequelae of bilateral gliding contusions.

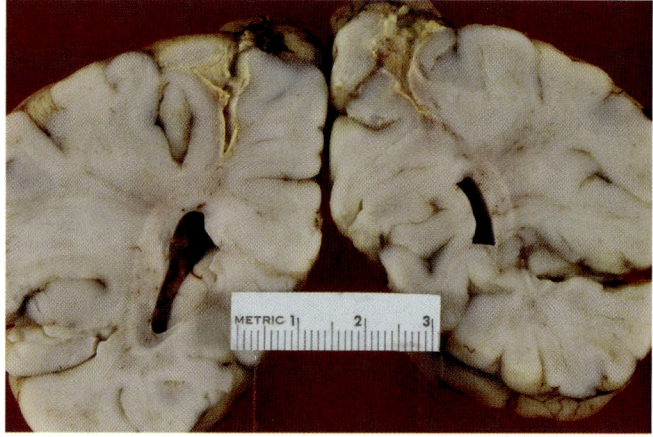

Fig. 7.3. Coronal section of parietal lesions at level of posterior trigone, demonstrating chronic sequelae of bilateral gliding contusions.

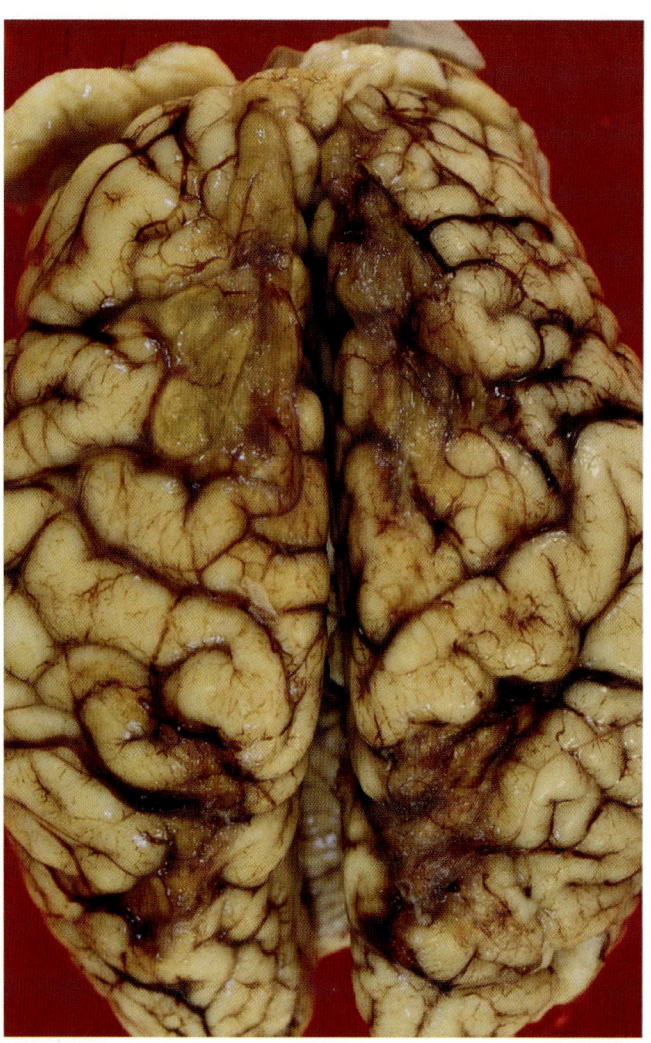

Fig. 7.1. Dorsal view of brain, with brownish-yellow discoloration and partial collapse of the cerebral cortex involving parasagittal areas of posterior frontal and parietal lobes bilaterally.

A useful generalization in evaluating possible child abuse cases is to avoid becoming extensively distracted by various hypothetical causes of a given single finding, to the extent that insufficient emphasis is placed on the totality of the findings and circumstances. Cause and manner of death are to be based on careful documentation of all evidence of injury, correlation of injuries with the mechanism of injury offered in the history provided, and any additional relevant data that can be collected (Case 7.2, Figs. 7.4–7.10).

Microscopic analysis of the aging/dating of lesions must take into consideration the possible unreliability of data provided by caretakers, and one should not be surprised that there are discrepancies between the provided history and what the gross and microscopic findings indicate. In addition, the fatal event is often not the sole traumatic event in abused children, and remote and recent injuries may coexist.[2] Sequelae of severe brain

Case 7.2. (Figs. 7.4–7.10)

A 4-month-old female infant was brought apneic to a medical center with a history of being "ill" for several days after immunization. She was pronounced dead shortly after admission. (Case courtesy of Thomas T. Noguchi, M.D.)

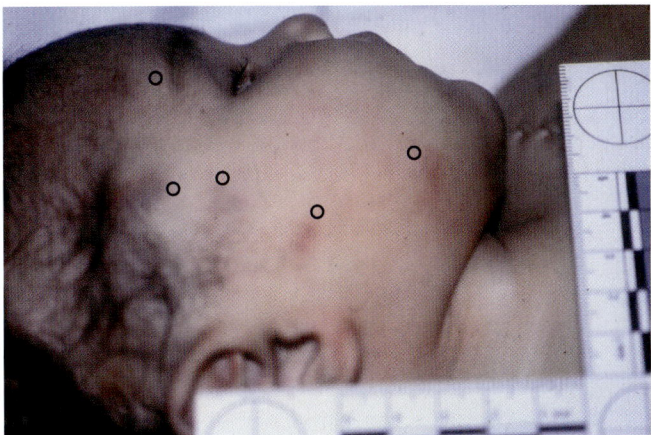

Fig. 7.4. Right side of face of infant shows 5-point skin bruises (adjacent to circles) in a curvilinear distribution from the forehead toward the chin, corresponding to the fingertips of a spread hand (thumb at forehead; greater spacing between the thumb and index finger and between 4th and 5th fingers; middle 3 fingers more closely spaced).

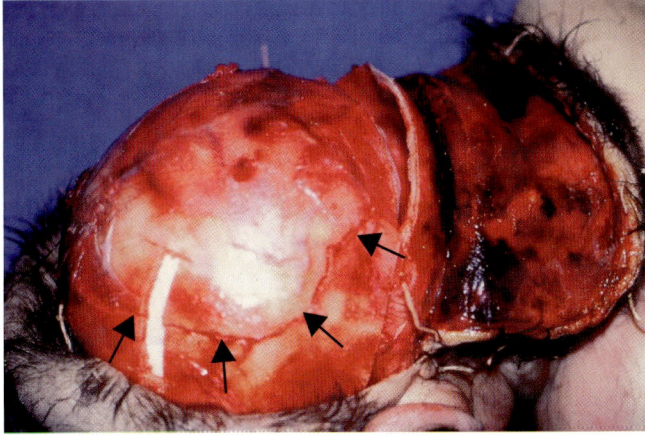

Fig. 7.5. Exposed calvarial surface showing a long curvilinear fracture (*arrows*) of frontal and parietal bones, and galeal and subperiosteal hemorrhage, evidence of trauma unaccounted for by the history given by caretakers.

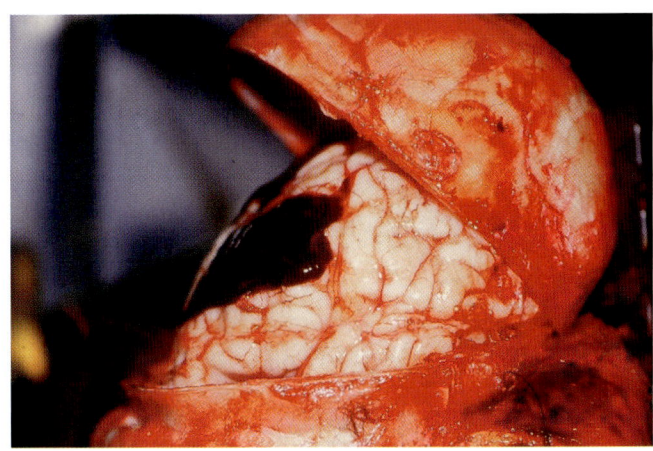

Fig. 7.6. Small acute subdural hematoma underlying the fracture.

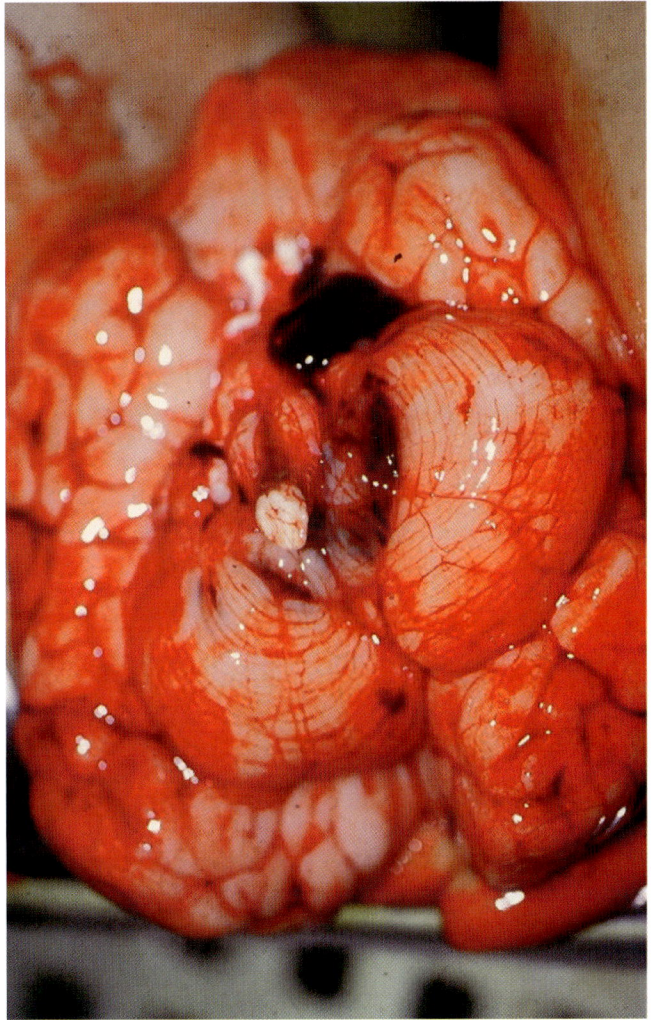

Fig. 7.7. Inferior aspect of cerebellum and lower brainstem in the fresh state, demonstrating acute contused, herniated tonsils, left greater than right, and hemorrhage over the biventer lobules corresponding to the margins of foramen magnum. Impact trauma against the edge of the foramen magnum, rather than simply herniation contusion, is suggested by the findings in Figs. 7.9–7.10.

The Suspected Child Abuse Case

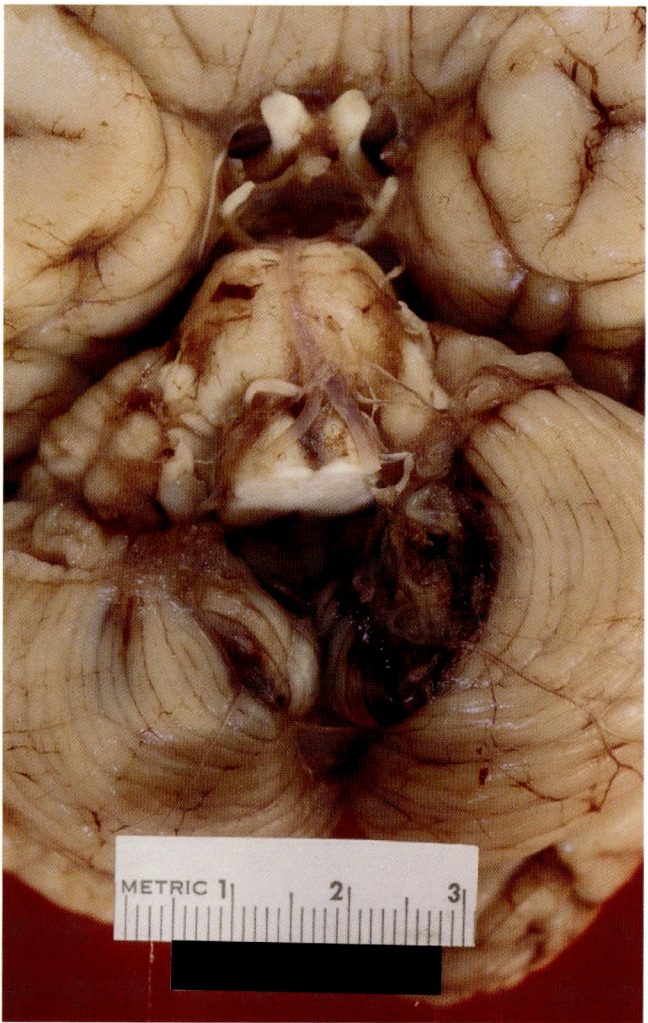

Fig. 7.8. Same area as in Fig. 7.7 postfixation and after removal of caudal medulla, demonstrating contused left (greater than right) tonsil and biventer.

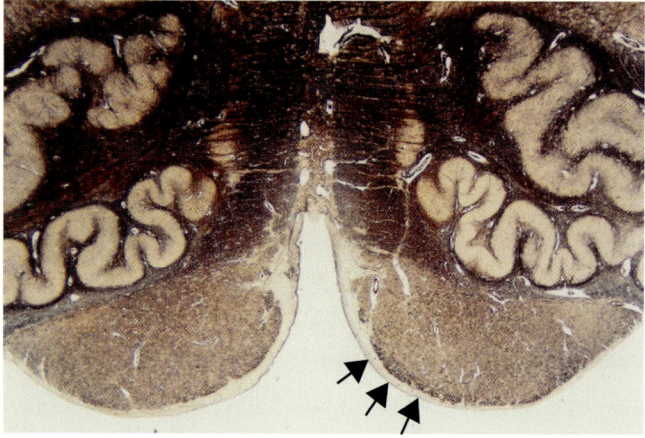

Fig. 7.9. Low-power micrograph of a transverse section of medulla. Arrows point to a faint layer of silver-positive deposits at the subpial surface of the pyramid unilaterally. (Bodian.)

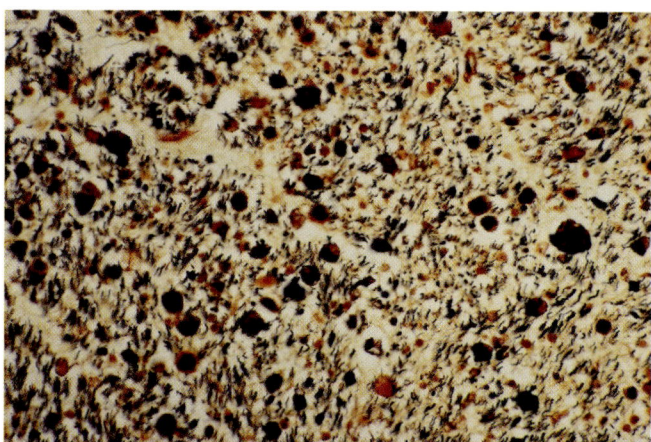

Fig. 7.10. Medium-power micrograph of the subpial area indicated in Fig. 7.9, showing in transverse section swollen axons. (Bodian.) (*Comment*: In another case of child abuse in a 3-month-old infant, this same finding led to the caretaker's confession that the incident leading to acute unresponsiveness of the decedent was vigorous downward compression by the palm over the anterior fontanel region.)

injury have been discussed in Chapter 6, and are applicable to the pediatric age group as well as adults.[59,60]

Writing one's report in suspected child abuse cases will typically follow the general departmental protocol pertaining to that jurisdiction, but the guidelines detailed so lucidly by David[19] are so important that it should be recommended reading. With regard to mechanical models of childhood head injury, our impression is that, to borrow a quote from Peterson,[83] "the real issue is that there is a crucial difference between a 'thing' and a mathematical model of the 'thing'." Given the claims generated by some of these model builders, this point deserves more attention.

Bergmann[7] noted that "advocacy is not a specialty" in medicine. One must use something other than claims of objectivity to assess conflicting viewpoints expressed in publications in this still very controversial subject of pediatric head injury. Some suggestions on how to evaluate the information generated by such diverse viewpoints follow.

Representative Examples of Questions to Consider in Child Abuse Cases

- Is there a post-traumatic lucid interval in infants who later die?
- Can rebleeding in subdural hemorrhage be abrupt and lethal?
- Can shaking alone produce the typical findings in "shaken baby syndrome," or is impact required?
- Is shaken baby syndrome really an entity, or should different terminology be used?
- Are short falls lethal?

- Which causes of natural death may be confused with child abuse?
- What are the ocular findings in child abuse, and are any of them pathognomonic?
- Can seizures, vomiting, or cardiopulmonary resuscitation cause retinal hemorrhages?
- Is diffuse axonal injury common in child abuse cases?
- What do biomechanical models tell us about child abuse cases?
- What are the neuropathologic findings in child abuse cases?

Information Sources

Texts and Monographs

In addition to chapters related to child abuse contained in most textbooks on forensic pathology, or monographs on forensic neuropathology cited elsewhere in this text, there are review series on forensic pathology topics that often contain chapters related to child abuse[12] as well as numerous texts and atlases more specific to the topics of accidental and inflicted trauma in infants and children. Among the latter are the well-known references on child abuse, such as that by Reese and Ludwig,[89] and on the radiologic examination in child abuse cases by Kleinman.[62] Natural causes of sudden unexpected death important in the differential diagnosis of child abuse in infants and older children, as well as chapters on accidental, inflicted, and suicidal deaths in children, are included in Byard's monograph on sudden unexpected death.[10] Publications of conference proceedings on topics such as inflicted childhood neurotrauma[90] are very useful. An inspection of bookseller displays at forensic pathology national meetings will reveal dozens of other relevant texts.

Conferences and Workshops

Several organizations sponsor workshops and/or conferences on various topics relevant to suspected child abuse cases. Platform papers, poster presentations, and workshops on child abuse topics are a regular component of the annual meetings of, for example, the American Academy of Forensic Sciences, the American Academy of Pediatrics, and the Annual North American Conference on Shaken Baby Syndrome sponsored by the National Center on Shaken Baby Syndrome. These and other related meetings are useful in updating one's knowledge of developments in the field and provide opportunities to discuss topics with researchers who have a special interest in this area.

Periodical Literature

Selected review articles are recommended to our trainees to be used initially as a brief overview of consensus opinions and areas of controversy in the child abuse literature. Those that we consider to be a minimum "window" into the vast literature on this topic are the following citations (which include one recent chapter review).* After an initial overview is obtained from such sources, original references cited in these reviews are recommended reading. Articles that propose a very novel theory or viewpoint often prompt letters to the editor in subsequent issues of that journal, and such letters can add information either supporting or rebutting the original article and generate additional data by the original authors. *The Quarterly Update*[99] is an essential component of reviewing the child abuse literature, since it contains many abstracts and critical reviews by specialists in the field. It abstracts a wide range of journals not readily accessible or routinely reviewed in our department.

Colleagues

Many forensic medicine departments have staff members with a special interest in the topic of child abuse. This can not only be an informal source of information based on extensive experience, but may also include individuals with extensive and ongoing contacts with other researchers interested in this topic worldwide. James Ribe, J.D., M.D., is an example of such an individual in our department who also has compiled an extensive, and periodically updated, annotated bibliography of the child abuse literature that is accessible to other staff members via an in-house website. Such individuals can also provide one with both medical (e.g., National Library of Medicine) and legal (e.g., FindLaw) websites that can be a source of useful information on specific child abuse topics.

Public Record Court Transcripts

For most medical examiners and consultants, court transcripts are a less frequently used source of information. However, at times they can provide very helpful information not readily obtained in any other fashion. Court transcript citations are not unusual in articles written by attorneys, but are also starting to appear somewhat more frequently in forensic medicine articles, particularly in the child abuse field. In our jurisdiction, a high percentage of suspected child abuse cases eventually require testimony by the medical examiner and/or consultant involved in the case. It is difficult to overstate the many valuable teaching lessons that can be acquired from a careful review of court transcripts, such as *R. v. L. Harris et al.*[88] Differences in the methods used by medical researchers, and those used by attorneys in an adversarial system such as a courtroom, can be more readily appreciated by reviewing such material. A better understanding of such differences in approach is also provided

*References 4, 5, 9, 11, 12, 14, 17, 23, 32, 50–53, 63, 64, 76–78, 100.

by Bird's overview of this topic.[8] The court's approach to the analysis of particularly controversial topics in the child abuse literature,[6,33–36,38,87,92,95] as displayed in such transcripts, can be used to improve the quality of expert witness testimony.

Tadd Syndrome

This term is sometimes used in the child abuse literature to indicate the talk and deteriorate and die sequence of events in which there is a varying degree of lucid interval between a head injury and subsequent deterioration and death. Issues such as how long and how truly asymptomatic such lucid intervals may be are presently areas of active investigation and debate in child abuse research (see reviews and controversy papers cited earlier). The main hindrance to prompt and definitive resolution to this issue continues to be that the only source of the clinical history in the most controversial cases is that provided by the caretaker(s).

Tin Ear Syndrome

"Tin ear syndrome" has been used to describe the triad of unilateral ear bruising, hemorrhagic retinopathy, and ipsilateral subdural hematoma and brain swelling in children, the suggested mechanism of which is rotational acceleration due to blunt force trauma to the ear.[46]

In other cases not fulfilling criteria for the above triad, signs of trauma in and around the ear that may have implications for intracranial injury include hemorrhage from the ear canal; hemorrhage behind or within the tympanic membrane; tympanic membrane disruption; pinna abrasions, lacerations, contusions, hematomas, and patterned injuries; partial or complete avulsion of the pinna; Battle's sign (often related to basal skull fracture); and burns.[97]

How Unique Is the Mechanism of Head Injury in Your Case?

Certain types of injuries in children may be unusual, but there is a good chance that others have described a similar event. Short-distance falls are discussed in several of the reviews cited previously. A few of the numerous examples available will illustrate the prudence of screening the literature for events related to any case:

Falls from windows.[96]
Falling off a bed.[81,98]
Dropped from caretaker's arms.[98]
Rolling off a couch.[98]
Baby-rocker injuries.[57]
Ice skating, skateboarding, roller skating and in-line skating injuries.[73]
Baby-bouncer injuries.[26]
Playground falls.[82,105]
Falls from bunk beds.[70]
Infant walker injuries.[5,91]
Highchair injuries.[71]
Stairway falls.[56]

Although injuries may occur to multiple body sites in each of these examples, head injuries often predominate.

Terson's Syndrome

Review articles on the topic of child abuse cited in this chapter sometimes include mention of Terson's syndrome.[9,64,100] Definitions, if provided, vary from "vitreous hemorrhage after intracranial hemorrhage" (seen in adults but not in children)[64] to "retinal hemorrhages with subdural hemorrhages and subarachnoid hemorrhages (SAHs), with or without trauma, blood in the optic nerve sheath, and increased intracranial pressure," "seen in adults."[9] Such nonuniformity in the use of this term warrants further comment.

The original description of vitreous hemorrhage with subarachnoid hemorrhage is ascribed to Litten in 1881, although the description by Terson in 1900 led to his name being applied to this syndrome.[72] Some subsequent reports do not consistently adhere to the strict limitation of this syndrome to vitreous hemorrhages associated with subarachnoid hemorrhage, and they have included subhyaloid, intraretinal, and/or subretinal hemorrhages, with or without associated vitreous hemorrhages, in association with any form of intracranial bleeding, or with other causes of increased intracranial pressure.[54,72,74,94]

The proposed mechanism of Terson's syndrome (using the strict definition of vitreous hemorrhage associated with subarachnoid hemorrhage) was originally thought to be that blood in the pericerebral subarachnoid space tracked along the optic nerve sheath subarachnoid space and entered the vitreous (with or without associated retinal hemorrhage).[74,84,94] In other words, the intraocular blood was from an intracranial source. Subsequent experimental and clinical observations suggest that such a mechanism rarely, if ever, occurs. Rather, the source of the intraocular blood is now believed to be ocular, and proposed mechanisms for this source of hemorrhage include a rapid rise in intracranial pressure, resultant venous congestion related to impaired venous drainage to the cavernous sinus and/or retinochoroidal connections and central retinal vein, leading to intraocular venous stretching, stasis, and hemorrhage;[16,72,74] increased intrathoracic pressure; direct impact trauma to the head; and ocular mechanical injury caused by shaking.[79]

The most convincing recent supportive evidence for an ocular, and not intracranial, source of the vitreous hemorrhage (with or without associated retinal hemorrhage) is the production of Terson's syndrome within minutes following rapid injection of lumbar or caudal epidural saline (with or without steroids).[39,80] In both of these reports, postinjection magnetic resonance scans revealed no evidence of intracranial blood. It was suggested that the cerebrospinal fluid acute pressure spike is the key causative event, and that fluid infusions into the spinal epidural space of 20 ml or more be administered slowly to reduce the risk of this complication.[80] The abrupt development of intracranial hypertension may be the factor that accounts for the inconsistent correlation between intracranial hypertension and ocular hemorrhages described in some studies.[79] Alternatively, ocular hemorrhage may result from more than one mechanism in different cranial injury cases. One study in normal adults, if applicable to infants, indicates that gravity inversion (i.e., the head-down vertical position) alone, for up to 30 minutes, is capable of producing orbital and conjunctival congestion, eyelid petechiae, subconjunctival hemorrhage, and excessive tearing, but not retinal hemorrhage.[29]

Terson's syndrome, as originally defined, does appear to be a phenomenon primarily seen in adults, and is associated with a higher fatality rate compared with individuals with subarachnoid hemorrhage without Terson's syndrome. Morad et al.[79] found only preretinal hemorrhages in 7%, and preretinal plus intraretinal hemorrhages in 8% of 75 cases diagnosed as shaken baby syndrome. The youngest patient in the series of Pfausler et al.[84] was in the "10–19" year age bracket. When the broader usage of Terson's syndrome is employed, encompassing retinal hemorrhages without the need for associated vitreous hemorrhage, children are found in other case series.[94] Further studies in children within the most common age group for child abuse, with precise descriptions of the location of the ocular hemorrhage(s) by ophthalmologists, are needed to clarify the age bracket and circumstances in which Terson's syndrome occurs. A potential therapeutic benefit from early ophthalmologic consultation in survivors of these intracranial catastrophies is the accumulating evidence that treatments such as pars plana vitrectomy or intravitreous injections of anti-Rh serum may significantly improve long-term visual acuity.[74]

Shaken Adult Syndrome

Although opinions to the contrary exist,[37] available data are consistent with rare cases of "shaken adult syndrome" occurring.[13,49,86]

Crush Injuries to the Head

Fatal childhood crush injuries to the head are much less frequently seen than the acceleration–deceleration type of head injuries.[24,101] A combination of mechanisms may occur in some cases, such as when a toddler pulls a television set over. The television set can impact a mobile head before both the television and the interposed head strike the floor for combined dynamic and static loading. Combined head and cervical spine injury is not unusual in such cases.

Subdural Hemorrhage in Childhood

This topic is discussed in several of the articles on consensus opinions and areas of controversy cited earlier in the section on periodical literature sources. Additional articles with a focus more specifically on subdural hemorrhages in children are also included in the references at the end of this chapter.[18,22,27,48,59,69,77,103,104] Subdural hemorrhage is also discussed elsewhere in this text, both with regard to sequelae of blunt force trauma (see Chap. 6) and dating/aging of subdural hemorrhage (see Chap. 3).

Folk Remedies and Customs

The multiethnic population in our jurisdiction produces cases in which folk medicine or other cultural practices may result in diagnostic issues that require familiarity with such customs. Most of these concern cutaneous lesions, such as those related to dermabrasion,[43] cupping, or focal burns.[47] Santeria, a religious cult of African origin that is practiced primarily by, for example, some immigrants from the Caribbean, can include rituals in which human skulls are decorated in a variety of ways.[106] Medical examiners usually encounter such cases when skeletal remains (often an admixture of human and animal parts) and various other paraphernalia are found in wilderness areas by hikers, and brought to the attention of authorities. Such cases are not infrequent in our jurisdiction.

Excluding potential poisoning from herbal remedies, there are at least two cultural practices that may lead to neuropathology questions. One is purposeful infant head molding, that may cause confusion with a variety of natural diseases that lead to malformed heads, or to misinterpretation as child abuse.[28] The second is Hispanic folk medicine measures used to treat "fallen fontanelle" (caida de mollera), probably simply a sign of dehydration.[102] One article claims that such treatment may include measures vigorous enough to produce intracranial hemorrhage.[44] As pointed out by Hansen,[47] such folk medicine therapy is typically gentle, and no other similar published case reports have appeared. On the basis of available information, it would appear reasonable to do a thorough evaluation of cases in which this alleged mechanism is suggested, in order to rule out other acci-

Spinal Injury in Child Abuse

Spinal column and spinal cord injury, alone or associated with head injury, is common enough in both accidental and inflicted trauma in all age groups to warrant specific examination of both anatomic areas at autopsy.[75] This is particularly relevant in cases in which there is clinical evidence of cord compromise or spinal column trauma by imaging studies; cases in which the injury to death interval is too brief to include imaging studies; or cases in which coma was present throughout the injury to death interval, so that a complete neurologic examination (including volitional movements, etc.) was not possible.

There is an extensive literature on spinal injury, both in adults and children, and selected aspects are reviewed elsewhere in this text (see discussion on neuroradiology studies in forensic neuropathology in Chapter 17).

Limiting our discussion here to cases of suspected child abuse, several of the publications already cited in this chapter provide an overview of this topic. A few additional comments follow. SCIWORA (spinal cord injury without radiologic abnormality) is a phenomenon emphasized in spinal injury in children and young adults primarily, presumably because certain anatomic features of the immature spine, such as more horizontally oriented facets and more soft tissue laxity, predispose to subluxations rather than fractures,[45] and/or because incomplete ossification masks more subtle fracture findings that would be apparent in a well-ossified, mature spine.

Spinal cord injury without head injury is relatively uncommon as a presenting symptom of child abuse, but it does occur.[21,31,67,85] Onset of symptoms following spinal trauma may be delayed for hours or days, the presumed mechanism being delayed vascular compromise and cord infarction.[1,15]

The mechanism of spinal injury in child abuse cases undoubtedly varies from case to case and, in most instances, is not observed by an objective and reliable unrelated witness. Impact injuries of various types occur, as revealed by accompanying contusions or other evidence of blunt force injury. In cases of cervical spinal cord injury (with or without associated brainstem injury) without evidence of impact, shaking of the child is often reported or suspected. The method of shaking usually cited is by grasping the child's shoulders or trunk to initiate the shaking motion. In another potential shaking mechanism, first brought to our attention by Maureen J. Frikke, M.D. (Office of the Medical Examiner, Salt Lake City, Utah), the perpetrator grasps the child by the head and shakes it back and forth. Mary E. Case, M.D., has also described this shaking method in symposium lectures attended by the authors. The cord injury from this method likely reflects the combined effects of vigorous back-and-forth motion together with spinal cord traction secondary to the suspended body weight. This injury mechanism may be suspected if, as in one of Dr. Frikke's cases, the decedent also demonstrates ocular injuries suspicious for direct globe blunt force trauma (e.g., dislocated lens, anterior chamber hemorrhage) due to the perpetrator's thumb(s) pressing into the orbits. Alternatively, infrequent cases occur in which the infant is grasped by the ankles and vigorously swung about, with or without evidence of head impact.[30]

More emphasis is currently being placed on examining the spine, associated blood vessels, and spinal cord, particularly in the cervical area, in addition to the brain, in cases of suspected child abuse.[17,42,76,93] As further studies that include spinal cord findings are reported, a more accurate determination of the frequency of spinal cord injury in nonaccidental neurotrauma and its potential role in causing post-traumatic apnea and other disturbances will follow.

Retinal Hemorrhage

Retinal hemorrhage is a frequent topic in the child abuse literature. The birth process is the most common cause of retinal hemorrhage in infants, as was briefly mentioned in Chapter 4. It occurs in about 33% of normal vaginal deliveries, 75% of vacuum-assisted deliveries, and 7% of infants delivered by cesarean section,[25] although figures vary considerably in different studies.[66] Incidence in forceps delivery is unclear due to the small number of cases studied. Most birth-related retinal hemorrhages are intraretinal dot-blot (i.e., larger than dot hemorrhages, but not strictly defined as to size) or flame hemorrhages.[66] Flame hemorrhages are within the nerve fiber layer, and dot-blot hemorrhages are within the deeper layers of the retina. White centers may or may not be present, are more frequent in larger hemorrhages, and are not pathognomonic for sepsis. Preretinal (including vitreous) and subretinal hemorrhages are rare in relationship to the birth process. Retinal hemorrhages may enlarge over a few days following their initial appearance.[100] Their distribution, whether by birth process or by subsequent accidental or nonaccidental trauma, may be anywhere in the retina from the posterior pole to the ora serrata, and they may be unilateral or bilateral. According to Levin,[66] all birth-related flame hemorrhages should be resolved (i.e., not visible by ophthalmoscopic examination by an ophthalmologist) within 1 week. Most disappear in 3 to 5 days, and they may resolve within 24 hours.[100] Dot-blot hemorrhages resolve by 6 weeks after birth.[66] Levin[66] notes one case of subinternal limiting membrane hemorrhage that took 1 month to resolve, a deep macular hemorrhage that took 6 weeks to resolve, and a large peripheral birth hemorrhage that took 6 weeks to resolve.

In a series of 149 newborns examined by Emerson et al.,[25] retinal hemorrhages were found in 34% of cases and were bilateral in 52% of these. Ninety percent of the hemorrhages (dot-blot and flame) detected at birth had resolved within 2 weeks, and none were seen by 4 weeks after birth. Only one subretinal hemorrhage was present in their series, and it resolved at 6 weeks after birth. Opthalmoscopy in this series was performed at 2-week intervals if initial examination revealed retinal hemorrhages.[25]

The Ophthalmology Child Abuse Working Party[100] indicated that mild retinal hemorrhages may clear rapidly, with birth-related flame hemorrhages resolving as early as within 24 hours; moderately severe retinal hemorrhages clear within a few weeks; and severe, widespread retinal hemorrhages and vitreous hemorrhages may take many months to clear and often leave evidence of residual damage.

Accurate dating/aging of retinal hemorrhage by appearance is not possible by current methods.[100] A generalization suggested is that brown-yellow ochre membranes are "consistent with injuries at least one month old," and eyes with severe intraocular hemorrhage develop a "greenish hue on slit-lamp or ophthalmoscopic examination."[100]

We are not aware of any large series of human studies on the speed of appearance of hemosiderin in retinal hemorrhage, or how long it persists. By analogy with studies on other tissues in humans, it probably takes 2 to 3 days following acute hemorrhage for hemosiderin to appear in retinal lesions, as is the case in other soft tissues and in subdural hematomas. It has been suggested that if hemosiderin persists in the retina of infants who are more than 2 to 3 months old, it is likely that this is from a hemorrhage occurring after birth, but data supporting this view were not cited.[100] One experimental study in adult rhesus monkeys, using central retinal vein occlusion to produce retinal hemorrhage, detected hemosiderin (based on positivity with Prussian blue stain) within 2 days' survival (which was the shortest survival period studied in this report), and persistence of hemosiderin for up to 16.8 months after hemorrhage induction (the longest survival period studied in this report).[41] This latter result is more consistent with our experience in human brain and dura, where hemosiderin may sometimes be present in lesions, which, by history and appearance, are many months to several years old. Variations undoubtedly occur based on tissue vascularity, amount of initial hemorrhage, and, very likely, based on age and other undetermined factors that remain to be studied.

It is preferable that funduscopic appearance be documented clinically by an experienced ophthalmologist prior to death. The ophthalmologist is best equipped by training to determine the nature of the retinal hemorrhage, including conditions such as retinopathy of prematurity or examination-induced retinal hemorrhage.[68] When this has not occurred, there may still exist some opportunity to obtain postmortem funduscopic information prior to any procedures such as vitreous aspiration, but the window of opportunity to do so by direct or indirect ophthalmoscopy[65] may be limited by advancing corneal clouding or by anterior chamber, lens, or vitreous opacities or infiltrates. Recently, ophthalmic endoscopy has been suggested as an approach that overcomes many of the previous problems encountered in retinal visualization at the time of autopsy, and forensic applications of this technology are discussed.[3,20]

Kevorkian,[61] for the purpose of determining retinal findings useful in estimating time of death, described the sequential changes in the retina by routine ophthalmoscopy for up to 15 to 20 hours after death, and he documented the retinal appearance by fundus photographs for up to 11 hours after death. He was able to accomplish this by frequent (every 30–40 seconds) moistening of the cornea with "a drop or two of water," and covering the cornea with the upper eyelid between examinations. Other retinal findings that may aid the estimation of time since death have been described.[20,55]

One study that provides details on the pace of postmortem corneal clouding is that of Wróblewski and Ellis.[107] They examined a series of 303 bodies at postmortem intervals of less than 15 minutes to 3 days. In 68% of cases, autopsy was performed within the first 2 hours after death. No corneal opacity was seen in less than 15 minutes after death. Opacity developed in 4 cases within 15 to 30 minutes; in 5 cases within 30 to 60 minutes; in 6 cases within 1 to 1½ hours; in 10 cases within 1½ to 2 hours; and in 71 additional cases beyond 2 hours' postmortem. The authors commented that corneal cloudiness may be prevented by continued moistening of the cornea with water or saline, but they could not confirm the clearing of a "hopelessly hazy cornea" by such methods, as claimed by some previous authors.[107]

It has been said that an ophthalmologist may still be able to determine the presence and distribution of retinal hemorrhages for up to 72 hours after death, but not after the eyeball has been removed.[66] Whether this refers to an occasional case with very delayed corneal clouding, or results from some combination of continued corneal hydration combined with techniques such as indirect ophthalmoscopy or slit-lamp examination, was not further specified.

Subsequent to eye removal, a description of hemorrhage location(s) within the globe and retinal layers can be documented by a specialty-trained ophthalmic pathologist, although practical considerations will limit the histologic sampling possible when there are multiple and widespread hemorrhages.

References

1. Ahmann PA, Smith SA, Schwartz JF, Clark DB. Spinal cord infarction due to minor trauma in children. Neurology 1975;25: 301–307.
2. Alexander R, Crabbe L, Sato Y, et al. Serial abuse in children who are shaken. Am J Dis Child 1990;144:58–60.
3. Amberg R, Pollak S. Postmortem endoscopy of the ocular fundus: A valuable tool in forensic postmortem practice. Forensic Sci Int 2001;124:157–162.
4. American Academy of Pediatrics, Committee on Child Abuse and Neglect. Distinguishing sudden infant death syndrome from child abuse fatalities. Pediatrics 2001;107:437–441.
5. American Academy of Pediatrics. Committee on Injury and Poison Prevention. Injuries associated with infant walkers. Pediatrics 2001;108:790–792.
6. Andrew T. Review of "Geddes JF, Taskert RC, Hackshaw AK, et al.: Dural hemorrhage in non-traumatic infant deaths: Does it explain the bleeding in 'shaken baby syndrome'"? Neuropathol Appl Neurobiol 2003;29:14–22." Child Abuse Quarterly Medical Update 2003;X:15–17.
7. Bergman AB. Advocacy is not a specialty. Arch Pediatr Adolesc Med 2005;159:892.
8. Bird SJ. Scientific certainty: Research versus forensic perspectives. J Forensic Sci 2001;46:978–981.
9. Block RW. Child abuse—Controversies and imposters. Curr Probl Pediatr 1999;29:253–272.
10. Byard RW. Sudden Death in Infancy, Childhood and Adolescence, ed 2. New York: Cambridge University Press, 2004.
11. Byard RW. Unexpected infant death: Lessons from the Sally Clark case. Med J Aust 2004;181:52–54.
12. Byard RW. Medicolegal problems with neonaticide. In Tsokos M (ed): Forensic Pathology Reviews. Totowa, NJ: Humana Press, 2004;1:171–185.
13. Carrigan TD, Walker E, Barnes S. Domestic violence: The shaken adult syndrome. J Accid Emerg Med 2000;17:138–139.
14. Case ME, Graham MA, Handy TC, et al. The National Association of Medical Examiners Ad Hoc Committee on Shaken Baby Syndrome: Position paper on fatal abusive head injuries in infants and young children. Am J Forensic Med Pathol 2001;22:112–122.
15. Choi J-U, Hoffman HJ, Hendrick EB, et al. Traumatic infarction of the spinal cord in children. J Neurosurg 1986;65:608–610.
16. Choudhari KA, Pherwani AA, Gray WJ. Terson's syndrome as the sole presentation of aneurysmal rupture. Br J Neurosurg 2003; 17:355–357.
17. Conway EE Jr. Nonaccidental head injury in infants: "The shaken baby syndrome" revisited. Pediatr Ann 1998;27:677–690.
18. Datta S, Stoodley N, Jayawant S, et al. Neuroradiological aspects of subdural haemorrhages. Arch Dis Child 2005;90:947–951.
19. David TJ. Avoidable pitfalls when writing medical reports for court proceedings in cases of suspected child abuse. Arch Dis Child 2004;89:799–804.
20. Davis NL, Wetli CV, Shakin JL. The retina in forensic medicine: Applications of ophthalmic endoscopy—the first 100 cases. Am J Forensic Med Pathol 2006;27:1–10.
21. Diamond P, Hansen CM, Christofersen MR. Child abuse presenting as a thoracolumbar spinal fracture dislocation: A case report. Pediatr Emerg Care 1994;10:83–86.
22. Duhaime A-C, Christian C, Armonda R, et al. Disappearing subdural hematomas in children. Pediatr Neurosurg 1996;25: 116–122.
23. Duhaime A-C, Christian CW, Rorke LB, Zimmerman RA. Nonaccidental head injury in infants—The "shaken baby syndrome." N Engl J Med 1998;338:1822–1829.
24. Duhaime A-C, Eppley M, Margulies S, et al. Crush injuries to the head in children. Neurosurgery 1995;37:401–407.
25. Emerson MV, Pieramici DJ, Stoessel KM, et al. Incidence and rate of disappearance of retinal hemorrhage in newborns. Ophthalmology 2001;108:36–39.
26. Farmakakis T, Alexe DM, Nicolaidou P, et al. Baby-bouncer-related injuries: An under-appreciated risk. Eur J Pediatr 2004; 163:42–43.
27. Feldman KW, Bethel R, Shugerman RP, et al. The cause of infant and toddler subdural hemorrhage: A prospective study. Pediatrics 2001;108:636–646.
28. FitzSimmons E, Prost JH, Peniston S. Infant head molding: A cultural practice. Arch Fam Med 1998;7:88–90.
29. Friberg TR, Weinreb RN. Ocular manifestations of gravity inversion. JAMA 1985;253:1755–1757.
30. Frikke MJ. Personal communication, 1997.
31. Gabos PG, Tuten HR, Leet A, Stanton RP. Fracture-dislocation of the lumbar spine in an abused child. Pediatrics 1998;101: 473–477.
32. Geddes J. Pediatric head injury. In Golden JA, Harding BN (eds): Pathology and Genetics: Developmental Neuropathology. Basel: ISN Neuropath Press, 2004, pp 184–191.
33. Geddes JF, Hackshaw AK, Vowles GH, et al. Neuropathology of inflicted head injury in children. I. Patterns of brain damage. Brain 2001;124:1290–1298.
34. Geddes JF, Tasker RC, Adams GGW, Whitwell HL. Violence is not necessary to produce subdural and retinal hemorrhage: A reply to Punt et al. Pediatr Rehabil 2004;7:261–265.
35. Geddes JF, Tasker RC, Hackshaw AK, et al. Dural hemorrhage in non-traumatic infant deaths: Does it explain the bleeding in "shaken baby syndrome"? Neuropathol Appl Neurobiol 2003;29: 14–22.
36. Geddes JF, Vowles GH, Hackshaw AK, et al. Neuropathology of inflicted head injury in children. II. Microscopic brain injury in infants. Brain 2001;124:1299–1306.
37. Geddes JF, Whitwell HL. Shaken adult syndrome revisited (letter to editor). Am J Forensic Med Pathol 2003;24:310–311.
38. Geddes JF, Whitwell HL. Inflicted head injury in infants. Forensic Sci Int 2004;146:83–88.
39. Gibran S, Mirza K, Kinsella F. Unilateral vitreous haemorrhage secondary to caudal epidural injection: A variant of Tersons's syndrome (letter to editor). Br J Ophthalmol 2002;86:353–354.
40. Gilbert-Barness E, Debich-Spicer DE. Handbook of Pediatric Autopsy Pathology. Totawa, NJ: Humana Press, 2005.
41. Gilliland MGF, Folberg R, Hayreh SS. Age of retinal hemorrhages by iron detection: An animal model. Am J Forensic Med Pathol 2005;26:1–4.
42. Gleckman AM, Kessler SC, Smith TW. Periadventitial extracranial vertebral artery hemorrhage in a case of shaken baby syndrome. J Forensic Sci 2000;45:1151–1153.
43. Golden SM, Duster MC. Hazards of misdiagnosis due to Vietnamese folk medicine. Clin Pediatr 1977;16:949–950.
44. Guarnaschelli J, Lee J, Pitts FW. "Fallen fontanelle" (caida de mollera): A variant of the battered child syndrome. JAMA 1972;222:1545–1546.
45. Hadley MN, Zabramski JM, Browner CM, et al. Pediatric spinal trauma: Review of 122 cases of spinal cord and vertebral injuries. J Neurosurg 1988;68:18–24.
46. Hanigan WC, Peterson RA, Njus G. Tin ear syndrome: Rotational acceleration in pediatric head injuries. Pediatrics 1987;80: 618–622.
47. Hansen KK. Folk remedies and child abuse: A review with emphasis on caide de mollera and its relationship to shaken baby syndrome. Child Abuse Negl 1997;22:117–127.

48. Hobbs C, Childs A-M, Wynne J, et al. Subdural haematoma and effusion in infancy: An epidemiological study. Arch Dis Child 2005;90:952–955.
49. Holmgren B. Review of "Geddes JF, Whitwell H. Shaken adult syndrome revisited. Am J Forensic Med Pathol 2003;24:310–311." Child Abuse Quarterly Medical Update 2004;XI(I):24–25.
50. Hymel KP. Small steps in the right direction: The ongoing challenge of research regarding inflicted traumatic brain injury. Child Abuse Negl 2005;29:945–947.
51. Hymel KP and the Committee on Child Abuse and Neglect of the American Academy of Pediatrics. Distinguishing sudden infant death syndrome from child abuse fatalities. Pediatrics 2006;118:421–427.
52. Hymel KP, Bandak FA, Partington MD, Winston KR. Abusive head trauma? A biomechanics-based approach. Child Maltreatment 1998;3:116–128.
53. Hymel KP, Jenny C, Block RW. Intracranial hemorrhage and rebleeding in suspected victims of abusive head trauma: Addressing the forensic controversies. Child Maltreatment 2002;7:329–348.
54. Iwase T, Tanaka N. Bilateral subretinal haemorrhage with Terson's syndrome. Graefes Arch Clin Exp Ophthalmol 2006;244:507–509.
55. Jaafar S, Nokes LDM. Examination of the eye as a means to determine the early postmortem period: A review of the literature. Forensic Sci Int 1994;64:185–189.
56. Joffe M, Ludwig S. Stairway injuries in children. Pediatrics 1988;82(Pt 2):457–461.
57. Jones MD, James DS, Cory CZ, et al. Subdural hemorrhage sustained in a baby-rocker? A biomechanical approach to causation. Forensic Sci Int 2003;131:14–21.
58. Judkins AR, Hood IG, Mirchandani HG, Rorke LB. Technical communication. Rationale and technique for examination of nervous system in suspected infant victims of abuse. Am J Forensic Med Pathol 2004;25:29–32.
59. Kemp AM. Investigating subdural haemorrhage in infants. Arch Dis Child 2002;86:98–102.
60. Kemp AM, Stoodley N, Cobley C, et al. Apnoea and brain swelling in non-accidental head injury. Arch Dis Child 2003;88:472–476.
61. Kevorkian J. The eye in death. Clin Symp 1961;13:51–62.
62. Kleinman PK. Diagnostic Imaging of Child Abuse, ed 2. St Louis: Mosby, 1998.
63. Krous HF, Byard RW. Shaken infant syndrome: Selected controversies. Pediatr Dev Pathol 1999;2:497–498.
64. Krous HF, Byard RW. Controversies in pediatric forensic pathology. Forensic Sci Med Pathol 2005;1:9–18.
65. Lantz PE, Adams GGW. Postmortem monocular indirect ophthalmoscopy. J Forensic Sci 2005;50:1450–1452.
66. Levin AV. Retinal hemorrhages and child abuse. In David TJ (ed): Recent Advances in Pediatrics, Vol 18. New York: Churchill Livingstone, 2000, pp 18:151–219.
67. Levin TL, Berdon WE, Cassell I, Blitman NM. Thoracolumbar fracture with listhesis—An uncommon manifestation of child abuse. Pediatr Radiol 2003;33:305–310.
68. Lim Z, Tehrani NN, Levin AV. Retinal hemorrhages in a preterm infant following screening examination for retinopathy of prematurity. Br J Ophthalmol 2006;90:799–800.
69. Maxeiner H. Demonstration and interpretation of bridging vein ruptures in cases of infantile subdural bleedings. J Forensic Sci 2001;46:85–93.
70. Mayr JM, Seebacher U, Lawrenz K, et al. Bunk beds—A still underestimated risk for accidents in childhood? Eur J Pediatr 2000;159:440–443.
71. Mayr JM, Seebacher U, Shimpl G, Fiala F. Highchair accidents. Acta Pediatr 1999;88:319–322.
72. McCarron MO, Alberts MJ, McCarron P. A systematic review of Terson's syndrome: Frequency and prognosis after subarachnoid hemorrhage. J Neurol Neurosurg Psychiatry 2004;75:491–493.
73. McGeehan J, Shields BJ, Smith GA. Children should wear helmets while ice-skating: A comparison of skating-related injuries. Pediatrics 2004;114:124–128.
74. Medele RJ, Stummer W, Mueller AJ, et al. Terson's syndrome in subarachnoid hemorrhage and severe brain injury accompanied by acutely raised intracranial pressure. J Neurosurg 1998;88:851–854.
75. Michael DB, Goyot DR, Darmody WR. Coincidence of head and cervical spine injury. J Neurotrauma 1989;6:177–189.
76. Minns RA. Shaken baby syndrome: Theoretical and evidential controversies. J R Coll Physicians Edinb 2005;35:5–16.
77. Minns RA. Subdural haemorrhages, haematomas, and effusions in infancy. Arch Dis Child 2005;90:883–884.
78. Minns RA, Busuttil A. Patterns of presentation of the shaken baby syndrome: Four types of inflicted brain injury predominate. BMJ 2004;328:766.
79. Morad Y, Kim YM, Armstrong DC, et al. Correlation between retinal abnormalities and intracranial abnormalities in the shaken baby syndrome. Am J Ophthalmol 2002;134:354–359.
80. Naseri A, Blumenkranz MS, Horton JC. Terson's syndrome following epidural saline injection. Neurology 2001;57:364.
81. Nimityongskul P, Anderson LD. The likelihood of injuries when children fall out of bed. J Pediatr Orthop 1987;7:184–186.
82. Norton C, Rolfe K, Morris S, et al. Head injury and limb fracture in modern playgrounds. Arch Dis Child 2004;89:152–153.
83. Peterson I. Flight of the bumblebee. Science News Online 2004;166 (11). Available at: http://www.sciencenews.org. Accessed 1/3/06.
84. Pfausler B, Belcl R, Metzler R, et al. Terson's syndrome in spontaneous subarachnoid hemorrhage: A prospective study in 60 consecutive patients. J Neurosurg 1996;85:392–394.
85. Piatt JH Jr, Steinberg M. Isolated spinal cord injury as a presentation of child abuse. Pediatrics 1995;96:780–782.
86. Pounder DJ. Shaken adult syndrome. Am J Forensic Med Pathol 1997;18:321–324.
87. Punt J, Bonshek RE, Jaspan T, et al. The "unified hypothesis" of Geddes et al. is not supported by the data. Pediatr Rehabil 2004;7:173–184.
88. *R v. Lorraine Harris, Raymond Charles Rock, Alan Barry, Joseph Cherry, Michael Ian Faulder*, Court of Appeal (2005) EWCA Crim 1980, pp 1–67.
89. Reece RM, Ludwig S (eds). Child Abuse: Medical Diagnosis and Management, ed 2. Philadelphia: Lippincott Williams & Wilkins, 2001.
90. Reece RM, Nicholson CE (eds). Inflicted Childhood Neurotrauma. Elk Grove Village, IL: American Academy of Pediatrics, 2003.
91. Ribe JK, Lopez T, Pena LA. Distinguishing infant walker injuries from fatal abusive head trauma (abstract). Proc Am Acad Forensic Sci 2001;7:232–233.
92. Ribe J. Review of "Geddes JF, Whitwell HL: Inflicted head injury in infants. Forensic Sci Int 2004;146:83–88." Child Abuse Quarterly Medical Update 2005;XII(2):10–11.
93. Rorke LB. Neuropathology of inflicted childhood neurotrauma. In Reece RM, Nicholson CE (eds): Inflicted Childhood Neurotrauma. Elk Grove Village, IL: American Academy of Pediatrics, 2003, pp 165–179.
94. Schloff S, Mullaney PB, Armstrong DC, et al. Retinal findings in children with intracranial hemorrhage. Ophthalmology 2002;109:1472–1476.
95. Spivack B. Review of "Geddes JF, Hackshaw AK, Vowles GH, et al.: Neuropathology of inflicted head injury in children. I. Patterns of brain damage. Brain 2001;124:1290–1298," and of "Geddes JF, Vowles GH, Hackshaw AK, et al: Neuropathology of inflicted head injury in children. II. Microscopic brain injury in children. Brain 2001;124:1299–1306." Child Abuse Quarterly Medical Update 2001;VIII(4):8–9.

96. Stone KE, Lanphear BP, Pomerantz WJ, Khoury J. Childhood injuries and deaths due to falls from windows. J Urban Health 2000;77:26–33.
97. Swift B, Rutty GN. The human ear: Its role in forensic practice. J Forensic Sci 2003;48:153–160.
98. Tarantino CA, Dowd MD, Murdock TC. Short vertical falls in infants. Pediatr Emerg Care 1999;15:5–8.
99. The Quarterly Update. Reviews of Current Child Abuse Medical Research. Available at: http://www.quarterlyupdate.org.
100. The Ophthalmology Child Abuse Working Party. Child abuse and the eye. Eye 1999;13:3–10.
101. Tortosa JG, Martínez-Lage JF, Poza M. Bitemporal head crush injuries: Clinical and radiological features of a distinctive type of head injury. J Neurosurg 2004;100:645–651.
102. Trotter RT II, de Montellano BO, Logan MH. Fallen fontanelle in the American Southwest: Its origin, epidemiology, and possible organic causes. Med Anthropol 1989;10:211–221.
103. Tzioumi D, Oates RK. Subdural hematomas in children under 2 years: Accidental or inflicted? A 10-year experience. Child Abuse Negl 1998;22:1105–1112.
104. Vinchon M, Noizet O, Defoort-Dhellemmes S, et al. Infantile subdural hematomas due to traffic accidents. Pediatr Neurosurg 2002;37:245–253.
105. Waltzman ML, Shannon M, Bowen AP, Bailey MC. Monkeybar injuries: Complications of play. Pediatrics 1999;103:e58 (p 1020).
106. Wetli CV, Martinez R. Forensic sciences aspects of Santeria, a religious cult of African origin. J Forensic Sci 1981;26:506–514.
107. Wróblewski B, Ellis M: Eye changes after death. Br J Surg 1970;57:69–71.

Injuries Due to Firearms and Other Missile-Launching Devices

INTRODUCTION 211
INFORMATION RESOURCES 211
WOUND BALLISTIC ISSUES 212
HEAD WOUNDS 213
SKULL WOUNDS 213
TANGENTIAL SKULL WOUNDS 215
ENTRANCE AND EXIT WOUNDS OF THE SKULL 215
INTERNAL RICOCHET 216
BRAIN WOUND CHARACTERISTICS 216
BACKSPATTER 220
"DISAPPEARING" BULLETS 220
INTERMEDIATE TARGETS 222
BULLET MIGRATION IN BODIES 223
SUICIDAL VERSUS HOMICIDAL GUNSHOT WOUNDS TO THE HEAD 226
TANDEM BULLETS 227
PROGNOSTIC FACTORS IN CRANIOCEREBRAL GUNSHOT WOUNDS 227

CENTERFIRE RIFLE WOUNDS 228
SHOTGUN WOUNDS 228
POTENTIAL COMPLICATIONS OF CRANIOSPINAL GUNSHOT WOUNDS 229
COMPLICATIONS OF RETAINED BULLETS AND OTHER MISSILES RELATED TO BULLET COMPOSITION 230
INJURY TO DEATH INTERVAL 231
SPEED OF INCAPACITATION ISSUES IN GUNSHOT WOUNDS 232
REACTION/RESPONSE TIME ISSUES 237
UNUSUAL MISSILE-LAUNCHING DEVICES AND AMMUNITION 239
FIREARM WOUND IMITATORS 243
BULLET/SHRAPNEL TRACE EVIDENCE 243
RADIOLOGY OF GUNSHOT WOUNDS 244
GUNSHOT WOUND AUTOPSY PROTOCOLS 246
REFERENCES 248

Introduction

A person is shot in the head and dies. Few circumstances in forensic pathology are more deceptive in their illusion of simplicity, or better illustrate the intricate interdependence that exists between various members of the forensic team in order to arrive at the most accurate answer to questions that may arise during subsequent investigation and adjudication.

The consulting neuropathologist involved in the study of firearm and other missile injuries will find this to be a complex and challenging area, and one in which some of the neuropathology questions posed are not presently answerable. The consultant with limited firearms knowledge may be asked to simply describe the neuroanatomic structures damaged in the bullet path. The consultant interested in providing a more comprehensive service to his colleagues, however, requires a broader knowledge base in this subspecialty.

Information Resources

The literature on gunshot wounds is extensive, as would be expected considering the frequency of such injuries and the interest of professionals in types of firearms and in the pathology and biophysics of the injuries they produce. Understanding the literature requires knowledge of its unique vocabulary. The consulting neuropathologist need not necessarily develop the level of expertise required of the medical examiner performing the autopsy to determine, for example, range of fire or to distinguish entrance from exit wounds by characteristics of the skin and subcutaneous tissue injury. Nevertheless, communication between consultant and referring physician is greatly facilitated by the former being familiar with topics such as the following:

- Types of firearms commonly used in civilian firearm fatalities.
- Firearm caliber or gauge designations.
- Firearm ballistic issues.
- The role of firearms examiners, trace evidence experts, and scene investigation experts.
- Ammunition types and components for each of the common firearms.
- Perforating (i.e., passing completely through) versus penetrating (entering an object, but not exiting) wounds.[54]
- Low-velocity and high-velocity wounds (best expressed as numerical values, since definitions vary).[70]
- Wound effects produced by the bullet, flame jet, grains of unburned smokeless gunpowder, and various combustion products that accompany the bullet.
- Characteristics of entrance versus exit wounds, and pitfalls in distinguishing one from the other (e.g., shored wounds, wounds on skin overlying bone, drying of wound edges, effects of therapeutic intervention, effects of postmortem decomposition on wounds).
- Pathology of typical missile wounds in skin, soft tissue, bone, and central nervous system (CNS) tissues.
- Varieties of atypical wounds, such as those due to unusual projectiles (frangible bullets, Glaser safety slugs, sabot slugs, duplex rounds, flechettes, etc.), the varied potential effects of intermediate targets, and of weapon–ammunition mismatch.[279]
- Powder stippling/tattooing versus imitators. Our department makes a distinction between stippling (skin abrasion due to impact by unburned gunpowder particles) and powder tattooing (presence of unburned powder granules embedded in skin).
- Range-of-fire definitions and methods of determination (e.g., contact, near-contact, intermediate range, and distant [or indeterminate] range).
- Temporary cavity formation and its effects.
- Bullet wipe versus lead fouling.
- Backspatter.
- Wound size and bullet caliber relationships.
- Radiology of missile wounds: usefulness and pitfalls.

Owing to the questions that arise in cases relating to one or more of these and many other firearms-related topics, it is useful to refer to chapters on gunshot wounds in several contemporary general forensic pathology textbooks and atlases, a few examples of which are included in the references at the end of this chapter.[58,59,243,268,298] Our most frequently used single source of information on this subject, however, is that by DiMaio.[54] It not only provides a general survey of the topic, but it also contains considerable information relevant to the neuropathology of missile wounds. Other examples of sources available include textbooks,[57,180] articles,[51] and *Wound Ballistics Review*. The latter, edited by Dr. Martin L. Fackler, is no longer published, but locating and reviewing past issues is well worth the effort. Many workshops and other presentations at national meetings, such as that of the American Academy of Forensics Sciences, are also valuable sources of data on gunshot wounds.

In this chapter, we focus on a few selected topics and cases viewed primarily from the vantage point of the neuropathology consultant, and suggest procedures that will aid the medical examiner in obtaining as much useful information as possible from such consultations. It is assumed that the reader is already, at a minimum, familiar with an overview of the subject comparable to the information in DiMaio's excellent monograph.[54] If certain points made later appear redundant in that they are cited in several of the references listed previously, it is because their reiteration appears to require reinforcement based on the frequency of questions received or because such sources contradict one another on certain points.

The initial portion of this chapter emphasizes handgun wounds, since they account for approximately 90% of firearms-related homicide cases and a slightly lower percentage of firearms-related suicide cases seen in our department. Discussion of other weapons follows. Unless otherwise specified, statements refer to wounds from ammunition using smokeless powder.

Wound Ballistic Issues

There are three mechanisms by which a bullet or other missile causes wounds in tissue. The first mechanism is the crushing and tearing of the tissue by the bullet as it passes through the tissue.[112] The second mechanism is the temporary cavity phenomenon that contributes to some wounds. This is due to displacement of the tissue away from the bullet surface during its passage through the tissue, and the degree to which it contributes to the amount of tissue destruction produced is influenced by a number of variables, such as bullet construction, mass, velocity, presence of yaw, and nature of the tissues struck.[112,183] The third mechanism of wound production is the displacement of tissues by gases generated by the burning powder, which both lead (to a lesser degree) and follow (to a greater degree) the bullet into the tissue. This mechanism of tissue injury occurs only in situations where the gun muzzle contacts, or nearly contacts, the skin surface when the bullet is fired.[72] The resultant permanent wound cavity reflects the consequences of the first mechanism of injury, together with whatever additional tissue injury the second and third potential mechanisms may add under the circumstances of that wound.

The reader will note that no mention of sonic wave, sound wave, or shock wave phenomena contributing to tissue damage is included in the mechanisms of wound production described earlier. This omission is purposeful because available evidence does not, in our opinion, provide support for this hypothetical fourth mechanism of presumably significant tissue injury in gunshot wounds.[54,70,72–74,112,180,182,183]

An appreciation of the factors influencing these various mechanisms of wounding is helpful in avoiding errors based on common misconceptions. For example, tissue damage caused by military rifle bullets before they yaw may be indistinguishable from that due to a handgun bullet.[70] Some handgun bullets with less than one-third the velocity of some rifle bullets may produce a much larger temporary cavity than that produced by the rifle bullet.[69] A 240-grain .44 Magnum bullet fired from a revolver with a 1-inch barrel has a velocity at 15 feet of 742 feet per second; the same bullet fired from the same revolver with an 18-inch barrel has a velocity of 1575 feet per second at a range of 15 feet.[48]

The wound ballistic literature relevant to human wounds is difficult to interpret at times due to differences in terminology and emphasis. Some authors include all available information on the weapon, ammunition, range, presence or absence of intermediate targets, witness statements, etc., together with their description of the wounds. Most do not, which compromises efforts to correlate the injury with weapon type or speed of incapacitation variables, for example. As in all other areas of forensic pathology, the data presented should be examined as objectively and critically as possible.

In addition to the general references cited in the Introduction, references more specifically concerning ballistic issues that the reader may find useful are included at the end of this chapter.[67–69,102,181] Karger[131] provides a lucid description of how these principles apply to cranial gunshot wounds.

Head Wounds

The forensic neuropathology consultant is rarely involved in the interpretation of gunshot wounds of the skin, subcutaneous tissue, oronasal cavities, or eye in various head and neck wounds. Components of missile wound determination such as range, direction of fire, etc., will already have been documented by the referring medical examiner at the time of the general autopsy, using the criteria extensively described and illustrated in references such as those cited in the Introduction to this

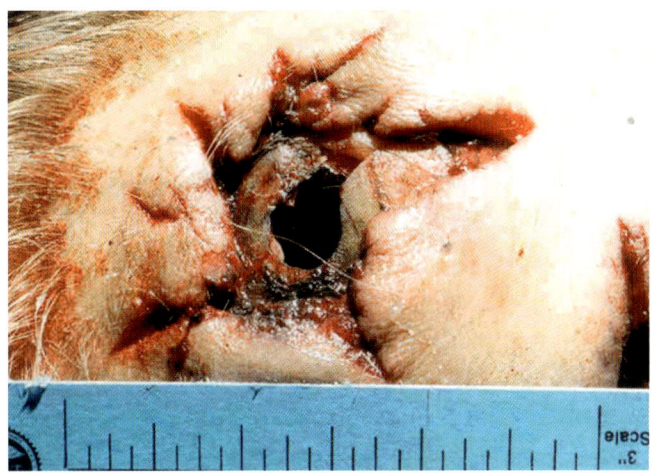

Fig. 8.1. Contact gunshot entrance wound of skull. Dark gray to black products of combustion stain skull and adjacent soft tissue. Focal external beveling is seen at upper margin of skull defect, but was much less extensive than the internal beveling. Stellate scalp laceration results from temporary cavity formation gases generated by burning gunpowder in this case, but scalp stellate wounds devoid of soot are also seen in gunshot wounds of sufficient mass and velocity at much greater ranges.

chapter, combined with the examiner's personal experience. As previously emphasized, the neuropathologist's familiarity with skin and subcutaneous tissue wound appearance, concepts, and terminology is essential for accurate interpretation of general autopsy data and for composing an accurate consultation report. The increased likelihood of stellate tight contact entrance wounds in skin closely overlying the skull (Fig. 8.1) (including facial structures overlying bone), the appearance of abrasion rings, graze (i.e., tangential, gutter) wounds, keyhole skull defects, and so on are familiar concepts that need not be reiterated here.

Skull Wounds

Somewhat less emphasized, and a source of occasional requests for a neuropathologist's opinion in the autopsy suite or later, are questions concerning wounds of the skull and its contents. The magnitude and number of entrance or exit wound skull fractures are not a reliable guide to direction of fire.[22] Skull exit wound fractures may be less than, greater than, or not significantly different in extent and number from those of entrance wounds. Factors that influence these variables include type of ammunition, range of fire, and location of the skull wound.[22] However, fracture patterns and presence or absence of beveled edges of fractures can be useful in the determination of direction of fire, sequence of wounds in cases of multiple gunshot wounds, and distinguishing gunshot wounds from blunt force cranial trauma.

The following description is summarized from selected articles and chapters,[20,107,259] verified by personal experience. It is recommended that these references be reviewed for additional details and helpful diagrams.

Cranial gunshot wounds produce primary (and sometimes secondary) radiating fracture lines that extend from the entrance wound outward. It has been established that these radiating fracture lines develop faster, and thus are already present, when a bullet exits the skull. Radiating fracture lines from the exit wound will, therefore, stop at the preexisting fracture lines caused by the entrance wound, allowing the entrance and exit wounds to be distinguished by this skull fracture pattern irrespective of beveling characteristics of the entrance and exit wounds. These radiating fractures are typically not beveled, but, rather, they are vertical or have a stair-step edge configuration.[259] Fracture line pattern will also be influenced by variations in skull anatomy (e.g., natural buttressed areas of skull; fused versus nonfused skull sutures). A similar analysis of radiating fracture line intersections can sometimes aid in the sequencing of multiple cranial gunshot wounds (Figs. 8.2–8.3).[230]

The distinction between gunshot wound and blunt force trauma to the skull may also be aided by fracture line examination if concentric (i.e., tertiary) fracture lines have also been produced by the injury.[20,107,259] Concentric fracture lines surround the entrance wound in roughly circular fashion, perpendicular to radiating fracture lines, and are typically beveled. They are most likely to occur in wounds due to low-velocity bullets, or in contact wounds due to high-velocity bullets.[252] They may, in wounds of sufficient magnitude, form on the opposite side of the skull (again, before the bullet impacts the exit site) and stop further propagation of radiating fractures from the exit wounds.

Concentric fractures can allow distinction between gunshot wounds (in which the concentric fracture beveling will be external, also referred to as heaving fractures due to the bone being lifted outward by increased intracranial pressure),[259] versus blunt force trauma (in which the concentric fracture line beveling will be inward due to the blunt force applied from outside the skull, compressing it inward).

Even when both radial and concentric fractures occur, however, occasional cases may develop atypical fractures that have features of both gunshot wounds and blunt force trauma.[259,260] Internal ricochet was a contri-

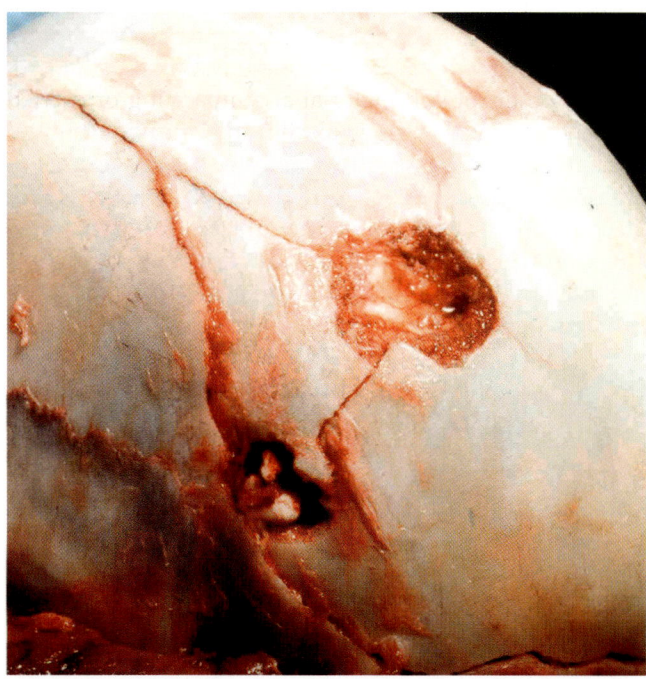

Fig. 8.2. Sequencing gunshot wounds. Right lateral skull with two exit wounds. The lower skull defect (1) is from the first bullet fired, with fracture lines radiating upward at 11 o'clock (A) and 1 o'clock. The upper skull defect is from the second bullet fired.

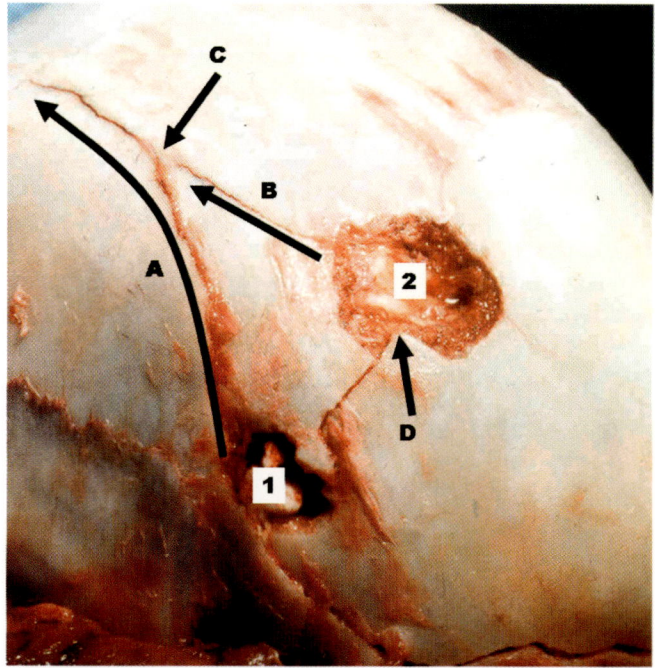

Fig. 8.3. The second exit wound (2) demonstrates a fracture line at approximately 9:30 o'clock (B), which stops where it meets the preexisting 11 o'clock fracture line from the first exit wound (C), and also demonstrates a step-off margin inferiorly where it intersected the 1:00 o'clock fracture line from the first exit wound (D).

buting factor in one of the two cases reported by Smith et al.[260] (see later). When the gunshot is a contact midline skull wound, the resulting fracture lines may be quite symmetric.[78] Basal skull fractures of various patterns, including hingelike fractures, may also occur due to direct or indirect effects of cranial gunshot wounds.[23]

Tangential Skull Wounds

A gunshot wound that penetrates the scalp and subcutaneous tissue may be angled sufficiently to graze or groove the skull but not enter it (Fig. 8.4). Such a wound has also been termed a tangential wound or gutter wound. It may involve the skull external table only; may produce, in addition, a linear nondisplaced inner table fracture; or may perforate the skull and cause fragments of bone to be displaced inward. A slightly deeper penetration may be accompanied by the deeper portion of the bullet being sheared off by the skull and entering the intracranial cavity, often with skull bone fragments, while the remainder of the bullet continues on an extracranial path. The latter situation may result in a keyhole skull fracture, with its characteristic combination of external and internal beveling of the wound edges. Rarely has a keyhole wound pattern been described in a skull exit gunshot wound.[56] When even a portion of the bullet perforates the skull, it is no longer classified as a tangential wound.[10,54]

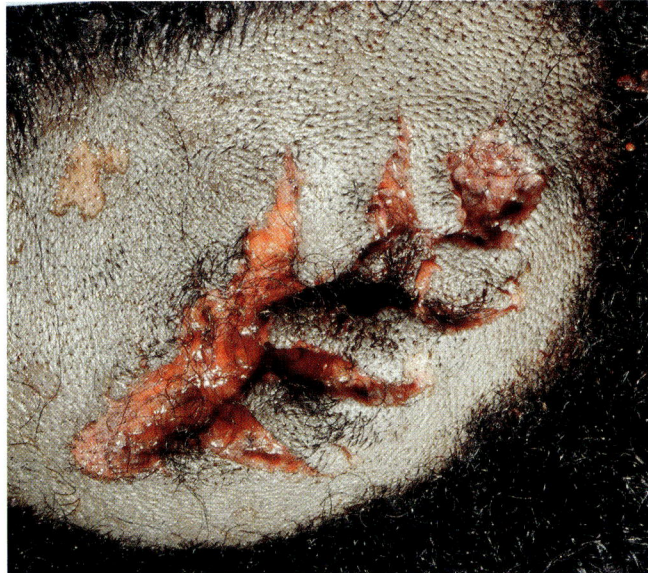

Fig. 8.4. Tangential gunshot wound of scalp. The direction of bullet travel was from lower left to upper right. The points of the torn scalp are oriented toward the lower left corner of the photo, indicating the location of the weapon muzzle.

Most tangential gunshot wounds do not result in fatality or serious intracranial consequences, but exceptions occur. Tangential gunshot wounds to the head can cause profuse extracranial hemorrhage,[64] epidural hematoma, subdural hematoma, subarachnoid hemorrhage, cerebral contusions, and intracerebral hemorrhages.[10,103,272]

Complicating the clinical evaluation of tangential skull gunshot wounds is the observation in one study that the determination by clinical history of loss of consciousness, a Glasgow Coma Scale score of less than 15 on emergency department presentation, presence of skull fracture, location of wound on the skull, or presence of extracranial bullet fragments, alone or in combination, did not reliably distinguish which cases would develop intracranial hemorrhage.[10]

Entrance and Exit Wounds of the Skull

Bullets do not "drill" their way into the body.[54,162] The distance required for one complete revolution of the bullet due to rotational effects imparted by the twist rate of barrel rifling (e.g., 1 × 8 inches, 1 × 12 inches) in comparison to the bullet's forward motion due to velocity is such that a more apt analogy is that bullets rapidly push their way into the body. Entrance wounds in thin portions of the skull, such as the temporal squama, may not demonstrate the internal beveling which assists one in distinguishing entrance versus exit (which typically demonstrate external beveling) wounds in thicker portions of the skull (Figs. 8.5–8.6). Soot, bullet wipe, lead fouling, smokeless powder, or black powder[173] granules, and/or other foreign material accompanying or carried into the wound by the projectile may provide helpful clues by their exclusive presence or preponderance at the entrance versus exit wound. Thermal damage to tissues is less common.[188] Unfortunately, many or all of these ancillary clues are less likely to be present with distant range wounds, with jacketed bullets, or in cases in which the skull is skeletonized or there is advanced decomposition.

The beveling patterns of skull wounds are sometimes misleading. External beveling of entrance wounds is more likely to occur in entrance wounds in which the bullet is undergoing yaw, where the defect involves a suture line or a preexisting fracture, or the defect is caused by a tight contact gunshot wound with a more powerful handgun.[252] Rifle wounds may also produce entrance wound external beveling.[219] Even in the presence of these uncommon atypical features, it is usually possible to determine entrance wounds because internal beveling is nearly always present, and it is more pronounced than the external beveling.[162,222] Ancillary clues such as abrasion rings, skull fracture patterns, and bullet fragmentation patterns are also often present.

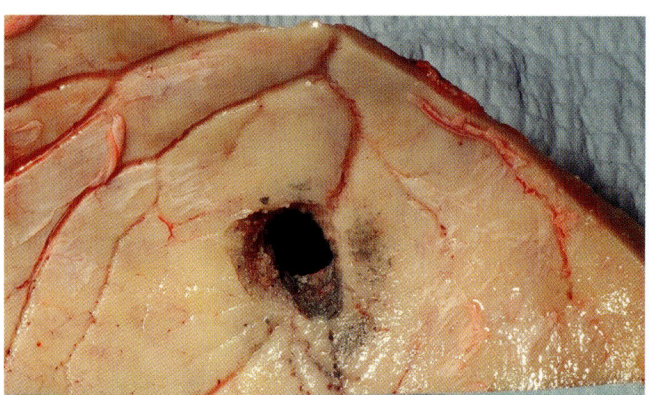

Fig. 8.5. Gunshot entrance wound as seen from inner surface of skull, with pronounced internal beveling. The dark gray soot staining is characteristic of a contact or near-contact wound.

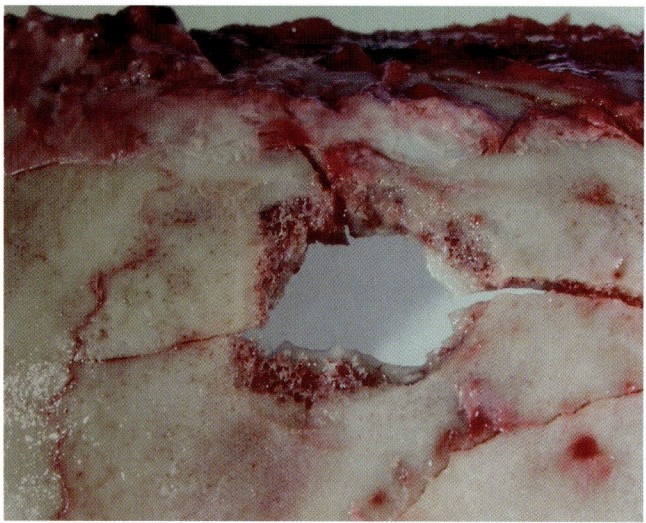

Fig. 8.6. Gunshot exit wound, as seen on external surface of skull, with pronounced external beveling. The radiating fracture line at approximately 8:30 o'clock stops at a cranial suture. Hemorrhagic tissue at top of photo is periosteum and temporalis muscle.

With rare exceptions, in the absence of intermediate targets the skull entrance wounds are smaller than the exit wounds.[233] Most skull entrance wounds are round to ovoid,[222,223] as are most exit wounds, but, if the wound is irregular in shape (i.e., square, triangular, or rectangular), it is more likely to be an exit wound.[223]

No attempt should be made to determine precise bullet caliber from entrance wound size, since the latter may be larger or, less often, smaller than the bullet producing it.[19,54] On rare occasions, head gunshot wound patterns may provide clues linking more than one victim to a single assailant.[83] In Frazer's report, both victims had two bullets fired through the same skull entrance wound.[83] The reverse phenomenon, with two separate entrance gunshot wounds converging to exit through a single exit wound, has been described in the thorax, but not, to our knowledge, in a head wound.[111]

Internal Ricochet

Upon entering the skull, a bullet may travel along the curvature of the inner table of the skull and come to rest external to the brain or it may travel along the inner table and then bounce off the inner table to enter the brain. It may also bounce off the inner table of the skull after penetrating the brain one or more times, with production of two or more intracerebral bullet paths at various angles to one another.[85] Intracranial bullet ricochet is not restricted to bony surfaces. It may also occur from other firm surfaces, such as the falx cerebri or tentorium cerebelli.[85] Exceptionally, the bullet may ricochet off the inner table of the skull and exit the calvarium through its entrance wound.[95] These various alternatives to a straight linear path of the bullet upon entering the skull are termed "internal ricochets." The internal ricochet site(s) may be identified on the dura mater by bullet wipe, lead fouling, abrasion, laceration, and/or hemorrhage (Case 8.1, Figs. 8.7–8.9). It may be identified on the skull by bullet wipe, lead fouling, and/or fracture, depending on variables such as bullet composition, construction, and velocity. Internal ricochet is one factor that, among others such as bullet migration (discussed later), makes speculation as to bullet path based solely on location of entrance wound and position of bullet on radiographs subject to error.

Brain Wound Characteristics

Inelastic tissues, such as brain, liver, spleen, and bone, are more vulnerable to temporary cavity effects than elastic tissues, such as muscle, lung, skin, or bowel.[70] An additional factor in brain injuries is the inelasticity of the restraining skull, against which brain thrust outward due to temporary cavity effects may impact and produce injury remote from the bullet path.[131] The effects of gunshot wounds to the brain *exclusive* of such remote effects are discussed in this section.

While passing through scalp, skull, and meninges into the brain, the bullet produces a permanent cavity with variable superimposed injuries due to temporary

Case 8.1. (Figs. 8.7–8.9)

A 28-year-old man was shot in the head. Although there was no immediate loss of consciousness, his condition deteriorated within minutes. On arrival at the hospital, lower greater than upper extremity paralysis, dysconjugate gaze, respiratory insufficiency requiring airway assistance, and seizure activity occurred, followed by coma. A computed tomography head scan revealed multiple intracranial metal fragments, cerebral swelling with right to left midline subfalcine herniation, and a small right subdural hemorrhage. Intracranial pressure ranged from 30 to 50 mm Hg, later rising into the 70s. Despite supportive care, he expired 6 days' postinjury.

Autopsy examination revealed the following bullet path. The entrance was at the left upper parietal area and there was no exit wound. Direction was left to right, slightly front to back, and slightly downward. Brain structures involved included the posterior aspect of the left superior frontal gyrus and the left pre- and postcentral gyri superiorly, the superior sagittal sinus, right pre- and postcentral gyri, and subcortical white matter. Exiting the brain, the bullet entered the dorsolateral right hemisphere subarachnoid space (Fig. 8.7). The bullet then was deflected by an internal ricochet from the inner table of the skull in the right inferolateral parietal/posterior lateral temporal lobe region (Figs. 8.8–8.9), reentering the brain substance and terminating in the immediate subcortical white matter of the junction between the right posterior temporal and occipital lobes laterally.

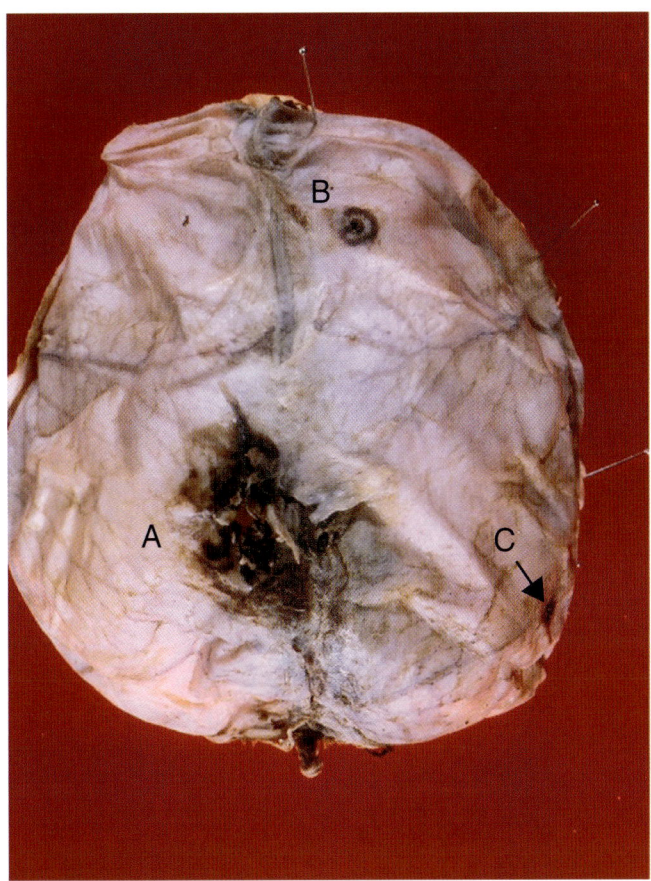

Fig. 8.8. Outer surface of cranial dura mater. A, Bullet path through superior sagittal sinus; B, intracranial pressure monitor site; C, hemorrhagic dura mater at internal ricochet site.

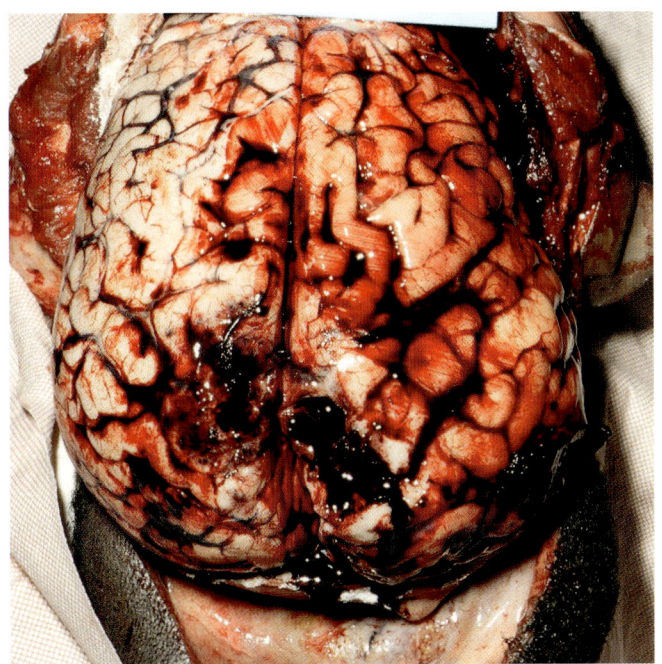

Fig. 8.7. Exposed brain with subarachnoid hemorrhage. The posterior hemispheres demonstrate more dense blood clot obscuring the bullet path in the region of the bilateral pre- and postcentral gyri.

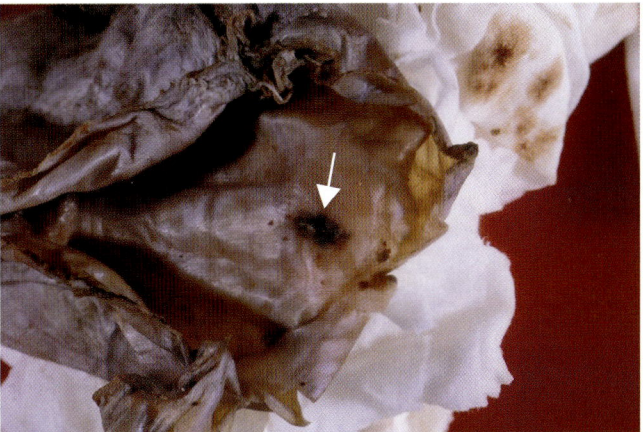

Fig. 8.9. Inner dura mater surface at internal ricochet site (*arrow*).

cavity effects and expanding gases accompanying the bullet (as discussed in the earlier section, "Wound Ballistic Issues"). Subarachnoid hemorrhage will be present, and surface contusions around the entrance and/or exit wound may be present.[54] Depending on whether or not the bullet undergoes fragmentation, and the presence and number of accessory missiles (such as fractured skull bone fragments or other intermediate target matter) entering the brain, secondary wound paths may occur that deviate from the main bullet permanent cavity, typically in an expanding cone-shaped pattern.

Studies of brain gunshot wounds have been performed in humans[154,212,213] and in animals.[139] Both types of studies have tended to focus on common handgun calibers (e.g., varying from .22 caliber to .45 caliber). In human studies[212,213] the permanent bullet cavity had a margin of necrotic tissue that was devoid of glial fibrillary acid protein (GFAP)-positive astrocytes and contained hemorrhages. Within a narrow zone, however, GFAP positivity in astrocytes abruptly returned as one moved further from the permanent cavity. Neutrophils were seen in all cases in the margin of the permanent cavity as single infiltrating cells or in aggregates within somewhat over 2 to 2.5 hours' survival of the victim. There was no clear relationship between numbers of neutrophils and survival times or presence or absence of clinical brain death in this report.[213] More difficult to reconcile with other known CNS injury timetables described in this text was the authors' description of infiltrating macrophages in the margin of the permanent wound cavity in this same time frame, that is, somewhat over 2 to 2.5 hours, based on the presence of cells stained with CD68 immunostain.[213] These cells were located peripheral to the necrotic area that contained neutrophil infiltration, in the zone of injured axons. Disruption of neurons, myelin sheaths, and axons extended into a more peripheral circumferential zone. The overall circumference of injured tissue was larger at the entrance than at the exit of the permanent wound cavity. The mean values of radial distance for each of these zones, measured from the border of the permanent cavity near the bullet entry (including wounds from revolvers, pistols, or rifles varying in caliber from .22 long to 9mm), was estimated as 3.2 ± 1.2mm for astrocyte destruction, 8.25 ± 5.08mm for hemorrhage, 17.7 ± 4.78 for axonal injury, and 18.47 ± 5.24mm for neuronal injury.[212] More widespread staining of axons for β-amyloid precursor protein (β-APP) has been described in portions of the brain remote from the gunshot wound permanent cavity, including the opposite cerebral hemisphere and brainstem. However, distinguishing the effect of the gunshot wound per se from secondary causes of β-APP positivity, such as hypoxic-ischemic injury, has been problematic[154,213] (Case 8.2, Figs. 8.10–8.13; Case 8.3, Figs. 8.14–8.15).

Case 8.2. (Figs. 8.10–8.13)

A 23-year-old man sustained a gunshot wound to the head while in his wheelchair in a gang-related drive-by shooting. He was a paraplegic following a motor vehicle accident a year before. Although found unresponsive, his vital signs were strong. However, he died 14 hours after the shooting without recovering consciousness.

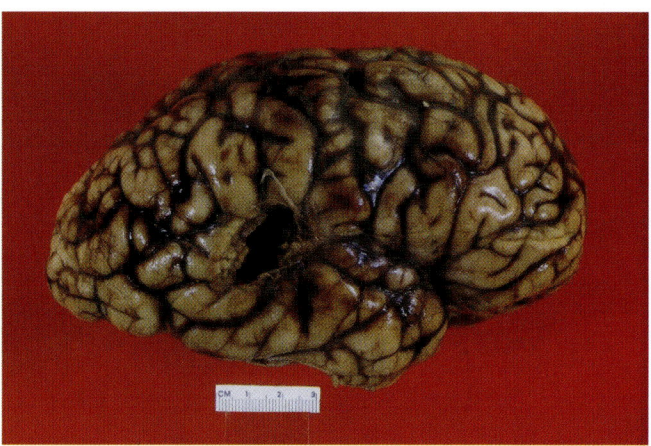

Fig. 8.10. Entry wound to right superior temporal gyrus at the level of postcentral gyrus. Margins of wound are relatively sharp with little maceration and subarachnoid hemorrhage.

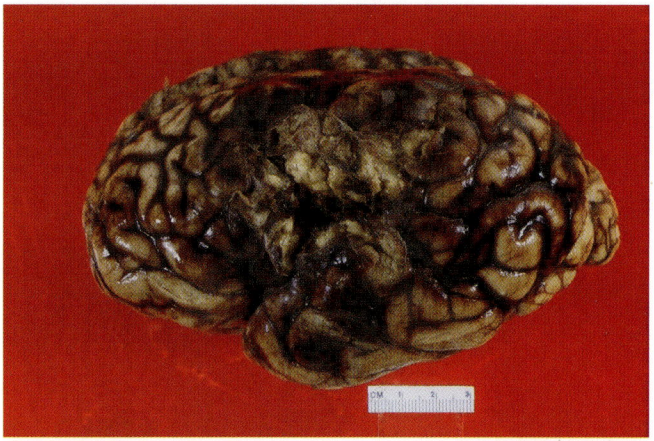

Fig. 8.11. Exit wound at left lateralmost paracentral region and Broca's area. Note broad zone of cortical maceration and wide area of subarachnoid hemorrhage.

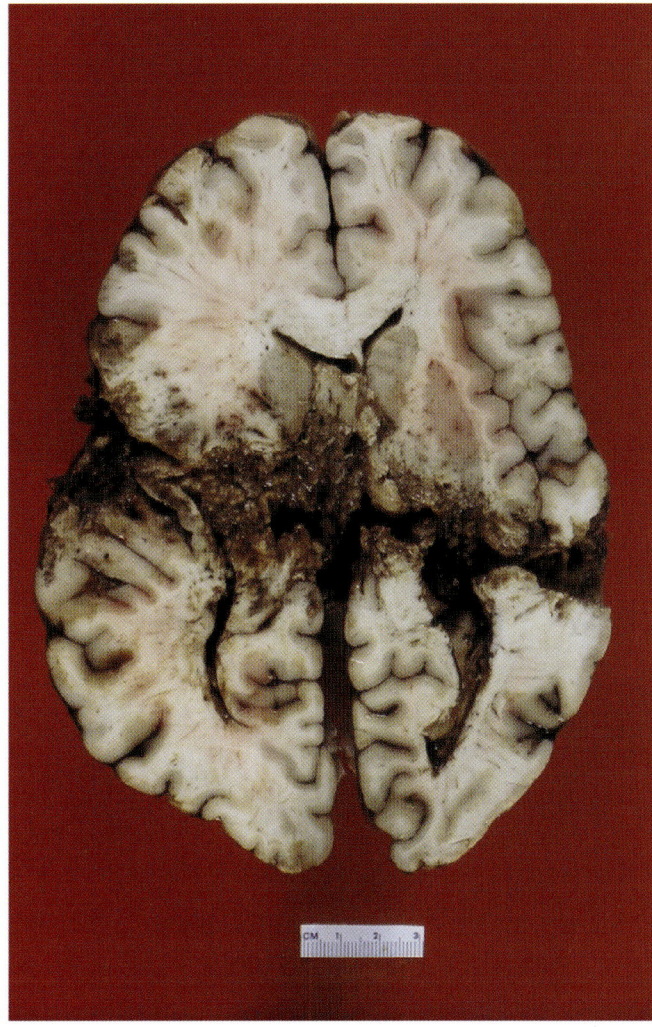

Fig. 8.12. Horizontal section connecting entry and exit wounds displays missile track crossing thalami. Exit wound is slightly larger and more hemorrhagic.

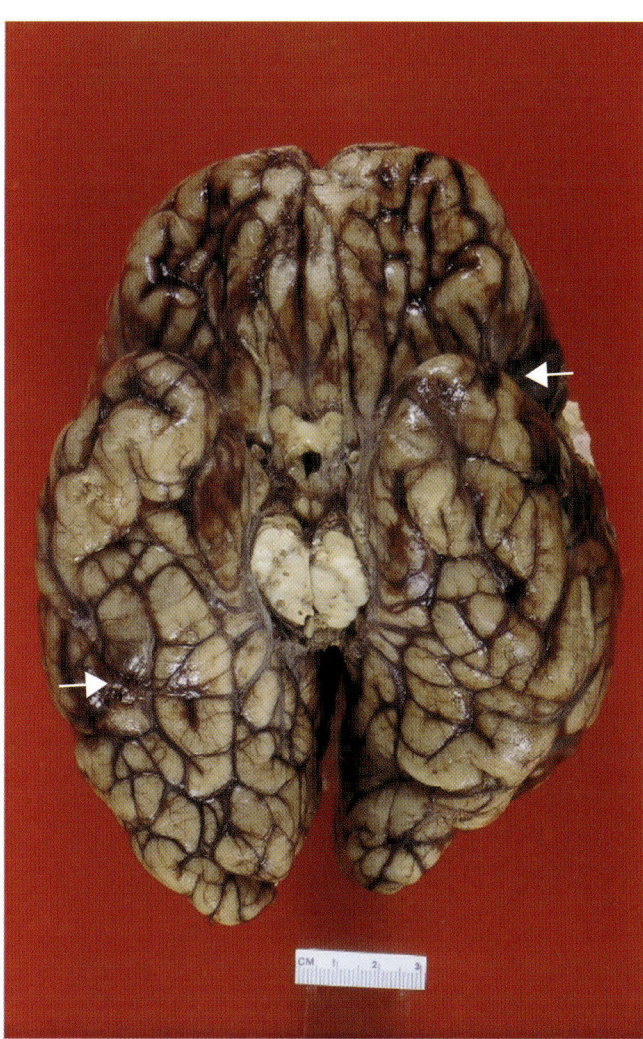

Fig. 8.13. Small remote contusions (*arrows*).

Case 8.3. (Figs. 8.14 and 8.15)

This adult male was shot in the head 4 years prior to death, and was in a vegetative state throughout this interval.

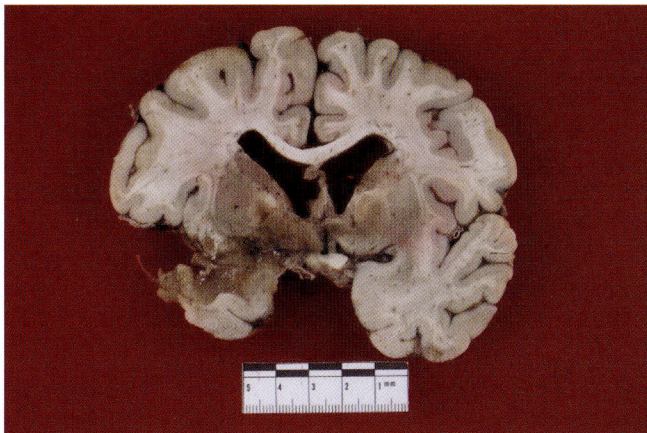

Fig. 8.14. Coronal section of brain with left superior temporal lobe and left basal ganglia tissue loss and brownish discoloration at level of gunshot entrance wound in left temple area.

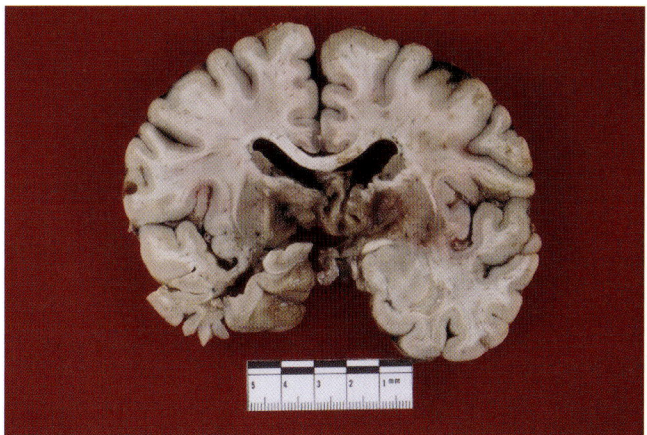

Fig. 8.15. Bilateral globus pallidus greater than putamen atrophy and brownish discoloration extending beyond the permanent wound cavity, most likely due to hypoxic-ischemic injury following the gunshot wound as a secondary complication.

Karger et al.,[139] in experimental brain wounds in calves produced by 9-mm pistol rounds, emphasized the frequency with which secondary bone missiles and intracranial pressure effects from the temporary cavity contributed to the wounding effects in brain. They discussed the value of radiologic imaging studies such as magnetic resonance imaging (MRI) and computed tomography (CT) scans, and they compared their findings with those described in selected human cases.

The sequence of histologic changes that occur in contusions and lacerations in nongunshot wound cases described elsewhere in this text has proved to be consistent with that seen in our office at similar survival times in gunshot wounds, and as described by others in a recently reported case.[185]

Using these various techniques, wound studies that attempt to correlate these findings with weapon caliber, bullet construction, range of fire, presence or absence of secondary overpressure effects (e.g., contusions remote from the gunshot wound path, skull fractures remote from the entrance and exit wound–related fractures, heaving fractures), and with reliable witness observations on speed of incapacitation would add considerable forensically useful information to our present knowledge of the effects of gunshot wounds to the head. Admittedly, such information is often difficult to obtain.

Backspatter

Backspatter has been demonstrated in both human gunshot wound cases and in animal experimentation, and it has been defined as "the ejection of biologic material from the entrance wound in a retrograde direction to the line of fire."[138] Gunshot wounds to the head may result in deposition of fragments of brain, fat, muscle, bone, skin, hair, or eyes on the weapon (even inside the barrel), on the shooter, and/or on other surfaces in the vicinity.[138] It is more likely to occur in cases of head wounds with contact or close range, with larger-caliber weapons, or with more powerful weapons.[54,138,271] The search for such evidence is primarily within the province of the scene investigator, criminalist, firearms examiner, and the medical examiner, but the neuropathologist should have some familiarity with its potential presence and relative frequency.

"Disappearing" Bullets

See Case 8.4 and following section, "Bullet Migration in Bodies," p. 223.

Case 8.4. (Figs. 8.16–8.23)

A 33-year-old woman was one of the victims in a triple homicide. All three victims were found bound and shot in the head. Bullets were recovered in the other two victims, but not in this woman. In this case there was an entrance wound in the left posterior parietal scalp, but no exit wound (Fig. 8.16). The bullet path included the underlying skull (which demonstrated internal beveling and bullet wipe), left parietal and temporal lobes, and left ventral pons, terminating in a fracture site at the left posterior clinoid process. Additional findings included laceration of pituitary gland and cavernous sinus, laceration of the right cerebral peduncle and adjacent hippocampus, subarachnoid hemorrhage, and diffuse brain swelling with bilateral uncal and cerebellar tonsillar herniations. No bullet was found in the victim's clothes or in the body (including by full body x-rays) (Fig. 8.17). Reexamination of the scene and items placed in evidence from the scene revealed a pillow with two soot-ringed defects on one side (Fig. 8.18). A slitlike tear and an adjacent blunt-tipped extrusion of blood-stained polyester pillow-fill material were found on the opposite side of the pillow (Figs. 8.19–8.20). Dissection of the latter (Fig. 8.21) revealed a bullet contained in its tip. The bullet was a round nose .38 or .357 caliber, 140-grain, nonjacketed lead bullet with two cannelures, rifled 6 left. A bone fragment and soft tissue was embedded in the bullet tip (Fig. 8.22). Examination of the polyester pillow-fill "cocoon" in which the bullet was held revealed broad zones of fusion of fibers forming a very firm sheet of synthetic material (Fig. 8.23). DNA analysis of the blood on the surface of the extruded pillow-fill material was compared with blood from the decedent, and the two samples were genetically indistinguishable.

Comment. Examination of this case excluded other potential mechanisms by which bullets can seemingly disappear from bodies, further discussed elsewhere in this chapter. Further analysis by LAPD Crime Laboratory expert Susan Brockbank[231] and her associates revealed that the polyester pillow-fill used in this particular pillow melted and fused at a temperature of 491°F, whereas the flame jet preceding a fired bullet may reach a temperature of approximately 1400°F.[55] The bullet rifling suggested a weapon manufactured by Colt's Manufacturing Company, and the pillow soot pattern near the entrance holes was consistent with a revolver cylinder gap stain approximately 3 inches from the near entrance wound and 6 inches from the far entrance wound on the pillow (see faint stain to left of entrance hole, A in Fig. 8.18). It was concluded that the pillow was used as an improvised silencer for two of the victims, and, in the present case, the circumstances were such that the flame jet accompanying the bullet fused sufficient polyester pillow-fill fibers to form a sleeve accompanying the bullet in its path intracranially in the victim.[231] This synthetic sleeve was sufficiently firm to extract the bullet it surrounded from the victim's head when the murderer pulled the pillow away and tossed it several feet from the victim. (Case courtesy of James K. Ribe. M.D.)

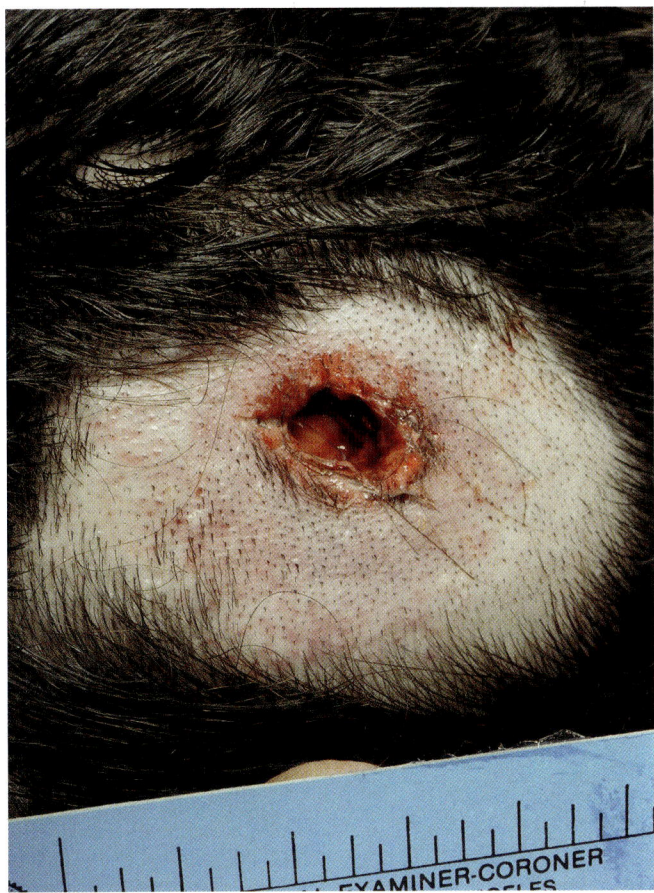

Fig. 8.16. Scalp entrance bullet wound.

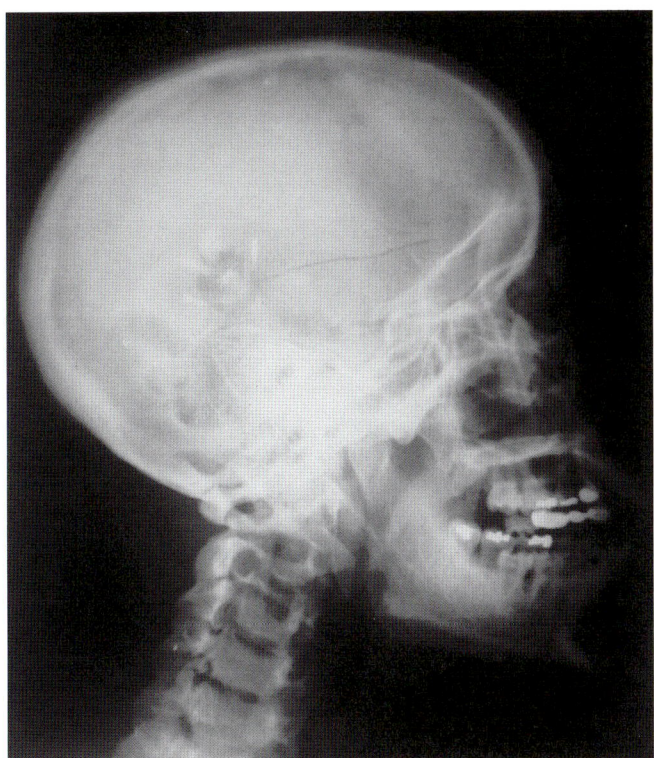

Fig. 8.17. Lateral skull radiograph reveals fractures, but no metallic opacity aside from dental fillings.

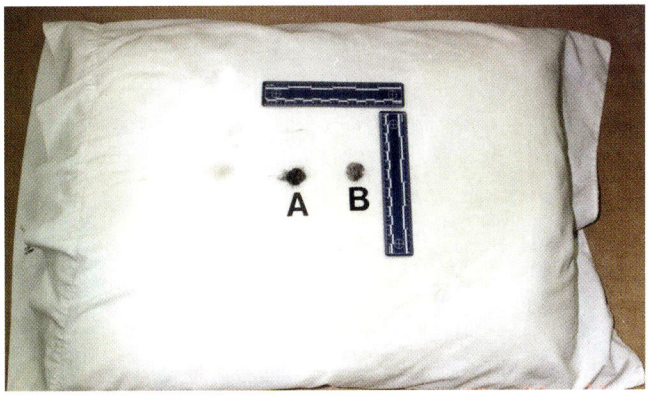

Fig. 8.18. Pillow with two soot-ringed gunshot entrance wounds (A and B).

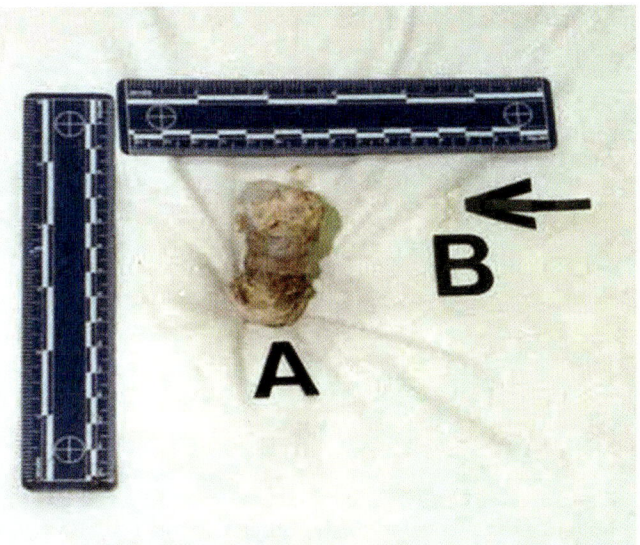

Fig. 8.19. Opposite side of pillow seen in Fig. 8.18. Label A indicates a blood-discolored protrusion through torn pillow case. B label indicates a small tear in pillowcase (*arrow*).

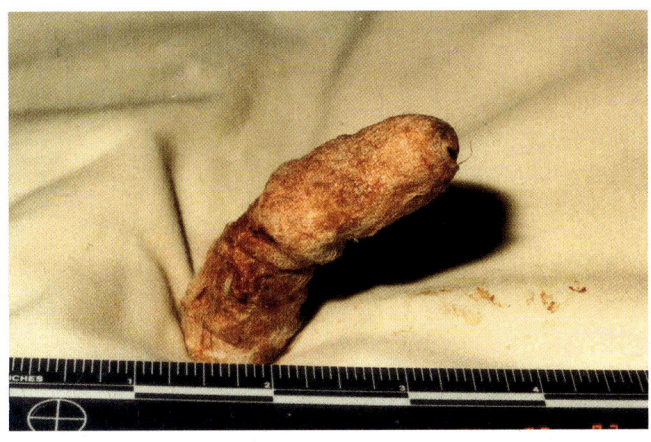

Fig. 8.20. Side view of blood-stained protrusion from pillow (designated "A" in Fig. 8.19).

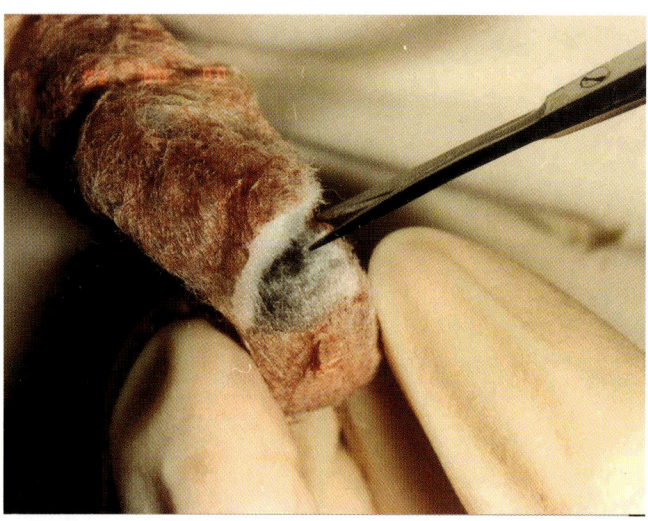

Fig. 8.21. Dissection of top of protrusion seen in Fig. 8.20 reveals bullet.

Fig. 8.23. Detail of inner aspect of pillow protrusion, demonstrating heat-fused polyester pillow-fill material.

Fig. 8.22. Detail of removed bullet, with bone and soft tissue embedded in deformed nose (anterior aspect) of bullet.

Intermediate Targets

The aforementioned unique case demonstrated bullet removal from a body by an intermediate target. More usual effects of an intermediate target include an altered bullet path by means of missile deflection or ricochet, altered clues of range of fire estimates such as blocking or deflection of gunshot residue products, bullet fragmentation, adding secondary missiles to the wound due to fragments of the intermediate target being carried into and/or around the surface of the wound created by the bullet (e.g., pseudostippling), and/or deforming or altering the appearance of the bullet itself due to contact of its surface with the intermediate target.[54] Pseudostippling can also occur in the absence of intermediate targets.[221] Revolver cylinder indexing errors may also fragment the bullet as it enters the barrel, producing atypical wounds.[125]

Thick head hair can prevent the appearance of typical indicators of a near-contact gunshot wound, the hair acting as a form of intermediate target.[119] Davis[47] cited a case in which an individual's dense, tightly coiled hair stopped a bullet. Some unusual intermediate targets, such as an empty cartridge case with intact primer[258] and metal coins,[186] have been carried into head wounds by bullets. In several of our cases of gunshot wounds to the head, an upper extremity of the victim had been an intermediate target.

Silencers (suppressors) may alter not only the loudness of the report, but may also increase or decrease bullet velocity, its tendency to yaw, or produce bullet deformity or fragmentation (if improperly constructed, or misalignment occurs).[140] They may cause atypical entrance wounds due to reduced mechanical and thermal effects at contact entrance wounds, reduce carbon monoxide in contact entrance wounds, produce an enlarged muzzle imprint in contact wounds, or add silencer-derived foreign material to the wound.[191] A foam-filled domestic pillow used as an improvised silencer was noted to be more effective in noise reduction than some metal silencers tested by the authors.[191]

Bullet Migration in Bodies

The preceding case report (Case 8.4) is an example of an unusual mechanism by which a bullet entering the body becomes difficult to locate. The mechanisms discussed later are also applicable to many other types of missiles (shrapnel, etc.) that can produce wounds.

After entering the body, the bullet may, due to deflection by bone or losing energy, enter a tissue plane, such as subcutaneous tissue of the scalp, and track along a path of least resistance rather than continuing in a straight line. This can result in its path terminating at a quite unexpected site (see Case 8.8 later). Tracking of a bullet along a muscle fascial plane has been described in the neck, for example.[304] In one unusual case, this path of least resistance was the spinal canal, with the bullet entry at the C6 level and destruction of the spinal cord along the bullet path between C6 and T10, where the bullet came to rest.[278]

Once entering the body and coming to rest, bullets may subsequently move, variously referred to as wandering or migrating bullets. Such bullet movement has been described in multiple sites, including in the pleural, peritoneal, and pericardial cavities; in the tracheobronchial tree; entering the gastrointestinal (GI) tract and passing out through it[112]; entering the genitourinary (GU) tract and passing out through it; and migrating within the central nervous system (CNS).[165] One remarkable case involved a gunshot wound to the gravid uterus of a 23-year-old woman in her 36th week of pregnancy.[32] The bullet entered and remained within the uterus. The newborn infant, delivered by urgent cesarean section, had sustained a puncture wound at the right nasolabial crease without other injury. Subsequent x-rays revealed the bullet in the GI tract of the infant, not in the mother. With the aid of laxatives, the infant passed the bullet through the rectum. Bizarre cases are clearly not limited to CNS gunshot wounds, although the latter is the focus in this chapter.

Significant intracranial or intraspinal bullet migration has occurred within a time frame as brief as 2 hours (during which a bullet moved from the left to right posterior fossa).[86] Such cases have prompted several authors to recommend immediately preoperative x-rays, or intraoperative x-rays with the patient in position for operation, to ensure final bullet position prior to surgery.[15] Bullets have been observed to move in the spinal canal under fluoroscopy,[263] and intraoperatively with a change in patient position.[100] More often, migration of bullets or other metallic foreign objects in the CNS occurs over a period of several days, weeks, months, or even several years.

Postulated mechanisms of bullet migration include movement of the bullet in brain parenchyma damaged by the bullet during its passage, and decreased brain tissue density in the presence of hemorrhage or abscess cavity. Missiles may gain access to intraventricular or subarachnoid spaces that facilitate missile movement with body motion and/or other speculated influencing factors such as brain pulsations and cerebrospinal fluid (CSF) circulation. The bullet may gain access to the circulatory system.[152,199,225] To paraphrase and extend the criteria suggested by Michelassi et al.,[198] possibilities to consider include bullet migration, bullet embolization, internal ricochet, and loss of bullet at the scene of injury or anywhere between the scene and the autopsy table. Injuries mimicking gunshot wounds (e.g., screwdriver wound) should be suspected when a head, neck, spine (or other body area) entrance wound exists without an exit wound, signs and symptoms prior to death do not correlate well with suspected missile path(s), and radiologic examination either fails to reveal a missile or reveals an unexpected type of missile located within the body. Head, neck, or spine bullet or other metallic foreign body migration patterns relevant to the neuropathologist include the following selected examples:

1. *Bullet migration within the head.* Bullet migration within the intracranial cavity occurred in 4.2% in one series of 213 cases in the absence of abscess formation.[227] In this series, fragments in the anterior fossa tended to migrate toward the sella turcica and those in the middle fossa and posterior hemispheres tended to migrate toward the torcular Herophili.[227] This migration pattern has not been consistent, however, and several other migration patterns are described in case reports:
   - A bullet located initially in the left posterobasal temporal lobe was later found in the right cerebellopontine angle near the inner surface of the skull, having found its way through the tentorial incisura.[152]
   - A bullet initially entering the left parietal bone came to rest in the right frontal lobe, and by day 7 postinjury had migrated to the right occipital lobe.[199]
   - In a left frontal area gunshot wound, the bullet came to rest in the left frontal lobe. Over a period of a week, it migrated to the left temporoparietal and finally to the left occipital region.[307]

- A bullet entering the right frontal region terminated in the cerebellum, where serial CT head scans revealed a tumbling or rotational movement of the bullet within the cerebellar hemisphere during the subsequent 7 weeks.[307]
- A gunshot wound in the left parietal region resulted in the bullet coming to rest in the left cerebellar hemisphere. Eight hours later the bullet had moved posteriorly approximately 1.0 cm. A left suboccipital craniotomy revealed contused cerebellar tissue but no bullet, and a postoperative CT scan (performed only 2 hours after the preoperative scan demonstrated a left posterior fossa location) revealed the bullet to be in the right posterior fossa. Just prior to the next planned surgery, skull radiographs revealed the bullet had returned to the left posterior fossa, and it was successfully removed from that site.[86]
- A bullet entering the cranium migrated via gunshot wound-related orbital fractures into the ipsilateral maxillary sinus, accompanied by infection.[232]

2. *Bullet migration from head to spinal canal.* Bullets have migrated from the intracranial cavity to spinal cord levels varying from cervical to sacral levels.[13,145,306] The intracranial to intraspinal route has also occurred with other projectiles, including BBs[284] and air rifle pellets.[116] Occasional fragments of shrapnel (8) and metallic surgical clips have also found their way from an intracranial to an intraspinal location.[116]

3. *Bullet migration within the spinal canal.* Bullets entering the spinal canal at one level may migrate to other levels of the spinal canal. Most published examples describe a rostral to caudal migration, as in the following cases:
   - T7-level bullet entry, with the bullet found at the S1 vertebral level (presumably by migration rather than internal ricochet).[305]
   - Bullet entry and termination at the T11–12 level, with subsequent migration to the L4–5 level within 1 day.[143]
   - C7-level entry wound and initial bullet position, with subsequent migration to the sacral level.[276]
   - Bullet entrance at the C1 level, migrating to the T6 level within 12 days' postinjury, and eventually to the S2 level on a radiograph taken 3 years' postinjury.[225]

In a few instances, bullets that initially migrated in a caudal direction were subsequently found in a more rostral location.[15,100,263] In one of these cases, the bullet was moved back down to the open operative site (at the S1–2 level, where the bullet location was documented by prior x-ray) from its new, intraoperatively discovered L3-level position, by elevating the head end of the operating table. The bullet was described as "lying freely in the thecal sac in between the nerve roots."[100]

4. *Bullet migration within blood vessels.* Movement of bullets that have entered the venous or arterial system is termed "bullet embolization." Michelassi et al.,[198] in a literature review published in 1990, found 153 cases of bullet emboli in the English language literature. Although most entry wounds in this series were in the truncal region, it is also possible for neck vein[288] and common carotid artery wounds to result in bullet embolization (Case 8.5, Fig. 8.24). Dural venous sinus wounds (superior sagittal sinus, transverse sinus, and sigmoid sinus) have also been sources of embolization of bullets to the extracranial venous circulation.[198,208] In one case series, 14.7% of venous bullet emboli moved in a retrograde fashion, and 10.4% of arterial emboli resulted from right heart or venous injury.[198] Retrograde embolization in some cases was ascribed

Case 8.5. (Fig. 8.24)

This young adult male was shot in the right midback, the bullet path traversing the liver and right thoracic cavity, exiting the right upper torso, reentering the right neck, and terminating in the right internal jugular vein. The bullet then embolized to its final location in the inferior vena cava (Fig. 8.24).

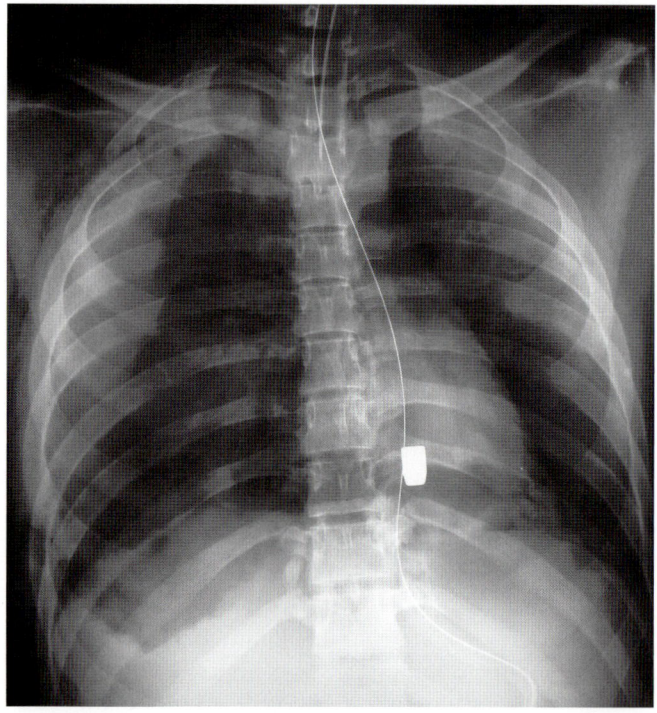

Fig. 8.24. Anteroposterior chest radiograph. Bullet-shaped metallic foreign body at the level of the distal esophagus partially overlies a nasogastric tube just proximal to the gastroesophageal junction. Bilateral pneumohemothoraces are present with subcutaneous emphysema in the right chest wall.

to body position, gravity, respiratory movements, and bullet caliber and weight. Paradoxical embolization could be ascribed to a patent foramen ovale or atrioventricular septum perforation by missile(s).[113,198] Other examples relevant to the CNS include:

- Shotgun wounds involving the carotid artery[303] and heart[144] have resulted in pellet emboli to the middle cerebral artery (Case 8.6, Figs. 8.25–8.27).
- An air gun pellet wound to the common carotid artery remained at that site for 3 days, but, by the 12th day postinjury, it had migrated to the ipsilateral internal carotid artery just below the level of the clinoid process, resulting in complete occlusion of this vessel.[216]
- Vascular injury in head or neck gunshot wounds may result in air embolism.
- Gunshot wounds or severe blunt force trauma of the brain may result in pulmonary embolization by cerebral tissue.[204] Parenthetically, CNS tissue embolization to edible portions of livestock slaughtered with the aid of either conventional cartridge-generated, or pneumatically operated, captive bolt guns has also been demonstrated.[11] This has raised concern in some quarters as to whether this could facilitate transmission of CNS prion diseases in such animals to humans.[11]

5. *Bullet migration within GI, respiratory, or GU tracts.* Migration within these systems is well described in the literature.[54,165] Wounds of the head or neck, particularly with skull base, nasal sinus, or palatal involvement, may allow the missile to gain access to the nasopharynx region, with subsequent descent into the upper GI tract or tracheobronchial tree. Loss of the missile from the body may then occur through vomiting,[54] expectoration after coughing,[238] or spontaneous passage per rectum after being swallowed.[205]

6. *Miscellaneous.* A review of how bullets may seem to disappear would not be complete without commenting briefly on bullets that are found at autopsy, and then vanish. Use of sink traps in the autopsy suite, and careful attention to uninterrupted possession of the bullet by the medical examiner or pathology consultant from the time it is retrieved to its packaging and submission to the evidence safe, will minimize

Case 8.6. (Figs. 8.25–8.27)

A 24-year-old security guard sustained a shotgun wound to the head. His initial Glasgow Coma Scale (GCS) score was 15 (normal), and he was without focal neurologic abnormalities despite x-ray evidence of multiple shotgun pellets in the head and neck area as caudal as the C3 vertebra. One day later he developed lethargy, and on day 2 postinjury he developed left hemiparesis, progressing to coma and death over the next few days.

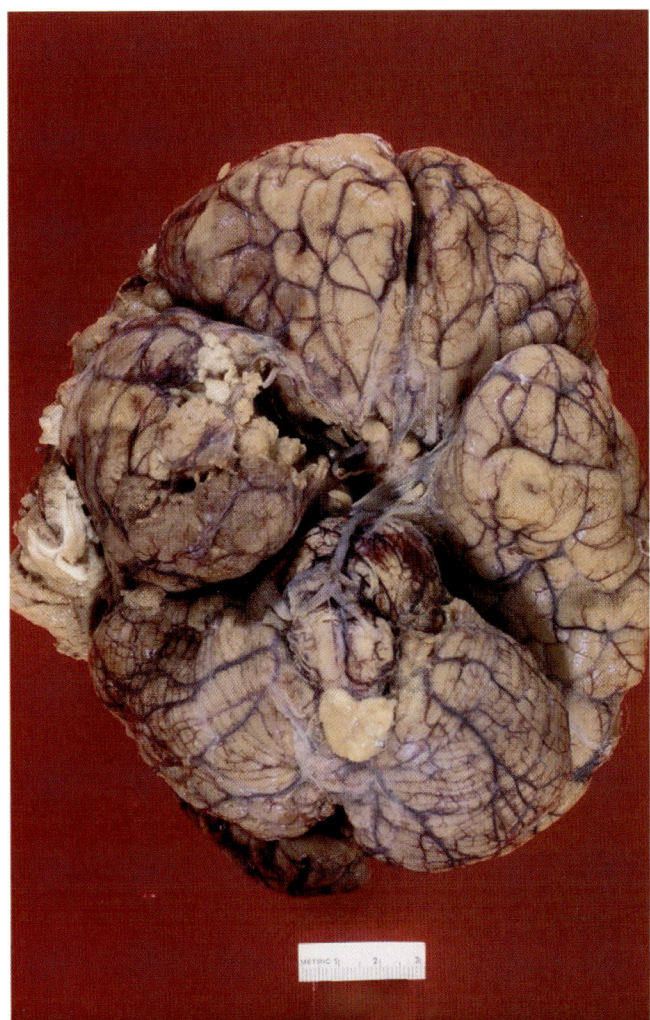

Fig. 8.25. External basal view of swollen, friable, and fragmented brain tissue in right middle cerebral artery territory, with right uncal herniation.

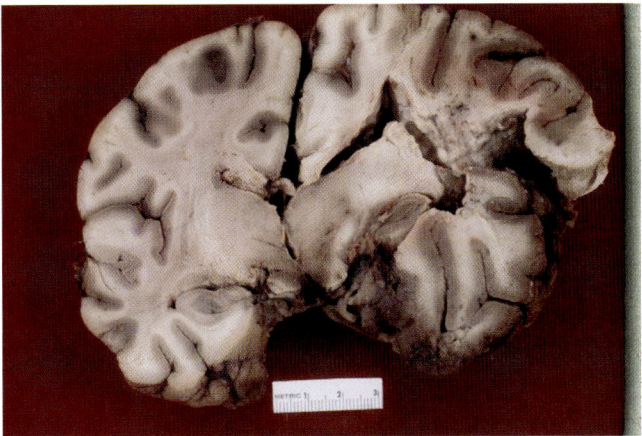

Fig. 8.26. Coronal section of brain from right middle cerebral artery territory infarction, with right to left midline shift and right uncal herniation.

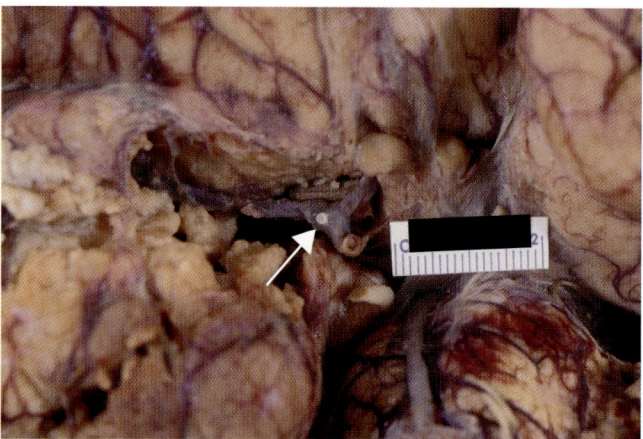

Fig. 8.27. Silver-colored shotgun pellet embolus (*arrow*) lodged in proximal right middle cerebral artery, and producing complete occlusion.

such incidents. On one occasion known to us, a defense attorney inferred in court that a bullet was lost due to incompetence by either our department or the law enforcement agency to whom the bullet was released. It was found that the bullet had been subpoenaed by, and was sequestered in, that attorney's office throughout the time it supposedly had "disappeared." Meticulous chain of evidence records will quickly resolve such situations.

Suicidal versus Homicidal Gunshot Wounds to the Head

This section emphasizes three major points:

1. There is no single location of craniocerebral gunshot wound that, by itself, reliably distinguishes suicidal from homicidal gunshot wounds.
2. Multiple suicidal gunshot wounds to the head do occur.[160]
3. It is the totality of the circumstances in a given case that is most likely to lead to the correct manner of death. In some cases, supplemental data from a "psychological autopsy," a detailed investigation of the personality, life circumstances, behaviors, etc. of the deceased individual by forensic psychiatrists, can be useful in aiding the medical examiner in determination of the most likely manner of death.

Some misconceptions seem to persist, especially in the nonscientific literature, that are sufficiently common to warrant mention here. Awareness of case examples such as the following can be useful in avoiding premature conclusions as to, for example, manner of death.

- Women not infrequently shoot themselves in the head when committing suicide, despite the potential for disfigurement.[271]
- Close range or contact wounds do not invariably result in backspatter on or in the weapon's barrel.[54]
- Although multiple gunshot wounds to the head are more common in homicide,[37] multiple suicidal gunshot wounds to the head may be the result of several possible circumstances. These include an individual shooting him- or herself simultaneously with two different weapons,[54,218,236,254] defective or low-energy ammunition that allows initial nonincapacitating wound(s), initial poor placement of the weapon with regard to vital areas, tandem bullets, multiple bullet loadings, sympathetic discharge of revolver rimfire cartridges, and weapon/bullet caliber mismatch.[54,91,115,117,160,252,254]
- Intraoral or submental gunshot wounds may be suicidal[37,85,134,160] or homicidal.[308]
- Both contact and noncontact wounds to the head may be either suicidal or homicidal.[271]
- Right-handed suicidal persons may shoot themselves in the left temple, and vice versa.[271]
- Entrance wounds in the posterior aspect of the head or neck, or in the eye occur both in suicides and in homicides.[37,54,134,153]
- Intermediate-range wounds can occur in both suicide and homicides[37] although an estimated 80 to 99% of suicides in several case series (compared with only 6–11% of homicides) demonstrate contact or near-contact wounds.[134]
- Complex suicides in which a gunshot wound to the head, for example, is combined with other mechanisms of suicide (e.g., hanging,[25,38] poisoning,[217] or drug overdose,[54] or by assuming a position ensuring the body will fall into the ocean subsequent to a cranial gunshot wound[38]), are also occasionally encountered. The combination of methods may be preplanned, or a result of an initial failed attempt. A recent review of the subject of complex suicides, many of which have a cranial gunshot wound component, is available.[28]
- Although the location of an entrance wound in the head, in isolation, is not reliable in distinguishing suicides from homicides, there is evidence that certain *angles* of the bullet's path *in certain locations* may increase the index of suspicion that homicide is more likely than suicide. For example, excluding cases of bullet ricochet, a downward and/or back-to-front angle in gunshot wounds to the temple, or a downward angle in gunshot wounds into the mouth, are quite unusual in suicidal gunshot wounds to the head.[134]
- When there are two or more gunshot wounds in cranial or high cervical areas in which each can be reliably considered to cause instant and permanent incapacitation, homicide is confirmed assuming one can exclude two weapons fired by the decedent simultaneously, and that tandem bullets can be excluded.[117,134,230] In cases where two or more separate

gunshot wounds to the head that can be sequenced occurred, and in which the first one can be reliably considered instantly and permanently incapacitating, homicide can be assumed if a nonautomatic weapon is responsible.[3]
- Suicides that initially mimic homicides are not always due to removal of the weapon from the scene by another. In one reported case an individual employed creative means to conceal the weapon following a self-inflicted fatal shot to the head by positioning himself so that the gun, to which a weight was tied, fell into a river, whereas his body fell on the river bank.[97]

Tandem Bullets

Multiple bullets simultaneously entering a single gunshot entrance wound when a gun is fired a single time are referred to as "tandem" (or "piggy-back") bullets in the forensic literature. This phenomenon can be recognized by a single entrance wound (usually at contact or near-contact range), plus two or more projectiles within the body in continuity with that entrance wound, plus the presence of characteristic deformities of the bullets that leave no doubt that one bullet was pushed out of the barrel of the weapon by a second, following bullet(s). The deformity of the base of the initial bullet will correspond to the deformed tip of the following bullet that forced it out of the barrel. Such bullet deformities exclude the infrequent case of multiple separate and independent bullets fired into the same entrance wound.[121]

Arbitrarily limiting our examples to a few of the many head wound cases in the tandem bullet literature, the cause of such events is most often ascribed to faulty ammunition[200,256,280] or to weapon/bullet mismatch.[54] Either of these situations may result in a bullet remaining in the weapon's barrel. The next bullet fired strikes the impacted first bullet or bullets, in the unusual instance of more than one bullet stuck in the barrel.[256] The resultant series of tip to base bullets exiting the barrel produce the single entrance wound. Bulging of the barrel may result, or the barrel may rupture. The latter can abort the continued progress of the missiles toward the intended target by allowing escape of propellant gases through the rupture.

Depending largely on the range of fire, atypical or even multiple entrance wounds may also result from tandem bullet discharge.[256] In one unusual case, the initial projectile pushed from a rifle barrel by the aftercoming bullet was a bore cleaning brush, both projectiles penetrating the victim's head.[63]

As Fackler noted, the types of cases described earlier, which dominate the forensic literature, are examples of accidentally produced tandem projectiles. Tandem bullets intentionally produced commercially or by handloaders may also be encountered by a medical examiner on rare occasions.[71] Multiple bullet loadings may use bullets of similar caliber but different weights,[71] and shotshells may contain pellets of varying size.[194] Such ammunition could provide an alternate explanation to, for example, the presence of multiple shooters and/or use of multiple weapons.

Prognostic Factors in Craniocerebral Gunshot Wounds

Patients who have sustained craniocerebral gunshot wounds tend to have poorer prognosis for morbidity and mortality if, at the time of initial examination, they demonstrated the following findings[60]:

- Glasgow Coma Scale (GCS) score of 3 to 5.
- Apnea or respiratory depression.
- Hypovolemia.
- Unequal or bilaterally fixed and dilated pupils.
- A central bihemispheric gunshot wound path (compared with an anterior bifrontal, or posterior biooccipital, gunshot wound path).
- A transventricular gunshot wound path.
- Scattered bone and metal fragments away from the bullet path.[207]

Additional unfavorable prognostic factors which appear somewhat less reliable as predictors of high mortality rate include:

- Increased intracranial pressure (ICP) within the first 72 hours' postinjury.[1]
- Multilobar bihemispheric gunshot wound path worse than multilobar unihemispheric gunshot wound path worse than unilobar gunshot wound path.[60]
- A posterior fossa gunshot wound path.[60]
- Associated epidural, subdural, subarachnoid, or intraventricular hematoma.[286]
- Penetrating or perforating gunshot wounds worse than shrapnel or tangential wounds.[60]

Most of these prognostic criteria would seem obvious, based on a commonsense approach. However, the many variables involved in a specific gunshot wound case support a cautious approach to predicting outcome. The best single indicator of outcome and mortality rate appears to be the presenting GCS score.[1,286] Even bihemispheric, multilobar gunshot wounds with a 2-day delay in treatment onset may, for example, demonstrate a remarkable degree of recovery in an isolated case.[240]

It may be concluded that some findings on admission are more important than others in determining prognosis in gunshot wounds and that individual variations that are difficult to predict can occur. The prognostic issues of gunshot wounds in testimony are primarily within the sphere of expertise of experienced clinicians who work with such cases and are intimately familiar with standards of care in the community, rather than pathologists.

Centerfire Rifle Wounds

The primary difference between handgun wounds, which receive the preponderance of attention in this chapter due to the frequency with which they are encountered, and centerfire rifle wounds is the potential for greater tissue destruction in the latter.[54] This increased destructiveness is often most obvious with contact wounds, and with head (Fig. 8.28), solid organ (e.g., liver, spleen), and bone wounds.

Basing conclusions regarding weapon type by wound appearance alone is unwise, however. For example, multiple pellet-containing rounds are manufactured in several handgun calibers. Handguns chambered for most rifle rounds, and rifles chambered for handgun rounds, are currently available. Several handguns presently manufactured, due to improvements in design and safety, are capable of firing rounds far more powerful than was the case several years ago (e.g., .475 Linebaugh, .454 Casull, .460 and .500 S&W Magnum).[196] Wounds from military rifle rounds before they yaw may be indistinguishable from handgun wounds.

As a general rule, however, compared with the commonly encountered handgun round wounds, one will find most centerfire rifle round wounds more destructive of tissue at a greater distance from the permanent wound track, in large part related to the ability of rifle rounds to produce a larger temporary cavity.[54] Permanent and temporary cavity effects with several common military rifle rounds are reviewed by Fackler.[66] Military centerfire FMJ rounds tend to be somewhat less destructive in soft tissue wounds compared with soft point or hollowpoint hunting rounds of the same caliber. The latter are also more likely to produce the so-called "lead snowstorm" x-ray appearance, in which widely scattered small dense opacities are present in and around the wound path due to bullet fragmentation.[54,193,194,195] The pattern of such bullet fragments, at least in truncal wounds, is considered an unreliable indicator of entry/exit direction, however.[273]

Shotgun Wounds

Most shotguns have a smooth inner barrel (bore) and fire shotgun shells that contain either pellets or a single shotgun slug. Some shotguns designed specifically for slugs have rifled barrels and riflelike sights in order to increase the range of slug accuracy compared with smooth bores. Shotshell and rifled or sabot[46] slugs are capable of causing highly destructive wounds, similar to those caused by some centerfire rifles, and either form of shotgun round may cause major loss of head tissue, including brain exenteration, with close range or contact wounds.[133,257]

While skin pellet wound patterns, together with soot, etc., are useful in range of fire determination (assuming absence of intermediate targets),[40] pellet patterns alone on x-rays are not, due to the so-called "billiard ball" effect. That is, at short ranges the multiple spherical pellets maintain a relatively tight aggregate in space. Pellets leading the aggregate slow first as they strike tissue. Thus, they themselves are struck from behind by aftercoming pellets that are still traveling at a higher velocity. The result is a wider cone of pellet dispersal in the tissue than would have resulted in the absence of such ricochet effects, the diameter of pellet dispersal in the body significantly exceeding that of the skin wound pattern.[195]

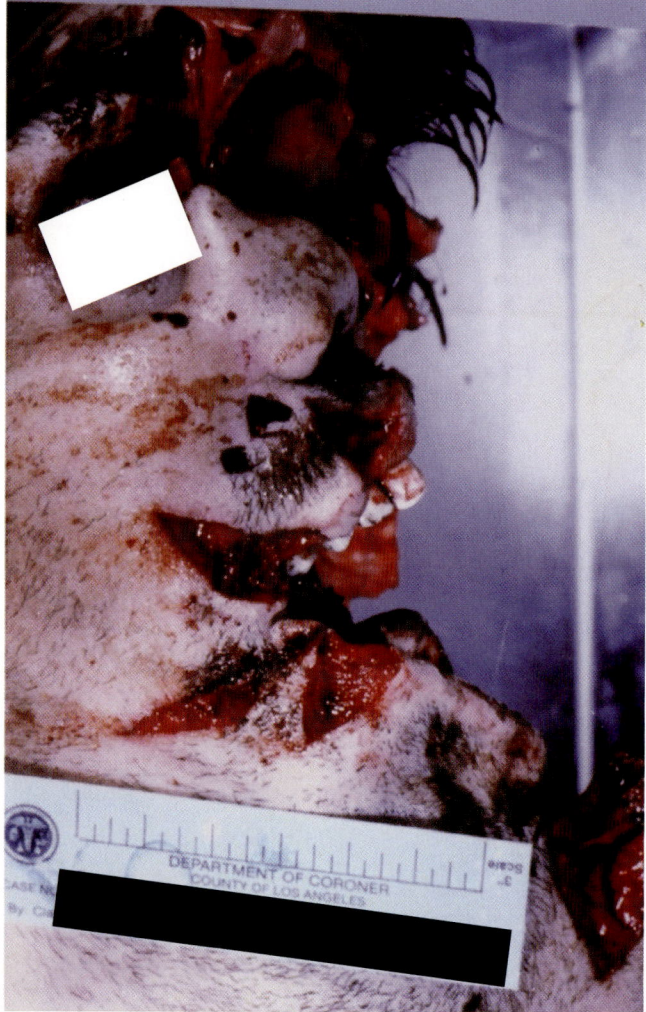

Fig. 8.28. Wound from an upward-directed 30-06 caliber rifle hunting round that entered beneath the left chin has largely avulsed the left face and cranium.

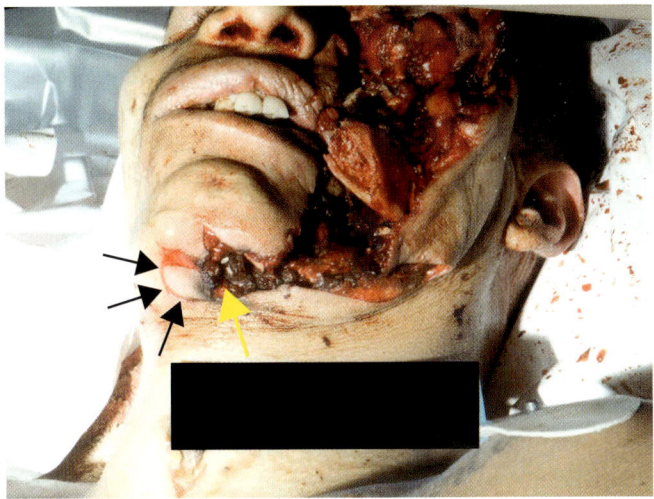

Fig. 8.29. Contact shotgun entrance wound under chin (*yellow arrow*), fired in upward direction causing severe destruction of left face and head. The black arrows outline an adjacent circular abrasion caused by entrance-wound gas pressure forcing the adjacent skin backward against the second (unfired) muzzle of the double-barreled shotgun that produced the wound.

Perhaps less widely known than the usual examples of entrance (Fig. 8.29) and exit (Fig. 8.30) shotgun wounds are reentry wounds by shotguns[99] (reentry wounds are more common in handgun and rifle wounds), or that a single "petal mark" from a plastic shot cup may occur from contact wounds.[75,76] Shotgun gauge estimation by wound appearance in suicidal contact wounds of the head has been studied, and, although 12-gauge rounds are generally more destructive, overlap in wound appearance can occur with other gauges (particularly wounds with 20-gauge and 16-gauge rounds).[106] As noted in that report, most suicidal gunshot wounds are contact or very close range and are often to the head, but more complicated techniques have been devised by individuals that permit a suicidal shotgun wound to be distant range.[62]

Shotshells are also handloaded by recreational shooters, typically with manufacturer-recommended components and component amounts, due to the significant reduction in cost when large numbers of rounds are involved. Occasionally, however, the medical examiner may encounter shotgun wounds in which the hand-loaded missiles consist of screws, bolts, glass fragments, or rock salt, sometimes referred to as "junk loads." Other unusual loads, such as pellets chained together by wires, have been reported.[177]

Potential Complications of Craniospinal Gunshot Wounds

Some of the following effects and complications of gunshot wounds are seen in virtually every case. Others

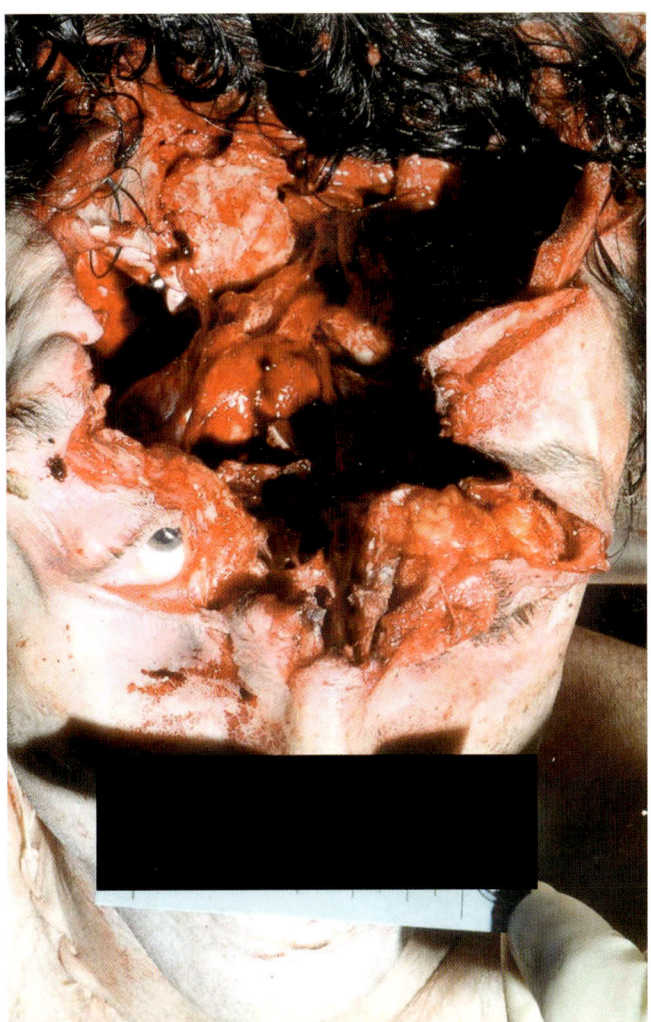

Fig. 8.30. Anterior head exit wound caused by upward-directed shotgun entrance wound in hard palate, with near-complete brain exenteration.

are infrequent to rare. Since it may be helpful for reports or for testimony in a given case to cite other examples of similar gunshot wound consequences, a few selected references are included. A more extensive list of references can be obtained from the articles cited, Internet data banks, and other sources.

- Exsanguination due to major vessel injury. Although described in basal skull fractures associated with cerebral blunt force trauma,[220] similar results may be expected if gunshot wounds directly or indirectly involve the internal carotid artery or other major head and neck arteries, veins, or dural sinuses.[18]
- Hemorrhage in brain parenchyma and/or other compartments, such as subcutaneous tissues including periorbital areas ("raccoon eyes") with orbital plate fractures; retroauricular (i.e., Battle's sign) with mastoid skull area or basilar skull fractures; and hem-

orrhage in subgaleal, epidural, subdural, subarachnoid, and intraventricular foci,[103] can all occur.

- Increased intracranial pressure due to mass effects of brain/cord swelling and/or hemorrhage, with clinical documentation by ICP monitors and/or by anatomic markers, such as brain swelling and herniation syndromes revealed by clinical imaging techniques, or discovered at autopsy.[21]
- Hydrocephalus.[286] There are at least two mechanisms by which gunshot wounds may cause hydrocephalus. As in subarachnoid hemorrhage of nontraumatic etiology, CSF absorption by arachnoid villi may be compromised, leading to communicating hydrocephalus. Alternatively, sufficient CNS injury may lead to cerebral atrophy with hydrocephalus *ex vacuo*, given sufficient survival time.
- Increased intracranial pressure due to tension pneumocephalus.[92]
- Air embolism.[118]
- Depressed skull fractures.
- Neurologic defects documented clinically in areas peripheral to the permanent wound cavity, and presumed secondary to temporary cavity/expanding gas effects or vascular compromise.[203]
- Infection (e.g., meningitis, brain abscess, osteomyelitis).[6,43,105,124,126,265] (*Comment*: Bullets do not become sterilized when fired.[54] However, infection is much more likely if bone or extracorporeal intermediate-target fragments are carried into the wound by the bullet.)
- Post-traumatic apnea (and other complications, as described for blunt force injury in Chapter 6).
- Cerebrospinal fluid leakage (providing a route for infection).
- Encephalocele.[206]
- Post-traumatic cerebral vasospasm.[309]
- Post-traumatic intracerebral or extracerebral carotid or vertebral artery pseudoaneurysm.[1,120,234]
- Post-traumatic major intracerebral artery thrombosis.[18,82]
- Hypopituitarism.[45]
- Acute or delayed-onset post-traumatic seizure disorder.[150]
- Complications indirectly related to immobility caused by gunshot wounds, or to decreased levels of consciousness (e.g., deep vein thrombosis, pneumonia, urinary tract infections).
- Bullet migration (as described earlier).
- Complications related to bullet composition (e.g., copper, lead; see following section).

Some complications are more specific to, or particularly unusual for, spinal gunshot wounds. These include:

- Diabetes insipidus.[161]
- Development of delayed spinal mass lesion effects due to shrapnel, granulomas, or fibrotic masses or presence of intraspinal epidermoids and lipomas.[8,159,233,257,269,302]
- Lumbar disk herniation.[187,233]
- Chronic arachnoiditis[233] or local adhesions.[87]
- Chronic pain.[233,293]
- Subarachnoid-pleural fistula.[242]
- Cerebrospinal fluid-lymphatic fistula.[155]
- Hypertension (rather than the more commonly encountered hypotension following spinal cord injury).[156,235]
- Arachnoid cyst.[87]
- Syringomyelia.[87]
- Predisposition to fractures secondary to reduced bone density in chronically paralyzed patients.[84]

Complications of Retained Bullets and Other Missiles Related to Bullet Composition

Copper

A few reports exist linking human brain gunshot wounds by copper-plated or copper-jacketed bullets to an increased incidence of local brain softening with necrosis or cyst formation, increased likelihood of bullet migration, and gliosis.[192] Adjacent necrotic brain tissue acquired a greenish stain within 9 months after a gunshot wound in a case reported by Messer and Cerza.[192] In another case, a copper-jacketed bullet that had migrated from the intracranial cavity to the cervical spinal canal over approximately 5 years was removed from the subarachnoid space with no fibrosis or other reactive changes described in that location.[306] Concern has been expressed as to whether bullets with copper surfaces should be removed prophylactically, or whether patients wounded with such missiles should simply be observed more frequently for possible bullet migration, cyst formation, inflammatory reactions (including sterile abscess), and excessive fibrosis.

Such concerns are based on occasional case reports[192,255] supported by experimental animal studies using implanted copper-coated gunshot pellets[255] or copper fragments obtained from commercially available bullets.[281] In cat brain, implanted copper-coated shotgun pellets produced severe tissue reactions, nickel-plated pellets produced a much milder local reaction, and lead pellets the least reactive changes.[255] Reactive changes provoked by copper-coated pellets included brain edema sufficient to produce herniation effects, greenish-discolored exudate in the necrotic trail of the migrating pellet, and acute and chronic inflammation. Other changes present were foamy and pigment-laden macrophages, neuronal chromatolysis, oligodendroglial swelling, engorged blood vessels, cyst formation, and leptomeningeal inflammation and fibrosis where the pellet came to rest in the basal meninges after migrating through the brain. More chronic lesions (4 weeks' survival) included perivascular cuffing, new blood vessel formation, and reactive astrocytes. At 16 weeks' survival, necrosis, prominent foamy macrophages, cyst formation, and astrocytosis remained conspicuous, and microglial and vascular changes (hyalinized small vessels, endothelial proliferation, and some occluded vessels) were noted. At 6 months' survival, inflammation was sparse and gliosis and fibrosis were prominent, with pellets contained within multiloculated cysts with fibrotic walls.[255]

In a spinal *extra*dural location in rabbits, implanted lead, aluminum, or copper bullet fragments alike produced no remarkable dural, pial, arachnoid, or spinal cord pathology.[281] In a spinal *intra*dural location in rabbits, however, copper produced significant local pia-arachnoid fibrosis and adjacent subpial spinal cord necrosis. A surrounding spinal cord zone demonstrated relative axon preservation but substantial myelin loss. Macrophages and inflammatory cells were sparse.[281] Intradural lead fragments in the rabbit produced local leptomeningeal fibrosis but only equivocal vacuolar change in spinal cord. Intradural aluminum fragments were relatively inert, producing no significant pia-arachnoid or cord changes.

Of related interest is a study of the effects of human decomposition on bullet striations (rifling marks). Copper-containing bullet surfaces, compared with aluminum, lead, or nylon-coated bullets, demonstrated the most severe degradation of rifling marks when embedded within a body undergoing decomposition. It was sufficient to prevent matching the bullet with the weapon used to fire it in all body regions tested.[261] Slightly less pronounced corrosion, but still sufficient to obscure bullet/weapon matching, occurred in lead bullets in muscle and in fat, but not in other body sites tested (i.e., head, chest, and abdomen).[261]

More attention by neuropathologists to reactive changes at the bullet recovery sites correlated with bullet composition and survival times would be helpful in determining the incidence and extent of such copper-related complications in human CNS wounds.

Lead

Retained lead bullets or shotgun pellets in soft tissues are usually surrounded by poorly vascularized scar tissue and do not become symptomatic enough to warrant surgical removal in most cases. Excluding accompanying infection, they also typically generate much less local inflammatory reaction in or near CNS tissues than, for example, copper (see earlier). Occasionally they will produce a soft tissue chronic inflammatory reaction or cystic changes,[113,297] and, in joint spaces, they may cause lead arthropathy[113,164] with or without associated systemic lead poisoning. Systemic lead poisoning is more likely to be present with retained lead missiles that are multiple, or that are located in joints, bones, or intervertebral disks, and rare fatal cases have been reported.[54,96,113,174]

Symptoms of systemic lead poisoning due to retained bullets may develop gradually or rather abruptly, within several weeks to a few months of the gunshot wound[89,174,249] or after a delay of one to many years.[77,122,148,274] Chronic occult lead poisoning may become symptomatic as a result of superimposed stresses such as febrile illnesses, metabolic acidosis, hyperthyroidism, pregnancy, prolonged immobilization, and possibly alcoholism.[77,174] In cases of pregnancy in women with prior gunshot wounds and retained lead fragments, lead levels may rise to clinically significant levels during the course of the pregnancy in both mother and fetus, and may adversely affect fetal outcome.[228,285]

Ingested lead missiles, including those ingested voluntarily by mentally disturbed persons,[190] by young children,[285] or by persons inadvertently consuming lead missiles while eating wild game downed by hunters using firearms,[42,101,163,245] may be asymptomatic[42,285] or may cause symptoms due to acute appendicitis (associated with intraluminal pellets,[163] abdominal pain and GI bleeding)[190] or systemic lead poisoning developing acutely or gradually.[101,179]

It is to be expected that inclusion of lead poisoning in the differential diagnosis list of clinical symptoms or death becomes less likely as the time interval between gunshot wound and morbidity or death increases. Several case reports indicate that elevated blood lead levels can be seen anywhere from days to up to several decades after wounds with retained missiles.[77] In addition, more subtle forms of occupational or environmental lead exposure may lead to clinically significant blood lead levels. Examples include employees, instructors, and shooters at indoor ranges. Indoor range shooters may be asymptomatic but show clinically significant elevated blood levels,[35,88] may develop relatively subtle symptoms,[81,299] or more serious lead systemic toxicity symptoms.[88] An increased index of suspicion of possible lead toxicity in cases in which the cause of death is initially obscure could prompt specific questioning of survivors concerning the decedent's recreational, as well as occupational, lead exposures.

Injury to Death Interval

Several studies have investigated the question of survival time of individuals subjected to various fatal injuries. In a following section, data are presented that deal more specifically with how quickly a person might become incapacitated, and what functions a person might be capable of performing during that injury to death interval. These questions are not infrequently asked of medical examiners during sessions with various case investigators, attorneys, etc. Davis[47] indicates that the brain contains approximately 5 to 10 seconds' worth of oxygen after complete cessation of circulation, such as the heart being burst by a projectile. DiMaio[54] described a case in which the victim's heart was shredded by a gunshot wound, yet was able to run 65 feet prior to his collapse. Experimental work has demonstrated that a person "can remain conscious for at least 10 to 15 seconds after complete occlusion of the carotid arteries"[54] (see also Chap. 10). Even wounds that produce instant incapacitation may not produce immediate death, survival periods in these circumstances varying from minutes, to hours, days, or longer, depending on many variables (e.g., location and nature of injury, speed of blood loss, extent of medical supportive care received).[54,167,267]

In gunshot wound cases, anatomic evidence exists, in the form of characteristic chronic lesions, which indicates that persons have survived cerebellar tonsillar herniation contusions or medullary contusion for extended periods,[85] survived gunshot wounds involving the hypothalamus and thalamus for many years,[85] and that an individual maintained consciousness for at least 2 hours following a gunshot wound involving bilateral caudate head anterior tips.[54] These and other reported cases demonstrate the problems of attempting to predict time of death following gunshot wounds to the head based solely on neuroanatomic structures involved.

Devoid of specific information regarding the exact neuroanatomic (gross and microscopic), cardiovascular system, and other autopsy findings in a given case not always detailed in published reports, the findings at the scene, eyewitness reports, and so on, the most accurate answer to the question of injury to death interval may be "sometime between when the victim was last seen alive by a reliable witness and when they were pronounced." Any other estimate requires additional supporting facts.

Speed of Incapacitation Issues in Gunshot Wounds

In Chapter 10, experimental work is cited which indicates that sudden cessation of blood flow to the brain due to heart failure or vascular occlusion in the neck does not necessarily result in instantaneous cessation of purposeful activity. As noted above, Davis estimated this still-functional interval to be in the range of 5 to 10 seconds.[47]

A dictionary definition of incapacitation is "the act of incapacitating or state of being incapacitated."[294] Incapacity is defined in the same source as "the quality or state of being incapable, especially lack of physical or intellectual power or natural or legal qualification."[294] A more practical definition for forensic purposes is suggested by Karger[131]: "Incapacitation is an early and unavoidable inability to perform complex and longer lasting movements." Karger's definition excludes reflexes or automatisms, and by the term "unavoidable" he refers to physiologic effects independent of any unpredictable psychological mechanisms.[131] In other words, the incapacitated person cannot perform complex interactions between him- or herself and the assailant, or between him- or herself and the environment, in any goal-directed fashion.

In offering expert testimony on the topic of speed of incapacitation, there is the assumption of obligation on the part of the witness to be familiar not only with anatomic and physiologic mechanisms underlying normal and ordinary functional capability, but also to have an appreciation of the potential influence of psychological factors, and an awareness of prior witnessed observations on the extent of wounding some persons may sustain while simultaneously retaining volitional activity to a degree that many would regard as extraordinary. As MacPherson[180] indicates, "Caliber and all its aspects is much less important than any of three other factors (shot placement, penetration, and the psychology of the individual being shot)."

The previous generation in this country included many individuals with military combat experience. A higher proportion of the present generation has neither military nor law enforcement officer experience, and are less likely to have been exposed to armed combat, lessons learned by those who have faced and survived life-threatening encounters, and lessons learned by the examination of cases in which individuals did not survive such incidents. The unfortunate result is that opinions on armed combat are more likely to be grounded on television and movie illusion rather than on reality. One responsibility of the medical examiner is not to add to such misinformation on gunshot wound cases.

Primarily Non-Central Nervous System Wounds

Cases in which multiple gunshot wounds in several areas of the body, including the head, might be expected to incapacitate very rapidly but did not, have been described in references cited earlier (see section, "Injury to Death Interval"). Many other published examples serve to reinforce this important point, such as cases reviewed by Adams et al.[2] and by Anderson.[9]

The necessary number of shots required to cause incapacitation varies from person to person.[209] Persons under the influence of drugs, who are psychotic,[68] or who have a certain mind-set, can be much more difficult to incapacitate short of a high cervical spinal cord or an intracranial gunshot wound to certain specific sites (see later).

An individual as described by General Julian H. Hatcher (quoted by MacPherson[180]) who is "intoxicated with excitement or rage," determined to "kill or fight clear" at any cost no matter the opposition, and "to do as much harm to the other fellow as possible" before death can be the most difficult of all to stop. This was not the opinion of a casual observer; General Hatcher was a lifelong student and recognized expert on firearms.[109] The risk is further increased if that individual is in good physical condition, well-trained in martial arts and in the use of weapons, or is simply having a lucky day. How many bullets are enough? More than 32 in some reported cases.[2]

To summarize, gunshot wounds that are in areas other than certain CNS sites proven sufficient to cause immediate incapacitation may cause rapid (i.e., over several seconds or minutes), but not immediate, incapacitation as defined earlier. Further, the degree of wounding that can be absorbed by some persons before volitional activity ceases may be regarded with skepticism or disbelief by the uninformed or misinformed. Medical examiners should consider such facts before concluding that a given shooting case represents an "unnecessary" number of shots or "overkill."

Central Nervous System Wounds

The functional effects of gunshot wounds in which the primary injury is anatomic or physiologic spinal cord transection, without major cardiovascular, pulmonary, or other solid organ involvement, can be reasonably predicted by knowledge of the spinal segment level of involvement. Clinical-pathologic correlation of spinal cord injuries has led to some accepted generalities. A complete transection at the C3 spinal segment or above, for example, can be expected to paralyze all respiratory function, and a complete lesion at C4 or above can be expected to paralyze all four extremities.[29] Attempts, however, to correlate the gunshot wound permanent cavity location at more caudal spinal cord levels with immediate incapacitation in this same fashion is not necessarily going to produce consistent results. The reason is that the permanent wound cavity anatomic location is not necessarily identical to the physiologic functional loss produced, since even incomplete spinal cord wounds may produce transient or longer-lasting physiologic functional impairment extending several segments above and below the anatomic lesion. Rapid death would prevent histologic studies from identifying nonhemorrhagic abnormalities that may require hours to days to become histologically apparent as distance from the permanent wound cavity increases.

With regard to intracranial CNS lesions, a careful analysis of the relationship between various head gunshot wounds and presence or absence of immediate incapacitation is that of Karger.[131,132] It is recommended reading for medical examiners and neuropathology consultants involved in such cases. With rare exceptions, our case material has been quite consistent with Karger's conclusions, which are briefly summarized later.

Signs of secondary cranial gunshot wound overpressure effects are to be sought at the time of autopsy, and include the following:

a. Tertiary concentric (i.e., heaving) skull fractures (see earlier section on skull gunshot wounds).
b. Skull fractures remote from primary (entrance/exit) fractures or secondary wound fractures (fractures radiating from entrance/exit defects). That is, intact bone separates the gunshot wound primary and secondary fractures from so-called remote fractures. The latter are most common in the orbital roof, ethmoid plate, and thinner tympanic roof sites, but they may occur elsewhere. Variation in skull thickness and strength will influence fracture propensity.
c. Remote brain contusions. That is, contusions that are separated from the direct brain trauma due to the permanent and temporary cavity effects by intact brain tissue. Remote contusions are most common, though not limited to, the same areas prone to contusions caused by blunt force trauma. These areas include the cerebellar tonsils, anterolateral or anteroinferior frontal and temporal lobes (Case 8.7, Figs. 8.31–8.32), medial temporal lobes, and posterolateral occipital poles.
d. Brain intraparenchymal hemorrhages separated from the direct gunshot wound path by intact brain tissue. These are most common in the basal ganglia, midbrain, pons, and cerebellum. Intraparenchymal hemorrhages due to secondary brain swelling and herniation syndromes are not included in this category.
e. Explosive, comminuted skull fractures with partial or complete brain exenteration can be considered an extreme example of secondary overpressure effects. Such wounds are most often seen with high-powered rifles, shotguns at close or contact range, or with some of the more powerful modern handgun loads at close or contact range.

The reason these secondary overpressure effects are emphasized by Karger[131,132] is that they have been found, together with anatomic structures involved in the wound path, to be useful in predicting whether or not a given cranial gunshot wound produced instant incapacitation. It is important that their presence or absence be documented at the time of the autopsy and neuropathologic examination. Unfortunately, many of the gunshot wound reports in the literature do not provide specific details regarding these important secondary overpressure effects, permanent cavity size, existence and distribution of secondary missile tracks (such as from bone or bullet fragments), weapon and ammunition type, and exact neuroanatomic structures damaged. Individual variations in craniocerebral relationships make extrapolation of exact neuroanatomic structures involved based solely on skull entrance and exit wound positions suboptimal. However, careful correlation of findings with published images of skull–brain relationships in various planes can aid in analysis in some cases[34] (also see Chap. 1).

Case 8.7. (Figs. 8.31 and 8.32)

A 23-year-old male was shot in the head. The entrance wound was in the left parietal lobe, the exit wound in the left upper frontal lobe, and the permanent cavity involved cortex and immediate subcortical white matter. Deep white matter and basal ganglia were grossly intact.

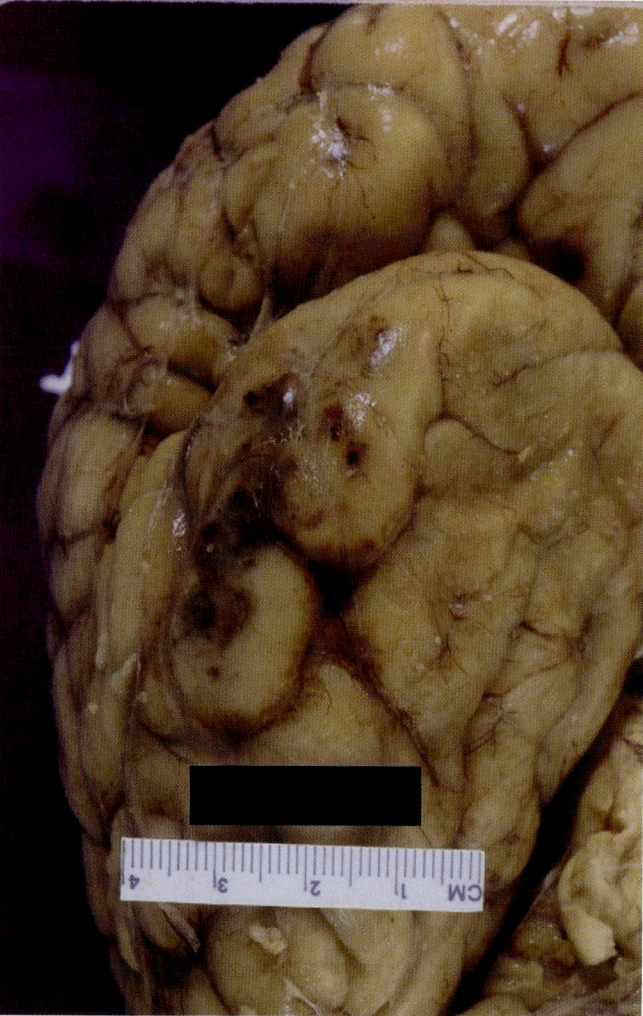

Fig. 8.32. Remote contusion in right anterolateral temporal lobe.

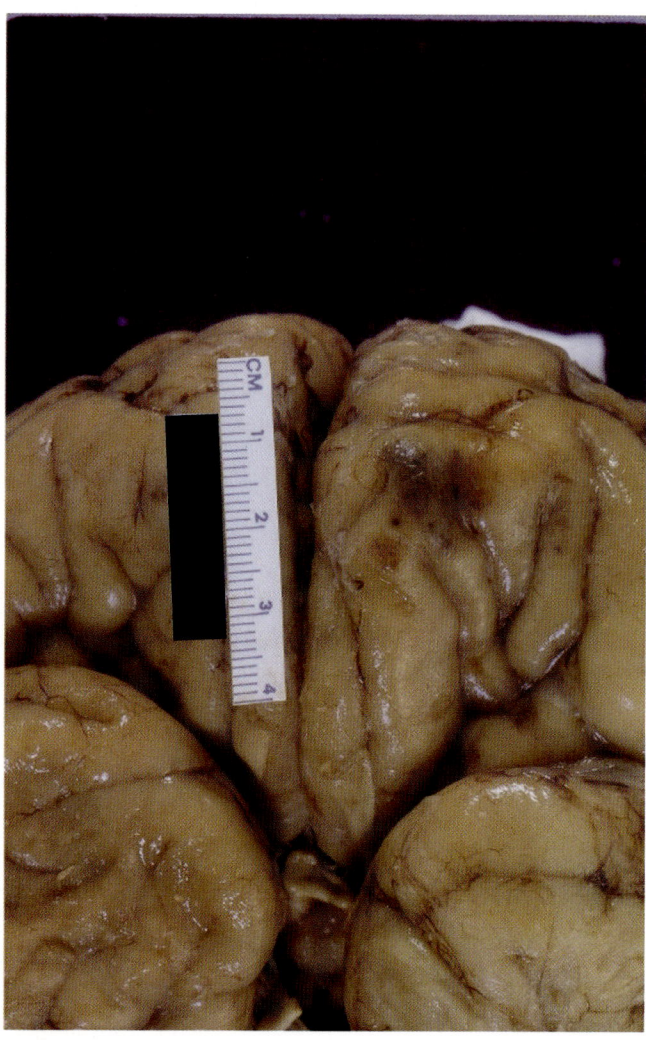

Fig. 8.31. Remote cortical contusion was present in left anterior orbitofrontal cortex.

Karger's summary of available data led to certain preliminary conclusions, including the observation that immediate incapacitation is very likely in the following head gunshot wound conditions:

- An intracerebral hit from a centerfire rifle or shotgun at close range.
- An intracerebral hit from modern firearms (and ammunition) from about 9 mm parabellum caliber (9 by 19 mm) destructive capability upward (unless the path intracranially is very short, or is limited to the anterior cranial fossa, either of which reduces the likelihood of immediate incapacitation).
- Evidence of increased intracranial overpressure effects as described earlier.
- The bullet path involves brainstem, diencephalon, cerebellum, major motor pathways, or central gray matter (excluding one or both of the most anterior caudate head regions) (Case 8.8).[54,132]

Case 8.8. (Figs. 8.33–8.38)

This 26-year-old male suspect in a bank robbery was crouched in the back seat of the getaway car driven by one of his several co-suspects as they were pursued by police. Gunshots were exchanged in both directions. The suspect was taken into custody "walking and talking," and was brought to a local emergency room for treatment of a small right scapular area wound interpreted by the emergency room (ER) physician to be a superficial laceration. A chest x-ray was negative. The wound was sutured and the suspect taken to the jail ward of a nearby hospital where, during continued questioning, the officer noted that the suspect was having increasing difficulty responding to questions. Due to the change in mental status, he was transferred to the intensive care unit where he shortly thereafter became comatose. He continued to deteriorate despite emergency supportive measures, and was pronounced approximately 1¾ hours after onset of coma and approximately 6½ hours after the car chase.

At autopsy, the right scapular wound was found to be an entrance gunshot wound (Figs. 8.33–8.34) with the subsequent bullet path entering the head (Figs. 8.35–8.37). In sequence, the bullet path involved the subcutaneous tissue overlying the right trapezius muscle, the deep posterior musculature of the right neck, the right posterior fossa of the occipital bone (lateral to midportion of foramen magnum and posterior to jugular foramen), entered the anterior margin of the inferior right cerebellar hemisphere just lateral to the tonsil, continued through the right cerebellar hemisphere into the posterior right hippocampal formation, right pulvinar of the thalamus, the splenium of the corpus callosum, and terminated in the subcortical white matter of the left anterior superior parietal lobule (Fig. 8.37). A medium-caliber jacketed, moderately deformed bullet was recovered from this site. The bullet path was thus upward, right to left, and back to front. Hemorrhage and swelling was present in the wound path, with compression and right to left midline shift of the brainstem without Duret's hemorrhages. Only minor subarachnoid and intraventricular hemorrhage was present. Minor cortical contusions involving the left frontal and parietal operculae, the midportions of the left middle and inferior temporal gyri, and the lateralmost left orbital cortex, were also present. Subsequent study confirmed that these remote cortical contusions were not acute, but rather subacute to early chronic lesions unrelated to the fatal gunshot wound. Examination also revealed shotgun pellets in the upper inner thighs and groin area from a remote healed wound (Fig. 8.38). Blood toxicology screen was positive for phencyclidine.

Comment. Lessons can be learned from such a case, including the following. It is incumbent upon an ER physician attending a patient who has just been in a gunfight and has an acute skin defect, and an initial x-ray that does not show bullets or bullet fragments, to promptly order an x-ray survey of the rest of the body to search for an unexpected location of a bullet. Psychotropic agents can alter an individual's response to injury, but the extent to which phencyclidine altered this individual's response to his wound is speculative.

This case also highlights the point that one needs to be cautious when attempting to infer clinical behavior from neuropathologic findings alone due to the numerous variables that may influence the situation. There was no specific statement recorded that the decedent remained mentally alert and intact after sustaining the gunshot wound in the getaway car and before being apprehended. It was known that the police found him to be "walking and talking" at the time of apprehension, and he was sufficiently intact to not raise any suspicion on the part of the ER physician regarding CNS injury a short time thereafter. Thus, this case meets criteria outlined by Karger for inclusion in his case series, in which "short initial periods of unconsciousness or incapacitation could not be excluded beyond any doubt. But the reconstruction of events . . . did not supply indicators of primary unconsciousness."[132] This case is also a reminder that some individuals who have sustained fatal gunshot wounds have been involved in prior nonfatal shootings, and bullets or a representative sampling of shotgun pellets from healed wounds should be removed and filed in evidence.

Possible discrepancies exhibited by this case, when compared with the preliminary criteria for likelihood of immediate incapacitation outlined by Karger,[132] include the following. The gunshot wound in our case involved both cerebellum and thalamus, either of which alone would have been expected to produce immediate incapacitation based on the criteria suggested. However, this case demonstrated no secondary overpressure effects. The only remote brain contusions present were found to be from a prior injury and not the fatal gunshot wound. No secondary, tertiary, or remote skull fractures were present, and no remote intraparenchymal hemorrhages were present. The relatively low velocity of the bullet by the time it entered the brain is indicated by its termination in the soft brain tissue, with neither exit wound nor evidence of an internal ricochet present. Intracranial temporary cavity effect would have been nil.

Perhaps the main lesson from this case is that, as Karger[132] suggests, case reports that detail the several variables influencing wound severity are relatively sparse, and more data are necessary in order to refine the suggested pathologic criteria for speed of incapacitation following CNS gunshot wounds.

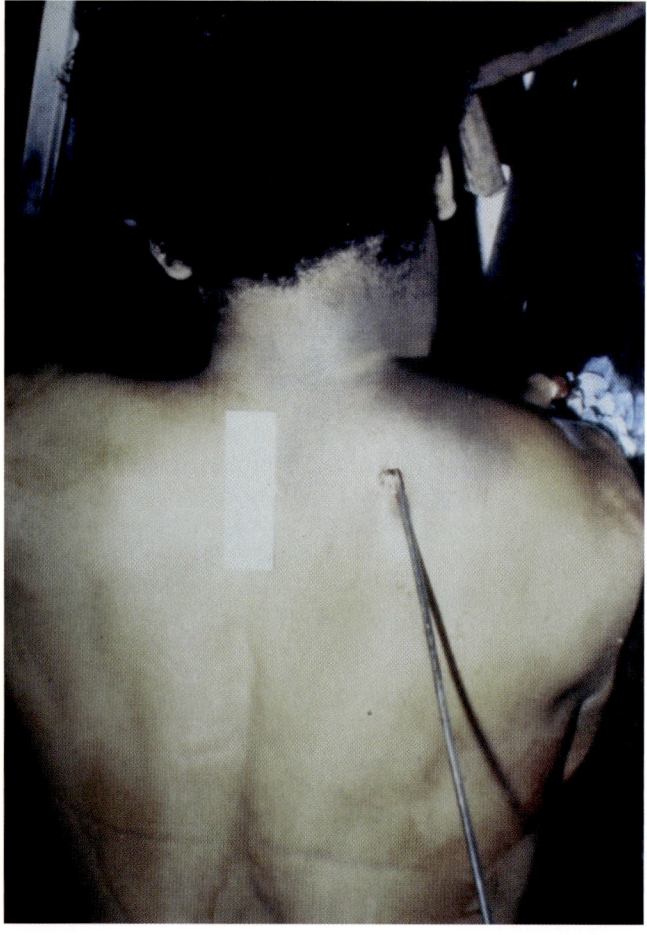

Fig. 8.33. Probe touches the skin entrance wound and has been angled to indicate the approximate bullet trajectory.

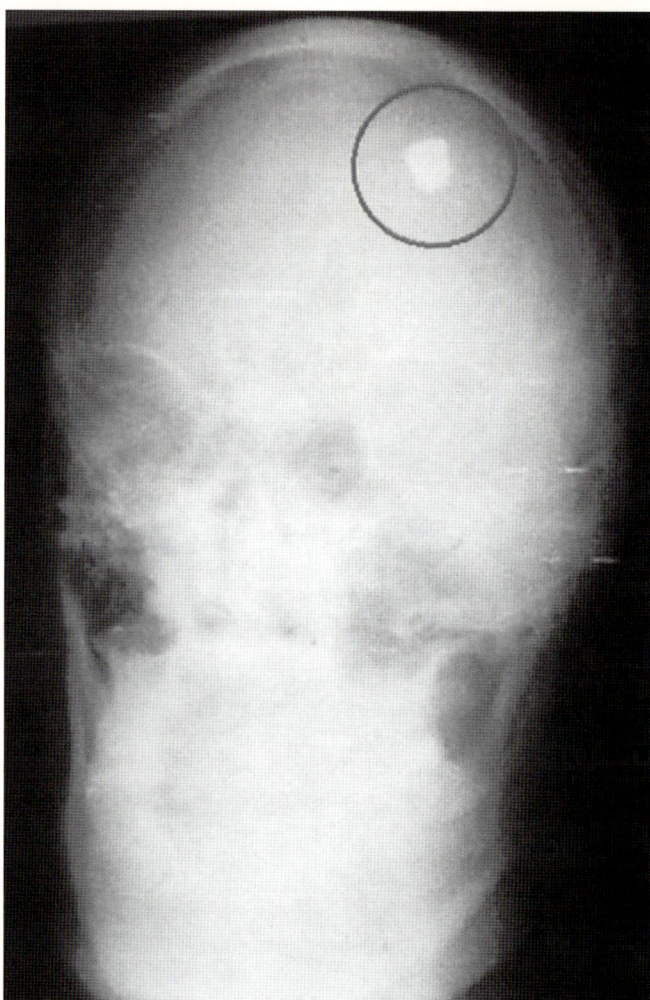

Fig. 8.35. Anteroposterior skull radiograph, radiopaque bullet circled.

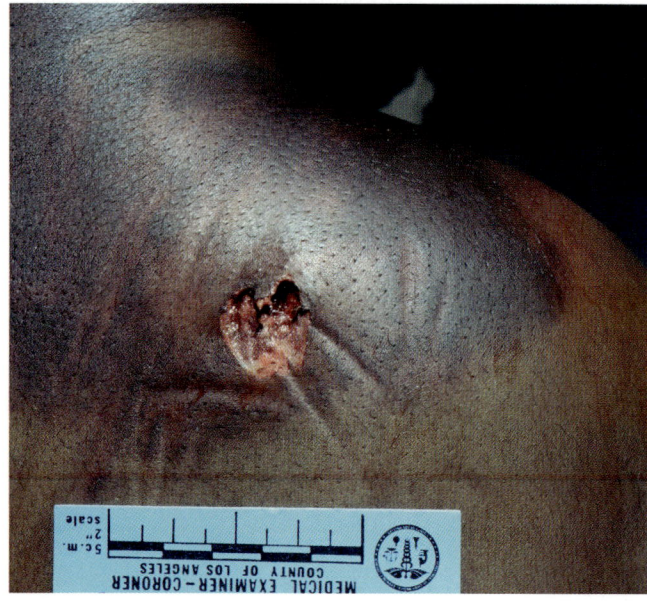

Fig. 8.34. Detail of gunshot entrance wound. Note broad zone of skin abrasion inferiorly and undermined wound edge superiorly.

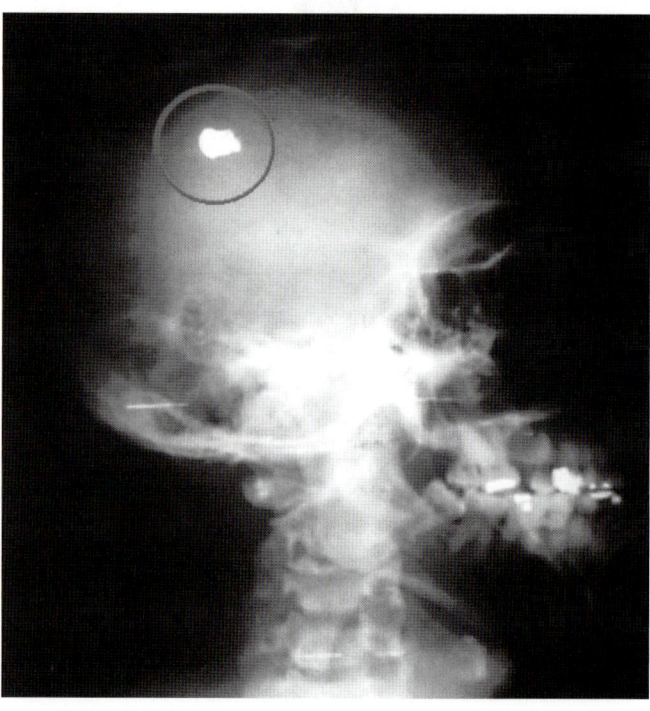

Fig. 8.36. Lateral skull radiograph, radiopaque bullet circled.

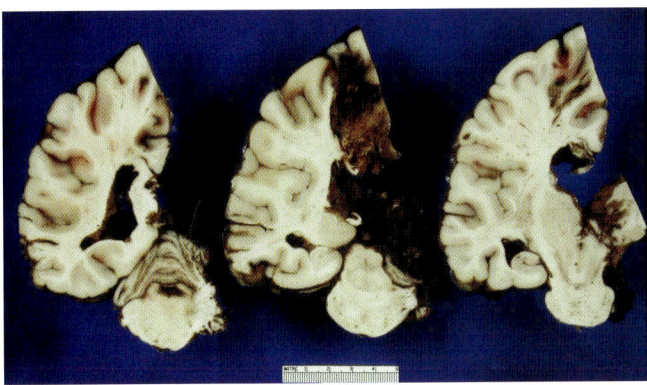

Fig. 8.37. Three coronal levels of left hemisphere with transverse sections of brainstem and cerebellum (see case report for description of wound path).

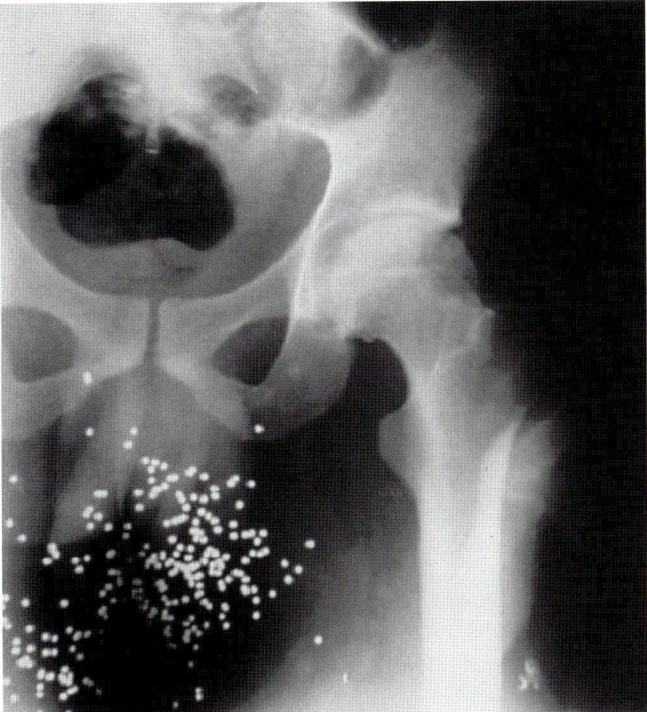

Fig. 8.38. Pelvic x-ray demonstrates gunshot pellets from remote, healed wound.

Parenthetically, it has also been noted that some individuals with minor, superficial wounds that do not involve any vital structure or major vessel may rapidly or immediately become incapacitated. This has usually been ascribed to the inability of such individuals to control fear, their lack of courage, or similar adverse psychological reactions to the situation. Some authors acknowledge the possibility that these adverse psychological responses may be accompanied by symptoms and signs that would be expected in an overwhelming catecholamine surge. We have wondered whether it is possible that persons with a genetically or situationally hyperactive autonomic nervous system (ANS) could develop such disabling (or even fatal cardiac arrhythmia) reactions to minor trauma without necessarily experiencing excessive psychological feelings of fear, etc. An analogy might be a willing mind inadvertently betrayed by one's own ANS. Most nurses and physicians experienced in performing phlebotomy quickly learn that the most likely group of patients to develop syncope during or shortly after the procedure are the young, strong, and healthy male athletes, often well-acquainted with continuing play despite sports injuries, not the "little old ladies from Pasadena." Well-documented experience has demonstrated that use of tactical breathing relaxation techniques in stressful situations, including armed combat (and court testimony), has provided many individuals with the means to control the unwelcome accompaniments of catecholamine surge, with resultant improved performance and self-control.[98] Could lack of this type of training, rather than lack of courage, have been a contributing factor in at least some of these otherwise inexplicable deaths? It remains unanswered, but it seems a potentially testable hypothesis.

Reaction/Response Time Issues

DiMaio[54] briefly comments on a study demonstrating that the time required for a trained police officer to fire a drawn handgun, once the decision is made to fire (mean time of 0.677 second with trigger finger outside the trigger guard, as required by many departmental policy statements), is longer than the time required for an individual facing the officer to turn around 90 degrees (mean time 0.310 second) or 180 degrees (mean time 0.676 second) so that his back is facing the officer.[282] Thus, it is possible for an officer who states he shot an individual facing him, with the entry wound located in the suspect's side or back, to be making a perfectly truthful statement. In the cited study, the mean time required to simply fire the weapon in response to an audible signal, with gun drawn, pointed at the target, and finger on the trigger, was 0.365 seconds. Alternatively, an individual with his back toward the officer can easily turn his head and bring his gun arm around, aiming and firing at the officer while his back continues to face the officer. Lewinski,[169] for example, describes 11 different gunfight scenarios in which an individual can be shot in the back while he is attempting to shoot an officer.

The preceding figures for officer reaction/response times in firing a handgun already drawn and pointed at a target do not take into consideration other aspects of real-world gunfights. Another study measured the time required for an officer to perceive a threat, evaluate it, decide whether or not to shoot, and shoot if justified by the scenario presented in an interactive shooting simulator with different types and complexities of threat viewed on a screen.[283] The starting point in this experiment had the officer with gun drawn, pointed at the screen, with finger on the trigger. The purpose was to measure the mean decision time (appearance of stimulus to shot minus the reaction/response time of 0.365 second established in the previous study).[282] The mean decision time was 0.211 second for a simple scenario and 0.895 second for a more complex scenario, the average of the two being 0.553 second. Thus, the total reaction time (0.365 second) plus the averaged decision time (0.553 second) results in a time of 0.918 second. This time applies to both the beginning and the end of the series of shots, since once the threat is perceived to have ended, a reaction time is required to stop shooting (Case 8.9). A free-falling body can descend 6 feet in about 0.6 second, so that once the officer perceives the suspect falling and decides to discontinue firing it is possible for one or two more shots to strike the body after it reaches the ground.[283] Even untrained individuals can fire up to approximately five shots in 1 second, providing some insight as to how critical even brief delays in response to a threat can be. Since these reaction/response time studies were testing trained, professional law enforcement personnel, it is

Case 8.9. (Fig. 8.39)

Officers from the DEA and a local SWAT team conducted a raid on a residence known to have received a shipment of imported drugs. The team, led by an officer behind a tactical bunker with a bulletproof window, cleared the kitchen and started to exit the kitchen to proceed down a hallway. As the officer with the bunker cleared the kitchen door, one of the residents appeared in the hallway, advancing toward the officers with a semiautomatic handgun in his right hand. The gun was held down by his right side. According to the officer, he and the resident locked eyes, the subject started to raise the handgun, and the officer fired immediately from his gun that was held in his right hand just off to the right side of the bunker. The officer fired three rounds at his assailant. Range tests indicated it would take approximately 1 second for the officer to react to a sound and fire three rounds. The subject did not fire a round at the officer.

The subject was fatally hit with a through-and-through round in his upper torso and right forearm. The round through the torso went from back to front, slightly upward, and very slightly to the left. The subject also had a graze wound on his left forearm.

The coroner opined that the subject received the graze wound on the left forearm as he was turning counterclockwise away from the officer. This particular wound, according to the coroner, was from the same round that had penetrated and exited his right forearm. The coroner said the subject was then shot in the back. A third round was found in the wall just slightly above where the subject's head would have been. The coroner could not identify whether it was the first, second, or third round that struck the wall.

For the coroner to be correct, the subject would have, at some point in his rapid turn, to have held both arms symmetrically and parallel to each other in the brief time that he turned his body 180 degrees and while he was rapidly trying to turn and run from this armed encounter.

Research. In research conducted by Dr. Lewinsky, subjects holding a gun in hand, without pointing that gun at an officer, but engaging in a turn to run away from an officer, can complete a 180-degree turn, on average, in just over half a second. Almost all the subjects used their arms to assist them as they turned. The use of the arm that held the gun caused it to initially be raised and pointed toward the officers as they were turning (Fig. 8.39). The nongun hand was generally flung to the side and backward. As subjects were about to complete the turn, the left hand, in right-handed subjects, was frequently but briefly passed across the chest before it came to a position on the left side of the body where it would assist the subject as he or she began to engage in running away from the officer. Male subjects in particular also dropped slightly as they gained compression from their legs to assist in the dynamic, rapid turn. They also leaned into the turn with their upper torso. Not a single subject held the arms even slightly parallel, and never did he or she hold them symmetrically positioned while turning.

Analysis. Because the subject in this case had not fired his loaded weapon at the officer and was able to complete a 180-degree turn in the time span it took the officer to fire twice, just over two-thirds of a second, it appears the subject was determined to run away from the encounter with this armed officer. Unfortunately, while engaging in the turning motion, he raised his handgun from his side as he used both arms to assist him in his rapid turn in flight, as was found typical in research subjects engaged in a rapid, dynamic turn. As he raised his gun and brought it across his body, he briefly but unintentionally created the threat that led the officer to perceive this subject was beginning to point his gun to shoot him (Fig. 8.39). The first round from the officer, which would have been fired approximately 4/10 of a second after the subject started his turn (most of the time in a turn is spent in developing momentum), went through the subject's forearm. The second round, fired approximately 7/10 of a second after the subject started the turn, was fired higher and entered the subject's back after he had almost completed his turn but before he could draw his hand away from his chest. This was the round that grazed his left forearm as it passed in front of his chest at this point. The third round from the officer was fired at an even higher point and struck the wall just above where the subject's head would have been.

Conclusion. The jury found the officer not guilty of all charges. Post-trial jury interviews revealed that jury members initially believed the officer lied, but became convinced of his honesty when the research data were presented.

(Case courtesy of Bill Lewinsky, Ph.D.)

Fig. 8.39. This research subject was specifically instructed *not* to point the gun, or mimic firing the gun, at the cameraman. He was instructed only to turn counter-clockwise and run away as rapidly as possible. As he did so, this freeze-frame from the movie shows the initial phase of the natural arm motions used to assist the dynamic turn, indistinguishable from intent to shoot in the direction of the cameraman.

likely such times would be longer in most civilian self-defense shootings.

Examples of a few of the several other reaction time studies relevant to forensic cases are included in the references for this chapter.[170–172] Another reference cited considers some common juror misconceptions about reaction/response time issues.[296] Such articles are not typically found in the forensic pathology literature, but interested individuals should be able to access these and other relevant studies through websites such as www.forcesciencenews.com or through contacts with law enforcement agencies.

Perhaps the most appropriate summary of this reaction/response time section is to remind ourselves that a medical examiner with sufficient experience in gunshot wounds might reasonably be considered a wound ballistic expert within a given spectrum of firearm types.

Such expertise does not automatically translate to one also being a firearms ballistic expert, or an expert in lethal combat scenario analysis, each of which is a distinct field with different knowledge requirements.[65] This is important to keep in mind when responding to questions involving various hypothetical scenarios during court testimony.

Unusual Missile-Launching Devices and Ammunition

This section briefly summarizes various gunpowder, spring tension, or pneumatic-propelled devices and missiles that have demonstrated a potential for, or actually caused, lethal CNS and other injuries. Its purpose is to facilitate an initial access to some of the relevant literature on these unusual injuries. The emphasis is on CNS-related injuries, and the list includes:

- Compressed gas cylinder from paintball gun.[104]
- Pellets fired from air guns.[7,201]
- BBs fired from air guns.[202]
- Pneumatic or cartridge-driven nail or stud guns[27,54,175,250] (Case 8.10, Figs. 8.40–8.41). *Comment*: Nail guns have also been reported to cause traumatic aneurysms of the middle[26] and posterior[229] cerebral arteries, a potential source of delayed fatality.
- Captive bolt (i.e., slaughterer's, butcher's) gun.[289,290]
- Pneumatic hammer.[50]
- Blasting caps.[5]
- Black powder, muzzle loading weapons.[142]
- A wide variety of improvised homemade, or commercially available, weapons exist, easily overlooked because they are designed to appear as motorcycle handlebars, belt buckles, batons, large bolts, canes, cigarette lighters, highway flares, knives, cell phones, key chains, and myriad other easily overlooked everyday items. Most of these fire standard bullets.[108,211]
- A wide variety of ammunition exists in addition to the more standard bullet types.[54] Examples that have caused, or potentially could cause, fatal craniocerebral wounds include Glaser rounds[52,124] and frangible ammunition.[93,129,247] Rarely, frangible bullets penetrating the skull do not undergo fragmentation.[252]
- Blank ammunition and compressed air-powered gun craniocerebral wounds may cause serious injury or fatality[14,90] (Case 8.11, Figs. 8.42–8.44).
- Certain unmodified blank weapons[224] and modified blank weapons[239] can also fire projectiles.
- Less lethal devices for law enforcement use may cause serious injury or death. In this category are included rubber, plastic, or ceramic bullets[54,292]; rubber or plastic baton rounds[151,184,197,270] (Case 8.12, Figs. 8.45–8.47); tear gas cartridges[39]; and so called "bean bag" projectiles. The last are typically 12-gauge shotgun bean bag

240 Injuries Due to Firearms and Other Missile-Launching Devices

Case 8.10. (Figs. 8.40 and 8.41)

A 29-year-old construction worker committed suicide with a nail gun fired into his right temple. The flange in the nail prevented deeper penetration. The nailer used an explosive charge to drive 16 penny nails through 2 × 4-inch boards and into underlying cement foundations in the course of house frame construction.

(Case courtesy of Lisa A. Scheinin, M.D.)

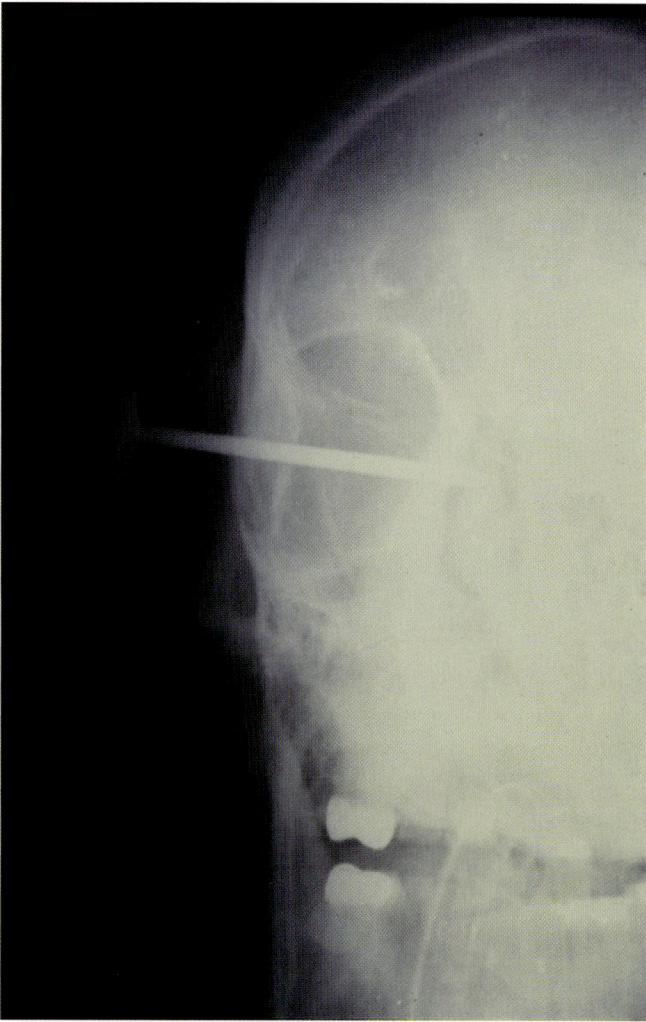

Fig. 8.41. Anteroposterior skull radiograph. Nail lodged in skull at level of right temple. The aluminum flange is radiolucent (see following section, "Radiology of Gunshot Wounds").

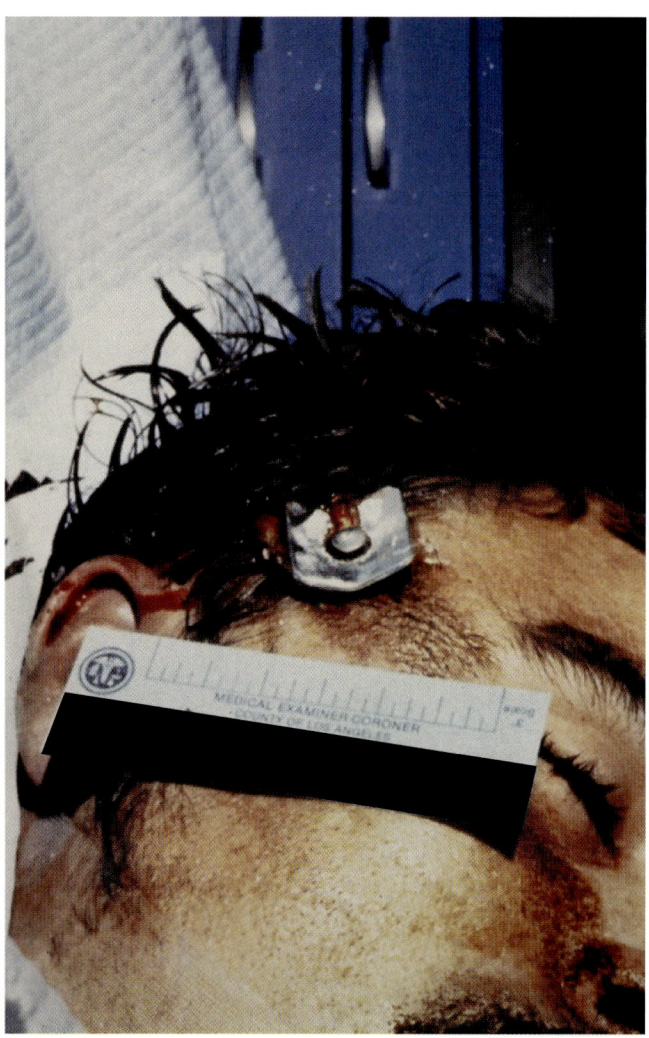

Fig. 8.40. Nail with aluminum flange in right temple of subject.

Case 8.11. (Figs. 8.42–8.44)

A 26-year-old man played Russian roulette in front of associates by loading three dummy rounds and two live blank cartridges in a .44 caliber revolver, spun the cylinder, and placed the barrel to the right temple. He pulled the trigger, and a blank cartridge fired, causing a wound to the temple. Craniotomy was performed to elevate a depressed skull fracture and remove bone fragments driven into the basal ganglia. Brain death was pronounced a few days after the injury.

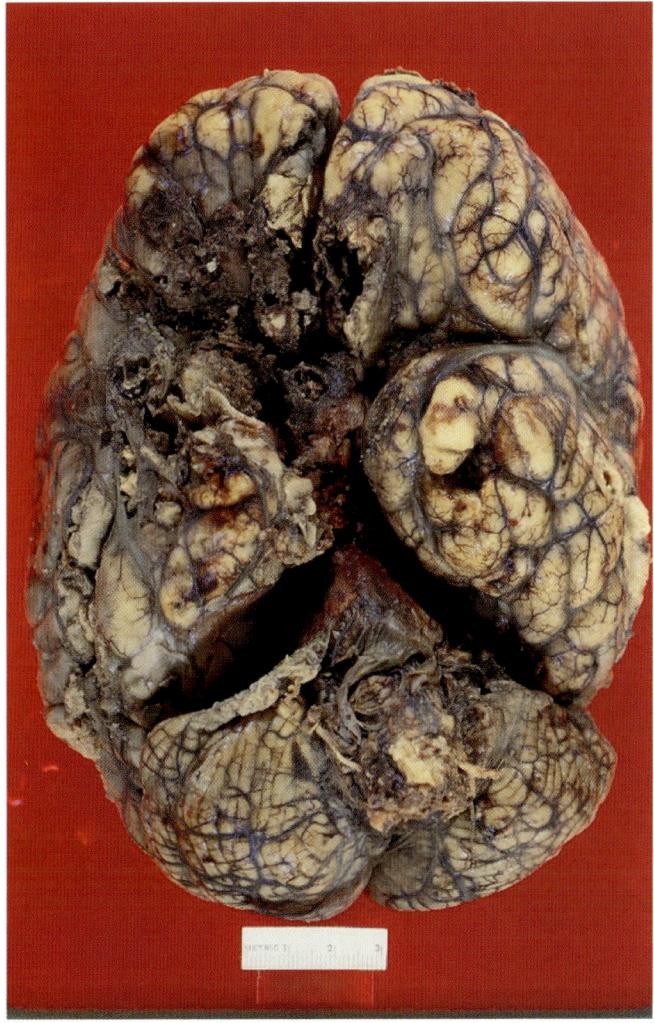

Fig. 8.42. Base view shows a dark hemorrhagic stain of right orbital frontal lobe and along anterior sylvian vallecula, on a background of respirator brain changes.

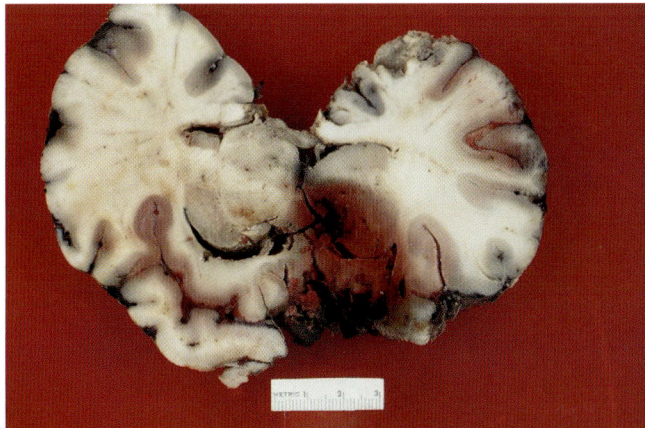

Fig. 8.43. Frontal lobes on coronal section show acute hemorrhagic stain of ventral half of anterior right striatum extending to the midline and orbital cortex. Note remote contusion of right middle frontal gyrus. Separation of superior interhemispheric fissure is due to artifact.

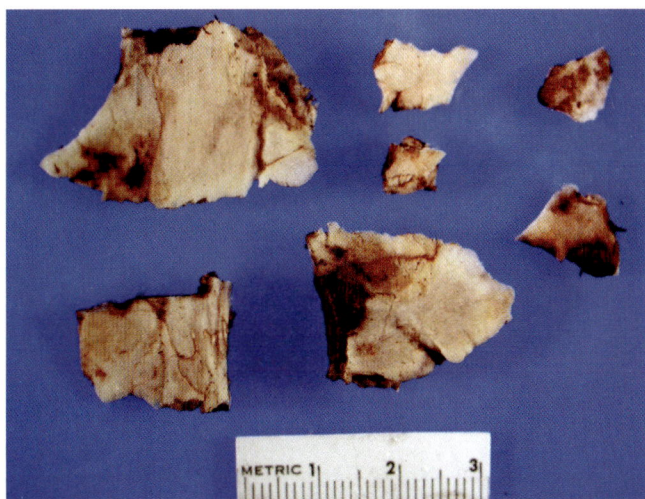

Fig. 8.44. Bone fragments that had been driven into the brain by blank cartridge discharge.

Case 8.12. (Figs. 8.45–8.47)

A wanted subject barricaded himself and engaged in a prolonged gun battle with police. In the early phase of the gunfight, police attempted to resolve the situation with less lethal, ARWEN-type plastic impact baton missiles while being fired upon by the suspect. These missiles were directed at the only exposed portion of the subject, his head. Despite the suspect being struck in the head by two such rounds, witnesses reported no interruption in the suspect's activity level or his ability to fire at police. The suspect died from exsanguination after later receiving a gunshot wound to the radial artery at the level of the wrist. Postmortem toxicology was positive for methamphetamine.

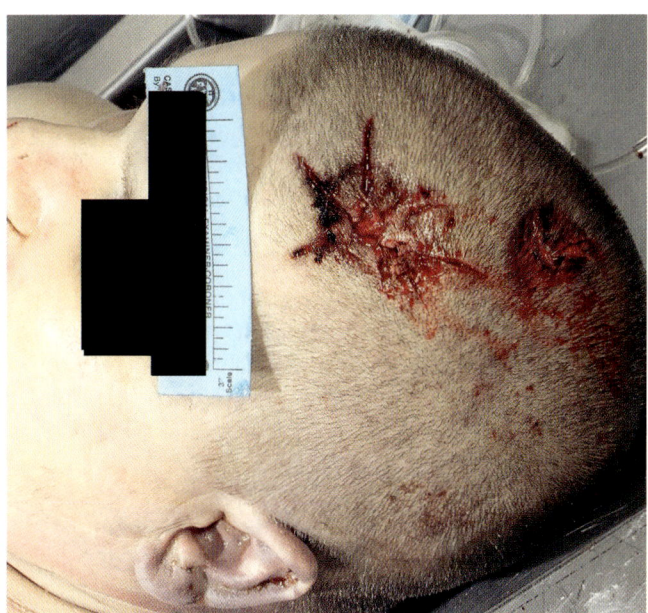

Fig. 8.45. Two scalp impact sites from ARWEN-type plastic baton rounds are evident.

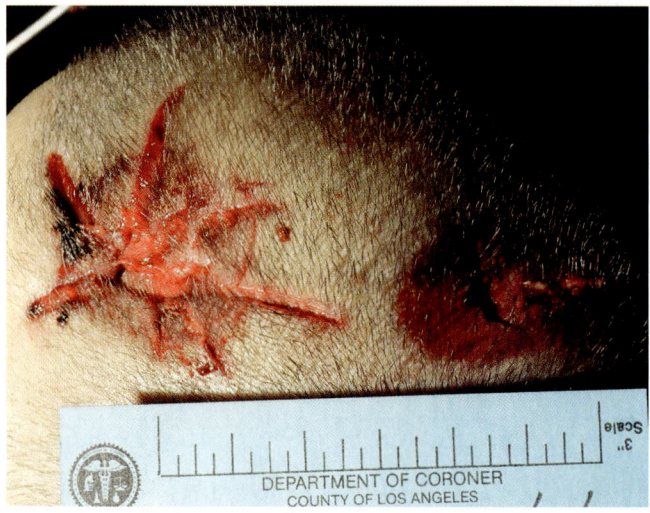

Fig. 8.46. Anterior wound was a stellate laceration with abrasions.

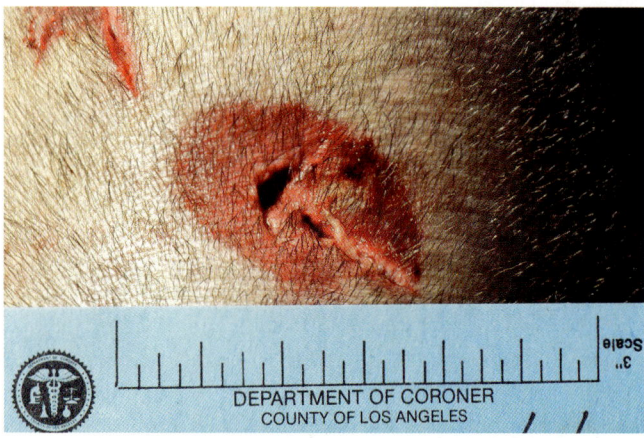

Fig. 8.47. Posterior baton impact site. Abrasion with milder laceration suggesting a more tangential impact. There was no skull fracture or intracranial abnormality in this decedent.

of two probes that can be launched from a hand-held device into the clothing of the subject from distances of up to approximately 15 to 21 feet (air taser). Other devices have a similar purpose, but they require the user and a subject to be within arm's length of one another (stun gun). Although effects on voluntary muscles are the goal, some critics have raised

> **Case 8.13.** (Figs. 8.48–8.50)
>
> This adult male attacked police with rocks and metal objects when they arrived in response to a citizen's call about the suspect creating a disturbance. His continued assault led to the officers firing two of the old-style stitched-fabric square "bean bag" rounds (flexible batons) at an estimated range of 25 to 27 feet. One round entered the left jaw area. The second round entered the chest, producing a fatal wound.
>
> (Case courtesy of W.E. Sherry, M.D.)

projectiles of two main types. The first is the older stitched-fabric squares containing lead birdshot (so-called flexible batons),[151,184,253] which have caused occasional serious and fatal injuries, with examples seen in our office (Case 8.13, Figs. 8.48–8.50). We have not seen any serious injury with the newer, drag-stabilized (bulb with tail) bean bag projectiles.[44]

- Air tasers and stun guns are an additional approach to the less-lethal equipment choices used by law enforcement to subdue dangerous and violent persons who are a risk to others and/or themselves. They produce an electrical discharge causing temporary voluntary muscle paralysis in many subjects by means

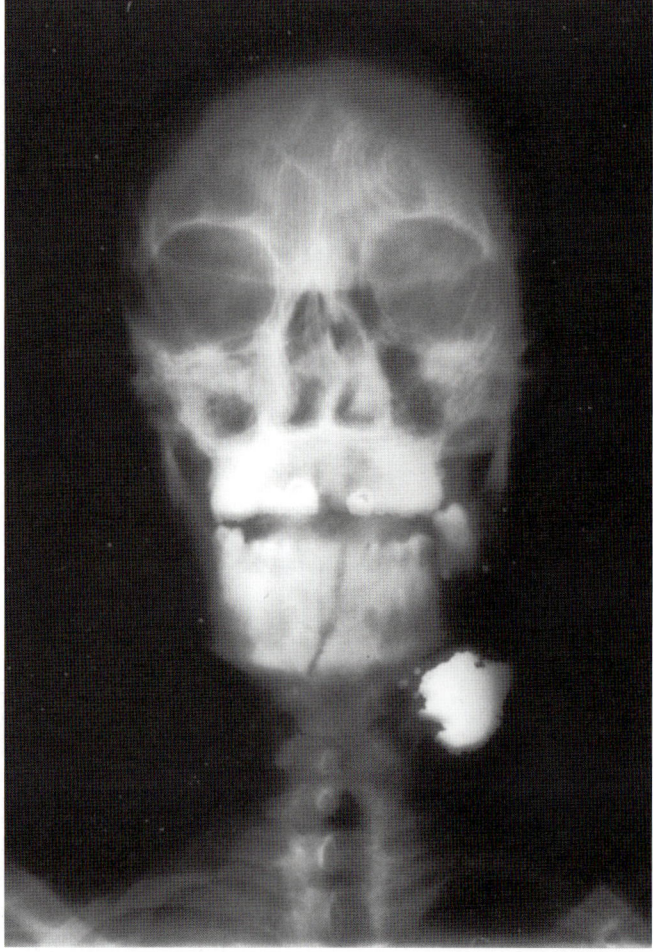

Fig. 8.48. Anteroposterior skull radiograph demonstrating BB-containing bean bag projectile in left submental area. Note central and left comminuted mandibular fractures, and loose BBs adjacent to bag.

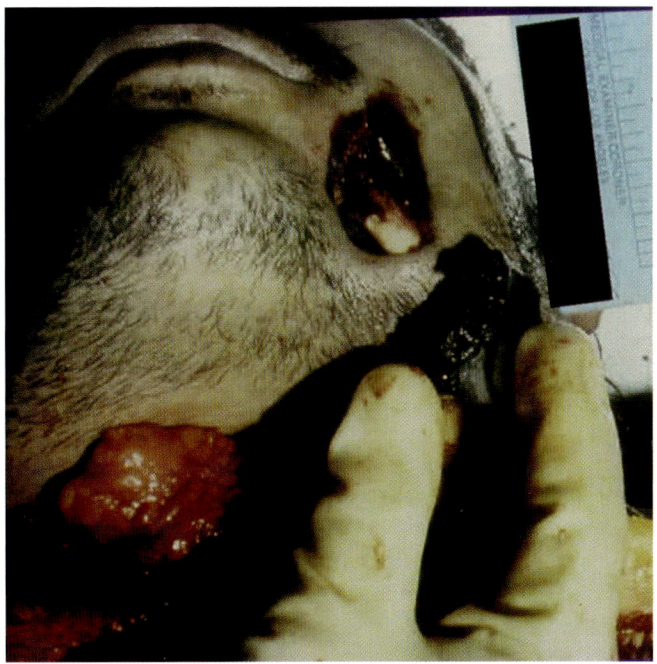

Fig. 8.49. Autopsy photo of removed bean bag held near cheek entrance wound. White object in entrance wound is a fractured, displaced mandibular tooth.

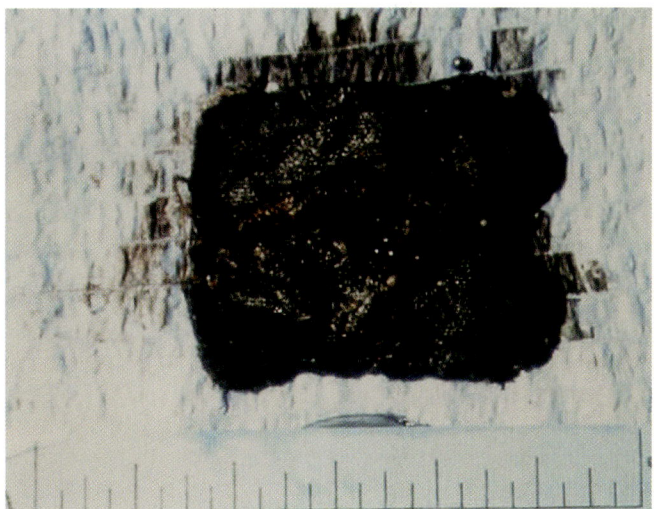

Fig. 8.50. Detail of blood-soaked bean bag, with loose BBs above, to right side, and on upper surface of bag due to fabric ruptures.

include information regarding toxicology studies, details of postevent cardiac evaluation, and other important considerations.[149] It suggests that further study is required to establish the relative risk of such devices. Use of such devices on infants have also been speculated as a cause of death.[287]

Death of persons subjected to these electrical devices is more likely to warrant consultation with a cardiologist than a neuropathologist. Current interest in this subject is reflected, in part, by nine abstracts addressing various issues and experiences with use of tasers in the Proceedings of the 2006 Meeting of the American Academy of Forensic Sciences.[4,157,176,244,246,262,275,295,301]

Firearm Wound Imitators

Wounds from several different tools and nonfirearm devices can superficially resemble gunshot wounds,[54] particularly when the wounding instrument has been hidden or removed from the scene. A few examples seen in our department and cited in the literature as sources of serious or fatal CNS wounds provide some indication of the wide spectrum of instruments used in attacks. Some examples include:

- Screwdrivers (including power screwdrivers)[130] (Figs. 8.51–8.52).
- Ice pick.
- Pen.
- Pencil.
- Tapered steel punch.
- Rebar.
- Awl.
- Homemade potato gun.[16]
- Slingshot.[24]
- Blasting caps.[5]
- Crossbow bolts/arrows,[49,226] depending on type of arrowhead.[33,61,158,215]
- Ballpoint pen point (propelled by pistol crossbow).[237]
- Nails.[53]

Bullet/Shrapnel Trace Evidence

As Davis[47] has stated, "Always consider the recovered bullet to contain trace evidence until proven otherwise." How compulsively this should be pursued, as with other time-consuming and expensive special studies, will depend on the specifics of the case and the judgment of the medical examiner assigned to the case. Briefly reviewed here are some of the types of trace evidence that bullets may provide. When such studies are contemplated, one needs to consult with other members of the forensic team to determine the most appropriate speci-

questions concerning whether the electrical stimulus is also capable of inducing fatal cardiac arrhythmia in otherwise normal adult persons, persons with pre-existing cardiac disease, children, and persons under the influence of certain drugs. A recent letter to the editor reports on a possible relationship between use of an air taser and life-threatening cardiac arrhythmias in rare subjects. This communication did not

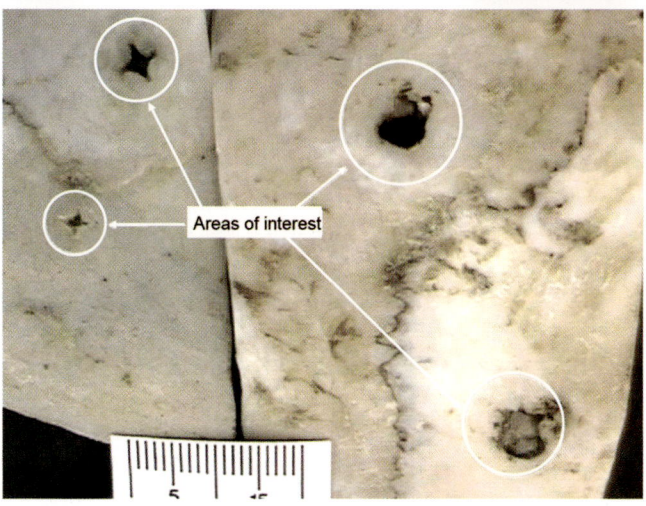

Figs. 8.51–8.52. A Phillips-type screwdriver (Fig. 8.51, *top*) will produce either a crisscross skull defect, or a round skull defect, depending on its depth of penetration (Fig. 8.52, *bottom*). A round defect may be misinterpreted as a gunshot wound. (Case courtesy of Steve Dowell, Los Angeles County Department of Coroner.)

men handling method(s) and sequencing of the planned studies. For example, in cases where matching a specific bullet to a specific victim is considered necessary, methods to preserve and collect DNA evidence should take precedence over collection of fiber or other foreign inert intermediate-target material present on the bullet.[136] This would also be one indication for immediate bullet retrieval rather than initial fixation prior to dissection, despite the fact that a detailed neuroanatomic examination is compromised by sectioning an unfixed, compared with well-fixed, brain or spinal cord. In other cases where no DNA or other trace evidence studies are necessary, blood and tissue elements can be gently cleaned from the recovered bullet according to department protocol and the dry bullet sent to the firearms examiner. In yet other cases, the type and sequential order of arrangement of tissues on a bullet might be examined by cytologic methods to aid in the determination of bullet path direction, with the understanding that some tissues adhere to bullets more than others.[210] In bullets passing into or through the brain, for example, there tends to be a disproportionate amount of vascular elements, ependyma, choroid plexus, dura, leptomeninges, and cranial muscle fragments.[210] Brain was the only tissue in this particular study that produced lengthy fragments of capillaries and other blood vessels. It was coincidentally also noted that the CNS-derived vessels were "cleanly stripped of nervous tissue."[210]

DNA analysis (perhaps combined with cytologic methods) might also be used to link a particular bullet (and thus weapon) to a particular wound in a victim, which could be useful in cases with multiple assailants and weapons, for example.[137,141]

Missiles requiring analysis for toxins are presently more likely to be encountered outside of the United States. Examples include bullets containing a toxin such as cyanide,[146] and anticoagulant-coated shrapnel from improvised explosive devices intended to produce death by exsanguination in victims who might otherwise survive the shrapnel-induced wounds.

Radiology of Gunshot Wounds

Clinicians, in contrast to pathologists, routinely employ a much broader spectrum of radiologic techniques in order to discover occult injury, localize bullets and secondary fragments, discover emerging complications at an early stage, disclose vascular injury, and aid therapeutic decisions in gunshot wound cases.[113]

They also must be aware, given the limitations of current equipment, of the possible need to avoid certain procedures. Weapons are to be safely secured outside of MRI suites, not only due to their metal components but because spontaneous discharge of loaded weapons can occur in the strong magnetic field.[17,128] MRI may be omitted in cases with intraocular metal fragments, cerebral aneurysm clips, cardiac pacemakers, or the presence of bullets with ferro- or paramagnetic materials, to avoid possible added injury or death due to magnet-induced projectile migration.[79,110,113,127,277] Such cases may be referred to the medical examiner because of periprocedural complications, but it is not clear whether projectile movement resulting from postmortem MRI examination would be sufficient to confuse wound path determinations in the CNS or in other body locations.

Modern imaging methods are capable of providing interactive three-dimensional reconstruction of a bullet path in a format that is easily visualized, and possibly more aesthetically acceptable (due to its medical illustration-like, bloodless appearance) to courtroom use.[214] Medical examiners can also be aided by supplemental use of modern imaging techniques such as CT or MRI, combined with current autopsy methods[118] (also see section on neuroradiology in Chap. 17). Due to budget limitations in most departments, such tools are almost exclusive to research settings at the present time.

In medical examiner cases, issues can arise that are not necessarily critical for treatment decisions in living patients, and may not be included in hospital radiology reports. However, they may influence wound interpretation. Davis[47] noted that hospital radiographs taken during initial examination in acute gunshot wounds may show tissue gas bubbles indicating not only that a given wound was an entrance wound, but also that it was a contact wound. Such information may not always be included in the formal report, and it is one more reason why review of all available data is important in case analysis.

A survival time allowing surgical treatment, especially if some healing has also occurred prior to death, may compromise later determination of bullet path direction. Hollerman et al.[113] also noted that air acutely visible between muscle bundles may represent temporary cavitation effect. At later stages, tissue air may represent subcutaneous or mediastinal emphysema, or wound infections with gas-forming organisms. With sufficient death to autopsy intervals, postmortem decomposition may also produce tissue gas.

In many cases it becomes important to correlate autopsy and radiologic findings, and account for any apparent discrepancies. Forensic radiology consultants play an essential role for this purpose. Imaging reports may appear inconsistent with postmortem anatomic findings under some circumstances, and such discrepancies should be explained whenever possible. Some examples are the following:

- Bullet-shaped opacities are not always bullets. Gravel, nonmissile metallic objects worn as jewelry or in clothing or hair, contaminated or double-exposed radiographs, bullets in the oral cavity region mistaken for dental fillings and, conversely, displaced dental fillings mimicking bullets (such as from blunt force trauma) are some examples.[36,47]
- Objects outside of the skull may appear to be inside of the skull on routine studies limited to anteroposterior and lateral radiographs.[36]
- Underexposed x-rays may cause bone density to obscure the bullet.[47]
- Some bullet (or shotshell) components may be nearly or completely radiolucent. Examples are aluminum jackets of bullets (Case 8.14; Figs. 8.53–8.56); synthetic filler granules of polyethylene or similar material, wadding materials, or other shotshell components; and the plastic cap of Glaser rounds and the plastic tip inserted in the nose of some hunting rounds.

Case 8.14. (Figs. 8.53–8.56)

This 66-year-old man with pancreatic cancer died from a single, self-inflicted gunshot wound that entered the hard palate and exited the vertex of the head.

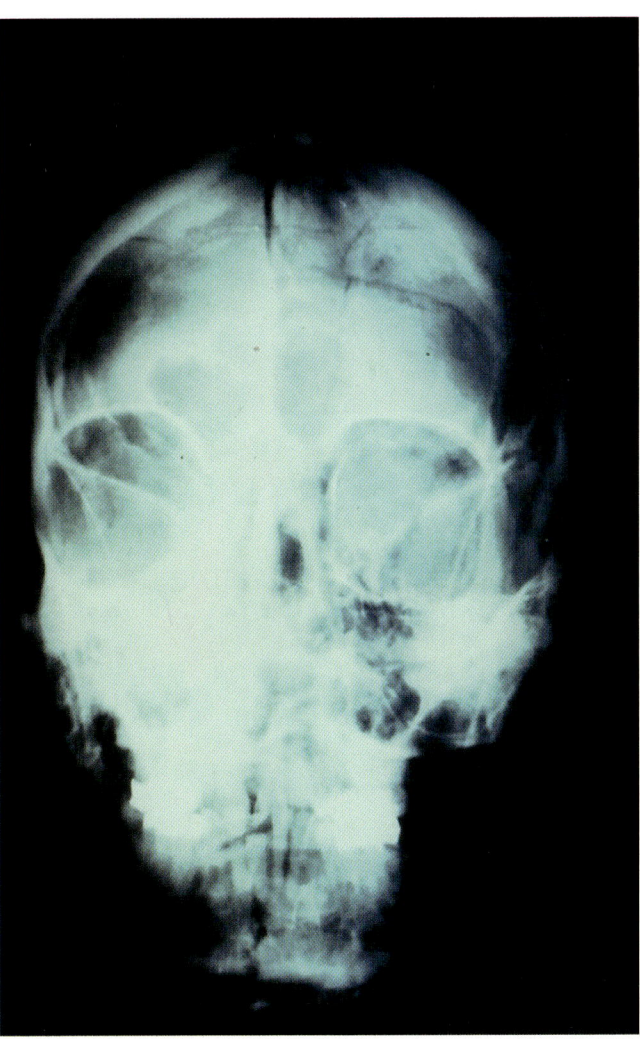

Fig. 8.53. Anteroposterior skull radiograph reveals no metallic fragments, but does demonstrate skull fractures.

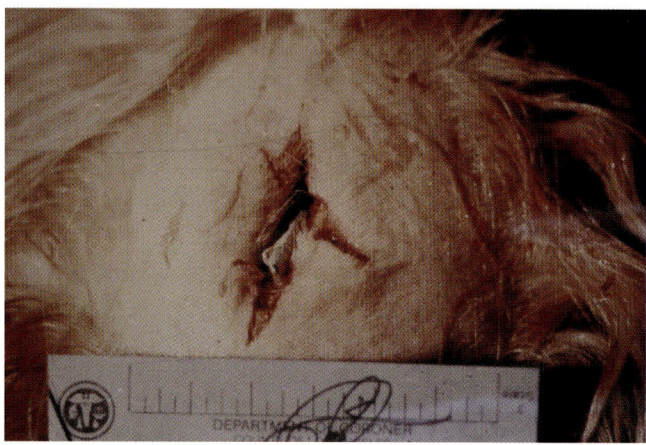

Fig. 8.54. A deformed, blood-stained metallic bullet jacket was partially exposed at the exit wound (*arrow*).

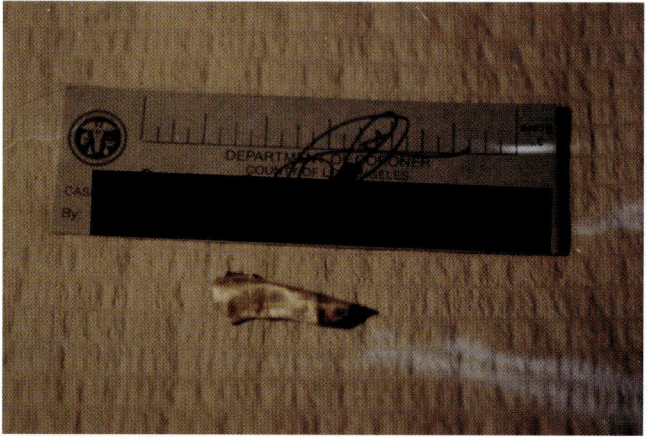

Fig. 8.55. Detail of the recovered blood-discolored radiolucent aluminum bullet jacket.

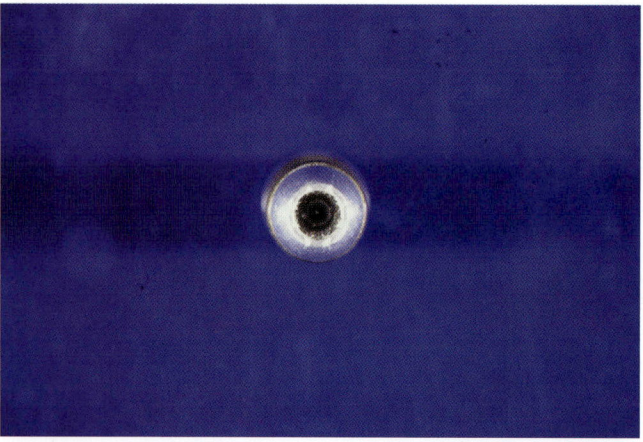

Fig. 8.56. Anterior view of a similar, intact aluminum-jacketed hollowpoint bullet. It is not unusual for the lighter-weight jacket components of bullets to remain within wounds, while heavier bullet-core components retain sufficient inertia to extend the depth of the wound or even exit the body.

Nyclad bullet coating material is also included in this category, but it is very firmly adherent to the underlying lead bullet and, in our experience, it is unlikely to separate from the lead core (unless it strikes thick bone, according to DiMaio[54]). Radiolucent plastic bullets exist, but they are rarely seen in wounds in the United States.[113] Radiolucent foreign material from intermediate targets may also be present in wounds. For example, wood splinters may appear as air shadows.[194] Careful search of the wound path will minimize overlooking such findings.

- Failure to do full body radiographs or screen with whole body fluoroscopy prior to autopsy increases the potential for bullets being present but remaining undiscovered. This is most likely in cases of bullet migration or bullet embolism, internal ricochet, old healed bullet wounds without conspicuous scars, and bullets in decomposed or burned bodies.
- A bullet-shaped opacity in long-buried skulls may persist after bullet retrieval, presumably due to adherent metallic salts derived from a bullet that was touching the inner skull for a prolonged period.[147]
- Radiographic magnification of deformed bullets or pellets may lead to incorrect caliber and missile type estimations, based solely on radiographs.

Modern imaging techniques have also been used to resolve long-standing disputes related to delayed neurologic defects from gunshot wounds to the head.[135]

Readers interested in sources that contain more numerous illustrations of radiographs of a wide spectrum of firearm wounds, as well as examples of other missile-type injury, are directed to the following citations.[30,31,113,193–195,291,292,300]

Gunshot Wound Autopsy Protocols

There are several published guidelines for the medical examiner suggesting which procedures can or should be performed during the general autopsy[80,178,251] and, more specifically in gunshot wound victims, what data to include in the gunshot wound autopsy report.[47,54,57,58,80,83,94,168,243,266] The reader may refer to these and similar references for details. In addition, medical examiner departments will often have their own individual guidelines based on the experience of that department. Our current Departmental Deputy Medical Examiner Procedural Manual devotes nine pages to gunshot wounds, including forms used to describe key features of each individual gunshot wound in a body. Clearly, many such cases do not warrant consultation by a neuropathologist (Fig. 8.57). In some cases it is appropriate to describe a closely associated cluster of gunshot wounds with overlapping paths as a group, rather than as individual wounds. For testimony purposes, an anatomic diagram devoid of all data except identification

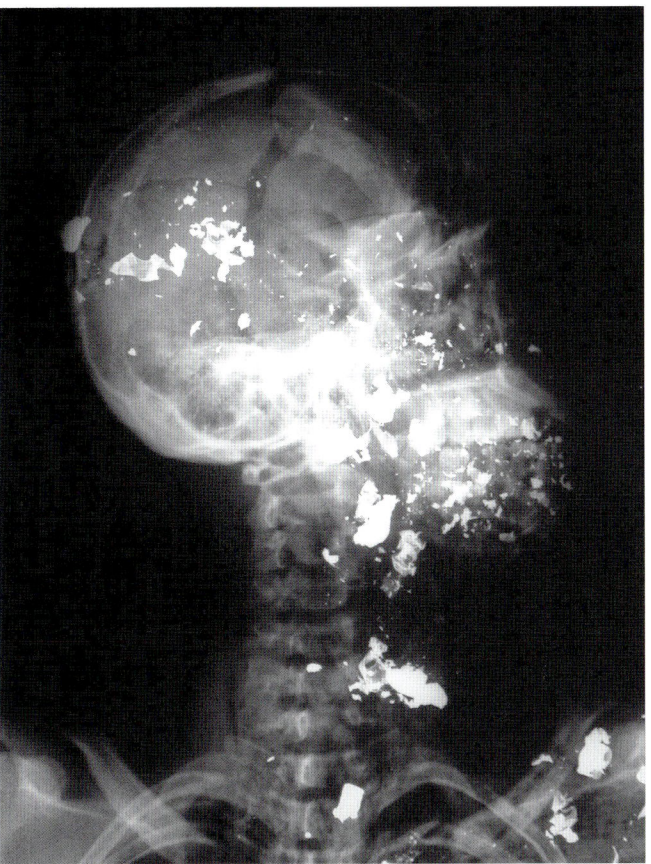

Fig. 8.57. Lateral skull/cervical spine radiograph. Extensive bullet fragments throughout the cranium, facial structures, and soft tissues of the neck and chest wall with highly comminuted fractures of the skull, skull base, and facial structures. Frontal pneumocranium is present. Numerous other gunshot wounds, with fractures and metallic fragments, were present in trunk and extremities. A neuropathology consultation would not contribute meaningful forensic information to the medical examiner in such cases.

information and numbered arrows corresponding to the path of the gunshot wounds present can be helpful in presenting a concise overview of injuries to the court.

Wound interpretations by hospital clinicians unfamiliar with the forensic aspects of gunshot wounds are often incorrect.[12,41] Discrepancies between conclusions described in the hospital record and the forensic pathologist's opinion are likely to be focused upon by attorneys in court, either to clarify the discrepancy or to impeach the testimony of one or more sources of the varied opinions. Such issues can be resolved by providing clear data from the forensically trained observer of gunshot wounds, who has the added advantage of exploring the entire wound, externally and internally, without simultaneously performing urgent and multiple procedures in an attempt to preserve the life of the victim.

The limited suggestions that follow are based on our experience concerning information we would find particularly helpful in answering neuropathology-related questions, such as those described in this chapter. Other less widely cited observations on examining gunshot wounds are suggested, in the event that they may be helpful to neuropathology consultants without prior forensic pathology experience.

In general, gunshot wound information already available to the forensic neuropathology consultant by the time the brain and/or spinal cord are adequately fixed in formalin and ready for gross examination usually includes the following:

- Gunshot wound(s) by type (e.g., bullet, shotgun pellet) and by number (with a comment that if multiple gunshot wounds are described the numbering system is arbitrary and not necessarily in the order the wounds were received, unless sequencing was specifically performed in that case).
- Description and size of entrance wound, including measured location from top of head, midline, or other major, easily understood landmarks; size; shape; presence of soot or stippling; pseudostippling; abrasion rings; carbon monoxide (CO) discoloration; bone beveling; gas disruption of underlying tissue; bullet wipe; lead fouling; and searing.
- Evidence of internal ricochet.
- Whether the projectile was recovered, and a general description of the projectile. In our department, the projectile description is usually restricted to whether it is metallic, plastic, etc.; color; jacket type if present; bullet configuration (round nose, hollowpoint, etc.); deformed or intact; and whether the caliber is small (approximately .22), medium (approximately 9 mm), or large (.45 or greater).
- Time and anatomic location where projectile was recovered, how processed or cleaned, how packaged, and when submitted to evidence safe or evidence technician for chain of custody issues, etc., are also specified. In our department, no additional marks are placed on the bullet or bullet fragments recovered.
- Direction (left/right, upward/downward, anterior/posterior) of wound path.
- Structures involved in bullet path in sequence, including wound size and appearance. In cases submitted for neuropathology consultation, the wound path description may simply indicate "brain," with the detailed brain examination deferred to the neuropathology consultant.
- Exit wound size, appearance (with details as noted for entrance wound, and given the same number as the respective entrance wound for that bullet).
- Whether the wound is judged to be lethal or nonlethal, and whether it was sustained antemortem or postmortem.

- Other injuries external to the nervous system.
- All wounds diagramed, photographed, x-rayed, and a brief description of projectiles on x-rays by the medical examiner (even if the x-rays are later to be submitted for formal forensic radiology consultation).
- Findings from the dissecting microscope examination, if performed.
- Whether or not toolmark analysis or gunshot residue examination was requested.
- Any effects on wound appearance of healing, therapeutic intervention, insect or predator activity, or decomposition.
- Whether any atypical features are present on the wound suggesting a muzzle pattern, intermediate target, shored exit wound, revolver cylinder gap, flash suppressor, muzzle brake, etc.
- Miscellaneous associated abnormalities noted by the medical examiner, such as location and volume of epidural or subdural hematomas, subarachnoid hemorrhage, etc.

The most common omissions we encounter in departmental or outside consultation cases submitted for neuropathology consultation include the following:

- Information from first responder, law enforcement, scene description of amount and distribution of blood, hospital records, and so on, which allow the development of a detailed timeline of events including witnessed activities, if any, carried out by the decedent following the injury.
- Absence of description of presence or absence of possible secondary overpressure effects. Were offset skull fractures present when the scalp was reflected and before removal of the skull cap (i.e., heaving fractures)? Were detailed descriptions/diagrams/photographs of entrance/exit wound; primary, secondary, and tertiary fractures made, including the presence or absence of remote fractures? Was there a search for subtle evidence of internal ricochet on the dura prior to adding dural artifacts by stripping of the dura, and was there examination of the skull for internal ricochet evidence after dural stripping?
- It is important for the consulting neuropathologist to know exactly why the brain was sent for consultation and the specific questions to be addressed.

The neuropathologist must also be familiar with the chain of custody protocol for the department, so that evidence discovered during the brain examination that is to be submitted for further evaluation is handled in the appropriate fashion. Special search may be indicated, for example, for radiolucent shotshell components in a brain wound. Bomb shrapnel components, radiopaque or radiolucent, are particularly important to collect in order to provide the bomb expert with the maximum amount of information concerning the nature and source of the device. Radiographs of the brain specimen prior to dissection can be helpful in such cases.

The decision as to when to recover a projectile must weigh the relative need for rapid bullet identification versus the detailed neuroanatomic data that might later be needed. In addition, one should consider avoiding damage to projectiles, such as rapid rusting of steel-surfaced projectiles which can compromise later analysis of rifling marks and weapon matching.[114]

The risk of injury to the examiner by sharp metal or other missile fragments is also minimized by a predissection review of x-rays taken either at autopsy or just prior to dissection. This can identify projectiles such as those with splayed pointed flanges produced upon tissue impact by some hollowpoint bullets. The bullet or shrapnel can be removed by rubber-tipped forceps (designed to prevent the addition of toolmark artifact to missiles).[241] In rare instances, there may be indications for other special techniques.[123]

Surgical removal of unexploded weapon projectiles from the head and other body areas has been accomplished. Technical advice from a local law enforcement agency or military bomb experts, and familiarity with published recommendations, is appropriately obtained before attempting such procedures in living patients or in an autopsy setting.[166,189,264] Some recent sociopolitical trends suggest that medical examiners would be prudent to review such information and establish contacts for the necessary technical advice, if not already in place. The main requirement in the medical examiner setting would be recognition of the presence of such an unexploded device prior to accidental detonation. Once recognized, appropriate safety precautions could be implemented, and the decision regarding how to render the device inactive (or attempt retrieval for its examination by the appropriate agencies) determined.

References

1. Aarabi B, Alden T, Chesnut RM, et al. Management and prognosis of penetrating brain injury. J Trauma 2001;51(Suppl 2): S1–S86.
2. Adams RJ, McTernan TM, Remsberg C. Street Survival: Tactics for Armed Encounters. Northbrook, IL: Calibre Press, 1980.
3. Al-Alousi L. Automatic rifle injuries: Suicide by eight bullets. Report of an unusual case and a literature review. Am J Forensic Med Pathol 1990;11:275–281.
4. Aleksander AK. Forensic engineering analysis of TASER™ issues and safety warnings (abstract). Proc Am Acad Forensic Sci 2006; 12:143–144.
5. Allaire MT, Manhein MH, Listi GA. Blasting caps: An alternative source of high velocity trauma in human skeletal remains (abstract). Proc Am Acad Forensic Sci 2005;11:298.
6. Ameen AA. Penetrating craniocerebral injuries: Observations in the Iraqi Iranian War. Milit Med 1987;152:76–79.
7. Amirjamshidi A, Abbassioun K, Roosbeh H. Air-gun pellet injuries to the head and neck. Surg Neurol 1997;47:331–338.
8. Amitani K, Tsuyuguchi Y, Hukuda S. Delayed cervical myelopathy caused by bomb shell fragment. J Neurosurg 1976;44: 626–627.

9. Anderson WF. Forensic Analysis of the April 11, 1986, FBI Firefight. Los Angeles: W. French Anderson, MD (Self-published), 1996.
10. Anglin D, Hutson HR, Luftman J, et al. Intracranial hemorrhage associated with tangential gunshot wounds to the head. Acad Emerg Med 1998;5:672–678.
11. Anil MH, Love S, Helps CR, et al. Jugular venous emboli of brain tissue induced in sheep by the use of captive bolt guns. Vet Rec 2001;148:619–620.
12. Apfelbaum JD, Shockley LW, Wahe JW, Moore EE. Entrance and exit gunshot wounds: Incorrect terms for the emergency department? J Emerg Med 1998;16:741–745.
13. Arasil E, Tascioglu AO. Spontaneous migration of an intracranial bullet to the cervical spinal canal causing Lhermitte's sign. J Neurosurg 1982;56:158–159.
14. Aslan S, Uzkeser M, Katirci Y, et al. Air guns: Toys or weapons? Am J Forensic Med Pathol 2006;27:260–262.
15. Avci SB, Acikgoz B, Gundogdu S. Delayed neurological symptoms from the spontaneous migration of a bullet in the lumbosacral spinal canal: Case report. Paraplegia 1995;33:541–542.
16. Barker-Griffith AE, Streeten BW, Abraham JL, et al. Potato gun ocular injury. Ophthalmology 1998;105:535–538.
17. Beitia AO, Meyers SP, Kanal E, Bartell W. Spontaneous discharge of a firearm in an MR imaging environment. AJR Am J Roentgenol 2002;178:1092–1094.
18. Benzel EC, Day WT, Kesterson L, et al. Civilian craniocerebral gunshot wounds. Neurosurgery 1991;29:67–72.
19. Berryman HE, Smith OC, Symes SA. Diameter of cranial gunshot wounds as a function of bullet caliber. J Forensic Sci 1995;40:751–754.
20. Berryman HE, Symes SA. Recognizing gunshot and blunt cranial trauma through fracture interpretation. In Reichs KJ (ed): Forensic Osteology: Advances in the Identification of Human Remains, ed 2. Springfield, IL: Charles C Thomas, 1998, pp 333–352.
21. Besenski N, Jandro-Santel D, Jelavic-Koic F, et al. CT analysis of missile head injury. Neuroradiology 1995;37:207–211.
22. Betz P, Stiefel D, Eisenmenger W. Cranial fractures and direction of fire in low velocity gunshots. Int J Legal Med 1996;109:58–61.
23. Betz P, Stiefel D, Hausmann R, Eisenmenger W. Fractures at the base of the skull in gunshots to the head. Forensic Sci Int 1997;86:155–161.
24. Bhootra BL, Bhana BD. An unusual missile-type head injury caused by a stone: Case report and medicolegal perspectives. Am J Forensic Med Pathol 2004;25:355–357.
25. Blanco-Pampin JM, Suárez-Peñaranda JM, Rico-Boquete R, Concheiro-Carro L. Planned complex suicide: An unusual suicide by hanging and gunshot. Am J Forensic Med Pathol 1997;18:104–106.
26. Blankenship BA, Baxter AB, McKahn GM II. Delayed cerebral artery pseudoaneurysm after nail gun injury. AJR Am J Roentgenol 1999;172:541–542.
27. Bock H, Neu M, Betz P, Seidl S. Unusual craniocerebral injury caused by a pneumatic nail gun. Int J Legal Med 2002;116:279–281.
28. Bohnert M. Complex suicides. In Tsokos M (ed): Forensic Pathology Reviews. Totowa, NJ: Humana Press, 2005;2:127–143.
29. Brazis PW, Masdeu JC, Biller J. Localization in Clinical Neurology, ed 3. Boston: Little, Brown, 1996.
30. Brogdon BG. Forensic Radiology. Boca Raton, FL: CRC Press, 1998.
31. Brogdon BG, Vogel H, McDowell JD. A Radiologic Atlas of Abuse, Torture, Terrorism, and Inflicted Trauma. Boca Raton, FL: CRC Press, 2003.
32. Buchsbaum HJ, Caruso PA. Gunshot wound of the pregnant uterus: Case report of fetal injury, deglutition of missile, and survival. Obstet Gynecol 1969;33:673–676.
33. Byard RW, Koszyca B, James R. Crossbow suicide: Mechanisms of injury and neuropathologic findings. Am J Forensic Med Pathol 1999;20:347–353.
34. Cahill DR, Orland MJ, Miller GM. Atlas of Human Cross-Sectional Anatomy: With CT and MRI Images, ed 3. New York: Wiley-Liss, 1995.
35. Centers for Disease Control and Prevention (CDC). Lead exposure from indoor firing ranges among students on shooting teams—Alaska, 2002–2004. MMWR Morb Mortal Wkly Rep 2005;54:577–579.
36. Cina SJ, Gelven PL, Nichols CA. A rock in a hard place: A brief case report. Am J Forensic Med Pathol 1995;16:333–335.
37. Cina SJ, Ward ME, Hopkins MA, Nichols CA. Multifactorial analysis of firearm wounds to the head with attention to anatomic location. Am J Forensic Med Pathol 1995;20:109–115.
38. Cingolani M, Tsakri D. Planned complex suicide: Report of three cases. Am J Forensic Med Pathol 2000;21:255–260.
39. Clarot F, Vaz E, Papin F, et al. Lethal head injury due to tear-gas cartridge gunshots. Forensic Sci Int 2003;137:45–51.
40. Coe JI, Austin N. The effects of various intermediate targets on dispersion of shotgun patterns. Am J Forensic Med Pathol 1992;13:281–283.
41. Collins KA, Lantz PE. Interpretation of fatal, multiple, and exiting gunshot wounds by trauma specialists. J Forensic Sci 1994;39:94–99.
42. Cox WM, Pesola GR. Buckshot ingestion. N Eng J Med 2005;353:e23.
43. Craig JB. Cervical spine osteomyelitis with delayed onset tetraparesis after penetrating wounds of the neck: A report of 2 cases. S Afr Med J 1986;69:197–199.
44. Dahlstrom DB, Prowley K, Fackler ML. Drag-stabilized (bulb with tail) 12 gauge shotgun bean bag projectile wound. Wound Ballistics Rev 2001;5(1):8–12.
45. D'Angelica M, Barba CA, Morgan AS, et al. Hypopituitarism secondary to transfacial gunshot wound. J Trauma 1995;39:768–771.
46. Davis GJ. An atypical shotgun entrance wound caused by a sabot slug: All is not as it seems. Am J Forensic Med Pathol 1993;14:162–164.
47. Davis JH. Forensic pathology in firearms cases. Wound Ballistics Rev 1998;3(4):5–15.
48. Davis WC Jr. Revolver barrel length. American Rifleman 2003;151:28.
49. deJongh K, Dohmen D, Salgado R, et al. "William Tell" injury: MDCT of an arrow through the head. AJR Am J Roentgenol 2004;182:1551–1553.
50. DeLetter EA, Piette MHA. An unusual case of suicide by means of a pneumatic hammer. J Forensic Sci 2001;46:962–965.
51. Denton JS, Segovia A, Filkins JA. Practical pathology of gunshot wounds. Arch Pathol Lab Med 2006;130:1283–1289.
52. deRoux S, Prendergast NC, Tamburri R. Wounding characteristics of Glaser safety ammunition: A report of three cases. J Forensic Sci 2001;46:160–164.
53. Devilliers JC. Stab wounds of the brain and skull. In Vinken PJ, Bruyn GW, Braakman R (eds): Handbook of Clinical Neurology. Injuries of the Brain and Skull, Part I. Amsterdam: North-Holland, 1975;23:477–503.
54. DiMaio VJM. Gunshot Wounds: Practical Aspects of Firearms, Ballistics, and Forensic Techniques, ed 2. Boca Raton, FL: CRC Press, 1999.
55. DiMaio VJM. Dana SE: Handbook of Forensic Pathology. Austin, TX: Landes Bioscience, 1998.
56. Dixon DS. Exit keyhole lesion and direction of fire in a gunshot wound of the skull. J Forensic Sci 1984;29:336–339.
57. Dodd MJ. Terminal Ballistics: A Text and Atlas of Gunshot Wounds. Boca Raton, FL: CRC Taylor and Francis, 2006.
58. Dolinak D, Matshes E. Medicolegal Neuropathology: A Color Atlas. Boca Raton, FL: CRC Press, 2002.

59. Dolinak D, Matshes EW, Lew EO. Forensic Pathology: Principles and Practice. Boston: Elsevier Academic Press, 2005.
60. Dosoglu M, Orakdogen M, Somay H, et al. Civilian gunshot wounds to the head. Neurochirurgie 1999;45:201–207.
61. Downs JCU, Nichols CA, Scala-Barnett D, Lifschultz BD. Handling and interpretation of crossbow injuries. J Forensic Sci 1994;39:428–445.
62. Durak D, Fedakar R, Turkmen N. A distant-range, suicidal shotgun wound of the back. J Forensic Sci 2006;51:131–133.
63. Ellis PSJ. Fatal gunshot injury caused by an unusual projectile—A barrel-cleaning brush as a tandem bullet. Am J Forensic Med Pathol 1997;18:168–171.
64. Elron M, Soustiel JF, Guilburd JN, et al. Profuse hemorrhage from cerebral vessels in tangential missile injuries. Acta Neurochir (Wien) 1998;140:255–259.
65. Fackler M. The Lee Clegg case: A study in self-deception. Wound Ballistics Rev 1999;4(3):8–20.
66. Fackler ML. Wound ballistics: A review of common misconceptions. JAMA 1988;259:2730–2736.
67. Fackler ML. Wound ballistics research of the past twenty years: A giant step backwards. Wound Ballistics Rev 1992;1(3):18–24.
68. Fackler ML. Police handgun ammunition selection. Wound Ballistics Rev 1992;1(3):32–37.
69. Fackler ML. Errors published in the *Journal of Trauma*. Wound Ballistics Rev 1995;2(1):40–47.
70. Fackler ML. Gunshot wound review. Ann Emerg Med 1996;28:194–203.
71. Fackler ML. Commentary on: "Simmons GT: Findings in gunshot wounds from tandem projectiles. J Forensic Sci 1997;42:678–681." J Forensic Sci 1997;42:1214.
72. Fackler ML. Civilian gunshot wounds and ballistics: Dispelling the myths. Emerg Med Clin North Am 1998;16:17–28.
73. Fackler ML. What's wrong with the wound ballistics literature, and why. Wound Ballistics Rev 2001;5(1):37–47.
74. Fackler ML, Peters CE. The "shock wave" myth. Wound Ballistics Rev 1991;1(1)38–40.
75. Fackler ML, Welch NE. Monica Dunn's suicide investigation: A study in tunnel vision. Wound Ballistics Rev 1996;2(3):29–36.
76. Fackler ML, Welch NE, Dahlstrom DB, Powley KD. Shotcup petal distortion separates contact from two foot distant shotgun wounds. Wound Ballistics Rev 1996;2(3):37–41.
77. Farrell SE, Vandevander P, Schoffstall JM, Lee DC. Blood lead levels in emergency department patients with retained lead bullets and shrapnel. Acad Emerg Med 1999;6:208–212.
78. Fenton TW, Stefan VH, Wood LA, Sauer NJ. Symmetrical fracturing of the skull from midline contact gunshot wounds: Reconstruction of individual death histories from skeletonized human remains. J Forensic Sci 2005;50:274–285.
79. Finitsis SN, Falcone S, Green BA. MR of the spine in the presence of metallic bullet fragments: Is the benefit worth the risk? AJNR Am J Neuroradiol 1999;20:354–356.
80. Finkbeiner WE, Ursell PC, Davis RL. Autopsy Pathology: A Manual and Atlas. Philadelphia: Churchill Livingston, 2004.
81. Fisher-Fischbein J, Fischbein A, Melnick HD, Bardin CW. Correlation between biochemical indicators of lead exposure and semen quality in a lead-poisoned firearms instructor. JAMA 1987;257:803–805.
82. Fitzgerald LF, Simpson RK, Trask T. Locked-in syndrome resulting from cervical spine gunshot wound. J Trauma 1997;42:147–149.
83. Frazer M. An unusual pattern of gunshot injury linking two homicides to the same assailant. J Forensic Sci 1987;32:262–265.
84. Freehafer AA, Mast WA. Lower extremity fractures in patients with spinal-cord injury. J Bone Joint Surg Am 1965;47:683–694.
85. Freytag E. Autopsy findings in head injuries from firearms: Statistical evaluation of 254 cases. Arch Pathol 1963;76:215–225.
86. Fujimoto Y, Cabrera HT, Pahl FH, et al. Spontaneous migration of a bullet in the cerebellum. Neurol Med Chir (Tokyo) 2001;41:499–501.
87. Gellad FE, Paul KS, Geisler FH. Early sequelae of gunshot wounds to the spine: Radiologic diagnosis. Radiology 1988;167:523–526.
88. George PM, Walmsley TA, Currie D, Wells JE. Lead exposure during recreational use of small bore rifle ranges. N Z Med J 1993;106:422–424.
89. Gerhardsson L, Dahlin L, Knebel R, Schütz A. Blood lead concentration after a shotgun wound. Environ Health Perspect 2002;110:115–117.
90. Giese A, Koops E, Lohmann F, et al. Head injury by gunshots from blank cartridges. Surg Neurol 2002;57:268–277.
91. Gilson T. Suicide involving multiple gunshot wounds. ASCP Check Sample Forensic Pathology No. FP 00-4 (FP-255). Chicago: ASCP Press, 2000;42:39–49.
92. Gönül E, Baysefer A, Erdogan E, et al. Tension pneumocephalus after frontal sinus gunshot wound. Otolaryngol Head Neck Surg 1998;118:559–561.
93. Graham JW, Petty CS, Flohr DM, Peterson WE. Forensic aspects of frangible bullets. J Forensic Sci 1966;11:507–515.
94. Graham MA, Hanzlick R. Forensic Pathology in Criminal Cases. Carlsbad, CA: Lexis Law Publishing, 1997.
95. Grey TC. The incredible bouncing bullet: Projectile exit through the entrance wound. J Forensic Sci 1993;38:1222–1226.
96. Grogan DP, Bucholz RW. Acute lead intoxication from a bullet in an intervertebral disc space. J Bone Joint Surg Am 1981;63:1180–1182.
97. Gross A, Kunz J. Suicidal shooting masked using a method described in Conan Doyle's novel. Am J Forensic Med Pathol 1995;16:164–167.
98. Grossman D, with Christensen LW. On Combat: The Psychology and Physiology of Deadly Conflict in War and in Peace. St Louis: PPCT Research Publications, 2004.
99. Gulmann C, Hougen HP. Entrance, exit, and reentrance of one shot with a shotgun. Am J Forensic Med Pathol 1999;20:13–16.
100. Gupta S, Senger RLS. Wandering intraspinal bullet. Br J Neurosurg 1999;13:606–607.
101. Gustavsson P, Gerhardsson L. Intoxication from an accidentally ingested lead shot retained in the gastrointestinal tract. Environ Health Perspect 2005;113:491–493.
102. Haag LC. Falling bullets: Terminal velocities and penetration studies. Wound Ballistics Rev 1995;2(1):21–26.
103. Hadas N, Schiffer J, Rogev M, Shperber Y. Tangential low-velocity missile wound of the head with acute subdural hematoma: Case report. J Trauma 1990;30:538–539.
104. Haikal NA, Harruff RC. Compressed gas cylinder related injuries: Case report of a fatality associated with a recreational paintball gun: Review of the literature and safety recommendations (abstract). Proc Am Acad Forensic Sci 2004;10:249.
105. Hales DD, Duffy K, Dawson EG, Delamarter R. Lumbar osteomyelitis and epidural and paraspinous abscesses: Case report of an unusual source of contamination from a gunshot wound to the abdomen. Spine 1991;16:380–383.
106. Harruff RC. Comparison of contact shotgun wounds of the head produced by different gauge shotguns. J Forensic Sci 1995;40:801–804.
107. Hart GO. Fracture pattern interpretation in the skull: Differentiating blunt force from ballistics trauma using concentric fractures. J Forensic Sci 2005;50:1276–1281.
108. Hartshorne NJ, Reay DT, Harruff RC. Accidental firearm fatality involving a hand-crafted pen gun: Case report. Am J Forensic Med Pathol 1997;18:92–95.

109. Hatcher JS. Hatcher's Notebook. Harrisburg, PA: Stackpole, 1962.
110. Hess U, Harms J, Schneider A, et al. Assessment of gunshot bullet injuries with the use of magnetic resonance imaging. J Trauma 2000;49:704–709.
111. Hiss J, Kahana T. Confusing exit gunshot wound—"Two for the price of one." Int J Legal Med 2002;116:47–49.
112. Hollerman JJ, Fackler ML, Coldwell DM, Ben-Menachem Y. Gunshot wounds. 1. Bullets, ballistics, and mechanisms of injury. AJR Am J Roentgenol 1990;155:685–690.
113. Hollerman JJ, Fackler ML, Coldwell DM, Ben-Menachem Y. Gunshot wounds. 2. Radiology. AJR Am J Roentgenol 1990;155:691–702.
114. Holley LS, Barnard JJ, Fletcher LA, et al. Special ammunition: Problems for pathologists and medical examiners (abstract). Proc Am Acad Forensic Sci 1999;5:186–187.
115. Hudson P. Multishot firearm suicide: Examination of 58 cases. Am J Forensic Med Pathol 1981;2:239–242.
116. Ilkko E, Reponen J, Ukkola V, Koivukangas J. Spontaneous migration of foreign bodies in the central nervous system. Clin Radiol 1998;53:221–225.
117. Jacob B, Barz J, Haarhoff K, et al. Multiple suicidal gunshots to the head. Am J Forensic Med Pathol 1989;10:289–294.
118. Jackowski C, Thali M, Sonnenschein M, et al. Visualization and quantification of air embolism structure by processing postmortem MSCT data. J Forensic Sci 2004;49:1339–1342.
119. Jason A. The effect of hair upon the deposition of gunshot residue (abstract). Proc Am Acad Forensic Sci 2001;7:225.
120. Jean WC, Barrett MD, Rockswold G, Bergman TA. Gunshot wound to the head resulting in a vertebral artery pseudoaneurysm at the base of the skull. J Trauma 2001;50:126–128.
121. Jentzen JM, Lutz M, Templin R. Tandem bullet versus multiple gunshot wounds. J Forensic Sci 1995;40:893–895.
122. John BE, Boatright D. Lead toxicity from gunshot wound. South Med J 1999;92:223–224.
123. Johnson AC, Kinard WD, Washington WD. Nondestructive recovery and examination of bullet fragments in brain tissue. J Forensic Sci 1980;25:297–301.
124. Jones AM, Reyna M Jr, Sperry K, Hock D. Suicidal contact gunshot wounds to the head with .38 Special Glaser safety slug ammunition. J Forensic Sci 1987;32:1604–1621.
125. Jones EG, Hawley DA, Thompson EJ. Atypical gunshot wound caused by cylinder index error. Am J Forensic Med Pathol 1993;14:226–229.
126. Jones RE, Bucholz RW, Schaefer SD, et al. Cervical osteomyelitis complicating transpharyngeal gunshot wounds to the neck. J Trauma 1979;19:630–634.
127. Kanal E: Response to: "Finitsis SN, Falcone S, Green BA. MR of the spine in the presence of metallic bullet fragments: Is the benefit worth the risk? AJNR Am J Neuroradiol 1999;20:354." AJNR 1999;20:355–356.
128. Kanal E, Shaibani A. Firearm safety in the MR imaging environment. Radiology 1994;193:875–876.
129. Kaplan J, Klose R, Fossum R, DiMaio VJM. Centerfire frangible ammunition: Wounding potential and other forensic concerns. Am J Forensic Med Pathol 1998;19:299–302.
130. Karabatsou K, Kandasami J, Rainov NG. Self-inflicted penetrating head injury in a patient with manic-depressive disorder. Am J Forensic Med Pathol 2005;26:174–176.
131. Karger B. Penetrating gunshots to the head and lack of immediate incapacitation. I. Wound ballistics and mechanisms of incapacitation. Int J Legal Med 1995;108:53–61.
132. Karger B. Penetrating gunshots to the head and lack of immediate incapacitation. II. Review of case reports. Int J Legal Med 1995;108:117–126.
133. Karger B, Banaschak S. Two cases of exenteration of the brain from Brenneke shotgun slugs. Int J Legal Med 1997;110:323–325.
134. Karger B, Billeb E, Koops E, Brinkmann B. Autopsy features relevant for discrimination between suicidal and homicidal gunshot injuries. Int J Legal Med 2002;116:273–278.
135. Karger B, Heindel W, Fechner G, Brinkmann B. Proof of a gunshot wound and its delayed effects 54 years post injury. Int J Legal Med 2001;115:173–175.
136. Karger B, Meyer E, DuChesne A. STR analysis on perforating FMJ bullets and a new VWA variant allele. Int J Legal Med 1997;110:101–103.
137. Karger B, Meyer E, Knudsen PJT, Brinkmann B. DNA typing of cellular material on perforating bullets. Int J Legal Med 1996;108:177–179.
138. Karger B, Nüsse R, Bajanowski T. Backspatter on the firearm and hand in experimental close-range gunshots to the head. Am J Forensic Med Pathol 2002;23:211–213.
139. Karger B, Puskas Z, Ruwald B, et al. Morphological findings in the brain after experimental gunshots using radiology, pathology and histology. Int J Legal Med 1998;111:314–319.
140. Karger B, Rand SP. Multiple entrance wounds from one bullet due to the use of a silencer. Am J Forensic Med Pathol 1998;19:30–33.
141. Karger B, Stehmann B, Hohoff C, Brinkmann B. Trajectory reconstruction from trace evidence on spent bullets. II. Are tissue deposits eliminated by subsequent impacts? Int J Legal Med 2001;114:343–345.
142. Karger B, Teige K. Fatalities from black powder percussion handguns. Forensic Sci Int 1998;98:143–149.
143. Karim NO, Nabors MW, Golocovsky M, Cooney FD. Spontaneous migration of a bullet in the spinal subarachnoid space causing delayed radicular symptoms. Neurosurgery 1986;18:97–100.
144. Kase CS, White RL, Vinson TL, Eichelberger RP. Shotgun pellet embolus to the middle cerebral artery. Neurology 1981;31:458–461.
145. Kerin DS, Fox R, Mehringer CM, et al. Spontaneous migration of a bullet in the central nervous system. Surg Neurol 1983;20:301–304.
146. Kerkhoff W, Lusthof KJ. A case of ballistic toxicology. Bull Int Assoc Forensic Toxicol 1999;29(1):4–5.
147. Kessler SC, Evans RJ, Mires AM. Demystifying the missing bullet—A pathologist's nightmare: A case report of a 17-year-old buried homicide skull with a radiopaque metal stain in the shape of a bullet (abstract). Proc Am Acad Forensic Sci 2001;7:226–227.
148. Kikano GE, Stange KC. Lead poisoning in a child after a gunshot injury. J Fam Pract 1992;34:498–504.
149. Kim PJ, Franklin WH. Ventricular fibrillation after stun-gun discharge (letter to editor). N Engl J Med 2005;353:958–959.
150. Kirkpatrick JB. Gunshots and other penetrating wounds of the central nervous system. In Leestma JE: Forensic Neuropathology. New York: Raven Press, 1988, pp 276–299.
151. Klinger D, Hubbs K. Citizen injuries from law enforcement impact munitions: Evidence from the field. Wound Ballistics Rev 2000;4(4):9–13.
152. Kocak A, Ozer MH. Intracranial migrating bullet. Am J Forensic Med Pathol 2004;25:246–250.
153. Kohlmeier RE, McMahan CA, DiMaio VJM. Suicide by firearms: A 15-year experience. Am J Forensic Med Pathol 2001;22:337–340.
154. Koszyca B, Blumbergs PC, Manavis J, et al. Widespread axonal injury in gunshot wounds to the head using amyloid precursor protein as a marker. J Neurotrauma 1998;15:675–683.
155. Kozic Z, Zingesser LH. Traumatic cerebrospinal fluid-lymphatic fistula: Case report. J Neurosurg 1978;49:607–609.
156. Krassioukov AV, Bunge RP, Pucket WR, Bygrave MA. The changes in human spinal sympathetic preganglionic neurons after spinal cord injury. Spinal Cord 1999;37:6–13.
157. Kroll MW, Sweeney J, Swerdlow CD. Theoretical considerations regarding the cardiac safety of law enforcement electronic control devices (abstract). Proc Am Acad Forensic Sci 2006;12:139–140.

158. Krukemeyer MG, Grellner W, Gehrke G, et al. Survived crossbow injuries. Am J Forensic Med Pathol 2006;27:274–276.
159. Kuijlen JM, Herpers MJ, Beuls EA. Neurogenic claudication, a delayed complication of a retained bullet. Spine 1997;22:910–914.
160. Kury G, Weiner J, Duval JV. Multiple self-inflicted gunshot wounds to the head: Report of a case and review of the literature. Am J Forensic Med Pathol 2000;21:32–35.
161. Kuzeyli K, Cakir E, Baykal S, Karaarslan G. Diabetes insipidus secondary to penetrating spinal cord trauma: Case report and literature review. Spine 2001;26: E510–E511.
162. Lantz PE. An atypical, indeterminate-range, cranial gunshot wound of entrance resembling an exit wound. Am J Forensic Med Pathol 1994;15:5–9.
163. Larsen AR, Blanton RH. Appendicitis due to bird shot ingestion: A case study. Am Surg 2000;66:589–591.
164. Learch TJ, Andrews CL, Dowel SJ, et al. Lead arthropathy following bullet injuries (abstract). Proc Am Acad Forensic Sci 2000;6:197.
165. Ledgerwood AM. The wandering bullet. Surg Clin North Am 1977;57:97–109.
166. Lein B, Holcomb J, Brill S, et al. Removal of unexploded ordinance from patients: A 50-year military experience and current recommendations. Mil Med 1999;164:163–165.
167. Levy V, Rao VJ. Survival time in gunshot and stab wound victims. Am J Forensic Med Pathol 1988;9:215–217.
168. Lew E, Dolinak D, Matshes E. Firearm injuries. In Dolinak D, Matshes EW, Lew EO (eds): Forensic Pathology: Principles and Practice. Amsterdam: Elsevier Academic Press, 2005, pp 162–200.
169. Lewinski B. Why is the suspect shot in the back? Police Marksman 2000;XXV(6):20–28.
170. Lewinski B. Biomechanics of lethal force encounters—Officer movements. Police Marksman 2002;XXVII(6):19–23.
171. Lewinski B, Hudson B. Time to start shooting? Time to stop shooting? The Tempe Study. Police Marksman 2003;XXVIII(5):216–229.
172. Lewinski B, Hudson B. The impact of visual complexity, decision making and anticipation. The Tempe Study. Experiments 3 & 5. Police Marksman 2003;XXVIII(6):24–27.
173. Lieske K, Janssen W, Kulle K-J. Intensive gunshot residues at the exit wound: An examination using a head model. Int J Legal Med 1991;104:235–238.
174. Linden MA, Manton WI, Stewart RM, et al. Lead poisoning from retained bullets: Pathogenesis, diagnosis, and management. Ann Surg 1982;195:305–313.
175. Litvack ZN, Hunt MA, Weinstein JS, West GA. Self-inflicted nail-gun injury with 12 cranial penetrations and associated cerebral trauma: Case report and review of the literature. J Neurosurg 2006;104:828–834.
176. Lucas WJ, Cairns JT. Lethality of TASERs—The Canadian experience (abstract). Proc Am Acad Forensic Sci 2006;12:141.
177. Luchini D, DiPaolo M, Morabito G, Gabbrielli M. Case report of a homicide by a shotgun loaded with unusual ammunition. Am J Forensic Med Pathol 2003;24:198–201.
178. Ludwig J. Handbook of Autopsy Practice, ed 3. Totowa, NJ: Humana Press, 2002.
179. Lyons JD, Filston HC. Lead intoxication from a pellet entrapped in the appendix of a child: Treatment considerations. J Pediatr Surg 1994;29:1618–1620.
180. MacPherson D. Bullet Penetration: Modeling the Dynamics and the Incapacitation Resulting from Wound Trauma. El Segundo, CA: Ballistic Publications, 1994.
181. MacPherson D. Wound ballistics misconceptions. Wound Ballistics Rev 1996;2(3):42–43.
182. MacPherson D. Shock wave. Wound Ballistics Rev 1999;4(1):9–12.
183. MacPherson D. The temporary wound cavity. Wound Ballistics Rev 1999;4(2):22–25.
184. MacPherson D. Comments on impact munitions. Wound Ballistics Rev 2000;4(4):14–15.
185. Malandrini A, Villanova M, Salvadori C, et al. Neuropathological findings associated with retained lead shot pellets in a man surviving two months after a suicide attempt. J Forensic Sci 2001;46:717–721.
186. Mallak C, Chute D, Smialek J. A penny (or peso) for your thoughts: An unusual intermediate target. Am J Forensic Med Pathol 1998;19:230–233.
187. Mariottini A, Delfini R, Ciappetta P, Paolella G. Lumbar disc hernia secondary to gunshot injury. Neurosurgery 1994;15:73–75.
188. Marty W, Sigrist T, Wyler D. Measurement of the skin temperature at the entry wound by means of infrared thermography: An investigation involving the use of .22- to .38-caliber handguns. Am J Forensic Med Pathol 1994;15:1–4.
189. Mavroudis C: Physicians and the Navy Cross: A treatise on courage. Surgery 1991;110:896–902.
190. McNutt TK, Chambers-Emerson J, Dethlefsen M. Bite the bullet: Lead poisoning after ingestion of 206 lead bullets. Vet Hum Toxicol 2001;43:288–289.
191. Menzies RC, Scroggie RJ, Labowitz DI. Characteristics of silenced firearms and their wounding effects. J Forensic Sci 1981;26:239–262.
192. Messer HD, Cerza PF. Copper jacketed bullets in the central nervous system. Neuroradiology 1976;12:121–129.
193. Messmer JM. Radiology of gunshot wounds. In Brogdon BG (ed): Forensic Radiology. Boca Raton, FL: CRC Press, 1998, pp 225–248.
194. Messmer JM, Brogdon BG. Pitfalls in the radiology of gunshot wounds. In Brogdon BG, Vogel H, McDowell JD (eds): A Radiologic Atlas of Abuse, Torture, Terrorism, and Inflicted Trauma. Boca Raton, FL: CRC Press, 2003, pp 179–194.
195. Messmer JM, Brogdon BG, Vogel H. Conventional weapons, including shotguns. In Brogdon BG, Vogel H, McDowell JD (eds): A Radiologic Atlas of Abuse, Torture, Terrorism, and Inflicted Trauma. Boca Raton, FL: CRC Press, 2003, pp 161–178.
196. Metcalf D. Today's hunting handgun. Guns & Ammo 2005;49:54–63.
197. Metress EK, Metress SP. The anatomy of plastic bullet damage and crowd control. Int J Health Serv 1987;17:333–342.
198. Michelassi F, Pietrabissa A, Ferrari M, et al. Bullet emboli to the systemic and venous circulation. Surgery 1990;107:239–245.
199. Milhorat TH, Elowitz EH, Johnson RW, Miller JI. Spontaneous movement of bullets in the brain. Neurosurgery 1993;32:140–143.
200. Miller FP III, Barnard JJ, Ross KF, et al. Wherever two or more are gathered: Suicidal gunshot wounds of the head involving tandem bullets with one entrance wound (abstract). Proc Am Acad Forensic Sci 1998;4:151–152.
201. Milroy CM, Clark JC, Carter N, et al. Air weapon fatalities. J Clin Pathol 1998;51:525–529.
202. Miner ME, Cabrera JA, Ford E, et al. Intracranial penetration due to BB air rifle injuries. Neurosurgery 1986;19:952–954.
203. Mirovsky Y, Shalmon E, Blankstein A, Halperin N. Complete paraplegia following gunshot injury without direct trauma to the cord. Spine 2005;30:2436–2438.
204. Miyaishi S, Moriya F, Yamamoto Y, Ishizu H. Massive pulmonary embolizations with cerebral tissue due to gunshot wound to the head. Brain Inj 1994;8:559–564.
205. Morrow JS, Haycock CE, Lazaro E. The "swallowed bullet" syndrome. J Trauma 1978;18:464–466.
206. Nader MA, Halabi S, Rachid B, et al. Neglected craniocerebral gunshot wound resulting in an encephalocele: Case report. Surg Neurol 2000;54:397–400.
207. Nagib MG, Rockswold GL, Sherman RS, Lagaard MW. Civilian gunshot wounds to the brain: Prognosis and management. Neurosurgery 1986;18:533–537.

208. Nehme AF. Intracranial bullet migrating to pulmonary artery. J Trauma 1980;20:344–346.
209. Newgard K. The physiological effects of handgun bullets: The mechanisms of wounding and incapacitation. Wound Ballistics Rev 1992;1(3):12–17.
210. Nichols CA, Sens MA. Cytologic manifestations of ballistic injury. Am J Clin Pathol 1991;95:660–669.
211. Nowicki EJ, Ramsey DA. Street Weapons. Powers Lake, WI: Performance Dimensions Publishing, 1991.
212. Oehmichen M, Meissner C, König HG: Brain injury after gunshot wounding: Morphometric analysis of cell destruction caused by temporary cavitation. J Neurotrauma 2000;17:155–162.
213. Oehmichen M, Meissner C, König HG: Brain injury after survived gunshot to the head: Reactive alterations at sites remote from the missile track. Forensic Sci Int 2001; 115:189–197.
214. Oliver WR, Chancellor AS, Soltys M, et al. Three-dimensional reconstruction of a bullet path: Validation by computed radiography. J Forensic Sci 1995;40:321–324.
215. Opeskin K, Burke M. Suicide using multiple crossbow arrows. Am J Forensic Med Pathol 1994;15:14–17.
216. Padar SC. Air gun pellet embolizing the intracranial internal carotid artery. J Neurosurg 1975;43:222–224.
217. Padosch SA, Schmidt PH, Madea B. Planned complex suicide by self-poisoning and a manipulated blank revolver: Remarkable findings due to multiple gunshot wounds and self-made wooden projectiles. J Forensic Sci 2003;48:1371–1378.
218. Parroni E, Caringi C, Ciallella C. Suicide with two guns represents a special type of combined suicide. Am J Forensic Med Pathol 2002;23:329–333.
219. Peterson BL. External beveling of cranial gunshot entrance wounds. J Forensic Sci 1991;36:1592–1595.
220. Pollanen MS, Blenkinsop B, Farkas EM. Fracture of temporal bone with exsanguination: Pathology and mechanism. Can J Neurol Sci 1992;19:196–200.
221. Prahlow JA, Allen SB, Spinder T, Poole RA. Pseudo-gunpowder stippling caused by fragmentation of a plated bullet. Am J Forensic Med Pathol 2003;24:243–247.
222. Quatrehomme G, Iscan MY. Analysis of beveling in gunshot entrance wounds. Forensic Sci Int 1998;93:45–60.
223. Quatrehomme G, Iscan MY. Gunshot wounds to the skull: Comparison of entries and exits. Forensic Sci Int 1998;94:141–146.
224. Rabl W, Riepert T, Steinlechner M. Metal pins fired from unmodified blank cartridge guns and very small calibre weapons—Technical and wound ballistic aspects. Int J Legal Med 1998;111:219–223.
225. Rajan DK, Alcantara AL, Michael DB. Where's the bullet? A migration in two acts. J Trauma 1997;43:716–718.
226. Randall B, Newby P. Comparison of gunshot wounds and field-tipped arrow wounds using morphologic criteria and chemical spot tests. J Forensic Sci 1989;34:579–586.
227. Rapp LG, Arce CA, McKenzie R, et al. Incidence of intracranial bullet fragment migration. Neurol Res 1999;21:475–480.
228. Raymond LW, Ford MD, Porter WG, et al. Maternal-fetal lead poisoning from a 15-year-old bullet. J Maternal-Fetal Neonatal Med 2002;11:63–66.
229. Rezai AR, Lee M, Kite C, et al. Traumatic posterior cerebral artery aneurysm secondary to an intracranial nail: Case report. Surg Neurol 1994;42:312–315.
230. Rhine JS, Curran BK. Multiple gunshot wounds of the head: An anthropological view. J Forensic Sci 1990;35:1236–1245.
231. Ribe JK, Brockbank SA, Edwards RL, Andrews JM. Bullet removal from victim's head by pillow used as silencer (abstract). Proceedings of the International Association of Forensic Science, 15th Triennial Meeting, August 22–28, 1999, p 183.
232. Rinaldi A, Gazzeri R, Conti L, et al. Cranio-orbital missile wound and bullet migration: Case report. J Neurosurg Sci 2000;44:107–112.
233. Robertson DP, Simpson RK, Narayan RK. Lumbar disc herniation from a gunshot wound to the spine: A report of two cases. Spine 1991;8:994–995.
234. Robinson NA, Flotte CT. Traumatic aneurysms of the carotid arteries. Am Surg 1974;40:121–124.
235. Roche WJ, Nwofia C, Gittler M, et al. Catecholamine-induced hypertension in lumbosacral paraplegia: Five case reports. Arch Phys Med Rehabil 2000;81:222–225.
236. Rogers DR. Simultaneous temporal and frontal suicidal gunshots. Am J Forensic Med Pathol 1989;10:338–339.
237. Rompen JC, Meek MF, vanAndel MV. A cause célèbre: The so-called "ballpoint murder." J Forensic Sci 2000;45:1144–1147.
238. Roth P. Spontaneously expectorated penetrating foreign body. Ann Emerg Med 1986;15:381–382.
239. Rothschild MA, Maxeiner H, Schneider V. Cases of death caused by gas or warning firearms. Med Law 1994;13:511–518.
240. Rothschild MA, Schneider V. Gunshot wound to the head with full recovery. Int J Legal Med 2000;113:349–351.
241. Russell MA, Atkinson RD, Klatt EC, Noguchi TT. Safety in bullet recovery procedures: A study of the Black Talon bullet. Am J Forensic Med Pathol 1995;16:120–123.
242. Sarwal V, Suri RK, Sharma OP, et al. Traumatic subarachnoid-pleural fistula. Ann Thoracic Surg 1996;62:1622–1626.
243. Saukko P, Knight B. Knight's Forensic Pathology, ed 3. London: Arnold, 2004.
244. Scheil AT, Collins KA. TASER-related fatalities: Case report and review of the literature (abstract). Proc Am Acad Forensic Sci 2006;12:238–239.
245. Schep LJ, Fountain JS. Lead shot in the appendix (letter to editor). N Engl J Med 2006;354:1757.
246. Schlosberg M. Stun gun fallacy: How the lack of TASER regulation endangers lives (abstract). Proc Am Acad Forensic Sci 2006;12:140–141.
247. Schyma C, Bittner M, Placidi P. The MEN frangible: Study of a new bullet in gelatin. Am J Forensic Med Pathol 1997;18:325–330.
248. Sekula-Perlman A, Tobin JG, Pretzler E, et al. Three unusual cases of multiple suicidal gunshot wounds to the head. Am J Forensic Med Pathol 1998;19:23–29.
249. Selbst SM, Henretig F, Fee MA, et al. Lead poisoning in a child with a gunshot wound. Pediatrics 1986;77:413–416.
250. Shakir A, Koehler SA, Wecht CH. A review of nail gun suicides and an atypical case report. J Forensic Sci 2003;48:409–413.
251. Sheaff MT, Hopster DJ. Post Mortem Technique Handbook. London: Springer, 2001.
252. Sherry WE. Personal communication, 2005.
253. Sherry WE, Bockhacker LE, Riley SL Jr, Sathyavagiswaran L. Use of bean bag ammunition—Injuries observed (abstract). Proc Am Acad Forensic Sci 2000;6:206.
254. Shields LBE, Hunsaker DM, Hunsaker JC III, Rolf CM. Multiple self-inflicted suicidal gunshot wounds of the head: A matter of timing and placement—simultaneous or sequential? ASCP Check Sample Forensic Pathology No. FP 03-2 (FP-283). Chicago, ASCP Press, 2003;45:17–34.
255. Sights WP, Bye RJ. The fate of retained intracerebral shotgun pellets: An experimental study. J Neurosurg 1970;33:646–653.
256. Simmons GT. Findings in gunshot wounds from tandem projectiles. J Forensic Sci 1997;42:678–681.
257. Simpson RK Jr, Venger BH, Fischer DK, et al. Shotgun injuries of the spine: Neurosurgical management of five cases. Br J Neurosurg 1988;2:321–326.
258. Skinker DM, Coyne CM, Lanham C, Hunsaker JC III. Chasing the casing: A .38 Special suicide. J Forensic Sci 1996;41:709–712.
259. Smith OC, Berryman HE, Lahren CH. Cranial fracture patterns and estimate of direction from low velocity gunshot wounds. J Forensic Sci 1987;32:1416–1421.

260. Smith OC, Berryman HE, Symes SA, et al. Atypical gunshot exit defects to the cranial vault. J Forensic Sci 1993;38:339–343.
261. Smith OC, Jantz L, Berryman HE, Symes SA. Effects of human decomposition on bullet striations. J Forensic Sci 1993;38:593–598.
262. Smith R. TASER® non-lethal weapons: Safety data and field results (abstract). Proc Am Acad Forensic Sci 2006;137–139.
263. Soges LJ, Kinnebrew GH, Limcaco OG. Mobile intrathecal bullet causing delayed radicular symptoms. AJNR Am J Neuroradiol 1988;9:610. (Also see erratum: AJNR Am J Neuroradiol 1988;9:890.)
264. Spencer JD. Accidental death by light anti-tank weapon: A dangerous autopsy? J Forensic Sci 1979;24:479–482.
265. Spitz DJ, Ouban A. Meningitis following gunshot wound of the neck. J Forensic Sci 2003;48:1369–1370.
266. Spitz WU (ed). Spitz and Fisher's Medicolegal Investigation of Death: Guidelines for the Application of Pathology to Crime Investigation, ed 3. Springfield, IL: Charles C. Thomas, 1993.
267. Spitz WU, Petty CS, Fisher RS. Physical activity until collapse following fatal injury by firearms and sharp pointed weapons. J Forensic Sci 1961;6:290–300.
268. Spitz WU, Spitz DJ (eds). Spitz and Fisher's Medicolegal Investigation of Death: Guidelines for the Application of Pathology to Crime Investigation, ed 4. Sprinfield, IL: Charles C Thomas, 2006.
269. Staniforth P, Watt I. Extradural "plumboma": A rare cause of acquired spinal stenosis. Br J Radiol 1982;55:772–774.
270. Steele JA, McBride SJ, Kelly J, et al. Plastic bullet injuries in Northern Ireland: Experiences during a week of civil disturbance. J Trauma 1999;46:711–714.
271. Stone IC. Characteristics of firearms and gunshot wounds as markers of suicide. Am J Forensic Med Pathol 1992;13:275–280.
272. Stone JL, Lichtor T, Fitzgerald LF, Gandhi YN. Civilian cases of tangenital gunshot wounds to the head. J Trauma 1996;40:57–60.
273. Straathof D, Bannach BG, Wilson AJ, Dowling GP. Radiography of perforating centerfire rifle wounds of the trunk. J Forensic Sci 2000;45:597–601.
274. Stromberg BV. Symptomatic lead toxicity secondary to retained shotgun pellets: Case report. J Trauma 1990;30:356–357.
275. Sweeney JD, Kroll MW, Panescu D. Analysis of electrical activation of nerve and muscle by TASERs (abstract). Proc Am Acad Forensic Sci 2006;12:142–143.
276. Tanguy A, Chabannes J, Deubelle A, et al. Intraspinal migration of a bullet with subsequent meningitis. J Bone Joint Surg Am 1982;64:1244–1245.
277. Teitelbaum GP, Yee CA, Van Horn DD, et al. Metallic ballistic fragments: MR imaging safety and artifacts. Radiology 1990;175:855–859.
278. Tekavcic I, Smrkolj VA. The path of a wounding missile along the spinal canal. Spine 1996;21:639–641.
279. Thogmartin JR, Start DA. Nine-millimeter ammunition used in a 40 caliber Glock pistol: An atypical gunshot wound. J Forensic Sci 1998;43:712–714.
280. Timperman J, Cnops L. Tandem bullet in the head in a case of suicide. Med Sci Law 1975;15:280–283.
281. Tindel NL, Marcillo AE, Tay BK-B, et al. The effect of surgically implanted bullet fragments on the spinal cord in a rabbit model. J Bone Joint Surg Am 2001;83:884–890.
282. Tobin EJ, Fackler ML. Officer reaction-response times in firing a handgun. Wound Ballistics Rev 1997;3(1):6–9.
283. Tobin EJ, Fackler ML. Officer decision time in firing a handgun. Wound Ballistics Rev 2001;5(2):8–12.
284. Traeger M, Wood BP. Radiologic cases of the month: The migrating BB and the medicine man. Am J Dis Child 1993;147:901–902.
285. Treble RG, Thompson TS. Elevated blood lead levels resulting from the ingestion of air rifle pellets. J Anal Toxicol 2002;26:370–373.
286. Tudor M. Prediction of outcome in patients with missile craniocerebral injuries during the Croatian war. Mil Med 1998;163:486–489.
287. Turner MS, Jumbelic ML. Stun gun injuries in the abuse and death of a seven-month-old infant. J Forensic Sci 2003;48:180–182.
288. Van Arsdell GS, Razzouk AJ, Fandrich BL, et al. Bullet fragment venous embolus to the heart: Case report. J Trauma 1991;31:137–139.
289. Ventura F, Blasi C, Celesti R. Suicide with the latest type of slaughterer's gun. Am J Forensic Med Pathol 2002;23:326–328.
290. Viola L, Costantinides F, DiNunno C, et al. Suicide with a butcher's bolt. J Forensic Sci 2004;49:595–597.
291. Vogel H, Brogdon BG. Air guns. In Brogdon BG, Vogel H, McDowell JD (eds): A Radiologic Atlas of Abuse, Torture, Terrorism, and Inflicted Trauma. Boca Raton, FL: CRC Press, 2003, pp 195–198.
292. Vogel H, Brogdon BG. Unconventional loads and weapons. In Brogdon BG, Vogel H, McDowell JD (eds): A Radiologic Atlas of Abuse, Torture, Terrorism, and Inflicted Trauma. Boca Raton, FL: CRC Press, 2003, pp 199–209.
293. Waters RL, Sie IH. Spinal cord injuries from gunshot wounds to the spine. Clin Orthop Relat Res 2003;408:120–125.
294. Webster's Third New International Dictionary of the English Language, Unabridged. Chicago: Encyclopaedia Britannica, Inc, 1986.
295. Wecht CH, Lee HC, Baden MM. The role of the forensic scientist in the investigation of police-related deaths—A current dilemma (abstract). Proc Am Acad Forensic Sci 2006;12:19.
296. Weeg J. What you need to tell the prosecutor in your next use-of-force case. Police Marksman 2002;XXVII(3):45–47.
297. Weinrach DM, Stickel AE, Diaz LK. Soft tissue cyst secondary to bullet retention. Arch Pathol Lab Med 2001;125:1391.
298. Wetli CV, Mittleman RE, Rao VJ. An Atlas of Forensic Pathology. Chicago: ASCP Press, 1999.
299. White SA, Narula AA. A complication of indoor pistol shooting. J Laryngol Otol 1996;110:663–664.
300. Wilson AJ. Gunshot injuries: What does a radiologist need to know? Radiographics 1999;19:1358–1368.
301. Wright A, LeMelle G. USA: Excessive and lethal force? Amnesty International's concerns about deaths and ill-treatment involving police use of TASERs (abstract). Proc Am Acad Forensic Sci 2006;12:141–142.
302. Wu WQ. Delayed effects from retained foreign bodies in the spine and spinal cord. Surg Neurol 1986;25:214–218.
303. Yaari R, Ahmadi J, Chang GY. Cerebral shotgun pellet embolism. Neurology 2000;54:1487.
304. Yetiser S, Kahramanyol M. High-velocity gunshot wounds to the head and neck: A review of wound ballistics. Mil Med 1998;163:346–351.
305. Yip L, Sweeny PJ, McCarroll KA. Spontaneous migration of an intraspinal bullet following a gunshot wound. Am J Emerg Med 1990;8:569–570.
306. Young WF Jr, Katz MR, Rosenwasser RH. Spontaneous migration of an intracranial bullet into the cervical canal. South Med J 1993;86:557–559.
307. Zafonte RD, Watanabe T, Mann NR. Moving bullet syndrome: A complication of penetrating head injury. Arch Phys Med Rehabil 1998;79:1469–1472.
308. Zietlow C, Hawley DA. Unexpectedly homicide: Three intraoral gunshot wounds. Am J Forensic Med Pathol 1993;14:230–233.
309. Zubkov AY, Pilkington AS, Bernanke DH, et al. Posttraumatic cerebral vasospasm: Clinical and morphological presentations. J Neurotrauma 1999;16:763–770.

Sudden Unexpected Death

INTRODUCTION 255
ACUTE SUBARACHNOID HEMORRHAGE 256
EPILEPSY 262
CYSTICERCOSIS 265
ACUTE INTRACEREBRAL HEMORRHAGE 268
VASCULAR MALFORMATION 271
MENINGITIS 274
HYDROCEPHALUS 275
TUMORS 277
MULTIPLE SCLEROSIS 281
REYE'S SYNDROME 281
REFLEXES AND SUDDEN UNEXPECTED DEATH 281
REFERENCES 284

Introduction

The heart and brain share the notoriety of organs giving rise to sudden unexpected natural (nontraumatic) death, with the heart being responsible for the larger share. In the definition of sudden unexpected death (SUD), no established general agreement exists as to the limits of survival time from onset of symptoms to death. In the forensic literature, the term "sudden" has been used to indicate a time interval from minutes to 24 hours. Sudden may be relatively sudden or near-sudden, in which the victim may reach the hospital emergency room and expire there, sometimes after a few hours. *Webster's Third New International Dictionary*[121] defines sudden death as "unexpected death that is instantaneous or occurs within minutes from any cause other than violence." In our practice, "sudden" is defined as occurring within 24 hours.

"Unexpected" (i.e., unforeseen or surprising, as defined in Webster's)[121] is also a relative and perhaps redundant term in that it is implied in the word *sudden*. Death may be unexpected in that the individual may have been considered to be in good health or may have had only minor symptoms not considered life-threatening even by a treating physician. Therefore, a sudden death may be exactly that or in some cases it might be better characterized as a relatively sudden, relatively unexpected death. Most of the cases in this chapter fall within this latter category. The circumstances under which a sudden, unexpected death might occur are usually: (1) the decedent has a witnessed collapse followed rapidly by death within 24 hours; (2) the decedent has an unwitnessed death after being seen alive within the previous 24 hours; (3) the decedent is found down with signs of life but dies quickly in the field or the emergency room; or (4) following sudden collapse of the decedent, the time interval to presence of signs of brain death is a matter of a few hours, but the decedent is maintained on life support. Abnormalities in a variety of organ systems other than the central nervous system (CNS) may result in SUD, and they are reviewed elsewhere.[15,16,23]

The neuropathologic processes in most cases falling into the aforementioned categories of SUD usually lend themselves to accurate determination of cause of death by gross examination alone, although the exact mechanism of death is not always clear. Vital to the determination of the mode in cases of SUD is the accurate recognition of several pathologic processes. Excluding drug overdose, intentional or otherwise, nondrug intoxications, and so-called reflex SUD, nervous system causes of SUD tend to group into limited categories of anatomically identifiable diseases. These most commonly include epilepsy, nontraumatic subarachnoid hemorrhage, spontaneous acute intracerebral hemorrhage (especially those

Sudden Unexpected Death

in the brainstem and cerebellum), intraventricular cysts, brain tumors, acute purulent meningitis, and hydrocephalus. Rare miscellaneous causes account for the remaining cases.[10,23] Other than epilepsy, the pathologic process is clearly identifiable on gross examination, and the obligatory microscopic examination is usually only confirmatory. Microscopy provides details that, with rare exceptions (such as occult neoplasia within a parenchymal hemorrhage), do not materially alter the diagnosis made by gross examination. That is not to say that microscopic examination can be dispensed with, because the unexpected does occur.

Acute Subarachnoid Hemorrhage

A ruptured berry (saccular) aneurysm is by far the major cause of potentially lethal, acute nontraumatic subarachnoid hemorrhage. Its incidence has been estimated to be 0.87% in a large autopsy series.[88] Only about 14% of those with subarachnoid hemorrhage died within 1 day in the Cooperative Study of Intracranial Aneurysms and Subarachnoid Hemorrhage[91,92] and fall within the purview of SUD. Despite statistics suggesting rarity of ruptured berry aneurysm causing SUD, several dozen cases come to the attention of our medical examiner's office annually. The large majority of berry aneurysms occur near the anterior portion of the circle of Willis, such as the bifurcations of the anterior cerebral and anterior communicating arteries (Case 9.1, Figs. 9.1–9.3), internal carotid and anterior cerebral arteries (Case 9.2, Figs. 9.4–9.5), and the horizontal portion of the middle cerebral artery (Case 9.3, Fig. 9.6). They are much less common at other sites, such as the apex of the basilar artery and on the posterior cerebral or vertebral arteries (Case 9.4, Figs. 9.7–9.8). The critical size of symptomatic aneurysms in most locations was 7 to 10 mm in diameter in the Cooperative Study. No symptomatic aneurysm was less than 3 mm in diameter. However, many ruptured aneurysms found in our SUD cases were less than 7 mm in diameter, with the anterior communicating artery location being the most prevalent. A recent study suggests that hypertension, relatively young age, and posterior circulation location are risk factors for rupture of small diameter (i.e., 7 mm or less) intracranial aneurysms.[78] The rupture site is usually at the apex of the aneurysmal

Case 9.1. (Figs. 9.1–9.3)

A 44-year-old man complained of "not feeling well" on the day prior to death. He was found unresponsive in bed in the morning, approximately 10 hours after having last been seen alive.

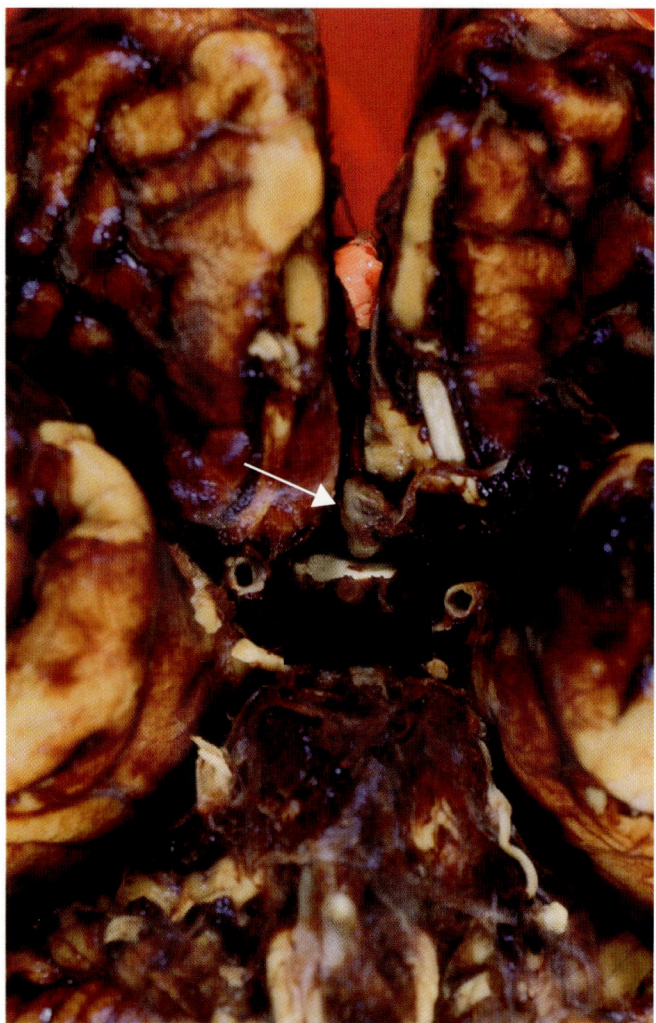

Fig. 9.1. Symmetric massive acute subarachnoid hemorrhage, primarily at the base of brain. A 0.9 × 0.8 × 0.4-cm multilobulated berry aneurysm arises from the anterior communicating artery (*arrow*). The hemorrhage dissected into left lateral ventricle through basal medial cortex, distending the entire ventricular system and forming a "blood cast."

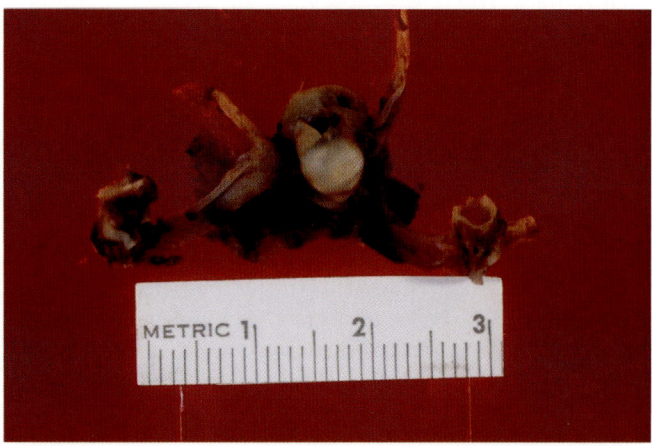

Fig. 9.2. Closeup view of aneurysm.

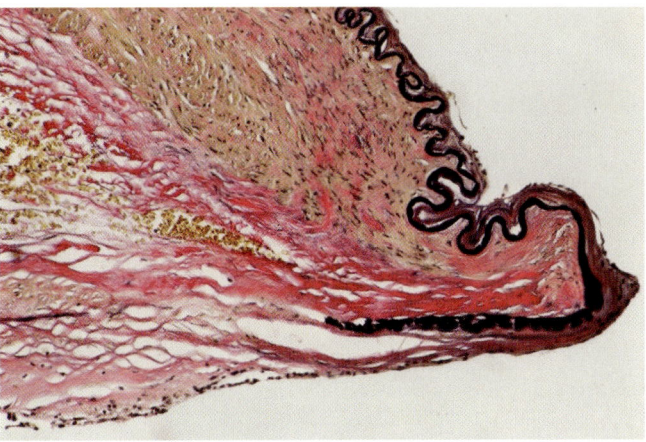

Fig. 9.3. Medium-power micrograph at aneurysm origin shows severe degeneration of elastic lamina with frayed disappearance at the neck of aneurysm as it folds back over the adventitia of the parent vessel. (Elastic van Gieson.)

Case 9.2. (Figs. 9.4 and 9.5)

A 54-year-old woman was found unresponsive on the bathroom floor and pronounced dead at the scene. There was no known medical illness, but she was overweight.

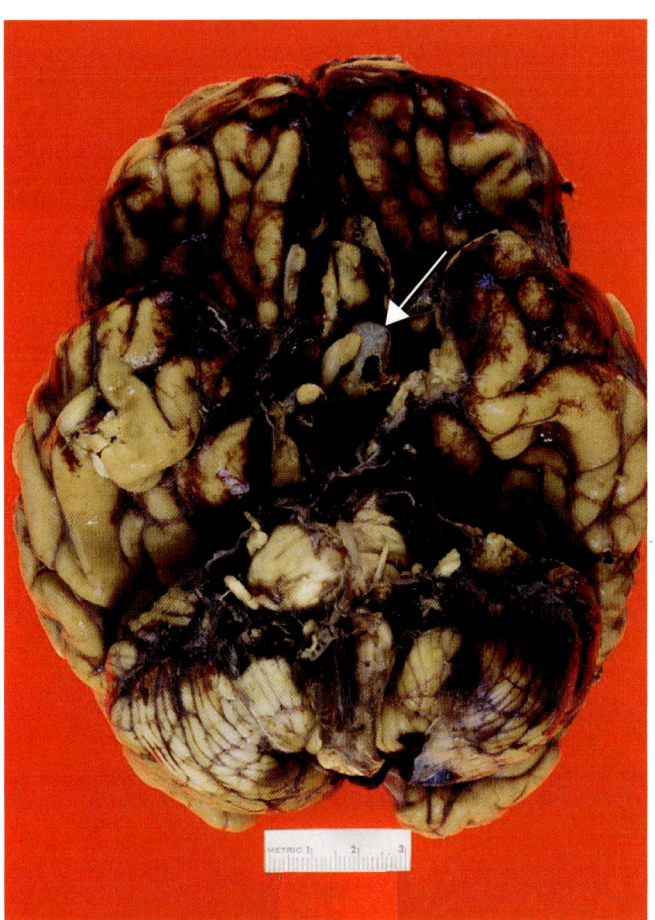

Fig. 9.4. Widely distributed but only moderate acute subarachnoid hemorrhage at the base of brain. Aneurysm dome is visible (*arrow*).

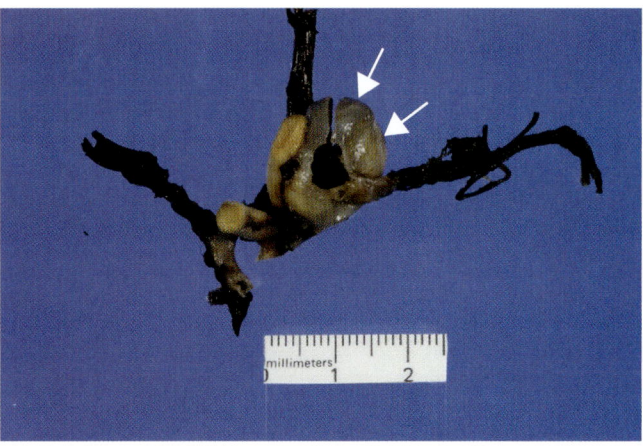

Fig. 9.5. Closeup view of the large saccular aneurysm measuring 1.8 × 1.5 × 1.5 cm is shown arising from the proximal left internal carotid artery at the bifurcation of middle and anterior cerebral arteries (*arrows*). The aneurysm uplifted and flattened adjacent left optic nerve. A large rupture defect is present at the dome apex.

Case. 9.3. (Fig. 9.6)

A 43-year-old man was found down by his wife upon returning home. He was supine and covered with a blanket without signs of life; he was pronounced dead at the scene.

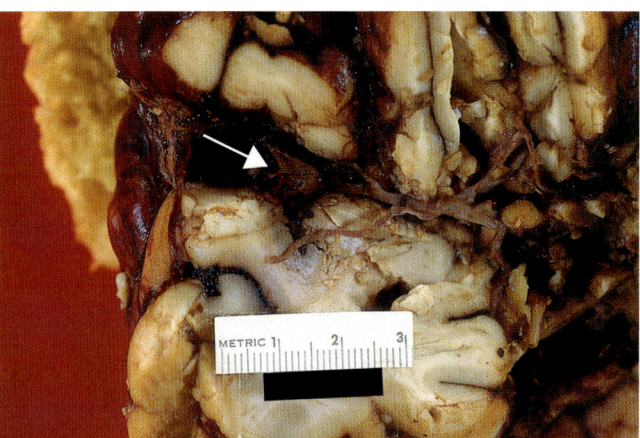

Fig. 9.6. A saccular aneurysm measuring 0.8 × 0.5 cm is shown at the distalmost portion of the horizontal right middle cerebral artery (*arrow*). A large rupture was found at the aneurysmal apex. A large amount of acute subarachnoid hemorrhage was found at the base of brain and extending into parasylvian regions, greater on the right.

258 Sudden Unexpected Death

> **Case 9.4.** (Figs. 9.7 and 9.8)
>
> A 28-year-old woman who was abusing cocaine began having "seizures." She was restrained by her boyfriend for 5 hours before 911 was called when she became unresponsive. Paramedics found the decedent comatose, hypotensive, and hypothermic. She subsequently fulfilled criteria for brain death and was pronounced dead about 3 days later.

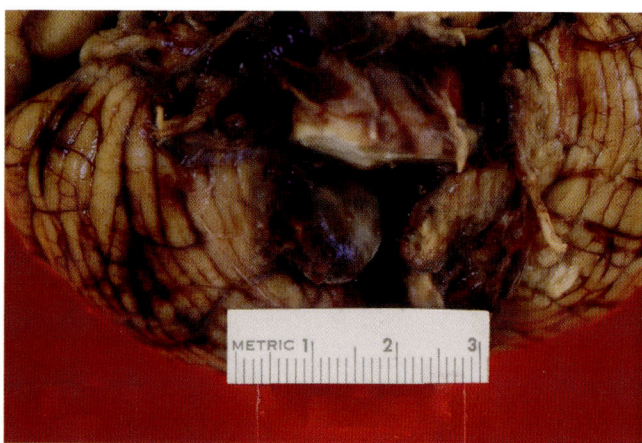

Fig. 9.7. Acute subarachnoid hemorrhage around the brainstem and a 1.4 × 1.2-cm berry aneurysm of right anterior inferior cerebellar artery.

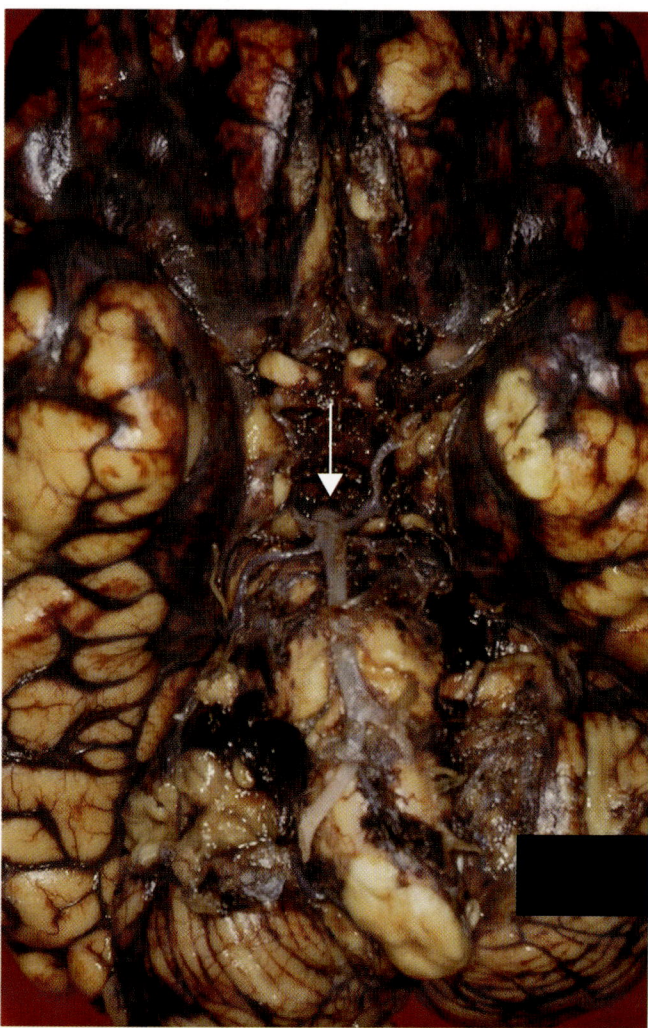

Fig. 9.9. Acute subarachnoid hemorrhage at the midline base of brain as a result of rupture of a small, barely visible berry aneurysm at the apex of the basilar artery (*arrow*).

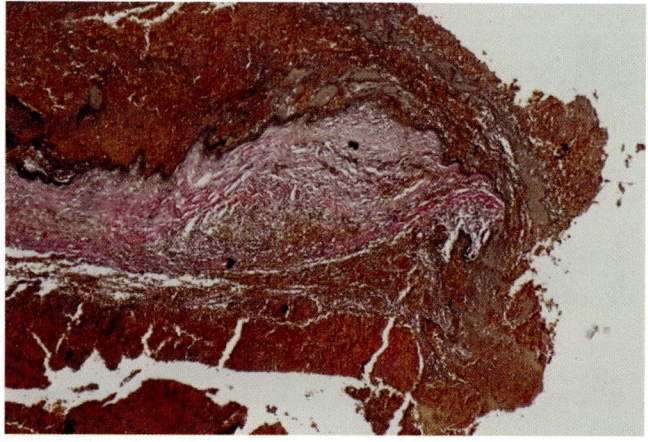

Fig. 9.8. Low-power photomicrograph shows thinning and degeneration of the aneurysmal wall at rupture site. (Elastic van Gieson.)

dome (66%) but may occur at the body (12%) or the neck (3%) of the aneurysm.[91,92] The resultant acute subarachnoid hemorrhage is usually massive (Fig. 9.9; Case 9.5, Fig. 9.10), sufficient to obscure the cortical surfaces, fill the basal cisterns, and migrate superiorly along the sylvian valleculae and onto the cerebral convexity to form a parasylvian "Mercury wing" pattern. When the rupture is at the anterior communicating artery site, the hemorrhage is symmetric, but when the rupture is heavily to one side, it gives a clue to aneurysmal rupture on the ipsilateral middle cerebral artery. Reflux of subarachnoid hemorrhage into the fourth ventricle and retrograde is usually minimal, but occasionally it is enough to partially fill the third and lateral ventricles.

Brain herniation is not usually a feature in cases of SUD following ruptured aneurysm unless an intracerebral hemorrhage occurs secondary to an arterial jet from the ruptured aneurysm dissecting through the cortex to

> **Case 9.5.** (Fig. 9.10)
>
> A 51-year-old woman slipped without falling or striking her head, and, for the next 3 days, she had head and neck pain before seeking medical attention. While in the emergency room, she lost consciousness, and a computed tomography scan revealed basal subarachnoid hemorrhage and hydrocephalus. She died 2 days later.

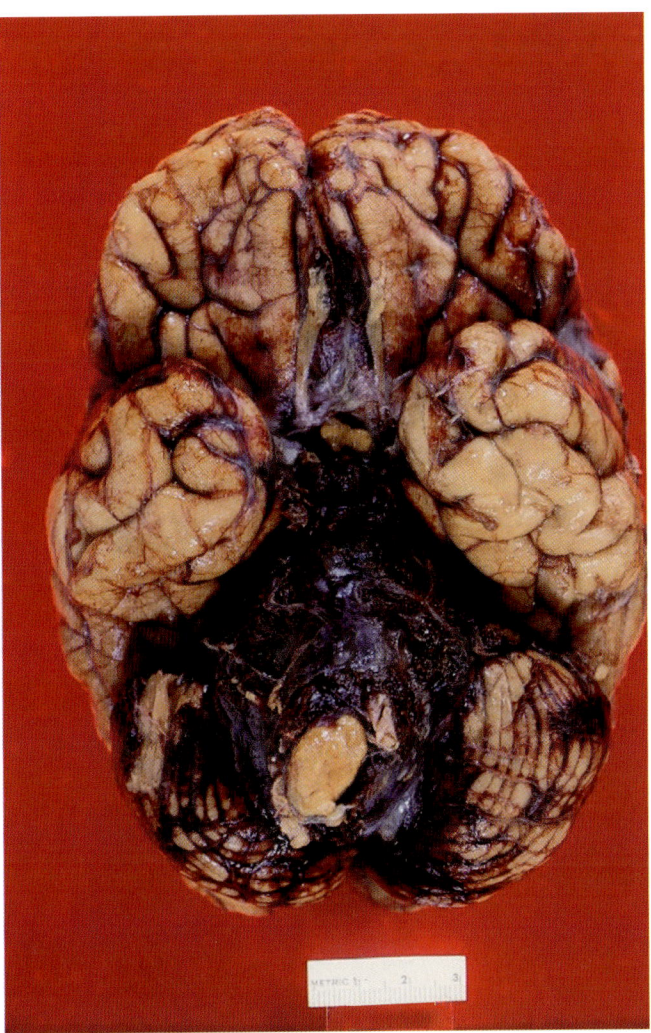

Fig. 9.10. Massive acute subarachnoid hemorrhage at midline base of brain. A ruptured berry aneurysm of the left vertebral artery was found, with a linear rupture site crossing the dome. Reflux hemorrhage into the fourth and third ventricles was present.

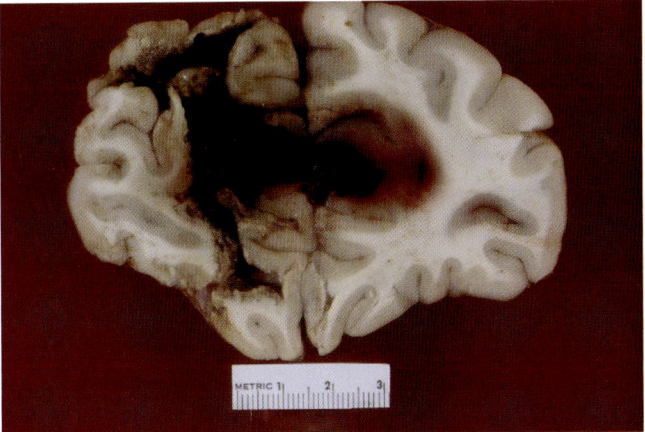

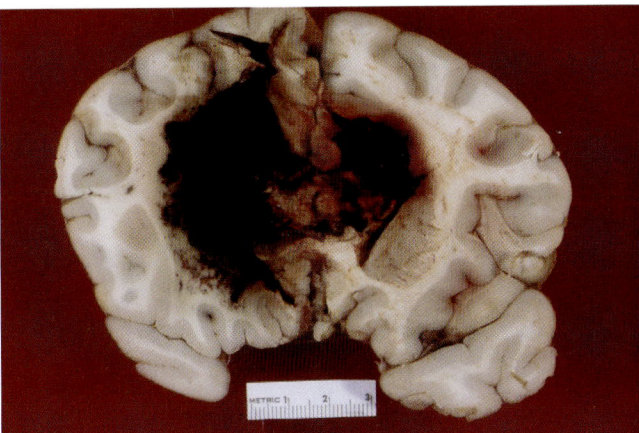

Figs. 9.11–9.12. *Top* (Fig. 9.11) and *bottom* (Fig. 9.12), bilateral acute cerebral hemorrhage at two anterior frontal levels originated from interhemispheric fissure where a traumatic aneurysm was located on pericallosal branch of anterior cerebral artery.

form a hematoma in the underlying basal ganglia or frontal white matter (Figs. 9.11–9.14). In the latter situation, herniation can be expected in the longer survivors as a result of mass effect of hematoma and secondary brain swelling.

Berry aneurysms may be multiple, but the offending one is usually identified by the location of the predominant subarachnoid hemorrhage and definitively by identification of the rupture site. Search for suspected aneurysms should be conducted at the autopsy table in the unfixed state. This is the best opportunity to inspect individual blood vessels at risk of the circle of Willis and its branches. The search for aneurysms is made exceedingly tedious and difficult after formalin fixation and hardening of subarachnoid hemorrhage. In an estimated 10% of cases, even the most careful search fails to reveal a source of hemorrhage.[118] The aneurysm may be too small to be detected by the naked eye or collapsed and obscured by the hemorrhage. It may be destroyed during the course of rupture. In addition, subarachnoid hemorrhage may have a nonaneurysmal source.

The previous discussion of ruptured berry aneurysm pertains to adults, but exceedingly rare cases occur in childhood as a cause of SUD. Many of the children presenting with a ruptured cerebral aneurysm had a prior history of headaches, and the many reports of aneurysm in infancy and childhood focus attention on neurosurgical treatment and not sudden death.[124]

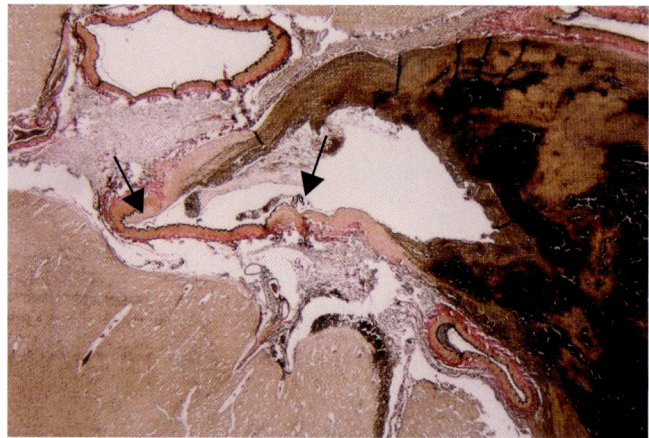

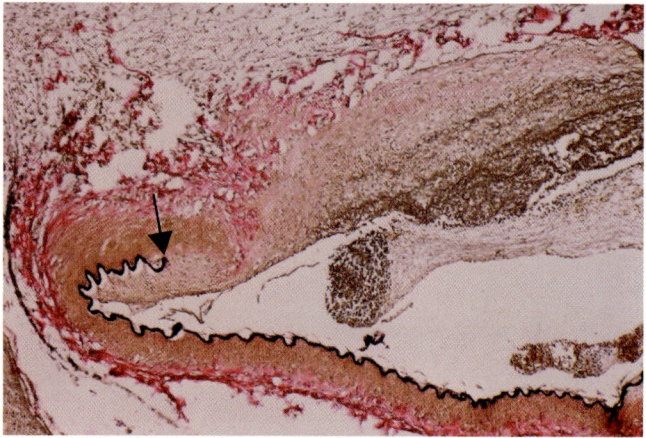

Fig. 9.13–9.14. Photomicrographs at low (Fig. 9.13, *top*) and medium (Fig. 9.14, *bottom*) power demonstrate the two ends of sharp presumed traumatic disruptions of elastic lamina (*arrows*). Recent hematoma extends to right from rent (elastic Van Gieson). From same case seen in Figs. 9.11–9.12.

In a forensic setting, an important differential diagnosis of acute subarachnoid hemorrhage, especially in the posterior fossa, should include an unsuspected traumatic mechanism such as that reported by Contostavlos[20] and Mant.[65] They described a mechanism by which a blow behind the ear resulted in a fracture of the transverse process of the atlas leading to a vertebral artery tear, resulting in a hemorrhagic dissection into the posterior fossa and consequent rupture into the subarachnoid space. The blow itself may be relatively minor and the cutaneous mark inconspicuous.

Craniocerebral trauma may produce a traumatic aneurysm,[42,45] which usually ruptures within several days to a few weeks. Rarely the delay between trauma and aneurysm (or pseudoaneurysm) rupture is much longer, with reported intervals of 14 months to 10 years prior to acute subarachnoid hemorrhage producing SUD.[106] The proximal portion of the pericallosal artery close to the inferior margin of the falx cerebri[98] and the internal carotid artery as it penetrates the parasellar dura[45] are likely sites for development of traumatic aneurysms (Figs. 9.11–9.14). Traumatic aneurysms also, it should be noted, may occur in scalp, meningeal, and immediately infracranial vessels, in vessels as they traverse bony canals within the skull, and in several other intracranial vessels in addition to those mentioned earlier, resulting in the hemorrhage not only in the subarachnoid space but also in other tissue compartments.[44]

Other uncommon natural causes of acute subarachnoid hemorrhage that may present as SUD include rupture of a mycotic aneurysm, superficially located cortical or dural angiomas, and atherosclerotic aneurysm. Giant "berry" aneurysms, reported as large as 7.8 by 7.0 by 6.2 cm,[109] arising at the base of the brain are more likely to present with mass effect rather than rupture. These large aneurysms tend to have lumens largely filled by laminated thrombus, and are consequently less likely to hemorrhage. Atherosclerotic aneurysms, typically situated on the basilar (Fig. 9.15; Case 9.6, Figs. 9.16–9.18)

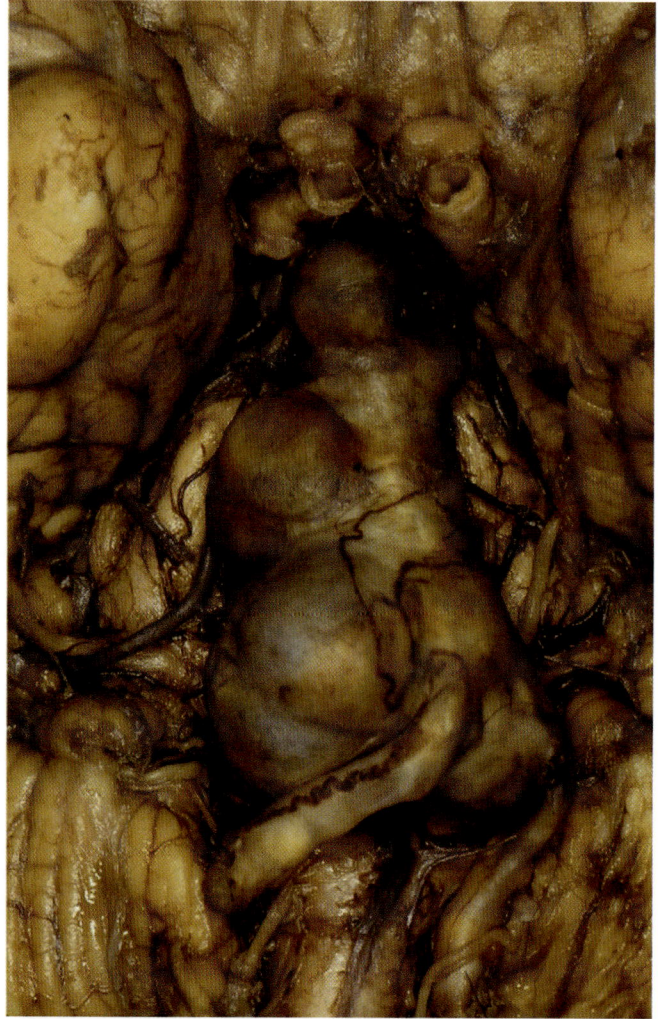

Fig. 9.15. Giant atherosclerotic aneurysm of basilar artery (unruptured).

Case 9.6. (Figs. 9.16–9.18)

A 37-year-old man with a history of hypertension and renal failure was found down apneic. He arrived at the hospital in cardiopulmonary arrest, and he could not be resuscitated.

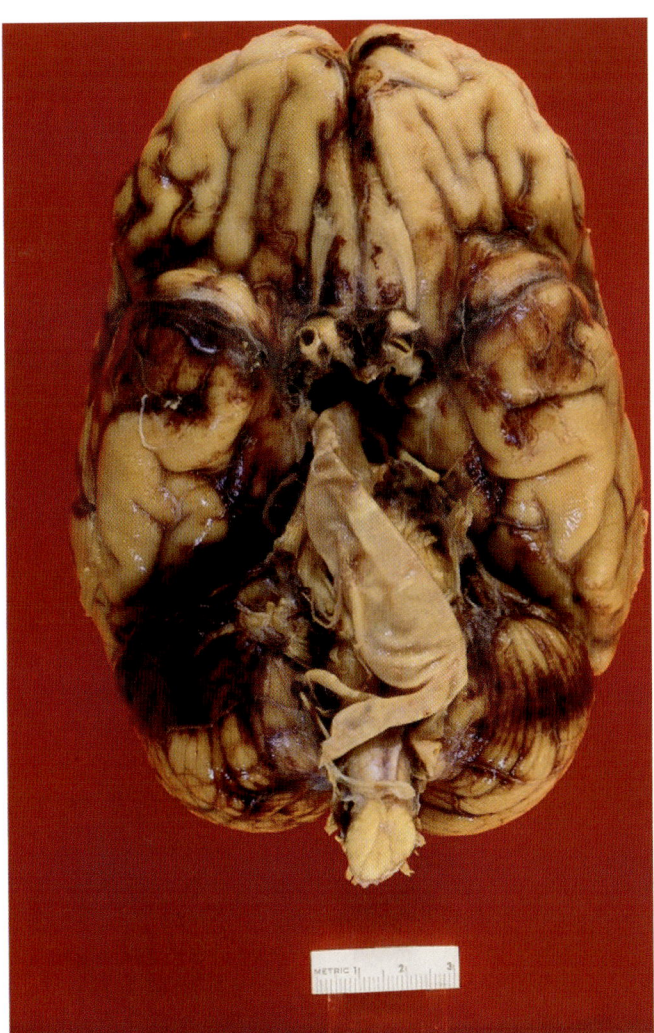

Fig. 9.16. A hugely ectatic atherosclerotic fusiform aneurysm measuring 7.0 cm in length and 2.2 cm in diameter replaced the entire basilar artery and extended into one vertebral artery.

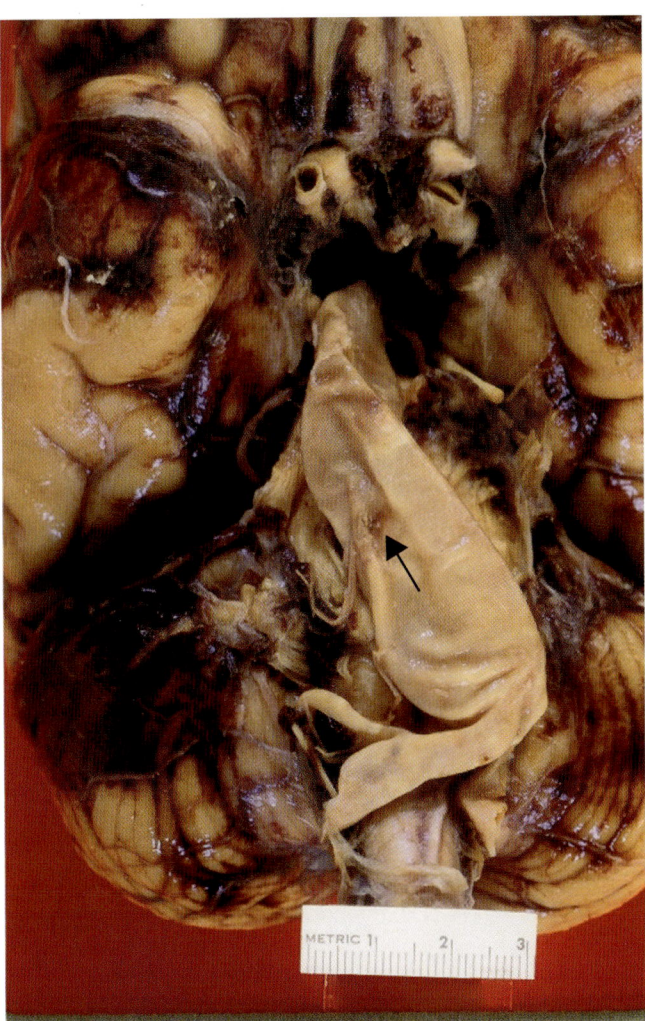

Fig. 9.17. A pinpoint rupture site was found at midlevel (*arrow*).

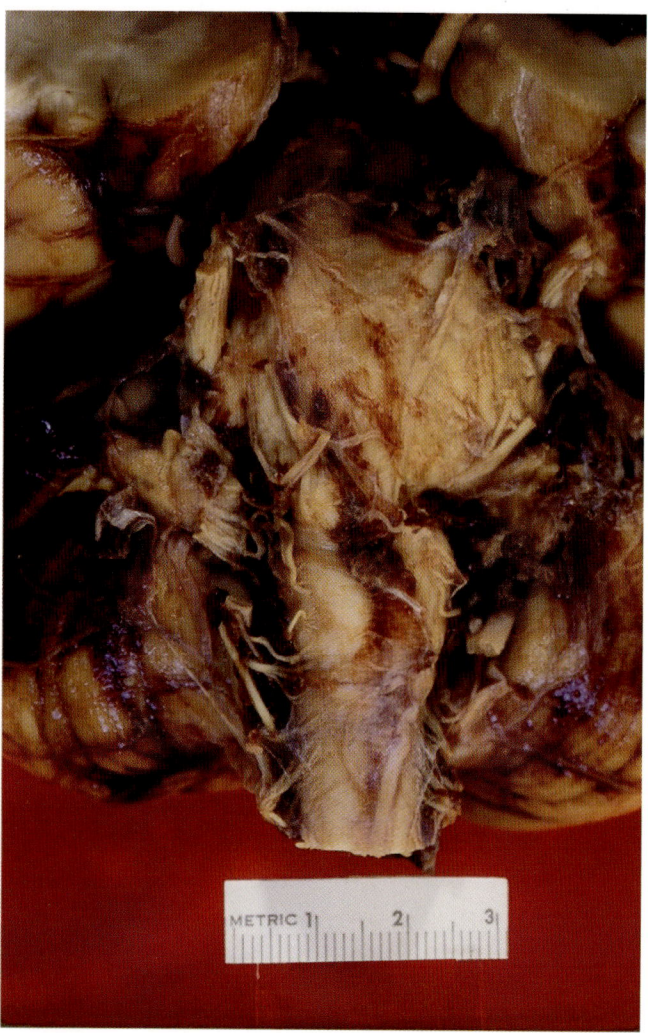

Fig. 9.18. Removal of the aneurysm discloses irregular impressions on the pons and medulla caused by the bulky aneurysm, apparently without neurologic deficit.

or internal carotid arteries, are infrequent sources of acute subarachnoid hemorrhage. Very rarely, cerebral arteries rupture at the site of a nonaneurysmal atheroma. One should be alerted to the presence of a mycotic aneurysm with the finding of acute subarachnoid hemorrhage localized to the parasylvian cerebral hemisphere, characteristically but not exclusively affecting the candelabra of arteries arising from the middle cerebral artery, in the context of a history of intravenous drug abuse. The demonstration of cardiac valvular vegetations in the case strongly supports the diagnosis of subarachnoid hemorrhage secondary to a ruptured mycotic aneurysm.

Epilepsy

Prior to the last few decades, SUD in epilepsy was not generally recognized as a potential outcome by the general practitioner or even the specialist. The reader is referred to concise reviews of this history given by Leestma and collegues.[46,58,59] Sudden unexpected death in an epileptic patient (SUDEP) is a regularly recurring event in our case material and in that of others.[3] Most often the decedent is found without signs of life in bed in the morning after retiring in his or her usual state of health. Scene investigation is usually devoid of indications of a major motor seizure having occurred. Bedding is not in disarray, and evidence of tongue biting[12] and urinary incontinence are usually absent. It would appear that death came quietly and peacefully. Occasionally, the victim is found on the floor next to the bed. Postmortem drug analyses frequently,[56] but not invariably,[97] indicate subtherapeutic levels of anticonvulsants, probably from noncompliance and not due to a metabolic complication. In this circumstance, and without evidence of positional asphyxia or other possible causes, the mechanism of death is largely a matter of exclusion. Available data favor the probable mechanism being an autonomic nervous system seizure leading to asystole and/or respiratory arrest.[21,42,44,48,50,60,61,79] In other circumstances, epileptics are subject to accidental death such as by drowning in a bathtub during a seizure, losing control of a motor vehicle leading to a lethal collision, and falling from a great height with a seizure. In one extraordinary case, a young epileptic boy with a history of an automatism of "running fits" was on a group hike when he bolted, ran into the deep end of a lake, and drowned.

Neuropathologic findings in such cases are usually devoid of an anatomic cause of a secondary (or symptomatic) seizure disorder, such as chronic cortical contusion, tumor, cerebrovascular disorders, arachnoid cyst (Case 9.7, Fig. 9.19), cerebral malformation, cysticercosis, vascular malformation, and vasculitides such as systemic lupus erythematosus, to name a few. All of these conditions have been seen in our case material. Other than the common and nonspecific focal findings of hippocampal sclerosis (Case 9.8, Figs. 9.20–9.23), cerebellar cortical atrophy (Figs. 9.22–9.23), and less constant findings in the cerebral cortex and amygdala/periamygdaloid cortex,[66] the brain is usually otherwise normal.

Hippocampal sclerosis may be unilateral (Case 9.8, Figs. 9.20–9.23) or bilateral (Case 9.9, Figs. 9.24–9.25), with the former being more common.[93] Aside from noticeable shrinkage of the hippocampal formation,

> **Case 9.7.** (Fig. 9.19)
>
> A 29-year-old man had a history of a major motor seizure disorder. However, he was apparently without medical attention. He was last seen alive the previous evening at 1900 hours, and he was found lifeless on his mattress in a closed bedroom at 0830 hours. The cause of the seizures was attributed to an arachnoid cyst over the pyriform cortex.

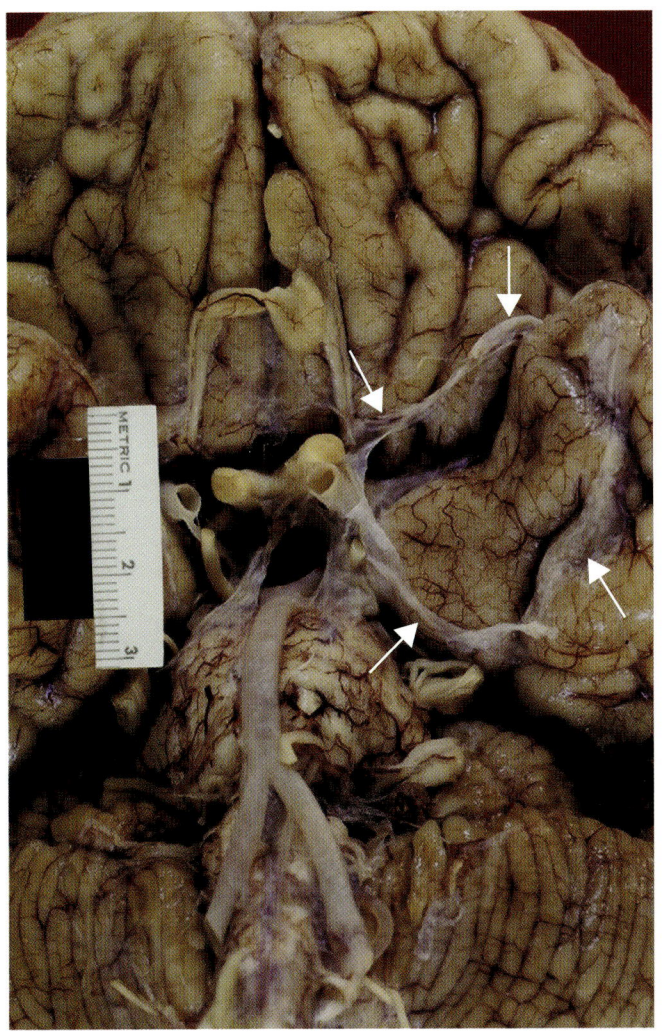

Fig. 9.19. An arachnoid cyst overlies the left pyriform area with displacement of the adjacent temporal pole. Margins of the now collapsed cyst are indicated by arrows.

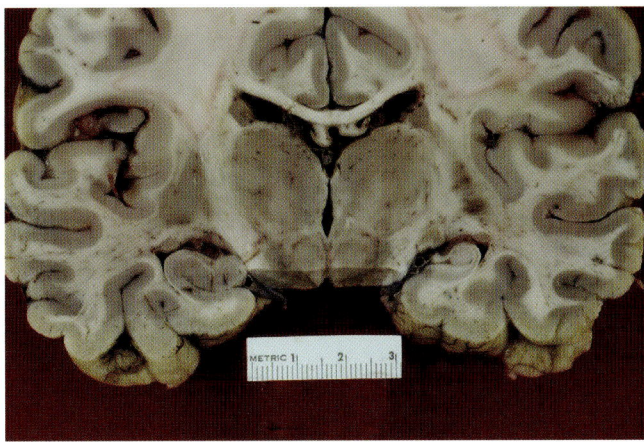

Fig. 9.20. Unilateral hippocampal atrophy on left.

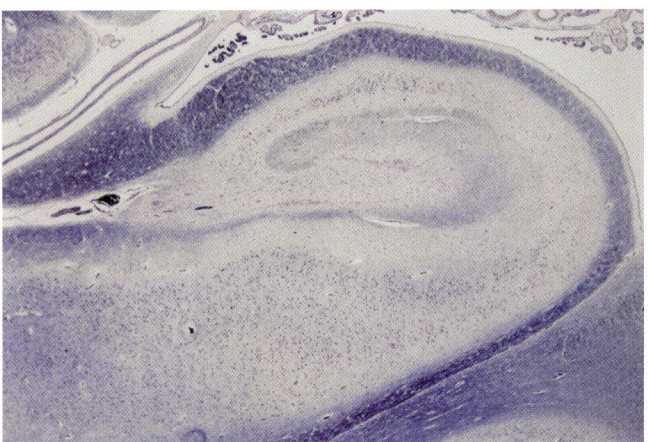

Fig. 9.21. Typical selective cell loss of hippocampal sclerosis is shown by microscopy with near-total loss of neurons in CA_1 and CA_4 with a few surviving in CA_2. Dentate fascia is rarefied. (Luxol fast blue/cresyl violet.)

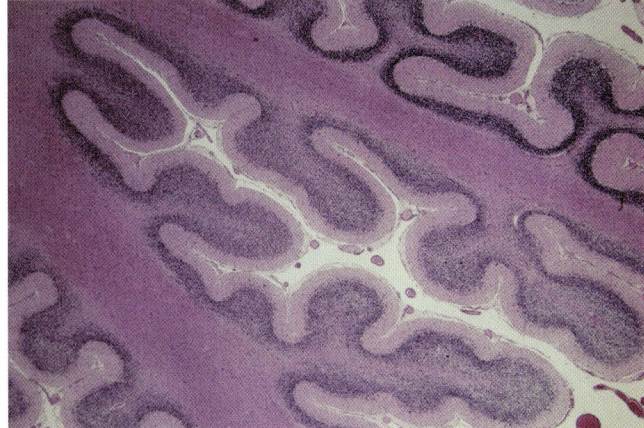

Fig. 9.22. Cerebellum shows widespread but not generalized cortical atrophy, appreciated at this low-power magnification by the thinning of the granule cell layer. (H&E.)

careful scrutiny with a magnifying lens usually demonstrates thinning of the gray band of the hippocampus proper beneath the alveus (Fig. 9.20). Microscopically, the well-described selective vulnerability of the neurons of the hippocampus proper is typically seen in the form of their depopulation or complete loss, with accompanying gliosis, in Sommer's sector (CA1 of Lorente de No,

Case 9.8. (Figs. 9.20–9.23)

A 44-year-old nurse's aide had not returned to duty after her break. She was discovered down on the floor of a locked hospital bathroom and pronounced dead at the scene. She had a history of major motor seizures. Postmortem levels of phenytoin and phenobarbital were in the therapeutic range.

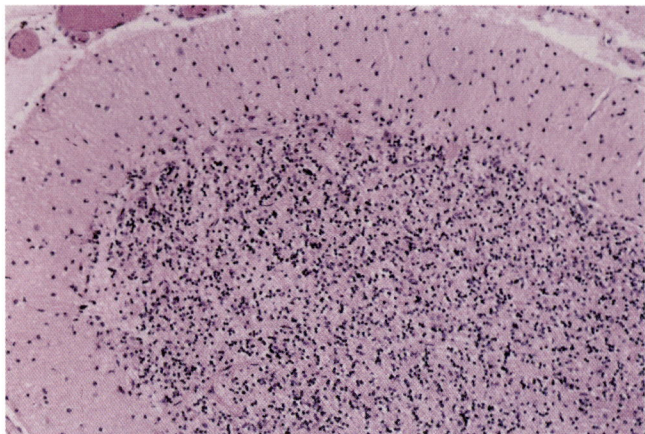

Fig. 9.23. Medium-power micrograph shows Purkinje cell loss and Bergmann's gliosis, the latter seen as radiating glial fibrils in molecular layer, along with atrophy of granule cell layer. (H&E.)

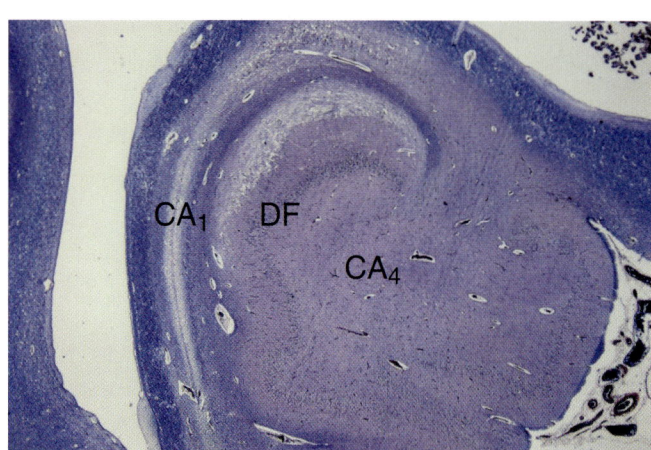

Fig. 9.25. Low-power micrograph confirmed severe bilateral hippocampal sclerosis with total loss of neurons of Sommer's sector (CA_1) and endfolium (CA_4) and severe loss of dentate fascia (DF). (Phosphotungstic acid hematoxylion.)

Case 9.9. (Figs. 9.24 and 9.25)

A 26-year-old woman had a history of frequently recurring seizures since the age of 8 years, but she refused to take prescribed anticonvulsant medications. She was found face down in a bathtub with the water running, and she was pronounced dead at the scene. Her last witnessed seizure was the day prior to death.

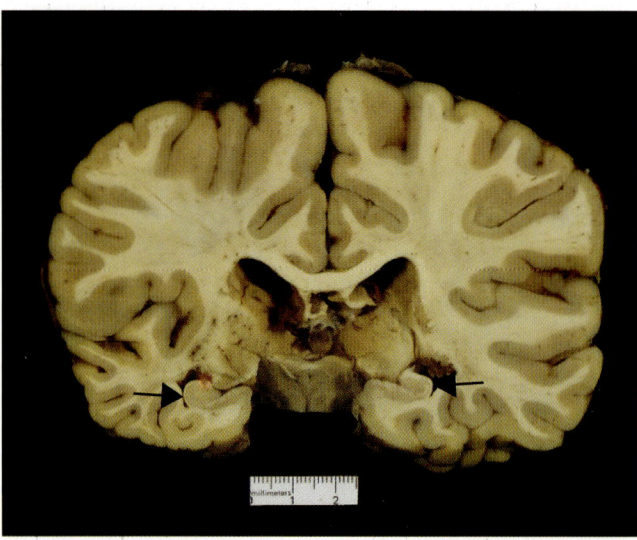

Fig. 9.24. Bilateral hippocampal sclerosis with noticeable thinning of the gray band of hippocampus proper (*arrows*).

or H1 of Rose) (Fig. 9.21). Less consistently, additional neuronal loss is noted in the endfolium (CA4) with relative sparing of the dorsal cell band (CA2) (Fig. 9.21). The dentate fascia is resistant, although it is often atrophied in severe cases (Fig. 9.25). Anterograde degeneration of the fornices is sometimes seen secondary to hippocampal sclerosis. Atrophy of the pyriform cortex may accompany the hippocampal loss.[30] Short of extensive sampling, isocortical loss attributed to seizures is more difficult to demonstrate.

Cerebellar cortical atrophy is widespread but not always diffuse, that is, atrophy has a greater propensity for the folia toward the depths of the sulci. It may be widespread in random areas of the hemisphere with relative sparing of adjacent zones (Fig. 9.22). There is a nonspecific dropout of Purkinje cells accompanied by hypertrophy and proliferation of Bergmann's astrocytes[103] (Fig. 9.23). In more severe cases, isomorphic radial gliosis is conspicuous in the molecular layer, with straight glial fibers perpendicular to the pial surface (Fig. 9.23). The granular cell layer is invariably rarefied in severe cases.

Status epilepticus is a medical emergency that leads to important brain damage without immediate treatment. The minimum duration of continuous seizures for the diagnosis of status epilepticus (SE) is a gray area. Various studies define SE as seizures lasting for periods of 30 minutes, 1 hour, or simply "prolonged."[120] Seizure type may be various, such as absence attacks and partial complex seizures, but major motor seizures (grand mal) present the greatest risk for brain damage. Mortality is said to be greatest when SE is due to underlying pathologic processes, such as brain tumor, cerebrovascular disorders, trauma, drug overdose, and some metabolic derangements including hypoxia.[120] About 50% of SE cases are said to occur in this group, and 50% occur with idiopathic seizure disorder. Mortality is directly related to duration of seizures. After 60 minutes of SE, irreversible damage can be expected.

Grossly, there is little change in the brain of those dying in SE other than acute brain swelling. Microscopically, there may be acute ischemic necrosis of neurons of Sommer's sector of the hippocampus proper, Purkinje cells of the cerebellum, the thalamus, and pseudolaminar necrosis of layers II and III of the cerebral cortex. Neuronal damage may also occur in the amygdaloid complex of nuclei, and in the periamygdaloid and entorhinal cortices.[30] The histopathology as seen by hematoxylin-eosin (H&E) stain is no different than that seen with acute hypoxic-ischemic injury.

Naturally occurring SE is a complex event with several simultaneously occurring systemic factors, such as hypoxemia, hyperemia, hypotension, hypoglycemia, and hyperpyrexia. Experiments with controlled systemic variables, in paralyzed and ventilated animals, have shown seizures alone produce selective neuronal damage by a mechanism of glutamate excitotoxic injury.[122] However, human cases of SE with stable systemic factors examined in detail at autopsy are few.[30] Cases of death during an episode of SE are only rarely seen at our office.

Focal cortical dysplasia and more subtle diffuse cerebral cortical cytoarchitectural anomalies, often with a history of mental retardation, may be encountered in some cases.[100]

Chronic cortical contusion is by far the most common cause of secondary epilepsy. The neurosurgical experience of finding focal cortical atrophy in brain tissue resected for the treatment of epilepsy has not been as commonly reported in autopsy series, possibly reflecting the usual selection criteria for neurosurgical intervention (e.g., more severe or frequent seizures, unresponsiveness to medical management). Both hippocampal sclerosis and cerebellar cortical atrophy are now considered to be sequelae of excitotoxic injury[30,69,70,75,111,112,122] rather than the long-hypothesized hypoxic-ischemic injury. Experimental studies are beginning to expose some of the complex mechanisms by which excitotoxicity leads to neuronal death.[81,85,107] Such evidence does not exclude a primary hippocampal lesion as an instigating cause of some forms of epilepsy, however.[19,69,70,101,111,112] In cases with severe cerebellar cortical atrophy, retrograde degeneration of the inferior olivary nuclei is invariably found.

Cysticercosis

Cysticercosis is a major public health problem in several countries and regions, including Mexico, Central America, and Asia. As a consequence, immigrants to the United States from such endemic areas may present with central nervous system (CNS) complications of previously undiagnosed cysticercosis.[37] CNS manifestations of this metazoic parasitic disorder can be classified anatomically as parenchymal, meningeal, intraventricular, and rarely spinal. *Taenia solium*, or pork tapeworm, develops in the intestine of humans following ingestion of larvae in incompletely cooked pork. Ova in proglottids released from the tapeworm and ingested via fecal-oral transfer (mostly through contaminated foods) penetrate the stomach wall, and are disseminated by blood vessels to various organs including the brain. Humans are an end or permanent host—that is, the cycle ends in humans. Larvae encyst in the brain and other organs and can remain viable up to several years,[63] but eventually die and calcify. The larva lies within a cystic membrane generated by itself. Microscopically, this trilaminar parasitic membrane (Fig. 9.30) is characteristic and can be useful in microscopic diagnosis after the larva has died. Cysts may attach to the dura mater (Case 9.10, Fig. 9.26), or pia-arachnoid (Case 9.11, Figs. 9.27–9.29). Within the brain parenchyma, the discrete cyst with its encysted opaque white larva is usually round and less than 1.0 cm in average diameter. Occasionally, the parasitic membrane may be profusely elaborated in the meningeal rac-

Case 9.10. (Fig. 9.26)

A 24-year-old man had an episode of severe headache with dizziness that was treated with medication by a physician 2 years before with no known recurrence. At about 2200 hours, the decedent came home from work as usual, went to his room and to bed. He was found unresponsive in bed at 0620 hours of the following morning. Paramedics pronounced the decedent at the scene.

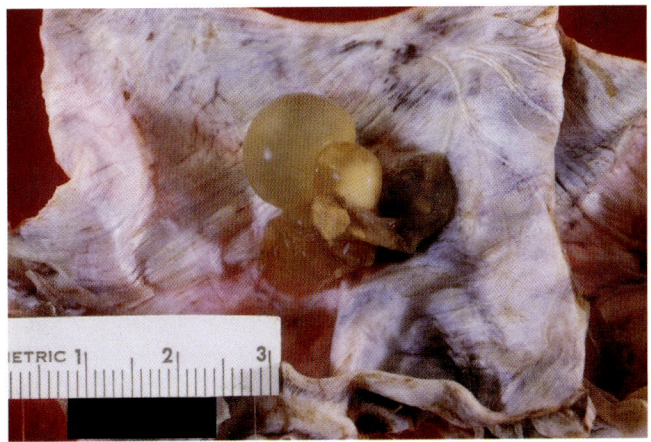

Fig. 9.26. Subdural surface of cranial dura mater with a partly racemose cysticercus cyst attached. The role of the cyst as a possible cause of death was not clear.

266 Sudden Unexpected Death

> **Case 9.11.** (Figs. 9.27–9.29)
> A previously healthy 26-year-old man was witnessed to have a seizure lasting about 5 minutes, followed by apnea. He was pronounced dead at the scene by arriving paramedics.

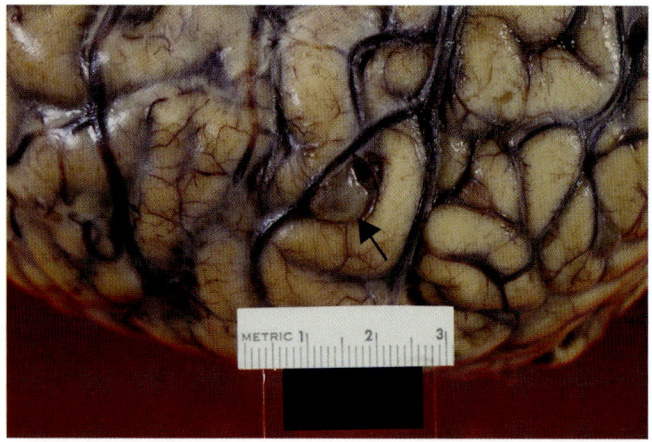

Fig. 9.27. Focal thickening of meninges marks location of a discrete meningeal cyst (*arrow*).

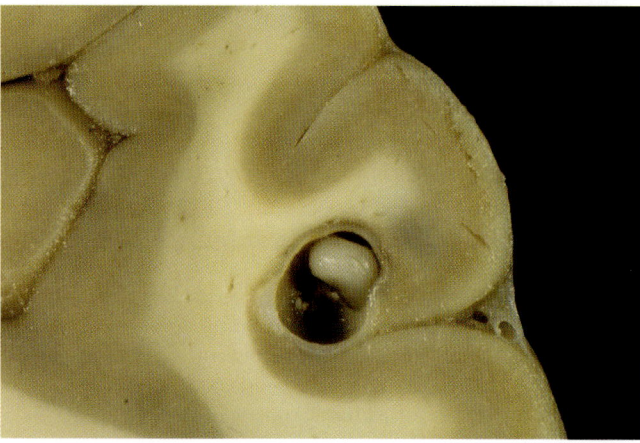

Fig. 9.28. Section of cortex reveals a cyst containig an opaque white larva.

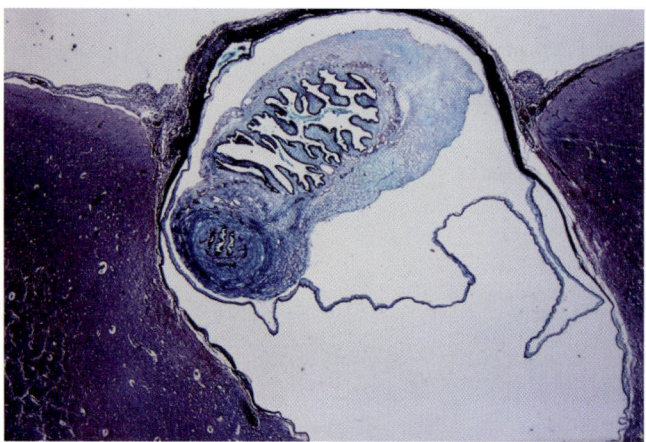

Fig. 9.29. Low-power photomicrograph illustrates a vital larva within a meningeal cyst. (Trichrome.)

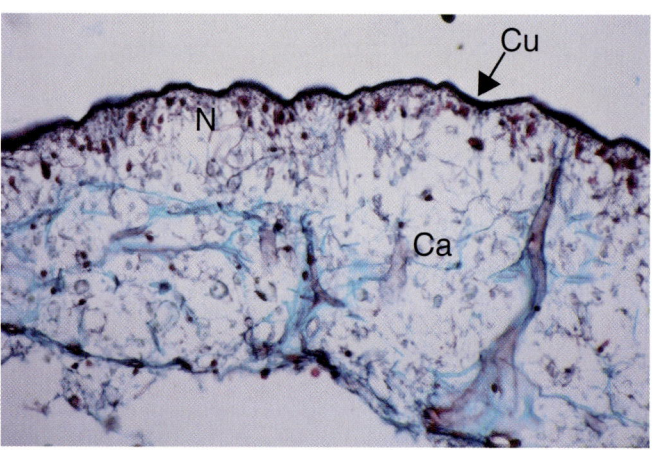

Fig. 9.30. High-power micrograph demonstrates features of typical trilaminar parasitic membrane of *Cysticercus cellulosae* (trichrome). Cu = cuticular layer; N = nuclear layer; Ca = canalicular layer.

emose form. The ventricular type may be racemose or may form discrete round cysts that can obstruct the various narrowings of the ventricular system (Case 9.12, Fig. 9.31), leading to noncommunicating hydrocephalus. The demise of a cysticercus cyst, spontaneous or following treatment, and its larva and membranes may lead to a host inflammatory reaction to the release of the degenerated materials. This fact is a matter of concern for those treating cerebral cysticercosis with trematodicides, such as praziquantel. Pathologically, depending on the timing after demise of the cyst, remnants of the larva and undulating ghost of the parasitic membrane may still be discerned. Later, positive morphologic identification may not be possible (Figs. 9.32–9.35).

Secondary seizure disorder is the most common presentation of parenchymal cysticercosis,[74] with its propensity for cerebral cortical encystment. Up to 10% of seizure cases in large urban hospital emergency rooms in California are due to this parasite.[63] Unexpectedly, the more common finding in our cases of SUD in cysticercosis is a cyst lodged in the third (see Fig. 9.31) or fourth ventricle without a history of seizure disorder. The precise mechanism of death in these cases is not always clear. The finding of hydrocephalus, apparently acute, and inflammatory changes along the walls of the ventricles (including the floor of the fourth ventricle) suggests a possible cardiorespiratory arrest secondary to an effect on the dorsal motor nucleus of the vagus and/or on the hypothalamus (adjacent to the walls of the distended third ventricle). Such a mechanism of death would not require a terminal seizure.

Parasitic infections causing SUD may also include rare cases of coenurosis presenting with a complex cyst membrane in the fourth ventricle (Case 9.13, Figs. 9.36–

Case 9.12. (Fig. 9.31)

A 16-year-old boy was awakened by a headache at 0100 hours; he took aspirin and returned to bed. In the morning, he was difficult to arouse, and paramedics transported him to a hospital. At 1100 hours, he developed respiratory arrest and subsequent brain death.

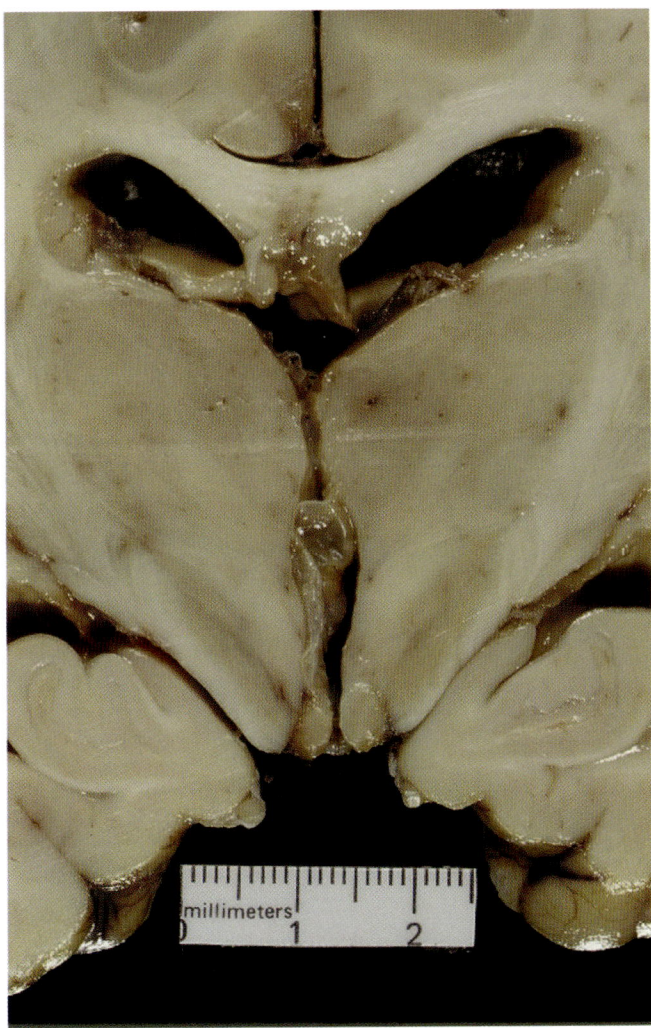

Fig. 9.31. Racemose cysticercus cyst in the third ventricle caused acute hydrocephalus.

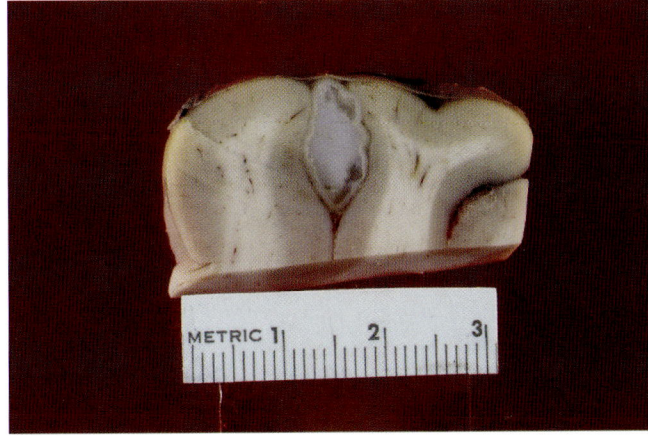

Fig. 9.32. Degenerated meningeal cysticercus cyst contains opacified colloid material.

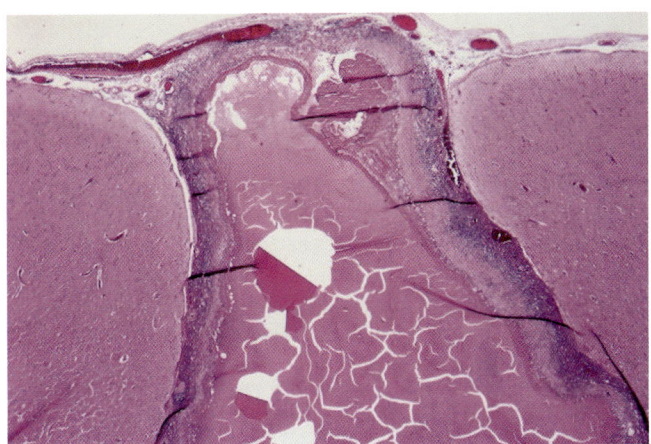

Fig. 9.33. Low-power micrograph illustrates a portion of same cyst as in Fig. 9.32 with homogeneous colloid remnant of a dead larva, artifactually cracked. The blue tint along the external surface of the cyst wall is an inflammatory cell reaction. (H&E.)

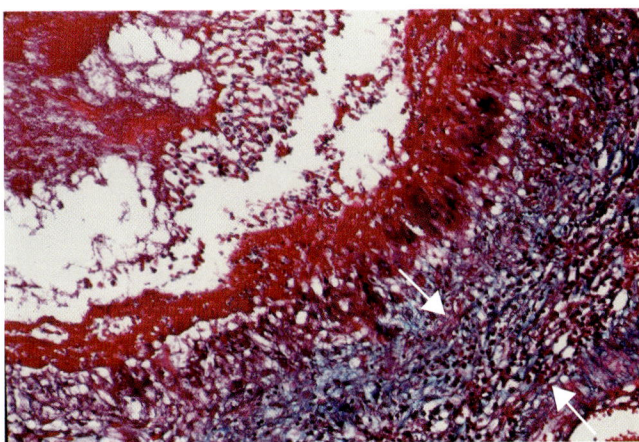

Fig. 9.34. Outer surface of the parasitic membrane (*arrows*) shows a band of florid inflammatory reaction. Bright red detritus is necrotic remnant of the parasitic membrane to which there is giant cell reaction. (Trichrome.)

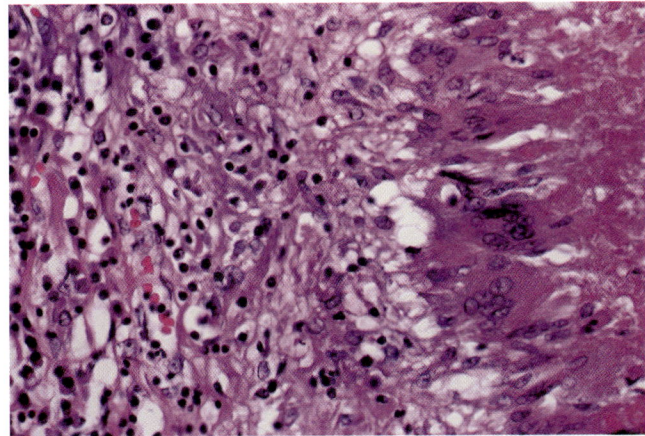

Fig. 9.35. Higher-power micrograph shows a zone of inflammation with foreign body giant cell reaction to a degenerated parasitic membrane. (H&E.)

Case 9.13. (Figs. 9.36–9.39)

A 38-year-old Korean woman developed abdominal symptoms with vomiting. The following morning she complained of abdominal pain and severe headache. At 1410 hours, the husband found the decedent incoherent, and herbal medicine was given. About 30 minutes later she was found unresponsive and apneic. Paramedics transported her to a hospital where she arrived comatose, with fixed dilated pupils, and apneic without assisted respirations. Computed tomography scan of the head showed hydrocephalus. Severe generalized brain swelling and moderately dilated ventricles were noted at autopsy.

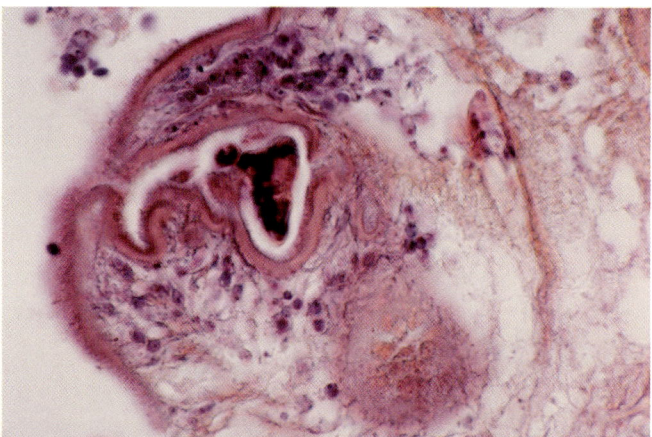

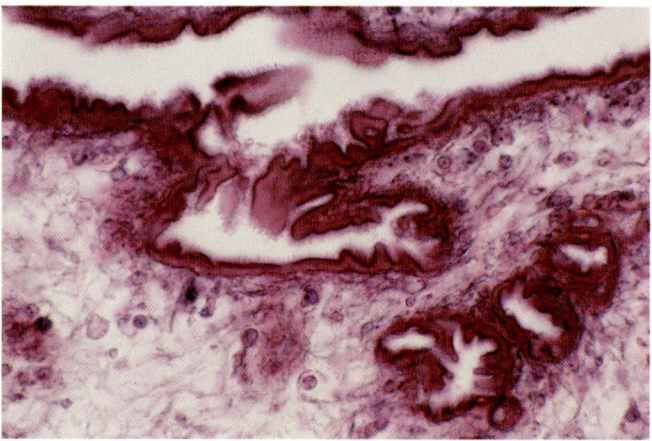

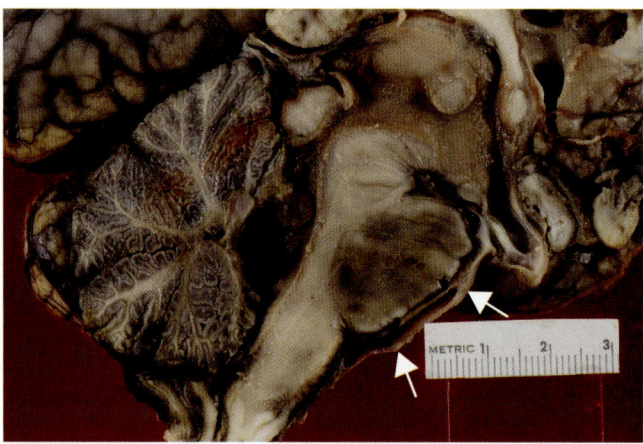

Fig. 9.36. Midsagittal section of brainstem and cerebellum demonstrates mild dilatation of fourth ventricle with a small clear parasitic cyst of *coenurus* impacted in the rostral fourth ventricle. Note thickened leptomeninges along the anterior brainstem (*arrows*).

Figs. 9.38–9.39. *Top* (Fig. 9.38; high-power) and *bottom* (Fig. 9.39; high-power) micrographs detail the special features of the *Coenurus* parasitic membrane. The membrane is trilaminar with a thick cuticular layer, but a distinguishing feature of the cyst membrane of *Coenurus cerebralis* from that of *Cysticercus cellulosae* are the many invaginated inlets. (H&E.)

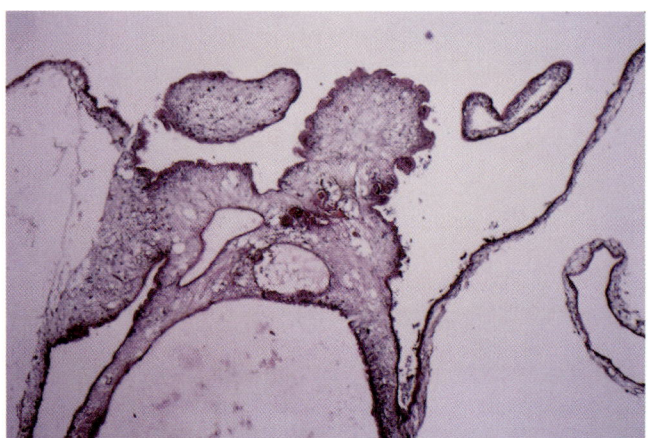

Fig. 9.37. Lower-power photomicrograph of the parasitic membrane. (Trichrome.)

9.39).[55,85] The cyst is the larval form of the dog tapeworm *Multiceps multiceps* and is called *Coenurus cerebralis*. A dog was in the household in this case.

Acute Intracerebral Hemorrhage

Of the various types of cerebral "strokes," intracerebral hemorrhages have a particularly high percentage of fatal outcome, some of which can be categorized as sudden.[72] This reflects the fact that hemorrhages more often involve basal ganglia (Case 9.14, Fig. 9.40), thalamus, pons, and cerebellum (Case 9.15, Fig. 9.41) (see also Chap. 11 on cardiovascular disease), and are often accompanied by rupture into the ventricular system. There are definite prognostic correlations with the size and location of intracerebral hemorrhages.[117] The underlying pathogenesis of the hemorrhage is not always identifiable, but the

Case 9.14. (Fig. 9.40)

A 93-year-old woman had a history of hypertension with circulatory problems. On the day prior to death she fell, sustaining a large bruise on her head. She was found dead on the floor the following day.

Case 9.15. (Fig. 9.41)

This 45-year-old woman with a history of hypertension complained of a headache at 2:30 AM. About 15 minutes later, she complained of nausea and stated she "never had a more severe headache." Her symptoms worsened and 911 was called, after which she slumped over. Resuscitative efforts at the scene failed, and she was pronounced dead at 4:52 AM.

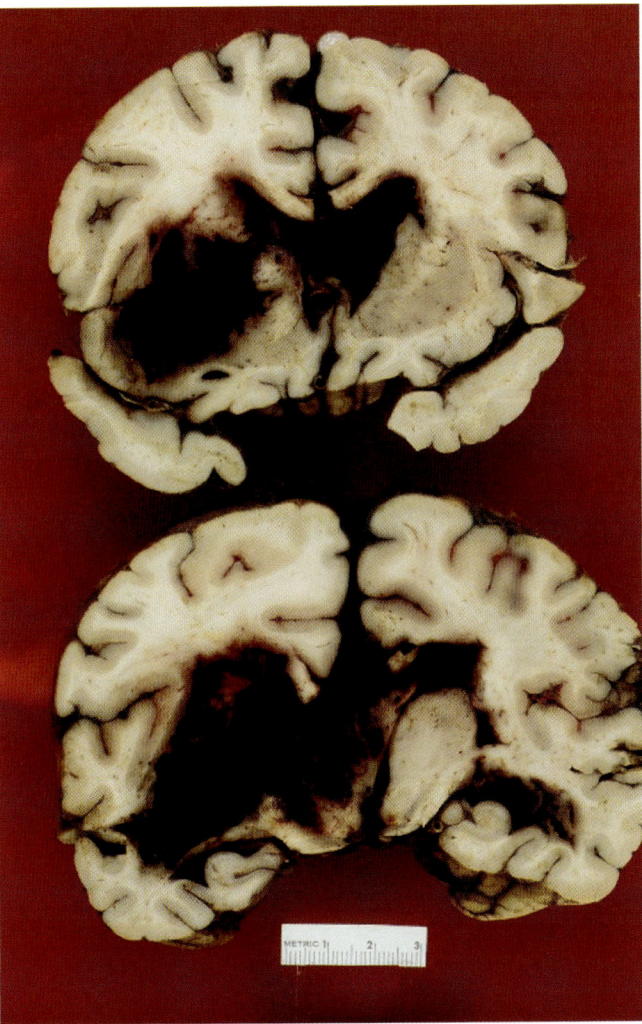

Fig. 9.40. Massive acute hemorrhage of the left basal ganglia with extension into the lateral and third ventricles and thalamus. Left uncal-parahippocampal herniation was accompanied by Duret's hemorrhage of midbrain and pons. The hemorrhage location is typical of patient with history of hypertension.

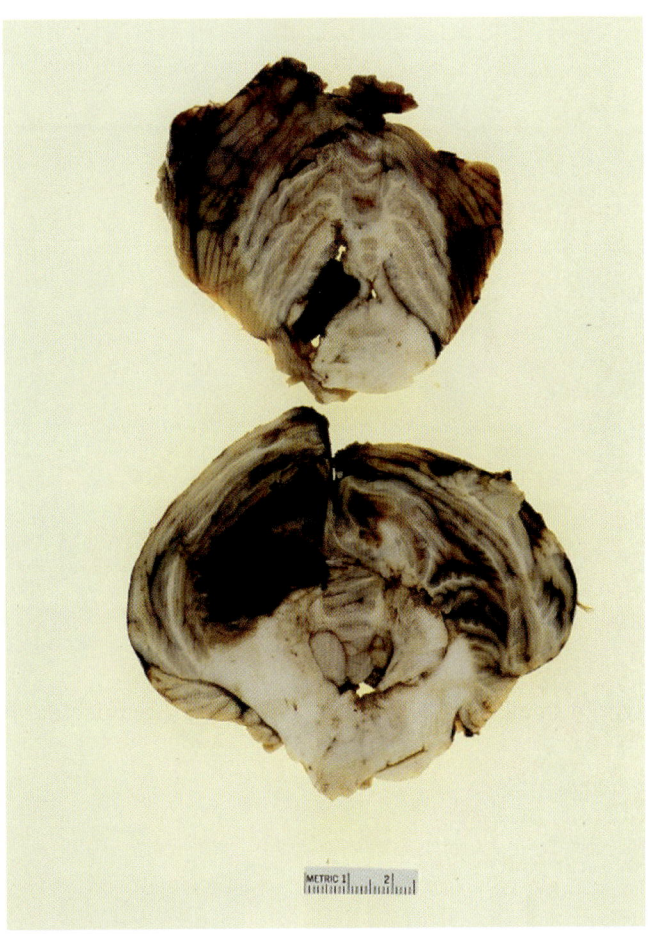

Fig. 9.41. Acute intracerebellar hemorrhage, with extension to, and compression of, left dorsolateral pontine tegmentum. Compression of pontine tegmentum was likely cause of arrest in vital functions.

majority of such cases have a history of systemic hypertension (Case 9.14, Fig. 9.40; Case 9.15, Fig. 9.41).[90] Microscopic examination often shows evidence of underlying hypertensive vascular disease, and fortuitous sections will sometimes demonstrate findings such as a Charcot-Bouchard aneurysm (see Chap. 11).

Other causes of intracerebral hemorrhage to consider include bleeding diatheses (Case 9.16, Figs. 9.42–9.44),

leukemias or lymphomas, vascular malformations, vasculitides, amyloid angiopathy in the elderly, neoplasms, and drug (especially amphetamine and cocaine) abuse. Hemorrhage accompanying amyloid angiopathy is characteristically located in the superficial cerebral cortex with subcortical extension as well as subarachnoid and sometimes subdural spread. Involvement shows a frontoparietal bias, although vascular changes may be most

270 Sudden Unexpected Death

> **Case 9.16.** (Figs. 9.42–9.44)
>
> A 50-year-old man had cleaned out a cesspool and septic tank using caustic soda flakes 3 days prior to death. On the following day, hemorrhagic lesions appeared on his body and legs. One day prior to death, he went on a 3-mile hike. On the day of death he felt hot, complained of headaches, and had blood in his vomitus, urine, and stool.

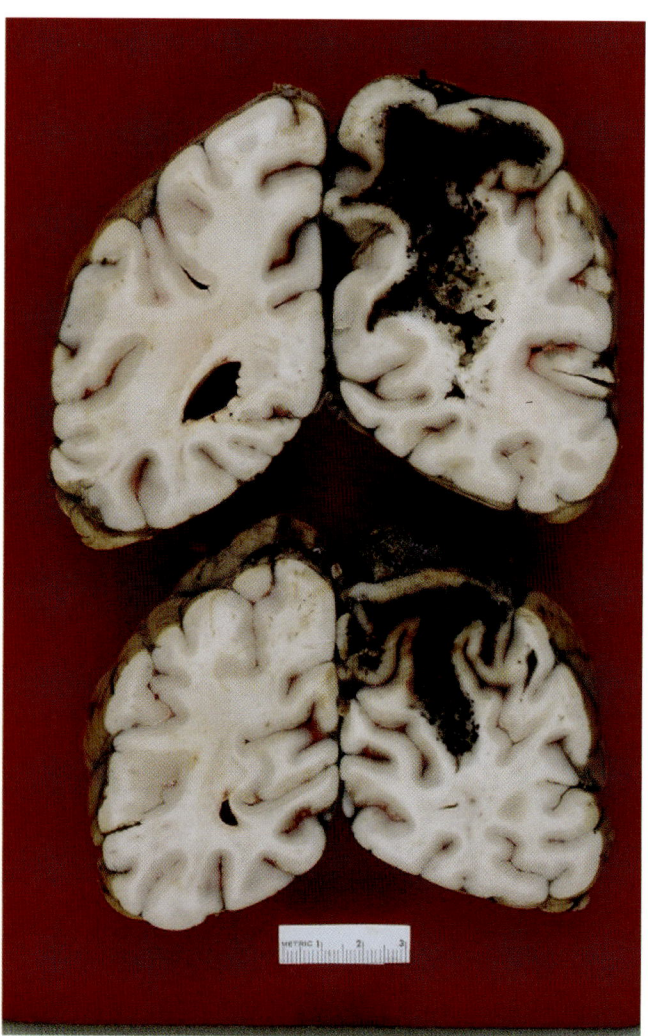

Fig. 9.42. Large acute hemorrhage destroyed subcortical white matter of the right superior parietal lobe, leaving cortical rind.

Fig. 9.43. Duret's hemorrhage of midbrain and pons.

in cerebral cortical/subcortical white matter. With rare exceptions, it is considered an incidental finding of no clinical importance. On the other hand, venous malformation (Case 9.17, Figs. 9.45–9.47) and AVM (Case 9.18, Figs. 9.48–9.50) are potentially dangerous lesions prone

Case 9.17. (Figs. 9.45–9.47)

A 7-year-old boy with asthma was found collapsed on the bathroom floor complaining of headache. He was put to bed where he was later found unresponsive, and he was transported to a hospital by paramedics but was pronounced dead 30 minutes after arrival.

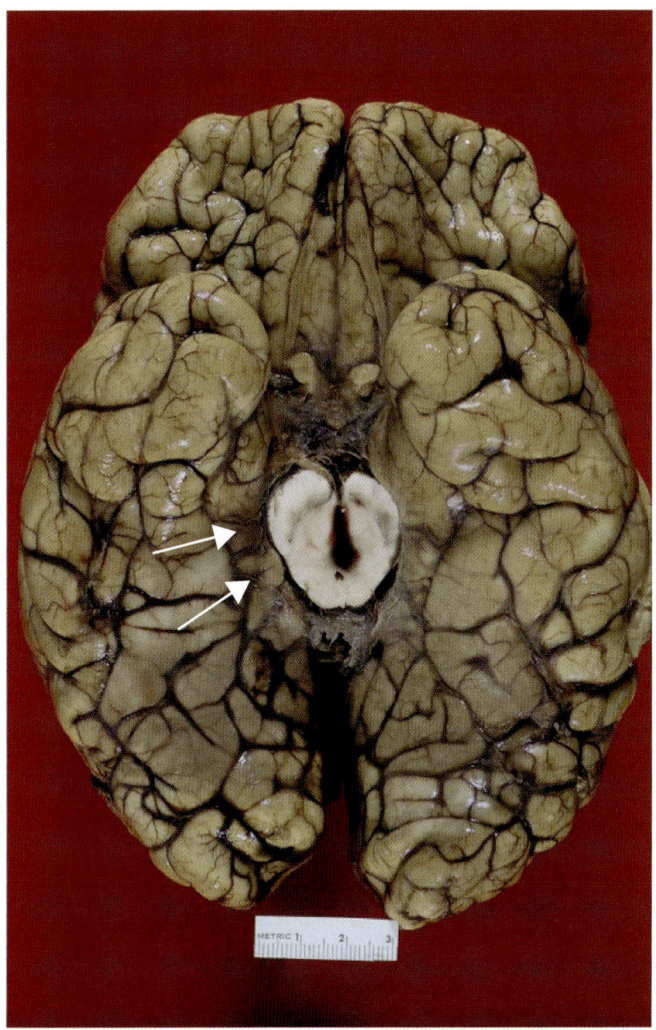

Fig. 9.44. Small inconspicuous parahippocampal herniation on right (*arrows*). Large intracerebral hemorrhage was thought to be part of a general bleeding diathesis.

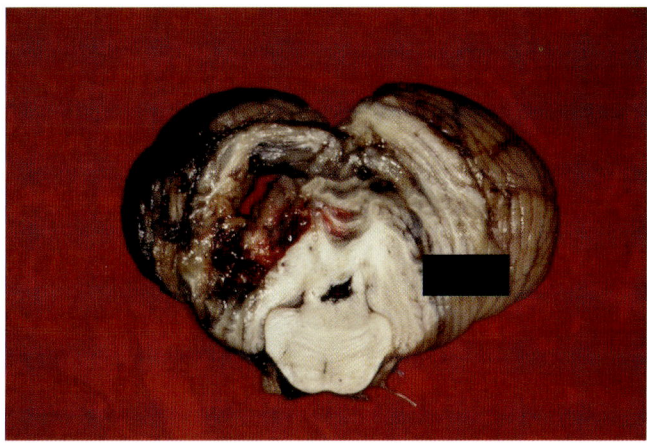

Fig. 9.45. Hemorrhagic cavity is seen on right (left in photo) cerebellar hemisphere where a clot has fallen out postmortem. Overlying acute subarachnoid hemorrhage is present.

prominent in the parieto-occipital region. Hemorrhage associated with amyloid angiopathy can be neurologically devastating, but not often the cause of SUD (see Chap. 11).

Vascular Malformation

Cerebral vascular malformations include arteriovenous malformation (AVM), venous malformation, cavernous hemangioma, varix, and capillary telangiectasia.[8] In addition, these types may occur in various combined forms, and several genetic syndromes such as Sturge-Weber-Dimitri syndrome and von Hippel-Lindau disease exist in which various types of vascular malformations occur. Capillary telangiectasia is a very common incidental finding at autopsy, most often in the basis pontis or

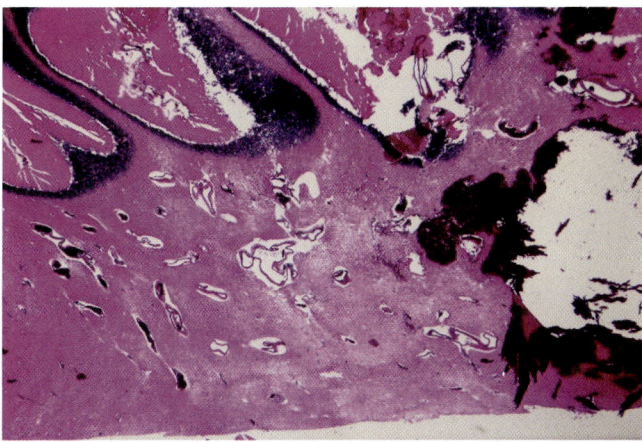

Fig. 9.46. Low-power photomicrograph shows abnormal medium size to large veins of a venous malformation in cerebellar white matter adjacent to hemorrhage. (H&E.)

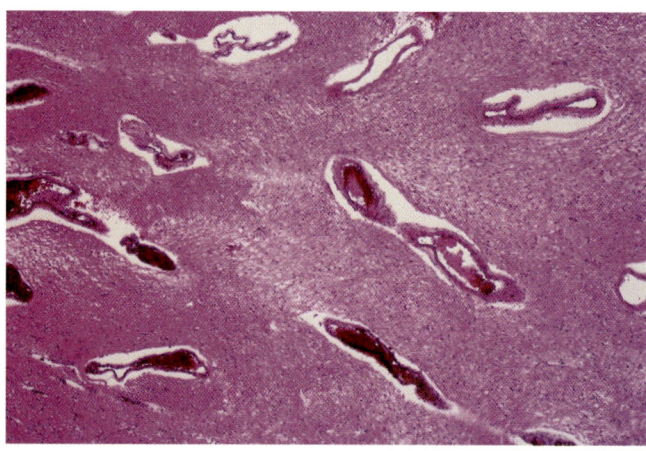

Fig. 9.47. Medium-power micrograph showing typical distribution of abnormal veins of malformation. No evidence of past hemorrhage or parenchymal atrophy is present. (H&E.)

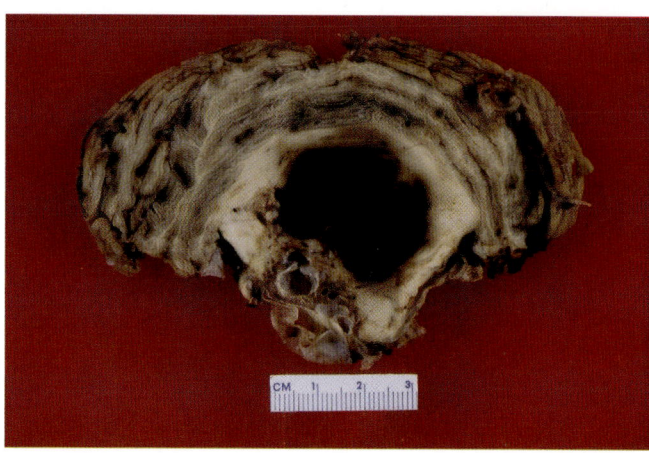

Fig. 9.49. Transverse section of caudal pons shows an arteriovenous malformation destroying right half of basis pontis and a massive hemorrhage confluent with a hugely expanded fourth ventricle, the borders of which are no longer discernible.

Case 9.18. (Figs. 9.48–9.50)

A 72-year-old man was found unresponsive in a shower at a retirement home. He was thought to be a victim of a fall. CT scan showed a "brain stem hemorrhage obstructing the fourth ventricle." A pressure monitor and ventriculostomy were placed, but he died the following day.

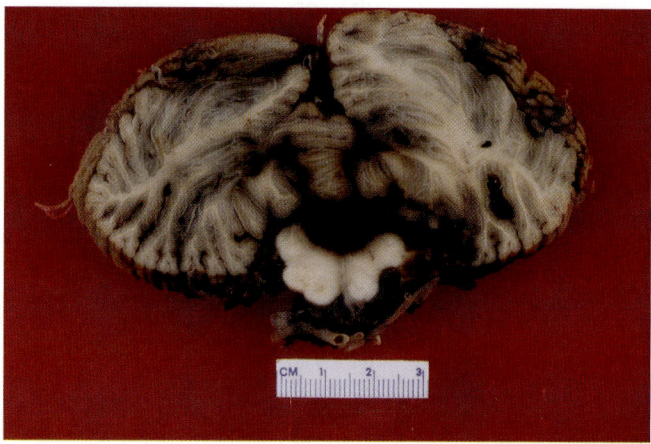

Fig. 9.50. Caudal fourth ventricle is distended by solid blood clot, and numerous small blood vessels of the arteriovenous malformation pack perimedullary cistern.

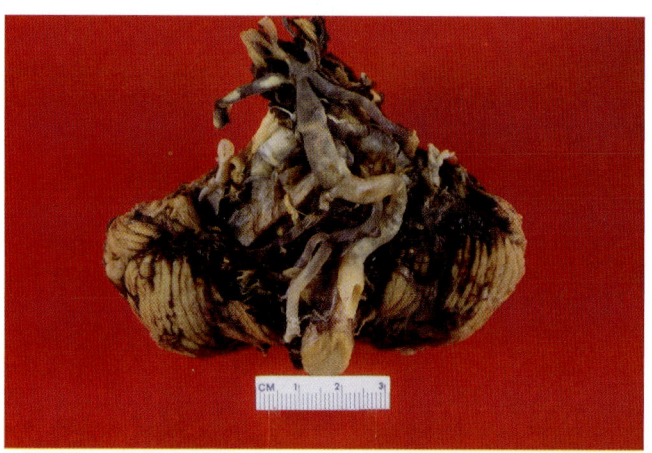

Fig. 9.48. Ventral view of brainstem and cerebellum shows a very tortuous vertebral-basilar artery. A tangle of large anomalous blood vessels characteristic of an arteriovenous malformation overlies right pons (left in photo). Cerebellar hemisphere is slightly smaller on that side.

to spontaneous rupture, seizures and, depending on location, an occasional dramatic cause of SUD (see Chap. 11 on cerebrovascular diseases for description of microscopic findings in the malformations mentioned earlier). As with AVM, cases of cavernous hemangioma (Case 9.19, Figs. 9.51–9.53; Figs. 9.54–9.55) usually have a history of neurologic symptoms having prompted prior medical attention. Most of our cases of AVM have had seizures or other neurologic symptoms prior to death, and death is not often unexpected, although it may be sudden. In other cases, the malformation is an incidental autopsy finding when located in a clinically silent area.

Case 9.19. (Figs. 9.51–9.53)

A 17-year-old previously healthy boy was found dead in bed.

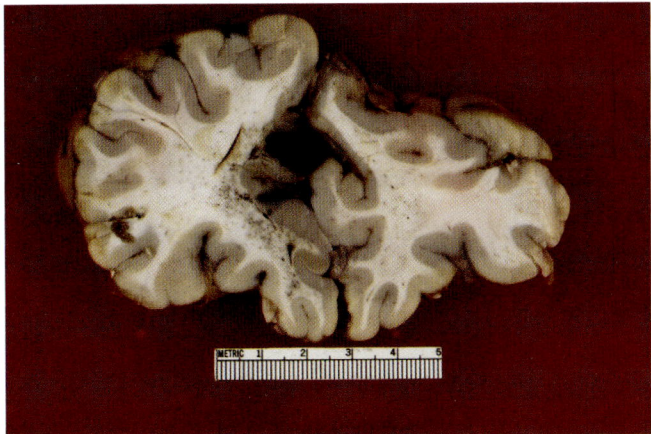

Fig. 9.51. Vascular malformation of medial-facing left superior frontal gyrus. Subjacent white matter over wide extent shows diffuse capillary telangiectasia.

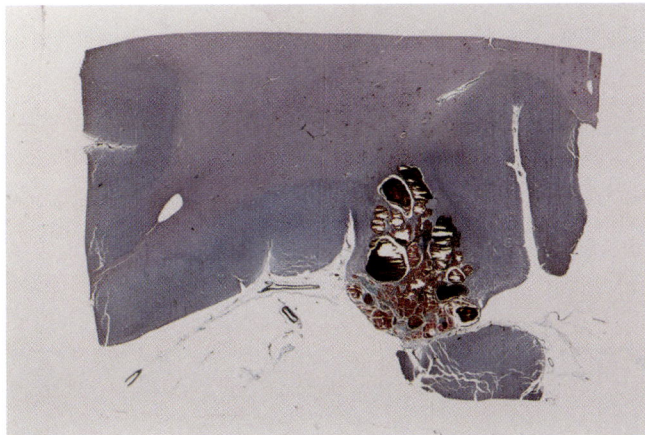

Fig. 9.52. Low-power micrograph shows a cavernous hemangioma extending into subarachnoid space. (Trichrome.)

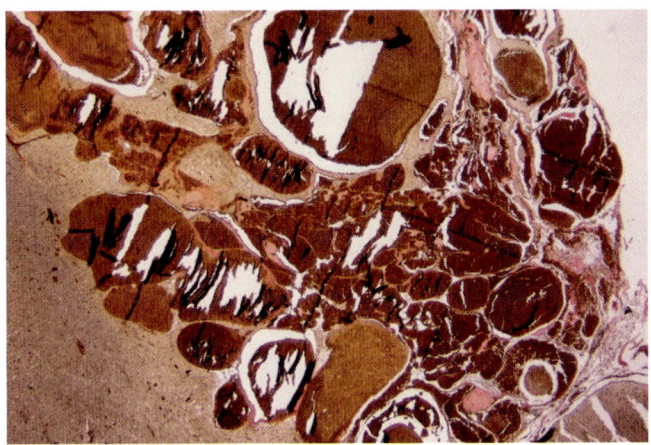

Fig. 9.53. Medium-power micrograph shows thin-walled vascular channels without media sharing party walls. Intervening parenchyma is severely atrophic, and decorated with hemosiderin. (Elastic van Gieson.)

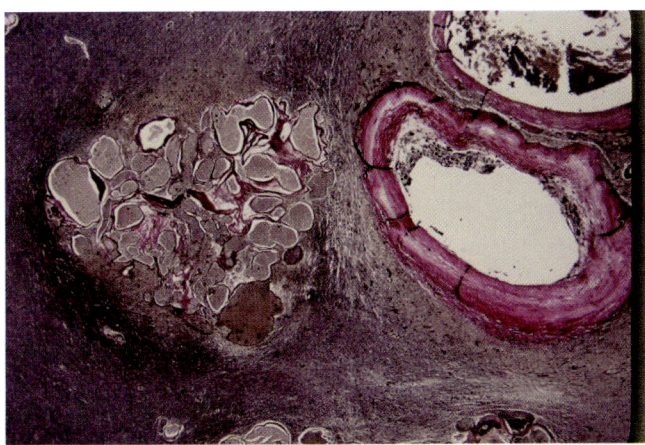

Fig. 9.54. Low-power micrograph from another case shows cavernous hemangioma of closely clustered dilated vascular channels. (Trichrome.)

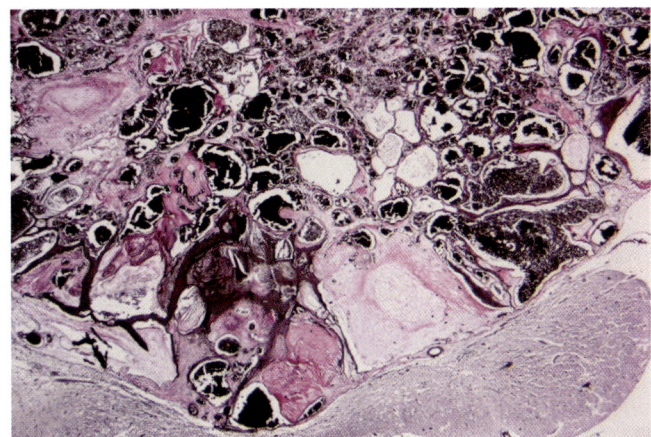

Fig. 9.55. Medium-power micrograph from case seen in Fig. 9.54 shows myriads of closely approximated, mostly thin-walled, fibrotic vascular channels with shared walls and lack of intervening parenchyma. (Trichrome.)

Meningitis

In the great majority of cases, symptoms accompanying infections of the CNS result in the patient visiting a physician before death. Unattended infectious death coming under the jurisdiction of the medical examiner-coroner is relatively unusual. SUD due to cysticercosis has been discussed earlier. Acute meningitis is a well-known but relatively uncommon cause of SUD. Hyperacute bacterial meningitis in the very young and the elderly due to organisms such as group B hemolytic *Streptococcus, Escherichia coli, Listeria monocytogenes, Neisseria meningitidis, Haemophilus influenzae,* and *Streptococcus penumoniae* can cause death within 24 hours of onset of symptoms. An elderly person with mild headache and malaise found dead in bed the following morning with acute bacterial meningitis is not altogether rare.

At autopsy a purulent exudate may not be apparent on gross inspection, or the leptomeninges may be only slightly clouded. Microscopic examination in untreated cases will invariably demonstrate subarachnoid spaces infiltrated by leukocytes and bacteria. On occasion, a person found dead may have the brain surfaces entirely obscured by purulent material.

Such patients surely must have been symptomatic for at least a few days prior to death, but for various reasons had not come to medical attention. Interview of family and friends will usually uncover several days of symptoms not brought to medical attention. Such cases might be properly categorized as sudden but unexpected only because of the absence of trained observers. Reactive, secondary changes such as early leptomeningeal fibrosis, vasculitis, ventriculitis, choroid plexitis, cerebritis, abscess (Case 9.20, Figs. 9.56–9.57), cortical infarction, and hydrocephalus indicate that the illness was likely symptomatic, but unattended for a period of time that can be estimated by the histologic changes present.

Meningitis can produce SUD through multiple possible mechanisms occurring singly or in combination. The three main mechanisms are (1) cerebral edema, (2) obstruction to the flow of cerebrospinal fluid (CSF), and (3) systemic collapse. Inflammatory bacterial mediators can produce a direct toxic or cytopathic effect, resulting in cytotoxic edema. In addition, bacterial toxins affecting neurons can increase metabolic demand, resulting in hypoxia. As an example, patients with higher levels of tumor necrosis factor (TNF) are more likely to die.[51] Injury to the blood–brain barrier results in vasogenic edema. All of these pathways have hypoxic/ischemic effect as the final common denominator. Obstruction to the flow of CSF by purulent exudate, filling and obstructing subarachnoid spaces and foramina and blocking absorption at the arachnoid villi, may produce acute hydrocephalus. This can lead to interstitial edema with its secondary sequelae.

The systemic effects of bacteremia may be concurrent with the acute meningitis, with both being factors in the development of SUD. This is most common with meningococcal and pneumococcal infection, in which disseminated intravascular coagulopathy, adrenal hemorrhage (Waterhouse-Friderichsen syndrome), and hepatic necrosis can occur, leading to septic shock and circulatory collapse as part of the systemic inflammatory response syndrome (SIRS) (see Chap. 6).

Case 9.20. (Figs. 9.56 and 9.57)

A 29-year-old man had a 1-month history of headache, vomiting, and weight loss. He was found dead at home 2 days after last being seen alive.

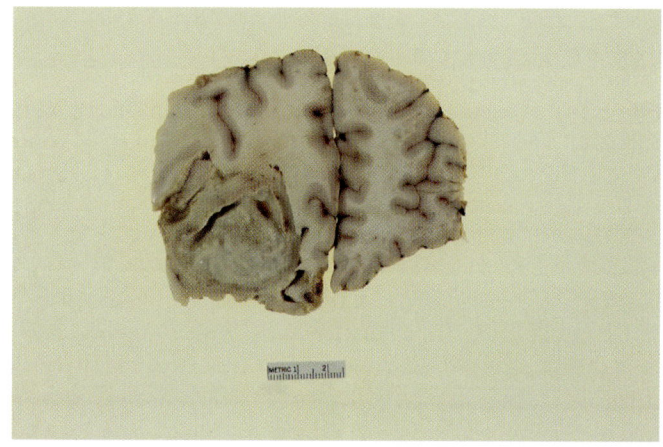

Fig. 9.56. A large solitary 3.5 × 4.0-cm abscess of left orbitofrontal lobe with surrounding discoloration due to edema. Ipsilateral uncal herniaton was present.

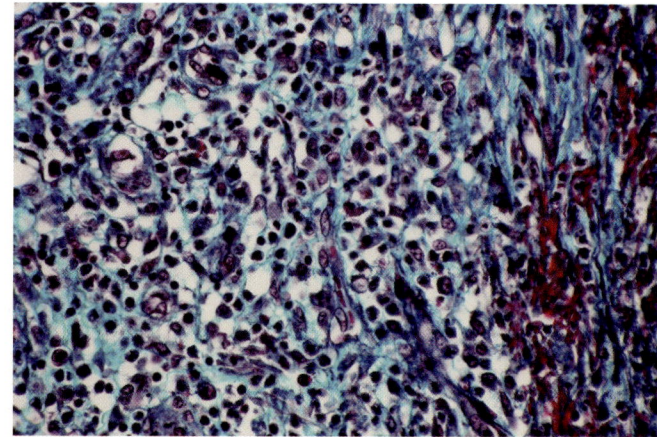

Fig. 9.57. High-power photomicrograph shows innumerable polymorphonuclear neutrophils and fibroblastic proliferation along abscess wall. Colonies of cocci were also demonstrated. (Trichrome.)

It is unclear how frequently SUD in meningitis may be related to focal intraparenchymal reactive changes. Cerebritis or abscess may occur, not only secondary to meningitis but also by direct spread from adjacent (e.g., purulent sinusitis) or remote (e.g., septic emboli from lung abscess) sites. This can lead to fatal seizures or to rupture of abscess content into the ventricular system. Ventricular purulent exudate may breach the ependymal lining and involve such vital areas as the hypothalamus adjacent to the third ventricle, and the dorsal vagal nuclei at the floor of the fourth ventricle.

Hydrocephalus

Infrequently, hydrocephalus is discovered at autopsy in an individual who died suddenly. The presence of cysticercosis or colloid cyst of the third ventricle (Case 9.21, Figs. 9.58–9.60) may provide an obvious etiology.[26] Other causes include congenital aqueductal stenosis, aqueductal atresia, and membrane of the aqueduct. Glioma of the aqueduct is one of the smallest tumors of the body that may be fatal (Case 9.22, Figs. 9.61–9.64). The mechanism by which such lesions cause sudden death is uncertain; brainstem compression secondary to herniation is rarely a feature. One assumption is that the acuteness of the onset of hydrocephalus, with resultant interstitial edema and impaired vascular perfusion in the face of increased intracranial pressure (ICP), leads to death before herniation can develop. In the presence of elevated intracranial pressure, brain autoregulatory compensatory mechanisms include (1) vasoconstriction and (2) development of decreased cerebrospinal fluid (CSF) volume. This latter mechanism includes (1) decreased CSF production and (2) increased CSF absorption[73] (e.g., "transependymal absorption").[76] These mechanisms may be in effect for many months, until a point is reached on the ICP-intracranial volume curve when the gradient rises steeply, compensation is no longer adequate, and clinical deterioration occurs quickly with what appears to be sudden death in what was, up to that point, a slowly evolving pathologic process. In the resulting acute obstructive

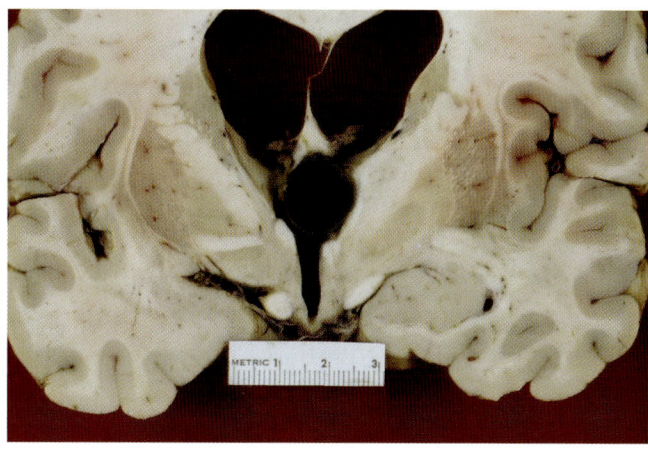

Fig. 9.58. Closeup of a colloid cyst impacted in anterior third ventricle, obstructing the interventricular foramina creating a large noncommunicating hydrocephalus and bilateral uncal-parahippocampal herniation. The cyst is unusually dark brown from past and recent hemorrhage.

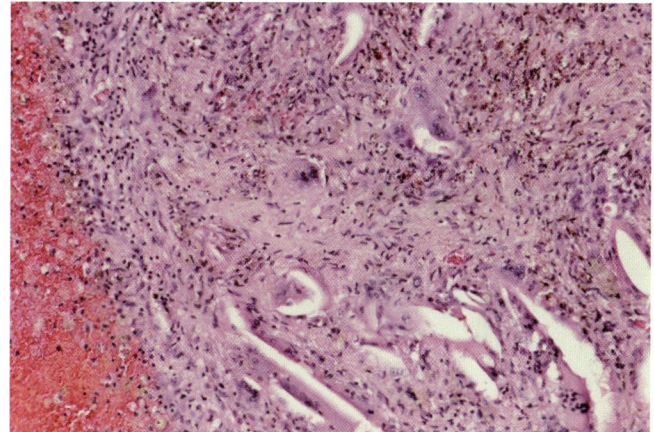

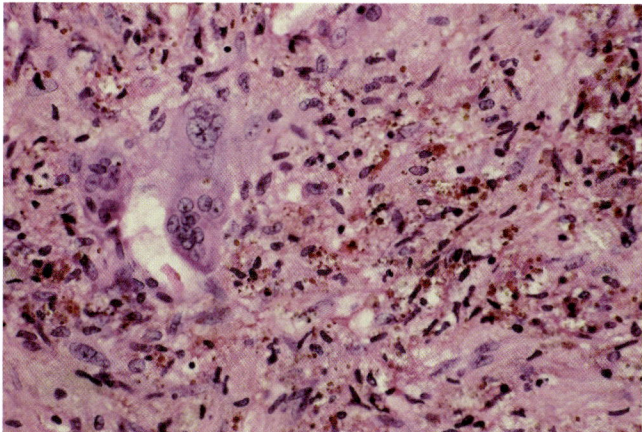

Figs. 9.59–9.60. *Top* (Fig. 9.59; medium-power) and *bottom* (Fig. 9.60; high-power) photomicrographs show organization of past hemorrhage within the cyst with hemosiderin, macrophages, foreign body giant cells, and cholesterol clefts. Recent hemorrhage is also present. (H&E.)

Case 9.21. (Figs. 9.58–9.60)

A 31-year-old man developed headaches 6 months prior to death. The cause of the headaches remained undetermined after medical examination, and the decedent was placed on analgesics with instructions to return if not better. He did not return. About 10 hours prior to death, he complained of "not feeling well." He then developed nausea, vomiting, and lightheadedness. He was last seen apparently asleep on the floor with heavy breathing, but he was later found unresponsive and pronounced dead at the scene.

hydrocephalus, the ependymal lining is damaged by stretching, with breakdown of the CSF–brain barrier. CSF is consequently forced into the periventricular white matter, producing interstitial edema.[29]

In infants, an additional intrinsic compensatory mechanism in increased ICP is provided by cranial sutures not yet being fused. With a sudden rise in intracranial

> **Case 9.22.** (Figs. 9.61–9.64)
>
> A 23-year-old woman had fallen off of bleachers at a high school basketball game at the age of 14 years, striking her head, with a resultant seizure disorder. She was poorly compliant with her medications. She was last seen alive while sleeping, but she was found without signs of life 10 hours later.

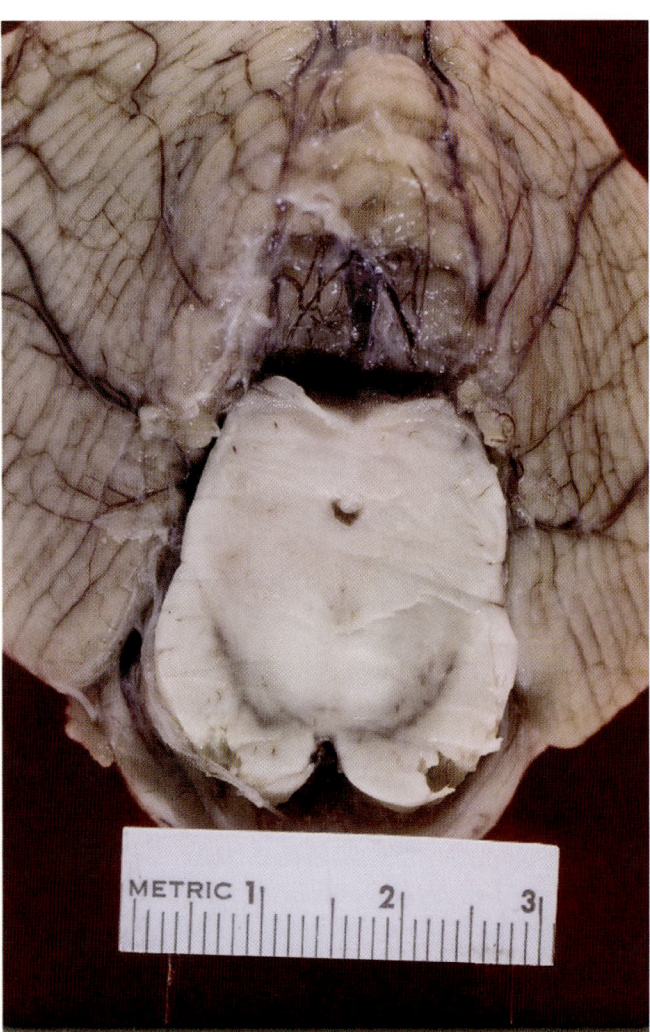

Fig. 9.62. Transverse section of midbrain shows partial blockage of aqueduct. Third ventricle and aqueduct proximal to the partial occlusion were dilated. Cerebellar tonsils were herniated. No remote trauma was found.

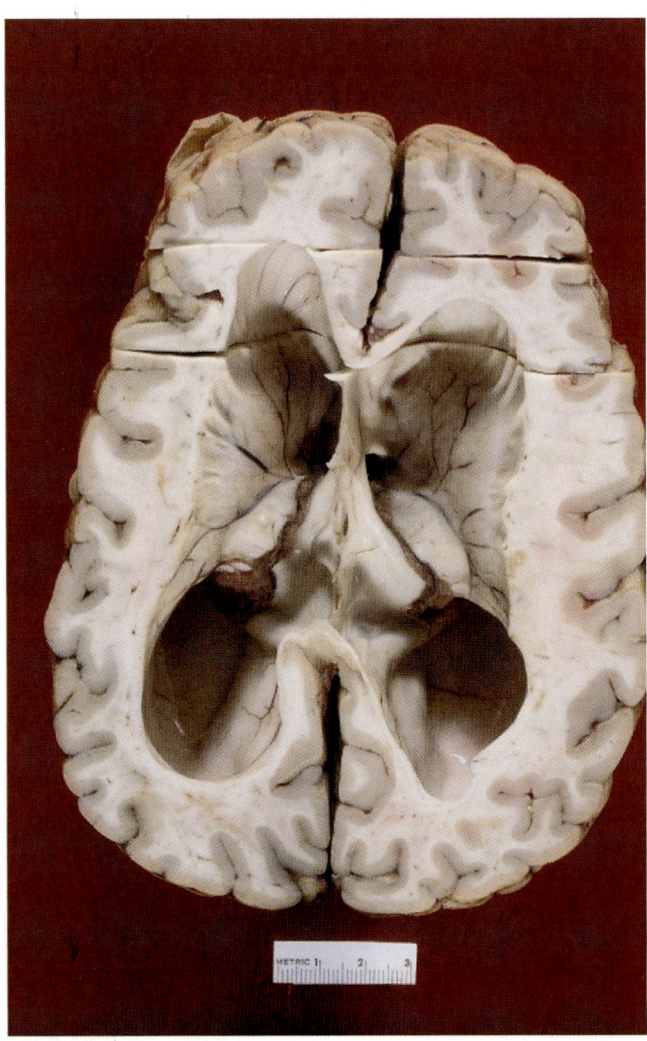

Fig. 9.61. Horizontal section of cerebral hemispheres shows massive symmetric enlargement of lateral ventricles.

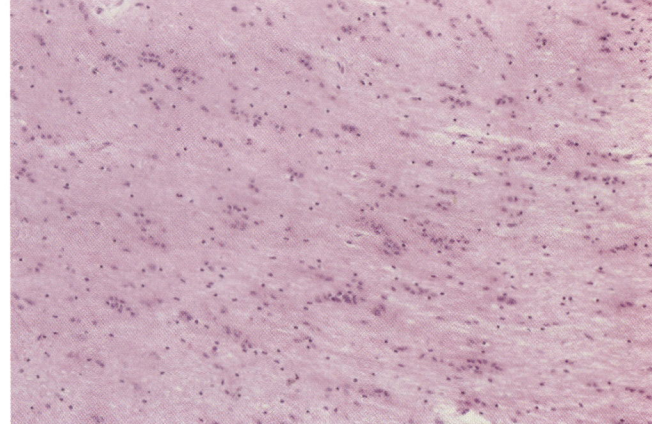

Fig. 9.63. Medium-power micrograph of the tumor shows sparsely clustered nuclei in field of glial fibers of a low-grade glioma of aqueduct. (H&E.)

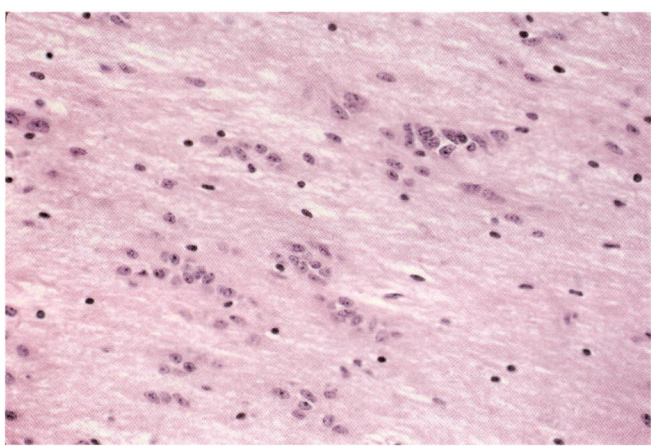

Fig. 9.64. High-power micrograph of same field seen in Fig. 9.63 shows cellular detail. (H&E.)

volume, sutures can expand, lessening the rate at which ICP rises. This can allow earlier detection of the problem as a result of, for example, bulging fontanels or an abnormally accelerated increase in cranial circumference, with possible life-saving intervention. In the elderly, preexisting cerebral atrophy may provide room for expansion of the brain into a preexisting dilated subarachnoid space, allowing the development of symptoms before life-threatening decompensation occurs.

In the special circumstance of tumors of the posterior fossa, there may be not only production of acute obstructive hydrocephalus, but also direct or indirect brainstem compression from cerebellar tonsillar and/or superior cerebellar herniations. This may lead to SUD by direct compromise of cardiac or respiratory centers, more often in the medulla. However, these patients are usually symptomatic with their disease, and death may be sudden but not entirely unexpected.

Tumors

Neoplasms of the brain or the cervical spinal cord can cause sudden death. In the vast majority of cases of brain tumor, the diagnosis has been made antemortem, and death is neither sudden nor unexpected. The tumors may be of dural, intraparencymal, or intraventricular origin. In most instances of SUD, the tumors are primary. Metastatic tumors more frequently have produced local or systemic symptoms leading to diagnosis long before death due to CNS metastases occurs. In our experience, SUD due to CNS spread from an occult primary tumor is uncommon.[104] In many such instances, questioning of the decedent's social contacts reveals a period of months or more of nonspecific symptoms, such as headache or blurred vision, without medical attention having been sought. Rarely are antemortem symptoms absent.

There are several mechanisms by which tumor-related SUD occurs. The first is via simple mass effect, in which the tumor finally reaches a critical size and/or volume of surrounding cerebral edema large enough to produce herniation with all of its secondary phenomena (Case 9.23, Figs. 9.65–9.67; Case 9.24, Figs. 9.68–9.71). A second mechanism is acute hemorrhage into the tumor, causing a sudden large increase in mass effect with herniation (Case 9.25, Figs. 9.72–9.73). A third mechanism is blockage of the ventricular system causing acute noncommu-

Case 9.23. (Figs. 9.65–9.67)

A 12-year-old girl had a 6- to 7-year history of epilepsy treated with phenobarbital and was seizure-free for 2 years. On the night prior to death, she complained of severe headaches and vomited. She was seen alive in the morning but was found dead shortly thereafter.

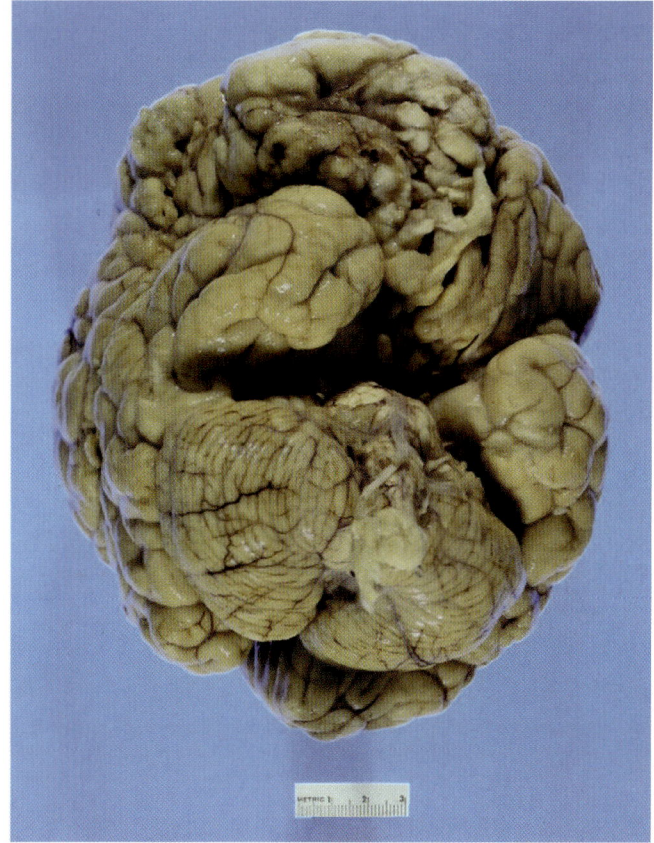

Fig. 9.65. Huge tumor protrudes from the right orbital frontal region, with a midline shift from right to left.

Sudden Unexpected Death

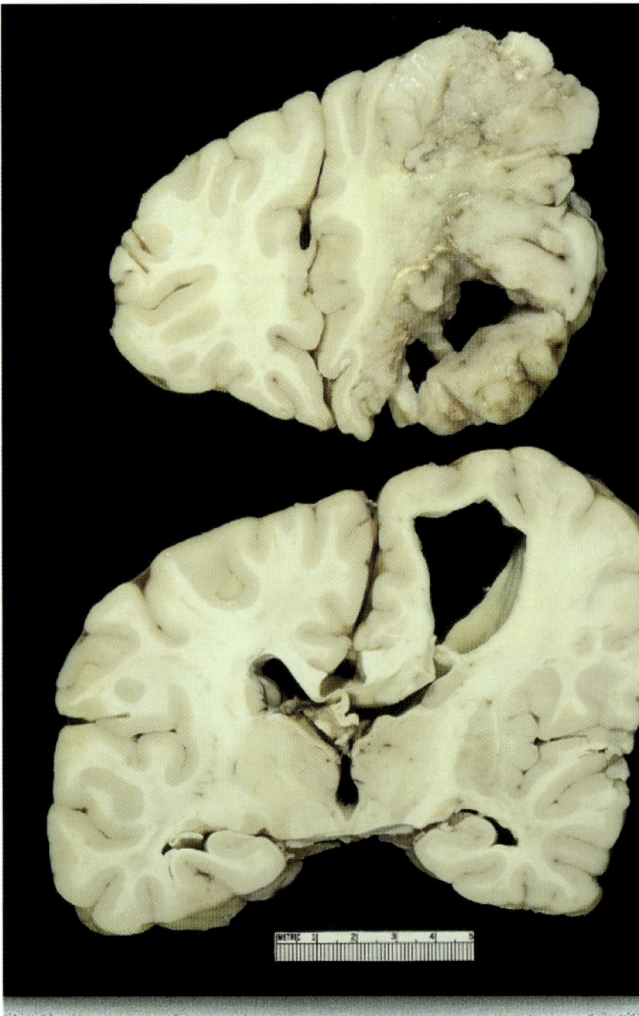

Fig. 9.66. Coronal sections show solid and cystic components of tumor replacing most of right frontal lobe. Extensive necrosis and calcification were present.

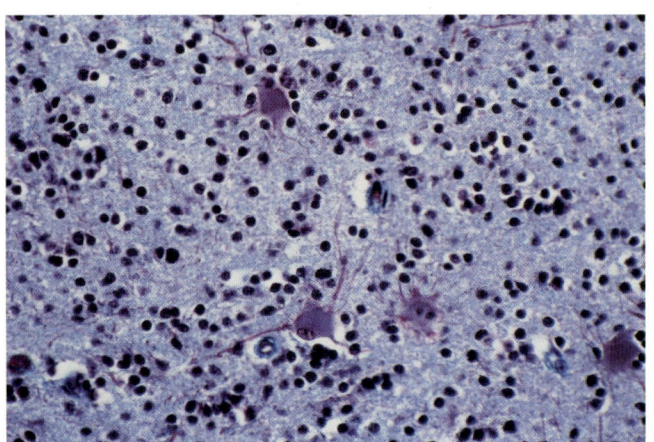

Fig. 9.67. High-power photomicrograph demonstrates a well-differentiated oligodendrioglioma. Hypertrophied, reactive astrocytes are shown scattered through the tumor. (PTAH).

Case 9.24. (Figs. 9.68–9.71)

A 29-year-old woman complained of a severe headache on the day of death, and she was given Demerol at the hospital and sent home. Approximately 1 hour after going to bed that evening, she was found in full arrest and could not be resuscitated.

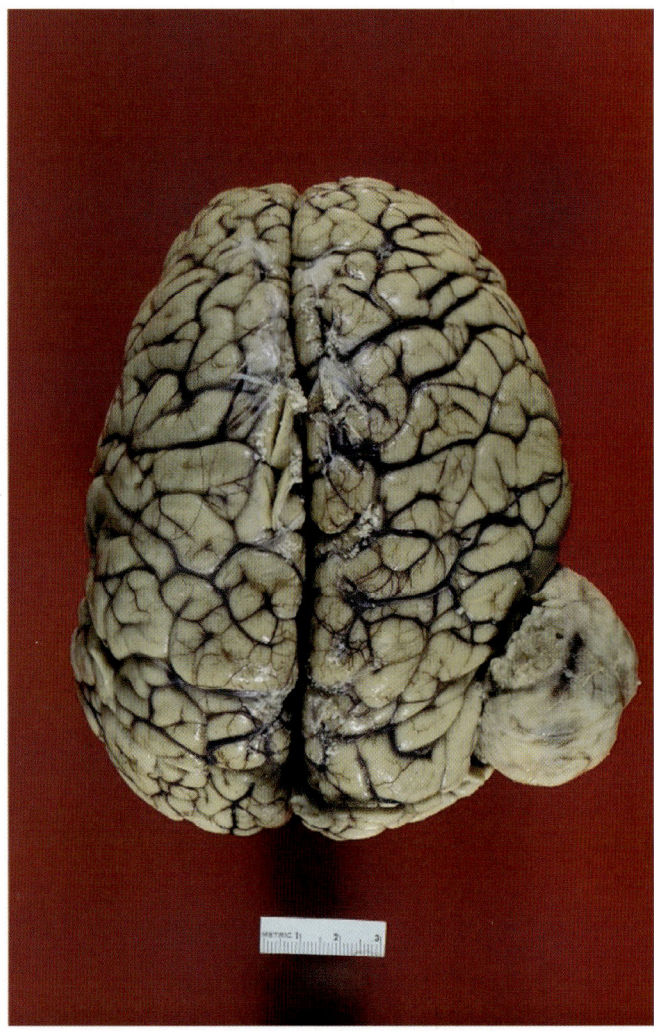

Fig. 9.68. Vertex view of brain shows massive tumor of right parietal lobe measuring 4.5 × 6.0 × 8.0 cm. The tumor weighed 110 g and fell away cleanly, revealing deeply indented brain. Marked brain swelling, with bilateral uncal herniation, compressed midbrain.

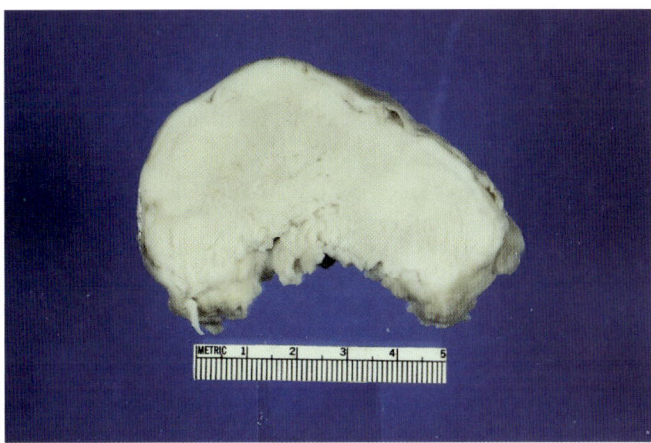

Fig. 9.69. On section, the meningioma was firm and uniformly white.

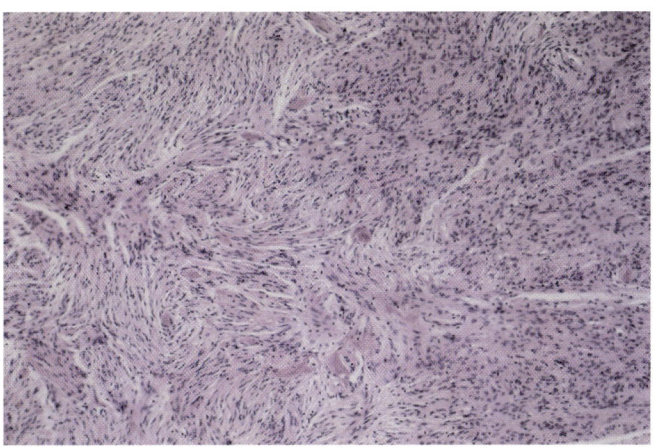

Fig. 9.70. Medium-power micrograph shows interlacing patterns of spindle tumor cells typical of fibroblastic meningioma. (H&E.)

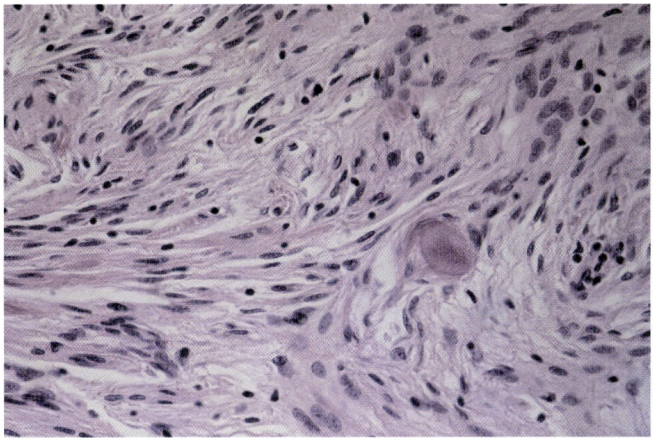

Fig. 9.71. High-power micrograph shows detail of fibroblastic pattern of tumor cells. (H&E.)

Case 9.25. (Figs. 9.72 and 9.73)

A 37-year-old salesman had "moderate to severe" headaches for approximately 12 days prior to death. Two days prior to death a physician prescribed Fiorinal with codeine, with instructions to call back if the headaches persisted after 48 hours. On the evening prior to death, his headaches, shortness of breath, and occasional chest pains were attributed to job stress. He was found dead the following morning.

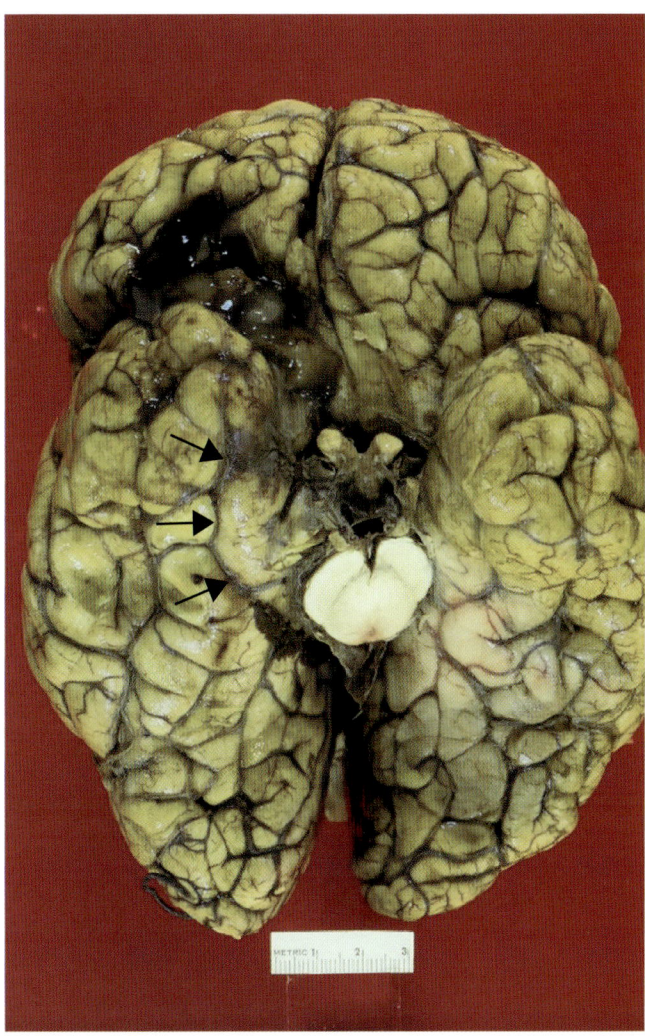

Fig. 9.72. A dark hemorrhagic mass protrudes from the right (left in photo) posterior orbital cortex, with right to left shift of the midline and an enlarged right temporal lobe. A large uncal-parahippocampal herniation is seen on the right (*arrows*), and midbrain is displaced.

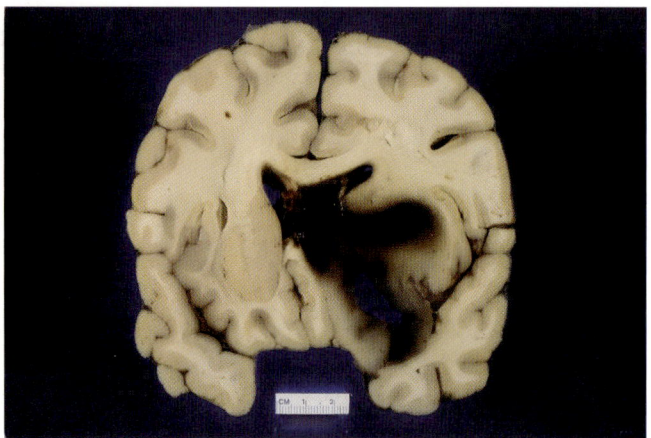

Fig. 9.73. Coronal section from Case 9.25 shows hemorrhage and a dark gelatinous, mucoid-textured area with cystic degeneration and indistinct boundaries seen along lower margin of the cyst of this oligodendroglioma.

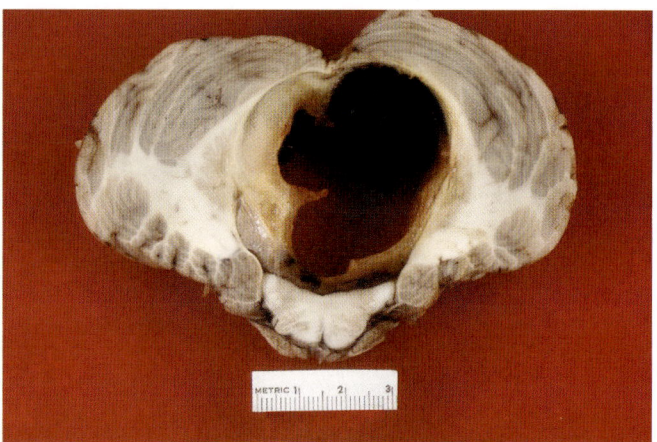

Fig. 9.74. Huge discrete cystic tumor with fresh hemorrhage shows a rim of gray-tan semitranslucent tissue replacing the cerebellar vermis. Tumor fills and massively expands the fourth ventricle, with severe compression of floor of pons and medulla. Recent cerebellar tonsillar herniation was present. Microscopically, the tumor was a pilocytic astrocytoma.

nicating hydrocephalus, with a rapid rise in intracranial pressure (Case 9.27, Fig. 9.75; Case 9.28, Fig. 9.76). A fourth mechanism is via compression of anatomic regions critical to cardiac or respiratory functions, such as the hypothalamus or medulla oblongata (Case 9.26, Fig. 9.74). A fifth mechanism is via direct neoplastic infiltration of these anatomic regions.[87] A sixth circumstance in which a tumor might theoretically cause sudden death is by precipitating an epileptic seizure with resultant apnea and/or fatal cardiac arrhythmia as discussed earlier. This mechanism might merit special consideration if prior seizures had been clinically documented, with or without knowledge of an underlying neoplasm, and no other anatomic cause of death is apparent.

A recent review summarizes many of the tumors included in the World Health Organization classification of tumors of the nervous system that had been a reported cause of SUD.[71] Additional cases that have been reported include brainstem glioma,[82] cases with multiple tumors,[83] and cases of leptomeningeal gliomatosis.[39] Phakomatoses, such as tuberous sclerosis, may produce SUD by a variety of different mechanisms, including intracranial or intracardiac tumor mass effects, fatal cardiac arrythmias, seizures, and exsanguination due to massive intratumoral hemorrhage.[16] Nonneoplastic infiltrative lesions such as sarcoidosis of the CNS are also rare causes of SUD, causing parenchymal damage by infiltration and/or mass effect.[31]

Case 9.26. (Fig. 9.74)

A 10-year-old child was reportedly well until she complained of a severe headache 3 days prior to death. She was taken to a clinic where a physician prescribed analgesics and antibiotics and sent her home. In the afternoon of the day prior to death, she was again seen by a physician and told to continue the medications. She was found dead the following morning, about 5 hours after last being seen alive.

Case 9.27. (Fig. 9.75)

A 34-year-old man was admitted to the hospital for a possible cocaine overdose. His symptoms fluctuated from agitation to lethargy, and he was thought to be acutely psychotic due to cocaine withdrawal. He became apneic, hypotensive, and comatose, and he was pronounced dead about 15 hours after admission. CT scan had shown a colloid cyst with hydrocephalus. Postmortem blood benzoylecgonine was positive.

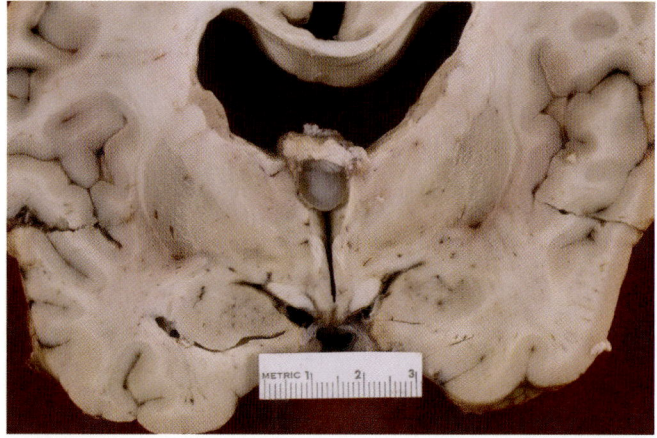

Fig. 9.75. Colloid cyst of anterior third ventricle blocks interventricular foramina with resultant hydrocephalus. Brain was markedly swollen, with bilateral parahippocampal herniation and midbrain compression.

Case 9.28. (Fig. 9.76)

A 17-year-old boy complained of intense headaches, nausea, vomiting, neck pain, and fever for 7 days prior to death, but he had not sought medical attention. On the day of death, he collapsed, and 911 was called, but he was pronounced dead at the scene.

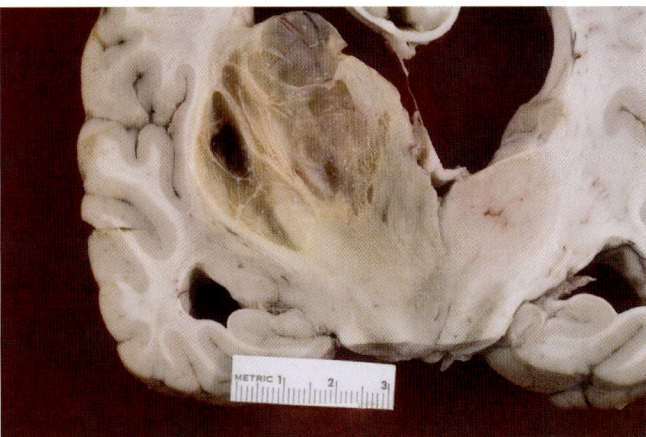

Fig. 9.76. A large and partially cystic tumor is centered in left thalamus and extended rostrally into basal ganglia, caudally into midbrain, and crossed midline at caudal thalamic level. Posterior third ventricle and aqueduct were occluded by tumor, resulting in severe hydrocephalus. Microscopic examination showed a grade II astrocytoma.

Multiple Sclerosis

In general, causes of death in multiple sclerosis (MS) are those expected in other chronic debilitating diseases.[89] Seizures attributable to MS occur, but are infrequent. Rarely, we have seen SUD cases in MS patients in which the terminal event was related to some activity that would predictably elevate body temperature, such as a hot tub bath or sunbathing.[41,53] A major contributing factor is considered to be a rapid increase in disability due to hyperthermia-exaggerated conduction impairment in demyelinated CNS nerve fibers in plaques—basically an extension of Uhthoff's phenomenon to CNS lesions other than those in visual pathways. The patient unexpectedly becomes too weak to extricate him- or herself from the heat source, leading to death before the individual can be assisted by others. Most of our cases have been found submerged in a hot tub, with anatomic findings consistent with drowning. Other mechanisms, such as heat shock proteins and other undetermined factors, may also contribute to death in such cases.[41,53]

Reye's Syndrome

This syndrome, also known as encephalopathy with fatty degeneration of the viscera, has become rare since salicylates are avoided in the treatment of childhood febrile illnesses. The few cases that do occur are more likely to be the result of inborn errors of metabolism. Most cases have a gradually increasing intensity of symptoms over several days, with high mortality rate, but rare instances of SUD are recorded.[125] The brain exhibits severe diffuse swelling and evidence of hypoxic-ischemic injury. Alzheimer type II cells may be found in sites such as the thalamus, basal ganglia, and cerebral cortex in some, but not all, cases.

Reflexes and Sudden Unexpected Death

In the introduction to this chapter, it was noted that in the category of SUD, most of the diagnosis in an individual case can be accurately determined by gross examination alone. Some exceptions to this general rule applicable to anatomically identifiable pathology have already been cited, and reflex causes of SUD now need mentioning. Most diagnoses of reflex causes of SUD are based primarily on circumstantial evidence, and diagnoses are mainly by exclusion. Clinical tests documenting, for example, a hyperactive carotid reflex or postural hypotension can only be performed on living patients.[116] In the absence of accurate witnesses to the terminal event, careful questioning of a decedent's contacts may sometimes elicit a history of prior symptoms that provide support for a diagnosis of one or another predisposing factor to reflexogenic death. Examples might include a history of episodes of dizziness, palpitations, or syncope with prolonged standing or with rapid change from a supine to standing position.[38,113] Evidence of cardiac disease may be a predisposing factor in some cases, even if insufficient by usual anatomic criteria to be considered a convincing cause of death. With these precautions in mind, the following selected reports relevant to refloxogenic death are reviewed.

Sudden Unexpected Death Related to Stressful Events

SUD has occurred closely related in time to a number of psychologically stressful events, representative examples of which are listed by Engel[27] and Saposnik et al.,[94] including after receipt of news of the collapse, death, or imminent loss of a person close to the decedent; during the following 3 weeks of grief, or upon the anniversary of loss of such a person; during an episode of extreme personal danger or threat of injury, or shortly after such an event has passed; and during the excitement of reunion, triumph, birthdays, or a happy ending. The additional presence of preexisting cardiovascular disease (Case 9.29), exertion, depression, or unusual or extreme fatigue appears to increase the risk of fatal cardiac arrhythmias associated with such emotional triggers.

Engel[27] cites evidence that favors a more complex set of reactions to such psychological stressors than a simple "vasovagal" reflex. These reactions may include an interplay between sequential or simultaneous sympathetic/parasympathetic neural reflexes (fight-flight versus conservation-withdrawal physiologic responses, respectively), particularly when the circumstances of the psychic stresses provide conflicting and uncertain stimuli as to which is the most appropriate response for the individual's survival. Such combined sympathetic and parasympathetic hyperactivity is considered the basis for the apparent contradictory physical findings of bradycardia in the presence of other findings of sympathetic hyperactivity, and may be more conducive to fatal arrhythmias than the presumably more common situation in which either sympathetic or parasympathetic reactions alone predominate.

Such studies preceded the more recent literature emphasizing the importance of catecholamine surge, for example, as a more indirect effect of sympathetic hyperactivity, or demonstrating myocardial lesions consistent with such a mechanism in certain assault victims with apparently nonfatal injuries who die.[18] Due to the circumstances of such deaths precluding careful clinical assessment in most cases, and the difficulties of converting data from animal experiments (especially in anesthetized animals) to the forensic setting, a recent report that supports the mechanism of myocardial stunning due to catecholamine surge in at least some such cases warrants the attention of medical examiners.[68,123] From a forensic perspective, the focus must also be on how the circumstances surrounding such deaths influence the determination of the manner of death.[95,115] The effect of stress on cardiac function is also believed to contribute to the increased incidence of cardiac deaths associated with mass disasters or even the threat of an impending mass disaster attack.[108]

Reflex Cardiac Arrest Secondary to Carotid Sinus Stimulation

In forensic practice, this situation more frequently arises when a neck hold is applied to an individual. The neck hold may be either the "arm bar" type of choke hold in which the midforearm (or other rigid instrument) presses directly against the anterior midline neck structures, or may be the "carotid sleeper" hold in which the tip of the elbow of the compressing arm is directly in line with the anterior midline of the neck such that flexion of the elbow causes compression of the anterolateral aspects of the neck bilaterally, rather than the anterior midline. Such holds have been regarded as a likely cause of cardiac arrest secondary to parasympathetic stimulation.[23,95] Reports such as that of Ali *et al.*,[4] which describes complete heart block with ventricular asystole in three patients during therapeutic left vagus nerve stimulation for epilepsy, tend to support this hypothesis. Such neck holds may be used to subdue violent and aggressive individuals by law enforcement personnel, when simple extremity and truncal restraint is insufficient. The violent person requiring restraint can exhibit extraordinary strength as a result of psychiatric illness and/or use of illicit drugs. Hand pressure against the neck has been another mechanism implicated in some cases,[25] as well as carotid stimulation used for diagnostic purposes.[36,40] Proper determination of cause and manner of death may be very challenging in these circumstances.

Consideration of the role of physiologic consequences of psychological factors may also be relevant to some deaths associated with apparent nonlethal wounds sustained in military or civilian combat settings, and is discussed further in Chapter 8.

"Café Coronary" Deaths

Although the mechanism of death in many cases included in this category is choking due to airway obstruction by a food bolus, others have indicated that there are instances in which the witnessed terminal circumstances

---

Case 9.29.

A 66-year-old man was assaulted, pushed and struck with a box held by the assailant, as he retreated from the attack and tried to fend off the attacker. He suffered a forehead contusion when struck by the box, but no loss of consciousness or fall. Witnesses described the decedent as "shaken up" by the incident after the assailant left the scene. The decedent walked to a chair, sat down, complained of "not feeling well," and shortly thereafter collapsed, unresponsive. The time period from cessation of the attack to collapse was estimated by witnesses to be a few minutes. Arriving paramedics found the decedent in cardiorespiratory arrest. He did not respond to resuscitative efforts and was pronounced dead shortly thereafter at the receiving hospital.

Autopsy revealed a 1½-inch-diameter midline contusion overlying the frontal sinus area, minor abrasions of both elbows, and severe multivessel coronary atherosclerosis (up to 90–95% occlusion with calcification) in a 600-g heart with areas of remote myocardial infarction. Neuropathology examination of the brain was unremarkable except for incidental findings of minimal cerebral atherosclerosis and mild to moderate basal ganglia calcification. The case was closed with cause of death listed as acute cardiac arrest due to cardiomegaly with severe coronary atherosclerosis, with other conditions listed as head trauma. Manner of death was determined to be homicide.

gave no hint of a choking or hypoxic mechanism.[95] They suggested a vasovagal mechanism of cardiac arrest secondary to stimulation of the laryngeal or pharyngeal mucosa by the food bolus as the mechanism of death in such cases.

Blunt Force Trauma to the Throat

Sudden deaths due to blows to the throat involving the hypopharynx, larynx, or carotid sinus area are said to cause reflex cardiac arrest as a result of a vagal (parasympathetic) reflex mechanism.[95] This could be applicable to cases in which evidence of excessive hemorrhage, or crushing and obstruction of the upper airway, is absent. Such a mechanism may also be theorized to play a role in some cases of strangulation deaths, whether by means of manual or ligature pressure, or by hanging.

Sudden Unexpected Death Due to Cold Water Immersion

The mechanism of sudden death associated with immersion, especially in cold water below 20°C, has been attributed to an exaggerated dive reflex.[24,80] The essential neural input consists of water touching the face, and reflex or voluntary inhibition of respiratory centers resulting in apnea.[34] Animal experiments suggest the afferent pathway to be the trigeminal and/or laryngeal nerves. The response appears to be a combination of sympathetic and parasympathetic influences, and it consists of bradycardia; peripheral vasoconstriction to a degree that reduces blood flow to all tissues except heart, lungs, and brain; and maintained or elevated blood pressure. It has been observed that either fear or alcohol ingestion potentiates this reflex, and the bradycardia may progress to cardiac arrest.[34,95] Fatal arrhythmias other than asystole may occur in some individuals with cool water exposure. Individual variation in cardiac status or autonomic responses may result in ventricular ectopic beats or extrasystoles upon immersion[33] or with water simply sprayed on the head (so that breathing could continue) in a subject immersed in water up to the neck.[49] A stimulus as mild as cold ocular irrigation under general anesthesia (which was performed prior to a planned left scleral buckle procedure) has produced a significant bradycardia that was attributed to the dive reflex.[5] Whether this mechanism accounts for some cases of "drowning without aspiration" is unclear.[77]

Burke et al.[14] reported the case of a 12-year-old boy who died suddenly after a cold drink. Ventricular fibrillation was observed during resuscitation efforts, followed by postdefibrillation asystole not responding to resuscitative efforts. At autopsy, a previously undiagnosed cardiac myxoma associated with myocardial scarring was believed to be a contributory factor to the "cold-induced reflex" arrhythmia. Saukko and Knight[95] theorized that a similar reflex mechanism (distinct from any pharmacologic effects) may be responsible for some very rapidly appearing SUDs in cases of inhalant abuse that use cooling gases sprayed directly into the pharynx and larynx. This may initiate sudden freezing temperatures into this sensitive area.[95] Endotracheal intubation alone has long been recognized by anesthesiologists as a potential trigger for cardiac arrhythmias, the predominant effect in most patients suggesting sympathetic hyperactivity (e.g., tachycardia, ventricular asystole associated with elevated blood pressure), although other patients develop bradycardias.[47]

The neuropathologist should also construct his or her opinion in such cases with an awareness that other mechanisms may play a role. For example swimming (i.e., immersion) has recently been proposed as a gene-specific arrhythmogenic trigger for inherited long QT syndrome.[1,2] Ventricular fibrillation following bradycardia is known to occur in cases of long QT syndrome, and may thus be another predisposing factor in certain immersion deaths.

The Oculocardiac and Trigeminocardiac Reflex Complex

This group of reflexes involves afferent nerve stimulation in any of the three divisions of the trigeminal nerve, and a vagus nerve efferent limb, resulting in sudden bradycardia or asystole with arterial hypotension.[32,35,84,105] When the site of stimulation involves the ophthalmic division of the trigeminal nerve, it has historically been termed the oculocardiac reflex, despite the recognition that it may involve stimuli in the orbital region in cases of prior enucleation or in cases of congenital anophthalmia,[119] or in forehead areas adjacent to the orbit.[102] The trigeminocardiac reflex is the terminology typically used in the literature for cases in which the stimulus is in the territory of the maxillary and/or mandibular divisions of the trigeminal nerve.[6,11,57,99]

The importance of these reflexes to the forensic neuropathologist and medical examiner is twofold. First, these reflexes have been observed in a variety of ophthalmic, maxillofacial, dental, and neurosurgical procedures, and in a periprocedural complication involving inadvertent intubation of the orbit during a nasal intubation procedure.[35] These reflexes also may occur in surgical procedures involving more central portions of the trigeminal nerve,[13,17] and they can mimic certain clinical signs suggestive of increasing intracranial pressure in cases of eye or orbital blunt force trauma.[110] Although now generally well recognized and usually controllable by prompt intervention by surgeons and anesthesiologists, the potential for fatality exists. Reports of three fatal cases exist in the ophthalmic surgery literature.[9,52,64] In a fourth intraoperative death occurring in the course of strabismus surgery, preexisting myocarditis was felt to be an important contributory factor.[28] It follows that such reflexes should be kept in mind when analyzing periprocedural deaths involving surgical procedures in the distribution of the trigeminal nerve.

Second, as noted earlier, the oculocardiac reflex has been described in cases of trauma in the region of the eye and orbit.[110] A recent report in the forensic literature calls attention to this reflex as a possible mechanism of death in a case of multiple stab wounds to one eye globe and bilateral periorbital regions.[62] Apparently nonfatal degrees of trauma in the second or third division of the trigeminal nerve might similarly be theorized to stimulate a trigeminal cardiac reflex, and be a possible mechanism of death. In assault cases involving these anatomic areas, the associated fear, excitement, and/or physical exertion might also predispose to the likelihood of a fatal cardiac arrhythmia[52,62] (also see previous section on psychologically stressful events and reflex cardiac arrhythmia).

Observations of oculocardiac and trigeminocardiac reflexes in clinical settings suggest that they are also more likely to occur with abrupt and sustained surgical traction of relevant structures or actual severing of a nerve, as opposed to smooth and gentle traction alone[17]; with manipulation of the medial, lateral, and superior rectus muscles but rarely with manipulation of the inferior oblique and inferior rectus muscles; and in the presence of hypoxemia, hypercarbia, pain, young age (children and adolescents are more sensitive), and possibly certain drugs.[6,7,32,35,52,105]

Sudden Unexpected Deaths Related to Miscellaneous Presumptive Reflex Mechanisms

Various authors postulate reflex cardiac inhibition as a possible mechanism leading to SUD occurring within seconds or minutes after varying types and degrees of ordinarily nonlethal peripheral trauma or stimulation such as blows to the back of the neck[22] or to the abdomen (not the precordium, in which the mechanism is regarded as commotio cordis), scrotal/testicular trauma, or cannulation of the uterine cervix.[95,107] Excess catecholamine levels secondary to associated fear or emotion is also suggested as a predisposing factor in such circumstances.

Final Comment

Reflexogenic SUD is a challenging problem for the medical examiner. Although the focus in this text is on the neuropathology and potential neurophysiologic mechanisms (cases that are often referred to the neuropathologist as a "last resort" in attempting to explain the cause of death after investigation of other possibilities have proved fruitless), these cases also serve as reminders that the medical examiner and neuropathology consultant depend on the expertise of the scene investigator, the witness interrogator, the toxicologist, and the molecular biologist, among others on the forensic team, for solutions to these cases. As one example, a case seen by us in the early days of the "rave party" fad was a young female who died in her sleep. Initial autopsy, toxicology, and medical history were negative. A skilled interrogator requestioned the decedent's family, who initially denied any drug use by the decedent, and extracted the new information that she had been abusing gamma hydroxybutyric acid (GHB) for the past year at numerous rave parties. Determination of GHB tissue levels (not a component of the routine drug screen previously performed) revealed levels exceeding those found in other reported fatalities ascribed to GHB. A search for possible rapid eye movement sleep-related reflexogenic mechanisms of death, and so on was obviously unnecessary.

Postmortem molecular analysis will also, we believe, prove useful in providing a more definitive diagnosis in some of these cases by exposing predisposing factors that spontaneously (or in some instances by various environmental triggers) may result in SUD. Gene mutations associated with long QT syndrome[1,2,24,114] and some forms of Brugada's syndrome,[54,67,86] for example, can now be identified by postmortem genetic screening. This can be helpful diagnostically, and may also result in medical intervention and prevention of deaths in similarly affected relatives of the decedent. Study of these gene mutations may also eventually provide a better understanding of how fever, sleep and awakening, fear, exercise, swimming and other yet-undefined factors can precipitate potentially fatal arrythmias.

References

1. Ackerman MJ, Tester DJ, Driscoll DJ. Molecular autopsy of sudden unexplained death in the young. Am J Forensic Med Pathol 2001;22:105–111.
2. Ackerman MJ, Tester DJ, Porter CJ. Swimming, a gene-specific arrythmogenic trigger for inherited long QT syndrome. Mayo Clin Proc 1999;74:1088–1094.
3. Annegers JF, Coan SP. SUDEP: Overview of definitions and review of incidence data. Seizure 1999;8:347–352.
4. Ali II, Pirzada NA, Kanjwal Y, et al. Complete heart block with ventricular asystole during left vagus nerve stimulation for epilepsy. Epilepsy Behav 2004;5:768–771.
5. Arndt GA, Stock MC. Bradycardia during cold ocular irrigation under general anaesthesia: An example of the diving reflex. Can J Anaesth 1993;40:511–514.
6. Bainton R, Barnard N, Wiles JR, Brice J. Sinus arrest complicating a bitemporal approach to the treatment of panfacial fractures. Br J Oral Maxillofac Surg 1990;28:109–110.
7. Bainton R, Lizi E. Cardiac asystole complicating zygomatic arch fracture. Oral Surg Oral Med Oral Pathol 1987;64:24–25.
8. Bebin J, Smith EE. Vascular malformations of the brain. In Smith RR, Haerer A, Russell WF (eds): Vascular Malformations. New York: Raven Press, 1982.
9. Bietti GB. Problems of anesthesia in strabismus surgery. Int Ophthal Clin 1966;6:727–737.
10. Black M, Graham DI. Sudden unexplained death in adults caused by intracranial pathology. J Clin Pathol 2002;55:44–50.
11. Blanc VF. Editorial CK. Trigeminocardiac reflexes. Can J Anaesth 1991;38:696–699.

12. Blisard KS, McFeeley PJ. The spectrum of neuropathologic findings in deaths associated with seizure disorders. J Forensic Sci 1988;33:910–914.
13. Braun JA, Preul MC, Nimr S. Trigeminocardiac reflexes (letter to editor, with reply by Lang SA, Blanc VF). Can J Anaesth 1992;39:303–305.
14. Burke AP, Afzal MN, Barnett DS, Virmani R. Sudden death after a cold drink: Case report. Am J Forensic Med Pathol 1999;20:37–39.
15. Byard RW. Sudden Death in Infancy, Childhood and Adolescence, ed 2. Cambridge: Cambridge University Press, 2004.
16. Byard RW, Blumbergs PC, James RA. Mechanisms of unexpected death in tuberous sclerosis. J Forensic Sci 2003;48:172–176.
17. Cha ST, Eby JB, Katzen JT, Shahinian HK. Trigeminocardiac reflex: A unique case of recurrent asystole during bilateral trigeminal sensory root rhizotomy. J Craniomaxillofac Surg 2002;30:108–111.
18. Cebelin MS, Hirsch CS. Human stress cardiomyopathy: Myocardial lesions in victims of homicidal assaults without internal injuries. Hum Pathol 1980;11:123–132.
19. Chang BS, Lowenstein DH. Epilepsy. N Engl J Med 2003;349:1257–1266.
20. Contostavlos DL. Massive subarachnoid hemorrhage due to laceration of the vertebral artery associated with fracture of the transverse process of the atlas. J Forensic Sci 1971;16:40–56.
21. Dasheiff RM, Dickinson LJ. Sudden unexpected death of epileptic patient due to cardiac arrhythmia after seizure. Arch Neurol 1986;43:194–196.
22. Davis GG, Glass JM. Case report of sudden death after a blow to the back of the neck. Am J Forensic Med Pathol 2001;22:13–18.
23. DiMaio VJ, DiMaio D. Forensic Pathology, ed 2. Boca Raton, FL: CRC Press, 2001.
24. DiPaolo M, Luchini D, Bloise R, Priori SG. Postmortem molecular analysis in victims of sudden unexplained death. Am J Forensic Med Pathol 2004;25:182–184.
25. Eisele JW, Berry GJ, Ackerman MJ, Tester DJ. Sudden death following brief compression of the neck (abstract). Proc Am Assoc Forensic Sci 2005;11:240–241.
26. Eisenstat J. Sudden death resulting from a colloid cyst of the third ventricle. Forensic Pathology No. FP 04-5 (FP-296). Chicago: ASCP, 2004;46:51–62.
27. Engel GL. Psychologic stress, vasodepressor (vasovagal) syncope, and sudden death. Ann Intern Med 1978;89:403–412.
28. Fayon M, Gauthier M, Blanc VF, et al. Intraoperative cardiac arrest due to the oculocardiac reflex and subsequent death in a child with occult Epstein-Barr virus myocarditis. Anesthesiology 1995;83:622–624.
29. Fishman RA. Brain edema. N Engl J Med 1975;293:706–711.
30. Fujikawa DG, Itabashi HH, Wu A, Shinmei SS. Status epilepticus-induced neuronal loss in humans without systemic complications of epilepsy. Epilepsia 2000;41:981–991.
31. Gleckman AM, Patalas ED, Joseph JT. Sudden unexpected death resulting from hypothalamic sarcoidosis. Am J Forensic Med Pathol 2002;23:48–51.
32. Gold RS, Pollard Z, Buchwald IP. Asystole due to the oculocardiac reflex during strabismus surgery: A report of two cases. Ann Ophthalmol 1988;20:473–477.
33. Goode RC, Duffin J, Miller R, et al. Sudden cold water immersion. Respir Physiol 1975;23:301–310.
34. Gooden BA. Drowning and the dive reflex in man. Med J Aust 1972;2:583–587.
35. Green JG, Wood JM, Davis LF. Asystole after inadvertent intubation of the orbit. J Oral Maxillofac Surg 1997;55:856–859.
36. Greenwood RJ, Dupler DA. Death following carotid sinus pressure. JAMA 1962;181:605–609.
37. Grisolia JS, Widerholt WC. CNS cysticercosis. Arch Neurol 1982;39:540–544.
38. Grubb BP. Neurocardiogenic syncope. N Engl J Med 2005;352:1004–1010.
39. Havlik DM, Becher MW, Nolte KB. Sudden death due to primary diffuse leptomeningeal gliomatosis. J Forensic Sci 2001;46:392–395.
40. Hilal H, Massumi R. Fatal ventricular fibrillation after carotid-sinus stimulation. N Engl J Med 1966;275:157–158.
41. Henke AF, Cohle SD, Cottingham SL. Fatal hyperthermia secondary to sunbathing in a patient with multiple sclerosis. Am J Forensic Med Pathol 2000;21:204–206.
42. Hirsch CS, Martin DL. Unexpected death in young epileptics. Neurology 1971;21:682–690.
43. Holmes B, Harbaugh RE. Traumatic intracranial aneurysms: A contemporary review. J Trauma 1993;35:855–860.
44. Howell SJL, Blumhardt LD. Cardiac asystole associated with epileptic seizures: A case report with simultaneous EEG and ECG. J Neurol Neurosurg Psychiatry 1989;52:795–798.
45. Jackson FE, Gleave JRW, Janon E. The traumatic cranial and intracranial aneurysms. In Vinken PJ, Bruyn GW, Braakman R (eds): Handbook of Clinical Neurology. Part II: Injuries of the Brain and Skull. Amsterdam: North-Holland, 1976;24:381–398.
46. Jay G, Leetsma J. Sudden death in epilepsy: A comprehensive review of the literature and proposed mechanisms. Acta Neurol Scand Suppl. 1981;82:1–66.
47. Katz RL, Bigger Jr JT. Cardiac arrhythmia during anesthesia and operation. Anesthesiology 1970;33:193–213.
48. Kawai M, Goldsmith IL, Verma A. Differential effects of left and right hemispheric seizure onset on heart rate. Neurology 2006;66:1279–1280.
49. Keating WR, Hayward MG. Sudden death in cold water and ventricular arrhythmia. J Forensic Sci 1981;26:459–461.
50. Kiok MC, Terrence CF, Fromm GH, Lavine S. Sinus arrest in epilepsy. Neurology 1986;36:115–116.
51. Kirkpatrick JB. Neurologic infections due to bacteria, fungi, and parasites. In Davis RL, Robertson DM (eds): Textbook of Neuropathology, ed 3. Baltimore: Williams & Wilkins, 1997.
52. Kirsch RE, Samet P, Kugel V, Axelrod S. Electrocardiographic changes during ocular surgery and their prevention by retrobulbar injection. Arch Ophthalmol 1957;58:348–356.
53. Kohlmeir RE, DiMaio VJM, Kagan-Hallet K. Fatal hyperthermia in hot baths in individuals with multiple sclerosis. Am J Forensic Med Pathol 2000;21:201–203.
54. Kum LCC, Fung JWH, Sanderson JE. Brugada syndrome unmasked by febrile illness. Pacing Clin Electrophysiol 2002;25:1660–1661.
55. Kuper S, Mendelow H, Proctor NSF. Internal hydrocephalus caused by parasitic cysts. Brain 1958;81:235–242.
56. Kutt H, Haynes J, McDowell F. Some causes of ineffectiveness of diphenylhydantoin. Arch Neurol 1966;14:489–492.
57. Lang S, Lanigan DT, van der Wal M. Trigeminocardiac reflexes: Maxillary and mandibular variants of the oculocardiac reflex. Can J Anaesth 1991;38:757–760.
58. Leestma JE. Forensic Neuropathology. New York: Raven Press, 1988.
59. Leestma JE, Kalelkar MB, Teas SS, et al. Sudden unexpected death associated with seizures: Analysis of 66 cases. Epilepsia 1984;25:84–88.
60. Liedholm LJ, Gudjonsson O. Cardiac arrest due to partial epileptic seizures. Neurology 1992;42:824–829.
61. Lim ECH, Lim S-H, Wilder-Smith E. Brain seizes, heart ceases: A case of ictal asystole. J Neurol Neurosurg Psychiatry 2000;69:557–559.
62. Lynch MJ, Parker H. Forensic aspects of ocular injury. Am J Forensic Med Pathol 2000;21:124–126.
63. Maguire JH. Tapeworms and seizures—Treatment and prevention. N Engl J Med 2004;350:215–217.

64. Mallinson FB, Coombes SK. A hazard of anesthesia in opththalmic surgery. Lancet 1960;1:574–575.
65. Mant AK. Traumatic subarachnoid hemorrhage following blows to the neck. J Forensic Sci Soc 1972;12:567–572.
66. Margerison JH, Corsellis JAN. Epilepsy and the temporal lobes: A clinical, electroencephalographic and neuropathological study of the brain in epilepsy, with particular reference to the temporal lobes. Brain 1966;89:499–530.
67. Marill KA, Ellinor PT. Case 37:2005: A 35-year-old man with cardiac arrest while sleeping. N Engl J Med 2005;353:2492–2501.
68. Maseri A, Kurisu S, Inoue I, et al. Myocardial stunning due to sudden emotional stress (letter to editor). N Engl J Med 2005;352:1923–1924; Wittstein IS, Champion HC (authors' reply). Ibid, pp 1924–1925.
69. Mathern GW, Adelson PD, Cahan LD, Leite JP. Hippocampal neuron damage in human epilepsy: Meyer's hypothesis revisited. Prog Brain Res 2002;135:237–251.
70. Mathern GW, Kuhlman PA, Mendoza D, Pretorius JK. Human fascia dentata anatomy and hippocampal neuron densities differ depending on the epileptic syndrome and age at first seizure. J Neuropathol Exp Neurol 1997;56:199–212.
71. Matschke J. Primary cerebral neoplasms as a cause of sudden unexpected death. In Tsokos M (ed): Forensic Pathol Rev 2005;2:45–58.
72. Matsumoto N, Whisnant JP, Kurland LT, Okazaki H. Natural history of stroke in Rochester, Minnesota, 1955 through 1969: An extension of a previous study, 1945 through 1954. Stroke 1973;4:20–29.
73. McComb JG. Recent research into the nature of cerebrospinal fluid formation and absorption. J Neurosurg 1983;59:369–383.
74. McCormick GF, Zee C-S, Heiden J. Cysticercosis cerebri: Review of 127 cases. Arch Neurol 1982;39:534–539.
75. Meldrum BS. Concept of activity-induced cell death in epilepsy: Historical and contemporary perspectives. Prog Brain Res 2002;135:3–11.
76. Miller JD, Ironside JW. Raised intracranial pressure, oedema and hydrocephalus. In Graham DI, Lantos PL (eds): Greenfield's Neuropathology, ed 6. London: Arnold, 1997.
77. Modell JH, Bellefleur M, Davis JH. Drowning without aspiration: Is this an appropriate diagnosis? J Forensic Sci 1999;44:1119–1123.
78. Nahed BV, DiLuna ML, Morgan T, et al. Hypertension, age, and location predict rupture of small intracranial aneurysms. Neurosurgery 2005;57:676–683.
79. Nashef L, Walker F, Allen P, et al. Apnoea and bradycardia during epileptic seizures: Relation to sudden death in epilepsy. J Neurol Neurosurg Psychiatry 1996;60:297–300.
80. Newman AB, Stewart RD. Submersion incidents. In Auerbach PS (ed): Wilderness Medicine: Management of Wilderness and Environmental Emergencies, ed 3. St. Louis: Mosby, 1995.
81. Olney JW, Sharpe LG, Geigin R. Glutamate-induced brain damage in infant primates. J Neuropathol Exp Neurol 1972;31:464–488.
82. Opeskin K, Ruszkiewicz A, Anderson RMcD. Sudden death due to undiagnosed medullary-pontine astrocytoma. Am J Forensic Med Pathol 1995;16:168–171.
83. Ortiz-Reyes R, Dragovic L, Eriksson A. Sudden unexpected death resulting from previously nonsymptomatic subependymoma. Am J Forensic Med Pathol 2002;23:63–67.
84. Palm E, Strömblad R. Respiratory and circulatory responses to manipulations of the eye. Acta Ophthalmol 1954;32:615–629.
85. Pau A, Perria C, Turtas S, et al. Long-term follow-up of the surgical treatment of intracranial coenurosis. Br J Neurosurg 1990;4:39–44.
86. Porres JM, Brugada J, Urbistondo V, et al. Fever unmasking the Brugada syndrome. Pacing Clin Electrophysiol 2002;25:1646–1648.
87. Rajs J, Rosten-Almqvist P, Nennesmo I. Unexpected death in two young infants mimics SIDS: Autopsies demonstrate tumors of medulla and heart. Am J Forensic Med Pathol 1997;18:384–390.
88. Richardson JC, Hyland HH. Intracranial aneurysms: A clinical and pathological study of subarachnoid and intracerebral hemorrhage caused by berry aneurysms. Medicine 1941;20:1–83.
89. Riudavets MA, Colegial C, Rubio A, et al. Causes of unexpected death in patients with multiple sclerosis: A forensic study of 50 cases. Am J Forensic Med Pathol 2005;26:244–249.
90. Russell DS. Discussion: The pathology of spontaneous intracranial haemorrhage. Proc R Soc Med 1954;47:689–704.
91. Sahs AL, Nibbenlink DW, Turner JC (eds): Report of the Cooperative Study. Aneurysmal Subarachnoid Hemorrhage. Baltimore-Munich: Urban & Schwarzenberg, 1981.
92. Sahs AL, Perret GE, Locksley HB, Nishioka H. Intracranial Aneurysms and Subarachnoid Hemorrhage: A Cooperative Study. Philadelphia: JB Lippincott, 1969.
93. Sano K, Malamud N. Clinical significance of sclerosis of cornu ammonis. Arch Neurol Psychiatry 1953;70:40–53.
94. Saposnik G, Baibergenova A, Dang J, Hachinski V. Does a birthday predispose to vascular events? Neurology 2006;67:300–304.
95. Saukko P, Knight B. Knight's Forensic Pathology, ed 3. London: Arnold, 2004.
96. Schwartz PJ, Zaza A, Locati E, Moss AJ. Stress and sudden death: The case of the long QT syndrome. Circulation 1991;83(4 Suppl):II-71–II-80.
97. Schwender LA, Troncoso JC. Evaluation of sudden death in epilepsy. Am J Forensic Med Pathol 1986;7:283–287.
98. Senegor M. Traumatic pericallosal aneurysm in a patient with no major trauma: Case report. J Neurosurg 1991;75:475–477.
99. Seo K, Takayama H, Araya Y, et al. A case of sinus arrest caused by opening the mouth under general anesthesia. Anesth Prog 1994;41:17–18.
100. Shields LB, Hunsaker DM, Hunsaker JC III, Parker Jr JC. Sudden unexpected death in epilepsy: Neuropathologic findings. Am J Forensic Med Pathol 2002;23:307–314.
101. Shorvon S. Does convulsive status epilepticus (SE) result in cerebral damage or affect the course of epilepsy—The epidemiological and clinical evidence? Prog Brain Res 2002;135:85–93.
102. Slade CS, Cohen SP. Elicitation of the oculocardiac reflex during endoscopic forehead lift surgery. Plast Reconstr Surg 1999;104:1828–1830.
103. Slemmer JE, Dezeeuw CI, Weber JT. Don't get too excited: Mechanisms of glutamate-mediated Purkinje cell death. Prog Brain Res 2005;148:368–390.
104. Somers GR, Smith CR, Perrin DG, et al. Sudden unexpected death in infancy and childhood due to undiagnosed neoplasia: An autopsy study. Am J Forensic Med Pathol 2006;27:64–69.
105. Sorenson EJ, Gilmore JE. Cardiac arrest during strabismus surgery. Am J Ophthal 1956;41:748–752.
106. Soria ED. Traumatic aneurysms of cerebral vessels: A case study and review of the literature. Angiology 1988;39:609–615.
107. Spitz WU (eds). Spitz and Fisher's Medicolegal Investigation of Death: Guidelines for the Application of Pathology to Crime Investigation; ed 3. Springfield, IL: Charles C Thomas, 1993.
108. Stalnikowicz R, Tsafrir A. Acute psychosocial stress and cardiovascular events. Am J Emerg Med 2002;20:488–491.
109. Stebbens WE. Intracranial arterial aneurysms. Austral Ann Med 1954;3:214–218.
110. Stortebecker TP. Posttraumatic oculocardiac syndrome from a neurosurgical point of view. J Neurosurg 1953;10:682–686.
111. Sutula TP, Pitkänen H. More evidence for seizure-induced neuron loss. Is hippocampal sclerosis both cause and effect of epilepsy? Neurology 2001;57:169–170.
112. Sutula T, Pitkänen A. Summary: Evidence for seizure-induced damage in human studies: Epidemiology, pathology, imaging, and clinical studies. Prog Brain Res 2002;135:315–317.

113. Thomas JE. Hyperactive carotid sinus reflex and carotid sinus syncope. Mayo Clin Proc 1969;44:127–139.
114. Towbin JA, Wang Z. Genotype and severity of long QT syndrome. Arch Pathol Lab Med 2001;125:116–121.
115. Turner SA, Barnard JJ, Spotswood SD, Prahlow JA. "Homicide by heart attack" revisited. J Forensic Sci 2004;49:598–600.
116. Varga E, Wórum F, Szabó Z, et al. Motor vehicle accident with complete loss of consciousness due to vasovagal syncope. Forensic Sci Int 2002;130;156–159.
117. Victor M, Ropper AH. Adams and Victor's Principles of Neurology, ed 7. New York: McGraw-Hill, 2001.
118. Walton JN. Subarachnoid Hemorrhage. Edingburgh: E&S Livingstone, 1956.
119. Ward B, Bass S. The oculocardiac reflex in a congenitally anophthalmic child. Pediatr Anaesth 2001;11:372–373.
120. Wasterlain CG, Fukijawa DG, Penix L, Sankar R. Pathophysiological mechanisms of brain damage from status epilepticus. Epilepsia 1993;34(Suppl 1):S37–S53.
121. *Webster's Third New International Dictionary.* Chicago: G&C Merriam, 1966.
122. Whetsell WO. Current concepts of excitotoxicity. J Neuropathol Exp Neurol 1996;55:1–13.
123. Wittstein IS, Thiemann DR, Lima JAC, et al. Neurohumoral features of myocardial stunning due to sudden emotional stress. N Engl J Med 2005;352:539–548.
124. Wojtacha M, Bazowski P, Mandera M, et al. Cerebral aneurysms in childhood. Child's Nerv Syst 2001;17:37–41.
125. Young TW. Reye's syndrome: A diagnosis occasionally first made at medicolegal autopsy. Am J Forensic Med Pathol 1992;13:21–27.

Responses of the Central Nervous System to Acute Hypoxic-Ischemic Injury and Related Conditions

10

INTRODUCTION 289
CLINICAL FEATURES 289
SELECTIVE VULNERABILITY OF THE CENTRAL NERVOUS SYSTEM TO HYPOXIC-ISCHEMIC INJURY: CELLULAR ASPECTS 291
UNUSUAL PATTERNS OF CENTRAL NERVOUS SYSTEM INJURY IN WHICH A COMPONENT OF HYPOXIC-ISCHEMIC INJURY APPEARS TO BE A MAJOR CONTRIBUTORY FACTOR 293
HEAT STROKE 295
EPILEPSY 297
ASPHYXIA 297
HYPOGLYCEMIA 297
RED NEURONS, DARK NEURONS, AND THE CATEGORY OF "I'M NOT SURE HOW TO INTERPRET THESE NEURONS" 298
REFERENCES 304

Introduction

The cerebrovascular diseases reviewed in Chapter 11 can injure the central nervous system (CNS) via mechanisms such as an altered cerebrospinal fluid (CSF) chemical milieu in intraventricular hemorrhage, mass effect in intracerebral hemorrhage, or obstruction of blood supply to a given area. However, a feature shared by all these conditions is some component of hypoxic-ischemic injury. This injury mechanism leads to numerous questions in forensic cases, and this chapter addresses several of those most frequently encountered in this office.

Clinical Features

Time Interval to Incapacitation

Rossen et al.,[80] using a cervical pressure cuff in 126 normal male subjects (age range, 17–31 years), demonstrated that rapid and essentially complete interruption of circulation to the brain by this means resulted in severe functional impairment within a range of 4.0 to 10.0 seconds and, in almost half the subjects, within 5.0 to 5.5 seconds. These experiments were not designed to determine whether such a narrow time range of loss of voluntary motor function could be lengthened by, for example, a reward system that would alter the subjects' motivation, and thus allow some extrapolation to situations where one's very physical survival is at stake. They do, however, provide some general reference value for what we would interpret as "speed of incapacitation" under these hypoxic-ischemic conditions.

When respiration ceases but circulation is uninterrupted, the time interval to incapacitation is much longer. Hong[48] reported that pearl divers of the Tuamotu Archipelago "make repetitive dives to a depth of 30 to 40 meters (each dive lasting about 1.5–2.5 minutes) for about 6 hours a day during the diving season." He commented that 10 to 30% of these divers developed symptoms by the end of the day that could include nausea, vertigo, partial or complete paralysis, temporary unconsciousness, and even fatal outcomes. These complications were speculated to be related to nitrogen retention.[48]

In 1992 an organization was established to promote education, competitions, and recognition of world records for free diving. Called AIDA (International Association for Development of Apnea) (Authors' note to reader: this is not a misprint), it lists the current unaided static apnea (i.e., holding breath face down in pool) world records for males as 8 minutes 58 seconds, and for females, 7 minutes and 30 seconds.[1] The world record for static apnea achieved with the aid of prebreathing 100% oxygen is reportedly 12 minutes 47 seconds.[30] Pre- and post-training blood gases, pulmonary function studies, serial tests of cognition, and other measurements would be of interest in these competitors, but have not come to our attention.

Time Interval to Loss of Consciousness

Rossen et al.[80] found that actual loss of consciousness (LOC) in the cervical pressure cuff study described earlier quickly followed conscious functional impairment. LOC in these subjects occurred within a range of approximately 4.5 to 11 seconds, with most subjects becoming unconscious after 6.0 to 6.5 seconds (average, 6.8 seconds). In other words, LOC generally followed onset of functional impairment within approximately 0.5 to 1.0 second. Other sources were quoted indicating that loss of consciousness after complete cardiac arrest usually occurred within 8 seconds, and regularly occurred within 12 seconds.[80]

Recovery of consciousness, if cervical cuff pressure was released immediately following LOC, required from 3.0 to 11.5 seconds and was more variable in a given subject than the interval from cuff inflation to loss of consciousness. If cuff pressure was maintained for up to 100 seconds, consciousness was regained in 30 to 40 seconds and the subject could walk from the room within 2 minutes after the procedure.[80]

Less complete compromise of circulation to the brain in the form of a judo choke hold applied by one individual to another is also efficient in producing loss of consciousness. A properly applied judo choke hold, termed "shime waza," produces unconsciousness in approximately 10 seconds (range, 8 to 14 seconds).[59] Vertebral artery flow is not compromised by this maneuver. The judo practitioner is taught to recognize the loss of consciousness and to release pressure on the neck immediately, which results in the subject regaining consciousness spontaneously within 10 to 20 seconds. Pressure of 250 mm Hg on the neck, or 5 kg (11 lb) of rope pressure on the neck, are reportedly sufficient to occlude carotid arteries.[59] Pressures of 4.4 lb will occlude jugular veins,[29] and estimates for the amount of pressure needed to collapse the airway vary from 33 lb (for trachea)[29] to 66 lb (for "airway").[59] Conversion of such varied pressure measurements to identical units would require detailed information on techniques used, surface area involved, and so on, not consistently provided in such sources.

Deaths related to choke holds in law enforcement settings are typically ascribed to improper technique resulting in injury to neck structures.[59] Confounding factors include attempted restraint of agitated combative individuals who may be under the influence of drugs known to increase pain tolerance and/or cause cardiac arrhythmias. Decedents among our cases include suspects who engaged in furious, directed attacks with intent to inflict serious or fatal injury upon law enforcement officers attempting nonlethal restraint. Such factors demand consideration in judicial decisions of suspect fatalities resulting from such encounters. A recent review of this general topic is available.[28]

In judicial hangings, loss of consciousness is reportedly immediate with proper technique, since the mechanism of death is related to injuries that include upper cervical fracture-dislocation with cord compression or transection, vertebral artery laceration, and intracranial hemorrhage[29] (also see Chap. 17). Sauvageau and Racette[82] described sequential agonal events in a 37-year-old man who filmed his suicide hanging, one in which the airway was not completely occluded. There was loss of consciousness at 13 seconds, convulsions at 15 seconds, decorticate posturing at 21 seconds, decerebrate posturing at 46 seconds, a second episode of decerebrate posturing and saliva flowing from the mouth at 1 minute 11 seconds, loss of muscle tone with a few isolated muscle movements at 1 minute 38 seconds, cessation of all respiratory movements at 2 minutes, and the last isolated muscle movement was noted at 4 minutes and 10 seconds.

Time Interval to Onset of Irreversible Central Nervous System Damage

Estimates of time required for irreversible cellular damage following CNS oxygen deprivation vary widely, with claims of evidence of microscopic CNS damage being evident within 30 to 40 seconds.[88] The authors are not aware of convincing data to support such rapid morphologic changes. It is clear that neuronal vulnerability to hypoxic-ischemic insults can vary with age and other circumstances.

Pregnant women are reportedly more vulnerable to cerebral anoxia than nonpregnant women, and can develop irreversible brain damage within 4 to 6 minutes after cardiac arrest.[56] The pregnant woman is also more vulnerable to cerebral anoxia than her unborn child. Katz et al.[56] surveyed the literature from 1900 to 1985, and found that perimortem cesarean sections performed within 5 minutes of the mother's death produced "normal" infants in 100% of 42 reported cases. After 6 to 10 minutes' delay, 7 of 8 infants were reported to be normal and 1 had mild neurologic sequelae. After 11 to 15 minutes' delay, 6 of 7 infants were normal and 1 had severe neurologic sequelae. After a delay of 16 to 20 minutes, 1 infant had severe neurologic sequelae, and after a 21+-minute delay between maternal death and delivery of the infant by cesarean section, 1 infant was normal and 2 had severe neurologic sequelae. They recommended cesarean delivery within 4 minutes of maternal cardiac arrest to obtain optimum infant survival. They also commented that resuscitation of the mother was more likely to be successful after such perimortem cesarean section than before the procedure, owing to improved venous return and possibly other factors. In contrast, the cellular hypoxia caused by carbon monoxide (CO) intoxication is more likely to cause fetal than maternal death, owing to the altered CO kinetics in fetal hemoglobin compared with maternal hemoglobin.[34]

Adults undergoing temporary arterial occlusion during cerebral aneurysm surgery provide a situation difficult to extrapolate to an awake, alert individual. Such cases do, however, underscore the degree of individual variation in resistance to localized hypoxic-ischemic brain injury that can occur. In this unique intraoperative setting, arterial occlusion lasting less than 14 minutes in an anesthetized normothermic patient produced no cerebral infarction in a series of 121 patients reported by Samson et al.[81] Arterial occlusion for 19 minutes was tolerated well by 95% of cases, and arterial occlusion for greater than 31 minutes resulted in infarction in 100% of cases.

Selective Vulnerability of the Central Nervous System to Hypoxic-Ischemic Injury: Cellular Aspects

The term "selective vulnerability" in the CNS is employed to refer to the fact that some cells within the CNS are more susceptible than others to irreversible injury with diffuse hypoxic-ischemic insults.[83] In mature brains, the most vulnerable cells are neurons, followed in turn by oligodendrocytes, astrocytes, and microglia. In the perinatal period, the order of vulnerability is not yet as clearly defined, and seems to vary with age group and other unidentified factors.[93]

Particularly in newborns, hypoxic-ischemic injury may lead not only to cellular necrosis but also to apoptosis. Apoptosis refers to spontaneous cell death that typically occurs as a physiologic event during development in many growing tissues, including the nervous system. It occurs as a result of activation of specific genes, and can also be precipitated by external factors such as physical and chemical agents, and hypoxic-ischemic injury. The histologic appearance of apoptosis is sufficiently variable and nonspecific, however, to cause confusion with cell necrosis (or nonapoptotic cell death). Differentiation of these two forms of cell death thus requires molecular biologic techniques, and routine light microscopy cannot reliably separate these processes in conditions such as neoplasia, ischemic insult, and viral infections.[58,61] Apoptosis of neurons may sometimes mimic necrosis of neurons, with nuclear karyolysis, pyknosis, karyorrhexis, cytoplasmic intense eosinophilia, and loss of cellular structure and fragmentation occurring in both conditions. One feature that may be helpful in distinguishing these two forms of cell death is that apoptotic cells tend to die as individuals, whereas necrotic cells tend to die in groups.[61] Apoptosis occurs rapidly ("in about 30 minutes,"[79] or is "completed in less than an hour"[46]).

Patterns of selective vulnerability to hypoxic-ischemic injury in the CNS vary with both known and unknown factors. Age, degree and duration of insult, adequacy of collateral circulation in cases of localized compromise of blood flow, general health status, and other unknown factors can influence the outcome of otherwise seemingly identical insults. Such factors can also complicate interpretations of microscopic findings, such as the relative immaturity of Sommer's sector neurons the of hippocampus proper prior to approximately 2 years of age (see Chap. 2). Nonetheless, certain sequelae of hypoxic-ischemic or other injury can be useful in dating the likely developmental period of insult, when found in brains of individuals who have succumbed months or years later (e.g., ulegyria and others; see Chaps. 4 and 5). *The hierarchy of cell vulnerability that we discuss subsequently is best viewed as a general guide in various areas of the nervous system, and difficult-to-predict combinations of injury in the brain are commonly encountered.*

Table 10.1 is an effort to compile a large amount of information into a practical format. The majority of the table refers to neuronal vulnerability, with anatomic areas of increased vulnerability listed in the left column. Also included are the age-related vulnerability of

CNS Responses to Acute Hypoxic-Ischemic Injury

Table 10.1. Areas Most Vulnerable to Hypoxic-Ischemic Injury (Varies with Brain Maturation[a])

Brain Region	Premature Infant	Term Infant and Early Childhood	After 2+ Years–Postnatal Brain
Cerebral neocortex	Some evidence suggests increased vulnerability of watershed areas.	Calcarine cortex; pre- and postcentral gyri; depths of sulci > crowns of gyri	Layer III > layer V/VI > layer II/IV.[13] Layers III, IV,V > I, II, VI; depths and sides of sulci > crests of gyri. Watershed areas > central zones of arterial supply; parietal/occipital > temporal/frontal poles.
Hippocampus Sommer's sector (CA1)	Single report of selective CA1 necrosis in 22-week gestation fetus[40]	More common than in premature infant	CA1 > CA3/4 > CA2 > dentate fascia or parahippocampal gyrus
Subiculum (often with ventral pons)	More common than in term infant		Rare
Entorhinal cortex		More common than in premature infant	Uncommon
Basal ganglia/ thalamus (as a group)	Slightly more common in prematures, but also seen in term		May occur
Caudate and putamen	—	Outer ½ caudate + outer ½ putamen > caudate inner ½; small neurons > large neurons	
Globus pallidus	—	Inner and outer segments > caudate and putamen	
Amygdala	—		Basolateral > corticomedial
Thalamus (excluding lateral geniculate body)	—	Anterior, dorsal, medial, and ventrolateral nuclei	Anterior nucleus > dorsomedial nucleus (parvocellular portion) > lateral angles of ventrolateral nucleus > centromedian nucleus > sensory relay, midline and intralaminar nuclei
Lateral and medial geniculate body	Equally common in premature and term infants		Lateral > medial geniculate body
Hypothalamus Supraoptic n. Paraventricular n. Lateral nuclei Mammillary body		More common than in premature infants	Rare in adult
Brainstem Cranial nerve nuclei III IV Motor nucleus V Motor nucleus VII Dorsal cochlear nucleus Vestibular nuclei Dorsal motor nucleus X Nucleus ambiguus	Equally common in prematures and terms infants (plus spinal nucleus V in term infants)		Not uncommon

Table 10.1. Continued

Brain Region	Premature Infant	Term Infant and Early Childhood	After 2 + Years–Postnatal Brain
Substantia nigra	Equally common in premature and term infants		Reticular zone > compact zone
Inferior colliculi		More common at term than in premature infants	May occur
Midbrain reticular formation and pontine reticular formation	Equally common in premature and term infants		May occur
Ventral pontine nuclei	More common in premature infants		Uncommon
Inferior olivary nucleus	More common in premature infants		Common
Cuneate and gracile nuclei		More common in term infants	Occasional
Cerebellum Purkinje cells Granule cells	Equally common in premature and term infants		Purkinje cells > granule cells > Golgi type II neurons, dentate nucleus, and other deep cerebellar nuclei
Spinal cord Anterior horn cells alone		More common in term infants	
Infarction (including anterior horn)	More common in premature infants		Not uncommon
Periventricular white matter injury	"		Uncommon
Periventricular germinal matrix hemorrhage	"		NA

Data from Armstrong[4]; Brierley and Graham[13]; Brierley et al.[14]; Garcia and Anderson[41]; Kinney and Armstrong[57,58]; Norenberg and Bruce-Gregorios[71]; and Volpe.[93]
[a]Premature and neonatal brain data predominantly from Volpe JJ. Neurology of the Newborn, ed 3. Philadelphia, WB Saunders, 1995 with permission.
NA, not applicable.

premature infant proliferating glial precursors in the periventricular area, and of blood vessel rupture in the subependymal germinal matrix. It is acknowledged that factors other than simply hypoxic-ischemic conditions may be contributory to these lesions (e.g., neurotransmitter-mediated excitotoxicity).[45] The varying vulnerabilities with increasing age are to be regarded as estimates based on isolated cases, or as generalizations. No large single study specifically addressing this issue in a wide age range of subjects exists, to our knowledge.

Events following more severe focal or diffuse hypoxic-ischemic injury sufficient to cause infarction, in which all cellular elements in a given zone die, are discussed in Chapter 11. The timing of the sequence of the cellular changes following CNS infarction is discussed in Chapter 3.

Unusual Patterns of Central Nervous System Injury in Which a Component of Hypoxic-Ischemic Injury Appears to Be a Major Contributory Factor

The following case demonstrates selective involvement of structures, especially in the brainstem, which only partially follows the general guidelines given in Table 10.1. In addition to individual variation in responses, the findings in this Case (Case 10.1) may reflect a borderline age group (i.e., beyond age 1 year, but not yet a mature brain).

Case 10.1.

A 17-month-old boy was found with his head submerged in a water-filled 5-gallon plastic bucket after being left unattended; the time submerged was estimated at between 4 and 10 minutes. Arriving paramedics found him apneic and asystolic. Resuscitation efforts restored a pulse after 20 minutes. The child remained comatose despite treatment efforts, and he was pronounced dead 11 days after being submerged in water.

Microscopic examination revealed widespread, although not universal, acute neuronal hypoxic-ischemic change ("red neurons") and neuronal loss in the cerebral neocortex, near-universal involvement of hippocampal neurons in the end-folium and dorsal cell band, necrotic Sommer's sector and much of the subiculum, and less involvement of the entorhinal cortex. Neurons of the dentate fascia were relatively intact. The lateral geniculate body was necrotic, as was the lateral (but not medial) segment of the globus pallidus and putamen. The nucleus basalis and hypothalamus were intact, but the dorsal thalamus showed severe neuronal loss with remaining neurons having acute hypoxic-ischemic necrosis. Scattered swollen axons were found within necrotic gray matter but not in white matter. The brainstem areas involved were limited to the lateral zone of the substantia nigra, Edinger-Westphal nucleus, and adjacent dorsal tegmental gray (severe), lateral cuneate nucleus, and inferior olivary nuclei (partial in the latter, with inferior lamellae largely spared). Other spared areas included cerebellar cortex, medial substantia nigra, oculomotor nucleus, tectum, peri-aqueductal gray matter, cerebral peduncles, dorsal vagal nuclei, and rostral spinal cord.

Case 10.2. (Figs. 10.1 and 10.2)

A 30-year-old woman was hospitalized with abdominal pain attributed to fulminant liver disease. She had intermittent fever but remained alert. Following bronchoscopy, she was lethargic and later grunting with eyes open. Later still in the same day, she was found unresponsive, and died that day. Terminal cardiovascular data were not available.

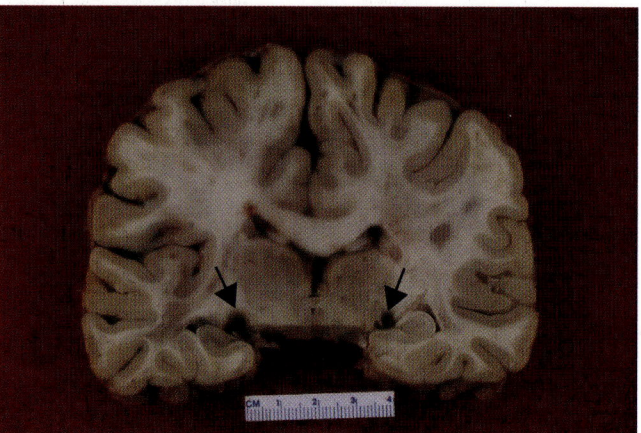

Fig. 10.1. Coronal section at level of lateral geniculate body shows discrete bilaterally symmetric dark brown hemorrhagic staining of lateral geniculate bodies (*arrows*) in absence of any other grossly detectable change.

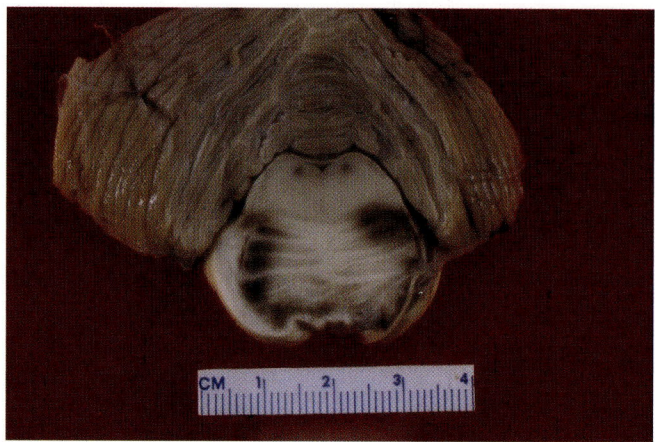

Fig. 10.2. Pons just rostral to middle cerebellar peduncles shows bilateral dark brown discoloration of hemorrhage in distribution of lateral pontine branches of basilar artery. Microscopically, stained areas contained acute ischemic necrosis of neurons bathed in acute hemorrhage. Cerebral cortex, basal ganglia, thalamus, hippocampal formation, and medial geniculate bodies were free of acute cell change.

Comment on Case 10.2. In some cases, distribution of hypoxic-ischemic injury is so unusual that individual variation in blood supply is suggested. The lateral one-half of the lateral geniculate body (LGB) typically receives its blood supply from the anterior choroidal artery (usually a branch of the internal carotid artery, but occasionally a branch of the middle cerebral artery), and the medial one-half of the LGB receives its blood supply from the thalamogeniculate artery (a branch of the posterior cerebral artery).[25] In this individual, possibly the entire LGB was supplied by the thalamogeniculate artery on each side, with a posterior circulation insufficiency thus accounting for both the LGB and basilar artery lateral pontine branch hemorrhagic infarction.

Other unusual patterns of injury may accentuate white matter rather than gray matter involvement (Case 10.3).

Predominantly white matter injury in association with hypoxic-ischemic injury in the perinatal period has been discussed elsewhere (see Chap. 4). Exceptionally, this reversal of the more common pattern in which gray matter is primarily damaged can occur in adults. The white matter degenerative changes may occur acutely, or develop following an interval of apparent partial or full recovery from the initial acute episode. In cases with acute white matter degeneration, it has been suggested that a combination of prolonged hypoxia, systemic hypotension, and metabolic acidosis may predispose to white

Case 10.3.

A 59-year-old man was found unresponsive with fixed, dilated pupils. Hospital evaluation revealed a large left-sided subdural hematoma with midline shift and herniation syndrome. Evacuation of the hematoma on the day of admission did not result in significant clinical improvement. He remained comatose until brain death was pronounced 21 days after admission.

Neuropathologic examination revealed bihemispheric chronic subdural neomembranes (estimated duration of several weeks), left greater than right side, status post burr holes and ventriculostomy, sequelae of left cingulate gyrus herniation with subacute infarction of the cingulate gyrus, merging with subacute herniation infarction in the territory of the left posterior cerebral artery secondary to uncal herniation, Kernohan's notch effect of the right midbrain, and small Duret hemorrhages. The most unusual finding consisted of widespread, seemingly random, multiple small and large foci of subacute necrosis involving such limited areas as the pathways of the hippocampal formation entering the alveus and small foci in the subcortical white matter of the parahippocampal and adjacent medial temporal isocortex. Other sites of necrosis involved a small zone in the dorsal thalamus, much of the central pontine tegmentum, degeneration of the pontocerebellar tracts, and extensive involvement of cerebellar folial white matter. The last was quite prominent, whereas cerebellar Purkinje cells and granular cells were spared. Only a mild to moderate increase in cerebellar deep white matter microglia and astrocytosis were present. The dentate fascia was virtually completely spared, as were pyramidal neurons of the hippocampus proper. Although the cause was undetermined, post–hypoxic-ischemic encephalopathy was suspected as the most likely basis for this unusual pattern.

---

matter-predominant degeneration.[42] This white matter-predominant pattern has been described in a variety of clinical settings,[42] including various drug and toxin exposures, cardiac arrest, head trauma, hypoglycemia, and other conditions.[16]

Delayed destruction of cerebral white matter, sometimes referred to as Grinker's myelinopathy, may follow an acute insult with a hypoxic-ischemic injury component. The precipitating event is most often CO poisoning,[64] but a similar syndrome may follow insults such as hypoglycemia, cardiac arrest, strangulation, or complications of surgical anesthesia.[75] The interval of improved or normal neurologic function between the acute insult (which is often sufficient to induce an initial coma for 1 to 2 days) and subsequent progressive neurologic deterioration may range from 4 to 14 days or more.[76] This delayed-onset neurologic deterioration may later stabilize, may rarely improve or even recover, or may progress to death. Although this unusual clinical syndrome demonstrates preponderant white matter pathology in most instances, its pathologic substrate in a few cases has instead been cavitating lesions of the caudate nucleus, putamen, and globus pallidus bilaterally, with sparing of the white matter.[32]

Under this general heading of unusual distributions of presumed hypoxic-ischemic injury may also be included such combinations as pallido-reticular preferential injury, preponderant combined involvement of the globus pallidus with the substantia nigra, and the basal ganglia and brainstem tegmental pattern, such as described by Opeskin and Burke,[73] both in children and adults. In our experience, cases in which these or other unusual patterns emerge are often accompanied by a clinical history of a circulatory insult, and reactive cellular events consistent in type, if not distribution, for a hypoxic-ischemic event.

Heat Stroke

Heat stroke may selectively involve neuroanatomic areas generally associated with vulnerability to hypoxic-ischemic injury. Clinical history of temperatures obtained at the scene investigation (i.e., environmental conditions and core or rectal body temperatures determined as soon as possible after discovery of a body), together with exclusion of other causes of death, may (with appropriate autopsy findings) be sufficient to provide the correct diagnosis. Neuropathologic findings, while nonspecific, can demonstrate some curious combinations of features.[68] The more nonspecific features include brain swelling, subarachnoid hemorrhage, progressive neuronal degeneration and gliosis corresponding in degree to the survival period examined, vascular congestion, edema, and parenchyma petechial hemorrhages. The latter occur particularly in the walls of the third ventricle, cerebral aqueduct, and fourth ventricle. Pituitary infarction has also been described.[22]

Somewhat more unusual is the rapid evolution of some of the microscopic findings. Certain cellular reactions in heat stroke are noted more rapidly than is often seen, for example, in most cases of hypoxic-ischemic encephalopathy uncomplicated by hyperpyrexia. Neurons may become swollen, chromatolytic, vacuolated, with pyknotic nuclei, and "ghosts" of neurons may be seen with survival times of between 4 and 11 hours. Within 24 hours, loss of neurons and early proliferation of glia, including rod cells, have been noted.[68] In the cerebellar cortex, rapid development of hypoxic-ischemic-like injury, with disintegration of some Purkinje cells and edema of the Bergmann cell layer, was present in cases surviving only 5½ hours. Virtually complete degeneration of Purkinje cells and Bergmann layer gliosis were present in cases with survival periods of only 72 hours. Within 12 days' postinsult, gliosis and neuronal loss were evident in the more resistant cerebellar cortex molecular layer and dentate nucleus, respectively. Changes were similar in the cerebellar hemispheres and in the vermis. In cases with basal ganglia involvement, the globus pallidus was least affected. The hypothalamus was largely spared, as were brainstem nuclei except for mild involvement of the quadrigeminal plate, inferior olivary nuclei, and reticular formation. This study did not specifically comment on changes in the hippocampus compared with other areas of the cerebral cortex,[68] but a more recent report documents sparing of Ammon's horn in heat stroke.[10]

In our experience, Purkinje cell injury of the degree described in hyperthermia cases within 3 days are not found in cases of hypoxic-ischemic injury unassociated with hyperthermia until approximately 6 to 7 days' postinsult, and a similar degree of Bergman's gliosis is not found in less than approximately 2 to 3 weeks, although one may encounter Bergmann glia nuclear enlargement within several days (Case 10.4, Figs. 10.3–10.4).

Malamud et al.[68] described only two cases of heat stroke in which widespread white matter involvement occurred. One of these cases with a 4-day survival was associated with brain purpura, and a case with 12-hour survival demonstrated prominent perivascular foci of rarefaction. Such cases may represent an underlying disseminated intravascular coagulation syndrome.[22]

The special vulnerability of the cerebellum in heat stroke appears to be reflected in some other forms of hyperpyrexia. For example, cerebellar degeneration has been described in early attempts at fever therapy,[43] neuroleptic malignant syndrome with[70] and without[66] associated lithium toxicity, and possibly serotonin syndrome.[37] More recent, improved methods of therapeutic hyperthermia that observe certain precautions regarding degree and duration of temperature elevation have not been associated with this complication.[91] It is also noted that some reported cases of neuroleptic malignant syndrome did not demonstrate prominent cerebellar changes,[49,53,62,69,72] and that clinical distinction between neuroleptic malignant syndrome and serotonin syndrome may be difficult in some cases.[21,36,51,78] Malignant hyperthermia can be distinguished, for example, by a history of anesthesia with certain agents, despite some pathologic and clinical features it shares with other hyperpyrexia syndromes.[21,23] A recent review article on the serotonin syndrome provides a useful summary of current clinical concepts.[12]

Heat stroke is not a "pure" hyperthermic insult, but rather is one in which other factors such as dehydration, redistribution of blood flow, reduced tissue oxygenation, hyperventilation with resultant respiratory alkalosis, electrolyte imbalance, and coagulopathy may occur.[84]

Perhaps one of the least complex situations in which effects of hyperthermia may be studied is in a setting of regional or whole body hyperthermia therapy, usually for cancer. Data from animal experiments indicate that

Case 10.4. (Figs. 10.3 and 10.4)

A 46-year-old man was witnessed standing outside his apartment on a summer day when he suddenly fell face down. Paramedics found him unresponsive, with a Glasgow Coma Scale (GCS) score of 3 and a rectal temperature of 108°F on arrival at hospital. CT head scan was negative. Clinical diagnoses included heat stroke, rhabdomyolysis, and sepsis due to Klebsiella pneumonia. He had a history of alcohol and drug abuse, and had been taking unspecified medications for severe depression (possible relationship of psychotropic medications to hyperthermia not mentioned in hospital records). Despite intensive care, he remained comatose, and died 20 days after admission.

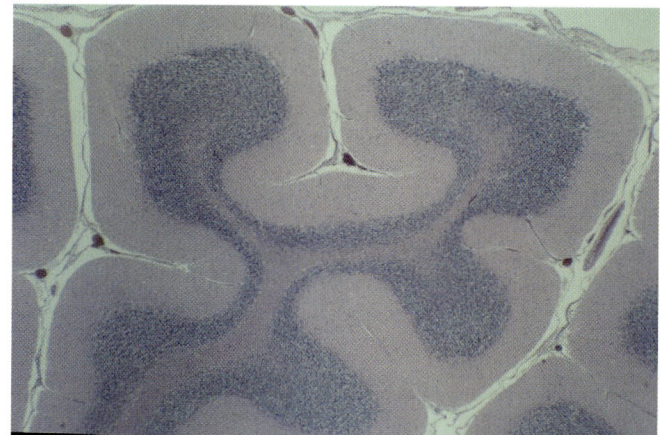

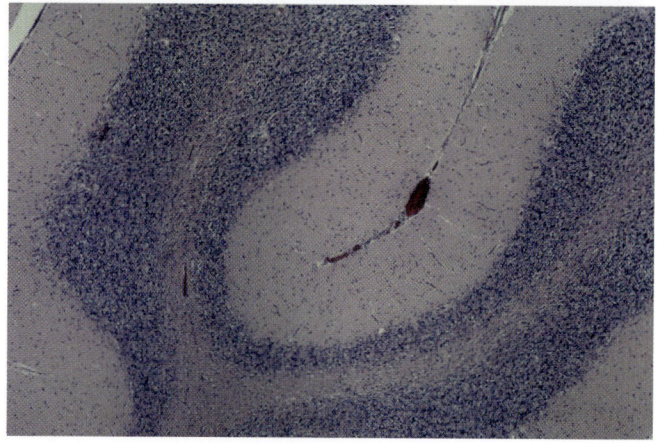

Figs. 10.3–10.4. Fig. 10.3 (*top*; low-power) and Fig. 10.4 (*bottom*; medium-power) micrographs show total dropout of Purkinje cells and mild diffuse increased cellularity in molecular layer. Granule cell layer is mildly autolyzed. Cerebral cortex, basal ganglia, and thalamus, among other areas, were remarkably intact except for prominence of astrocytic nuclei in thalamus suggestive of Alzheimer's type II cells. The pattern of CNS involvement was consistent with changes reported in fatal cases of heat stroke (see text). (H&E.)

neural tissue is more vulnerable to hyperthermia than other body tissues, and CNS tissue is more vulnerable than peripheral nervous system tissue.[92] From the standpoint of determining thresholds of heat tolerance in neural tissues, data from humans even within this relatively limited therapeutic category remain difficult to evaluate. For example, therapeutic hyperthermia often combines whole body hyperthermia with local cooling of the head to minimize CNS complications; combines use of hyperthermia with radiotherapy, resulting in enhanced radiosensitivity of tissues; or combines hyperthermia with chemotherapy, where hyperthermia appears to enhance cytotoxicity of certain drugs such as BCNU (carmustine) and metronidazole. Another confounding factor is the potential tendency of hyperthermia to elevate intracranial pressure and/or peritumoral edema in the brain.[84,92]

Available data indicate that examination of the CNS alone, without supporting data, provides no findings diagnostic of hyperthermia, but a disproportionate degree of cerebellar involvement compared with other areas of the brain as described earlier might warrant further exploration of the possibility of hyperthermia in difficult cases in which this possibility was not previously considered. Also, primary CNS lesions situated in locations from the upper pons to the hypothalamus, as well as hydrocephalus, may cause hyperthermia as a secondary effect.[74]

Epilepsy

Neuropathologic findings in cases of chronic recurrent convulsive disorders, and in status epilepticus, are discussed in Chapter 9.

Asphyxia

Neuropathologic changes in the various forms of asphyxia are often unremarkable aside from cerebrovascular congestion, owing to the brief survival periods of most cases precluding development of abnormalities apparent by routine histologic methods. More prolonged survivals provide us with no features that could allow reliable distinction from other types of hypoxic-ischemic CNS injury such as cardiac arrest. Fortunately, most cases provide clues at the scene investigation or general autopsy examination that will alert the first responders or medical examiner to the possibility of asphyxia (e.g., industrial toxic gas or oxygen-free environmental exposures, ligature marks, abnormalities in deeper tissues encountered during layer-by-layer neck dissection, and so on).[50]

Hypoglycemia

Opinion differs as to whether study of the CNS in itself can lead to a presumptive diagnosis of hypoglycemia rather than to hypoxic-ischemic injury. Some, probably the majority, consider the CNS changes to be nonspecific and essentially indistinguishable in these two conditions.[11,65,95] Others suggest that some topographic or qualitative differences exist in the neuropathology of hypoglycemia when compared with hypoxic-ischemic insults.[60]

The literature on hypoglycemia effects in the brain is difficult to evaluate, since findings may be described without always making a clear distinction between results in humans versus animal experiments. In other instances, what may have been initially a relatively isolated hypoglycemic insult is later complicated by a variety of ensuing medical problems including recurrent seizures, hyperpyrexia, and multisystem failure prior to death, rendering the determination of the relative contribution of each category of insult to the subsequent CNS findings problematic. If any patterns exist that might lead one to suspect hypoglycemia rather than hypoxic-ischemic injury as the major determinant of the CNS damage in adults, it seems to us that the following references apply.[5,6,9,31,55,63]

In mature brains following fatal hypoglycemic injury, the neocortical injury tends to be more extensive and widespread throughout the frontal, insular, and to varying degrees, temporal, parietal, and less commonly, occipital areas, without showing any preponderance of watershed area involvement. The neocortical injury may be pancortical or even associated with necrosis, but more often tends to be without necrosis and with a superficial laminar preponderance. Basal ganglia injury in several reported cases of hypoglycemia have tended to focus on the caudate and putamen, with generally less involvement of the globus pallidus and thalamus. The hippocampus is variably affected, typically in CA1 and CA3-4 greater than CA2 but with the unusual tendency in some cases to heavily involve the dentate fascia, which is usually spared in hypoxic-ischemic injury. Cerebral and cerebellar subcortical white matter involvement typically includes gliosis, but myelin stains may be unremarkable. Cerebellar cortex, dentate nucleus, and brainstem tend to be spared compared with severe neuronal injury elsewhere, which is perhaps the most emphasized discrepancy between hypoglycemic injury and the more commonly encountered hypoxic-ischemic forms of injury, such as postresuscitated cardiopulmonary arrest. Auer et al.[6] place more emphasis on prominent injury to the dentate fascia in cases with cerebellar sparing as a guide to hypoglycemia rather than hypoxic-ischemic injury, but consider other topographic distinctions described earlier to be less reliable. In cases with sufficient survival time, gliosis of involved areas tends to be very conspicuous, and microglial response is also typically prominent.

Some of our experience with atypical lesion distribution in what appears to be predominant hypoxic-ischemic injury leads us to be cautious in interpreting these changes, although it is certainly appropriate to raise the question of hypoglycemia, or describe the findings as "consistent with" those described in adult cases of known hypoglycemic injury, when the topographic patterns are those described earlier. The complexities involved have also been briefly summarized by Auer and Sutherland.[9] Using β-amyloid precursor protein immunostaining, Dolinak et al.[31] described evidence of widespread axonal injury in hypoglycemia cases sufficient to be considered in the differential diagnosis of such changes together with diffuse traumatic axonal injury and hypoxia.

The situation differs in neonatal hypoglycemia. Anderson et al.[2] found no significant differences between brain changes in fatal cases of neonatal hypoglycemia and those due to severe hypoxia in neonates (estimated gestational age of cases reported 25–39 weeks), except for noting that the classic "ischemic cell change" of neuronal eosinophilic cytoplasm with an irregular, pyknotic nucleus was frequent in hypoxic cases and rare in hypoglycemia in this age group. Neuronal injury in hypoglycemia tended to be more widespread and severe. It is of interest that despite the widespread nature of the changes in these neonatal cases (which were throughout the CNS including cerebrum, cerebellum, brainstem, and spinal cord), they noted sparing of the subependymal germinal zones. Also, one of their hypoglycemia cases (estimated hypoglycemia of 38 hours' duration) with less severe CNS injury demonstrated changes most evident in the occipital cortex, putamen, caudate, and internal granular layer of the cerebellum. As Armstrong[4] noted, however, immature brains often demonstrate nuclear pyknosis or karyorrhexis rather than cytoplasmic eosinophilia typical of more mature brains even in response to hypoxic-ischemic insults, rendering this suggested distinction by Anderson et al.[2] less reliable as a guide to hypoglycemia in this age group. Larroche[65] also described no histologic distinguishing features between neonatal hypoglycemia and severe asphyxia, and mentioned that hypoglycemia is one of the precursors of periventricular leukomalacia. Armstrong[4] noted that conditions that may accompany asphyxia in the neonate include not only hypoglycemia but also seizures, hypotension, kernicterus, and idiopathic respiratory distress syndrome, thus emphasizing the need for careful review of medical records in the analysis of such cases.

With regard to the time required for irreversible neuronal injury to occur, it appears a longer duration is necessary for severe hypoglycemia to cause irreversible change than that required for hypoxic-ischemic injury. In adults, it has been commented that hypoglycemic coma for up to 1 to 3 hours is "usually associated with complete recovery, while longer periods are dangerous."[63] In neonates, cases with significant hypoglycemia (i.e., blood glucose well below 20 mg/100 ml) for up to an estimated 9 hours have been successfully treated, and two other cases in this same report died only after hypoglycemia had been present for 38 to 40 hours, during which time other medical complications evolved.[2]

Red Neurons, Dark Neurons, and the Category of "I'm Not Sure How to Interpret These Neurons"

As demonstrated in this text, dating/aging of cellular changes in several types of brain lesions (e.g., infarction, contusion) are not uniform in the published literature. Considerable variation exists, for example, in estimates of the first appearance of red neurons in these lesions. Several general reasons for such discrepancies were stated in the introduction to Chapter 3, and factors potentially relevant to this topic are discussed in further detail later. The frequency with which questions are raised concerning red neurons and dark neurons warrants additional attention to this topic.

As was stated by Johnson,[52] "The origin of dark, or hyperchromatic, neurons has been debated for nearly a century." Exactly when a red, or acidophilic, neuron can be interpreted as irreversibly on its way to cell death is also a matter of some disagreement. The interpretation of the appearance of neurons by the standard hematoxylin-eosin (H&E) stain as it relates to its timing with death is a constant challenge. Occasional cases are encountered in which this conundrum is presented after many years of practice.

The most frequent questions can be answered in part by reviewing selected animal experimental findings and the human neuropathology literature. The findings are summarized here with a proposed approach to this issue.

Many of the original descriptions of neuronal hypoxic-ischemic injury were made with the Nissl stain rather than H&E. Terms used to describe these changes can be confusing. They have included pallor, swelling, coagulation necrosis, ischemic cell change, homogenizing cell change, liquefaction necrosis, Nissl's acute cell disease, Spielmeyer's acute swelling, and Nissl's chronic cell degeneration. More recently published observations tend to describe changes as observed in H&E stains.

Brown[15] described ischemic cell change, irrespective of cause, and the reader is referred to his article for details. He described a rather stereotyped sequence of change, indicated that it is a more rapid process in small neurons than in large neurons, and that increased severity of hypoxic-ischemic injury is reflected only in the numbers of neurons involved, not the nature or sequence of cellular changes. The pattern of neuronal involvement is in those areas selectively vulnerable to hypoxic injury. The sequence of change may be briefly summarized as follows. Light microscopically, the earliest change consists of microvacuolization (0.16–2.5 μm vacuole diameter) with the nucleus normal or slightly shrunken, normal nucleolus, cell shrinkage seen primarily in the neocortex, cytoplasm normal to slightly basophilic, and some neurons demonstrating perineuronal spaces. This change is said to occur in human subjects within 1 hour after cardiac arrest. Ultrastructurally, the analogous changes are swollen mitochondria, and some dilated tubules and cisternae of the endoplasmic reticulum. There is an increase in free ribosomes, and some neurons are surrounded by swollen astrocytic processes. As we have previously noted in this text, we cannot distinguish this state from autolysis in our human autopsy material.

"Classic" ischemic cell change[15] is the first change which can be recognized with confidence in human autopsy material. The cell body is variably shrunken, stains more or less intensely with aniline dyes, and the cytoplasm tint varies from vivid pink to reddish with H&E stains (Figs. 10.5–10.9). Nissl substance appears finely granular and dispersed. The nucleus is shrunken, often triangular, and dark with aniline dyes. Cytoplasmic vacuoles may be less obvious. This stage is said to persist in humans for at least 6 hours. The associated ultrastructural changes include vacuolar change in the Golgi complex and endoplasmic reticulum, persistence of free ribosomes and ribosomes located on expanded rough endoplasmic reticulum membranes, some breakdown of granular endoplasmic reticulum, and some neurons demonstrating increased cytoplasmic and nuclear density. An irregular cell contour is present in some cells, related to swollen astrocyte processes irregularly indenting the neuronal cell body and dendrites ("incrustations"). These changes continue to evolve with

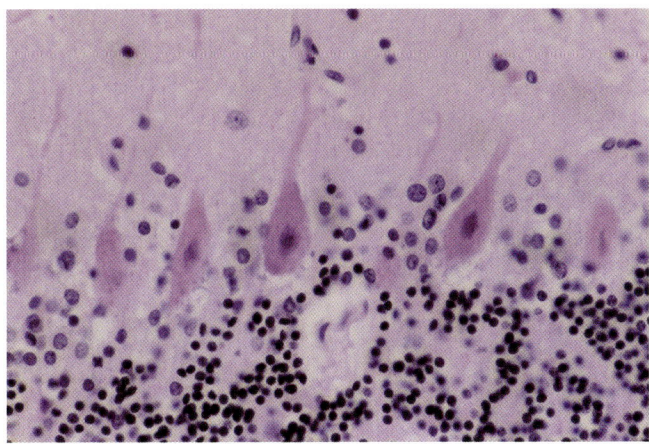

Fig. 10.6. Well-developed red neuron changes in Purkinje cells (cytoplasmic eosinophilia, cell shrinkage, nuclear pyknosis with indistinct nucleoli) and mild Bergmann's gliosis (nuclei of Bergmann glia are approximately two to three times the diameter of granule cell nuclei, with fine granular chromatin). (H&E.)

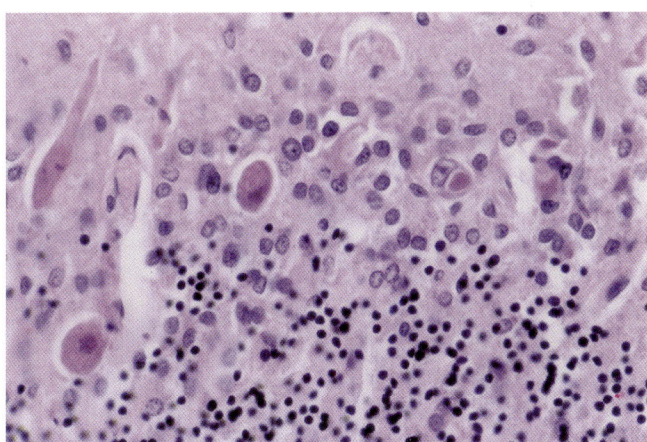

Fig. 10.7. Later stage of hypoxic-ischemic insult to cerebellar cortex. Further cell shrinkage with some Purkinje cell loss, decreased basophilia of nuclei, more conspicuous Bergmann's gliosis, and some granule cell depopulation. (H&E.)

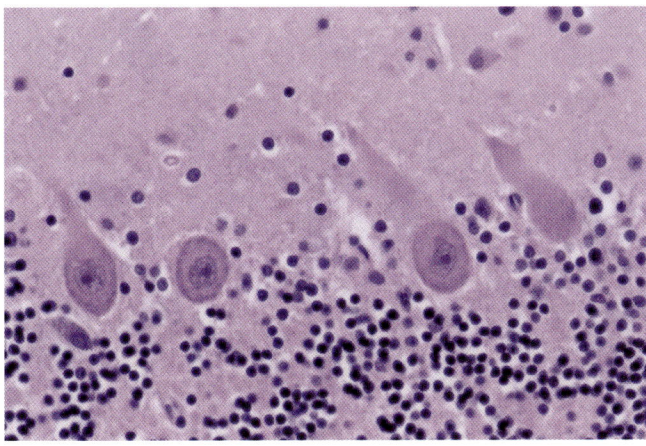

Fig. 10.5. Normal cerebellar cortex Purkinje cell layer, for comparison with Figs. 10.6–10.7. Bergmann glia are inconspicuous. (H&E.)

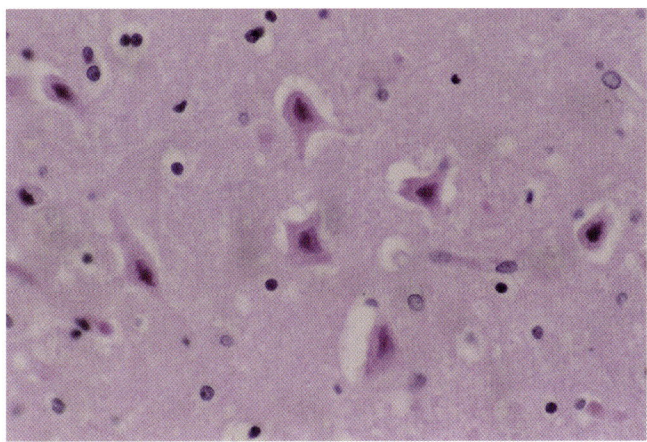

Fig. 10.8. Thalamus with red neuron change. (H&E.)

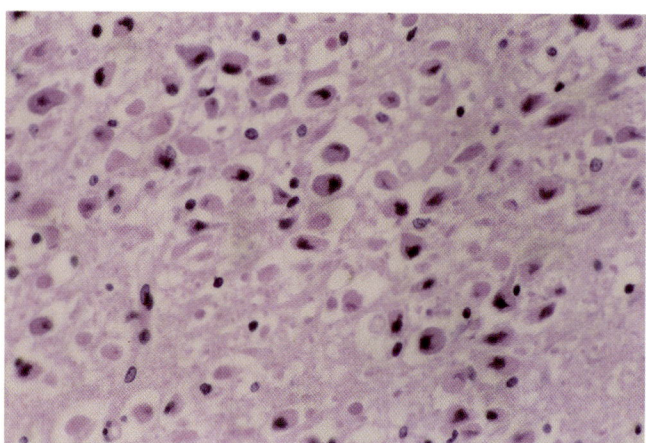

Fig. 10.9. CA1 sector of hippocampus with well-developed red neuron change and neuropil edema. (H&E.)

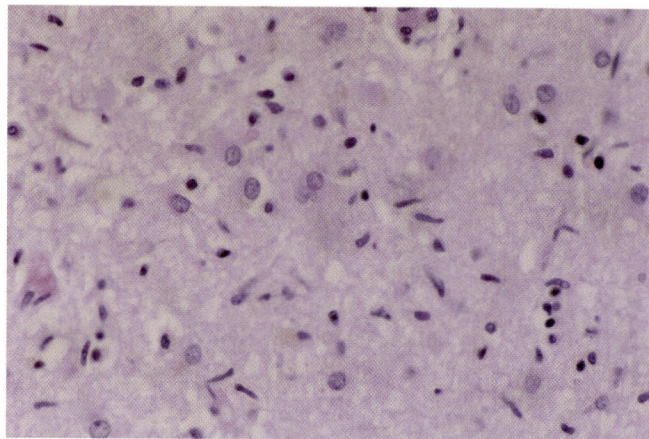

Fig. 10.11. CA1 sector of hippocampus with chronic state of hypoxic-ischemic injury. Absence of neurons, with presence of astrocytosis and increased rod cells. (H&E.)

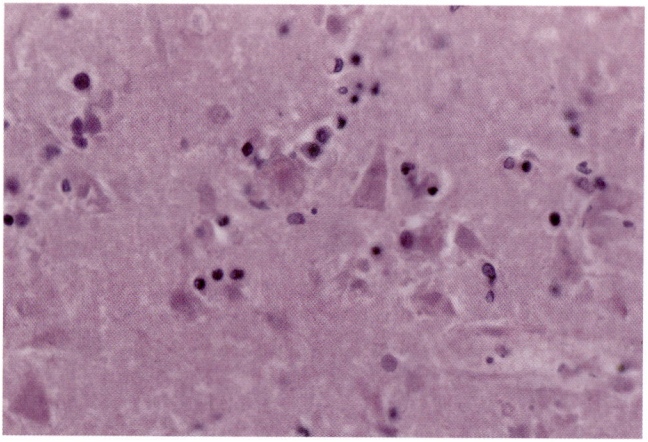

Fig. 10.10. Homogenizing cell change in neocortex, demonstrating cells with indistinct cell membranes and organelles. (H&E.)

further shrinkage of the neuronal nucleus and cytoplasm, and continued pink staining of the cytoplasm. Evolution of ultrastructural changes consists of a more electron-dense nuclear and cytoplasmic appearance, intranuclear clusters of dense osmiophilic granular material, decreased cytoplasmic volume, vacuolar remnants of endoplasmic reticulum and mitochrondria, and presence of autophagic vacuoles. Some free ribosomes persist.

The stage of "homogenizing cell change" is characterized by disappearance of incrustations, increased homogenation and loss of staining intensity of cytoplasm, and fragmentation of the shrunken triangular nucleus in some cells by light microscopy. This change is typically seen after 24 hours' survival (Figs. 10.9–10.10). Ultrastructurally, there is a general decrease in electron density of both nucleus and cytoplasm, and organelles show progressive deterioration, although a few dense mitochondria may still be visible. Profiles suggesting phagocytosis of neuronal cytoplasm by astrocytes are also described.[15]

The final stage is a "ghost cell" appearance, with cytoplasm reduced to absent around a shrunken dark-staining nucleus, and eventual cell loss. Gliosis and an increase in rod cells typically appear in areas of neuronal loss (Fig. 10.11).

Brown[15] described a second type of hypoxic-ischemic neuronal injury seen in experimental animals which may be a reversible change and is referred to as a "scalloped cell." In contrast to the previously described neuronal changes, scalloped cells are slightly dense, with mildly distorted cell outline and nucleus, and without cytoplasmic vacuoles or nuclear pyknosis. They are also said to occur in "edematous" human cortex. It is difficult to convincingly distinguish this change from the earliest stage of classic ischemic cell change, as described earlier, in humans.

Brown[15] described two separate and distinct types of neuronal artifacts to be distinguished from the aforementioned cell changes. The first is the dark or hyperchromatic neuron (subsequently referred to as dark neuron). It often has an irregular and indented soma, a corkscrew-like apical dendrite, and a shrunken, darkly stained nucleus and cytoplasm in which Nissl bodies are obscured (Fig. 10.12). Clear perinuclear halos may be seen (due to a swollen Golgi complex). The nucleolus is normal in size but may be difficult to see due to the nuclear shrinkage and hyperchromicity. Cytoplasmic vacuolization, other than the perinuclear vacuoles, and neuronal incrustations are typically not present.

The second type of artifact described by Brown[15] was called the "hydropic cell." It is a neuron that is pale, swollen, and vacuolated. This is a commonly encountered artifact in human autopsy material secondary to

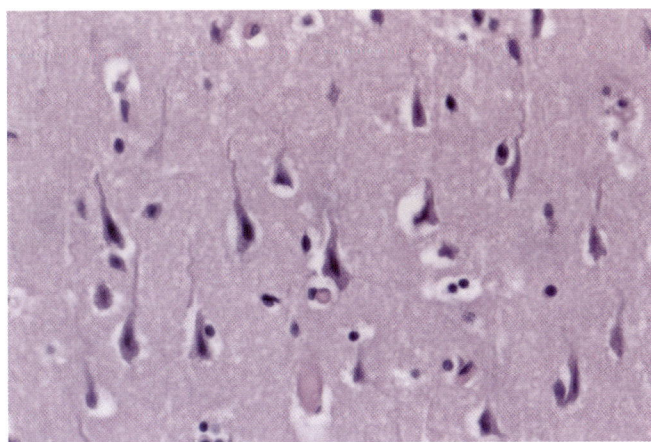

Fig. 10.12. Dark neuron change in neocortex. Note corkscrew-like apical dendrites in shrunken neurons with dark red to basophilic cytoplasm. (H&E.)

autolysis. Both of these artifacts (dark neuron and hydropic cell) are said to be unrelated to areas of selective vulnerability, or to the intensity or time course of the hypoxic-ischemic injury.

This 1977 description of cellular changes by Brown,[15] at both the light and electron microscopic levels, might lead one to consider the distinction between dark neurons and red neurons (i.e., Brown's ischemic cell change) to be relatively straightforward. At that time, most neuropathologists were in agreement with this viewpoint, largely influenced by the elegant studies of Cammermeyer.[17–20] Cammermeyer studied several species of experimental animals, primarily using the periodic acid-Schiff and gallocyanin-chromealum stains. There was general agreement that the dark neurons demonstrated by this method were equivalent to dark neurons using H&E, a variety of Nissl stains, neurons that were argyrophilic by certain methods,[91] and the dense osmiophilic neurons encountered in ultrastructural studies. Cammermeyer proposed that dark neurons were an artifact produced experimentally in CNS tissue in the following instances:

- Surgical biopsy.
- Immersion fixation.
- Inadequate perfusion fixation.
- Trauma to brain tissue prior to optimal perfusion fixation.
- Trauma to brain tissue if it is removed too quickly after optimal perfusion fixation (various times, depending largely on the fixative employed, but requiring a fixative to autopsy delay of at least 10 hours, and sometimes greater than 24 hours with formalin fixation, in order to minimize the development of dark neuron change).[17]

Cammermeyer noted that dark neuron change was more easily produced in neurons with abundant Nissl material. He also indicated that it should be distinguished from the diffuse neuronal basophilia that can occur in neurons undergoing acute retrograde neuronal degeneration. He seemed to emphasize the corkscrew apical dendrite as a more reliable guide to dark neuron change than other more subtle tinctorial or morphologic changes.[20]

Brierley et al.[14] regarded dark neurons as artifact, but indicated they are infrequently found in human autopsy material because the death to autopsy interval is typically greater than 10 hours, after which time this artifact is unlikely to occur as a consequence of brain tissue removal and immersion fixation.

The dark neuron as artifact viewpoint prevailed for the next several years in neuropathology texts, and it was also noted that it could be produced by perfusion with hyperosmolar solutions.[33,41,44] Gradually, evidence accumulated suggesting that dark neurons could also be the result of disease. Hirano[47] indicated that dark neurons could be found in pathologic conditions such as Alzheimer's disease, amyotrophic lateral sclerosis, and other chronic human diseases. Loberg and Tovik[67] noted classic dark neurons developing within minutes in human brain contusions, and described what seemed to be an evolution of dark neurons to similarly shaped eosinophilic shrunken cells within 2 to 3 days' survival. They suggested that dark neurons were an early stage of traumatic neuronal death. A human traumatic brain lesion study by Anderson and Opeskin[3] also described dark neurons in traumatic lesions with less than 1-hour survival, and persisting in cases with greater than 48-hour survival. They concluded that dark neurons can be artifact, but are seen more commonly in cases of brain trauma.

Several animal experimental studies were also beginning to collectively lend support to the concept that dark neurons were not exclusively due to artifact. Dark, atrophic neurons were noted in transneuronal degeneration in rats[52]; in normal adult cats in which extensive precautions were taken to avoid the artifact-inducing pitfalls noted by Cammermeyer[24]; in bicuculline-induced epilepsy in rats[85]; in hypoxic-ischemic injury in rats[26]; and in a transgenic mouse model of Huntington's disease.[90] The latter study interpreted the dark neurons as degenerating cells in this light microscopic (using thionin as a Nissl stain) and ultrastructural study. They also found these dark neurons to be TUNEL-negative, suggesting that the mechanism was other than apoptosis. Excluding the filamentous neuronal intranuclear inclusions present, it is difficult to distinguish their Figure 4D in a human case of Huntington's disease from the ultrastructural example of an acidophilic neuron undergoing necrosis in Figure 5.21 by Auer and Sutherland,[9] but the limited number and relatively low-power ultrastructural illustrations preclude reliable comparison.

Gallyas et al.[38,39,91] used a silver stain to demonstrate dark neurons in both experimental animals and in human brains exposed to localized trauma. Based on these studies, they proposed a hypothesis that allows for the development of dark neurons either by antemortem neuronal disease or injury, or by postmortem artifact. They suggested that stored mechanical energy remains for a time within incompletely fixed/degraded neurofilaments, and could lead to neuronal "collapse" with the production of dark neurons.[91]

In a detailed light and electron microscopic study on the effects of hypoglycemia in the rat cerebral cortex, it was found that classic dark neurons developed within 30 to 60 minutes.[7] Subsequently, some dark neurons evolved back to morphologically normal neurons within 6 hours' survival time. Other dark neurons evolved to typical acidophilic neuronal change within 4 to 6 hours and continued on to unequivocal cell death within the next several days. The fate of dark neurons was determined largely by their location, with reversible changes seen in cortical laminae 4-6 and irreversible injury with cell death occurring in cortical laminae 2-3. Neither dark neuron change nor mitochondrial swelling was an early ultrastructural indicator of reversible change versus cell death. The following aggregate of ultrastructural changes was required before eventual cell death could be reliably predicted: amorphous cytoplasm devoid of ribosomes or Golgi complex; flocculent densities in mitochondria; thick, electron-dense chromatin clumps; cytorrhexis; and karyorrhexis.[7] They also noted that the details and pace of cellular changes varied, being different in hippocampus[8] and in caudoputamen[54] than in cerebral cortex.[7]

Implications Based on the Experimental Data Cited Previously

1. What are referred to as dark neurons are sometimes produced by artifact, and sometimes they are produced by disease. If in the latter category, sometimes they represent a reversible change and sometimes they progress to cell death. That is, sometimes they become red neurons. It follows that there may be stages within this latter evolution that made it difficult for the light microscopist to distinguish dark neurons from red neurons, and that we should no longer assume that dark neurons in human material are invariably artifact.
2. The pace of development of dark neurons and red neurons varies from species to species, and in different brain areas of the same species.
3. Neuronal death following brain lesions, whether traumatic[89] or ischemic,[35] need not be synchronous, but may be asynchronous, occurring in successive waves due to secondary changes at the lesion margin. This can potentially lead to error in the interpretation of the duration of persistence of certain cell changes, such as red (or dark) neurons.

Conclusions Based on Human Neuropathologic Studies Cited Previously, and Personal Observations

1. It is likely that more is asked of our routine stains on autopsy tissue than our current methods can provide. Many medical examiner departments work primarily at the light microscopy level of resolution with immersion-fixed human autopsy tissue without the luxury of a battery of special stains, and often with death to autopsy intervals that can extend to a few days. Animal experimental studies indicate that investigator opinions differ even in material optimally fixed by perfusion methods, and examined not only with a variety of light microscopic stains but also by electron microscopy.
2. Not infrequently, typical dark neurons are encountered in human autopsy material where the death to autopsy interval exceeds 10 hours. Such delays probably cannot prevent the appearance of dark neurons.
3. Age, general health, and abruptness of the terminal illness may influence the appearance of red neurons (see discussion of cerebral infarction dating/aging in Chap. 3).
4. All H&E stains are not the same. Anyone practicing anatomic pathology is aware that H&E stains from different histology laboratories, and from time to time from the same laboratory (e.g., due to turnover of histotechnologists, purchasing officers deciding to obtain eosin or hematoxylin from another source, and the like). Quality assurance screening also varies from time to time, resulting in variations in stain characteristics that can alter how "red" an ischemic neuron appears, or how "dark" a dark neuron appears, leading to a "gray" zone of interpretation.
5. Postmortem autolysis can result in not only a diffuse loss of tissue basophilia, but it can also reduce detail as a result of neuropil breakdown that compromises interpretation in human material.
6. There could be other explanations for instances in which experienced and careful observers' opinions differ. It is now established that subtle variations in red-green color vision occur even in individuals considered to have normal color vision on routine screening tests.[27] Restricting comments regarding congenital color vision deficiencies to only the medical profession, studies indicate that color vision deficiency can interfere with the proper interpretation of patient physical examination findings (e.g., ophthalmoscopy, otoscopy, recognition of jaundice, cherry red color of CO poisoning, pallor of anemia, inflammation, skin bruises), chemical color change endpoints, distinguishing blood from bile, and interpreting blood and

urine test strips.[86,87] More specifically, with regard to histopathologists and histotechnologists, there is evidence that those with color vision deficiency have more difficulty in finding acid-fast bacteria and gram-negative bacteria, more difficulty interpreting Congo red and several other stains, and make significantly more errors in test slide interpretation than those without color vision deficiencies.[77,86,94] Studies have revealed an incidence of color vision deficiency in an aggregate of histopathologists, cytologists, and clinical laboratory workers varying from 10.3% (of 23 subjects) to 13% (of 133 subjects), and the majority of subjects were unaware of their color vision problem.[86] These data do not include the estimated larger number of individuals with acquired forms of color vision deficiency. The degree, if any, to which color vision deficiency has contributed to the discrepant opinions regarding distinguishing red and dark neurons, or aging/dating of red neurons in experimental animal and human neuropathologic material, is unknown.

Experimental studies are decoding the complex chemical events related to, for example, cerebral hypoxic-ischemic neuronal injury.[35] It seems reasonable to anticipate that histochemical or immunocytochemical markers providing a more accurate interpretation of the nature of what we now crudely classify as red or dark neurons will be forthcoming. In the meantime, we use the following approach with H&E-stained material.

1. Dark neurons should display most or all of the criteria described earlier.[15] Prominent shrinkage, basophilia, and corkscrew apical dendrites in many such cells are convincing in the aggregate.
2. Red neurons should display most or all of the criteria described earlier.[15] A bright red to pink cytoplasm (not magenta), pyknotic or disintegrating nucleus, indistinct Nissl bodies and nucleolus, and prominent cell shrinkage are key points, together with the distribution of such changes.

Red neuron change should occur in a group of cells, not simply in a rare isolated cell interspersed with normal-appearing neurons. It should be appreciated that metabolic differences between cells may allow some to develop this change earlier than others, and that, for example, isolated cells surrounded by normal cells can still be valid dark neurons.[20] A more convincing totality of findings is demanded for forensic purposes. It is reassuring to see some associated changes in the neuropil, such as locally increased pallor and/or vacuoles compared with nearby areas, in early stages of change, and other cellular reactive changes at later stages. Subtle cytoplasmic increased eosinophilia or magenta tint (Fig. 10.13) to the cytoplasm in a minority of neurons should not be considered sufficient for a diagnosis of definite hypoxic-ischemic (red) neuron injury. As in other types

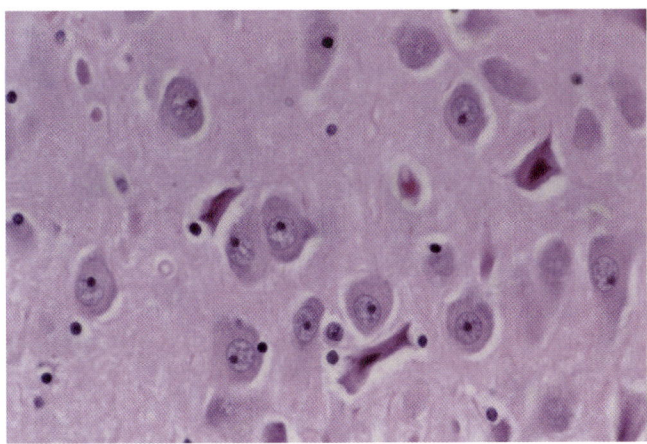

Fig. 10.13. CA1 sector of hippocampus. A few shrunken neurons with magenta, rather then bright red cytoplasm in a field of otherwise normal neurons. A problematic case, particularly when devoid of nuclear abnormalities, as is often the case. These magenta neurons do show nuclear pyknosis, but were very sparse in an otherwise normal-appearing CA1 sector and were not seen elsewhere in the brain (see text). (H&E stain.)

of abnormality, the color change is also judged by the company it keeps.

Occasionally the cellular changes are classic for red neurons, but in an atypical distribution compared with the usual sites of selective vulnerability. For example, classic red neurons in the dentate fascia can sometimes be seen in the company of normal-appearing hippocampal pyramidal cells. Red neurons are occasionally limited to either the superior, or only the inferior portion of the inferior olivary nucleus. Red neurons often take longer to disappear in the neocortex (an average of 5–7 days, rarely up to 2 weeks) than in cerebellar Purkinje cells (often less than 4 days). In children less than 1 year of age, apparently global hypoxic-ischemic change may show classic red neurons in the dorsal thalamus but not in Purkinje cells. By 12 months of age, a distribution of red neurons can be seen similar to that of adults in analogous settings. The variety of selective vulnerability patterns needs to be interpreted in the context of age group, nature of insult, and other variables, as discussed earlier in this chapter.

Auer and Benveniste stated that necrotic neurons, whatever the cause of the necrosis, decorated with acid dyes, and "the term neuronal acidophilia would appear more descriptive, accurate and inclusive."[5] The analogous term "red neuron" is appropriate, since it is so commonly found in the forensic literature. The same is not necessarily true at the ultrastructural level where, for example, differences between hypoxic-ischemic injury and hypoglycemic injury can be appreciated in at least some experimental animal neuronal populations.[8]

References

1. International Association for Development of Apnea (AIDA). Available at: http://www.aida-international.org/. Accessed 8/4/06.
2. Anderson JM, Milner RDG, Strich SJ. Effects of neonatal hypoglycemia on the nervous system: A pathological study. J Neurol Neurosurg Psychiatry 1967;30:295–310.
3. Anderson RMcD, Opeskin K. Timing of early changes in brain trauma. Am J Forensic Med Pathol 1998;19:1–9.
4. Armstrong DD. Neonatal encephalopathies. In Duckett S (ed): Pediatric Neuropathology: Baltimore: Williams & Wilkins, 1995, pp 334–351.
5. Auer RN, Benveniste H. Hypoxia and related conditions. In Graham DI, Lantos PL (eds): Greenfield's Neuropathology, ed 6. London: Arnold, 1997;I:263–314.
6. Auer RN, Hugh J, Cosgrove E, Curry B. Neuropathologic findings in three cases of profound hypoglycemia. Clin Neuropathol 1989;8:63–68.
7. Auer RN, Kalimo H, Olsson Y, Siesjö BK. The temporal evolution of hypoglycemic brain damage. I. Light- and electron-microscopic findings in the rat cerebral cortex. Acta Neuropathol (Berl) 1985;67:13–24.
8. Auer RN, Kalimo H, Olsson Y, Siesjö BK. The temporal evolution of hypoglycemic brain damage. II. Light- and electron-microscopic findings in the hippocampal gyrus and subiculum of the rat. Acta Neuropathol (Berl) 1985;67:25–36.
9. Auer RN, Sutherland GR. Hypoxia and related conditions. In Graham DI, Lantos PL (eds): Greenfield's Neuropathology, ed 7. London: Arnold, 2002;I:233–280.
10. Bazille C, Megarbane B, Bensimhon D, et al. Brain damage after heat stroke. J Neuropathol Exp Neurol 2005;64:970–975.
11. Becker LE, Prior TW, Yates AJ. Metabolic diseases. In Davis RL, Robertson DM (eds): Textbook of Neuropathology, ed 3. Baltimore: Williams & Wilkins, 1997, pp 407–509.
12. Boyer EW, Shannon M. The serotonin syndrome. N Engl J Med 2005;352:1112–1120.
13. Brierley JB, Graham DI. Hypoxia and vascular disorders of the central nervous system. In Adams JH, Corsellis JAN, Duchen LW (eds): Greenfield's Neuropathology, ed 4. New York: John Wiley & Sons, 1984, pp 125–207.
14. Brierley JB, Meldrum BS, Brown AW. The threshold and neuropathology of cerebral "anoxic-ischemic" cell change. Arch Neurol 1973;29:367–374.
15. Brown AW. Structural abnormalities in neurons. J Clin Pathol Suppl (R Coll Pathol) 1977;11:155–169.
16. Brucher JM. Neuropathological problems posed by carbon monoxide poisoning and anoxia. In Bour H, Ledingham I McA (eds): Carbon Monoxide Poisoning. Progress in Brain Research, Vol 24. Elsevier, Amsterdam, 1967, pp 75–100.
17. Cammermeyer J. The post-mortem origin and mechanism of neuronal hyperchromatosis and nuclear pyknosis. Exp Neurol 1960;2:379–405.
18. Cammermeyer J. The importance of avoiding "dark" neurons in experimental neuropathology. Acta Neuropathol 1961;1:245–270.
19. Cammermeyer J. "Ischemic neuronal disease" of Spielmeyer. A reevaluation. Arch Neurol 1973;29:391–393.
20. Cammermeyer J. Is the solitary dark neuron a manifestation of postmortem trauma to the brain inadequately fixed by perfusion? Histochemistry 1978;56:97–115.
21. Carbone JR. The neuroleptic malignant and serotonin syndromes. Emerg Med Clin North Am 2000;18:317–325.
22. Chao TC, Sinniah R, Pakiam JE. Acute heat stroke deaths. Pathology 1981;13:145–156.
23. Christiansen LR, Collins KA. Pathologic findings in malignant hyperthermia: A case report and review of the literature. Am J Forensic Med Pathol 2004;25:327–333.
24. Cohen EB, Pappas GD. Dark profiles in the apparently-normal central nervous system: A problem in the electron microscopic identification of early anterograde axonal degeneration. J Comp Neurol 1969;136:375–396.
25. Crosby EC, Humphrey T, Lauer EW. Correlative Anatomy of the Nervous System. New York: Macmillan, 1962.
26. Czurko A, Nishino H. "Collapsed" (argyrophilic, dark) neurons in rat model of transient focal cerebral ischemia. Neurosci Lett 1993;162:71–74.
27. Deeb SS. The molecular basis of variation in human color vision. Clin Genet 2005;67:369–377.
28. DiMaio TG, DiMaio VJM. Excited Delirium Syndrome: Cause of Death and Prevention. Boca Raton, FL: CRC Taylor and Francis, 2006.
29. DiMaio VJ, DiMaio D. Forensic Pathology, ed 2. Boca Raton: CRC Press, 2001.
30. Divertnet. Available at: http://www.divernet.com/news/stories/breath170603.shtml. Accessed 8/4/06.
31. Dolinak D, Smith C, Graham DI. Hypoglycaemia is a cause of axonal injury. Neuropathol Appl Neurobiol 2000;26:448–453.
32. Dooling EC, Richardson EP Jr,. Delayed encephalopathy after strangling. Arch Neurol 1976;33:196–199.
33. Duchen LW. General pathology of neurons and neuroglia. In Adams JH: Corsellis JAN, Duchen LW (eds): Greenfield's Neuropathology, ed 4. New York: John Wiley & Sons, 1984, pp 1–52.
34. Farrow JR, Davis GJ, Roy TM, et al. Fetal death due to nonlethal maternal carbon monoxide poisoning. J Forensic Sci 1990;35:1448–1452.
35. Ferrer I, Planas AM. Signaling of cell death and cell survival following focal cerebral ischemia: Life and death struggle in the penumbra. J Neuropathol Exp Neurol 2003;62:329–339.
36. Foster DW, Rubenstein AH. Hypoglycemia. In Wilson JD, Braunwald E, Isselbacher KJ, et al (eds): Harrison's Principles of Internal Medicine, ed 12. New York: McGraw-Hill, 1991, pp 1759–1765.
37. Fujino Y, Tsuboi Y, Shimoji E, et al. Progressive cerebellar atrophy following acute antidepressant intoxication. Rinso Shinkeigaku 2000;40:1033–1037. (Article in Japanese; reviewed in abstract form only.)
38. Gallyas F, Zoltay G, Dames W. Formation of "dark" (argyrophilic) neurons of various origin proceeds with a common mechanism of biophysical nature (a novel hypothesis). Acta Neuropathol (Berl) 1992;83:504–509.
39. Gallyas F, Zoltay G, Horváth Z. Light microscopic response of neuronal somata, dendrites and axons to post-mortem concussive head injury. Acta Neuropathol (Berl) 1992;83:499–503.
40. Galloway PG, Roessmann U. Neuronal karyorrhexis in Sommer's sector in a 22-week stillborn. Acta Neuropathol (Berlin) 1986;70:343–344.
41. Garcia JH, Anderson ML. Circulatory disorders and their effects on the brain. In Davis RL, Robertson DM (eds): Textbook of Neuropathology, ed 3. Baltimore: Williams & Wilkins, 1997, pp 715–822.
42. Ginsberg MD, Hedley-White ET, Richardson EP Jr. Hypoxic-ischemic leukoencephalopathy in man. Arch Neurol 1976;33:5–14.
43. Gore I. The pathology of hyperpyrexia: Observations at autopsy in 17 cases of fever therapy. Am J Pathol 1949;25:1029–1059.
44. Hardman JM. Microscopy of traumatic central nervous system injuries. In Perper JA, Wecht CH (eds): Microscopic Diagnosis in Forensic Pathology. Springfield, IL: Charles C Thomas, 1980, pp 268–326.
45. Hayes RL, Jenkins LW, Lyeth BG. Neurotransmitter-mediated mechanisms of traumatic brain injury: Acetylcholine and excitatory amino acids. J Neurotrauma 1992;9(Suppl 1):S173–S187.
46. Hetts SW. To die or not to die: An overview of apoptosis and its role in disease. JAMA 1998;279:300–307.
47. Hirano A. Neurons and astrocytes. In Davis RL, Robertson DM (eds): Textbook of Neuropathology, ed 3. Baltimore: Williams & Wilkins, 1997, pp 1–109.

48. Hong SK. The physiology of breath-hold diving. In Strauss RH (ed): Diving Medicine. New York: Grune & Stratton, 1976, pp 269–286.
49. Horn E, Lach B, Lapierre Y, Hrdina P. Hypothalamic pathology in the neuroleptic malignant syndrome. Am J Psychiatry 1988;145:617–620.
50. Hunter S, Ballinger WE Jr, Greer M. Nitrogen inhalation in the human. Acta Neuropathol 1985;68:115–121.
51. John L, Perreault MM, Tao T, Blew PG. Serotonin syndrome associated with nefazodone and paroxetine. Ann Emerg Med 1997;29:287–289.
52. Johnson JE Jr. The occurrence of dark neurons in the normal and deafferentated lateral vestibular nucleus in the rat: Observations by light and electron microscopy. Acta Neuropathol 1975;31:117–127.
53. Jones EM, Dawson A. Neuroleptic malignant syndrome: A case with post-mortem brain and muscle pathology. J Neurol Neurosurg Psychiatry 1989;52:1006–1009. (Comment on: J Neurol Neurosurg Psychiatry 1990;53:271; Author reply 271).
54. Kalimo H, Auer RN, Siesjö BK. The temporal evolution of hypoglycemic brain damage. III. Light and electron microscopic findings in the rat caudoputamen. Acta Neuropathol (Berl) 1985;67:37–50.
55. Kalimo H, Olsson Y. Effects of severe hypoglycemia on the human brain: Neuropathological case reports. Acta Neurol Scand 1980;62:345–356.
56. Katz VL, Dotters DJ, Droegemueller W. Perimortem cesarean delivery. Obstet Gynecol 1986;68:571–576.
57. Kinney HC, Armstrong DD. Perinatal neuropathology. In Graham DI, Lantos PL (eds): Greenfield's Neuropathology, ed 6. London: Arnold, 1998,I:535–599.
58. Kinney HC, Armstrong DD. Perinatal neuropathology. In Graham DI, Lantos PL (eds): Greenfield's Neuropathology, ed 7. London: Arnold, 2002,I:519–606.
59. Koiwai EK. Deaths allegedly caused by the use of "choke holds" (shime waza). J Forensic Sci 1987;32:419–432.
60. Koskinen PJ, Nuutinen HMJ, Laaksonen H, et al. Importance of storing emergency serum samples for uncovering murder with insulin. Forensic Sci Int 1999;105:61–66.
61. Kreutzberg GW, Blakemore WF, Graeber MB. Cellular pathology of the central nervous system. In Graham DI, Lantos PL (eds): Greenfield's Neuropathology, ed 6. New York: Oxford University Press, 1997,I:85–156.
62. Lannas PA, Pachar JV. A fatal case of neuroleptic malignant syndrome. Med Sci Law 1993;33:86–88.
63. Lawrence RD, Meyer A, Nevin S. The pathological changes in the brain in fatal hypoglycemia. Q J Med 1942;11:181–202.
64. Lapresle J, Fardeau M. The central nervous system and carbon monoxide poisoning. II. Anatomical study of brain lesions following intoxication with carbon monoxide (22 cases). In Bour H, Ledingham I McA (eds): Carbon Monoxide Poisoning. Progress in Brain Research, Vol 24. Amsterdman: Elsevier, 1967, pp 31–74.
65. Larroche J-C. Developmental Pathology of the Neonate. Amsterdam: Excerpta Medica, 1977.
66. Lee S, Merriam A, Kim T-S, et al. Cerebellar degeneration in neuroleptic malignant syndrome: Neuropathologic findings and review of the literature concerning heat-related nervous system injury. J Neurol Neurosurg Psychiatry 1989;52:387–391.
67. Løberg EM, Torvik A. Brain contusions: The time sequence of the histological changes. Med Sci Law 1989;29:109–115.
68. Malamud N, Haymaker W, Custer RP. Heat stroke: A clinicopathologic study of 125 fatal cases. Mil Surg 1946;99:397–449.
69. Martin DT, Swash M. Muscle pathology in the neuroleptic malignant syndrome. J Neurol 1987;235:120–121.
70. Naramoto A, Koizumi N, Itoh N, Shigematsu H. An autopsy case of cerebellar degeneration following lithium intoxication with neuroleptic malignant syndrome. Acta Pathol Jpn 1993;43:55–58.
71. Norenberg MD, Bruce-Gregorios J. Nervous system manifestations of systemic disease. In Davis RL, Robertson DM (eds): Textbook of Neuropathology, ed 3. Baltimore: Williams & Wilkins, 1997, pp 547–625.
72. Norman ES, Winston DC. Forensic Pathology No. FP-01–9 (FP-270). Check Sample. Forensic Pathology. Chicago: American Society of Clinical Pathology 2001;43:105–118.
73. Opeskin K, Burke MP. Hypotensive hemorrhagic necrosis in basal ganglia and brainstem. Am J Forensic Med Pathol 2000;21:406–410.
74. Parvizi J, Damasio AR. Neuroanatomical correlates of brainstem coma. Brain 2003;126:1524–1536.
75. Plum F, Posner JB. The Diagnosis of Stupor and Coma, ed 3. New York: Oxford University Press, 2000.
76. Plum F, Posner JB, Hain RF. Delayed neurological deterioration after anoxia. Arch Int Med 1962;110:56–63.
77. Poole CJM, Hill DJ, Christie JL, Birch J. Deficient colour vision and interpretation of histopathology slides: Cross sectional study. BMJ 1997;315:1279–1281.
78. Reeves RR, Mack JE, Beddingfield JJ. Neurotoxic syndrome associated with risperidone and fluvoxamine. Ann Pharmacother 2002;36:440–443. (Comment in: Ann Pharmacother 2002;36:1293; Author reply 1294).
79. Ross DW. Apoptosis. Arch Pathol Lab Med 1997;121:83.
80. Rossen R, Kabat H, Anderson JP. Acute arrest of cerebral circulation in man. Arch Neurol Psychiatry 1943;50:510–528.
81. Samson D, Batjer HH, Bourman G, et al. A clinical study of the parameters and effects of temporary arterial occlusion in the management of intracranial aneurysms. Neurosurgery 1994;34:22–28.
82. Sauvageau A, Racette S. Agonal sequences in a filmed suicidal hanging: Analysis of respiratory and movement responses to asphyxia by hanging (abstract). Proc Am Acad Forensic Sci 2006;12:226–227.
83. Schade JP, McMenemey WH (eds). Selective Vulnerability of the Brain in Hypoxaemia. Philadelphia: FA Davis, 1963.
84. Sminia P, van der Zee J, Wondergem J, Haveman J. Effect of hyperthermia on the central nervous system: A review. Int J Hyperthermia 1994;10:1–30.
85. Söderfeldt B, Kalimo H, Olsson Y, Siesjö BK. Bicuculline-induced epileptic brain injury: Transient and persistent cell changes in rat cerebral cortex in the early recovery period. Acta Neuropathol (Berl) 1983;62:87–95.
86. Spalding JAB. Colour vision deficiency in the medical profession. Br J Gen Pract 1999;49:469–475.
87. Spalding JAB. Confessions of a colour blind physician. Clin Exp Optom 2004;87:344–349.
88. Spitz W. Asphyxia. In Spitz W (ed): Spitz and Fisher's Medicolegal Investigation of Death. Springfield, IL: Charles C Thomas, 1993, p 444.
89. Stein TD, Fedynyshyn JP, Kalil RE. Circulating autoantibodies recognize and bind dying neurons following injury to the brain. J Neuropathol Exp Neurol 2002;61:1100–1108.
90. Turmaine M, Raza A, et al. Nonapoptotic neurodegeneration in a transgenic mouse model of Huntington's disease. Proc Natl Acad Sci 2000;97:8093–8097.
91. Van Den Pol AN, Gallyas F. Trauma-induced Golgi-like staining of neurons: A new approach to neuronal organization and response to injury. J Comp Neurol 1990;296:654–673.
92. van der Zee J. Heating the patient: A promising approach? Ann Oncol 2002;13:1173–1184.
93. Volpe JJ. Neurology of the Newborn. ed 3. Philadelphia: WB Saunders, 1995.
94. Vorster BJ, Milner LV. Colour-blind laboratory technologists. Lancet 1979;2:1295.
95. Winston DC. Suicide via insulin overdose in nondiabetics: The New Mexico experience. Am J Forensic Med Pathol 2000;21:237–240.

Vascular Diseases of the Central Nervous System

11

INTRODUCTION 307
CEREBRAL THROMBOSIS 307
CEREBRAL EMBOLISM 311
CEREBRAL HEMORRHAGE 311
SECONDARY CHANGES OF CEREBRAL
    INFARCTION 316
DURAL SINUS THROMBOSIS 317
CEREBRAL VASCULAR MALFORMATIONS 317

RUPTURED CEREBRAL ANEURYSMS 323
CEREBRAL AMYLOID ANGIOPATHY 327
DRUG ABUSE-RELATED CARDIOVASCULAR
    DISEASE 331
STATES OF ALTERED CONSCIOUSNESS AND THEIR
    ANATOMIC SUBSTRATES 331
REFERENCES 333

Introduction

With the very high prevalence of strokes and other cerebro-vascular diseases (CVD) in our population, it is inevitable that a significant percentage of cases referred for forensic investigation have CVD as the primary or contributory cause of death. Some of the more common reasons for referral of suspected CVD cases for forensic investigation follow:

Stroke leading to death within 24 hours of onset, or found dead at scene.
Sudden unexpected death from ruptured cerebral aneurysm.
Sudden unexpected death from hemorrhage of vascular malformation.
Embolization from remote sites.
Rupture of mycotic aneurysm.
Stroke leading to lethal motor vehicle accident.
Traumatic injury leading to a stroke.
Rupture of traumatic aneurysm (see Chap. 9).
Traumatic injury alleged as stroke.
Periprocedural vascular injury (see Chap. 16).

Cerebral Thrombosis

Stroke has been defined as "an abrupt onset of focal or global neurologic symptoms caused by ischemia or hemorrhage."[62] By definition, the symptoms must last longer than 24 hours and be the result of irreversible brain damage, thus distinguishing them from transient ischemic attacks (TIAs). Fortunately, modern emergent thrombolytic therapy has rescued many stroke victims from debilitating handicaps and even death. However, cerebro-vascular disease is still by far the most common neurologically disabling and lethal disorder in the United States, with atherosclerosis and hypertension as the major predisposing causes of stroke. Strokes are the fourth leading cause of death in the United States. Nevertheless, depending on circumstances, only a relatively small fraction of strokes become the subject of forensic investigation. The innumerable factors contributing to the risk for atherosclerosis leading to cerebral infarction need not be repeated here. For a review of the risk factors in the pathogenesis of stroke, the reader is referred to a modern textbook such as *Greenfield's Neuropathology*.[27] Although by far the most common cause, atherosclerosis is but one of several causes of occlusive vascular disease.

Relevant to the forensic pathologist for the determination of the mechanism of death in stroke are the following:

1. The elucidation of sites of atherosclerosis, or other vascular disease, and thrombosis. Some of the more common sites of thrombosis and embolic sources with atherosclerosis are the internal carotid artery just distal to the bifurcation of the common carotid artery, the internal carotid artery at its bifurcation into the middle and anterior cerebral arteries, the horizontal portion of the middle cerebral artery (Case 11.1, Figs. 11.1–11.4), the apex of the basilar artery, the body of the basilar artery, and the vertebral artery at its origin at the subclavian artery. Severe systemic atherosclerosis does not necessarily find a corresponding degree of involvement in the cerebral arteries. Atherosclerosis of the cerebral arteries may be surprisingly mild in the presence of severe systemic disease. An exception may be in the presence of diabetes mellitus, where cerebral and systemic arteries may show a similar degree of involvement.

Case 11.1. (Figs. 11.1–11.4)

A 65-year-old man who had undergone lobectomy for lung cancer a year earlier was treated with an anticoagulant for deep vein thrombosis in his leg. Several months later, he developed severe headaches and difficulty speaking. Examination disclosed an expressive aphasia with anomia. Difficulty with comprehension and apraxia were suspected. CT scan revealed a hypodense left posterior parietal region. He died within a week of onset of symptoms. General autopsy revealed a patent foramen ovale.

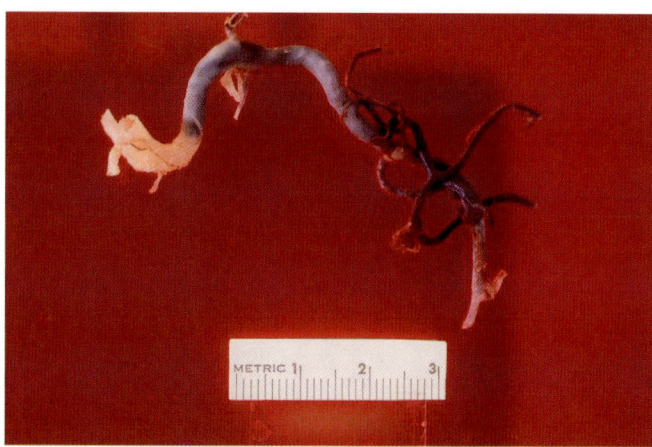

Fig. 11.2. Left middle cerebral artery with open internal carotid artery at the left of the photo and distal arteries to the right.

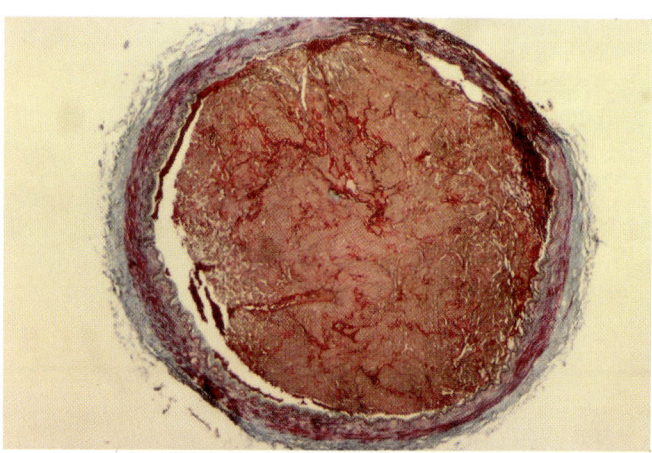

Fig. 11.3. Section of horizontal segment of left middle cerebral artery shows pure fibrin embolus distending a normal artery. (Trichrome.)

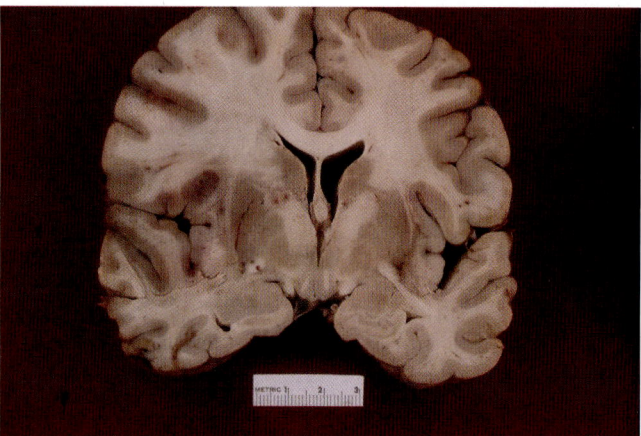

Fig. 11.1. Faintly hemorrhagic acute infarction is seen involving left middle and deep inferior frontal gyri, insula, and superior and middle temporal gyri. Slight midline shift to right is due to swollen left hemisphere, also causing narrowing of left lateral ventricle.

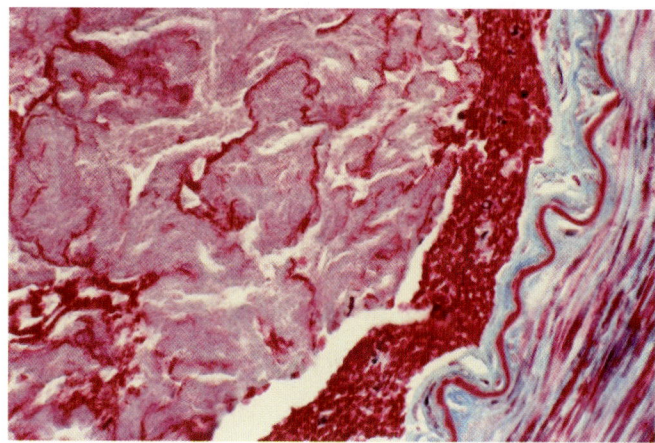

Fig. 11.4. Higher-power micrograph shows detail of embolus and luminal wall of artery. Folds of elastic lamina are slightly straightened due to distention of the artery by the embolus. (Trichrome.)

2. Anatomic extent of cerebral cortical infarction, whether acute or chronic, may be visible or palpable within a defined vascular territory and should be clearly described and diagramed (Figs. 11.5–11.6).

3. Secondary effects of edema should be described in terms of asymmetry of cerebral hemispheres; number of millimeters of shift across the midline; and presence or absence of cingulate, orbital frontal, uncal-parahippocampal, superior cerebellar vermis, and tonsillar herniations. A description of the overall shape and symmetry of the midbrain, shape of the aqueduct, and measurement of the lateral diameter of the midbrain will lend support to the determination of cause of death as midbrain compression. Uncal-parahippocampal herniation inevitably leads to asymmetric side-to-side compression of the midbrain with narrowing of the lateral diameter, the narrower half on the side opposite to the herniation. Of course, a good set of photographs will obviate the need for an overly detailed description. When present, Duret's hemorrhage of the midbrain and rostral pons is important to document.

4. Whenever timing of the pathologic process is important, several samples for microscopy from the margins of the infarction can be compared with several available histologic timetables (see Chap. 3). A note of caution is that histologic timetables provide only a range. Reliance on tables for precise timing inevitably leads to controversy, as emphasized in Chapter 3.

Including all age groups, 71% of strokes were due to cerebral thrombosis in one study.[43] Resulting cerebral infarctions are acute, subacute, or chronic, with loosely defined and overlapping divisions in histologic features and dating. The potential problem areas are in the acute–subacute and subacute–chronic transitions. Identifying early acute and late chronic infarcts is self-evident. In the former, grossly limited areas of cortical softening in a partial or total vascular territory are accompanied by histologic changes of acute ischemic necrosis of neurons (see Chap. 10) in the absence of vascular hyperplasia or macrophagic infiltration (see Chap. 3). At the other extreme, a cystic remnant within a vascular territory with little residual inflammation indicates a chronic infarct (Figs. 11.5–11.6). Subacute infarct spans the histologic period of coagulative to liquefactive necrosis in which there is ever-increasing infiltration by macrophages, eventually leading to a wall-to-wall sea of macrophages with endothelial hypertrophy, hyperplasia, and astrogliosis at the margins. However, both gross and microscopic appearances may be very misleading, as shown in a case of cerebral infarction, historically 4½ months old, with a persistent picture of a coagulative necrosis and large numbers of macrophages (Case 11.2, Figs. 11.7–11.10). The case highlights the importance of carefully investigated historical data which, fortunately, correlate well with the pathologic findings most of the time.

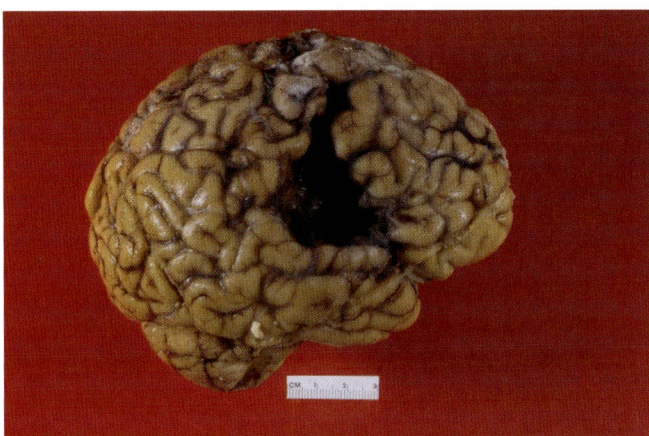

Fig. 11.5. Lateral view of right cerebral hemisphere shows a cavitary remnant of an old infarct with discrete loss of lateral portions of pre- and postcentral gyri, with anterior cavity margin at precentral sulcus and posterior margin at supramarginal gyrus.

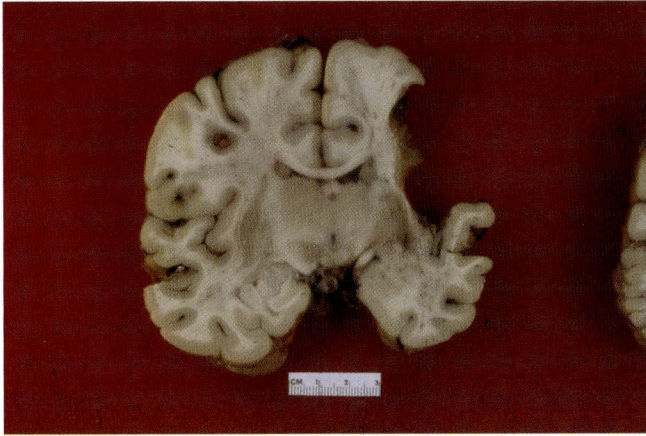

Fig. 11.6. Coronal section at midthalamic level shows total loss of cortex and subcortical white matter to the lateral margin of putamen. Superior temporal gyrus is spared.

Case 11.2. (Figs. 11.7–11.10)

An 85-year-old woman was assaulted in a robbery attempt, and sustained bilateral mandibular fractures. She drove herself to an emergency room where she lost consciousness and was found to have a left hemiparesis. Following open reduction and fixation of the fractures, she remained in a coma. She was thought to be in coma vigil. She died 4½ months after the attack.

Autopsy revealed marked concentric atherosclerosis of the right internal carotid artery, leaving a pinpoint lumen, but the right A_1 segment of the anterior cerebral artery was widely patent, and a large anterior communicating artery was present.

310 Vascular Diseases of the Central Nervous System

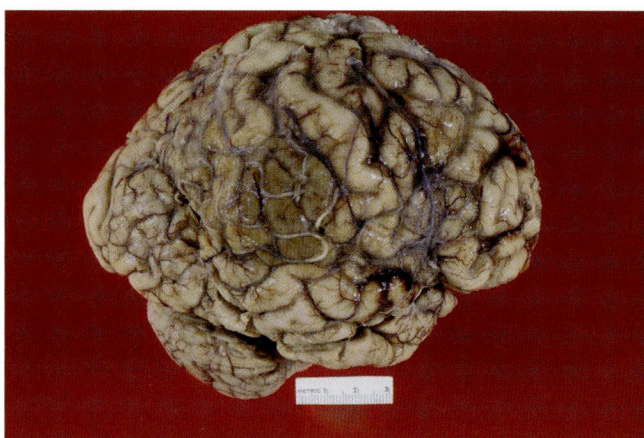

Fig. 11.7. Right lateral cerebral hemisphere shows dark discoloration of inferior parietal lobule just behind postcentral sulcus, and partial cortical collapse extending posteriorly into occipital lobe. A Stryker saw cut passes through middle of area.

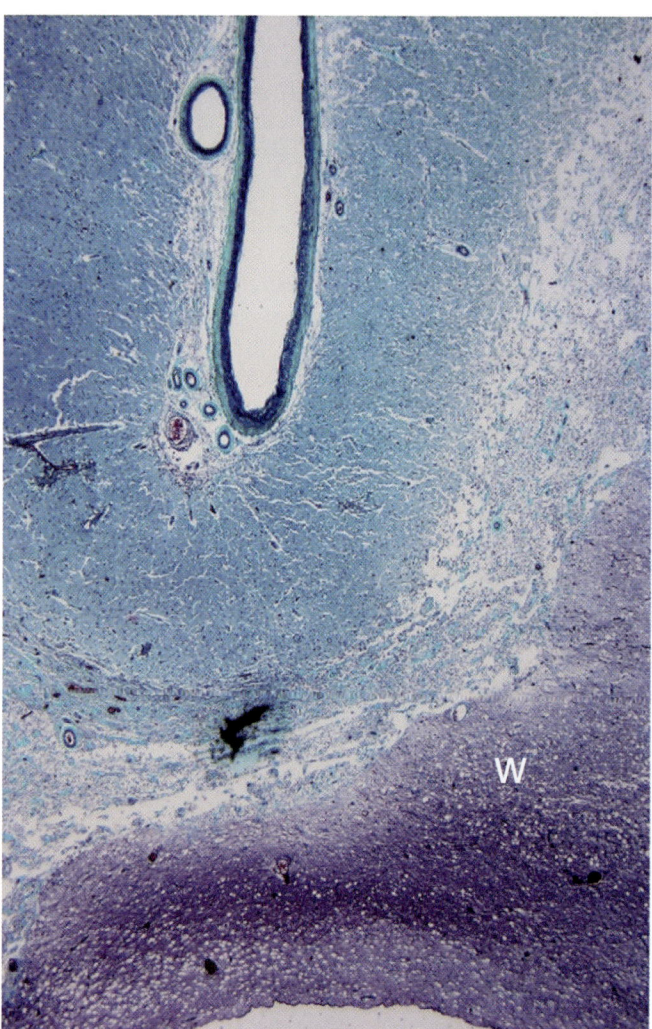

Fig. 11.9. Low-power micrograph shows infarcted cerebral cortex along the depth of a sulcus and partially sparing subcortical white matter (W). (Gomori trichrome.)

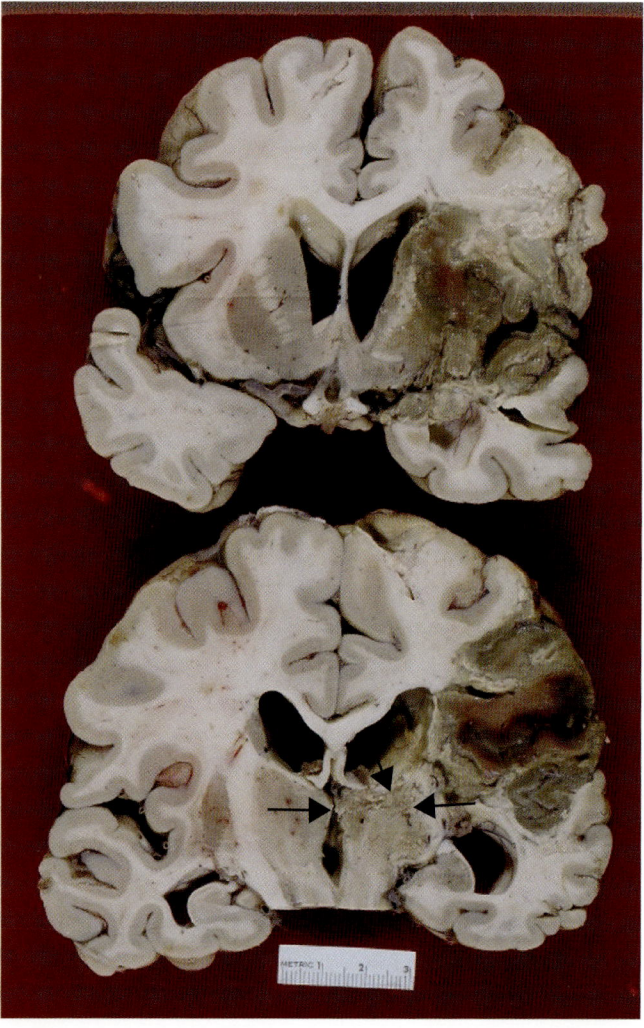

Fig. 11.8. Coronal sections show necrotic remnants of chronic infarct in right middle cerebral artery territory, including levels of anterior and posteriormost basal ganglia. Infarct had transected anterior limb and all of posterior limb of internal capsule. In the territory of posterior circulation, thalamus was partially infarcted (*arrows*).

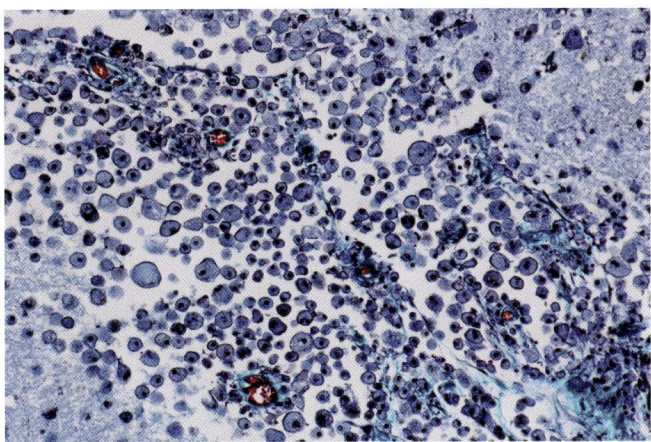
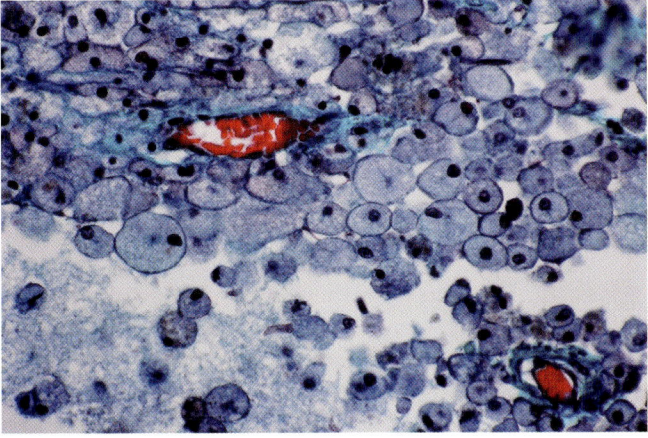

Fig. 11.10. *Top*, medium-power micrograph shows a sea of foamy macrophages in a liquefying area of infarct of 4½ months' age. *Bottom*, higher-power detail. (Gomori trichrome.)

Cerebral Embolism

Cerebral embolism comprises a relatively small percentage of causes of stroke. Strokes due to embolization from extracranial arteries, that is, artery-to-artery, are said to comprise only 3.8% of all strokes. However, emboli of cardiac origin may cause as many as 13.6% of strokes.[27] Demonstration of an artery occluded and distended by a thrombus, with infarction distal to that point, in the absence of disease at the point of obstruction and absence of intracranial vascular disease in general, leaves little option other than the diagnosis of embolization. The precise source of the embolus is not always easily determined, but heart and atheromatous plaque at the carotid bifurcation are most often suspects. Paradoxical embolization from a deep leg vein thrombosis through a patent foramen ovale occasionally occurs (see Case 11.1, Figs. 11.1–11.4). A large assortment of types of emboli have caused cerebral infarctions that are often hemorrhagic. Emboli of forensic importance might include a shotgun pellet or bullet fragment (see Chap. 8); bone fragments following traumatic fracture or surgical procedure; and intentional therapeutic emboli gone astray, such as those used to occlude cerebral aneurysms; fat; and tumor, among others. Traumatic vascular occlusion and traumatic embolization are discussed in Chapter 6. Gas and air embolization and amniotic fluid embolization are discussed in Chapter 4. Mycotic embolization secondary to cardiac valvular vegetation is not uncommonly encountered in a forensic setting. An unusual cause of sudden unexpected death from embolization to the brain resulted from food embolizing to the brain following the development of an esophageal-cardiac fistula in a mentally retarded man with a habit of self-induced vomiting.[30]

A thrombotic infarct, such as that due to atherosclerosis causing occlusion by gradual compromise of the vessel lumen, is mainly anemic. Although not always the case, cerebral thromboembolization often leads to local reperfusion and conversion of an anemic to a hemorrhagic infarction of proximal areas, such as the basal ganglia (Case 11.3, Fig. 11.11). This mechanism of hemorrhagic infarction is thus based on the concept of flow-reflow.

A different pattern of limited hemorrhagic infarction may be seen in the border zones between occluded and nonoccluded vascular territories, that is, by perfusion of infarcted parenchyma by collateral blood supply from adjacent unaffected vascular territories. The histologic dating of CVD infarcts is discussed in Chapter 3.

Cerebral Hemorrhage

The list of causes of nontraumatic intracerebral hemorrhage is long, and a few more commonly encountered causes are listed in Table 11.1. By far the most common of those causes in decedents coming to the attention of the medical examiner is chronic systemic hypertension. It is well established that hypertensive intracerebral hemorrhage is more common in blacks versus whites and even more common in Asians. It is more frequent in males. The diagnosis of hypertension may have been made in life, but some individuals are found dead at the scene with intracerebral hemorrhage without a prior history of hypertension. In that event, autopsy findings of cardiomegaly, left ventricular hypertrophy, nephro-

Case 11.3. (Fig. 11.11)

A 22-year-old man, found down at work 4 days prior to death, had complained of headaches. CT head scan 3 days after admission showed a left cerebral hemorrhage. He remained comatose until death. Ten months prior to death, his vehicle was rear-ended. Any complaints resulting from that accident were unknown.

Vascular Diseases of the Central Nervous System

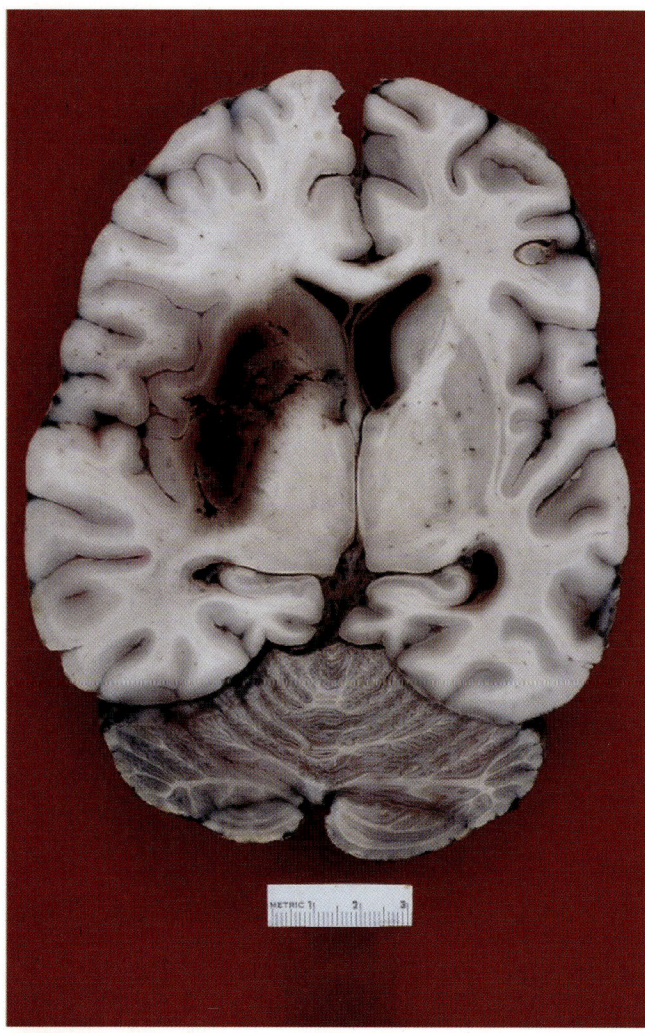

Fig. 11.11. Brain sectioned in CT plane shows an acute hemorrhagic infarct of basal ganglia in left middle cerebral artery territory, while cortical areas are relatively anemic (seen as subtle blurring of cortical/white matter boundaries contrasted with distinct cortical ribbon on right).

Table 11.1. Some Nontraumatic Causes of Intracerebral Hemorrhage

Systemic hypertension
Vascular malformation
Cerebral amyloid angiopathy
Neoplasms (primary and secondary)
Disseminated intravascular coagulation
Thrombotic thrombocytopenia purpura (Moschcowitz's disease)
Drug related
    Anticoagulant therapy
    Contraceptive use
    Cocaine, methamphetamine, heroin, ephedrine

sclerosis, and vascular hypertrophy and sclerosis in the brain, if present, will support the diagnosis of hypertension as the likely cause of the hemorrhage.

Hypertensive cerebral hemorrhages have their favored anatomic sites. Common sites are putamen and thalamus, lobar white matter, cerebellum, and pons. In an autopsy series, 42% involved striatum; 16%, pons; 15%, thalamus; 12%, cerebellum; 10%, cerebral white matter; and 5%, other sites.[24] Hemorrhage at each locus presents its relatively characteristic clinical signs. The most common putaminal hemorrhage presents primarily as a contralateral hemiplegia and forced adversive eye movements, with the eyes directed toward the hemisphere containing the hemorrhage. Thalamic hemorrhage may manifest as hemianesthesia at onset, quickly followed by hemiplegia and coma when hemorrhage extends caudally into the midbrain. Cerebellar hemorrhage presents as ipsilateral ataxia, vomiting, forced lateral gaze, often progressing to brainstem compression and coma. Pontine hemorrhage presents as coma at the onset, with quadriparesis and dysconjugate gaze. Hemorrhage in lobar white matter has a presentation similar to that of infarct within that vascular territory.

Clearly, the volume of hemorrhage is an important, but not the only, determinant of outcome, and those cases referred to the medical examiner usually have large hemorrhages. "Large" is a relative term, but some attempts have been made to quantify the volume. For example, a hemorrhage greater than 3.0 cm in diameter in the cerebral hemisphere, 2.0 cm in the cerebellum, and 1.0 cm in the brainstem has been considered large.[25] Aside from volume of hemorrhage, anatomic site of involvement is critical. Other secondary determinants of lethal outcome discussed elsewhere in this text are (1) generalized cerebral swelling and edema assessed by increased brain weight, flattened cortical surfaces, and reduced ventricular volume; (2) asymmetric brain edema and mass effect, with midline shift leading to uncal-parahippocampal and cingulate herniations; and (3) consequent caudal displacement of the rostral brainstem and brainstem compression leading to Kernohan's notch and Duret's hemorrhage (see Fig. 11.23). Survivors of Duret's hemorrhage do occur but are infrequent.

Putaminal hemorrhages have been further described as subinsular to emphasize their characteristic lateral localization sparing insular cortex as demonstrated by computed tomography (CT) and magnetic resonance (MR) scans in smaller, less overwhelming, hemorrhages (Case 11.4, Fig. 11.12). Such hemorrhages chronically resolve as slitlike subinsular cysts parallel to the claustrum with cystic replacement of the lateral putamen, sparing the globus pallidus (Fig. 11.13). Follow-up CT scan studies of intracerebral hemorrhages have shown that even large hemorrhages are found to have mostly resolved in survivors.[23] Even large hemorrhages usually spare overlying insular cortex while destroying the

Case 11.4. (Fig. 11.12)

A 44-year-old man was found down in his secured residence during a welfare check. Autopsy revealed a contusion of parietal scalp. Toxicologic survey was negative.

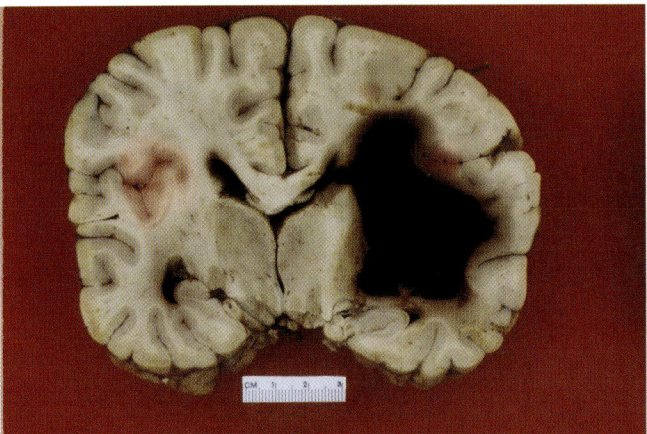

Fig. 11.12. Coronal section at midthalamic level shows massive acute subinsular hemorrhage destroying basal ganglia. Hemorrhage extended from pregenu frontal white matter, through basal ganglia, and into post-splenial parietal white matter, including part of optic radiation. Ipsilateral uncal groove was only slightly increased.

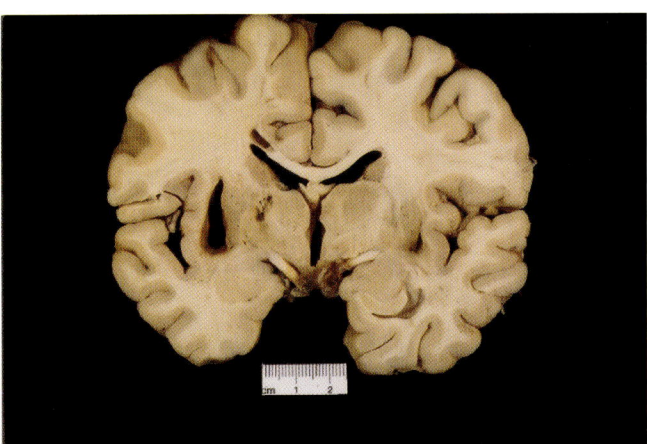

Fig. 11.13. Coronal section at level of anterior thalamus shows an example of a characteristic vertical subinsular chronic cystic lesion of left putamen due to a resolved hemorrhage (and smaller cystic/malacic lesions in left corpus callosum and internal capsule regions).

putamen, internal capsule, and portions of the caudate, with eventual dissection into the lateral ventricle. The results of attempted surgical evacuation of basal ganglia hematomas have been dismal, with 75% mortality. There are a greater number of survivors with conservative therapy.[48] As might be anticipated, some 20 years later surgical mortality was down to 28% in a small series, although still worse than with conservative treatment.[71,72]

Large thalamic hemorrhages invariably involve nuclei subserving sensation such as posterior ventral nuclei and adjacent internal capsule, causing a clinical constellation of contralateral hemianesthesia and hemiplegia (Case 11.5, Fig. 11.14). As in other "hypertensive" hemorrhages, the precise point of vascular rupture is virtually never identified. On a rare occasion, a potential source is serendipitously found in the form of a Charcot-Bouchard microaneurysm (Case 11.6, Figs. 11.15–11.18). Such aneurysms have been demonstrated by injection/corrosion

Case 11.5. (Fig. 11.14)

A 50-year-old woman had a history of chronic cocaine and alcohol abuse. Paramedics transported her to a hospital when she was found with altered consciousness. A diagnosis of "intracranial bleed, status post ingestion of cocaine" was made when she tested positive for cocaine. She remained comatose and expired 5 days after admission. Cocaine was not detectable in blood taken 9 days after admission to the hospital, but levels of benzoylecgonine were detected at less than $0.3\,\mu g/ml$.

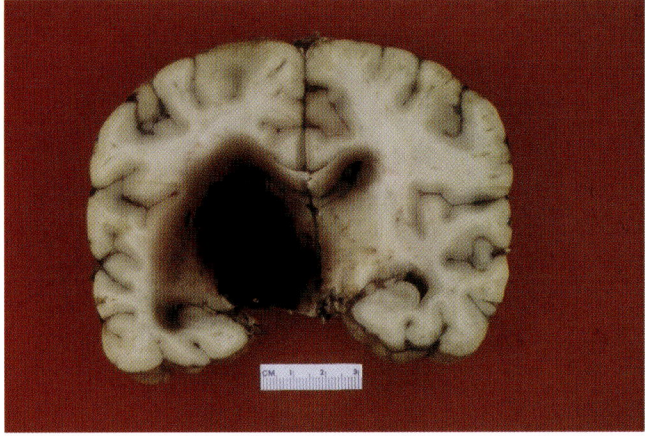

Fig. 11.14. A massive left thalamic hemorrhage also involves the posterior limb of internal capsule, with prominent displacement of adjacent structures. The hemorrhage extended caudally through midbrain and into ipsilateral pontine tegmentum.

Case 11.6. (Figs. 11.15–11.18)

A 97-year-old woman reportedly had a stroke 8 months earlier that left her bedridden, aphasic, and on gastric tube feeding. A diagnosis of parkinsonism was also considered. Autopsy revealed a heart weight of 370g with slight ventricular hypertrophy. État lacunaire was also present.

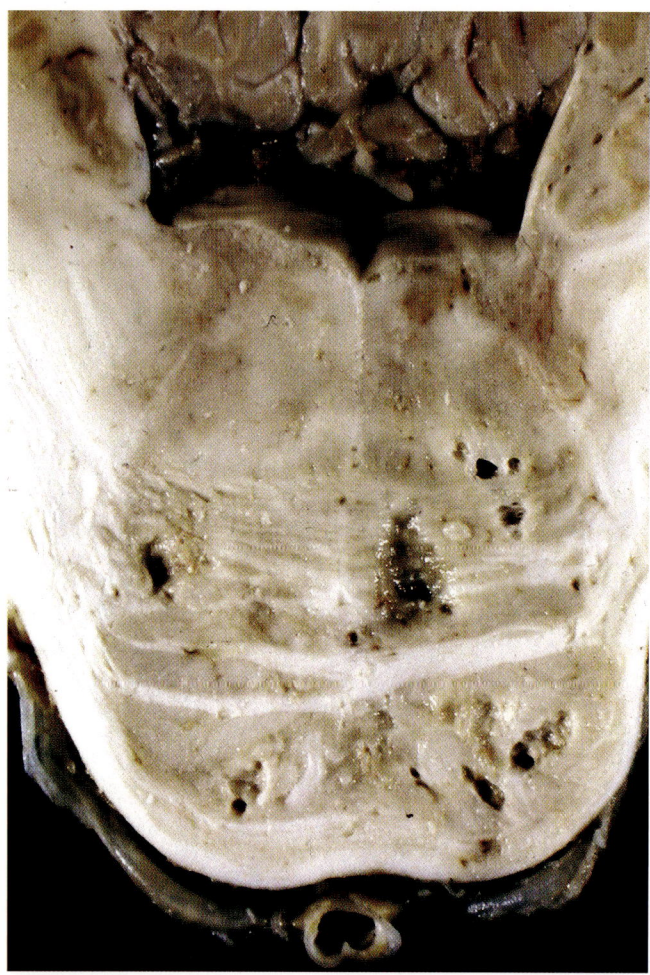

Fig. 11.15. Closeup view of état lacunaire of pons.

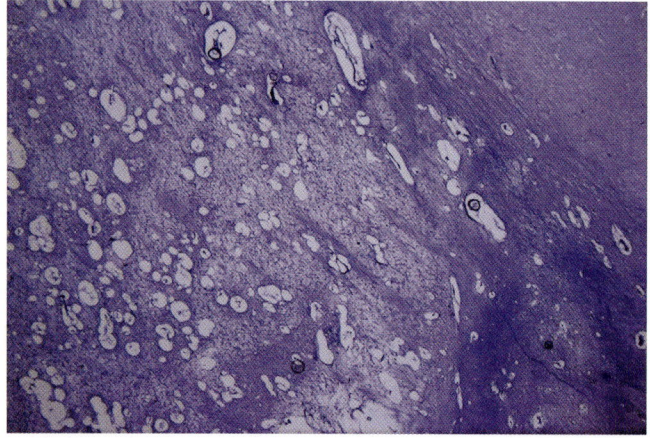

Fig. 11.16. Lower-power micrograph of putamen shows extensive perivascular atrophy of état lacunaire. (Holzer.)

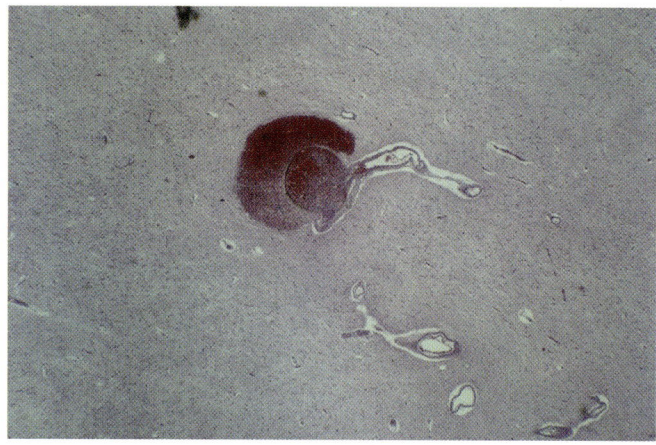

Fig. 11.17. Low-power photomicrograph shows a Charcot-Bouchard aneurysm of thalamus. (H&E stain.)

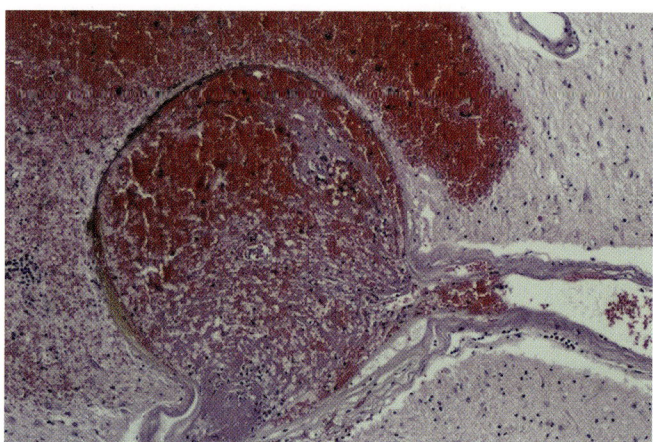

Fig. 11.18. Medium-power micrograph shows details of the aneurysm seen in Fig. 11.17. Small perianeurysmal hemorrhage is present, but rupture site is not evident in this plane of section. (H&E.)

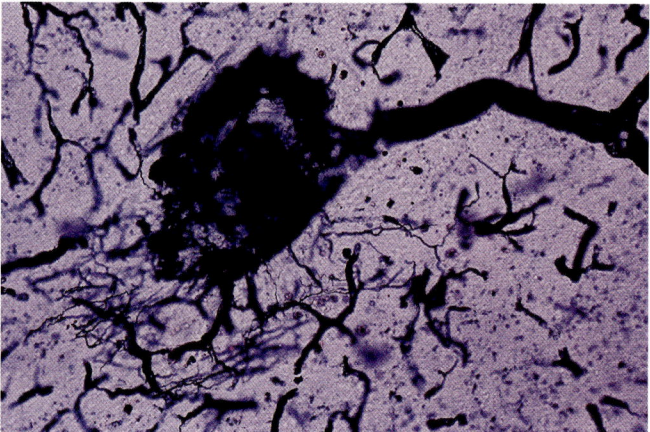

Fig. 11.19. Medium-power micrograph of Charcot-Bouchard aneurysm in thalamus in another case. (India ink perfusion.)

techniques in the past (Fig. 11.19), but another explanation for the "aneurysm" has been proposed.[14] Using a new technique, these authors provided evidence that hemorrhages originate in tortuous arterioles mimicking aneurysms. Moreover, even the relationship of Charcot-Bouchard aneurysm as a cause of intracerebral hemorrhage has been challenged.[14]

Cerebellar hemorrhage in a hypertensive setting is most often centered in central white matter in or near the dentate nucleus, suggesting that a penetrating branch of the superior cerebellar artery is the source. Branches of the anterior inferior cerebellar artery may be another source. The small capacity of the posterior fossa leaves little room for expansion before the hemorrhage mass leads to tonsillar herniation, medullary compression, and cardiorespiratory arrest (Case 11.7, Fig. 11.20). Upward herniation of the superior cerebellum through the incisura tentorii may also occur, with loss of the pupillary light reflex and presence of bilateral extensor rigidity.[27] As opposed to cerebral hemispheric hemorrhages for which surgical intervention is usually inadvisable or outcome not significantly greater than conservative treatment,[33,48] timely evacuation of cerebellar hemorrhage may be life-saving with good outcome.[33]

Although less frequent, hypertensive pontine hemorrhage causes a most devastating clinical presentation with coma at the onset, and it has a very high mortality of about 72%.[35] The hemorrhage is usually located in the midline basis pontis at midpontine level, and it frequently expands into the tegmentum and dissects into the fourth ventricle (Case 11.8, Fig. 11.21). A major hemorrhage into the tegmentum is almost always incompatible with life due to interruption of autonomic systems regulating cardiorespiratory functions (Case 11.9, Fig. 11.22). The source of pontine hemorrhages is found in

Case 11.8. (Fig. 11.21)

A 79-year-old man was heard arguing with his common law wife followed by pounding noises coming from the couple's trailer, heard by neighbors. The wife was overheard saying, "Are you dead yet?" The man was found dead at the scene with his head in a pool of blood. A blood-stained claw hammer and a pipe wrench were next to the body. Cabinet and appliances in the room were blood spattered.

Case 11.7. (Fig. 11.20)

A 67-year-old man was found down on the street, reportedly having fallen without loss of consciousness, and he was taken to the hospital where initial vital signs were normal. Alcohol was detected on his breath, and his blood alcohol was 0.183%. He used abusive language, and refused to give a history. Physical examination was refused. On the following morning, 15 hours after found down, he was discovered pulseless, and resuscitative efforts failed.

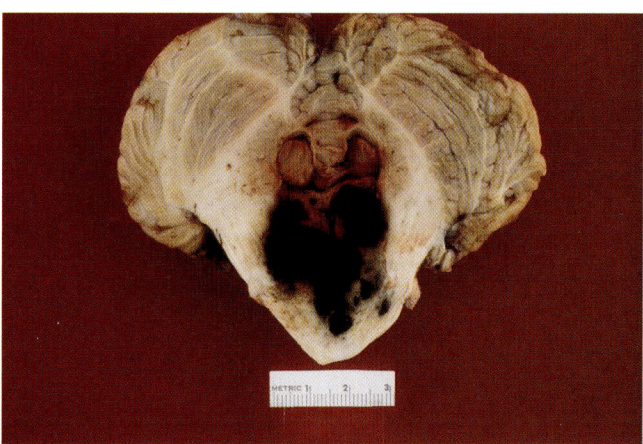

Fig. 11.21. Transverse section at midpontine level reveals massive hemorrhage destroying tegmentum and extending into basis pontis. Microscopy failed to expose a cause. Dura mater and cerebrum were free of traumatic injury. Mode was homicide, but the mechanism may have been related to catecholamine surge effects accompanying the assault.

Case 11.9. (Fig. 11.22)

A 47-year-old male was discovered unresponsive on a sidewalk. He had complained of right temporal pain and left-sided numbness prior to his collapse. His blood pressure recorded in the field was 300/130. He was admitted in coma and in full cardiopulmonary arrest. CT head scan was interpreted showing a "posterior fossa bleed and brain stem infarct." He was pronounced dead a few hours after admission.

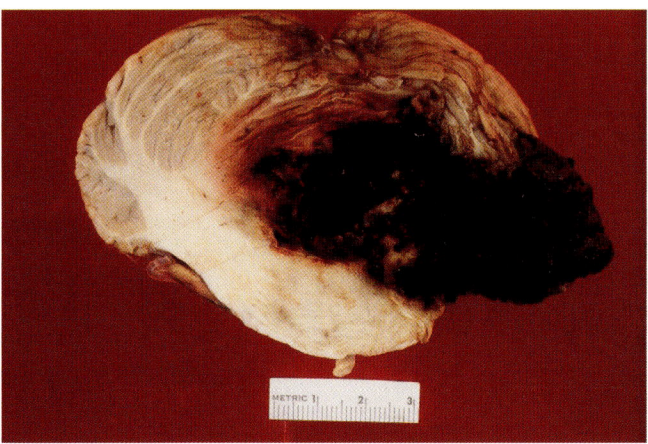

Fig. 11.20. Horizontal section through midpontine level shows a massive acute cerebellar hemorrhage with partial erosion of pontine tegmentum. The cause could not be determined microscopically.

Vascular Diseases of the Central Nervous System

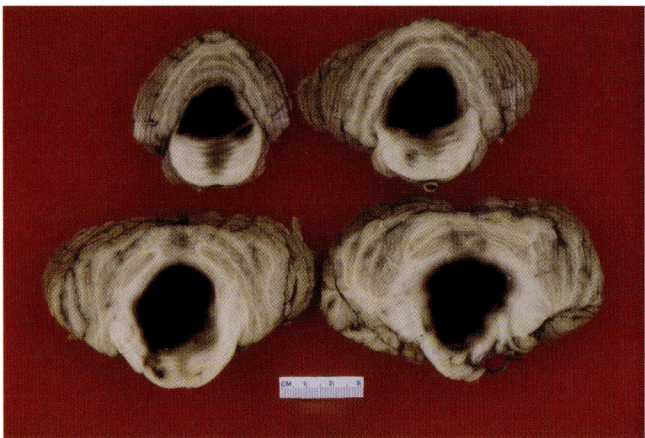

Fig. 11.22. A massive pontine hemorrhage extends through several levels, primarily involving tegmentum. Postmortem toxicologic study detected a blood benzoylecgonine level of 0.7 μg/ml.

midline and paramedian perforators arising from the basilar artery.

Secondary Changes of Cerebral Infarction

Secondary changes of cerebral infarction due to thrombosis or embolization (and cerebral hemorrhage) become as important as primary lesions in determination of outcome. Without assigning any priority, these changes include cerebral edema (vasogenic); ischemic-hypoxic injury (cytotoxic edema); spread of secondary infarction around the primary lesion and at remote sites; uncal-parahippocampal, tonsillar, upward transtentorial cerebellar vermis, cingulate, and orbital frontal herniations; and secondary vascular compression, often with hemorrhagic infarction and most commonly in the territory of the posterior cerebral artery. Unilateral mass effect of hemorrhage, infarction, and associated edema may result in uncal-parahippocampal herniation leading to a disastrous Duret's hemorrhage in the midbrain and pons, interrupting vital hypothalamobulbar pathways and centers subserving respiration and cardiac function (Case 11.10, Fig. 11.23). Predictably, the presence of Duret's hemorrhage correlates with poor outcome. The volume and location of cerebral hemorrhage, territory and extent of infarction, and the volume and location of subarachnoid hemorrhage vary considerably from case to case, and it would be a great challenge indeed to collect a series with a measured volume of hemorrhage and infarction to correlate with the presence or absence of Duret's hemorrhage. In one large autopsy series, however, the incidence of Duret's hemorrhage was determined in cerebral hemorrhage (45%), ruptured

Case 11.10. (Fig. 11.23)

A 50-year-old male was in good health when he cleaned out a cesspool and a septic tank using caustic soda flakes. The following day, he fell ill with nausea and hemorrhagic lesions on his body and legs. Two days later, he awoke complaining of headaches and feeling feverish. Hematemesis followed a half-hour later, along with hematochezia and hematuria. He died shortly thereafter.

Fig. 11.23. Photo shows typical pattern of secondary brainstem Duret hemorrhage, beginning rostrally in midbrain as an anterior midline hemorrhage, expanding into pontine tegmentum, with less in basis pontis. The primary lesion was an acute hemorrhage in right superior parietal lobule, with secondary brain swelling and herniation syndrome (also see Chap. 9).

cerebral aneurysm (36%), and infarction (29%).[52] These secondary effects are more fully covered in Chapter 6. Brain death is discussed in Chapter 3.

Dural Sinus Thrombosis

Dural sinus thrombosis occurs in several recurring locations, leading to disparate clinical syndromes. Thrombosis of the cavernous sinus most often has an infectious etiology, such as paranasal sinusitis, dental abscess of the maxillary teeth, or skin infection around the nose and eyes. Involvement of the third, fourth, and sixth cranial nerves and the ophthalmic division of the trigeminal nerve leads to ocular palsies and sensory symptoms of the upper face, unilateral when the thrombosis is confined to one side. Extension into the ophthalmic veins can result in chemosis and proptosis of the eye. Spread of thrombosis into the superior petrosal sinus may eventuate in a transverse sinus thrombosis, a more dangerous situation. Only rarely does our experience include a case referred to our office of an individual who had died with cavernous sinus thrombosis.

The most frequent dural sinus thrombosis involves the superior sagittal sinus,[2] and one or two such cases are encountered yearly in our office. The circumstances under which thrombosis occurs include, for example, the woman in her postpartum period or taking oral contraceptives, malnutrition, dehydration, presence of remote malignancy, and hematologic disorders including hypercoagulable states. The clinical presentation is almost always initially headaches, hemiplegia, and seizures followed by signs of increased intracranial pressure.

The pathology is immediately grossly apparent upon reflection of the dura, which exposes parasagittal hemorrhagic cortical infarction and distended thrombosed superficial cortical veins draining into the superior sagittal sinus (Figs. 11.24–11.26). The sinus itself will be firm and full in appearance. The hemorrhage may be unilateral (Figs. 11.24–11.25), or bilateral (Fig. 11.26).

Microscopy usually adds no further information other than what may be important for timing if that is an issue (Figs. 11.27–11.28). Nevertheless, microscopy is a necessary procedure to rule out possible inflammatory, neoplastic, and other etiologic factors.

Cerebral Vascular Malformations

Vascular malformations of the brain and spinal cord are, in order of their clinical importance, arteriovenous malformation, cavernous hemangioma, venous angioma (malformation), mixed angioma, and capillary telangiectasia.[6,31] Only rare cases of *capillary telangiectasia* of clinical importance have been reported, such as the case involving the medulla and pons in a patient with lymphosarcoma.[21] One of our cases of capillary telangiectasia included a sudden unexpected death in a 21-year-old man with a telangiectasia occupying bilaterally symmetric paramedian medulla that included the medial lemnisci and medial medullary reticular formation and adjacent nuclei and tracts (Case 11.11, Figs. 11.29–11.32). Capillary telangiectasia is a common, focal, and usually small incidental finding in the pons (Fig. 11.33), most frequent in the basis pontis, cerebral cortex/subcortical white matter, less frequent in the basal ganglia and cerebellum, and more rarely in such locations as the corpus callosum (Figs. 11.34–11.35). Capillary telangiectasia may accompany other types of malformations such as cavernous hemangioma. It appears as a faint pink to reddish stain in the fresh state and pale gray to a darker blush

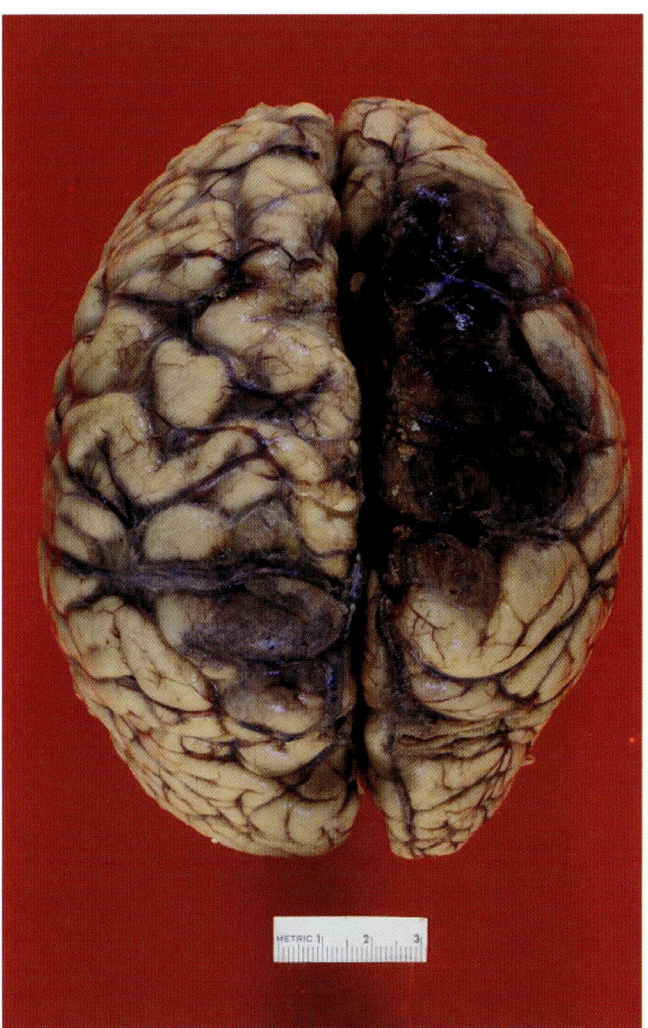

Fig. 11.24. Vertex view of cerebral hemispheres shows distended thrombosed superficial cortical veins secondary to thrombosis of superior sagittal sinus. Darkened cortex indicates acute hemorrhagic infarction and associated subarachnoid hemorrhage.

318 Vascular Diseases of the Central Nervous System

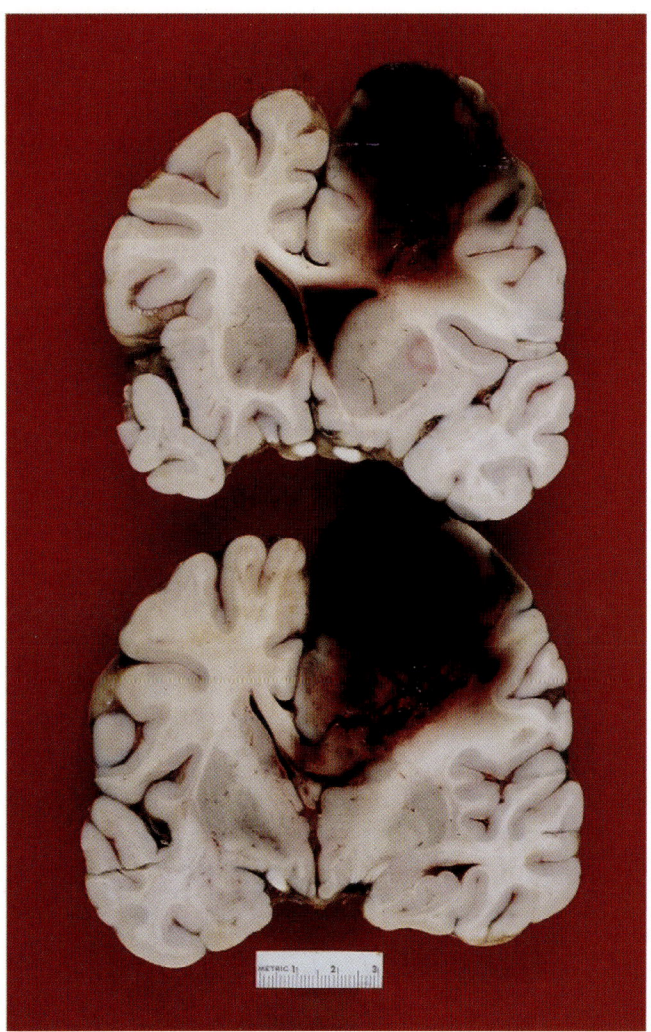

Fig. 11.25. Coronal section shows a large hemorrhage in the superior sagittal sinus drainage territory.

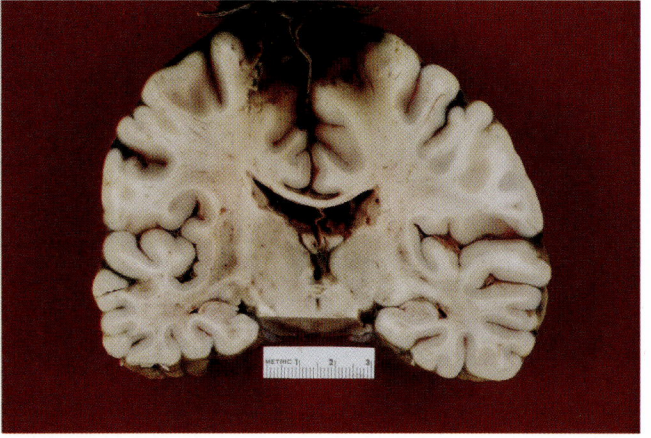

Fig. 11.26. Hemorrhage secondary to superior sagittal sinus thrombosis may be bilateral.

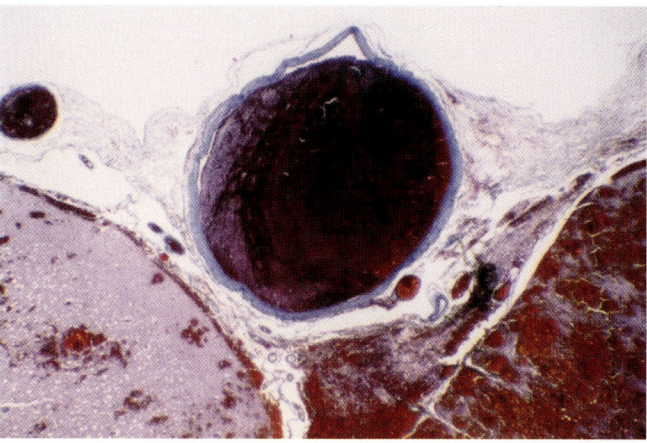

Fig. 11.27. Low-power micrograph of subarachnoid cortical vein distended by acute thrombus in a case of dural sinus thrombosis. Note underlying hemorrhagic cortical infarction. (Trichrome.)

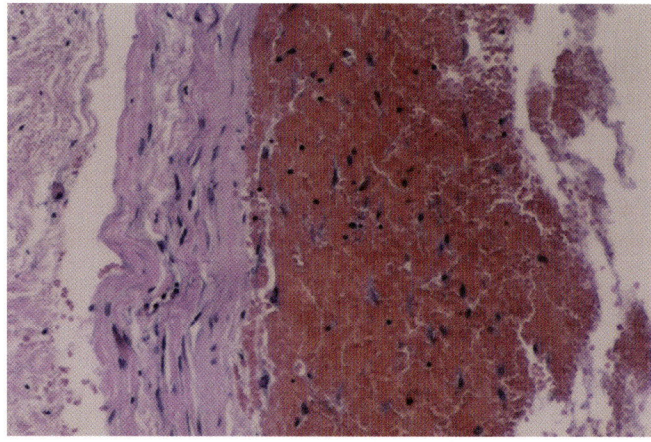

Fig. 11.28. Medium-power micrograph of the vein seen in Fig. 11.27 shows sparse scattered fibroblasts in the thrombus consistent with a few days' survival. Vein wall shows little change. (H&E.)

Case 11.11. (Figs. 11.29–11.32)

A 21-year-old man came home complaining of "not feeling well" at 8 PM, and went to lie down. Two and a half hours later, he was heard "gagging" in his sleep. Paramedics transported the decedent to the hospital where he was pronounced dead a half hour later. General autopsy revealed pulmonary edema and congestion, but no clear cause of death. Neuropathologic examination revealed a capillary telangiectasia in the right frontal white matter and in the medulla. Neither had hemorrhaged.

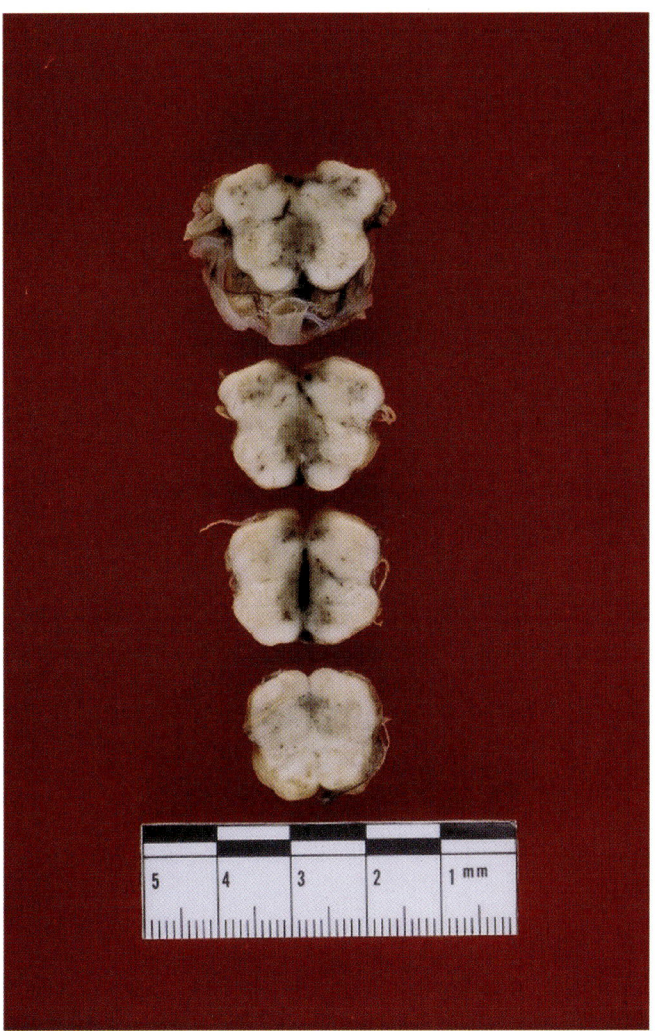

Fig. 11.29. Gross neuropathologic examination revealed a vascular "blush" symmetrically involving medullary paramedian structures, including medial lemnisci, medial reticular formations, dorsal vagal nuclei, and parasolitary nuclei.

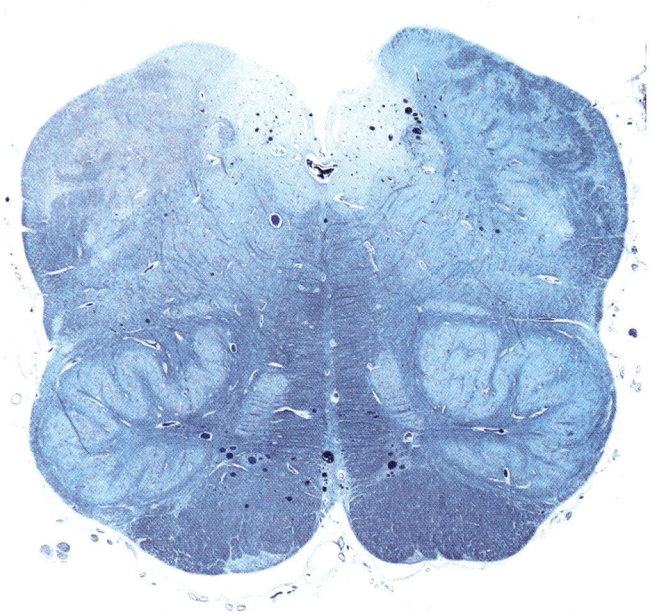

Fig. 11.30. Whole mount of medulla demonstrates dilated vascular channels, primarily at lower medial lemnisci and floor of fourth ventricle at this level. (Luxol fast blue.)

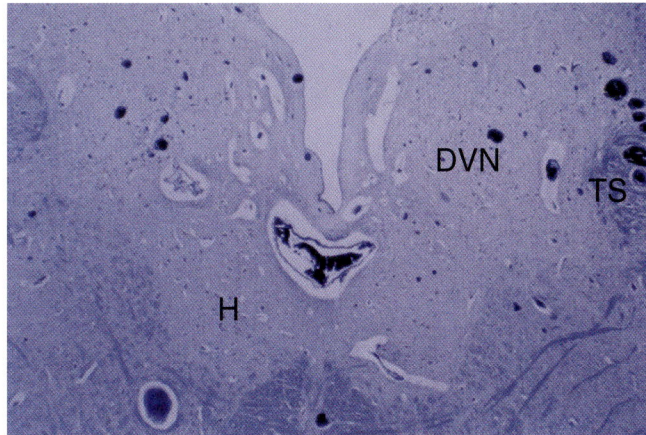

Fig. 11.31. Low-power micrograph of floor of fourth ventricle at level seen in Fig. 11.30. An excess of capillary-sized vessels is present bilaterally (and was even more florid at other section levels). H, hypoglossal nucleus; DVN, dorsal vagal nucleus; TS, tractus solitarius. (Luxol fast blue.)

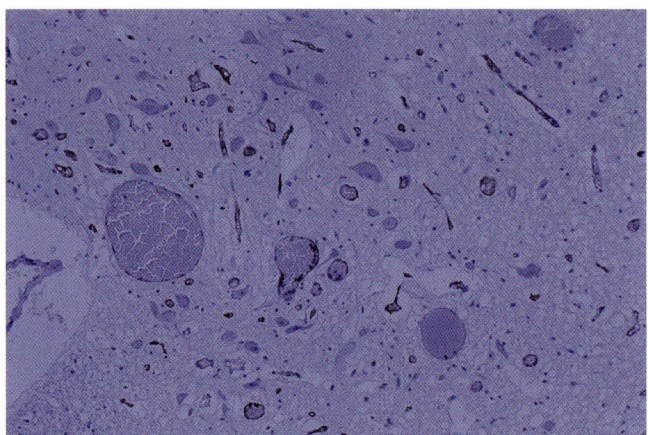

Fig. 11.32. Right dorsal vagal nucleus of Case 11.11 demonstrates dilated and normal-sized capillaries with brown-stained endothelium in the midst of normal neurons and neuropil. (CD34 immunostain.)

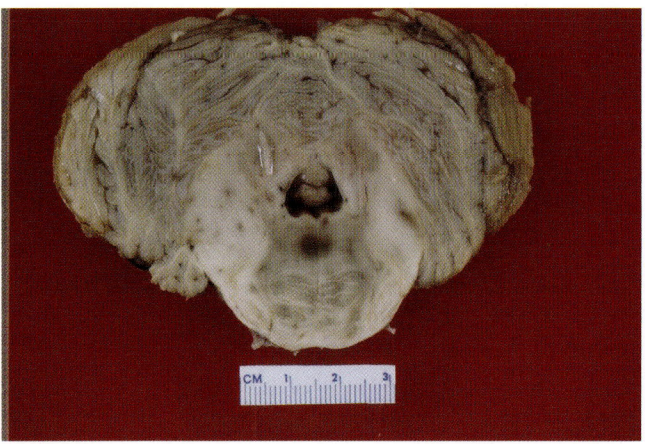

Fig. 11.33. Transverse section at midpontine level of another case shows a tan to brown blush (postfixation) characteristic of capillary telangiectasias at the midline. They occur more frequently in basis pontis, but can be located in tegmentum, as in this case.

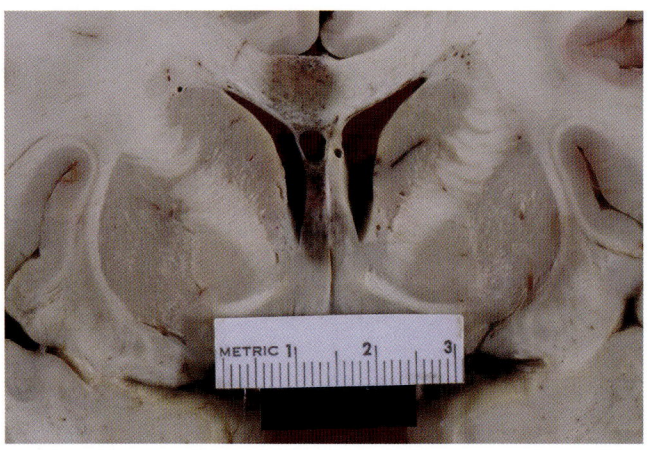

Fig. 11.34. Closeup view of corpus callosum at the level of the anterior commissure shows a hemorrhagic-appearing blush of a capillary telangiectasia to left of midline.

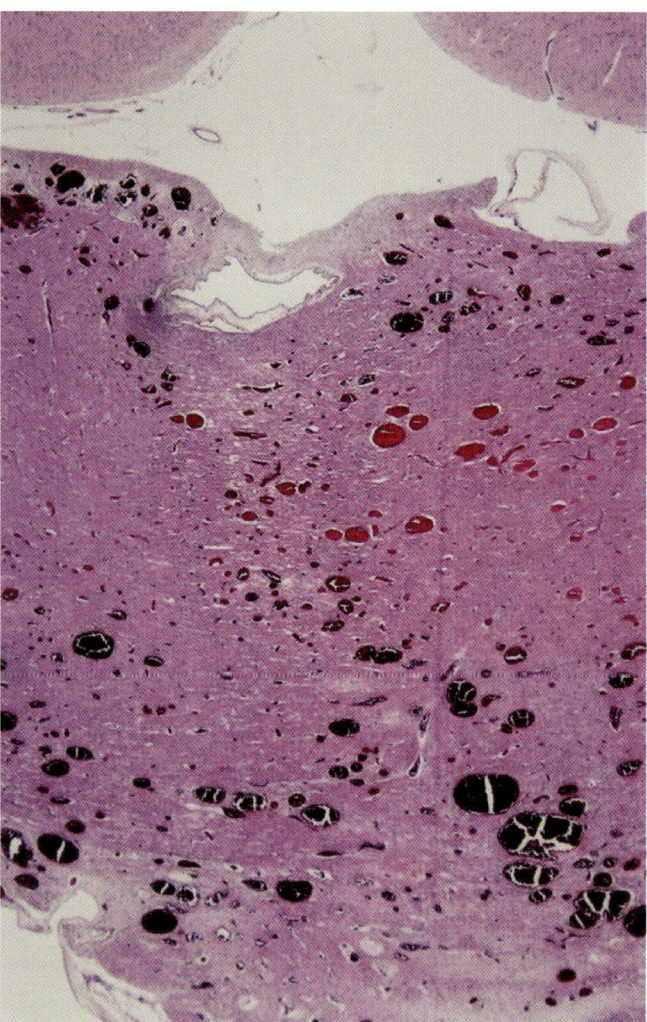

Fig. 11.35. Low-power micrograph of the telangiectasia seen in Fig. 11.34 shows a field of dilated capillaries of various size in undamaged white matter of corpus callosum. (H&E.)

after formalin fixation. Its favorite site is the midpontine basis pontis, occasionally unilateral and sharply demarcated at the midline, but more often bilaterally symmetric. It is composed of a closely juxtaposed collection of dilated capillaries of otherwise normal appearance, with intervening parenchyma normal and devoid of perivascular atrophy or evidence of prior hemorrhage.

Arteriovenous malformations. Arteriovenous malformations (AVMs), compared with other CNS vascular malformations, are usually the largest and most dangerous. Decedents with larger AVMs are often clinically symptomatic with seizures, headaches, and focal neurologic deficits. Most are readily imaged by CT and MR scans although some remain "occult" (see later). Decedents with AVMs usually have had neurologic/neurosurgical attention, some with prior surgery except where the anatomic site of involvement made the lesion inoperable. Such cases may have had neuroradiologic interventional therapy.[64] The majority of AVMs are supratentorial, with only a small percentage in the posterior fossa. Few involve the spinal cord, where clinical presentation may include that of transverse myelitis.[8] Involved sites are most often cerebral cortical and/or meningeal (Case 11.12, Figs. 11.36–11.37; Case 11.13, Figs. 11.38–11.39). AVMs primarily affecting the dura have been reported.[15]

> **Case 11.12.** (Figs. 11.36 and 11.37)
>
> A 39-year-old man had a seizure while riding on a bus, and was taken to the hospital where he was given an injection of phenobarbital. After an electrocardiogram was taken, he left against medical advice. Three hours later, he was found unresponsive in a park and was pronounced deal at the scene.

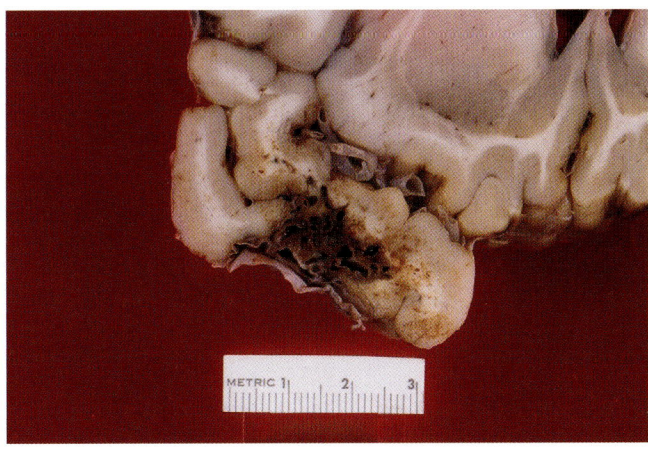

Fig. 11.37. Closeup view of coronal section of temporal pole exposes the arteriovenous malformation, composed of closely grouped irregular vascular channels destructive of brain.

Grossly, AVMs usually declare their presence by their obviously enlarged draining veins and feeding arteries, directing attention to their location (Fig. 11.40). Small, so-called cryptic, AVMs may not expose their presence at autopsy until sections are taken. Microscopically, AVMs are composed mainly of enlarged blood vessels of highly variable size and thickness, and with vessel walls a caricature of normal arteries and veins. In both, the walls of vessels vary from extreme fibromuscular thickening to sectors of markedly attenuated walls of fibrous connective tissue, vulnerable to rupture (Case 11.13, Fig. 11.39).[44] The elastic lamina of the arterial components is often interrupted and frayed beneath a zone of intimal fibrosis. Affected blood vessels are often focally calcified. Brain parenchyma between vascular channels shows severe atrophy and gliosis, with hemosiderin deposits as evidence of past hemorrhage. The designation "cryptic" or "occult" vascular angioma refers to small AVMs and venous angiomas that first announce their presence by spontaneous hemorrhage. They are now more frequently discovered without hemorrhage, as incidental findings on CT and MR studies performed for other reasons.[7]

Cavernous hemangiomas occur most commonly in cerebral subcortical white matter, measuring up to a few

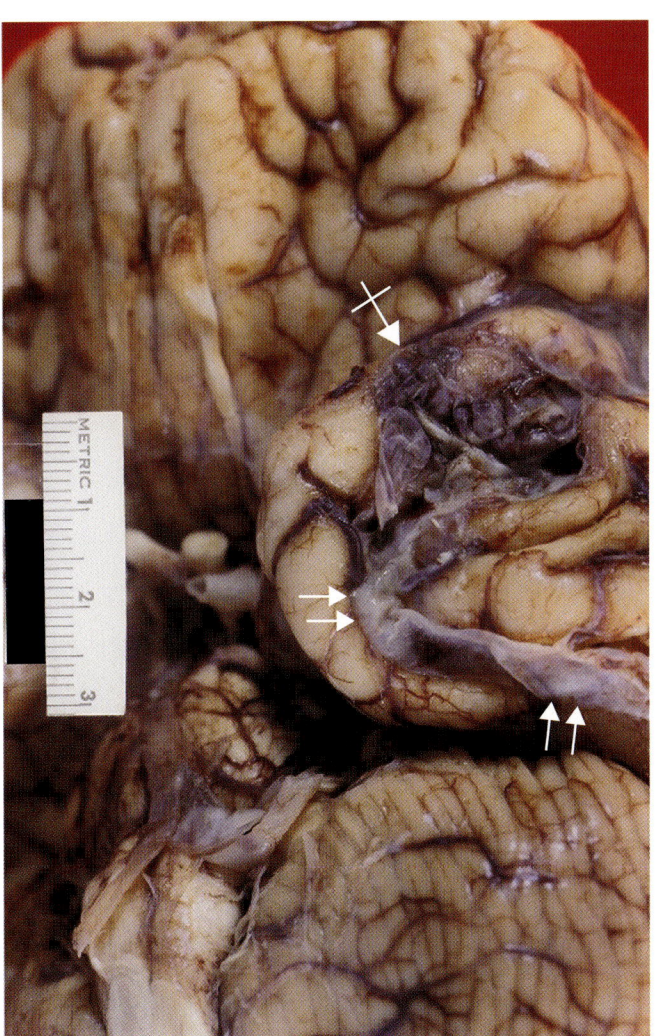

Fig. 11.36. Closeup of base view of brain, with olfactory and optic nerves and brainstem to the left, shows a tangle of large abnormal blood vessels at temporal pole (*crossed arrow*). A large feeding artery (*double arrows*) from posterior circulation enters the arteriovenous malformation.

> **Case 11.13.** (Figs. 11.38 and 11.39)
>
> A 37-year-old man had a seizure disorder for 19 years. On the day before death, he had two major motor seizures and was left in bed alive. About 18 hours later, he was found dead on the floor next to his bed. Autopsy showed an arteriovenous malformation of the left frontal lobe without recent hemorrhage. Unilateral mild hippocampal sclerosis and moderate cerebellar cortical atrophy were present.

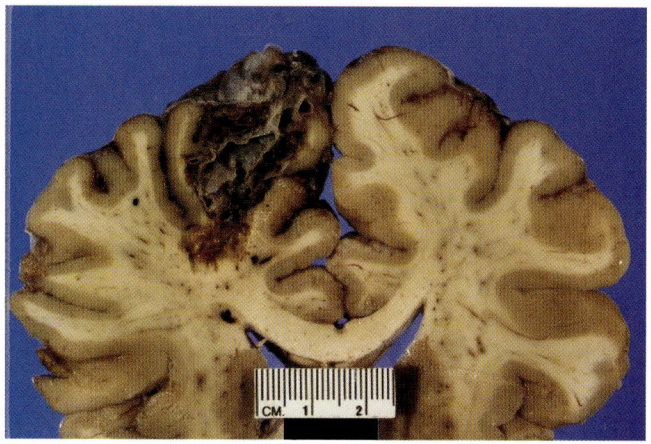

Fig. 11.38. Arteriovenous malformation involves left superior frontal gyrus and overlying meninges.

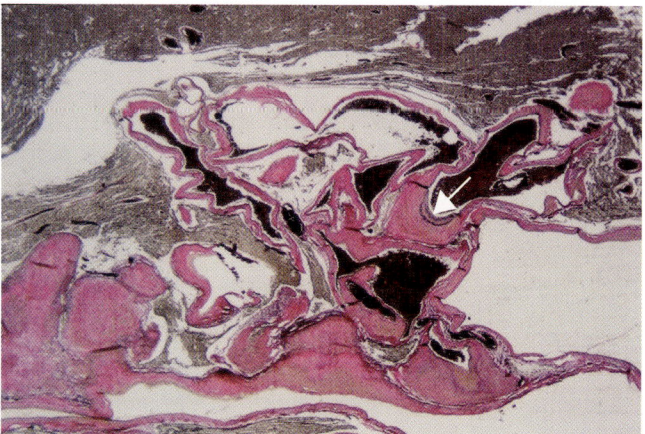

Fig. 11.39. Low-power micrograph of arteriovenous malformation shows characteristic abnormally large vascular channels, with walls varying in thickness from extremely thick to markedly attenuated. A short segment of elastic lamina is still visible in thickened sector of an artery (*arrow*). (Elastic van Gieson.)

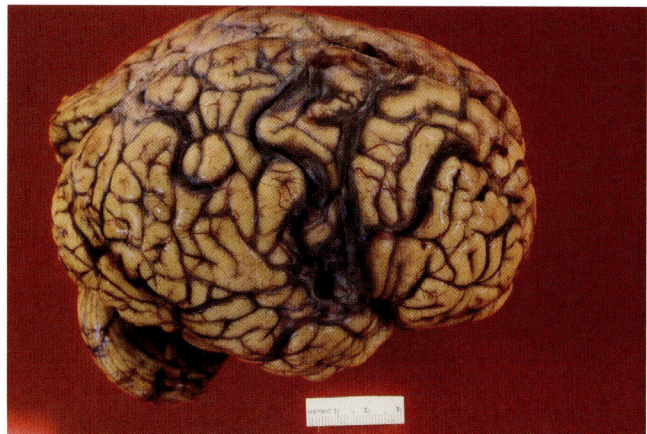

Fig. 11.40. Greatly distended superficial cortical veins drain from nidus of an arteriovenous malformation in the anterior sylvian fissure.

centimeters in diameter. Decedents with a history of seizures, found dead at the scene with or without hemorrhage in a cavernous hemangioma, are an occasional finding at our office. Their numbers may be underestimated as a cause of intracerebral hemorrhage.[41] The frontal lobe is the most common site, but, in one series of symptomatic patients, a distinct bias (86%) was to the temporal lobe.[20] Cavernous hemangiomas may be multiple; one case seen in our office had 43 separate lesions. Grossly, hemangiomas are usually not large, measuring a few millimeters to 1 to 2 cm in diameter, well-circumscribed, and dark red to black (Figs. 11.41–11.42). Occasional larger ones occur. Microscopically, cavernous hemangioma consists of a multitude of closely approximated small to large vascular channels with, in start contrast to AVMs, still relatively thin fibrous walls devoid of muscularis (Fig. 11.42). Apposed vessels some-

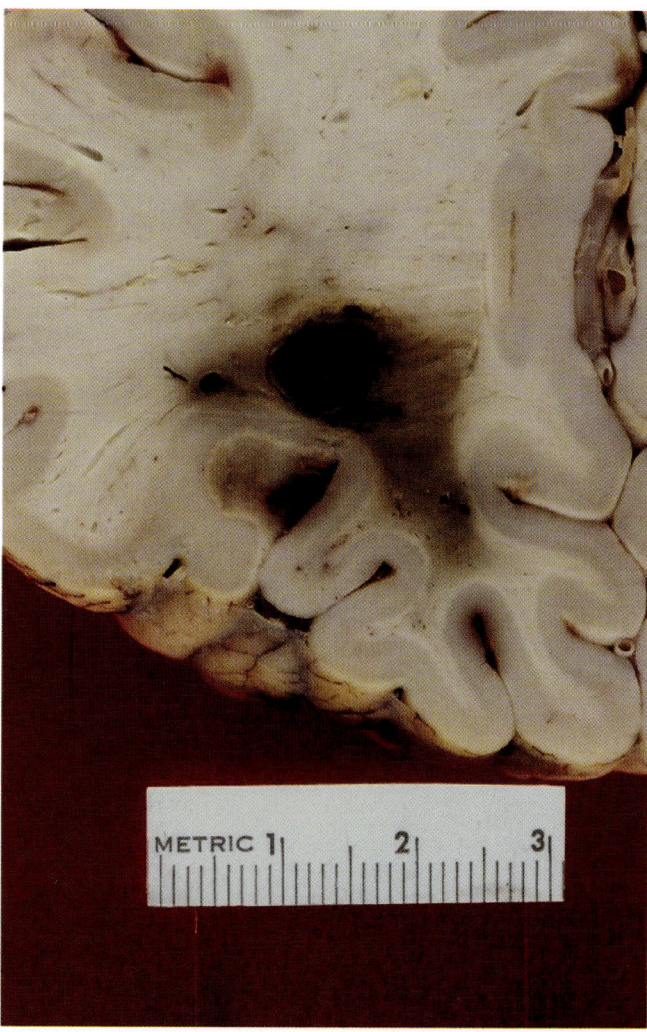

Fig. 11.41. Closeup view of left orbital frontal lobe shows a small fresh subcortical hemorrhage, marking the site of a cavernous hemangioma.

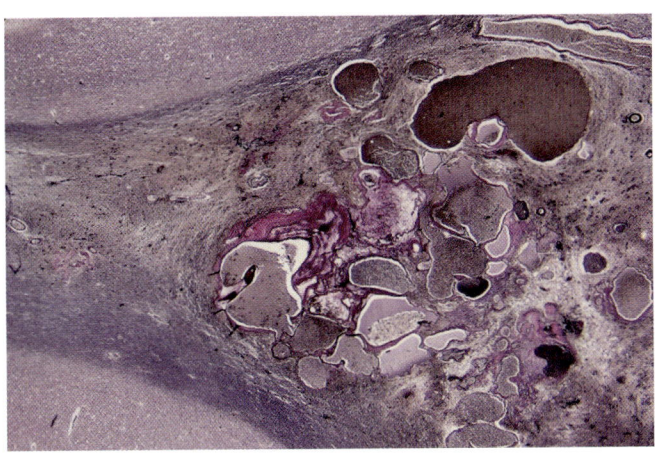

Fig. 11.42. Low-power micrograph of an area taken just adjacent to hemorrhage shown in Fig. 11.41 shows a cavernous hemangioma with its tangle of relatively thin-walled dilated blood vessels, some sharing walls, lying in atrophic brain parenchyma. (Elastic van Gieson.)

times appear to share a common wall. Brain parenchyma between vessels is invariably atrophic with hemosiderin deposits as evidence of past hemorrhage. Calcification may be present. Familial occurrences have been reported.[58]

Our collection of vascular malformations includes a single case of cranial dural cavernous angioma (Figs. 11.43–11.45). In addition to the two cases of dural angiomas found by McCormick and Boulter[47] in a review of 500 vascular malformations, only three other cases acceptable to them were found on review of the literature.

Venous malformations (angiomas) are found in white matter of the cerebrum and cerebellum, among other sites, and in the meninges of the spinal cord. The malformation is not an uncommon incidental finding at autopsy. Increasingly encountered on CT[61] and MR studies, and considered benign in some reports, the authors have found this malformation as a source of hemorrhage in cases of sudden unexpected death, especially in the cerebellum of the young (see Chap. 9, Case 9.17, Figs. 9.45–9.47). Of four cases quoted by Rosen *et al.*[59] of death due to cerebellar hemorrhage, three were diagnosed as due to "vascular malformation," not further classified. The fourth was diagnosed as an AVM. As an incidental finding, venous malformation may simply appear microscopically as a focal grouping of engorged, enlarged veins mimicking congestion.

Microscopically, the malformation appears as a scattering of not very impressive, but obviously enlarged and mildly thickened, veins with walls of uniform thickness separated by normal-appearing white matter (see Chap. 9, Figs. 9.46–9.47). Mixed vascular malformations are occasionally found composed of various combinations of the malformation described earlier.[5]

Angiomas of the spinal cord giving rise to *Foix-Alajouanine syndrome* are classified as venous malformations. However, they have special features not found with cerebral angiomas. The angioma itself is meningeal, occupying the subarachnoid space, mainly on the dorsal aspect of the cord (Figs. 11.46–11.48). Although largely composed of veins, a minority of abnormal blood vessels displays arterial features. Aminoff *et al.*[3] hypothesized that an arteriovenous shunt bypassing the capillary bed leads to elevated intramedullary venous pressure, causing stagnation and dilatation of the capillary bed of the cord with consequent fibrotic thickening of capillaries and arterioles and hypoxic-ischemic damage to the cord (Fig. 11.48).

Sturge-Weber-Dimitri disease (encephalotrigeminal angiomatosis) and telangiectatic disorders such as Osler-Weber-Rendu disease and ataxia-telangiectasia, or Louis-Barr syndrome, are infrequently encountered in a forensic setting, and are discussed in standard textbooks on neuropathology.

Ruptured Cerebral Aneurysms

Of the nontraumatic intracranial hemorrhages, acute subarachnoid hemorrhage secondary to ruptured congenital berry aneurysm is by far the most common cause of death in cases referred to the medical examiner. Clinical presentations vary, and can include cases of sudden unexpected death (see Chap. 9).

Subarachnoid hemorrhage secondary to ruptured aneurysm is primarily a disease of middle-aged and young adults, with the highest incidence in women in their 50s, and less than 1% occurs in children. However, death from a ruptured aneurysm is reported from time to time in children.[49,55] Subarachnoid hemorrhage secondary to the rupture of a mycotic aneurysm is also discussed in Chapter 9. Other causes of nontraumatic subarachnoid hemorrhage are blood dyscrasias, anticoagulant and thrombolytic therapies, and leukemias. Coagulopathy of hepatic failure due to chronic alcoholism can be included. In all of these circumstances, a majority survive the initial subarachnoid hemorrhage, only to succumb within several days to recurrent hemorrhage or life-threatening complications of the initial hemorrhage. Complications include vasospasm with resultant ischemia and infarction in vital territories, edema, and herniations and their consequences. Various vasoconstrictive agents released by hemorrhage have been proposed as a cause of vasospasm, but the pathogenesis remains uncertain.[36] Vasospasm following subarachnoid hemorrhage, demonstrated by angiography in life, has also been demonstrated postmortem.[34] Increased intracranial pressure secondary to failed absorption of cerebrospinal fluid (CSF) due to blockage by ventricular and subarachnoid hemorrhages contrib-

324 Vascular Diseases of the Central Nervous System

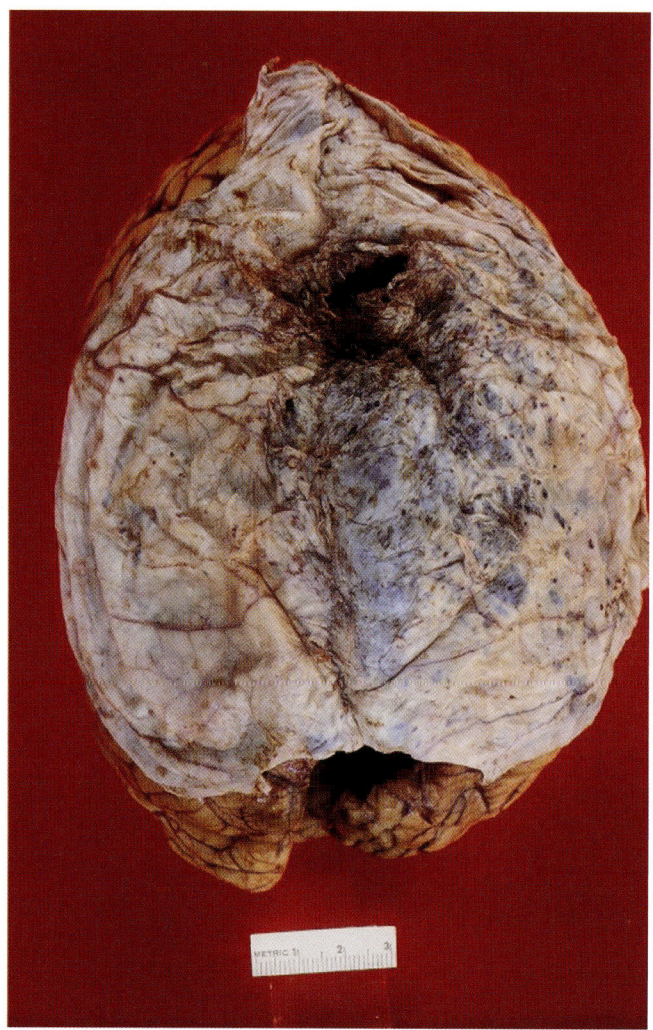

Fig. 11.43. Convexity view of dura mater overlying cerebral hemispheres shows gray darkening of an angioma of the right parietal parasagittal region, sharply demarcated at midline.

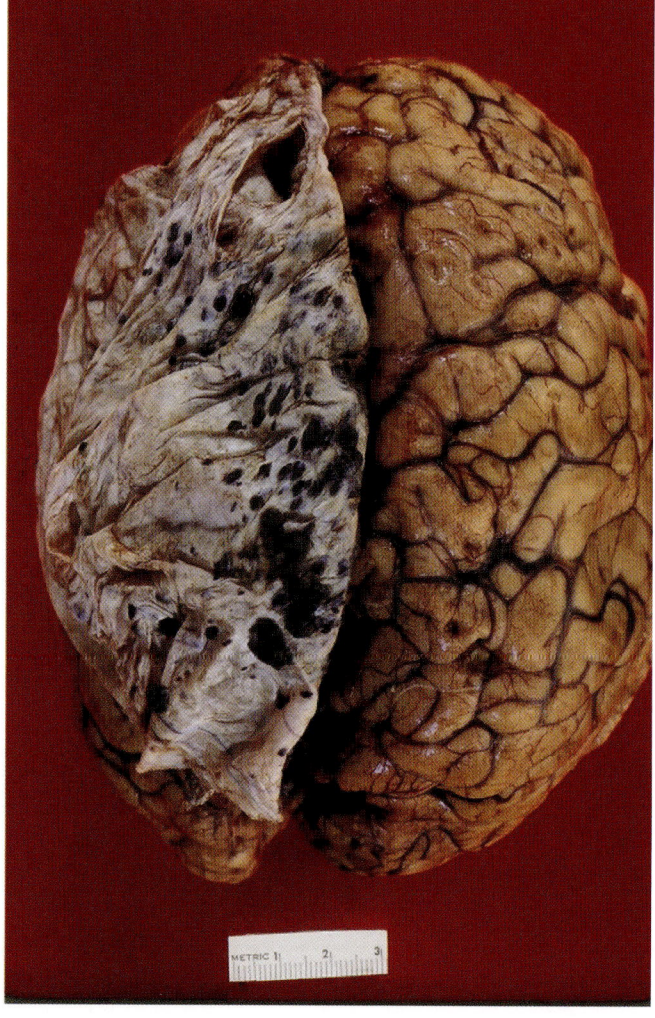

Fig. 11.44. Reflection of right dural covering onto the left hemisphere exposes a widespread intradural vascular lesion on subdural side.

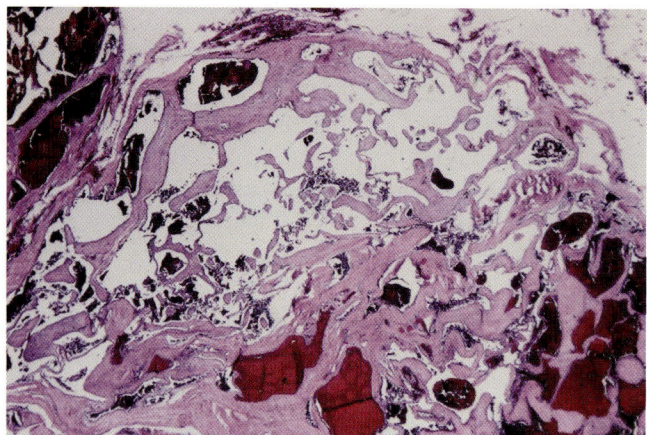

Fig. 11.45. Low-power micrograph of the lesion seen in Figs. 11.43 and 11.44 reveals a chaotic complex of vascular channels best characterized as a cavernous angioma. (H&E.)

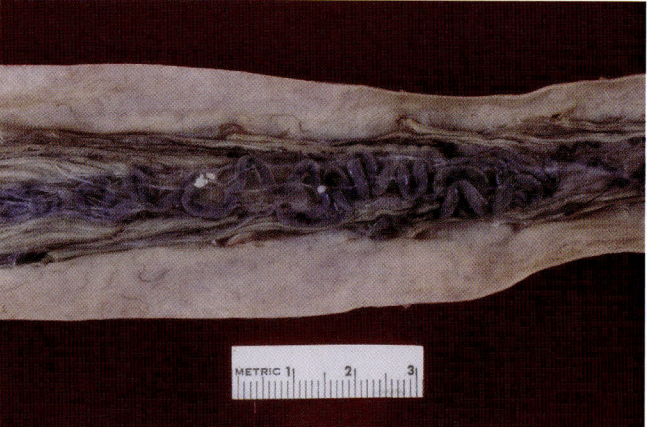

Fig. 11.46. Closeup view of tangle of tortuous blood vessels of spinal angioma on the dorsal aspect of cord. A few small incidental leptomeningeal white plaques are also present (see Chap. 1, Figs. 1.15–1.17).

utes to mortality.²⁹ The finding of acute hydrocephalus on follow-up scan after subarachnoid hemorrhage from a ruptured aneurysm may indicate an ominous prognosis for fatal infarction.⁶⁸ In the presence of primary intraventricular hemorrhage, that is, where the presence of ventricular hemorrhage is not an extension of a parenchymal hemorrhage or reflux from a subarachnoid hemorrhage, hemorrhage of a vascular malformation of the choroid plexus should be considered.¹⁹ Such malformations are rare.

Ruptured Cerebral Aneurysm in a Roller Coaster Ride

The death of a roller coaster rider with acute subarachnoid hemorrhage was referred to our office. As anticipated, the case evoked considerable controversy (Case 11.14, Figs. 11.49–11.54). The manner in which the pathologic findings were thoroughly documented is a model on how careful description, photography, and diagrams can forestall wild allegations and obviate controversy, and the case is briefly presented.

Case 11.14. (Figs. 11.49–11.54)

A young woman, allegedly without past neurologic symptoms, was unresponsive at the end of an amusement park roller coaster ride, and she did not respond to resuscitative measures. Autopsy findings included cardiomegaly (410 g), acute subarachnoid hemorrhage (Fig. 11.49), and a solitary left posterior communicating artery saccular aneurysm, measuring up to 1.6 cm in diameter, with a rupture site in the body of the aneurysm (Figs. 11.50–11.51).⁴⁶ Neuropathologic consultation was sought after fixation of the brain. This case generated considerable interest from the press, attorneys, a regulatory agency, and others.

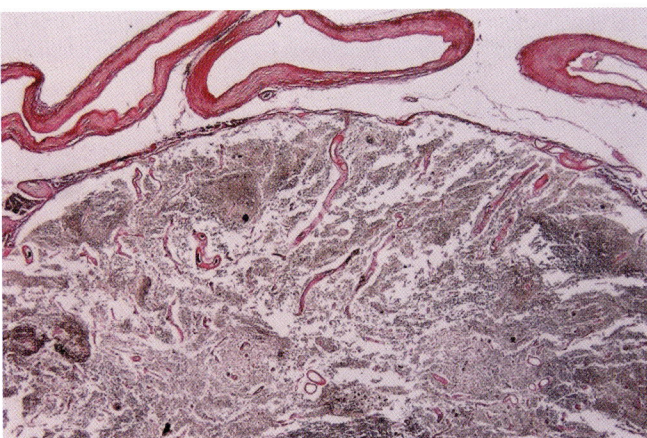

Fig. 11.47. Low-power micrograph shows abnormally large subarachnoid veins of angioma and small sclerotic vessels in atrophic cord. Fragmentation of cord is an artifact. (van Gieson.)

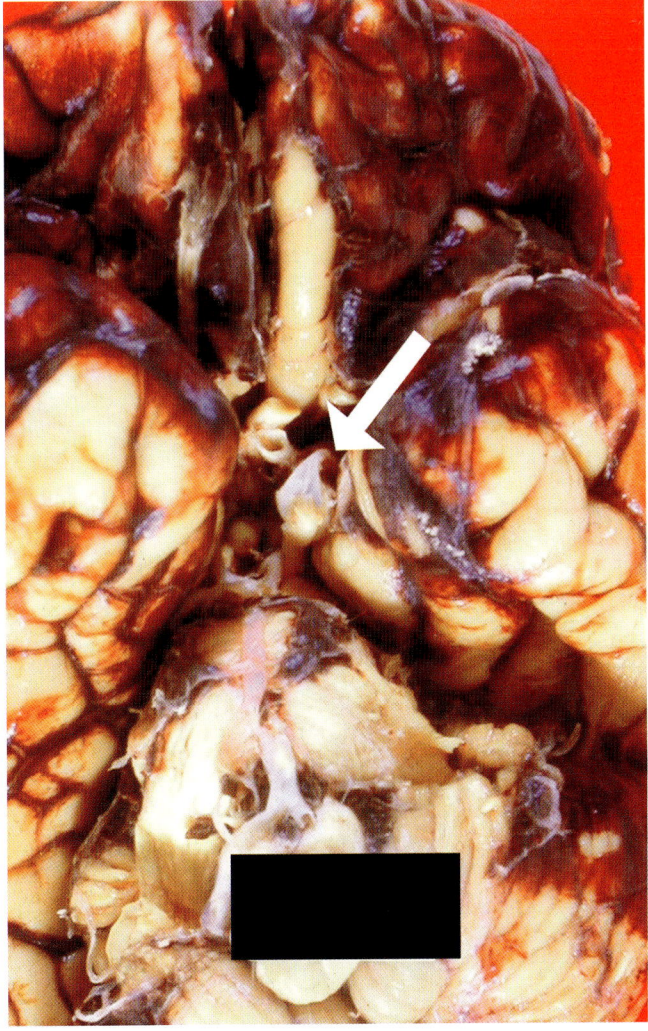

Fig. 11.49. Base view of cerebrum shows acute subarachnoid hemorrhage, some of which has been washed away at autopsy to expose the ruptured left communicating artery saccular aneurysm.

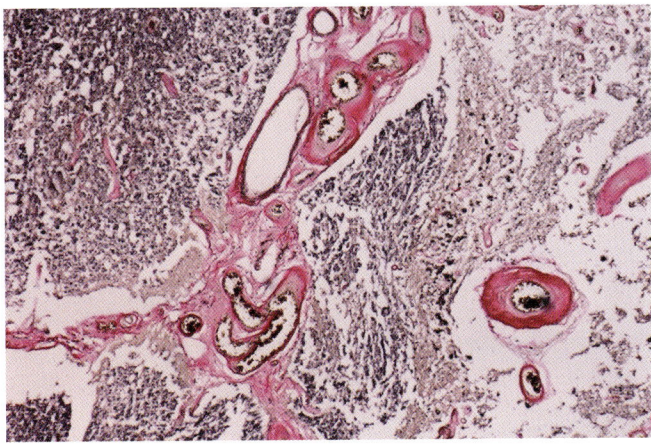

Fig. 11.48. Higher power micrograph shows detail of sclerotic small blood vessels within abnormal cord parenchyma. (van Gieson.)

326 Vascular Diseases of the Central Nervous System

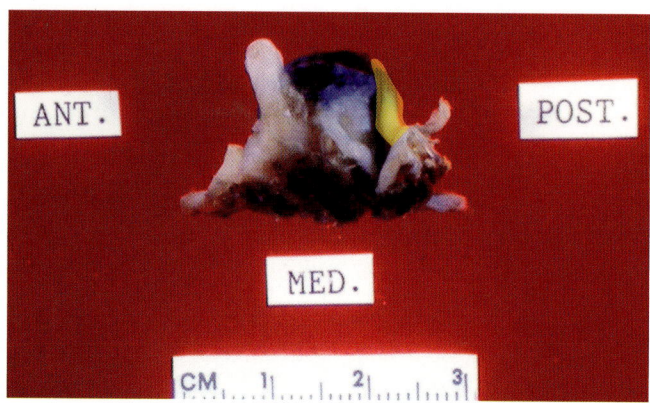

Fig. 11.50. Closeup view of aneurysm dissected from the circle of Willis with internal carotid artery and left oculomotor nerve attached. Dome of aneurysm is painted blue.

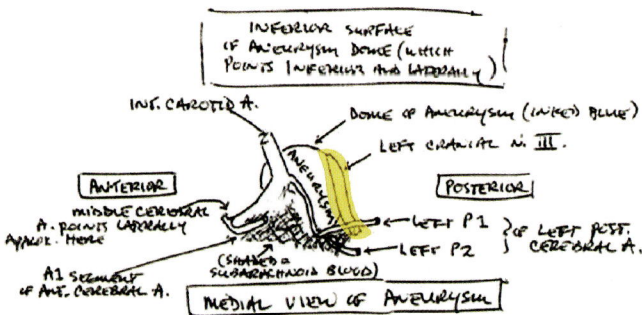

Fig. 11.51. One of several diagrams of specimen shown in Fig. 11.50 made at the time of the gross neuropathologic examination, with labeling of anatomic components. The yellow-colored structure in Figs. 11.50 and 11.51 is the left third cranial nerve, displaced and stretched as it courses over the dome of the aneurysm.

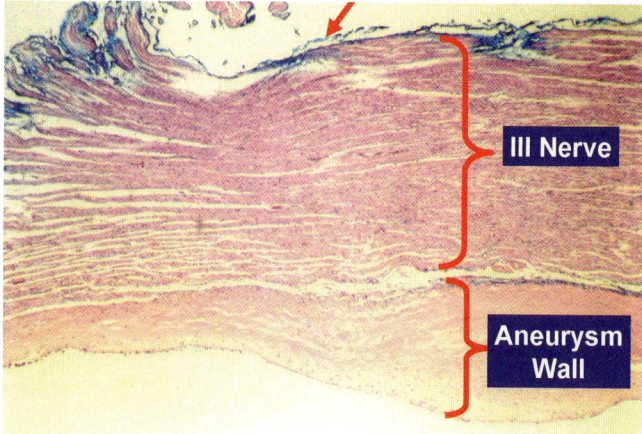

Fig. 11.52. Medium-power micrograph shows atrophic left oculomotor nerve on the surface of aneurysm wall. Arrow indicates perineurium stained by blue ink used to paint aneurysm dome (also seen in Fig. 11.54). (H&E.)

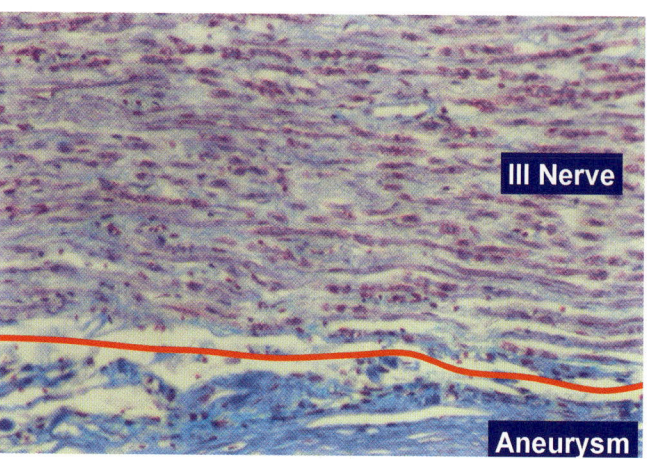

Fig. 11.53. Higher-power micrograph shows detail of atrophic oculomotor nerve immediately adjacent to the surface of aneurysm. The myelin sheaths are fragmented and reduced in number (magenta-colored). Endoneurial connective tissue (*light blue*), situated between individual nerve fibers, is increased. Red line is the nerve–aneurysm wall boundary. (Trichrome.)

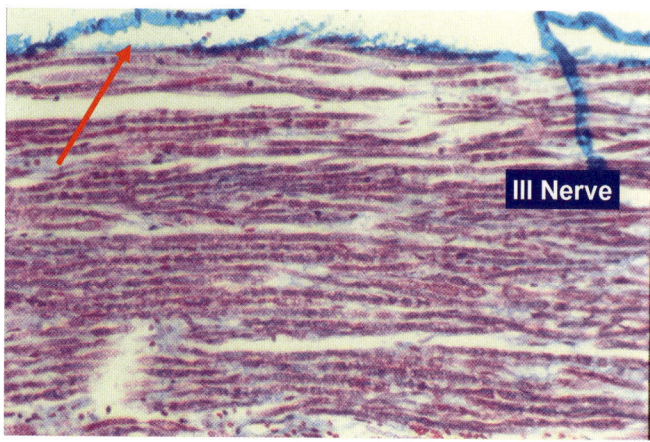

Fig. 11.54. High-power micrograph of the more normal-appearing, outer portion of the same oculomotor nerve (i.e., most distant from the aneurysm wall to which the nerve adheres) for comparison with Fig. 11.53. Magnification of Figs. 11.53 and 11.54 is identical. Arrow points to perineurium, stained by the blue ink used to paint the aneurysm dome. (Trichrome.)

Case 11.14. cont'd

Speculations as to cause and manner of death in Case 11.14 were rampant from many sources. Two examples will suffice. The neuropathology consultant on this case was asked to "disprove" the opinion of an opposing expert witness that the decedent had a normal cerebrovascular system before the ride, but the ride caused the aneurysm to develop, grow to its final dimensions, and burst due entirely to the ride. It was also opined by others that the aneurysm rupture site being at the aneurysm body rather than at the dome "proved" that the roller coaster ride caused the rupture, since spontaneous aneurysm ruptures "always" occur at the dome. Prior studies, however, have demonstrated that spontaneous aneurysmal ruptures occur at the dome in 66%, the body in 12%, the neck in 3%, and at an unknown site in 19% of cases.[63]

Certain precautions were taken during analysis of the case to minimize later problems. Photographic and diagrammatic documentation both before and after dye was placed on the outer aneurysm surface; dissection of the aneurysm followed by additional multiangled photographs and corresponding diagrams (Figs. 11.50–11.51); written and verbal instructions to the histotechnologists as to embedding and orientation for sectioning; numbering and mounting all serial sections; and staining of every 10th section with hematoxylin-eosin (H&E) stain were among the precautions taken. Furthermore, initially stained sections were reviewed and specific adjacent levels selected for special stains, and photomicrographs of key findings were made and labeled with a slide number and the stain used. These precautions were designed to anticipate possible questions and special requests by interested parties months and years later. This approach allowed identification of the aneurysm type (saccular), the anatomic source of each protuberance on the specimen (aneurysm with oculomotor nerve, etc.), identification of the exact site of aneurysm rupture, evidence of chronicity (macrophages, calcification and hyalinization of aneurysmal wall, degenerative changes of oculomotor nerve (Figs. 11.52–11.54), and other features.

Other issues to which attention was directed by interested parties in this case included the gravitational forces involved in roller coaster rides. Relevant published reports included those suggesting a possible causal relationship between some neuropathologic conditions and amusement park rides, other activities in which patients were engaged when aneurysm rupture occurred, data on traumatic aneurysm cases, and suggested criteria for linking aneurysm rupture to a traumatic event. Selected references on several of these topics are cited for the interested reader.[4,11,13,22,26,45,63,65]

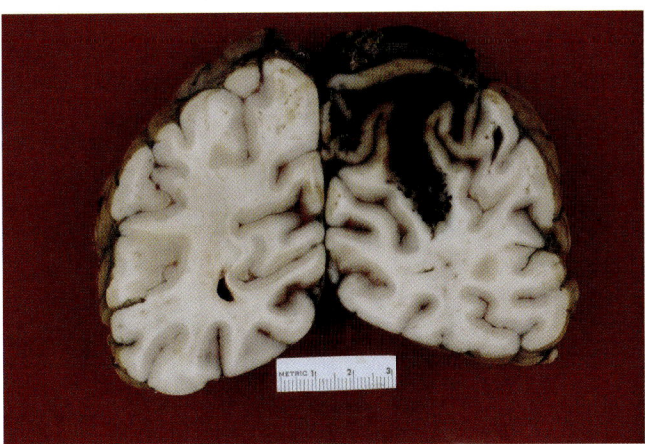

Fig. 11.55. Coronal section at posterior parietal level in a case of cerebral amyloid angiopathy shows large, mainly subcortical, fresh hemorrhage undercutting cortex.

Case 11.15. (Figs. 11.56–11.58)

An 89-year-old woman was admitted to a convalescent home after a fall at home in which she sustained a frontal hematoma. She was alert but forgetful. Three months before death, she fell and struck her head again. X-rays were negative, but her condition gradually deteriorated, eventuating in a coma.

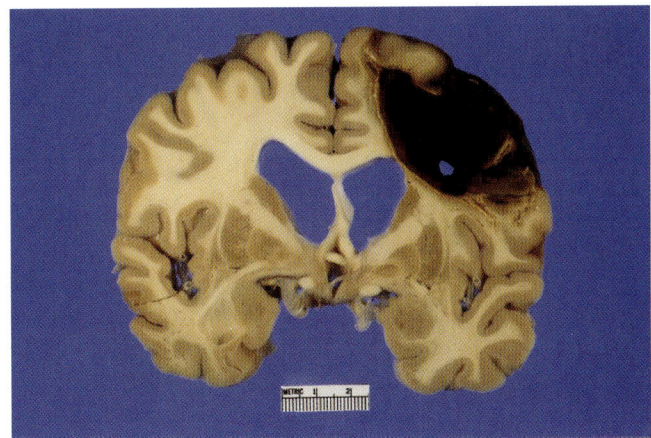

Fig. 11.56. A large chronic and superimposed recent cortical/subcortical hemorrhage is shown involving the right frontal lobe at the level of anterior striatum. The very thin margin of necrosis and gliosis around the central recent blood clot indicates its older age. Other than slight distortion of dorsolateral angle of the frontal horn of ventricle, displacement of adjacent structures is not conspicuous.

Cerebral Amyloid Angiopathy

Cerebral amyloid angiopathy is a disorder of the elderly, although cases may occur in individuals in their 60s.[53] The condition may present dramatically with a large acute lobar intracerebral hemorrhage.[53] Typically, the hemorrhage is superficially located, arising from intracortical or subarachnoid blood vessels of any lobe (Fig. 11.55; Case 11.15, Figs. 11.56–11.58). Affected cortical blood vessels are small arterioles, microscopically seen as amyloid-positive thickened "donuts" (Case 11.16, Figs. 11.59–11.61).[70] Polarizing microscopy demonstrates

Vascular Diseases of the Central Nervous System

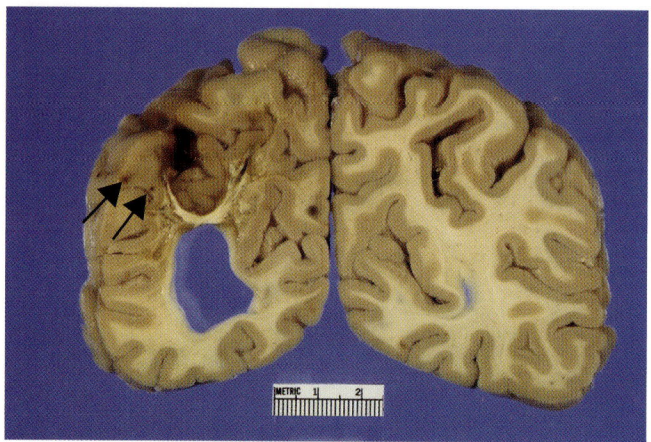

Fig. 11.57. Two cortical petechiae (*arrows*) are characteristic but nonspecific for intracortical hemorrhage of cerebral amyloid angiopathy in the setting of an elderly person. An adjacent, more deeply placed, lesion is also present.

Case 11.16. (Figs. 11.59–11.65)

A 64-year-old man had a spontaneous right intracerebral hemorrhage that was evacuated and diagnosed as an occult arteriovenous malformation. Two months later, he was found down but alert with no sign of trauma. Over 1 to 2 hours, there was progressive deterioration of responsiveness. By the time of admission to the hospital, he was comatose with spontaneous decerebrate posturing. CT scan showed a large left parietooccipital intracerebral hemorrhage. He died a week later.

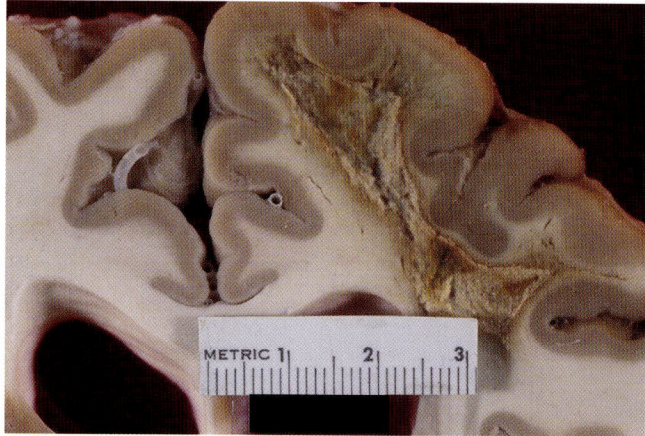

Fig. 11.58. Closeup view of another region from the same case, showing chronic sequelae of a subcortical lobar hemorrhage in cerebral amyloid angiopathy.

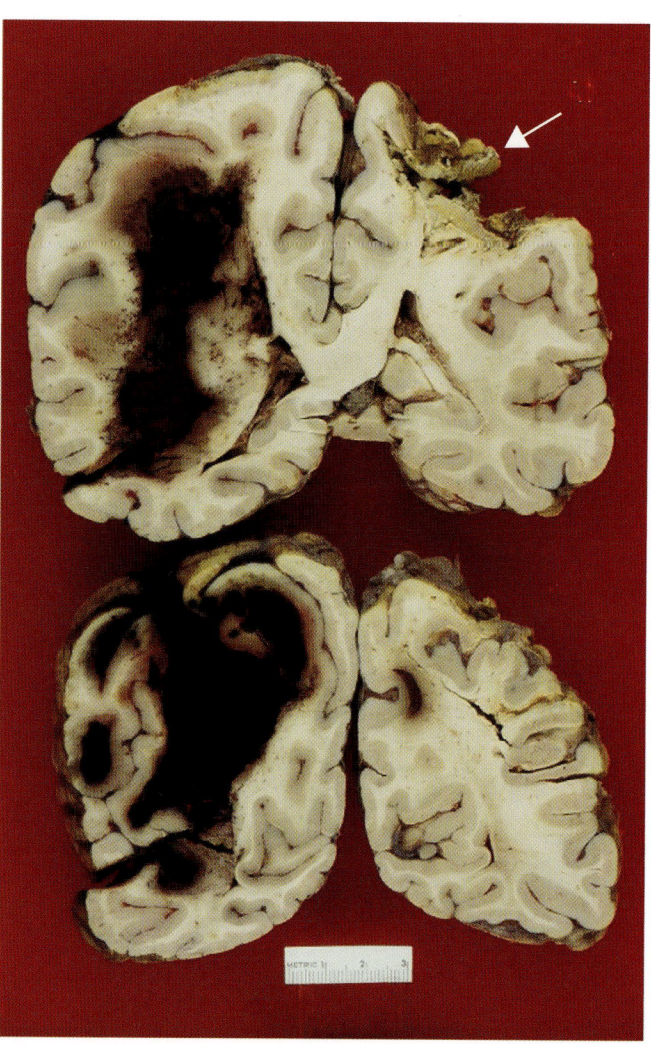

Fig. 11.59. Massive cortical/subcortical hemorrhage occupies left parietooccipital white matter at splenial and postsplenial levels. A chronic necrotic cortical remnant of hemorrhage evacuated 2 months previously can be seen at right parietal convexity (*arrow*).

apple green birefringence in vessel walls on Congo red stain (Figs. 11.62–11.63). Hemosiderin-laced cortical microinfarcts are sometimes found nearby. Overlying subarachnoid arteries may show circumferential intramural lucencies, giving a "double-barreled," or a vessel within a vessel, appearance (Figs. 11.64–11.65).[70] These arteries are amyloid positive, but the changes are typically patchy. From the location of the abnormality, hemorrhages often extend into the overlying subarachnoid space and sometimes beyond, developing to a spontaneous subdural hematoma that can mislead one to implicate a traumatic cause. Hemorrhages directed deeply may enter the lateral ventricle. Decedents with congophilic cerebral amyloid angiopathy may have a CVD

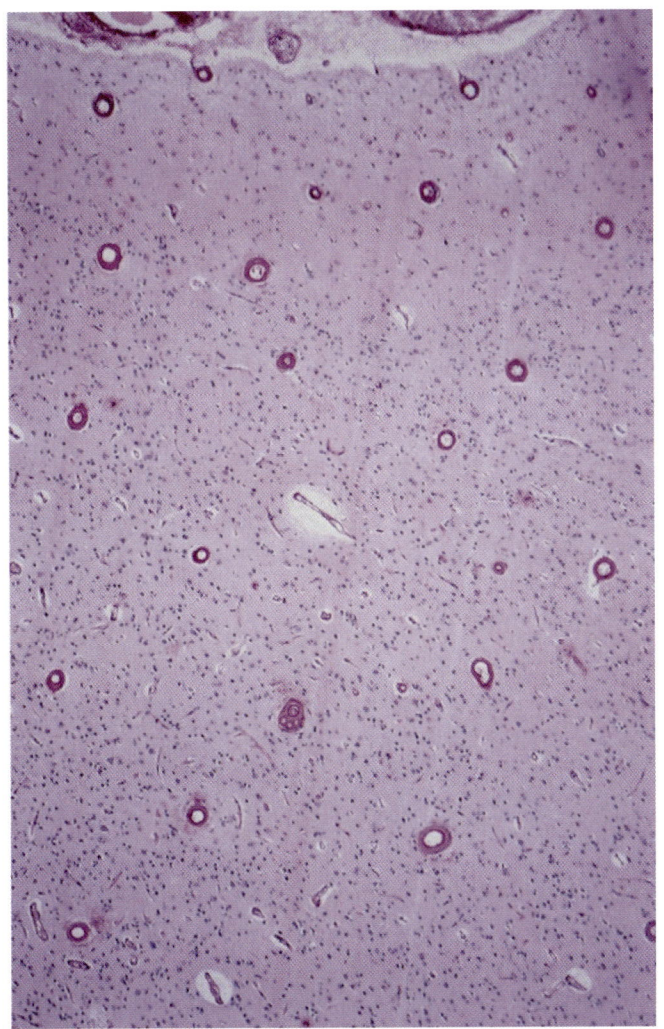

Fig. 11.60. Medium-power micrograph of cortex shows numerous PAS-positive thickened small arteries and arterioles. (Periodic-acid-Schiff.)

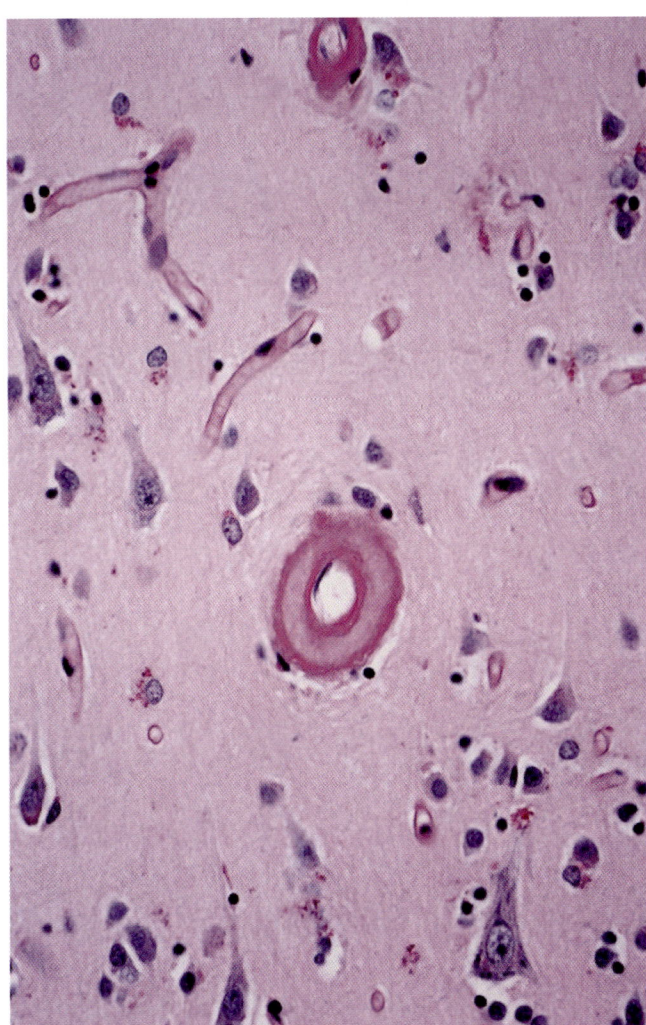

Fig. 11.61. High-power micrograph shows detail of one thickened arteriole. (Periodic-acid-Schiff.)

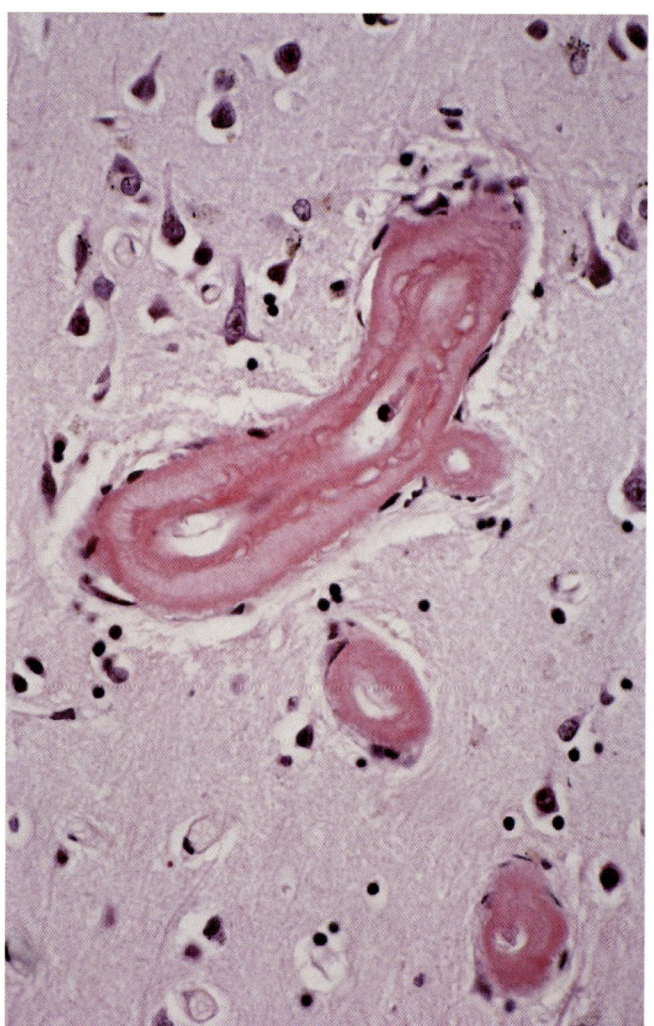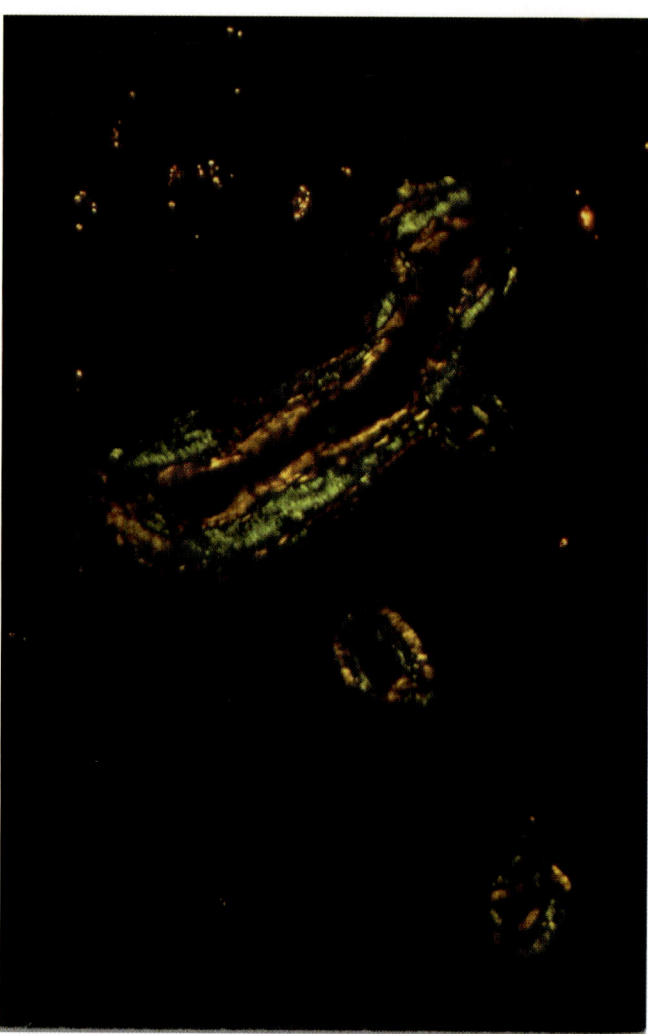

Fig. 11.62. High-power micrograph of cerebral cortex shows concentrically thickened small arteries and arterioles of cerebral amyloid angiopathy. Note radiating fine fibrils replacing media. (Congo red.)

Fig. 11.63. Same section as in Fig. 11.62 viewed by polarization shows vessel walls with apple green birefringence of amyloid. (Congo red.)

history of repeated small strokes and dementia.[42] Microscopic examination commonly reveals senile plaques, especially at the sites of vascular abnormality, suggesting Alzheimer's disease. However, differences have been shown in the plaques of cerebral amyloid angiopathy versus those of Alzheimer's disease.[69]

Drug Abuse-Related Cardiovascular Disease

The majority of victims of cocaine-induced deaths have no morphologic changes in the brain. Seizures have been suggested as the cause of death in some cases. A relatively small number of cases of cocaine abuse, however, have presented with cerebral and cerebellar hemorrhage, cerebral infarction, and subarachnoid hemorrhage.[18] Some, if not most, of the hemorrhages were found to have been from an underlying vascular abnormality, such as previously asymptomatic AVM, berry aneurysm,[50] or hypertensive vascular disease.[38] A few others were purported to have been secondary to a vasculitis attributed to a cocaine effect.[10,51] The sites of cocaine-induced hemorrhage in an autopsy series, which included subjects with findings consistent with hypertensive vascular disease, were predictably found in the putamen and the pons. Locations of hemorrhage in those without features of hypertension were lobar. Other abused drugs associated with brain hemorrhage and infarction, apparently uncommonly lethal, are phencyclidine,[9] methamphetamine,[60] and heroin,[12] at times associated with vasculitis of uncertain pathogenesis, although drug-induced hypersensitivity has been suspected.[12,17]

States of Altered Consciousness and Their Anatomic Substrates

CVD cases referred to the medical examiner often come with clinically catastrophic histories, and a comatose state is a common outcome. Therefore, the anatomic substrates of several altered states of consciousness are briefly reviewed.

In accordance with the definition for various states of altered consciousness as outlined in Plum and Posner,[57] and confining our attention to those states that are most likely to come to the attention of the forensic pathologist, the following summary of clinical-anatomic correlates may be helpful. Aside from CVD, these clinical states may develop as a result of trauma, toxin exposures, or progressive neurodegenerative diseases, for example. Their inclusion in this chapter reflects the fact that most cases are caused by hypoxic-ischemic injury resulting from cerebrovascular disease, incomplete recovery following resuscitation post-cardiac arrest, and so on.

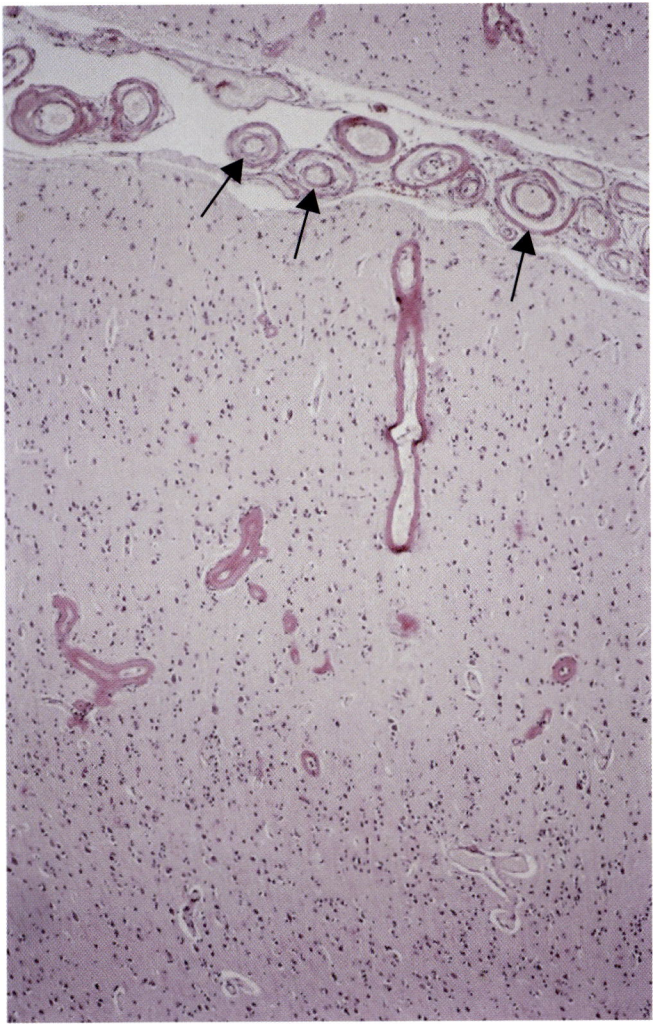

Fig. 11.64. Medium-power of cerebral cortex shows "double barrel" arteries in subarachnoid space, better described as "vessel in vessel," or mural lucency due to loss of media. (Congo red.)

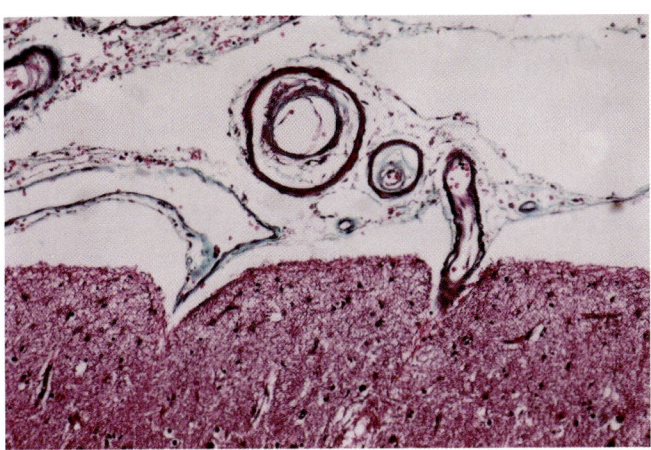

Fig. 11.65. Medium-power micrograph of small arteries shows detail of characteristic mural lucency of media (double-barreled artery) in cerebral amyloid angiopathy. (Trichrome.)

Coma

The comatose patient is unresponsive and unarousable, eyes closed, with no apparent response to external stimulus or inner need.[57,73] Lesions that produce coma include extensive lesions of the cerebral hemispheres bilaterally; lesions of the paramedian diencephalon and of the brainstem; or lesions of the cerebral hemispheres and brainstem together (Case 11.17).[57] The most discrete single lesion producing coma is probably a rostral pontine tegmental lesion, with or without a contiguous caudal midbrain tegmental lesion.[54] The latter authors indicated that coma is of shorter duration with unilateral than with bilateral lesions, and that the pontine tegmental nuclei most consistently involved include the raphe complex, locus ceruleus, and tegmental nucleus, among others (using the terminology of Paxinos and Huang).[56] An incidental observation in this study was that hyperthermia (of apparently central origin) was present in four of their cases, all of whom had involvement of the rostral pontine tegmentum.[54]

Locked-in State

These patients, following brainstem lesions that abolish descending control of voluntary movement below the level of the lesion, demonstrate paralysis of lower cranial nerves and are quadriplegic, but without impairment of consciousness. Communication indicating their awareness of self and environment can be established by means of vertical eye movement and blinking in most cases.[57] Lesion sites reported to produce this syndrome include the pontine basis and ventral midbrain.[28,37,57] Only rarely is the syndrome secondary to a lesion rostral to the midbrain, such as the single case reported with bilateral infarcts of the internal capsule demonstrated by CT scan,[16] but without autopsy confirmation.

Akinetic Mutism

These patients demonstrate a mute, or nearly mute, alert-appearing immobility with intact sleep-wake cycles but no evidence of mental activity or spontaneous motor activity. Zeman[73] described it as a state of "profound apathy with evidence of preserved awareness, characterized by attentive visual pursuit and an unfulfilled 'promise of speech'." This clinical state may be produced by bilateral deep medial frontal lobe lesions that typically include the cingulate gyri, septal area, varying portions of the frontal poles, and with variable inclusion of the hypothalamus, thalamus, or lesions elsewhere in the brain.[57]

Vegetative State

The patient in this state demonstrates a return of apparent awareness after a period of coma, but without evidence of cognitive function. The patient's eyes typically open spontaneously in response to verbal stimuli or to noxious stimuli, sleep-wake cycles occur, and cardiorespiratory functions are normal. No discrete localized motor responses, obeying of commands, or comprehensible speech is seen.[57,73] This state has also been described as "wakefulness without awareness."[73] The Multi-Society Task Force on PVS (persistent vegetative state) defines it as a clinical condition of complete unawareness of self and environment, accompanied by sleep-wake cycles, with either complete or partial preservation of hypothalamic or brainstem autonomic functions, no evidence of sustained, reproducible, purposeful, or voluntary behavioral responses to various stimuli, no evidence of language comprehension or expression, presence of bowel and bladder incontinence, and variably preserved cranial nerve and spinal reflexes.[66,67]

If this clinical state exceeds 1 month in duration, it is referred to as a persistent vegetative state.[73] The term "permanent" vegetative state is not recommended,

---

Case 11.17.

A 45-year-old man was assaulted with a baseball bat, including blows to the head area, and found unconscious at the scene. He remained comatose until death 5 months later.

Neuropathologic findings included bilateral chronic subdural hematomas involving virtually all compartments (although minimal in the posterior fossa); cerebral and cerebellar cortical atrophy with partial sparing of the temporal isocortex (brain weight, 1030 g); soft and partially collapsed brain parenchyma, including white matter and deep gray matter; and moderate panventricular enlargement. Brainstem showed pale and softened cerebral peduncles bilaterally and pallor of the medullary pyramids and inferior olivary nuclei. Microscopic examination of the neocortex, basal ganglia, dorsal thalamus, hippocampus, brainstem, and cerebellum showed extensive neuronal loss, patchy necrosis, profuse astrogliosis, and macrophage infiltration. Neocortical involvement was widespread but variable in severity. The bulk of the cerebral cortex was severely involved, however. Cerebral white matter demonstrated severe degeneration, with astrogliosis and macrophage infiltration. The basal ganglia, thalamus, and hippocampus showed severe neuronal loss with areas of necrosis and the reactive changes noted previously. Midbrain demonstrated marked loss of substantia nigra neurons and necrosis of cerebral peduncles, primarily in the midzone (corticospinal tract), with relative preservation of corticopontine tracts. The pons demonstrated severe necrosis of both superior cerebellar peduncles and medial lemnisci with relative sparing of the central tegmental bundles. Corticospinal tracts of the basis pontis were gliotic, with macrophage infiltration, as were the medullary pyramids. The cerebellum showed loss of Purkinje cells, striking Bergmann's gliosis, and patchy persistence of the granule cell layer. The microscopic appearance of the subdural neomembranes was consistent with the 5-month survival period.

owing to rare cases with delayed partial recovery of function sufficient to make the previous definition inapplicable.

Anatomic lesions in such cases involve forebrain areas, often with cortical laminar necrosis, and associated with basal ganglia and hippocampal lesions. Brainstem involvement is not a consistent finding, but loss of cerebellar Purkinje cells is common.[57] Emphasis has been placed on the presence of severe bilateral thalamic lesions in some of these cases.[39,40] Three major topographic patterns of cerebral injury, or some combination thereof, may result in a persistent vegetative state: (1) widespread and bilateral cerebral cortical damage; (2) extensive bilateral involvement of intra- and subcortical connections of the cerebral hemispheric white matter; and/or (3) bilateral severe thalamic involvement.[1,32,39]

References

1. Adams JH, Jennett B, McLellan DR, et al. The neuropathology of the vegetative state after head injury. J Clin Pathol 1999;52:804–806.
2. Ameri A, Bousser MG. Cerebral venous thrombosis. Neurol Clin 1992;10:87–111.
3. Aminoff MJ, Barnard RO, Logue V. The pathophysiology of spinal vascular malformations. J Neurol Sci 1974;23:255–263.
4. Asari S, Nakamura S, Yamada O, et al. Traumatic aneurysm of peripheral cerebral arteries: Report of two cases. J Neurosurg 1977;46:795–803.
5. Awad IA, Robinson JR, Mohanty S, Estes ML. Mixed vascular malformations of the brain: Clinical and pathogenetic considerations. Neurosurgery 1993;33:179–188.
6. Bebin J, Smith EE. Vascular malformations of the brain. In Smith RR, Haerer A, Russell WF (eds): Vascular Malformations. New York: Raven Press, 1982, pp 13–29.
7. Becker DH, Townsend JJ, Kramer RA, Newton TH. Occult cerebrovascular malformations: A series of 18 histologically verified cases with negative angiography. Brain 1979;102:249–287.
8. Bergstrand A, Höök O, Lindvall H. Vascular malformations of the spinal cord. Acta Neurol Scand 1964;40:169–183.
9. Bessen HA. Intracerebral hemorrhage associated with phencyclidine abuse. JAMA 1982;248:585–586.
10. Bostwick DG. Amphetamine induced cerebral vasculitis. Hum Pathol 1981;11:1031–1033.
11. Braksierk RJ, Roberts DJ. Amusement park injuries and deaths. Ann Emerg Med 2002;39:65–72.
12. Brust JCM, Richter RW. Stroke associated with addiction to heroin. J Neurol Neurosurg Psychiatry 1976;39:194–199.
13. Burton C, Velasco F, Dorman J. Traumatic aneurysm of a peripheral cerebral artery: Review and case report. J Neurosurg 1968;28:468–474.
14. Challa VR, Moody DM, Bell MA. The Charcot-Bouchard aneurysm controversy: Impact of new histologic technique. J Neuropathol Exp Neurol 1992;51:264–271.
15. Challa VR, Moody DM, Brown WR. Vascular malformations of the central nervous system. J Neuropathol Exp Neurol 1995;54:609–621.
16. Chia L-C. Locked-in state with bilateral internal capsule infarcts. Neurology 1984;34:1365–1367.
17. Citron BP, Halpern M, McCarron M, et al. Necrotizing angiitis associated with drug abuse. N Engl J Med 1970;283:1003–1011.
18. Cregler LL, Mark H. Medical complications of cocaine abuse. N Engl J Med 1986;315:1495–1500.
19. Doe FD, Shuanghoti S, Netsky MG. Cryptic hemangioma of the choroid plexus: A cause of intraventricular hemorrhage. Neurology 1972;22:1232–1239.
20. Farmer J-P, Cosgrove GR, Villemure J-G, et al. Intracerebral cavernous angiomas. Neurology 1988;38:1699–1704.
21. Farrell DF, Forno LS. Symptomatic capillary telangiectasis of the brain stem without hemorrhage: Report of an unusual case. Neurology 1970;20:341–346.
22. Fleischer AS, Patton JM, Tindall GT. Cerebral aneurysms of traumatic origin. Surg Neurol 1975;4:233–239.
23. Franke CL, van Swieten JC, van Gijn J. Residual lesions on computed tomography after intracerebral hemorrhage. Stroke 1991;22:1530–1533.
24. Freytag E. Fatal hypertensive intracerebral haematomas: A summary of the pathological anatomy of 393 cases. J Neurol Neurosurg Psychiatry 1968;31:616–620.
25. Garcia JH. Circulatory disorders and their effects on the brain. In Davis RL, Robertson DM (eds): Textbook of Neuropathology. Baltimore: Williams & Wilkins, 1985.
26. Gonsoulin M, Barnard JJ, Prahlow JA. Death resulting from ruptured cerebral artery aneurysm. Am J Forensic Med Pathol 2002;23:5–14.
27. Graham DI, Lantos P (eds). Greenfield's Neuropathology. London: Arnold, 1997.
28. Hawkes CH. "Locked-in" syndrome: Report of seven cases. BMJ 1974;4:379–382.
29. Heros RC. Acute hydrocephalus after subarachnoid hemorrhage. Stroke 1989;20:715–717.
30. Itabashi HH, Granada LO. Cerebral food embolism secondary to esophageal-cardiac perforation. JAMA 1972;219:373–375.
31. Jellinger K. Vascular malformations of the central nervous system: A morphological overview. Neurosurg Rev 1986;9:177–216.
32. Jennett B, Adams JH, Murray LS, Graham DI. Neuropathology in vegetative and severely disabled patients after head injury. Neurology 2001;56:486–490.
33. Kanno T, Sano H, Shimomiya Y, et al. Role of surgery in hypertensive intracerebral hematoma: A comparative study of 305 nonsurgical and 154 surgical cases. J Neurosurg 1984;61:1091–1099.
34. Karhunen PJ, Servo A. Sudden fatal or non-operable bleeding from ruptured intracranial aneurysm: Evaluation by post-mortem angiography with vulcanizing contrast medium. Intl J Legal Med 1993;106:55–59.
35. Kase CS, Maulsby GO, Mohr JP. Partial pontine hematomas. Neurology 1980;30:652–655.
36. Kassell NF, Sasaki T, Colohan ART, Nazar G. Cerebral vasospasm following aneurysmal subarachnoid hemorrhage. Stroke 1985;16:562–572.
37. Keane JR, Itabashi HH. Locked-in syndrome due to tentorial herniation. Neurology 1985;35:1647–1649.
38. Kibayashi K, Mastri AR, Hirsch CS. Cocaine induced intracerebral hemorrhage: Analysis of predisposing factors and mechanisms causing hemorrhagic stroke. Hum Pathol 1995;26:659–663.
39. Kinney HC, Samuels MA. Neuropathology of the persistent vegetative state: A review. J Neuropathol Exp Neurol 1994;53:548–558.
40. Kinney HC, Korein J, Panigrahy A, et al. Neuropathological findings in the brain of Karen Ann Quinlan: The role of the thalamus in the persistent vegetative state. N Engl J Med 1994;330:1469–1475.
41. Malik GM, Morgan JK, Boulos RS, Ausman JI. Venous angiomas: An underestimated cause of intracerebral hemorrhage. Surg Neurol 1988;30:350–358.
42. Mandybur TI. Cerebral amyloid angiopathy: The vascular pathology and complications. J Neuropathol Exp Neurol 1986;45:79–90.
43. Matsumoto N, Whisnant JP, Kurland LT, Okazaki H. Natural history of stroke in Rochester, Minnesota, 1955 through 1969: An

43. extension of a previous study, 1945 through 1954. Stroke 1973;4: 20–29.
44. McCormick WF. The pathology of vascular ("arteriovenous") malformations. J Neurosurg 1966;24:807–816.
45. McCormick WF. The relationship of closed-head trauma to rupture of saccular intracranial aneurysms. Am J Forensic Med Pathol 1980;1:223–226.
46. McCormick WF, Acosta-Rua GJ. The size of intracranial saccular aneurysms: An autopsy study. J Neurosurg 1970;33:422–427.
47. McCormick WF, Boulter TR. Vascular malformations ("angiomas") of the dura mater: Report of two cases. J Neurosurg 1966;25: 309–311.
48. McKissock W, Richardson A, Taylor J. Primary intracerebral haemorrhage: A controlled trial of surgical and conservative treatment in 180 unselected cases. Lancet 1961;2:221–226.
49. Meldgaard K, Vesterby A, Ostergaard JR. Sudden death due to rupture of a saccular intracranial aneurysm in a 13-year-old boy. Am J Forensic Med Pathol 1997;18:342–344.
50. Mody CK, Miller BL, McIntyre HB, et al. Neurologic complications of cocaine abuse. Neurology 1988;38:1189–1193.
51. Morrow PL, McQuillen JB. Cerebral vasculitis associated with cocaine abuse. J Forensic Sci 1993;38:732–738.
52. Nedergaard M, Klinken L, Paulson OB. Secondary brain stem hemorrhage in stroke. Stroke 1983;14:501–505.
53. Okazaki H, Reagan TJ, Campbell RJ. Clinicopathologic studies of primary cerebral amyloid angiopathy. Mayo Clin Proc 1979; 54:22–31.
54. Parvizi J, Damasio AR. Neuroanatomical correlates of brainstem coma. Brain 2003;126:1524–1536.
55. Patel AN, Richardson AE. Ruptured intracranial aneurysms in the first two decades of life: A study of 58 patients. J Neurosurg 1971; 35:571–576.
56. Paxinos G, Huang XF. Atlas of the Human Brainstem. San Diego: Academic Press, 1995.
57. Plum F, Posner JB. The Diagnosis of Stupor and Coma, ed 3. New York: Oxford University Press, 2000.
58. Rigamonti D, Hadley MN, Drayer BP, et al. Cerebral cavernous malformations: Incidence and familial occurrence. N Engl J Med 1988;319:343–347.
59. Rosen RS, Armbrustmacher V, Sampson BA. Spontaneous cerebellar hemorrhage in children. J Forensic Sci 2003;48:177–179.
60. Rothrock JF, Rubenstein R, Lyden PD. Ischemic stroke associated with methamphetamine inhalation. Neurology 1988;38:589–592.
61. Rothfus WE, Albright AL, Casey KF, Latchaw RE, Roppolo MN. Cerebellar venous angioma: "Benign" entity? AJNR Am J Neuroradiol 1984;5:61–66.
62. Sacco RL. Frequency and determinants of stroke. In Fisher M (ed): Clinical Atlas of Cerebrovascular Disorders. London: Wolfe, 1994, pp 1.2–1.16.
63. Sahs AL, Perret GE, Locksley HB, Nishioka H. Intracranial Aneurysms and Subarachnoid Hemorrhage: A Cooperative Study. Philadelphia: JB Lippincott, 1969.
64. Schweitzer JS, Chang BS, Madsen P, et al. The pathology of arteriovenous malformations of the brain treated by embolotherapy. II. Results of embolization with multiple agents. Neuroradiology 1993;35:468–474.
65. Smith DH, Meaney DF. Roller coasters, g forces, and brain trauma: On the wrong track? J Neurotrauma 2002;19:1117–1120.
66. The Multi-Society Task Force on PVS. Medical aspects of the persistent vegetative state (First of two parts). N Engl J Med 1994; 330:1499–1508.
67. The Multi-Society Task Force on PVS. Medical aspects of the persistent vegetative state (Second of two parts). N Engl J Med 1994;330:1572–1579.
68. Van Gijn J, Hijdra A, Wijdicks EFM, et al. Acute hydrocephalus after aneurysmal subarachnoid hemorrhage. J Neurosurg 1985; 63:355–362.
69. Verbeek MM, Eikelenboom P, de Waal RMW. Differences between the pathogensis of senile plaques and congophilic angiopathy in Alzheimer disease. J Neuropathol Exp Neurol 1997;56: 751–761.
70. Vinters HV. Cerebral amyloid angiopathy: A critical review. Stroke 1987;18:311–324.
71. Waga S, Miyazaki M, Okada M, et al. Hypertensive putaminal hemorrhage: Analysis of 182 patients. Surg Neurol 1986;26: 159–166.
72. Waga S, Yamamoto Y. Hypertensive putaminal hemorrhage: Treatment and results. Is surgical treatment superior to conservative one? Stroke 1983;14:480–485.
73. Zeman A. Consciousness. Brain 2001;124:1263–1289.

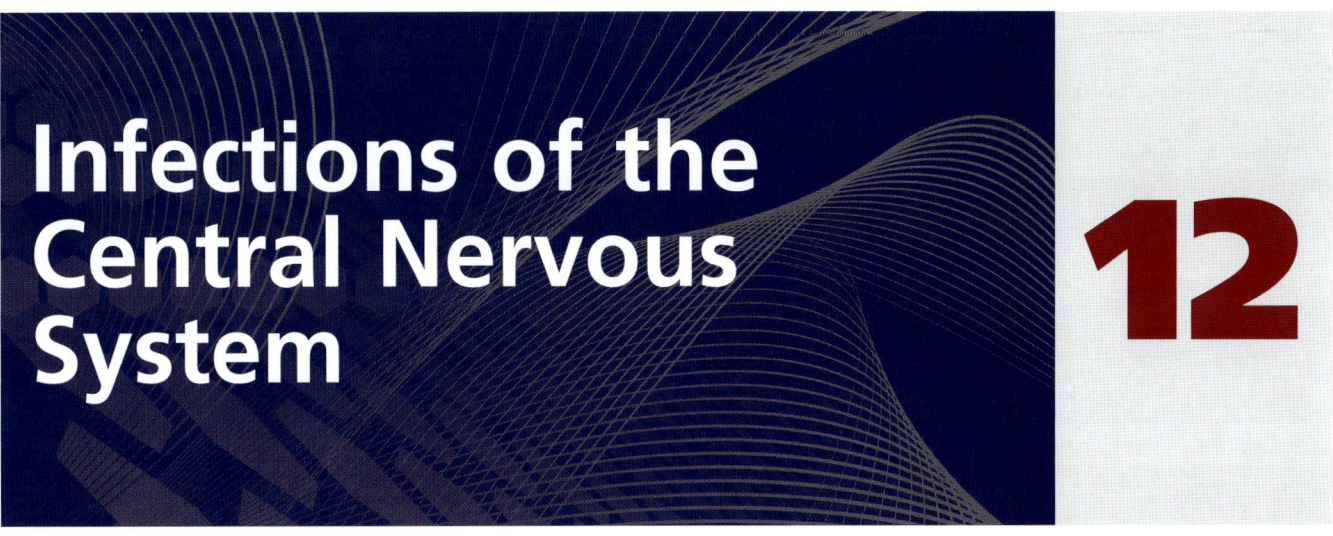

Infections of the Central Nervous System

INTRODUCTION 335
PERINATAL INFECTIONS 335
BACTERIAL INFECTIONS: SPREAD FROM PERICRANIAL PRIMARY FOCI 336
EPIDURAL ABSCESS 336
SUBDURAL EMPYEMA 336
MENINGITIS 337
BRAIN ABSCESS 339
RICKETTSIAL INFECTIONS 339

CHRONIC BACTERIAL INFECTIONS 341
FUNGAL INFECTIONS 342
VIRAL INFECTIONS 345
TRANSMISSIBLE SPONGIFORM ENCEPHALOPATHIES (PRION DISEASES) 351
PARASITIC INFECTIONS 353
BIOLOGIC TERRORISM 355
REFERENCES 355

Introduction

Sources of cases referred for disposition by the medical examiner include a relatively high percentage of individuals who, for a wide variety of reasons, have suboptimal nutrition, a history of insufficient medical evaluation and treatment for chronic illness including immunocompromised states, or have had inadequate preventive medicine screening for conditions such as routine prenatal visits, and so on. It follows that infectious diseases are often among the mechanisms leading to the terminal state. The purpose of this chapter is to provide a brief overview of the wide spectrum of infectious diseases encountered in a medical examiner setting, and a few selected examples of such cases seen at this office. More comprehensive reviews of central nervous system (CNS) infections are available in several general textbooks on neuropathology and in monographs devoted to this subject.[36,48]

Whenever indicated, and when circumstances allow, microbiologic cultures and other special studies should be used to confirm preliminary morphologic diagnoses.[14,64,65,71,72] Antemortem microbiologic studies performed with appropriate collection techniques are correlated with gross and microscopic findings.[73] In the absence of antemortem microbiologic studies in cases in which infection is suspected, it is especially important to correlate the clinical history, gross and microscopic autopsy findings, and results of postmortem microbiologic studies, since estimates of contaminants or false positives in postmortem cultures have varied from approximately 13 to 76%, and vary with the collection technique used.[27,30,73]

Perinatal Infections

Transcervical ascending infections provide a source that can spread to infect amniotic fluid, or be inhaled by the infant as he or she passes through an infected birth canal at parturition. Furthermore, intrauterine infections may result in preterm birth due to prostaglandin release from neutrophils. Inhalation of infected amniotic fluid can lead to neonatal pneumonia and meningitis. Neonatal meningitis, within 4 to 5 days' postbirth, is usually due to group B streptococcus or *Escherichia coli* species. *Listeria*, *Klebsiella-Enterobacter*, *Staphylococcus aureus*, *Staphylococcus epidermidis*, and *Candida* sp. are more likely sources of late-onset sepsis.[38]

Listeria and *Treponema* both can be spread from mother to child through the chorionic villi. *Toxoplasma*, rubella, cytomegalovirus (CMV), and herpes simplex are the main components of the TORCH syndrome, and can result in various combinations of CNS infection, pneumonia, chorioretinitis, myocarditis, hemolytic anemia, and skin lesions.[22]

Bacterial Infections: Spread from Pericranial Primary Foci

Intracranial suppuration and parameningeal infection may occur by direct extension from a pericranial focus. Scalp infections may progress to subcutaneous or subgaleal abscesses, cranial osteomyelitis, and intracranial spread via diploic emissary veins. Acute or chronic sinusitis, mastoiditis, otitis media, or petrous osteomyelitis (**Diagram 12.1**) may also be facilitated by diploic and emissary vein spread. Cerebral veins and dural sinuses are valveless, allowing blood to flow in either direction and facilitating spread of infection from nearby local sources. Dehydration and hypercoagulable states, including the presence of circulating antiphospholipid antibodies, predispose to cerebral venous thromboses, and can be additional factors complicating infections.

The roof of the frontal and ethmoid sinuses form a portion of the anterior cranial fossa, so sinusitis can result in orbital or retroorbital cellulitis, frontal epidural abscess, dural sinus thrombosis, subdural empyema, or frontal lobe abscess, although brain abscess is not invariably within the area of the CNS in closest proximity to a given sinus. Sphenoid sinusitis may spread locally to cause cavernous sinus thrombosis, meningitis, temporal lobe abscess, or superior orbital fissure syndrome. The latter is characterized by orbital pain, exophthalmos, and ophthalmoplegia secondary to involvement of the third, fourth, sixth, and ophthalmic division of the fifth cranial nerve.

Maxillary sinus infections can lead to osteomyelitis of facial bones with retroorbital cellulitis, proptosis, and ophthalmoplegia. Direct intracranial extension of maxillary sinusitis is rare except in rhinocerebral mucormycosis.[15] Nasotracheal intubation has been described as causing sinusitis in 2 to 5% of patients.[15]

Middle ear mastoid infections may extend into the middle fossa to involve the temporal lobe, or into the posterior fossa to involve the cerebellum or brainstem. *Pseudomonas aeruginosa* is one of the more serious causes of malignant otitis externa in diabetics, and may spread to the temporal or occipital bones and to intracranial sites, leading to multiple cranial nerve palsies.

Epidural Abscess

Epidural abscess may, as with subdural empyema, develop as a complication of craniotomy, or due to compound skull fracture that results in spread of infection from an adjoining sinus, middle ear, or orbit. Usually it is secondary to osteomyelitis or from hematogenous seeding of the epidural space. Both the epidural and subdural space are potential spaces and not actual compartments. Infections causing epidural abscess must dissect the dura away from the skull as they spread. Epidural abscesses, in contrast to subdural empyema or subgaleal infections, tend to be restricted by dural attachments at cranial suture lines. Organisms involved in epidural abscess formation are similar to those causing subdural empyema (see later).

Subdural Empyema

Subdural empyema can occur from direct inoculation of bacteria into the subdural space as a complication of neurosurgical procedures, but more often it spreads from local infections such as paranasal sinusitis, mastoiditis, or cranial osteomyelitis. The subdural space is a large compartment with little mechanical barrier, allowing for rapid spread of infection. The condition calls for immediate surgical drainage and aggressive antibiotic therapy. Organisms more likely in neurosurgical procedure-related subdural empyema include *Staphylococcus* and gram-negative bacilli. Aerobic and microaerobic streptococci and anaerobic bacteria are typically involved in sinusitis-associated subdural empyema. Subdural effu-

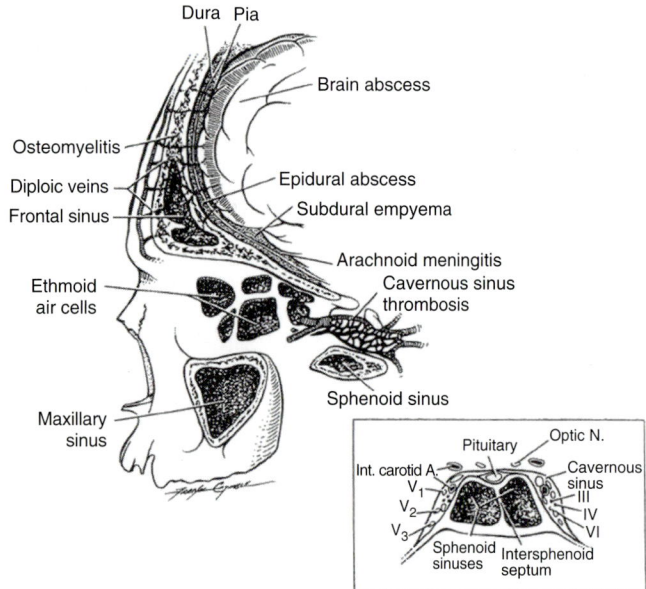

Diagram 12.1. From Chow AW. Life-threatening infections of the head and neck. Clin Infect Dis 1992; 14:991–1004.

sions can also complicate bacterial meningitis in children, and consist of sterile collections of protein-rich fluid that result from increased permeability of thin-walled capillaries and veins of the inner dura (Case 12.1, Figs. 12.1–12.2; Fig. 12.3).

> ### Case 12.1. (Figs. 12.1 and 12.2)
>
> A 26-year-old man was recently released from the emergency room after convervative treatment for a respiratory infection, but he was readmitted 3 days later unconscious following a seizure. CT scan showed bilateral maxillary opacification and a fluid level, and a fluid collection of the left parietal area. Shortly after admission, he was pronounced brain dead, and life support was withdrawn 48 hours later.

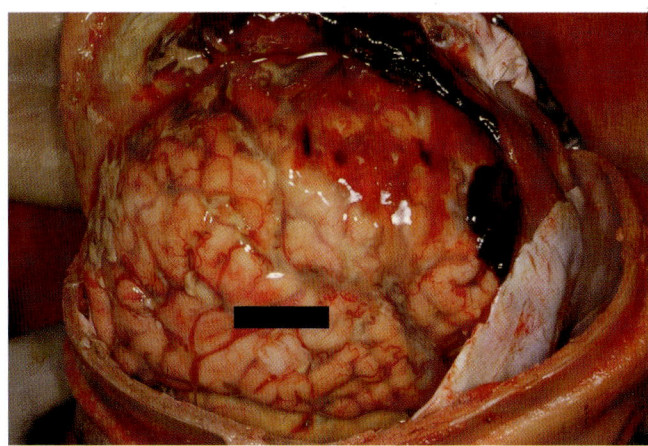

Fig. 12.3. View of right lateral cerebral hemisphere *in situ* at autopsy of another case (frontal right) shows fresh sanguinopurulent subdural empyema layered over frontotemporoparietal lobe. Underlying acute purulent leptomeningitis is also noticeable at margins of empyema.

Meningitis

Acute bacterial meningitis is the most common suppurative intracranial infection (Case 12.2, Figs. 12.4–12.5). In adults, organisms commonly found include *Streptococcus pneumoniae* (~ 50%), *Neisseria meningitidis* (~ 25%), group B *Streptococcus* (~ 10%), and *Listeria monocytogenes* (~ 10%). *Haemophilus influenzae* meningitis, often associated with subdural empyema, was not uncommon in the past[16,55] but has declined due to the Hib vaccine introduced in 1987. *Pneumococcus* and *Meningococcus* colonize the nasopharynx by attaching to nasopharyngeal epithelial cells. They are transported across the epithelial cells in membrane-bound vacuoles to reach the intravascular space, or invade by creating separations in apical tight junctions of columnar epithelial cells. Bacteria with polysaccharide capsules are able to resist phagocytosis as they enter the bloodstream, choroid plexus, and eventually spinal fluid, where rapid multiplication of organisms occurs due to lack of host-mediated immune defense. Complications of bacterial meningitis can be due to the immune response rather than direct bacterial-induced injury. Hence, progression of disease can continue after bacteria are killed by antibiotic treatment. Release of cell wall-lysed bacteria may be the initial step in inducing the inflammatory response leading to formation of subarachnoid purulent exudate. Bacterial cell wall components induce meningeal inflammation by production of cytokines and chemokines derived from inflammatory cells, astrocytes, and endothelial cells. The subarachnoid exudate of proteinaceous material and leukocytes may become sufficiently florid to obstruct exit of spinal fluid from the ventricular system, pack and block flow of cerebrospinal fluid (CSF) in subarachnoid spaces (Fig. 12.6), and diminish the capacity of arachnoid granu-

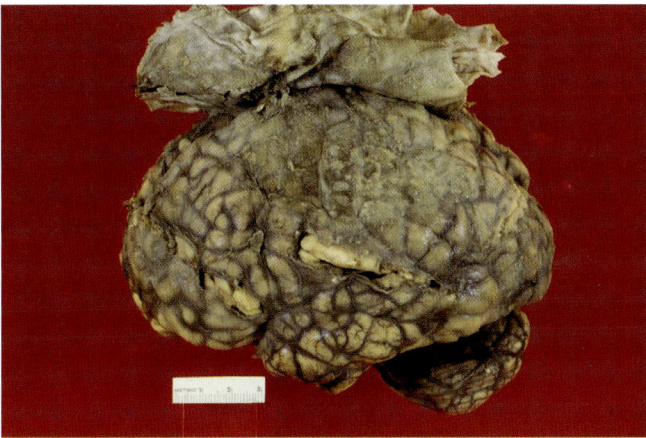

Fig. 12.1. Left frontoparietal convexity and overlying subdural surfaces are covered with a dirty gray empyema. The gray discoloration is characteristic of a respirator brain.

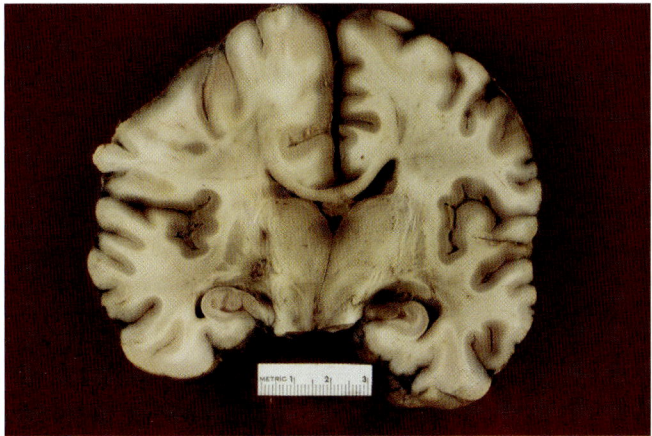

Fig. 12.2. Shallow-dished compression of the left hemisphere created by the empyema is accompanied by ipsilateral cerebral edema.

Infections of the Central Nervous System

Case 12.2. (Figs. 12.4 and 12.5)

A 59-year-old man was found down in an alley in the morning with lacerations and contusions of his forehead and scalp. He was unresponsive in the emergency room and found to have fractured ribs and a flail chest. Fever and leukocytosis of 19,200 mm^3 were found. Blood culture grew out *Streptococcus pneumoniae*. No abnormality was noted in the CT head scan. He remained febrile despite aggressive antibiotic therapy and died on the fifth hospital day.

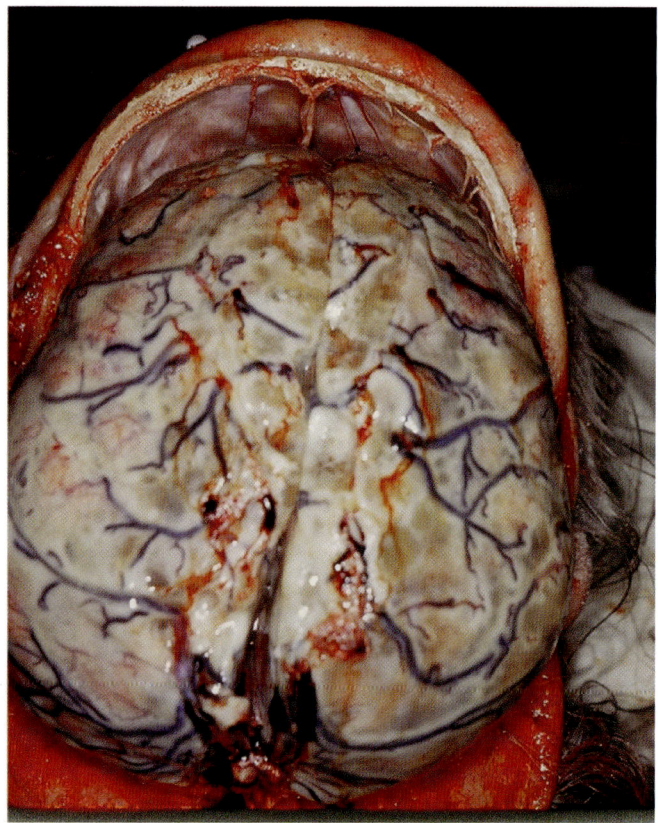

Fig. 12.5. Vertex view of another case of severe acute purulent meningitis in the fresh state.

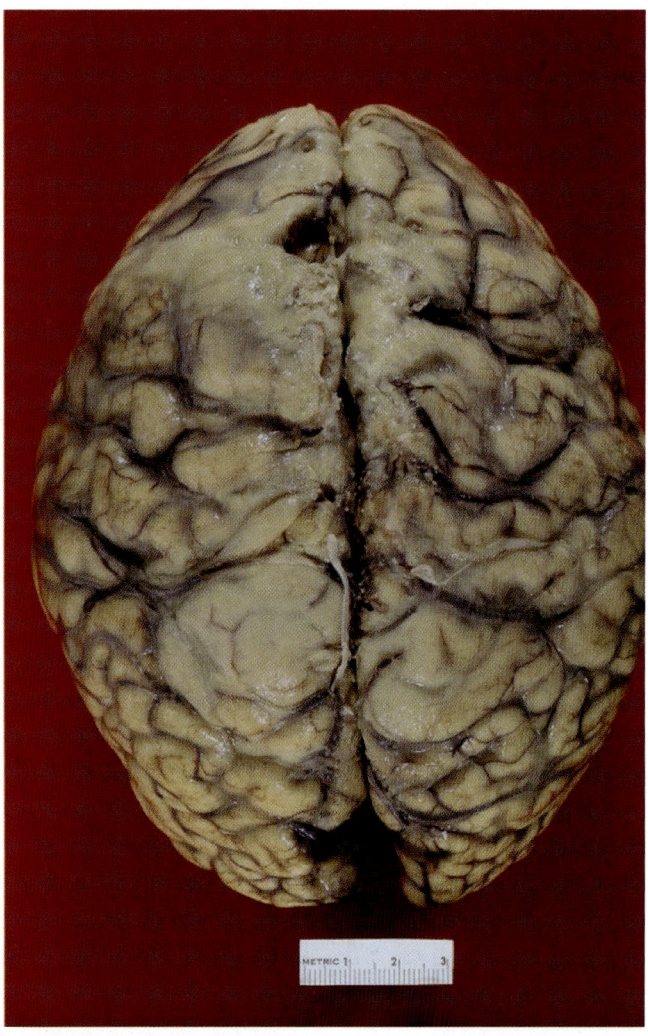

Fig. 12.4. Vertex view of cerebral hemispheres after formalin fixation shows pockets of pus over frontoparietal parasagittal regions. No evidence of recent traumatic injury was found in brain. Mild to moderate brain swelling was present without herniation.

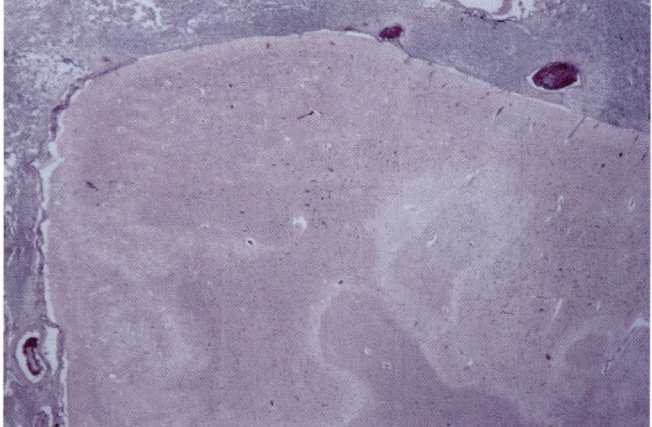

Fig. 12.6. Low-power micrograph shows acute purulent leptomeningitis with leukocytes packing and distending subarachnoid space. Mottled appearance of cortex is from acute infarction. (H&E.)

lations to filter CSF into the venous system, causing obstructive and/or communicating hydrocephalus and other secondary effects resulting from the increased intracranial pressure. Involvement of intracranial vessels in the inflammatory process can progress to vasculitis, thrombophlebitis, major venous sinus thrombosis, and consequences of the secondary hypoxic-ischemic injury to the neural parenchyma. Surgical procedures, such as cochlear implants in children, may predispose to bacterial meningitis,[60] particularly in cases with other inner ear abnormalities that facilitate communication to the intracranial space.[4,61]

Brain Abscess

Brain abscess occurs when sepsis leads to local vascular inflammation, as described earlier, which results in ischemic or necrotic brain tissue that is a focus of easily colonized parenchyma due to local breakdown of the blood–brain barrier. Abscesses are of hematogenous origin or spread from local infection, such as from the paranasal sinuses and ears. Those of hematogenous origin are most common in the territory of the middle cerebral artery and at the cortical/subcortical white matter junction. Once bacteria have established parenchymal infection, the abscess evolves through a stage of acute cerebritis (~ days 1–3) characterized by perivascular infiltration of inflammatory cells surrounding a central core of coagulative necrosis, which in turn is surrounded by edema. In the late cerebritis stage (~ days 4–9) pus formation leads to an enlarged necrotic center bordered by increasing numbers of macrophages and fibroblasts. Fibroblasts then begin to form a thin capsule. By the third stage (~ days 10–13) capsule formation is seen more prominently on the outer aspect of the lesion, which correlates with the ring enhancement seen on imaging studies (Case 12.3, Figs. 12.7–12.8). By approximately day 14, a well-developed typical abscess cavity is usually seen, consisting of a central necrotic area with inflammatory exudates; a middle zone of macrophages, lymphocytes, and plasma cells; and a peripheral fibrotic capsule with associated gliosis[38] (Figs. 12.9–12.10). Combinations of aerobic and anaerobic bacteria may occur. Anaerobes involved include streptococci, *Haemophilus* sp. (facultative), *Bacteroides* (non-*fragilis*), and *Fusobacterium*. In contrast, brain abscess secondary to bacterial endocarditis is more commonly due to a single organism, such as *S. aureus* in acute bacterial endocarditis. Uncommonly, CNS abscesses are caused by *Actinomyces* spp.[62] Currently, a significant percentage of brain abscesses are caused not by classic pyogenic bacteria, but rather by *Toxoplasma gondii*, *Aspergillus*, *Nocardia* (Figs. 12.11–12.12), *Mycobacterium*, and fungi such as *Cryptococcus neoformans*, and occur in immunocompromised individuals.[74,75]

Rickettsial Infections

Rickettsia are obligate intracellular bacteria with the morphology of short bacilli or coccobacilli. CNS rickettsial infections, such as Rocky Mountain spotted fever and typhus, are typically characterized by vasculitis and perivasculitis, as well as parenchymal microglial nodules.

Case 12.3. (Figs. 12.7 and 12.8)

A 26-year-old woman with a long history of intravenous drug abuse developed an infection around an injection site on her arm. While in the hospital for treatment of the infection, an arterial occlusion developed in her arm and she underwent an embolectomy. Following this, mitral valvular disease was discovered, and she underwent a valve replacement. Postoperatively, embolization to the brain was noted and treated, but she died 5 weeks after the valve replacement.

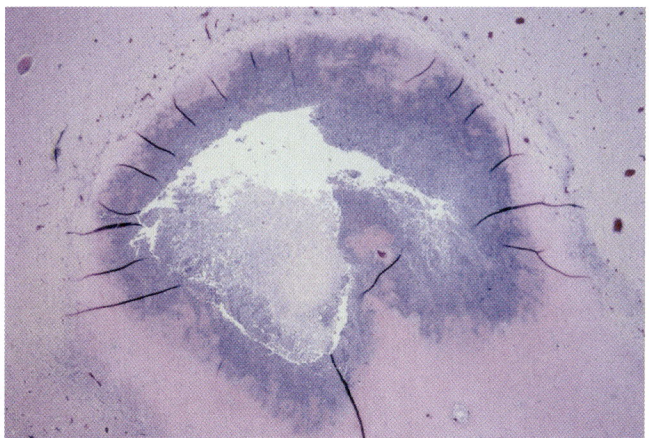

Fig. 12.7. Low-power micrograph shows a focus of septic infarction, now an abscess, with a central zone of necrosis. (H&E.)

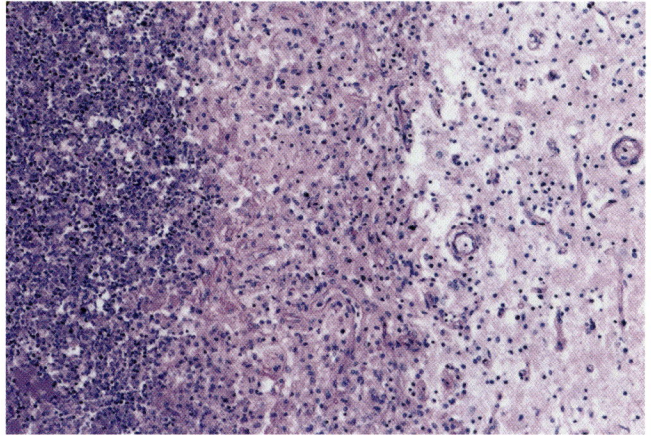

Fig. 12.8. Medium-power micrograph demonstrates early formation of abscess wall of granulation tissue. (H&E.)

Infections of the Central Nervous System

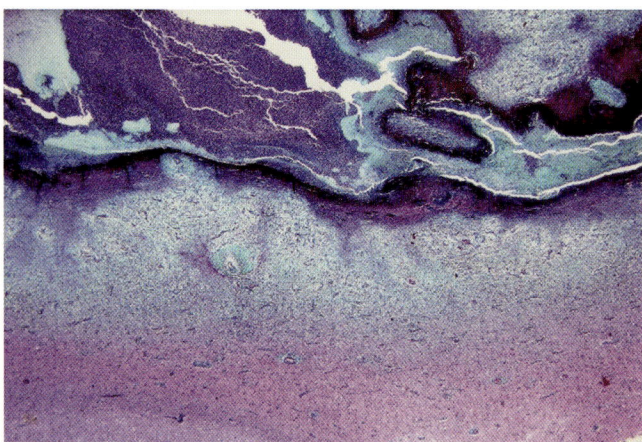

Fig. 12.9. Low-power micrograph of another case shows margin of a subacute abscess wall in the middle zone of the image. Abscess cavity is at top, containing dark necrotic content, and pink brain is below. (Trichrome.)

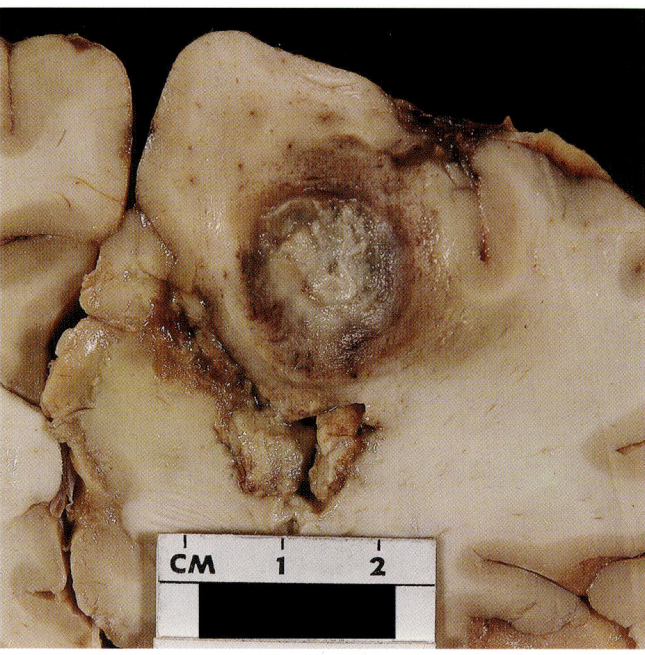

Flg. 12.11. Closeup view of *Nocardia* abscess at cerebral cortex–white matter border.

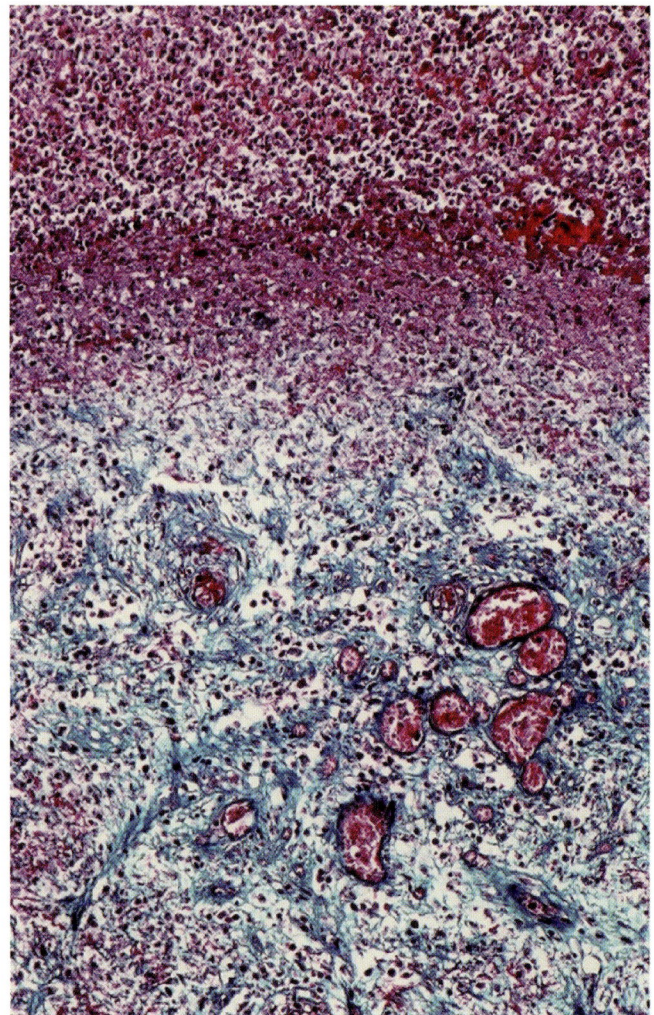

Fig. 12.10. Medium-power micrograph shows early formation of abscess wall with fibrovascular proliferation in granulation tissue. Collagen is green. (Trichrome.)

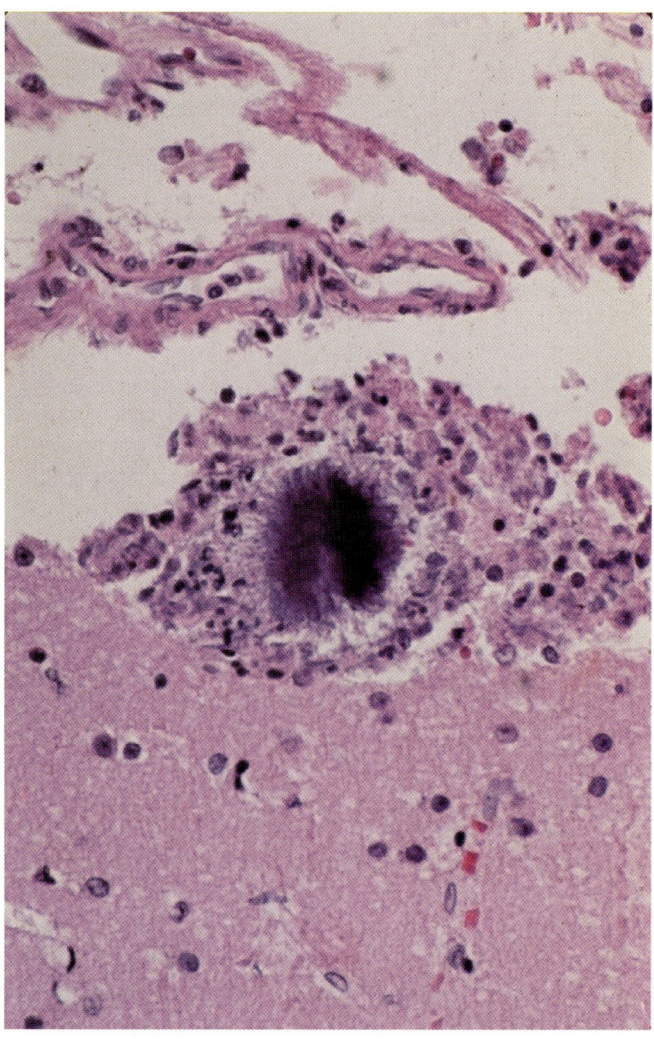

Fig. 12.12. High-power micrograph shows a colony of *Nocardia* hyphae in pia. (H&E.)

There is direct invasion of the vascular endothelium by *Rickettsia*, and secondary microinfarction of brain tissue can occur.

Chronic Bacterial Infections

Tuberculosis may produce a variety of CNS lesions. One form is meningoencephalitis with varying degrees, mild to severe, of granulomatous inflammation characterized by epithelioid cells, caseous necrosis, giant cells, and a predominantly lymphocytic infiltrate. Obliterative endarteritis often accompanies this form of the disease. A second, and much less common, form consists of a neutrophil-rich acute meningitis picture, usually with readily found acid-fast tuberculous bacilli present (Case 12.4, Figs. 12.13–12.15). Such cases emphasize the need for a battery of routine stains in diagnosing acute meningitis. A third form of tuberculous CNS infection consists of intraparenchymal tuberculomas, which are mass lesions up to several centimeters in diameter, microscop-

Case 12.4. (Figs. 12.13–12.15)

A 34-year-old man was found sleeping naked on a public lawn on Christmas Day. He had an altered mental state and he was taken to the hospital where a temperature of 99.4°F, hyponatremia, and pleural effusion were found. CT scan revealed enlargement of all ventricles. Basilar meningitis was suspected, and ventriculostomy was performed. CSF contained 370 red blood cells per mm^3, WBC's 99, 33% polymorphonuclears and 66% mononuclears. Total protein was 103 mg/dl and glucose 31. Tuberculosis and fungal meningitis were considered. Repeat CSF on the fourth hospital day showed a WBC 1155, protein 319, and glucose 39. Pleural biopsy was negative for acid-fast bacillus and fungus. He died 3 weeks after admission.

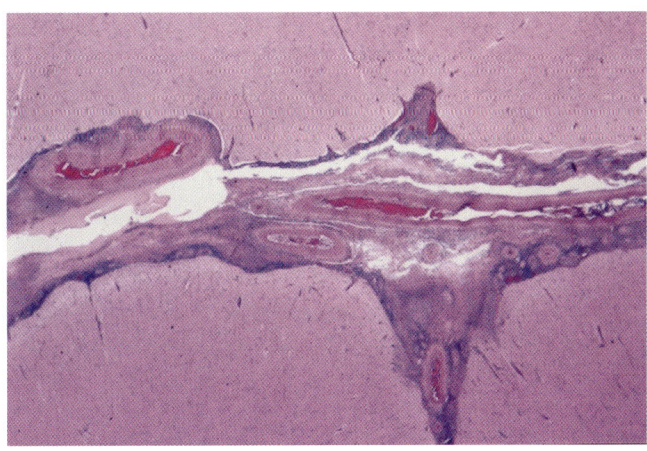

Fig. 12.14. Florid leptomeningitis in cerebral sulcus and vasculitis of smaller blood vessels of subarachnoid space (*arrows*). (H&E.)

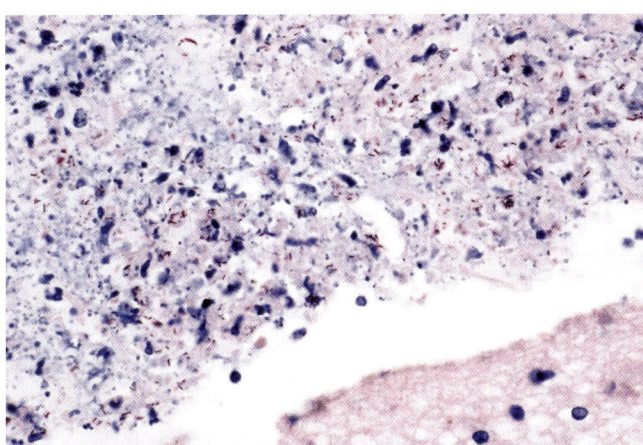

Fig. 12.15. High-power micrograph reveals large numbers of acid-fast bacilli. (Ziehl-Neelsen.)

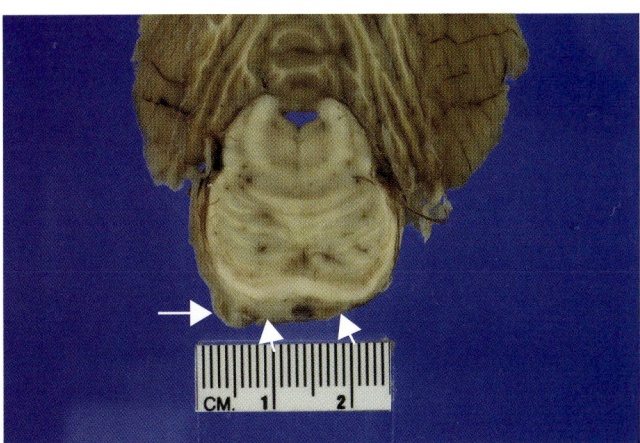

Fig. 12.13. Pons shows thick adherent leptomeninges of prepontine cistern (*arrows*). Rostral fourth ventricle is normal.

ically characterized by central caseous necrosis with a typical peripheral tuberculous granulomatous reaction including Langhans and foreign body-type giant cells. As the number of immunocompromised patients increases, so also do CNS infections with organisms not usually expected in this site, such as nontuberculous acid-fast organisms of the *Mycobacterium avium* complex.[26]

Syphilis is caused by *Treponema pallidum*, a spirochete that produces a variety of CNS lesions. Acute aseptic meningitis may occur in the secondary stage. In secondary or tertiary stages, possible lesions include meningovascular syphilis characterized by granulomatous arachnoiditis, often accompanied by an obliterative endarteritis known as Heubner's arteritis, leading to stroke syndrome. The arteritis is characterized by a distinctive perivascular inflammatory reaction rich in plasma cells and lymphocytes. A form of neurosyphilis more characteristic of the tertiary stage consists of cerebral (and rarely meningeal) gummas, characterized by circumscribed mass lesions with central areas of coagulative necrosis and a more peripheral zone of fibrosis, epithelioid cells, multinucleated giant cells, lymphocytes, and plasma cells. Paretic neurosyphilis or general paresis of the insane (GPI) is a tertiary-stage form of neurosyphilis characterized by thickened meningeal chronic inflammation and cortical atrophy, often prominent in the frontal lobe as well as other areas of the isocortex. Microscopic findings include loss of neurons, perivascular chronic inflammation, proliferation of microglia, and gliosis. Microglia are unusually elongated and aligned along the axes of dendrites. Iron deposits are demonstrated by the Prussian blue reaction, and occur in perivascular areas as well as in the neuropil. Spirochetes are demonstrated by a Warthin-Starry-Levaditi silver stain, and have a corkscrew appearance. A spinal form of late-developing neurosyphilis is tabes dorsalis, in which the focus of damage is on sensory nerves. It may occur in combination with GPI (taboparesis). Microscopically it is characterized by loss of both axons and myelin in dorsal roots, and spinal cord dorsal column atrophy (Fig. 12.16). An infrequently mentioned clue to the possibility of congenital syphilis, in addition to CNS findings similar to adult forms of neurosyphilis and findings in other organ systems, is the presence of binucleated Purkinje cells.

Neuroborreliosis, in the form caused by the spirochete *Borrelia burgdorferi*, produces Lyme disease. Potential CNS effects can include meningitis, cranial neuritis, radiculoneuritis, perivascular or vasculitic inflammation, and multifocal areas of periventricular demyelination.[24] One case with parkinsonism-like clinical symptoms, pathologic features of striatonigral degeneration, and antibody and polymerase chain reaction evidence of *Borrelia burgdorferi* infection was reported, but the possibility of chance association of the two disorders was not excluded.[12]

Fungal Infections

Gomori methenamine silver (GMS) and periodic-acid Schiff (PAS) stains, as well as acid-fast bacillus and Giemsa stains, are routine components in the initial stain panel used in CNS infection cases where bacterial or fungal organisms are suspected based on preliminary hematoxylin-eosin (H&E) screening. Fungal meningitis or meningoencephalitis occurs with a variety of organisms.[70]

Cryptococcal infection, due to *Cryptococcus neoformans*, is the most common form of fungal meningitis in many parts of the world.[44] Masses of these organisms accumulate in subarachnoid and Virchow-Robin spaces, sometimes forming microcysts called "soapsuds cysts" in the parenchyma in areas such as the cerebral cortex ventral aspects and basal ganglia, with occasional loss of a clear relationship to a vessel (Figs. 12.17–12.19). Grossly, there may be a faintly visible meningeal exudate that produces a slippery brain surface, as if the brain had been dipped in clear liquid detergent. Parenchymal lesions consist of aggregates of organisms within spaces surrounded by little or no inflammatory reaction. Well-formed granulomas, or cryptococcomas, are seen but rarely. Organisms are approximately 5 to 10 µm yeast forms with budding and a characteristic wide capsule readily demonstrated by PAS or mucicarmine stains.

Cases are occasionally brought to our office for consultation in which corpora amylacea are mistaken for cryp-

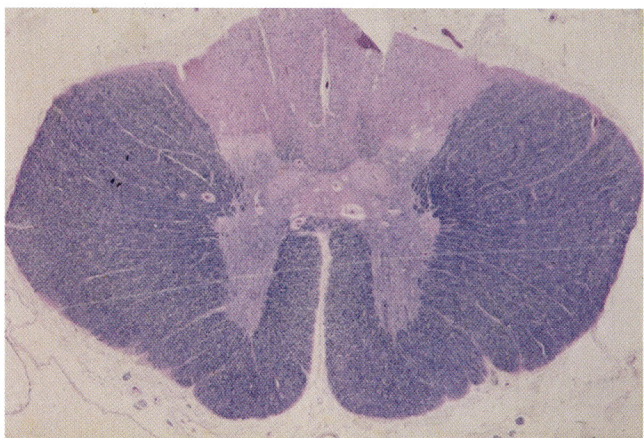

Fig. 12.16. Micrograph of transverse section of thoracic level of spinal cord demonstrates atrophy of posterior columns in tabes dorsalis. The anatomic finding itself is suggestive of tabes, but nevertheless nonspecific. The diagnosis of tabes requires both its neurologic findings and serologic confirmation. (Luxol fast blue/H&E.)

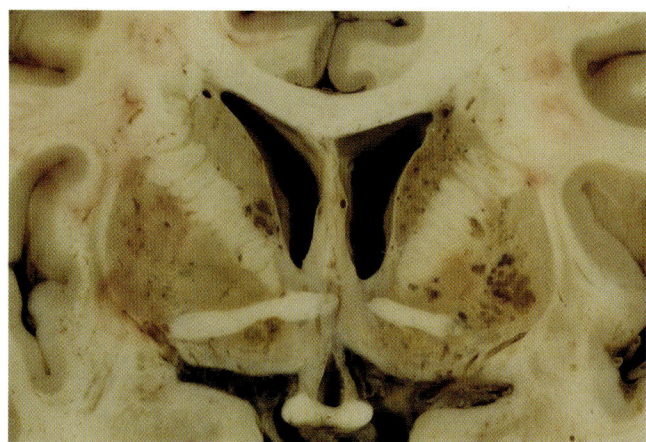

Fig. 12.17. Closeup view of coronal section at anterior commissure shows typical microcysts ("soapsuds cysts") in right putamen, left body of caudate, and more subtle ones beneath anterior commissure. Cerebral cryptococcosis.

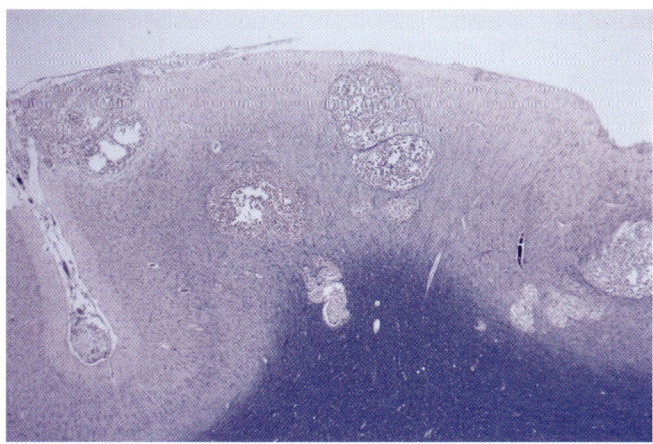

Figs. 12.18. Low-power micrograph of cerebral cortex shows mirocysts filled with fungi evoking no noticeable inflammation. Cerebral cryptococcosis. (Luxol fast blue/H&E.)

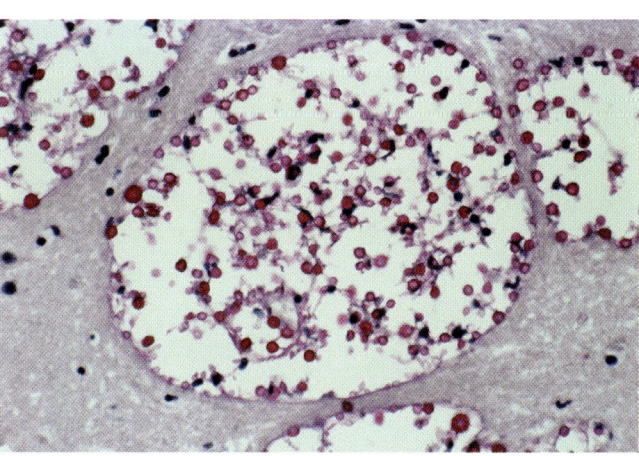

Fig. 12.20. High-power micrograph shows yeast forms of *Cryptococcus neoformans* in a microcyst. (Mucicarmine.)

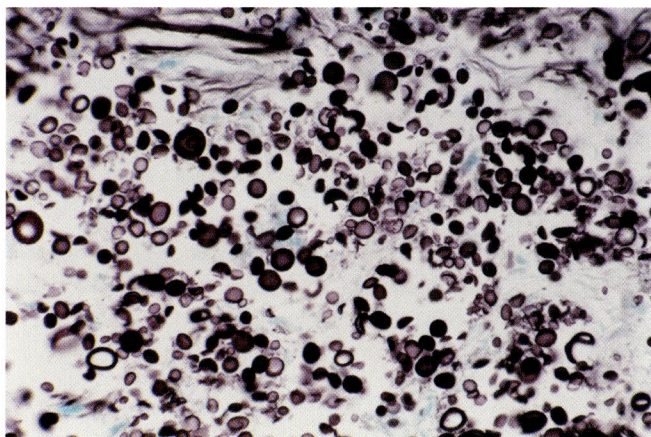

Figs. 12.19. High-power micrograph shows yeast forms of *Cryptococcus neoformans*. (Methanamine silver.)

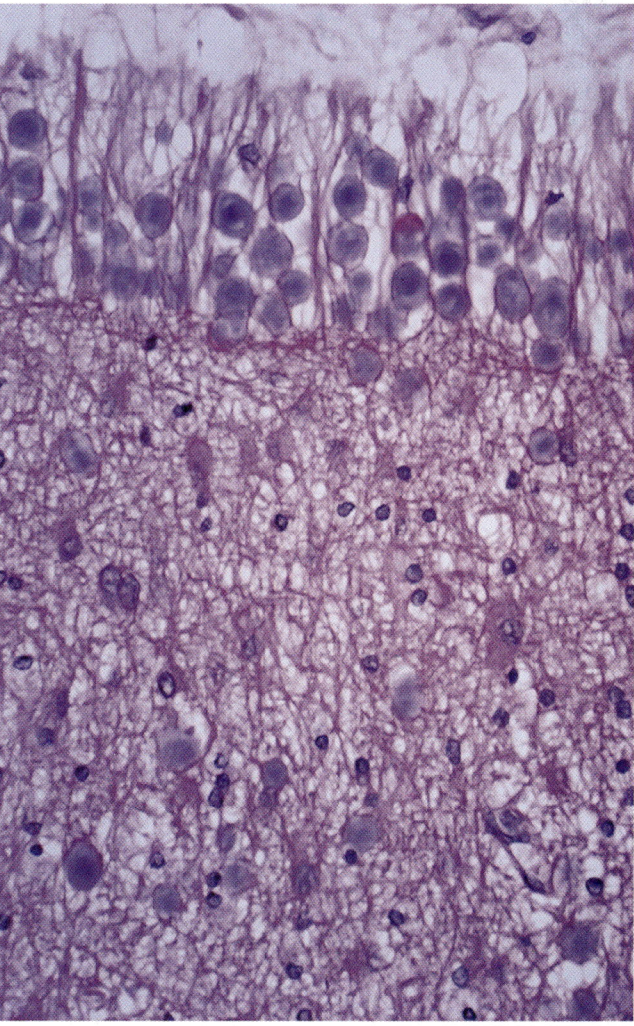

Fig. 12.21. High-power micrograph shows collections of subpial corpora amylacea at a higher power than with the yeast forms seen in Fig. 12.20. (H&E.)

tococcal organisms (Figs. 12.20–12.21), and this potential pitfall has been noted by others.[20] Both corpora amylacea and cryptococci stain with Alcian blue, PAS, and GMS, but they differ in their structural detail and the "company they keep." Corpora amylacea cluster in subpial, subependymal, and perivascular areas, increase with age, are within astrocytic processes and not lying free within neuropil cysts or the subarachnoid space, and do not demonstrate a conspicuous mucicarmine-positive capsule.

Histoplasma microorganisms may produce a meningoencephalitis, are intracellular, and show narrow-necked budding profiles. They are small, in the range of approximately 1 to 5 μm.

Coccidioides immitis causes a granulomatous meningitis, especially at the base of the brain and around the spinal cord, seen as opacified thickening (Figs. 12.22–12.24). The organism can be found as a spherule of approximately 20 to 60 μm, with many endospores inside

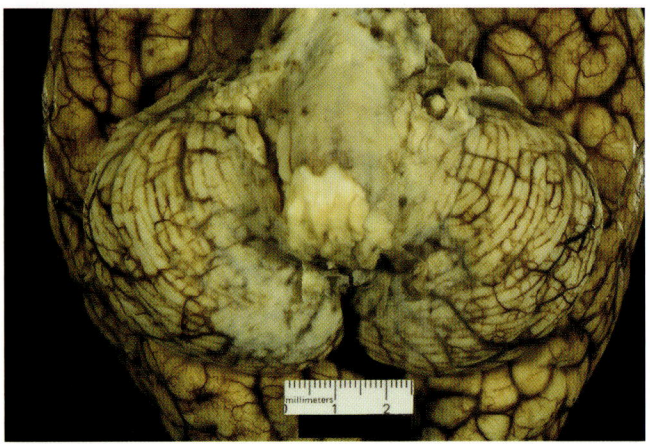

Fig. 12.22. Base view of brainstem and cerebellum shows opacified thickening of leptomeninges over belly of pons and similar patches over cerebellum in a case of coccidioidomycosis.

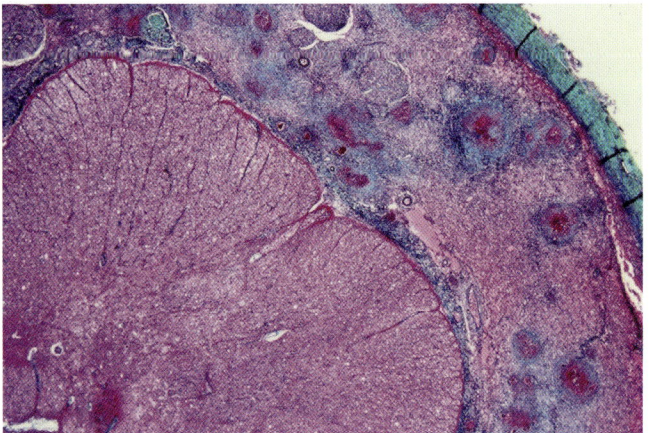

Fig. 12.23. Low-power micrograph of spinal cord shows subarachnoid space distended by granulomatous meningitis. Granulomas can be distinguished even at this power by their red centers, separating them from nerve roots. (Trichrome.) Coccidioidomycosis.

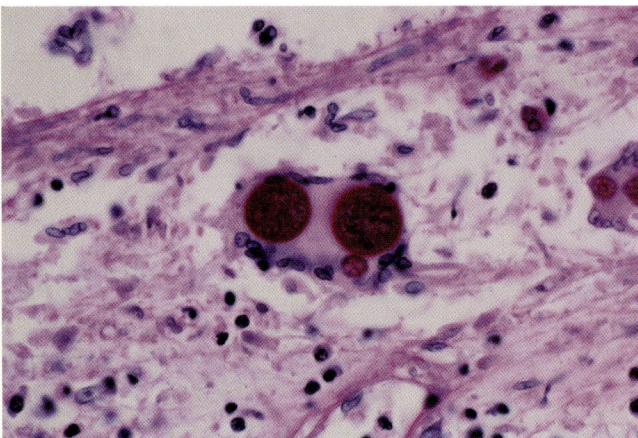

Fig. 12.24. Higher-power micrograph shows two spherules of *Coccidioides immitis* within a single giant cell. (H&E.)

a multinucleated giant cell, or as individual yeast forms approximately 4 to 10 µm in diameter. They are more easily identified with a PAS stain. GMS stains the endospores, but not the outer spherule wall.

North American blastomycosis is caused by *Blastomyces dermatitidis*, and also causes a granulomatous meningitis, a fibrinopurulent meningitis, and multiple abscesses. This organism has typical yeast cell features with a thick double-layered cell wall, measuring up to approximately 25 µm, and demonstrates broad-based budding.

Brain involvement by *Aspergillus* sp. usually occurs in immunocompromised individuals, such as patients with human immunodeficiency virus (HIV) infection or who have been on chemotherapy. Pathologic findings are characterized by hemorrhagic necrosis of white matter due to infarction resulting from angiocentric fungal organism invasion, with resultant vasculitis and secondary thrombosis. The *Aspergillus* organism is characterized by septated hyphae with approximately 45-degree angle branching. It is well demonstrated by both PAS and GMS stains (Case 12.5, Figs. 12.25–12.26), and occasionally a fruiting body may be seen. The organism also extensively invades brain parenchyma.

Disseminated candidiasis, also most commonly seen in immunocompromised individuals, typically produces

> **Case 12.5.** (Figs. 12.25 and 12.26)
>
> A 60-year-old woman had a complicated medical history that included hypertension, chronic obstructive pulmonary disease, diabetes mellitus, and strokes. She was hospitalized because of slurred speech, and an extensive diagnostic workup lasted for a month before she died. The case was referred to the coroner because of alleged poisoning. None was found. Neuropathologic examination revealed multiple acute to subacute abscesses.

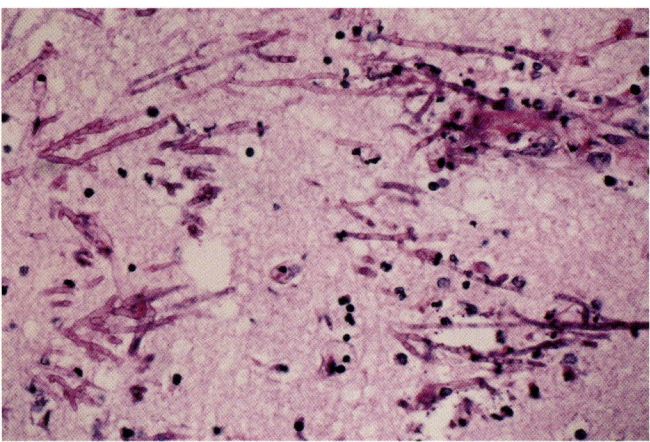

Fig. 12.25. High-power micrograph of an abscess shows, on close examination, septation and acute branching of *Aspergillus* sp. (H&E.)

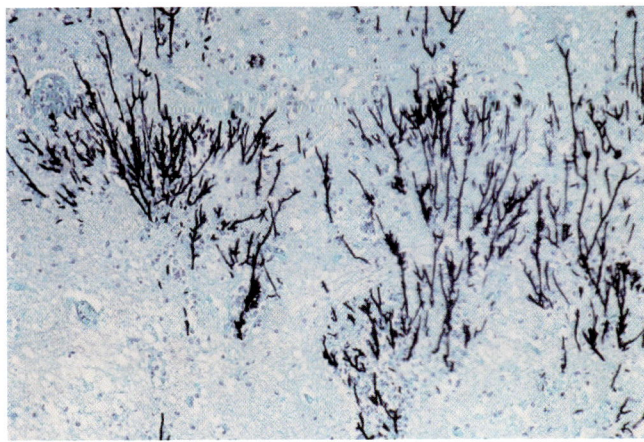

Fig. 12.26. Medium-power micrograph of same abscess as in Fig. 12.25 shows pattern of growth of this organism. (Methenamine silver.)

acute microabscesses with or without giant cell granuloma formation. Thin, branching pseudohyphae and small, approximately 2- to 4-μm yeast forms are the most common morphologic forms seen, but atypical morphologic forms occasionally occur and can be misleading.[1] Rhinocerebral mucormycosis is a progressive destructive infection of the paranasal sinuses caused by the Mucoraceae fungal family. It occurs primarily in debilitated immunocompromised patients, particularly in the context of uncontrolled diabetes and ketoacidosis, in profoundly dehydrated children, and in neutropenic patients receiving cytotoxic therapy. Infection is initially in the nose and nasopharynx, spreads to the sinuses, and thence into the orbit leading to panophthalmitis (Fig. 12.27) by extension through the nasolacrimal duct, or into the CNS. The latter route may extend through the cribriform plate to involve the meninges, adjacent frontal lobe, and cranial nerves. *Mucor*, like *Aspergillus* spp., is angiocentric, involving the walls of arteries with resultant thrombosis and infarction. Grossly, black necrotic lesions often can be seen on the nasal mucosa and soft palate. Diagnosis is aided by the presence of large nonseptate hyphae with irregular, wide-angle branching.

Viral Infections

Enteroviruses account for 75 to 80% of aseptic meningitis cases. Herpes simplex virus DNA is frequently amplified from the CSF of patients with herpes simplex encephalitis type 1, and recurrent and lymphocytic meningitis due to herpes simplex virus type 2. Polymerase chain reaction techniques are now available for the diagnosis of CNS infections due to varicella-zoster virus, Epstein-Barr virus, CMV, Coxsackievirus, Echovirus, polio, and enterovirus.

Arbovirus meningoencephalitis infections occur in the summer months, have clear geographic locations, and may occur in epidemics.[67] This reflects their ecology of transmission through infected insect vectors. Eastern encephalitides, Western encephalitis, St. Louis encephalitis, and California encephalitis are types of arboviral encephalitides seen in the United States. Characteristically, there is a lymphocyte-predominant perivascular inflammatory cell response, with neutrophils and other elements being less conspicuous. Multifocal necrotic foci in gray and white matter are found, and neuronophagia is often present. Viral antigens can be detected in neurons by immunohistochemical methods. Necrotizing vasculitis with hemorrhage may occur. In some cases, cortical involvement predominates; in others, basal ganglia, brainstem, or spinal cord bear the brunt of the disease.

Herpes simplex virus type 1 (HSV-1) is the most common form of nonepidemic encephalitis in otherwise healthy adults, and such cases usually involve the inferomedial temporal lobes including the amygdala and hippocampus, orbital frontal lobes, insula cortex, and cingulate gyrus (Case 12.6, Figs. 12.28–12.30; Case 12.7, Figs. 12.31–12.32). HSV-1 is a necrotizing hemorrhagic infection, microscopically characterized by perivascular inflammatory cell infiltrates that are lymphocyte-rich, and Cowdry type A intranuclear eosinophilic "glassy" viral inclusions in neurons and glia. After about a month or so from the onset of infection, chances of finding inclusion bodies is very small even with diligent search. However, evidence of the virus may still be detectable by *in situ* hybridization. In children, it may cause a chronic granulomatous form of encephalitis.[47] In immunocompromised patients, rather than the typical limbic form described earlier, one may see clinically and pathologically atypical forms of herpes simplex encephalitis, such as a more diffuse nonnecrotizing encephalitis of the hemispheres and brainstem, a more isolated brainstem encephalitis, a noninflammatory pseudoischemic histo-

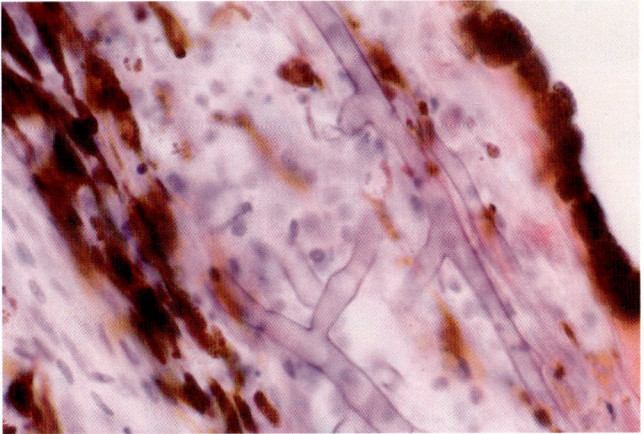

Fig. 12.27. High-power photomicrograph shows large nonseptate hyphae of irregular diameter with right-angle branching of mucormycosis in retina. (H&E.)

346 Infections of the Central Nervous System

Case 12.6. (Figs. 12.28–12.30)

A 14-year-old boy was hospitalized with a diagnosis of encephalitis, not otherwise specified. He remained hospitalized for 4 months and was sent home in the care of a nurse, and lived for another 4 months before he died. Neuropathologic examination showed gross features typical of herpes simplex encephalitis.

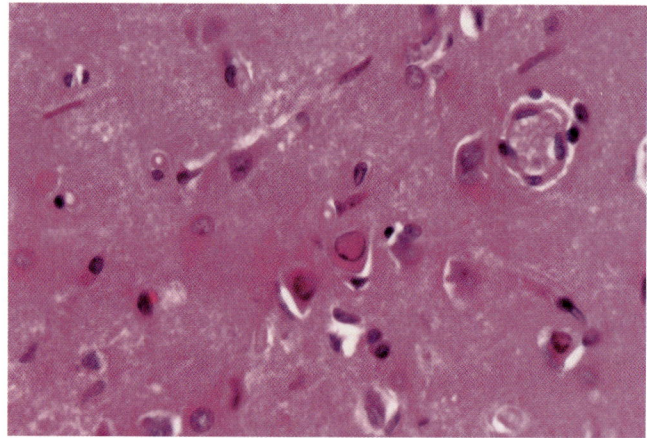

Fig. 12.30. High-power micrograph from another case shows a neuron in center field with a large Cowdry type A nuclear inclusion body. (H&E.)

Case 12.7. (Figs. 12.31 and 12.32)

A 41-year-old man had a seizure disorder of unknown duration and cause. Neurologic examination was reportedly negative.

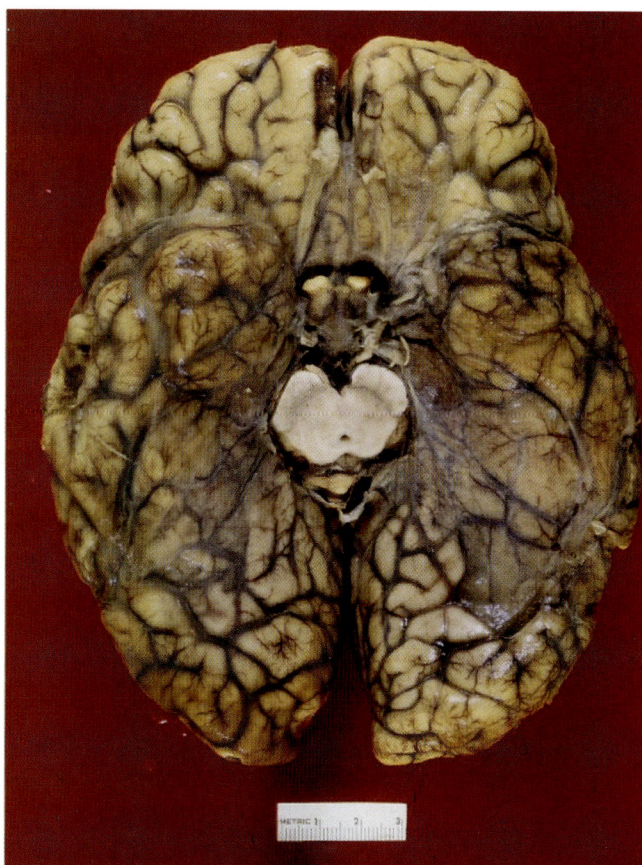

Fig. 12.28. Base view of brain shows bilateral symmetric gray-tan discoloration of necrosis of temporal lobe from the pole to occipital lobe in this case of herpes simplex encephalitis.

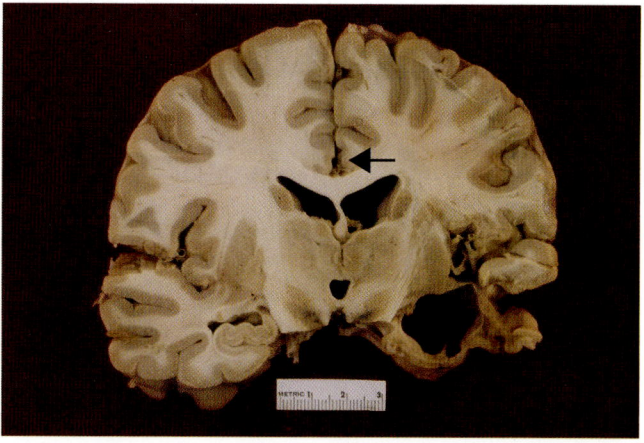

Fig. 12.31. Chronic cystic destructive remnant of entire right temporal lobe leaves no trace of hippocampal formation, amygdala rostrally, and insula cortex. Ipsilateral cingulate gyrus is atrophic (*arrow*). On morphologic basis alone, the findings are highly suggestive of herpes simplex encephalitis.

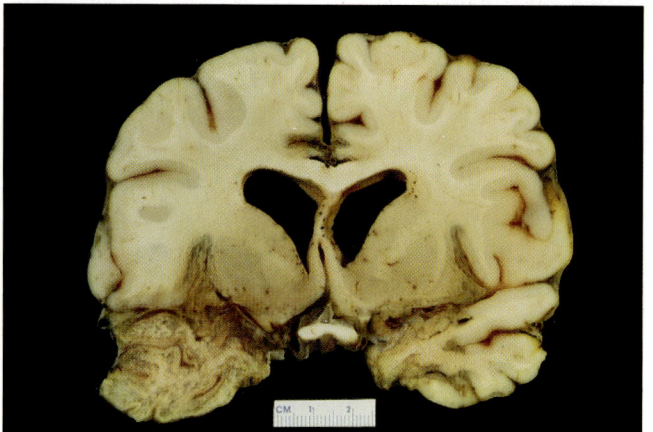

Fig. 12.29. Coronal section shows chronic necrosis of temporal lobe, insula cortex, prepyriform area, and cingulate gyri characteristic of anatomic distribution of necrosis of herpes simplex encephalitis.

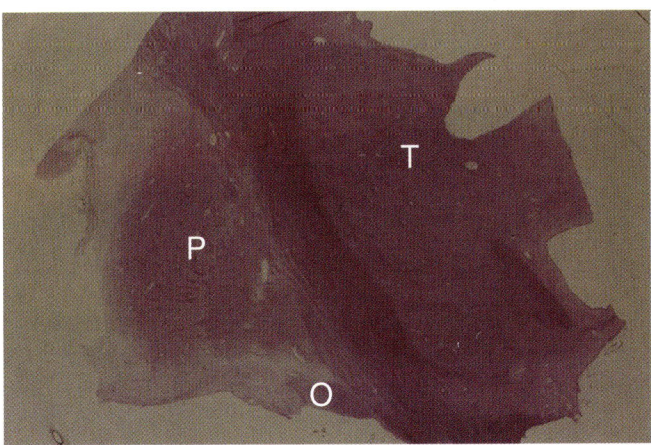

Fig. 12.32. Micrograph of basal ganglia and thalamus (T) shows loss of insula cortex and subjacent areas of superficial putamen (P). Base of lenticular nucleus shows minor loss, leaving optic tract (O) intact. (H&E.)

logic pattern, or herpes simplex virus type 2-associated meningoencephalitis.[34,63]

Herpes simplex virus type 2 (HSV-2) has also been the reported cause of infection affecting up to 50% of neonates born by vaginal delivery to women with primary HSV genital infections, and is a cause of neonatal encephalitis. Microscopically one sees "glassy" clear to reddish nuclear inclusions, often within multinucleated cells, in H&E, Giemsa, or Wright stains.

In herpes zoster (shingles), due to reactivation of latent varicella-zoster virus in the sensory ganglia, ganglia reveal a dense, predominantly mononuclear, infiltrate with herpetic intranuclear inclusions within neurons and their supporting cells. As with other viruses described earlier, it may produce atypical CNS involvement in immunocompromised individuals, such as CNS large or medium vessel vasculopathy leading to ischemic or hemorrhagic infarction, or a deep white matter necrotic and/or demyelinating lesion, or ependymal/subependymal lesions due to varying topographic patterns of small vessel vasculopathy.[40]

Human herpesvirus-6 (HHV-6) is one of several possible agents responsible for an acute febrile illness in childhood and sudden death. The brain is edematous, and atypical mononucleated cells are present in vessels.[29]

Cytomegalovirus (CMV) causes a subacute encephalitis in immunocompromised individuals, with CMV inclusion-bearing cells. The virus tends to localize to ependymal and subependymal regions, at times leading to a hemorrhagic and necrotizing encephalitis, ventriculitis, and choroid plexitis (Case 12.8, Figs. 12.33–12.35). The organism can be demonstrated by *in situ* hybridization techniques.[59] Acute and chronic inflammation, often with involvement of endothelium, is seen. Inclusions may be intranuclear or intracytoplasmic, and generally can be readily visualized with H&E stain or immunomethods (Figs. 12.34–12.35).

Poliomyelitis involves anterior horn cells of the spinal cord predominantly, but may involve bulbar areas. In acute cases, one classically sees mononuclear cell perivascular cuffing and neuronophagia. Chronic survivors show neuronal depopulation, gliosis, and atrophy of

> **Case 12.8.** (Figs. 12.33–12.35)
>
> A 41-year-old man was admitted to the hospital alert but confused and disoriented after being found down on a sidewalk. CT scan showed a mass lesion in the left frontoparietal parasagittal area. He was HIV positive. *Toxoplasma* serology was also positive, and he was treated for toxoplasmosis without response. He died of pneumonia and sepsis after several weeks of hospitalization. Neuropathologic examination revealed cytomegalovirus.

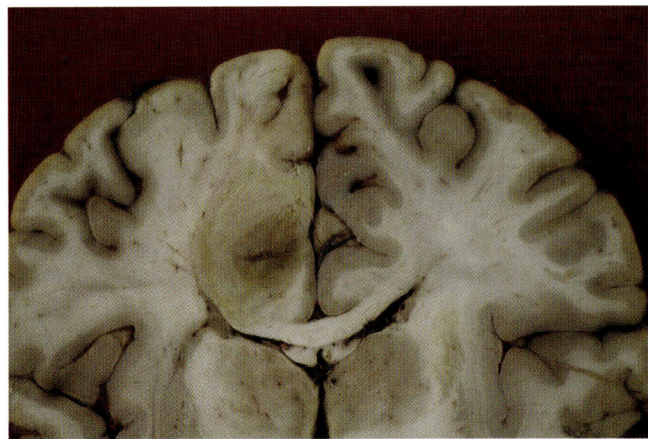

Fig. 12.33. Closeup view of coronal section at midthalamic level shows focal cortical/subcortical necrosis of left medial frontal/limbic lobe due to CMV.

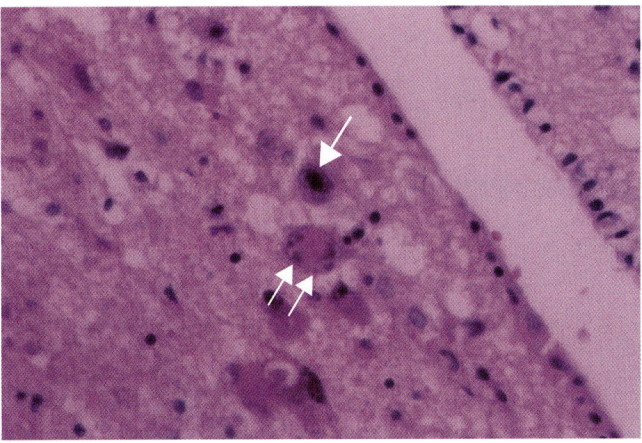

Fig. 12.34. High-power micrograph of periventricular gray matter shows a dark inclusion body in an unidentified cell (*arrow*). A Creutzfeldt cell is next to it (*double arrow*). (H&E.)

Infections of the Central Nervous System

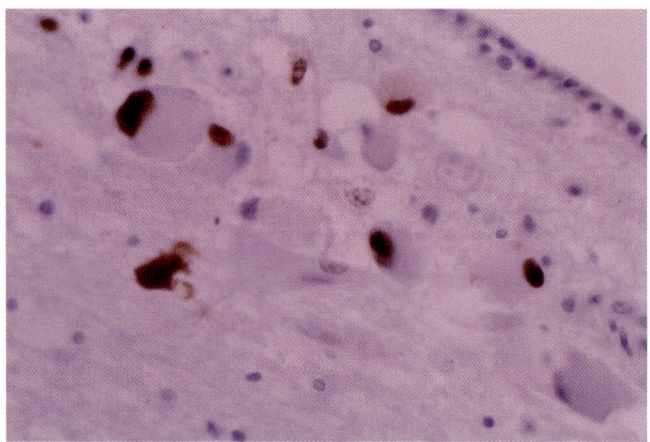

Fig. 12.35. High-power micrograph of the same area as Fig. 12.34 shows immunoperoxidase stain positive for cytomegalovirus.

Case 12.9. (Figs. 12.37–12.39)

A 22-year-old male immigrant developed nausea, vomiting, and low back pain diagnosed as renal stone and urinary tract infection and treated with analgesics and antibiotics. Shortly later, he presented to an emergency room with flank pain and throat tightness. He was agitated, combative, and having excessive salivation. Fever and leukocytosis were noted. A infectious disease consultation included rabies in the differential diagnosis, but no public health report was entered. The patient expired 1 week after his initial visit. Autopsy findings were diagnostic of rabies encephalitis.

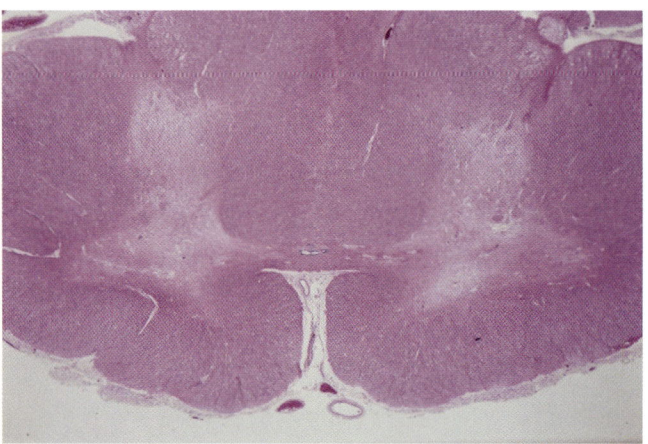

Fig. 12.36. Low-power micrograph of transverse section of cervical spinal cord of a long-term survivor of poliomyelitis shows pallor of both gray columns, partial collapse of anterior horns, and total loss of motor neurons bilaterally. (H&E.)

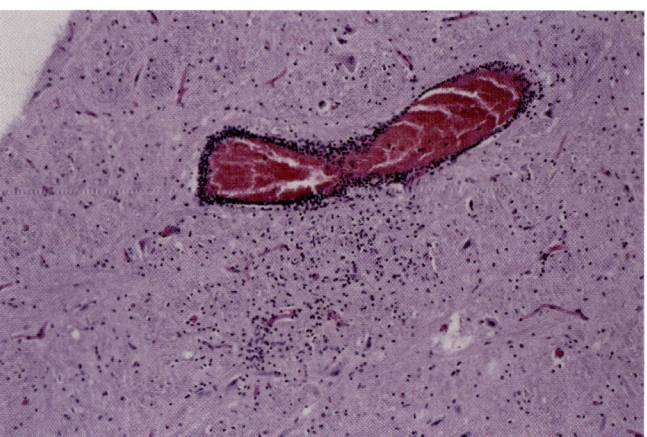

Fig. 12.37. Medium-power micrograph shows a perivascular cuff of chronic inflammatory cells, a nonspecific finding, and areas of parenchymal chronic inflammatory cell infiltrate. (H&E.)

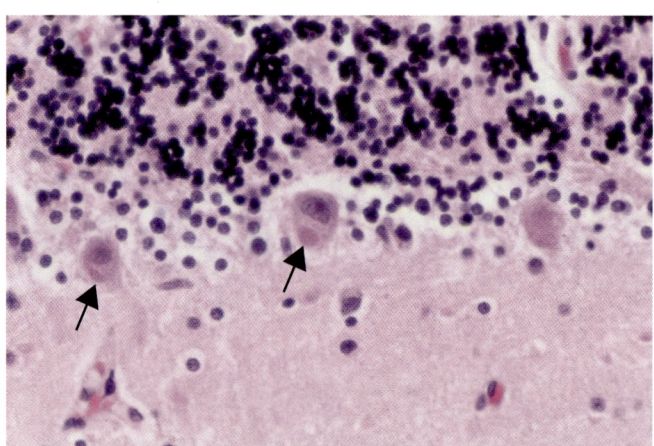

Fig. 12.38. High-power micrograph of cerebellum shows pathognomonic Negri body (*arrows*) in otherwise normal-appearing Purkinje cells. Note lack of local inflammation. Numerous such cytoplasmic inclusion bodies were present. (H&E.)

anterior horns and roots, occasionally with some sparse residual inflammation (Fig. 12.36).

Rabies, an arbovirus, generally is caused by the bite of a rabid animal, usually a dog or bat, and spreads to the CNS via peripheral nerves. Grossly, the brain is edematous and congested. Microscopically, neuronal degeneration and inflammation with nonspecific perivascular cuffs of chronic inflammatory cells is widespread, but most severe in the basal ganglia, midbrain, and floor of the fourth ventricle. The characteristic Negri bodies are intracytoplasmic round to oval eosinophilic inclusions, most consistently found in the pyramidal neurons of the hippocampus and Purkinje cells of the cerebellum. Perhaps counter to usual expectation, Negri body-rich sites typically show little or no inflammation (Case 12.9, Figs. 12.37–12.39). Rabies virus can be demonstrated by

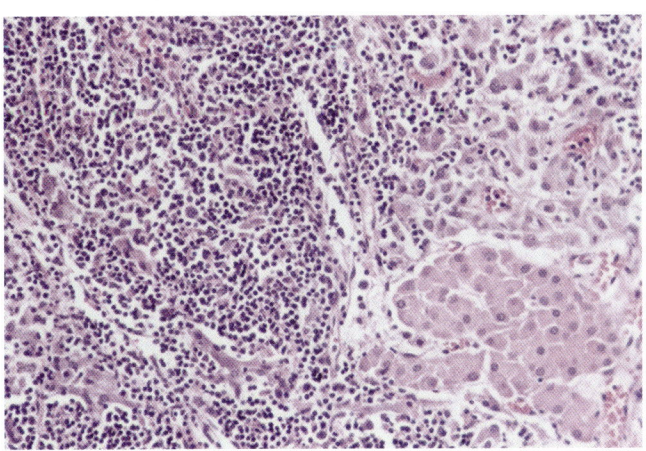

Fig. 12.39. Medium-power micrograph shows adrenal gland with medullitis, another finding in this case of rabies infection. (H&E.)

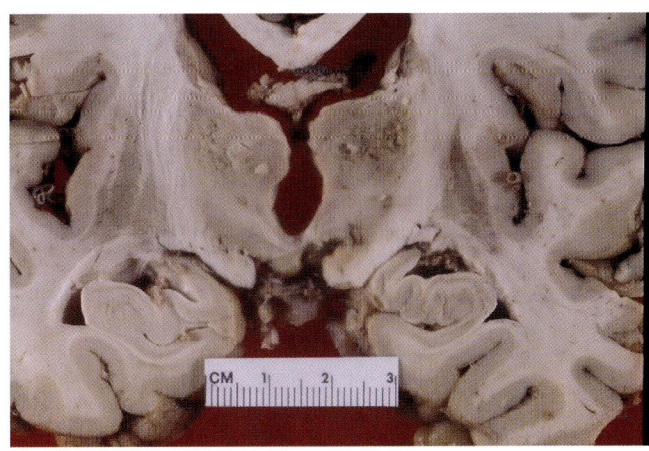

Fig. 12.40. Coronal section of cerebral hemispheres shows focal necrosis of dorsal thalamus bilaterally in a long survivor of West Nile virus encephalitis.

ultrastructural and immunohistochemical examination of Negri bodies.

A clue to the presence of rabies brought to our attention by our most recent case was the presence of lymphoid hypophysitis and lymphoid adrenal medullitis (Fig. 12.39). Such changes in the pituitary gland can be seen in a variety of disorders in which an autoimmune mechanism is suspected[45,66]; in the adrenal gland in a presumably normal adrenal medulla[11]; in association with retroperitoneal inflammatory disorders[43,66]; and in type 1 diabetes mellitus.[7] However, rabies viral antigen has been demonstrated not only in the CNS but also in several extracranial sites, including the adrenal medulla.[32,33] It is recommended, therefore, in cases of encephalopathy of unknown etiology in which there is adrenal medullary or pituitary chronic inflammation, these findings should be considered additional indications to screen known sites of predilection for Negri bodies. Adrenal medullitis is not to be confused with neuroblastic cells that are normally seen during fetal life and can occasionally persist in the newborn and into early infancy.[69]

West Nile virus has become a public health concern in this country following its initial diagnosis in New York cases in 1999,[19] and as it has subsequently spread to other parts of the United States, including California.[58] In several cases seen in our office the findings were consistent with other reported cases, including the tendency of lesions to predominate in the brainstem and spinal cord.[19,53] Most cases of West Nile virus infection are asymptomatic or have nonspecific flulike symptoms, but the likelihood of CNS involvement increases with age.[10] Transmission is usually by the bite of infected mosquitos, but transmission by organ transplantation, blood transfusion, breast milk, intrauterine infection, and laboratory-acquired infection has been reported.[2,31,51]

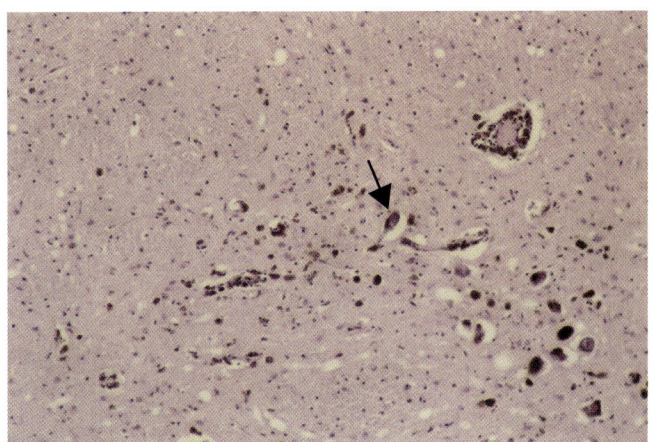

Fig. 12.41. Medium-power micrograph of atrophic substantia nigra shows a lone surviving neuron (*arrow*). The other pigment-containing cells to the right are macrophages. West Nile virus encephalitis. (H&E.)

These latter observations have led to screening for the virus in donated blood.[57]

CNS findings in West Nile virus encephalitis are consistent with several other viral encephalitides, including the presence of perivascular lymphocytic infiltrates, parenchymal microglial clusters, neuronophagia, and variable neuronal loss. These abnormalities may be widespread in the brain, and may include meningitis and cranial and spinal neuritis, but West Nile virus should be particularly suggested if the predominant inflammatory changes are in the brainstem and spinal cord gray matter.[5,25]

The autopsy evidence of West Nile virus encephalitis in long-term survivors may show only nonspecific findings of focal necrosis and atrophy in areas such as the brainstem and thalamus, but not in the cerebral cortex, as in a case studied in this office (Figs. 12.40–12.43). The

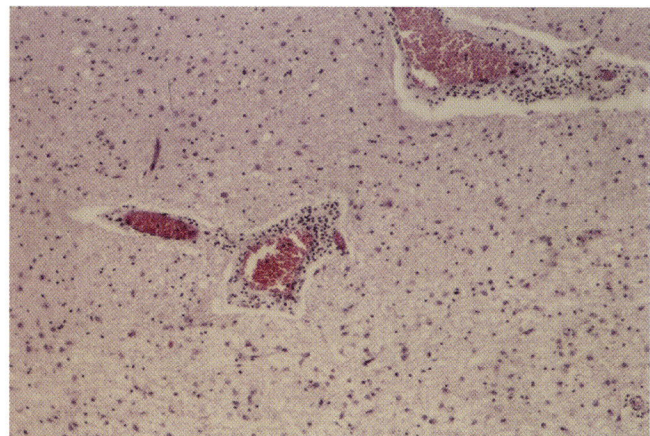

Fig. 12.42. Medium-power micrograph of severe atrophy of dorsal thalamus shows an area of total loss of neurons and astrogliosis. West Nile virus encephalitis. (H&E.)

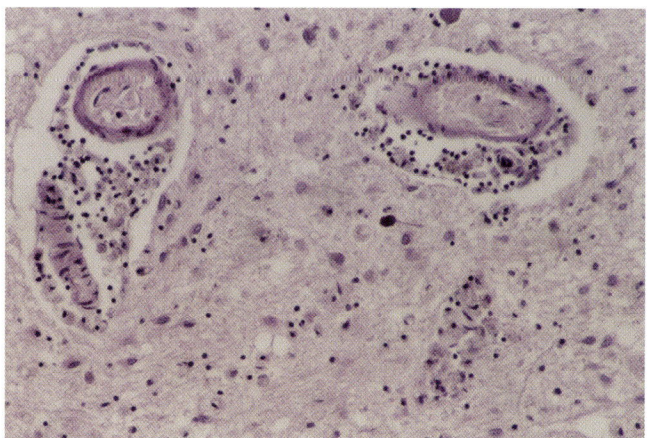

Fig. 12.43. High-power micrograph in the same vicinity as Fig. 12.42 shows sparse perivascular lymphocytes and macrophages. (H&E.)

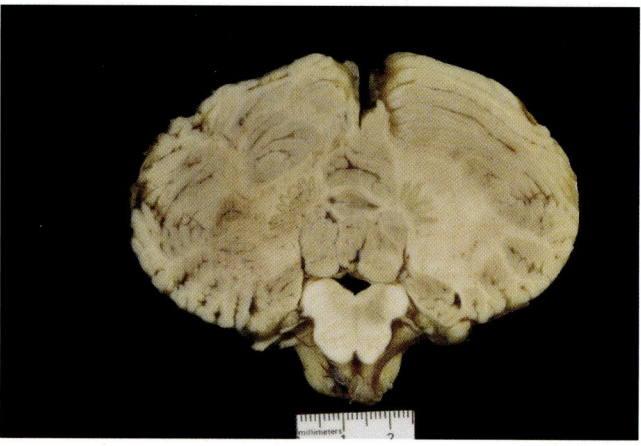

Fig. 12.44. Transverse section of cerebellum and medulla shows granular necrosis of white matter of progressive multifocal leukoencephalopathy on the left. Faint discoloration of a lesion is seen in right medulla.

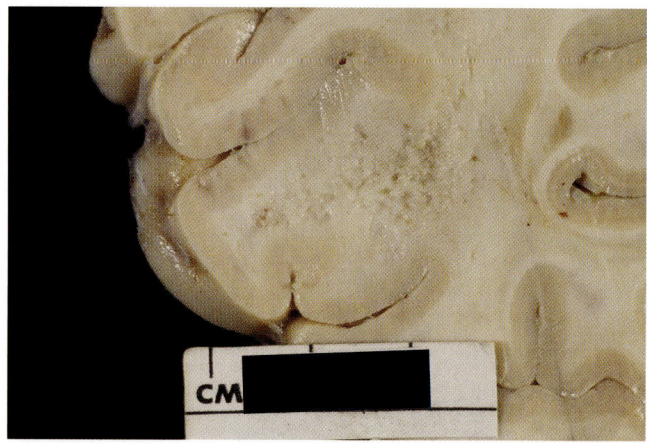

Fig. 12.45. Closeup view of necrosis of white matter in cerebrum in another case of progressive multifocal leukoencephalopathy.

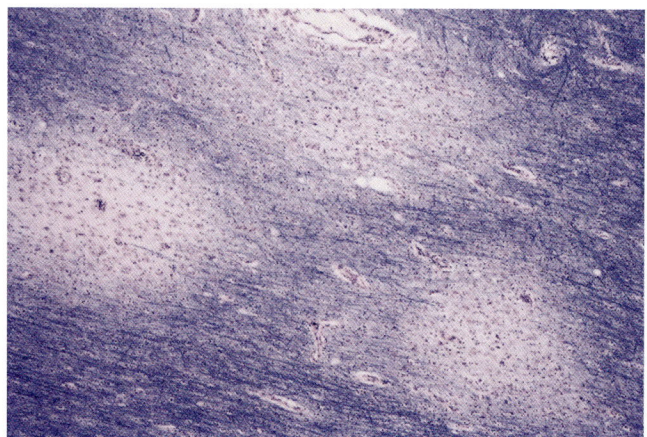

Fig. 12.46. Low-power micrograph of cerebral white matter shows typical multifocal white matter necrosis in progressive multifocal leukoencephalopathy. (Luxol fast blue/H&E.)

findings are nonspecific, and techniques such as *in situ* hybridization are required to prove the diagnosis.

HIV-1 typically causes a subacute meningoencephalitis, but several less common clinical-pathologic syndromes have been reported.[6,9,39] In subacute meningoencephalitis, the meningeal reaction is often unimpressive, but neural parenchyma may show a patchy widespread chronic inflammatory cell reaction with microglial nodules, necrosis, and gliosis. Macrophage-derived small multinucleated "giant cells" associated with the microglial nodules are highly suspicious for HIV-1.

Progressive multifocal leukoencephalopathy (PML) is caused by polyomavirus, and typically occurs in immunosuppressed individuals.[3,28,42] The typical gross and microscopic findings include patchy, irregular areas of demyelination (Figs. 12.44–12.46); oligodendrocyte nuclear inclusions that are acidophilic or amphophilic, may occasionally be granular, and often press nuclear

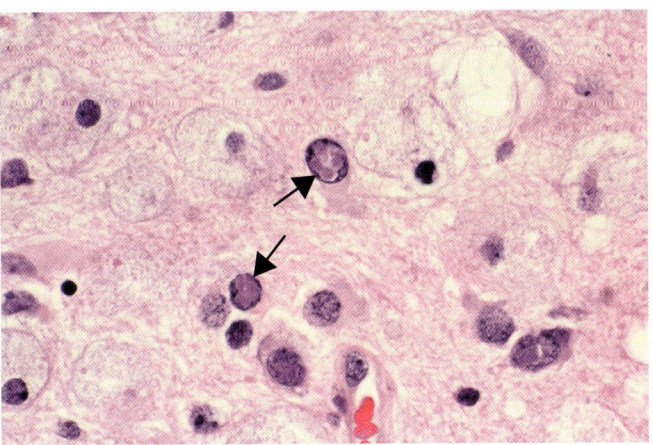

Fig. 12.47. High-power micrograph shows intranuclear inclusion bodies of progressive multifocal leukoencephalopathy in oligodendrocytes (*arrows*). Note the large macrophages in the field. (H&E.)

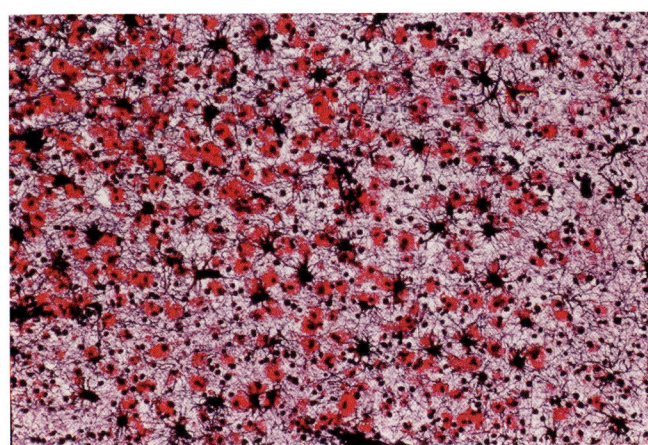

Fig. 12.49. Medium-power micrograph shows lipid-laden macrophages (*red*) commingled with hypertrophied astrocytes (*black*) in a case of progressive multifocal leukoencephalopathy. (Silver carbonate/oil red O.)

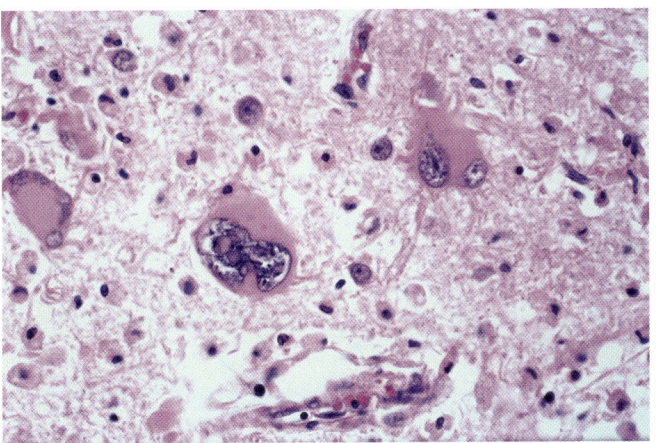

Fig. 12.48. High-power micrograph shows hypertrophied bizarre astrocytes typically found in progressive multifocal leukoencephalopathy. (H&E.)

chromatin to the periphery of the nucleus (Fig. 12.47); and hypertrophied and bizarre astrocytes (Fig. 12.48). The patchy white matter lesions range from a few millimeters to extensive involvement of nearly an entire lobe of the brain (Figs. 12.45–12.46). Scattered and sometimes profuse infiltration of lipid-laden macrophages mixed with hypertrophied astrocytes and variable axonal loss are seen (Fig. 12.49). *In situ* hybridization techniques can demonstrate JC virus in the brain white matter in PML.[52]

Subacute sclerosing panencephalitis is a condition thought to represent persistent infection of the CNS by altered measles virus, which, after a prolonged quiescent period, produces progressive illness. Microscopically, the typical picture is one of widespread neuronal degeneration and gliosis; and Cowdry type A, and less often Cowdry type B, viral inclusions are present within the nuclei of neurons and oligodendroglia. There is variable inflammation of white and gray matter, and neurofibrillary tangles may occur in long-term survivors. Ultrastructural studies have demonstrated that inclusions contain nucleocapsids characteristic of measles virus, and which are positive with measles virus antigen immunohistochemistry.[46]

Transmissible Spongiform Encephalopathies (Prion Diseases)

This group of diseases includes Creutzfeldt-Jakob disease, fatal insomnia syndromes, Gerstmann-Sträussler-Scheinker syndrome, and variant Creutzfeldt-Jakob disease.[17,35] As the class name implies, a dominant feature in many of the conditions within this class consists of spongiform change caused by intracellular vacuoles in neurons and glia. Spongiform change is conspicuous in many, but not all, forms of prion disease, and is not to be confused with status spongiosis, a more nonspecific finding seen in several diseases characterized by severe neuronal loss and gliosis.[50]

Creutzfeldt-Jakob disease (CJD) occurs in several sporadic and familial forms with varied clinical features, and is associated with an abnormal form of specific protein, termed prion protein, that is presently held to be the transmissible agent. Normal prion protein is a 30-kd cellular protein present in neurons. Disease is believed to result from a prion protein alteration from its alpha-helix isoform to an abnormal beta-pleated sheet isoform that is resistant to digestion by proteases. Accumulation of the beta-pleated sheet isoform prion protein in neural tissue is considered to be the cause of pathology in these diseases, but the exact mechanism of neuronal death is unknown. When disease progresses rapidly, the brain may show little gross evidence of atrophy. Grossly, the brain usually appears normal, but it may show mild cerebral cortical, striatal, or cerebellar atrophy. Microscopically, typical findings include a spongiform transformation of the cerebral cortex, including the hippocampal formation, and deep gray matter structures such as the corpus striatum (Figs. 12.50–12.53). Spongiform degeneration is characterized by randomly distributed clear vacuoles up to approximately 20 to 25 μm in diameter within the neuropil, with conspicuous associated gliosis. Expansion of vacuolated areas into larger cystlike spaces has been referred to as status spongiosis. Inflammatory infiltrates are not seen. Electron microscopy shows the vacuoles to be intracytoplasmic and membrane bound, and contained within both neurons and glial cells.

In kuru, in contrast, the aggregated abnormal protein is more typically extracellular, PAS positive, and prominent in the cerebellum. In fatal familial insomnia cases there is no spongiform pathology, but neuronal loss and reactive gliosis are seen in anterior ventral and dorsal medial nuclei of the thalamus, as well as some neuronal loss in the olivary nuclei.

A variant Creutzfeldt-Jakob disease was noted initially in the United Kingdom in 1995. Young adults presented with behavioral problems and a progressive neurologic syndrome that evolved more slowly than that usually observed in sporadic CJD. Autopsy in these cases showed neuropathologic findings similar to Creutzfeldt-Jakob disease, suggesting a close relationship of the two illnesses. Evidence linking the development of variant Creutzfeldt-Jakob disease to ingestion of food products derived from cattle with bovine spongiform encephalopathy has led to governmental regulations

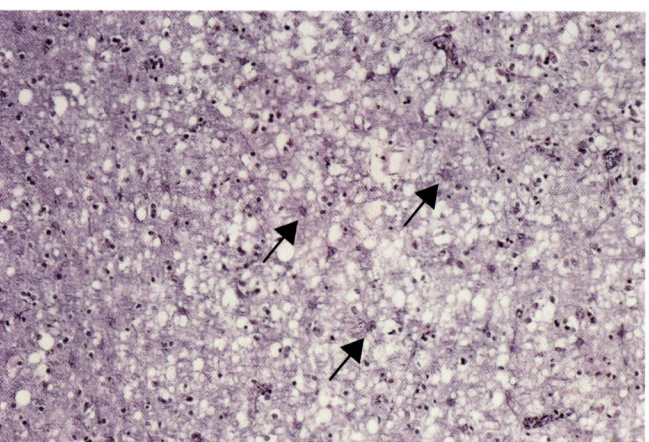

Fig. 12.51. Medium-power micrograph of cerebral cortex in Creutzfeldt-Jakob disease shows a characteristic very fine vacuolization unaccompanied by inflammation. Hypertrophic astrocytes are scattered in the field (*arrows*). (Phosphotungstic acid hematoxylin.)

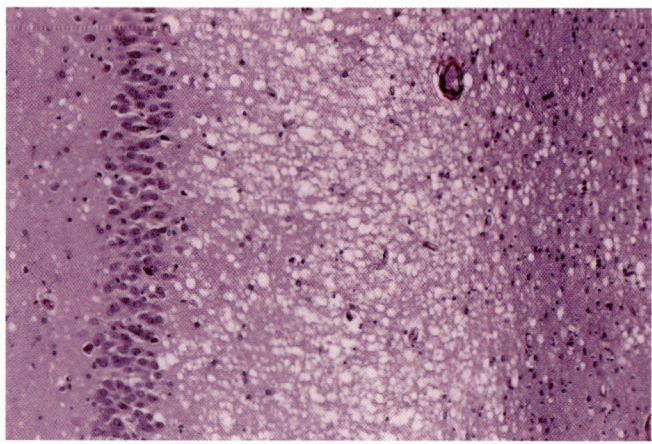

Fig. 12.52. Medium-power micrograph shows selective spongiform degeneration in molecular layer of dentate fascia. (H&E.)

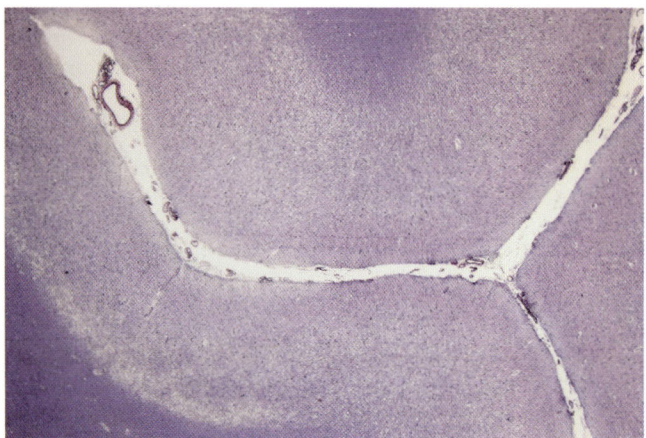

Fig. 12.50. Low-power micrograph of cerebral cortex in Creutzfeldt-Jakob disease shows widespread, not necessarily uniform, fine spongy change barely perceptible at this power. (Phosphotungstic acid hematoxylin.)

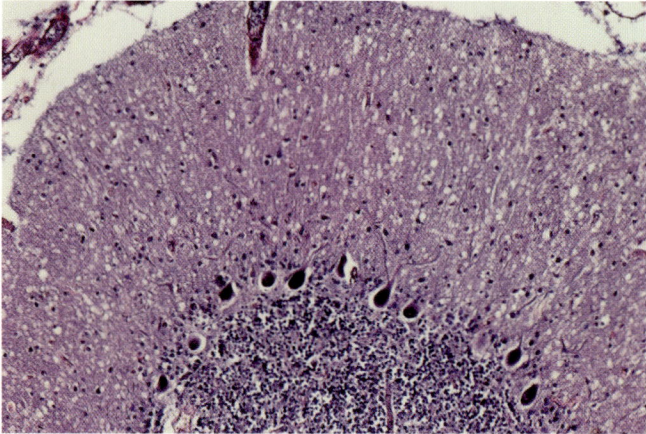

Fig. 12.53. Medium-power micrograph shows spongy degeneration of molecular layer of cerebellum. Purkinje cells are normal in number and appearance. (H&E.)

designed to minimize this potential source of the disease in humans. Variant CJD neuropathology differs from classic sporadic CJD in that one of the striking neuropathologic features is prion protein amyloid plaque formation extensively distributed in the cerebrum and cerebellum.

Gerstmann-Sträussler-Scheinker syndrome, in one of its more common presentations, results in cerebellar ataxia followed by progressive dementia. Amyloid-rich plaques in the cerebral cortex, white matter, thalamus, basal ganglia, and cerebellar cortex are characteristic, but spongiform change is not seen in all cases.

Iatrogenic transmission of prion diseases has been reported but is now waning.[8] For further details of the neuropathologic features distinguishing the various prion diseases, the reader is referred to recent general neuropathology texts and monographs on the subject.[23]

Parasitic Infections

Other CNS infections involve protozoal diseases including malaria, toxoplasmosis, amebiasis, and trypanosomiasis, and metazoal infections such as cysticercosis and echinococcosis.

Plasmodium falciparum is the most common cause of cerebral malaria,[18,56] and can cause sudden unexpected death. Grossly, the brain shows congestion, edema, and petechial hemorrhages. Microscopically, one sees parasitized erythrocyte sequestration in cerebral microvessels, often within ring hemorrhages. Hemozoin pigment deposition within the erythrocyte is derived from erythrocyte hemoglobin degradation by the parasite; it is birefringent.[56] Foci of neuronal hypoxic-ischemic lesions, at times with a microglial reaction, may occur. The microglial nodules are sometimes referred to as "Dürck" granulomas. Patients with cerebral malaria have increased amounts of intercellular adhesion molecule-1 thrombospondin receptor and CD46 on cerebral endothelial cells, perhaps activated by cytokines such as tumor necrosis factor, to which malarial-infected erythrocytes bind.

Toxoplasmosis is caused by a coccidian parasite most commonly occurring in the CNS in newborns, or in later life in the immunosuppressed. The congenital form often appears as bilaterally symmetric areas of subependymal necrosis and calcification, particularly in the region of the caudate nucleus. Manifestations may include chorioretinitis, microcephaly, and hydrocephalus. More widespread cerebral calcification may also be seen. In immunocompetent adults, the uncommon CNS involvement may include meningoencephalitis.

In cerebral toxoplasmosis as seen in immunosuppressed individuals, such as those with HIV infection, abscesses are frequently multiple and primarily involve the cortical medullary junction and deep gray nuclei.[70]

Lesions are less common in the cerebellum and brainstem, and rare in the spinal cord. Acute lesions consist of a central focus of necrosis surrounded by acute and chronic inflammation and vascular proliferation (Case 12.10, Figs. 12.54–12.56). Free tachyzoites and bradyzoites in pseudocysts can be found in the periphery of necrotic foci (Fig. 12.56), often visible on H&E and Giemsa stains, but are more reliably diagnosed with appropriate immunostains. Blood vessels surrounding lesions may show intimal thickening, fibrinoid necrosis, or even frank vasculitis.

Cysticercosis, the most common metazoal CNS parasite in the United States,[54] is due to infection with *Taenia solium* and is acquired by ingestion of eggs present in contaminated human feces. Cysticercal cysts can be located in the meninges, gray matter, white matter, cerebral aqueduct, and ventricular foramina (Figs. 12.57–12.58). The cysts are ovoid, white opalescent, rarely exceeding 1.5 cm, and contain an invaginated scolex with hooklets that are bathed in clear cyst fluid (Fig. 12.58). The trilaminar cyst wall is more than 100 μm thick and is rich in glycoprotein (Fig. 12.59). It evokes little host reaction when the organism is viable, but when the organism dies it evokes an inflammatory reaction leading

Case 12.10. (Figs. 12.54–12.56)

A 32-year-old woman known to be HIV positive developed progressive weakness, anorexia, weight loss, and obtundation over 3 weeks. She was found to have a pancytopenia of 2200 per mm³, with 46% segmented neutrophils. CSF total protein was 126, glucose 61 mg/100 ml, and a cell count 4 WBC's per mm³. CT scan showed a 2.5 × 3-cm lesion in the right thalamus interpreted as an early abscess. She died 1 week after admission.

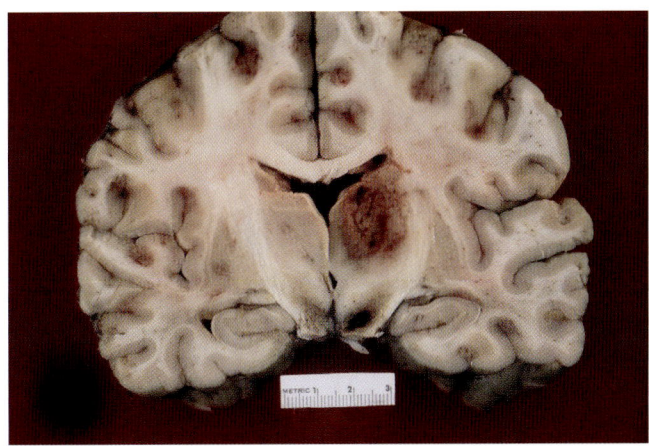

Fig. 12.54. Coronal section through thalamus shows necrosis of nearly the entire right thalamus, and a few less conspicuous lesions elsewhere in this photo. Similar, but smaller, disseminated lesions involved basal ganglia, brainstem, and cerebellum. Cerebral toxoplasmosis.

Infections of the Central Nervous System

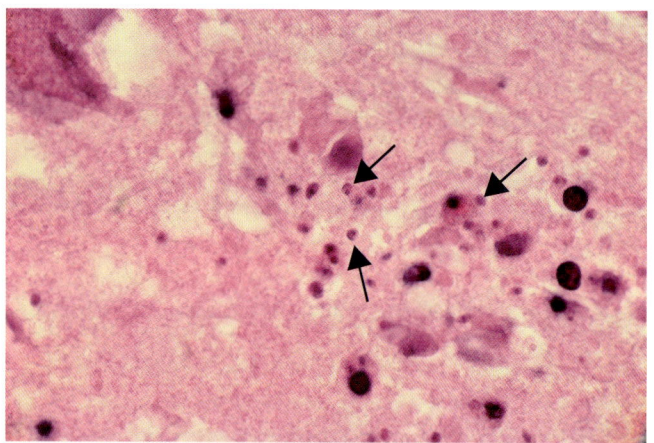

Fig. 12.55. High-power micrograph at the margin of necrosis shows several tachyzoites closely grouped (*arrows*). Cerebral toxoplasmosis. (H&E.)

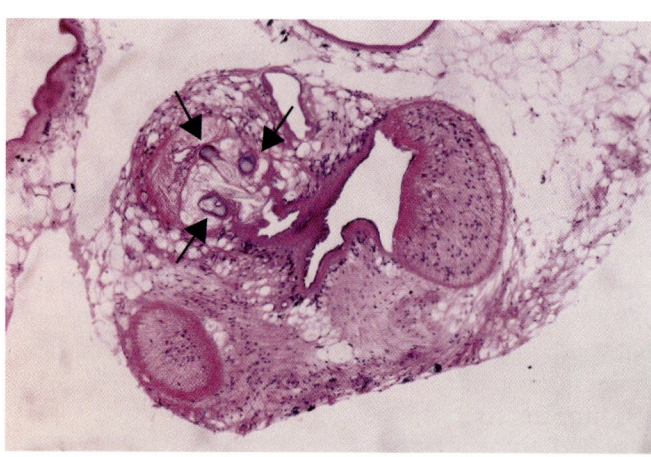

Fig. 12.58. Low-power micrograph of section through scolex of *Cysticercus cellulosae* shows its several features, including hooklets (*arrows*). (H&E.)

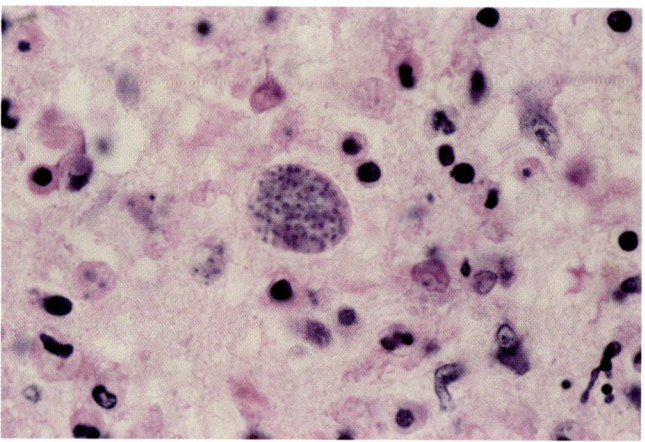

Fig. 12.56. High-power micrograph shows a pseudocyst bearing bradyzoites. Cerebral toxoplasmosis. (H&E.)

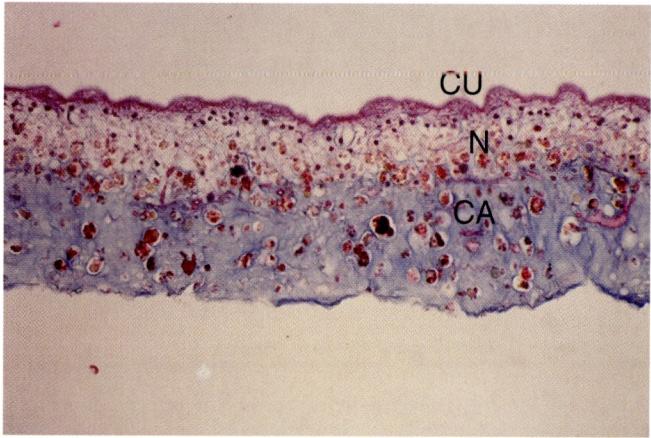

Fig. 12.59. Medium-power micrograph of parasitic membrane of cysticercus cyst shows the typical trilaminar structure consisting of cuticular (CU), nuclear (N), and canalicular (CA) layers. Cuticular layer is faintly ciliated. (Gomori trichrome.)

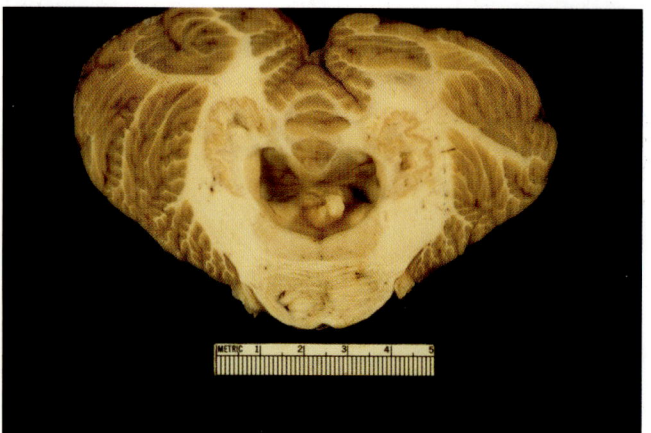

Fig. 12.57. Transverse section at midpontine level shows moderate distention of fourth ventricle due to a cysticercus cyst attached to floor of fourth ventricle and blocking lateral recesses and foramen of Magendie.

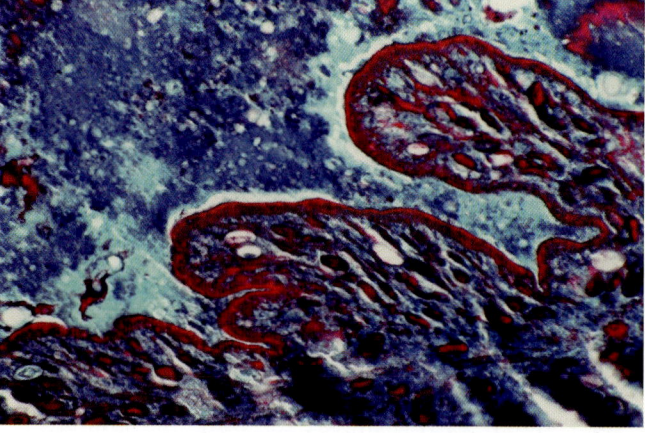

Fig. 12.60. In a degenerated lesion suspected of being that of a cysticercus cyst, trichrome stain is useful to demonstrate a residual cuticular layer seen as an undulating bright red line within a necrotic mass.

to scarring and calcification. Parasitic membranes may still be identifiable within the necrotic debris of the cyst by application of a trichrome stain (Fig. 12.60). Additional case examples are discussed in Chapter 9.

Coenurus cerebralis cyst membranes are similar to those of cysticercus, but a distinguishing features is the many invaginated inlets and more numerous scolices in *Coenurus* cysts[68] (see Chap. 9).

The spectrum of CNS amebic infections currently includes an acute primary amebic meningoencephalitis due to *Naegleria fowleri* (Fig. 12.61), a more chronic granulomatous amebic encephalitis due to *Acanthamoeba* sp and *Balamuthia mandrillaris*, and a more recently described case in which a newly recognized human pathogenic ameba, *Sappinia diploidea*, produced a tumor-like solitary mass in the temporal lobe of a previously healthy person.[21,37,49]

Amebae can be difficult to distinguish from histiocytes on histopathology. Methenamine silver (GMS) and PAS stains are helpful in visualizing the organisms, although definitive identification depends on immunofluorescent studies and culture. *N. fowleri* primary acute amebic meningoencephalitis follows aspiration of (usually stagnant) water contaminated with trophozoites or cysts, entering the brain via the olfactory nerves,[21,41] or by inhalation of contaminated dust that leads to invasion of the olfactory neuroepithelium by amebae and production of an intense acute hemorrhagic necrotizing meningoencephalitis of basal cerebral cortex and posterior fossa (Fig. 12.61).

Earliest possible treatment depends on a good clinical history, a high index of suspicion, and detection of motile trophozoites in wet mounts of fresh cerebrospinal fluid.

Acanthamoeba infection is more common, occurring in immunosuppressed patients who, for example, are chronically ill, debilitated, have lymphoproliferative disorders, have been on steroid therapy, or have the acquired immunodeficiency syndrome (AIDS). Amebae reach the CNS by hematogenous routes from primary infection sites in sinuses, skin, or lungs. In addition to meningitis, typical lesions include embolic infarcts, edema, and the presence of trophozoites and cysts on CSF wet mount, or in autopsy specimens. Fluorescein-labeled antiserum is available from the Centers for Disease Control and Prevention for definitive diagnosis of the protozoal species in biopsy specimens.

Both forms of African trypanosomiasis, or sleeping sickness (i.e., West African due to *Trypanosoma brucei gambiense*, and East African due to *Trypanosoma brucei rhodesiense*) cause a meningoencephalitis responsible for much of the morbidity and mortality of the disease. The West African form is a chronic disease lasting many months, whereas the East African type is more acute, with death from toxemia within a few weeks after onset of symptoms. In both forms, the final stage of encephalitis is characterized microscopically by perivascular mononuclear cells, including lymphocytic and plasma cell infiltrates, microglial nodules, and gliosis. There are also characteristic Mott cells, a distinctive form of plasma cell with multiple eosinophilic cytoplasmic inclusions composed of immunoglobulin.[70] Both Mott cells and trypanosomes can be seen in the CSF.

Other protozoal infections involving the brain seen, for example, in AIDS patients, may include unusual organisms such as *Trachipleistophora hominis*.[64,65]

Biologic Terrorism

Agents considered most likely to be potentially used in biologic warfare scenarios include many in which CNS manifestations are primary, or are an important component of the resultant clinical syndrome. Although beyond the scope of this discussion, a recent review of this topic is available.[13]

References

1. Alasio TM, Lento PA, Bottone EJ. Giant blastoconidia of *Candida albicans*: A case report and review of the literature. Arch Pathol Lab Med 2003;127:868–871.
2. Batalis NI, Galup L, Zaki SR, Prahlow JA. West Nile virus encephalitis. Am J Forensic Med Pathol 2005;26:192–196.
3. Berger JR, Koralnik IJ. Progressive multifocal leukoencephalopathy and natalizumab—Unforeseen consequences. N Engl J Med 2005;353:414–416.
4. Bluestone CD. Bacterial meningitis in children with cochlear implants (letter to editor). N Engl J Med 2003;349:1772.
5. Bouffard J-P, Riudavets MA, Holman R, Rushing EJ. Neuropathology of the brain and spinal cord in human West Nile virus infection. Clin Neuropathol 2004;23:59–61.
6. Brew BJ: HIV Neurology. New York: Oxford University Press, 2001.

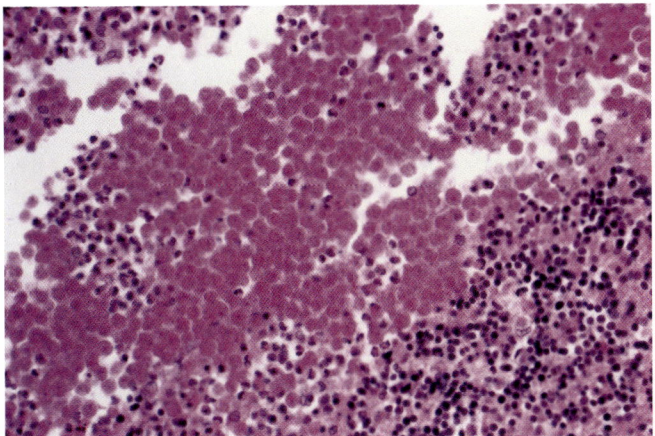

Fig. 12.61. High-power micrograph shows a swarm of trophozoites of *Naegleria fowleri* in cerebellum. Small nuclei in the corner of field are those of cerebellar granule cells. (H&E.)

7. Brown FM, Smith AM, Longway S, Rabinowe SL. Adrenal medullitis in type I diabetes. J Clin Endocrinol Metab 1990;71:1491–1495.
8. Brown P, Brandel J-P, Preese M, Sato T. Iatrogenic Creutzfeldt-Jakob disease: The waning of an era. Neurology 2006;67:389–393.
9. Büttner A, Weis S. HIV-1 infection of the central nervous system. In Tsokos M (ed): Forensic Pathology Reviews. Totowa, NJ: Humana Press, 2005;3:81–134.
10. Campbell GL, Marfin AA, Lanciotti RS, Gubler DJ. West Nile virus. Lancet Infect Dis 2002;2:519–529.
11. Carney JA. Adrenal gland. In Sternberg SS (ed): Histology for Pathologists. New York: Raven Press, 1992, pp 321–346.
12. Cassarino DS, Quezado MM, Ghatak NR, Duray PH. Lyme-associated parkinsonism: A neuropathologic case study and review of the literature. Arch Pathol Lab Med 2003;127:1204–1206.
13. Centers for Disease Control and Prevention: Medical Examiners, Coroners, and Biologic Terrorism: A Guidebook for Surveillance and Case Management. MMWR 2004;53(No. RR-8):1–36.
14. Chandler FW. Infectious disease pathology: Morphologic and molecular approaches to diagnosis. J Histotechnol 1995;18:183–186.
15. Chow AW. Life-threatening infections of the head and neck. Clin Infect Dis 1992;14:991–1004.
16. Curless RG. Subdural empyema in infant meningitis: Diagnosis, therapy, and prognosis. Childs Nerv Syst 1985;1:211–214.
17. Dearmond SJ, Kretzschmar HA, Prusiner SB. Prion diseases. In Graham DI, Lantos PL (eds): Greenfield's Neuropathology, ed 7. London: Arnold, 2002;2:273–323.
18. Ette HY, Koffi K, Botti K, et al. Sudden death caused by parasites: Postmortem cerebral malaria discoveries in the African endemic zone. Am J Forensic Med Pathol 2002;23:202–207.
19. Fratkin JD, Leis A, Stokic DS, et al. Spinal cord neuropathology in human West Nile virus infection. Arch Pathol Lab Med 2004;128:533–537.
20. Fuller GN, Burger PC. Central nervous system. In Sternberg SS (ed): Histology for Pathologists. New York, Raven Press, 1992, pp 145–167.
21. Gelman BB, Popov V, Chaljub G, et al. Neuropathological and ultrastructural features of amebic encephalitis caused by *Sappinia diploidea*. J Neuropathol Exp Neurol 2003;62:990–998.
22. Grafe MR. Intrauterine infections. In Golden JA, Harding BN (eds): Pathology and Genetics: Developmental Neuropathology. Basel: ISN Neuropath Press, 2004, pp 360–366.
23. Graham DI, Lantos PL (eds): Greenfield's Neuropathology, ed 7. London: Arnold, 2002.
24. Gray F, Alonso J-M. Bacterial infections of the central nervous system. In Graham DI, Lantos PL (eds): Greenfield's Neuropathology, ed 7. London: Arnold, 2002;2:151–193.
25. Guarner J, Shieh W-J, Hunter S, et al. Clinicopathologic study and laboratory diagnosis of 23 cases with West Nile virus encephalomyelitis. Hum Pathol 2004;35:983–990.
26. Gyure KA, Prayson RA, Estes ML, Hall GS. Symptomatic *Mycobacterium avium* complex infection of the central nervous system: A case report and review of the literature. Arch Pathol Lab Med 1995;119:836–839.
27. Haden KH, Lester LL, Elkins SK, Cole M. Postmortem microbiology: Friend or foe? (abstract). Proc Am Acad Forensic Sci 2002;8:175.
28. Hair LS, Nuovo G, Powers JM, et al. Progressive multifocal leukoencephalopathy in patients with human immunodeficiency virus. Hum Pathol 1992;23:663–667.
29. Hoang MP, Ross KF, Dawson DB, et al. Human herpesvirus-6 and sudden death in infancy: Report of a case and review of the literature. J Forensic Sci 1999;44:432–437.
30. Hove M, Pencil SD. Effect of postmortem sampling technique on the clinical significance of autopsy blood cultures. Hum Pathol 1998;29:137–139.
31. Iwamoto M, Jernigan DB, Guasch A, et al. Transmission of West Nile virus from an organ donor to four transplant recipients. N Engl J Med 2003;348:2196–2203.
32. Jackson AC, Ye H, Phelan CC, et al. Extraneural organ involvement in human rabies. Lab Invest 1999;79:945–951.
33. Jogai S, Radotra BD, Banerjee AK. Rabies viral antigen in extracranial organs: A post-mortem study. Neuropathol Appl Neurobiol 2002;28:334–338.
34. Johnson M, Valyi-Nagy T. Expanding the clinicopathologic spectrum of Herpes simplex encephalitis. Hum Pathol 1998;29:207–209.
35. Johnson RT, Gonzalez RG, Frosch MP. Case 27–2005: An 80-year-old man with fatigue, unsteady gait, and confusion. N Engl J Med 2005;353:1042–1050.
36. Johnson RT. Viral Infections of the Nervous System. Philadelphia: Lippincott-Raven, 1998.
37. Jung S, Schelper RL, Visvesvara GS, Chang HT. *Balamuthia mandrillaris* meningoencephalitis in an immunocompetent patient: An unusual clinical course and a favorable outcome. Arch Pathol Lab Med 2004;128:466–468.
38. Keohane C. Perinatal and postnatal infections. In Golden JA, Harding BN (eds): Pathology and Genetics: Developmental Neuropathology. Basel: ISN Neuropath Press, 2004, pp 367–379.
39. Kibayashi K, Mastri AR, Hirsch CS. Neuropathology of human immunodeficiency virus infection at different disease stages. Hum Pathol 1996;27:637–642.
40. Kleinschmidt-DeMasters BK, Amlie-Lefond C, Gilden DH. The patterns of Varicella zoster virus encephalitis. Hum Pathol 1996;27:927–938.
41. Knoblock RJ, Townsend JJ, Klatt EC. Pathologic quiz case: Fatal central nervous system lesions in an immunosuppressed patient. Arch Pathol Lab Med 2002;126:1247–1249.
42. Koralnik IJ, Schellingerhout D, Frosch MP. Case 14–2004: A 66-year-old man with progressive neurologic deficits. N Engl J Med 2004;350:1882–1893.
43. Lack EE. Tumors of the Adrenal Gland and Extra-Adrenal Paraganglia. Atlas of Tumor Pathology. Third Series. Fascicle 19. Washington, DC: Armed Forces Institute of Pathology, 1997.
44. Lee SC, Dickson DW, Casadevall A. Pathology of cryptococcal meningoencephalitis: Analysis of 27 patients with pathogenetic implications. Hum Pathol 1996;27:839–847.
45. Lloyd RV. Non-neoplastic pituitary lesions, including hyperplasia. In Lloyd RV (ed): Surgical Pathology of the Pituitary Gland. Major Problems in Pathology, Volume 27. Philadelphia: WB Saunders, 1993, pp 25–33.
46. Love S, Wiley CA. Viral diseases. In Graham DI, Lantos PL (eds): Greenfield's Neuropathology, ed 7. London: Arnold, 2002;2:1–105.
47. Love S, Koch P, Urbach H, Dawson TP. Chronic granulomatous Herpes simplex encephalitis in children. J Neuropathol Exp Neurol 2004;63:1173–1181.
48. Marra CM (ed). Central nervous system infections. Neurol Clin 1999;17:675–942.
49. Martinez AJ, Garcia CA, Halks-Miller M, Arce-Vela R. Granulomatous amebic encephalitis presenting as a cerebral mass lesion. Acta Neuropathol (Berl) 1980;51:85–91.
50. Masters CL, Richardson EP Jr. Subacute spongiform encephalopathy (Creutzfeldt-Jakob disease): The nature and progression of spongiform change. Brain 1978;101:333–344.
51. Morse DL. West Nile virus—Not a passing phenomenon. N Engl J Med 2003;348:2173–2174.
52. Naber SP. Molecular pathology—Diagnosis of infectious disease. N Engl J Med 1994;331:1212–1215.
53. Nichter CA, Pavlakis SG, Shaikh U, et al. Rhombencephalitis caused by West Nile fever virus. Neurology 2000;55:153.
54. Oeberst JL, Barnard JJ, Bigio EH, Prahlow JA. Neurocysticercosis. Am J Forensic Med Pathol 2002;23:31–35.

55. Ogilvy CS, Chapman PH, McGrail K. Subdural empyema complicating bacterial meningitis in a child: Enhancement of membranes with gadolinium on magnetic resonance imaging in a patient without enhancement on computed tomography. Surg Neurol 1992;37:138–141.
56. Peoc'h MY, Gyure KA, Morrison AL. Postmortem diagnosis of cerebral malaria. Am J Forensic Med Pathol 2000;21:366–369.
57. Peterson LR, Epstein JS. Problem solved? West Nile virus and transfusion safety. N Engl J Med 2005;353:516–517.
58. Peterson LR, Hayes EB. Westward Ho?—The spread of West Nile virus. N Engl J Med 2004;351:2257–2259.
59. Procop GW, Wilson M. Infectious disease pathology. Clin Infect Dis 2001;32:1589–1601.
60. Reefhuis J, Honein MA, Whitney CG, et al. Risk of bacterial meningitis in children with cochlear implants. N Engl J Med 2003; 349:435–445.
61. Reefhuis J, Mann EA, Whitney CG. Bacterial meningitis in children with cochlear implants (letter to editor). Author's reply. N Engl J Med 2003;349:1772–1773.
62. Roy S, Ellenbogen JM. Seizures, frontal lobe mass, and remote history of periodontal abscess. Arch Pathol Lab Med 2005;129: 805–806.
63. Schiff D, Rosenblum MK. Herpes simplex encephalitis (HSE) and the immunocompromised: A clinical and autopsy study of HSE in the settings of cancer and human immunodeficiency virus-type 1 infection. Hum Pathol 1998;29:215–222.
64. Schwartz DA. Emerging and reemerging infections: Progress and challenges in the subspecialty of infectious disease pathology. Arch Pathol Lab Med 1997;121;776–784.
65. Schwartz DA, Bryan RT, Hughes JM. Pathology and emerging infections—Quo vadimus? Am J Pathol 1995;147:1525–1533.
66. Sobrinho-Simões M, Brandão A, Paiva ME, et al. Lymphoid hypophysitis in a patient with lymphoid thyroiditis, lymphoid adrenalitis, and idiopathic retroperitoneal fibrosis. Arch Pathol Lab Med 1985;109;230–233.
67. Solomon T. Current concepts. Flavivirus encephalitis. N Engl J Med 2004;351:370–378.
68. Sparks AK, Naefie RC, Connor DH. Coenurosis. In Binford CH, Conner DH (eds): Pathology of Tropical and Extraordinary Diseases. Washington, DC: Armed Forces Institute of Pathology, 1976;2(Section 11):543–545.
69. Turkel SB, Itabashi HH. The natural history of neuroblastic cells in the fetal adrenal gland. Am J Pathol 1974;76:225–244.
70. Turner G, Scaravilli F. Parasitic and fungal diseases. In Graham DI, Lantos PL (eds): Greenfield's Neuropathology, ed 7. London: Arnold, 2002;2:107–150.
71. Watts JC. Surgical pathology and the diagnosis of infectious diseases (editorial). Am J Clin Pathol 1994;102:711–712.
72. Watts JC, Chandler FW. The surgical pathologist's role in the diagnosis of infectious diseases. J Histotechnol 1995;18:191–193.
73. Wilson SJ, Wilson ML, Reller LB. Diagnostic utility of postmortem blood cultures. Arch Pathol Lab Med 1993;117:986–988.
74. Woods CR Jr. Brain abscess and other intracranial suppurative complications. Adv Pediatr Infect Dis 1995;10:41–79.
75. Yorke RF, Rouah E. Nocardiosis with brain abscesses due to an unusual species, *Nocardia transvalensis*. Arch Pathol Lab Med 2003;127:224–226.

Brain Tumors

13

INTRODUCTION 359
ASTROCYTIC TUMORS 359
OLIGODENDROGLIOMA 370
EPENDYMOMA 370
MEDULLOBLASTOMA 374
GANGLIOGLIOMA 375
MIXED GLIOMAS 377
MENINGIOMA 377
SCHWANNOMAS 382
TUMORS OF MALDEVELOPMENT 383
CHORDOMA 386
ARACHNOID CYST 387
LIPOMA 388
PITUITARY TUMORS 388
METASTATIC NEOPLASMS 390
PRIMARY MALIGNANT LYMPHOMAS 391
EXTRAMURAL METASTASIS 393
CAUSES OF BRAIN TUMOR 393
REFERENCES 393

Introduction

Brain neoplasms are sometimes seen in cases referred to the medical examiner, and they generally fall into the following circumstances: sudden unexpected death,[9] diagnostic failure (oversight), therapeutic misadventure, failure to recognize postoperative complications, and patient's refusal of therapy, among other unusual special situations. Some tumors are incidental findings, for example, small meningiomas, lipomas, and subependymomas. The incidence of the types of brain tumors encountered, in our experience, is probably not much different from that of published large series of the past, such as those of Courville,[8] Zimmerman,[49] Zülch,[50] and Graham and Lantos,[15] among others. Though all forms of primary gliomas have been seen, meningiomas and metastatic neoplasms are more frequent. Most of the remaining tumors are encountered uncommonly.

Astrocytic Tumors

Astrocytic neoplasms are, by far, the most common glial tumors, accounting for more than 60% of all primary brain tumors.[26] Although sharing a common cell of origin, astrocytomas are highly variable in their gross and microscopic appearance, propensity for anatomic sites, and, thereby, their clinical expression. Based on histopathologic features, astrocytomas are graded from I to IV according to the World Health Organization (WHO) classification.[26] Low-grade diffuse astrocytomas are designated grade II; anaplastic astrocytomas, grade III; and glioblastoma multiforme, grade IV. The pilocytic astrocytomas, with their restricted anatomic sites and indolent behavior, are grade I.[26]

For the clinical presentations of astrocytomas of the various grades, the reader is referred to any number of excellent textbooks on neurology and neurosurgery.

Diffuse Astrocytomas

The diffuse astrocytomas (Grade II, WHO) make up 10 to 15% of all astrocytic brain tumors. They occur in all age groups, with a mean age of 34 years, 70% occurring between 20 and 45 years of age.[26] Frontal and temporal lobes of the cerebrum are favored sites followed by the brainstem and spinal cord, whereas a cerebellar location is uncommon. Due to its diffusely infiltrative nature, this tumor has imperceptible boundaries on gross examination, and microscopy is required for determination of its true extent (Case 13.1, Figs. 13.1–13.2). Its characteristic gross appearance is a pale uniform expansion of tissue with blurring of gray-white anatomic borders (Case 13.2, Figs. 13.3–13.4). The tumor texture may vary from firm and rubbery to gelatinous. Where infiltrated by tumor, anatomic features of such areas as the thalamus and basal ganglia may be generally retained except for their enlargement. Low-grade astrocytomas are generally nonhemorrhagic, but biopsy sometimes may lead to a false impression of a hemorrhagic tumor (Case 13.3, Figs. 13.5–13.8).

Microscopically, the tumor is characterized by a diffuse field of differentiated astrocytes of relatively low cellularity with some atypia, although mitoses are rare. Within the same grade, cytomorphology may vary considerably from fibrillary (Fig. 13.4) to protoplasmic (Figs. 13.7–13.8) to a more aesthetically pleasing gemistocytic type (Figs. 13.9–13.10). It may also be microcystic, especially with protoplasmic forms (Fig. 13.8).

Confluence of cysts may eventually result in a largely cystic tumor. Tumor cells of low-grade astrocytomas mimic reactive astrocytes, and they are betrayed by their glial processes in fibrillary types. However, cell type may not be obvious in protoplasmic forms, as processes may be inconspicuous. A positive reaction to glial fibrillary acidic protein (GFAP) immunostain is helpful in such cases.

Case 13.1. (Figs. 13.1 and 13.2)
A 31-year-old woman, who had epilepsy since the age of 7 years, fell and struck her head on the ground. Her last known seizure occurred 2 days before death. She was without complaint at 2315 hours, but she was found dead the next morning face down on a pillow.

Case 13.2. (Figs. 13.3 and 13.4)
A 38-year-old woman with a 5-month history of headaches was seen by three different physicians without a diagnosis being made. CT head scan was read as normal. She was unexpectedly found dead.

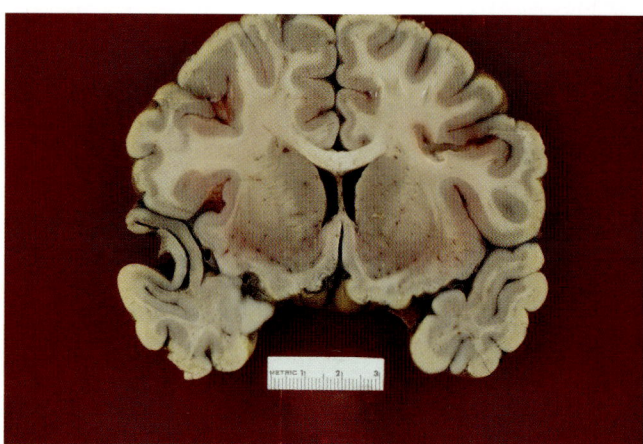

Fig. 13.1. What appears to be a chronic contusion of anterior left superior temporal gyrus is a cystic tumor.

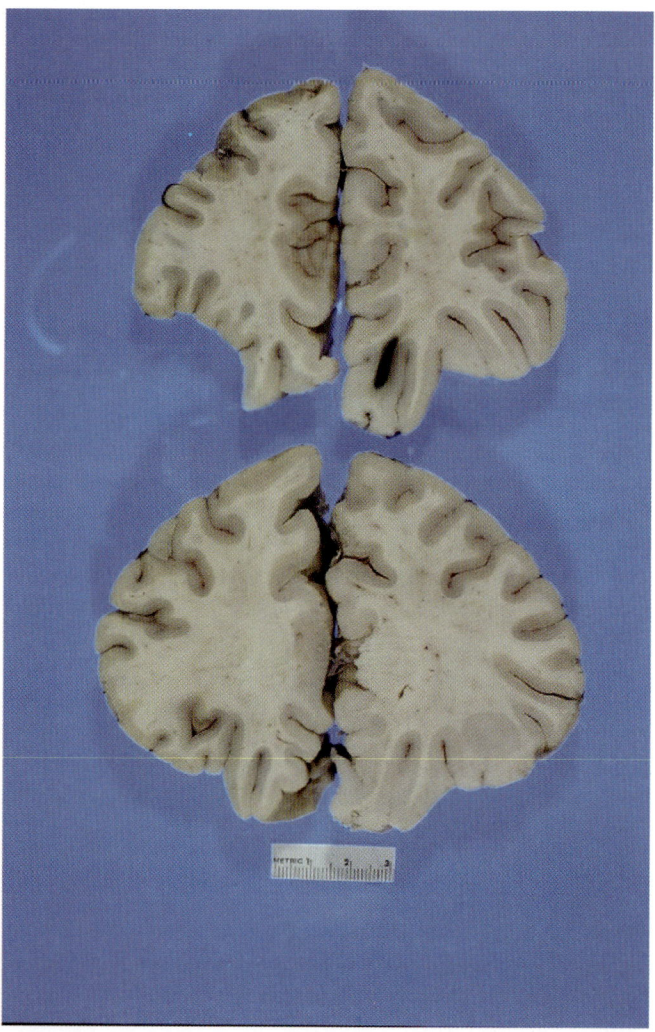

Fig. 13.3. Coronal sections of anterior frontal lobes show tumor forming focal enlargement of right rectus and orbital cortex and subcortex with a small fresh hemorrhage. Margins suggestive of a neoplasm are imperceptible.

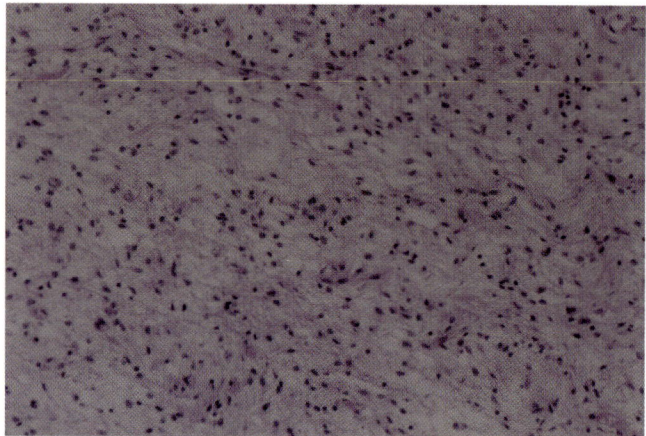

Fig. 13.2. Medium-power micrograph shows relatively sparsely cellular field of small tumor astrocytes. (H&E.)

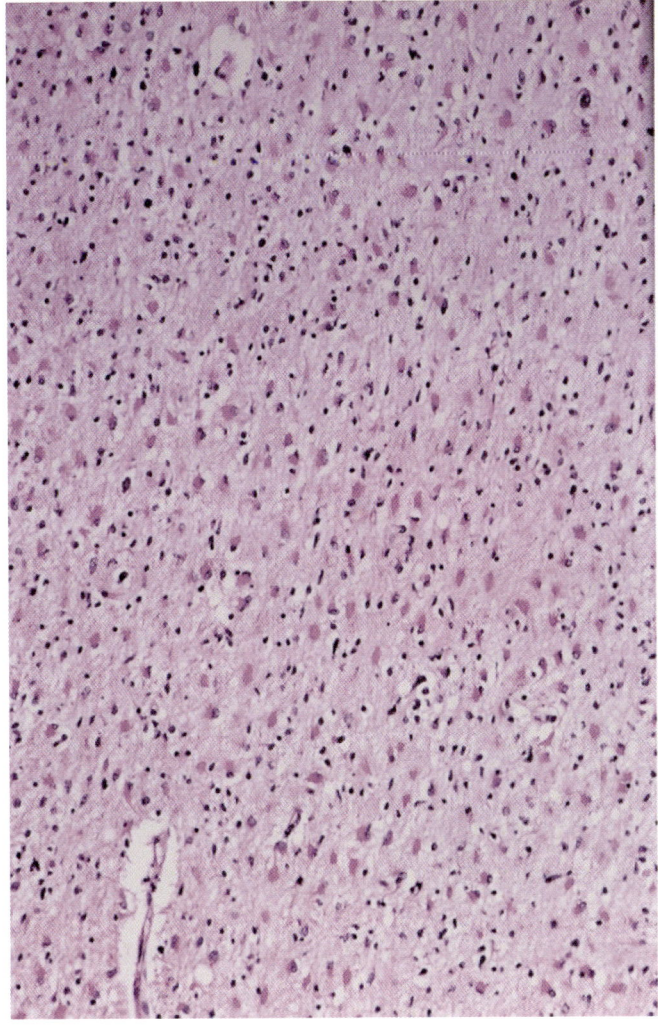

Fig. 13.4. Medium-power micrograph of same lesion as in Fig. 13.3 shows a diffuse uniform field of small fibrillary tumor astrocytes. (H&E.)

Case 13.3. (Figs. 13.5–13.8)

A 52-year-old man, while in custody for robbery, developed seizures and was discovered to have a brain tumor. Still in custody in the hospital while receiving chemotherapy, he became involved in a struggle with correctional officers, collapsed, became unresponsive, and died at the scene.

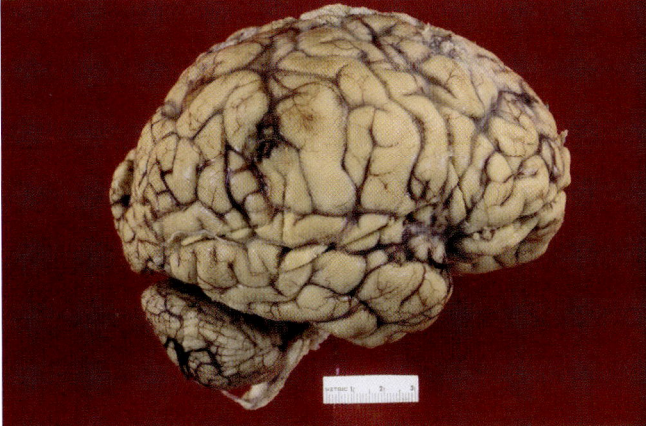

Fig. 13.5. Lateral view of right cerebral hemisphere shows broad, swollen, and flattened cortex of paracentral gyri and parietal lobe. The small hemorrhagic lesion in supramarginal gyrus is a biopsy site.

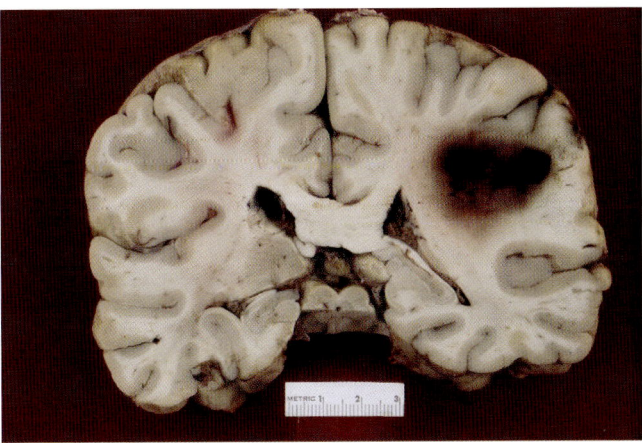

Fig. 13.6. Coronal section at the splenium shows a hemorrhagic tumor whose margins are indefinable. Hemorrhage may be from biopsy.

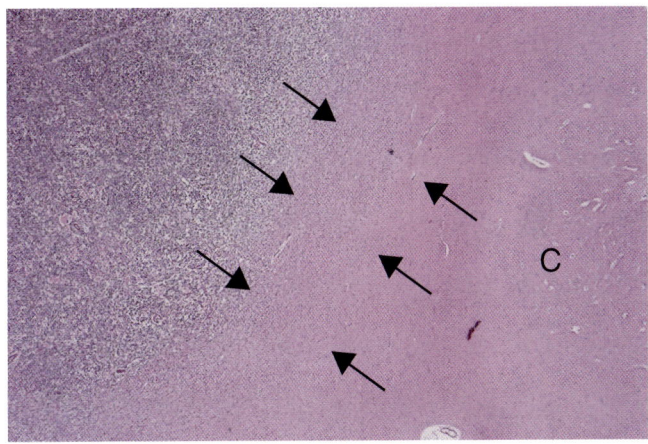

Fig. 13.7. Low-power micrograph shows densely cellular basophilic tumor to the left and less cellular infiltrating tumor (*arrows*) to the right. C, cortex. (H&E.)

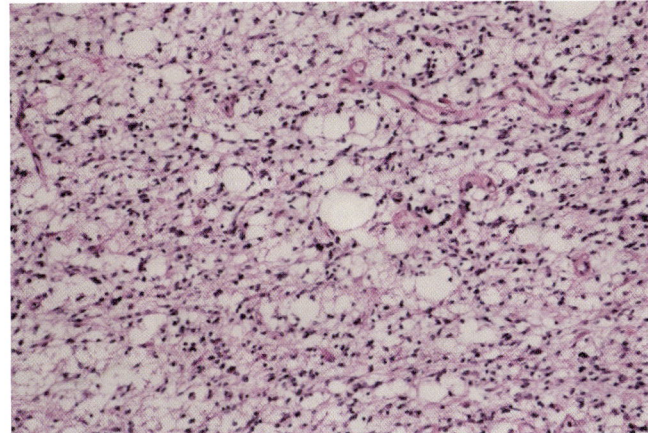

Fig. 13.8. Medium-power micrograph of protoplasmic astrocytoma showing microcystic degeneration. (H&E.)

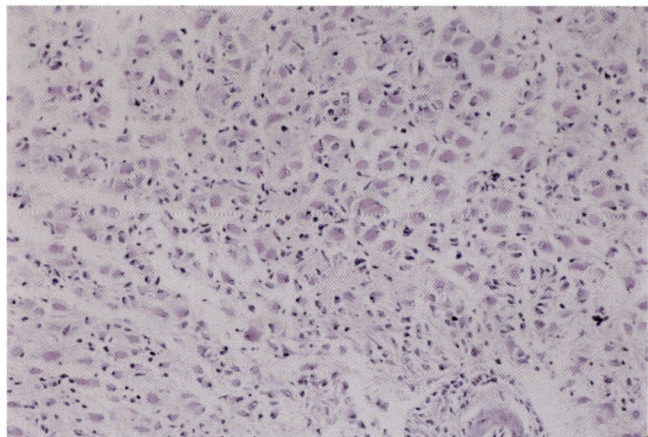

Fig. 13.9. Medium-power micrograph of gemistocytic astrocytoma shows a moderately cellular tumor composed of plump cells with homogeneous "glassy" acidophilic cytoplasm and eccentric nuclei, lying in loose, finely fibrillar glial matrix. (H&E.)

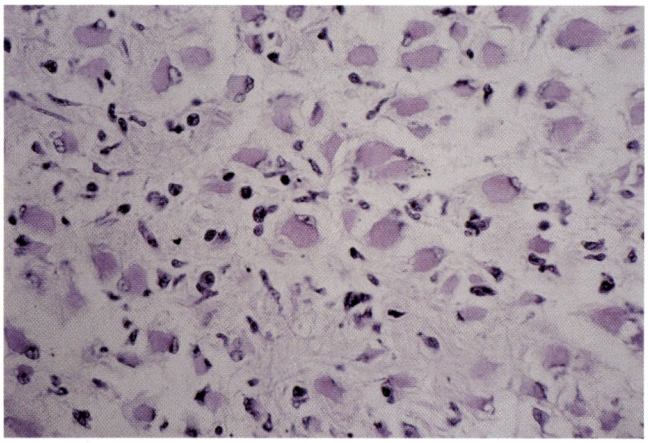

Fig. 13.10. Higher-power micrograph of same field as in Fig. 13.9. (H&E.)

Case 13.4. (Fig. 13.11)
A 10-year-old child was reportedly well until she complained of a severe headache 3 days prior to death. She was taken to a clinic where a physician prescribed analgesics and antibiotics and sent her home. In the afternoon of the day prior to her death, she was again seen by a physician and told to continue the medications. She was found dead the following morning, about 5 hours after last being seen alive.

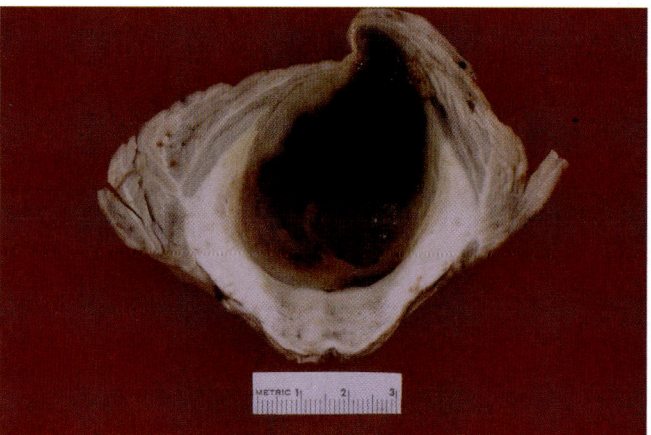

Fig. 13.11. Large discrete cystic tumor with fresh hemorrhage shows a rim of gray-tan semitranslucent tissue replacing the cerebellar vermis. The tumor and hemorrhage fill and massively expand the fourth ventricle, with severe compression of the floor of pons. Recent cerebellar tonsillar herniation was present. Microscopically, the tumor was a pilocytic astrocytoma.

Pilocytic Astrocytoma
Pilocytic astrocytoma (Grade I, WHO) is the most common glioma in children. These tumors have preferential anatomic localization in such areas as the optic nerve, optic chiasm, hypothalamus, brainstem, spinal cord, and cerebellum, with the greatest number occurring in the cerebellum (see Chapter 9). Some may present in the basal ganglia and thalamus. Grossly, the tumor is soft to gelatinous and remains relatively circumscribed (Case 13.4, Fig. 13.11). It has a tendency to form large cysts lined by tumor, characteristically in the cerebrum and cerebellum. Another cystic tumor of the cerebellum, hemangioblastoma, may present a similar appearance on gross examination. With astrocytoma, the entire cyst wall may be lined by neoplasm. However, with hemangioblastoma, tumor is limited to a mural nodule, with

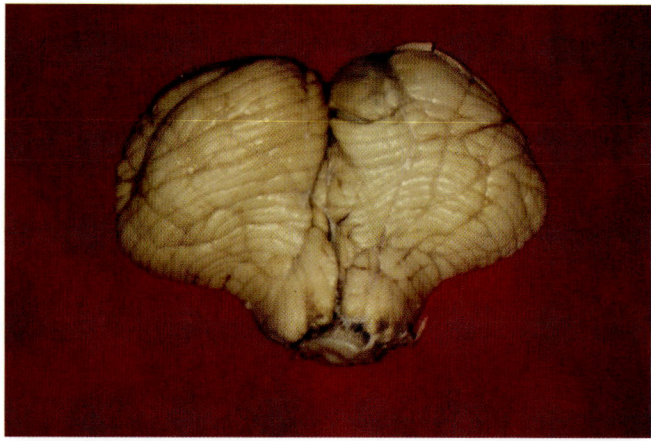

Fig. 13.12. Cerebellar hemangioblastoma presents at surface of right medial hemisphere effacing folia. Note molded tonsils and biventers around medulla.

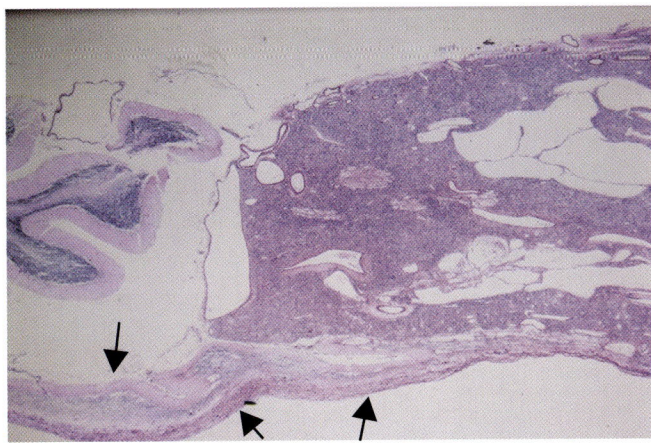

Fig. 13.13. Low-power photomicrograph shows markedly attenuated and atrophic folia at the bottom (*arrows*), above which is partly cystic hemangioblastoma forming a mural nodule. (H&E.)

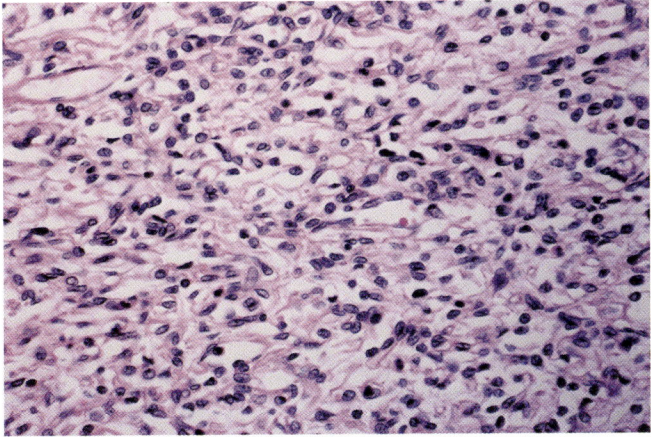

Fig. 13.14. High-power photo of tumor shows a cellular variant of hemangioblastoma. (H&E.)

the remainder of the cyst wall free of tumor (Figs. 13.12–13.14).

Microscopically, pilocytic astrocytoma exhibits a biphasic pattern of isomorphic alignment of bipolar piloid cells accompanied by Rosenthal fibers, on the one hand, and a loose microcystic area of multipolar cells on the other. GFAP stain is not helpful toward identifying Rosenthal fibers since the fibers are GFAP negative. Occasional microvascular glomeruloid formation may be found in the tumor, but such findings are not to be misinterpreted as indication of a higher-grade neoplasm. Although usually confined within the pial barrier, pilocytic astrocytomas may become exophytic, such as in pontine gliomas, and spread along the subarachnoid space can occur.

Anaplastic Astrocytoma

Anaplastic astrocytoma (Grade III, WHO) occurs mostly in an age group between those of grade II and grade IV tumors, with the average age at diagnosis being around 40 years.[26] Although anaplastic astrocytoma may be the initial presentation, in many instances it may result from malignant transformation of a previously low-grade tumor. Surgical biopsy diagnoses may be misleading, as it may be a fortuitous nonrepresentative sampling of a low-grade area in an already high-grade tumor.

Grossly, tumor margins are only slightly more distinguishable in sectors characterized by higher cellularity, vascularity, and hemorrhage that add a gray to reddish-tan color. Infiltration at the margins may still be indistict and impossible to define grossly, but ill-defined massive expansion of involved structures draws attention to the tumor's location. For the size of the tumor, mass effect may be disproportionately greater than that seen with low-grade tumors due to presumed greater vasogenic edema. White matter surrounding the tumor may take on a faint yellow to greenish cast from breakdown of the blood–brain barrier, chronic edema, and accompanying gliosis.

Microscopically, central features of anaplasia are hypercellularity clearly exceeding that seen with grade II astrocytomas, pleomorphism, hyperchromatic nuclei, and frequent mitoses with aberrant forms. Focal necrosis, when present, is small.

Glioblastoma Multiforme

Glioblastoma multiforme (Grade IV, WHO) is a highly malignant, rapidly growing neoplasm. It constitutes about 12 to 15% of all intracranial tumors and approximately 50 to 60% of all astrocytic neoplasms.[26] It is seen in an older age group with the peak incidence between 45 and 75 years, 70% falling within this range.[26] In one study, 8.8% of glioblastoma multiforme was found in children.[11] In adults, glioblastoma is a tumor of the cerebrum involving white matter and the overlying cortex of all lobes, but with preference for the temporal and frontoparietal regions. Deep infiltration of the basal ganglia and spread to the opposite hemisphere through the corpus callosum is common with tumors arising in the frontal and parietal lobes. The clinical course is usually short, measured in months, with seizures, symptoms of increasing intracranial pressure, and rapidly progressing neurologic signs.

Gross examination reveals asymmetrically swollen cerebral hemispheres with flattened cortical surfaces and midline shift. Focal broad expansion of gyri may indicate the tumor's location (Case 13.5, Figs. 13.15–13.16). At the base of the brain, the uncus ipsilateral to the larger hemisphere may be herniated along with the parahippocampal gyrus, with shift and rotation of the upper brainstem (Fig. 13.16). Subfalcine cingulate herniation and herniation of the posterior orbital cortex over the margin of the lesser wing of the sphenoid bone may be present when tumor mass effect is primarily in a frontal location

(Case 13.6, Fig. 13.17). On section, the tumor typically displays broad variegated areas of mixed yellow to dark red to brown foci reflecting deposits of myelin debris, macrophagic infiltration, necrosis, and various stages of hemorrhage (Fig. 13.17). Tumor margins may be deceptively sharp but are more extensively infiltrating than appears. Combined tumor bulk and edema almost always cause significant mass effect. Infiltrated gray matter such as the cortex and basal ganglia may be enlarged (Case 13.7, Figs. 13.18–13.25). In some cases, a large hemorrhage within necrotic zones may precipitate the terminal event (Fig. 13.17). In addition to spread through the corpus callosum (Fig. 13.20), invasive tumor may be grossly evident along tracts such as the fornices, posterior limb of the internal capsule, and the cerebral peduncles. Metastasis may occur within the subarachnoid space,[12] and rare distant extraneural metastases to lung, bone, and other sites are known.[38] The latter usually follow, but occasionally antedate, surgical intervention.

Histopathologic features are extremely variable in the individual participating cells, their growth patterns, vascular response, incorporation and participation of nonglial elements, and mode of spread. Typical features include areas of densely cellular undifferentiated small cells, hyperchromatic nuclei, and sparse cytoplasm (Fig. 13.21). Mitoses are present but may be difficult to detect in postmortem material. Tumor cells stacked along margins of irregular areas of necrosis produce characteristic pseudopalisades (Figs. 13.22–13.23). Another important diagnostically supportive finding is a proliferative vascular change called glomeruloid formation, or vascular endothelial proliferation, usually found as several clusters aligned adjacent to areas of necrosis (Figs. 13.24–13.25). Areas within and adjacent to undifferentiated cells may contain fields of a relatively pure population of tumor cells readily recognizable as astrocytic, such as foci of gemistocytes, fibrillary astrocytes, and monster giant cells with bizarre multinucleated and hyperchromatic nuclei and expansive cytoplasm. Gemistocytes are plump cells with eccentric nuclei and a moderate amount of homogeneous pale-pink cytoplasm and sparse blunt process (see Figs. 13.9–13.10). Astrocytomas and other glial neoplasms occasionally track along certain planes such as the subpial, perivascular, perineurial, perifascicular, and/or subependymal spaces, forming so-called

> **Case 13.5.** (Figs. 13.15 and 13.16)
>
> A 32-year-old female accepted no medical care, due to her religious beliefs, despite a severe seizure disorder of over 7 years' duration. A progressive neurologic syndrome began 1 year before death characterized by right facial numbness and "impairment" of the right side of her body, progressing to difficulty in swallowing. After a short period of gasping respirations, she was found dead at home.

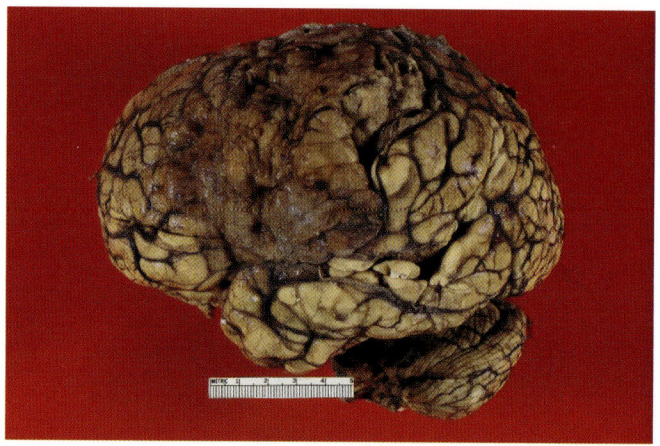

Fig. 13.15. Glioblastoma multiforme involving the surface of brain distorts the left paracentral cortex.

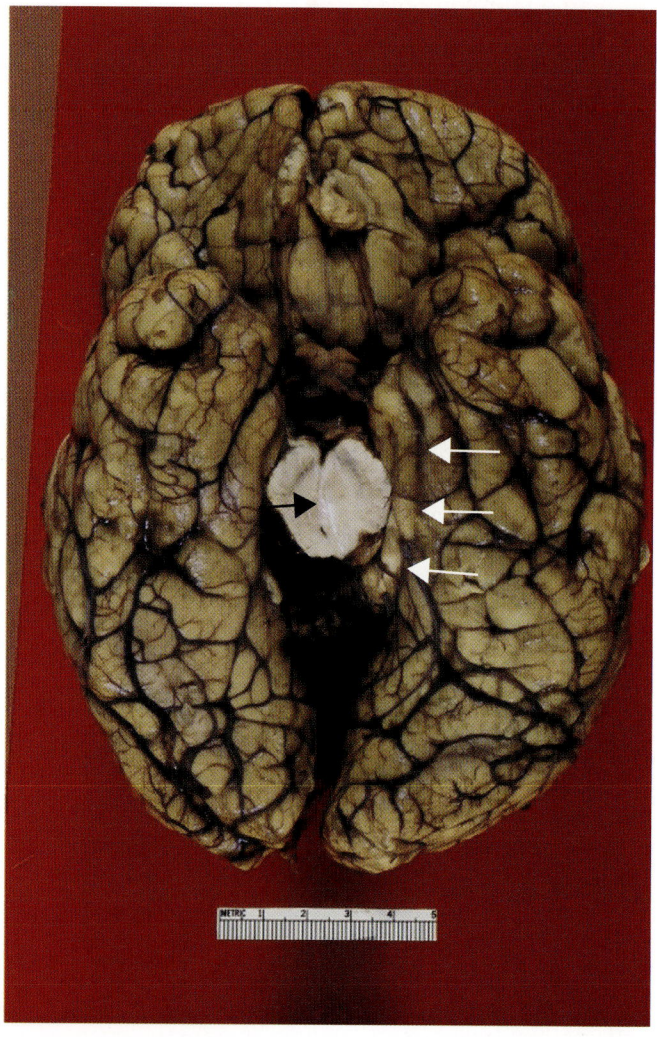

Fig. 13.16. Base of brain shows a left uncal-parahippocampal herniation (*white arrows*) accompanied by asymmetric midbrain, with midline bowed to the contralateral side (*black arrow*). Despite the shift, note that contralateral half of midbrain is narrower (due to compression against the margin of tentorium).

Case 13.6. (Fig. 13.17)

A 51-year-old woman slipped and fell in the bathroom, striking her head, about 3½ months before death. Following the fall, she lost some mobility, and, in succeeding months, lost her ability to manage her life and became cachectic. She was placed in a sanitarium for Christian Scientists and received nursing care. Two months later, she was found in bed without signs of life.

Case 13.7. (Figs. 13.18–13.25)

A 34-year-old woman had a history of head trauma sustained in an auto accident 15 years before, resulting in coma lasting 3 to 4 days. No information was available as to neurologic impairment. Five years before death, she developed seizures attributed to a brain tumor. An exploratory craniotomy was performed. Shortly before death, the decedent exhibited unusual behavior, such as sitting on the toilet for 3 hours, watching a water faucet drip, and staring at a roadmap for 5 hours. While sleeping in a motel room, she started screaming out names. When an attempt was made to awaken her a few hours later, she was unresponsive, and paramedics pronounced her dead at the scene.

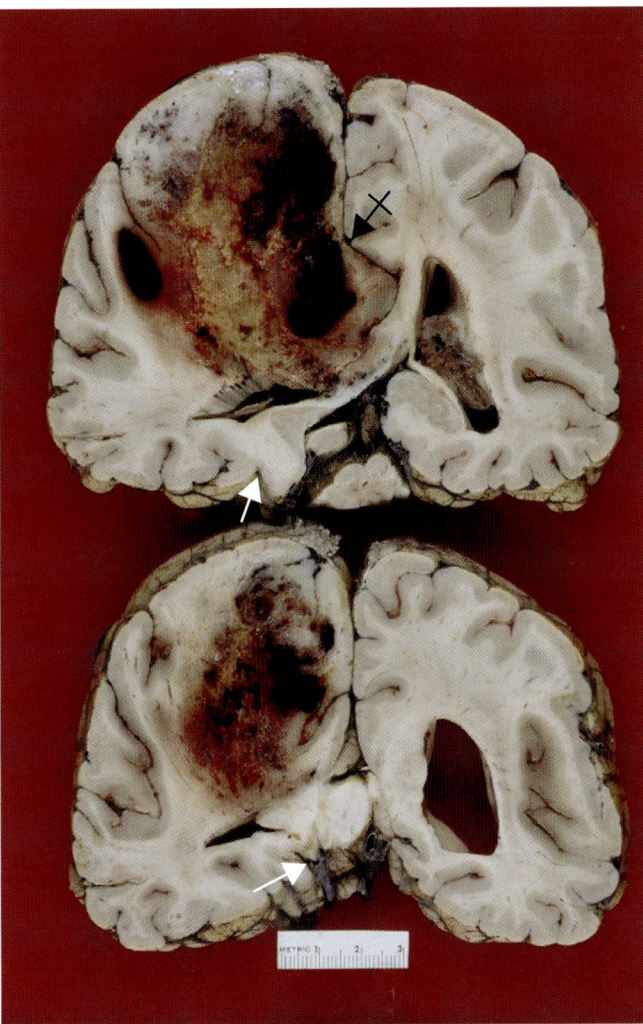

Fig. 13.17. Parietal coronal sections show a huge necrotic and hemorrhagic tumor causing downward displacement of splenium of corpus callosum. Sharp indentation of uncal herniation is noted on the left (*white arrows*). Note subfalcial cingulate herniation exaggerated by tumor at level of splenium (*crossed black arrow*). Glioblastoma multiforme.

Scherer's secondary structures.[43] Infrequently, islands within glioblastoma composed of isomorphically aligned groups of larger spindle cells are found, considered to be sarcomatous and originating from the vascular mesenchyme, termed gliosarcoma. Cases of glioblastoma currently invariably have a fatal outcome.

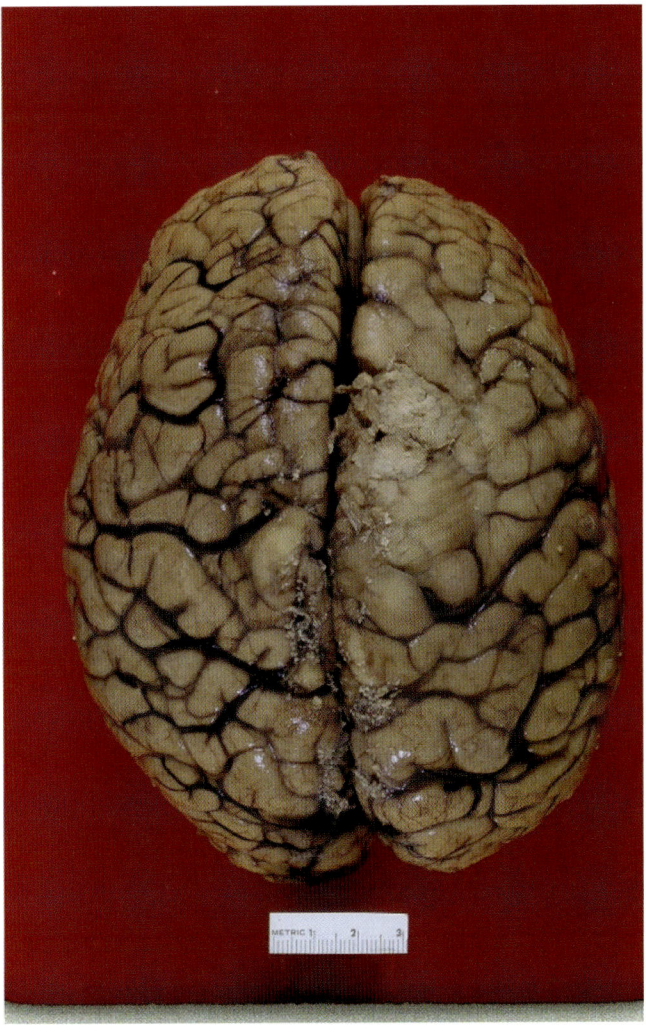

Fig. 13.18. Vertex view of cerebrum harboring glioblastoma multiforme shows flattening of cortical surfaces without noticeable asymmetry.

Brain Tumors

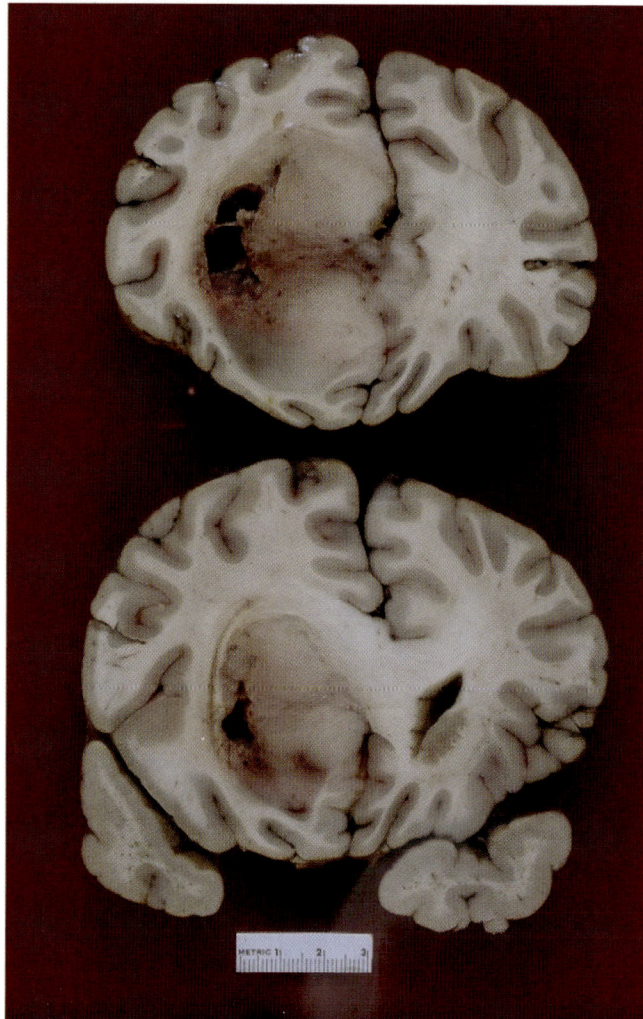

Fig. 13.19. Frontal coronal sections at genu and pregenu levels show a large tumor occupying white matter and filling the frontal horn. The cavity along the lateral margin of the tumor is postmortem artifact left by dropout of necrotic tumor. The relatively discrete margins of the tumor belie diffuse infiltration of surrounding white matter.

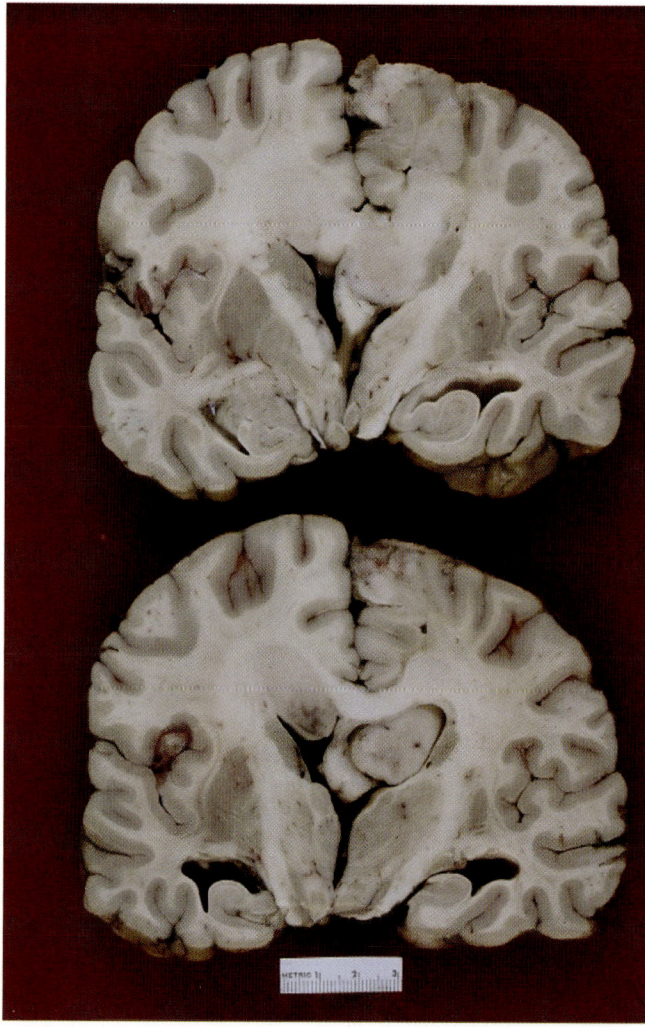

Fig. 13.20. Nodular tumor expansion of corpus callosum extends posteriorly as intraventricular growth.

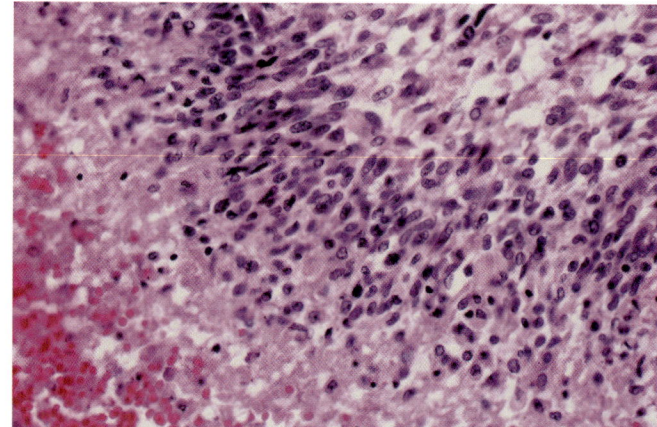

Fig. 13.21. High-power micrograph of glioblastoma multiforme shows small, ovoid, undifferentiated tumor cells with hyperchromatic nuclei at margin of necrosis. (H&E.)

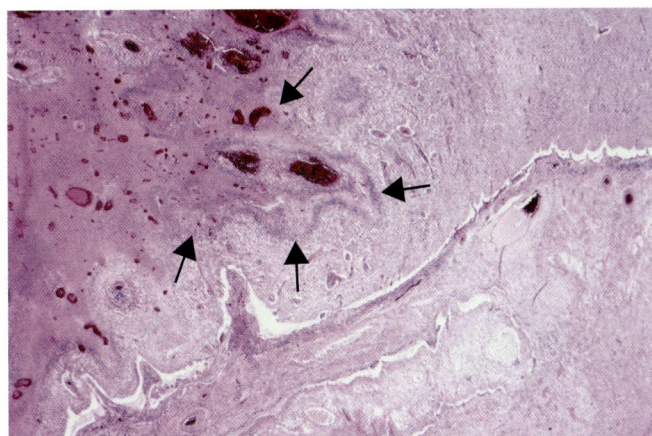

Fig. 13.22. Low-power micrograph of pseudopalisading shows a convoluted dense line of undifferentiated cells aligned along margin of necrosis (*arrows*). (H&E.)

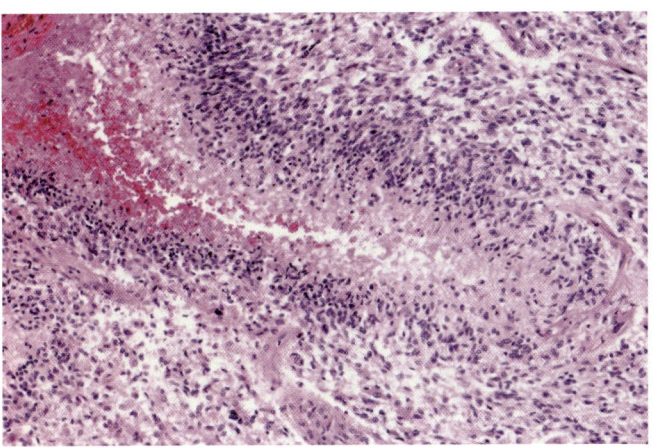

Fig. 13.23. Medium-power micrograph shows detail of pseudopalisading of tumor cells at the margin of necrosis. (H&E.)

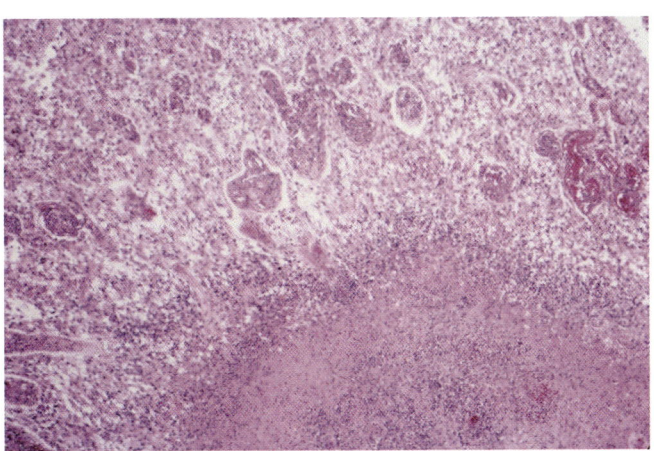

Fig. 13.24. Medium-power micrograph shows characteristic glomeruloid microvascular proliferation of glioblastoma multiforme at margin of necrosis. (H&E.)

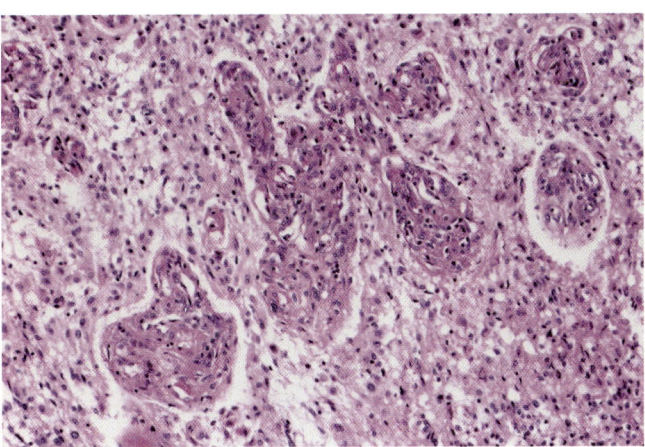

Fig. 13.25. High-power micrograph of the same field as in Fig. 3.24 shows detail of glomeruloid vascular change. (H&E.)

Pleomorphic Xanthoastrocytoma

Pleomorphic xanthoastrocytoma, described by Kepes et al.,[25] is a relatively circumscribed tumor occurring in childhood and in young adults, preferentially involving the superficial cortex of the temporal and parietal lobes. As originally noted, overlying leptomeninges may be locally infiltrated, and distant seeding is known.[26,36] Subjects usually have a relatively favorable prognosis due to the tumor's circumscribed growth pattern.

On section, the tumor is grossly variegated in color with yellow zones reflecting lipidized areas. The tumor is usually solid, but it may be partly cystic, containing xanthochromic serous fluid. Microscopically, tumor cells form mixed fields of large plump to multinucleated giant astrocytes with eccentric nuclei and fields of interlacing elongated fibrillar forms (Figs. 13.26–13.27). In areas, cells may be lipidized giving rise to their yellow color (Fig. 13.27). Silver stain will unveil an extensive reticulin

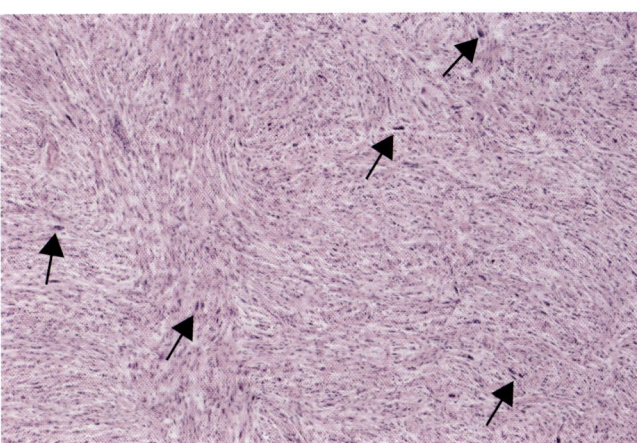

Fig. 13.26. Low-power micrograph of this case of pleomorphic xanthoastrocytoma shows a compact field of interlacing fibrillary astrocytes interspersed with large forms (*arrows*). (H&E.)

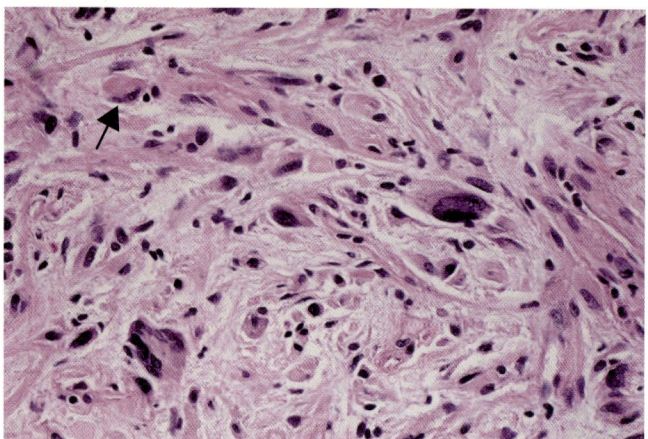

Fig. 13.27. High-power micrograph of a site pointed out at low power in Fig. 13.26 shows giant and multinucleated forms of astrocytes and a lipidized cell (arrow). (H&E.)

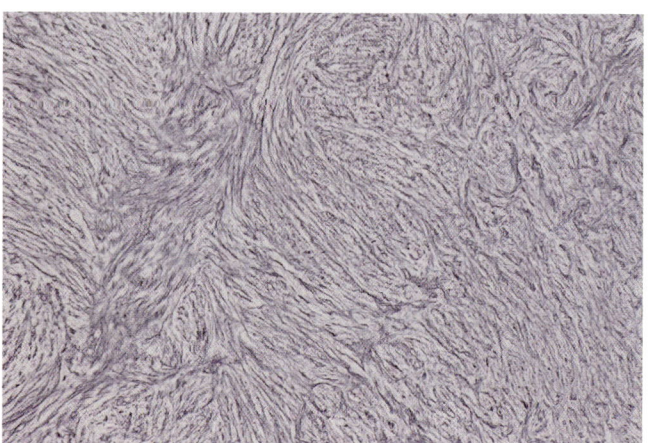

Fig. 13.28. Low-power micrograph of Wilder-stained section shows an intense reticulin deposition paralleling interlacing pattern seen on H&E stain in Fig. 13.26.

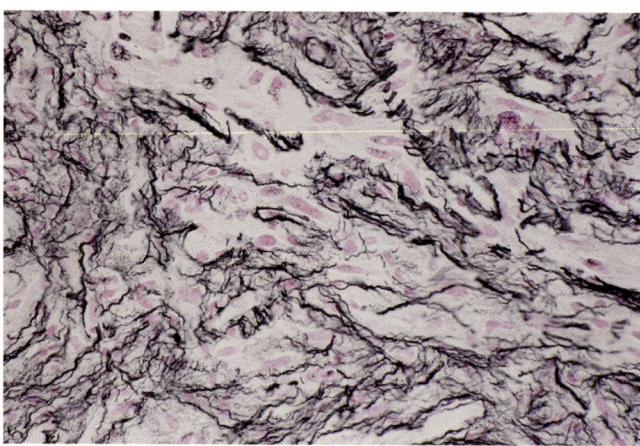

Fig. 13.29. Higher-power micrograph of this pleomorphic xanthoastrocytoma shows reticulin closely encircling tumor cells. (Wilder.)

network that often outlines individual cells (Figs. 13.28–13.29). Mitoses are rare. Highly vascularized forms have been described.

Subependymoma

Subependymoma is one of the more commonly encountered brain tumors in our office, usually as an incidental finding. Most are found in the fourth ventricle, and occasional ones occur on the lateral ventricular surface close to the interventricular foramina (Case 13.8, Fig. 13.30). They are found primarily in middle and older age groups. Large tumors fill and obstruct the fourth ventricle or occlude the interventricular foramen to produce symptomatic hydrocephalus.

Grossly, the tumor forms a firm, uniformly pale, solid, sometimes gritty, nodular or multinodular growth (Fig. 13.31). Cysts and hemorrhage in the tumor may be present but are infrequent. Larger tumors of the fourth ventricle sometimes extend through the lateral recess to appear at the cerebellopontine angle, mimicking an acoustic schwannoma, or grow out through the roof of the fourth ventricle (foramen of Magendie) to appear in the cisterna magna (Fig. 13.31). Smaller tumors lie loosely in the fourth ventricle, attached at the rhombic lip (located at the base of the attachment of the choroid plexus at the lateral margin of the fourth ventricle) (Fig. 13.32). Larger tumors flatten the floor of the fourth ventricle and compress floor nuclei, such as the dorsal vagal, hypoglossal, and vestibular nuclei (Fig. 13.33).

Microscopically, clustered groups of subependymal tumor cells, similar to those normally found buried in glia at the base of the attachment of the choroid plexus at the rhombic lip of the fourth ventricle, are distributed in a very dense fibrillary glial matrix (Fig. 13.34). Once encountered, it is a pattern not easily forgotten. Mitoses

Case 13.8. (Fig. 13.30)

A 67-year-old woman with Alzheimer's disease was found slumped to the floor. She was last seen alive seated in an adjacent chair.

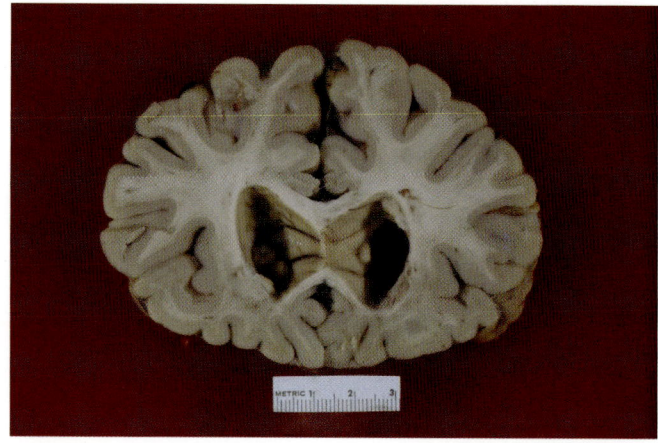

Fig. 13.30. Multiple small subependymomas on the walls of frontal horns of lateral ventricles measure about up to 4 mm in diameter. The tumors were incidental findings.

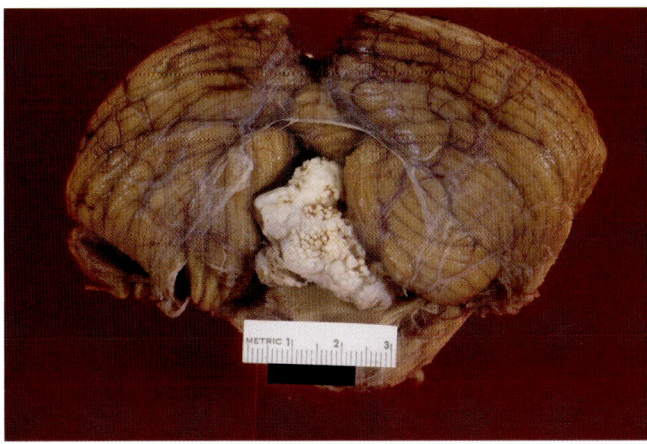

Fig. 13.31. Inferior view of cerebellum shows an unusually large white irregular tumor with verrucous surface expanding cisterna magna and laterally displacing biventers. Microscopically, it was a subependymoma.

are absent to rare. Calcification may be present in the tumor (Fig. 13.35).

Gliomatosis Cerebri

Gliomatosis cerebri is a rare, diffusely and widely infiltrating neoplasm of uncertain histogenesis, but most show features of anaplastic astrocytoma.[7] Most occur in the middle decades of life, but notable cases have been reported in the first and second decades.[42] By definition, the tumor involves more than two cerebral lobes and may continue into the brainstem and beyond, and even into the subarachnoid space. The tumor regularly crosses the midline. Imaging studies would suggest existence of separate multicentric sites, but microscopic connectivity cannot be entirely ruled out by these techniques. The tumor expands affected anatomy with imperceptible margins, similar to diffuse low-grade astrocytoma, except for its much wider extent. Expectedly, prognosis is poor.

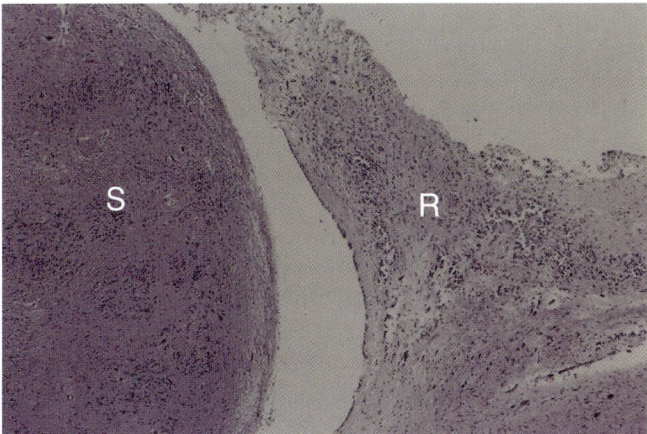

Fig. 13.32. Low-power micrograph of subependymoma (S) is shown just adjacent to rhombic lip (R) of fourth ventricle. Note similarity of tumor and tissue of rhombic lip with its small clusters of ependymal cells in a glial matrix. (H&E.)

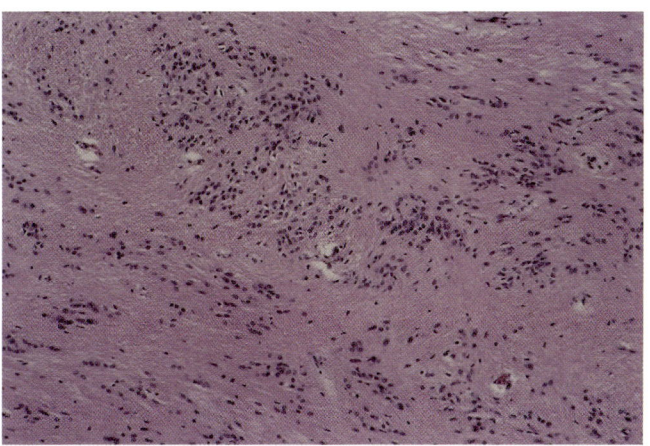

Fig. 13.34. Medium-power micrograph shows detail of subependymoma with clusters of ependymal cells in dense glial matrix. (H&E.)

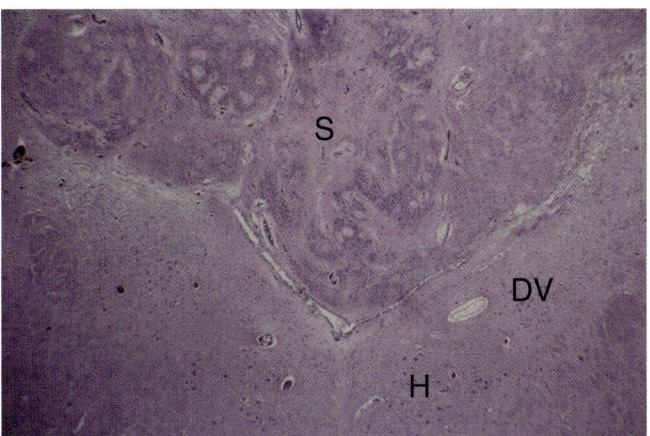

Fig. 13.33. Low-power micrograph shows a subependymoma (S) compressing floor nuclei of fourth ventricle. H, hypoglossal nucleus; DV, dorsal vagal nucleus. (H&E.)

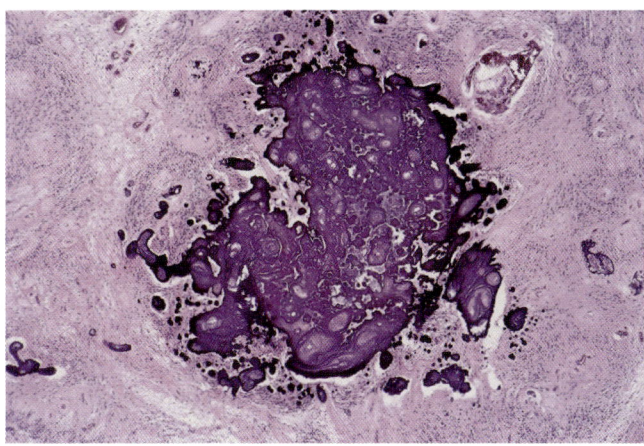

Fig. 13.35. Low-power micrograph shows an atypically large focus of calcification within the tumor. (H&E.)

Oligodendroglioma

Oligodendrogliomas make up approximately 4 to 5% of all intracranial neoplasms and 10 to 17% of all intracranial gliomas.[37] The peak incidence is in the fifth and sixth decades, and about 6% occur in infancy and childhood.[37] White matter of cerebral hemisphere, preferentially in the frontal lobe, is the usual site of origin. Infiltration of the overlying cortex and subarachnoid space is not uncommon. The tumor is found in the brainstem, cerebellum, and spinal cord much less frequently. Seizures may herald presence of the tumor. Externally, asymmetric swelling of the cerebral hemispheres usually gives a clue to the presence of underlying tumor (Fig. 13.36), and herniation may or may not be present depending on the

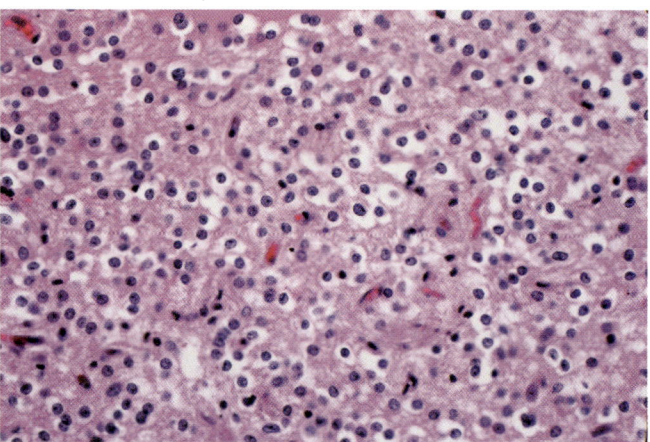

Fig. 13.37. High-power micrograph of sample taken from margin of cystic cavity of case seen in Fig. 13.36 shows characteristic "fried egg" appearance of oligodendroglioma. (H&E.)

bulk of tumor mass. On section, the tumor is characteristically a relatively discrete, pale, pink-gray soft granular mass. It is sometimes cystic, with tumor lining the cyst wall (Fig. 13.36). Small hemorrhages may arise from the distinctive vascularity of proliferative capillaries at marginal zones of the tumor. Larger hemorrhages within the tumor may be a cause of sudden death (Fig. 13.36). Gritty resistance to the knife may be encountered due to calcification.

The typical microscopic appearance of the tumor is a "honeycombed" or "fried egg" pattern of tumor cells with regular, small, round nuclei with clear cytoplasm bound in groups by open, thin fibrovascular trabeculae (Fig. 13.37). Tumor cells of atypical forms present a diagnostic challenge. For example, tumor cells may take on elongate or signet ring forms. Sometimes, cells form a "marching soldiers" pattern. Anaplastic oligodendrogliomas present further diagnostic challenge needing expert consultation. Granular microcalcification is often found scattered within the tumor (Figs. 13.38–13.39) and at the marginal zones of tumor along with glomeruloid vascular endothelial proliferation (Figs. 13.40–13.41), a potential source of hemorrhage.[34] Mitoses are usually not found. Mixed oligodendroglioma and astrocytoma, the so-called oligoastrocytoma, is uncommon, accounting for 1.8% of gliomas in one study.[26] The latter tumor is classified as grade II by WHO. Tumor cells may be intermixed, or groups of the two types of tumor cells may be juxtaposed.

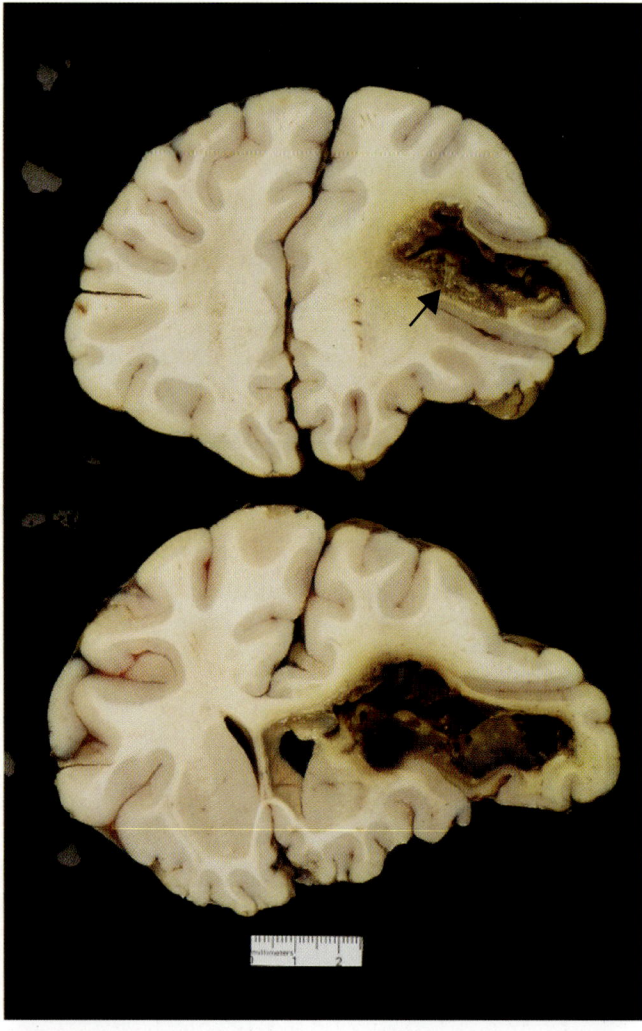

Fig. 13.36. Coronal sections of frontal lobes show artifactually partially collapsed hemorrhagic cystic tumor causing massive expansion of affected hemisphere and shift of midline to the left. Tumor is difficult to identify on gross inspection, but a small amount of necrotic-appearing tissue along a sector of cyst margin is tumor (shown microscopically in Fig. 13.37) (arrow).

Ependymoma

Ependymomas are most often found in children and young adults, but older age groups are not exempt. They comprise about 30% of intracranial tumors in children younger than 3 years.[26,32] In the spinal cord, ependymomas make up 50 to 60% of gliomas.[26] As might be expected, the location of ependymoma is in relation to the ventricles and central canal of the spinal cord. The posterior fossa, in and around the fourth ventricle, is a common location of the tumor.[21] Spinal cord ependymomas are found mostly in adults, and they may be found at all levels. Some find cervical and cervicothoracic levels as preferential locations.[26] The myxopapillary type is almost always located at the clonus medullaris and cauda equina. At its common site of involvement around the fourth ventricle, ependymoma may arise from the roof or floor of the ventricle as a solid soft gray granular discrete mass filling the ventricle (Case 13.9, Figs. 13.42–13.43). Larger growths may protrude from the lateral

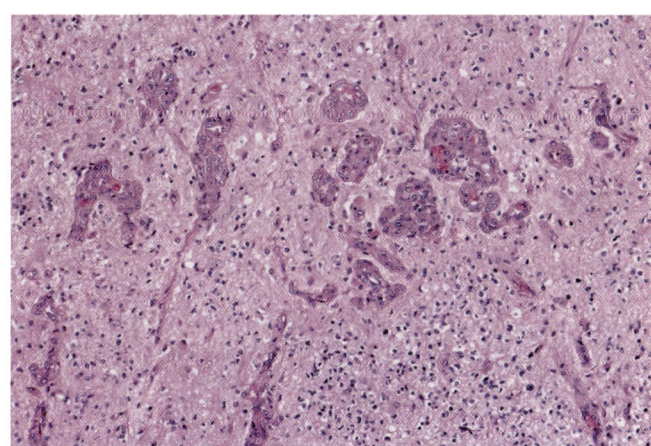

Fig. 13.40. Medium-power micrograph of capillary endothelial proliferation in glomeruloid form does not herald malignant tendency in oligodendroglioma as it does in glioblastoma. The change is most often found along marginal zones of tumor. (H&E.)

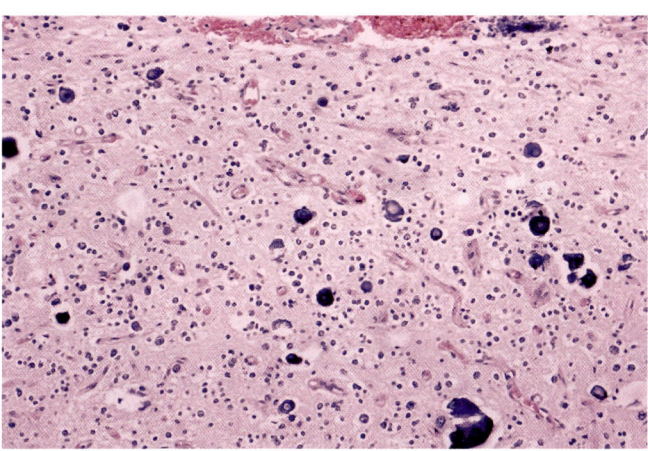

Fig. 13.38. Medium-power micrograph of another case of oligodendroglioma shows sparse monotonous field of cells typical of this tumor with small round nuclei centered in clear cytoplasm, giving rise to the "fried egg" moniker. Sparse open capillary network and granular calcification are also characteristic. (H&E.)

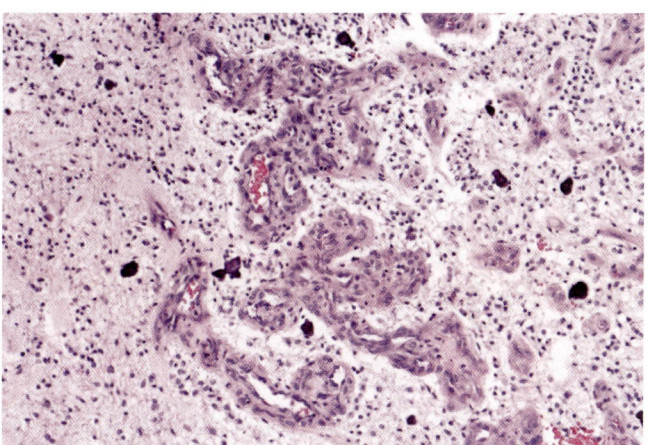

Fig. 13.41. Medium-power micrograph of proliferative vascular change in another oligodendroglioma, similar to that seen in Fig. 13.40, with calcification. (H&E.)

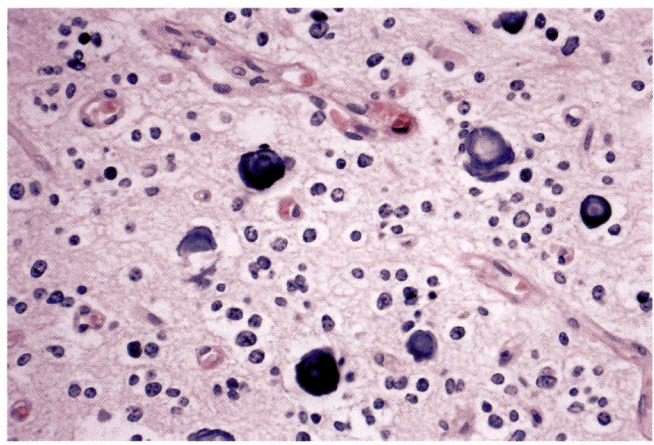

Fig. 13.39. High-power micrograph from tumor in Fig. 13.38 illustrates tumor cells and calcification in detail. (H&E.)

Case 13.9. (Figs. 13.42 and 13.43)

A 17-year-old boy in apparent good health excused himself from his classroom stating that he felt sick, went to the restroom, and vomited. About 5 minutes later, he was discovered in the restroom incoherent and unresponsive to questioning. When paramedics arrived, he was able to state that he exercised that morning, ate lunch, and was attending class when he was struck by a severe headache radiating from the back of his head. CT head scan revealed "possible pineal tumor vs bleed." No surgical procedure was performed other than a ventriculostomy, and he died 6 days after the onset.

Brain Tumors

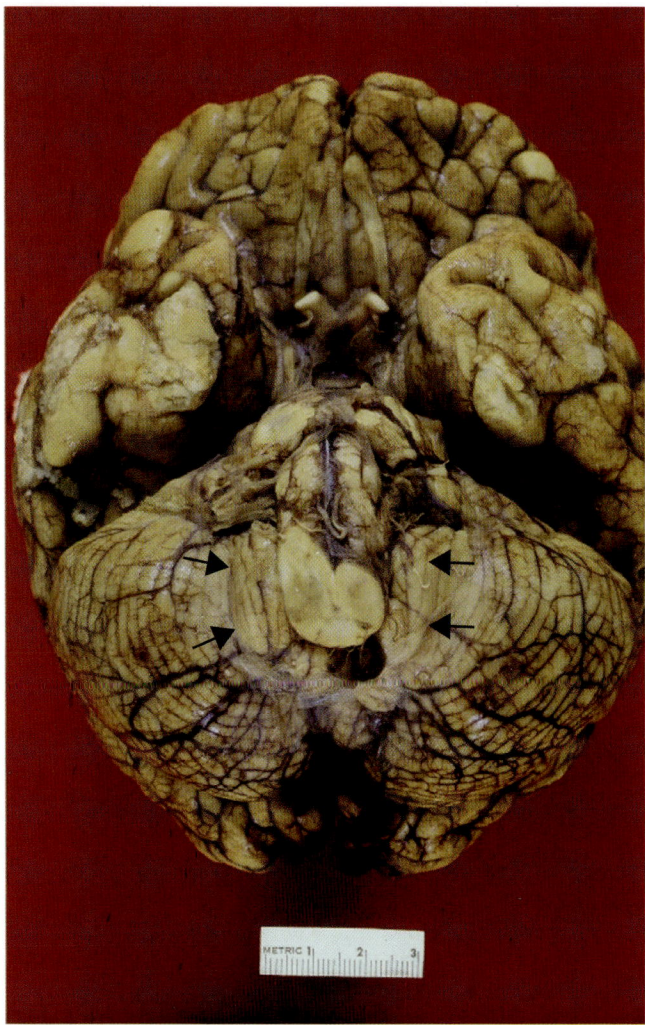

Fig. 13.42. Base view of brainstem and cerebellum of ependymoma of fourth ventricle. Note molded biventers (*arrows*) compressing medulla on (left in photo).

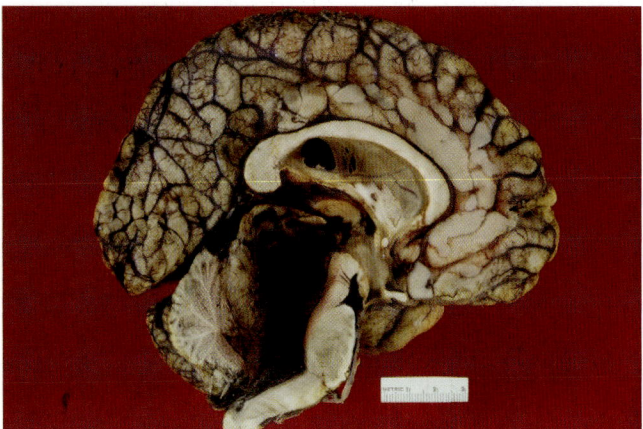

Fig. 13.43. Ependymoma with hemorrhage seen on midsagittal section massively expands fourth ventricle, rostrally involves midbrain tectum, and extends into caudal third vent-ricle. Note hydrocephalus and fenestrated septum pellucidum.

recess or through the roof of the fourth ventricle into the cisterna magna. The tumor may infiltrate the brainstem and cerebellum locally. Ependymomas of the spinal cord may occur around the central canal as a discrete growth, or over several segments. Those of the cerebral hemisphere arise from, and frequently maintain connection with, the lateral ventricular wall. Others may be isolated from the ventricle. Those arising from the third ventricle are rare. Some find preferential location to be white matter of the trigone region of the lateral ventricle.[29]

Microscopically, the more common cellular subtype of this tumor is moderately cellular and consists of a fairly uniform population of cells of cohesive appearance with uniformly round nuclei and moderate pale cytoplasm, occasionally forming diagnostic ependymal rosettes. A circumferential perivascular arrangement of tumor cells dropping fibrillar processes to a blood vessel wall to form a pseudorosette aids in the diagnosis, but it is not pathognomonic, since perivascular pseudorosettes occur in several other neural neoplasms (Figs. 13.44–13.46). Papillary, clear cell, tanycytic, and other histologic subtypes and variants of ependymomas present more of a challenge in differential diagnosis.

Expanding tumors of the fourth ventricle with increasing noncommunicating hydrocephalus cause headaches and other symptoms of increased intracranial pressure and, on occasion, sudden unexpected death. In most cases caretakers are alerted to the need for medical attention long before risk of sudden death. Tumor filling the fourth ventricle may also cause ataxia, nystagmus, and gaze palsies by involvement of overlying cerebellum and its connections with brainstem centers. Cerebral masses may produce focal neurologic signs depending on location, or may present with nonspecific signs of increased intracranial pressure.

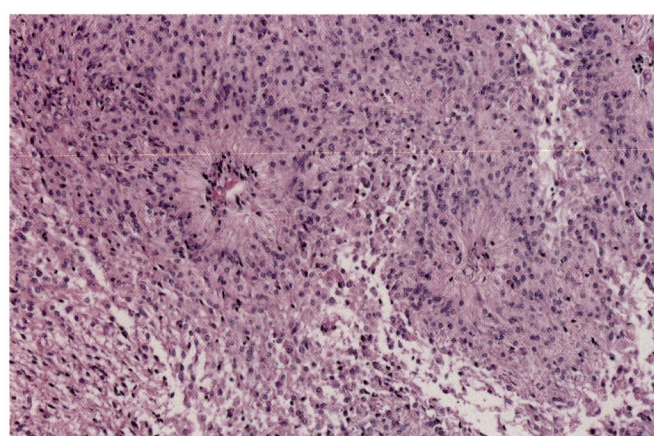

Fig. 13.44. Medium-power micrograph of cellular ependymoma shows two typical perivascular pseudorosettes with fibrillary processes converging on central small blood vessels. (H&E.)

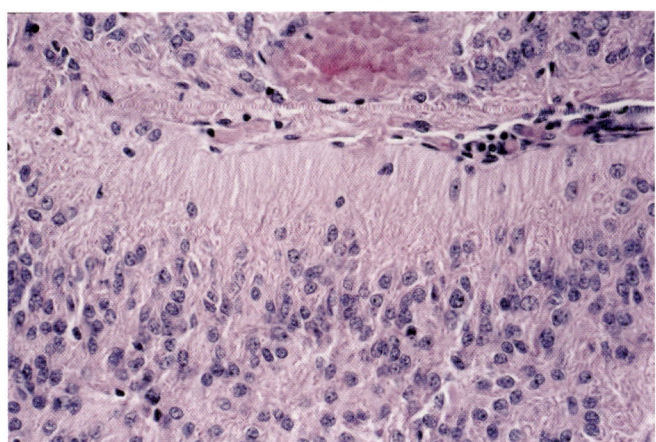

Fig. 13.45. High-power micrograph of same tumor as Fig. 13.44, with tumor fibrils directed to blood vessel wall. (H&E.)

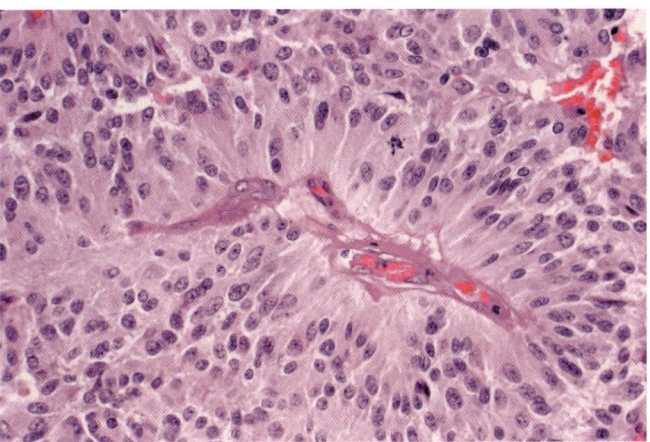

Fig. 13.46. High-power micrograph of an ependymoma with perivascular pseudorosette. (H&E.)

Fig. 13.47. Midsagittal section of brainstem and cerebellum shows slightly enlarged basis pontis containing a few small hemorrhages, and small amounts of grey tissue in interpeduncular fossa and layered on dorsal aspect of medulla, which are tumor metastases from an ependymoblastoma.

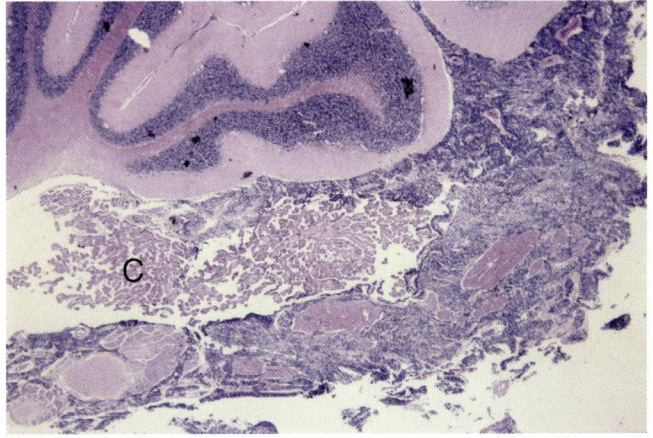

Fig. 13.48. Low-power micrograph of cerebellum from case seen in Fig. 13.47 demonstrates subarachnoid metastasis. C, choroid plexus. (H&E.)

Ependymoblastoma

Ependymoblastoma is a rare tumor of childhood with the median age at diagnosis of 2 years.[10] The majority occur supratentorially. Other sites are the fourth ventricle and pons[40] (Figs. 13.47–13.51). The tumor is relatively circumscribed, frequently reaching massive size in the cerebral hemisphere adjacent to a ventricular surface. The tumor aggressively infiltrates leptomeninges (Fig. 13.48), and distant metastases may occur. The ependymoblastoma example shown in Figs. 13.47–13.51 was centered in the basis pontis with extensive leptomeningeal spread with desmoplasic features.

Microscopically, the tumor is highly cellular in anastomosing cords, comprised of uniform small- to medium-sized cells with sparse cytoplasm and round to spindle-shaped hyperchromatic nuclei with frequent mitoses. The diagnostic feature is the ependymal rosette with an irregularly clustered ring of nuclei at the periph-

Brain Tumors

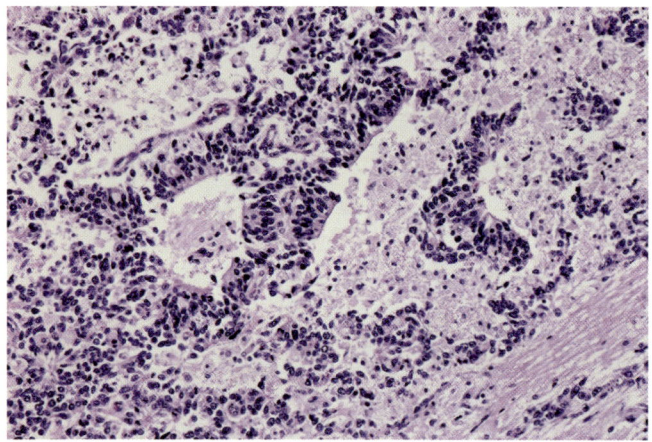

Fig. 13.49. Medium-power micrograph of tumor in basis pontis from case seen in Fig. 13.47 shows a focus of viable tumor cells within area of necrosis, containing a few rosettes. (H&E.)

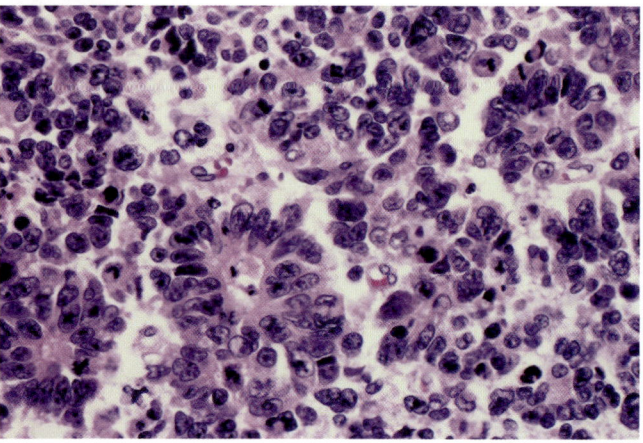

Fig. 13.50. High-power micrograph shows a typical ependymal rosette with a central lumen. Tumor nuclei are hyperchromatic and pleomorphic in this ependymoblastoma. (H&E.)

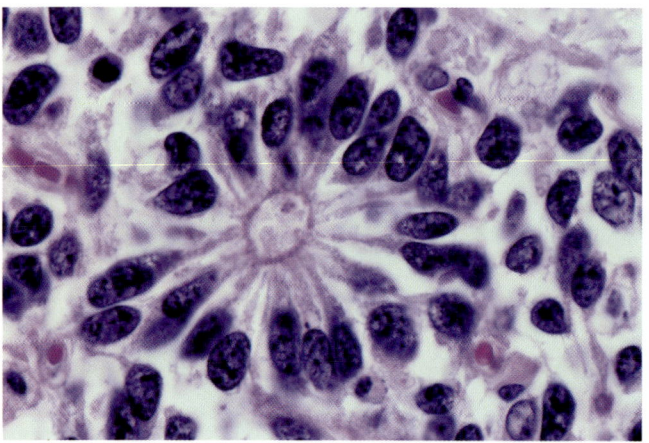

Fig. 13.51. Ependymoblastoma. Very high power of a diagnostic ependymal ("true") rosette consists of columnar circular grouping of unipolar cells converging on a central lumen (or tubule). Condensations around lumen are blepharoplasts. (H&E.)

ery and processes converging on a central lumen lined by a limiting membrane (see Fig. 13.51). Juxtaluminal blepharoplasts are demonstrable by appropriate stains such as phosphotungstic acid hematoxylin (PTAH). These true rosettes are relatively frequent and readily discovered at low power.

Prognosis is poor, with fatal outcome in a matter of months.

Medulloblastoma

Medulloblastoma is found mainly in children and less frequently in young adults. It is a highly malignant neoplasm known for its propensity to seed along subarachnoid and ventricular spaces. Most commonly originating in the cerebellar vermis, the tumor grows into the fourth ventricle and invades the subarachnoid space (Case 13.10, Fig. 13.52). It is thought to arise from the embryonic external granular cell layer of developing cerebellar cortex.[22] In older age groups, the tumor tends to occur in the lateral hemisphere and is usually of desmoplastic subtype.

Grossly, the tumor is gray to pink, soft, friable, and relatively demarcated. Microscopically, the neoplasm consists of a densely compact array of small cells with round, oval, to irregular hyperchromatic nuclei (Figs. 13.53–13.55), often described as carrot-shaped, with sparse, mostly imperceptible cytoplasm. When present, neuroblastic (Homer Wright) rosettes are highly supportive of the diagnosis of medulloblastoma. The rosette is composed of a small circular array of tumor nuclei, often carrot-shaped, with thin cell processes converging to a central point devoid of lumen. Its presence is highly variable, from frequent to rare. The tumor invading the subarachnoid space may excite a fibroblastic response, which is to be distinguished from the reticulin-rich desmoplasia of the desmoplastic variant of medulloblastoma (Figs. 13.54–13.55).

Desmoplastic Variant of Medulloblastoma

Desmoplastic variant of medulloblastoma usually affects young adults and has a preference for the lateral cerebel-

> **Case 13.10.** (Fig. 13.52)
>
> A 22-year-old man allegedly sustained a "flexion-extension cervical sprain injury" at work while lifting a barrel. X-ray of the neck was negative, and he was treated with medications. He complained that his neck was "locking" accompanied by chills. For this, he was placed in a bathtub of warm water, but he was found shuddering about a half-hour later. Still complaining of neck pain, he was admitted to the hospital on the same day. Two days later, he was found on the hospital floor aware enough to apologize for not requesting help, but 5 minutes later was in respiratory arrest with fixed pupils. Life support was withdrawn two days later.

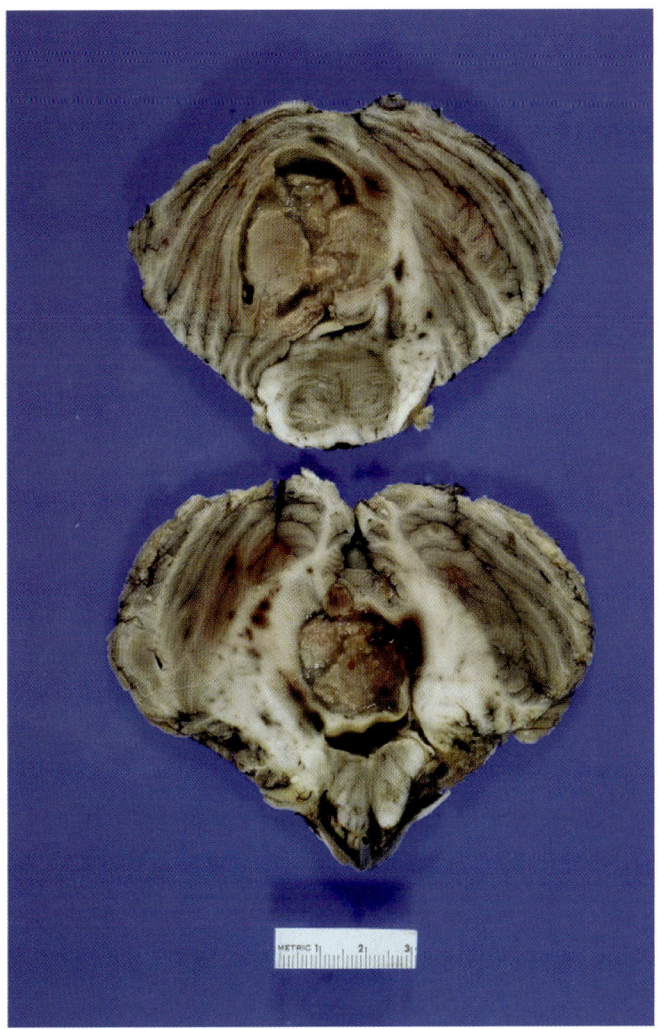

Fig. 13.52. Medulloblastoma centered at anterior vermis of cerebellum and encroaching into fourth ventricle. Fragmented and necrotic appearance due partly to postmortem change.

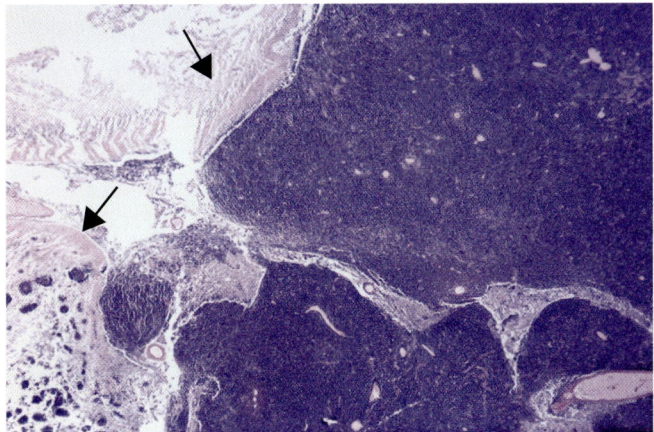

Fig. 13.53. Low-power micrograph of medulloblastoma shows a compact, densely cellular basophilic field of nuclear-rich tumor that is destructive of cerebellum (seen in mechanically disrupted fragments) (*arrows*). (H&E.)

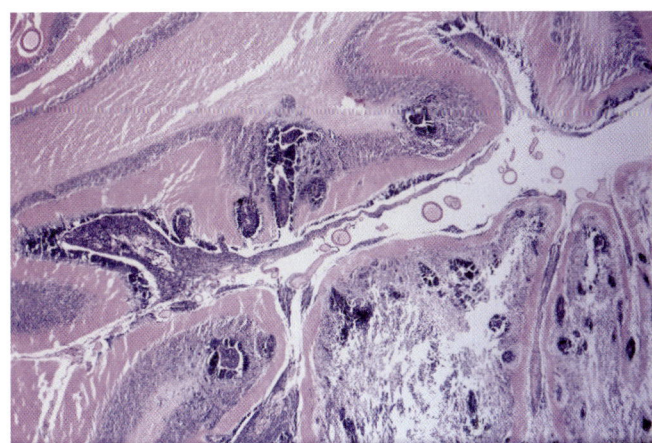

Fig. 13.54. Low-power micrograph shows extensive metastases in cerebellar parenchyma and subarachnoid space. (H&E.)

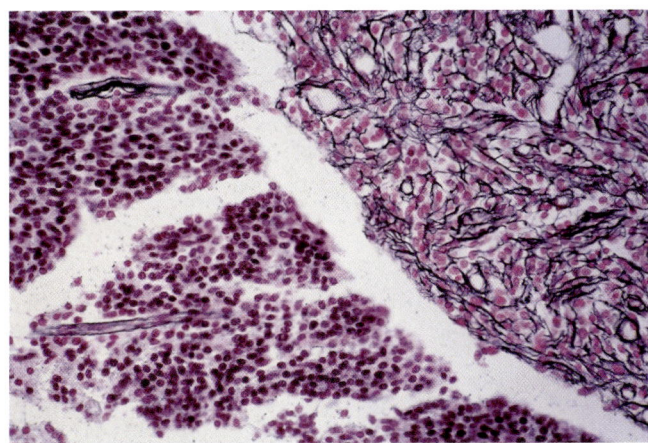

Fig. 13.55. High-power micrograph shows densely cellular tumor with high nuclear to cytoplasmic ratio (*left*) and tumor in subarachnoid space to the right exciting reticulin-rich desmoplasia. (Wilder.)

lar hemisphere (Case 13.11, Figs. 13.56–13.57). Microscopically, its most distinguishing feature is round nodular islands of uniform cells of less dense cellularity amid a background of a densely cellular field of hyperchromatic cells (Figs. 13.58–13.59). The cells of the island show hints of neuronal and astrocytic differentiation enmeshed in a rich reticular network. Survival with this variant was noted to be better than with the classic form.[5] Clinically, patients typically present with unrelenting vomiting, stumbling, falls, and then eventually with signs such as nystagmus and ocular palsies. The tumor is rapidly growing, and the clinical course is usually short.

Ganglioglioma

Gangliogliomas are slow-growing tumors found in a wide age range but with age at diagnosis in most cases

Brain Tumors

> **Case 13.11.** (Figs. 13.56–13.57)
>
> A 27-year-old woman developed progressively severe headaches 2 months before death, followed by daily nausea and vomiting. There was increasing somnolence for several days before admission to the hospital. Other symptoms were "weakness" of the left arm, dizziness, and lightheadedness. She died before diagnostic tests could be completed. An anonymous telephone call indicated possible poisoning.

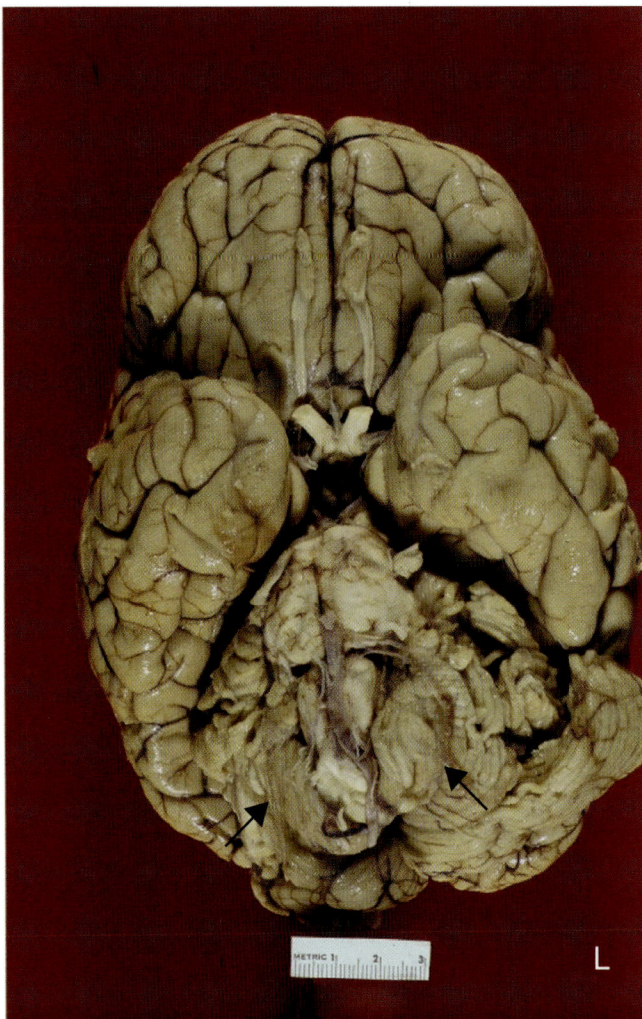

Fig. 13.56. Base view of brainstem and cerebellum shows an enlarged left cerebellar hemisphere and herniation of biventers, left greater (arrows indicate indentations created by margins of foramen magnum). Biventers compress lateral aspects of medulla.

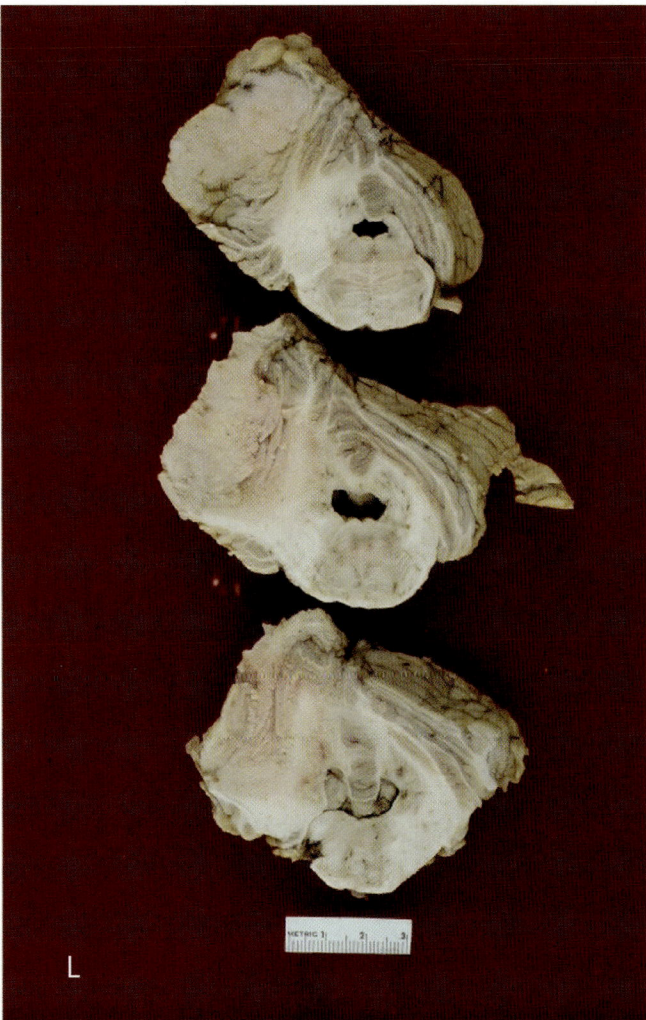

Fig. 13.57. Pale pink-gray medulloblastoma replaces the left superior cerebellar hemisphere.

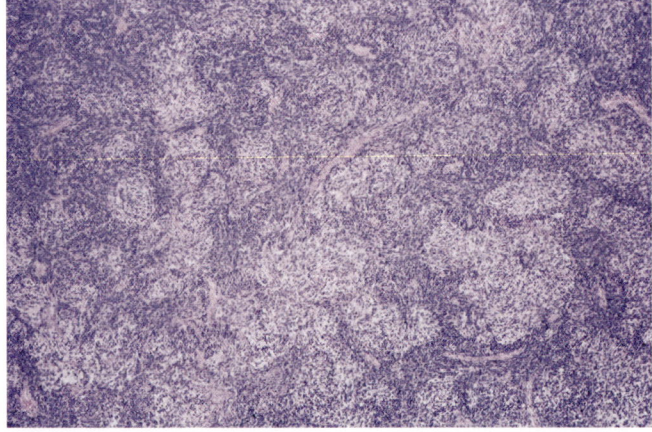

Fig. 13.58. Low-power micrograph of desmoplastic medulloblastoma shows a distinctive biphasic pattern of predominant irregularly rounded islands of pale tumor amid darker zone of denser cellularity. (H&E.)

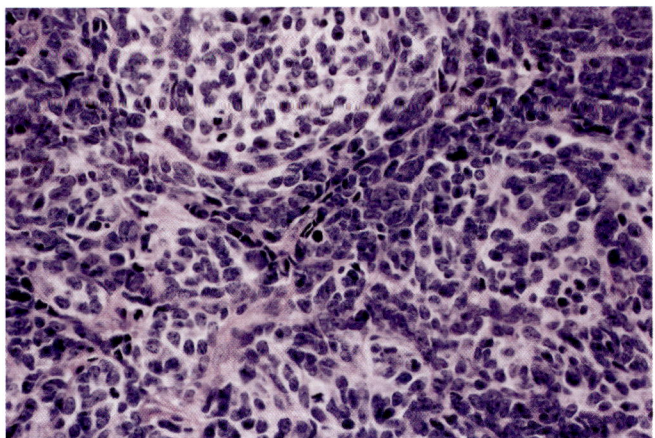

Fig. 13.59. High-power micrograph of tumor seen in Fig. 13.58 shows details of a narrow band of dense growth between lighter zones. (H&E.) The zone of dense cellularity is reticulin-rich, not easily recognized by H&E stain, but clearly demonstrated by reticulin stain.

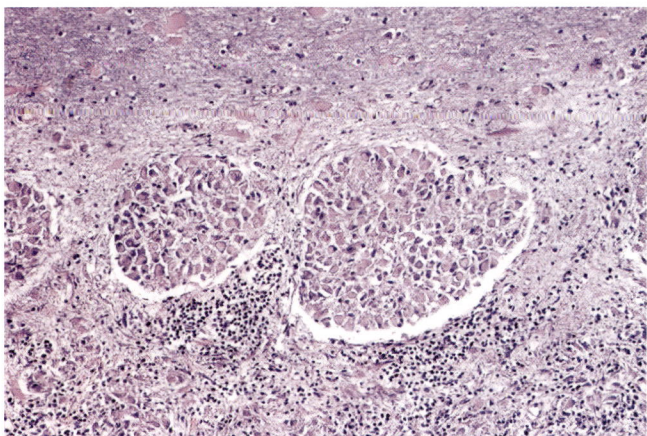

Fig. 13.60. Medium-power micrograph of ganglioglioma shows nests of neoplastic neurons surrounded by tumor astrocytes and "lymphocytic" infiltration. (H&E.)

between 8 and 25 years.[47] It has a predilection for the temporal lobe (74%), and its presence is commonly heralded by partial complex seizure.[47] Its presence is confirmed by computed tomography (CT) or magnetic resonance (MR) scans, although the findings are nonspecific. Grossly, the tumor is fairly well demarcated, and it may be variably cystic. Calcification of the tumor is common, and may be detected by CT scan and by its grittiness at autopsy.

Microscopically, irregular groupings of dysmorphic neurons of variable size are dispersed in a field of neoplastic astrocytes (Figs. 13.60–13.61) enmeshed in a reticulin network. Binucleated neurons (Fig. 13.61) and large to giant forms with large vesicular nuclei and thick processes of random orientation are hallmarks of the neoplasm. Groups of small neurons and astrocytes of varied forms such as gemistocytes may be present, along with Rosenthal fibers. What appears to be lymphocytic infiltration (Fig. 13.60) has been interpreted in at least some cases as neuronal precursors.[41]

On rare occasions, the astrocytic component may display anaplasia. Gangliogliomas of the temporal lobe, because of their location, the tumor's indolent behavior, and its relative discreteness, lend themselves to successful resection that is curative in most cases.[23]

Mixed Gliomas

Mixed gliomas, mostly oligoastrogliomas, are not rare.[17] An incidence of 3.4% was reported in a very large series of brain tumors.[48] Individual cases of mixed glioma with astrocytic and ependymal components have been described.[26] Cavanaugh[4] reported eight cases of epilepsy with "small tumors of the temporal lobe" collected during the course of temporal lobectomy for treatment of

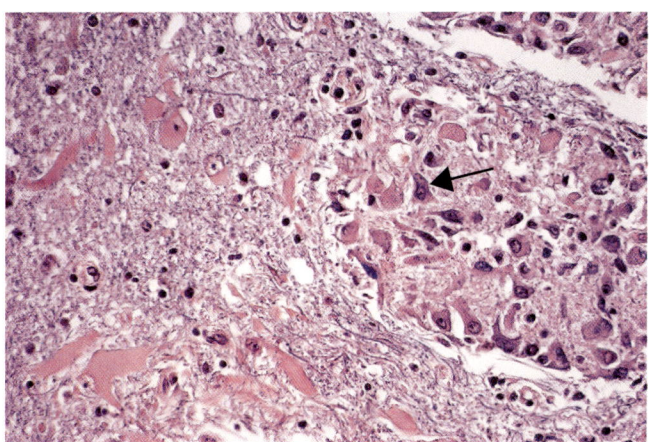

Fig. 13.61. High-power micrograph of same case seen in Fig. 13.60 demonstrates a binucleated neuron (*arrow*) and details of neoplastic nerve cells and astrocytes. (H&E.)

epilepsy. Pathologically, the tumors were considered hematomatous with astroglial, oligodendroglial, spongioblastic, and neuronal elements in various combinations, mostly oligoastroglial. A rare example of a similar case is illustrated (Figs. 13.62–13.65). A well-circumscribed mixed ependymoma-astrocytoma of the parietal cortex was reported as a unique case in a 38-year-old woman.[28] Biopsy findings suggested a diagnosis of subependymoma in an atypical location. A tumor of similar appearance of the frontal lobe was encountered in our autopsy material in a 21-year-old male who died in status epilepticus (Case 13.12, Figs. 13.66–13.67).

Meningioma

Meningiomas are said to comprise 14 to 15% of all primary brain tumors in the United States in autopsy

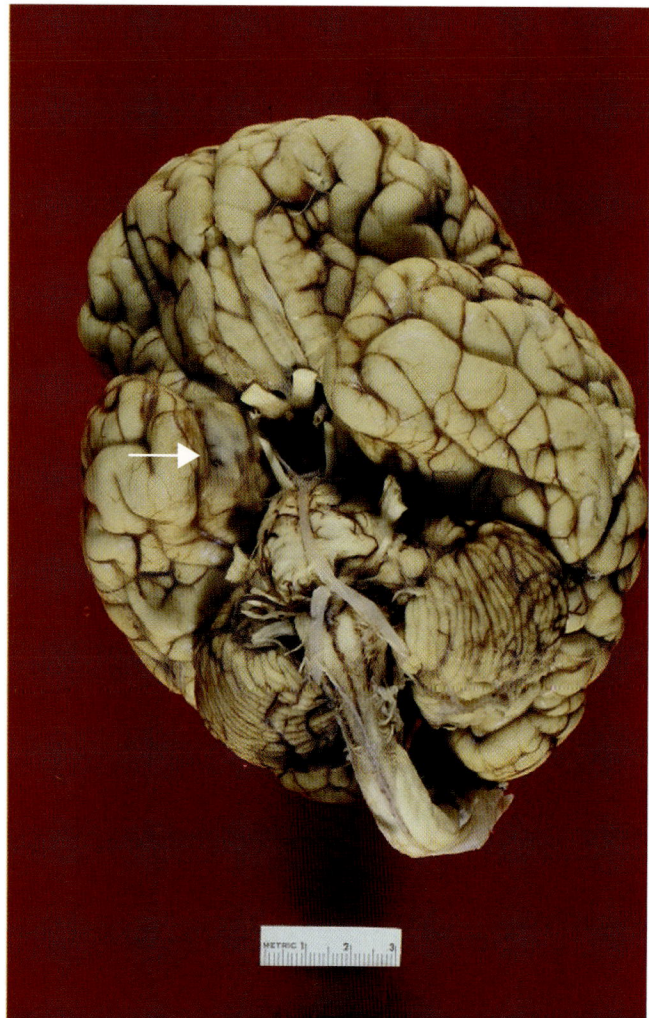

Fig. 13.62. Tilted base view of brain to display the right pyriform area shows a slight cortical bulge of tumor medial to rhinal fissure (*arrow*).

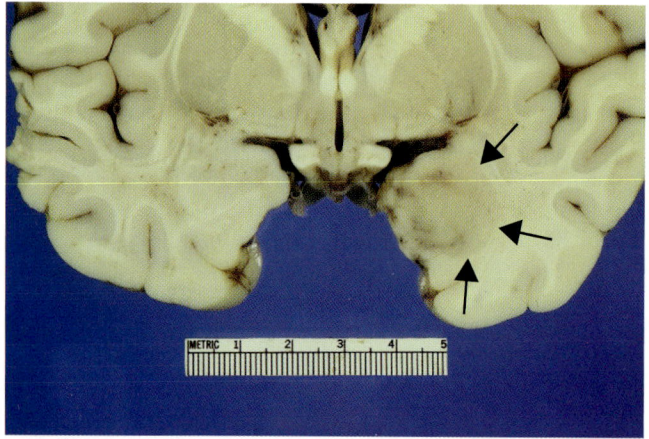

Fig. 13.63. Closeup view of coronal section through amygdaloid complex shows enlargement on the right by a tumor with relatively sharp margins (*arrows*). Tumor is typical of that described by Cavanaugh[4] as "small tumours of the mesial temporal lobe" excised during surgical treatment of epilepsy.

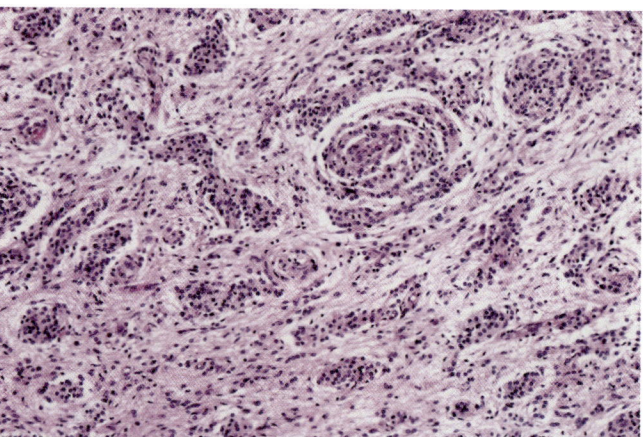

Fig. 13.64. Medium-power micrograph of the tumor shows a mixed glioma composed of ependymal tumor cells in small groups in a field of low-grade astrocytoma, an ependymoastrocytoma.

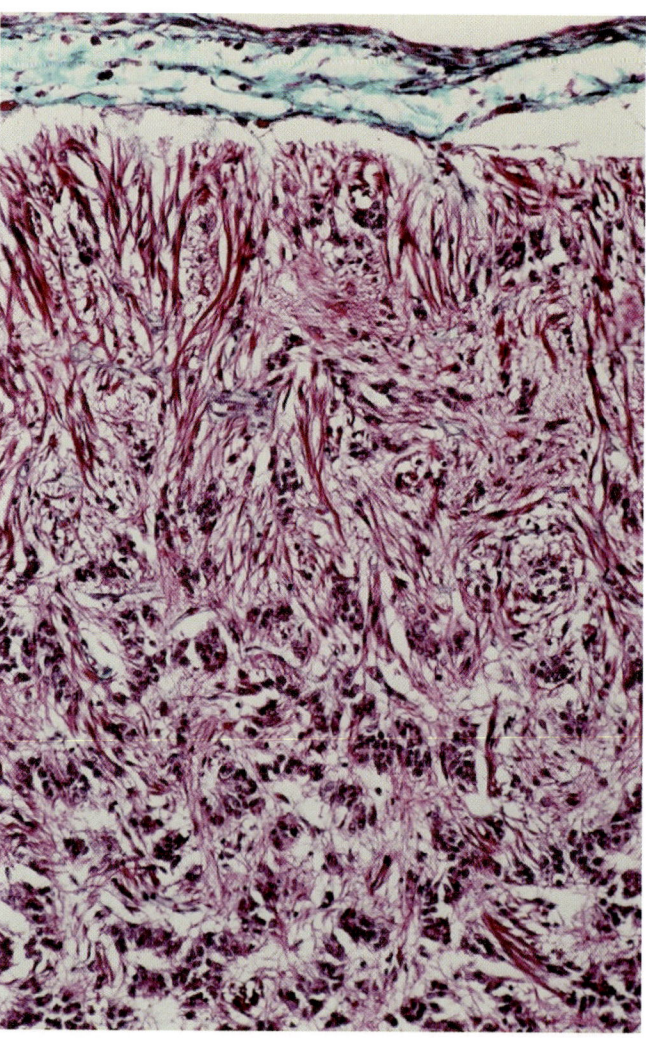

Fig. 13.65. High-power micrograph of tumor at pial (*green*) surface of cortex shows interlacing fibrillar astroglial component and a number of small clustered cells of ependymal appearance in this mixed glioma. (Trichrome.)

Case 13.12. (Figs. 13.66 and 13.67)

A 21-year-old man had a seizure disorder incompletely controlled with medications so that he was unable to obtain a vehicle operator license. He sought help in alternative medicine where he was told to stop all medications. Three days after having done so, seizures became increasingly frequent until they were occurring every few minutes. Unable to eat properly, he was losing weight. Returning to his attending, he was told the seizures would soon stop, and he was told to sleep on the floor to avoid falling out of bed. His brother observed continuous seizures until breathing became labored, and paramedics were summoned. He was pronounced dead at the scene. Attending was indicted for practicing medicine without a license and involuntary manslaughter.

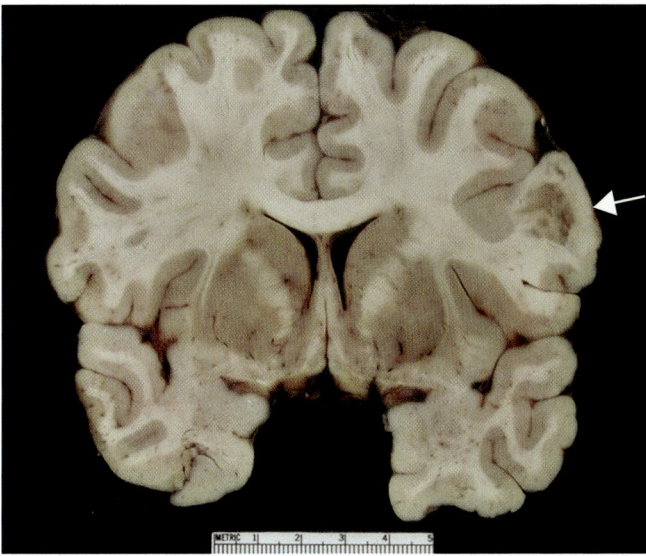

Fig. 13.66. Brain was externally "unremarkable". Coronal section at level of anterior striatum and amygdala shows a partially cystic whitish-tan tumor replacing right inferior frontal gyrus (*arrow*).

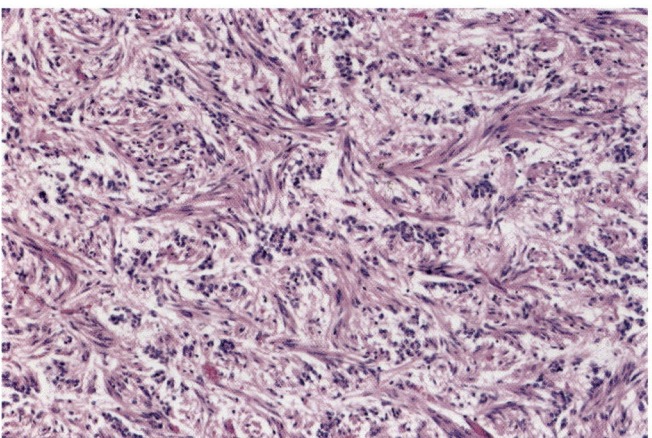

Fig. 13.67. Medium-power micrograph shows low-grade fibrous astrocytoma with interspersed aggregates of ependymal cells. (H&E.)

series,[24] and they form one of the more common neoplasms encountered in our office. Of 760 patients who died of brain tumors in Minnesota from 1958 through 1962, 6.8% died of meningiomas.[6] Meningoma is primarily a tumor of adulthood, peaking in the sixth and seventh decades, but it may occur in childhood infrequently, 1.1% of all meningiomas in one review.[45] The higher incidence of meningiomas in women has been long recognized. In one reported series of meningiomas, the female to male ratio was 2.5:1 in the intracranial space and 9:1 in a spinal location.[24] In our experience, meningiomas have presented in the individuals with sudden unexpected death, postsurgical complications, as the cause of symptoms but not the cause of death, and as incidental findings. The cell of origin in meningiomas is controversial, but the most likely candidate is arachnoid cap cells. The variety of histologic types that meningiomas manifest may present a diagnostic challenge. The more common varieties, such as meningothelial and psammomatous types, replicate normal arachnoid cap cells and should present no problem. Expectedly, the tumor arises wherever arachnoid cap cells occur, and this includes the glomus of the choroid plexus where meningothelial elements are always present, often with psammoma bodies, frequently accounting for the "calcification" seen in the glomus on x-ray. The calcification is said to occur with aging, but it may be found even in the young. It should be no surprise, therefore, that intraventricular meningiomas are seen from time to time and, with rare exception, they are found in the region of the atrium of the lateral ventricle. The most common sites for this tumor include the cerebral convexity, falx cerebri, olfactory groove, and parasellar region.

Grossly, the tumor is firm, globular, lobulated, or a sessile mass attached to subdural surfaces, compressing and displacing the brain and spinal cord. Three examples of cerebral tumors are shown (Figs. 13.68–13.74; Case 13.13, Figs. 13.75–13.76). An occasional tumor may present as a scalp mass by erosion through the skull. Growing as a discrete localized tumor at vital anatomic sites on the surface of the brain, it can cause focal clinical signs such as unilateral anosmia, uniocular visual field cut, rarely total visual loss, or unilateral facial palsy with deafness. Seizures are common. Meningiomas of the foramen magnum, however, give rise to a complex set of symptoms and signs with a difficult differential diagnosis.

More common histologic types of meningiomas are nicely illustrated in the WHO reference.[26] The common meningothelial type forms a lobulated syncytium of uniformly ovoid nuclei, some with characteristic clear centers described as nuclear inclusions, that are actually deep cytoplasmic invaginations of the nuclear membrane (Fig. 13.70). In reality, the syncytium is a complex interdigitation of thin cell processes unresolved by light microscopy. Transitional type (or mixed type) is characterized by small whorls that are composed of tumor cells

380 Brain Tumors

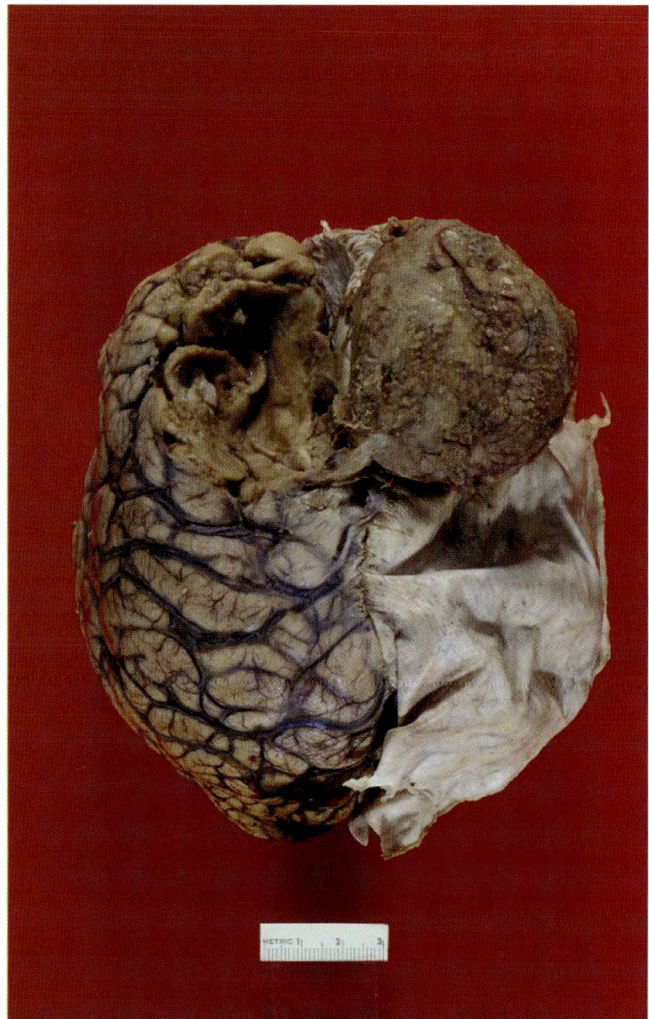

Fig. 13.68. Vertex view of brain shows a massive, discrete, globular tumor attached to left paramedian subdural surface, creating a huge excavation of left frontal lobe. Left dural leaf is turned over to the right, lifting the tumor from its bed. Cortex at cavity base is atrophic. In small meningiomas, cortex or other structures are simply pushed aside.

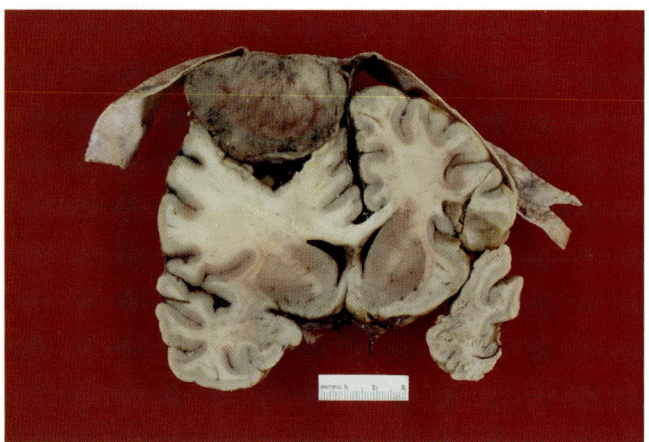

Fig. 13.69. Coronal section at anterior striatum displays the tumor deeply indenting the dorsal half of hemisphere. White matter is expanded by edema.

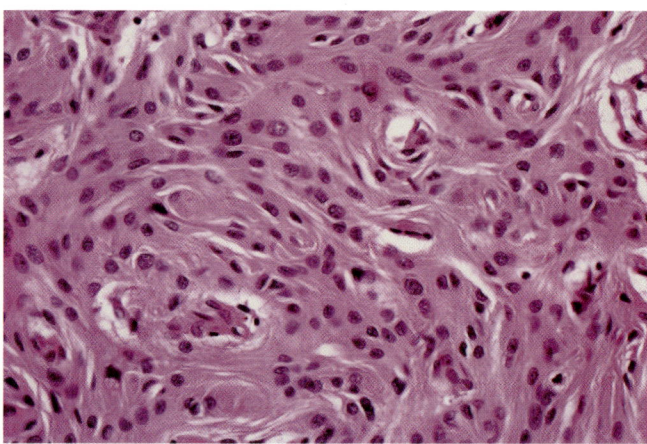

Fig. 13.70. High-power micrograph of meningioma shows typical round to oval nuclei of meningothelial cells in a pink "syncytium" composed of complex interdigitation of cell processes. (H&E.)

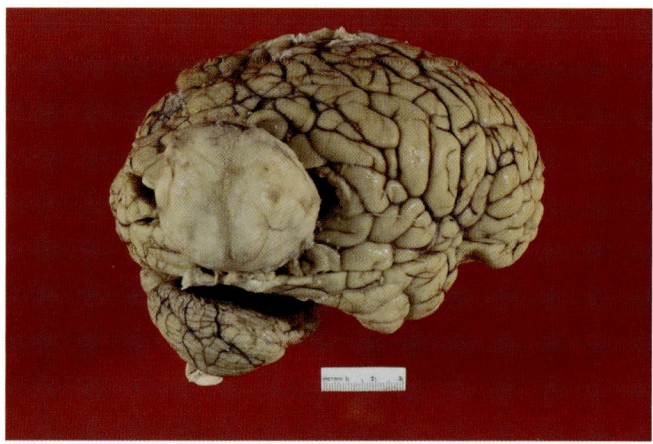

Fig. 13.71. A large discrete globular meningioma still covered with dura mater, the rest of surrounding dura having been cut away, occupies junction of posterior parieto/occipito/temporal lobes. Remaining cortical surfaces show generalized flattening due to brain swelling.

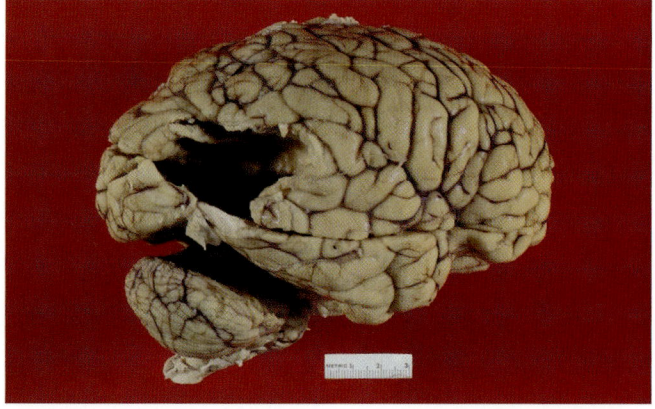

Fig. 13.72. Deep cavitation of cortex and subcortex created by tumor seen in Fig. 13.71 is exposed upon its removal.

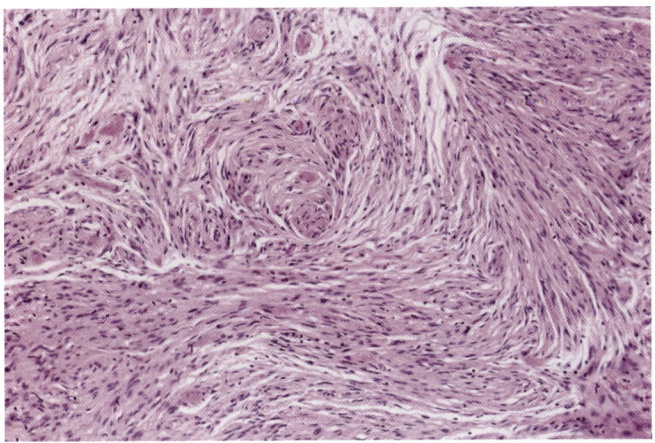

Fig. 13.73. Medium-power micrograph shows swirling isomorphic growth of spindle cell fibroblastic meningioma. Centrally, small whorls of meningothelial pattern are present. (H&E.)

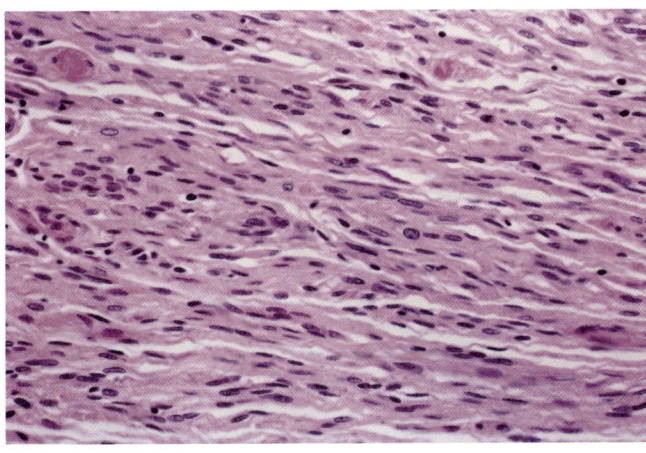

A

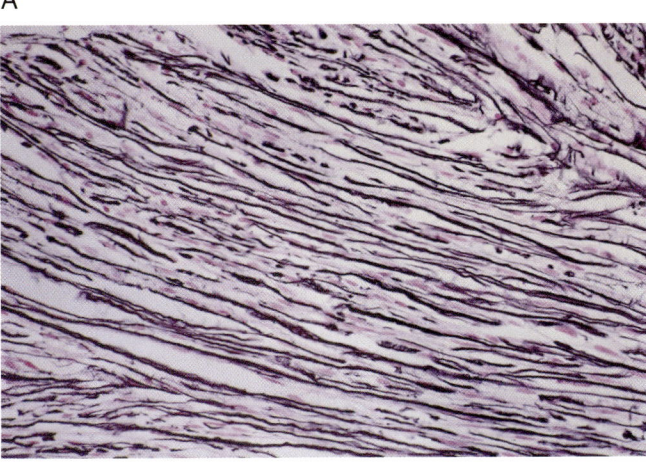

B

Fig. 13.74. High-power micrograph (A) shows detail of tumor seen in Fig. 13.73 with mostly bipolar cells possessing nuclei with blunt ends (H&E.) Special stain (B) demonstrates a rich reticulin network between tumor cells. (Wilder.)

Case 13.13. (Figs. 13.75 and 13.76)

A 53-year-old man became involved in an altercation over a traffic accident, and was struck on the head with a wrench. Although not rendered unconscious, he complained of a headache. Just prior to admission to the hospital, he developed an "exploding" headache with loss of vision. While in the emergency room, he had a seizure, became comatose, and died the following day.

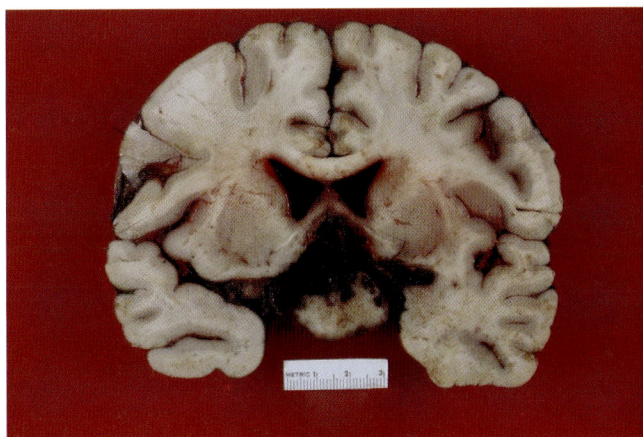

Fig. 13.75. Meningioma with fresh hemorrhage dorsally elevates floors of lateral ventricles with upward tilt of left anterior commissure at level of optic chiasm. Chiasm is buried in fresh hemorrhage.

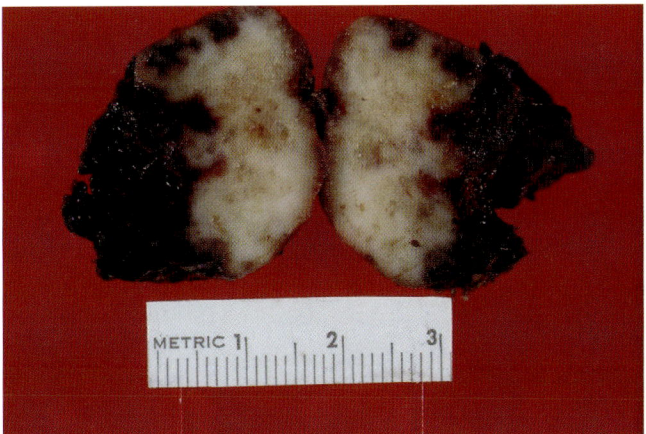

Fig. 13.76. Globular, solid tumor with fresh hemorrhage is yellowish-white.

Brain Tumors

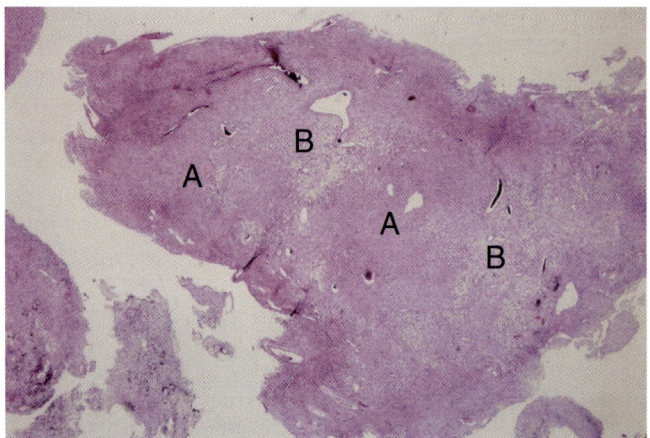

Fig. 13.77. Lower-power micrograph of fragment of acoustic Schwannoma tumor with a biphasic pattern of darker solid Antoni A (A) and lighter Antoni B (B) areas. (H&E.)

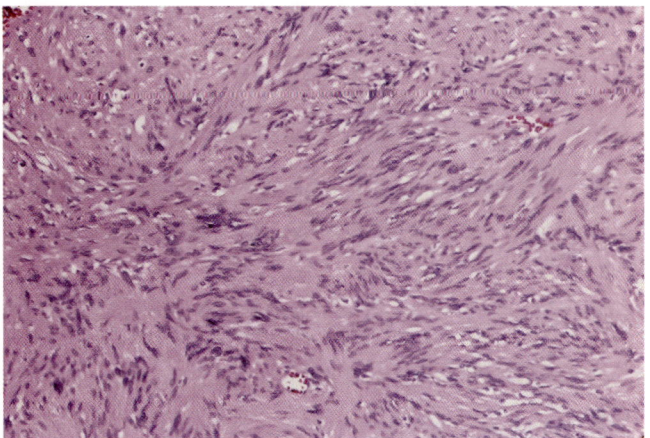

Fig. 13.78. Medium-power micrograph of the Schwannoma seen in Fig. 13.77 shows Antoni A area as a solid field of spindle cells with elongate clustered nuclei with intervening anuclear fibrillar zone suggesting Verocay body. Verocay bodies with well-organized waves of nuclear palisades are not as common in the acoustic tumors as they are in peripheral nerve schwannomas. (H&E.)

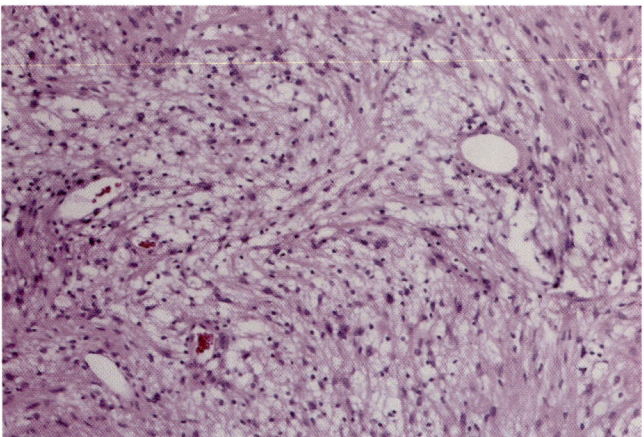

Fig. 13.79. Medium-power micrograph from another portion of tumor seen in Fig. 13.77 shows more loosely arranged Antoni B area. (H&E.)

concentrically tightly wrapped around themselves and around blood vessels. Calcification of the whorl is termed a psammoma body, and a tumor made up of large numbers of such bodies is a psammomatous meningioma, a type more commonly found in the spinal canal. Fibroblastic meningiomas form isomorphic growths of elongated spindle cells forming interlacing bundles (Figs. 13.73–13.74). Nuclei tend to maintain meningothelial character, with blunt ends, as distinguished from elongated eel-like tapered nuclei of fibroblasts. The amount of intercellular collagen may vary considerably. Angiomatous meningioma is composed predominantly of vascular channels of various caliber with intervening meningothelial cells. Hemorrhage may occur in meningiomas but not related to any specific histologic type, to our knowledge (Case 13.13; Figs. 13.75–13.76). Variants of meningioma not described earlier include myxomatous, secretory, papillary, clear cell, chordoid, and other less common subtypes, some of which demonstrate a more aggressive clinical course, as discussed in the reference cited.[26]

Schwannomas

Schwannomas are tumors arising from Schwann cells of peripheral nerves and peripheral portions of cranial nerves, acoustic tumors being the most common of the intracranial sites. Schwannomas comprise about 8% of all primary intracranial tumors.[26] For variable distances beyond the brainstem surface, cranial nerves remain as CNS tissue supported by oligodendroglia. The acoustic/vestibular nerve has the longest segment of oligodendroglia-supported myelinated nerve fibers beyond its exit from the brainstem before its sharp transition to Schwann cells, the Obersteiner-Redlich zone, as supportive cells. This transition occurs at around the internal auditory meatus (porus acusticus) for the acoustic/vestibular nerve, and it is assumed that schwannomas arise at or near this transition. Enlargement of the porus by a tumor demonstrated on x-ray is almost always diagnostic of an acoustic schwannoma.

Grossly, schwannomas are discrete, encapsulated, globular masses, white to tan on section. Areas may show yellowish discoloration, hemorrhage, and cyst formation. Microscopically, the neoplasm is characteristically biphasic with histologically differentiated areas called Antoni A and B (Figs. 13.77–13.79). The Antoni A area is composed of compact isomorphically aligned elongate tumor cells in interlacing bundles. Occasionally, tumor nuclei palisade in waves with rows separated by anuclear fibrillar zones into a structure termed a Verocay body (Fig. 13.80). The Verocay body is more frequently found in peripheral nerve schwannomas and may not appear in acoustic tumors. Antoni B areas are relatively lucent zones of loosely and randomly arranged

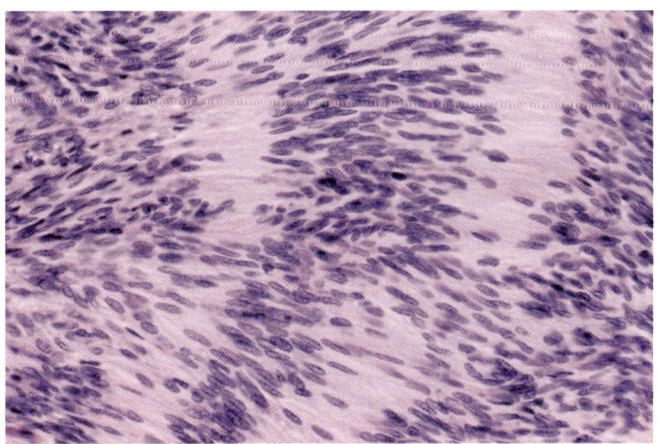

Fig. 13.80. Verocay body in acoustic schwannoma. (H&E.)

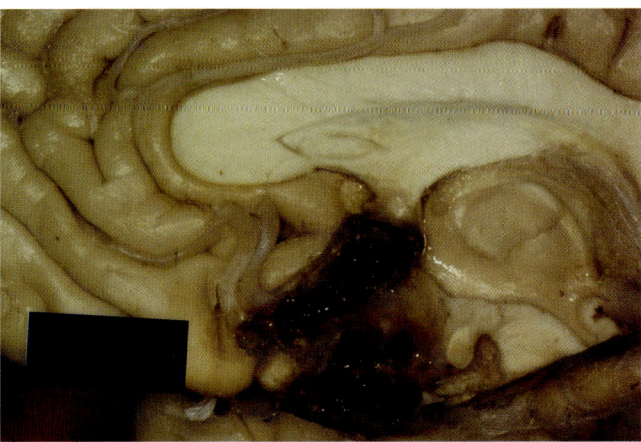

Fig. 13.81. Multilobulated cystic craniopharyngioma filled with hemorrhage invades anterior hypothalamus, anterior column of fornix on one side, and parolfactory gyri.

tumor cells. The two areas are juxtaposed and not intermixed. Especially in Antoni B areas, schwannomas have a tendency to take on variegated appearances. Microcystic areas may be present and are thought to become confluent to form larger cysts. Mucinous and xanthomatous areas of the tumor may be encountered along with areas of increased vascularity with hyalinized blood vessels. Schwannomas may contain not only these degenerative changes, but also cells with large, hyperchromatic nuclei. Cases with these features are sometimes referred to as "ancient schwannomas." These findings, as well as occasional mitoses in more cellular variants of schwannoma, are not in isolation to be interpreted as signs of malignancy.[41]

Occupying the cerebellopontine angle, acoustic schwannomas are clinically manifested by hearing loss, tinnitus, vertigo, and facial weakness. If left unattended, the tumor may attain sufficient size to compress the brainstem, produce hydrocephalus and its consequent increased intracranial pressure, and possibly death. Other tumors to be considered in this location are meningioma, epidermoid cyst, glomus jugulare tumor, metastatic tumor, and arachnoid cyst. Schwannomas are also found in the spinal canal and produce compression syndromes, but they are almost always incidental findings in the coroner's office. Malignant change is exceptionally rare.

Tumors of Maldevelopment

Malformative intracranial tumors, arguably neoplasms, form a small but important group of tumors encountered in the coroner's office. Because of their generally slow growth, tumors may attain a very large size, accommodated up to a point before catastrophic consequences, or they may produce symptoms when still small due

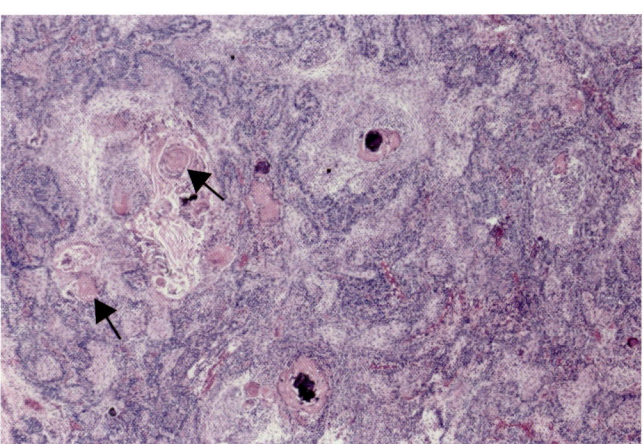

Fig. 13.82. Low-power micrograph of craniopharyngioma shows a chaotic field of squamous epithelium marked by sinuous basophilic lines representing nuclear basal layer. Arrows point to "wet keratin" nodules. Foci of calcification are common in this tumor. (H&E.)

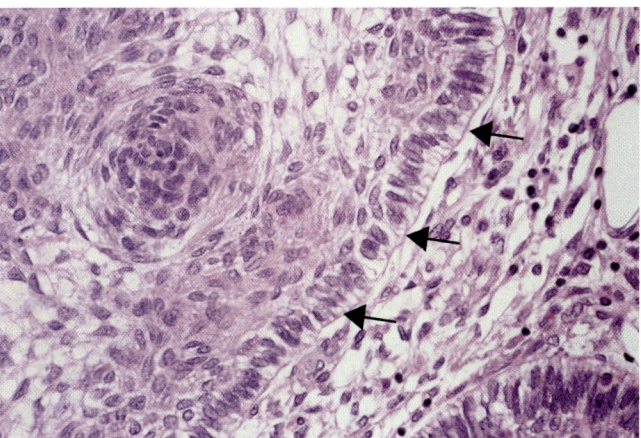

Fig. 13.83. High-power micrograph of tumor seen in Fig. 13.82 shows squamous epithelium in detail, with basal columnar cell layer on basal lamina (*arrows*) overlying more loosely organized stellate reticulum. (H&E.)

to involvement of vital sites. *Craniopharyngioma*, or Rathke's pouch tumor, is of this class of neoplasms, and is therapeutically challenging because of its suprasellar location and its tendency to infiltrate the hypothalamus (Fig. 13.81). More than half of these tumors are found in children and adolescents, which is to say fewer may be found in later life, some quite late, suggesting a slow rate of growth of these tumors.[41] In its suprasellar location, the tumor compresses the optic chiasm on its anterior border, elevates the floor and potentially obstructs the third ventricle, points posteriorly over the mammillary bodies, and sometimes extends through the diaphragma sellae into the pituitary fossa. Upon contact with the overlying hypothalamus, infiltrating fingers of tumor and resultant intense reactive gliosis make total surgical removal unlikely. The tumor is solid to cystic with machinery oil-like content, including cholesterol crystals that can be readily identified on a slide by polarizing microscopy as rainbow-colored parallelograms. The solid portion of the tumor is comprised of keratinizing stratified squamous epithelium with a columnar basal layer, sitting on a basal laminar membrane overlying a more loosely organized microcystic zone termed the stellate reticulum (Figs. 13.82–13.83).

Epidermoid Cyst

Epidermoid cyst, or cholesteatoma, makes up less than 1% of all intracranial neoplasms.[41] It is another slow-growing tumor, usually occurring at the cerebellopontine angle and parapituitary locations, causing predictable clinical syndromes of tumors in those locations. Some occupy midline anterior cerebral hemispheres, reach massive size, and present with psychiatric symptoms and sometimes sudden unexpected death.

Grossly, the tumor is a well-circumscribed encapsulated lobulated mass whose characteristic surface appearance led to its description as a "pearly tumor" because of its mother-of-pearl luster (Case 13.14, Figs. 13.84–13.86). The cyst wall is comprised of usually thin, flattened keratinizing stratified squamous epithelium on an outer collagenous membrane (Fig. 13.85). Cyst content is made up of hairlike desquamating parakeratotic keratin and grumous amorphous debris. When its contents are spilled by spontaneous or surgical rupture into the intracranial space, an acute inflammatory and giant cell reaction ensues, resulting in a picture of acute meningitis (Fig. 13.86).

Colloid Cyst

Colloid cyst is a benign neuroepithelial tumor found typically in the anterior third ventricle, a critical location for potential blockage of the interventricular foramina of Monro and CSF with resultant hydrocephalus. The cyst is said to arise from the paraphysis cerebri, an embryonic organ in the rudimentary choroidal fold hanging between the interventricular foramina.[1] The possibility of a choroidal epithelial origin has also been suggested.[31] Paraphyseal cuboidal epithelium is devoid of cilia, whereas choroidal epithelium is ciliated. Electron microscopy of the colloid cyst lining has shown both ciliated and nonciliated cells, voiding ciliation as a differential point. Further confounding electron microscopic details have led some to conclude that the epithelium is of endodermal origin, most likely from respiratory epithelium.[18]

> ### Case 13.14. (Figs. 13.84–13.86)
> A 33-year-old man with unconfirmed report of a brain tumor was found down lying prone in bed. He had a history of seizures, allegedly secondary to head injury sustained while in the military.

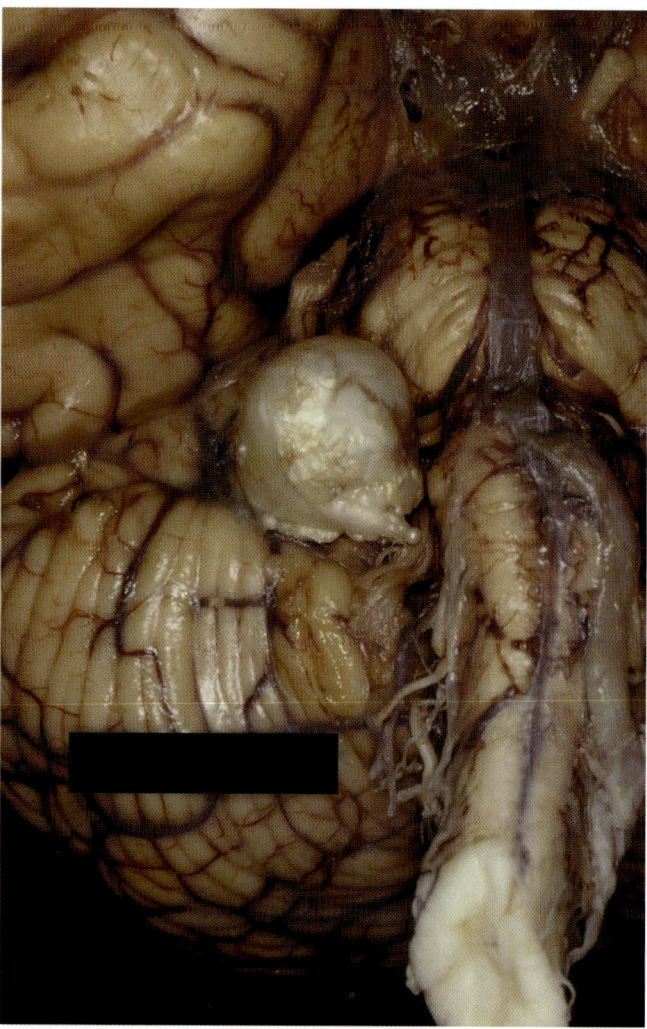

Fig. 13.84. Base view of brain shows a pearly tumor characteristic of epidermoid cyst in right cerebellopontine angle.

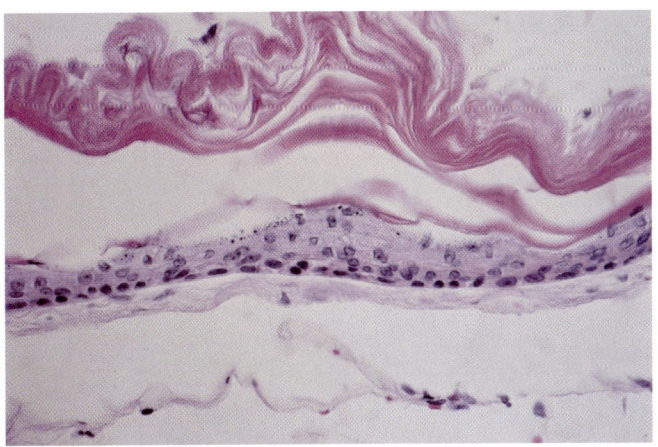

Fig. 13.85. High-power micrograph of epidermoid cyst wall shows thin squamous epithelium with keratohyaline granules lying upon thin layer of connective tissue. Desquamated keratin within the cyst lies above. (H&E.)

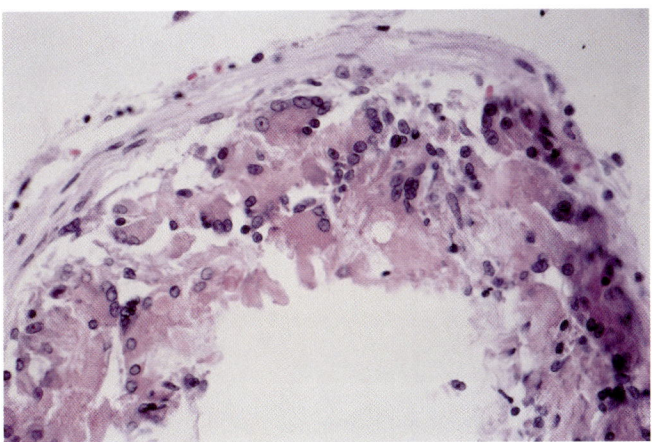

Fig. 13.86. High-power micrograph shows foreign body giant cell reaction where cyst wall has been disrupted. (H&E.)

How such epithelium finds its location in the anterior third ventricle is unexplained.

Grossly, the cyst is a smooth globe measuring about 2 cm in diameter on average in symptomatic cases, and attached to the roof of the third ventricle (Case 13.15, Fig. 13.87). Rare ones may attain larger size, and others may be a small incidental finding. The cyst is soft and usually pale tan, but it can be dark when hemorrhagic (Fig. 13.88) (see Chap. 9). Uncomplicated cysts are composed of a homogeneous colloid content usually lined by a single layer of cuboidal epithelium with secretory vacuoles oriented toward the cyst, supported by an outer layer of thin fibrous connective tissue (Fig. 13.89). A third ventricular cysticercus cyst may have a similar appearance, but it is most often unattached and never hemorrhagic (see Chap. 12). Colloid cyst is occasionally encountered as an incidental finding at autopsy, but it is also seen as

> **Case 13.15.** (Fig. 13.87)
>
> A 22-year-old man and his friends were involved in an altercation with other males, following which he sustained a gunshot wound to the right temple region as he was leaving the scene in a vehicle. CT head scan revealed skull fracture and intracerebral hemorrhage underlying a graze gunshot wound. Coincidentally, a "cyst on the roof of the third ventricular junction" was noted in the scan. Scalp wound was débrided and bullet fragments removed, and the decedent was released asymptomatic on the ninth hospital day. Some 3 months later, paramedics were summoned to the decedent's home because of headache, vomiting, and seizures. He was pronounced dead later that day.
>
> Autopsy confirmed a mild depressed fracture of the right temporal bone and chronic cortical contusion of the right temporal parietal region.

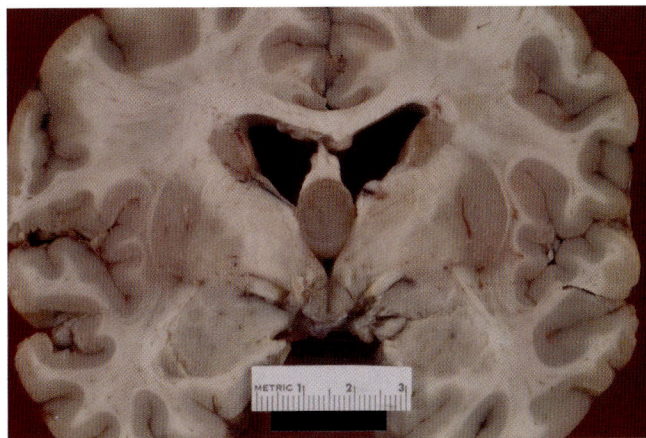

Fig. 13.87. Colloid cyst obstructs interventricular foramina and elevates fornices. Lateral ventricles are mildly dilated.

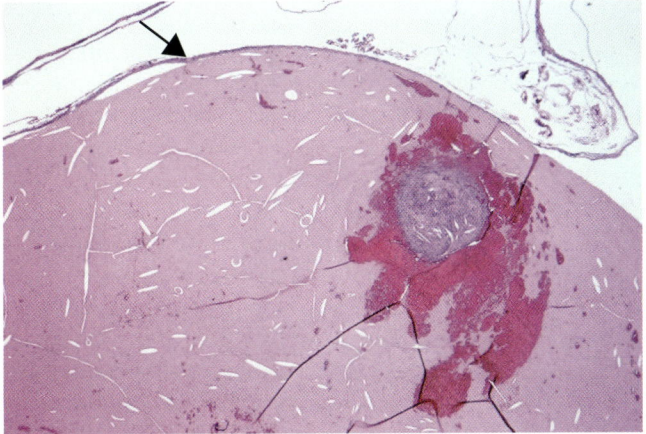

Fig. 13.88. Low-power micrograph of a colloid cyst from another case shows a thin cyst wall (*arrow*) and "colloid" content with scattered cholesterol clefts, a focus of recent hemorrhage, and a central zone of inflammation. (H&E.)

Brain Tumors

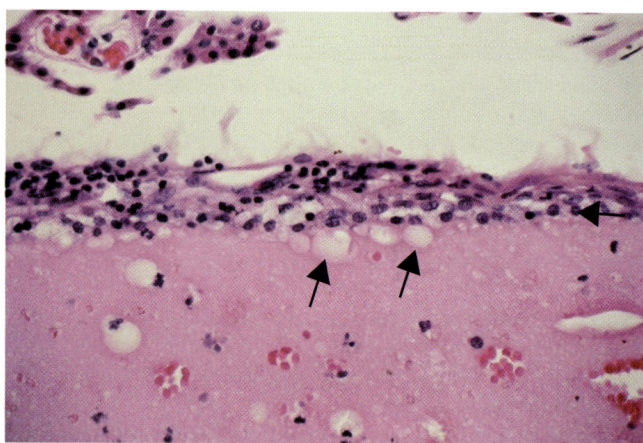

Fig. 13.89. High-power micrograph of colloid cyst shows a single layer (*horizontal arrow*) of cuboidal epithelium with small dark round nuclei supported by a thin layer of connective tissue. Secretory vacuoles arise from the epithelium (*vertical arrows*). (H&E.)

a cause of sudden unexpected death[2,35] (see Chap. 9). A careful review of the history may reveal that the decedent experienced severe intermittent headaches altered by change in posture. Intermittent vomiting, dimming of vision, or drop attacks have been described in some cases. In its deep-seated location, colloid cyst of the third ventricle presents a difficult surgical extraction when large. Intentional or unintentional section of the fornices to obtain access to the tumor results in serious impairments, such as loss of immediate recall memory.[44]

Neuroepithelial cysts may occur intraspinally, and in other intracranial locations such as the fourth ventricle, subarachnoid space, and, in rare instances, intracerebrally.[14]

Chordoma

Chordomas are rare tumors classified as intracranial and vertebral. Those involving the spinal column are more frequent, with most occurring in the sacrococcygeal region. The vertebral form is rarely a concern in the medical examiner's office. Intracranial chordomas consistently occur at the clivus of the sphenoid bone, are believed to arise from notochordal remnants, and typically penetrate the dura, infiltrate lower cranial nerves, compress the brainstem and sometimes the cerebellum, and extend rostrally into parasellar and orbital regions.

Grossly, the tumor is semitranslucent, gray, soft, and gelatinous, eroding the clivus and forming variable mass effect on ventral aspects of the brainstem and beyond, often surrounding the basilar artery (Fig. 13.90). Microscopically, a characteristic feature of the tumor is the physaliforous cell with its bubbly cytoplasm filled with clear vacuoles (Figs. 13.91–13.92). The cell, rich in glyco-

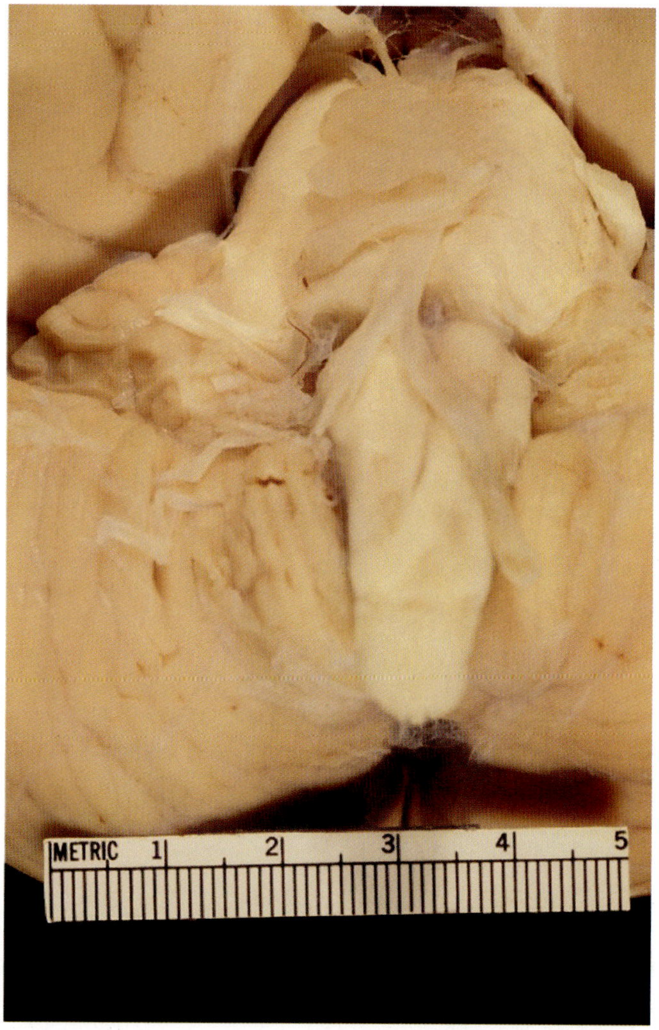

Fig. 13.90. Chordoma is seen as pale tan tissue partially covering basilar artery and belly of pons.

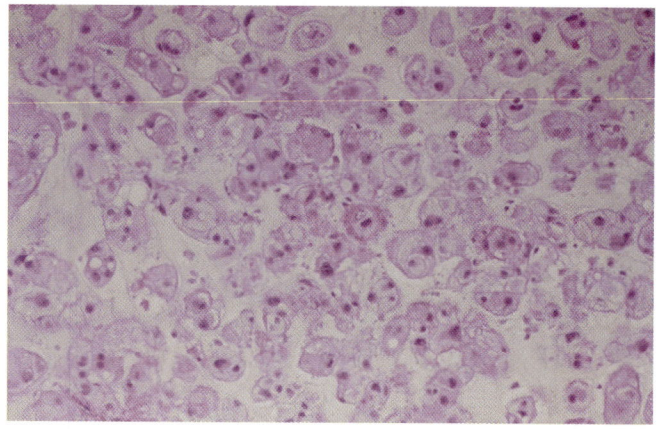

Fig. 13.91. Medium-power micrograph of chordoma shows irregularly round tumor cells, a few coherent, with bubbly cytoplasm in a mucinous matrix. (H&E.)

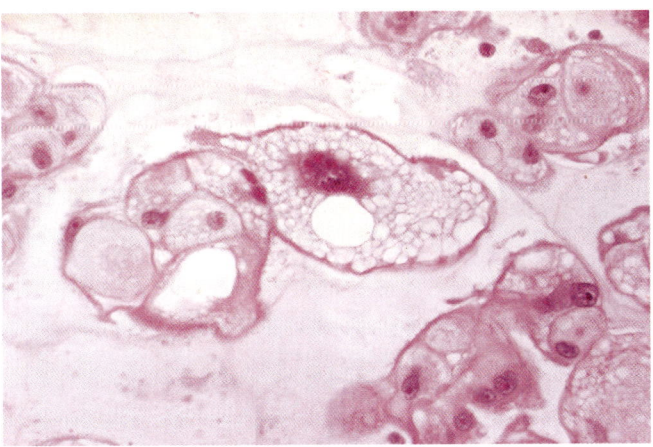

Fig. 13.92. High-power micrograph of the same tumor as seen in Fig. 13.91 better displays physaliforous cells. (H&E.)

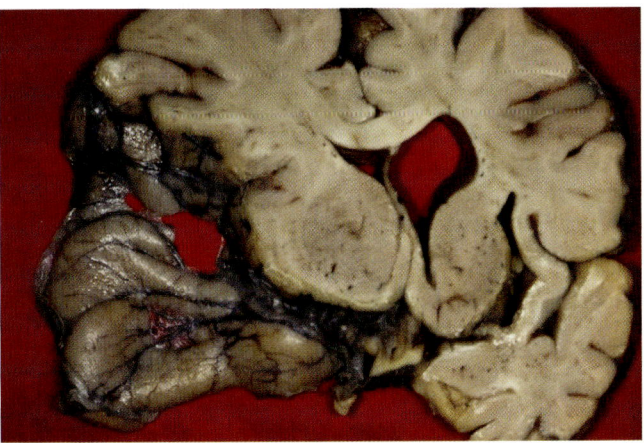

Fig. 13.94. Coronal section of brain seen in Fig. 13.93 at level of anterior striatum demonstrates massive displacement from right (left side of image) to left with partial effacement of ipsilateral, and moderate enlargement of contralateral, frontal horns. Note cortex bordering cyst is not atrophic. Subinsular vertical cyst (right side of image) is unrelated residual of an old hemorrhage of basal ganglia, typical in location and form of hypertensive hemorrhage.

gen, is periodic-acid schiff (PAS) positive. Thin fibrous septae may create a lobulated grouping of tumor cells that are embedded in a homogeneous background of mucinous matrix. The term *ecchordosis physalifora* is applied to a small intradural gelatinous globule, composed of cells not unlike those of chordoma, usually lying on the arachnoid surface of the midline belly of the pons. Small lesions are often loosely attached to the arachnoid and easily lost if not carefully handled. The finding is a relatively common incidental finding that is said to occur in about 2% of autopsies.[41] It is thought not to be a source of chordomas.

Arachnoid Cyst

Arachnoid cyst, strictly speaking, is not a neoplasm, but it may present as a slowly expanding mass causing compression and, at times, massive displacement of the brain (Figs. 13.93–13.94; Case 13.16, Fig. 13.95). The cyst is formed by splitting of the malformed arachnoid membrane supported by a layer of collagen.[13] The outer surface of the cyst is comprised of the normal interface layer of the dura and arachnoid.[13] By far the most frequent location of the cyst is the anterior sylvian vallecula (Case 13.16, Fig. 13.95). Other sites include the cerebellopontine angle, quadrigeminal plate cistern, cisterna magna, and other unusual locations.[13]

Grossly, the cyst wall consists of a semitransparent membrane that is usually collapsed due to rupture at autopsy, and one only becomes aware of its presence by the distorted anatomy (Figs. 13.93–13.95). The cyst mem-

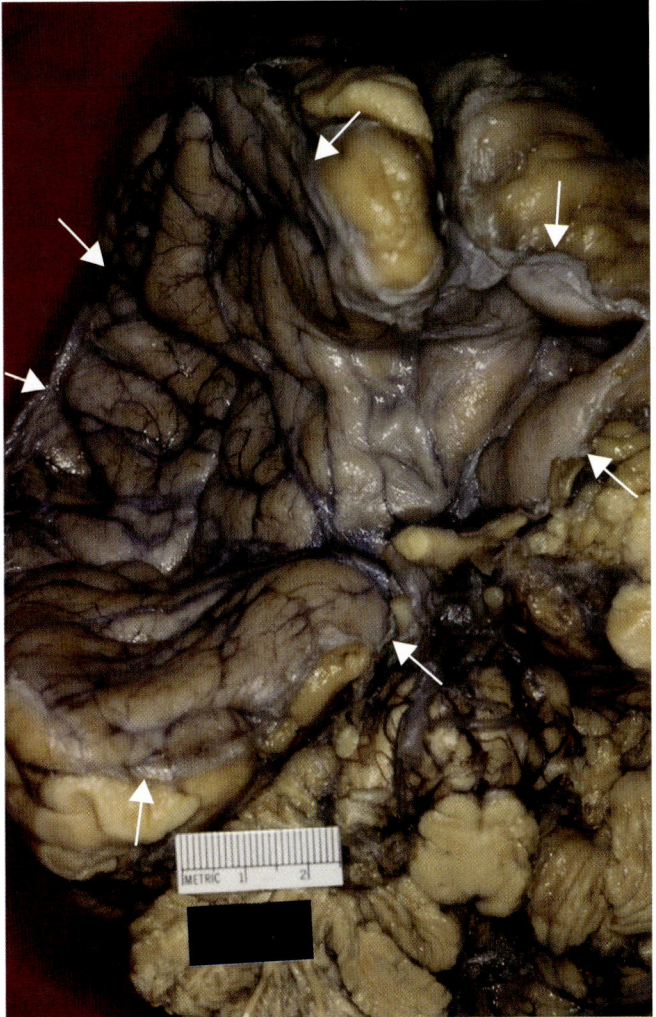

Fig. 13.93. Base view of brain shows a large arachnoid cyst that compresses right inferior frontotemporal region. Usual profile of temporal pole is obliterated by extreme compression and posterior displacement. Membrane covering cyst is torn and retracted to cyst margins (*arrows*).

Case 13.16. (Fig. 13.95)

A 31-year-old man's roommate was awakened by a noise in the early morning, and he discovered that the decedent was missing from the couch where he had been sleeping. A window was open, and he was found on the pavement three floors down. Resuscitative efforts continued for over an hour at the hospital but were unsuccessful.

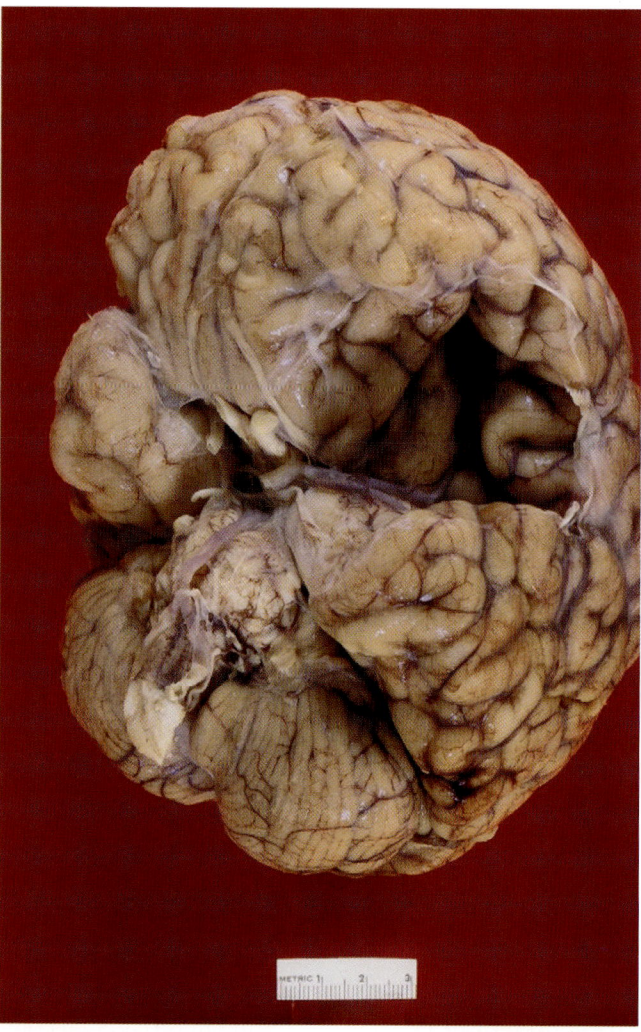

Fig. 13.95. A tilted view of the left lateral cerebral hemisphere shows a large arachnoid cyst that opens left sylvian vallecula by displacement of temporal pole and frontal lobe to expose insula cortex. Cortex in bed of cyst appears normal.

brane is best demonstrated under water. Microscopic sampling of the margin of the cyst membrane may demonstrate the point of splitting of the arachnoid membrane. An occasional arachnoid cyst exhibits thicker, whitish opaque areas in its wall. The cyst may become clinically evident at any age. In infancy, macrocrania may result. Development of increased intracranial pressure or seizures are other potential clinical presentations.

Lipoma

Intracerebral and intraspinal lipomas are benign tumors that, depending on their location, may cause important clinical syndromes. In general, they are found along the midline in such typical sites as the corpus callosum, floor of the third ventricle, quadrigeminal plate, and cerebellopontine angle. Tumors involving the quadrigeminal plate can compress and occlude the aqueduct or rostral fourth ventricle and produce hydrocephalus (Case 13.17, Fig. 13.96). Troublesome spinal lipomas typically involve the region of the conus medullaris and cauda equina, and their infiltration of rootlets can compromise attempted surgical removal. Large lipomas of the corpus callosum are often accompanied by agenesis of the body of that structure. A fibrous capsule is poorly organized at best, and the cingulate cortex is usually infiltrated.

Plain skull x-ray may expose the tumor as a lucent globe with peripheral calcification. Microscopically, the tumor usually differs little from mature fat, but occasional examples contain admixtures of neural, muscle, cartilaginous, and other tissue components.[41] Incidental dural lipomas, if one can dignify them as such, are a very common occurrence on the sub-dural surfaces of dorsal convexity regions and the falx cerebri. They usually measure no more than 0.2 to 0.4 cm in diameter as a flat round lesion. Occasional ones are larger, but not large enough to cause difficulty.

Pituitary Tumors

Pituitary tumors comprise about 4 to 8% of all intracranial tumors in general autopsy series, occurring mainly in adults. Only about 10% occur in children.[41] The tumors are classified as microadenoma (<10 mm) and macroadenoma (>10 mm). Further definitions in gradings rely on neuroimaging,[30] radiologic study,[16] and neurosurgical findings,[46] based upon features such as sellar erosion, suprasellar and parasellar extension (Case 13.18, Fig. 13.97–13.99), compression of the optic chiasm, impressions on the third ventricle, and evidence of metastases. The clinical classification of pituitary adenomas is based on endocrine effects and visual field defects. Less frequent are ocular palsies from compression of cavernous sinus walls and contained cranial nerves, CSF rhinorrhea, and endocrine disorders from impingement on the hypothalamus, on the adjacent pituitary gland, or due to endocrine activity by tumor cells.

Pathologically, the histologic classification of acidophil, basophil, and chromophobe tumors has become obsolete, replaced by a classification based mainly on immunohistochemistry and electron microscopic findings. The histologic pattern may be diffuse, sinusoidal, or papillary, but the architecture has no strict correlation with hormonal types. Two or more patterns may occur in the same tumor (Figs. 13.98–13.99). The combination

Case 13.17. (Fig. 13.96)

A 68-year-old man with a history of inoperable brain tumor sustained bilateral subdural hematomas when his pickup truck collided with a parked semi-truck, apparently after he suffered a stroke while driving. His Glasgow Coma Scale score was 3 in the emergency room, and he expired shortly thereafter.

Autopsy revealed an estimated 200-ml bilateral acute subdural hematoma, right side greater, with right uncal herniation without Duret's hemorrhage. Underlying the acute hematoma, bilateral thick chronic subdural hematomas were present.

Case 13.18. (Figs. 13.97–13.99)

A 46-year-old woman had a "stroke" 13 years earlier with resultant right-sided paralysis, but she had not seen a physician in 2 years. On the day of death, she complained of pain in her throat and was found without respirations about an hour later. She was DOA at the hospital. Autopsy findings included a large chronic cystic infarction of the left frontoparietotemporal lobe and a pituitary tumor.

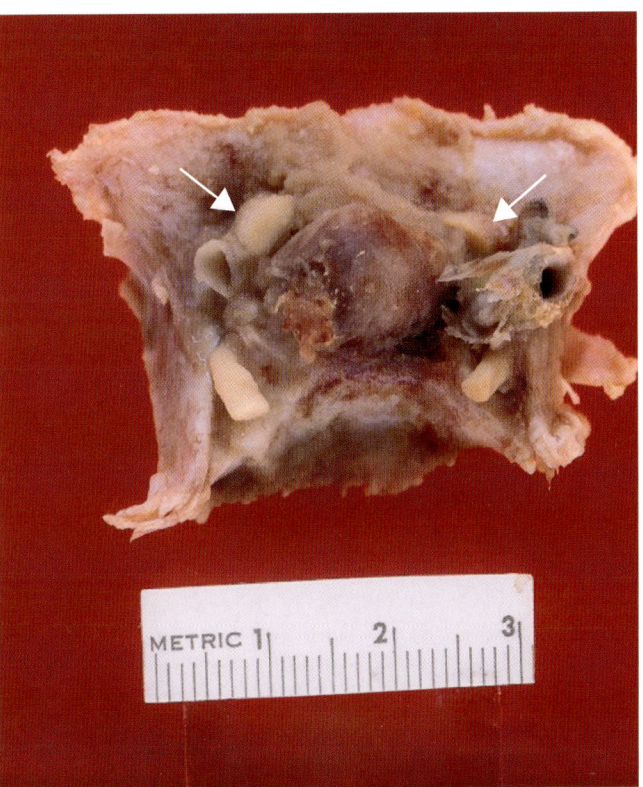

Fig. 13.97. Dome of suprasellar globular pituitary adenoma lies between internal carotid arteries. Optic nerves (*arrows*) are asymmetric, thinner on the right, although the tumor shows no greater mass on that side.

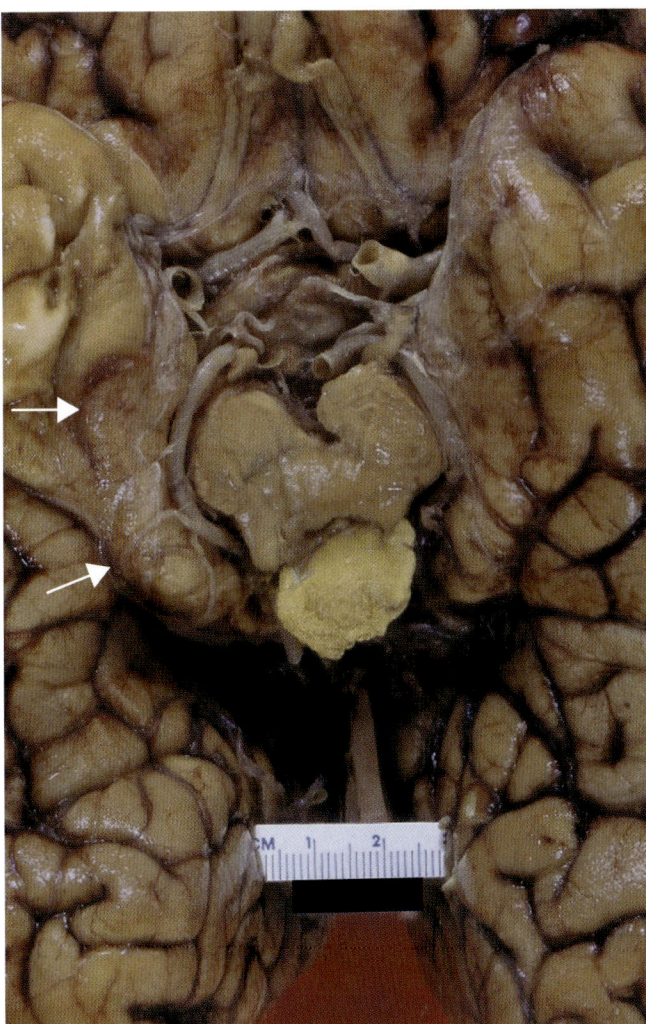

Fig. 13.96. Photo shows a large lipoma of quadrigeminal plate that caused partial occlusion of the cerebral aqueduct and rostral fourth ventricle. A large noncommunicating hydrocephalus accompanied partial "agenesis" of septum pellucidum and corpus callosum. "Agenesis" may have actually been lysis secondary to chronic hydrocephalus beginning at an early age. Note mild right parahippocampal herniation (*arrows*). Decedent was status post ventriculostomy at an unknown age.

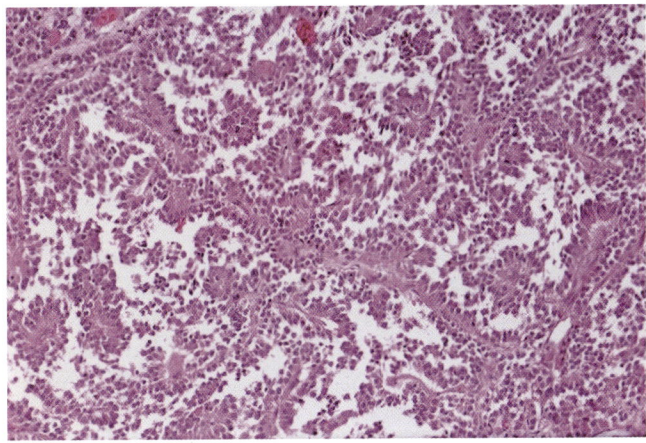

Fig. 13.98. Medium-power micrograph of pituitary adenoma exhibiting papillary pattern. Empty spaces are artifact. (H&E.)

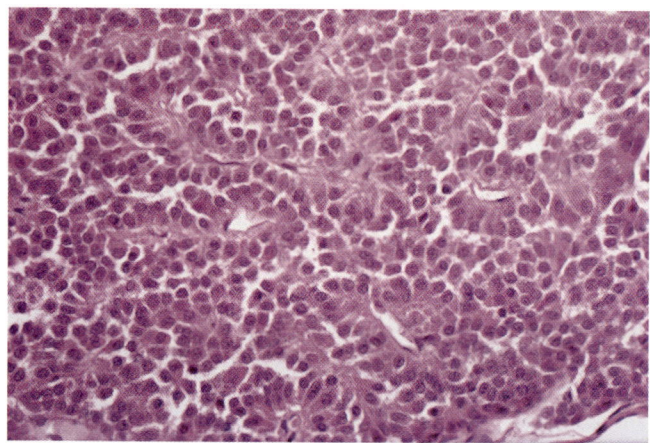

Fig. 13.99. High-power of same tumor as Fig. 13.98 shows monomorphic cells in sinusoidal pattern. (H&E.)

of six hormones and features of fine structure provides differentiation of some 16 different types of adenomas.[20] In forensic practice, tumor categorization to this degree is impractical and not critical in most situations.

Metastatic Neoplasms

Metastatic neoplastic disease of the CNS occurs primarily through hematogenous routes from distant primaries. Local metastases from malignant neoplasms of nasopharynx, nasal sinuses, and ear are rare. Statistics of metastatic tumors vary widely according to types of institutions from which they are compiled. Hospitals for the terminally ill and cancer hospitals have high figures of CNS metastases, followed by neurosurgical and community hospitals. In one large autopsy series from New York City of 3849 cases of malignant neoplasms collected over 25 years, 26% had intracranial metastases, including those involving the dura mater, leptomeninges, and pituitary gland. By contrast, only 8% had metastases in the spinal canal.[19] There is a general agreement that lung is the most common primary site of metastatic brain tumor in men and breast the most common primary site in women.[39] Other common sites of origin are kidney and gastrointestinal tract. A high percentage of malignant melanoma, testicular seminoma, and choriocarcinoma metastasize to the brain.

The majority of cases of metastatic brain tumor presenting to the office of the medical examiner have been symptomatic and had received medical attention for their primary tumor. Less often, the brain metastasis was the first manifestation of disease while the primary remained unknown.

The purpose of the forensic autopsy in this circumstance is substantiation of the primary, anatomic localization of metastatic tumors, and developing a clinical pathologic correlation relative to possible mechanism(s) of cause of death. Not surprisingly, the order of frequency of metastatic tumor sites is the cerebrum, cerebellum, and brainstem, in that order. Metastasis to the brainstem is likely to be the cause of death by destruction of centers and tracts subserving vital functions of respiration and cardiac rhythm (Case 13.19, Fig. 13.100). Metastases to the cerebrum and cerebellum may have lethal outcome secondary to mass effect, such as uncal-parahippocampal herniation and Duret's hemorrhage in the midbrain and pontine tegmentum, or pontine and medullary compression from cerebellar metastases (Case 13.20, Fig. 13.101). Small metastases in the cerebral and cerebellar hemispheric parenchyma tend to be multiple, although solitary ones are found and may be surgically extirpated. They tend to be spherical, sharply marginated (Case 13.21, Fig. 13.102), and preferentially located at the cortical/subcortical white matter boundary, but may be found at almost any site. Large metastases, although still sharply demarcated, tend to have irregular geographic margins, central necrosis, and hemorrhage. Necrosis, however, may also be a feature of small metastases. Some small metastases generate a disproportionate degree of edema, with resultant mass effect and herniation syndromes.

As previously mentioned, it is not unusual for some primary gliomas to metastasize within intracranial and intraspinal spaces. An uncommon example is metastasis from a choroid plexus carcinoma arising from the glomus of the choroid plexus (Fig. 13.103). Usually found in childhood, case reports of this tumor occurring in adults have been contested as more likely metastasis from occult adenocarcinoma from other loci such as the bronchus.[41] The microscopic characteristics of individual metastatic neoplasms can be found in standard reference books on pathology. An important caveat is being cognizant of similarities of appearances between some primary and metastatic neoplasms. For example, the stromal cell area of a cerebellar capillary hemangioblastoma might be confused with a metastatic clear cell carcinoma of the kidney. A complete autopsy should resolve most questions.

> **Case 13.19.** (Fig. 13.100)
>
> A 69-year-old woman was found to have gallbladder disease, and, during cholecystectomy, carcinoma of the sigmoid colon was discovered with positive lymph nodes and resected. In the ensuing 3 months, loss of brainstem functions included loss of sensation on the left side of the face, diminished left corneal reflex, vocal cord paralysis, loss of gag reflex, and central respiratory failure, along with decreasing mental status.

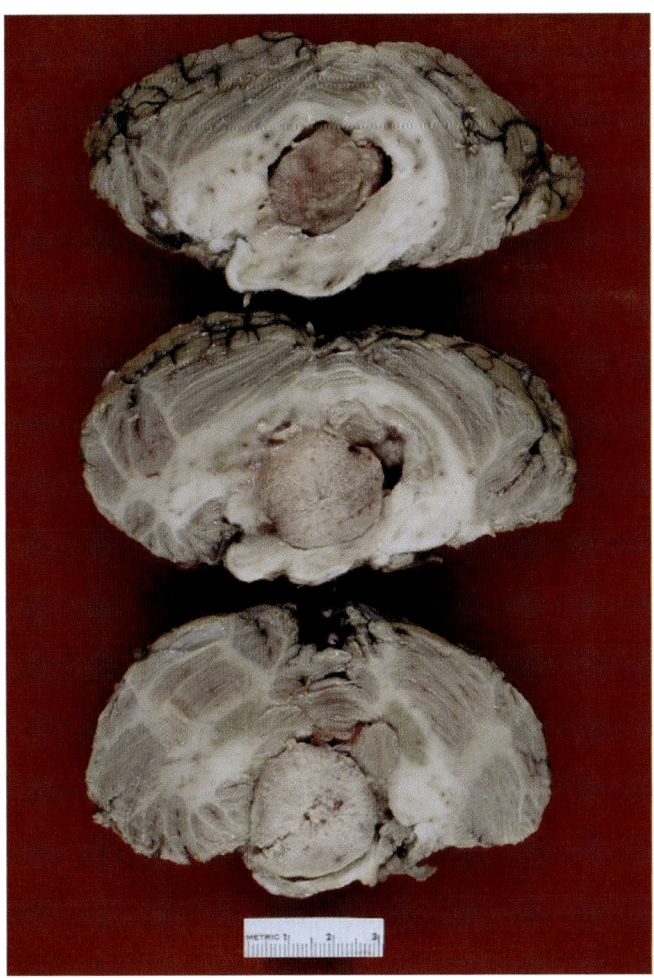

Fig. 13.100. Solitary globular intraventricular metastatic tumor expands the fourth ventricle with extreme attenuation of lower brainstem. Medulla is flattened to a mere ribbon in lowest section.

Case 13.20. (Fig. 13.101)
A 80-year-old man had a history of carcinoma of the large intestine.

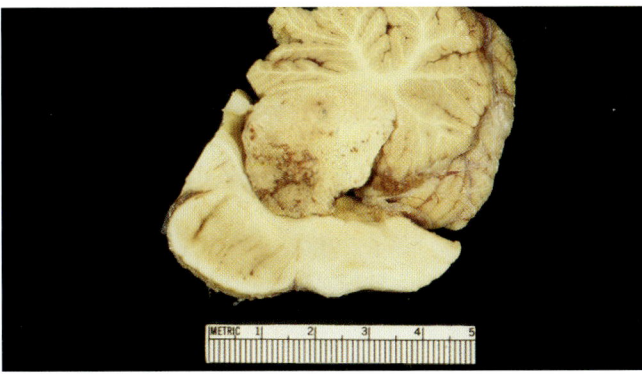

Fig. 13.101. Large solitary metastatic tumor arises from anterior vermis to fill fourth ventricle, compressing floor structures of pons and medulla.

Case 13.21. (Fig. 13.102)
A 41-year-old man had abused heroin and was on a methadone program. His history also included treatment for "liver and brain cancer." He was found dead at his residence a week and a half after discharge from the hospital.

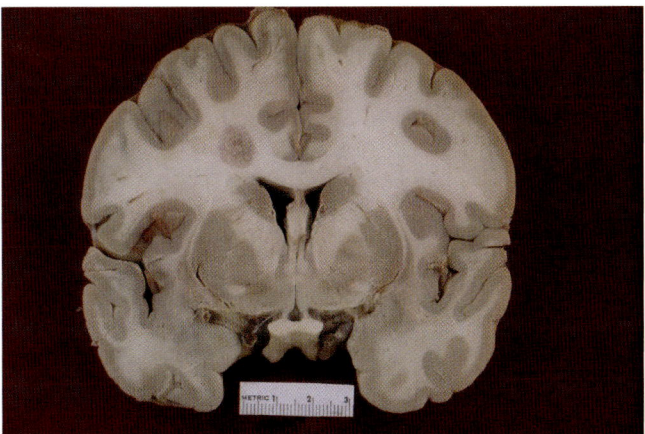

Fig. 13.102. Metastatic carcinoma of typically discrete round profile is shown in left frontal white matter lateral to corpus callosum. Only two other smaller metastases were found in the right frontal and left temporal white matter.

Primary Malignant Lymphomas

Primary malignant lymphomas of the brain were rare until the advent of immunosuppression associated with transplantation surgery and with human immunodeficiency virus (HIV) infection. The term excludes lymphomas that accompany extraneural involvement. Tumors are of B- or T-cell types, with B-cell lymphomas predominating by far (95%). Some 2 to 12% of patients with HIV infection develop this malignant neoplasm. It may rarely occur in the absence of therapeutic immunosuppression or HIV infection. The majority of tumors are found in the cerebrum with the frontal lobe being the most frequent site. It is rare in the spinal cord (1%).[26]

Grossly, the tumor size can vary widely, but some may attain considerable size. Margins of tumor may be diffuse to relatively sharply demarcated (Case 13.22, Fig. 13.104). The process may be multicentric, especially with HIV infection. Tumor is typically pink to gray with a granular texture, and larger tumors may be centrally necrotic or hemorrhagic (Case 13.23, Fig. 13.105). The gross appearance is quite variable, and some cases may be grossly indistinguishable from a malignant glioma. Microscopically, at low power, angiocentricity of tumor

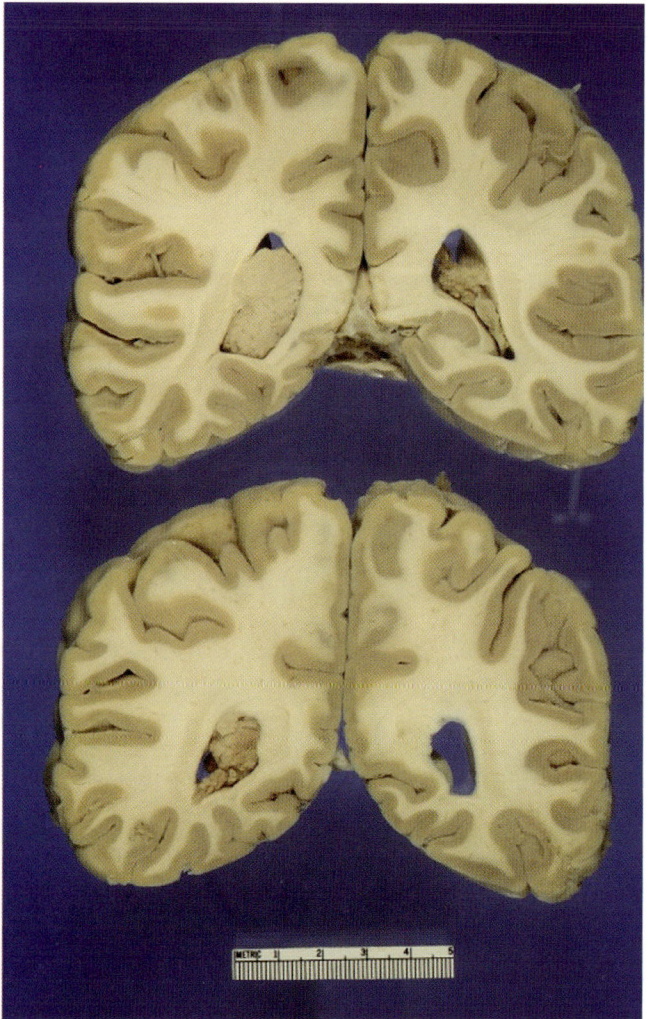

Fig. 13.103. Choroid plexus carcinoma overgrows left glomus of choroid plexus, with metastasis involving right glomus and tumor seeding the subarachnoid space.

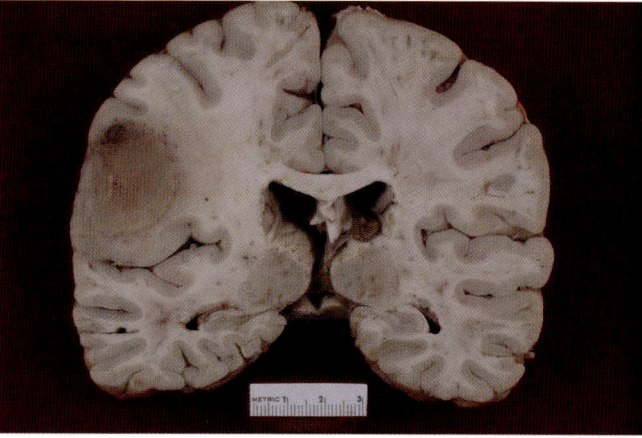

Fig. 13.104. Primary lymphoma in left inferior parietal lobule appears as pale pink-tan relatively homogeneous mass with indistinct borders obliterating cortical ribbon. No evidence of toxoplasmosis was found.

Case 13.22. (Fig. 13.104)

A 35-year-old man with acquired immunodeficiency syndrome was in custody for battery. With development of confusion, he was transferred to a hospital jail ward. MR scan revealed a left parietal gadolinium-enhancing ring lesion. Toxoplasmosis was suspected, and he was treated for it. Two weeks into his hospitalization, the decedent was awake and responding to verbal commands, but he suddenly died the following day.

Case 13.23. (Fig. 13.105)

A 23-year-old male homosexual and polydrug abuser was admitted to the hospital because of weight loss, generalized weakness, and difficulty speaking. CT head scan disclosed a mass in the left frontotemporal region. Craniotomy and partial lobectomy were performed, and the surgical pathology diagnosis was lymphoma. He died on the ninth hospital day. Tumor was limited to the brain at general autopsy.

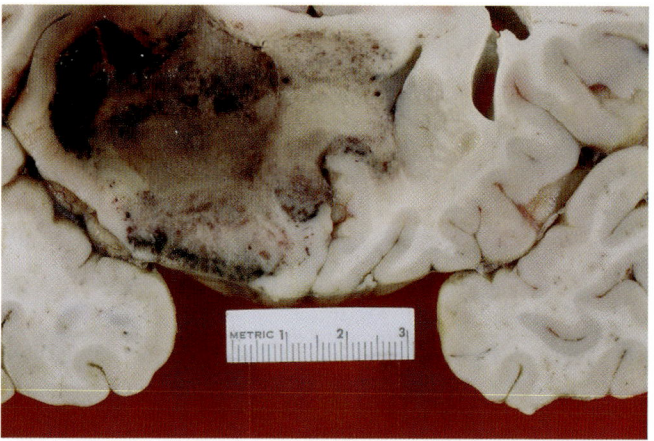

Fig. 13.105. Massive primary malignant hemorrhagic and necrotic lymphoma of left frontal lobe crosses midline through genu of corpus callosum.

cells is distinctive, and expansion of vessel walls infiltrated by tumor cells is highly characteristic. The finding can be demonstrated to advantage with a reticulin stain. Tumor cells are often a mixed population of large B lymphocytes, small lymphocytes, and occasional plasmacytoid cells. Mitoses are common, and large bizarre multinucleated forms may be present.

Extramural Metastasis

Extramural metastasis of brain tumors is a rare occurrence. Glioblastoma multiforme is by far the most common source, followed by medulloblastoma.[38] Other tumors known to metastasize include astrocytomas, ependymomas, and oligodendrogliomas. Meningiomas are also known to metastasize to extraneural sites, but those that do usually exhibit some features of malignancy such as high cellularity, mitoses, necrosis, and invasion of the brain.[24] Kepes[24] reviewed 85 published cases of extraneural metastases, and nearly all had undergone prior surgery. However, it is not to be assumed that all followed surgery.

Causes of Brain Tumor

Other than genetic risk, etiologies of brain neoplasms remain controversial. Workers in all manner of occupations including white collar, health professions, farming, mining, and manufacturing have claimed being placed at risk for brain tumor, presumably by exposure to some carcinogen or carcinogenic factor in individual circumstances. An excess of brain tumors has been also alleged in persons exposed to environmental factors such as electromagnetic fields of various sources (e.g., cell phones and power lines).[33]

Brain tumors discovered following head trauma and diagnostic x-rays generate recurring claims of causal relationships.[24] As to trauma predisposing to meningioma, strict criteria for acceptance of a possible causal relationship have been stipulated by Auster.[3] Less controversial is the possible relationship of brain tumors arising within the target field of therapeutic x-irradiation.[27]

References

1. Ariëns Kappers J. The paraphysis cerebri. In Crosby EC, Humphrey T, Lauer EW (eds): Correlative Anatomy of the Nervous System. New York: Macmillan, 1962.
2. Aronica PA, Ahdab-Barmada M, Rozin L, Wecht CH. Sudden death in an adolescent boy due to a colloid cyst of the third ventricle. Am J Forensic Med Pathol 1998;19:119–122.
3. Auster LS. The role of trauma in oncogenesis: A judicial consideration. JAMA 1961;175:946–950.
4. Cavanaugh JB. On certain small tumours encountered in the temporal lobe. Brain 1958;81:389–405.
5. Chatty EM, Earle KM. Medulloblastoma: A report of 201 cases with emphasis on the relationship of histologic variants to survival. Cancer 1971;28:977–983.
6. Choi NW, Schuman LM, Gullen WH. Epidemiology of primary central nervous system neoplasms. I. Mortality from primary central nervous system neoplasms in Minnesota. Am J Epidemiol 1970;91:238–259.
7. Couch JR, Weiss SA. Gliomatosis cerebri: Report of four cases and review of the literature. Neurology 1974;24:504–511.
8. Courville CB. Pathology of the Central Nervous System, ed 3. Mountain View, CA: Pacific Press, 1950.
9. DiMaio SM, DiMaio VJ, Kirkpatrick JB. Sudden unexpected deaths due to primary intracranial neoplasms. Am J Forensic Med Pathol 1980;1:29–45.
10. Dohrman GJ, Farwell JR, Flannery JT. Ependymomas and ependymoblastomas in children. J Neurosurg 1976;45:273–283.
11. Dohrman GJ, Farwell JR, Flannery JT. Glioblastoma multiforme in children. J Neurosurg 1976;44:442–448.
12. Erlich SS, Davis RL. Spinal subarachnoid metastasis from primary intracranial glioblastoma multiforme. Cancer 1978;42:2854–2864.
13. Friede RL. Developmental Neuropathology, ed 2. Berlin-Heidelberg: Springer-Verlag, 1989.
14. Friede RL, Yasargil MG. Supratentorial intracerebral epithelial (ependymal) cysts: Review, case reports, and fine structure. J Neurol Neurosurg Psychiatry 1977;40:127–137.
15. Graham DI, Lantos PL (eds): Greenfield's Neuropathology, ed 7. London: Arnold, 2002.
16. Hardy J, Vezina JL. Transsphenoidal neurosurgery of intracranial neoplasm. Adv Neurol 1976;15:261–274.
17. Hart MN, Petito CK, Earle KM. Mixed gliomas. Cancer 1974;33:134–140.
18. Ho KL, Garcia JH. Colloid cysts of the third ventricle: Ultrastructural features are compatible with endodermal derivation. Acta Neuropathol 1992;83:605–612.
19. Hojo S, Llena J, Hirano A, Zimmerman HM. Metastatic tumors in the central nervous system. J Neuropathol Exp Neurol 1978;37:628.
20. Horvath E, Scheithauer BW, Kovacs K, Lloyd RV. Regional neuropathology: Hypothalamus and pituitary. In Graham DI, Lantos PL (eds): Greenfield's Neuropathology, ed 6. London: Arnold, 1997.
21. Ingraham FD, Matson DD. Neurosurgery of Infancy and Childhood. Springfield, IL: Charles C Thomas, 1954.
22. Kadin ME, Rubinstein LJ, Nelson JS. Neonatal cerebellar medulloblastoma originating from the fetal external granular layer. J Neuropathol Exp Neurol 1970;29:583–600.
23. Kalyan-Raman UP, Olivero WC. Gangliogliomas: A correlative clinicopathological and radiological study of ten surgically treated cases with follow-up. Neurosurgery 1987;20:428–433.
24. Kepes JJ. Meningomas: Biology, Pathology, and Differential Diagnosis. New York: Masson Publishing USA, 1982.
25. Kepes JJ, Rubinstein LJ, Eng LF. Pleomorphic xanthoastrocytoma: A distinctive meningocerebral glioma of young subjects with relatively favorable prognosis; a study of 12 cases. Cancer 1979;44:1839–1852.
26. Kleihues P, Cavenee WK (eds). World Health Organization Classification of Tumours. Pathology and Genetics of Tumours of the Nervous System. Lyon: International Agency for Research on Cancer Press, 2000.
27. Kleinschmidt-DeMasters BK, Kang JS, Lillehei KO. The burden of radiation-induced central nervous system tumors: A single institution's experience. J Neuropathol Exp Neurol 2006;65:204–216.
28. Kondziolka D, Bilbao JM. Mixed ependymoma-astrocytoma (subependymoma?) of the cerebral cortex. Acta Neuropathol 1988;76:633–637.
29. Koos WT, Miller MH. Intracranial Tumors of Infants and Children. St Louis: CV Mosby, 1971.
30. Kovacs K, Horvath E. Tumors of the Pituitary Gland. Atlas of Tumor Pathology. Second Series, Fascicle 21. Washington, DC: Armed Forces Institute of Pathology, 1986.
31. Lach B, Scheithauer BW, Gregor A, Wick MR. Colloid cyst of the third ventricle: A comparative immunohistochemical study of neuraxis cysts and choroid plexus epithelium. J Neurosurg 1993;78:101–111.
32. Liu HM, Boggs J, Kidd J. Ependymomas of childhood. I. Histological survey and clinicopathological correlation. Child's Brain 1976;2:92–110.

33. Lin RS, Dischinger PC, Conde J, Farrell KP. Occupational exposure to electromagnetic fields and the occurrence of brain tumors. J Occup Med 1985;27:413–419.
34. Liwnicz BH, Wu SZ, Tew JM Jr. The relationship between the capillary structure and hemorrhage in gliomas. J Neurosurg 1987;66:536–541.
35. McDonald JA. Colloid cyst of the third ventricle and sudden death. Ann Emerg Med 1982;11:365–367.
36. McLean CA, Jellinek DA, Gonzales MF. Diffuse leptomeningeal spread of pleomorphic xanthoastrocytoma. J Clin Neurosci 1998;5:223–233.
37. Mørk SJ, Lindegaard K-F, Halvorsen TB, et al. Oligodendroglioma: Incidence and biologic behavior in a defined population. J Neurosurg 1985;63:881–889.
38. Pasquier B, Pasquier D, N'Golet A, et al. Extraneural metastases of astrocytomas and glioblastomas: Clinicopathological study of two cases and review of the literature. Cancer 1980;45:112–125.
39. Posner JB. Management of brain metastases. Rev Neurol 1992;148:477–487.
40. Queiroz LS, Lopes de Faria J, Cruz Neto JN. An ependymoblastoma of the pons. J Pathol 1975;115:207–210.
41. Russell DS, Rubinstein LJ. Pathology of Tumours of the Nervous System, ed 5. Baltimore: Williams & Wilkins, 1989.
42. Rubinstein LJ. Tumors of the Central Nervous System. Atlas of Tumor Pathology. Second Series, Fascicle 6. Washington, DC: Armed Forces Institute of Pathology, 1972.
43. Scherer HJ. The forms of growth in gliomas and their practical significance. Brain 1940;63:1–35.
44. Sweet WH, Talland GA, Ervin FR. Loss of recent memory following section of fornix. Trans Am Neurol Assoc 1959;84:76–82.
45. Taptas JN. Intracranial meningioma in a four-month-old infant simulating subdural hematoma. J Neurosurg 1961;18:120–121.
46. Wilson CB. A decade of pituitary microsurgery. The Herbert Olivecrona lecture. J Neurosurg 1984;61:814–833.
47. Wolf HK, Müller MB, Spänle M, et al. Ganglioglioma: A detailed histopathological and immunohistochemical analysis of 61 cases. Acta Neuropathol 1994;88:166–173.
48. Zimmerman HM. Brain tumors—their incidence and classification in man and their experimental production. Ann N Y Acad Sci 1969;159:337–359.
49. Zimmerman HM. Introduction to tumors of the central nervous system: Neoplasms. In Minckler J (ed): Pathology of the Nervous System, Vol 2. New York: McGraw-Hill, 1971.
50. Zülch KJ. Brain Tumours: Their Biology and Pathology. New York: Springer-Verlag, 1957.

Neurodegenerative Disorders 14

INTRODUCTION 395
ALZHEIMER'S DISEASE 395
PICK'S DISEASE 400
DIFFUSE LEWY BODY DISEASE 402
PARKINSON'S DISEASE 402
HUNTINGTON'S DISEASE 404

PROGRESSIVE SUPRANUCLEAR PALSY 405
MULTIPLE SYSTEM ATROPHY 405
FRIEDREICH'S ATAXIA 408
MOTOR NEURON DISEASE 409
REFERENCES 410

Introduction

"Neurodegenerative disorders" are a heterogeneous collection of well-recognized conditions of unknown etiology having in common a clinical history of a set of characteristic neurologic deficits with pathoanatomic correlation. Recent classifications emphasize clinical presentations such as dementing, movement, and neuromuscular disorders. A few of the group are readily distinguished by characteristic features, such as the senile plaque in Alzheimer's disease and the Lewy body in Parkinson's disease. Others may be identified only by extensive and expensive immunocytochemical and biochemical studies. In our office, by far, the most common disorder encountered is Alzheimer's disease (AD). Some cases with these chronic conditions are seen by the medical examiner due to an unexpectedly sudden death. For example, asphyxia may result from aspiration of food or foreign objects in persons with dementia and/or neuromuscular disorders.[30] More usual circumstances include death at home with evidence of neglect by a caretaker, the possibility of elder abuse of the demented, and those killed by auto and other accidents while wandering. Decedents with non-dementing disorders under medical management, such as those with Parkinson's disease, are seen much less frequently. An exception is deaths that occur in government institutions, all of which are referred to the medical examiner's office.

Alzheimer's Disease

AD is the most common of the neurodegenerative disorders, comprising about two-thirds (42–81%) of dementia cases.[46] In the year 2000, an estimated 4.5 million persons had Alzheimer's disease in the United States, and this number will increase to 13.2 million by 2050 with the increasing population of advanced age.[25] In this same report, 7% of those with AD were between 65 and 74 years of age, 53% were between 75 and 84 years, and 40% were 85 years and older. Age of onset is considered early before the age of 60 years. The mean age of onset of dementia, probably of mixed etiology, was said to be 82.3 years in France.[26] AD leads to a high of risk of death, with an average survival of about 4.5 years from onset. Prevalence is about 1% in those 60 to 69 years of age, increasing to 40 to 50% in those 95 years of age and older.[31] In the Framingham study, prevalence of AD was zero for men and women before age 65 years but rose to 37 per 1000 for men and 88.4 per 1000 for women for the age range of 80 to 84 years.[4] The clinical diagnosis of AD is made after exclusion of treatable and other untreatable dementing disorders, and definitive diagnosis requires autopsy.[7]

The gross anatomic features of the brain of a case of AD, even those submitted with a history of dementia, may show surprisingly minor deviations from the norm. That is, brain weight, cortical fullness, sulcal openings, ventricular size, and hippocampal formation/amygdala may show little noticeable alterations from norms for age. At the other extreme, the brain may be small, weighing less than 1000 g, show almost knife-blade cortical atrophy, have thickened leptomeninges, marked ventricular dilatation, and shrunken hippocampal formations. In the majority of cases of AD, the findings fall someplace in between (Case 14.1, Figs. 14.1–14.2). Focal cortical atrophy, when present, of the pyriform cortex and parahippocampal gyri may push the diagnosis toward AD. Although gross findings of AD are nonspecific, microscopic features, with qualifications, are diagnostic. These are the senile plaque (Figs. 14.3–14.4), neurofibrillary tangle (Fig. 14.5), and neuropil thread. Ancillary findings are granulovacuolar degeneration (Fig. 14.5) and the Hirano body (Fig. 14.6). Although once thought unique to AD, senile plaques have been reported in other conditions, such as dementia pugilistica and progressive supranuclear palsy (PSP).[22] All of the classic pathologic features, including senile plaques,

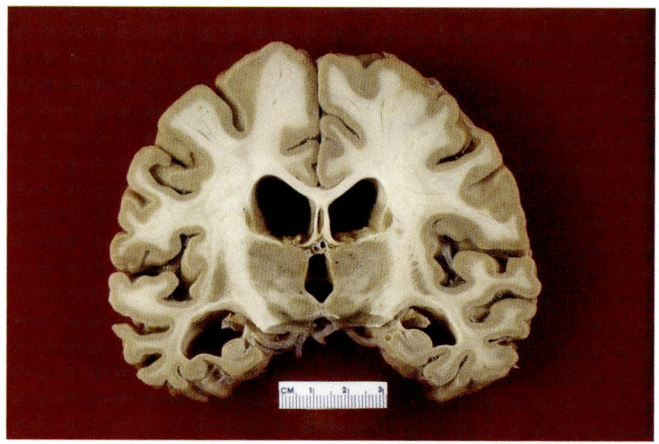

Fig. 14.2. Coronal section at midthalamic level exposes moderate lateral ventricular dilatation including temporal horns, with mild atrophy of hippocampal formations. Sulci are abnormally open, but obvious thinning of cortical ribbon is limited and most apparent in temporal lobes.

Case 14.1. (Figs. 14.1 and 14.2)

A 94-year-old woman who had resided in a board and care facility for about 5 years, the reason for which was not given, fell at home and sustained a hip fracture. She underwent open reduction and internal fixation hip surgery. Until then, she was reportedly ambulatory and in "fairly good health." No history of dementia was mentioned. She was transferred to a skilled nursing facility 3 days' postoperatively. About 4 months later, she was returned for treatment to an acute care hospital because of lethargy and possible urinary tract infection. About a week later, she was transferred to a chronic care facility where she was found to have stage III–IV decubitus ulcers. Two days later, she was found dead.

Fig. 14.3. Low-power micrograph of cerebral cortex shows numerous senile plaques seen as diffusely distributed minute specks. (Bielschowsky.)

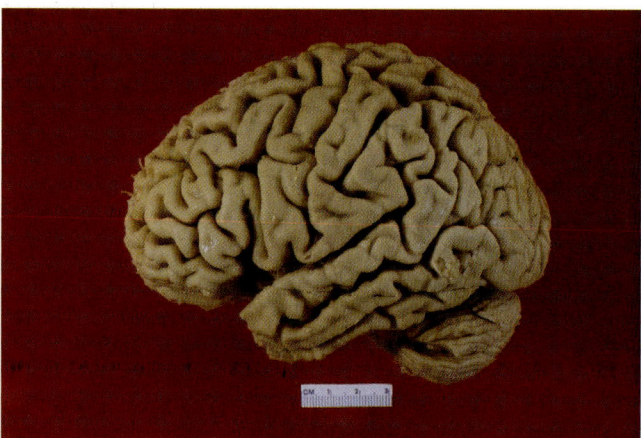

Fig. 14.1. Brain stripped of leptomeninges shows diffuse cerebral cortical atrophy seen about equally in frontal, temporal, and parietal lobes in this case of Alzheimer's disease (AD). Atrophy is characteristic but not specific for AD.

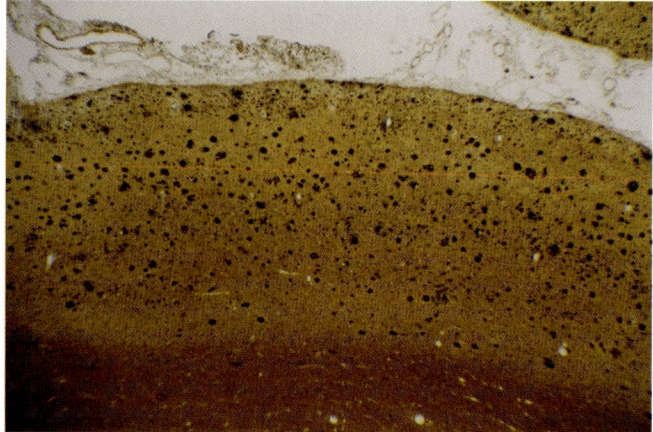

Fig. 14.4. Senile plaques are shown at slightly higher power than in Fig. 14.3. Note deeper layers of cortex have fewer plaques. (Bielschowsky.)

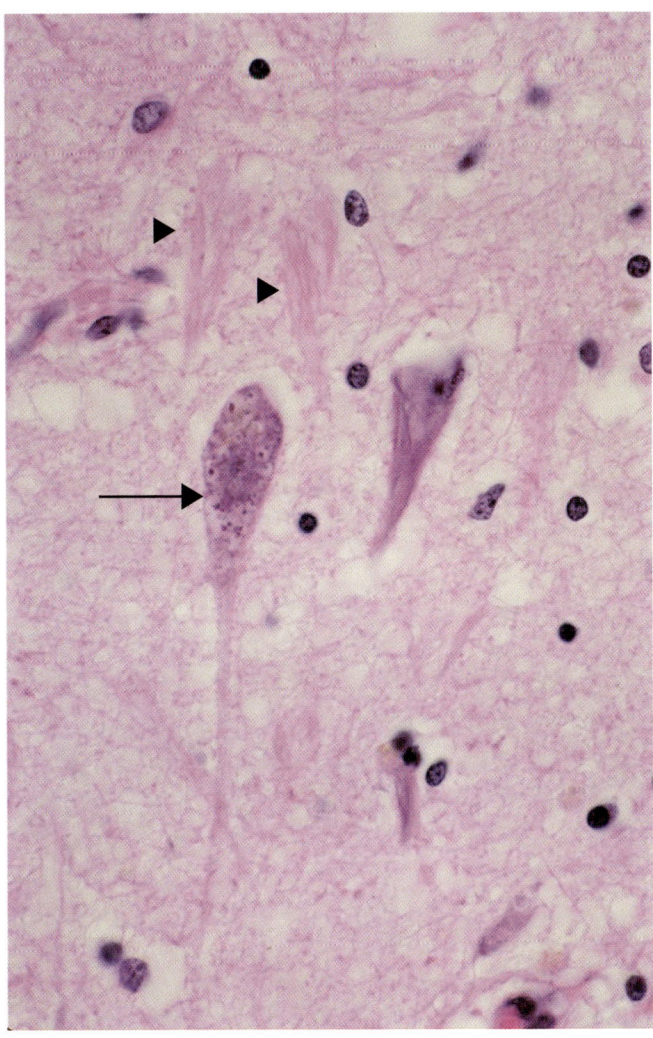

Fig. 14.5. High-power micrograph of CA2 of hippocampus proper shows granulovacuolar degeneration in a case of Alzheimer's disease (*arrow*) and a neurofibrillary tangle to its right. Two ghost neurofibrillary tangles are above (*arrowheads*). (H&E.)

are also found in Down syndrome.[49] Survivors of Down syndrome beyond the age of 30 to 40 years invariably show changes of AD indistinguishable from those with AD. Senile plaques are round bodies measuring 5 to 200 μm in diameter,[22] found mainly in the cerebral cortex with selective distribution in a few subcortical sites such as the amygdaloid complex of nuclei (Fig. 14.7), the striatum (Figs. 14.8–14.9), and rarely the molecular layer of the cerebellum. Distribution of senile plaques in the hippocampus proper tends to occur in greater numbers in CA1 into the subiculum than in CA2 or the endfolium (CA4). In the dentate fascia, plaques are aligned in the molecular layer (Fig. 14.10). Morphologic subtypes of senile plaques are neuritic and diffuse (Fig. 14.11). Neuritic plaques, which carry a higher correlation with dementia, have a central amyloid core circled by debris of neuritic degeneration, rare astrocytes, and microglia (Fig. 14.12). As the name indicates, diffuse plaques are composed of a small to large, irregular to round deposit of fine amyloid fibrils, and their presence in normal aging has been suggested as an indicator of presymptomatic AD.[43] Diffuse plaques are said to eventually form neuritic plaques.[15] Depending on the quality of the hematoxylin-eosin (H&E) stain, plaques may be difficult to detect, and sections stained in this manner cannot be relied on as the basis for semiquantitative plaque counts used for diagnosis. Stains recommended for quantitative evaluation are mentioned later. The neurofibrillary tangle (NFT) is an unmistakable neuronal cytoplasmic marker second in importance in AD. Its light microscopic appearance, however, is nonspecific, as NFTs may be found in other widely disparate conditions. Some of these are dementia pugilistica, PSP, end-stage subacute sclerosing panencephalitis (SSPE), postencephalitic parkinsonism

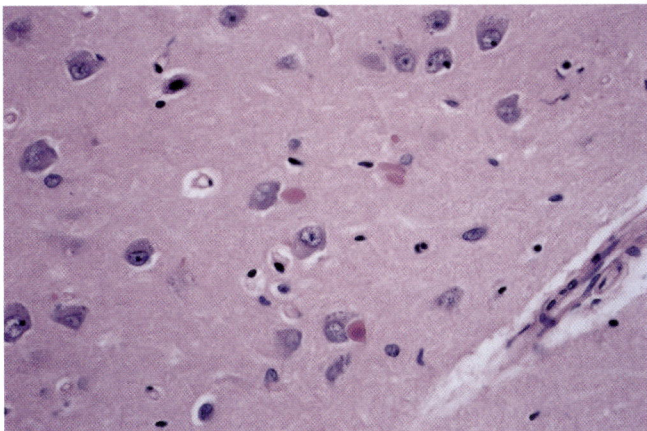

Fig. 14.6. High-power micrograph shows perineuronal brightly acidophilic bodies, most often rod-shaped, called Hirano bodies. (H&E.)

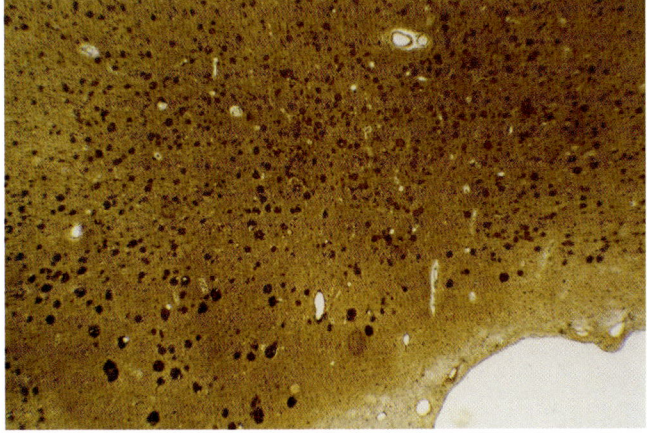

Fig. 14.7. Medium-power micrograph shows myriads of senile plaques seen as dark spots in basomedial nucleus of the amygdaloid complex. (Bielschowsky.)

Neurodegenerative Disorders

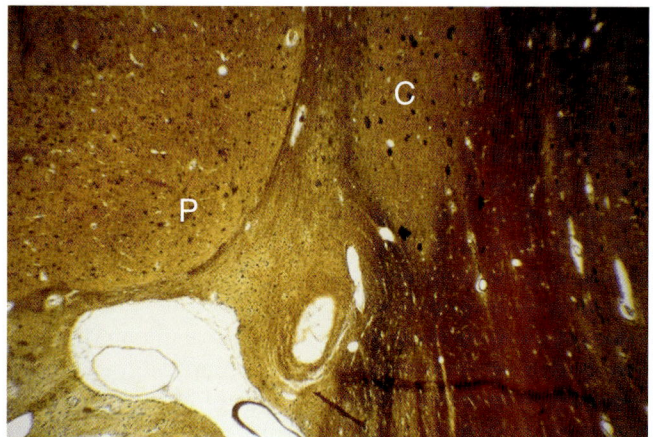

Fig. 14.8. Senile plaques are seen as small dark spots in claustrum (C) and ventral putamen (P). (Bielschowsky.)

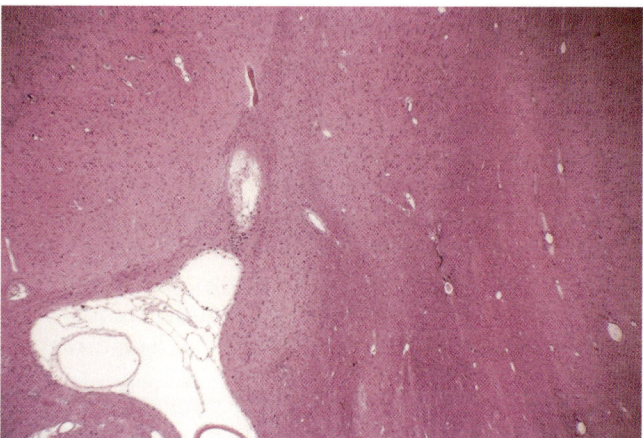

Fig. 14.9. The H&E-stained section of the same field stained by Bielschowsky stain in Fig. 14.8 is shown to emphasize the difficulty in detecting senile plaques with this stain at this, and even higher-power, microscopy. (H&E.)

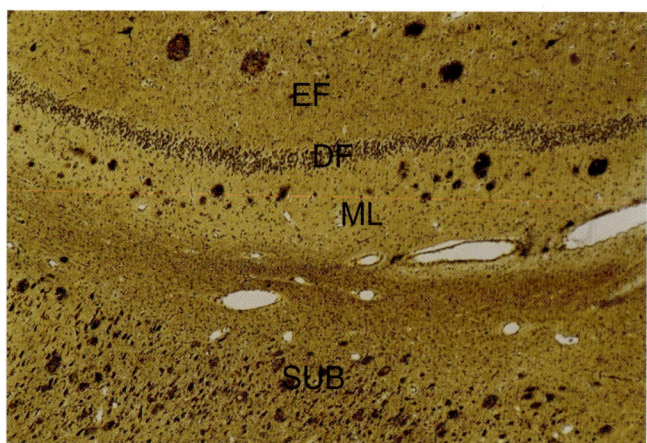

Fig. 14.10. Medium-power micrograph of dentate fascia (DF) and part of adjacent subiculum (SUB) shows plaques arranged in molecular layer (ML) of dentate fascia. EF, endfolium. (Bielschowsky.)

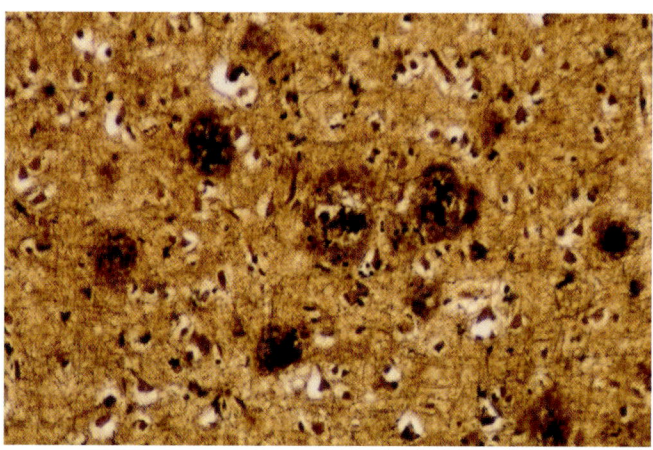

Fig. 14.11. High-power micrograph stained with silver method shows several senile plaques. The one in the middle with a central condensation (amyloid) is a neuritic plaque. Diffuse plaques are less organized and form a diffuse nest of fibrils. (Bielschowsky.)

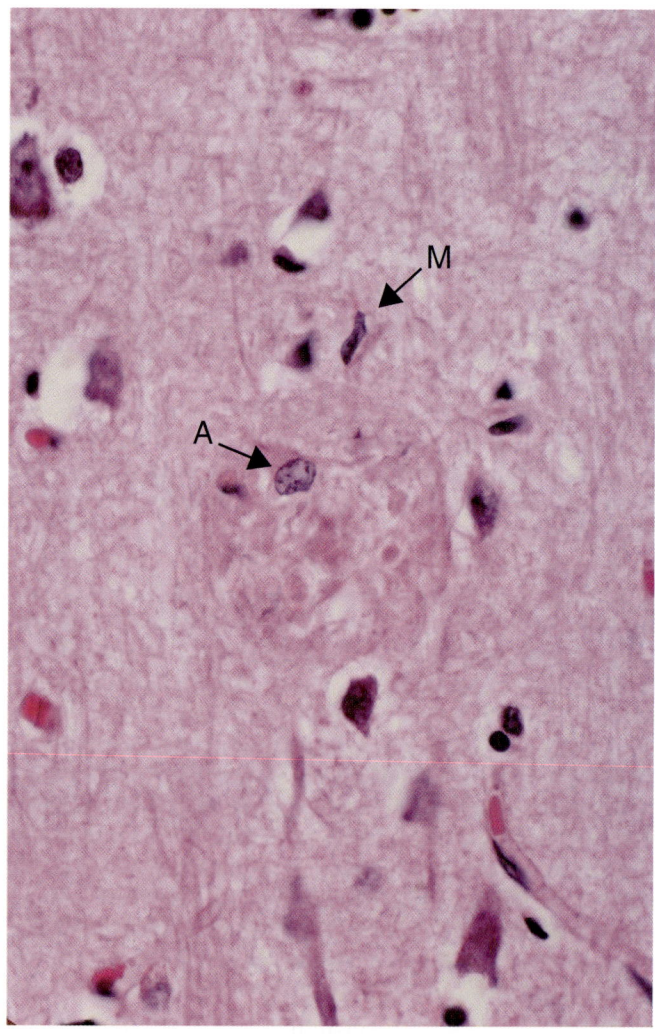

Fig. 14.12. High-power micrograph of senile plaque composed of a donut of debris, a faint central amyloid core, astrocyte (A), and microglia (M). (H&E.)

(PEP), parkinsonism-dementia complex of Guam, Down syndrome, Hallervorden-Spatz disease, and Gerstmann Sträussler-Scheinker disease, among others. Survivors of Down syndrome beyond the age of 30 years invariably show changes of AD indistinguishable from those who have Alzheimer's disease. Tangles by light microscopy appear as thickened rods, flame-shaped fibrils (Fig. 14.5), or skein-wool globose cytoplasmic inclusions of neurons that are usually basophilic by H&E stain (Fig. 14.13). NFTs in the neocortex typically occur in the shape of a question mark or hook, and those in the pyramidal neurons of the hippocampus proper are usually flame-shaped (Fig. 14.5). NFTs form more complex crossing fibrillar patterns in the entorhinal cortex. A neurofibrillary remnant following death of the cell is called a ghost tangle, and they are sometimes found in large numbers in such locations as CA2 of the hippocampal formation. As ghosts, tangles turn faintly acidophilic. Although readily detectable in optimally stained H&E-stained sections, NFTs are brought out in relief with silver methods (Fig. 14.14) and with Congo red-stained sections viewed by polarizing microscopy. They are strongly birefringent. Globose forms are found in the neurons of the subcortical gray such as the nucleus basalis (Fig. 14.13), hypothalamus, tegmental nuclei of the midbrain, locus ceruleus, and pontine raphe. By electron microscopy, the tangles are composed of bundles of paired helical microtubules giving a twisted appearance at 80-nm intervals.[34,60] The microtubules contain highly phosphorylated abnormal tau protein for which immunostains are commercially available.

Only a microscopic examination will provide a definitive diagnosis of AD. For practical reasons, in institutions such as the coroner's office, a minimum number of tissue blocks that will provide a consistently secure diagnosis is desirable. As yet, there is no unanimity in recommendation for the number of blocks needed and specific anatomic sites to be sampled, nor is there total agreement on the requirements of histologic findings for diagnosis. Several groups of investigators have published age-related semiquantitative histologic criteria based on numbers of senile plaques correlated with presence or absence of dementia.[42,45] One such guideline is that proposed by the Consortium to Establish a Registry for Alzheimer's Disease, or CERAD.[42] A minimum of five (5) anatomic regions were selected for microscopic examination, including the middle frontal gyrus, a block including the superior and middle temporal gyri, the inferior parietal lobule, the hippocampal formation with entorhinal cortex, and the midbrain with substantia nigra. Paraffin-embedded blocks are sectioned at 6- to 8-μm thickness and stained with H&E and Bielschowsky methods. Plaques can also be visualized by Congo red, periodic acid-Schiff (PAS), methenamine silver, and Bodian methods, but these stains are not recommended for plaque counts.[42] Other silver methods have been offered.[10]

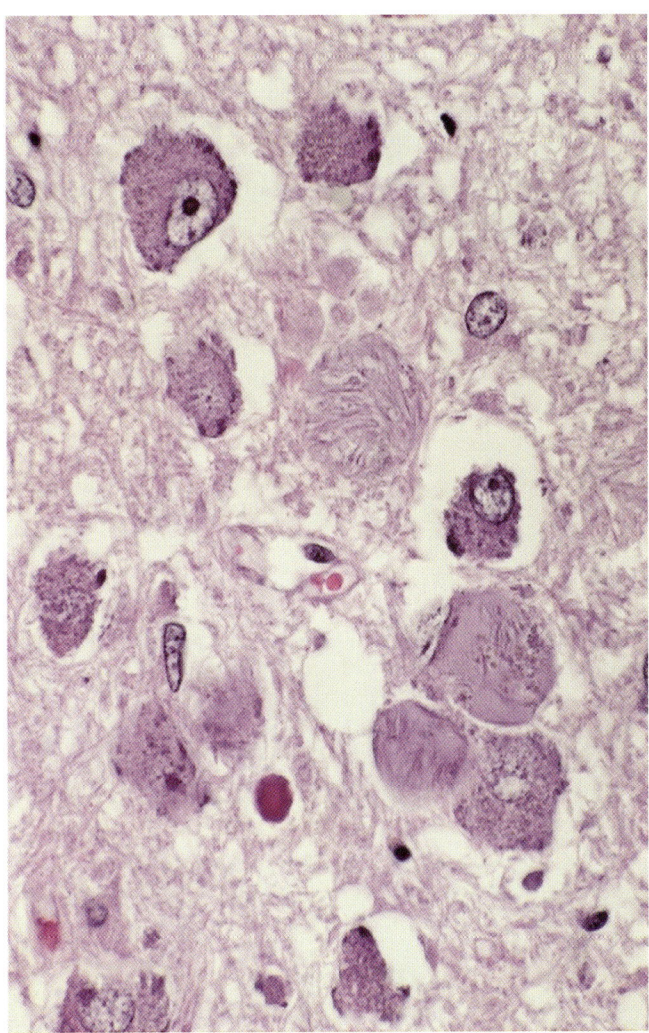

Fig. 14.13. High-power micrograph of nucleus basalis in Alzheimer's disease shows examples of globose type neurofibrillary tangles. (H&E.)

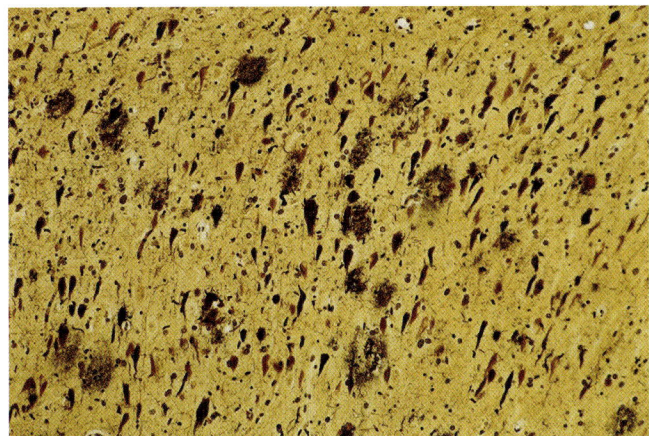

Fig. 14.14. Medium-power micrograph of subiculum in Alzheimer's disease shows numerous neurofibrillary tangles amid small senile plaques. (Bielschowsky.)

Plaques can also be stained with thioflavin S and viewed by fluorescent microscopy. Counts are made at 100× magnification and assessed as "sparse, moderate, or frequent" matched with illustrated diagrams of plaque fields.[42] Plaque score is rated according to the age of the decedent, and a final diagnosis of "definite, probable, and possible" AD is made depending on the presence or absence of a history of dementia after eliminating other possible causes of dementia. The CERAD diagnostic criteria are stated to be based on assessment of frequency of neocortical plaques of the neuritic type. However, at the same time, the statement is made that the protocol does not differentiate diffuse from neuritic plaques. Exclusion of diffuse plaques in plaque count is not clearly stated. Some have downplayed the role of diffuse plaques and amyloid load in the dementia of AD.[20] Omission of diffuse plaques from the equation would leave a large group of demented individuals needing a cause other than AD to explain their dementia. Clinical correlation with dementia of the two subtypes of senile plaques remains controversial, and a determinant measure may be the amyloid load.[12] Parkinson's disease (23%), cerebrovascular disease (34%), and a host of other conditions including diffuse Lewy body disease, tumors, vascular malformations, and hippocampus sclerosis, for example, may accompany AD, adding to and complicating the clinical features of AD.[42] Forensic pathologists, sometimes working with a limited history, should interpret with caution senile plaques found in the elderly decedent. It is a common experience to find sparse, and even moderate (CERAD designations) numbers of senile plaques in decedents with no history of dementia.[24] The CERAD protocol does not address the question of correlation between plaque frequency and degree of dementia. Rather than frequency of senile plaques in the neocortex, others have found better correlation of severity of dementia with location and counts of neurofibrillary tangles. The theory is that degree and progression of dementia are the consequence of the extent of disconnectivity of integration between the entorhinal cortex, hippocampal formation, and amygdala with the neocortex and subcortical targets. This is thought to be better gauged by measure of numbers of NFTs at the said sites rather than by senile plaques.[2] Neuritic plaque counts in the neocortex may be an early indicator of AD rather than the numbers of NFTs, but NFTs may play a larger role in the later stages of AD.[62] Some authors have found the numbers and distribution of NFTs and neuropil threads a better measure of staging AD than that of amyloid load and senile plaques,[8,20] amyloid and neuritic plaque distribution being too variable between individuals exhibiting dementia.

Granulovacuolar degeneration (GVD) is a neuronal marker of aging seen in increasing numbers with advancing age from about 60 years in intellectually normal individuals, but often more numerous in the company of the other changes in cases of AD.[5,64] It is readily seen with H&E stain in the pyramidal neurons of the hippocampus proper, especially from the lateral half of CA2 through CA1 and into the subiculum (see Fig. 14.5). Its significance in the normal aging process is unknown.[64] The name "granulovacuolar degeneration" is descriptive of minute cytoplasmic clear round vacuoles measuring 3 to 5 µm in diameter, each containing a single basophilic granule of 0.5 to 1.5 µm in diameter.[64] Vacuoles may be single or multiple in a single nerve cell. GVD changes accompany NFTs and senile plaques in AD, but their role in AD, if any, is not known. Although increasing numbers of GVD parallel the severity of AD, the appearance of NFTs precedes the presence of GVD.[64] The relationship may be one of accidental juxtaposition of changes of AD and aging. GVD has been described with many disorders other than AD. Aside from the hippocampus proper, rare GVD in single affected cells is also found in the amygdala, various neocortical areas, third nerve nucleus, and mammillary body.[64]

The Hirano body is a brightly eosinophilic paraneuronal or intraneuronal rodlike body (a dot about the size of red blood cells when cut in transverse section) that was first noted in parkinsonism-dementia complex by Hirano *et al.* but has since been found in other conditions.[27,28] The bodies are commonly found in or adjacent to pyramidal neurons of CA1 into the subiculum, and less commonly in CA2 (see Fig. 14.6). Hirano bodies can be found in both demented and nondemented elderly accompanying changes of AD.

A recurrent topic in the pathoanatomic study of the diagnosis of AD is finding a consistent measure in the differentiation of demented versus nondemented subjects.[3,11,13,33,40] No consensus has yet been reached. Perhaps a generally applicable set of histopathologic criteria as to what constitutes earliest changes of AD will never be reached because of individual variability in what has been called neurocognitive reserve. What constitutes such reserve is itself composed of a host of complex factors. Of importance for those who work in the forensic arena is avoidance of overinterpretation of pathologic changes of AD in the absence of a history of dementia.

Other than AD and cerebrovascular disease (CVD), causing multi-infarct dementia, disorders causing dementia encountered in the Los Angeles County Department of the Coroner are uncommon. Several suggested protocols for the pathologic evaluation of such cases have been published.[37,38] Among non-AD causes of dementia encountered at this office are Pick's disease, frontotemporal dementia, and diffuse Lewy body disease.

Pick's Disease

Pick's disease makes up a very small percentage (about 2%) of the cases of dementia.[22] The clinical onset overlaps with that of AD, mostly between 45 and 64 years of age, but onset is uncommon after age 75 in contrast with AD. The gross features of Pick's disease are highly characteristic when fully developed, and the term "lobar atrophy" is descriptive. Brain weight is usually moderately to greatly reduced, with weights less than 1000g not unusual when the disease has run its clinical course. Marked brain weight loss is reflected in severe cortical atrophy of such extent as to be described as "knife blade atrophy" (Case 14.2, Fig. 14.15). Frontal and temporal lobes are predominantly involved and, in the latter, the posterior two-thirds of the superior temporal gyrus is typically spared. The parietal lobe may be affected in some cases with precentral and postcentral gyri left relatively intact, leaving an island of spared cortex surrounded by atrophy (Case 14.3, Fig. 14.16). The diagnosis of Pick's disease can be almost assured when these gross features are present. A number of cases show asymmetry of the previously mentioned features, and, when this occurs, the findings are said to be more prevalent on the left. Lateral ventricles are enlarged (hydrocephalus *ex vacuo*) due to atrophy of subcortical white matter of the frontal and temporal lobes and caudate. The atrophy of the latter may mimic that of Huntington's disease in the setting of dementia. Other than the substantia nigra, brainstem and spinal cord are unaffected, and the cerebellum may show only minor changes.

Case 14.3. (Figs. 14.16–14.18)

A 68-year-old woman had a diagnosis of Alzheimer's disease, and she sustained a fall and hip fracture that was successfully treated. While in a recumbent position for removal of surgical staples, she became unresponsive with gasping respirations. Paramedics found partial airway obstruction that was cleared, but further efforts at resuscitation failed.

Case 14.2. (Fig. 14.15)

A 67-year-old woman stopped talking and was cared for by relatives. Taken to the hospital, she was dehydrated and malnourished. The family reported that the decedent was bedridden for 2 to 3 months and had stopped eating 2 weeks ago. Pelvic and heel decubiti were present. She died 2 weeks into hospitalization.

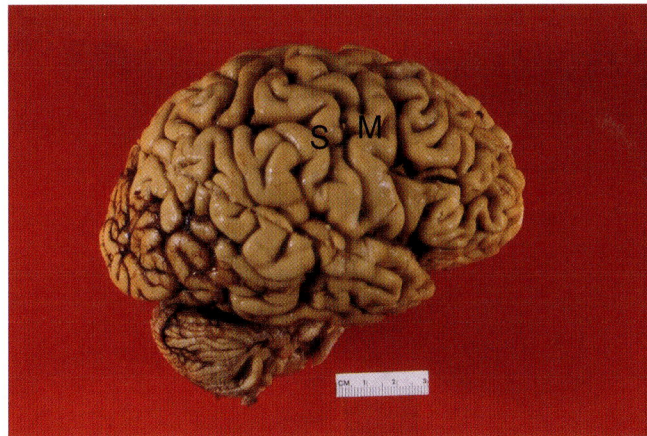

Fig. 14.16. Right cerebral hemisphere stripped of meninges shows frontotemporal cortical atrophy in Pick's disease, more prominent in frontal lobe. Precentral (M) and postcentral (S) gyri are relatively preserved.

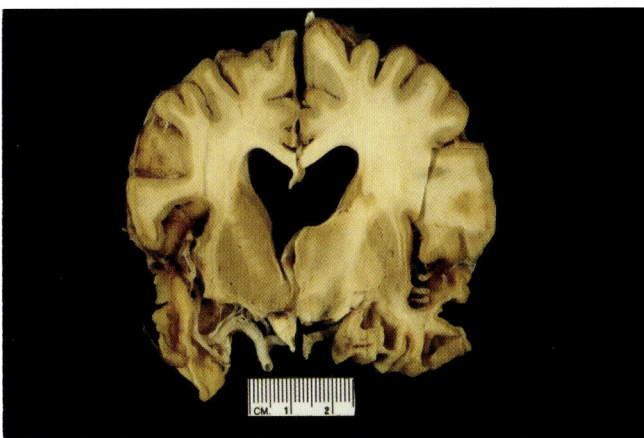

Fig. 14.15. Coronal section at anterior striatum on the left and anterior thalamus on the right (hemispheres sectioned separately) show severe selective "knife blade" (cut at right angle) cortical atrophy of temporal lobes and insula with gross sparing of remaining areas. Lateral ventricles are moderately enlarged. Posterior half of superior temporal gyrus was selectively preserved along with pre- and postcentral gyri. Microscopy of temporal cortex showed severe diffuse atrophy and scattered ballooned neurons with eccentric nuclei and cytoplasmic inclusion typical of Pick's disease. Motor cortex was intact.

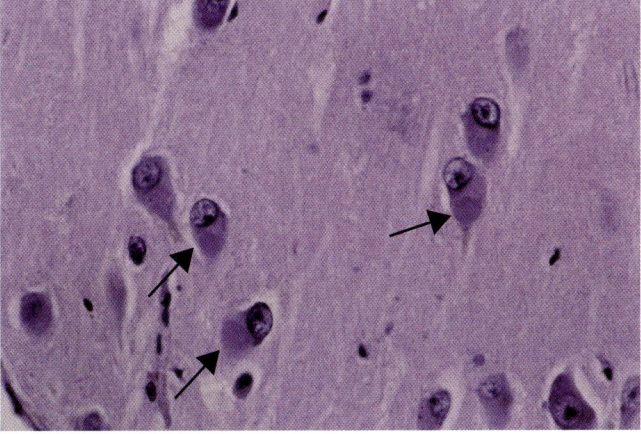

Fig. 14.17. High-power micrograph of pyramidal neurons of subiculum demonstrates discrete round cytoplasmic inclusions, or Pick bodies (*arrows*). (H&E.)

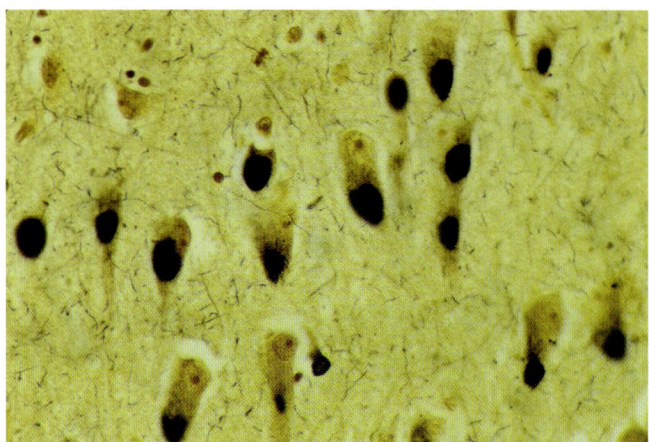

Fig. 14.18. High-power micrograph of Case 14.3 from the same region seen in Fig. 14.17 stained with silver method shows numerous Pick bodies. (Bielschowsky.)

Microscopically, advanced cases show severe, sometimes near-total, loss of cortical neurons with loss of laminar architectural pattern. Layer III is said to be most affected in cases of partial involvement. Severely involved cerebral cortex may show only end-stage, nonspecific atrophy with near-total loss of neurons, and a loosely expanded interstitium with diffuse astrogliosis. Subcortical white matter shows marked loss of myelin and diffuse gliosis. When accompanied by the previously described gross findings, the presence of Pick cells and bodies is diagnostic (Fig. 14.17). The Pick cell, by H&E stain, is a ballooned neuron with nuclei eccentrically displaced by a pale eosinophilic to amphophilic cytoplasmic inclusion. Pick bodies are neuronal cytoplasmic round inclusions, 10 to 30 μm in diameter, and strongly argyrophilic (Fig. 14.18). Pick bodies are positive by immunostains for ubiquitin and tau protein. Pick cells may be rare or not found at all in the cortex in end-stage atrophy, and the cell should be sought for in the cortex with lesser degree of atrophy. Occasionally, Pick bodies are grouped in large numbers in such areas as the subiculum (Fig. 14.18). Cases with dementia mimicking the gross pathoanatomic features of Pick's disease, but lacking in the distinctive histopathologic features, further complicate the classification of any case when Pick cells are rare or cannot be found. The nosologic placement of those cases remains to be settled.

Diffuse Lewy Body Disease

Diffuse Lewy body disease (DLBD) is said to be the second leading cause of cognitive impairment in the elderly,[55] occurring with or without changes of Alzheimer's disease and/or Parkinson's disease.[41] A guideline for the clinical and pathologic diagnosis has been recommended by an international consortium[41] and by others.[21]

When clinical signs of Parkinson's disease are present, autopsy usually reveals classic Lewy bodies, described later. Unless accompanied by changes of AD, which sometimes occurs, the brain may show only mild nonspecific atrophy and ventricular enlargement. Lewy bodies in DLBD are found in neurons of the deeper layers of the medial temporal, cingulate, and insula cortex. Compared with the classic Lewy body, they are relatively inconspicuous cytoplasmic bodies, round to irregular in shape and pale-staining with H&E without a halo.[48] They are positive for ubiquitin but tau protein negative.

Parkinson's Disease

Parkinsonism is a term applied to a clinical syndrome that includes rigidity and movement disorder as essential features.[52] Tremors may or may not be present. A number of conditions other than Parkinson's disease may exhibit these signs, such as multiple system atrophy, PSP, Wilson's disease, AD, side effects of neuroleptic drugs,[53] and postencephalitic parkinsonism. Parkinson's disease (PD) is the second most common neurodegenerative disorder after AD. Mean age of onset of PD is said to be 60 years,[52] and its prevalence is 30 to 190 per 100,000 (or 0.5–1% of those 65–69 years and 1–3% in those 80 years and older).[59] In our material, cases exhibiting pathologic changes of PD may or may not be accompanied by a clinical diagnosis of PD. Those cases without a history of signs of PD may only be an artifact of an incomplete medical history, or they may represent preclinical cases. The brain in PD is usually devoid of distinguishing external features. In cases suspected of PD, our midbrain examination routine is to make the first cut an exact transverse section of the brainstem at the superior pontine sulcus, and then obtain two 3-mm-thick sections of the exposed midbrain rostral to the first cut. Typically, the substantia nigra is depigmented in PD, especially in its lateral half.[29] Some residual gross pigmentation may remain toward the interpeduncular fossa. The locus ceruleus may also have lost pigmentation, but the remainder of the brainstem sections displays no change. Cerebral hemispheres are usually unremarkable, including the basal ganglia.

Microscopically, depopulation of pigmented neurons of the substantia nigra is accompanied by melanin "incontinence," in which neurons have dropped out, leaving extracellular and phagocytosed melanin pigment, along with mildly increased cellularity of relatively inconspicuous astrogliosis. The pathognomonic finding is presence of the classic Lewy body, which is a round cytoplasmic inclusion found in pigmented nerve cells, measuring about 10 to 30 μm in diameter. By H&E stain, a clear halo surrounds a bright eosinophilic core that is occasionally targetoid (Fig. 14.19). Similar bodies may be found in other widespread anatomic sites, including the dorsal tegmental nucleus of the locus ceruleus, dorsal nucleus of the raphe, dorsal vagal nucleus, posterior hypothalamus, and nucleus basalis of Meynert. They may also be found in neurons of the intermediolateral cell column of the spinal cord and in sympathetic and parasympathetic ganglia.[58] PD may coexist with AD.[16] However, dementia in PD is not always accounted for by histologic changes of AD nor by cortical Lewy body disease.

A condition entering into the differential diagnosis with idiopathic Parkinson's disease is postencephalitic parkinsonism (PEP), now mostly of historical interest, which followed the influenza pandemic between 1915 and 1927. In some, signs of parkinsonism developed many decades later. The anatomic sites of pathologic involvement, including the substantia nigra, are shared by the two conditions, but PEP is characterized by neurofibrillary tangles in neurons in place of Lewy bodies (Case 14.4; Figs. 14.20–14.22). Differential diagnosis also includes progressive supranuclear palsy (PSP), which also affects the substantia nigra and shares neurofibrillary tangle pathology with PEP. However, with PSP, midbrain atrophy is more widespread with distinct shrinkage and narrowing of the distance from the interpeduncular fossa to the aqueduct and tectal surface. A condition of historical interest, which may still recur, is the permanent parkinsonian syndrome that followed

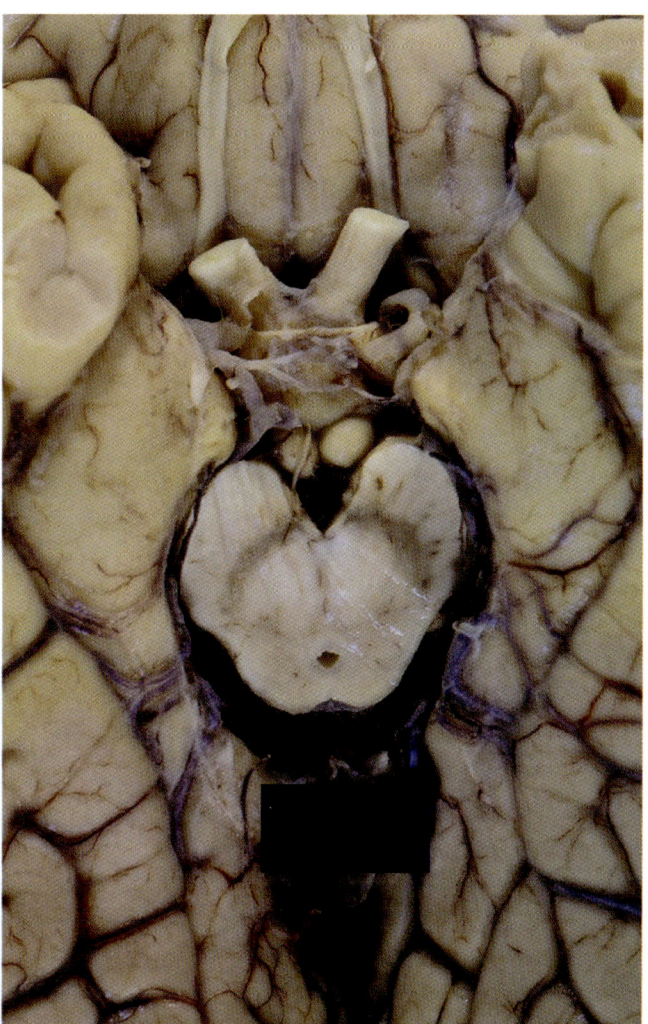

Fig. 14.20. Closeup view of midbrain transected at level of decussation of brachium conjunctivum shows light brown-colored substantia nigra lacking normal black pigmentation.

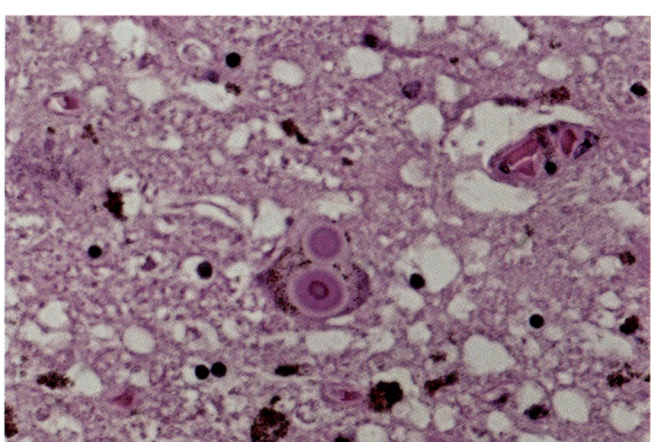

Fig. 14.19. High-power micrograph shows a pair of Lewy bodies in a single neuron in Parkinson's disease showing characteristic concentric lamination. (H&E.)

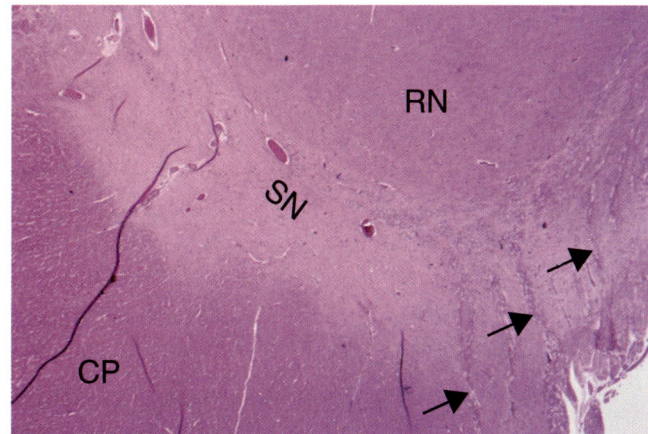

Fig. 14.21. Low-power micrograph shows a portion of anterolateral midbrain with substantia nigra (SN) depleted of pigmented neurons. RN, red nucleus; CP, cerebral peduncle; arrows point to third nerve fascicles. (H&E.)

Case 14.4. (Figs. 14.20–14.22)
A 61-year-old man had a diagnosis of Parkinson's disease and angina. He was found at the bottom of his home pool with a pool skimmer in his hands. His PD was described as severe, and he had fallen before on several occasions because of poor balance. He was dead on arrival at the hospital.

Neurodegenerative Disorders

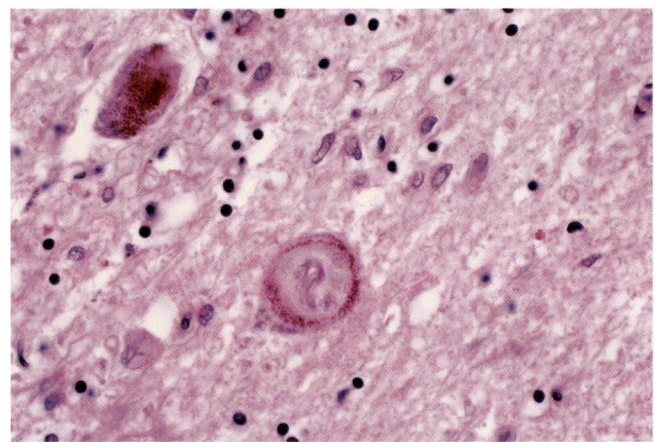

Fig. 14.22. High-power micrograph shows neurofibrillary tangle in a pigmented neuron of substantia nigra in this case of postencephalitic parkinsonism. (H&E.)

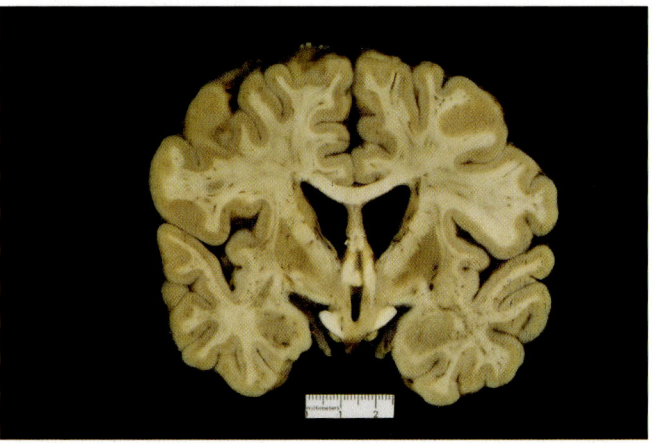

Fig. 14.23. Huntington's disease. Note marked symmetric shrinkage of body of caudate nucleus and small lenticular nuclei. Ventricles are mildly enlarged. Cortex and white matter appear unremarkable.

intentional human abuse of MPTP (1-methyl-4-phenyl-1,2,3,6,-tetra-hydropyridine), which caused necrosis of substantia nigra neurons.[6,35] Inclusion bodies similar to, but not identical to, Lewy bodies were found with experimental exposure of MPTP in primates.[19]

Huntington's Disease

Huntington's disease (HD) is a progressive hereditary disorder of autosomal dominant transmission clinically manifested by a triad of chorea, personality change, and eventual dementia. The usual age of onset is about 35 to 40 years, with childhood forms developing as early as 4 years of age. The onset may be as late as 70 years. Its prevalence has been given as between 4 and 8 per 100,000 in the United States. Gross examination of the brain shows nonspecific mild to moderate cortical atrophy and associated reduced weight. Coronal sections of advanced cases, however, demonstrate bilaterally symmetric atrophy of the head of the caudate nucleus with characteristic straightened or concave contour in coronal sections of the frontal horns of the lateral ventricles, in contrast to its normal convex contour (Figs. 14.23–14.25). As a consequence, the lateral ventricles, especially the frontal horns, show moderate to marked *ex vacuo* dilatation. The putamen is also atrophic, and the globus pallidus appears shrunken (Fig. 14.23). Thinning of the cerebral cortical ribbon may be noticeable in some, but not in all, cases.

Microscopically, depending on the state of the disease, the striatum shows loss of small and medium-sized neurons, with dorsal sectors more severely affected than the better-preserved ventral striatum. Normally, smaller numbers of large neurons remain relatively preserved. The areas of atrophy show prominent diffuse proliferation of small fibrillary astrocytes (Figs. 14.24–14.25). Based on a large experience, a five-grade (0–4) system

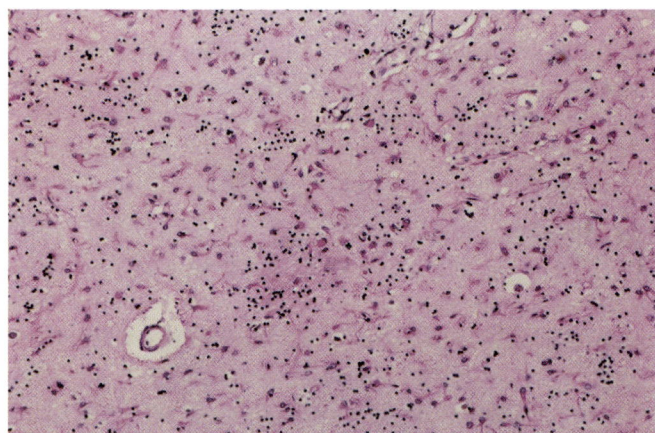

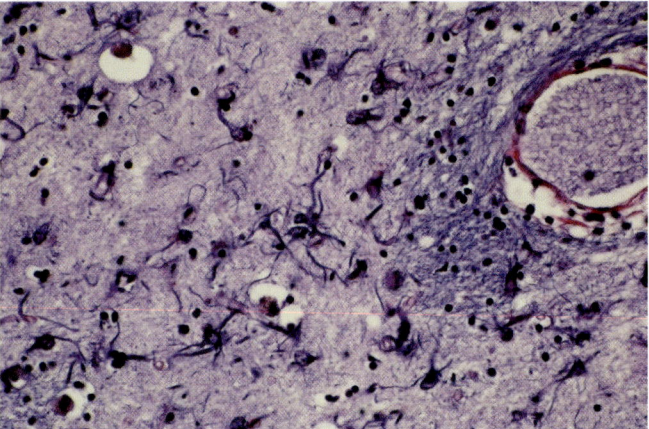

Figs. 14.24–14.25. Medium-power micrograph (*top*, Fig. 14.24) shows near-total dropout of neurons of striatum and diffuse gliosis of small fibrillary astrocytes. (H&E.) Special stain demonstrates detail of reactive astrocytes at high power (*bottom*) (Fig. 14.25). (Phosphotungstic acid hematoxylin.)

has been proposed defining the severity of Huntington's disease, useful as a pathologic shorthand to correlate with the clinical state.[44,65] Even in the presence of clinical features of HD, some cases may show little or no gross or microscopic abnormality (grade 0). Outside of predominant loss in the striatum and cerebral cortex, nerve cells may be diminished in the hippocampus proper, hypothalamus, substantia nigra, and nucleus centrum medianum.[14]

Progressive Supranuclear Palsy

Progressive supranuclear palsy (PSP), or Steele-Richardson-Olszewski syndrome, is manifested by a constellation of signs and symptoms that may be mimicked by different histopathologic causes. The central clinical features as first described included supranuclear ophthalmoplegia (gaze palsy), akinesia, rigidity, nuchal dystonia, pseudobulbar palsy, and cognitive decline leading to dementia.[57] The clinical expression may be variable, as noted in later case reports. The disorder is uncommon but not rare, and it is said to constitute 1 to 8% of cases with clinical features of parkinsonism. Its prevalence is not well documented. The median age of onset is about 64 years.[39] On gross examination, the midbrain and pontine tegmentum are atrophic, and atrophy of the globus pallidus may be noticeable.[57] Characteristic atrophy of the midbrain may be detected by imaging studies as shortening of the midsagittal anterior-posterior diameter from the depth of the interpeduncular fossa to the tectum in PSP versus controls,[66] and by reduced area of the midbrain.[47] The substantia nigra and locus ceruleus are abnormally pale. Cerebral and cerebellar cortices are usually unremarkable. Microscopically, areas of subcortical gray matter show widespread neuronal loss, astrogliosis, and NFTs in surviving neurons. Affected areas include the substantia nigra, locus ceruleus, globus pallidus, subthalamic nucleus, tectum, periaqueductal gray, and dentate nucleus. Granulovacuolar degeneration was noted in neurons in the red nucleus, nuclei basis pontis, and other nuclei. Immunostain of affected neurons with NFTs are tau protein positive, but the protein is different from that of AD.[18]

Multiple System Atrophy

Three conditions previously regarded as separate entities are presently included under the term multiple system atrophy.[23,36] They are olivopontocerebellar atrophy (OPCA), Shy-Drager syndrome, and striatonigral degeneration of Adams et al.[1] The true incidence of these conditions is unknown. They are uncommon if not rare, as in striatonigral degeneration. The age of onset is approximately in the fourth and sixth decades. Symptoms and signs are variable depending on the predominant anatomic involvement.

Clinically, OPCA is characterized by ataxia beginning in the legs, and progresses rostrally to eventually involve the bulbar musculature. The salient feature of Shy-Drager syndrome is orthostatic hypotension accompanied by other autonomic and extrapyramidal signs. Striatonigral degeneration masquerades as Parkinson's disease without tremor or other involuntary movements, but it is often more resistant to the usual pharmacologic treatments for PD. Mental function is preserved.

On gross examination, OPCA is characterized by an atrophic peaked belly of the pons, which is normally rounded but for the midline impression of the basilar artery. The basis pontis is shrunken and the tegmentum is relatively preserved (Fig. 14.26). Middle cerebellar pe-

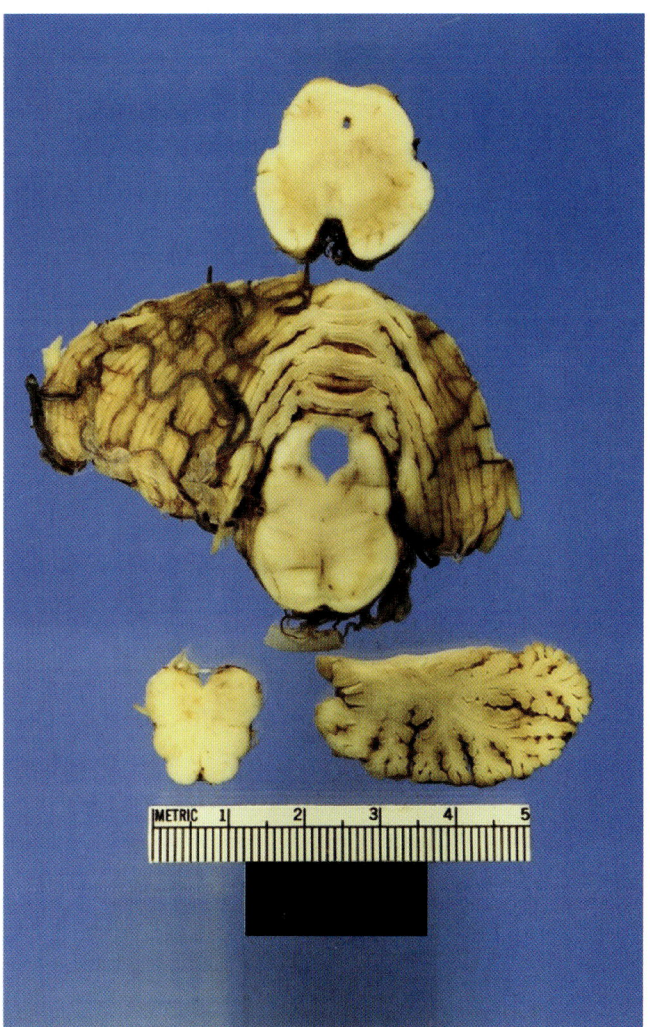

Fig. 14.26. Transverse sections of caudal midbrain, rostral pons, medulla, and cerebellum in a case of olivopontocerebellar atrophy (OPCA) show depigmented substantia nigra, disproportionately small atrophic basis pontis, pale inferior olives, and cerebellar cortical atrophy with expanded sulci. Cerebellar white matter is also diffusely atrophic.

duncles are shrunken, and olives are flattened. Atrophy of the cerebellar hemispheres may be present, but not invariably (Figs. 14.26–14.30). Microscopically, there is extensive loss of neurons of the nuclei basis pontis and severe loss of pontocerebellar fibers, while corticospinal fascicles are preserved (Fig. 14.31). The middle cerebellar peduncles and cerebellar white matter are severely atrophic, while the dentate nucleus remains intact (Fig. 14.32). There is variable, and sometimes severe, loss of Purkinje cells (Fig. 14.33). Purkinje cell loss and preservation of basket cells give rise to "empty baskets," or ramifications of basket cell endings where Purkinje cells dropped out (Fig. 14.34). The pontine tegmentum is generally unaffected. Nonspecific, although characteristic of OPCA (Figs. 14.35–14.36), is the development of an axonal swelling just distal to the axon hillock of Purkinje cells, called a torpedo.[22]

Striatonigral degeneration[1] may show little on external examination. However, on section, the caudate and putamen are atrophic and discolored a slate gray to pale green bilaterally and symmetrically. The substantia nigra and locus ceruleus are depigmented. Neuronal loss in the striatum is severe with near-total disappearance of small neurons and some preservation of large neurons. Diffuse astrogliosis accompanied by deposits of various pigments accounts for the discolored gross appearance. Lewy bodies and neurofibrillary tangles are absent in the atrophic substantia nigra and locus ceruleus.

Shy-Drager syndrome may show little gross abnormality in the spinal cord. The principal abnormality correlating with orthostatic hypotension is the loss of neurons of the preganglionic intermediomedial and intermediolateral cell columns of the thoracic and upper

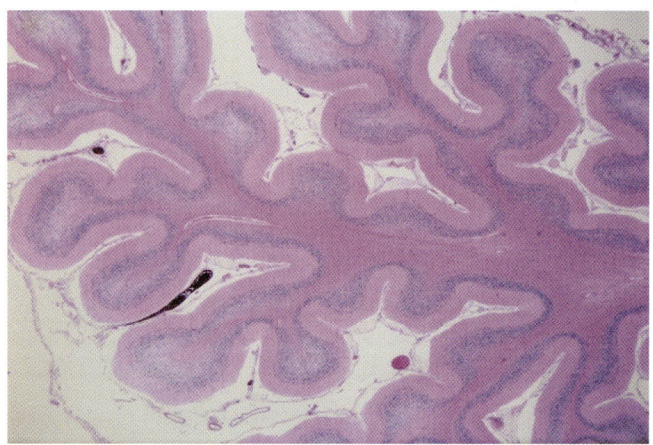

Fig. 14.27. Diffuse cerebellar cortical atrophy with enlarged sulci. OPCA. (H&E.)

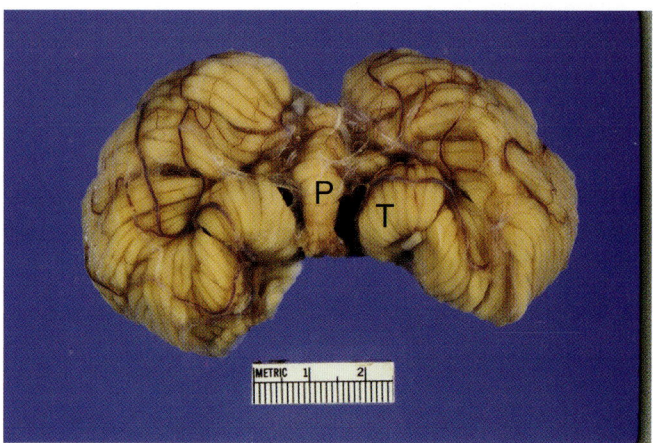

Fig. 14.29. View of inferior cerebellum in OPCA shows moderate atrophy with gaping subarachnoid spaces between tonsils (T) and posterior vermis (P).

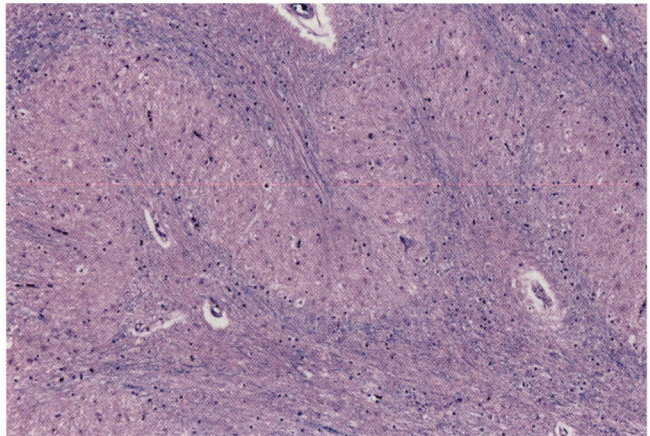

Fig. 14.28. Medium-power micrograph of inferior olivary nucleus in OPCA shows complete loss of nerve cells. Sparse myelinated fibers remain in hilus. (Luxol fast blue/H&E.)

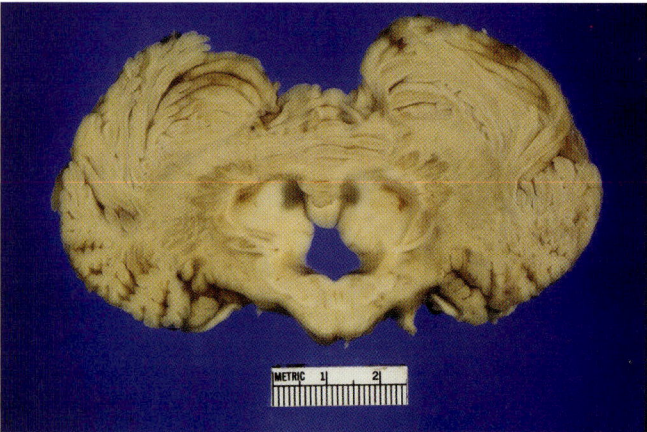

Fig. 14.30. Severe atrophy of pons, at level of middle cerebellar peduncles, and of cerebellar white matter around dentate nuclei. Fourth ventricle is markedly dilated. OPCA.

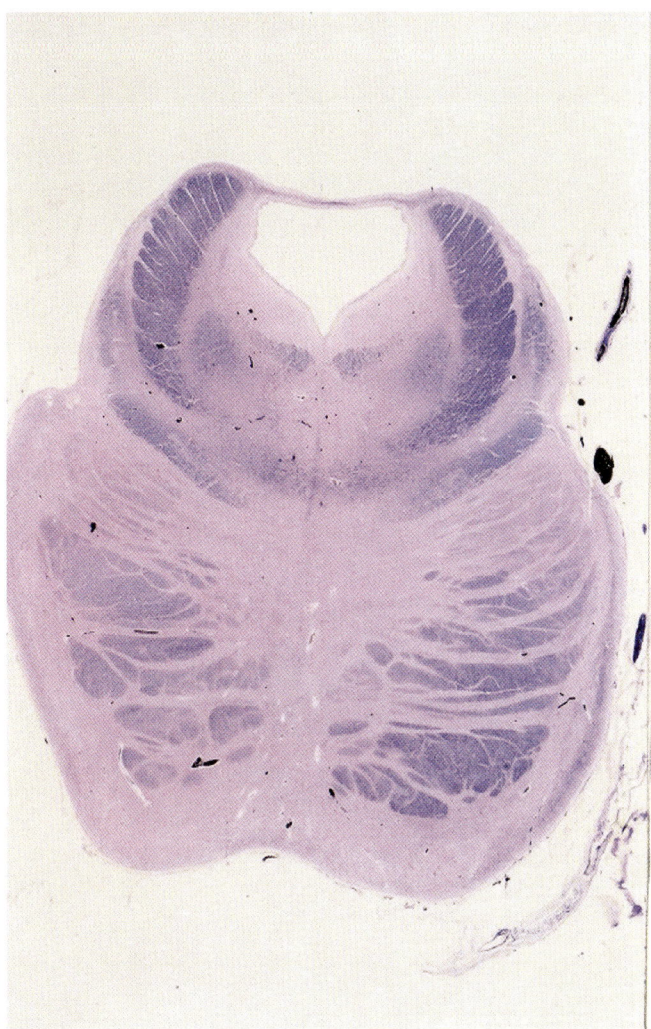

Fig. 14.31. Low-power micrograph of transverse section of rostral pons shows intact bundles of corticospinal tracts in a field of total loss of transversely crossing pontocerebellar fibers, seen as loss of myelin staining. Loss of nuclei basis pontis neurons cannot be appreciated at this power. OPCA. (Luxol fast blue/H&E.)

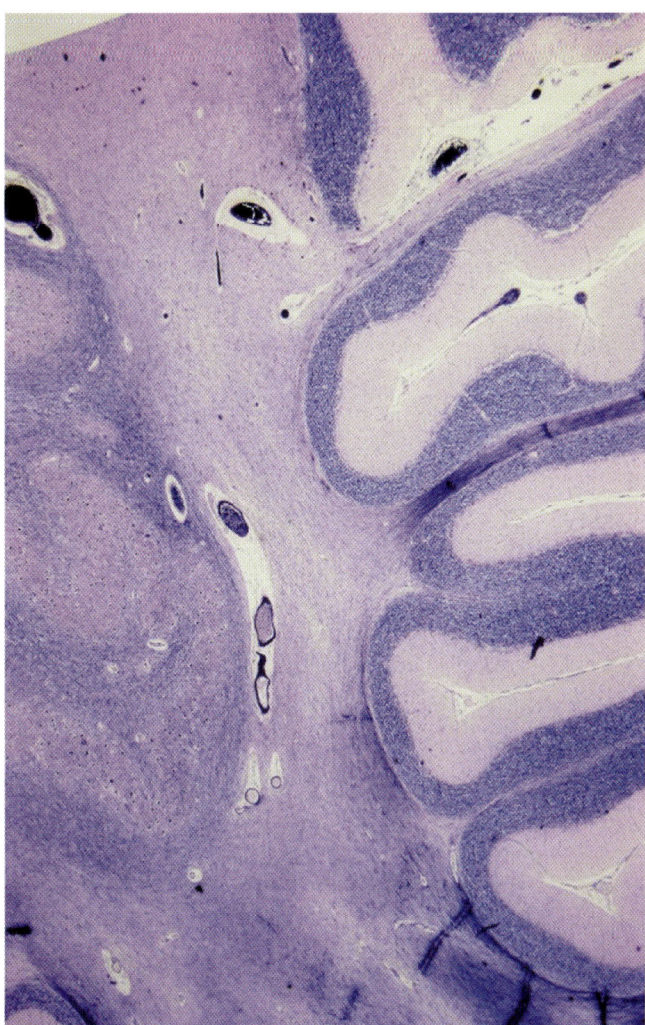

Fig. 14.32. Low-power micrograph of central cerebellum shows severe atrophy and loss of myelin staining of white matter except for some preservation around dentate nuclei. OPCA. (Luxol fast blue/H&E.)

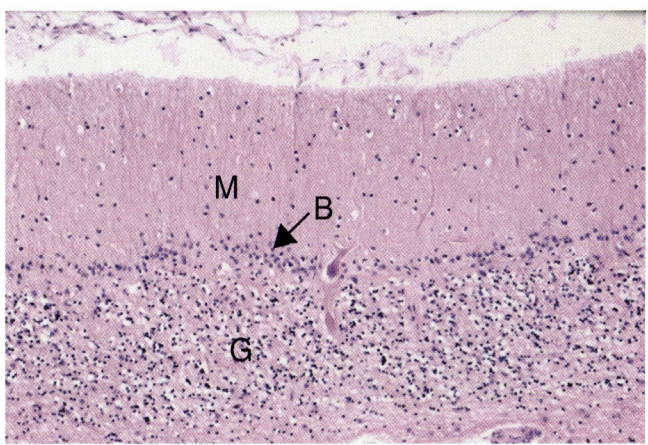

Fig. 14.33. Medium-power micrograph shows severe cerebellar cortical atrophy. Note gliotic molecular layer (M), severe loss of Purkinje cells, proliferated nuclei of Bergmann's astrocytes (B), and rarefied granule cell layer (G). OPCA. (H&E.)

lumbar spinal cord. Neuronal loss may also be appreciated in widespread anatomic sites in the brain, including the putamen, globus pallidus, substantia nigra, locus ceruleus, and dorsal vagal nuclei.[36] Lewy bodies are found in some cases.

The unifying histopathologic feature of these three formerly apparently disparate conditions is a glial cytoplasmic inclusion present in oligodendrocytes that is undetected with H&E stain but demonstrable by silver stains such as Bodian and modified Bielschowsky methods.[36,50] The inclusion is variably reactive with immunostains for ubiquitin and tau protein. Other associated inclusions include neuronal cytoplasmic inclusions and neuronal and glial nuclear argentophilic inclusions. The glial cytoplasmic inclusions are found in very widespread distribution from the motor cortex to

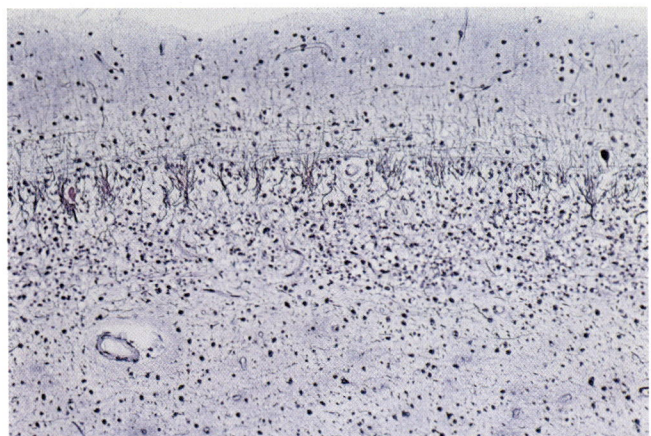

Fig. 14.34. Medium-power micrograph of silver-stained atrophic cerebellar cortex shows empty baskets where Purkinje cells dropped out. OPCA. (Bodian.)

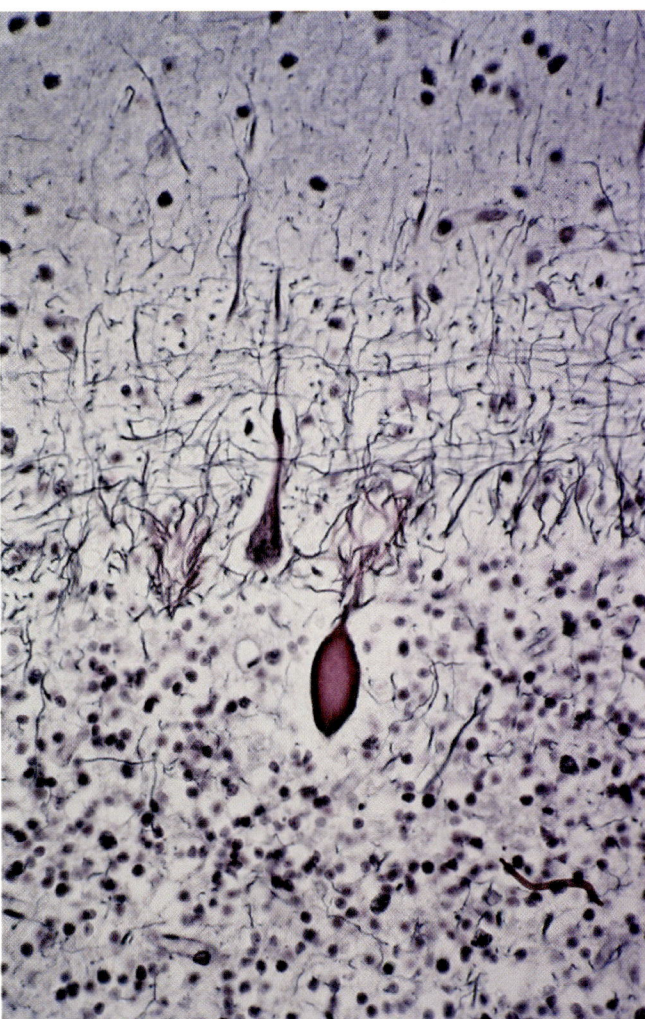

Fig. 14.36. High-power micrograph of silver stain of empty baskets and torpedo (swelling of Purkinje cell proximal axon). OPCA. (Bodian.)

the intermediolateral cell column of the spinal cord as well as in white matter.[51]

Friedreich's Ataxia

Friedreich's ataxia is the prototype of the heterogeneous group of spinocerebellar degenerative diseases. It is a rare autosomal recessive disorder with a reported prevalence of 1 to 2 per 100,000. The age of onset is usually between 8 and 15 years and before 20 years. Later onset is uncommon. Its clinical presentation is ataxia of gait with posterior column signs, advancing to dysarthria and weakness with upper motor neuron signs. Visual impairment may occur in some cases. An abnormal electrocardiogram is usually present due to associated cardiomyopathy. Scoliosis and pes cavus are typical features. Gross examination shows striking selective atrophy

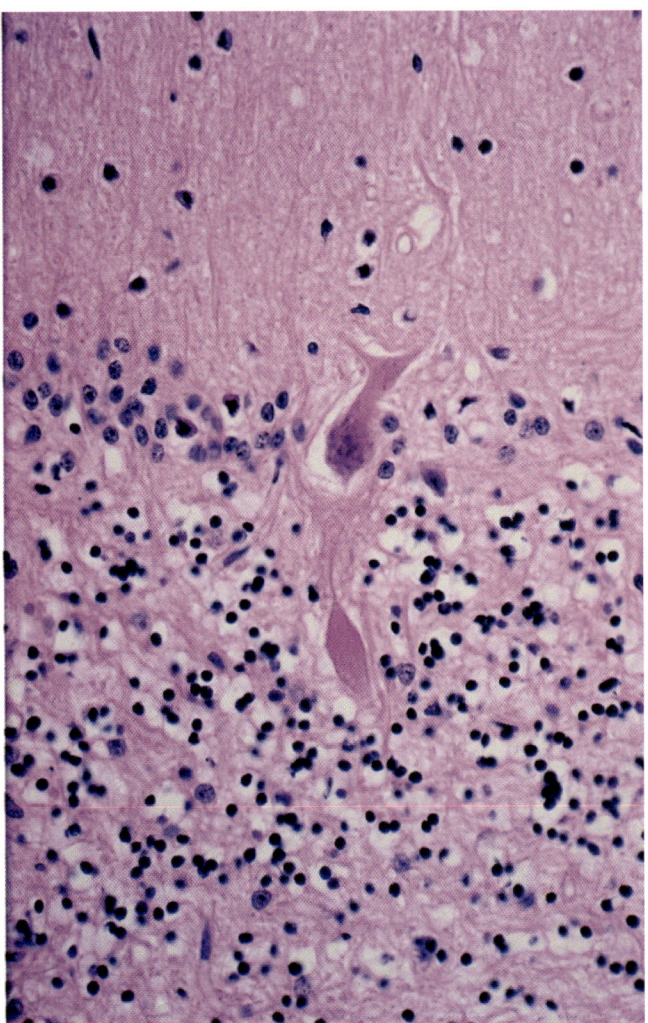

Fig. 14.35. High-power micrograph shows details of a surviving Purkinje cell with axonal swelling called a torpedo. Note straight, finely fibrillar gliosis of molecular layer and depopulation of granule cell layer. OPCA. (H&E.)

of posterior spinal rootlets and flattening of the dorsal aspect of the cord. The cerebrum remains unremarkable except for most cases showing mild optic atrophy.[22] Microscopically, the posterior columns show severe atrophy, fasciculus gracilis more than cuneatus. Lateral corticospinal tracts and adjacent spinocerebellar tracts show atrophy of somewhat lesser degree than the posterior columns. There is loss of neurons of Clarke's column, while the anterior horn remains unscathed. Atrophy of the dentate nucleus is accompanied by secondary degeneration of the superior cerebellar peduncles. Brainstem nuclei such as the cochlear and vestibular nuclei and superior olives may show atrophy. The cerebellar cortex usually remains intact.

Motor Neuron Disease

Motor neuron diseases (MND) are a group of progressive neuromuscular disorders that have in common loss of motor neurons at selective sites in the cerebral cortex, in certain brainstem nuclei, and in the anterior horn of the spinal cord. Depending on the particular type of MND, there may or may not be accompanying upper motor neuron degeneration. The various conditions are distinguished by their different clinical expressions, their predominant anatomic sites of involvement, age of onset, and, in some cases, genetic determinants inferred by recent molecular biology research. Disorders of various other disease categories such as infections and, forensically important, toxic disorders may masquerade as motor neuron disease. Of the neurotoxins mimicking MND, lead causing motor neuropathy and *Lathyrus sativus* (chickling pea) causing spastic paraparesis can be mentioned.[56] The most common of the MNDs is amyotrophic lateral sclerosis (ALS). Its prevalence is about 5 per 100,000 with a mean age at onset of about 60 years and mean survival of about 3 to 5 years. Most cases are sporadic, but about 5 to 10% are familial with an earlier age of onset and often a more malignant course. Clinically, muscle weakness with gradual loss of muscle mass typically begins in the legs on one or both sides and progresses to arms and bulbar muscles, accompanied by fasciculations and upper motor neuron signs. The common mode of death is aspiration pneumonia. Diagnostic accuracy of ALS is said to be more than 95%.[54]

Parenthetically, the demonstration of the anatomic features of MND requires a slight modification of the autopsy procedure. Prior to sectioning, a minimum of two 3-mm thick blocks of motor cortex are obtained by cuts at a right angle to the central sulcus, including precentral and postcentral gyri, from parasagittal mid-dorsolateral levels. After the brainstem and cerebellum are removed by transverse section through the superior pontine sulcus and a midbrain section is removed, the cerebrum is horizontally sectioned to expose the internal capsule. A block including the immediate postgenu internal capsule and adjacent posterior limb is removed. Transverse sections of the brainstem at the levels of the middle cerebellar peduncle just rostral to its midlevel, and caudal pons just rostral to the inferior pontine sulcus, will include the trigeminal motor nucleus and facial nucleus, respectively. Transverse section of the medulla at the midlevel will include nucleus ambiguus, hypoglossal nucleus, and the pyramids. If carefully examined grossly, hypoglossal nerves as they arise from the anterolateral sulcus may appear thin and gray. Anterior and posterior rootlets of the cauda equina are separated, and bundles of each are submitted in separate cassettes. Rootlets of the cervical enlargement are usually included in transverse sections of the spinal cord from C5, C6, and C7 levels and from three levels of lumbosacral enlargement. At the general autopsy, samples of moderately atrophic skeletal muscle are removed from upper and lower extremities and the diaphragm. Sampling of severely atrophic muscle is less likely to be diagnostically helpful.

Gross brain examination in ALS is usually unremarkable with the possible exception of selective atrophy of the motor cortex, rarely noticeable, and thinning of the hypoglossal nerves. The general diameter of the cord is reduced, and the normally thin anterior spinal rootlets are even more so, appearing gray compared with normally thicker and opaque posterior rootlets. Microscopically, the essential feature is the loss of motor neurons at several characteristic anatomic sites.[9] There is loss of Betz cells and pyramidal neurons of the fifth layer of the motor cortex. Motor neuron loss may also occur in the motor nucleus of cranial nerve V, facial nucleus, nucleus ambiguus, and hypoglossal nucleus (Fig. 14.40), and almost always there is loss of motor neurons of the anterior horn of the spinal cord and gliosis (Fig. 14.39). Loss of neurons in the oculomotor and abducens nuclei may occur but less conspicuously. Surviving neurons may contain Bunina bodies, which are small irregular cytoplasmic eosinophilic inclusions. Lateral and ventral corticospinal tracts show myelin loss, particularly more distally, with sometimes difficult to detect astrogliosis and macrophage infiltration (Fig. 14.37). In some cases, degeneration at the pyramids may be difficult to appreciate by H&E stain alone (Fig. 14.38). In advanced cases, changes in the corticospinal tract may be appreciated as rostral as the internal capsule.[9]

One of several rare neuro-degenerative disorders is *Hallervorden-Spatz disease* which is characterized by progressive rigidity, choreoathetosis, progressive mental deterioration, and less often torsion dystonia.[32] Cases are grouped by age of onset: late infantile (3 months–6 years), classic (7–15 years), and rare late or adult (24–64 years). Pathognomonic for the disease is an expanded rusty brown substantia nigra and globus pallidus. The discoloration in the pars reticularis of the substantia nigra and

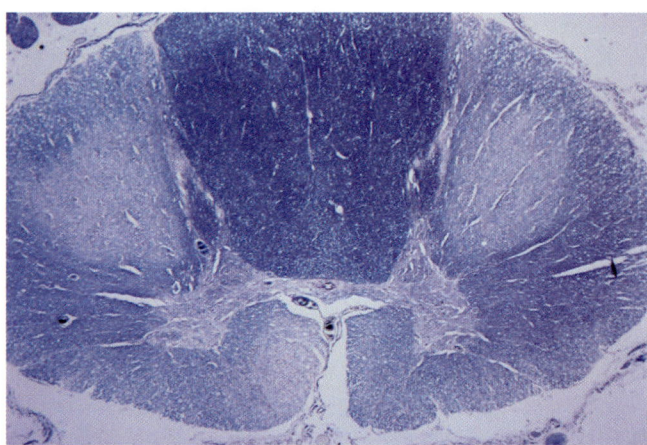

Fig. 14.37. Low-power micrograph of a transverse section of lower cervical spinal cord in a case of amyotrophic lateral sclerosis (ALS) demonstrates loss of myelin staining in atrophic lateral corticospinal tract bilaterally and ventral corticospinal tract predominantly on one side. (Luxol fast blue/cresyl violet.)

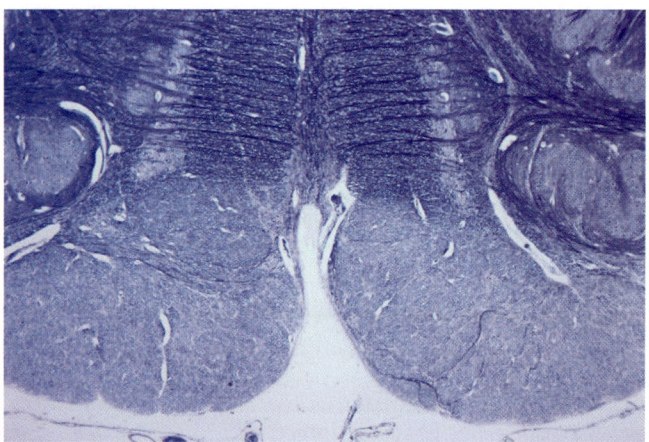

Fig. 14.38. Low-power of ventral medulla shows atrophy of pyramids seen as loss of myelin/staining next to normally-stained medial lemnisci. ALS. (Luxol fast blue/cresyl violet.)

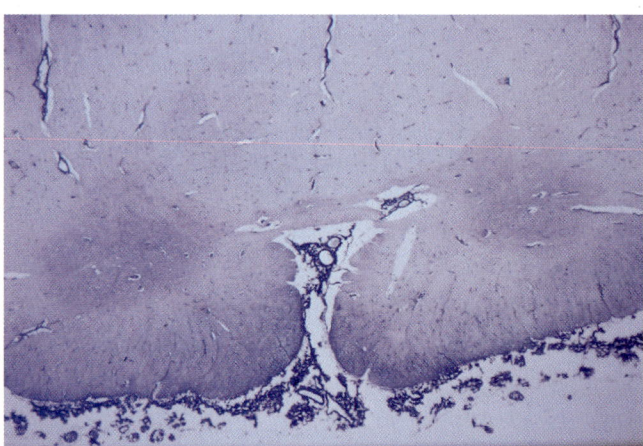

Fig. 14.39. Low-power micrograph of transverse section of spinal cord shows blue-stained gliosis in anterior horns in ALS. (Holzer.)

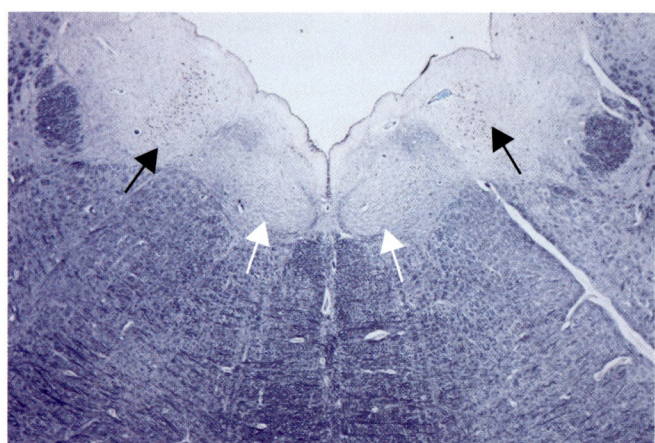

Fig. 14.40. Low-power micrograph of medulla in ALS shows loss of neurons in hypoglossal nuclei (*white arrows*) with preservation of dorsal vagal nuclei (*black arrows*). (Luxol fast blue/cresyl violet.)

globus pallidus is caused by various deposits including iron, lipid, and melanin. Axonal spheroids are present among microgliosis and astrogliosis. Neurofibrillary tangles may be found in some cases in the brainstem, hippocampus, and neocortex.[17]

References

1. Adams RD, Van Bogaert L, Vander Eecken M. Striatonigral degenerations. J Neuropathol Exp Neurol 1964;23:584–608.
2. Arriagada PV, Growdon JH, Hedley-Whyte ET, Hyman BT. Neurofibrillary tangles but not senile plaques parallel duration and severity of Alzheimer's disease. Neurology 1992;42:631–639.
3. Arriagada PV, Marzloff K, Hyman BT. Distribution of Alzheimer-type pathologic changes in nondemented elderly individuals matches the pattern in Alzheimer's disease. Neurology 1992;42:1681–1688.
4. Backman DL, Wolf PA, Linn R, et al. Prevalence of dementia and probable senile dementia of the Alzheimer type in the Framingham study. Neurology 1992;42:115–119.
5. Ball MJ, Lo P. Granulovacuolar degeneration in the ageing brain and in dementia. J Neuropathol Exp Neurol 1977;36:474–487.
6. Ballard PA, Tetrud JW, Langston JW. Permanent human parkinsonism due to l-methyl-4-phenyl-1,2,3,6-tetrahydropyridine (MPTP): Seven cases. Neurology 1985;35:949–956.
7. Boller F, Lopez OL, Moosy J. Diagnosis of dementia: Clinicopathologic correlations. Neurology 1989;39:76–79.
8. Braak H, Braak E. Neuropathological staging of Alzheimer-related changes. Acta Neuropathol 1991;82:239–259.
9. Brownell B, Oppenheimer DR, Hughes JT. The central nervous system in motor neurone disease. J Neurol Neurosurg Psychiatry 1970;33:338–357.
10. Campbell SK, Switzer RC, Martin TL. Alzheimer's plaques and tangles: A controlled and enhanced silver-staining method. Soc Neurosci Abstr 1987;13:678.
11. Crystal H, Dickson D, Fuld P, et al. Clinico-pathologic studies in dementia: Nondemented subjects with pathologically confirmed Alzheimer's disease. Neurology 1988;38:1682–1687.
12. Cummings BJ, Pike CJ, Shankle R, Cotman CW. Amyloid deposition and other measures of neuropathology predict cognitive status of Alzheimer's disease. Neurobiol Aging 1996;17:921–933.

13. Davis DG, Schmit FA, Wekstein DR, Markesbery WR. Alzheimer neuropathologic alterations in aged cognitively normal subjects. J Neuropathol Exp Neurol 1999;58:376–388.
14. de la Monte SM, Vonsattel J-P, Richardson EP Jr. Morphometric demonstration of atrophic changes in the cerebral cortex, white matter, and neostriatum in Huntington's disease. J Neuropathol Exp Neurol 1988;47:516–525.
15. Dickson DW. The pathogenesis of senile plaques. J Neuropathol Exp Neurol 1997;56:321–339.
16. Ditter SM, Mirra SS. Neuropathologic and clinical features of Parkinson's disease in Alzheimer's disease patients. Neurology 1987;37:754–760.
17. Eidelberg D, Sotrel A, Joachim C, et al. Adult onset Hallervorden-Spatz disease with neurofibrillary pathology. Brain 1987;110:993–1013.
18. Flament S, Delacourte A, Verny M, et al. Abnormal tau proteins in progressive supranuclear palsy: Similarities and differences with the neurofibrillary degeneration of the Alzheimer's type. Acta Neuropathol 1991;81:591–596.
19. Forno LS, Langston JW, DeLanney LE, et al. Locus ceruleus lesions and eosinophilic inclusions in MPTP-treated monkeys. Ann Neurol 1986;20:449–455.
20. Giannakopoulos P, Herrmann FR, Bussiere T, et al. Tangle and neuron numbers, but not amyloid load, predict cognitive status in Alzheimer's disease. Neurology 2003;60:1495–1500.
21. Gibb WRG, Esiri MM, Lee AJ. Clinical and pathological features of diffuse cortical Lewy body disease (Lewy body dementia). Brain 1987;110:1131–1153.
22. Graham DI, Lantos PL (eds). Greenfield's Neuropathology, ed 6. London: Arnold, 1997.
23. Graham J, Oppenheimer D. Orthostatic hypotension and nicotine sensitivity in a case of multiple system atrophy. J Neurol Neurosurg Psychiatry 1969;32:28–34.
24. Haroutunian V, Perl DP, Purohit DP, et al. Regional distribution of neuritic plaques in the nondemented elderly and subjects with very mild Alzheimer disease. Arch Neurol 1998;55:1185–1191.
25. Hebert LE, Scherr PA, Bienias JL, et al. Alzheimer disease in the US population: Prevalence estimates using the 2000 census. Arch Neurol 2003;60:1119–1122.
26. Helmer C, Joly P, Letenneur L, et al. Mortality with dementia: Results from a French prospective community-based cohort. Am J Epidemiol 2001;154:642–648.
27. Hirano A. Hirano bodies and related neuronal inclusions. Neuropathol Appl Neurobiol 1994;20:3–11.
28. Hirano A, Dembitzer HM, Kurland LT, Zimmerman HM. The fine structure of some intraganglionic alterations. J Neuropathol Exp Neurol 1968;27:167–182.
29. Hughes AJ, Daniel SE, Blankson S, Lees AJ. A clinicopathologic study of 100 cases of Parkinson's disease. Arch Neurol 1993;50:140–148.
30. Hunsaker DM, Hunsaker JC III. Therapy-related café coronary deaths: Two case reports of rare asphyxial deaths in patients under supervised care. Am J Forensic Med Pathol 2002;23:149–154.
31. Hy LX, Keller DM. Prevalence of AD among whites: A summary by levels of severity. Neurology 2000;55:198–204.
32. Jellinger K. Neuroaxonal dystrophy: Its natural history and related disorders. In Zimmerman HM. Progress in Neuropathology, Vol II. New York: Grune & Stratton, 1973, pp 158–160.
33. Jellinger KA. Neuropathological staging of Alzheimer-related lesions: The challenge of establishing relations to age. Neurobiol Aging 1997;18:369–375.
34. Kidd M. Alzheimer's disease—An electron microscopic study. Brain 1964;87:307–320.
35. Langston JW, Ballard P, Tetrud JW, Irwin I. Chronic parkinsonism in humans due to a product of meperidine-analog synthesis. Science 1983;219:979–980.
36. Lantos PL. The definition of multiple system atrophy: A review of recent developments. J Neuropathol Exp Neurol 1998;57:1099–1111.
37. Love S. Post mortem sampling of the brain and other tissues in neurodegenerative disease. Histopathology 2004;44:309–317.
38. Lowe J. Establishing a pathological diagnosis in degenerative dementias. Brain Pathol 1998;8:403–406.
39. Maher ER, Lee AJ. The clinical features and natural history of the Steele-Richardson-Olszewski syndrome (progressive supranuclear palsy). Neurology 1986;36:1005–1008.
40. McKeel DW Jr, Price JL, Miller JP, et al. Neuropathologic criteria for diagnosing Alzheimer disease in persons with pure dementia of Alzheimer type. J Neuropathol Exp Neurol 2004;63:1028–1037.
41. McKeith IG, Galasko D, Kosaka K, et al. Consensus guidelines for the clinical and pathologic diagnosis of dementia with Lewy bodies (DLB): Report of the consortium on DLB international workshop. Neurology 1996;47:1113–1124.
42. Mirra SS, Heyman A, McKeel D, et al. The consortium to establish a registry for Alzheimer's disease (CERAD). Part II. Standardization of the neuropathologic assessment of Alzheimer's disease. Neurology 1991;41:479–486.
43. Morris JC, Storandt M, McKeel DW Jr, et al. Cerebral amyloid deposition and diffuse plaques in "normal" aging: Evidence for presymptomatic and very mild Alzheimer's disease. Neurology 1996;46:707–719.
44. Myers RH, Vonsattel JP, Stevens TJ, et al. Clinical and neuropathologic assessment of severity in Huntington's disease. Neurology 1988;38:341–347.
45. National Institute on Aging and Reagan Institute Working Group on Diagnostic Criteria for the Neuropathological Assessment of Alzheimer's Disease. Consensus recommendations for the postmortem diagnosis of Alzheimer's disease. Neurobiol Aging 1997;18(Suppl 4):1–15.
46. Nussbaum RL, Ellis CE. Alzheimer's disease and Parkinson's disease. N Engl J Med 2003;348:1356–1364.
47. Oba H, Yagishita A, Terada H, et al. New and reliable MRI diagnosis for progressive supranuclear palsy. Neurology 2005;64:2050–2055.
48. Okazaki H, Lipkin LE, Aronson SM. Diffuse intracytoplasmic ganglion inclusion (Lewy body) associated with progressive dementia and quadriparesis in flexion. J Neuropathol Exp Neurol 1961;20:237–244.
49. Olson MI, Shaw CM. Presenile dementia and Alzheimer's disease in mongolism. Brain 1969;92:147–156.
50. Papp MI, Kahn JE, Lantos PL. Glial cytoplasmic inclusions in the CNS of patients with multiple system atrophy (striatonigral degeneration, olivopontocerebellar atrophy and Shy-Drager syndrome). J Neurol Sci 1989;94:79–100.
51. Papp MI, Lantos PL. The distribution of oligodendroglial inclusions in multiple system atrophy and its relevance to clinical symptomatology. Brain 1994;117:235–243.
52. Rajput A. Clinical features and natural history of Parkinson's disease (special consideration of aging). In Calne D (ed): Neurodegenerative Diseases. Philadelphia: WB Saunders, 1994, pp 555–571.
53. Rajput AH, Rozdilsky B, Hornykiewicz O, et al. Reversible drug-induced parkinsonism: Clinicopathologic study of two cases. Arch Neurol 1982;39:644–646.
54. Rowland LP, Shneider NA. Amyotrophic lateral sclerosis. N Engl J Med 2001;344:1688–1700.
55. Samuel W, Alford M, Hofstetter R, Hansen L. Dementia with Lewy bodies versus Alzheimer disease: Differences in cognition, neuropathology, cholinergic dysfunction, and synapse density. J Neuropathol Exp Neurol 1997;56:499–508.
56. Spencer PS. Lathyrism. In Vinken PJ, Bruyn GW (eds): Handbook of Clinical Neurology. deWolfe FA (ed). Vol 21(65): Intoxications of the Nervous System, Part II. Amsterdam: Elsevier 1995, pp 1–20.

57. Steele JC, Richardson JC, Olszewski J. Progressive supranuclear palsy: A heterogeneous degeneration involving the brain stem, basal ganglia and cerebellum with vertical gaze and pseudobulbar palsy, nuchal dystonia and dementia. Arch Neurol 1964;10:333–359.
58. Takeda S, Yamazaki K, Miyakawa T, Arai H. Parkinson's disease with involvement of the parasympathetic ganglia. Acta Neuropathol 1993;86:397–398.
59. Tanner CM, Goldman SM. Epidemiology of Parkinson's disease. Neurol Clin 1996;14:317–335.
60. Terry RD, Gonatas NK, Weiss M. Ultrastructural studies in Alzheimer's presenile dementia. Am J Pathol 1964;44:269–297.
61. Terry RD, Peck A, DeTeresa R, et al. Some morphometric aspects of the brain in senile dementia of the Alzheimer type. Ann Neurol 1981;10:184–192.
62. Tiraboschi P, Hansen LA, Thal LJ, Corey-Bloom J. The importance of neuritic plaques and tangles to the development and evolution of AD. Neurology 2004;62:1984–1989.
63. Tomlinson BE, Blessed G, Roth M. Observations on the brains of demented old people. J Neurol Sci 1970;11:205–242.
64. Tomlinson BE, Kitchener D. Granulovacuolar degeneration of hippocampal pyramidal cells. J Pathol 1972;106:165–185.
65. Vonsattel J-P, Myers RH, Stevens TJ, et al. Neuropathological classification of Huntington's disease. J Neuropathol Exp Neurol 1985;44:559–577.
66. Warmuth-Metz M, Naumann M, Csoti I, Solymosi L. Measurement of the midbrain diameter on routine magnetic resonance imaging: A simple and accurate method of differentiating between Parkinson disease and progressive supranuclear palsy. Arch Neurol 2001;58:1076–1079.

Demyelinating Disorders

15

MULTIPLE SCLEROSIS 413
LEUKODYSTROPHY 418

REFERENCES 421

Multiple Sclerosis

Multiple sclerosis (MS) is one of a large number of diseases of varied pathogenesis in which central nervous system (CNS) white matter is focally or diffusely the main target. (Those of infectious cause are covered in Chapter 12.) For a detailed review of disorders of white matter, the reader is referred to a reference such as the chapter on "Demyelinating Diseases" in *Greenfield's Neuropathology*.[6] Among demyelinating disorders, cases of MS constitute, by far, the most frequently encountered in our office. Others only rarely encountered include conditions such as progressive multifocal leukoencephalopathy and vacuolar encephalopathy associated with human immunodeficiency virus (HIV) infection; genetic/metabolic leukodystrophy such as metachromatic leukodystrophy, Krabbe's disease, and spongiform leukodystrophy (Canavan's disease); and postinfectious perivenous encephalomyelitis. Central pontine myelinolysis is relatively frequently found and usually localized only to the basis pontis, although similar lesions can sometimes be found in the medulla and cerebrum. Necrosis of white matter found with carbon monoxide intoxication and/or ischemic encephalopathy is described in Chapter 10.

The prevalence of MS varies considerably with geographic location, but it is distinctly higher toward the northern latitudes of the northern hemisphere. For example, the prevalence is 175 per 100,000 in northern Scotland[6] and 30 per 100,000 in San Francisco.[10] With a population of approximately 10 million, Los Angeles County may have a very roughly estimated 2500 individuals with MS. However, only a few of these patients become subjects of medical examiner investigation each year, due to some medicolegal special circumstance. Of 50 cases of MS examined in the Office of the Chief Medical Examiner of Maryland over 22 years (about 2.3 per year), the manner of death was natural in 52%, accidental in 26%, suicide in 8%, homicide in 6%, and undetermined in 8%.[16] Death from suicide and accidents is not unexpected in any chronic debilitating disease,[14,17] but death from elevation of body temperature such as from sun exposure[1] or drowning in a hot tub bath may be unique to MS because of the well-known worsening of symptoms by elevated body temperature (discussed in Chap. 9).[3,20] In a Danish study, the risk of suicide in MS was twice that expected in that population.[2]

Multiple sclerosis has a mean age of onset of about 30 years, but the disease is uncommon in early childhood[18] and uncommon after 50 years.[6] Its epidemiologic and genetic factors are rarely relevant in a forensic setting. Clinical manifestations are quite variable in space and time, and virtually the entire spectrum of neurologic symptoms and signs can occur at different times. The reader is referred to references such as that of McAlpine et al.[9,10] for a full clinical discussion. Certain favored sites of anatomic involvement, such as the optic nerves and medial longitudinal fasciculus, lead to characteristic signs, namely, sudden loss of visual acuity with central scotoma and internuclear ophthalmoplegia with nystagmus, respectively. A clinical course of relapses and remissions with stepwise incremental progression is also a characteristic occurrence. The cerebrospinal fluid (CSF) profile of mild elevation of lymphocyte count and presence of oligoclonal bands on electrophoresis is supportive but not diagnostic of MS.[12] Magnetic resonance imaging is an increasingly important tool in the diagnosis due to its ability to detect plaques.[13]

The gross anatomic pathology of the plaque of MS, not the disease, was first recognized and illustrated by Carswell (1828) and Cruveilhier (1835–42) in their publications of anatomic pathology[9] as patchy alterations on the surface of the belly of the pons, medulla, and spinal cord, locations where white matter appears at the surface (Figs. 15.1–15.3). The optic nerve is another site where plaque can be recognized prior to sectioning along with the corpus callosum which, however, is mostly hidden from view. Plaques are rarely detected by the naked eye on the surface of the cerebral cortex, but they can be demonstrated in cortex on stained sections (Fig. 15.3).

The essential and diagnostic pathognomonic lesion of MS is the plaque, which is a smoothly contoured and sharply marginated zone of true demyelination (Fig. 15.4) in which axons are left stripped of myelin (Figs. 15.5–15.6). Plaques occur at all levels of the brain and spinal cord with a distinct propensity for the periventricular white matter, especially deep to the dorsolateral angle of the lateral ventricle (Case 15.1, Figs. 15.7–15.8). They are variable in size from a few millimeters to several centimeters or larger in both white and gray matter, respecting no anatomic boundary and usually easily detected. Newer plaques are pink in the fresh state and turn gray with formalin fixation (Fig. 15.8). In a case of suicide by carbon monoxide intoxication, the plaques were dark red (Fig. 15.7). Smaller plaques appear as sharply punched-out perivenous zones of demyelination (Fig. 15.9). Plaques in gray matter, such as in the thalamus, may grossly mimic acute/subacute infarcts (Fig. 15.10). The boundaries of plaques entering the cerebral cortex are difficult to detect and require a myelin stain to demonstrate (Figs. 15.11–15.12). The so-called shadow plaque is one with less dense staining of myelin compared with normal, thought likely to be a result of incomplete remyelination (Fig. 15.13). In both fresh and fixed brain and spinal cord, plaques may become more distinct after brief exposure to air, and reexamination of sections a few minutes after initial inspection may be rewarded

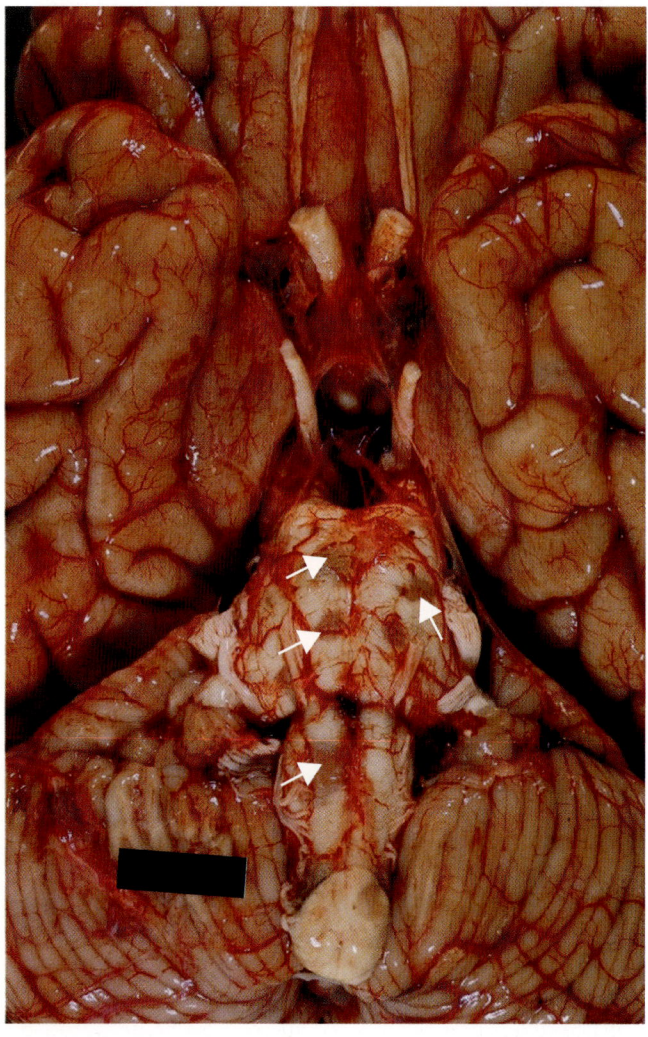

Fig. 15.1. Closeup view of base of brain and brainstem shows multiple sclerosis plaques as discrete small tan discolorations of surface white matter (*arrows*).

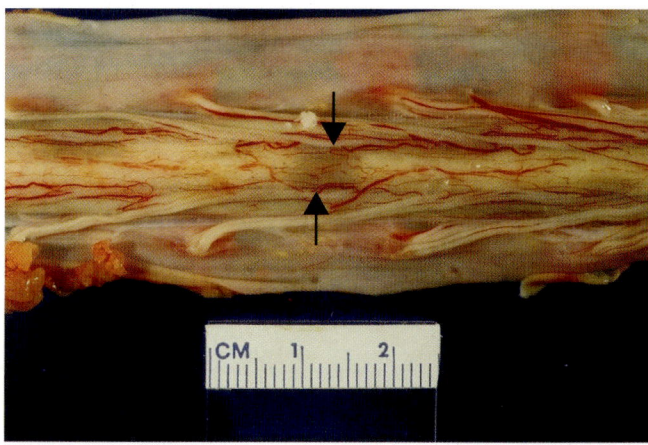

Fig. 15.2. Closeup view of dorsal aspect of spinal cord shows a tan focal discoloration of multiple sclerosis plaque (*arrows*).

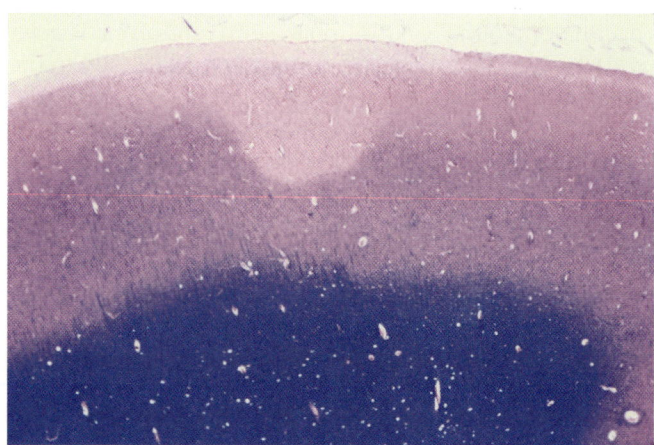

Fig. 15.3. Low-power photomicrograph of cerebral cortex shows a multiple sclerosis plaque as a sharply marginated zone of paling of demyelination based at the pia (*arrows*). (Luxol fast blue/H&E.)

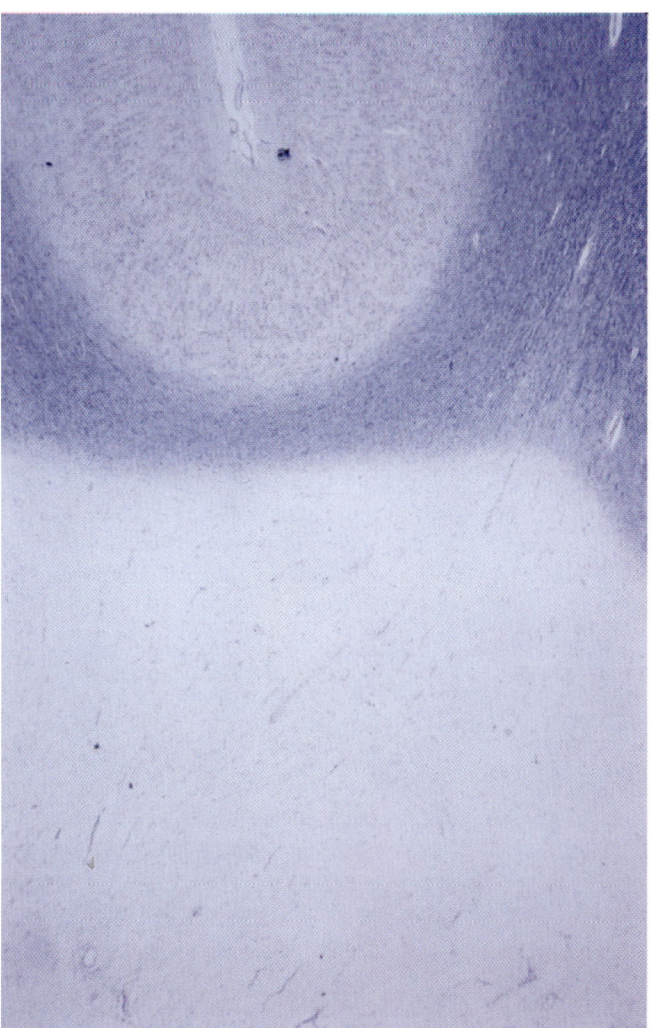

Fig. 15.4. Low-power micrograph shows a multiple sclerosis plaque as a large pale zone below with smoothly marginated sharp loss of blue staining myelin. Cortex is above. (Luxol fast blue/cresyl violet.)

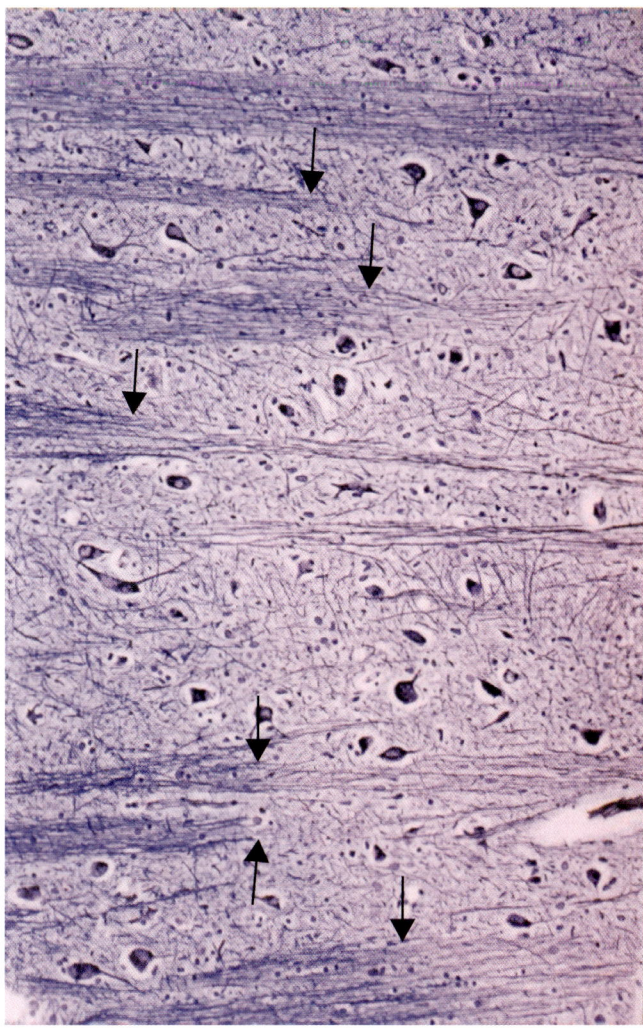

Fig. 15.5. Medium-power micrograph of the thalamus demonstrates sharp transition (*arrows*) from myelinated (*left*) to demyelinated nerve fibers at margin of a plaque. (Luxol fast blue/eosin/Bodian.)

by discovery of a surprising number of previously undetected plaques. Relatively small plaques in critical locations such as the pyramids or the spinal cord may cause devastating neurologic impairment (Figs. 15.14–15.15). On the other hand, the finding of an MS plaque, even in a critical site such as the spinal cord, may have been clinically asymptomatic.[5] Small plaques of MS may be an unexpected finding at autopsy.[15]

Examples of several variants of MS include neuromyelitis optica (Devic's disease) with devastating visual loss and transverse myelitis (Fig. 15.16); Balo's type with characteristic concentric zones of demyelination; and Schilder's type with giant plaques that are usually bilaterally symmetric and should be differentiated from X-linked adrenoleukodystrophy (ALD) (Figs. 15.17–15.18). A leukodystrophy with demyelination similar in appearance and histologic findings to ALD, including trilaminar bodies by electron microscopy and periodic-acid Schiff-positive macrophages, may occur with chronic organic solvent vapor inhalation.[7]

Histologically, inactive chronic plaques are zones of myelin loss and astrogliosis in which a few axons remain, and inflammation is scant, if present at all. Chronic active plaques show a narrow zone of inflammation at the plaque margin composed of lymphocytes, microglia, and variable numbers of lipid-laden macrophages (Fig. 15.19). Usually rather scant, perivascular lymphocytic infiltrates can be found within the plaque zone (Fig. 15.20) and in immediately adjacent white matter.

Some cases of multiple sclerosis are referred to the medical examiner because of sudden unexpected death, as discussed in Chapter 9.

A massive volume of literature has been compiled in human and animal experimental work on the pathogen-

Demyelinating Disorders

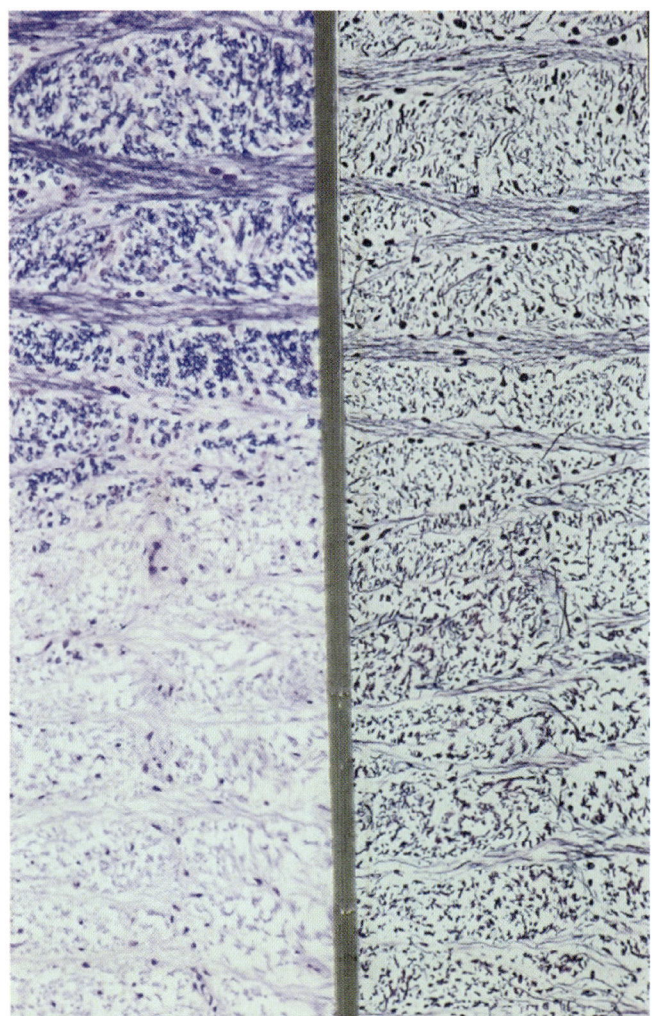

Fig. 15.6. Medium-power micrograph of medial lemniscus in medulla shows serial sections stained by Luxol fast blue/H&E (*left*) and Bodian for axons (*right*). Axons are uniformly normal in the image on the right. On the other hand, lemniscus is demyelinated in the same zone of preserved axons in the image on the left.

Case 15.1. (Figs. 15.7 and 15.8)

A 53-year-old man with multiple sclerosis and heart disease was found dead seated in his automobile parked in his garage with the doors closed. The ignition key was turned off, but his skin was noted to be "cherry red." The officer initially on the scene did not notice any fumes. His MS had been progressively worse, and he had threatened suicide in the past.

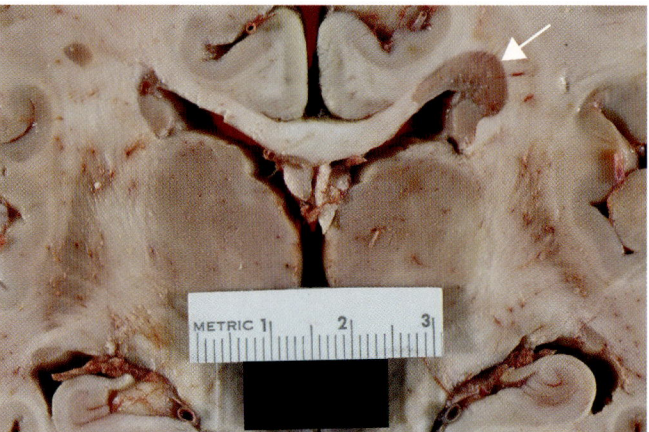

Fig. 15.7. Closeup view of coronal section of cerebral hemisphere at midthalamic level shows a sharply demarcated, rounded-contour, periventricular plaque of multiple sclerosis involving corpus callosum (*arrow*). The plaque is cherry red in a person dying of carbon monoxide intoxication.

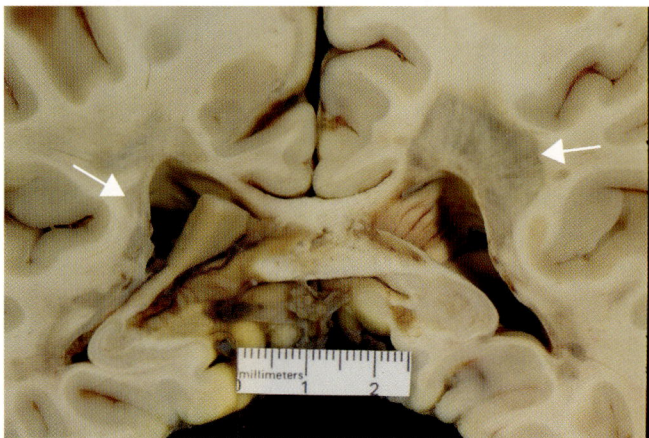

Fig. 15.8. Closeup view of periventricular plaque involving the tapetum bilaterally, a location typical and common for multiple sclerosis plaques. The plaques are gray post-formalin fixation (*arrows*).

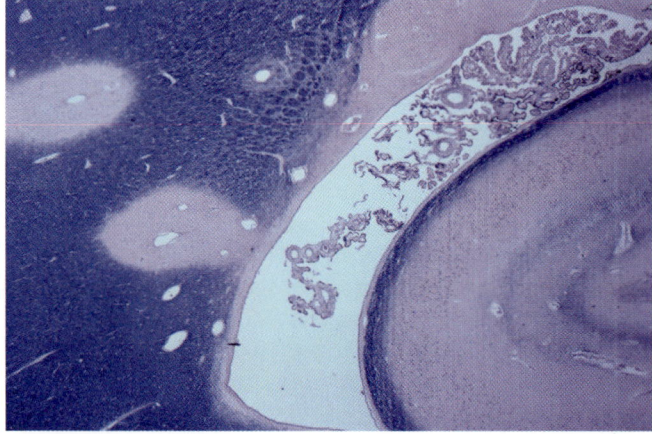

Fig. 15.9. Low-power micrograph shows two small perivenous plaques of multiple sclerosis in white matter adjacent to temporal horn of lateral ventricle. Note their smoothly rounded shape around small central veins. (Luxol fast blue/H&E.)

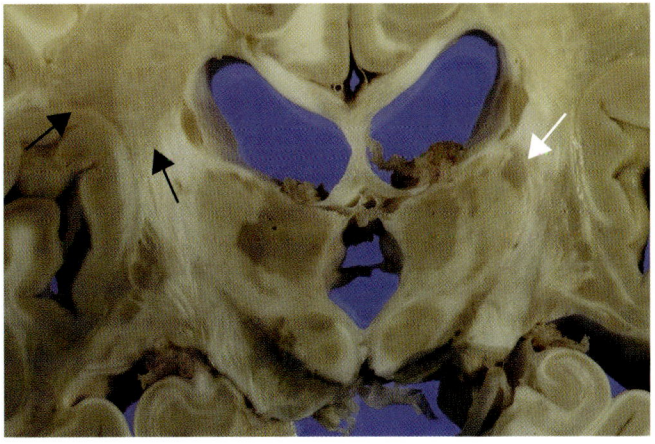

Fig. 15.10. Closeup view of coronal section of cerebral hemisphere shows plaques of multiple sclerosis in both thalami; the smaller plaque is present on the right (*white arrow*). Larger plaques are also in the left subcortical white matter and capsule (*black arrows*), among others.

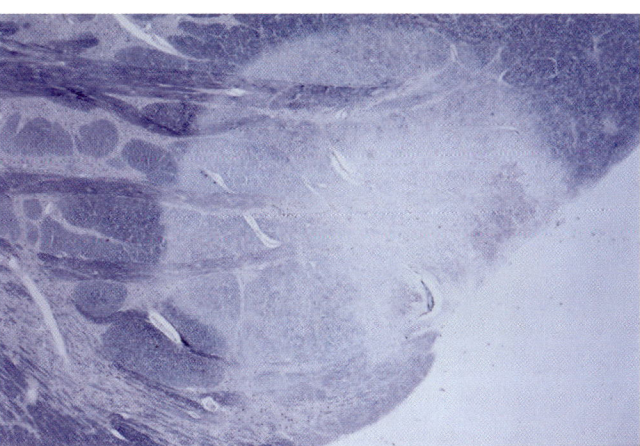

Fig. 15.13. Low-power micrograph of basis pontis demonstrates a shadow plaque of MS with smooth, rounded-contour zone of pale-staining myelin based on pia surface. (Luxol fast blue/cresyl violet.)

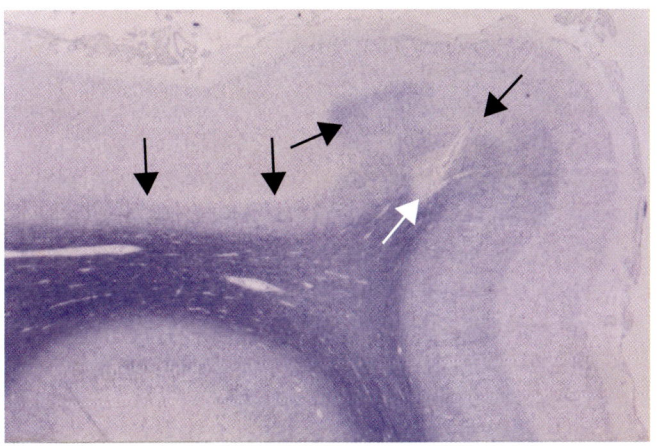

Fig. 15.11. Low-power micrograph of cerebral cortex shows areas of normally stained myelin architecture within cortex to the right and bottom of the image. Sharp cutoff of stained myelin in cortex at the top marks an intracortical MS plaque with margins indicated by black arrows. A small pale plaque is present at cortical/subcortical border (*white arrow*). (Luxol fast blue/cresyl violet.)

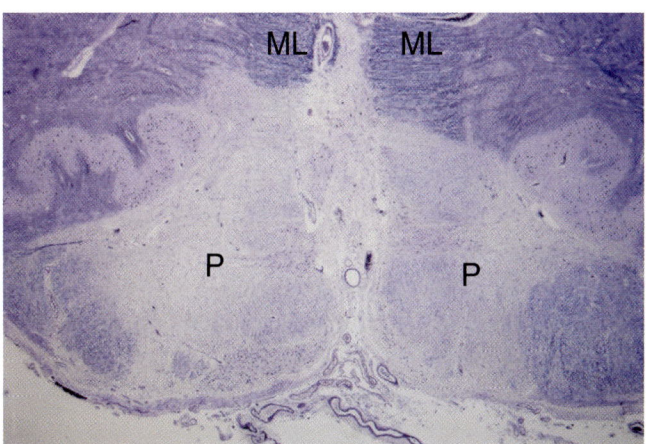

Fig. 15.14. Low-power micrograph of medulla shows a multiple sclerosis plaque sharply cutting across medial lemnisci (ML) and involving most of pyramids (P). (Luxol fast blue/cresyl violet.)

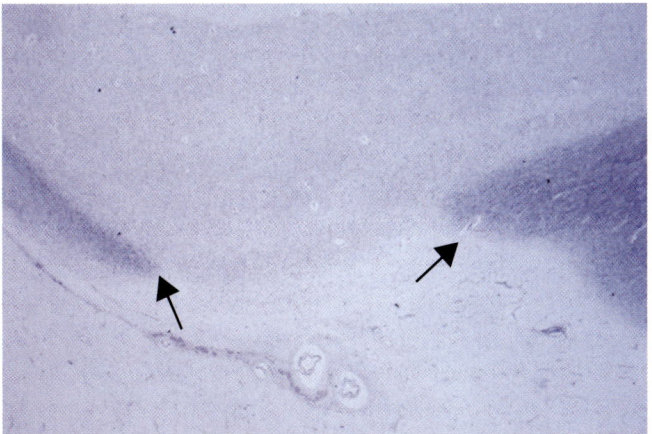

Fig. 15.12. Low-power micrograph of cerebral cortex and subcortical white matter shows cortex above and a large pale demyelinated MS plaque below that involves the U fiber (*arrows*). Intracortical portion of plaque cannot be discerned. (Luxol fast blue/cresyl violet.)

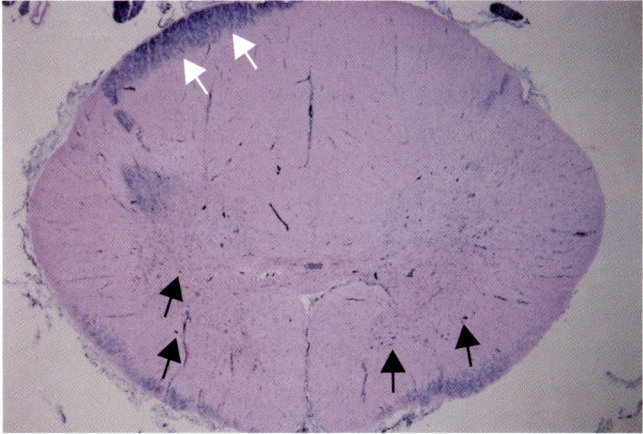

Fig. 15.15. Transverse section of spinal cord shows near-total demyelination with only thin rims of myelinated fibers remaining at the margins (*white arrows*). A few motor neurons of anterior horn can be discerned (*black arrows*). Despite the loss of myelin, note the preservation of normal cord profile. (Luxol fast blue/H&E.)

Demyelinating Disorders

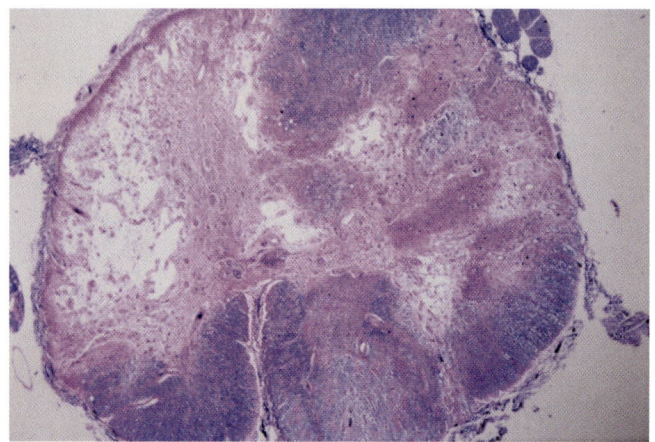

Fig. 15.16. Low-power micrograph shows transverse section of spinal cord demonstrating almost complete necrotizing transverse myelitis of Devic's disease. (Luxol fast blue/H&E.)

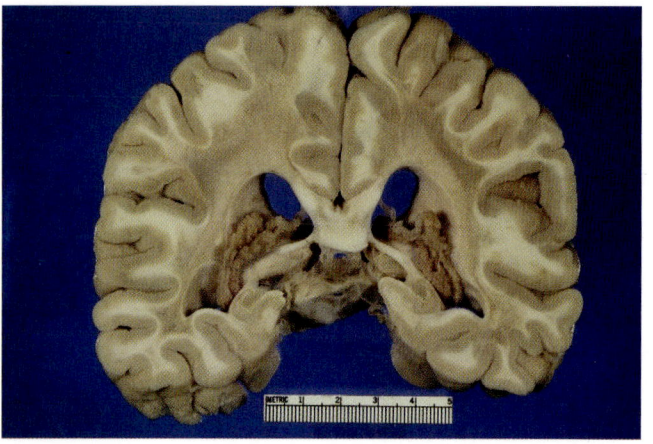

Fig. 15.17. Coronal sections at splenium of corpus callosum show bilaterally symmetric huge zones of demyelination in Schilder's type of multiple sclerosis.

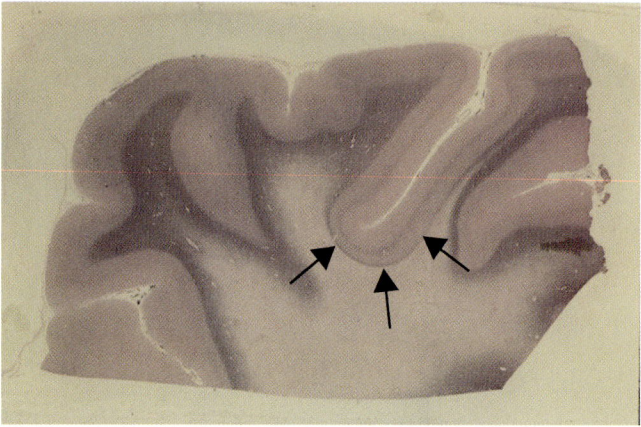

Fig. 15.18. Low-power micrograph of Schilder's type of multiple sclerosis shows broad zone of paling of demyelination of white matter with only a very fine line of **U** fiber-sparing at one sector (*arrows*). (Weil/eosin.)

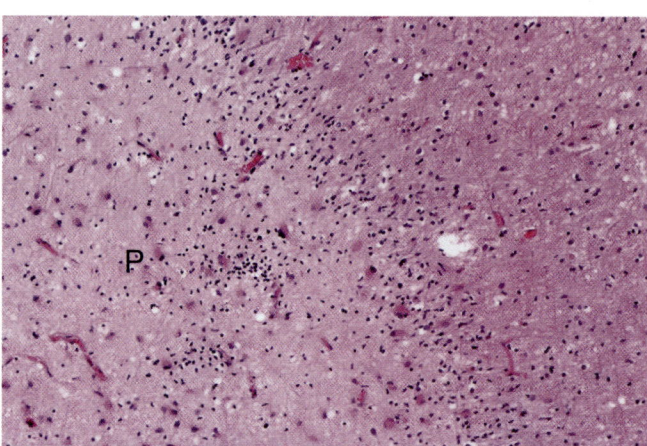

Fig. 15.19. Medium-power micrograph shows a margin of a chronic active plaque of MS with lymphocytes and microglia visible along margin and within plaque (P). Hypertrophied astrocytes are seen within plaque. (H&E.)

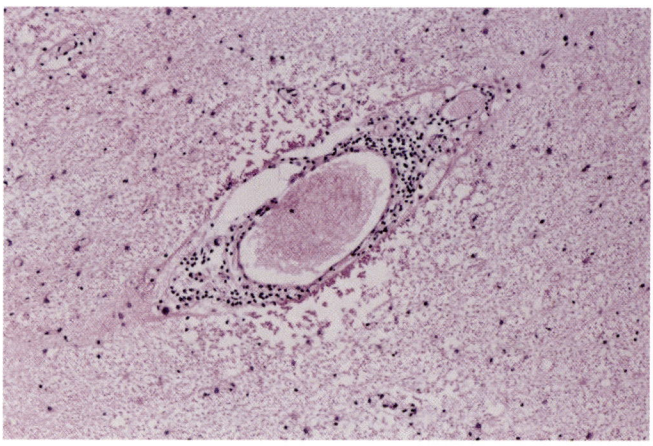

Fig. 15.20. Medium-power micrograph shows a typically scant perivascular cuff of chronic inflammatory cells within an MS plaque zone showing chronic gliosis. (H&E.)

esis of MS since it was defined as a clinical entity. For a recent review of the subject, the reader is referred to the paper by Ludwin[8] and by Frohman *et al.*[4]

Leukodystrophy

The leukodystrophies are a heterogeneous group of hereditary disorders of infancy and childhood primarily causing diffuse degenerative changes of white matter. Sporadic cases are known to occur. The conditions are rare, and still rarer adult forms are known. Over time, long-suspected molecular bases for the disorders have been and are being elucidated. A brief review of the biochemical defects can be found in resources such as *Greenfield's Neuropathology*.[6] The group includes, among others, metachromatic leukodystrophy, adrenoleukodystrophy, Krabbe's disease (globoid leukodystrophy), and Canavan's disease (spongy degeneration of van Bogaert and Bertrand). Alexander's disease, Pelizaeus-Merzbacher disease, and the various sudanophilic leukodystrophies have the gross characteristic of degeneration of white matter with various but identifiable histologic features but with incomplete explanation of pathogenesis. It is an unexpected occurrence when any of the aforementioned conditions are encountered during a forensic autopsy without a history of prior, and usually extensive, diagnostic workup. On rare occasions, cases of leukodystrophy are undiagnosed or misdiagnosed and come to the attention of the medical examiner.[19]

The clinical manifestations of each of the leukodystrophies may present differently depending on the type and age of onset. Most share motor disability such as gait disorder, ataxia, incoordination, loss of previously gained motor function, and progressive spasticity. Behavior disorder, irritability, nystagmus, and visual and auditory impairments are other signs of disturbed CNS functions, albeit nonspecific.[11]

Grossly, brain weights of the various leukodystrophies are variable, reduced in Pelizaeus-Merzbaucher disease and increased in the early stages of Canavan's disease, globoid leukodystrophy, and Alexander's disease.[6] Sections reveal a nonspecific diffuse change in white matter from gray discoloration to cavitation.[6] Microscopy provides the first clues to the diagnosis, and histochemistry, electron microscopy, and biochemical analysis usually lead to the eventual correct diagnosis. Some of the salient gross and microscopic features seen in our experience include cases of Pelizaeus-Merzbacher's disease (Case 15.2, Figs. 15.21–15.24), metachromatic leukodystrophy (Figs. 15.25–15.26), Canavan's disease (Figs. 15.27–15.28), and globoid leukodystrophy (Krabbe's disease) (Fig. 15.29).

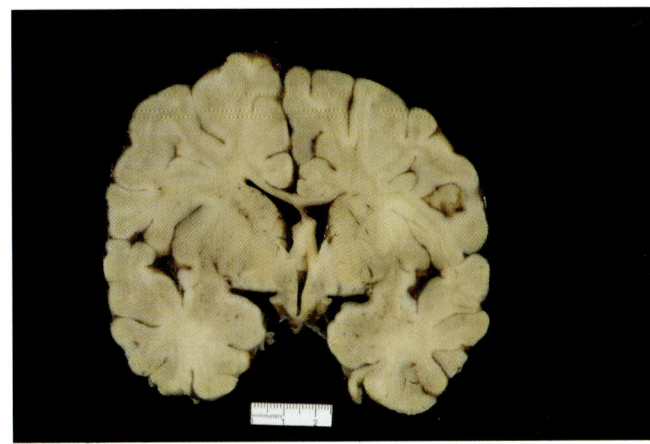

Fig. 15.21. Coronal section at frontal level in this case of Pelizaeus-Merzbacher disease shows diffuse pale tan discoloration of white matter with preservation of cortical ribbon.

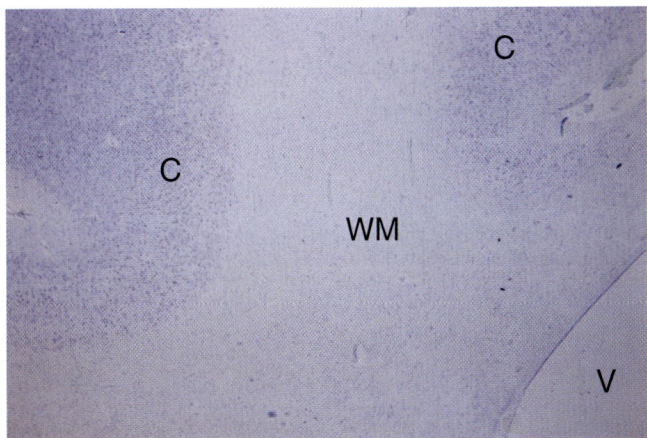

Fig. 15.22. Low-power micrograph of white matter (WM) reveals total absence of myelin. C, cerebral cortex; V, ventricle. (Luxol fast blue/cresyl violet.) PMD.

Case 15.2. (Figs. 15.21–15.24)

An 11-year-old boy was delivered at full term after an apparently normal pregnancy. At about 1 month of age, he developed severe respiratory problems that required a tracheotomy for management. Along with failure to thrive, he suffered recurrent episodes of pneumonitis, bronchiolitis, and seizures. At the age of 6 years, he was committed to a state hospital for the developmentally impaired with diagnoses of severe mental retardation, spastic quadriplegia, seizure disorder, neurogenic dysphagia, and tracheal stenosis.

Autopsy findings included a brain weight of 1100 g, abnormally pale tan and semitranslucent optic nerves of normal diameter, and normal oculomotor nerves. Spinal cord on section was uniformly gray and semitranslucent, whereas nerve roots appeared normal. Pelizaeus-Merzbacher's disease (PMD).

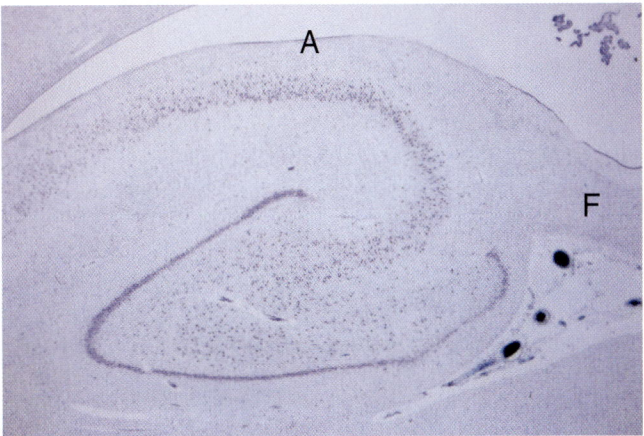

Fig. 15.23. Low-power micrograph of hippocampal formation stained with Luxol fast blue shows absence of myelin in alveus (A) and fimbria (F). PMD.

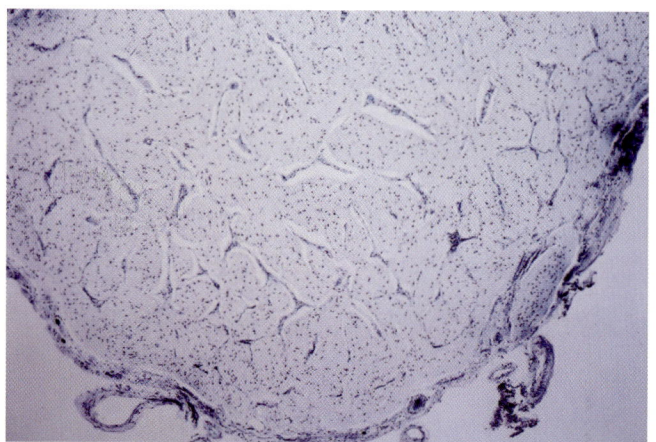

Fig. 15.24. Low-power micrograph of optic nerve demonstrates absence of myelin. (Luxol fast blue/cresyl violet.) PMD.

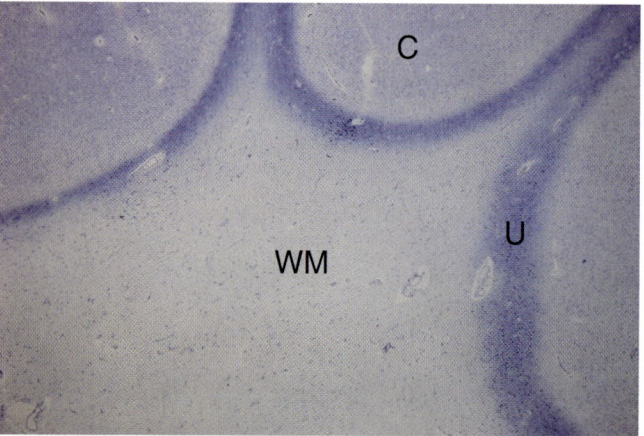

Fig. 15.25. Low-power micrograph of cerebrum degenerated from a case of metachromatic leukodystrophy shows pale broad area of white matter (WM) with sparing of **U** fibers (U). C, cerebral cortex. (Luxol fast blue/cresyl violet.)

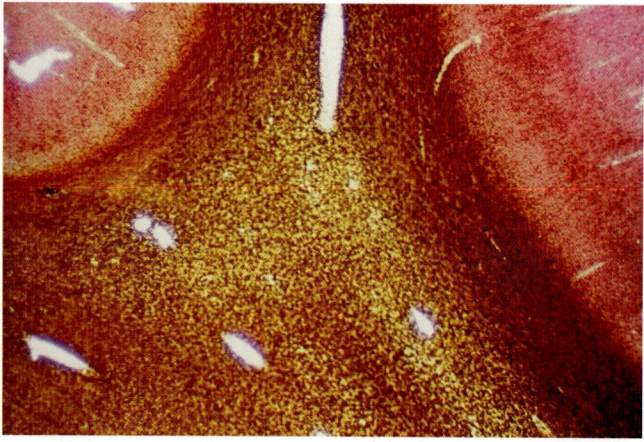

Fig. 15.26. Low-power micrograph of similar area of brain as in Fig. 15.25 shows brown metachromasia of white matter. (Toluidine blue method for metachromasia.)

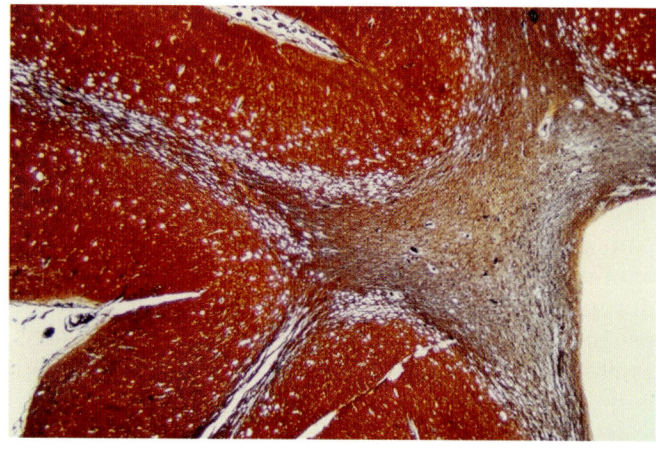

Fig. 15.27. Low-power micrograph of brain from a case of Canavan's disease shows pale atrophic white matter and characteristic spongy change mostly at cortical/subcortical white matter boundaries. (Modified trichrome stain.)

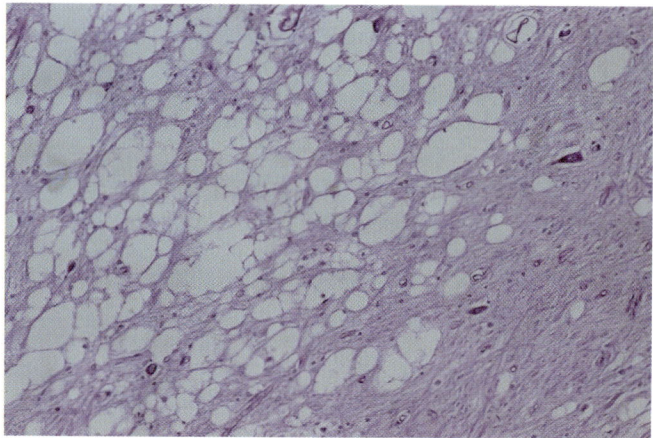

Fig. 15.28. Medium-power micrograph of cortical/white matter border in Canavan's disease showing spongy empty vacuolization in an area of gliosis. (H&E.)

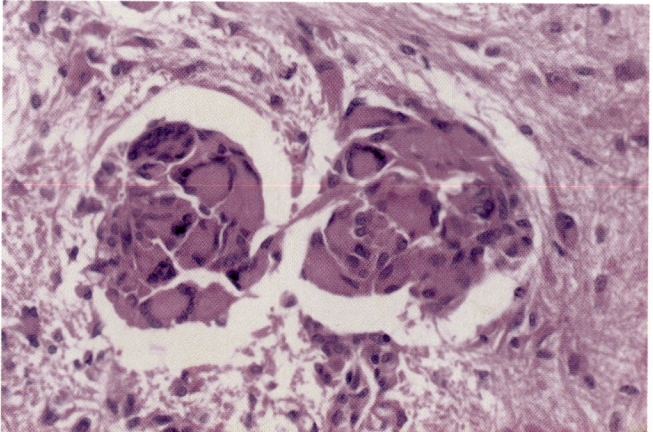

Fig. 15.29. High-power micrograph of white matter shows globoid macrophages with periodic acid-Schiff-positive content, characteristic of globoid leukodystrophy (Krabbe's disease).

References

1. Avis SP, Pryse-Phillips WM. Sudden death in multiple sclerosis associated with sun exposure: A report of two cases. Can J Neurol Sci 1995;22:305–307.
2. Brønnum-Hansen H, Stenager E, Nylev Stenager E, Koch-Henriksen N. Suicide among Danes with multiple sclerosis. J Neurol Neurosurg Psychiatry 2005;76:1457–1459.
3. Edmound J, Fog T. Visual and motor instability in multiple sclerosis. Arch Neurol Psychiatry 1955;73:316–323.
4. Frohman EM, Racke MK, Raine CS. Multiple sclerosis—The plaque and its pathogenesis. N Engl J Med 2006;354:942–955.
5. Ghatak NR, Hirano A, Lijtmaer H, Zimmerman HM. Asymptomatic demyelinated plaque in the spinal cord. Arch Neurol 1974;30:484–486.
6. Graham DI, Lantos PL (eds): Greenfield's Neuropathology, ed 6, Vol 1. London: Arnold, 1997.
7. Kornfeld M, Moser AB, Moser HW, et al. Solvent vapor abuse leukodystrophy: Comparison to adrenoleukodystrophy. J Neuropathol Exp Neurol 1994;53:389–398.
8. Ludwin SK. The pathogenesis of multiple sclerosis: Relating human pathology to experimental studies. J Neuropathol Exp Neurol 2006;65:305–318.
9. McAlpine D, Compston ND, Lumsden CE. Multiple Sclerosis. Edinburgh: E&S Livingstone, 1955.
10. McAlpine D, Lumsden CE, Acheson ED. Multiple Sclerosis: A Reappraisal. Edinburgh: E&S Livingstone, 1965.
11. Menkes JH, Sarnat HB (eds). Child Neurology, ed 6. Philadelphia: Lippincott Williams & Wilkins, 2000.
12. Moulin D, Paty DW, Ebers GC. The predictive value of cerebrospinal fluid electrophoresis in 'possible' multiple sclerosis. Brain 1983;106:809–816.
13. Paty DW, Oger JJM, Kastrukoff LF. Magnetic resonance imaging (MRI) in the diagnosis of multiple sclerosis: A prospective study with comparison of clinical evaluation, evoked potential, oligoclonal banding, and computer tomography. Neurology 1988;38:180–185.
14. Phadke JG. Survival pattern and cause of death in patients with multiple sclerosis: Results from an epidemiological survey in northeast Scotland. J Neurol Neurosurg Psychiatry 1987;50:523–531.
15. Phadke JG, Best PV. Atypical and clinically silent multiple sclerosis: A report of 12 cases discovered unexpectedly at necropsy. J Neurol Neurosurg Psychiatry 1983;46:414–420.
16. Riudavets MA, Colegial C, Rubio A, et al. Causes of unexpected death in patients with multiple sclerosis: A forensic study of 50 cases. Am J Forensic Med Pathol 2005;26:244–249.
17. Sadovnick AD, Eisen K, Ebens GC, Paty DW. Cause of death in patients attending multiple sclerosis clinics. Neurology 1991;41:1193–1196.
18. Salguero LF, Itabashi HH, Gutierrez NB. Childhood multiple sclerosis with psychotic manifestations. J Neurol Neurosurg Psychiatry 1969;32:572–579.
19. Shields LBE, Handy TC, Parker JC Jr, Burns C. Postmortem diagnosis of leukodystrophies. J Forensic Sci 1998;43:1068–1071.
20. Simons DJ. A note on the effect of heat and of cold upon certain symptoms of multiple sclerosis. Bull Neurol Inst N Y 1937;6:385–386.

Periprocedural Complications

16

INTRODUCTION 423
DEFINITIONS 423
CATEGORIZATION OF THE PROBLEM 424
MECHANISMS OF INJURY 424
FOREIGN MATERIALS 433
SURGICAL INSTRUMENTS, HARDWARE, AND ABLATIVE PROCEDURES 433
MECHANICAL FORCES 436
CHALLENGES FOR THE FORENSIC PATHOLOGIST 438
DETERMINATION OF MODE 438
FINAL COMMENT 439
REFERENCES 439

Introduction

The neuropathology of complications of medical procedures includes a broad spectrum of conditions. The reader is referred to several reviews of this subject.[14,26,40,97,107,116,128] These complications generally fall into two groups, those that are an expected risk, and those that are unexpected. The latter group includes complications viewed as therapeutic accidents or misadventures. These cases have several medicolegal implications. For example, the medical examiner will determine whether or not the final mode should be natural, as when the findings are those of an expected complication such as infection, or whether the mode is accident, homicide, or undetermined. The determination of final mode may require considerable study and analysis. There can be many shades of gray between an anticipated disease complication and true therapeutic negligence.[19] In addition, these cases often lead to prolonged civil litigation irrespective of whether or not treatment error is determined by medical experts to have occurred, for whatever reason is considered sufficient by the plaintiff and their attorney. The case examples presented in this chapter include the circumstances, complications, and neuropathology of the lethal outcome, and the issue of individual liability is intentionally omitted.

There are several purposes of the present chapter. First, to help determine whether or not a case in which someone claims the results of therapy to be suboptimal is a coroner's case. Second, to suggest special procedures that might be helpful prior to, during, or following examination and dissection of the formalin-fixed brain and spinal cord. For example, in a rare case, examination of peripheral nerves and muscle may be germane to a complete study. Third, to help determine whether or not neuropathology, surgery, anesthesiology, and/or other specialty consultation is needed to resolve questions. Fourth, to give some guidelines concerning the determination of appropriate mechanism and mode of death. It is hoped that these purposes will be served through selected citations and case examples representative of several of the major categories of iatrogenic neuropathologic complications.

Definitions

A periprocedural death is "a death that is known or suspected as having resulted in whole or in part from diagnostic, therapeutic, or anesthetic procedures."[52] Iatrogenic indicates any adverse condition in a patient occurring as the result of treatment by a physician. *Iatro* is from the Greek, meaning physician. *Genic* is also from the Greek ("gennan"), meaning to produce. Literally, therefore, iatrogenic could be interpreted as "something that creates physicians," analogous to "thrombogenic" or "ketogenic." A more precise term would be *iatric* or *iatrical*, also from the Greek ("iatrikos"), meaning pertaining to medicine or to a physician,[38] but the terminology already entrenched in common usage will be used herein. In determining whether or not a death is in fact periprocedural, an approach is to ask whether the procedure caused, contributed to, or hastened the death of a patient.

Categorization of the Problem

It has been estimated that 44,000 to 98,000 Americans die each year due to medical errors, and preventable adverse events cost $17 to $27 billion per year.[65,118] One report indicated that drug errors in hospitals, nursing homes, and doctors' offices injure over 1.5 million Americans per year.[7] These deaths are due to a very wide range of procedures.[127] There are various ways to classify these deaths, such as by medical specialty, device, mechanism, intent, or outcome. There are also classifications based on circumstances, or by types of drugs administered.[14,52,75]

Some medical specialties involve use of therapeutic procedures more likely to lead to neuropathologic complications. Intraoperative or postoperative complications resulting from neurosurgical procedures are an obvious example.[122] An older but detailed review is that of Horowitz and Rizzoli.[58] Favre et al.[42] analyzed 361 stereotactic neurosurgical procedures and found an overall risk for hematoma formation of 1.7%. Invasive diagnostic or therapeutic radiologic procedures,[113,114] such as those used in interventional neuroradiology and cardiology, include potential problems from contrast media injection for imaging studies,[12] and from the introduction of stents as well as therapeutic introduction of embolic materials for aneurysms, vascular malformations, or arteriovenous fistulae. Cardiovascular procedures with potential neurologic complications include arteriography, catheterization, angioplasty, and cardiac valve replacement.[1]

Other procedure-linked deaths include those related to anesthesiology,[10,131] general surgery,[44,55] and general medical and neurologic procedures,[103] including those associated with accidental introduction of air emboli during placement of central venous catheter,[109] total parenteral nutrition (TPN), or thrombolysis for coronary thrombi. Also included are deaths from dental (Case 16.1, Fig. 16.1) and chiropractic procedures.[79,91,115] The neuropathology of deaths from obstetric procedures is discussed in Chapter 4. In his analysis of cases ruled therapeutic misadventure over an 11-year period, Murphy[86] found an incidence of 0.46%.

The arbitrary classification used in this chapter for case examples is primarily by type of device. These include metal implants and emboli, surgical and ablative instruments, and mechanical forces. The etiology of some complications can be difficult to categorize, such as those due to needles that are also injecting drugs or plastic tubes containing foreign material used in TPN.

Case 16.1. (Fig. 16.1)

An 18-year-old man underwent extraction of an infected left molar tooth, followed by antibiotic therapy. He initially improved, but, about 2 weeks later, he became barely arousable and was admitted to the hospital. CT scan showed cerebral edema and left to right midline shift. A ventriculostomy was placed, and CSF was noted to be crystal clear. He suffered a cardiopulmonary arrest, and he was pronounced dead approximately 1½ weeks following hospital admission.

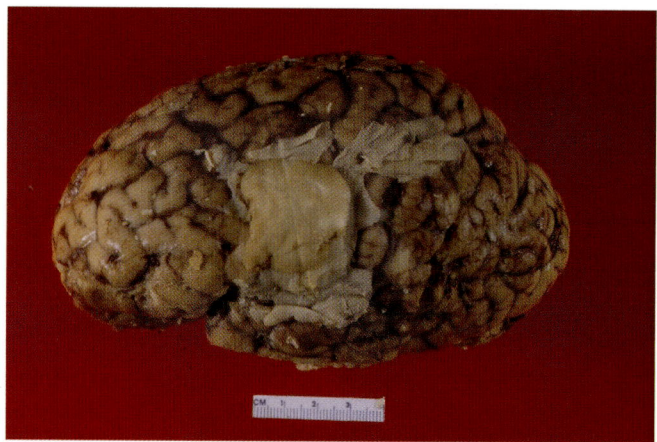

Fig. 16.1. Left cerebral hemisphere shows a large subdural pyogenic coagulum in the paracentral region. The left convexity dura mater showed more of the subdural empyema, consisting of a pale yellow film. Brain was asymmetrically swollen with left parahippocampal herniation and midbrain compression. Microscopic examination confirmed subdural empyema, and there was also acute superficial cerebral cortical necrosis. Cause of death was attributed to subdural empyema secondary to dental abscess.

Mechanisms of Injury

There are several ways in which neuropathologic deaths are produced by these various devices. Perforation of the dura, entering the subarachnoid space,[47] or entering cerebral ventricles can occur purposely or inadvertently from introduction of tubes and needles. Injury to blood vessels, nerves, or central nervous system (CNS) parenchyma can be produced by surgical instruments. Introduction of foreign materials such as non-CNS tissue implants, chemicals, air, vascular stents or therapeutic emboli, and so on can cause vascular obstruction.[18] Ablation of specific brain tissue is the goal of frontal leukotomy, but results may be suboptimal for a variety of reasons. Accidental application of mechanical forces such as excessive compression, hyperextension, or flexion may include poor positioning during surgery leading to peripheral neuropathies or other sequelae, or forceful neck movement during manipulation therapy leading to radiculopathy, myelopathy, or to brain embolism from dislodgement of emboli from extracranial carotid artery atheromatous foci.

Devices

Metal implants and emboli include coils, wires, clips, cages, and electrodes.[3,50,63] Coiling was approved by the Food and Durg Administration (FDA) in 1995. Cerebral aneurysms can be seeded with platinum coils via a femoral catheter (Case 16.2, Figs. 16.2–16.4). Complications from embolization of detachable coils include aneurysm perforation, arterial vasospasm, parent artery occlusion, cerebral embolization, and coil breakage and migration.[9] Fusiform, complex wide-necked, and giant aneurysms have a poor prognosis without treatment.[57] In addition to acute complications, wire embolization can lead to chronic sequelae such as thrombosis of the venous sinuses (Case 16.3, Fig. 16.5).

Clips are routinely used to isolate cerebral aneurysms from the vessel of origin, and rarely may become detached. Brain electrodes implanted at various targets such as the globus pallidus, ventral lateral thalamus, and subthalamic nucleus have been used for therapy of several conditions, including alleviation of symptoms of Parkinson's disease, but they are a potential source of life-threatening complications.

Plastic Tubes

Various kinds of neuropathologic injury can result from device misplacement, or even properly placed plastic tubes, such as ventriculostomy or ventriculoperitoneal shunt tubing, catheters, and stents.[64] In the intracranial compartment, ventriculostomy tubes can spontaneously migrate or be placed incorrectly, passing through the lateral ventricle into the caudate nucleus, thalamus, or entering the opposite hemisphere (Case 16.4, Figs. 16.6–16.9). Following endovascular aneurysm treatment[104] or in patients with a coagulopathy, the site of the ventriculostomy tract may show excessive hemorrhage, occasionally with rupture of blood into the lateral ventricle. Multiple tracts can indicate previous unsuccessful attempts at proper placement. Occasional intracerebral

Case 16.2. (Figs. 16.2–16.4)

A 50-year-old woman underwent left external carotid artery angiography and attempted embolization for uncontrolled epistaxis. Three coils inadvertently entered the left supraclinoid internal carotid artery (ICA). Two hours later, she developed right-sided weakness and obtundation. Radiologic studies demonstrated occlusion of the left ICA and a large left middle cerebral artery infarct. She was pronounced dead 27 hours following the procedure.

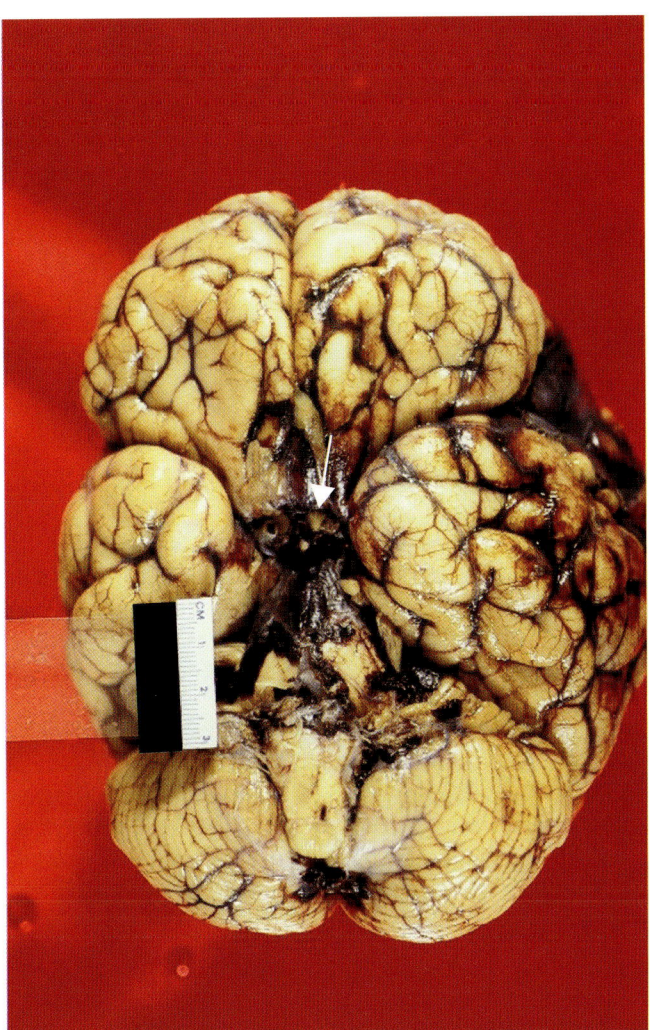

Fig. 16.2. Base of brain shows patchy subarachnoid hemorrhage, and sectioned end of left internal carotid artery is filled with firm, tan-gray material (*arrow*).

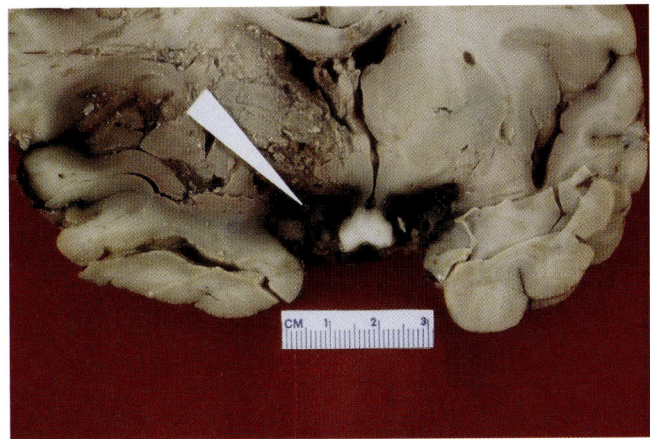

Fig. 16.3. Brain shows putrefactive changes and marked swelling, left greater than right. A 3-mm-long sliver of metal is embedded in the basal subarachnoid hemorrhage in the region of the junction of left internal carotid artery and middle cerebral artery (*arrow*).

426 Periprocedural Complications

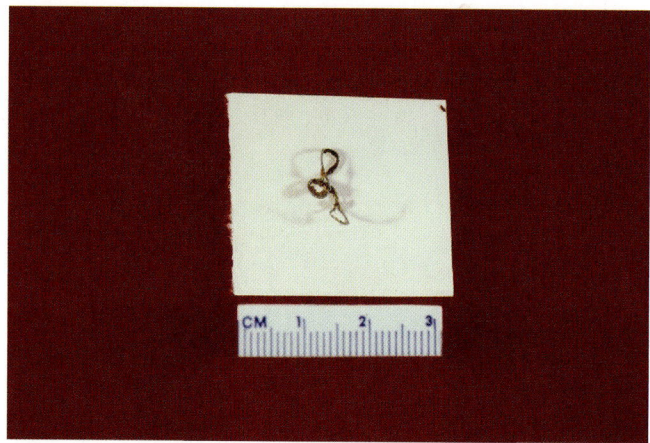

Fig. 16.4. In the process of sectioning, a highly coiled metallic thread fell from the region of the arrowhead tip in Fig. 16.3. Microscopic examination showed an acute thrombus with focal rupture of walls of the left internal carotid artery and middle cerebral artery.

Case 16.3. (Fig. 16.5)

A 4-year-old boy developed hydrocephalus secondary to a great cerebral vein of Galen aneurysm, diagnosed at the age of 5 months. He underwent placement of a ventriculoperitoneal shunt and embolization of the aneurysm.

Fig. 16.5. Transverse section of this great vein of Galen giant aneurysm shows a thick, fibrotic wall. Contents consist of numerous coils of thin wires amid grumous material. Straight sinus is at the dorsal midline. Hydrocephalus and thrombosis of the superior sagittal and transverse sinuses were present, with minimal residual sinus lumens.

Case 16.4. (Figs. 16.6–16.9)

A 50-year-old man with a history of alcohol abuse and cirrhosis developed a large acute subdural hematoma following an assault. He underwent a craniotomy for evacuation of the hematoma and died 2 days later.

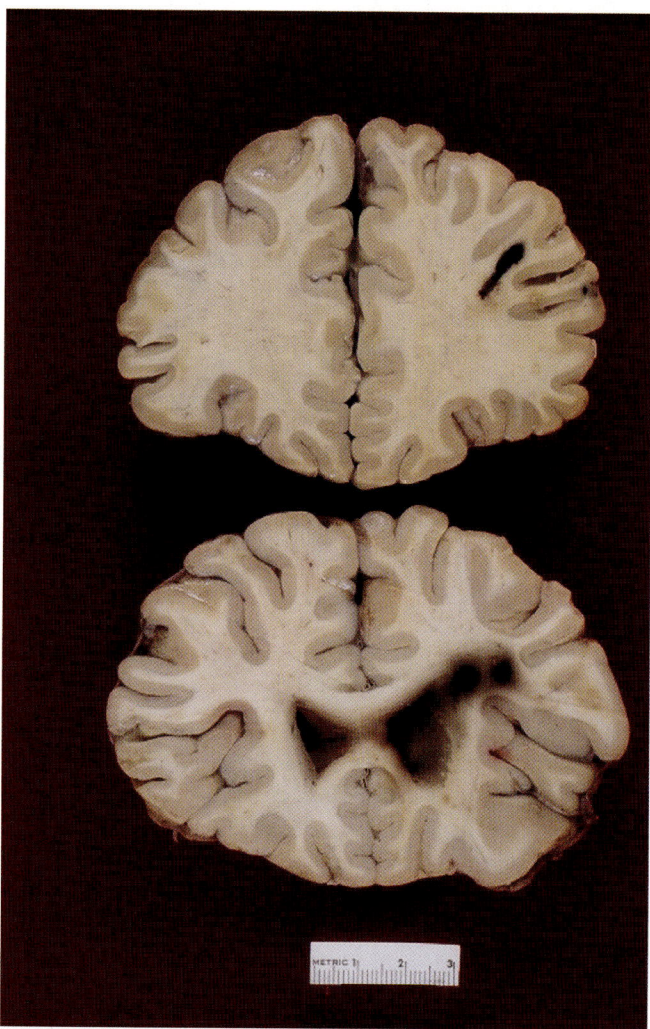

Fig. 16.6. Coronal section shows an initial single ventriculostomy tract beginning in the right middle frontal gyrus, which diverges into two tracts.

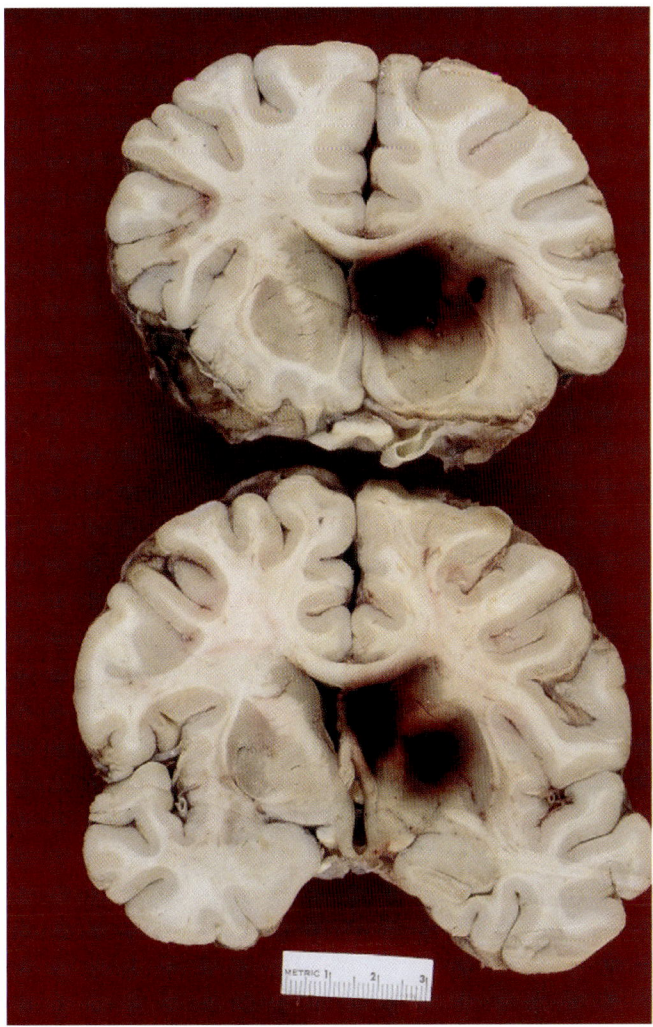

Fig. 16.7. These two tracts pass into the head of caudate nucleus and lateral ventricle, respectively, with hemorrhage in the right anterior horn.

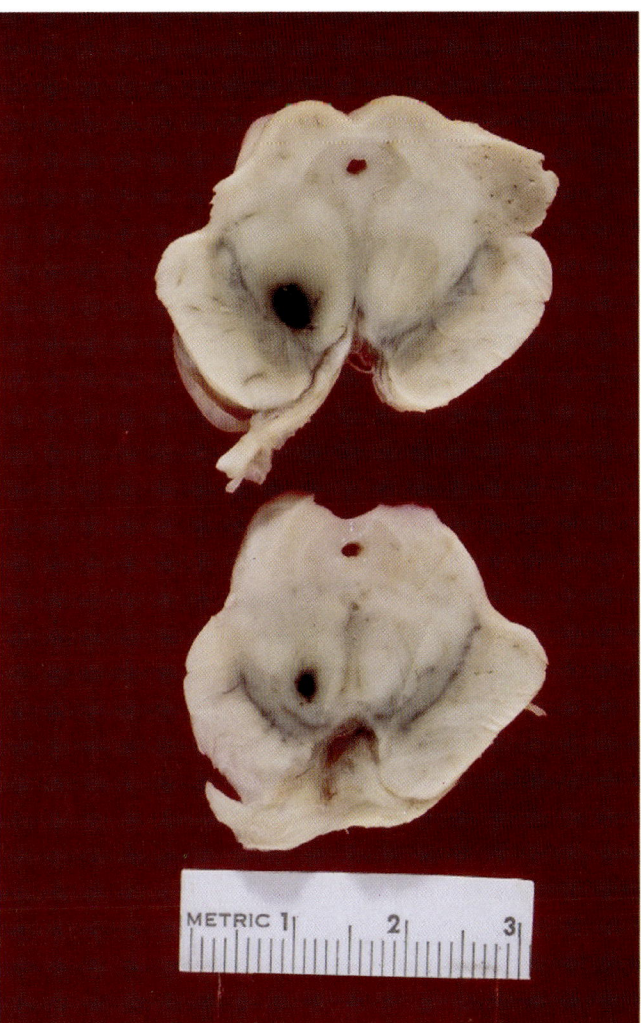

Fig. 16.9. Continuing path of one of ventriculostomy tracts ends in right red nucleus (left in image).

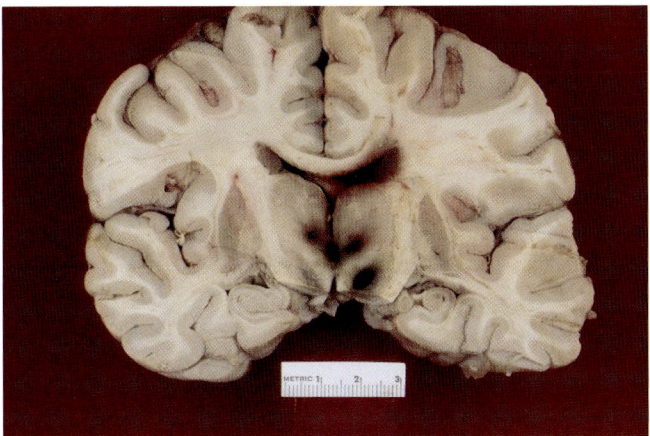

Fig. 16.8. Hemorrhages are also present in the right ventral thalamus and rostral midbrain.

hemorrhages have a tractlike shape, suggesting a surgical etiology. The finding of ventricular adhesions may be indicative of ventriculitis resulting from various procedures. Plugging, excessive drainage with slit ventricle syndrome, and other complications of ventriculoperitoneal shunting are well known.[21,70,85,88,89]

Various procedures are used to treat intracranial atherosclerosis.[49] Stents have been used in the treatment of fusiform and wide-necked aneurysms[94] and in vertebrobasilar angioplasty,[77] with complications including new aneurysm development and thrombosis leading to stroke.

In the intraspinal compartment, plastic catheters from morphine pumps, intended for drug delivery into the epidural space, may instead enter subdural or subarachnoid compartments. With careful dissection so as not to dislodge the tubing, the catheter tip can be localized to its actual termination site (Case 16.5, Figs. 16.10–16.11). Various complications can result after refilling of intrathecal infusion pump reservoirs, including catheter failure, abscess, meningitis, or instillation of an incorrect drug.[25,65]

Plastic tubes intended for extracranial sites can be inadvertently placed intracranially, with disastrous sequelae.[2,23] For example, nasogastric tubes can veer upward, entering the intracranial cavity via a fractured cribriform plate region (Case 16.6, Figs. 16.12–16.13). TPN tubes have inadvertently entered the common carotid artery, leading to stroke from intracarotid nutrient infusion.[18] Meningitis or subdural empyema may also occur (Case 16.7, Figs. 16.14–16.16).

Case 16.5. (Figs. 16.10 and 16.11)

A 71-year-old man had a history of bronchogenic carcinoma and severe lower back pain. He underwent revision of an intrathecal catheter of an implantable morphine pump. Approximately 12 hours later, he suffered a cardiac arrest.

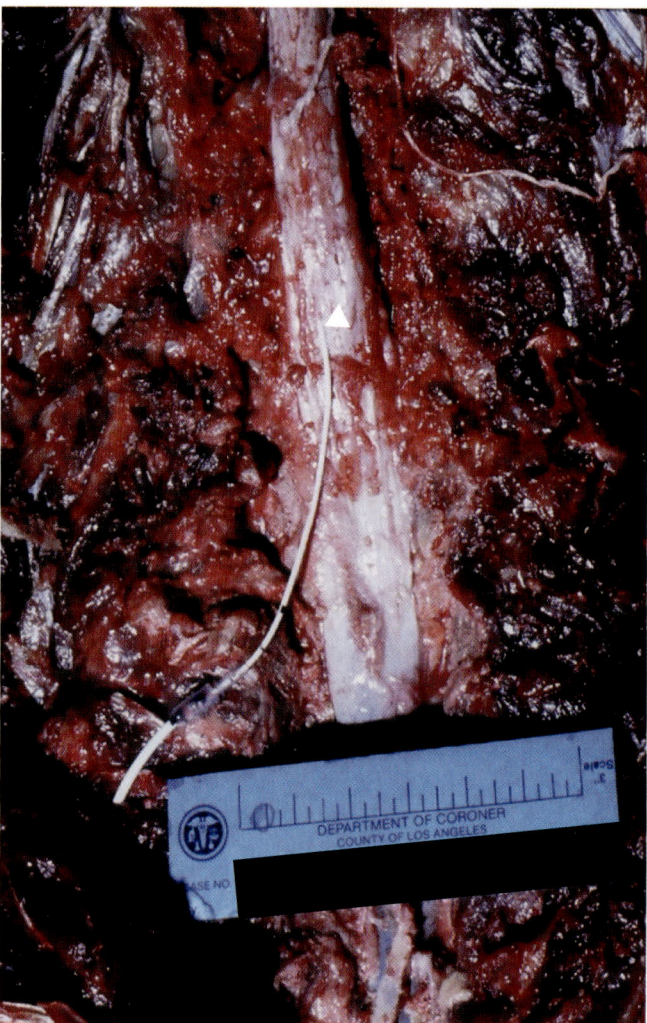

Fig. 16.11. The thinner catheter leading from the larger catheter enters the midline L1 spinal dura mater (*arrowhead*). Tip of the catheter was found at the anterior midline T12 level, lying on the anterior spinal artery, not where it was supposed to be. Minimal soft tissue hemorrhage is seen.

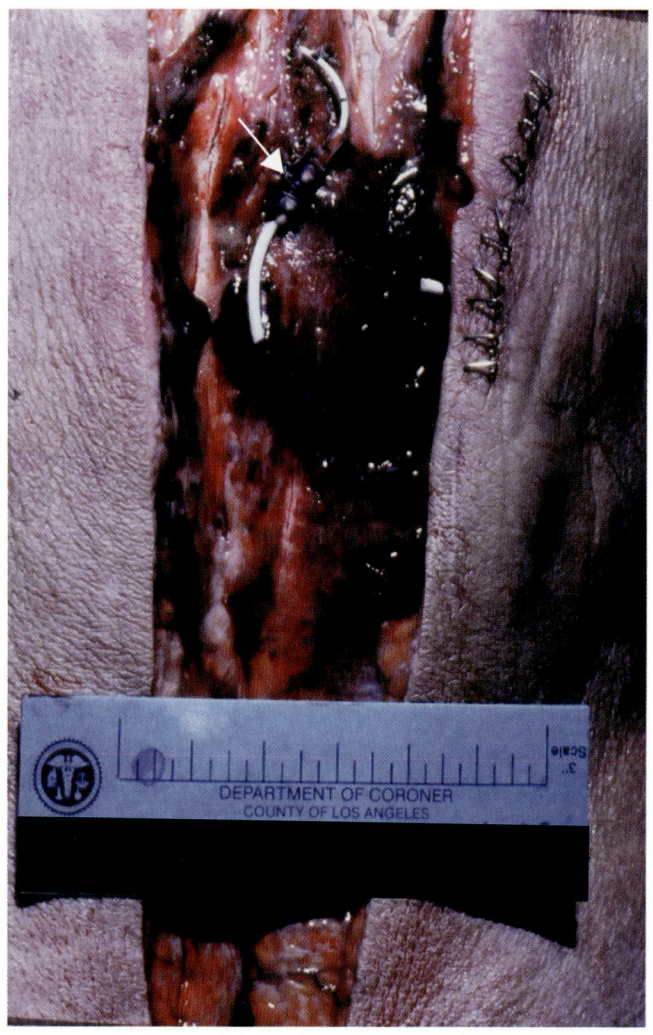

Fig. 16.10. The outer larger white plastic catheter is seen sutured onto the posterior L2 vertebral fascia (*arrow*).

Case 16.6. (Figs. 16.12 and 16.13)

A 20-year-old man suffered massive fractures of the base of the skull with extension into the calvarium from a motorcycle accident. There was no intracranial collection of blood.

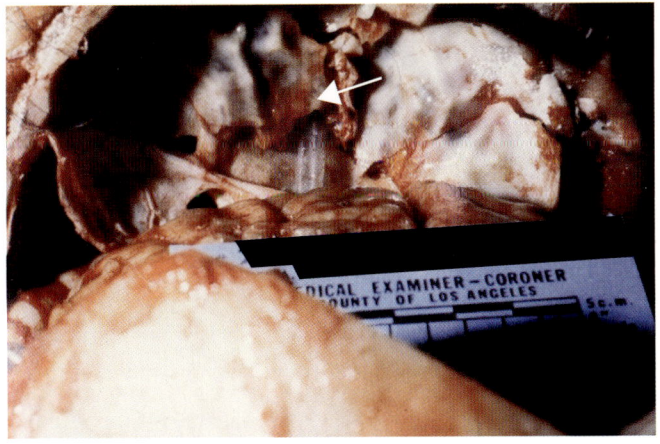

Fig. 16.12. Forty inches of nasogastric tubing had been inserted into the nasal cavity and intracranially, through the fracture of ethmoid bone (*arrow*).

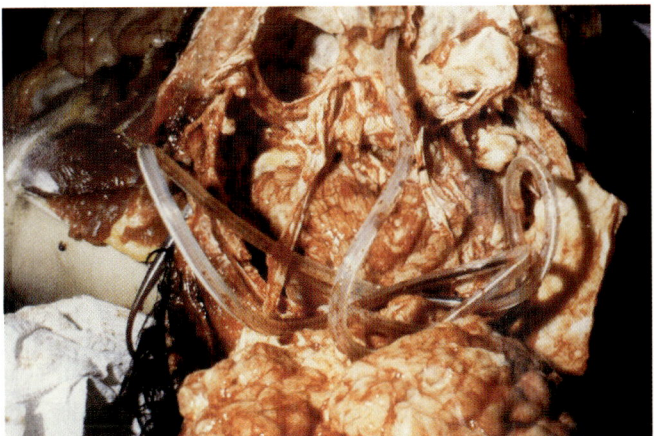

Fig. 16.13. Through lacerations of the tentorium cerebelli, several loops of tubing reached the posterior fossa.

Case 16.7. (Figs. 16.14–16.16)

A 71-year-old woman initially did well following gynecologic oncology surgery. Two and one-half weeks later, she developed seizures and lapsed into a coma. Lumbar puncture yielded milky cerebrospinal fluid with high glucose and protein. She died 3 days later.

The probable mechanism for the presence of intrathecal material consistent with infused TPN was reverse migration of intravenous TPN into subarachnoid space, although the access route between venous and CSF compartments was not identified.

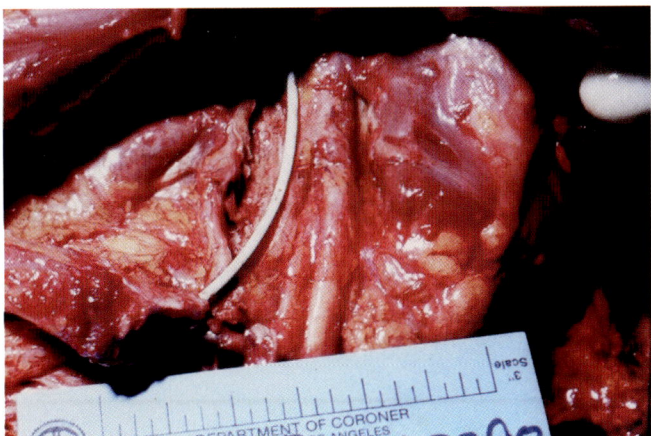

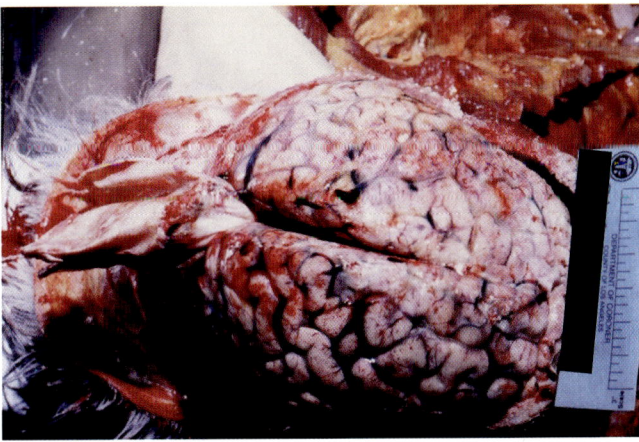

Fig. 16.15. Acute purulent leptomeningitis was found as well as acute pachymeningitis and acute subdural empyema.

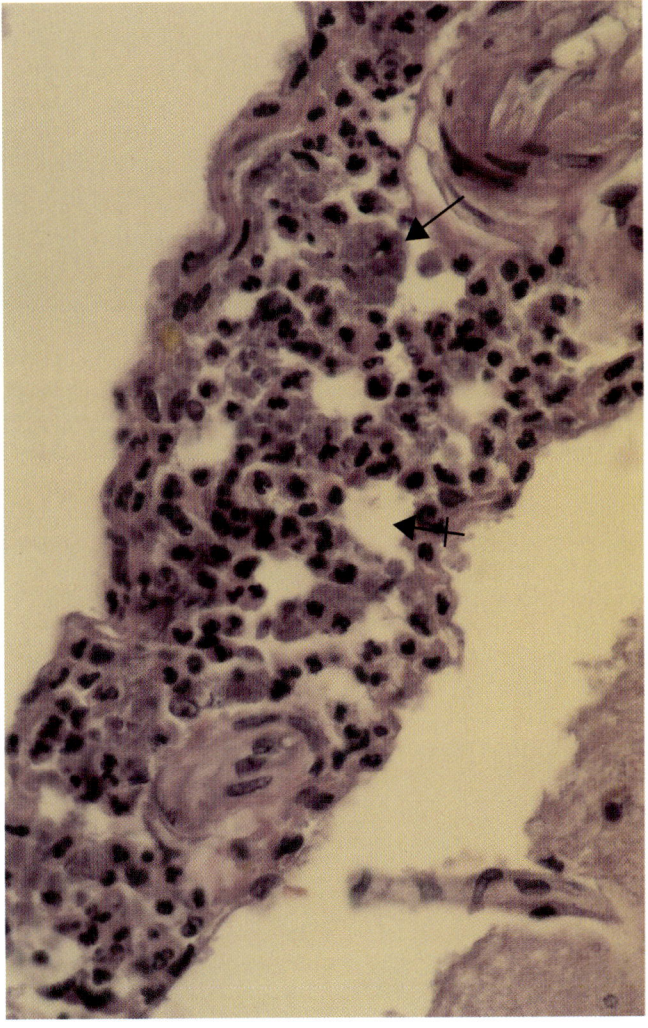

Fig. 16.16. High-power micrograph shows subarachnoid acute inflammatory infiltrate along with large amounts of ill-defined debris (*arrow*) and oil-red-O-positive material (seen as lucent globules with H&E) (*crossed arrow*).

Fig. 16.14. At autopsy, the total parenteral nutrition tube was seen still in the jugular vein.

Needles and Drugs

Disastrous neuropathologic sequelae can result from the inadvertent perforation of the brain or spinal cord during cervical myelography[117] or cisternal puncture procedures employed for other purposes.[68] With care during neuropathologic examination, these often minute defects can be identified. When lumbar puncture cannot be performed, such as when severe spinal degenerative changes are present, cisternal puncture without radiologic guidance may be resorted to, with rare fatal consequences such as massive subarachnoid hemorrhage or penetration of the medulla (Case 16.8, Figs. 16.17–16.21). Inadvertent removal of the meninges during withdrawal of a spinal needle has been reported.[108]

Incorrect drugs can be injected into a tissue compartment, and the correct drug can be injected into the wrong compartment. Injected anticoagulants and thrombolytics can lead to CNS hemorrhage.[34,35,54,119] Adverse idiosyncratic reactions to general and spinal anesthetics which have been properly administered may also occur.[8,30,59] Intrathecal injection of the wrong chemotherapeutic

> **Case 16.8.** (Figs. 16.17–16.21)
>
> A 73-year-old woman fell at the beach 10 years prior to death, suffering compression fractures and disk herniations for which she had decompressive laminectomy. Two months prior to death, laminectomy was performed from L3 to the sacrum for spinal stenosis and severe scoliosis. Postoperatively, she developed *Pseudomonas aeruginosa* meningitis and a lumbar abscess. Three days prior to death, she developed brainstem and intraventricular hemorrhage, having undergone several preceding cisternal taps for the administration of intrathecal antibiotics. A ventriculostomy was performed to relieve hydrocephalus, but her condition continued to deteriorate and death occurred soon thereafter.

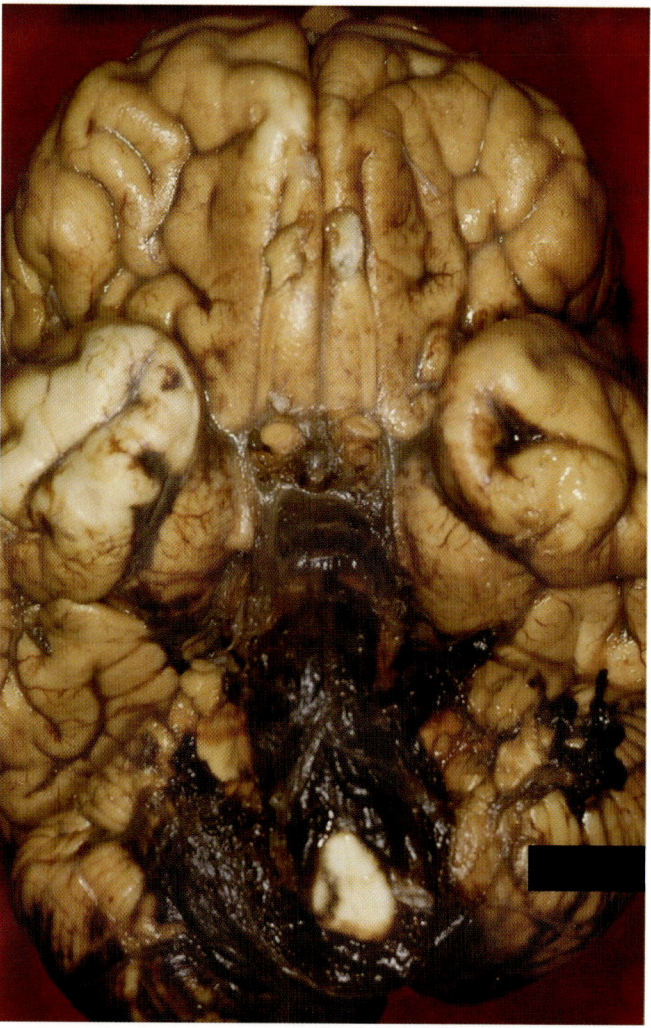

Fig. 16.17. Acute subarachnoid hemorrhage surrounds the ventral surface of brainstem, filling prepontine, cerebellopontine angle, and paramedullary cisterns.

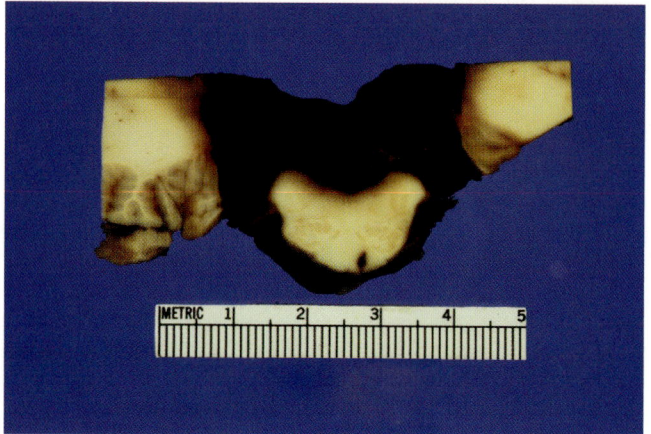
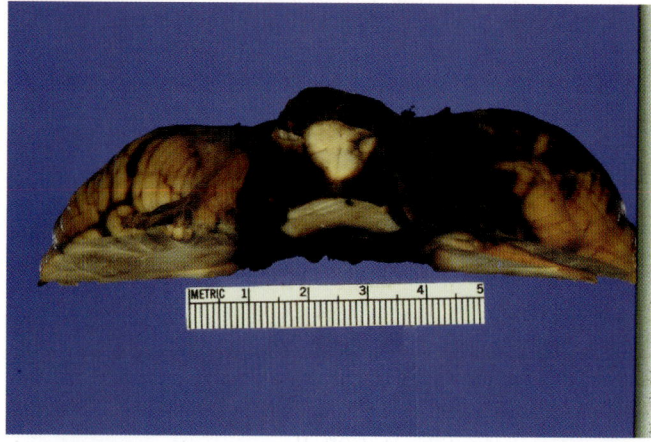

Figs. 16.18–16.19. *Left* (Fig. 16.18) and *right* (Fig. 16.19). A slender hemorrhagic track is present within caudal medulla, including the left pyramid. Branches of the basilar and vertebral arteries were embedded in blood clot, which also filled the basal cisterns and fourth ventricle. Comment: the presence of chronic arachnoiditis may have contributed to rupture of fixed subarachnoid blood vessel(s) during the course of cisternal punctures.

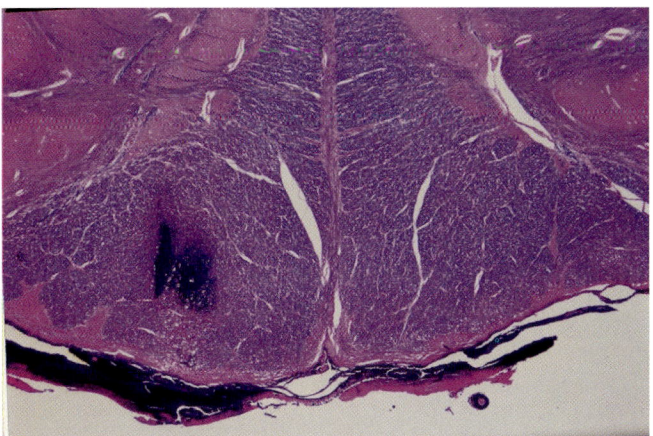

Fig. 16.20. Low-power micrograph of this cisternal puncture tract shows acute hemorrhage in the medullary pyramid. (Luxol fast blue/H&E.)

> **Case 16.9.** (Figs. 16.22–16.24)
>
> A 22-year-old man with a history of lymphoma was admitted to the hospital for a second cycle of chemotherapy with intrathecal methotrexate. He received instead intrathecal vincristine through a lumbar puncture (LP). The error was immediately noted, CSF drained through the LP, and ventriculostomy placed for CSF lavage. Despite these efforts, over the next several days he developed leg paresthesias, progressing to leg paralysis, arm incoordination, urinary retention, deafness, confusion, and somnolence. He died 8 days later. Gross examination showed moderate brain swelling and normal spinal cord with coverings.

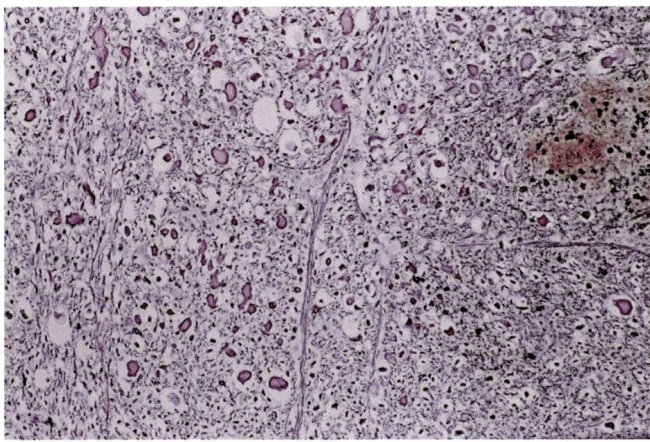

Fig. 16.21. Medium-power micrograph shows numerous axonal spheroids along the needle track margin. (Bodian.)

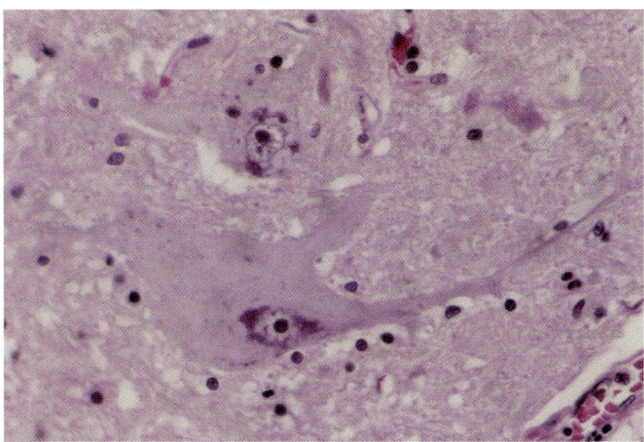

Fig. 16.22. High-power micrograph of anterior horn neurons of lumbosacral spinal cord shows cytoplasmic swelling, coarse clustering of Nissl substance, and swollen cell processes. These anterior horn changes were most prominent at this spinal cord level, diminishing rostrally. (H&E.)

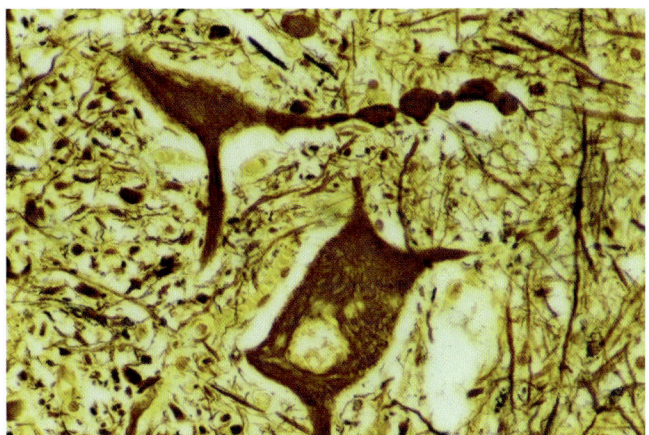

Fig. 16.23. High-power micrograph shows an anterior horn neuron (*upper*) with axonal swellings having a beadlike appearance. The other neuron is engorged with silver-positive intracytoplasmic material. (Bielschowsky.)

agent, such as vincristine or vindesine rather than methotrexate,[5,111,124] can result in death. Spinal cords in such circumstances may show enlarged motor neurons due to accumulation of neurofilaments (Case 16.9, Figs. 16.22–16.24) or other lesions, depending on the specific agent involved. Rare catastrophic mishaps may occur from accidental injection of agents, such as the subarachnoid injection of 5% glutaraldehyde instead of the intended reinjection of original CSF, resulting in death due to intravital brainstem and spinal cord fixation.[32] Inadvertent spinal cord penetration (Case 16.10, Figs. 16.25–16.27), and perforation of the vertebral artery[105] have occurred during cervical nerve root block.

Numerous drugs can lead to neurologic complications,[28] including death.[37,39] Following anticoagulant and

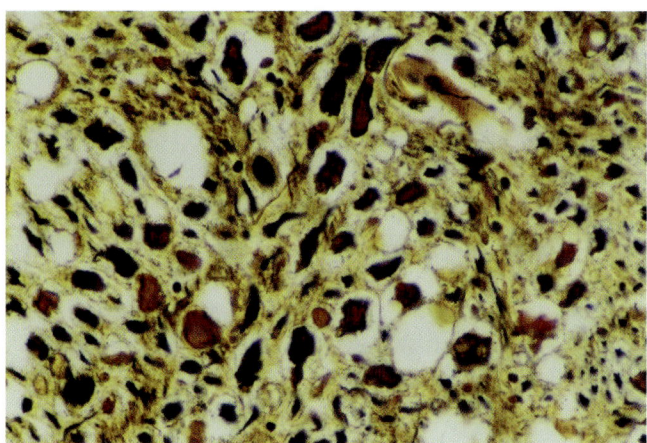

Fig. 16.24. High-power micrograph shows axonal thickenings and spheroids in posterior root entry zone. (Bielschowsky.)

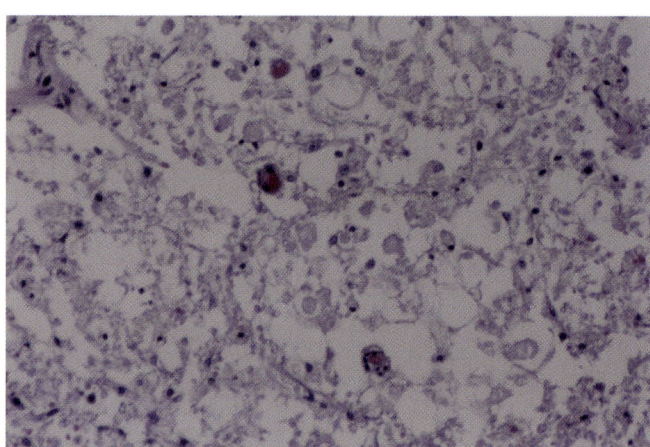

Fig. 16.26. High-power micrograph shows adjacent posterior column necrosis, expanded interstitial spaces, and a few scattered macrophages. (H&E.)

Case 16.10. (Figs. 16.25–16.27)

A 71-year-old woman had a history of myeloradiculopathy from cervical spine degenerative disease. Past surgery consisted of a cervical fusion and two cervical decompressions. She was reported to be slowly becoming paralyzed. While undergoing therapeutic injection of lidocaine and cortisone between the C6 and C7 vertebrae for her pain, she had a cardiac arrest. She was resuscitated, but remained comatose and died 20 days later. Cause of death was determined to be sequelae of hypoxic-ischemic encephalopathy, due to cardiac arrest, due to intraspinal injection of therapeutic agents. The mode was accident.

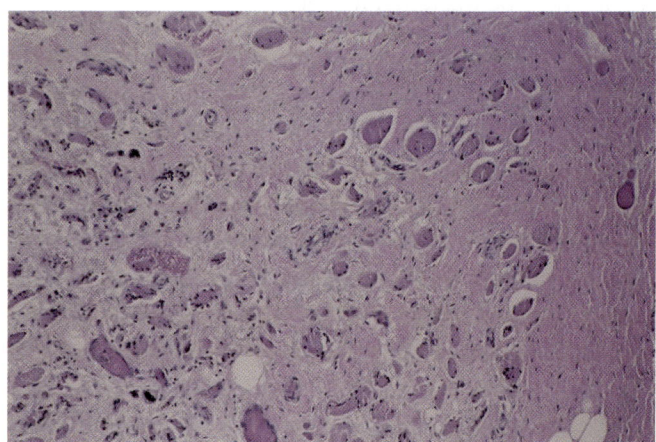

Fig. 16.27. Low-power micrograph shows atrophic muscle fibers within dense fibrous connective tissue in the thick dorsal epidural fibromuscular pannus that was centered at C6 level, compressing the cord. There was thick dura mater underlying the pannus, that contained aggregates of large foreign body type giant cells (not shown). (H&E.)

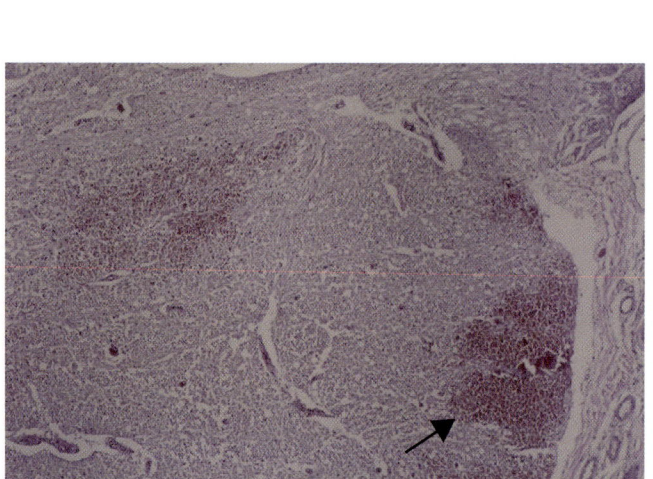

Fig. 16.25. Low-power micrograph shows subpial hemorrhage in the right cervical posterior column (*arrow*). Smaller foci of hemorrhage straddle the adjacent root entry zone and posterior horn. These foci represent the probable needle tract. (H&E.)

antithrombolytic therapy for cerebral ischemia or other reasons, there is a risk of intracerebral hemorrhage.[48,66,123]

Antiplatelet therapy has also been implicated in clinical deterioration following intracerebral hemorrhage.[125] The combination of heparin and tissue plasminogen activator for acute myocardial infarction together with sudden elevations in blood pressure can lead to lethal intracerebral hemorrhage (Case 16.11, Fig. 16.28). Anticoagulant therapy can also be a factor in the development of hemorrhage in other intracranial compartments, such as subdural hematoma or subarachnoid hemorrhage.[67,80]

Errors may occur during anesthesia[116] as a result of inadequate attention to vital signs, causing acute hypoxic-ischemic CNS injury. Idiosyncratic or other adverse reactions to spinal anesthetics can lead to cord ischemia or transverse myelopathy. The rate of drug or fluid administration may also be a factor in the development of neurologic injury, a well-known example being central pontine myelinolysis following too-rapid correction of hyponatremia.[73]

Foreign Materials

Many types of foreign materials can be introduced into (via blood supply) or around the brain or spinal cord, either inadvertently or deliberately.[47,95] These substances include fat,[133] TPN liquid,[18] silicone,[41,90,93,99,110] materials used for arteriovenous malformation embolization, Gelfoam, adrenal and embryonic tissue,[45,69,81] carbon dioxide,[74] and air.[4,24,60] Hemostatic substances can produce symptomatic, radiologically apparent mass lesions,[102] which may be mistaken for recurrent tumor or radiation necrosis. Gelfoam embolus of the middle cerebral artery with massive acute infarction has occurred following attempted external carotid artery embolization (Case 16.12, Figs. 16.29–16.32). A patent foramen ovale can provide a mechanism for an unexplained stroke following surgery.[72] Illicit cosmetic procedures have resulted in silicone embolization to lung and brain (Case 16.13, Figs. 16.33–16.34).

Surgical Instruments, Hardware, and Ablative Procedures

Procedures in which surgery is used to ablate tissue range from carotid angioplasty and endarterectomy[20,27,71] to procedures for psychiatric disorders, such as frontal

Case 16.12. (Figs. 16.29–16.32)

A 5-year-old girl underwent a left carotid angiogram for a left parotid mass. Using Gelfoam under fluoroscopic guidance, the external carotid was embolized just distal to the facial artery. All previously identified feeders to the tumor appeared to be occluded on follow-up angiogram. Ten hours later, she became lethargic, initially interpreted as oversedation. Eight hours later, a cerebral blood flow study showed nonperfusion. She died 2 days later.

Case 16.11. (Fig. 16.28)

A 79-year-old woman with a history of hypertension developed a myocardial infarction and she was placed on heparin and tissue plasminogen activator. She experienced a sudden spike in blood pressure 1 day later, and became unresponsive. CT scan showed extensive brainstem hemorrhage. She died 1 day later.

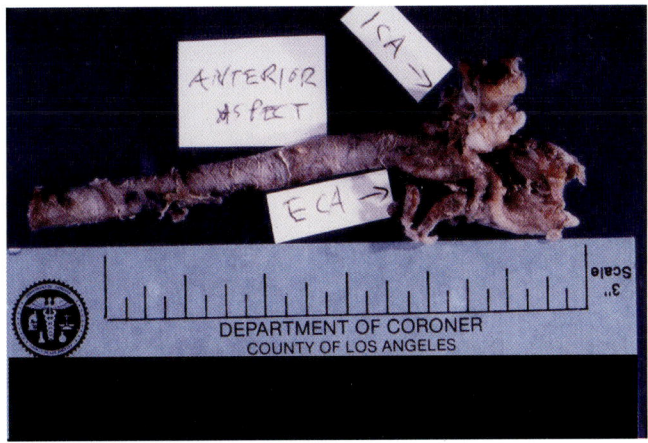

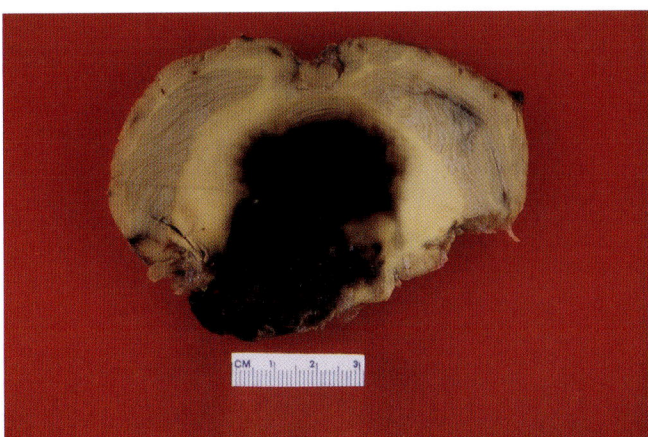

Fig. 16.28. Transverse section of pons and cerebellum shows massive hemorrhage extending into fourth ventricle and cerebellopontine angle cistern. Rostrally, this hemorrhage extended into midbrain, ventral thalamus, and third ventricle. Caudally, it entered rostral medulla. Brainstem blood vessels showed acute fibrinoid necrosis.

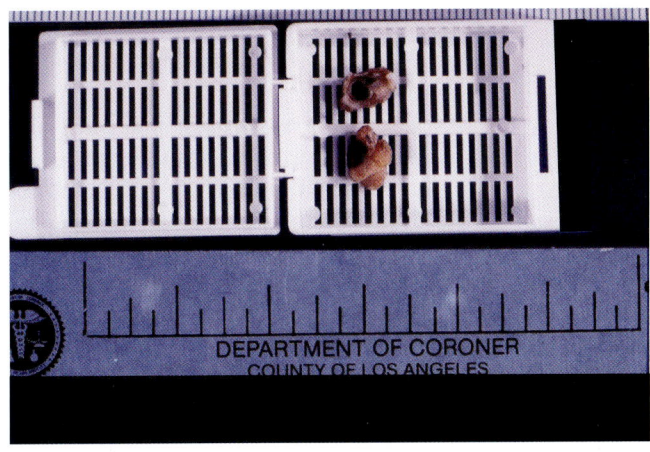

Figs. 16.29–16.30. *Top* (Fig. 16.29) and *bottom* (Fig. 16.30). Autopsy showed the left external and internal carotid arteries distended by brown material.

434 Periprocedural Complications

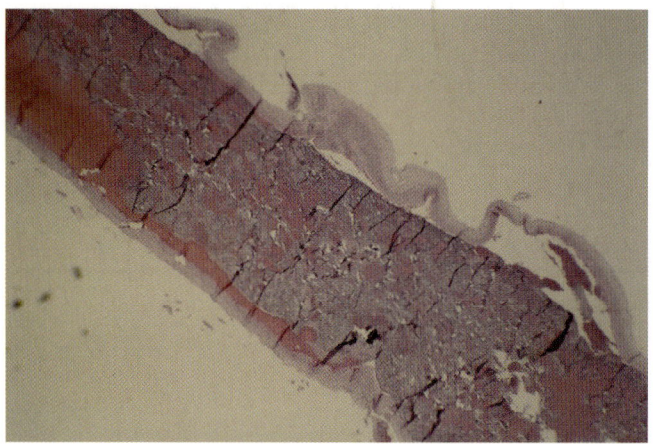

Fig. 16.31. Low-power micrograph shows an embolus filling and distending the horizontal portion of the left middle cerebral artery. (H&E.)

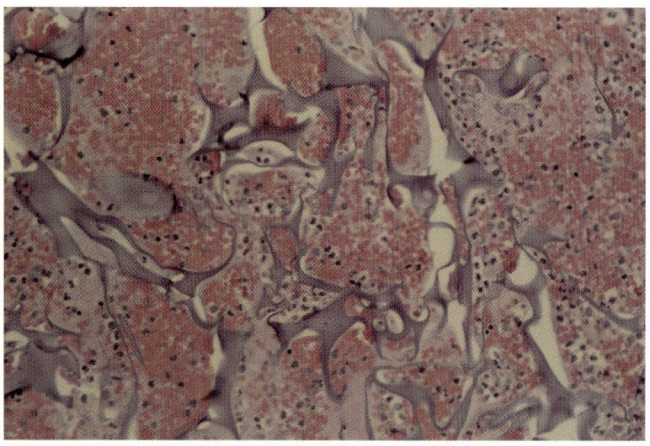

Fig. 16.32. High-power micrograph shows the embolus to consist of erythrocytes entrapped in a fine mesh of basophilic trabeculae, characteristic of Gelfoam. There was a massive acute anemic infarct in the territory of the left middle cerebral artery. (H&E.)

Case 16.13. (Figs. 16.33 and 16.34)

A 32-year-old woman underwent silicone injection into her buttock as part of a cosmetic procedure at a black market underground clinic at a residence. Approximately 50 cc had been injected when she developed chest tightness. Shortness of breath continued and she was hospitalized the next day. She failed to improve, underwent a cardiac arrest, and was pronounced dead 4 days following the injection. Autopsy revealed silicone emboli with petechial hemorrhages scattered throughout multiple organs, including the brain.

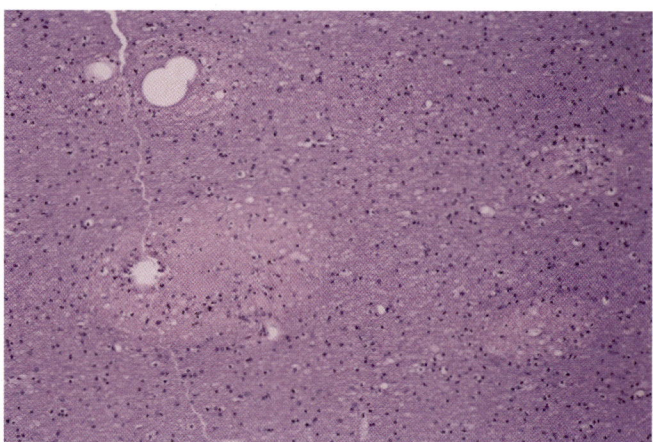

Fig. 16.33. Low-power photomicrograph shows multiple small petechiae consisting of ghost erythrocytes in cerebral white matter. Silicone microemboli are seen as empty round vacuoles. (H&E.)

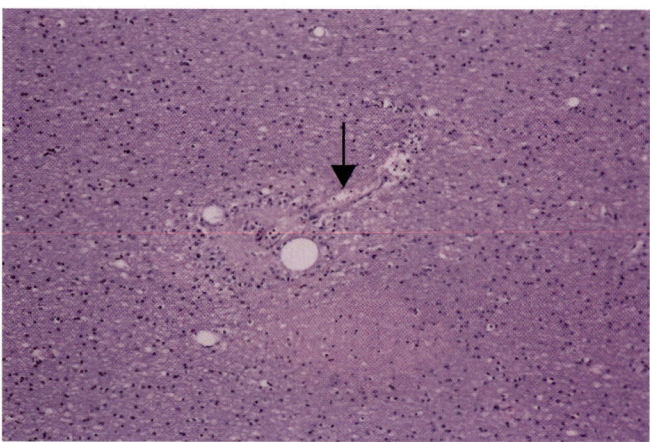

Fig. 16.34. Low-power photomicrograph shows a focal perivascular petechial hemorrhage surrounding a longitudinally sectioned capillary (*arrow*). (H&E.)

lobotomy and cingulotomy.[16] For neurologic disorders, callosotomy and other "ectomies" have been used to control seizures, thalamotomy used for the control of some types of tremors, and pallidotomy for treatment of Parkinson's disease.[43,112] Psychiatric, as well as surgical hardware, complications may follow deep brain stimulation.[82,87,120] Abscess[92] or hemorrhage (Case 16.14, Figs. 16.35–16.36) may occur from intracranial penetration of halo brace pins. Radiotherapy, chemotherapy, and immunoablation for malignant neoplasms can be considered forms of nonsurgical, selective tissue ablation.[6,28] CNS lymphoma has been reported following iatrogenic immunosuppression,[96] and prior irradiation has been implicated in the development of certain brain tumors (see Chap. 13).

Various procedures used to treat cervical and intracranial atherosclerosis can lead to complications. Although the rate of adverse neurologic consequences is low, carotid angioplasty can dislodge plaque fragments, resulting in atheroemboli to the brain.[100] Fatal exsanguination from suture line disruption following carotid endarterectomy has also been reported.[83]

With the advent of modern psychotherapeutic drugs, frontal or prefrontal lobotomy or leukotomy are procedures of the past.[76] These cases, while not complications of a procedure but rather the intended consequences for therapeutic purposes, must be recognized as such and not misidentified as, for example, consequences of vascular accidents or foul play. These cases are now rarely encountered (Case 16.15, Figs. 16.37–16.39). Cingulotomy has been performed to treat severe depression (Case 16.16, Fig. 16.40), and callosotomy to control seizures (Case 16.17, Figs. 16.41–16.42). In one large study,[33] uni-

> **Case 16.14.** (Figs. 16.35 and 16.36)
>
> A 39-year-old man with a history of childhood seizures sustained an unstable C1 vertebral fracture following an assault. He underwent halo brace fixation, and remained neurologically intact. Three months later he suddenly developed right-sided weakness and speech difficulty, and, while attempting to call 911, he fell. At the hospital, he was found to have bilateral penetration of the halo brace pins into the parietal lobes. Toxicology was positive for opiates and cocaine. The halo brace was removed. Brain death was pronounced 10 days later. The cause of death was determined to be sequelae of cervical trauma, with the mode homicide.

> **Case 16.15.** (Figs. 16.37–16.39)
>
> A 75-year-old man was involved in a motor vehicle accident in which his clothing became caught on a truck, and he was dragged approximately 70 feet. He sustained severe abrasions. The chief complaint on that admission was "dementia." Two days later, he was found unresponsive at home. Resuscitative efforts failed, and he was pronounced dead.

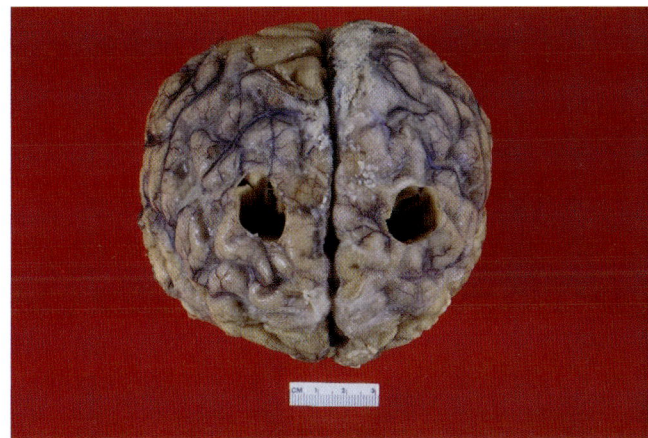

Fig. 16.37. Dorsal aspect of frontal lobes shows two symmetric round surgical cortical defects of approach to leukotomy. There were also corresponding sites of dural thinning.

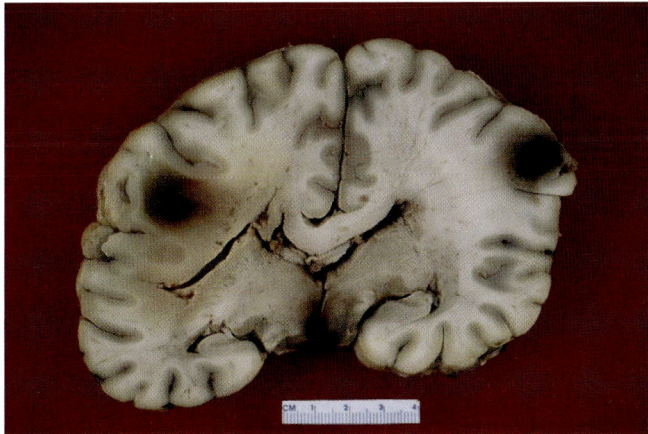

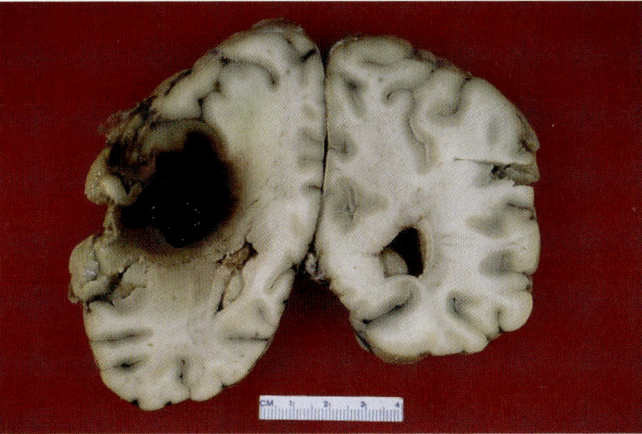

Figs. 16.35–16.36. *Top* (Fig. 16.35) and *bottom* (Fig. 16.36) show hemorrhage and edema in cortex and subcortical white matter. Duret hemorrhage is also seen in this severely swollen and herniated respirator brain.

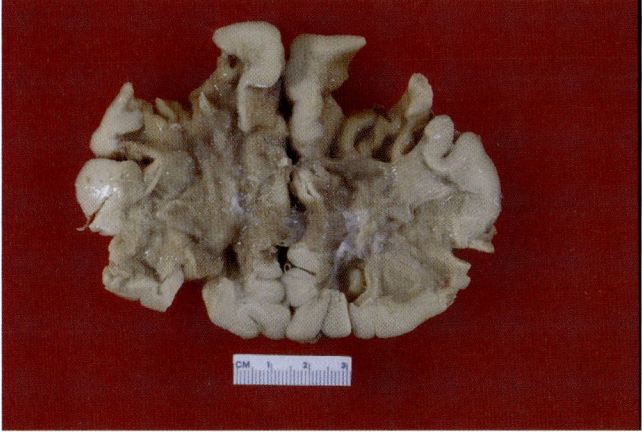

Fig. 16.38. Frontal coronal sections through cortical openings show massive cystic destruction of pregenu white matter of leukotomy.

Periprocedural Complications

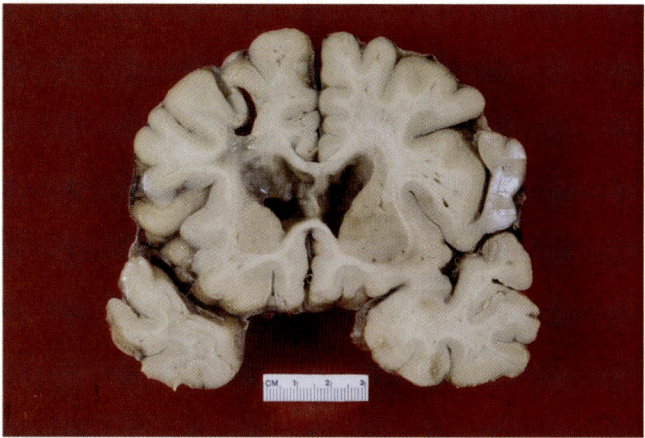

Fig. 16.39. Frontal horns of lateral ventricles are secondarily enlarged. Secondary degeneration of anterior limbs of internal capsule is present. Bilateral degeneration of dorsomedial nucleus of thalamus and frontopontine tracts was also present.

Case 16.16. (Fig. 16.40)

A 39-year-old woman had a history of seizure disorder, depression since the age of 18 years, and several suicide attempts. Approximately 3 years prior to death, she underwent a cingulotomy. She was unexpectedly found dead at her residence.

Case 16.17. (Figs. 16.41 and 16.42)

A 5-year-old boy had a history of seizure disorder since age 7 months. He had undergone two operations for section of corpus callosum a month prior to death. He was last known alive at 0230 hours on the day of death, and found unresponsive approximately 4½ hours later.

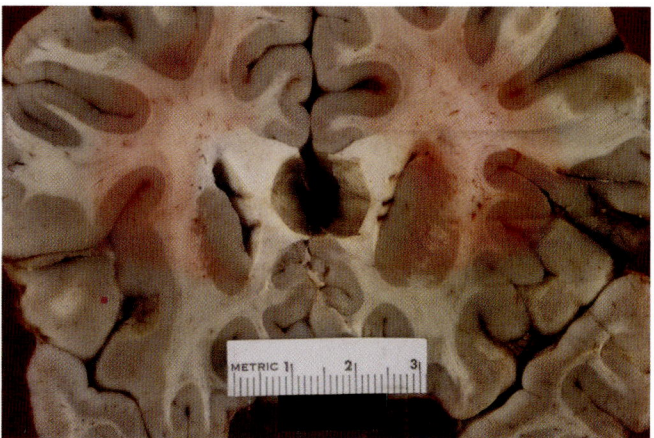

Fig. 16.41. Complete midline surgical transection of callosotomy was present from genu (shown) through splenium. Central pink discoloration is due to incomplete fixation.

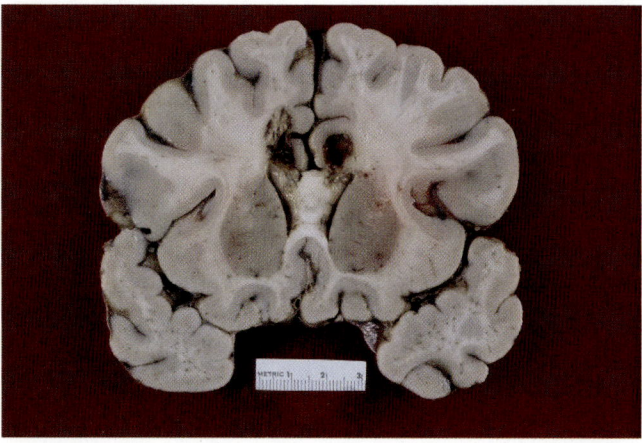

Fig. 16.40. Symmetric, mostly subcortical, necrosis of cingulate gyri bilaterally, centered at the level of genu of corpus callosum and extending 1 cm anteriorly and posteriorly.

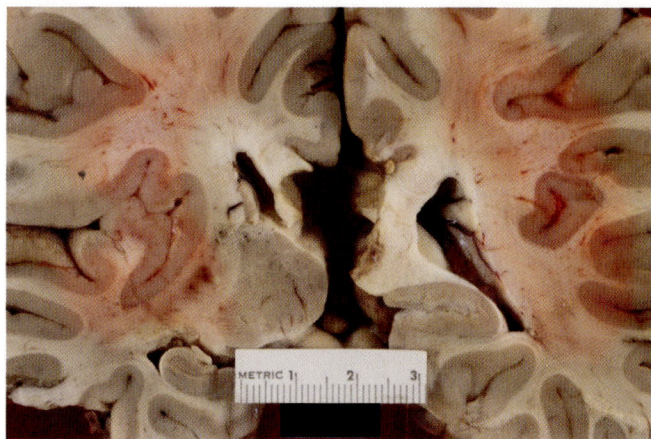

Fig. 16.42. Transection at level of splenium.

lateral pallidotomy led to permanent adverse effects in 14% of patients. Evidence for these and other neurosurgical procedures, such as skull irregularities and dural scarring or calcification consistent with prior burr hole placement, may sometimes be found even when specific history is lacking (Case 16.18, Figs. 16.43–16.46).

Radiation necrosis and/or the development of radiation-induced secondary brain tumors can follow radiation therapy for extra- or intracranial neoplasms (see Chap. 13). Radiation necrosis can occur acutely, or be delayed in onset as much as several decades following the therapy.[13,132]

Mechanical Forces

Manual spinal manipulation occurs in, for example, chiropractic treatment.[62,101,106,121] Spinal manipulation has occasionally led to neurologic complications including paralysis, disk herniation, vertebral artery dissection,[29] and vertebrobasilar ischemia with resultant infarction,[22,130] including locked-in syndrome.[15] In India, neck manipulation by barbers is common following a haircut, and subsequent persistent quadriparesis has been reported.[84] One literature review[17] indicated that 49% of cases of vertebral artery dissection were associated with minor trauma or neck manipulation. Under rare circumstances, it might be necessary to exclude self-sustained

> **Case 16.18.** (Figs. 16.43–16.46)
>
> A 55-year-old man was found unresponsive and pronounced dead at the scene. Details of medical history were lacking, but he was known to have spastic dystonia and previous neurosurgery.

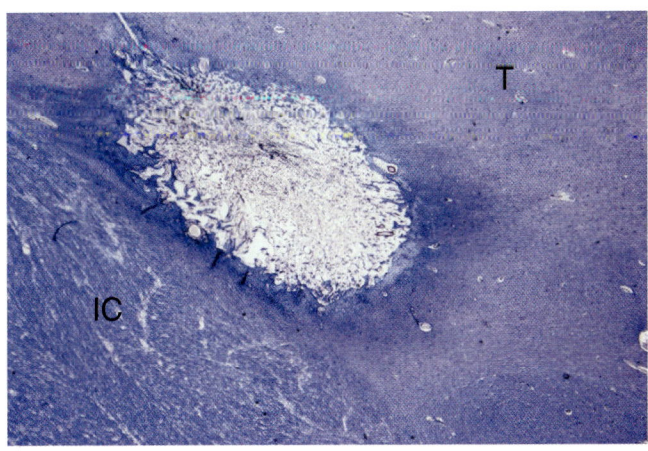

Fig. 16.44. Higher-power micrograph of lesion seen in Fig. 16.43 shows the targeted parenchymal destruction of ventral thalamus. (Luxol fast blue/cresyl violet.)

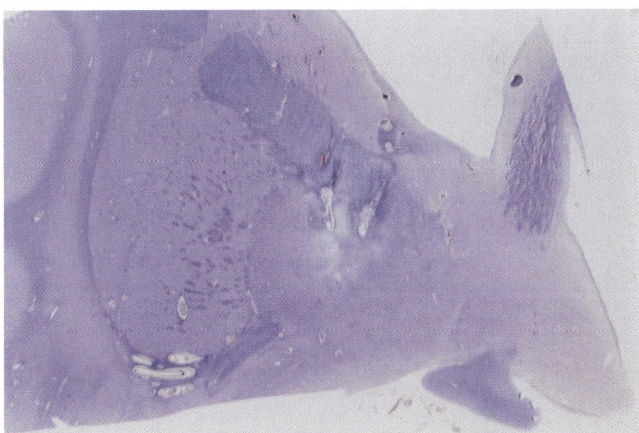

Fig. 16.45. Low-power micrograph shows two discrete lesions of pallidotomy in left globus pallidus, similar to contralateral thalamic lesion. (Luxol fast blue/cresyl violet.)

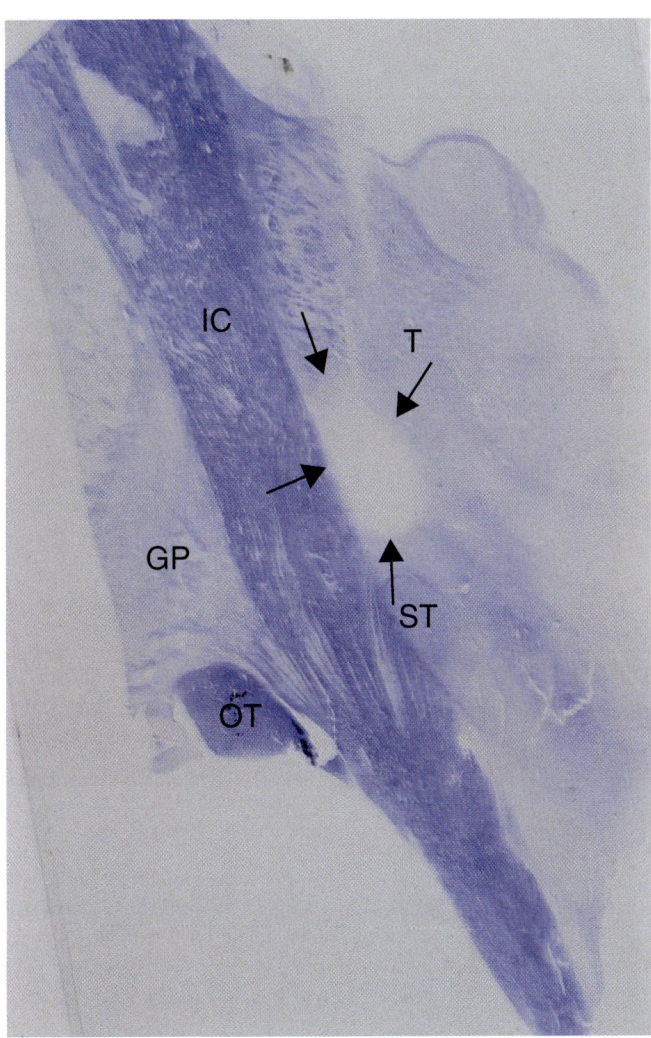

Fig. 16.43. Low-power micrograph shows a pale oval therapeutic lesion destroying ventral area of lateral ventral nucleus abutting posterolateral ventral nucleus of thalamus (*arrows*). The lesion consisted of loose gliosis. T, thalamus; IC, internal capsule, GP, globus pallidus; OT, optic tract; ST, subthalamic nucleus. (Luxol fast blue/cresyl violet.)

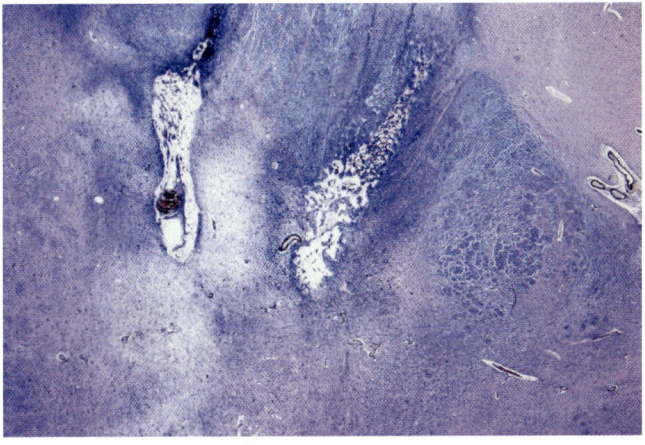

Fig. 16.46. Detail of therapeutic lesions seen in Fig. 16.45, partially involving internal capsule. (Luxol fast blue/cresyl violet.)

injury resulting from violent movements,[36] or to exclude self-injection of drugs.[61]

Anesthetic and neurosurgical procedures may require the patient to be placed in positions other than the more common prone, supine, or lateral positions. Prolonged head flexion, for example, may rarely result in myelopathy such as a midcervical tetraplegia.[10,56,78,126,129]

Challenges for the Forensic Pathologist

Periprocedural deaths require a detailed review of all available medical records by the medical examiner. Consultation with specialists and relevant clinical disciplines such as surgery, anesthesiology, or medical device equipment consultants may be needed. There is also the matter of deciding whether or not standard of care issues are in question, and whether notification of the appropriate authorities and/or medical boards is indicated. Careful documentation of all autopsy findings and a clear presentation of conclusions will best serve all involved parties, whether in the context of physician education, care review committees, or legal proceedings.

Determination of Mode

The determination of mode in periprocedural deaths can be difficult. These cases have been organized by some into five categories.[98] Suggestions for classifications of cases and of approaches to determining mode or manner of death, including those by experienced medical examiners, are available.[11,51–53,98] "Therapeutic complication" has been suggested as a sixth manner of death (i.e., in addition to natural, accident, suicide, homicide, and undetermined), although, as a recent review of this issue indicates,[46] the medical examiner is limited by laws in each jurisdiction as to manner of death terminology options allowed. Factors that are taken into account include age (young versus frail elderly), nature of the medical condition, presence of other associated disease, and so on.[14] A case classified as natural would have a complication of the natural disease being treated that was anticipated or expected. A therapeutic "complication" does not imply that a mistake was made; development of a predictable complication of indicated therapy is one example. A death classified as accident might have a complication that was an adverse reaction or condition unrelated to any intentional or criminal act, with an example being an order given for the incorrect drug. The category of gross negligence would include failure to meet the standard of care considered appropriate by qualified medical peers. Only rare periprocedural deaths qualify as homicide. In occasional cases, the mode remains undetermined (Case 16.19, Fig. 16.47) despite thorough and objective review.

Case. 16.19. (Fig. 16.47)

A 5-year-old boy was shot in the abdomen by his father, and he subsequently developed short bowel syndrome, sepsis, and liver failure. He had received total parenteral nutrition. He died almost 2 months following the incident.

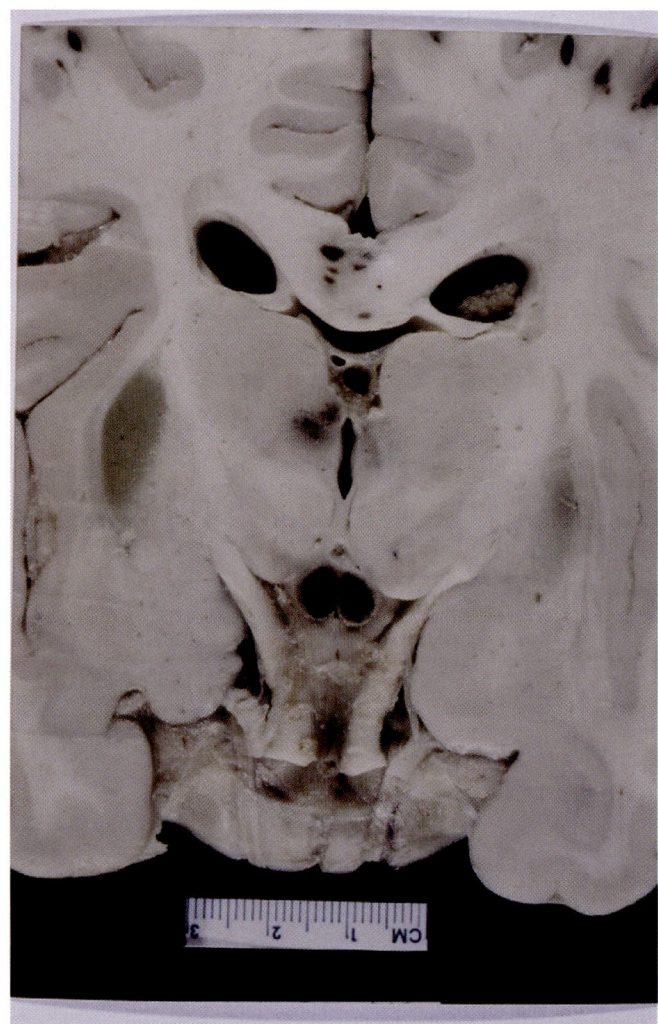

Fig. 16.47. Photo shows striking dark focal hemorrhages in mammillary bodies, periventricular thalamus, and posterior body of corpus callosum. Green staining of putamen is seen bilaterally, due to jaundice (with some blood–brain barrier breakdown). Hemorrhages in subcortical white matter and inferior colliculi were also present. Microscopically, capillary endothelial hypertrophy was present within areas of mammillary hemorrhage. There were acute periventricular petechiae. The findings were consistent with acute Wernicke's encephalopathy.

Final Comment

This brief outline of periprocedural errors and complications relevant to neuropathology would be incomplete without including a comment on diagnostic and procedural errors made by pathologists, including forensic pathologists, which can set in motion a chain of events leading to errors in subsequent clinical or judicial decisions. Error analysis and avoidance in pathology is discussed in several other sections of this text, perhaps to somewhat greater extent in Chapters 1, 7, 8, and Appendix E, with selected references cited. In this context, attention is also directed to the thoughtful comments of Cramer [31] concerning error analysis in pathology.

References

1. Adams HP. Neurologic complications of cardiovascular procedures. In Biller J (ed): Iatrogenic Neurology. Boston: Butterworth-Heinemann, 1998.
2. Adler JS, Graeb DA, Nugent RA. Inadvertent intracranial placement of a nasogastric tube in a patient with severe head trauma. Can Med Assoc J 1992;147:668–669.
3. Ahn JY, Han IB, Yoon PH, et al. Clipping vs coiling of posterior communicating artery aneurysms with third nerve palsy. Neurology 2006;66:121–123.
4. Akhtar N, Jafri W, Mozaffar T. Cerebral artery air embolism following an esophagogastroscopy: A case report. Neurology 2001;56:136–137.
5. Andrews JM, Itabashi HH, Chinwah O, et al. Intrathecal vincristine administration—A lethal therapeutic error (abstract). Proc Amr Acad Forensic Sci 2000;6:197–198.
6. Antunes NL, Souweidane MM, Lis E, et al. Methotrexate leukoencephalopathy presenting as Kluver-Bucy syndrome and uncinate seizures. Pediatr Neurol 2002;26:305–308.
7. Aspden P, Wolcott J, Bootman JL, Cronenwett LR (eds). Preventing Medication Errors. Washington, DC: National Academies Press, 2006.
8. Auroy Y, Narchi P, Messiah A, et al. Serious complications related to regional anesthesia. Anesthesiology 1997;87:479–486.
9. Baltsavias GS, Byrne JV, Halsey J, et al. Effects of timing of coil embolization after aneurysmal subarachnoid hemorrhage on procedural morbidity and outcomes. Neurosurgery 2000;47:1320–1331.
10. Barash PG, Cullen BF, Stoelting RK (eds). Clinical Anesthesia. Philadelphia: Lippincott, Williams & Wilkins, 2001.
11. Batalis NI, Harley RA, Collins KA. Iatrogenic deaths following treatment for hypertrophic obstructive cardiomyopathy: Case reports and an approach to the autopsy and death certification. Am J Forensic Med Pathol 2005;26:343–348.
12. Bender A, Elstner M, Paul R, Straube A. Severe symptomatic aseptic chemical meningitis following myelography. Neurology 2004;63:1311–1313.
13. Bigner DD, McLendon RE, Bruner JM (eds). Russell and Rubinstein's Pathology of Tumors of the Nervous System. New York: Oxford University Press, 1998.
14. Biller J. Iatrogenic Neurology. Boston: Butterworth-Heinemann, 1998.
15. Bindal RK, Nelson PB. Neurologic complications of spinal manipulation. In Biller J (ed): Iatrogenic Neurology. Boston: Butterworth-Heinemann, 1998.
16. Binder DK, Iskandar BJ. Modern neurosurgery for psychiatric disorders. Neurosurgery 2000;47:9–23.
17. Bin Saeed A, Shuaib A, Al-Sulaiti G, Emery D. Vertebral artery dissection: Warning symptoms, clinical features and prognosis in 26 patients. Can J Neurol Sci 2000;27:292–296.
18. Bohlega S, McLean DR. Hemiplegia caused by inadvertent intracarotid infusion of total parenteral nutrition. Clin Neurol Neurosurg 1997;99:217–219.
19. Brennan TA, Gawande A, Thomas E, Studdert D. Accidental deaths, saved lives, and improved quality. N Engl J Med 2005;355:1405–1409.
20. Buhk JK, Wellmer A, Knauth M. Late in-stent thrombosis following carotid angioplasty and stenting. Neurology 2006;66:1594–1596.
21. Byard RW, Koszyca B, Qiao M. Unexpected childhood death due to a rare complication of ventriculoperitoneal shunting. Am J Forensic Med Pathol 2001;22:207–210.
22. Cagnie B, Barbaix E, Vinck E, et al. Atherosclerosis in the vertebral artery: An intrinsic risk factor in the use of spinal manipulation? Surg Radiol Anat 2006;28:129–134.
23. Castiglione AG, Bruzzone E, Burrello C, et al. Intracranial insertion of a nasogastric tube in a case of homicidal trauma. Am J Forensic Med Pathol 1998;19:329–334.
24. Caulfield AF, Lansberg MG, Marks MP, et al. MRI characteristics of cerebral air embolism from a venous source. Neurology 2006;66:945–946.
25. Chaney MA. Side effects of intrathecal and epidural opioids. Can J Anaesth 1995;42:891–903.
26. Chang SM. Complications of medical therapy. In Bernstein M, Berger MS (eds): Neuro-Oncology: The Essentials. New York: Thieme Medical Publishers, 2000.
27. Chaturvedi S, Caplan LR. Angioplasty for intracranial atherosclerosis. Is the treatment worse than the disease? Neurology 2003;61:1647–1648.
28. Chen JT, Collins DL, Atkins HL, et al. Brain atrophy after immunoablation and stem cell transplantation in multiple sclerosis. Neurology 2006;66:1935–1937.
29. Christian MD, Detsky AS. A twist of fate? N Engl J Med 2004;351:69–73.
30. Cote CJ, Karl HW, Notterman DA, et al. Adverse sedation events in pediatrics: Analysis of medications used for sedation. Pediatrics 2000;106:633–644.
31. Cramer SF. Judging mistakes in pathology—Res ipse non loquitur (letter to editor). Arch Pathol Lab Med 2006;130:1430–1432.
32. Davis JH, Mittleman RE. In-vivo glutaraldehyde fixation of the brain stem and spinal cord after inadvertent intrathecal injection. J Forensic Sci 1998;43:1232–1236.
33. De Bie RMA, de Haan RJ, Schuurman PR, et al. Morbidity and mortality following pallidotomy in Parkinson's disease: A systematic review. Neurology 2002;58:1008–1012.
34. Demchuk AM, Tanne D, Hill MD, et al. The Multicentre tPA Stroke Survey Group. Predictors of good outcome after intravenous tPA for acute ischemic stroke. Neurology 2001;57:474–480.
35. Derex L, Nighoghossian N, Perinetti M, et al. Thrombolytic therapy in acute ischemic stroke patients with cardiac thrombus. Neurology 2001;57:2122–2125.
36. Dobbs M, Berger JR. Cervical myelopathy secondary to violent tics of Tourette's syndrome. Neurology 2003;60:1862–1863.
37. Donaghy M. Assessing the risk of drug-induced neurologic disorders: Statins and neuropathy. Neurology 2002;58:1321–1322.
38. Dorland's Medical Dictionary, ed 26. Philadelphia: WB Saunders, 1981.
39. Dukes PD, Robinson GM, Thomson KJ, Robinson BJ. Wellington coroner autopsy cases 1970–89: Acute deaths due to drugs, alcohol and poisons. N Z Med J 1992;105:25–27.
40. Evans RW (ed). Iatrogenic Disorders. Neurologic Clinics. Philadelphia: WB Saunders, 1998.

41. Fangtian D, Rongping D, Lin Z, Weihong Y. Migration of intraocular silicone into the cerebral ventricles. Am J Opthalmol 2005;140:156–158.
42. Favre J, Taha JM, Burchiel KJ. An analysis of the respective risks of hematoma formation in 361 consecutive morphological and functional stereotactic procedures. Neurosurgery 2002;50:48–57.
43. Fine J, Duff J, Chen R, et al. Long-term follow-up of unilateral pallidotomy in advanced Parkinson's disease. N Engl J Med 2000;342:1708–1714.
44. Foster D, Falah M, Kadom N, Mandler R. Wernicke encephalopathy after bariatric surgery: Losing more than just weight. Neurology 2005;65:1987.
45. Freed CR, Greene PE, Breeze RE, et al. Transplantation of embryonic dopamine neurons for severe Parkinson's disease. N Engl J Med 2001;344:710–719.
46. Gill JR, Goldfeder LB, Hirsch CS. Use of "therapeutic complication" as a manner of death. J Forensic Sci 2006;51:1127–1133.
47. Gladstone JP, Nelson K, Patel N, Dodick DW. Spontaneous CSF leak treated with percutaneous CT-guided fibrin glue. Neurology 2005;64:1818–1819.
48. Graham DI, Lantos PL (eds). Greenfield's Neuropathology, ed 7. New York: Arnold, 2002.
49. Gupta R, Schumacher HC, Mangla S, et al. Urgent endovascular revascularization for symptomatic intracranial atherosclerotic stenosis. Neurology 2003;61:1729–1735.
50. Hadjivassiliou M, Tooth CL, Romanowski CAJ, et al. Aneurysmal SAH: Cognitive outcome and structural damage after clipping or coiling. Neurology 2001;56:1672–1677.
51. Hanzlick R (ed). The Medical Cause of Death Manual. Northfield, Ill.: College of American Pathologists, 1994.
52. Hanzlick RL (ed). Cause-of-Death Statements and Certification of Natural and Unnatural Deaths: Protocol and Options. Northfield, Ill.: College of American Pathologists, 1998.
53. Hanzlick RL. Medical certification of death and cause-of-death statements. In Froede RC (ed): Handbook of Forensic Pathology, ed 2. Northfield, Ill.: College of American Pathologists, 2003.
54. Hartmann A, Rundek T, Mast H, et al. Mortality and causes of death after first ischemic stroke. Neurology 2001;57:2000–2005.
55. Hinkle DA, Raizen DM, McGarvey ML, Liu GT. Cerebral air embolism complicating cardiac ablation procedures. Neurology 2001;56:792–794.
56. Hitselberger WE, House WF. A warning regarding the sitting position for acoustic tumor surgery. Arch Otolaryngol 1980;106:69.
57. Hoh BL, Putman CM, Budzik RF, et al. Combined surgical and endovascular techniques of flow alteration to treat fusiform and complex wide-necked intracranial aneurysms that are unsuitable for clipping or coil embolization. J Neurosurg 2001;95:24–35.
58. Horowitz NH, Rizzoli HV. Postoperative Complications in Neurosurgical Practice: Recognition, Prevention, and Management. Baltimore: Williams & Wilkins, 1967.
59. Houten JK, Errico TJ. Paraplegia after lumbosacral nerve root block: Report of three cases. Spine J 2002;2:70–75.
60. Huber M, Litz RJ, von Kummer R, Albrecht DM. Intrathecal air following spinal anaesthesia (letter to editor). Anaesthesia 2002;57:307.
61. Hwang W, Ralph J, Marco E, Hemphill JC. Incomplete Brown-Séquard syndrome after methamphetamine injection into the neck. Neurology 2003;60:2015.
62. Inamasu J, Guiot BH. Iatrogenic vertebral artery injury. Acta Neurol Scand 2005;112:349–357.
63. Molyneux AJ, Kerr RS, Yu LM, et al. International Subarachnoid Aneurysm Trial (ISAT) Collaborative Group. International subarachnoid aneurysm trial (ISAT) of neurosurgical clipping versus endovascular coiling in 2143 patients with ruptured intracranial aneurysms: A randomised comparison of effects on survival, dependency, seizures, rebleeding, subgroups, and aneurysm occlusion. Lancet 2005;366:809–817.
64. Jiang WJ, Srivastava T, Gao F, et al. Perforator stroke after elective stenting of symptomatic intracranial stenosis. Neurology 2006;66:1868–1872.
65. Jones TF, Feler CA, Simmons BP, et al. Neurologic complications including paralysis after a medication error involving implanted intrathecal catheters. Am J Med 2002;112:31–36.
66. Kase CS, Furlan AJ, Wechsler LR, et al. Cerebral hemorrhage after intra-arterial thrombolysis for ischemic stroke. Neurology 2001;57:1603–1610.
67. Kavcic A, Meglic B, Pecaric Meglic N, et al. Asymptomatic huge calcified subdural hematoma in a patient on oral anticoagulant therapy. Neurology 2006;66:758.
68. Keane JR. Cisternal puncture complications. Calif Med 1973;119:10–15.
69. Kelly PJ, Ahlskog JE, van Heerden JA, et al. Adrenal medullary autograft transplantation into the striatum of patients with Parkinson's disease. Mayo Clin Proc 1989;64:282–290.
70. Kestle JRW. Pediatric hydrocephalus: Current management. Neurol Clin North Am 2003;21:883–895.
71. Kistler JP, Furie KL. Carotid endarterectomy revisited. N Engl J Med 2000;342:1743–1745.
72. Kizer JR, Devereux RB. Patent foramen ovale in young adults with unexplained stroke. N Engl J Med 2005;353:2361–2372.
73. Kleinschmidt-DeMasters BK, Rojiani AM, Filley CM. Central and extrapontine myelinolysis: Then . . . and now. J Neuropathol Exp Neurol 2006;65:1–11.
74. Lantz PE, Smith JD. Fatal carbon dioxide embolism complicating attempted laparoscopic cholecystectomy—Case report and literature review. J Forensic Sci 1994;39:1468–1480.
75. Lau G. Iatrogenic injury: A forensic perspective. In Tsokos M (ed): Forensic Pathology Reviews, Vol 3. Totowa, NJ: Humana Press, 2005.
76. Lerner BH. Last-ditch medical therapy—Revisiting lobotomy. N Engl J Med 2005;353:119–121.
77. Levy EI, Horowitz MB, Koebbe CJ, et al. Transluminal stent-assisted angioplasty of the intracranial vertebrobasilar system for medically refractory, posterior circulation ischemia: Early results. Neurosurgery 2001;48:1215–1223.
78. Levy LM. An unusual case of flexion injury of the cervical spine. Surg Neurol 1982;17:255–259.
79. Lifschultz BD, Kenney JP, Sturgis CD, Donoghue ER. Fatal intracranial hemorrhage following pediatric oral surgical procedure. J Forensic Sci 1995;40:131–133.
80. Lopez AE, Barnard JJ, White CL, et al. Motor-vehicle collision-related death due to delayed-onset subarachnoid hemorrhage associated with anticoagulant therapy. J Forensic Sci 2004;49:807–808.
81. Lopez-Lozano JJ, Bravo G, Abascal J, et al. Clinical outcome of cotransplantation of peripheral nerve and adrenal medulla in patients with Parkinson's disease. J Neurosurg 1999;90:875–882.
82. Lyons KE, Wilkinson SB, Overman J, Pahwa R. Surgical and hardware complications of subthalamic stimulation: A series of 160 procedures. Neurology 2004;63:612–616.
83. Melinek J, Lento P, Moalli J. Postmortem analysis of anastomotic suture line disruption following carotid endarterectomy. J Forensic Sci 2004;49:1077–1081.
84. Misra UK, Kalita J, Khandelwal D. Consequences of neck manipulation performed by a non-professional. Spinal Cord 2001;39:112–113.
85. Moza K, McMenomey SO, Delashaw JB. Indications for cerebrospinal fluid drainage and avoidance of complications. Otolaryngol Clin North Am 2005;38:577–582.
86. Murphy GK. Therapeutic misadventure: An 11-year study from a metropolitan coroner's office. Am J Forensic Med Pathol 1986;7:115–119.
87. Olanow CW, Brin MF, Obeso JA. The role of deep brain stimulation as a surgical treatment for Parkinson's disease. Neurology 2000;55(Suppl):S60–S66.

88. Olson S. The problematic slit ventricle syndrome. Pediatr Neurosurg 2004;40:264–269.
89. Omuro AMP, Lallana EC, Bilsky MH, DeAngelis LM. Ventriculoperitoneal shunt in patients with leptomeningeal metastasis. Neurology 2005;64:1625–1627.
90. Orenstein JM, Sato N, Aaron B, et al. Microemboli observed in deaths following cardiopulmonary bypass surgery: Silicone antifoam agents and polyvinyl chloride tubing as sources of emboli. Hum Pathol 1982;13:1082–1090.
91. Page C, Lehmann P, Jeanjean P, et al. Intracranial abscess and empyemas from E.N.T. origin. Ann Otolaryngol Chir Cervicofac 2005;122:120–126.
92. Papagelopoulos PJ, Sapkas GS, Kateros KT, et al. Halo pin intracranial penetration and epidural abscess in a patient with a previous cranioplasty: Case report and review of the literature. Spine 2001;26:E463–E467.
93. Papp A, Toth J, Kerenyi T, et al. Silicone oil in the subarachnoid space—A possible route to the brain? Pathol Res Pract 2004;200:247–252.
94. Phatouros CC, Sasaki TYJ, Higashida RT, et al. Stent-supported coil embolization: The treatment of fusiform and wide-neck aneurysms and pseudoaneurysms. Neurosurgery 2000;47:107–115.
95. Platt MS, Kohler LJ, Ruiz R, et al. Deaths associated with liposuction: Case reports and review of the literature. J Forensic Sci 2002;47:205–207.
96. Podolsky DK, Gonzalez RG, Hasserjian RP. Case 8–2006: A 71-year-old woman with Crohn's disease and altered mental status. N Engl J Med 2006;354:1178–1184.
97. Prahlow JA. Therapy-related deaths. American Society of Clinical Pathologists Check Sample FP-272. Chicago: ASCP, 2002;44:1–17.
98. Prahlow JA, McClain JL. Deaths due to medical therapy. In Froede RC (ed): Handbook of Forensic Pathology, ed 2. Northfield, Ill.: College of American Pathologists, 2003.
99. Price EA, Schueler H, Perper JA. Massive systemic silicone embolism: A case report and review of the literature. Am J Forensic Med Pathol 2006;27:97–102.
100. Rapp JH, Pan XM, Sharp FR, et al. Atheroemboli to the brain: Size threshold for causing acute neuronal cell death. J Vasc Surg 2000;32:68–76.
101. Restuccia D, Rubino M, Valeriani M, et al. Cervical cord dysfunction during neck flexion in Hirayama's disease. Neurology 2003;60:1980–1983.
102. Ribalta T, McCutcheon IE, Neto AG, et al. Textiloma (gossypiboma) mimicking recurrent intracranial tumor. Arch Pathol Lab Med 2004;128:749–758.
103. Ropper AH, Brown RH (eds). Adams and Victor's Principles of Neurology, ed 8. New York: McGraw-Hill, 2005, pp 11–34.
104. Ross IB, Dhillon GS. Ventriculostomy-related cerebral hemorrhages after endovascular aneurysm treatment. Am J Neuroradiol 2003;24:1528–1531.
105. Rozin L, Rozin R, Koehler SA, et al. Death during transforaminal epidural steroid nerve root block (C7) due to perforation of the left vertebral artery. Am J Forensic Med Pathol 2003;24:351–355.
106. Sakaguchi M, Kitagawa K, Hougaku H, et al. Mechanical compression of the extracranial vertebral artery during neck rotation. Neurology 2003;61:845–847.
107. Saukko P, Knight B. Knight's Forensic Pathology, ed 3. New York: Oxford University Press, 2004, pp 480–487.
108. Scheller MS, Sarnat AJ, Astarita RW. Unintentional removal of meningeal tissue with a 25-gauge spinal needle during spinal anesthesia: A case report. Anesthesiology 1984;61:593–594.
109. Schlimp CJ, Loimer T, Rieger M, et al. The potential of venous air embolism ascending retrograde to the brain. J Forensic Sci 2005;50:906–909.
110. Schmid A, Tzur A, Leshko L, Krieger BP. Silicone embolism syndrome: A case report, review of the literature, and comparison with fat embolism syndrome. Chest 2005;127:2276–2281.
111. Schochet SS, Lampert PW, Earle KM. Neuronal changes induced by intrathecal vincristine sulfate. J Neuropathol Exp Neurol 1968;27:645–657.
112. Schupbach M, Gargiulo M, Welter ML, et al. Neurosurgery in Parkinson disease: A distressed mind in a repaired body? Neurology 2006;66:1811–1816.
113. Scott JA, DeNardo AJ. Complications of interventional neuroradiology. In Biller J (ed): Iatrogenic Neurology. Boston: Butterworth-Heinemann, 1998, pp 39–49.
114. Segal AZ, Abernethy WB, Palacios IF, et al. Stroke as a complication of cardiac catheterization: Risk factors and clinical features. Neurology 2001;56:975–977.
115. Sherman MR, Smialek JE, Zane WE. Pathogenesis of vertebral artery occlusion following cervical spine manipulation. Arch Pathol Lab Med 1987;111:851–853.
116. Silverstein A (ed). Neurologic Complications of Therapy. Mount Kisco, NY: Futura, 1982.
117. Simon SL, Abrahams JM, Sean Grady M, et al. Intramedullary injection of contrast into the cervical spinal cord during cervical myelography: A case report. Spine 2002;27:E274–E277.
118. Sirota RL. The Institute of Medicine's report on medical error: Implications for pathology. Arch Pathol Lab Med 2000;124:1674–1678.
119. Sloan MA: Neurologic complications of thrombolytic therapy. In Biller J (ed): Iatrogenic Neurology. Boston: Butterworth-Heinemann, 1998, pp 335–378.
120. Smeding HMM, Speelman JD, Koning-Haanstra M, et al. Neuropsychological effects of bilateral STN stimulation in Parkinson disease: A controlled study. Neurology 2006;66:1830–1836.
121. Smith WS, Johnston SC, Skalabrin EJ, et al. Spinal manipulative therapy is an independent risk factor for vertebral artery dissection. Neurology 2003;60:1424–1428.
122. Strowitzki M, Moringlane JR, Steudel WI. Ultrasound-based navigation during intracranial burr hole procedures: Experience in a series of 100 cases. Surg Neurol 2000;54:134–144.
123. Torn M, Algra A, Rosendaal FR. Oral anticoagulation for cerebral ischemia of arterial origin: High initial bleeding risk. Neurology 2001;57:1993–1999.
124. Tournel G, Bécart-Robert A, Courtín P, et al. Fatal accidental intrathecal injection of vindesine. J Forensic Sci 2006;51:1166–1168.
125. Toyoda K, Okada Y, Minematsu K, et al. Antiplatelet therapy contributes to acute deterioration of intracerebral hemorrhage. Neurology 2005;65:1000–1004.
126. Warner MA, Martin JT. Patient positioning. In Barash PG, Cullen BF, Stoelting RK (eds): Clinical Anesthesia, ed 4. Philadelphia: Lippincott Williams & Wilkins, 2001.
127. Warnick RE. Complications of surgery. In Bernstein M, Berger MS (eds): Neuro-Oncology: The Essentials. New York: Thieme Medical Publishers, 2000, pp 148–157.
128. Wijdicks EFM. Neurologic Complications of Critical Illness, ed 2. Contemporary Neurology Series. New York: Oxford University Press, 2002.
129. Wilder BL. Hypothesis: The etiology of midcervical quadriplegia after operation with the patient in the sitting position. Neurosurgery 1982;11:530–531.
130. Williams LS, Biller J. Vertebrobasilar dissection and cervical spine manipulation: A complex pain in the neck. Neurology 2003;60:1408–1409.
131. Woolley EJ. Neurologic complications of anesthesia. In Silverstein A (ed): Neurological Complications of Therapy. Mount Kisco, NY: Futura, 1982, pp 199–204.
132. Yang SY, Wang KC, Cho BK, et al. Radiation-induced cerebellar glioblastoma at the site of a treated medulloblastoma: Case report. J Neurosurg 2005;102:417–422.
133. Yoo KM, Yoo BG, Kim KS, et al. Cerebral lipiodol embolism during transcatheter arterial chemoembolization. Neurology 2004;63:181–183.

Miscellaneous Topics 17

INTRODUCTION 443
FORENSIC NEURORADIOLOGY 443
NEUROPATHOLOGIC CONSEQUENCES OF
   CONTEMPORARY JUDICIAL EXECUTIONS 444
SHARP FORCE INJURY 445

FORENSIC ANTHROPOLOGY 448
TOXICOLOGY 451
NEUROPATHOLOGIC COMPLICATIONS OF SYSTEMIC
   DISEASES 452
REFERENCES 460

Introduction

Topics reviewed in this chapter include some disease categories infrequently encountered in forensic neuropathology referrals, or circumstances in which the neuropathologist is more likely to interact with other subspecialty forensic consultants. Topics such as the effects of therapeutic radiation on the body[44] are, for the most part, omitted unless they are frequent sources of questions in a forensic setting. Some of these topics will be very familiar to neuropathologists in a nonforensic setting, others less so. A few representative cases are included for emphasis of certain points in this brief overview.

Forensic Neuroradiology

The application of radiologic techniques to forensic sciences, and more specifically to forensic pathology, has become standard procedure. Several recent publications provide an overview of such applications.[6,17,18,33,34,76,83,126,151] For the neuropathologist seeking information on various systemic diseases, syndromes, metabolic disorders, and skeletal dysplasias that may exhibit a specific neuroradiologic abnormality (e.g., basal ganglia calcification), the "Gamuts" section of Taybi and Lachman's text[140] provides an excellent starting point.

Routine x-ray, computed tomography (CT), ultrasonography, magnetic resonance imaging (MRI), and positron emission tomography (PET) reports have become familiar components of hospital records that are reviewed in forensic cases. As technical developments continue, terms such as turbo spin-echo T_2-weighted images, fluid attenuated inversion recovery (FLAIR), and echo-planar diffusion-weighted images will become more common. Virtual autopsy methods by CT and/or MRI are increasingly being used as a tool for investigating civilian forensic cases, victims of mass disasters, and military casualties. New techniques under development will also change current protocols concerning the preferred technique(s) applicable to, for example, pediatric trauma cases.[2,3,33,34,151]

Occasional errors in interpretation of imaging studies are unavoidable.[24,59,99,147] Averting such errors is an ongoing challenge in all areas of medicine, given the complexities of modern medical practice. This point has led some to argue for greater use of the autopsy for quality control. Unfortunately, the autopsy also has its limitations, including varying degrees of experience and skill by the autopsy pathologist, and differing opinions concerning the interpretation of findings. Radiographic methods and the autopsy are best viewed as complementary and not competitive procedures, as each may disclose abnormalities undiscovered by the other,[62,96] and neither approach has yet reached a degree of accuracy sufficient to replace the other for postmortem examination.[54,66,72,106,122,141,142] Postmortem imaging studies are appearing in the periodical literature with increasing frequency.

Neuropathologists and radiologists who consult on forensic cases can occasionally encounter situations in which there is an apparent discrepancy between antemortem radiologic findings and autopsy findings. In court proceedings, such apparent discrepancies may be exploited by interested parties in an effort to discredit the testimony of the consulting experts. The imaging techniques used each have their strengths and weaknesses; if this were not the case, only one technique would be used for all applications. As innovations in imaging techniques continue to progress, radiologic-pathologic correlation will continue to improve. Currently the following areas are some examples of potential interpretation problems:

- Normal stages of incomplete myelination in young infants may be misinterpreted as abnormalities in brain density.[46]
- Normal germinal matrix remnants in subependymal areas of the frontal horns of the lateral ventricles in premature infants can produce imaging alterations in ultrasonography or MRI that may be misinterpreted as pathology.[146]
- Thin blood collections directly under the inner table of the skull, or in parafalcine areas, are difficult to identify by CT scan.[46]
- It may be difficult to distinguish between subdural and subarachnoid blood on CT scans.[50,80,147]
- The terms *hyperacute*, *acute*, and *chronic*, as variously applied to subdural hematomas by radiologists, may not be consistent with the use of these terms by pathologists.[147] Mixed density fluid in subdural hematomas, as seen on CT scans, can be present shortly after the precipitating trauma,[129] and hypodense subdural hematomas can be seen "within a few days."[147] Subdural taps, if performed prior to imaging studies, may also produce fresh bleeding in an older subdural fluid collection and thus alter the preexisting imaging findings.[147]
- Estimation of the quantity of subarachnoid hemorrhage by imaging studies may not correlate well with autopsy findings.[50]
- Subarachnoid hemorrhage may be diagnosed in its absence by the radiologist ("pseudosubarachnoid hemorrhage"), particularly in the presence of purulent meningitis or severe hypoxic-ischemic encephalopathy, using nonenhanced CT scans.[25]
- Opinions as to whether CT or MRI is more sensitive in the detection of acute intracerebral hemorrhage vary,[46,80] but MRI appears to be preferable for imaging pathology in the posterior fossa and anterior temporal area.[46]
- Hyperintense white matter foci on MRI often demonstrate at least subtle abnormalities on pathologic examination, but occasionally both gross and microscopic examination of the area of interest is negative.[14,15]
- Lesions can evolve over time, and findings from earlier imaging studies may change significantly due to an increase or decrease in initial abnormalities, or by superimposition of secondary complications.[33,34,46,147]
- Spinal cord injury without radiographic abnormality (SCIWORA) occurs both in adults and in children[68,101,107,139] (also see comment on SCIWORA in Chapter 7).
- Formalin fixation of the brain alters the appearance of MRI images, an important consideration if postmortem MRI results are to be correlated with antemortem MRI studies.[14,112]

As a final comment, the medical examiner should also be aware of potential periprocedural complications unique to the MRI laboratory environment. Patients with implanted therapeutic devices, such as electrical stimulation units, shunts for hydrocephalus, and so on may experience complications related to exposure to the magnetic fields generated by the MRI.[1,67,95,120] Potential problems related to retained bullets and to firearms within the MRI suite are discussed in Chapter 8.

Neuropathologic Consequences of Contemporary Judicial Executions

Hanging and electrocution are briefly discussed, primarily for the purpose of providing a few citations for initial exploration of these subjects by the interested reader. Lethal gas, lethal injection, and firing squad executions are not relevant for neuropathologic purposes. A description of present-day and recent past execution methods is available.[81]

Autopsy findings in judicial hangings, as opposed to most suicidal and accidental hangings, are described in several currently published forensic pathology textbooks, to which the reader is referred for details.[35,37] Although the so-called classic hangman's fracture-dislocation through the C2 pedicles with resultant rostral spinal cord contusion may occur, findings can vary from case to case to include a variety of fractures, such as those of the hyoid bone, thyroid cartilage, styloid process, occipital bone, odontoid process, and various types of vertebral fractures of C1, C3, C5, and C6; spinal column dislocations; ligamentous tears; disk herniations; and soft tissue injuries (hemorrhage, carotid and/or vertebral artery lacerations; spinal epidural, subdural, and subarachnoid hemorrhage; and complete spinal cord transection).*

Judicial electrocutions have yielded few reports with detailed descriptions of neuropathologic findings. Exclusive of annular head burns from the scalp electrode, and a single case of first-degree frontal scalp and facial burns, the only remarkable neuropathologic findings in 37 cases studied by autopsy from a case series executed by elec-

*References 35, 37, 63, 73, 81, 105, 118, 131, 136, 148.

trocution methods employed in Florida from 1983 to 1999 were epidural hematoma, induration of the dura mater, and brownish discoloration of the cerebral surface.[88] Another author commented on the presence of cerebral edema in 61% of 21 cases of judicial execution examined in Alabama since 1983, although 11% had "coagulation of any viscera," not otherwise specified.[39]

Hassin[64] examined brains of judicial execution cases that revealed "rents" in neural tissue, and elastic membrane and muscularis ruptures in large vessels including basilar, vertebral, and cerebellar arteries. He also described myelin sheath injury, neuronal chromatolysis, and other findings that he retrospectively may have considered less persuasive, as they were not emphasized in the comments in a subsequent article.[65]

Historical reviews of autopsies performed on judicial execution subjects during the initial implementation of this form of capital punishment, using technology bearing little resemblance to modern equipment, contained descriptions of scalp, calvarial, and brain lesions[74] that include some lesions now seen virtually exclusively in fatalities due to accidental high-voltage line electrocutions, lightning strikes, or prolonged heat exposure (e.g., bodies in burning buildings; boiling in water). Neuropathologic findings in autopsies in the latter instances have included subarachnoid hemorrhage; "wrinkling" of the pia-arachnoid and cerebral cortex surface; gray discoloration of brain; increased brain firmness and brittleness; skull fracture; scalp, skull, and brain charring and/or calcination (i.e., reducing to powder by heat); pulpified brain; and extrusion of pulpified brain through a dural defect to enter the epidural space.[5,108]

In some cases of accidental electrocution in which the patient survived, unusual immediate or delayed spinal cord syndromes have occurred, with clinical features of an ascending paralysis reminiscent of Guillian-Barré syndrome, transverse myelitis, or a motor neuron disease-like syndrome.[111] Pathologic findings in such cases are relatively limited, and pathogenesis, particularly for the delayed-onset forms, is unclear.

Kohr[84] has noted that, despite existence of legal precedent on the performance of autopsy in executed prisoners, legal challenges to this precedent have led to inconsistencies in court rulings, and awareness of current local jurisdictional protocol in such cases is advisable.

Sharp Force Injury

Fatalities resulting from sharp force injuries are included in the cases referred for medical examiner evaluation at this department. Most involve stab wounds to the heart, or large vessels in the neck, thorax, and/or extremities and do not require neuropathologic consultation. Homicidal wounds with sharp instruments are more reliably delivered to vital organs that are unprotected by the skull or spine, but, if the skull is the target, it is most likely to be penetrated through the eye and orbital roof or through the temporal squama.

When sharp force fatal injuries do involve the central nervous system (CNS), homicide[32] is more likely to be the manner of death than either suicide[45,91] or accident.[115] Sharp instruments used in producing wounds can be quite varied, and include swords, knives (Case 17.1, Figs. 17.1–17.2; Case 17.2, Figs. 17.3–17.5), sewing needles, nails, wires, metal spikes, sickles, pencils, pens, crochet hooks, umbrella points, car radio antennae, pitchforks, iron rods (such as rebar), bicycle wheel spokes, table forks, scissors, broom handles, spear shafts, axes, hatchets, machetes, tomahawks, drills, glass bottles,[94] and others. Knives are responsible for the vast majority of contemporary cases.[11,32] Broken knife tips from past wounds may be incidentally found at autopsy.[28] Delayed onset of symptoms or progression of symptoms from knife wounds, either cranial or spinal, usually is due to a retained portion of knife blade causing subsequent additional injury or infection, or infectious or other complications in the absence of a retained foreign body.[32,86]

The most common cause of a rapid death from a cranial stab wound is the development of an intracerebral hematoma with mass effect.[32] Other causes include exsanguination from scalp or deeper head wounds; rapidly developing epidural, subdural, or intraventricular hemorrhage; or direct penetration of vital brain centers. Later-developing complications include cavernous sinus thrombosis, brain swelling, pneumocephalus (see Case 17.1), infection such as meningitis or brain abscess, cranial nerve palsies, epilepsy, false (traumatic) aneurysm, carotid-cavernous fistulae, and cerebrospinal fluid (CSF) leakage via the wound path. Intracranial infection is more likely when there is retained foreign material introduced by the wounding instrument, and when the wound path involves the nasal cavity or paranasal sinuses (Case 17.3, Figs. 17.6–17.7).

Gross examination of spinal cord sharp force injury, due to the relatively short distances involved, may be aided by application of a dye that will permeate the injury cavity, but not penetrate through its walls, prior to obtaining cross-sections of the spinal cord (Case 17.4, Figs. 17.8–17.9). This allows a graphic demonstration of

Case 17.1. (Figs. 17.1 and 17.2)

A 29-year-old man was stabbed in the head during a fight, and he reportedly stood up, walked a few steps, and then slowly collapsed to the ground unresponsive. Imaging studies in the hospital revealed the tip of the embedded knife at the left foramen of Monro, subarachnoid hemorrhage, hemorrhage along the knife path, intraventicular hemorrhage, pneumocephalus, and cerebral edema. He remained comatose and died approximately 12 hours after the injury.

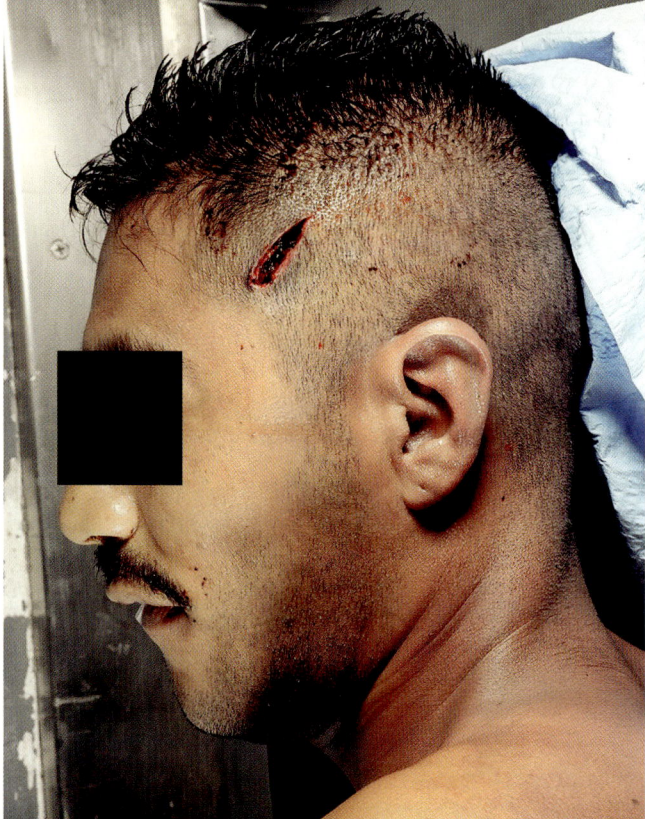

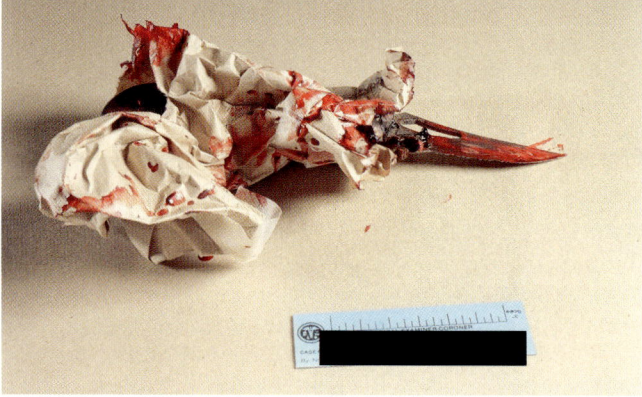

Fig. 17.1. *A,* Scalp knife wound. *B,* Knife blade producing the wound: handle covered with paper prior to fingerprint analysis.

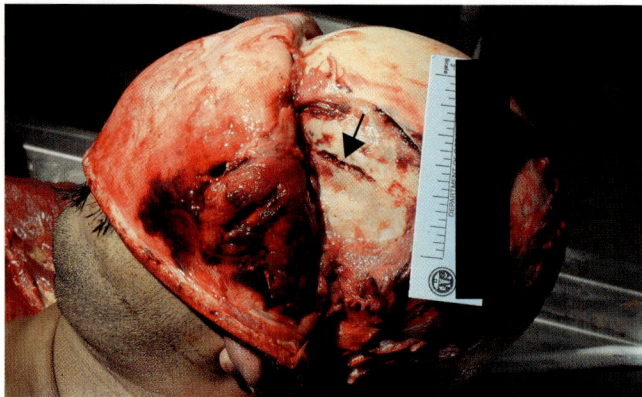

Fig. 17.2. Skull is readily penetrated in the thinner squama of temporal bone (*arrow*).

> **Case 17.2.** (Figs. 17.3–17.5)
> A 56-year-old man died of unrelated cause 27 years after sustaining a stab wound to the head, with chronic neurologic sequelae of seizure disorder and right leg paresis.

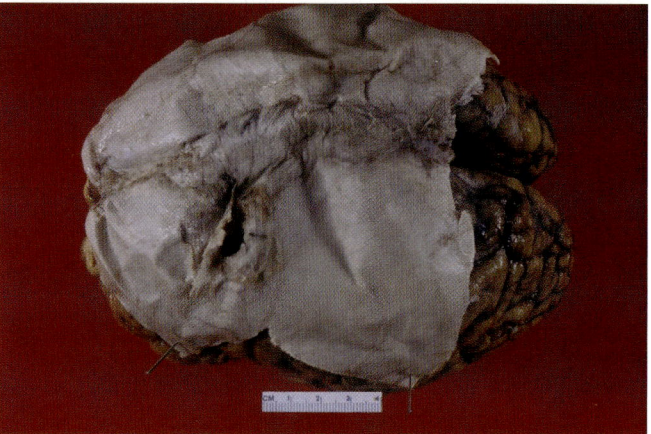

Fig. 17.3. Dural defect at prior wound site resulted from tear of dura during removal (artifact) secondary to adhesions. A surgical clip was present within the dural scar.

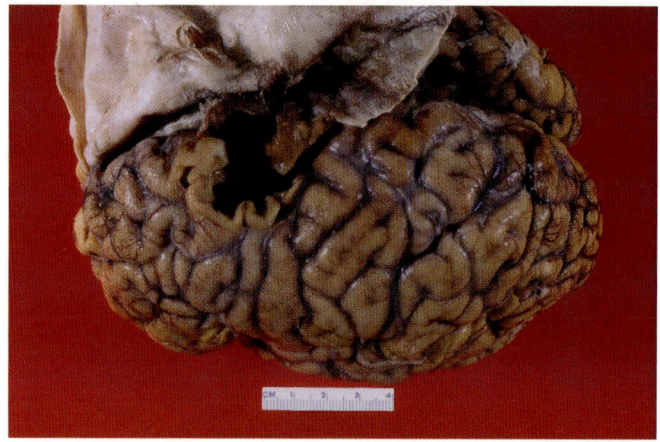

Fig. 17.4. The left convexity dura is reflected over the right hemisphere, exposing a left frontal cortical/subcortical cavitary defect 3.0 × 3.0 × 3.0 cm.

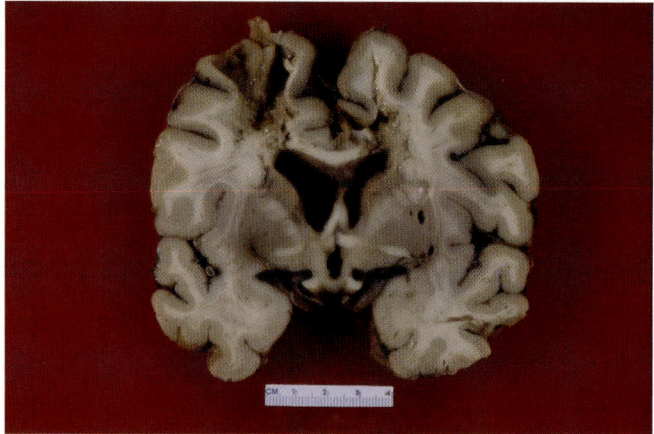

Fig. 17.5. Chronic sequelae of stab wound. Coronal section of brain demonstrates healed lesion far broader than might be anticipated in a stab wound. The lesion involved undercutting of the white matter of the medial half of the ipsilateral precentral gyrus, and partially transected corpus callosum. It also involved the contralateral medial hemisphere. Lateral ventricles are enlarged, left greater than right side.

Case 17.3. (Figs. 17.6 and 17.7)

A 31-year-old man was assaulted, stabbed, and thrown from a moving vehicle, sustaining multiple sharp and blunt force injuries. He was comatose and hypotensive on admission to the hospital. Complications ensued despite intensive care, and he died 9 days' postinjury.

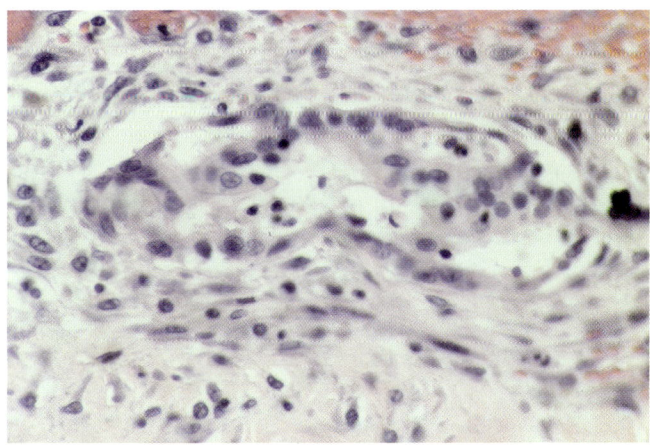

Fig. 17.7. High-power micrograph of epidural tissue at hematoma–dura interface revealed ciliated respiratory epithelium of frontal paranasal sinus origin displaced intracranially by knife blade. (H&E.)

Case 17.4. (Figs. 17.8 and 17.9)

A 42-year-old woman was attacked in a hotel room, sustaining multiple sharp and blunt force injuries including a stab wound at the C4 vertebral level. She was pronounced dead at the scene.

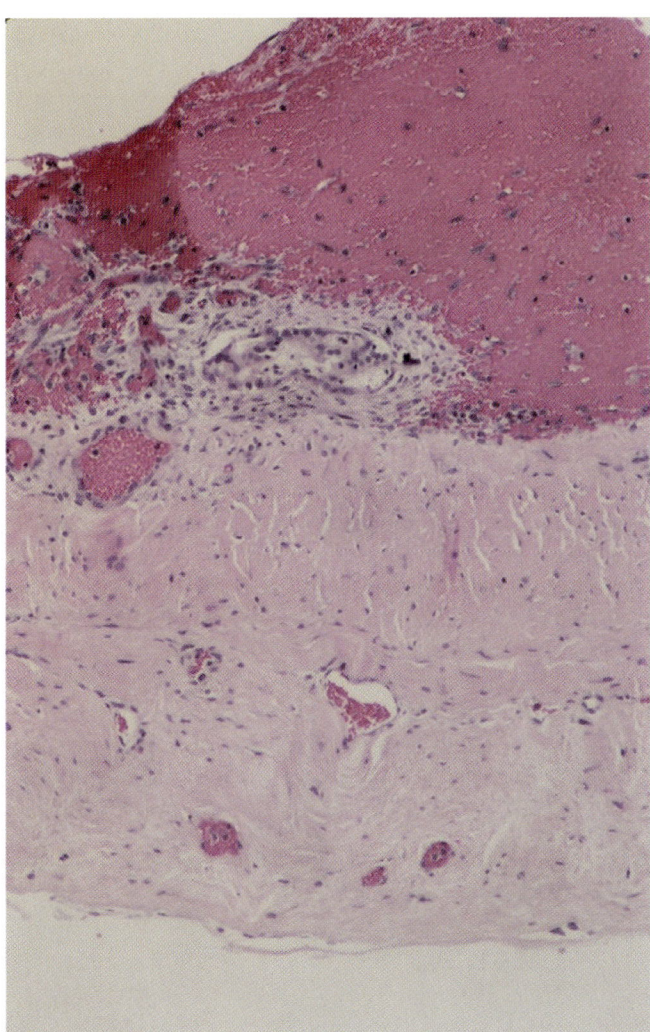

Fig. 17.6. Low-power micrograph. Among the multiple wounds was a stab wound in the left lower forehead that produced a subfrontal epidural hematoma and entered the brain. Dura (lower portion of photo) with epidural hematoma (upper portion of photo), and tissue with epithelium at dura–hematoma interface. (H&E.)

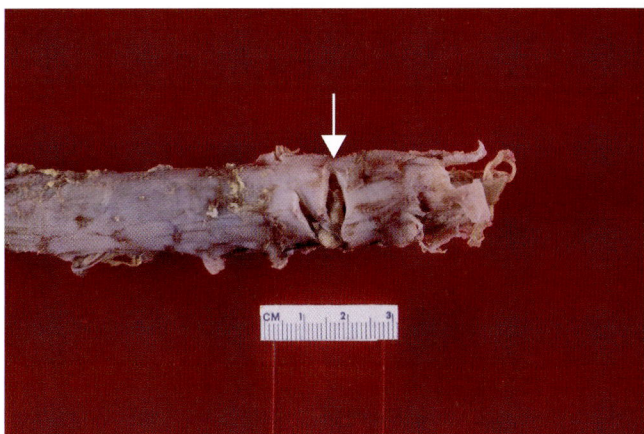

Fig. 17.8. Stab wound of spinal cord at C4 vertebral level. The **V**-shaped dural defect is consistent with twisting of the blade between entry and exit. The wound in the spinal cord corresponds to the more vertical dural defect (*arrow*).

448 Miscellaneous Topics

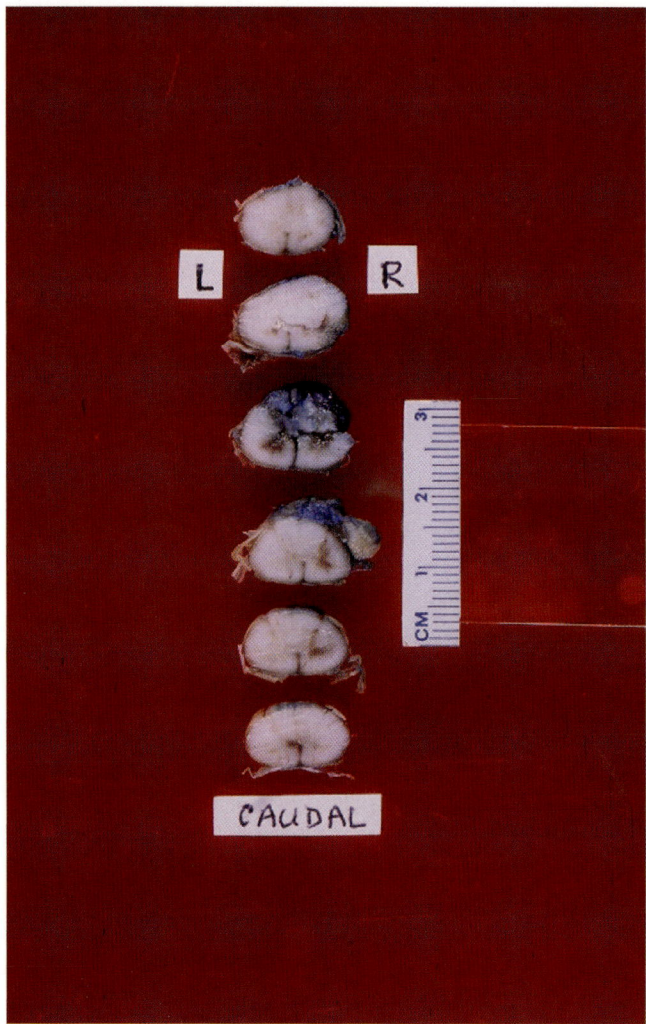

Fig. 17.9. Blue dye swabbed on the entry wound and allowed to "set" prior to making transverse sections of cord facilitates demonstration of acute wound margins in photos, which can be supplemented with diagrams for clinical-anatomic correlation.

the depth and location of cord injury in cases where survival periods are too brief to allow histologic reactive changes to occur at wound margins. Correlation of wound location with spinal pathways involved, either by clinical neurology or neurosurgery consultation and/or by use of clinical-anatomic reference sources,[16] can be helpful in answering questions later raised by interested parties, with the caveat that functional impairment due to acute circulatory or neurophysiologic wound effects could involve areas beyond the zone of disrupted tissue.

Speed of incapacitation, as well as survival periods, are as difficult to predict for stab wounds as they are in cases of gunshot wounds (see Chapter 8). Some generalizations based on available data are the following:

- Stab wounds are more likely to be multiple than single.
- Statistically, individuals sustaining fatal stab wounds are more likely to continue volitional physical activity longer, and are more likely to survive for a longer period, compared with individuals sustaining fatal gunshot wounds.
- Stab wounds to the heart or great vessels, or cases with multiple stab wounds, tend to result in shorter survival times than cases with a single or a few stab wounds, or involvement of other abdominal, thoracic, or neck organs. Exceptions do occur, such as cases with heart stab wounds that have survived up to 42 hours.[143]
- One historical review of continuing activity and survival periods in individuals sustaining dueling sharp force injuries caused by edged and pointed weapons also emphasized the unpredictability of both of these postinjury parameters.[92]

There are only limited data on the degree of activity and survival period after stab wounds to the head or neck. Most fatal neck wounds are due to large vessel involvement.[87,138]

Forensic Anthropology

A few comments on this forensic subspecialty are included herein due to its critical role in many medical examiner cases, the fact that certain material of general anatomic and neuroanatomic interest is almost exclusively found in publications in the anthropology or forensic (rather than neuropathology) literature, and because this material is rarely, if ever, presented in any detail in neuropathology training programs. The forensic neuropathology consultant may find it useful to become more familiar with some general references on topics in which a forensic anthropologist may be the best initial source of expertise.[57,58,119] A few examples of such topics follow:

- Recovery of buried remains and estimation of time since death.
- Use of skull and other skeletal elements in determining sex, age, body identification and other characteristics.[70,75,124]
- Determination of perimortem versus postmortem injuries.[144]
- Patterned injury analysis in soft tissues and bone, including various blunt and sharp force injuries and missile injuries, identification of tools used in dismemberment cases, and so on.
- Effects of heat (including fire and electrical injury)[5,114] (Case 17.5, Figs. 17.10–17.13) and cold (e.g., freezing, producing diastatic fractures of the skull).
- Distinguishing fatal animal attacks[22,102] (Case 17.6, Figs. 17.14–17.15), and postmortem animal predation injury or other pseudotrauma, from human-inflicted or accidental injury.[13] An example would be suture

Case 17.5. (Figs. 17.10–17.13)

The dismembered body of a 73-year-old man was found in his home. The head and upper neck were in a cooking pot, having previously been boiled in water for several hours (details of this case have been previously published by Andrews et al.[5]).

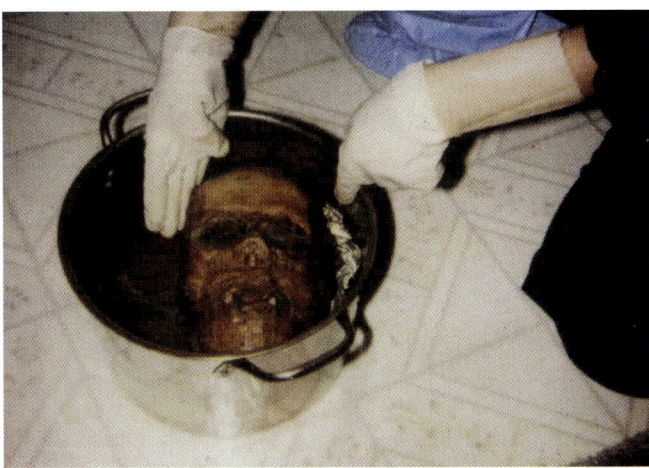

Fig. 17.10. Dismembered head as it was found in cooking pot.

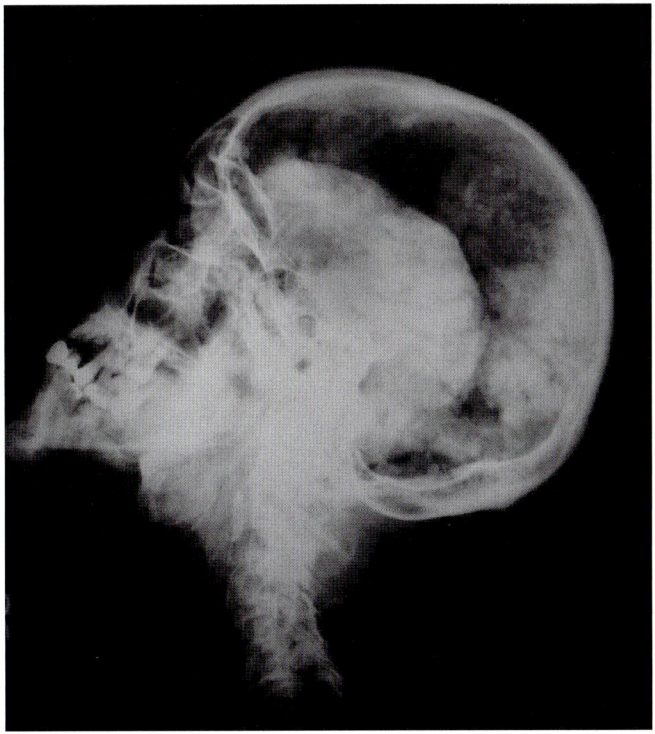

Fig. 17.11. X-ray of dismembered head, demonstrating large epidural space due to heat-induced shrinkage of dura, separating dura from inner skull surface.

Fig. 17.12. Curdlike fragments of brain filled the enlarged epidural space created by the dural shrinkage and its separation from inner skull.

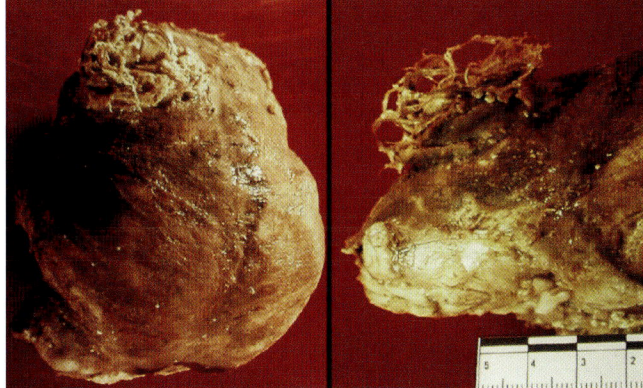

Fig. 17.13. Two views of shrunken dura, with dural tear in left frontal area marked by protruding cerebral blood vessels. As the heat-induced shrinkage of dura squeezed the essentially nonshrinking brain through the sievelike vessels in the dural defect, the now-fragmented brain tissue entered the epidural space.

Case 17.6. (Figs. 17.14 and 17.15)

A 6½-week-old infant was placed on a mattress on the floor while the caretaker was performing tasks in the garage. The caretaker returned after hearing the infant cry to find the family dog holding the infant by the head within its jaws, and immediately separated the infant from the dog. First responders found the infant bleeding profusely from the head. Despite emergency treatment, the infant died approximately 1½ hours later. At autopsy, multiple scalp and skull puncture wounds and skull fractures were present.

450 Miscellaneous Topics

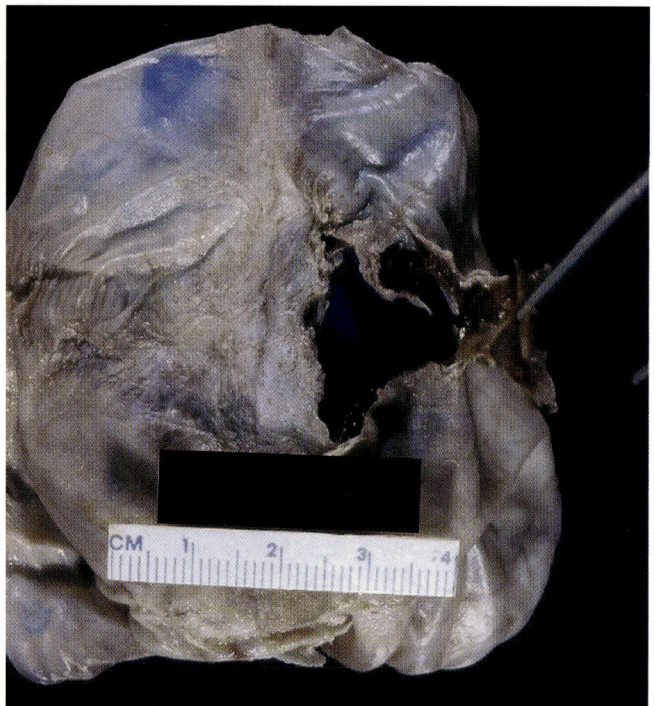

Fig. 17.14. One of several defects in cranial dura produced by dog bite, in this instance lacerating the right wall of the superior sagittal sinus.

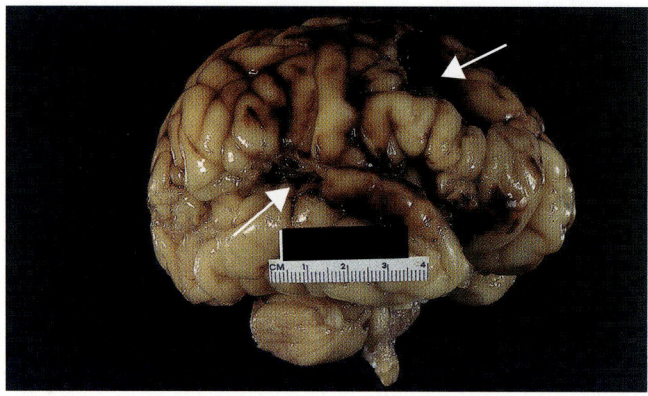

Fig. 17.15. Lateral view of right side of brain, with multiple puncture wounds (*arrows*) and subarachnoid hemorrhage.

Case 17.7. (Figs. 17.16 and 17.17)

An off-road vehicle operator discovered bones, identified by authorities as human, in 1997, including skull fragments insufficient to allow identification or cause of death. A handgun was also found in the vicinity. Additional search of the vicinity yielded no further bones at that time. Approximately 1½ years later, a hiker in the same general area found bones later identified as human in a shallow rain wash area beneath some brush. The additional fragments were found to match the previously discovered skull fragments, and allowed a positive identification of the decedent. The decedent had a history of suicidal ideation prior to his disappearance.

(Case courtesy of Steve Dowell, B.S., Research Criminalist, Department of Coroner, Los Angeles County.)

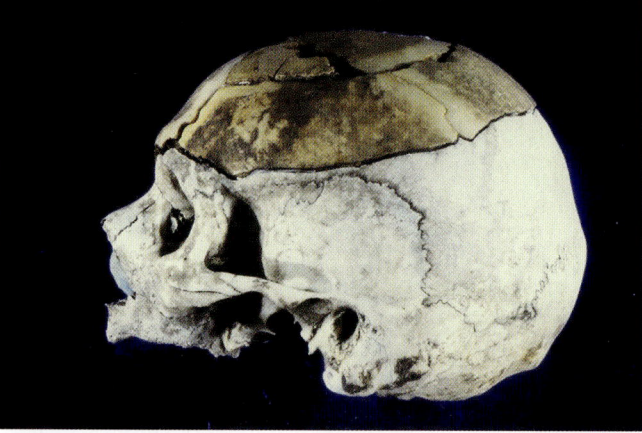

Fig. 17.16–17.17. Skull lateral view (Fig. 17.16; *top*) and vertex view (Fig. 17.17; *bottom*) demonstrating that the less-weathered, tan to brown, first-discovered (1997) skull bone fragments fit perfectly with the more-weathered, bleached white skull fragments (discovered in 1999) upon skull reconstruction.

separation in the immature skull due to weathering and warping, and not the result of antemortem increased intracranial pressure.[26]
- Distinguishing animal from human skeletal remains.[145]
- Analysis of skull fracture lines in the determination of gunshot versus blunt force trauma.[10]
- Individuation of fatal animal attack source by bite mark analysis (usually in conjunction with forensic dental consultation).[31,102]
- Effects of desiccation, mummification, and prolonged immersion (e.g., burial at sea) on the body in general and the CNS in particular.[19,41,69,71,89,103,116]
- Linkage of partial human remains found at different times and locations[56] (Case 17.7, Figs. 17.16–17.17).

Toxicology

Neurotoxicology is a complex and important subject, as reflected in the frequency with which it is addressed in textbooks and periodicals on toxicology, general and forensic pathology, neurology, and adult and pediatric neuropathology. Representative general sources are cited in the references for this section.* Many cases will not differ significantly from those encountered by the neuropathologist in a nonforensic setting (e.g., CNS complications of chemotherapeutic agents). Additional recommended information sources include departmental forensic toxicologists, Internet databases, local and national poison control centers, public health departments, workshops at national meetings of organizations representing the disciplines indicated earlier, and departments of parks and recreation (for local poisonous plants and animal species).

The usual role of the consulting forensic neuropathologist in toxicology cases is threefold. First, to be alert to the possibility of CNS complications arising from the introduction of newer therapeutic agents, so that proper studies can be designed to prove or disprove such associations.[150] Second, to determine whether CNS, peripheral nervous system (PNS), or muscle findings are, or are not, consistent with intoxication from an agent already suspected or identified by the medical examiner who performed the autopsy, or by the forensic toxicologist who analyzed the specimens provided. Third, and far

> **Case 17.8.** (Figs. 17.18 and 17.19)
>
> A 41-year-old man was found at home confused, with mild dyspnea and difficulty standing. Hospital evaluation revealed severe metabolic acidosis, hypertension, hyperkalemia, and acute renal failure. He became comatose, developed seizures and respiratory failure, and died on the second hospital day. A urinalysis was reported to show "hippuric acid crystals." There was no history of antifreeze, solvent, or other oxalate-containing materials having been ingested.

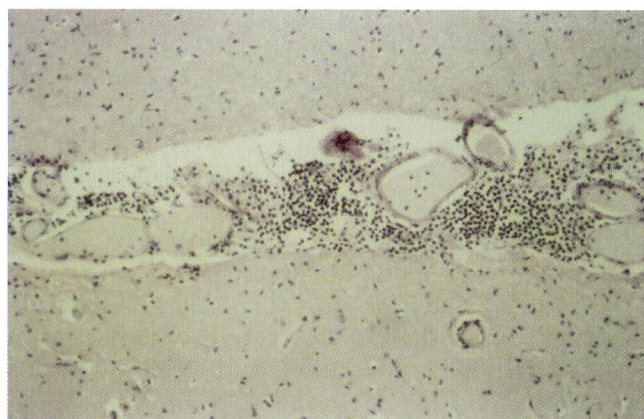

Fig. 17.18. Low-power micrograph shows acute meningitis, with neutrophils and less prominent mononucleated cells in subarachnoid space. Stains for infectious agents were negative. (H&E.)

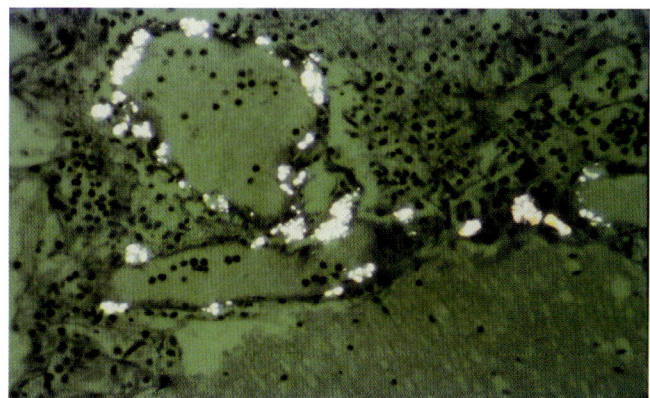

Fig. 17.19. Medium-power micrograph shows birefringment calcium oxalate crystals in leptomeningeal birefringent vessel wall. Similar crystals were subsequently found to be prominent in the kidney. (H&E with polarization.)

*References 8, 12, 21, 27, 29, 30, 36, 38, 40, 42, 43, 51, 53, 60, 77, 125, 130, 137.

less common in our experience, is to call attention to CNS findings of a previously unsuspected poisoning. This may occur, for example, when the offending agent is not revealed on routine toxicology screens, and the CNS microscopic examination is performed before the medical examiner has examined slides from other organ systems (Case 17.8, Figs. 17.18–17.19). An interesting finding in Case 17.8 was the conspicuous presence of calcium oxalate crystals in the CNS in slides stained by hematoxylin-eosin (H&E), Giemsa, acid-fast, and Gram stains, but no calcium oxalate crystals in the slide stained by the Gomori methenamine silver (GMS) method. This would indicate that tissue processing methods may influence persistence of such crystals, since calcium oxalate crystals may be seen in acid or neutral environments but tend to disappear in alkaline environments and are soluble in dilute hydrochloric acid.

It is not difficult to find materials that, when directly introduced into the CSF compartment, evoke an inflammatory or toxic response in the CNS or its coverings. Rather, the challenge has been to discover therapeutic, anesthetic, or diagnostic agents that cause no adverse effects when placed in the CSF. Case 17.8 is of interest in that, to our knowledge, ethylene glycol (or oxalosis secondary to ingestion of oxalate-containing plants)[128] is the only systemically ingested toxin category that can produce a chemical meningitis. The resultant birefringent calcium oxalate crystals may not be easily seen without polarization, and can persist even in severely autolyzed tissues. A single article reported encephalitic reactions to polyol-containing infusions,[109] but subsequent reports confirming this have not come to our attention.

Neuropathologic complications of drugs of abuse are common in medical examiner departments, and are reviewed in depth by Karch.[77] They are discussed briefly in Chapter 11.

Medical complications of alcohol abuse are commonly found in forensic settings. Fatal acute alcohol intoxication is not a likely referral to the consulting neuropathologist. Cases of acute ethanol intake combined with blunt force head trauma, or of chronic alcoholism, in contrast, are frequently referred to evaluate possibilities such as the sequelae of recent[78,79,98,117,152] or remote blunt force head trauma (including intracranial hemorrhage, post-traumatic apnea, or epilepsy), hepatic encephalopathy, cerebellar degeneration, central pontine myelinolysis, Marchiafava-Bignami disease, and Wernicke's encephalopathy.[135] (See section on neuropathologic complications of systemic diseases, later.)

In some instances, neurotoxic substances may produce neuropathologic findings that mimic degenerative or metabolic CNS disorders, such as CNS degenerations, manifested primarily by leukoencephalopathy. Included in this category are individuals who have chronically abused inhaled solvents or other substances, including toluene and heroin.[48,85] Problems in interpretation can arise in cases of polydrug abusers, since the neuropathologic lesion(s) may represent combined effects of abused agents.

Blood from intracranial hematomas may be useful for detection of various drugs taken prior to death, although interpretation of results must consider the possibility of therapeutic agents disseminating into a preexisting hematoma if medical treatment preceded death.[97,100] Brain tissue can provide evidence of drug or toxin exposure, even if severely decomposed, when it is not necessarily the usual organ emphasized for sampling of a given agent (such as thallium),[132] or in cases where the only available tissue is a decapitated head. Thallium may, incidentally, also be the only known homicidal poisoning that can be confirmed after cremation of a body.[130]

Despite the extensive information available on neurotoxicology, cases are encountered that raise yet-unanswered questions, as in Case 17.9 (Fig. 17.20).

In summary, a large percentage of cases referred for forensic neuropathology consultation at our office have positive toxicology screens, and in some cases neuropathologic complications are an important contributor to death. The literature on this subject cited earlier precludes the need for further review here. The reader is referred to those sources for more detailed information.

Neuropathologic Complications of Systemic Diseases

In textbooks on neuropathology, chapters dealing with this general topic vary considerably in their content. This is not unexpected, given the fact that one cannot present other than a very selective review of the subject of general medicine covering countless conditions of each organ system in which neuropathologic changes may occur. Natural disease cases comprise the majority of cases seen in most forensic pathology departments. The information available is vast and continually changing with the rapid advances in so many related medical fields.[121] This section reviews a very few of the myriad common and less common CNS manifestations of systemic diseases encountered in a forensic office.

Coagulopathies

Cases in which multiple or single hemorrhages in the brain are in locations atypical for hemorrhage as seen in hypertensive vascular disease (see Chap. 11), cases with unusually widespread hemorrhages that seem disproportionate to the external evidence of trauma, or cases in which there are systemic findings suspicious for coagu-

Case 17.9. (Fig. 17.20)

A 20-year-old man and four co-workers were cleaning the interior of a tanker that had carried 90% formic acid. He collapsed while working, and died in the hospital 1½ hours after his collapse. One of his co-workers, age 28 years, also became ill and died the same day.

At autopsy, the brain surface of the 20-year-old was not discolored when first removed, but over the subsequent 2 to 3 minutes after exposure to air assumed a turquoise blue hue. Coronal sections revealed an initially normal color on the cut surface, which then turned turquoise blue within a few minutes after exposure to air. The color change was most conspicuous in gray matter, including the basal ganglia, with minimal involvement of white matter. The brain tissue of the 28-year-old co-worker who also died did not demonstrate any atypical color change upon removal, sectioning, and exposure to air. The turquoise color in the 20-year-old persisted in formalin-fixed brain tissue (Fig.17.20). No pigment deposits or other significant brain pathology was seen histologically except for the presence of Alzheimer type II glia in the globus pallidus, putamen, and some cortical areas. Histologic organ survey, including liver, was unremarkable. Toxicology screen was positive for formic acid (2 ± 0.2μg) but negative for hydrogen sulfide. An electron probe analysis of the trachea, liver, and brain tissue performed by Steven Dowell (Research Criminalist, Los Angeles County Department of Coroner) was negative for excess copper or other trace elements.

Comment. Formic acid exposure in industrial settings may occur through oral, dermal, or inhalation routes, and can produce acidosis, hypotension, renal failure, apnea, shock, and death.[123] We have not found reports of brain discoloration due to formic acid poisoning, whether by direct formic acid exposure or by indirect exposure, such as by metabolism of methanol to formic acid in cases of acute methanol intoxication.[47] Formic acid treatment of tissue blocks is one of several methods used to decontaminate brain material with suspected prion diseases, and does not produce this color change. Tissue analysis for copper, some compounds of which have a propensity to develop a bluish color, was also negative, as noted earlier. This is the only case with this specific dramatic color change in the brain encountered in over 20,000 neuropathology consultations performed in this department over the last quarter century.

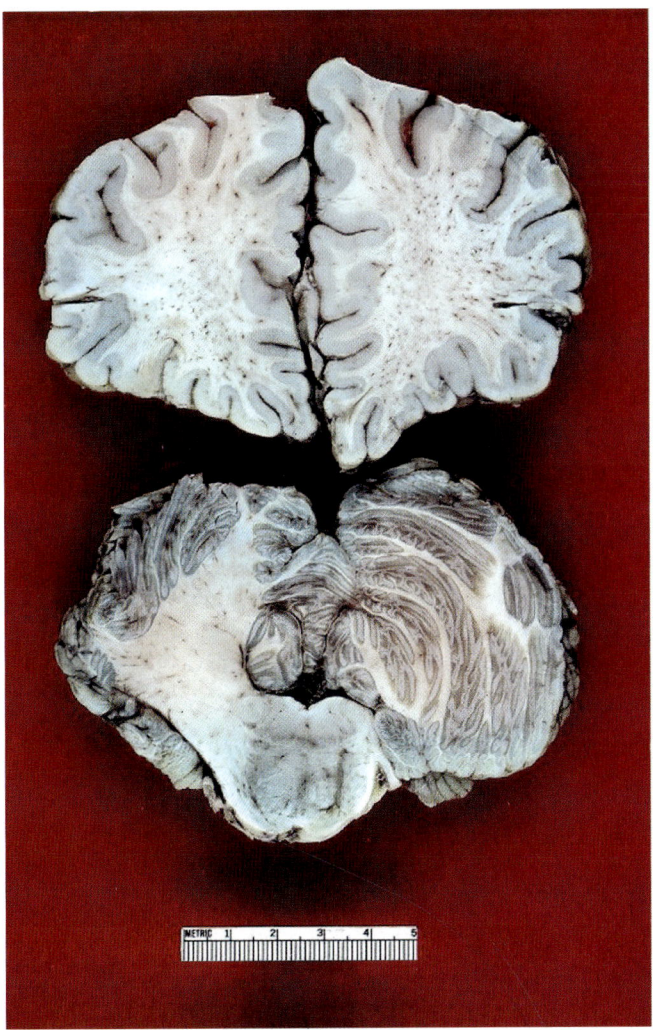

Fig. 17.20. Coronal section of frontal lobes and transverse section of pons and cerebellum, demonstrating bright turquoise discoloration.

lopathy in skin or other organs, warrant extensive efforts to exclude coagulopathy. A history of long-standing coagulation disorder in a decedent will obviate further search for a cause of hemorrhage in the nervous system. Disseminated intravascular coagulation is a commonly encountered complication in persons dying of other causes such as sepsis, and, in the forensic setting, major trauma. An intracerebral hemorrhage found in a decedent with a history of chronic alcoholism and liver cirrhosis upon autopsy is highly suggestive of a coagulopathy as the cause, whose precise pathogenesis usually remains unexplored. Other possible causes of hemorrhage, such as trauma, are always considered (see Chap. 6).

Focal Hemorrhages

In any spontaneous hemorrhage, the margins of the clot and the CNS parenchyma are sampled and also an area within the clot is sampled. Although such sections unusually reveal that the clot consists of blood elements only, this routine will occasionally lead to a diagnosis of hemorrhage due to metastatic or primary neoplasm, discovery of an occult vascular malformation, or discovery of vasculitis, infectious disease, or other disorders.

Intravascular Blood Elements

In every case, the slide screening routine should include specific attention to intravascular blood elements in the CNS. Presumptive (and subsequently confirmed) diagnoses of sickle cell trait or disease have first been made by the neuropathology consultant at our office by follow-

Case 17.10. (Fig. 17.21)

A 13-month-old girl had cyanosis and cardiac murmur shortly after birth, and evaluation revealed severe complex cyanotic congenital heart disease that included severe pulmonary stenosis, absent ductus arteriosus, significant right ventricle sinusoids and a small right ventricle, and a right ventricular-dependent coronary circulation. Several cardiac surgical procedures were performed in an attempt to improve cardiac function, but were unsuccessful. She remained cyanotic, required continuous oxygen supplementation, developed congestive heart failure, and died. Autopsy findings included the postoperative cardiac alterations, severe myocardial fibrosis, and moderate epicardial fibrosis. Neuropathologic examination revealed severe cerebrovascular congestion, brain swelling without herniation syndrome, and a thin, chronic right cerebral convexity subdural neomembrane with profound vascularity.

ing this routine, since histologic screening of other organs is problem-oriented rather than routine, and may have been deferred while awaiting CNS findings. Vessel content scrutiny may also reveal other disorders, such as intravascular malignant lymphomatosis, and so on.

Subdural Neomembrane Vascularity

In certain cases, as discussed elsewhere in this text, findings raise questions that, to our knowledge, have not been explored. For example, the most hypervascular subdural hemorrhage neomembrane encountered in our office, with numerous capillaries and sinusoids, occurred in a case with congenital heart disease (Case 17.10, Fig. 17.21).

It was theorized that chronic hypoxemia due to the eventually fatal congenital heart disease in Case 17.10 led to increased expression of vascular endothelial growth factor (VEGF) and resultant stimulation of angiogenesis. A VEGF stain was not significantly different in this case compared with a control, which neither supports nor excludes such a possibility, and such cases do raise one's curiosity as to the degree to which yet-undiscovered systemic factors might influence individual variations that occur in the histologic appearance of reactive changes to what seem to be essentially similar precipitating events.

Extramedullary Hematopoiesis

In another case of congenital heart disease (Case 17.11, Figs. 17.22–17.24), autopsy revealed bilateral and rather symmetric epidural blood collections, initially leading to a question of occult trauma. Histologic examination, however, revealed extramedullary hematopoiesis, emphasizing the importance of histologic verification of gross impressions. Extramedullary hematopoiesis, more common in cases with chronic hypoxemia or chronic anemia, may present with various neurologic symptoms, and is in the differential diagnostic list for lesions that usually produce (spinal more frequently than cranial) epidural, dural, and subdural (including falx cerebri) apparent blood collections or mass lesions. Less often, leptomeningeal involvement may produce masses indenting the brain surface, or there may be involvement of the

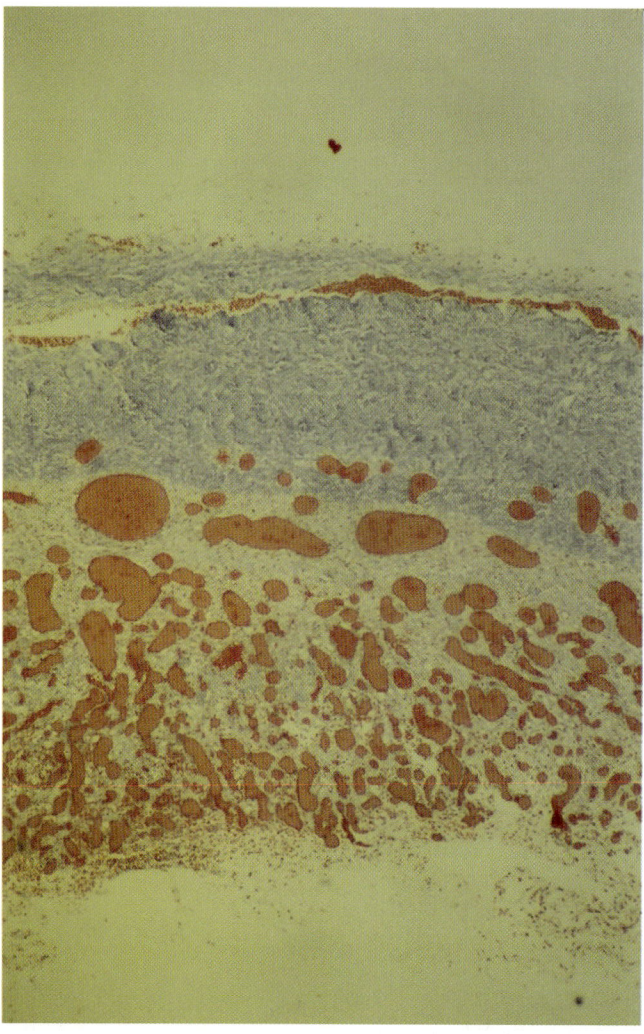

Fig. 17.21. Subdural neomembrane. Native dura is the darker blue layer in upper portion of photo. Hypervascular neomembrane is the thicker, very pale blue lower layer with numerous blood vessels. (Trichrome.)

Case 17.11. (Figs. 17.22–17.24)

A 38-week estimated gestational age male infant was born by uneventful vaginal delivery. At 13 days' postpartum he was found unresponsive by the mother, and did not respond to resuscitative efforts. Autopsy revealed congenital heart disease (ventricular septal defect, patent foramen ovale, aortic valve stenosis, aortic arch hypoplasia, and patent ductus arteriosus).

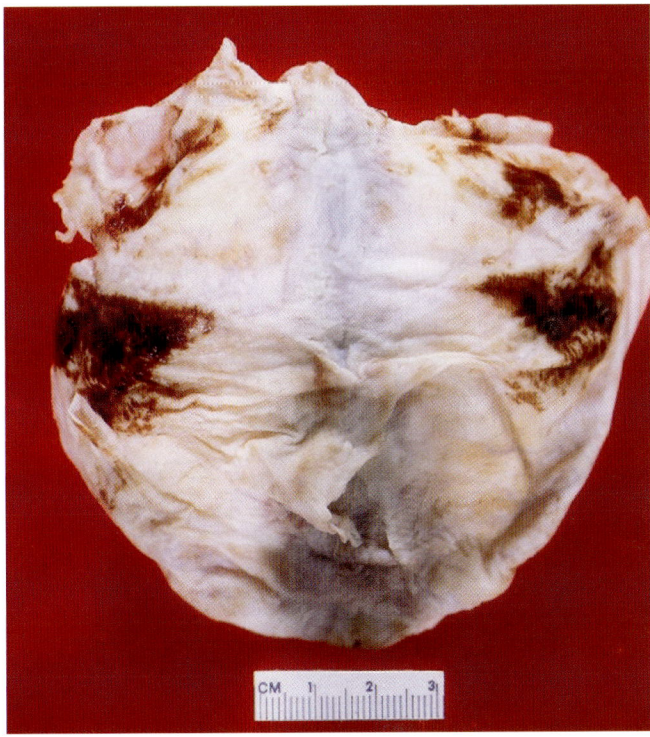

Fig. 17.22. In the fixed state, reddish-brown epidural material resembling zones of hemorrhage were rather symmetrically situated in dorsolateral frontal regions.

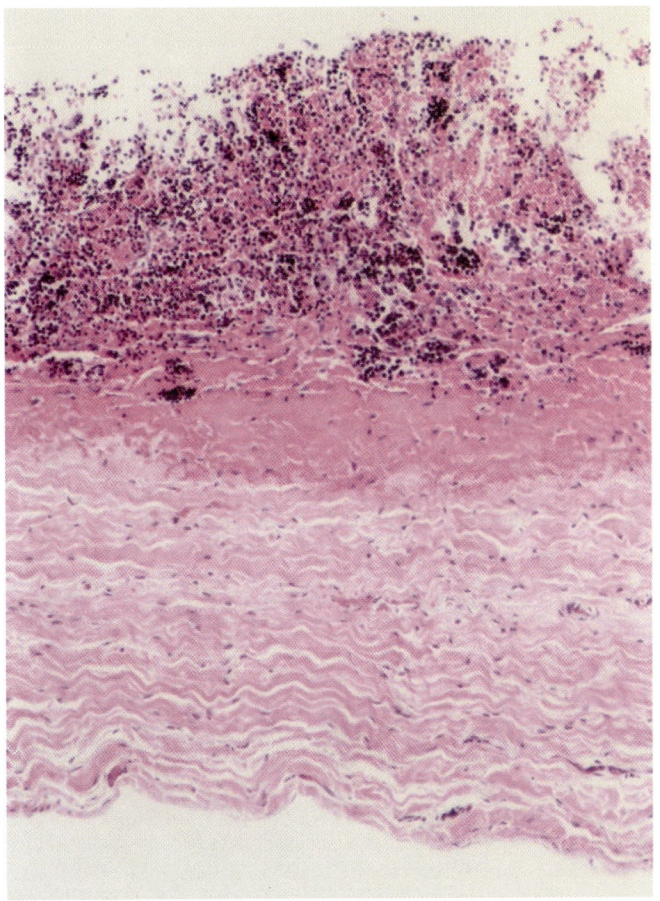

Fig. 17.23. Low-power micrograph. Nucleated cell-rich blood elements in epidural space (upper portion of photo). Native dura is below. (H&E.)

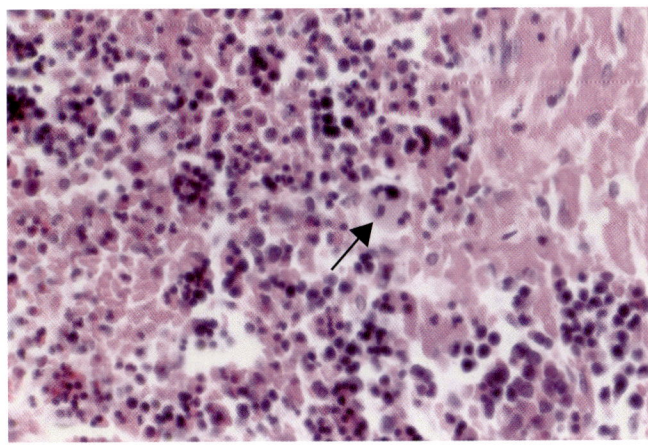

Fig. 17.24. Medium-power micrograph reveals immature erythrocyte and leukocyte precursors. A megakaryocyte is also present (*arrow*). (H&E.)

brain parenchyma or choroid plexus.[4,20,23,49,90,113,127,134] Symptoms may result from hemorrhage or hematopoietic tissue proliferation producing local mass effect. An unusual case of central diabetes insipidus due to posterior pituitary involvement by extramedullary hematopoiesis has been reported.[7]

Starvation

A review of statements concerning starvation in several contemporary books and articles would suggest that the brain is somehow protected from the loss of size and weight that affects nonneural tissues and organs.[35,93,130,133] It is understandable that this false impression persists. First, starvation is not a single entity, and may consist of varying degrees and combinations of insufficient intake of, for example, total calories, proteins, vitamins, and/or minerals. Many such cases also may have secondary complications such as decubiti and infections. Second, it can affect the immature brain differently than the mature brain. Third, the reduction in size and weight of the brain, while measurable, is not as dramatic as that seen in certain other organs. These factors, combined with the wide range of "normal" brain weights for a given sex, height, and age group that have been published, could easily lead to misinterpretation (see discussion of brain weights in Chap. 2). Fourth, it is theoretically possible that terminal events could lead to some degree of brain swelling, masking the preexisting size and weight reduction in some fatal cases.

Considerable clinical and experimental animal evidence has been accumulated over many years which clearly demonstrates that both gray and white matter in the CNS undergo reduction in size and weight in the starvation conditions studied. Emphasis in these studies has largely been on forms of chronic protein-calorie malnutrition such as kwashiorkor or marasmus in children.[61,104] In the case of fatal marasmus in a 3-year-old girl reported by Nagao et al.,[104] the brain weight was 980g (normal range cited was 1234 ± 110g). Imaging studies have not only described cortical atrophy and ventricular enlargement of brains of 11- to 21- year-old girls with anorexia nervosa, but have also demonstrated reversal of these abnormalities with nutritional rehabilitation treatment,[52] analogous to results in several imaging studies performed in starved children who have undergone nutritional therapy.[61]

Neuropathologic abnormalities in malnourished children have reportedly included neuronal lipofuscin accumulation in the cortex, basal ganglia, and inferior olives, and some neuronal atrophy in the midcortical layers.[61] In conclusion, the brain in severely starved persons undergoes reduction in size and weight that is sufficient to be detectable on gross examination and appears, at least in several cases studied thus far, to revert to normal size and weight when the protein-caloric deficiency is corrected. The details of the neuropathologic substrate for these changes, or for the cognitive deficits that may persist in some anorexia nervosa patients following nutritional treatment, are not fully understood.[52,149]

Sarcoidosis

Sarcoidosis, a granulomatous disease of unknown etiology, can involve multiple systems including the nervous system and its blood vessels and meninges. Most cases seen at our office have been diagnosed during life due to the frequent presence of symptomatic pulmonary involvement. Cases referred for neuropathology consultation may also include those with focal neurologic signs, such as diabetes insipidus (usually due to hypothalamic involvement). Much less frequently, the initial diagnosis of occult sarcoidosis is made by brain examination. Recent cases at our office have included involvement of the cerebellum (Figs. 17.25–17.27), occipital pole, or meninges at the foramina of Luschka and Magendie leading to hydrocephalus. CNS involvement may also be clinically asymptomatic, and subtle on gross examination, perhaps consisting only of focal meningeal thickening and/or granularity. A recent review of neurosarcoidosis is available.[55]

Scuba Deaths

The proximity of our office to the seacoast results in referral of deaths related to scuba diving, with detailed protocols for such cases including joint investigations with other agencies such as local law enforcement emergency services detail, diving safety officer's examination of equipment, and data sharing with representatives of the Department of the Navy. Additional consultation

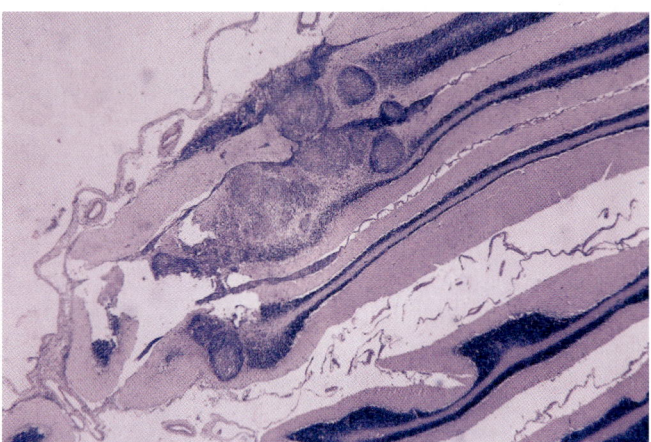

Fig. 17.25. Low-power micrograph of sarcoidosis involving cerebellar cortex. Granulomatous inflammation expands subarachnoid space and invades cerebellar parenchyma. (H&E.)

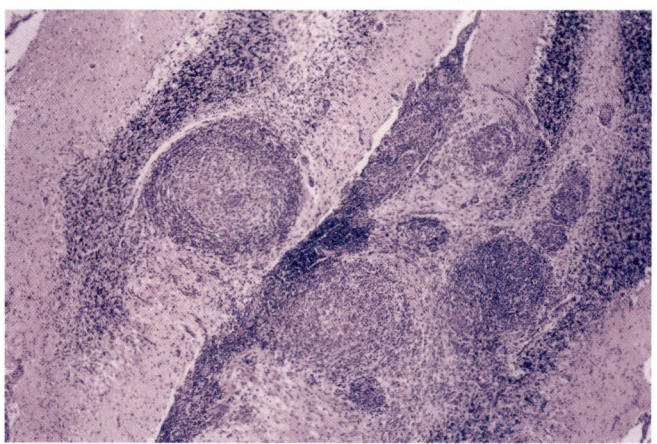

Fig. 17.26. Medium-power micrograph of noncaseating sarcoid granulomas. (H&E.)

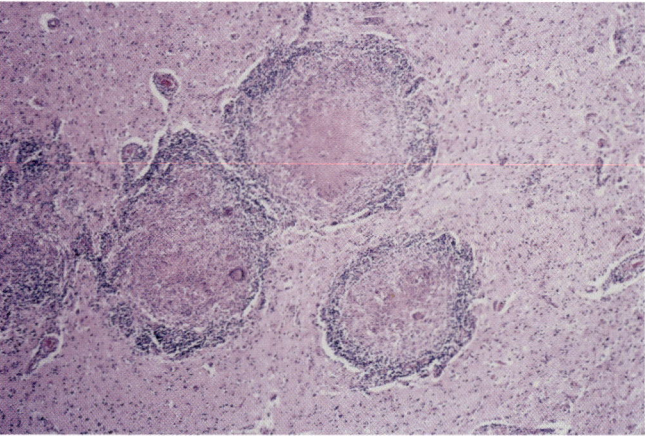

Fig. 17.27. Medium-power micrograph of caseating granulomas from a case of tuberculosis, for comparison with Fig. 17.26. (H&E.)

may be required depending on factors such as the presence or absence of marine animal predation, tissue injury due to impaction against coastline rocks, and so on. Suggested protocols are available through the abovementioned agencies as well as the Divers Alert Network and workshops at local and national forensic meetings. Medical aspects of scuba diving are also available in most forensic pathology texts and other sources.[82] Neuropathologic consultation is a routine part of the protocol in our office, with findings that show considerable case-to-case diversity. Some scuba deaths are related to sudden cardiac or intracranial catastrophe, and others are consequent to drowning or to barotrauma. In delayed deaths from drowning or barotrauma, the findings vary from those of acute hypoxic-ischemic injury or brain death to consequences of disseminated intravascular coagulation,[130] to subacute and chronic sequelae of hypoxic-ischemic injury, depending on survival time.

In rapid barotrauma deaths, one may encounter pneumocephalus[9] and air bubbles in arteries and capillaries (Case 17.12, Figs. 17.28–17.31). It is necessary, in the latter case, to exclude air bubbles limited to veins, since air bubbles in cerebral veins do not have the same significance. Venous air bubbles are a common artifact accompanying brain removal techniques, and with delayed death to autopsy intervals it is important to exclude vessel or parenchymal clear spaces due to postmortem overgrowth of gas-forming organisms. One also may see vessel-centered CNS petechiae (present in cases of less than 3 hours' survival encountered in our office), and with longer survival, infarction may appear.

Chronic Alcoholism

Neuropathologic conditions that are associated with, but not exclusive to, chronic alcoholics include sequelae of head trauma (see Chap. 6), central pontine myelinolysis, and Marchiafava-Bignami disease. Alcoholic peripheral neuropathy cases have not been seen in consultation; the cause of manner of death has not been directly attributed to these conditions in cases seen at our office. As noted by others, however, alcohol-related autonomic neuropathy is a theoretical basis for cardiac dysfunction in some cases of sudden expected death in chronic alcoholics.[61]

Case 17.12. (Figs. 17.28–17.31)

A 55-year-old woman had been scuba diving from a dive boat, and, when found missing at a routine roll call of divers, a search found her unresponsive at 200-feet depth without a regulator in her mouth. Resuscitative efforts included a hyperbaric chamber, but were unsuccessful and death was pronounced about 3 hours later. Neuropathologic abnormalities were limited to numerous intravascular vacuoles that often distended capillary-sized and larger vessels, and sparse petechiae.

Fig. 17.28. Transverse section of cerebellum and pons, demonstrating petechiae concentrated in basis pontis and largely sparing tegmentum.

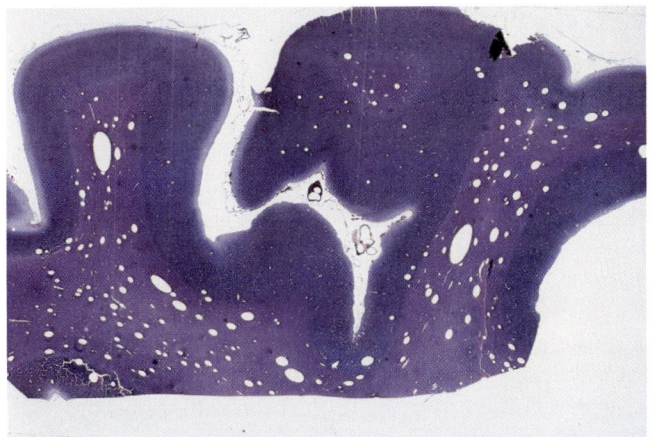

Fig. 17.29. Low-power micrograph of purposely overstained thick section to emphasize vessel-centered air bubbles. (Luxol fast blue/cresyl violet.)

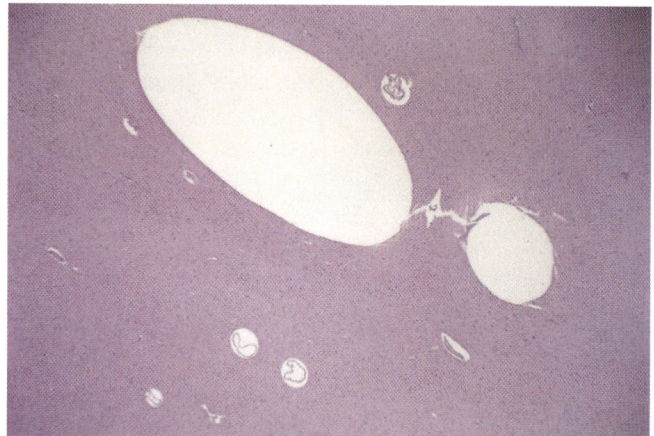

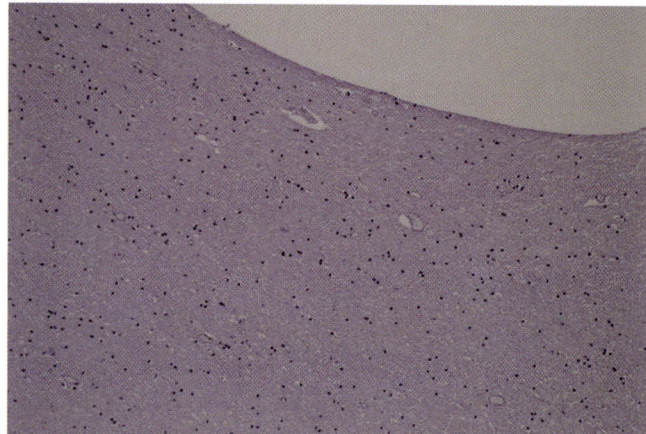

Fig. 17.30–17.31. Low-power (Fig. 17.30) and medium-power (Fig. 17.31) micrographs demonstrate vessel-centered clear vacuoles in scuba death. (H&E.)

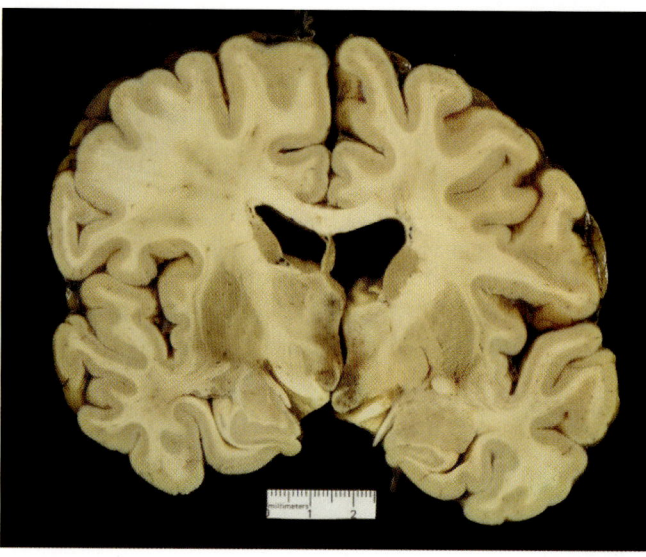

Fig. 17.32. Coronal section of brain in acute Wernicke's encephalopathy shows petechiae in walls of third ventricle and in mammillary bodies.

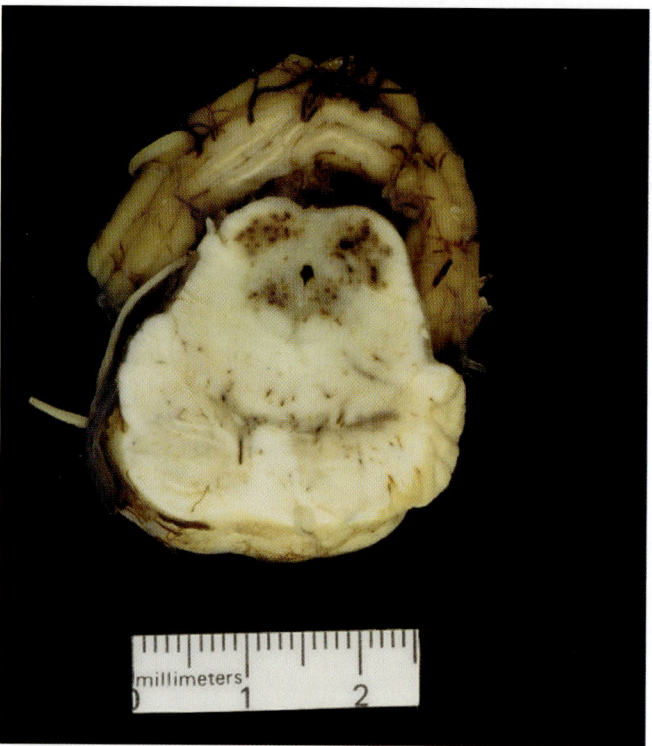

Fig. 17.33. Transverse section of pontomesencephalic junction in acute Wernicke's encephalopathy, with petechiae in periaqueductal region and inferior colliculi.

Wernicke's encephalopathy is not exclusive to chronic alcoholics, but is included here due to its frequent occurrence in the forensic case population and the fact that its acute or chronic CNS manifestations continue to generate questions, such as "Why are the mammillary bodies small and brownish in this case?" This is particularly likely in cases without a documented medical history available.

In acute Wernicke's encephalopathy, gross abnormalities may or may not be present (Figs. 17.32–17.33), but microscopic findings can include edema, petechiae, endothelial cell hypertrophy or hyperplasia, vascular proliferation, reactive astrocytes, loss of myelin and axons, and even necrosis in the mammillary bodies, but with an apparently incongruous degree of mammillary body neuronal perikaryon sparing (Figs. 17.34–17.36). Indeed, dense crowding of hypertrophied neurons of mammillary body neurons may occur.[110]

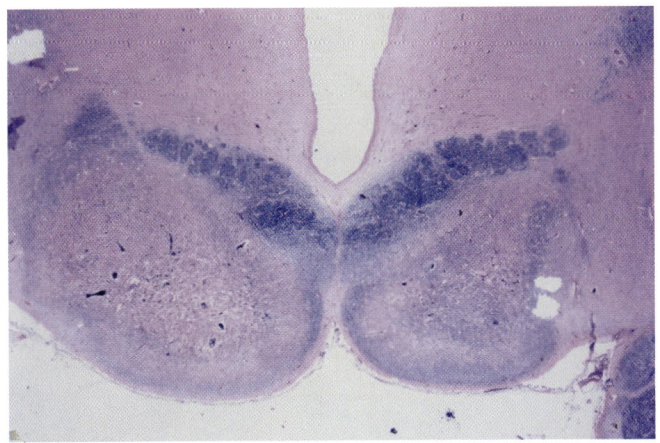

Fig. 17.34. Low-power micrograph of mammillary bodies in acute Wernicke's encephalopathy showing some vascular congestion, edema, and increased cellularity. (Luxol fast blue/cresyl violet.)

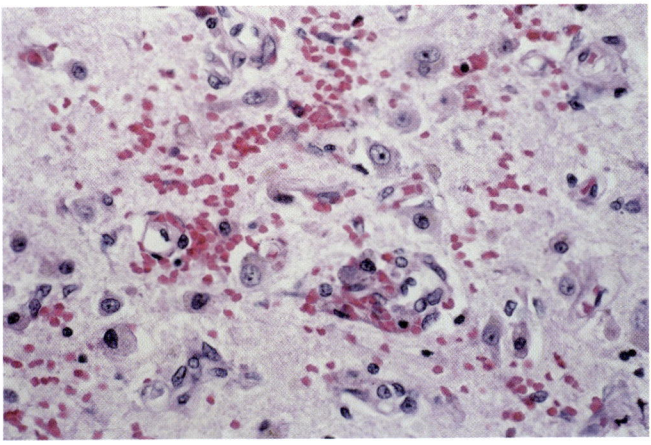

Fig. 17.35. Medium-power micrograph of mammillary body in acute Wernicke's encephalopathy, with capillary endothelial hypertrophy and hyperplasia, macrophages, interstitial hemorrhage, and intact neurons. (H&E.)

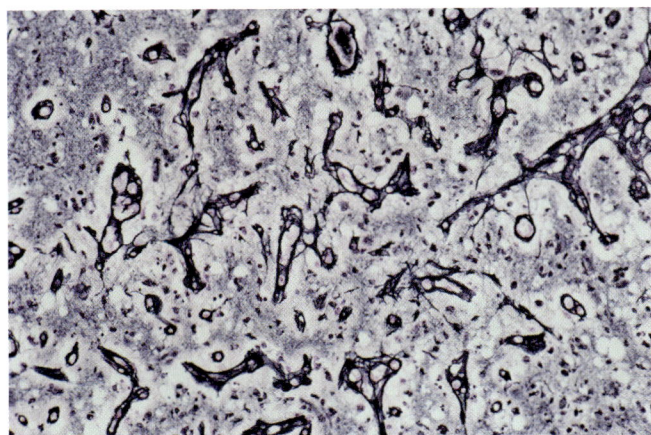

Fig. 17.36. Medium-power micrograph. A marked increase in capillary prominence in mammillary bodies is a useful marker for Wernicke's encephalopathy. Keeping a reticulin-stained section of normal mammillary body for comparison is helpful in questionable cases. (Reticulin.)

As survival time increases beyond a few days, one sees more conspicuous astrocytosis, microcystic change, variable neuronal depopulation, and hemosiderin-laden macrophages. The endpoint seen on gross examination is mammillary bodies that are atrophic and have a tan to brown discoloration. Although mammillary bodies demonstrate the most conspicuous changes, similar acute and chronic microscopic changes can occur in the walls of the third ventricle, dorsomedial nucleus of the thalamus, periaqueductal region, and floor of the fourth ventricle. The reader is referred to standard textbooks for additional variations in detail seen in some cases.[61] As Harper and Butterworth[61] note, combined acute and chronic changes may occur in individuals who have recovered from previous episodes. Thiamine deficiency is considered a key, although not necessarily the only, nutritional deficiency leading to the development of Wernicke's encephalopathy, and is also instrumental in the development of alcoholic circumscribed cerebellar degeneration and peripheral neuropathy.

A reticulin stain can be very useful in establishing the presence of the characteristic mammillary body vascular proliferation that occurs in Wernicke's encephalopathy, since it not only develops within a few days in acute cases but persists in chronic cases, and can be a convincing diagnostic feature in equivocal cases (Fig. 17.36).

In alcoholic circumscribed cerebellar cortical degeneration, the anterior vermis and anterior lobe bear the brunt of the injury, and suspicion of this condition is one of the few indications for midsagittal (rather than transverse) sectioning of the cerebellum at our office. The gross examination is quite characteristic (Fig. 17.37), and histologically one sees Purkinje cell loss and Bergmann's gliosis in involved areas (Case 17.13, Figs. 17.38–17.41). Although quantitative studies have demonstrated that

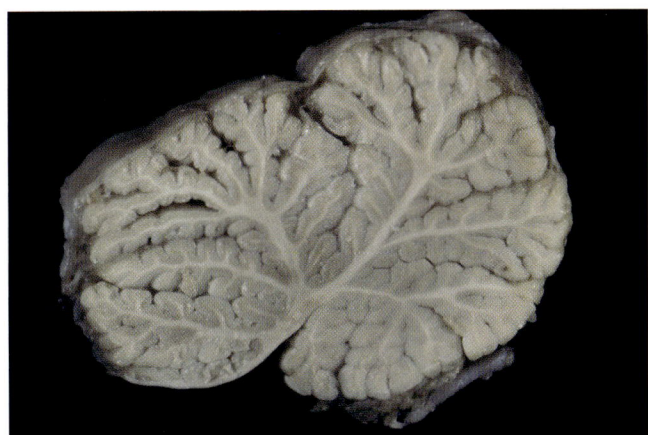

Fig. 17.37. Midsagittal section of cerebellum demonstrates anterior vermis atrophy in a chronic alcoholic, characterized by increased width of sulci between folia, most obvious in upper left portion of section.

Case 17.13. (Figs. 17.38–17.41)

A 33-year-old man was wheelchair-bound for 2 years for unstated cause. Ten years before, he was struck on the head with a baseball bat, and he went to the hospital but was released the same day. He later developed a seizure disorder, the details of which were unknown, and later became wheelchair-bound. He was unexpectedly found dead on his bedroom floor 10 hours after last seen apparently well. Although no history of alcoholism was offered, several empty beer cans were found on his nightstand and several more full cans were present in a box nearby. Autopsy revealed no traumatic brain injury or sequelae of chronic seizure disorder. Atrophy of anterior vermis of cerebellum was noted. It is likely that leg and truncal ataxia led to use of wheelchair.

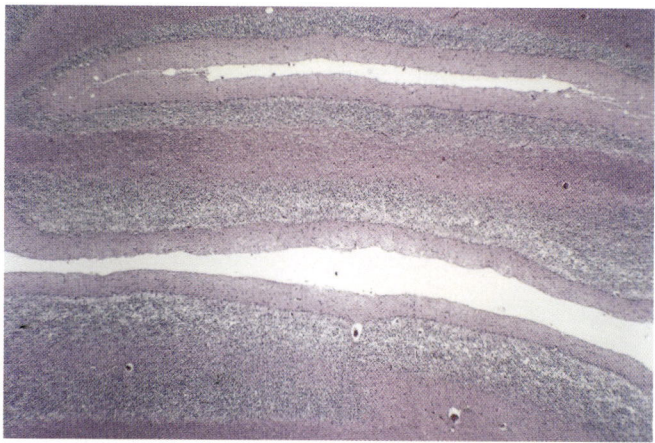

Fig. 17.40–17.41. Low-power (Fig. 17.40; *top*) and medium-power (Fig. 17.41; *bottom*) micrographs of anterior vermis (at same magnification as Figs. 17.38 and 17.39, respectively) show severe folial atrophy with total loss of Purkinje cells and marked rarefaction of granule cell layer. (H&E.)

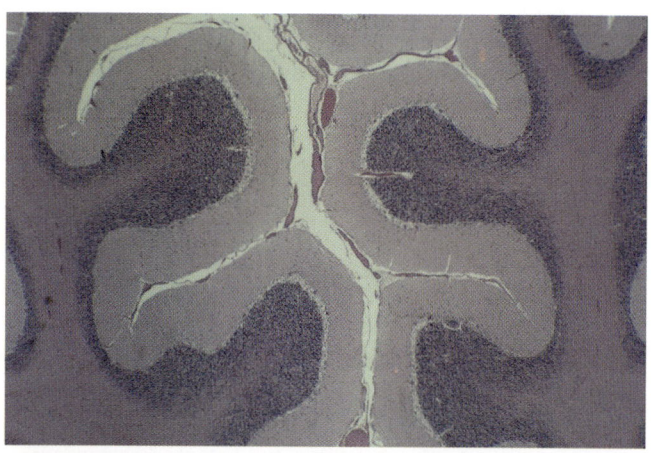

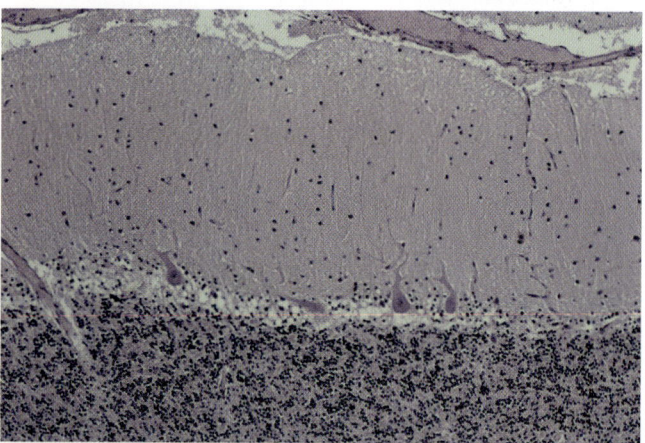

Figs. 17.38–17.39. Low-power (Fig. 17.38; *top*) medium-power (Fig. 17.39; *bottom*) micrographs of inferior cerebellum in this case reveals normal histology. (H&E.)

Purkinje cell dropout is more widespread in the cerebellum, it is too subtle to be appreciated apart from the anterior vermis in routine study of cases uncomplicated by, for example, prior trauma and/or hypoxic-ischemic episodes.

References

1. Akbar M, Stippich C, Aschoff A. Author's reply (letter to editor). N Engl J Med 2006;354:532.
2. American Academy of Pediatrics, Section on Radiology: Diagnostic imaging of child abuse. Pediatrics 2000;105:1345–1348.
3. American College of Radiology. ACR practice guideline for skeletal surveys in children. 2002. Available at: http://www.acr.org.
4. Anderson C, Duggan C, Kealy WF. Extramedullary haemopoiesis in the central nervous system: An unusual cause of epilepsy (letter to editor). J Neurol Neurosurg Psychiatry 1987;50:640–641.

5. Andrews JM, Gutstadt JP, Itabashi HH, et al. Central nervous system consequences of an unusual body disposal strategy: Case report and brief experimental investigation. J Forensic Sci 2003; 48:1153–1157.
6. Anslow P. Radiological assessment of the child with cerebral palsy and its medicolegal implications. In Squier W (ed): Acquired Damage to the Developing Brain: Timing and Causation. London: Arnold, 2002, pp 193–208.
7. Badon SJ, Ansell J, Smith TW, et al. Diabetes insipidus caused by extramedullary hematopoiesis. Am J Clin Pathol 1985;83:509–512.
8. Baselt RC. Disposition of Toxic Drugs and Chemicals in Man, ed 6. Foster City, CA: Biomedical Publications, 2002.
9. Bell MD. Drowning. In Dolinak D, Matshes EW, Lew EO (eds): Forensic Pathology: Principles and Practice. Amsterdam: Elsevier Academic Press, 2005, pp 227–237.
10. Berryman HE, Symes SA. Recognizing gunshot and blunt cranial trauma through fracture interpretation. In: Reichs KJ (ed): Forensic Osteology: Advances in the Identification of Human Remains, ed 2. Springfield, IL: Charles C. Thomas, 1998, pp 333–352.
11. Bhootra BK. An unusual penetrating head wound by a yard broom and its medicolegal aspects. J Forensic Sci 1985;30:567–571.
12. Blain PG, Harris JB (eds). Medical Neurotoxicology: Occupational and Environmental Causes of Neurological Dysfunction. London: Arnold, 1999.
13. Boglioli LR, Taff ML, Turkel SJ, et al. Unusual infant death. Dog attack or postmortem mutilation after child abuse? Am J Forensic Med Pathol 2000;21:389–394.
14. Braffman BH, Zimmerman RA, Trojanowski JQ, et al. Brain MR: Pathologic correlation with gross and histopathology. 1. Lacunar infarction and Virchow-Robin spaces. AJNR Am J Neuroradiol 1998;9:621–628.
15. Braffman BH, Zimmerman RA, Trojanowski JQ, et al. Brain MR: Pathologic correlation with gross and histopathology. 2. Hyperintense white-matter foci in the elderly. AJNR Am J Neuroradiol 1988;9:629–636.
16. Brazis PW, Masdeu JC, Biller J. Localization in Clinical Neurology, ed 3. Boston: Little, Brown, 1996.
17. Brogdon BG. Forensic Radiology. Boca Raton, FL: CRC Press, 1998.
18. Brogdon BG, Vogel H, McDowell JD. A Radiologic Atlas of Abuse, Torture, Terrorism, and Inflicted Trauma. Boca Raton, FL: CRC Press, 2003.
19. Brothwell D, Gill-Robinson H. Taphonomic and forensic aspects of bog bodies. In Haglund WD, Sorg MH (eds): Advances in Forensic Taphonomy. Method, Theory, and Archaeological Perspectives. Boca Raton, FL: CRC Press, 2002, pp 119–132.
20. Brown JA, Gomez-Leon G. Subdural hemorrhage secondary to extramedullary hematopoiesis in postpolycythemic myeloid metaplasia. Neurosurgery 1984;14:588–591.
21. Brust JCM. Neurological Aspects of Substance Abuse. Boston: Butterworth-Heinemann, 1993.
22. Bux RC, McDowell JD. Death due to attack from chow dog. Am J Forensic Med Pathol 1992;13:305–308.
23. Cameron WR, Ronnert M, Brun A. Extramedullary hematopoiesis of CNS in postpolycythemic myeloid metaplasia (letter to editor). N Engl J Med 1981;305:765.
24. Chung S, Schamban N, Wypij D, et al. Skull radiograph interpretation of children younger than two years: How good are pediatric emergency physicians? Ann Emerg Med 2004;43:718–722.
25. Chute DJ, Smialek JE. Pseudo-subarachnoid hemorrhage of the head diagnosed by computer axial tomography: A postmortem study of ten medical examiner cases. J Forensic Sci 2002;47:360–365.
26. Crist TAJ, Washburn A, Park H, et al. Cranial bone displacement as a taphonomic process in potential child abuse cases. In Haglund WD, Sorg MH (eds): Forensic Taphonomy: The Postmortem Fate of Human Remains. Boca Raton, FL: CRC Press, 1997, pp 319–336.
27. Dart RC (ed). Medical Toxicology, ed 3. Philadelphia: Lippincott Williams & Wilkins, 2004.
28. Davis NL, Kahana T, Hiss J. Souvenir knife: A retained transcranial knife blade. Am J Forensic Med Pathol 2004;25:259–261.
29. Davis RL, Robertson DM (eds). Textbook of Neuropathology, ed 3. Baltimore: Williams & Wilkins, 1997.
30. Delgado-Escueta AV, Janz D, Beck-Mannagetta G (suppl eds). Pregnancy and teratogenesis in epilepsy. Neurology 1992;42 (Suppl 5):1–160.
31. DeMunnynck K, Van de Voorde W. Forensic approach of fatal dog attacks: A case report and literature review. Int J Legal Med 2002;116:295–300.
32. deVilliers JC. Stab wounds of the brain and skull. In Vinken PJ, Bruyn GW, Braakman R (eds): Handbook of Clinical Neurology. Injuries of the Brain and Skull. Part I. Amsterdam: North-Holland, 1975;23:477–503.
33. Dias MS. Radiographic evaluation of inflicted childhood neurotrauma—Response. In Reece RM, Nicholson CE (eds): Inflicted Childhood Neurotrauma. Elk Grove Village, IL: American Academy of Pediatrics, 2003, pp 93–121.
34. Dias MS, Backstrom J, Falk M, Li V. Serial radiography in the infant shaken impact syndrome. Pediatr Neurosurg 1998;29:77–85.
35. DiMaio VJ, DiMaio D. Forensic Pathology, ed 2. Boca Raton, FL: CRC Press, 2001.
36. Dobbs MR, Wicklund MP (guest eds). Toxic and environmental neurology. Neurol Clinic 2005;23(No. 2):307–654.
37. Dolinak D, Matshes E. Asphyxia. In Dolinak D, Matshes EW, Lew EO: Forensic Pathology: Principles and Practice. Amsterdam: Elsevier, 2005, pp 201–225.
38. Dorman DC, Brenneman KA, Bolon B. Nervous system. In Haschek WM, Rousseaux CG, Wallig MA (eds): Handbook of Toxicologic Pathology, ed 2. San Diego: Academic Press, 2002;2:509–538.
39. Downs JCU. Judicial execution in Alabama (abstract). Proc Am Acad Forensic Sci 2001;7:224–225.
40. Duckett S (ed). Pediatric Neuropathology. Baltimore: Williams & Wilkins, 1995.
41. Eklektos N, Dayal MR, Manger PR. A forensic case study of a naturally mummified brain from the bushveld of South Africa. J Forensic Sci 2006;51:498–503.
42. Ellenhorn MJ (ed). Ellenhorn's Medical Toxicology: Diagnosis and Treatment of Human Poisoning, ed 2. Baltimore: Williams & Wilkins, 1997.
43. Ellison D, Love S, Chimelli L, et al. Neuropathology: A Reference Text of CNS Pathology. London: Mosby, 1998.
44. Fajardo LF, Berthrong M, Anderson RE. Radiation Pathology. Oxford: Oxford University Press, 2001.
45. Fekete JF, Fox AD. Successful suicide by self-inflicted multiple stab wounds of the skull, abdomen and chest. J Forensic Sci 1980;25:634–637.
46. Feldman KW, Brewer DK, Shaw DW. Evolution of the cranial computed tomography scan in child abuse. Child Abuse Negl 1995;19:307–314.
47. Ferrari LA, Arado MG, Nardo CA, Giannuzzi L. Post-mortem analysis of formic acid disposition in acute methanol intoxication. Forensic Sci Int 2003;133:152–158.
48. Filley CM, Halliday W, Kleinschmidt-DeMasters BK. The effects of toluene on the central nervous system. J Neuropathol Exp Neurol 2004;63:1–12.
49. Fucharoen S, Tunthanavatana C, Sonakul D, Wasi P. Intracranial extramedullary hematopoiesis in B°-thalassemia/hemoglobin E disease. Am J Hematol 1981;10:75–78.

50. Geddes J. Pediatric head injury. In Golden JA, Harding BN (eds): Pathology and Genetics: Developmental Neuropathology. Basel: ISN Neuropath Press, 2004, pp 184–191.
51. Gill JR. Practical toxicology for the forensic pathologist. In Tsokos M (ed): Forensic Pathology Reviews. Totowa, NJ: Humana Press, 2005;2:243–269.
52. Golden NH, Ashtari M, Kohn MR, et al. Reversibility of cerebral ventricular enlargement in anorexia nervosa, demonstrated by quantitative magnetic resonance imaging. J Pediatr 1996;128:296–301.
53. Graham DI, Lantos PL (eds). Greenfield's Neuropathology, ed 7. London: Arnold, 2002.
54. Griffiths PD, Variend D, Evans M, et al. Postmortem MR imaging of the fetal and stillborn central nervous system. AJNR Am J Neuroradiol 2003;24:22–27.
55. Gullapalli D, Phillips LH II. Neurologic manifestations of sarcoidosis. Neurol Clin 2002;20:59–83.
56. Haglund WD, Reay DT. Problems of recovering partial human remains at different times and locations: Concerns for death investigators. J Forensic Sci 1993;38:69–80.
57. Haglund WD, Sorg MH (eds). Forensic Taphonomy: The Postmortem Fate of Human Remains. Boca Raton, FL: CRC Press, 1997.
58. Haglund WD, Sorg MH (eds). Advances in Forensic Taphonomy: Method, Theory, and Archaeological Perspectives. Boca Raton, FL: CRC Press, 2002.
59. Halsted MJ, Kumar H, Paquin JJ, et al. Diagnostic errors by radiologic residents in interpreting pediatric radiographs in an emergency setting. Pediatr Radiol 2004;34:331–336.
60. Hardman JG, Limbird LE (eds). Goodman & Gilman's The Pharmacological Basis of Therapeutics. New York: McGraw-Hill, 1996.
61. Harper C, Butterworth R. Nutritional and metabolic disorders. In Graham DI, Lantos PL (eds): Greenfield's Neuropathology, ed 7. London: Arnold, 2002;1:607–652.
62. Hart BL, Dudley MH, Zumwalt RE. Postmortem cranial MRI and autopsy correlation in suspected child abuse. Am J Forensic Med Pathol 1996;17:217–224.
63. Hartshorne NJ, Reay DT. Judicial hanging (letter to editor). Am J Forensic Med Pathol 1995;16:87.
64. Hassin GB. Changes in the brain in legal execution. Arch Neurol Psychiatry 1933;30:1046–1060.
65. Hassin GB. Cerebral contusion in accidental electrocution: Pathologic study of a case. J Neuropathol Exp Neurol 1950;9:311–321.
66. Hayakawa M, Yamamoto S, Motani H, et al. Does imaging technology overcome problems of conventional postmortem examination? A trial of computed tomography imaging for postmortem examination. Int J Legal Med 2006;120:24–26.
67. Henderson JM, Thach J, Phillips M, et al. Permanent neurological deficit related to magnetic resonance imaging in a patient with implanted deep brain stimulation electrodes for Parkinson's disease: Case report. Neurosurgery 2005;57:1063.
68. Hendey GW, Wolfson AB, Mower WR, Hoffman JR. Spinal cord injury without radiographic abnormality: Results of the National Emergency X-Radiography Utilization Study in blunt cervical trauma. J Trauma 2002;53:1–4.
69. Hess MW, Klima G, Pfaller K, et al. Histological investigations on the Tyrolean ice man. Am J Phys Anthropol 1998;106:521–532.
70. Hogge JP, Messmer JM, Fierro MF. Positive identification by postsurgical defects from unilateral lambdoid synostectomy: A case report. J Forensic Sci 1995;40:688–691.
71. Hönigschnabl S, Schaden E, Stichenwirth M, et al. Discovery of decomposed and mummified corpses in the domestic setting—A marker of social isolation? J Forensic Sci 2002;47:837–842.
72. Jackowski C, Sonnenschein M, Thali MJ, et al. Virtopsy: Postmortem minimally invasive angiography using cross section techniques—Implementation and preliminary results. J Forensic Sci 2005;50:1175–1186.
73. James R, Nasmyth-Jones R. The occurrence of cervical fractures in victims of judicial hanging. Forensic Sci Int 1992;54:81–91.
74. Jones GRN. Judicial electrocution and the prison doctor. Lancet 1990;335:713–714.
75. Kahana T, Birkby WH, Goldin L, Hiss J. Estimation of age in adolescents—The basilar synchondrosis. J Forensic Sci 2003;48:504–508.
76. Kahana T, Hiss J. Forensic radiology. Forensic Pathol Rev 2005;3:443–460.
77. Karch SB. The Pathology of Drug Abuse, ed 2. Boca Raton, FL: CRC Press, 1996.
78. Kelly DF. Alcohol and head injury: An issue revisited. J Neurotrauma 1995;12:883–890.
79. Kelly DF, Lee SM, Pinanong PA, Hovda DA. Paradoxical effects of acute ethanolism in experimental brain injury. J Neurosurg 1997;86:876–882.
80. Kemp AM. Investigating subdural haemorrhage in infants. Arch Dis Child 2002;86:98–102.
81. Khan A, Leventhal RM. Medical aspects of capital punishment executions. J Forensic Sci 2002;47:847–851.
82. Kizer KW. Scuba diving and dysbarism. In Auerbach PS (ed): Wilderness Medicine: Management of Wilderness and Environmental Emergencies, ed 3. St Louis: Mosby, 1995, pp 1176–1208.
83. Kleinman PK. Diagnostic Imaging of Child Abuse, ed 2. St Louis: Mosby, 1998.
84. Kohr RM. Autopsies on executed federal prisoners (letter to editor). J Forensic Sci 2003;48:1203.
85. Kriegstein AR, Shungu DC, Millar WS, et al. Leukoencephalopathy and raised brain lactate from heroin vapor inhalation ("chasing the dragon"). Neurology 1999;53:1765–1773.
86. Kulkarni AV, Bhandari M, Stiver S, Reddy K. Delayed presentation of spinal stab wound: Case report and review of the literature. J Emerg Med 2000;18:209–213.
87. Levy V, Rao VJ. Survival time in gunshot and stab wound victims. Am J Forensic Med Pathol 1988;9:215–217.
88. Li M, Hamilton W. Review of autopsy findings in judicial electrocutions. Am J Forensic Med Pathol 2005;26:261–267.
89. London MR, Krolikowski FJ, Davis JH. Burials at sea. In Haglund WD, Sorg MH (eds): Forensic Taphonomy: The Postmortem Fate of Human Remains. Boca Raton, FL: CRC Press, 1997, pp 615–622.
90. Lund RE, Aldridge NH. Computed tomography of intracranial extramedullary hematopoiesis. J Comput Assist Tomogr 1984;8:788–790.
91. Lunetta P, Öhberg A, Sajantila A. Suicide by intracerebellar ballpoint pen. Am J Forensic Med Pathol 2002;23:334–337.
92. Lurz F, Morey MK, Gaugler WM. Dueling outcomes: Sharp force injury and continued physical activity (abstract). Proc Am Acad Forensic Sci 1998;4:167.
93. Madea B. Death as result of starvation: Diagnostic criteria. Forensic Pathol Rev 2005;2:3–23.
94. Madea B, Schmidt PH, Lignitz E, Padosch SA. Skull injuries caused by blows with glass bottles. Forensic Pathol Rev 2005;2:27–41.
95. Mauge C, Lilienfeld S. Magnetic resonance imaging and cerebrospinal fluid valves (letter to editor). N Engl J Med 2006;354:531–532.
96. McGraw EP, Pless JE, Pennington DJ, White SJ. Postmortem radiography after unexpected death in neonates, infants, and children: Should imaging be routine? AJR Am J Radiol 2002;178:1517–1521.
97. McIntyre IM, Hamm CE, Sherrard JL, et al. The analysis of an intracerebral hematoma for drugs of abuse. J Forensic Sci 2003;48:680–682.
98. Milovanovic AV, DiMaio VJM. Death due to concussion and alcohol. Am J Forensic Med Pathol 1999;20:6–9.

99. Minns RA. Subdural haemorrhages, haematomas, and effusions in infancy. Arch Dis Child 2005;90:883–884.
100. Moriya F, Hashimoto Y. Medicolegal implications of drugs and chemicals detected in intracranial hematomas. J Forensic Sci 1998;43:980–984.
101. Mower WR, Hendey GW, Hoffman JR, Wolfson AB. SCIWORA in childhood—The author's reply (letter to editor). J Trauma 2002;53:1198–1199.
102. Murmann DC, Brumit PC, Schrader BA. A comparison of animal jaws and bite mark patterns. J Forensic Sci 2006;51:846–860.
103. Murphy WA Jr, zur Nedden D, Gostner P, et al. The iceman: Discovery and imaging. Radiology 2003;226:614–629.
104. Nagao M, Maeno Y, Koyama H, et al. Estimation of caloric deficit in a fatal case of starvation resulting from child neglect. J Forensic Sci 2004;49:1073–1076.
105. Nokes LDM, Roberts A, James DS. Biomechanics of judicial hanging: A case report. Med Sci Law 1999;39:61–64.
106. Oehmichen M, Gehl H-B, Meissner C, et al. Forensic pathological aspects of postmortem imaging of gunshot injury to the head: Documentation and biometric data. Acta Neuropathol 2003;105:570–580.
107. Pang D, Wilberger Jr JE. Spinal cord injury without radiographic abnormalities in children. J Neurosurg 1982;57:114–129.
108. Panse F. Electrical trauma. In Vinken PJ, Bruyn GW, Braakman R (eds): Handbook of Clinical Neurology. Injuries of the Brain and Skull: Part I. Amsterdam: North Holland, 1975;23:683–729.
109. Peiffer J, Danner E, Schmidt PF. Oxalate-induced encephalitic reactions to polyol-containing infusions during intensive care. Clin Neuropathol 1984;3:76–87.
110. Pena CE. Wernicke's encephalopathy: Report of seven cases with severe nerve cell changes in the mammillary bodies. Am J Clin Pathol 1969;51:603–609.
111. Petty PG, Parkin G. Electrical injury to the central nervous system. Neurosurgery 1986;19:282–284.
112. Pfefferbaum A, Sullivan EV, Adalsteinsson E, Garrick T, Harper C. Postmortem MR imaging of formalin-fixed human brain. NeuroImage 2004;21:1585–1595.
113. Polliack A, Rosenmann E. Extramedullary hematopoietic tumors of the cranial dura mater. Acta Haematol 1969;41:43–48.
114. Pope EJ, Smith OC. Identification of traumatic injury in burned cranial bone: An experimental approach. J Forensic Sci 2004;49:431–440.
115. Prahlow JA, Ross KF, Lene WJW, Kirby DB. Accidental sharp force injury fatalities. Am J Forensic Med Pathol 2001;22:358–366.
116. Radanov S, Stoev S, Davidov M, et al. A unique case of naturally occurring mummification of human brain tissue. Int J Legal Med 1992;105:173–175.
117. Ramsay DA, Shkrum MJ. Homicidal blunt head trauma, diffuse axonal injury, alcoholic intoxication, and cardiorespiratory arrest: A case report of a forensic syndrome of acute brainstem dysfunction. Am J Forensic Med Pathol 1995;16:107–114.
118. Reay DT, Cohen W, Ames S. Injuries produced by judicial hanging: A case report. Am J Forensic Med Pathol 1994;15:183–186.
119. Reichs KJ (ed). Forensic Osteology: Advances in the Identification of Human Remains, ed 2. Springfield, IL: Charles C Thomas, 1998.
120. Rezai AR, Baker KB, Tkach JA, et al. Is magnetic resonance imaging safe for patients with neurostimulation systems used for deep brain stimulation? Neurosurgery 2005;57:1056–1062.
121. Riggs JE (ed). Neurologic manifestations of systemic disease. Neurol Clin 2002;20:1–289.
122. Roberts ISD, Benbow EW, Bisset R, et al. Accuracy of magnetic resonance imaging in determining cause of sudden death in adults: Comparison with conventional autopsy. Histopathology 2003;42:424–430.
123. Robles H. Formic acid. In Weyler P (ed): Encyclopedia of Toxicology. San Diego: Academic Press, 1998;2:35–36.
124. Rogers TL, Allard TT. Expert testimony and positive identification of human remains through cranial suture patterns. J Forensic Sci 2004;49:203 207.
125. Rorke LB. Lesions induced by toxins. In Golden JA, Harding BN (eds): Pathology and Genetics: Developmental Neuropathology. Basel: ISN Neuropath Press, 2004, pp 209–212.
126. Rutherford MA. Magnetic resonance imaging of injury to the immature brain. In Squier W (ed): Acquired Damage to the Developing Brain: Timing and Causation. London: Arnold, 2002, pp 166–192.
127. Rutman JY, Meidinger R, Keith JI. Unusual radiologic and neurologic findings in a case of myelofibrosis with extramedullary hematopoiesis. Neurology 1972;22:567–570.
128. Sanz P, Reig R. Clinical and pathological findings in fatal plant oxalosis: A review. Am J Forensic Med Pathol 1992;13:342–345.
129. Sargent S, Kennedy JG, Kaplan JA. "Hyperacute" subdural hematoma: CT mimic of recurrent episodes of bleeding in the setting of child abuse. J Forensic Sci 1996;41:314–316.
130. Saukko P, Knight B. Knight's Forensic Pathology, ed 3. London: Arnold, 2004.
131. Schneider RC, Livingston KE, Cave AJE, Hamilton G. "Hangman's fracture" of the cervical spine. J Neurosurg 1965;22:141–154.
132. Sharma AN, Nelson LS, Hoffman RS. Cerebrospinal fluid analysis in fatal thallium poisoning: Evidence for delayed distribution into the central nervous system. Am J Forensic Med Pathol 2004;25:156–158.
133. Simmons GT. Evaluation of homicidal starvation of a child. Forensic Pathology Check Sample No. FP 04-9 (FP-300). Chicago: American Society of Clinical Pathology. 2004;46(9):107–119.
134. Sitton JE, Reimund EL. Extramedullary hematopoiesis of the cranial dura and anhidrotic ectodermal dysplasia. Neuropediatrics 1992;23:108–110.
135. Skullerud K, Andersen SN, Lundevall J. Cerebral lesions and causes of death in male alcoholics. Int J Leg Med 1991;104:209–213.
136. Spence MW, Shkrum J, Ariss A, Regan J. Craniocervical injuries in judicial hangings: An anthropologic analysis of six cases. Am J Forensic Med Pathol 1999;20:309–322.
137. Spencer PS, Schaumburg HH. Experimental and Clinical Neurotoxicology, ed 2. New York: Oxford University Press, 2000.
138. Spitz WU, Petty CS, Fisher RS. Physical activity until collapse following fatal injury by firearms and sharp pointed weapons. J Forensic Sci 1961;6:290–300.
139. Spivack B. SCIWORA in childhood (letter to editor). J Trauma 2002;53:1198.
140. Taybi H, Lachman RS. Radiology of Syndromes, Metabolic Disorders, and Skeletal Dysplasias, ed 4. St Louis: Mosby, 1996.
141. Thali MJ, Braun M, Buck U, et al. VIRTOPSY—scientific documentation, reconstruction and animation in forensics: Individual and real 3-D data-based geometric approach including optical body/object surface and radiological CT/MRI scanning. J Forensic Sci 2005;50:428–442.
142. Thali MJ, Schweitzer W, Yen K, et al. New horizons in forensic radiology. The 60-second "digital autopsy"-full-body examination of a gunshot victim by multislice computed tomography. Am J Forensic Med Pathol 2003;24:22–27.
143. Thoresen SØ, Rognum TO. Survival time and acting capability after fatal injury by sharp weapons. Forensic Sci Int 1986;31:181–187.
144. Ubelaker DH, Adams BJ. Differentiation of perimortem and postmortem trauma using taphonomic indicators. J Forensic Sci 1995;40:509–512.
145. Ubelaker DH, Berryman HE, Sutton TP, Ray CE. Differentiation of hydrocephalic calf and human calvariae. J Forensic Sci 1991;36:801–812.

146. van Wezel-Meijler G, van der Knaap MS, Sie LTL, et al. Magnetic resonance imaging of the brain in premature infants during the neonatal period: Normal phenomena and reflection of mild ultrasound abnormalities. Neuropediatrics 1998;29:89–96.
147. Vinchon M, Noizet O, Defoort-Dhellemmes S, et al. Infantile subdural hematomas due to traffic accidents. Pediatr Neurosurg 2002;37:245–253.
148. Wallace SK, Cohen WA, Stern EJ, Reay DT. Judicial hanging: Postmortem radiographic, CT, and MR imaging features with autopsy confirmation. Radiology 1994;193:263–267.
149. Yager J, Anderson AE. Anorexia nervosa. N Engl J Med 2005; 353:1481–1488.
150. Yousry TA, Major EO, Ryschkewitsch C, et al. Evaluation of patients treated with Natalizumab for progressive multifocal leukoencephalopathy. N Engl J Med 2006;354:924–933.
151. Zimmerman RA. Radiographic evaluation of inflicted childhood neurotrauma. In Reece RM, Nicholson CE (eds): Inflicted Childhood Neurotrauma. Elk Grove Village, IL: American Academy of Pediatrics, 2003, pp 83–91.
152. Zink BJ, Feustel PJ. Effects of ethanol on respiratory function in traumatic brain injury. J Neurosurg 1995;82:822–828.

Case Sources for Neuropathology Consultation

Appendix A

All nonnatural in-custody deaths
All high profile cases
All child or elder abuse/neglect suspected cases
All suspected sudden infant death syndrome (SIDS) cases
All suspected pediatric homicides with head or spine trauma
All blunt head trauma cases from chronic care facilities
Question of fall versus blow to the head
Suspected homicide due to remote head or spinal trauma
Pediatric blunt head trauma, believed to be an accident
Epilepsy of traumatic or nontraumatic etiology
Rare accidents (e.g., dog bites to child's brain, and so on)
Cases with previously diagnosed or suspected neurologic disease (e.g., Alzheimer's disease, multiple sclerosis, sporadic or familial mental retardation)
History of significant psychiatric disorder (e.g., schizophrenia, paranoia)
Any unexplained neurologic signs and symptoms, including coma
Pediatric suicide with history of mental retardation or autism
Adult or child found down with no known explanation
Suicide by unusual chemical (e.g., ethylene glycol)
Subarachnoid hemorrhage, berry aneurysm not found at autopsy
Cases in which the functional significance of various brain lesions is likely to become an issue
Cases with circadian rhythm abnormality or major sleep disorder (e.g., narcolepsy)

Appendix B: The Glasgow Coma Scale

The Glasgow Coma Scale (GCS) is the most widely used clinical standard to assess level of consciousness worldwide (Table B.1).[4] Some have used it to categorize severity of head injury by GCS scores as mild (total score, 13–15), moderate (total score, 9–12), or severe (total score, 8 or less), with cases in the severe category generally requiring intubation to prevent aspiration, allow airway assistance as indicated, and so forth.[4]

Despite its widespread use, the medical examiner may find that GCS scores in some cases seem inconsistent with the subsequent clinical course and outcome. Among potential sources of such apparent discrepancies are inconsistencies in GCS scoring based on the experience of the clinician[3]; questions as to whether the GCS is less useful when first performed at the scene versus later in the emergency room[4]; questions as to its applicability in intubated patients[4]; and questions concerning its applicability in children 2 years of age or younger.[1,2] As noted earlier, however, it is currently the most widely encountered consciousness assessment scale in clinical records reviewed in this department, and there is not yet a consensus on an alternative scale for the preverbal child.

Table B.1. The Glasgow Coma Scale

Eyes Open	
Never	1
To pain	2
To verbal stimuli	3
Spontaneously	4
Best Verbal Responses	
No response	1
Incomprehensible sounds	2
Inappropriate words	3
Disoriented and converses	4
Oriented and converses	5
Best Motor Response	
No response	1
Extension (decerebrate rigidity)	2
Flexion abnormal (decorticate withdrawal)	3
Flexion withdrawal	4
Localizes pain	5
Obeys	6
Total score:	3–15

References

1. Durham SR, Clancy RR, Leuthardt E, et al. CHOP infant coma scale ("Infant Face Scale"): A novel coma scale for children less than two years of age. J Neurotrauma 2000;17:729–737.
2. Holmes JF, Palchak MJ, MacFarlane T, Kuppermann N. Performance of the pediatric Glasgow Coma Scale in children with blunt head trauma. Acad Emerg Med 2005;12:814–819.
3. Riechers RG II, Ramage A, Brown W, et al. Physician knowledge of the Glasgow Coma Scale. J Neurotrauma 2005;11:1327–1334.
4. Sternbach GL. The Glasgow Coma Scale. J Emerg Med 2000;19:67–71.

The Apgar Score

Appendix C

Table C.1. The Apgar Score*

Heart rate	
Absent	0
Below 100	1
Over 100	2
Respiratory effort	
Absent	0
Slow, irregular	1
Good, crying	2
Muscle tone	
Limp	0
Some flexion of extremities	1
Active motion	2
Response to catheter in nostril (tested after oropharynx is clear)	
No response	0
Grimace	1
Cough or sneeze	2
Color	
Blue, pale	0
Body pink, extremities blue	1
Completely pink	2
Total score:	0–10

*Tested at 1 and 5 minutes after delivery.

The Apgar Score is the most frequently used clinical tool worldwide to assess the condition of the newborn term infant, primarily from the standpoint of the infant's possible need for support of vital functions and the potential for survival (Table C.1).[2–4] A score of 7 or higher at 5 minutes after delivery suggests that the infant's condition is good to excellent, and lower scores call attention to the possible need for supportive intervention. Low scores may lead to additional scoring at 10, 15, and 20 minutes after delivery. The Apgar Score does not diagnose asphyxia or predict future neurodevelopmental deficits.[1,4] The Apgar Score can be influenced by a variety of factors such as presence of ongoing resuscitative efforts, prematurity, maternal medications, or concurrent cardiovascular or neurologic abnormalities. Such variables have led to suggestions for use of an expanded Apgar Score form, which the medical examiner may encounter in future medical record reviews.[1]

References

1. ACOG Committee Opinion. Number 333, May 2006 (Replaces Number 174, July 1996): The Apgar Score. Obstet Gynecol 2006; 107:1209–1212.
2. Casey BM, McIntire DD, Leveno KJ. The continuing value of the Apgar score for the assessment of newborn infants. N Engl J Med 2001;344:467–471.
3. Kliegman RM. The newborn infant. In Berhman RE, Kliegman RM, Arvin AM (eds): Nelson Textbook of Pediatrics, ed 15. Philadelphia: WB Saunders, 1996, pp 433–440.
4. Papile L-A. The Apgar score in the 21st century (editorial). N Engl J Med 2001;344:519–520.

Intracranial Pressure Monitoring

Appendix D

Brain swelling and increased intracranial pressure (ICP) are frequent components of the terminal state in various brain insults, including catastrophic cerebrovascular disorders, infections, mass lesions, and traumatic brain injury. Many of the case reports and illustrations in this text emphasize this association. Medical records in such cases often include data derived from ICP monitors, typically measured in mm Hg (torr). The data that follow are intended as very general guidelines only, and cited references should be reviewed for further details.

In order of decreasing accuracy of measurement, ICP can be monitored from the lateral ventricle; brain parenchyma; or subarachnoid, subdural, or epidural space. Measurement by lumbar puncture, tympanic membrane displacement, or transcranial Doppler is not considered adequate, and the lumbar route may even increase patient risk in the presence of brain swelling.[4]

ICP in normal individuals varies with age, and values given from different sources differ to some extent.[1,3] Values as suggested by Steiner and Andrews[4] are as follows:

- Adults:
  — 7–15 mm Hg supine.
  — ~ –10 mm Hg vertical (but not less than –15 mm Hg).
- Term infants: 1.5–6.0 mm Hg.
- Children: 3–7 mm Hg.
- In hydrocephalus: > 15 mm Hg is abnormal.
- In head-injured patients:
  — > 20–25 mm Hg is considered the range "above which outcome will be affected negatively."[4]
  — > 25 mm Hg is the threshold for active intervention in many care centers.
  — Consensus regarding ICP levels requiring intervention in children is less clear than in adults.[4]

The degree to which assessment of cerebral blood flow, cerebral perfusion pressure, and other measurable parameters, in conjunction with ICP, influences treatment decisions varies from institution to institution.[4] If conversion of units is needed, 1 mm Hg = ~ 13.5 mm H_2O (or cerebrospinal fluid) (extrapolated from data provided by Fishman)[2].

References

1. Czosnyka M, Pickard JD. Monitoring and interpretation of intracranial pressure. J Neurol Neurosurg Psychiatry 2004;75:813–821.
2. Fishman RA. Cerebrospinal Fluid in Diseases of the Nervous System, ed 2. Philadelphia: WB Saunders, 1992.
3. Jones PA, Andrews PJD, Easton VJ, Minns RA. Traumatic brain injury in childhood: Intensive care time series data and outcome. Br J Neurosurg 2003;17:29–39.
4. Steiner LA, Andrews PJD. Monitoring the injured brain: ICP and CBF. Br J Anaesth 2006;97:26–38.

Legal Testimony

Appendix E

Medicolegal issues have been briefly discussed elsewhere in this text (particularly in Chapters 1, 7, 8, and 16) and selected information sources cited therein. Additional selected readings are provided at the end of this appendix, and numerous additional sources of information on this subject are available (including publishers primarily oriented toward the legal profession, such as Seak, Inc., and Lawyers and Judges Publishing Co., Inc.).

The neuropathologist acting as a consultant in forensic cases may be called to testify in court or in depositions regarding a consultation case, even if the likelihood of such testimony was considered remote at the time the consultation was originally requested. For those already experienced in consultation work and/or giving testimony as an expert witness, such an unanticipated sequel to a consultation request is not problematic. For others, whose main interest in forming an alliance with a medical examiner or coroner's office was to broaden their spectrum of access to neuropathologic material for educational or public service purposes, and who are unaccustomed to the role of testifying as an expert witness, the arrival of a subpoena may be regarded as an unwelcome complication in an already busy schedule.

The neuropathologist unacquainted with commonly accepted rules of courtroom etiquette and demeanor, the importance of avoiding distracting mannerisms while on the stand, what is meant by "reasonable medical certainty," and common attorney interview strategies would do well to become familiar with such basics prior to encountering them in a very public adversarial setting (see Brodsky 1991, 1999 in Suggested Additional Readings). One's reputation, perhaps developed over many years, may be vulnerable in such settings unless scrupulous attention to preparation, honesty, objectivity, and the "rules of the game" in which one must function in our legal system are understood and observed. Anyone who doubts that some attorneys would purposely attempt to destroy the reputation of a medical witness whose testimony they know to be truthful might find it enlightening to read Ethics Opinion 1990-1 from the San Diego County Bar Association. One would hope that the conclusions in this opinion do not represent those of the vast majority of attorneys, any more than the negative behavior of an occasional medical examiner reflects the moral and ethical standards of most forensic pathology professionals.

Just as trainees in forensic pathology are given the opportunity to observe actual court testimony by senior forensic pathologists in the course of their training, we would encourage neuropathology consultants to avail themselves of similar opportunities and the medicolegal expertise that their forensic pathology colleagues will be pleased to share. Topics to be considered might reasonably include the differences between percipient (factual) witness versus expert witness; cause versus manner versus mechanism of death; chain of custody; a brief glossary of common legal terms encountered by forensic pathologists; and the differences between preliminary trial, bench trial, jury trial, depositions, and Grand Jury testimony.

Suggested Additional Readings

Brodsky SL. Testifying in Court: Guidelines and Maxims for the Expert Witness. Washington, DC: American Psychological Association, 1991.

Brodsky SL. The Expert Expert Witness: More Maxims and Guidelines for Testifying in Court. Washington, DC: American Psychological Association, 1999.

College of American Pathologists, Forensic Pathology Committee, The Pathologist in Court. Northfield, IL: College of American Pathologists, 2003.

Davis GG. Malpractice in pathology: What to do when you are sued. Arch Pathol Lab Med 2006;130:975–978.

Davis GG. The art of attorney interaction and courtroom testimony. Arch Pathol Lab Med 2006;130:1305–1308.

Finkel E. Expert witness guidelines: Pathologists given new order in the courtroom. CAP Today 2006;20:(7):5–8.

Fletcher CDM. Symposium on errors, error reduction, and critical values in anatomic pathology. Arch Pathol Lab Med 2006;130:602–603 (and continuing to p 653).

Foucar E. Pathology expert witness testimony and pathology practice: A tale of 2 standards. Arch Pathol Lab Med 2005;129:1268–1276.

Noguchi TT. Conflicts and challenges for the medical examiner. J Forensic Sci 1987;32:829–835.

Index

A

Abortion, illegal, 134
Abrasion rings, 247
Abscess, epidural, 336
Abuse. *See* Child abuse; Drug abuse-related cardiovascular disease
Acanthamoeba, 355
Acquired immunodeficiency syndrome (AIDS), 355
Actinomyces, 339
Acute amebic meningoencephalitis, 355
Acute cerebral cortical contusions, 176
Acute hypoxic-ischemic injury and related conditions, CNS responses to
   asphyxia, 297
   clinical features, 289–291
      time interval to incapacitation, 289–290
      time interval to loss of consciousness, 290
      time interval to onset of irreversible CNS damage, 290–291
   epilepsy, 297
   heat stroke, 295–297
   hypoglycemia, 297–298
   overview, 289
   red neurons and dark neurons, 298–302
   selective vulnerability of CNS to hypoxic-ischemic injury, 291–293
   unusual patterns, 293–295
Acute intracerebral hemorrhage, 268–271
Acute spinal cord injury, 84
Acute subarachnoid hemorrhage, 256–262
Acute subdural hemorrhage, 176
Acylcarnitine profile, 9
AD (Alzheimer's disease), 107, 194, 331, 395–400
Adrenoleukodystrophy (ALD), 415, 418–419
African trypanosomiasis, 355
Aging. *See* Dating/aging
AI. *See* Axonal injury
AIDA (International Association for Development of Apnea), 290
AIDS (acquired immunodeficiency syndrome), 355
Air bubbles in arteries and capillaries, 457
Air embolism, 129
Air tasers, 242

Akinetic mutism, 332
Alcoholic circumscribed cerebellar degeneration, 459
ALD (adrenoleukodystrophy), 415, 418–419
Alexander's disease, 419
ALS (amyotrophic lateral sclerosis), 301, 409
Altered consciousness, states of, 331–333
   akinetic mutism, 332
   coma, 332
   locked-in state, 332
   vegetative state, 332–333
Alzheimer's disease (AD), 107, 194, 331, 395–400
American Spinal Injury Association (ASIA) Impairment Scale, 86
Amniotic fluid embolism, 128–129
Amphophilic cytoplasmic inclusion, 402
Amphophilic globules, 32
Amyloid angiopathy, 269
Amyotrophic lateral sclerosis (ALS), 301, 409
Anaerobic bacteria, 336
Anaplastic astrocytoma, 363
Anaplastic oligodendrogliomas, 370
Anatomic lesions, 332
Anemia, 454
Anesthetic procedures, 438
Aneurysms, 256, 259–260, 325–327, 331
Angiomatous meningioma, 382
Animal horn cesarean section, 134
Anoxic-ischemic encephalopathy, 91
Anterior choroidal artery, 294
Anterior fontanel, 29
Anterior rectus gyri, 176
Anterior spinal rootlets, 409
Anterograde degeneration, 105, 264
Anthropology, forensic, 448–451
Anticoagulant-coated shrapnel, 244
Anticoagulants, 430
Antiplatelet therapy, 432
Apoptosis, 291
Arachnoid cells, 43, 64
Arachnoid cyst, 387–388
Arachnoid granulation fibrosis, 176
Arachnoid membrane, 387
Arbovirus meningoencephalitis, 345
Argentophilic tangles, 105

Arm bar hold, 282
Arnold-Chiari malformation, 28
Arrhythmias, 283
Arteries, air bubbles in, 457
Arteriovenous malformations, 124
Arteriovenous malformations (AVMs), 271, 320
AS (axonal spheroids), 51, 73, 74, 75, 185, 188, 410
ASIA (American Spinal Injury Association) Impairment Scale, 86
Aspergillus, 339, 343, 345
Asphyxia, 128, 297, 395
Assisted delivery, 133–134
Astrocytes, 35, 51
   terminology, 107–108
   wallerian degeneration, 103
Astrocytic neoplasms, 359
Astrocytic tumors, 369
   diffuse astrocytomas, 359–363
   glioblastoma multiforme, 363–365
   gliomatosis cerebri, 369
   pleomorphic xanthoastrocytoma, 367
   subependymoma, 367, 369
Asystole, 283
Atheromatous plaque, 311
Atherosclerosis, 308
Atherosclerotic aneurysm, 260
Audiovestibular nerve injuries, 191
Autolysis, 19
AVMs (Arteriovenous malformations), 271, 320
Axonal beading, 100
Axonal injury (AI), 50, 176
   diffuse (DAI), 74–75, 183–190
   traumatic, 77–79
Axonal spheroids (AS), 51, 73, 74, 75, 185, 188, 410
Axons
   axonal spheroids, 74
   β-Amyloid Precursor Protein (β-APP) immunostains, 75–77
   diffuse axonal injury (DAI), 74–75, 183–190
   traumatic axonal injury (AI), 77–79
   wallerian degeneration, 101–102
Axoplasm, 100

B

β-amyloid precursor protein (β-APP), 51, 75–77, 188, 218
B-cell lymphomas, 391
Babinski sign, 86
Backspatter, 220
Bacterial infections, spread from pericranial primary foci, 336
Bacteroides, 339
Balamuthia mandrillaris, 355
Barotrauma deaths, 457
Basal epidermal layer regeneration, 53
Basal ganglia, 32, 297, 362
Basal laminar membrane, 384
Basal skull fractures, 215
Basilar artery, 386
Basophil invasion, 43
Bergmann glia, 35–36, 107, 295
Bergmann's gliosis, 35, 296, 459
Berry aneurysms, 256, 259, 331
Bielschowsky method, 407
Bilateral burr holes, 55
Billiard ball effect, 228
Binucleated neurons, 377
Biologic terrorism, 355
Birefringent calcium oxalate crystals, 452
Biventer lobule, 35
Blastomyces dermatitidis, 343
Blood cells, 108–109
Blunt force head injury, 167–198
    boxing, 192–194
    brain contusions, lacerations, and hematomas, 177–183
    brainstem avulsion, 190–191
    consequences of, 195–197
        primary effects, 195
        secondary effects, 195–197
    diffuse axonal injury (DAI), 183–190
    epidural hematomas (EDHs), 169–170
    locked-in syndrome, 192
    overview, 167
    scalp abrasion, contusion, and laceration, 167–168
    skull fractures, 168–169
    sports, 192–194
    subarachnoid hemorrhage (SAH), 174–176
    subcutaneous scalp hemorrhage, 168
    subdural hematomas (SDHs), 170–174
        chronic subdural hematoma, 172–174
        dating/aging, 172
        source of neomembrane, 172
    subgaleal hemorrhage, 168
    traumatic cranial neuropathies, 191–192
    ventricular hemorrhage, 191
Blunt force trauma to throat, 283
Bodian method, 407
Bone demineralization, 61
Borrelia burgdorferi, 342
Boxing, 192–194
Brachial plexus, 132, 133
Bradycardia, 283
Brain. See entries beginning with Brain and entries beginning with Cerebral; Traumatic brain injury
Brain abscess, 339
Brain contusions, lacerations, and hematomas, 72–74, 177–183

Brain death, 92–98
    examination, 92–95
    microscopy, 95–97
    spontaneous/reflex movements, 97–98
Brain electrodes, 425
Brain embolism, 425
Brain herniation, 258
Brain intraparenchymal hemorrhages, 233
Brain necrosis-associated coagulopathy, 92
Brain parenchyma, 36, 195, 323
Brain pulsations, 223
Brain size, 27–28
Brain tumors, 359–394
    arachnoid cyst, 387–388
    astrocytic tumors, 359–370
        diffuse astrocytomas, 359–363
        glioblastoma multiforme, 363–365
        gliomatosis cerebri, 369
        pleomorphic xanthoastrocytoma, 367
        subependymoma, 367, 369
    causes of, 393
    chordoma, 386–387
    ependymoma, 370, 372–374
    extramural metastasis, 393
    ganglioglioma, 375–377
    lipoma, 388
    of maldevelopment, 383–386
        colloid cyst, 384–386
        epidermoid cyst, 384
    medulloblastoma, 374–375
    meningioma, 377–382
    metastatic neoplasms, 390
    mixed gliomas, 377
    oligodendroglioma, 370
    overview, 359
    pituitary tumors, 388
    primary malignant lymphomas, 391–392
    schwannomas, 382–383
Brain wounds, characteristics of firearms injuries, 216–220
Brainstem avulsion, 190–191
Brainstem diameter, 13
Brainstem glioma, 280
Brainstem injuries, in newborns, 132
Brainstem nuclei, 12
Bridging veins, 171
Bronchopneumonia, 87
Brugada's syndrome, 284
Bullets
    "disappearing," 220
    migration, 223–226
    tandem, 227
    trace evidence, 243–244
Bunina bodies, 409
Burns, 128

C

"Café coronary" deaths, 282–283
Caida de mollera, 205
Cajal's gold sublimate method, 79
California Department of Health Services SIDS protocol, 8–9
Callosotomy, 435
Calvarial bone, 58
Canavan's disease, 419
Candida, 335
Capillaries, air bubbles in, 457

Capillary buds, 53
Capillary telangiectasia, 271, 317
Caput succedaneum, 131
Carbon monoxide (CO), 128, 247, 291, 414
Cardiac arrest, 282, 297
Cardiac arrhythmias, 77, 134, 243, 282
Cardiac valvular vegetations, 262
Cardiopulmonary arrest, 87
Cardiorespiratory arrest, 315
Cardiovascular disease. See Vascular diseases, of central nervous system
Carotid sinus stimulation, 282
Carotid sleeper hold, 282
Cartilaginous callous, 55
Catecholamine surge, 196
Cavernous hemangioma, 271, 272, 317, 321–322
Cavum septi pellucidi, 36, 193
Cavum vergae, 36
CD68 immuno stain, 218
Cell extravascular migration, 53
Centerfire rifle wounds, 228
Central nervous system (CNS)
    blood cells in, 108–109
        erythrocytes (red blood cells), 108
        lymphocytes, 109
        macrophages, 108–109
        monocytes, 108–109
        neutrophils (polymorphonuclear leukocytes), 108
        other leukocytes, 109
    congenital lesions, 151–165
    development, 29–30
        cavum septi pellucidi, 36
        cavum vergae, 36
        cerebellum, 32–36
        cerebral cortex, 30–31
        cerebral ventricles, 41–42
        cerebrospinal fluid production, 41–42
        choroid plexus, 42–43
        hippocampus, 36–40
        myelination, 40–41
        noncortical regions, 31–32
        pituitary gland, 43–44
        reactive cells, 36
    developmental disorders, 151
    infections of, 335–357
        bacterial infections spread from pericranial primary foci, 336
        biologic terrorism, 355
        brain abscess, 339
        chronic bacterial infections, 341–342
        epidural abscess, 336
        fungal infections, 342–345
        meningitis, 337, 339
        overview, 335
        parasitic infections, 353–355
        perinatal infections, 335–336
        rickettsial infections, 339, 341
        subdural empyema, 336–337
        transmissible spongiform encephalopathies (prion diseases), 351–353
        viral infections, 345, 347–351
    malformations, 151–165
    responses of to acute hypoxic-ischemic injury and related conditions

asphyxia, 297
clinical features, 289–291
epilepsy, 297
heat stroke, 295–297
hypoglycemia, 297–298
overview, 289
red neurons and dark neurons, 298–302
selective vulnerability of CNS to hypoxic-ischemic injury, 291–293
unusual patterns, 293–295
vascular diseases of, 307–334
cerebral amyloid angiopathy, 327–331
cerebral embolism, 310–316
cerebral thrombosis, 307–309
drug abuse-related cardiovascular disease, 331
dural sinus thrombosis, 317–325
overview, 307
ruptured cerebral aneurysms, 325–327
secondary changes of cerebral infarction, 316–317
states of altered consciousness, 331–333
wallerian degeneration, 101–103
astrocytes, 103
axons, 101–102
microglia, 102–103
myelin sheath, 102
Central pontine myelinolysis, 452
Cephalhematomas, 132
Cephalopelvic disproportion, 131, 132
CERAD (Consortium to Establish a Registry for Alzheimer's Disease), 399–400
Cerebellar capillary hemangioblastoma, 390
Cerebellar cortex, 19, 32
Cerebellar cortical atrophy, 264, 265
Cerebellar cortical degeneration, 459
Cerebellar cortical heterotopias, 33
Cerebellar degeneration, 296, 452, 459
Cerebellar hemorrhage, 136, 312, 315, 331
Cerebellar herniations, 277
Cerebellar neuronal heterotopias, 33
Cerebellar pressure cone, 35
Cerebellar tonsillar herniations, 35, 277
Cerebellar vermis, 374
Cerebellum, 32–36, 41
Cerebral. See entries beginning with Cerebral and entries beginning with Brain; Traumatic brain injury
Cerebral amyloid angiopathy, 327–331
Cerebral aneurysms, 316, 325–327, 425
Cerebral anoxia, 291
Cerebral arteriosclerosis, 50
Cerebral atrophy, 277
Cerebral contusions, 72–74, 132
Cerebral cortex development, 30–31
Cerebral cortical contusions, 176, 181
Cerebral cortical dysplasia, 123
Cerebral cortical infarction, 309
Cerebral cysticercosis, 266
Cerebral edema, 87, 170, 176, 272, 316
Cerebral embolism, 310–316
Cerebral hemorrhage, 111, 331
Cerebral hypoxic-ischemic neuronal injury, 303
Cerebral infarction, 81
adult, 88–91
secondary changes of, 316–317

Cerebral palsy, 123
Cerebral parenchyma, 81
Cerebral parenchymal injury, 197
Cerebral peduncles, 185
Cerebral thrombosis, 307–309
Cerebral ventricles, 41–42
Cerebritis, 275
Cerebrospinal fluid (CSF), 63, 97, 136, 272, 275, 325, 413
accumulation, 28
appearance in subarachnoid hemorrhage, 68
circulation, 223
cytology in stroke, 98–99
leakage, 445
postmortem examination, 70–71
production, 41–42
pulsations, 55
rhinorrhea, 168
Cesarean section, 134
Charcot-Bouchard aneurysm, 269, 313, 315
Chemotherapeutic agent, 430–431
Chemotherapy, 434
Child abuse, 199–210
crush injuries, 205
folk remedies and customs, 205–206
information sources, 203–204
overview, 199–203
representative examples, 202–203
retinal hemorrhage, 206–208
shaken adult syndrome, 205
spinal injury, 206
subdural hemorrhage, 205
talk and deteriorate and die (Tadd) syndrome, 204
Terson's syndrome, 204–205
tin ear syndrome, 204
uniqueness of injury, 204
Childhood neurotrauma, 203
Cholesteatoma, 384
Chordoma, 386–387
Choroid plexus, 42–43, 71, 136, 390
Chronic bacterial infections, 341–342
Chronic cortical contusion, 265
Chronic epidural hematomas (EDH), 63
Chronic inflammatory cells, 30
Chronic protein-calorie malnutrition, 456
Chronic subdural hematoma, 172–174
Chronic systemic hypertension, 311
Chronic traumatic encephalopathy (CTE), 192
Cingulotomy, 435
CJD (Creutzfeldt-Jakob disease), 351–352
CMV (cytomegalovirus), 336, 347
CNS. See Central nervous system
CO (carbon monoxide), 128, 247, 291, 414
Coagulopathy, 190, 197
Coccidioides immitis, 343
Cochlear implants, 339
Coenurus cerebralis, 268, 355
Cold-induced reflex arrhythmia, 283
Cold water immersion, death due to, 283
Colloid cysts, 384–386
Coma, 332
Complications. See Periprocedural complications

Computed tomography (CT), 56, 168, 220, 312, 377, 443
Concentric fractures, 214
Congenital lesions, 151–165
Congo red stain, 194
Congophilic cerebral amyloid angiopathy, 328
Consciousness. See Altered consciousness, states of
Consortium to Establish a Registry for Alzheimer's Disease (CERAD), 399–400
Conspicuous meningeal arteries, 10
Contralateral hemianesthesia, 313
Contralateral hemiplegia, 312
Contusions
brain, 177–183
scalp, 167–168
Cooperative Study of Intracranial Aneurysms and Subarachnoid Hemorrhage, 256
Copper-coated shotgun pellets, 230
Corkscrew apical dendrite, 301
Corpora amylacea, 107, 343
Corpus callosum, 41, 153, 183, 364
Cortical architecture, 28
Cortical atrophy, 142, 404
Cortical dysplasia, 30
Cortical/subcortical hematomas, 181
Cortical/subcortical white matter junction, 339
Cortical thickness, 31
Cortical undercutting, 31
Corticospinal tract, 103, 104, 188
Coup cortical contusion, 182
Cranial dura mater, 10
Cranial gunshot wounds, 214
Cranial nerve injury, 132
Cranial stab wounds, 445
Craniocerebral relationships, 10
Craniopharyngioma, 384
Craniotomy, 336
Creutzfeldt cell, 107, 108
Creutzfeldt-Jakob disease (CJD), 351–352
Crush injuries, 205
Cryptic vascular angioma, 321
Cryptococcus neoformans, 339, 342
CSF. See Cerebrospinal fluid
CT (computed tomography), 56, 168, 220, 312, 377, 443
CTE (chronic traumatic encephalopathy), 192
Cuboidal epithelium, 385
Cuff inflation, 290
Cushing's syndrome, 129
CVD (cerebrovascular disease), 400
Cystic-malacic lesions, 138
Cystic tumor, 360
Cysticercosis, 265–266, 353
Cysts
arachnoid, 387–388
colloid, 384–386
epidermoid, 384
Cytoarchitectonic development, 30
Cytomegalovirus (CMV), 336, 347
Cytoplasmic vacuoles, 105, 299
Cytotoxic edema, 274
Cytotoxic therapy, 345

D

DAI (diffuse axonal injury), 51, 74–78, 183–190
Dandy-Walker malformation, 153
Dark neurons, 298–302
Dating/aging, 49–122
    adult cerebral infarction, 88–91
    astrocyte terminology, 107–108
    axonal injury (AI)
        axonal spheroids, 74
        β-Amyloid Precursor Protein (β-APP) immunostains, 75–77
        diffuse (DAI), 74–75
        traumatic, 77–79
    blood cells in CNS lesions, 108–109
    brain contusions, 72–74
    brain death, 92–98
        examination, 92–95
        microscopy, 95–97
        spontaneous/reflex movements, 97–98
    cerebrospinal fluid cytology in stroke, 98–99
    denervation of muscle, 112–114
    epidural hematomas (EDH), 62–63
    heterotopic ossification, 60–62
    hypoxic-ischemic injury, 91–92
    intracerebral hemorrhage, 81–82
    malignant cerebral edema, 99–100
    mineralization, 58–59
    overview, 49–52
    peripheral nerve regeneration, 103
    postmortem cerebrospinal fluid examination, 70–71
    puncture wounds, 79–81
    reactive cells changes, 109–112
    reinnervation of muscle, 114
    respirator brain, 92–98
        examination, 92–95
        microscopy, 95–97
        spontaneous/reflex movements, 97–98
    scalp injuries, 52–54
        antemortem versus postmortem, 53–54
        color changes in skin bruises, 52
    Schwannosis, 84
    skull fractures, 54–58
    spinal concussion, 87–88
    spinal cord infarction, 92
    spinal cord traumatic lesions, 82–84
    spinal shock, 84–87
    subarachnoid hemorrhage, 67–70
        cerebrospinal fluid (CSF) appearance, 68
        microscopy, 68–70
    subdural hematomas (SDH), 63–67, 172
        anatomy, 64
        size, 63–64
        spontaneous resolution, 66–67
    transneuronal degeneration, 104–107
        dentatoolivary system, 105–107
        trigeminal system, 104–105
        visual system, 104
    wallerian degeneration, 100–103
        central nervous system (CNS), 101–103
        peripheral nervous system (PNS), 100–101
Death. *See* Sudden unexpected death
Dehydration, 336
Delivery, 130–134
    assisted, 133–134
    illegal abortion, 134
    vaginal, 131–133
Dementia, 395, 400
Demyelinated CNS nerve fibers, 281
Demyelinating disorders, 413–421
    leukodystrophy, 418–419
    multiple sclerosis, 413–418
Denervation of muscle, 112–114
Dental abscess, 317
Dentate fascia, 397
Dentate nucleus, 107
Dentatoolivary system, 105–107
Depressed skull fracture, 169
Desquamating parakeratotic keratin, 384
Developmental abnormalities, 151–165
Devic's disease, 415
Diagnostic ependymal rosettes, 372
Diagnostic procedure artifacts, 19
Diastatic fracture, 169
Diffuse astrocytomas, 359–363
Diffuse axonal injury (DAI), 51, 74–78, 183–190
"Disappearing" bullets, 220
Disintegrating neutrophils, 108
Disk herniation, 436
Disseminated intravascular coagulation, 453, 457
Dive reflex, 283
Divers Alert Network, 457
Dorsal vagal nuclei, 176
Dot-blot hemorrhages, 137, 206
Down syndrome, 153, 397, 399
Drug abuse-related cardiovascular disease, 331
Dural border cells, 64, 171
Dural capillaries, 65
Dural sinus thrombosis, 317–325
Dural venous sinus wounds, 224
Duret's hemorrhages, 99, 170, 312, 316
Dysmorphic neurons, 377
Dysplastic cortex, 31

E

Ecchordosis physalifora, 387
Eclampsia, 128
ECMO (extracorporeal membrane oxygenation), 146
Ectopic calcification, 60
Edema, 73, 99, 309
"Edematous" human cortex, 300
EDH (epidural hematomas), 62–63, 66, 169–170
EDTA (ethylenediaminetetraacetic acid)-treated glass containers, 71
EGA (estimated gestational age), 21
Elastic lamina, 321
Electric shock, 126–128
Electrocution, 444–445
Electroencephalogram, 193
Electromagnetic fields, 393
Elongate sarcolemmal nuclei, 113
Embalming procedures, 15
Emboli, 311
Encephaloclastic lesions, 153
Encephalopathy, 281
Endoplasmic reticulum, 299
Endothelial hypertrophy, 65
Endotracheal intubation, 283
Eosinophilic corpuscle, 106
Eosinophilic cytoplasmic inclusion, 402
Eosinophils, 66
Ependymal cells, 33, 71
Ependymoma, 370, 372–374
Epicranial hematomas, 63
Epidermoid cyst, 384
Epidural abscess, 336
Epidural anesthetic procedures, 130–131
Epidural hematomas (EDH), 62–63, 66, 169–170
Epidural space, 336
Epilepsy, 255, 297
    pregnant women, 124, 126
    sudden unexpected death as result of, 262–265
Erythrocytes (red blood cells), 67, 70, 108
Erythrophagocytosis, 53, 69, 73, 97, 108
Escherichia coli, 335
Esophageal-cardiac fistula, 311
Esoteric degenerative disorders, 37
Estimated gestational age (EGA), 21
Ethylene glycol, 452
Ethylenediaminetetraacetic acid (EDTA)-treated glass containers, 71
Excitotoxicity-induced acute, 197
Executions, 444
Exsanguination, 229
External hydrocephalus, 28–29
Extracerebral fluid, 29
Extracorporeal membrane oxygenation (ECMO), 146
Extracranial bone fractures, 168
Extramedullary hematopoiesis, 454–455
Extramural metastasis, 393
Extravascular neutrophil infiltrates, 54
Extrinsic trauma, 132

F

Fallen fontanelle, 205
Fatal insomnia syndromes, 351
Ferruginated neurons, 58
Fetal artifacts, 129
Fetal bilateral subdural hematoma, 130
Fiber atrophy, 113
Fibrillary astrocytes, 404
Fibrillary glial matrix, 369
Fibrillary gliosis, 105
Fibrin, 53, 65
Fibrinopurulent meningitis, 343
Fibroblastic meningiomas, 382
Fibroblasts, 53, 65, 339
Fibromuscular thickening, 321
Fibrovascular stroma, 43
Fibrovascular trabeculae, 370
Final Neuropathologic Diagnosis, 4
Firearms injuries, 211–254
    autopsy protocols, 246–248
    backspatter, 220
    brain wound characteristics, 216–220
    bullet migration, 223–226
    bullet/shrapnel trace evidence, 243–244
    centerfire rifle wounds, 228
    complications of retained bullets, 230–231
    "disappearing" bullets, 220
    entrance and exit wounds, 215–216

firearm wound imitators, 243
head wounds, 213
information resources, 211–212
injury to death interval, 231–232
intermediate targets, 222–223
internal ricochet, 216
overview, 211
potential complications of craniospinal wounds, 229–230
prognostic factors, 227–228
radiology, 244–246
reaction/response time issues, 237–239
shotgun wounds, 228–229
skull wounds, 213–215
speed of incapacitation issues, 232–237
suicidal versus homicidal wounds, 226–227
tandem bullets, 227
tangential skull wounds, 215
unusual missile-launching devices and ammunition, 239–243
wound ballistic issues, 212–213
FLAIR (fluid attenuated inversion recovery), 443
Flame hemorrhages, 137, 206
Fluid attenuated inversion recovery (FLAIR), 443
Focal cortical dysplasia, 265
Foix-Alajouanine syndrome, 323
Fontanels, 29
Foramen magnum, 379
Forceps-assisted delivery, 133
Forensic anthropology, 448–451
Forensic neuroradiology, 443–444
Formalin fixation, 4, 14, 22, 320
Freeman-Watts landmark, 10
Friedreich's ataxia, 408–409
Full body radiographs, 246
Fungal infections, 342–345
Fusobacterium, 339

G

GA (gestational age), 19
Galactose, 9
Galen vein, 63
Gamma hydroxybutyric acid (GHB), 284
Ganglioglioma, 375–377
Gas-forming organisms, 245
Gastrointestinal (GI) tract, 223
GCS (Glasgow Coma Scale), 215, 225, 227
Gemistocytes, 364
Gemistocytic astrocytes, 51, 74
General paresis of the insane (GPI), 342
Genetic/metabolic leukodystrophy, 413
Genitourinary (GU) tract, 223
Germinal matrix (subependymal) hemorrhage, 129, 136
Gerstmann-Sträussler-Scheinker syndrome, 351, 353, 399
Gestational age (GA), 19
GFAP (glial fibrillary acidic protein) immunostain, 360
GFAP-positive processes, 102
GHB (gamma hydroxybutyric acid), 284
Ghost cells, 300
GI (gastrointestinal) tract, 223
Giemsa stain, 66
Glasgow Coma Scale (GCS), 215, 225, 227
Glial cell necrosis, 73
Glial cytoplasmic inclusions, 407
Glial fibrillary acidic protein (GFAP) immunostain, 360
Gliding contusion, 182
Glioblastoma multiforme, 363–365, 393
Gliomas, 275, 370
Gliomatosis cerebri, 369
Global hypoxic-ischemic brain insults, 51
Globoid leukodystrophy, 419
Globose forms, 399
Glomus, 42–43
Glutamate excitotoxic injury, 265
GMS (gomori methenamine silver), 66, 342, 452
Golgi complex, 299, 302
Gomori methenamine silver (GMS), 66, 342, 452
GPI (general paresis of the insane), 342
Gram stain, 66
Granule cells, 71
Granulocyte karyorrhexis, 73
Granulocytes, 52, 97
Granulomatous disease, 456
Granulomatous inflammation, 341
Granulomatous meningitis, 343
Granulovacuolar degeneration (GVD), 396, 400, 405
Gravity-related congestion, 17, 19
Grinker's myelinopathy, 295
Gross Impressions, 4
Grumous amorphous debris, 384
GU (genitourinary) tract, 223
Guillian-Barré syndrome, 445
Gunshot wounds. *See* Firearms injuries
GVD (granulovacuolar degeneration), 396, 400, 405

H

H&E (hematoxylin-eosin) stain, 8, 50, 101, 265, 327, 397, 452
Haemophilus influenzae, 337, 339
Hallervorden-Spatz disease, 399, 409
Hangings, 444
HD (Huntington's disease), 301, 401, 404–405
Head injury
 from blunt force, 167–198
  boxing, 192–194
  brain contusions, 177–183
  brainstem avulsion, 190–191
  diffuse axonal injury (DAI), 183–190
  epidural hematomas (EDHs), 169–170
  locked-in syndrome, 192
  overview, 167
  scalp abrasion, 167–168
  skull fractures, 168–169
  subarachnoid hemorrhage (SAH), 174–176
  subcutaneous scalp hemorrhage, 168
  subdural hematomas (SDHs), 170–174
  traumatic cranial neuropathies, 191–192
  ventricular hemorrhage, 191
 from firearms, 213
Head size, 27–28
Heart. *See entries beginning with* Cardiac
Heat stroke, 295–297
HELLP (hemolysis, elevated liver enzymes and low platelets) syndrome, 128
Hematoxylin-eosin (H&E) preparations, 172
Hematomas
 brain, 177–183
 epidural (EDH), 62–63, 169–170
  chronic, 63
  dating/aging, 63
  size, 62–63
  spontaneous resolution, 63
 subdural (SDH), 63–67, 170–174
  anatomy, 64
  chronic, 172–174
  dating/aging, 64–66, 172
  size, 63–64
  source of neomembrane, 172
  spontaneous resolution, 66–67
Hematoxylin-eosin (H&E) stain, 8, 50, 265, 327, 397, 452
Hemimegalencephaly, 28
Hemiplegia, 313
Hemolysis, elevated liver enzymes and low platelets (HELLP) syndrome, 128
Hemorrhage
 acute intracerebral, 268–271
 acute subarachnoid, 256–262
 acute subdural, 176
 cerebellar, 136, 312, 315, 331
 intracerebral, 81–82
 intraventricular (IVH), 136
 retinal, 137, 206–208
 subarachnoid hemorrhage (SAH), 19, 81, 129, 132, 174–176, 204, 205, 218, 262, 325
  cellular responses, 50
  cerebrospinal fluid (CSF) appearance, 68
  microscopy, 68–70
  in pregnant women, 124
 subcutaneous scalp, 168
 subdural, 67, 132, 204, 205
 subependymal (germinal matrix), 136
 subgaleal, 168
 subperiosteal, 168
 subretinal, 206
 ventricular, 191
Hemorrhagic lesions, 82, 109, 111–112, 136
Hemosiderin, 91, 207
Hemosiderin-decorated astrocytes, 182
Hemosiderin granules, 182
Hemosiderin-laced cortical microinfarcts, 328
Hemosiderin-laden macrophages, 65, 66, 112, 174
Hepatic encephalopathy, 452
Hepatomegaly, 87
Herniation contusion, 183
Herniation syndromes, 195
Heroin, 331
Herpes simplex virus type 1 (HSV-1), 345
Herpes simplex virus type 2 (HSV-2), 347
Herpes zoster, 347
Herpetic intranuclear inclusions, 347
Heterotopic calcification, 60, 61
Heterotopic ossification, 60–62
Heubner's arteritis, 341
HHV-6 (Human herpesvirus-6), 347
High-risk autopsy, 4–5
Hippocampal formation, 396

Hippocampal neurons, 40
Hippocampal pyramidal cells, 303
Hippocampal sclerosis, 36, 262, 265
Hippocampus, 4, 30, 36–40
Hirano body, 396, 400
Histopathologists, 303
Histoplasma microorganisms, 343
Histotechnologists, 64, 303
HIV (human immunodeficiency virus) infection, 343, 391, 413
Holoprosencephaly, 153
Holzer-positive fibrils, 108
Homicidal wounds, 226–227, 445
Homogenizing cell change, 300
Horner's syndrome, 132
Hortega's silver impregnation method, 79
HSV-1 (Herpes simplex virus type 1), 345
HSV-2 (Herpes simplex virus type 2), 347
Human herpesvirus-6 (HHV-6), 347
Human immunodeficiency virus (HIV) infection, 343, 391, 413
Huntington's disease (HD), 301, 401, 404–405
Hyaline membrane, 174
Hydrocephalus, 28–29, 230, 266, 275–277, 383, 456
Hydropic cell, 300
Hydropic meningothelial islands, 43
Hygroma, 28
Hyperactive carotid reflex, 281
Hyperacute cerebral edema, 99
Hyperchromatic cells, 375
Hyperchromatic nuclei, 364, 383
Hypercoagulable states, 336
Hyperextension, 190
Hypermyelination, 142
Hyperoxygenation, 145
Hyperpyrexia, 295, 296
Hypertensive cerebral hemorrhages, 312
Hypertensive vascular disease, 331, 452
Hyperthermia, 296, 332
Hyperthyroidism, 29
Hypervascular subdural hemorrhage neomembrane, 454
Hyperventilation, 296
Hypoglycemia, 197, 297–298
Hypoglycemic injury, 297–298
Hypokinesis, 196
Hypothalamus, 169, 183, 275
Hypoxemia, 454
Hypoxia, 52
Hypoxic-ischemic encephalopathy, 96, 133, 188
Hypoxic-ischemic injury, 51, 91–92, 139, 176, 185, 188, 196, 218, 281, 291, 293, 457
Hypoxic-ischemic lesions, 59, 109, 110–111, 123, 142
Hypoxic-ischemic neuron injury, 303
Hypoxic versus ischemic events, 52

I

Iatrogenic, 423
ICP (increased intracranial pressure), 227, 230, 275, Appendix D
Idiopathic benign external hydrocephalus, 28
Idiosyncratic reactions, 433
Imaging methods, 245
Immune disorders, 129
Immunoablation, 434
Immunocompromised states, 335
Immunocytochemical method, 188
Immunohistochemical method, 194
In-custody deaths, 5
Increased intracranial pressure (ICP), 227, 230, 275, Appendix D
Infantile atlantooccipital instability, 146
Infants
  newborn, 135–146
    cerebellar hemorrhage, 136
    extracorporeal membrane oxygenation (ECMO), 146
    infantile atlantooccipital instability, 146
    intraventricular hemorrhage (IVH), 136
    kernicterus, 145
    multicystic encephalopathy, 139
    neonatal sinovenous thrombosis, 136–137
    occipital osteodiastasis, 145–146
    paracentral cortical atrophy, 142, 145
    perinatal telencephalic leukoencephalopathy, 137
    periventricular leukomalacia, 137–138
    pontosubicular necrosis, 145
    retinal hemorrhage, 137
    skull congenital anomalies, 145
    status marmoratus, 142
    subependymal (germinal matrix) hemorrhage, 136
    ulegyria, 139, 142
  stillborn, 129–130
Infarction
  cerebral, 88–91
  spinal cord, 92
Infections, of central nervous system (CNS), 335–357
  bacterial infections spread from pericranial primary foci, 336
  biologic terrorism, 355
  brain abscess, 339
  chronic bacterial infections, 341–342
  epidural abscess, 336
  fungal infections, 342–345
  meningitis, 337, 339
  overview, 335
  parasitic infections, 353–355
  perinatal infections, 335–336
  rickettsial infections, 339, 341
  subdural empyema, 336–337
  transmissible spongiform encephalopathies (prion diseases), 351–353
  viral infections, 345, 347–351
Inferior olivary hypertrophy, 105, 106
Inflammatory bacterial mediators, 274
Inflammatory cells, 231, 339
Injury to death interval, 231–232
Intermediary coup, 182
Internal ricochet, 214–216
International Association for Development of Apnea (AIDA), 290
International Cerebral Palsy Task Force, 124
Intracerebral hematoma, 445
Intracerebral hemorrhages, 67, 81–82, 111, 289, 311
Intracerebral lipomas, 388
Intracortical hypoxic-ischemic injury, 139
Intracranial gliomas, 370
Intracranial hemorrhages, 62, 129, 146, 195, 452
Intracranial hypertension, 205
Intracranial infections, 168
Intracranial metastases, 390
Intracranial pressure (ICP), 227, 230, 275, Appendix D
Intracranial tumors, 128, 363
Intracranial vascular disease, 311
Intradiploic meningoencephalocele, 56
Intramuscular arteries, 113
Intramuscular nerve trunks, 113
Intraosseous carotid artery dissection, 5
Intraparenchymal CNS calcification, 59
Intraparenchymal hemorrhages, 92
Intraparenchymal tuberculomas, 341
Intrapartum trauma, 129
Intraretinal dot-blot Hemorrhage, 206
Intraspinal bullet migration, 223
Intraspinal lipomas, 388
Intrauterine infections, 335
Intrauterine subdural hemorrhages, 130
Intravascular coagulation syndrome, 296
Intravascular malignant lymphomatosis, 454
Intraventricular hemorrhage (IVH), 129, 132, 136, 191
Intravital brainstem fixation, 431
Ischemic cell change, 298, 299
Ischemic circulatory lesions, 72
Ischemic-hypoxic injury, 316
Ischemic lesions, 112
IVH (intraventricular hemorrhage), 129, 132, 136, 191

J

Joint ankylosis, 60
Juxtaluminal blepharoplasts, 374

K

Karyorrhexis, 108
Kernicterus, 145
Kernohan's notch, 312
Knife blade atrophy, 401
Krabbe's disease, 413, 419
Kwashiorkor, 456

L

Labor, 130–134
  assisted delivery, 133–134
  illegal abortion, 134
  vaginal delivery, 131–133
Lacerations
  brain, 177–183
  scalp, 167–168
Laidlaw's stain, 70
Lamina desiccans, 33
Langhans, 341
Laryngeal nerve injury, 146
Lateral folia, 33
Lateral geniculate body (LGB), 104, 294
Lathyrus sativus, 409
LCA (leukocyte common antigen), 102
Lead poisoning, 231
Leg vein thrombosis, 311

Leptomeningeal gliomatosis, 280
Leptomeningeal melanosis, 17
Leptomeningeal vessels, 10
Leptomeninges, 36, 373, 396
Lethal gas, 444
Lethal injection, 444
Leukocyte common antigen (LCA), 102
Leukocyte immunohistochemical stains, 71
Leukocytes, 53
    central nervous system (CNS) lesions, 109
    polymorphonuclear (neutrophils), 108
Leukocytosis, 69
Leukodystrophy, 415, 418–419
Leukoencephalopathy, 452
Leukophagocytosis, 97
Lewy body disease, 400, 404
LGB (lateral geniculate body), 104, 294
Lipoma, 388
Listeria monocytogenes, 335–337
Lithium toxicity, 296
Lobar intracerebral hemorrhage, 327
Locked-in state, 332
Locked-in syndrome, 192, 437
Locus ceruleus, 145
Louis-Barr syndrome, 325
LP (lumbar puncture), 68
Lucid period, 170
Lumbar puncture (LP), 68
Lumbosacral plexus injury, 133
Luschka, foramen of, 456
LV (left ventricle), 42
Lyme disease, 342
Lymphocytes, 66, 69, 71, 97, 109, 339
Lymphocytic cuffing, 12
Lymphoid adrenal medullitis, 349
Lymphoid hypophysitis, 349

M
Macrencephaly, 27
Macroadenoma, 388
Macrocephaly, 27
Macrophages, 52, 53, 65, 73, 74, 100, 108–109, 231, 309, 339
Magendie, foramen of, 456
Magnetic resonance imaging (MRI), 66, 132, 171, 220, 377, 443
Maldevelopment, tumors of, 383–386
    colloid cyst, 384–386
    epidermoid cyst, 384
Malformation syndromes, 29
Malignant cerebral edema, 99–100
Malignant cerebral swelling, 99, 195
Malignant hyperthermia, 296
Malignant middle cerebral artery edema syndrome, 99
Mammillary bodies, 459
Marasmus, 456
Marchi method, 112
Marchiafava-Bignami disease, 452, 457
Maternal diabetes mellitus, 129
Meconium-stained amniotic fluid, 129
Medial lemnisci, 185
Medial parahippocampal gyrus, 183
Medulla, 40
Medulloblastoma, 374–375, 393
Megalencephaly, 27, 28
Megalocephaly, 27, 28

Meningeal arteries, 169
Meningioma, 377–382, 393
Meningitis, 274–275, 337, 339
Meningococcus, 337
Mercury wing pattern, 258
Mesencephalic nucleus, 12
Mesothelial cell, 70
Metachromatic leukodystrophy, 413, 419
Metastatic brain tumor, 390
Metastatic neoplasms, 390
Metastatic tumors, 277
Methamphetamine, 331
Micrencephaly, 28
Microadenoma, 388
Microaerobic streptococci, 336
Microcephaly, 28
Microencephaly, 28
Microglia, 79, 102–103, 342
Micronodular mineralization, 32
Microscopy
    brain contusions, 72–74
    subarachnoid hemorrhage, 68–70
Microvascular glomeruloid formation, 363
Midbrain, 40
Midline rostral orbitofrontal gyri, 152–153
Mineralization
    dating/aging, 58–59
    reactive cell changes, 112
Missle-launching devices. *See* Firearms injuries
Mitosis, 108
Mixed gliomas, 377
MND (motor neuron diseases), 409–410
MODS (multiple-organ dysfunction syndrome), 197
Monocytes, 71, 108–109
Monocytoid cells, 108
Motor neuron diseases (MND), 409–410
MRI (magnetic resonance imaging), 66, 132, 171, 220, 377, 443
MS (multiple sclerosis), 281, 413–418
Mucor, 345
Müller cells, 107
Multi-Society Task Force on PVS (persistent vegetative state), 332
Multicystic encephalopathy, 139
Multiple-organ dysfunction syndrome (MODS), 197
Multiple sclerosis (MS), 281, 413–418
Multiple system atrophy, 405–408
Muscle atrophy, 113
Muscle denervation, 112–114
Muscle reinnervation, 114
Mycobacterium, 339, 341
Mycotic aneurysm, 260
Mycotic embolization, 311
Myelin degeneration, 102
Myelin sheath, 102
Myelin sheath injury, 445
Myelin staining, 105, 297
Myelinated axons, 100
Myelination, 40–41
Myelination gliosis, 41
Myelopathy, 425
Myocardial fibers, 59
Myocarditis, 283
Myxopapillary ependymomas, 371

N
Naegleria fowleri, 355
Necrotic cerebellar fragments, 97
Neisseria meningitidis, 337
Neomembrane
    hyalinization of, 66
    source of, 172
Neonatal encephalopathy, 135
Neonatal necrotic brain lesion, 59
Neonatal retinal hemorrhage, 137
Neonatal sinovenous thrombosis, 136–137
Neovascularization, 91
Nerve injuries, 191
Neuroborreliosis, 342
Neurodegenerative disorders, 395–412
    Alzheimer's disease (AD), 395–400
    Friedreich's ataxia, 408–409
    Huntington's disease (HD), 404–405
    motor neuron disease (MND), 409–410
    multiple system atrophy, 405–408
    overview, 395
    Parkinson's disease, 402–404
    Pick's disease, 400–402
    progressive supranuclear palsy, 405
Neuroepithelial cysts, 386
Neuroepithelial tumor, 384
Neurofibrillary tangles (NFTs), 193, 194, 397
Neurofilament immunohistochemical staining, 101
Neurofilament-positive elements, 102
Neurogenic pulmonary edema, 196
Neuroleptic malignant syndrome, 296
Neurologic deterioration, 295
Neuromelanin-containing cells, 12
Neuromyelitis optica, 415
Neuronal chromatolysis, 445
Neuronal encrustation, 73
Neuronal eosinophilic cytoplasm, 298
Neuronal ferrugination, 59
Neuronal hypoxic-ischemic lesions, 353
Neuronal incrustations, 91
Neuronal lipofuscin accumulation, 12
Neuronal migration, 30
Neuronal soma, ferrugination of, 112
Neuronophagia, 73
Neurons, ferrugination of, 58, 74
Neuropathologic complications of systemic diseases, 452–460
    chronic alcoholism, 457–460
    coagulopathies, 452–453
    extramedullary hematopoiesis, 454–455
    focal hemorrhages, 453
    intravascular blood elements, 453–454
    sarcoidosis, 456
    scuba deaths, 456–457
    starvation, 455–456
    subdural neomembrane vascularity, 454
Neuropathologic consequences of contemporary judicial executions, 444–445
Neuroradiology, forensic, 443–444
Neurosurgical procedures, 438
Neurotoxicology, 451
Neutrophilia, 97
Neutrophils (polymorphonuclear leukocytes), 53, 65, 69, 73, 91, 108, 112

Newborn infants, 135–146
    cerebellar hemorrhage, 136
    extracorporeal membrane oxygenation (ECMO), 146
    infantile atlantooccipital instability, 146
    intraventricular hemorrhage (IVH), 136
    kernicterus, 145
    multicystic encephalopathy, 139
    neonatal sinovenous thrombosis, 136–137
    occipital osteodiastasis, 145–146
    paracentral cortical atrophy, 142, 145
    perinatal telencephalic leukoencephalopathy, 137
    periventricular leukomalacia, 137–138
    pontosubicular necrosis, 145
    retinal hemorrhage, 137
    skull congenital anomalies, 145
    status marmoratus, 142
    subependymal (germinal matrix) hemorrhage, 136
    ulegyria, 139, 142
NFTs (neurofibrillary tangles), 193, 194, 397
Nissl bodies, 300
Nissl stains, 301
Nissl's acute cell change, 298
Nissl's chronic cell change, 298
Nocardia, 339
Noncortical regions, 31–32

O

Obersteiner-Redlich zone, 382
Occipital osteodiastasis, 136, 145–146
Occult vascular angioma, 321
Ocular hemorrhages, 205
Oculocardiac and trigeminocardiac reflex complex, 283–284
Oculocardiac reflex, 284
Olfactory neuroepithelium, 355
Oligodendroglioma, 370
Oligohydramnios, 129
Olivary hypertrophy, 105, 106, 107
Olivopontocerebellar atrophy (OPCA), 13, 405
OPCA (olivopontocerebellar atrophy), 13, 405
Ophthalmology Child Abuse Working Party, 207
Opthalmoscopy, 207
Optic atrophy, 104
Orbitofrontal cortex, 4
Orthostatic hypotension, 406
Ossification, 55, 60
Osteoblasts, 54
Osteoclasts, 54
Osteomyelitis, 336

P

Palatal myoclonus, 105
Paracentral cortical atrophy, 142, 145
Paradoxical embolization, 225, 311
Paranasal sinusitis, 317
Paraplegia, 60, 86
Parasitic infections, 266, 353–355
Parasylvian cerebral hemisphere, 262
Parasympathetic hyperactivity, 282
Parenchymal microcalcifications, 43
Parkinsonism-dementia complex, 399, 400

Parkinson's disease, 395, 400, 402–404, 434
Pars nervosa, 44
PAS (periodic-acid schiff), 342, 387, 399
Pediatric autopsy methods, 199
Pediatric heterotopic ossification, 61
Pelizaeus-Merzbacher disease, 419
Penfield's piloid astrocyte, 107
PEP (postencephalitic parkinsonism), 397–399, 403
Perimortem cesarean section, 134
Perinatal hypoxic-ischemic insults, 123
Perinatal infections, 335–336
Perinatal telencephalic leukoencephalopathy, 137
Periodic-acid schiff (PAS), 342, 387, 399
Periorbital ecchymosis, 168
Periosteal hemorrhage, 168
Peripheral blood leukocytes, 71
Peripheral nerve wallerian degeneration, 103
Peripheral nervous system (PNS), 100, 451
    regeneration, 103
    wallerian degeneration, 100–101
Peripheral neuropathy, 459
Periprocedural complications, 423–441
    categorization of, 424
    challenges for forensic pathologist, 438
    definitions, 423–424
    determination of mode, 438
    foreign materials, 433
    mechanical forces, 436–438
    mechanisms of injury, 424–433
        devices, 425
        needles and drugs, 430–433
        plastic tubes, 425–428
    overview, 423
    surgical instruments, hardware, and ablative procedures, 433–436
Peritoneal macrophages, 109
Perivascular aggregates, 30
Perivascular neutrophils, 53
Perivascular pseudorosettes, 372
Periventricular gliosis, 138
Periventricular leukomalacia (PVL), 110, 137–138
Persistent vegetative state (PVS), 332
Pes cavus, 407
PET (positron emission tomography), 443
Petal mark, 229
Petechiae, 132
Petrous ridge fracture, 168
Phakomatoses, 280
Phencyclidine, 331
Phlebotomy, 237
Phosphotungstic acid hematoxylin (PTAH)-positive fibrils, 108, 374
Phrenic nerve injury, 133
Physiologic conduction abnormality, 77
Pick's disease, 400–402
Piggy-back bullets, 227
Pigment-laden macrophages, 66
Pigmented osteoblasts, 64
Pilocytic astrocytoma, 362–363
Piloid astrocytes, 81, 107
Pineal gland, 22
Pituitary adenomas, 128, 388
Pituitary gland, 43–44
Pituitary infarction, 295

Pituitary tumors, 388
Plaques, 400, 414
Plasmodium falciparum, 353
Pleomorphic xanthoastrocytoma, 367
PML (progressive multifocal leukoencephalopathy), 350–351
Pneumocephalus, 69, 457
Pneumococcal meningitis, 56
Pneumococcus, 337
PNS. *See* Peripheral nervous system
Poisonings, pregnant women, 128
Poliomyelitis, 347
PolyHEMA sponge material, 59
Polymorphonuclear leukocytes (neutrophils), 53, 65, 69, 73, 91, 108, 112
Polyol-containing infusions, 452
Pontine hemorrhage, 312
Pontine myelinolysis, 457
Pontocerebellar tracts, 185, 188, 406
Pontomedullary avulsion, 191
Pontosubicular necrosis, 145
Pork tapeworm, 265
Positron emission tomography (PET), 443
Post-traumatic apnea, 206
Postcentral gyri, 409
Postdefibrillation asystole, 283
Postencephalitic parkinsonism (PEP), 397–399, 403
Posterior fontanel, 29
Posterior fossa, 176, 371
Posterior orbital cortex, 176
Posterior rootlets, 14
Post–head injury apnea, 196
Postinfectious perivenous encephalomyelitis, 413
Postresuscitated cardiopulmonary arrest, 297
Postural hypotension, 281
Precentral gyri, 409
Predominantly lymphocytic infiltrate, 341
Pregnancy, 123–149
    labor and delivery, 130–134
        assisted delivery, 133–134
        illegal abortion, 134
        vaginal delivery, 131–133
    newborn infants, 135–146
        cerebellar hemorrhage, 136
        extracorporeal membrane oxygenation (ECMO), 146
        infantile atlantooccipital instability, 146
        intraventricular hemorrhage (IVH), 136
        kernicterus, 145
        multicystic encephalopathy, 139
        neonatal sinovenous thrombosis, 136–137
        occipital osteodiastasis, 145–146
        paracentral cortical atrophy, 142, 145
        perinatal telencephalic leukoencephalopathy, 137
        periventricular leukomalacia, 137–138
        pontosubicular necrosis, 145
        retinal hemorrhage, 137
        skull congenital anomalies, 145
        status marmoratus, 142
        subependymal (germinal matrix) hemorrhage, 136
        ulegyria, 139, 142
    overview, 123–124
    pregnant women, 124–129

air embolism, 129
amniotic fluid embolism, 128–129
asphyxia, 128
burns, 128
eclampsia, 128
electric shock, 126–128
epilepsy, 124, 126
hemolysis, elevated liver enzymes and low platelets (HELLP) syndrome, 128
intracranial tumors, 128
poisonings, 128
stroke, 124
tocolytic agents, 129
stillborn infants, 129–130
comparison of lesions with liveborn infants, 129–130
death to delivery interval, 129
fetal artifacts, 129
Prepartum trauma, 129
Preretinal hemorrhages, 206
Presumptive reflex mechanisms, death related to, 284
Primary malignant lymphomas, 391–392
Primitive neuroectodermal tumors, 129
Prion diseases (transmissible spongiform encephalopathies), 351–353
Progressive multifocal leukoencephalopathy (PML), 350–351
Progressive supranuclear palsy (PSP), 396, 403, 405
Proliferative capillaries, 370
Protein immunohistochemistry, 102
Protoplasmic, 107
Protoplasmic astrocytes, 74
Prussian blue stain, 59
Psammomatous meningioma, 382
Pseudopalisades, 364
Pseudostippling, 222, 247
PSP (progressive supranuclear palsy), 396, 403, 405
PTAH (phosphotungstic acid hematoxylin)-positive fibrils, 108, 374
Punch drunk syndrome, 192
Puncture wounds, 79–81
Pupillary inequality, 19
Purkinje cells, 12, 33, 35–36, 71, 145, 194, 264, 265, 295, 296, 303, 332, 342, 406, 459, 460
Putaminal hemorrhage, 312
PVL (periventricular leukomalacia), 110, 137–138
PVS (persistent vegetative state), 332
Pyknosis, 40, 74
Pyramidal cell layer sectors hippocampus, 37
Pyriform cortex, 183, 396

Q
Quadriplegia, 86

R
Rabies virus, 348–349
Raccoon eyes, 168
Radiating fracture lines, 214
Radiation necrosis, 436
Radiculopathy, 425
Radiographs, 248, 443

Radiologic procedures, 424, 443
Radiology, 244–246
Radiolucent plastic bullets, 246
Radiotherapy, 434
Rathke's pouch tumor, 384
Reaction/response time issues, firearms, 237–239
Reactive astrocyte, 51
Reactive cells
appearance in developing brain, 36
changes, 109–112
hemorrhagic lesions, 111–112
hypoxic-ischemic lesions, 110–111
mineralization, 112
Rectus-orbital cortex, 191
Red blood cells (erythrocytes), 67, 70, 108
Red neurons, 51, 73, 112, 298–302
Redundant gyration, 28
Reflex movements, brain death, 97–98
Reflexes, and sudden unexpected death, 281–284
blunt force trauma to throat, 283
"café coronary" deaths, 282–283
death due to cold water immersion, 283
death related to miscellaneous presumptive reflex mechanisms, 284
death related to stressful events, 281–282
oculocardiac and trigeminocardiac reflex complex, 283–284
reflex cardiac arrest secondary to carotid sinus stimulation, 282
Reflexogenic death, 281
Reinnervation of muscle, 114
Remyelination, 414
Respirator brain, 92–98
examination, 92–95
microscopy, 95–97
spontaneous/reflex movements, 97–98
Reticulin stain, 459
Retinal hemorrhage, 137, 206–208
Retrograde embolization, 224
Retrohyperextension, 190
Reye's syndrome, 281
Rhinocerebral mucormycosis, 345
Ribosomes, 302
Rickettsial infections, 339, 341
Rifle wounds, 228
Ring fracture, 169
Rocky Mountain spotted fever, 339
Rod cells, 96
Rootlets, 409
Rosenthal fibers, 107, 363
Rostral pontine tegmentum, 185
Rostrum orbitale, 152
Rubber-tipped forceps, 248
Rubella, 336
Ruptured cerebral aneurysms, 176, 262, 325–327

S
Sabot slugs, 228
SAH. See Subarachnoid hemorrhage
Santeria, 205
Sappinia diploidea, 355
SAS (subarachnoid space), 28
Scalp abrasion, contusion, and laceration, 167–168

Scalp injuries, 52–54
antemortem versus postmortem, 53–54
color changes in skin bruises, 52
histologic dating/aging, 52–53
Schiff-positive macrophages, 415
Schwann cells, 100, 382
Schwannomas, 382–383
Schwannosis, 84
Scoliosis, 407
Scrotal trauma, 284
SDH. See Subdural hematomas
SE (status epilepticus), 264
Seizures, 123, 124, 370
Senile plaques, 397
Serotonin syndrome, 296
Shaken adult syndrome, 205
Shaken baby syndrome, 171
Sharp force injury, 445–448
Shime waza, 290
Shotgun wounds, 228–229
Shotshell, 228
Shrapnel, 224
Shy-Drager syndrome, 405
Sickle cell trait, 453
Siderophages, 70
SIDS (sudden infant death syndrome), 8–9, 71, 151
Silencers, 222–223
Silent chronic subdural hematomas, 174
Silver stains, 101
Sinuses
carotid sinus stimulation, reflex cardiac arrest secondary to, 282
dural sinus thrombosis, 317–325
dural venous sinus wounds, 224
paranasal sinusitis, 317
SIRS (systemic inflammatory response syndrome), 196, 197
Skin bruises, 52
Skull congenital anomalies, 145
Skull fractures, 54–58, 168–169
Skull wounds, 213–215
Slit ventricle syndrome, 427
Soft tissue chronic inflammatory reaction, 231
Soft tissue ossification, 60
Sommer's sector, 263, 291
Sonography, 42
Spielmeyer's acute swelling, 298
Spinal anesthetic procedures, 130–131
Spinal arachnoid diverticulae, 10
Spinal canal, 15
Spinal concussion, 87–88
Spinal cord dorsal column atrophy, 342
Spinal cord fixation, 431
Spinal cord infarction, 92
Spinal cord injuries, 233
acute, 84
in newborns, 132
traumatic lesions, 82–84
without radiologic abnormality (SCIWORA), 206, 444
Spinal lipomas, 388
Spinal manipulation, 436
Spinal meningeal ossification, 60
Spinal shock, 84–87
Spinal stenosis, 87–88

Spirochetes, 342
Spongiform leukodystrophy, 413
Spongioblast cells, 30
Spontaneous movements, and brain death, 97–98
Spontaneous subependymal/intraventricular hemorrhage, 111
Sports, 192–194
SSPE (subacute sclerosing panencephalitis), 351, 397
Stab wounds, 448
Staphylococcus, 336
Staphylococcus aureus, 335
Staphylococcus epidermidis, 335
Static apnea, 290
Status epilepticus (SE), 264
Status marmoratus, 142
Steele-Richardson- Olszewski syndrome, 405
Stellate, 169
Stenogyria, 28
Stillborn infants, 20, 129–130
Streptococcus, 337
Streptococcus pneumoniae, 337
Stressful events, sudden unexpected death related to, 281–282
Striatonigral degeneration, 406
Stroke
    cerebrospinal fluid (CSF) cytology, 98–99
    pregnant women, 124
Stroma, 64
Stromal cell, 390
Stun guns, 242
Sturge-Weber-Dimitri disease, 271, 323
Subacute sclerosing panencephalitis (SSPE), 351, 397
Subaponeurotic hemorrhages, 131
Subarachnoid hemorrhage (SAH), 19, 81, 129, 132, 174–176, 204, 205, 218, 262, 325
    cellular responses, 50
    cerebrospinal fluid (CSF) appearance, 68
    microscopy, 68–70
    in pregnant women, 124
Subarachnoid purulent exudate, 337
Subarachnoid space (SAS), 28
Subcutaneous scalp hemorrhage, 168
Subdural empyema, 336–337
Subdural hematomas (SDH), 19, 28, 192
    anatomy, 64
    chronic, 172–174
    chronic subdural hematoma, 172–174
    dating/aging, 64–66, 172
    size, 63–64
    source of neomembrane, 172
    spontaneous resolution, 66–67
Subdural hemorrhage, 67, 132, 204, 205
Subdural hygromas, 63
Subdural neomembranes, 56, 64
Subependymal (germinal matrix) hemorrhage, 136
Subependymal germinal zones, 298
Subependymal germinolysis, 138
Subependymal gliosis, 32
Subependymal tumor cells, 369
Subependymoma, 367, 369
Subfalcine cingulate herniation, 363
Subgaleal hemorrhage, 168

Subperiosteal hemorrhage, 168
Subpial calcification, 59
Subpial gliosis, 176
Subretinal hemorrhage, 206
Subsarcolemmal nuclei, 113
SUD. *See* Sudden unexpected death
Sudanophilic leukodystrophies, 419
Sudden infant death syndrome (SIDS), 8–9, 71, 151
Sudden unexpected death (SUD), 8–9, 255–287
    acute intracerebral hemorrhage, 268–271
    acute subarachnoid hemorrhage, 256–262
    cysticercosis, 265–266
    epilepsy, 262–265
    in epileptic patient (SUDEP), 262
    hydrocephalus, 275–277
    meningitis, 274–275
    multiple sclerosis, 281
    and reflexes, 281–284
        blunt force trauma to throat, 283
        "café coronary" deaths, 282–283
        death due to cold water immersion, 283
        death related to miscellaneous presumptive reflex mechanisms, 284
        death related to stressful events, 281–282
        oculocardiac and trigeminocardiac reflex complex, 283–284
        reflex cardiac arrest secondary to carotid sinus stimulation, 282
    Reye's syndrome, 281
    tumors, 277–280
    vascular malformation, 271–272
Suicidal wounds, 226–227
Supranuclear palsy, 13
Sutural diastasis, 62
Sympathetic/parasympathetic neural reflexes, 282
Syphilis, 341
Systemic heparinization, 146
Systemic inflammatory response syndrome (SIRS), 196, 197

T

Tadd (talk and deteriorate and die) syndrome, 204
Taenia choroidea, 191
Taenia solium, 265, 353
Talk and deteriorate and die (Tadd) syndrome, 204
TBI. *See* Traumatic brain injury
Temporary cavity phenomenon, 212
Terrorism, biologic, 355
Terson's syndrome, 204–205
Testicular trauma, 284
Thalamic hemorrhages, 313
Thalamus, 313, 362
Therapeutic complications, 438
Therapeutic emboli, 311
Therapeutic hyperthermia, 296–297
Throat, blunt force trauma to, 283
Thrombogenic, 423
Thrombolytics, 307, 430
Thrombosis, 92, 317

TIAs (transient ischemic attacks), 307
Tin ear syndrome, 204
Tissue mineralization, 59
Tissue oxygenation, 296
TNF (tumor necrosis factor), 274
Tocolytic agents, 129
Tonsillar herniation, 35
Torcular Herophili, 223
Total parenteral nutrition (TPN), 424
Toxicologic screening, 1
Toxicology, 451–452
Toxoplasma, 336
Toxoplasma gondii, 339
Toxoplasmosis, 353
TPN (total parenteral nutrition), 424
Trace evidence studies, 244
Trachipleistophora hominis, 355
Transcervical ascending infections, 335
Transient ischemic attacks (TIAs), 307
Transmissible spongiform encephalopathies (prion diseases), 351–353
Transneuronal degeneration, 104–107
    dentatoolivary system, 105–107
    trigeminal system, 104–105
    visual system, 104
Transverse myelitis, 445
Transverse myelopathy, 433
Traumatic axonal injury (AI), 77–79
Traumatic brain injury (TBI), 167–198
    from boxing, 192–194
    brain contusions, lacerations, and hematomas, 177
    brainstem avulsion, 190–191
    consequences of, 195–197
    diffuse axonal injury (DAI), 183–190
    epidural hematomas (EDHs), 169–170
    locked-in syndrome, 192
    overview, 167
    scalp abrasion, contusion, and laceration, 167–168
    skull fractures, 168–169
    sports, 192–194
    subarachnoid hemorrhage (SAH), 174–176
    subcutaneous scalp hemorrhage, 168
    subdural hematomas (SDHs), 170–174
        chronic subdural hematoma, 172–174
        dating/aging, 172
        source of neomembrane, 172
    subgaleal hemorrhage, 168
    traumatic cranial neuropathies, 191–192
    ventricular hemorrhage, 191
Traumatic cranial neuropathies, 191–192
Traumatic splenic rupture, 126
Traumatic subcutaneous hemorrhage, 168
Trematodicides, 266
Treponema, 336
Treponema pallidum, 341
Trigeminal cardiac reflex, 284
Trigeminal system, 104–105
Trigeminocardiac reflex, 283, 284
Trilaminar cyst, 353
Trypanosoma brucei gambiense, 355
Tumor necrosis factor (TNF), 274
Tumors. *See also* Brain tumors
    intracranial, 128
    of maldevelopment, 383–386

colloid cyst, 384–386
epidermoid cyst, 384
sudden unexpected death as result of, 277–280
TUNEL-negative, 301
Turnbull blue reaction stains, 59
Typhus, 339

U
Ulegyria, 139, 142
Ultrasonography, 443
Uncal herniation, 191
Uncal-parahippocampal herniation, 316
Unexpected death. *See* Sudden unexpected death
Uterine cervix, 284
Uterine rupture, 126

V
Vacuum-assisted delivery, 131, 133–134
Vaginal delivery, 131–133
Varicella-zoster virus, 345, 347
Varix, 271
Vascular congestion, 73
Vascular diseases, of central nervous system, 307–334
  cerebral amyloid angiopathy, 327–331
  cerebral embolism, 310–316
  cerebral thrombosis, 307–309
  drug abuse-related cardiovascular disease, 331
  dural sinus thrombosis, 317–325
  overview, 307
  ruptured cerebral aneurysms, 325–327
  secondary changes of cerebral infarction, 316–317
  states of altered consciousness, 331–333
    akinetic mutism, 332
    coma, 332
    locked-in state, 332
    vegetative state, 332–333
Vascular endothelial growth factor (VEGF), 454
Vascular endothelium, 341
Vascular hyperplasia, 309
Vascular malformation, 271–272, 325
Vasculitis, 339
Vasoconstriction, 275
Vasospasm, 325
Vasovagal reflex, 282
Vegetative state, 332–333
VEGF (vascular endothelial growth factor), 454
Veins, bridging, 171
Vena terminalis, 191
Venous malformation, 271, 323
Ventricular fibrillation, 283
Ventricular hemorrhage, 191
Ventricular purulent exudate, 275
Ventriculoperitoneal shunt tubing, 425
Ventriculostomy, 425
Vertebral artery dissection, 436, 437
Vertebral column, 15
Vertebrobasilar ischemia, 437
Viral infections, 345, 347–351
Virchow-Robin space, 12
Virtual subdural space, 171
Visual system, transneuronal degeneration, 104
Vitamin K levels, 124
Vitreous hemorrhages, 204, 207
Von Hippel-Lindau disease, 271
Von Kossa stain, 59
Von Recklinghausen's disease, 84

W
Wallerian degeneration, 100–103
  central nervous system (CNS), 101–103
    astrocytes, 103
    axons, 101–102
    microglia, 102–103
    myelin sheath, 102
  peripheral nervous system (PNS), 100–101
Warthin-Starry-Levaditi silver stain, 342
Water immersion, death due to, 283
Weber-Rendu disease, 325
Wernicke's encephalopathy, 105, 452, 458
West Nile virus, 349–350
Whiplash injury, 171
White leptomeningeal plaques, 10, 12

X
Xanthochromia, 68

PROPERTY OF
MARICOPA COUNTY
FORENSIC SCIENCE CENTER